MANUAL
OF STEEL
CONSTRUCTION

LOAD &
RESISTANCE
FACTOR
DESIGN

Volume I

Structural Members,
Specifications,
& Codes

Second Edition

FOREWORD

The American Institute of Steel Construction, founded in 1921, is the non-profit technical specifying and trade organization for the fabricated structural steel industry in the United States. Executive and engineering headquarters of AISC are maintained in Chicago, Illinois.

The Institute is supported by three classes of membership: Active Members totaling 400 companies engaged in the fabrication and erection of structural steel, Associate Members who are allied product manufacturers, and Professional Members who are individuals or firms engaged in the practice of architecture or engineering. Professional members also include architectural and engineering educators. The continuing financial support and active participation of Active Members in the engineering, research, and development activities of the Institute make possible the publishing of this Second Edition of the *Load and Resistance Factor Design Manual of Steel Construction*.

The Institute's objectives are to improve and advance the use of fabricated structural steel through research and engineering studies and to develop the most efficient and economical design of structures. It also conducts programs to improve product quality.

To accomplish these objectives the Institute publishes manuals, textbooks, specifications, and technical booklets. Best known and most widely used are the *Manuals of Steel Construction,* LRFD (Load and Resistance Factor Design) and ASD (Allowable Stress Design), which hold a highly respected position in engineering literature. Outstanding among AISC standards are the *Specifications for Structural Steel Buildings* and the *Code of Standard Practice for Steel Buildings and Bridges.*

The Institute also assists designers, contractors, educators, and others by publishing technical information and timely articles on structural applications through two publications, *Engineering Journal* and *Modern Steel Construction.* In addition, public appreciation of aesthetically designed steel structures is encouraged through its award programs: Prize Bridges, Architectural Awards of Excellence, Steel Bridge Building Competition for Students, and student scholarships.

Due to the expanded nature of the material, the Second Edition of the LRFD Manual has been divided into two complementary volumes. The present Volume I contains the LRFD Specification and Commentary, tables, and other design information for structural members, and a new Part 2: Essentials of LRFD. All of the information on connections is in an accompanying Volume II. Like the LRFD Specification upon which they are based, both volumes of this LRFD Manual apply to buildings, not bridges.

The Committee gratefully acknowledges the contributions of Roger L. Brockenbrough, Louis F. Geschwindner, Jr., and Cynthia J. Zahn to this Manual.

By the Committee on Manuals, Textbooks, and Codes,

William A. Thornton, Chairman

Barry L. Barger, Vice Chairman

Horatio Allison	Mark V. Holland	David T. Ricker
Robert O. Disque	William C. Minchin	Abraham J. Rokach
Joseph Dudek	Thomas M. Murray	Ted W. Winneberger
William G. Dyker	Heinz J. Pak	Charles J. Carter, Secretary
Ronald L. Hiatt	Dennis F. Randall	

REFERENCED SPECIFICATIONS, CODES, AND STANDARDS

Part 6 (Volume I) of this LRFD Manual contains the full text of the following:

American Institute of Steel Construction, Inc. (AISC)
Load and Resistance Factor Design Specification for Structural Steel Buildings,
December 1, 1993
Specification for Load and Resistance Factor Design of Single-Angle Members,
December 1, 1993
Seismic Provisions for Structural Steel Buildings, June 15, 1992
Code of Standard Practice for Steel Buildings and Bridges, June 10, 1992

Research Council on Structural Connections (RCSC)
Load and Resistance Factor Design Specifications for Structural Joints Using ASTM A325 or A490 Bolts, June 8, 1988

Additionally, the following other documents are referenced in Volumes I and II of the LRFD Manual:

American Association of State Highway and Transportation Officials (AASHTO)
AASHTO/AWS D1.5–88

American Concrete Institute (ACI)
ACI 349–90

American Iron and Steel Institute (AISI)
Load and Resistance Factor Design Specification for Cold-Formed Steel Structural Members, 1991

American National Standards Institute (ANSI)
ANSI/ASME B1.1–82ANSI/ASME B18.2.2–86
ANSI/ASME B18.1–72ANSI/ASME B18.5–78
ANSI/ASME B18.2.1–81

American Society of Civil Engineers (ASCE)
ASCE 7-88

American Society for Testing and Materials (ASTM)

ASTM A6–91b	ASTM A490–91	ASTM A617–92
ASTM A27–87	ASTM A500–90a	ASTM A618–90a
ASTM A36–91	ASTM A501–89	ASTM A668–85a
ASTM A53–88	ASTM A502–91	ASTM A687–89
ASTM A148–84	ASTM A514–91	ASTM A709–91
ASTM A153–82	ASTM A529–89	ASTM A770–86
ASTM A193–91	ASTM A563–91c	ASTM A852–91
ASTM A194–91	ASTM A570–91	ASTM B695–91
ASTM A208(A239–89)	ASTM A572–91	ASTM C33–90
ASTM A242–91a	ASTM A588–91a	ASTM C330–89
ASTM A307–91	ASTM A606–91a	ASTM E119–88
ASTM A325–91c	ASTM A607–91	ASTM E380–91
ASTM A354–91	ASTM A615–92b	ASTM F436–91
ASTM A449–91a	ASTM A616–92	

American Welding Society (AWS)

AWS A2.4–93	AWS A5.25–91
AWS A5.1–91	AWS A5.28–79
AWS A5.5–81	AWS A5.29–80
AWS A5.17–89	AWS B1.0–77
AWS A5.18–79	AWS D1.1–92
AWS A5.20–79	AWS D1.4–92
AWS A5.23–90	

PART 1

DIMENSIONS AND PROPERTIES

OVERVIEW

To facilitate reference to Part 1, the locations of frequently used tables are listed below.

STRUCTURAL STEELS

Availability

Section A3.1 of the AISC *Load and Resistance Factor Design Specification for Structural Steel Buildings* lists fifteen ASTM specifications for structural steel approved for use in building construction.

Five of these steels are available in hot-rolled structural shapes, plates, and bars. Two steels, ASTM A514 and A852, are available only in plates. Table 1-1 shows five groups of shapes and eleven ranges of thickness of plates and bars available in the various minimum yield stress* and tensile strength levels afforded by the seven steels. For complete information on each steel, reference should be made to the appropriate ASTM specification. A listing of shape sizes included in each of the five groups follows in Table 1-2, corresponding with the groupings given in Table A of ASTM Specification A6.

Seven additional grades of steel, other than those covering hot-rolled shapes, plates, and bars, are listed in Section A3.1a of the LRFD Specification. These steels cover pipe, cold- and hot-formed tubing, and cold- and hot-rolled sheet and strip.

The principal producers of shapes listed in Part 1 of this Manual are shown in Table 1-3. Availability and the principal producers of structural tubing are shown in Tables 1-4 through 1-6. For additional information on availability and classification of structural steel plates and bars, refer to the separate discussion beginning on page 1-133.

Space does not permit inclusion in Table 1-3, or in the listing of shapes and plates in Part 1 of this Manual, of all rolled shapes or plates of greater thickness that are occasionally used in construction. For such products, reference should be made to the various producers' catalogs.

To obtain an economical structure, it is often advantageous to minimize the number of different sections. Cost per square foot can often be reduced by designing this way.

Selection of the Appropriate Structural Steel

Steels with 50 ksi yield stress are now widely used in construction, replacing ASTM A36 steel in many applications. The 50 ksi steels listed in Section A3.1a of the LRFD Specification are ASTM A572 high-strength low-alloy structural steel, ASTM A242 and A588 atmospheric-corrosion-resistant high-strength low-alloy structural steels, and ASTM A529 high-strength carbon-manganese structural steel. Yield stresses above 50 ksi can be obtained from two grades of ASTM A572 steel as well as ASTM A514 and A852 quenched and tempered structural steel plate. These higher-strength steels have certain advantages over 50 ksi steels in certain applications. They may be economical choices where lighter members, resulting from use of higher design strengths, are not penalized because of instability, local buckling, deflection, or other similar reasons. They may be used in tension members, beams in continuous and composite construction where deflections can be minimized, and columns having low slenderness ratios. The reduction of dead load and associated savings in shipping costs can be significant factors. However, higher strength steels are not to be used indiscriminately. Effective use of all steels depends on thorough cost and engineering analysis. Normally, connection material is specified as ASTM A36. The connection tables in this Manual are for A36 steel.

---

*As used in the AISC LRFD Specification, "yield stress" denotes either the specified minimum yield point (for those that have a yield point) or specified minimum yield strength (for those steels that do not have a yield point).

With appropriate procedures and precautions, all steels listed in the AISC Specification are suitable for welded fabrication. To provide for weldability of ASTM A529 steel, the specification of a maximum carbon equivalent is recommended.

ASTM A242 and A588 atmospheric-corrosion-resistant, high-strength, low-alloy steels can be used in the bare (uncoated) condition in most atmospheres. Where boldly exposed under such conditions, exposure to the normal atmosphere causes a tightly adherent oxide to form on the surface which protects the steel from further atmospheric corrosion. To achieve the benefits of the enhanced atmospheric corrosion resistance of these bare steels, it is necessary that design, detailing, fabrication, erection, and mainte-nance practices proper for such steels be observed. Designers should consult with the steel producers on the atmospheric-corrosion-resistant properties and limitations of these steels prior to use in the bare condition. When either A242 or A588 steel is used in the coated condition, the coating life is typically longer than with other steels. Although A242 and A588 steels are more expensive than other high-strength, low-alloy steels, the reduction in maintenance resulting from the use of these steels usually offsets their higher initial cost.

Brittle Fracture Considerations in Structural Design

As the temperature decreases, an increase is generally noted in the yield stress, tensile strength, modulus of elasticity, and fatigue strength of the structural steels. In contrast, the ductility of these steels, as measured by reduction in area or by elongation, and the toughness of these steels, as determined from a Charpy V-notch impact test, decrease with decreasing temperatures. Furthermore, there is a temperature below which a structural steel subjected to tensile stresses may fracture by cleavage,* with little or no plastic deformation, rather than by shear,* which is usually preceded by a considerable amount of plastic deformation or yielding.

Fracture that occurs by cleavage at a nominal tensile stress below the yield stress is commonly referred to as brittle fracture. Generally, a brittle fracture can occur in a structural steel when there is a sufficiently adverse combination of tensile stress, tem-perature, strain rate, and geometrical discontinuity (notch) present. Other design and fabrication factors may also have an important influence. Because of the interrelation of these effects, the exact combination of stress, temperature, notch, and other conditions that will cause brittle fracture in a given structure cannot be readily calculated. Conse-quently, designing against brittle fracture often consists mainly of (1) avoiding conditions that tend to cause brittle fracture and (2) selecting a steel appropriate for the application. A discussion of these factors is given in the following sections.

Conditions Causing Brittle Fracture

It has been established that plastic deformation can occur only in the presence of shear stresses. Shear stresses are always present in a uniaxial or biaxial state-of-stress. How-ever, in a triaxial state-of-stress, the maximum shear stress approaches zero as the principal stresses approach a common value, and thus, under equal triaxial tensile stresses, failure occurs by cleavage rather than by shear. Consequently, triaxial tensile stresses tend to cause brittle fracture and should be avoided. A triaxial state-of-stress can result from a uniaxial loading when notches or geometrical discontinuities are present.

---

*Shear and cleavage are used in the metallurgical sense (macroscopically) to denote different fracture mechanisms.

Increased strain rates tend to increase the possibility of brittle behavior. Thus, structures that are loaded at fast rates are more susceptible to brittle fracture. However, a rapid strain rate or impact load is not a required condition for a brittle fracture.

Cold work and the strain aging that normally follows generally increase the likelihood of brittle fracture. This behavior is usually attributed to the previously mentioned reduction in ductility. The effect of cold work that occurs in cold forming operations can be minimized by selecting a generous forming radius and, thus, limiting the amount of strain. The amount of strain that can be tolerated depends on both the steel and the application.

The use of welding in construction increases the concerns relative to brittle fracture. In the as-welded condition, residual stresses will be present in any weldment. These stresses are considered to be at the yield point of the material. To avoid brittle fracture, it may be required to utilize steels with higher toughness than would be required for bolted construction. Welds may also introduce geometric conditions or discontinuities that are crack-like in nature. These stress risers will additionally increase the requirement for notch toughness in the weldment. Avoidance of the intersection of welds from multiple directions reduces the likelihood of triaxial stresses. Properly sized weld-access holes prohibit the interaction of these various stress fields. As steels being welded become thicker and more highly restrained, welding procedure issues such as preheat, interpass temperature, heat input, and cooling rates become increasingly important. The residual stresses present in a weldment may be reduced by the use of fewer weld passes and peening of intermittent weld layers. In most cases, weld metal notch toughness exceeds that of the base materials. However, for fracture-sensitive applications, notch-tough base and weld metal should be specified.

The residual stresses of welding can be greatly reduced through thermal stress relief. This reduces the driving force that causes brittle fracture, but if the toughness of the material is adversely affected by this thermal treatment, no increase in brittle fracture resistance will be experienced. Therefore, when weldments are to be stress relieved, investigation into the effects on the weld metal, heat-affected zone, and base material should be made.

Selecting a Steel To Avoid Brittle Fracture

The best guide in selecting a steel that is appropriate for a given application is experience with existing and past structures. A36 and Grade 50 (i.e., 50 ksi yield stress) steels have been used successfully in a great number of applications, such as buildings, transmission towers, transportation equipment, and bridges, even at the lowest atmospheric temperatures encountered in the U.S. Therefore, it appears that any of the structural steels, when designed and fabricated in an appropriate manner, could be used for similar applications with little likelihood of brittle fracture. Consequently, brittle fracture is not usually experienced in such structures unless unusual temperature, notch, and stress conditions are present. Nevertheless, it is always desirable to avoid or minimize the previously cited adverse conditions that increase the susceptibility of the steel to brittle fracture.

In applications where notch toughness is considered important, it usually is required that steels must absorb a certain amount of energy, 15 ft-lb or higher (Charpy V-notch test), at a given temperature. The test temperature may be higher than the lowest operating temperature depending on the rate of loading. See Rolfe and Barsom (1986) and Rolfe (1977).

Lamellar Tearing

The information on strength and ductility presented in the previous sections generally pertains to loadings applied in the planar direction (longitudinal or transverse orientation) of the steel plate or shape. It should be noted that elongation and area reduction values may well be significantly lower in the through-thickness direction than in the planar direction. This inherent directionality is of small consequence in many applications, but does become important in the design and fabrication of structures containing massive members with highly restrained welded joints.

With the increasing trend toward heavy welded-plate construction, there has been a broader recognition of the occurrence of lamellar tearing in some highly restrained joints of welded structures, especially those using thick plates and heavy structural shapes. The restraint induced by some joint designs in resisting weld deposit shrinkage can impose tensile strain sufficiently high to cause separation or tearing on planes parallel to the rolled surface of the structural member being joined. The incidence of this phenomenon can be reduced or eliminated through greater understanding by designers, detailers, and fabricators of (1) the inherent directionality of construction forms of steel, (2) the high restraint developed in certain types of connections, and (3) the need to adopt appropriate weld details and welding procedures with proper weld metal for through-thickness connections. Further, steels can be specified to be produced by special practices and/or processes to enhance through-thickness ductility and thus assist in reducing the incidence of lamellar tearing. Steels produced by such practices are available from several producers. However, unless precautions are taken in both design and fabrication, lamellar tearing may still occur in thick plates and heavy shapes of such steels at restrained through-thickness connections. Some guidelines in minimizing potential problems have been developed (AISC, 1973). See also Part 8 in Volume II of this LRFD Manual and ASTM A770, Standard Specification for Through-Thickness Tension Testing of Steel Plates for Special Applications.

Jumbo Shapes and Heavy Welded Built-up Sections

Although Group 4 and 5 W-shapes, commonly referred to as jumbo shapes, generally are contemplated as columns or compression members, their use in non-column applications has been increasing. These heavy shapes have been known to exhibit segregation and a coarse grain structure in the mid-thickness region of the flange and the web. Because these areas may have low toughness, cracking might occur as a result of thermal cutting or welding (Fisher and Pense, 1987). Similar problems may also occur in welded built-up sections. To minimize the potential of brittle failure, the current LRFD Specification includes provisions for material toughness requirements, methods of splicing, and fabrication methods for Group 4 and 5 hot-rolled shapes and welded built-up cross sections with an element of the cross section more than two inches in thickness intended for tension applications.

FIRE-RESISTANT CONSTRUCTION

Fire-resistant steel construction may be defined as structural members and assemblies which can maintain structural stability for the duration of building fire exposure and, in some cases, prevent the spread of fire to adjacent spaces. Fire resistance of a steel member is a function of its mass, its geometry, the load to which it is subjected, its structural support conditions, and the fire to which it is exposed.

Many steel structures have inherent fire resistance through a combination of the above factors and do not require additional insulation from the effects of fire. However, in many situations, building codes specify the use of fire-rated steel assemblies. In this case, ASTM Specification E119, Standard Methods of Fire Tests of Building Construction and Materials, outlines the procedures of fire testing of structural elements.

Structural fire resistance is a major consideration in the design of modern buildings. In general, building codes define the level of fire protection that is required in specific applications and structural fire protection is typically implemented in design through code compliance. In the United States, with a few notable exceptions, the majority of cities and states now enforce one of the following model codes:

- National Building Code, published by the Building Officials and Code Administrators International.
- Standard Building Code, published by the Southern Building Code Congress International.
- Uniform Building Code, published by the International Conference of Building Officials.

Building codes specify fire-resistance requirements as a function of building occupancy, height, area, and whether or not other fire protection systems (e.g., sprinklers) are provided.

Fire-resistance requirements are specified in terms of hourly ratings based upon tests conducted in accordance with ASTM E119. This test method specifies a "standard" fire for evaluating the relative fire-resistance of construction assemblies (i.e., floors, roofs, beams, girders, and columns). Specific end-point criteria for evaluating the ability of assemblies to prevent the spread of fire to adjacent spaces and/or to continue to sustain superimposed loads are included. In effect, ASTM E119 is used to evaluate the length of time that an assembly continues to perform these functions when exposed to the standard fire. Thus, code requirements and fire-resistance ratings are specified in terms of time (i.e., one hour, two hours, etc.). The design of fire-resistant buildings is typically accomplished in a very prescriptive fashion by selecting tested designs that satisfy specific building code requirements. Listings of fire-resistant designs are available from a number of sources including:

- Fire-Resistance Directory, Underwriters Laboratories.
- Fire-Resistance Ratings, American Insurance Services Group.
- Fire-Resistance Design Manual, Gypsum Association.

In general, due to the very prescriptive nature of fire-resistant design, changes in tested assemblies can be difficult to justify to the satisfaction of code officials and listing agencies. In the case of structural steel construction, however, the basic heat transfer and structural principles are well defined. As a result, relatively simple analytical techniques have been developed that enable designers to use a variety of different structural steel shapes in conjunction with tested assemblies. These analytical techniques are specifically recognized by North American building code authorities and are described in a series of booklets published by the American Iron and Steel Institute (AISI):

Designing Fire Protection for Steel Columns (1980)
Designing Fire Protection for Steel Beams (1984)
Designing Fire Protection for Steel Trusses (1981)

Since fire-resistant design is currently based on the use of tested assemblies, an important consideration is the degree to which a test assembly is "representative" of actual building construction. In reality, this consideration poses a number of technical difficulties due to the size of available testing facilities, most of which can only accommodate floor or roof specimens in the range of 15 ft by 18 ft in area. As a result, a test assembly represents a relatively small sample of a typical floor or roof structure. Most floor slabs and roof decks are physically, if not structurally, continuous over beams and girders. Beam and girder spans are often much larger than can be accommodated in available laboratory furnaces. A variety of connection details are used to frame beams, girders, and columns. In short, given the cost of testing, the complexity and variety of modern structural systems, and the size of available test facilities, it is unrealistic to assume that test assemblies accurately model real construction systems during fire exposure.

In recognition of the practical difficulties associated with laboratory scale testing, ASTM E119 includes two specific test conditions, "restrained" and "unrestrained." From a structural engineering standpoint, the choice of these two terms is unfortunate since the "restraint" that is contemplated in fire testing is restraint against the thermal expansion, not structural rotational restraint in the traditional sense. The "restrained" condition applies when the assembly is supported or surrounded by construction which is "capable of resisting substantial thermal expansion throughout the range of anticipated elevated temperatures." Otherwise, the assembly should be considered free to rotate and expand at the supports and should be considered "unrestrained." Thus, a floor system that is simply supported from a structural standpoint will often be "restrained" from a fire-resistance standpoint. In order to provide guidance on the use of restrained and unrestrained ratings, ASTM E119 includes an explanatory Appendix. It should be emphasized that most common types of steel framing can be considered "restrained" from a fire-resistance standpoint.

The standard fire test also includes other arbitrary assumptions. The specific fire exposure, for example, is based on furnace capabilities with continuous fuel supply and does not model real building fires with exhaustible fuel. Also, the test method assumes that assemblies are fully loaded when a fire occurs. In reality, fires are infrequent, random events and their design requirements should be probability based. Rarely will design structural loads occur simultaneously with fire. In addition, many structural elements are sized for serviceability (i.e., drift, deflection, or vibration) rather than strength, thereby providing an additional reserve strength during a fire. As a result of these and other considerations, more rational engineering design standards for structural fire protection are now being developed (*International Fire Engineering Design for Steel Structures: State-of-the-Art,* International Iron and Steel Institute). Although not yet standardized or recognized in North American building codes, similar design methods have been used in specific cases, based on code variances.

One such method has been developed by AISI for architecturally exposed structural steel elements on the exterior of buildings. In effect, ASTM E119 assumes that structural elements are located within a fire compartment and does not realistically characterize the fire exposure that will be seen by exterior structural elements. *Fire-Safe Structural Steel: A Design Guide* (American Iron and Steel Institute, 1979) defines a step-by-step analytical procedure for determining maximum steel temperatures, based on realistic fire exposures for exterior structural elements.

Occasionally, structural engineers will be called upon to evaluate fire-damaged steel structures. Although it is well known that the prolonged exposure to high temperatures can affect the physical and metallurgical properties of structural steel, in most cases steel members that can be straightened in place will be suitable for continued use (Dill, 1960). Special attention should be given to heat-treated or cold-formed steel elements and high-strength bolts and welds.

Effect of Shop Painting on Spray-Applied Fireproofing

Spray-applied fireproofing has excellent adhesion to unpainted structural steel. Mechanical anchorage devices, bonding agents, or bond tests are not required to meet Underwriters Laboratories, Inc. (UL) guidelines. In fact, moderate rusting enhances the adhesion of the fireproofing material, providing the uncoated steel is free of loose rust and mill scale. Customarily, any loose rust or mill scale as well as any other debris which has accumulated during the construction process is removed by the fireproofing application contractor. In many cases, this may be as simple as blowing it off with compressed air.

This ease of application is not realized when fireproofing is applied over painted steel. In order to meet UL requirements, bond tests in accordance with the ASTM E736 must be performed to determine if the fireproofing material has adequate adherence to the painted surface. Frequently, a bonding agent must be added to the fireproofing material and the bond test repeated to determine if the minimum bond strength can be met. Should the bond testing still not be satisfactory, mechanical anchorage devices are required to be applied to the steel before the fireproofing can be applied. The erected steel must still be cleaned free of any construction debris and scaling or peeling paint before the fireproofing may be applied.

Once it is determined that the bond tests are adequate, UL guidelines require that if fireproofing is spray-applied over painted steel, the steel must be wrapped with steel lath or mechanical anchorage devices must be applied to the steel if the structural shape exceeds the following dimensional criteria:

- For beam applications, the web depth cannot exceed 16 inches and the flange cannot exceed 12 inches.
- For column applications, neither the web depth nor the flange width can exceed 16 inches.

A significant number of structural shapes do not meet these restrictions.

The use of primers under spray-applied fireproofing significantly increases the cost of the steel and the preparation for and the application of the fireproofing material. In an enclosed structure, primer is insignificant in either the short- or long-term protection of the steel. LRFD Specification Section M3.1 states that structural steelwork need not be painted unless required by the contract. For many years, the AISC specifications have not required that steelwork be painted when it will be concealed by interior building finish or will be in contact with concrete. The use of primers under spray-applied fireproofing is strongly discouraged unless there is a compelling reason to paint the steel to protect against corrosion.

It is suggested that the designer refer to the UL Directory *Fire Resistance*—Volume 1, 1993, "Coating Materials," for more specific information on this topic.

EFFECT OF HEAT ON STRUCTURAL STEEL

Short-time elevated-temperature tensile tests on the structural steels permitted by the AISC Specification indicate that the ratios of the elevated-temperature yield and tensile strengths to their respective room-temperature values are reasonably similar in the 300° to 700°F range, except for variations due to strain aging. (The tensile strength ratio may increase to a value greater than unity in the 300° to 700°F range when strain aging occurs.) Below 300°F the strength ratios decrease only slightly. Above 700°F the ratio of elevated-temperature to room-temperature strength decreases more rapidly as the temperature increases.

The composition of the steels is usually such that the carbon steels (ASTM A36 and A529) exhibit strain aging with attendant reduced notch toughness. The high-strength low-alloy steels (ASTM A242, A572, and A588) and heat-treated alloy steels (ASTM A514 and A852) exhibit less-pronounced or little strain aging. As examples of the decreased ratio levels obtained at elevated temperature, the yield strength ratios for carbon and high-strength low-alloy steels are approximately 0.77 at 800°F, 0.63 at 1,000°F, and 0.37 at 1,200°F.

Coefficient of Expansion

The average coefficient of expansion for structural steel between 70°F and 100°F is 0.0000065 for each degree. For temperatures of 100°F to 1,200°F the coefficient is given by the approximate formula:

$$\varepsilon = (6.1 + 0.0019t) \times 10^{-6}$$

in which ε is the coefficient of expansion (change in length per unit length) for each degree Fahrenheit and t is the temperature in degrees Fahrenheit. The modulus of elasticity of structural steel is approximately 29,000 ksi at 70°F. It decreases linearly to about 25,000 ksi at 900°F, and then begins to drop at an increasing rate at higher temperatures.

Use of Heat to Straighten, Camber, or Curve Members

With modern fabrication techniques, a controlled application of heat can be effectively used to either straighten or to intentionally curve structural members. By this process, the member is rapidly heated in selected areas; the heated areas tend to expand, but are restrained by adjacent cooler areas. This action causes a permanent plastic deformation or "upset" of the heated areas and, thus, a change of shape is developed in the cooled member.

"Heat straightening" is used in both normal shop fabrication operations and in the field to remove relatively severe accidental bends in members. Conversely, "heat cambering" and "heat curving" of either rolled beams or welded girders are examples of the use of heat to effect a desired curvature.

As with many other fabrication operations, the use of heat to straighten or curve will cause residual stresses in the member as a result of plastic deformations. These stresses are similar to those that develop in rolled structural shapes as they cool from the rolling temperature; in this case, the stresses arise because all parts of the shape do not cool at the same rate. In like manner, welded members develop residual stresses from the localized heat of welding.

In general, the residual stresses from heating operations do not affect the design strength of structural members. Any reduction in strength due to residual stresses is incorporated in the provisions of the LRFD Specification.

The mechanical properties of steels are largely unaffected by heating operations, provided that the maximum temperature does not exceed 1,100°F for quenched and tempered alloy steels (ASTM A514 and A852), and 1,300°F for other steels. The temperature should be carefully checked by temperature-indicating crayons or other suitable means during the heating process.

EXPANSION JOINTS

Although buildings are typically constructed of flexible materials, expansion joints are required in roofs and the supporting structure when horizontal dimensions are large. The maximum distance between expansion joints is dependent upon many variables including ambient temperature during construction and the expected temperature range during the lifetime of the building. An excellent reference on the topic of thermal expansion in buildings and location of expansion joints is the Federal Construction Council's Technical Report No. 65, Expansion Joints in Buildings.

Taken from this report, Figure 1-1 provides a guide based on design temperature change for maximum spacing of structural expansion joints in beam-and-column-framed buildings with hinged-column bases and heated interiors. The report includes data for numerous cities and gives five modification factors which should be applied as appropriate:

1. If the building will be heated only and will have hinged-column bases, use the maximum spacing as specified;
2. If the building will be air-conditioned as well as heated, increase the maximum spacing by 15 percent provided the environmental control system will run continuously;
3. If the building will be unheated, decrease the maximum spacing by 33 percent;
4. If the building will have fixed column bases, decrease the maximum spacing by 15 percent;

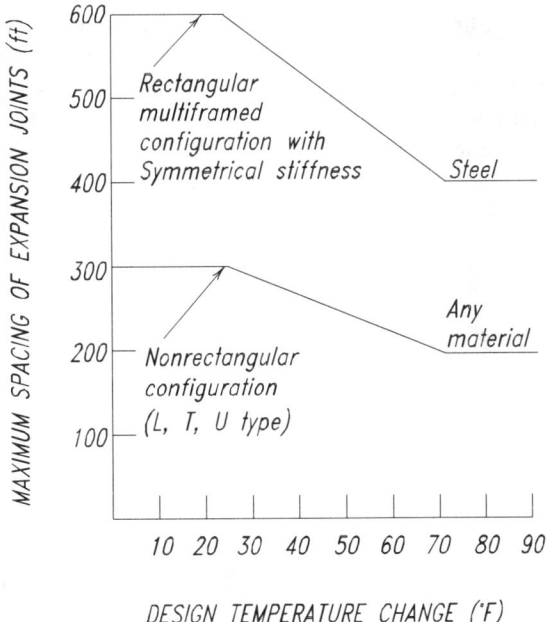

Fig. 1-1. Expansion joint spacing.

5. If the building will have substantially greater stiffness against lateral displacement in one of the plan dimensions, decrease the maximum spacing by 25 percent.

When more than one of these design conditions prevail in a building, the percentile factor to be applied should be the algebraic sum of the adjustment factors of all the various applicable conditions.

Additionally, most building codes include restrictions on location and spacing of fire walls. Such fire walls often become locations for expansion joints.

The most effective expansion joint is a double line of columns which provides a complete and positive separation. When expansion joints other than the double-column type are employed, low-friction sliding elements are generally used. Such systems, however, are never totally free and will induce some level of inherent restraint to movement.

COMPUTER SOFTWARE

AISC Database

The AISC Database contains the properties and dimensions of structural steel shapes, corresponding to Part 1 of this LRFD Manual. LRFD-related properties such as X1 and X2, as well as torsional properties, are included.

Two versions, one in U.S. customary units and one in metric units, are available.

Dimensions and properties of W, S, M, and HP shapes, American Standard Channels (C), Miscellaneous Channels (MC), Structural Tees cut from W, M, and S shapes (WT, MT, ST), Single and Double Angles, Structural Tubing, and Pipe are listed in ASCII format. Also included are: a BASIC read/write program, a sample search routine, and a routine to convert the file to Lotus *.PRN file format.

AISC for AutoCAD *

The program will draw the end, elevation, and plan views of W, S, M, and HP shapes, American Standard Channels (C), Miscellaneous Channels (MC), Structural Tees cut from W, M, and S shapes (WT, MT, ST), Single and Double Angles, Structural Tubing, and Pipe to full scale corresponding to data published in Part 1 of this LRFD Manual.

Version 2.0 runs in AutoCAD Release 12 only; Version 1.0 runs in AutoCAD Releases 10 and 11.

---

*AutoCAD is a registered trademark in the US Patent and Trademark Office by Autodesk, Inc. AISC for AutoCAD is copyrighted in the US Copyright Office by Bridgefarmer and Associates, Inc.

Table 1-1.
Availability of Shapes, Plates, and Bars According to
ASTM Structural Steel Specifications

Steel Type	ASTM Designation	F_y Minimum Yield Stress (ksi)	F_u Tensile Stress[a] (ksi)	Shapes — Group per ASTM A6					Plates and Bars										
				1[b]	2	3	4	5	To 1/2" incl.	Over 1/2" to 3/4" incl.	Over 3/4" to 1 1/4" incl.	Over 1 1/4" to 1 1/2" incl.	Over 1 1/2" to 2" incl.	Over 2" to 2 1/2" incl.	Over 2 1/2" to 4" incl.	Over 4" to 5" incl.	Over 5" to 6" incl.	Over 6" to 8" incl.	Over 8"
Carbon	A36	32	58–80																
		36	58–80[c]																
	A529[f] Grade 42	42	60–85																
	A529 Grade 50	50	70–100										d						
High-Strength Low-alloy	A572 Grade 42	42	60																
	A572 Grade 50	50	65																
	A572 Grade 60	60	75																
	A572 Grade 65	65	80																
Corrosion Resistant High-strength Low-alloy	A242	42	63																
		46	67																
		50	70																
	A588	42	63																
		46	67																
		50	70																
Quenched & Tempered Alloy	A852[e]	70	90–110																
Quenched & Tempered Low-Alloy	A514[e]	90	100–130																
	A514[e]	100	110–130																

[a] Minimum unless a range is shown.
[b] Includes bar-size shapes
[c] For shapes over 426 lb / ft minimum of 58 ksi only applies.
[d] Plates to 1 in. thick, 12 in. width; bars to 1 1/2 in.
[e] Plates only.
[f] To improve the weldability of A529 steel, the specification of a maximum carbon equivalent (per ASTM Supplementary Requirement S78) is recommended.

▨ Available

☐ Not Available

Table 1-2. Structural Shape Size Groupings for Tensile Property Classification					
Struc- tural Shapes	Group 1	Group 2	Group 3	Group 4	Group 5
W shapes	W24×55, 62	W44×230, 262	W44×290, 335	W40×466 to 593 incl.	W36×848
	W21×44 to 57 incl.	W40×149 to 264 incl.	W40×431	W40×392	W14×605 to 808 incl.
	W18×35 to 71 incl.	W36×135 to 210 incl.	W40×277 to 372 incl.	W36×328 to 798 incl.	
	W16×26 to 57 incl.	W33×118 to 152 incl.	W36×230 to 300 incl.	W33×318 to 354 incl.	
	W14×22 to 53 incl.	W30×90 to 211 incl.	W33×169 to 291 incl.	W30×292 to 477 incl.	
	W12×14 to 58 incl.	W27×84 to 178 incl.	W30×235 to 261 incl.	W27×307 to 539 incl.	
	W10×12 to 45 incl.	W24×68 to 162 incl.	W27×194 to 258 incl.	W24×250 to 492 incl.	
	W8×10 to 48 incl.	W21×62 to 147 incl.	W24×176 to 229 incl.	W18×211 to 311 incl.	
	W6×9 to 25 incl.	W18×76 to 143 incl.	W21×166 to 201 incl.	W14×233 to 550 incl.	
	W5×16,19	W16×67 to 100 incl.	W18×158 to 192 incl.	W12×210 to 336 incl.	
	W4×13	W14×61 to 132 incl.	W14×145 to 211 incl.		
		W12×65 to 106 incl.	W12×120 to 190 incl.		
		W10×49 to 112 incl.			
		W8×58, 67			
M Shapes	all				
S Shapes	to 35 lb/ft incl.	over 35 lb/ft			
HP Shapes		to 102 lb/ft incl.	over 102 lb/ft		
American Standard Channels (C)	to 20.7 lb/ft incl.	over 20.7 lb/ft			
Miscellane- ous Channels (MC)	to 28.5 lb/ft incl.	over 28.5 lb/ft			
Angles (L)	to ½-in. incl.	over ½- to ¾-in. incl.	over ¾-in.		

Notes:
Structural tees from W, M, and S shapes fall into the same group as the structural shapes from which they are cut. Group 4 and Group 5 shapes are generally contemplated for application as columns or compression compo- nents. When used in other applications (e.g., trusses) and when thermal cutting or welding is required, special material specification and fabrication procedures apply to minimize the possibility of cracking (see Part 6, LRFD Specification, Sections A3.1c, J1.5, J1.6, J2.8, and M2.2, and corresponding Commentary sections).

Structural Steel Shape Producers

Bayou Steel Corp.
P.O. Box 5000
Laplace, LA 70068
(800) 535-7692

Bethlehem Steel Corp.
301 East Third St.
Bethlehem, PA 18016-7699
(800) 633-0482

British Steel Inc.
475 N. Martingale Road #400
Schaumburg, IL 60173
(800) 542-6244

Chaparral Steel Co.
300 Ward Road
Midlothian, TX 76065-9501
(800) 527-7979

Florida Steel Corp.
P.O. Box 31328
Tampa, FL 33631
(800) 237-0230

Northwestern Steel & Wire Co.
121 Wallace St.
P.O. Box 618
Sterling, IL 61081-0618
(800) 793-2200

North Star Steel Co.
1380 Corporate Center Curve
Suite 215
P.O. Box 21620
Eagan, MN 55121-0620
(800) 328-1944

Nucor Steel
P.O. Box 126
Jewett, TX 75846
(800) 527-6445

Nucor-Yamato Steel
P.O. Box 1228
Blytheville, AR 72316
(800) 289-6977

Roanoke Electric Steel Corp.
P.O. Box 13948
Roanoke, VA 24038
(800) 753-3532

SMI Steel, Inc.
101 South 50th St.
Birmingham, AL 35232
(800) 621-0262

TradeARBED
825 Third Ave.
New York, NY 10022
(212) 486-9890

Structural Tube Producers

Steel Tube Institute
8500 Station St.
Suite 270
Mentor, OH 44060
(216) 974-6990

Acme Roll Forming Co.
812 North Beck St.
Sebewaing, MI 48759-0706
(800) 937-8823

Bull Moose
57540 SR 19 S
P.O. Box B-1027
Elkhart, IN 46515
(800) 325-4467

Copperweld Corp.
7401 South Linder Ave.
Chicago, IL 60638
(800) 327-8823

Dallas Tube & Rollform
P.O. Box 540873
Dallas, TX 75354-0873
(214) 445-0234

Eugene Welding Co.
P.O. Box 249
Marysville, MI 48040
(800) 336-3926

EXLTUBE, Inc.
905 Atlantic
North Kansas City, MO 64116
(800) 892-8823

Hanna Steel Corp.
3812 Commerce Ave.
P.O. Box 558
Fairfield, AL 35064
(800) 633-8252

Independence Tube Corp.
6226 West 74th St.
Chicago, IL 60638
(708) 496-0380

IPSCO Steel, Inc.
P.O. Box 1670, Armour Road
Regina, Saskatchewan S4P 3C7
CANADA
(306) 934-7700

UNR-Leavitt, Div. of UNR Inc.
1717 West 115th St.
Chicago, IL 60643
(800) 532-8488

Valmont Industries, Inc.
P.O. Box 358
Valley, NE 68064
(800) 331-3002

Welded Tube Co. of America
1855 East 122nd St.
Chicago, IL 60633
(800) 733-5683

Steel Pipe Producers

National Association of Steel Pipe
    Distributors, Inc.
12651 Briar Forest Dr., Suite 130
Houston, TX 77077
(713) 531-7473

Table 1-3.
Principal Producers of Structural Shapes

B—Bethlehem Steel Corp.	I—British Steel	S—North Star Steel	W—Northwestern Steel & Wire
C—Chaparral Steel	M—SMI Steel Inc.	T—TradeARBED	Y—Bayou Steel Corp.
F—Florida Steel Corp.	N—Nucor-Yamato Steel	U—Nucor Steel	
	R—Roanoke Steel		

Section, Weight per ft	Producer Code	Section, Weight per ft	Producer Code
W44×all	T	W24×103	B,W
		W24×84-94	B,I,N,W
W40×321-593	T	W24×55-76	B,C,I,N,W
W40×297	N		
W40×278	T	W21×182-201	I,W
W40×277	N,T	W21×166	B,I,W
W40×264	B,T	W21×83-147	B,I,N,W
W40×249	N,T	W21×44-73	B,C,I,N,W
W40×235	B,T		
W40×215	N,T	W18×258-311	B
W40×211	B,T	W18×175-234	B,W
W40×199	N,T	W18×130-158	B,N,W
W40×183	B,I,N,T	W18×76-119	B,N,W
W40×174	T	W18×65-71	B,I,N,W
W40×149-167	B,I,N,T	W18×35-60	B,C,I,N,W
W36×439-848	T	W16×67-100	B,N,W
W36×393		W16×...-...	B,I,N,W
W36×328-359	B,I,T	W16×26-50	B,C,I,N,W
W36×260-300	B,I,N,T		
W36×256		W14×...	
W36×245	B,I,N,T	W14×342-730	B,I,T
W36×232	B,I	W14×...-...	B,I,T,W
W36×135-230	B,I,N,T	W14×90-...	B,I,N,T,W
		W14×82	B,N,W
W33×263-354		W14×...	B,I,N,W
W33×201-241	B,N,T	W14×61-68	B,C,I,N,W
W33×169	B,T	W14×43-53	B,C,I,N,W
W33×118-152	B,I,N,T	W14×38	B,I,N,W
		W14×22-34	B,C,I,N,W
W30×391-477	T		
W30×261-326	B,T	W12×252-336	B
W30×173-235	B,I,N,T	W12×210-230	B,T
W30×148	B,I,T	W12×170-190	B,I,T,W
W30×99-132	B,I,N,T	W12×65-152	B,I,N,T,W
W30×90	B,N	W12×50-58	B,C,I,N,W
		W12×16-45	B,C,N,W
W27×307-539	T	W12×14	B,C,W
W27×258	N,T		
W27×235	N		
W27×146-217	B,N,T	W10×88-112	B,I,N,W
W27×129	B,I,T,W	W10×49-77	B,C,I,N,W
W27×84-114	B,I,N,T,W	W10×33-45	B,C,N,W
		W10×22-30	B,C,I,N,W
W24×279-492	T	W10×15-19	B,C,I,W
W24×250	B,N,W	W10×12	B,C,W
W24×229	B,N,T,W		
W24×207	B,N,W	W8×31-67	B,C,I,N,W
W24×192	B,I,N,T,W	W8×18-28	B,C,N,W
W24×104-176	B,I,N,T,W	W8×15	B,C,W,Y

Notes:
For the most recent list of producers, please see the latest January or July issue of the AISC magazine *Modern Steel Construction*.
Maximum lengths of shapes obtained vary with producer, but typically range from 60 ft to 75 ft. Lengths up to 100 ft are available for certain shapes. Please consult individual producers for length requirements.

Table 1-3 (cont.).
Principal Producers of Structural Shapes

B—Bethlehem Steel Corp.	I—British Steel	S—North Star Steel	W—Northwestern Steel & Wire
C—Chaparral Steel	M—SMI Steel Inc.	T—TradeARBED	Y—Bayou Steel Corp.
F—Florida Steel Corp.	N—Nucor-Yamato Steel	U—Nucor Steel	
	R-Roanoke Steel		

Section, Weight per ft	Producer Code	Section, Weight per ft	Producer Code
W8×10-13	B,C,M,W,Y	MC18×42.7-58	B,N
		MC13×31.8-50	B,N
W6×20-25	B,C,I,N,W	MC12×31-50	B,N
W6×16	B,C,W,Y	MC12×10.6	S,N
W6×15	B,C,I,N,W	MC10×22-41.1	B
W6×12	B,C,W,Y	MC10×8.4	S
W6×9	B,C,N,W,Y	MC9×23.9-25.4	B
		MC8×18.7-22.8	B,S
W5×16-19	B	MC8×8.5	M
		MC7×19.1-22.7	B
W4×13	B,C,M,Y	MC6×18	B
		MC6×12-16.3	B,S
M12×10.8-11.8	C		
M10×8-9	C		
M8×6.5	C		
M5×18.9	B		

Section by Leg Length & Thickness		Producer Code
L8×8×	$1\frac{1}{8}$	B
	1	B,S
	$\frac{7}{8}$	B,S
	$\frac{3}{4}$	B,S
	$\frac{9}{16}$	B,S
	$\frac{1}{2}$	B,S
L6×6×	1	B,U,Y
	$\frac{7}{8}$	B,Y
	$\frac{3}{4}$	B,M,U,Y
	$\frac{5}{8}$	B,M,U,Y
	$\frac{9}{16}$	B,M,U,Y
	$\frac{1}{2}$	B,M,S,U,Y
	$\frac{7}{16}$	B,M,U,Y
	$\frac{3}{8}$	B,M,S,U,Y
	$\frac{5}{16}$	M,U,Y
L5×5×	$\frac{7}{8}$	B,U,Y
	$\frac{3}{4}$	B,M,U,Y
	$\frac{5}{8}$	B,M,U,Y
	$\frac{1}{2}$	B,M,U,W,Y
	$\frac{7}{16}$	B,M,U,Y
	$\frac{3}{8}$	B,M,U,W,Y
	$\frac{5}{16}$	B,M,U,W,Y
L4×4×	$\frac{3}{4}$	M,U,Y
	$\frac{5}{8}$	M,U,Y
	$\frac{1}{2}$	F,M,R,U,W,Y
	$\frac{7}{16}$	F,M,U,Y
	$\frac{3}{8}$	F,M,R,U,W,Y
	$\frac{5}{16}$	F,M,R,U,W,Y
	$\frac{1}{4}$	F,M,R,U,W,Y

Remainder of left column (Section, Weight per ft | Producer Code):

Section, Weight per ft	Producer Code
S24×80-121	
S20×66-96	B,W
S18×54.7-70	B,W
S15×42.9-50	B,W
S12×31.8-50	B,W
S10×25.4-35	B,S
S8×18.4-23	B,C,
S6×12.5-17.25	C,S,Y
S5×10	C
S4×9.5	C
S4×7.7	C,Y
S3×7.5	C,Y
S3×5.7	C,M,Y
HP14×73-117	B,I,N,W
HP12×53-84	B,I,N,W
HP10×42-57	B,C,I,N,W
HP8×36	B,C,I,N,W
C15×33.9-50	B,N,W
C12×30	B,W
C12×20.7-25	B,C,S,W
C10×25-30	B,S,W
C10×15.3-20	B,C,S,W
C9×20	B
C9×13.4-15	B,S
C8×18.75	S,W,Y
C8×11.5-13.75	C,M,S,U,W,Y
C7×12.25	S,U,W
C7×9.8	M,S,U,W
C6×13	M,S,U,W,Y
C6×10.5	C,M,S,U,W,Y
C6×8.2	C,F,M,U,W,Y,
C5×9	M,U,W,Y
C5×6.7	F,M,U,W,Y
C4×5.4-7.25	F,M,U,W,Y
C3×6	M,U,W,Y
C3×4.1-5	F,M,R,U,W,Y

Table 1-3 (cont.).
Principal Producers of Structural Shapes

B—Bethlehem Steel Corp. | I—British Steel | S—North Star Steel | W—Northwestern Steel & Wire
C—Chaparral Steel | M—SMI Steel Inc. | T—TradeARBED | Y—Bayou Steel Corp.
F—Florida Steel Corp. | N—Nucor-Yamato Steel | U—Nucor Steel
R—Roanoke Steel

Section by Leg Length and Thickness		Producer Code	Section by Leg Length and Thickness		Producer Code
L3½×3½×	1/2	F,M,R,U,W,Y	L6×3½×	1/2	M,U,W,Y
	7/16	U,Y		3/8	B,M,U,W,Y
	3/8	F,M,R,U,W,Y		5/16	B,M,U,W,Y
	5/16	F,M,R,U,W,Y			
	1/4	F,M,R,U,W,Y	L5×3½×	3/4	M,U,Y
				5/8	M,U,Y
L3×3×	1/2	F,M,U,W,Y		1/2	M,U,W,Y
	7/16	U,Y		3/8	M,U,W,Y
	3/8	F,M,R,S,U,W,Y		5/16	M,U,W,Y
	5/16	F,M,R,S,U,W,Y		1/4	M,U,W,Y
	1/4	F,M,R,S,U,W,Y			
	3/16	F,M,R,U,W,Y	L5×3×	1/2	F,M,U,W,Y
				7/16	F,Y
L2½×2½×	1/2	F,U		3/8	F,M,U,W,Y
	3/8	F,S,U		5/16	F,M,U,W,Y
	5/16	F,S,U		1/4	F,M,U,W,Y
	1/4	[illegible]			
	3/16	F,U	L4×3½×	1/2	F,M,U,W
				3/8	F,M,R,U,W
L2×2×	3/8	F,S,U		5/16	F,M,R,U,W
	5/16	F,S,U		1/4	F,M,R,U,W
	1/4	F,S,U			
	3/16	F,S,U	L4×3×	5/8	M,U,Y
	1/8	F,S,U		1/2	F,M,U,W,Y
				[illegible]	[illegible]
L8×6×	1	B,S		3/8	F,M,R,U,W,Y
	7/8	B		5/16	F,M,R,U,W,Y
	3/4	B,S		1/4	F,M,R,U,W,Y
	5/8	B			
	9/16	B,S	L3½×3×	1/2	U,W
	1/2	B,S		3/8	M,U,W
	7/16	B,S		5/16	M,U,W
				1/4	M,U,W
L8×4×	1	B,S			
	7/8	B,S	L3½×2½×	1/2	U
	3/4	B,S		3/8	U
	5/8	B,S		1/4	U
	9/16	B,S			
	1/2	B,S	L3×2½×	1/2	U
	7/16	B,S		3/8	U,W
				5/16	U,W,Y
L7×4×	3/4	B,Y		1/4	R,U,W
	5/8	B,Y		3/16	U
	1/2	B,S,Y			
	7/16	B,Y	L3×2×	1/2	F
	3/8	B,S,Y		3/8	F,S,U
				5/16	F,S,U
L6×4×	7/8	B		1/4	F,R,S,U
	3/4	B,M,S,U,W,Y		3/16	F,R,U
	5/8	B,M,S,U,W,Y			
	9/16	B,M,S,U,W,Y	L2½×2×	3/8	R,S,U
	1/2	B,M,S,U,W,Y		5/16	S,U
	7/16	B,U,Y		1/4	R,S,U
	3/8	B,M,S,U,W,Y		3/16	R,S,U
	5/16	B,M,S,U,W,Y			

STRUCTURAL SHAPES: TABLES

Table 1-4.
Availability of Steel Pipe and Structural Tubing
According to ASTM Material Specifications

Steel	ASTM Specification	Grade	F_y Minimum Yield Stress (ksi)	F_u Minimum Tensile Stress (ksi)	Shape Round	Shape Square & Rectangular	Availability
Electric-Resistance Welded	A53 Type E	B	35	60	■		Note 3
Seamless	Type S	B	35	60	■		Note 3
Cold Formed	A500	A	33	45	■		Note 1
		B	42	58	■		Note 1
		C	46	62	■		Note 1
		A	39	45		■	Note 1
		B	46	58		■	Note 2
		C	50	62		■	Note 1
Hot Formed	A501	—	36	58	■		Note 1
High-Strength Low-Alloy	A618	I	50	70	■		Note 1
		II	50	70	■		Note 1
		III	50	65	■		Note 1

Notes:
1. Available in mill quantities only; consult with producers.
2. Normally stocked in local steel service centers.
3. Normally stocked by local pipe distributors.

■ Available

☐ Not Available

Table 1-5.
Principal Producers of Structural Tubing (TS)

A—Acme Rolling Forming Co.	D—Dallas Tube & Rollform	I—Independence Tube Corp.	V—Valmont Industries, Inc.
B—Bull Moose Tube Co.	E—Eugene Welding Co.	P—IPSCO Steel	W—Welded Tube Co. of America
C—Copperweld Corp.	H—Hanna Steel Corp.	U—UNR-Leavitt, Div. of UNR, Inc.	X—EXLTUBE

Nominal Size and Thickness	Producer Code	Nominal Size and Thickness	Producer Code
30×30×5/8	V*	4½×4½×3/8, 5/16	I,P,W
28×28×5/8	V*	4½×4½×1/4, 3/16	A,B,C,D,I,P,W,X
26×26×5/8	V*	4½×4½×1/8	A,B,C,P,I,W
24×24×5/8, 1/2, 3/8	V*		
22×22×5/8, 1/2, 3/8	V*	4×4×1/2	B,C,P,U,W
20×20×5/8, 1/2, 3/8	V*	4×4×3/8, 5/16	A,B,C,D,E,I,P,U,W
18×18×5/8, 1/2, 3/8	V*	4×4×1/4, 3/16, 1/8	A,B,C,D,E,I,P,U,V,W,X
16×16×5/8	V*	3½×3½×5/16	I,P,W
16×16×1/2, 3/8, 5/16	V*,W	3½×3½×1/4, 3/16, 1/8	A,B,C,D,E,I,P,U,W,X
14×14×5/8	V*	3×3×5/16	I,P,W
14×14×1/2, 3/8	V*,W	3×3×1/4, 3/16	A,B,C,D,E,I,P,U,W,X
14×14×5/16	W	3×3×1/8	A,B,C,D,E,I,P,U,W
12×12×5/8	B	2½×2½×5/16	
12×12×1/2, 3/8	B,V*,W	2½×2½×1/4, 3/16	A,B,C,D,E,I,P,U,V,W,X
12×12×5/16, 1/4		2½×2½×1/8	A,B,C,D,E,I,P,U,V,W
10×10×5/8	B,C		I,V
10×10×1/2, 3/8, 5/16, 1/4	B,C,P,U,W	2×2×1/4	A,B,C,D,I,U,V,W,X
10×10×3/16	B,C,P,W	2×2×3/16, 1/8	A,B,C,D,E,I,P,U,V,W,X
8×8×5/8	B,C	1½×1½×3/16	B,E,P,U,V
8×8×1/2	B,C,P,U,W		
8×8×3/8, 5/16, 1/4, 3/16	B,C,D,P,U,W	30×24×1/2, 3/8, 5/16	V*
		28×24×1/2, 3/8, 5/16	V*
7×7×5/8	B	26×24×1/2, 3/8, 5/16	V*
7×7×1/2	B,C,P,U,W	24×22×1/2, 3/8, 5/16	V*
7×7×3/8, 5/16, 1/4, 3/16	B,C,D,P,U,W	22×20×1/2, 3/8, 5/16	V*
6×6×5/8	B	20×18×1/2, 3/8, 5/16	V*
6×6×1/2	B,C,P,U,W	20×12×1/2, 3/8, 5/16	W
6×6×3/8, 5/16	B,C,D,I,P,U,W	20×8×1/2, 3/8, 5/16	W
6×6×1/4, 3/16	A,B,C,D,I,P,U,W,X	20×4×1/2, 3/8, 5/16	W
6×6×1/8	A,B,C,I,P		
		18×12×1/2, 3/8, 5/16	V*
5½×5½×3/8, 5/16, 1/4, 3/16, 1/8	B,I	18×6×1/2, 3/8, 5/16	B,W
		18×6×1/4	B
5×5×1/2	B,C,P,U,W		
5×5×3/8, 5/16	B,C,D,I,P,U,W	16×12×1/2, 3/8, 5/16	V*,W
5×5×1/4	A,B,C,D,I,P,U,W,X	16×8×1/2, 3/8, 5/16	B,W
5×5×3/16	A,B,C,D,I,P,U,V,W,X	16×4×1/2, 3/8, 5/16	B,W
5×5×1/8	A,B,C,I,P,V,W		

*Size is manufactured by Submerged Arc Welding (SAW) process and is not stocked by steel service centers (contact producer for specific requirements). All other sizes are manufactured by Electric Resistance Welding and are available from steel service centers. For the most recent list of producers, please see the latest January or July issue of the AISC magazine *Modern Steel Construction*.

Table 1-5 (cont.).
Principal Producers of Structural Tubing (TS)

A—Acme Rolling Forming Co.	D—Dallas Tube & Rollform	I—Independence Tube Corp.
B—Bull Moose Tube Co.	E—Eugene Welding Co.	P—IPSCO Steel
C—Copperweld Corp.	H—Hanna Steel Corp.	U—UNR-Leavitt, Div. of UNR, Inc.
		V—Valmont Industries, Inc.
		W—Welded Tube Co. of America
		X—EXLTUBE

Nominal Size and Thickness	Producer Code	Nominal Size and Thickness	Producer Code
14×12×1/2, 3/8	V*	7×5×1/2	B,C,P,U,W
14×10×1/2, 3/8, 5/16	B,W	7×5×3/8, 5/16	B,C,I,P,U,W
14×6×5/8	B	7×5×1/4, 3/16	A,B,C,H,I,P,U,W
14×6×1/2, 3/8, 5/16, 1/4	B,W	7×5×1/8	A,B,C,I,P
14×4×5/8	B	7×4×3/8, 5/16	B,C,D,H,I,P,U,W
14×4×1/2, 3/8, 5/16, 1/4	B,W	7×4×1/4, 3/16	A,B,C,D,H,I,P,U,W
14×4×3/16	B	7×4×1/8	A,B,C,H,I,P
		7×3×3/8, 5/16	B,C,D,H,I,P,W
12×10×1/2, 3/8, 5/16, 1/4	B	7×3×1/4, 3/16	A,B,C,D,H,I,P,W,X
12×8×5/8	B	7×3×1/8	A,B,C,D,H,I,P
12×8×1/2, 3/8, 5/16, 1/4	B,C,U,W		
12×8×3/16	B,C,W	6×4×1/2	B,C,P,U,W
12×6×5/8	B	6×4×3/8, 5/16	B,C,D,H,I,P,U,W
12×6×1/2, 3/8, 5/16, 1/4	B,C,U,W	6×4×1/4	A,B,C,D,H,I,P,U,W,X
12×6×3/16		6×4×3/16	A,B,C,D,H,I,P,U,V,W,X
12×4×5/8		6×4×1/8	A,B,C,D,H,I,P,V,W
12×4×1/2, 3/8, 5/16, 1/4, 3/16	B,U,W	6×3×1/2	P,U
12×3×5/16, 1/4, 3/16		6×3×3/8, 5/16	B,C,D,H,I,P,U
12×2×1/4, 3/16	B,U	6×3×1/4	A,B,C,D,H,I,P,U,X
		6×3×3/16	A,B,C,D,H,I,P,U,W,X
10×8×1/2, 3/8, 5/16, 1/4, 3/16	B,C,U,W	6×3×1/8	A,B,C,D,H,I,P,W
10×6×1/2	B,C,U,W	6×2×3/8	H
10×6×3/8, 5/16, 1/4, 3/16		6×2×5/16	I,P,W
10×5×3/8, 5/16, 1/4, 3/16	B,C,D	6×2×1/4, 3/16	A,B,C,D,E,H,I,P,U,W,X
10×4×1/2	B,C,P,U,W	6×2×1/8	A,B,C,D,E,H,I,P,U,W
10×4×3/8, 5/16, 1/4, 3/16	B,C,D,P,U,W		
10×3×3/8, 5/16	D	5×4×3/8, 5/16	I,P,W
10×3×1/4, 3/16	B,D	5×4×1/4, 3/16	B,C,D,I,P,U,W
10×2×5/16	D,P,W	5×3×1/2	C,P,U
10×2×1/4, 3/16	B,D,P,U,W	5×3×3/8, 5/16	B,C,D,H,I,P,U,W
		5×3×1/4, 3/16	A,B,C,D,E,H,I,P,U,W,X
8×6×1/2	B,C,P,U,W	5×3×1/8	A,B,C,D,E,H,I,P,U,W
8×6×3/8, 5/16, 1/4, 3/16	B,C,D,P,U,W	5×2×5/16	I,P,W
8×4×5/8	B	5×2×1/4, 3/16	A,B,C,D,E,H,I,P,U,W,X
8×4×1/2	B,C,P,U,W	5×2×1/8	A,B,C,D,E,H,I,P,U,W
8×4×3/8, 5/16	B,C,D,H,I,P,U,W		
8×4×3/16	A,B,C,D,H,I,P,U,W,X	4×3×5/16	B,I,P,W
8×4×1/8	A,B,D,I,P	4×3×1/4, 3/16	A,B,C,D,E,H,I,P,U,W,X
8×3×1/2	C,P,U	4×3×1/8	A,B,C,D,E,H,I,P,U,W
8×3×3/8, 5/16	B,C,D,I,P,U,W	4×2×3/8	H
8×3×1/4, 3/16	A,B,C,D,I,P,U,W	4×2×5/16	H,I,P,W
8×3×1/8	A,B,C,D,I,P	4×2×1/4, 3/16	A,B,C,D,E,H,I,P,U,W,X
8×2×3/8	H	4×2×1/8	A,B,C,E,H,I,P,U,W
8×2×5/16	H,I,P,W		
8×2×1/4, 3/16	A,B,D,I,P,U,W	3×2×5/16	I
8×2×1/8	A,B,D,I,P	3×2×1/4, 3/16	A,B,C,D,E,H,I,P,U,V,W,X
		3×2×1/8	A,B,C,D,E,H,I,P,U,V,W
		2 1/2×1 1/2×1/4, 3/16	H,X

Refer to most recent January or July issue of Modern Steel Construction Magazine

Table 1-6.
Principal Producers of Steel Tubing (Round)

C—Copperweld Corp. P—IPSCO	U—UNR-Leavitt, Div. of UNR, Inc.	V—Valmont Industries, Inc.	W—Welded Tube Co. of America X—EXLTUBE

Outside Diameter and Thickness	Producer Code	Outside Diameter and Thickness	Producer Code
20.000×.500,.375,.250	P*,W	6.626×.250,.188	P,U,V,W
		6.625×.125	P,V,W
18.000×.500,.375,.250	P*,W		
		6.000×.500,.375,.312	W
16.000×.500	P*,W	6.000×.280	X
16.000×.375,.250	P,W	6.000×.250,.188,.125	V,W
16.000×.188	P,V*		
16.000×.125	V*	5.563×.375	P,U
		5.563×.258	P,U,V,W
14.000×.500,.438,.375,.250	P,W	5.563×.134	P,V,W
14.000×.188	P,V*		
14.000×.125	V*	5.000×.500,.375,.312	P,C,W
		5.000×.258	P,X
12.750×.500,.406,.375	P,W	5.000×.250,.188	C,P,U,V,W
12.750×.188×.125	P,V*	5.000×.125	P,U,V,W
10.750×.500,.365,.250		4.500×.337	P,U,V,W
10.000×.625,.500,.375,.312	C	4.000×.337,.237	X
10.000×.250,.188		4.000×.226,.250,.188	P,V,W
10.000×.125	V		
		3.500×.337	X
9.625×.500	C,U	3.500×.300	P,W
9.625×.375,.312,.250,.188	C,P,U	3.500×.254,.203,.188,.125	P,U,V,W
		3.500×.221	
8.625×.500	C,P,U		
8.625×.375,.322	C,P,U,W		
8.625×.250,.188	C,P,U,V,W	3.000×.300,.216	X
8.625×.125	P,V,W		
		2.875×.276	W
7.000×.500	C,P,U	2.875×.250,.203,.188,.125	P,U,V,W
7.000×.375,.312,.250	C,P,U,W		
7.000×.188	C,P,U,V,W	2.375×.250,.218,.188	P,V,W
7.000×.125	C,P,V,W	2.375×.154,.125	P,U,V,W
6.625×.500,.432	P,U		
6.625×.375,.312,.280	P,U,W		

*Size is manufactured by Submerged Arc Welding (SAW) Process and is typically not stocked by steel service centers. Other sizes are manufactured by Electric Resistance Welding and typically are available from steel service centers. For more information contact the manufacturer or the American Institute for Hollow Structural Sections.

Also, other sizes and wall thicknesses may be available. Contact an individual manufacturer for more details.

Steel Pipe: For availability contact the National Association of Steel Pipe Distributors, Inc.

STRUCTURAL SHAPES

Designations, Dimensions, and Properties

The hot rolled shapes shown in Part 1 of this Manual are published in ASTM Specification A6/A6M, *Standard Specification for General Requirements for Rolled Steel Plates, Shapes, Sheet Piling, and Bars for Structural Use.*

W shapes have essentially parallel flange surfaces. The profile of a W shape of a given nominal depth and weight available from different producers is essentially the same except for the size of fillets between the web and flange.

HP bearing pile shapes have essentially parallel flange surfaces and equal web and flange thicknesses. The profile of an HP shape of a given nominal depth and weight available from different producers is essentially the same.

American Standard Beams (S) and American Standard Channels (C) have a slope of approximately 17 percent (2 in 12 inches) on the inner flange surfaces. The profiles of S and C shapes of a given nominal depth and weight available from different producers are essentially the same.

The letter M designates shapes that cannot be classified as W, HP, or S shapes. Similarly, MC designates channels that cannot be classified as C shapes. Because many of the M and MC shapes are only available from a limited number of producers, or are infrequently rolled, their availability should be checked prior to specifying these shapes. They may or may not have slopes on their inner flange surfaces, dimensions for which may be obtained from the respective producing mills.

The flange thickness given in the table from S, M, C, and MC shapes is the average flange thickness.

In calculating the theoretical weights, properties, and dimensions of the rolled shapes listed in Part 1 of this Manual, fillets and roundings have been included for all shapes except angles. Because of differences in fillet radii among producers, actual properties of rolled shapes may vary slightly from those tabulated. Dimensions for detailing are generally based on the largest theoretical-size fillets produced.

Equal leg and unequal leg angle (L) shapes of the same nominal size available from different producers have profiles which are essentially the same, except for the size of fillet between the legs and the shape of the ends of the legs. The k distance given in the tables for each angle is based on the theoretical largest size fillet available. Availability of certain angles is subject to rolling accumulation and geographical location, and should be checked with material suppliers.

W SHAPES
Dimensions

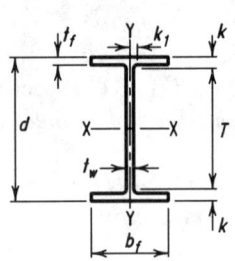

Desig-nation	Area A in.²	Depth d in.		Web Thickness t_w in.		$\dfrac{t_w}{2}$ in.	Flange Width b_f in.		Flange Thickness t_f in.		Distance T in.	k in.	k_1 in.
W44×335	98.3	44.02	44	1.020	1	1/2	15.950	15 3/4	1.770	1 3/4	38 7/16	2 9/16	1 5/16
×290	85.8	43.62	43 5/8	0.870	7/8	7/16	15.830	15 7/8	1.580	1 9/16	38 7/16	2 3/8	1 1/4
×262	77.2	43.31	43 3/8	0.790	13/16	3/8	15.750	15 3/4	1.420	1 7/16	38 7/16	2 3/16	1 3/16
×230	67.7	42.91	42 7/8	0.710	11/16	3/8	15.750	15 3/4	1.220	1 1/4	38 7/16	2	1 1/8
W40×593*	174	42.99	43	1.790	1 13/16	1	16.690	16 3/4	3.230	3 1/4	34 3/16	4 7/16	2 1/16
×503*	148	42.05	42 1/16	1.540	1 9/16	3/4	16.420	16 7/16	2.760	2 3/4	34 3/16	3 15/16	1 15/16
×431	127	41.26	41 1/4	1.340	1 5/16	11/16	16.220	16 1/4	2.360	2 3/8	34 3/16	3 9/16	1 7/8
×372	109	40.63	40 5/8	1.160	1 3/16	9/16	16.060	16 1/16	2.050	2 1/16	34 3/16	3 1/4	1 3/4
×321	94.1	40.08	40 1/16	1.000	1	1/2	15.910	15 7/8	1.770	1 3/4	34 3/16	2 15/16	1 11/16
×297	87.4	39.84	39 7/8	0.930	15/16	1/2	15.825	15 7/8	1.650	1 5/8	34 3/16	3 1/16	1 11/16
×277	81.3	39.69	39 3/4	0.830	13/16	7/16	15.830	15 7/8	1.575	1 9/16	34 3/16	2 3/4	1 5/8
×249	73.3	39.38	39 3/8	0.750	3/4	3/8	15.750	15 3/4	1.420	1 7/16	34 3/16	2 5/8	1 9/16
×215	63.3	38.98	39	0.650	5/8	5/16	15.750	15 3/4	1.220	1 1/4	34 3/16	2 3/8	1 1/2
×199	58.4	38.67	38 5/8	0.650	5/8	5/16	15.750	15 3/4	1.065	1 1/16	34 3/16	2 1/4	1 1/2
×174	51.1	38.20	38 1/4	0.650	5/8	5/16	15.750	15 3/4	0.830	13/16	34 3/16	2	1 1/2
W40×466*	137	42.44	42 7/16	1.67	1 11/16	13/16	12.640	12 5/8	2.950	2 15/16	34 3/16	4 1/8	2
×392*	115	41.57	41 9/16	1.42	1 7/16	11/16	12.360	12 3/8	2.520	2 1/2	34 3/16	3 11/16	1 7/8
×331	97.6	40.79	40 13/16	1.22	1 1/4	5/8	12.170	12 3/16	2.130	2 1/8	34 3/16	3 5/16	1 13/16
×278	81.8	40.16	40 3/16	1.02	1	1/2	11.970	12	1.810	1 13/16	34 3/16	3	1 11/16
×264	77.6	40.00	40	0.960	1	1/2	11.930	12	1.730	1 3/4	34 3/16	2 15/16	1 11/16
×235	68.9	39.69	39 3/4	0.830	13/16	7/16	11.890	11 7/8	1.575	1 9/16	34 3/16	2 3/4	1 5/8
×211	62.0	39.37	39 3/8	0.750	3/4	3/8	11.810	11 3/4	1.415	1 7/16	34 3/16	2 5/8	1 9/16
×183	53.7	38.98	39	0.650	5/8	5/16	11.810	11 3/4	1.220	1 1/4	34 3/16	2 3/8	1 1/2
×167	49.1	38.59	38 5/8	0.650	5/8	5/16	11.810	11 3/4	1.025	1	34 3/16	2 3/16	1 1/2
×149	43.8	38.20	38 1/4	0.630	5/8	5/16	11.810	11 3/4	0.830	13/16	34 3/16	2	1 1/2
W36×848*	249	42.45	42 1/2	2.520	2 1/2	1 1/4	18.130	18 1/8	4.530	4 1/2	31 1/8	5 11/16	2 1/4
×798*	234	41.97	42	2.380	2 3/8	1 3/16	17.990	18	4.290	4 5/16	31 1/8	5 7/16	2 3/16
×650*	190	40.47	40 1/2	1.970	2	1	17.575	17 5/8	3.540	3 9/16	31 1/8	4 11/16	2
×527*	154	39.21	39 1/4	1.610	1 5/8	13/16	17.220	17 1/4	2.910	2 15/16	31 1/8	4 1/16	1 3/4
×439*	128	38.26	38 1/4	1.360	1 3/8	11/16	16.965	17	2.440	2 7/16	31 1/8	3 9/16	1 5/8
×393*	115	37.80	37 3/4	1.220	1 1/4	5/8	16.830	16 7/8	2.200	2 3/16	31 1/8	3 5/16	1 5/8
×359*	105	37.40	37 3/8	1.120	1 1/8	9/16	16.730	16 3/4	2.010	2	31 1/8	3 1/8	1 9/16
×328*	96.4	37.09	37 1/8	1.020	1	1/2	16.630	16 5/8	1.850	1 7/8	31 1/8	3	1 1/2
×300	88.3	36.74	36 3/4	0.945	15/16	1/2	16.655	16 5/8	1.680	1 11/16	31 1/8	2 13/16	1 1/2
×280	82.4	36.52	36 1/2	0.885	7/8	7/16	16.595	16 5/8	1.570	1 9/16	31 1/8	2 11/16	1 1/2
×260	76.5	36.26	36 1/4	0.840	13/16	7/16	16.550	16 1/2	1.440	1 7/16	31 1/8	2 9/16	1 1/2
×245	72.1	36.08	36 1/8	0.800	13/16	7/16	16.510	16 1/2	1.350	1 3/8	31 1/8	2 1/2	1 7/16
×230	67.6	35.90	35 7/8	0.760	3/4	3/8	16.470	16 1/2	1.260	1 1/4	31 1/8	2 3/8	1 7/16

*Group 4 or Group 5 shape. See Notes in Table 1-2.

W SHAPES
Properties

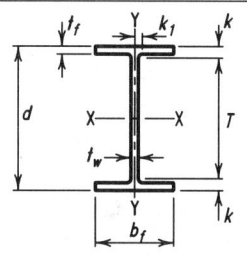

Nom-inal Wt. per ft	Compact Section Criteria					Elastic Properties						Plastic Modulus	
						Axis X-X			Axis Y-Y				
ft	$\frac{b_f}{2t_f}$	$\frac{h}{t_w}$	F_y'''	X_1	$X_2 \times 10^6$	I	S	r	I	S	r	Z_x	Z_y
lb			ksi	ksi	$(1/ksi)^2$	in.^{4}	in.^{3}	in.	in.^{4}	in.^{3}	in.	in.^{3}	in.^{3}
335	4.5	38.1	44	2430	5110	31100	1410	17.8	1200	150	3.49	1620	236
290	5.0	44.7	32	2140	8220	27100	1240	17.8	1050	133	3.50	1420	206
262	5.5	49.2	26	1930	12300	24200	1120	17.7	927	118	3.46	1270	183
230	6.5	54.8	21	1690	21200	20800	969	17.5	796	101	3.43	1100	157
593	2.6	19.1	—	4790	337	50400	2340	17.0	2520	302	3.81	2760	481
503	3.0	22.2	—	4110	620	41700	1980	16.8	2050	250	3.72	2300	394
431	3.4	25.5	—	3550	1100	34800	1690	16.6	1690	208	3.65	1950	327
372	3.9	29.5	—	3100	1860	29600	1460	16.4	1420	177	3.60	1670	277
321	4.5	34.2	55	2690	3240	25100	1250	16.3	1190	150	3.56	1420	234
297	4.8	36.8	47	2500	4240	23200	1170	16.3	1090	138	3.54	1330	215
277	5.0	41.2	38	2350	5370	21900	1100	16.4	1040	132	3.58	1250	204
249	5.5	45.6	31	2120	7940	19500	992	16.3	926	118	3.56	1120	182
215	6.5	52.6	23	1830	14000	16700	858	16.2	796	101	3.54	963	156
199	7.4	52.6	23	1690	20300	14900	769	16.0	695	88.2	3.45	868	137
174	9.5	52.6	23	1500	36000	12200	639	15.5	541	68.8	3.26	715	107
466	2.1	20.5	—	4560	473	36300	1710	16.3	1010	160	2.72	2050	262
392	2.5	24.1	—	3920	851	29900	1440	16.1	803	130	2.64	1710	212
331	2.9	28.0	—	3360	1560	24700	1210	15.9	646	106	2.57	1430	172
278	3.3	33.5	57	2860	2910	20500	1020	15.8	521	87.1	2.52	1190	140
264	3.4	35.6	50	2720	3510	19400	971	15.8	493	82.6	2.52	1130	132
235	3.8	41.2	38	2430	5310	17400	874	15.9	444	74.6	2.54	1010	118
211	4.2	45.6	31	2200	7890	15500	785	15.8	390	66.1	2.51	905	105
183	4.8	52.6	23	1900	13700	13300	682	15.7	336	56.9	2.50	781	89.6
167	5.8	52.6	23	1750	20500	11600	599	15.3	283	47.9	2.40	692	76.0
149	7.1	54.3	22	1610	31400	9780	512	14.9	229	38.8	2.29	597	62.2
848	2.0	12.5	—	7100	71	67400	3170	16.4	4550	501	4.27	3830	799
798	2.1	13.2	—	6720	87	62600	2980	16.4	4200	467	4.24	3570	743
650	2.5	16.0	—	5590	175	48900	2420	16.0	3230	367	4.12	2840	580
527	3.0	19.6	—	4630	365	38300	1950	15.8	2490	289	4.02	2270	454
439	3.5	23.1	—	3900	704	31000	1620	15.6	1990	235	3.95	1860	367
393	3.8	25.8	—	3540	1040	27500	1450	15.5	1750	208	3.90	1660	325
359	4.2	28.1	—	3240	1470	24800	1320	15.4	1570	188	3.87	1510	292
328	4.5	30.9	—	2980	2040	22500	1210	15.3	1420	171	3.84	1380	265
300	5.0	33.3	58	2720	2930	20300	1110	15.2	1300	156	3.83	1260	241
280	5.3	35.6	51	2560	3730	18900	1030	15.1	1200	144	3.81	1170	223
260	5.7	37.5	46	2370	5100	17300	953	15.0	1090	132	3.78	1080	204
245	6.1	39.4	41	2230	6430	16100	895	15.0	1010	123	3.75	1010	190
230	6.5	41.4	37	2100	8190	15000	837	14.9	940	114	3.73	943	176

W SHAPES
Dimensions

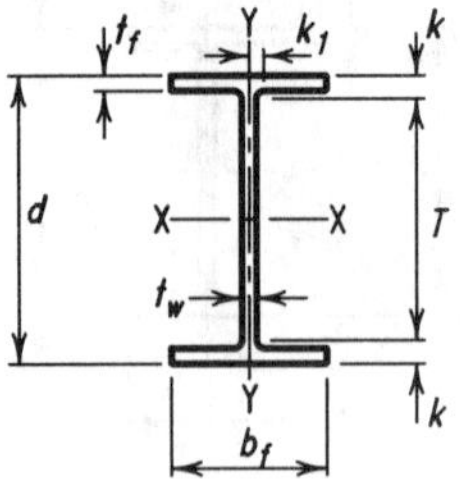

Designation	Area A	Depth d		Web Thickness t_w		Web $t_w/2$	Flange Width b_f		Flange Thickness t_f		Distance T	Distance k	Distance k_1
	in.^{2}	in.		in.		in.	in.		in.		in.	in.	in.
W36×256	75.4	37.43	37 3/8	0.960	1	1/2	12.215	12 1/4	1.730	1 3/4	32 1/8	2 5/8	1 5/16
×232	68.1	37.12	37 1/8	0.870	7/8	7/16	12.120	12 1/8	1.570	1 9/16	32 1/8	2 1/2	1 1/4
×210	61.8	36.69	36 3/4	0.830	13/16	7/16	12.180	12 1/8	1.360	1 3/8	32 1/8	2 5/16	1 1/4
×194	57.0	36.49	36 1/2	0.765	3/4	3/8	12.115	12 1/8	1.260	1 1/4	32 1/8	2 3/16	1 3/16
×182	53.6	36.33	36 3/8	0.725	3/4	3/8	12.075	12 1/8	1.180	1 3/16	32 1/8	2 1/8	1 3/16
×170	50.0	36.17	36 1/8	0.680	11/16	3/8	12.030	12	1.100	1 1/8	32 1/8	2	1 3/16
×160	47.0	36.01	36	0.650	5/8	5/16	12.000	12	1.020	1	32 1/8	1 15/16	1 1/8
×150	44.2	35.85	35 7/8	0.625	5/8	5/16	11.975	12	0.940	15/16	32 1/8	1 7/8	1 1/8
×135	39.7	35.55	35 1/2	0.600	5/8	5/16	11.950	12	0.790	13/16	32 1/8	1 11/16	1 1/8
W33×354*	104	35.55	35 1/2	1.160	1 3/16	5/8	16.100	16 1/8	2.090	2 1/16	29 3/4	2 7/8	1 3/8
×318*	93.5	35.16	35 1/8	1.040	1 1/16	9/16	15.985	16	1.890	1 7/8	29 3/4	2 11/16	1 5/16
×291*	85.6	34.84	34 7/8	0.960	1	1/2	15.905	15 7/8	1.730	1 3/4	29 3/4	2 9/16	1 1/4
×263*	77.4	34.53	34 1/2	0.870	7/8	7/16	15.805	15 3/4	1.570	1 9/16	29 3/4	2 3/8	1 3/16
×241	70.9	34.18	34 1/8	0.830	13/16	7/16	15.860	15 7/8	1.400	1 3/8	29 3/4	2 3/16	1 3/16
×221	65.0	33.93	33 7/8	0.775	3/4	3/8	15.805	15 3/4	1.275	1 1/4	29 3/4	2 1/16	1 3/16
×201	59.1	33.68	33 5/8	0.715	11/16	3/8	15.745	15 3/4	1.150	1 1/8	29 3/4	1 15/16	1 1/8
W33×169	49.5	33.82	33 7/8	0.670	11/16	3/8	11.500	11 1/2	1.220	1 1/4	29 3/4	2 1/16	1 1/8
×152	44.7	33.49	33 1/2	0.635	5/8	5/16	11.565	11 5/8	1.055	1 1/16	29 3/4	1 7/8	1 1/8
×141	41.6	33.30	33 1/4	0.605	5/8	5/16	11.535	11 1/2	0.960	15/16	29 3/4	1 3/4	1 1/16
×130	38.3	33.09	33 1/8	0.580	9/16	5/16	11.510	11 1/2	0.855	7/8	29 3/4	1 11/16	1 1/16
×118	34.7	32.86	32 7/8	0.550	9/16	5/16	11.480	11 1/2	0.740	3/4	29 3/4	1 9/16	1 1/16
W30×477*	140	34.21	34 1/4	1.630	1 5/8	13/16	15.865	15 7/8	2.950	3	26 3/4	3 3/4	1 9/16
×391*	114	33.19	33 1/4	1.360	1 3/8	11/16	15.590	15 5/8	2.440	2 7/16	26 3/4	3 1/4	1 7/16
×326*	95.7	32.40	32 3/8	1.140	1 1/8	9/16	15.370	15 3/8	2.050	2 1/16	26 3/4	2 13/16	1 5/16
×292*	85.7	32.01	32	1.020	1	1/2	15.255	15 1/2	1.850	1 7/8	26 3/4	2 5/8	1 1/4
×261	76.7	31.61	31 5/8	0.930	15/16	1/2	15.155	15 1/8	1.650	1 5/8	26 3/4	2 7/16	1 3/16
×235	69.0	31.30	31 1/4	0.830	13/16	7/16	15.055	15	1.500	1 1/2	26 3/4	2 1/4	1 1/8
×211	62.0	30.94	31	0.775	3/4	3/8	15.105	15 1/8	1.315	1 5/16	26 3/4	2 1/8	1 1/8
×191	56.1	30.68	30 5/8	0.710	11/16	3/8	15.040	15	1.185	1 3/16	26 3/4	1 15/16	1 1/16
×173	50.8	30.44	30 1/2	0.655	5/8	5/16	14.985	15	1.065	1 1/16	26 3/4	1 7/8	1 1/16
W30×148	43.5	30.67	30 5/8	0.650	5/8	5/16	10.480	10 1/2	1.180	1 3/16	26 3/4	2	1
×132	38.9	30.31	30 1/4	0.615	5/8	5/16	10.545	10 1/2	1.000	1	26 3/4	1 3/4	1 1/16
×124	36.5	30.17	30 1/8	0.585	9/16	5/16	10.515	10 1/2	0.930	15/16	26 3/4	1 11/16	1
×116	34.2	30.01	30	0.565	9/16	5/16	10.495	10 1/2	0.850	7/8	26 3/4	1 5/8	1
×108	31.7	29.83	29 7/8	0.545	9/16	5/16	10.475	10 1/2	0.760	3/4	26 3/4	1 9/16	1
×99	29.1	29.65	29 5/8	0.520	1/2	1/4	10.450	10 1/2	0.670	11/16	26 3/4	1 7/16	1
×90	26.4	29.53	29 1/2	0.470	1/2	1/4	10.400	10 3/8	0.610	9/16	26 3/4	1 5/16	1

*Group 4 or Group 5 shape. See Notes in Table 1-2.

W SHAPES
Properties

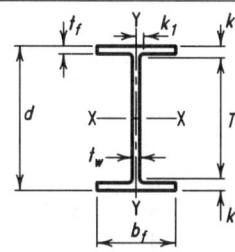

Nominal Wt. per ft	Compact Section Criteria			X_1	$X_2 \times 10^6$	Elastic Properties						Plastic Modulus	
						Axis X-X			Axis Y-Y				
	$\frac{b_f}{2t_f}$	$\frac{h}{t_w}$	F_y'''			I	S	r	I	S	r	Z_x	Z_y
lb			ksi	ksi	$(1/\text{ksi})^2$	in.^{4}	in.^{3}	in.	in.^{4}	in.^{3}	in.	in.^{3}	in.^{3}
256	3.5	33.8	56	2840	2870	16800	895	14.9	528	86.5	2.65	1040	137
232	3.9	37.3	46	2580	4160	15000	809	14.8	468	77.2	2.62	936	122
210	4.5	39.1	42	2320	6560	13200	719	14.6	411	67.5	2.58	833	107
194	4.8	42.4	36	2140	8850	12100	664	14.6	375	61.9	2.56	767	97.7
182	5.1	44.8	32	2020	11300	11300	623	14.5	347	57.6	2.55	718	90.7
170	5.5	47.8	28	1900	14500	10500	580	14.5	320	53.2	2.53	668	83.8
160	5.9	50.0	26	1780	18600	9750	542	14.4	295	49.1	2.50	624	77.3
150	6.4	52.0	24	1680	24200	9040	504	14.3	270	45.1	2.47	581	70.9
135	7.6	54.1	22	1520	38000	7800	439	14.0	225	37.7	2.38	509	59.7
354	3.8	25.8	—	3540	1030	21900	1230	14.5	1460	181	3.74	1420	282
318	4.2	28.8	—	3200	1530	19500	1110	14.4	1290	161	3.71	1270	250
291	4.6	31.2	—	2940	2130	17700	1010	14.4	1160	146	3.69	1150	226
263	5.0	34.5	54	2670	3100	15800	917	14.3	1030	131	3.66	1040	202
241	5.7	36.1	49	2430	4590	14200	829	14.1	932	118	3.63	939	182
221	6.2	38.7	43	2240	6440	12800	757	14.1	840	106	3.59	855	164
201	6.8	41.9	36	2040	9390	11500	684	14.0	749	95.2	3.56	772	147
169	4.7	44.7	32	2160	8150	9290	549	13.7	310	53.9	2.50	629	84.4
152	5.5	47.2	29	1940	12900	8160	487	13.5	273	47.2	2.47	559	73.9
141	6.0	49.6	26	1800	17800	7450	448	13.4	246	42.7	2.43	514	66.9
130	6.7	51.7	24	1660	25100	6710	406	13.2	218	37.9	2.39	467	59.5
118	7.8	54.5	22	1510	37700	5900	359	13.0	187	32.6	2.32	415	51.3
477	2.7	16.6	—	5420	193	26100	1530	13.7	1970	249	3.75	1790	390
391	3.2	19.9	—	4510	386	20700	1250	13.5	1550	198	3.68	1430	310
326	3.7	23.7	—	3860	735	16800	1030	13.2	1240	162	3.61	1190	252
292	4.1	26.5	—	3460	1110	14900	928	13.2	1100	144	3.58	1060	223
261	4.6	29.0	—	3110	1690	13100	827	13.1	959	127	3.54	941	196
235	5.0	32.5	61	2820	2460	11700	746	13.0	855	114	3.52	845	175
211	5.7	34.9	53	2510	3950	10300	663	12.9	757	100	3.49	749	154
191	6.3	38.0	44	2280	5840	9170	598	12.8	673	89.5	3.46	673	138
173	7.0	41.2	38	2070	8540	8200	539	12.7	598	79.8	3.43	605	123
148	4.4	41.5	37	2310	6180	6680	436	12.4	227	43.3	2.28	500	68.0
132	5.3	43.9	33	2050	10500	5770	380	12.2	196	37.2	2.25	437	58.4
124	5.7	46.2	30	1930	13500	5360	355	12.1	181	34.4	2.23	408	54.0
116	6.2	47.8	28	1800	17700	4930	329	12.0	164	31.3	2.19	378	49.2
108	6.9	49.6	26	1680	24200	4470	299	11.9	146	27.9	2.15	346	43.9
99	7.8	51.9	24	1560	34100	3990	269	11.7	128	24.5	2.10	312	38.6
90	8.5	57.5	19	1430	47000	3620	245	11.7	115	22.1	2.09	283	34.7

W SHAPES
Dimensions

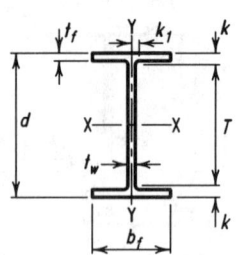

Desig-nation	Area A in.^{2}	Depth d in.		Web Thickness t_w in.		$\dfrac{t_w}{2}$ in.	Flange Width b_f in.		Flange Thickness t_f in.		Distance T in.	k in.	k_1 in.
W27×539*	158	32.52	$32\frac{1}{2}$	1.970	2	1	15.255	$15\frac{1}{4}$	3.540	$3\frac{9}{16}$	24	$4\frac{1}{4}$	$1\frac{5}{8}$
×448*	131	31.42	$31\frac{3}{8}$	1.650	$1\frac{5}{8}$	$\frac{13}{16}$	14.940	15	2.990	3	24	$3\frac{11}{16}$	$1\frac{1}{2}$
×368*	108	30.39	$30\frac{3}{8}$	1.380	$1\frac{3}{8}$	$\frac{11}{16}$	14.665	$14\frac{5}{8}$	2.480	$2\frac{1}{2}$	24	$3\frac{3}{16}$	$1\frac{5}{16}$
×307*	90.2	29.61	$29\frac{5}{8}$	1.160	$1\frac{3}{16}$	$\frac{5}{8}$	14.445	$14\frac{1}{2}$	2.090	$2\frac{1}{16}$	24	$2\frac{13}{16}$	$1\frac{1}{4}$
×258	75.7	28.98	29	0.980	1	$\frac{1}{2}$	14.270	$14\frac{1}{4}$	1.770	$1\frac{3}{4}$	24	$2\frac{1}{2}$	$1\frac{1}{8}$
×235	69.1	28.66	$28\frac{5}{8}$	0.910	$\frac{15}{16}$	$\frac{1}{2}$	14.190	$14\frac{1}{4}$	1.610	$1\frac{5}{8}$	24	$2\frac{5}{16}$	$1\frac{1}{8}$
×217	63.8	28.43	$28\frac{3}{8}$	0.830	$\frac{13}{16}$	$\frac{7}{16}$	14.115	$14\frac{1}{8}$	1.500	$1\frac{1}{2}$	24	$2\frac{3}{16}$	$1\frac{1}{16}$
×194	57.0	28.11	$28\frac{1}{8}$	0.750	$\frac{3}{4}$	$\frac{3}{8}$	14.035	14	1.340	$1\frac{5}{16}$	24	$2\frac{1}{16}$	1
×178	52.3	27.81	$27\frac{3}{4}$	0.725	$\frac{3}{4}$	$\frac{3}{8}$	14.085	$14\frac{1}{8}$	1.190	$1\frac{3}{16}$	24	$1\frac{7}{8}$	$1\frac{1}{16}$
×161	47.4	27.59	$27\frac{5}{8}$	0.660	$\frac{11}{16}$	$\frac{3}{8}$	14.020	14	1.080	$1\frac{1}{16}$	24	$1\frac{13}{16}$	1
×146	42.9	27.38	$27\frac{3}{8}$	0.605	$\frac{5}{8}$	$\frac{5}{16}$	13.965	14	0.975	1	24	$1\frac{11}{16}$	1
W27×129	37.8	27.63	$27\frac{5}{8}$	0.610	$\frac{5}{8}$	$\frac{5}{16}$	10.010	10	1.100	$1\frac{1}{8}$	24	$1\frac{13}{16}$	$\frac{15}{16}$
×114	33.5	27.29	$27\frac{1}{4}$	0.570	$\frac{9}{16}$	$\frac{5}{16}$	10.070	$10\frac{1}{8}$	0.930	$\frac{15}{16}$	24	$1\frac{5}{8}$	$\frac{15}{16}$
×102	30.0	27.09	$27\frac{1}{8}$	0.515	$\frac{1}{2}$	$\frac{1}{4}$	10.015	10	0.830	$\frac{13}{16}$	24	$1\frac{9}{16}$	$\frac{15}{16}$
×94	27.7	26.92	$26\frac{7}{8}$	0.490	$\frac{1}{2}$	$\frac{1}{4}$	9.990	10	0.745	$\frac{3}{4}$	24	$1\frac{7}{16}$	$\frac{15}{16}$
×84	24.8	26.71	$26\frac{3}{4}$	0.460	$\frac{7}{16}$	$\frac{1}{4}$	9.960	10	0.640	$\frac{5}{8}$	24	$1\frac{3}{8}$	$\frac{15}{16}$
W24×492*	144	29.65	$29\frac{5}{8}$	1.970	2	1	14.115	$14\frac{1}{8}$	3.540	$3\frac{9}{16}$	21	$4\frac{5}{16}$	$1\frac{9}{16}$
×408*	119	28.54	$28\frac{1}{2}$	1.650	$1\frac{5}{8}$	$\frac{13}{16}$	13.800	$13\frac{3}{4}$	2.990	3	21	$3\frac{3}{4}$	$1\frac{3}{8}$
×335*	98.4	27.52	$27\frac{1}{2}$	1.380	$1\frac{3}{8}$	$\frac{11}{16}$	13.520	$13\frac{1}{2}$	2.480	$2\frac{1}{2}$	21	$3\frac{1}{4}$	$1\frac{1}{4}$
×279*	82.0	26.73	$26\frac{3}{4}$	1.160	$1\frac{3}{16}$	$\frac{5}{8}$	13.305	$13\frac{1}{4}$	2.090	$2\frac{1}{16}$	21	$2\frac{7}{8}$	$1\frac{1}{8}$
×250*	73.5	26.34	$26\frac{3}{8}$	1.040	$1\frac{1}{16}$	$\frac{9}{16}$	13.185	$13\frac{1}{8}$	1.890	$1\frac{7}{8}$	21	$2\frac{11}{16}$	$1\frac{1}{8}$
×229	67.2	26.02	26	0.960	1	$\frac{1}{2}$	13.110	$13\frac{1}{8}$	1.730	$1\frac{3}{4}$	21	$2\frac{1}{2}$	1
×207	60.7	25.71	$25\frac{3}{4}$	0.870	$\frac{7}{8}$	$\frac{7}{16}$	13.010	13	1.570	$1\frac{9}{16}$	21	$2\frac{3}{8}$	1
×192	56.3	25.47	$25\frac{1}{2}$	0.810	$\frac{13}{16}$	$\frac{7}{16}$	12.950	13	1.460	$1\frac{7}{16}$	21	$2\frac{1}{4}$	1
×176	51.7	25.24	$25\frac{1}{4}$	0.750	$\frac{3}{4}$	$\frac{3}{8}$	12.890	$12\frac{7}{8}$	1.340	$1\frac{5}{16}$	21	$2\frac{1}{8}$	$\frac{15}{16}$
×162	47.7	25.00	25	0.705	$\frac{11}{16}$	$\frac{3}{8}$	12.955	13	1.220	$1\frac{1}{4}$	21	2	$1\frac{1}{16}$
×146	43.0	24.74	$24\frac{3}{4}$	0.650	$\frac{5}{8}$	$\frac{5}{16}$	12.900	$12\frac{7}{8}$	1.090	$1\frac{1}{16}$	21	$1\frac{7}{8}$	$1\frac{1}{16}$
×131	38.5	24.48	$24\frac{1}{2}$	0.605	$\frac{5}{8}$	$\frac{5}{16}$	12.855	$12\frac{7}{8}$	0.960	$\frac{15}{16}$	21	$1\frac{3}{4}$	$1\frac{1}{16}$
×117	34.4	24.26	$24\frac{1}{4}$	0.550	$\frac{9}{16}$	$\frac{5}{16}$	12.800	$12\frac{3}{4}$	0.850	$\frac{7}{8}$	21	$1\frac{5}{8}$	1
×104	30.6	24.06	24	0.500	$\frac{1}{2}$	$\frac{1}{4}$	12.750	$12\frac{3}{4}$	0.750	$\frac{3}{4}$	21	$1\frac{1}{2}$	1
W24×103	30.3	24.53	$24\frac{1}{2}$	0.550	$\frac{9}{16}$	$\frac{5}{16}$	9.000	9	0.980	1	21	$1\frac{3}{4}$	$\frac{13}{16}$
×94	27.7	24.31	$24\frac{1}{4}$	0.515	$\frac{1}{2}$	$\frac{1}{4}$	9.065	$9\frac{1}{8}$	0.875	$\frac{7}{8}$	21	$1\frac{5}{8}$	1
×84	24.7	24.10	$24\frac{1}{8}$	0.470	$\frac{1}{2}$	$\frac{1}{4}$	9.020	9	0.770	$\frac{3}{4}$	21	$1\frac{9}{16}$	$\frac{15}{16}$
×76	22.4	23.92	$23\frac{7}{8}$	0.440	$\frac{7}{16}$	$\frac{1}{4}$	8.990	9	0.680	$\frac{11}{16}$	21	$1\frac{7}{16}$	$\frac{15}{16}$
×68	20.1	23.73	$23\frac{3}{4}$	0.415	$\frac{7}{16}$	$\frac{1}{4}$	8.965	9	0.585	$\frac{9}{16}$	21	$1\frac{3}{8}$	$\frac{15}{16}$
W24×62	18.2	23.74	$23\frac{3}{4}$	0.430	$\frac{7}{16}$	$\frac{1}{4}$	7.040	7	0.590	$\frac{9}{16}$	21	$1\frac{3}{8}$	$\frac{5}{16}$
×55	16.2	23.57	$23\frac{5}{8}$	0.395	$\frac{3}{8}$	$\frac{3}{16}$	7.005	7	0.505	$\frac{1}{2}$	21	$1\frac{5}{16}$	$\frac{15}{16}$

* Group 4 or Group 5 shape. See Notes in Table 1-2.

W SHAPES
Properties

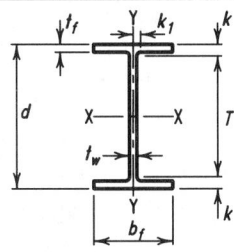

Nom-inal Wt. per ft	Compact Section Criteria			X_1	$X_2 \times 10^6$	Elastic Properties Axis X-X			Axis Y-Y			Plastic Modulus	
	$\dfrac{b_f}{2t_f}$	$\dfrac{h}{t_w}$	F_y'''			I	S	r	I	S	r	Z_x	Z_y
lb			ksi	ksi	$(1/\text{ksi})^2$	in.^{4}	in.^{3}	in.	in.^{4}	in.^{3}	in.	in.^{3}	in.^{3}
539	2.2	12.3	—	7160	66	25500	1570	12.7	2110	277	3.66	1880	437
448	2.5	14.7	—	6070	123	20400	1300	12.5	1670	224	3.57	1530	351
368	3.0	17.6	—	5100	243	16100	1060	12.2	1310	179	3.48	1240	279
307	3.5	20.9	—	4320	463	13100	884	12.0	1050	146	3.42	1020	227
258	4.0	24.7	—	3670	873	10800	742	11.9	859	120	3.37	850	187
235	4.4	26.6	—	3360	1230	9660	674	11.8	768	108	3.33	769	168
217	4.7	29.2	—	3120	1640	8870	624	11.8	704	99.8	3.32	708	154
194	5.2	32.3	61	2800	2520	7820	556	11.7	618	88.1	3.29	628	136
178	5.9	33.4	57	2550	3740	6990	502	11.6	555	78.8	3.26	567	122
161	6.5	36.7	47	2320	5370	6280	455	11.5	497	70.9	3.24	512	109
146	7.2	40.0	40	2110	7900	5630	411	11.4	443	63.5	3.21	461	97.5
129	4.5	39.7	41	2390	5340	4760	345	11.2	184	36.8	2.21	395	57.6
114	5.4	42.5	35	2100	9220	4090	299	11.0	159	31.5	2.18	343	49.3
102	6.0	47.0	29	1890	14000	3620	267	11.0	139	27.8	2.15	305	43.4
94	6.7	49.4	26	1740	19900	3270	243	10.9	124	24.8	2.12	278	38.8
84	7.8	52.7	23	1570	31100	2850	213	10.7	106	21.2	2.07	244	33.2
492	2.0	10.9	—	7950	43	19100	1290	11.5	1670	237	3.41	1550	375
408	2.3	13.1	—	6780	79	15100	1060	11.3	1320	191	3.33	1250	300
335	2.7	15.6	—	5700	156	11900	864	11.0	1030	152	3.23	1020	238
279	3.2	18.6	—	4840	297	9600	718	10.8	823	124	3.17	835	193
250	3.5	20.7	—	4370	436	8490	644	10.7	724	110	3.14	744	171
229	3.8	22.5	—	4020	605	7650	588	10.7	651	99.4	3.11	676	154
207	4.1	24.8	—	3650	876	6820	531	10.6	578	88.8	3.08	606	137
192	4.4	26.6	—	3410	1150	6260	491	10.5	530	81.8	3.07	559	126
176	4.8	28.7	—	3140	1590	5680	450	10.5	479	74.3	3.04	511	115
162	5.3	30.6	—	2870	2260	5170	414	10.4	443	68.4	3.05	468	105
146	5.9	33.2	58	2590	3420	4580	371	10.3	391	60.5	3.01	418	93.2
131	6.7	35.6	50	2330	5290	4020	329	10.2	340	53.0	2.97	370	81.5
117	7.5	39.2	42	2090	8190	3540	291	10.1	297	46.5	2.94	327	71.4
104	8.5	43.1	34	1860	12900	3100	258	10.1	259	40.7	2.91	289	62.4
103	4.6	39.2	42	2400	5280	3000	245	9.96	119	26.5	1.99	280	41.5
94	5.2	41.9	37	2180	7800	2700	222	9.87	109	24.0	1.98	254	37.5
84	5.9	45.9	30	1950	12200	2370	196	9.79	94.4	20.9	1.95	224	32.6
76	6.6	49.0	27	1760	18600	2100	176	9.69	82.5	18.4	1.92	200	28.6
68	7.7	52.0	24	1590	29000	1830	154	9.55	70.4	15.7	1.87	177	24.5
62	6.0	50.1	25	1700	25100	1550	131	9.23	34.5	9.80	1.38	153	15.7
55	6.9	54.6	21	1540	39600	1350	114	9.11	29.1	8.30	1.34	134	13.3

W SHAPES
Dimensions

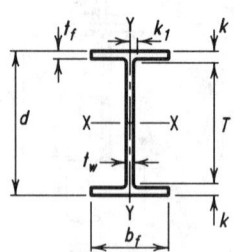

Desig-nation	Area A in.^{2}	Depth d in.	Web Thickness t_w in.	$\frac{t_w}{2}$ in.	Flange Width b_f in.	Flange Thickness t_f in.	T in.	k in.	k_1 in.
W21×201	59.2	23.03 23	0.910 $^{15}/_{16}$	$^1/_2$	12.575 $12^5/_8$	1.630 $1^5/_8$	$18^1/_4$	$2^3/_8$	1
×182	53.6	22.72 $22^3/_4$	0.830 $^{13}/_{16}$	$^7/_{16}$	12.500 $12^1/_2$	1.480 $1^1/_2$	$18^1/_4$	$2^1/_4$	1
×166	48.8	22.48 $22^1/_2$	0.750 $^3/_4$	$^3/_8$	12.420 $12^3/_8$	1.360 $1^3/_8$	$18^1/_4$	$2^1/_8$	$^{15}/_{16}$
×147	43.2	22.06 22	0.720 $^3/_4$	$^3/_8$	12.510 $12^1/_2$	1.150 $1^1/_8$	$18^1/_4$	$1^7/_8$	$1^1/_{16}$
×132	38.8	21.83 $21^7/_8$	0.650 $^5/_8$	$^5/_{16}$	12.440 $12^1/_2$	1.035 $1^1/_{16}$	$18^1/_4$	$1^{13}/_{16}$	1
×122	35.9	21.68 $21^5/_8$	0.600 $^5/_8$	$^5/_{16}$	12.390 $12^3/_8$	0.960 $^{15}/_{16}$	$18^1/_4$	$1^{11}/_{16}$	1
×111	32.7	21.51 $21^1/_2$	0.550 $^9/_{16}$	$^5/_{16}$	12.340 $12^3/_8$	0.875 $^7/_8$	$18^1/_4$	$1^5/_8$	$^{15}/_{16}$
×101	29.8	21.36 $21^3/_8$	0.500 $^1/_2$	$^1/_4$	12.290 $12^1/_4$	0.800 $^{13}/_{16}$	$18^1/_4$	$1^9/_{16}$	$^{15}/_{16}$
W21×93	27.3	21.62 $21^5/_8$	0.580 $^9/_{16}$	$^5/_{16}$	8.420 $8^3/_8$	0.930 $^{15}/_{16}$	$18^1/_4$	$1^{11}/_{16}$	1
×83	24.3	21.43 $21^3/_8$	0.515 $^1/_2$	$^1/_4$	8.355 $8^3/_8$	0.835 $^{13}/_{16}$	$18^1/_4$	$1^9/_{16}$	$^{15}/_{16}$
×73	21.5	21.24 $21^1/_4$	0.455 $^7/_{16}$	$^1/_4$	8.295 $8^1/_4$	0.740 $^3/_4$	$18^1/_4$	$1^1/_2$	$^{15}/_{16}$
×68	20.0	21.13 $21^1/_8$	0.430 $^7/_{16}$	$^1/_4$	8.270 $8^1/_4$	0.685 $^{11}/_{16}$	$18^1/_4$	$1^7/_{16}$	$^7/_8$
×62	18.3	20.99 21	0.400 $^3/_8$	$^3/_{16}$	8.240 $8^1/_4$	0.615 $^5/_8$	$18^1/_4$	$1^3/_8$	$^7/_8$
W21×57	16.7	21.06 21	0.405 $^3/_8$	$^3/_{16}$	6.555 $6^1/_2$	0.650 $^5/_8$	$18^1/_4$	$1^3/_8$	$^7/_8$
×50	14.7	20.83 $20^7/_8$	0.380 $^3/_8$	$^3/_{16}$	6.530 $6^1/_2$	0.535 $^9/_{16}$	$18^1/_4$	$1^5/_{16}$	$^7/_8$
×44	13.0	20.66 $20^5/_8$	0.350 $^3/_8$	$^3/_{16}$	6.500 $6^1/_2$	0.450 $^7/_{16}$	$18^1/_4$	$1^3/_{16}$	$^7/_8$
W18×311*	91.5	22.32 $22^3/_8$	1.520 $1^1/_2$	$^3/_4$	12.005 12	2.740 $2^3/_4$	$15^1/_2$	$3^7/_{16}$	$1^3/_{16}$
×283*	83.2	21.85 $21^7/_8$	1.400 $1^3/_8$	$^{11}/_{16}$	11.890 $11^7/_8$	2.500 $2^1/_2$	$15^1/_2$	$3^3/_{16}$	$1^3/_{16}$
×258*	75.9	21.46 $21^1/_2$	1.280 $1^1/_4$	$^5/_8$	11.770 $11^3/_4$	2.300 $2^5/_{16}$	$15^1/_2$	3	$1^1/_8$
×234*	68.8	21.06 21	1.160 $1^3/_{16}$	$^5/_8$	11.650 $11^5/_8$	2.110 $2^1/_8$	$15^1/_2$	$2^3/_4$	1
×211*	62.1	20.67 $20^5/_8$	1.060 $1^1/_{16}$	$^9/_{16}$	11.555 $11^1/_2$	1.910 $1^{15}/_{16}$	$15^1/_2$	$2^9/_{16}$	1
×192	56.4	20.35 $20^3/_8$	0.960 1	$^1/_2$	11.455 $11^1/_2$	1.750 $1^3/_4$	$15^1/_2$	$2^7/_{16}$	$^{15}/_{16}$
×175	51.3	20.04 20	0.890 $^7/_8$	$^7/_{16}$	11.375 $11^3/_8$	1.590 $1^9/_{16}$	$15^1/_2$	$2^1/_4$	$^7/_8$
×158	46.3	19.72 $19^3/_4$	0.810 $^{13}/_{16}$	$^7/_{16}$	11.300 $11^1/_4$	1.440 $1^7/_{16}$	$15^1/_2$	$2^1/_8$	$^7/_8$
×143	42.1	19.49 $19^1/_2$	0.730 $^3/_4$	$^3/_8$	11.220 $11^1/_4$	1.320 $1^5/_{16}$	$15^1/_2$	2	$^{13}/_{16}$
×130	38.2	19.25 $19^1/_4$	0.670 $^{11}/_{16}$	$^3/_8$	11.160 $11^1/_8$	1.200 $1^3/_{16}$	$15^1/_2$	$1^7/_8$	$^{13}/_{16}$
W18×119	35.1	18.97 19	0.655 $^5/_8$	$^5/_{16}$	11.265 $11^1/_4$	1.060 $1^1/_{16}$	$15^1/_2$	$1^3/_4$	$^{15}/_{16}$
×106	31.1	18.73 $18^3/_4$	0.590 $^9/_{16}$	$^5/_{16}$	11.200 $11^1/_4$	0.940 $^{15}/_{16}$	$15^1/_2$	$1^5/_8$	$^{15}/_{16}$
×97	28.5	18.59 $18^5/_8$	0.535 $^9/_{16}$	$^5/_{16}$	11.145 $11^1/_8$	0.870 $^7/_8$	$15^1/_2$	$1^9/_{16}$	$^7/_8$
×86	25.3	18.39 $18^3/_8$	0.480 $^1/_2$	$^1/_4$	11.090 $11^1/_8$	0.770 $^3/_4$	$15^1/_2$	$1^7/_{16}$	$^7/_8$
×76	22.3	18.21 $18^1/_4$	0.425 $^7/_{16}$	$^1/_4$	11.035 11	0.680 $^{11}/_{16}$	$15^1/_2$	$1^3/_8$	$^{13}/_{16}$
W18×71	20.8	18.47 $18^1/_2$	0.495 $^1/_2$	$^1/_4$	7.635 $7^5/_8$	0.810 $^{13}/_{16}$	$15^1/_2$	$1^1/_2$	$^7/_8$
×65	19.1	18.35 $18^3/_8$	0.450 $^7/_{16}$	$^1/_4$	7.590 $7^5/_8$	0.750 $^3/_4$	$15^1/_2$	$1^7/_{16}$	$^7/_8$
×60	17.6	18.24 $18^1/_4$	0.415 $^7/_{16}$	$^1/_4$	7.555 $7^1/_2$	0.695 $^{11}/_{16}$	$15^1/_2$	$1^3/_8$	$^{13}/_{16}$
×55	16.2	18.11 $18^1/_8$	0.390 $^3/_8$	$^3/_{16}$	7.530 $7^1/_2$	0.630 $^5/_8$	$15^1/_2$	$1^5/_{16}$	$^{13}/_{16}$
×50	14.7	17.99 18	0.355 $^3/_8$	$^3/_{16}$	7.495 $7^1/_2$	0.570 $^9/_{16}$	$15^1/_2$	$1^1/_4$	$^{13}/_{16}$
W18×46	13.5	18.06 18	0.360 $^3/_8$	$^3/_{16}$	6.060 6	0.605 $^5/_8$	$15^1/_2$	$1^1/_4$	$^{13}/_{16}$
×40	11.8	17.90 $17^7/_8$	0.315 $^5/_{16}$	$^3/_{16}$	6.015 6	0.525 $^1/_2$	$15^1/_2$	$1^3/_{16}$	$^{13}/_{16}$
×35	10.3	17.70 $17^3/_4$	0.300 $^5/_{16}$	$^3/_{16}$	6.000 6	0.425 $^7/_{16}$	$15^1/_2$	$1^1/_8$	$^3/_4$

*Group 4 or Group 5 shape. See Notes in Table 1-2.

W SHAPES
Properties

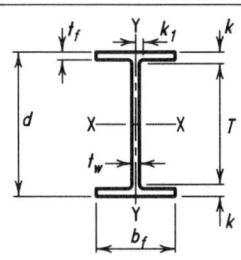

Nom-inal Wt. per ft	Compact Section Criteria			X_1	$X_2 \times 10^6$	Elastic Properties						Plastic Modulus	
						Axis X-X			Axis Y-Y				
ft	$\dfrac{b_f}{2t_f}$	$\dfrac{h}{t_w}$	F_y'''	X_1	$X_2 \times 10^6$	I	S	r	I	S	r	Z_x	Z_y
lb			ksi	ksi	$(1/\text{ksi})^2$	in.^{4}	in.^{3}	in.	in.^{4}	in.^{3}	in.	in.^{3}	in.^{3}
201	3.9	20.6	—	4290	453	5310	461	9.47	542	86.1	3.02	530	133
182	4.2	22.6	—	3910	649	4730	417	9.40	483	77.2	3.00	476	119
166	4.6	24.9	—	3590	904	4280	380	9.36	435	70.1	2.98	432	108
147	5.4	26.1	—	3140	1590	3630	329	9.17	376	60.1	2.95	373	92.6
132	6.0	28.9	—	2840	2350	3220	295	9.12	333	53.5	2.93	333	82.3
122	6.5	31.3	—	2630	3160	2960	273	9.09	305	49.2	2.92	307	75.6
111	7.1	34.1	55	2400	4510	2670	249	9.05	274	44.5	2.90	279	68.2
101	7.7	37.5	45	2200	6400	2420	227	9.02	248	40.3	2.89	253	61.7
93	4.5	32.3	61	2680	3460	2070	192	8.70	92.9	22.1	1.84	221	34.7
83	5.0	36.4	48	2400	5250	1830	171	8.67	81.4	19.5	1.83	196	30.5
73	5.6	41.2	38	2140	8380	1600	151	8.64	70.6	17.0	1.81	172	26.6
68	6.0	43.6	34	2000	10900	1480	140	8.60	64.7	15.7	1.80	160	24.4
62	6.7	46.9	29	1820	15900	1330	127	8.54	57.5	13.9	1.77	144	21.7
57	5.0	46.3	30	1960	13100	1170	111	8.36	30.6	9.35	1.35	129	14.8
50	6.1	49.4	26	1730	22600	984	94.5	8.18	24.9	7.64	1.30	110	12.2
44	7.2	53.6	22	1550	36600	843	81.6	8.06	20.7	6.36	1.26	95.4	10.2
311	2.2	10.6	—	8160	38	6960	624	8.72	795	132	2.95	753	207
283	2.4	11.5	—	7520	52	6160	564	8.61	704	118	2.91	676	185
258	2.6	12.5	—	6920	71	5510	514	8.53	628	107	2.88	611	166
234	2.8	13.8	—	6360	97	4900	466	8.44	558	95.8	2.85	549	149
211	3.0	15.1	—	5800	140	4330	419	8.35	493	85.3	2.82	490	132
192	3.3	16.7	—	5320	194	3870	380	8.28	440	76.8	2.79	442	119
175	3.6	18.0	—	4870	274	3450	344	8.20	391	68.8	2.76	398	106
158	3.9	19.8	—	4430	396	3060	310	8.12	347	61.4	2.74	356	94.8
143	4.2	21.9	—	4060	557	2750	282	8.09	311	55.5	2.72	322	85.4
130	4.6	23.9	—	3710	789	2460	256	8.03	278	49.9	2.70	291	76.7
119	5.3	24.5	—	3340	1210	2190	231	7.90	253	44.9	2.69	261	69.1
106	6.0	27.2	—	2990	1880	1910	204	7.84	220	39.4	2.66	230	60.5
97	6.4	30.0	—	2750	2580	1750	188	7.82	201	36.1	2.65	211	55.3
86	7.2	33.4	57	2460	4060	1530	166	7.77	175	31.6	2.63	186	48.4
76	8.1	37.8	45	2180	6520	1330	146	7.73	152	27.6	2.61	163	42.2
71	4.7	32.4	61	2680	3310	1170	127	7.50	60.3	15.8	1.70	145	24.7
65	5.1	35.7	50	2470	4540	1070	117	7.49	54.8	14.4	1.69	133	22.5
60	5.4	38.7	43	2290	6080	984	108	7.47	50.1	13.3	1.69	123	20.6
55	6.0	41.2	38	2110	8540	890	98.3	7.41	44.9	11.9	1.67	112	18.5
50	6.6	45.2	31	1920	12400	800	88.9	7.38	40.1	10.7	1.65	101	16.6
46	5.0	44.6	32	2060	10100	712	78.8	7.25	22.5	7.43	1.29	90.7	11.7
40	5.7	51.0	25	1810	17200	612	68.4	7.21	19.1	6.35	1.27	78.4	9.95
35	7.1	53.5	22	1590	30300	510	57.6	7.04	15.3	5.12	1.22	66.5	8.06

W SHAPES
Dimensions

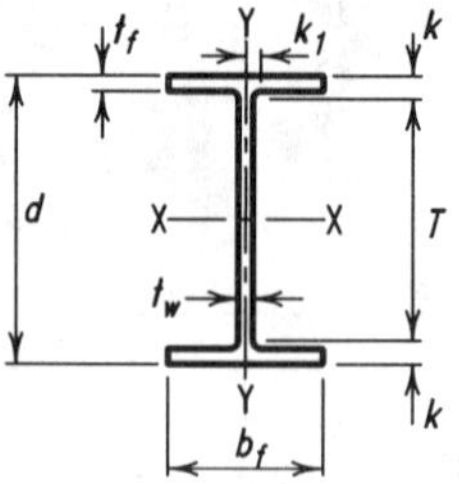

Desig-nation	Area A	Depth d		Web Thickness t_w		Web $\frac{t_w}{2}$	Flange Width b_f		Flange Thickness t_f		Distance T	Distance k	Distance k_1
	in.^{2}	in.		in.		in.	in.		in.		in.	in.	in.
W16×100	29.4	16.97	17	0.585	9/16	5/16	10.425	10 3/8	0.985	1	13 5/8	1 11/16	15/16
×89	26.2	16.75	16 3/4	0.525	1/2	1/4	10.365	10 3/8	0.875	7/8	13 5/8	1 9/16	7/8
×77	22.6	16.52	16 1/2	0.455	7/16	1/4	10.295	10 1/4	0.760	3/4	13 5/8	1 7/16	7/8
×67	19.7	16.33	16 3/8	0.395	3/8	3/16	10.235	10 1/4	0.665	11/16	13 5/8	1 3/8	13/16
W16×57	16.8	16.43	16 3/8	0.430	7/16	1/4	7.120	7 1/8	0.715	11/16	13 5/8	1 3/8	7/8
×50	14.7	16.26	16 1/4	0.380	3/8	3/16	7.070	7 1/8	0.630	5/8	13 5/8	1 5/16	13/16
×45	13.3	16.13	16 1/8	0.345	3/8	3/16	7.035	7	0.565	9/16	13 5/8	1 1/4	13/16
×40	11.8	16.01	16	0.305	5/16	3/16	6.995	7	0.505	1/2	13 5/8	1 3/16	13/16
×36	10.6	15.86	15 7/8	0.295	5/16	3/16	6.985	7	0.430	7/16	13 5/8	1 1/8	3/4
W16×31	9.12	15.88	15 7/8	0.275	1/4	1/8	5.525	5 1/2	0.440	7/16	13 5/8	1 1/8	3/4
×26	7.68	15.69	15 3/4	0.250	1/4	1/8	5.500	5 1/2	0.345	3/8	13 5/8	1 1/16	3/4
W14×808*	237	22.84	22 7/8	3.740	3 3/4	1 7/8	18.560	18 1/2	5.120	5 1/8	11 1/4	5 13/16	2 1/2
×730*	215	22.42	22 3/8	3.070	3 1/16	1 9/16	17.890	17 7/8	4.910	4 15/16	11 1/4	5 9/16	2 3/16
×665*	196	21.64	21 5/8	2.830	2 13/16	1 7/16	17.650	17 5/8	4.520	4 1/2	11 1/4	5 3/16	2 1/16
×605*	178	20.92	20 7/8	2.595	2 5/8	1 5/16	17.415	17 3/8	4.160	4 3/16	11 1/4	4 13/16	1 15/16
×550*	162	20.24	20 1/4	2.380	2 3/8	1 3/16	17.200	17 1/4	3.820	3 13/16	11 1/4	4 1/2	1 13/16
×500*	147	19.60	19 5/8	2.190	2 3/16	1 1/8	17.010	17	3.500	3 1/2	11 1/4	4 3/16	1 3/4
×455*	134	19.02	19	2.015	2	1	16.835	16 7/8	3.210	3 3/16	11 1/4	3 7/8	1 5/8
W14×426*	125	18.67	18 5/8	1.875	1 7/8	15/16	16.695	16 3/4	3.035	3 1/16	11 1/4	3 11/16	1 9/16
×398*	117	18.29	18 1/4	1.770	1 3/4	7/8	16.590	16 5/8	2.845	2 7/8	11 1/4	3 1/2	1 1/2
×370*	109	17.92	17 7/8	1.655	1 5/8	13/16	16.475	16 1/2	2.660	2 11/16	11 1/4	3 5/16	1 7/16
×342*	101	17.54	17 1/2	1.540	1 9/16	13/16	16.360	16 3/8	2.470	2 1/2	11 1/4	3 1/8	1 3/8
×311*	91.4	17.12	17 1/8	1.410	1 7/16	3/4	16.230	16 1/4	2.260	2 1/4	11 1/4	2 15/16	1 5/16
×283*	83.3	16.74	16 3/4	1.290	1 5/16	11/16	16.110	16 1/8	2.070	2 1/16	11 1/4	2 3/4	1 1/4
×257*	75.6	16.38	16 3/8	1.175	1 3/16	5/8	15.995	16	1.890	1 7/8	11 1/4	2 9/16	1 3/16
×233*	68.5	16.04	16	1.070	1 1/16	9/16	15.890	15 7/8	1.720	1 3/4	11 1/4	2 3/8	1 3/16
×211	62.0	15.72	15 3/4	0.980	1	1/2	15.800	15 3/4	1.560	1 9/16	11 1/4	2 1/4	1 1/8
×193	56.8	15.48	15 1/2	0.890	7/8	7/16	15.710	15 3/4	1.440	1 7/16	11 1/4	2 1/8	1 1/16
×176	51.8	15.22	15 1/4	0.830	13/16	7/16	15.650	15 5/8	1.310	1 5/16	11 1/4	2	1 1/16
×159	46.7	14.98	15	0.745	3/4	3/8	15.565	15 5/8	1.190	1 3/16	11 1/4	1 7/8	1
×145	42.7	14.78	14 3/4	0.680	11/16	3/8	15.500	15 1/2	1.090	1 1/16	11 1/4	1 3/4	1

*Group 4 or Group 5 shape. See Notes in Table 1-2.

W SHAPES
Properties

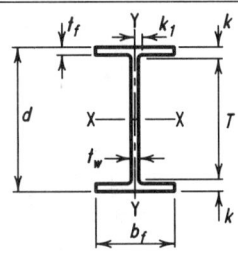

Nom-inal Wt. per ft	Compact Section Criteria					Elastic Properties						Plastic Modulus	
						Axis X-X			Axis Y-Y				
	$\dfrac{b_f}{2t_f}$	$\dfrac{h}{t_w}$	F_y'''	X_1	$X_2 \times 10^6$	I	S	r	I	S	r	Z_x	Z_y
lb			ksi	ksi	$(1/\text{ksi})^2$	in.^{4}	in.^{3}	in.	in.^{4}	in.^{3}	in.	in.^{3}	in.^{3}
100	5.3	24.3	—	3450	1040	1490	175	7.10	186	35.7	2.51	198	54.9
89	5.9	27.0	—	3090	1630	1300	155	7.05	163	31.4	2.49	175	48.1
77	6.8	31.2	—	2680	2790	1110	134	7.00	138	26.9	2.47	150	41.1
67	7.7	35.9	50	2350	4690	954	117	6.96	119	23.2	2.46	130	35.5
57	5.0	33.0	59	2650	3400	758	92.2	6.72	43.1	12.1	1.60	105	18.9
50	5.6	37.4	46	2340	5530	659	81.0	6.68	37.2	10.5	1.59	92.0	16.3
45	6.2	41.2	38	2120	8280	586	72.7	6.65	32.8	9.34	1.57	82.3	14.5
40	6.9	46.6	30	1890	12900	518	64.7	6.63	28.9	8.25	1.57	72.9	12.7
36	8.1	48.1	28	1700	20800	448	56.5	6.51	24.5	7.00	1.52	64.0	10.8
31	6.3	51.6	24	1740	20000	375	47.2	6.41	12.4	4.49	1.17	54.0	7.03
26	8.0	56.8	20	1470	40900	301	38.4	6.26	9.59	3.49	1.12	44.2	5.48
808	1.8	3.4	—	18900	1.45	16000	1400	8.21	5510	594	4.82	1834	927
730	1.8	3.7	—	17500	1.90	14300	1280	8.17	4720	527	4.69	1660	816
665	2.0	4.0	—	16300	2.50	12400	1150	7.98	4170	472	4.62	1480	730
605	2.1	4.4	—	15100	3.20	10800	1040	7.80	3680	423	4.55	1320	652
550	2.3	4.8	—	14200	4.20	9430	931	7.63	3250	378	4.49	1180	583
500	2.4	5.2	—	13100	5.50	8210	838	7.48	2880	339	4.43	1050	522
455	2.6	5.7	—	12200	7.30	7190	756	7.33	2560	304	4.38	936	468
426	2.8	6.1	—	11500	8.90	6600	707	7.26	2360	283	4.34	869	434
398	2.9	6.4	—	10900	11.0	6000	656	7.16	2170	262	4.31	801	402
370	3.1	6.9	—	10300	13.9	5440	607	7.07	1990	241	4.27	736	370
342	3.3	7.4	—	9600	17.9	4900	559	6.98	1810	221	4.24	672	338
311	3.6	8.1	—	8820	24.4	4330	506	6.88	1610	199	4.20	603	304
283	3.9	8.8	—	8120	33.4	3840	459	6.79	1440	179	4.17	542	274
257	4.2	9.7	—	7460	46.1	3400	415	6.71	1290	161	4.13	487	246
233	4.6	10.7	—	6820	64.9	3010	375	6.63	1150	145	4.10	436	221
211	5.1	11.6	—	6230	91.8	2660	338	6.55	1030	130	4.07	390	198
193	5.5	12.8	—	5740	125	2400	310	6.50	931	119	4.05	355	180
176	6.0	13.7	—	5280	173	2140	281	6.43	838	107	4.02	320	163
159	6.5	15.3	—	4790	249	1900	254	6.38	748	96.2	4.00	287	146
145	7.1	16.8	—	4400	348	1710	232	6.33	677	87.3	3.98	260	133

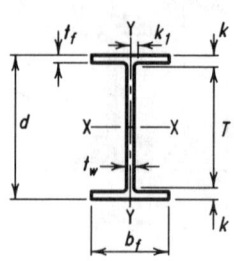

W SHAPES
Dimensions

Desig-nation	Area A	Depth d		Web			Flange				Distance		
				Thickness t_w		$\dfrac{t_w}{2}$	Width b_f		Thickness t_f		T	k	k_1
	in.^{2}	in.		in.		in.	in.		in.		in.	in.	in.
W14×132	38.8	14.66	14⅝	0.645	⅝	5⁄16	14.725	14¾	1.030	1	11¼	1¹¹⁄16	15⁄16
×120	35.3	14.48	14½	0.590	9⁄16	5⁄16	14.670	14⅝	0.940	15⁄16	11¼	1⅝	15⁄16
×109	32.0	14.32	14⅜	0.525	½	¼	14.605	14⅝	0.860	⅞	11¼	1⁹⁄16	⅞
×99	29.1	14.16	14⅛	0.485	½	¼	14.565	14⅝	0.780	¾	11¼	1⁷⁄16	⅞
×90	26.5	14.02	14	0.440	7⁄16	¼	14.520	14½	0.710	11⁄16	11¼	1⅜	⅞
W14×82	24.1	14.31	14¼	0.510	½	¼	10.130	10⅛	0.855	⅞	11	1⅝	1
×74	21.8	14.17	14⅛	0.450	7⁄16	¼	10.070	10⅛	0.785	13⁄16	11	1⁹⁄16	15⁄16
×68	20.0	14.04	14	0.415	7⁄16	¼	10.035	10	0.720	¾	11	1½	15⁄16
×61	17.9	13.89	13⅞	0.375	⅜	3⁄16	9.995	10	0.645	⅝	11	1⁷⁄16	15⁄16
W14×53	15.6	13.92	13⅞	0.370	⅜	3⁄16	8.060	8	0.660	11⁄16	11	1⁷⁄16	15⁄16
×48	14.1	13.79	13¾	0.340	5⁄16	3⁄16	8.030	8	0.595	⅝	11	1⅜	⅞
×43	12.6	13.66	13⅝	0.305	5⁄16	3⁄16	7.995	8	0.530	½	11	1⁵⁄16	⅞
W14×38	11.2	14.10	14⅛	0.310	5⁄16	3⁄16	6.770	6¾	0.515	½	12	1¹⁄16	⅝
×34	10.0	13.98	14	0.285	5⁄16	3⁄16	6.745	6¾	0.455	7⁄16	12	1	⅝
×30	8.85	13.84	13⅞	0.270	¼	⅛	6.730	6¾	0.385	⅜	12	15⁄16	⅝
W14×26	7.69	13.91	13⅞	0.255	¼	⅛	5.025	5	0.420	7⁄16	12	15⁄16	9⁄16
×22	6.49	13.74	13¾	0.230	¼	⅛	5.000	5	0.335	5⁄16	12	⅞	9⁄16

W SHAPES
Properties

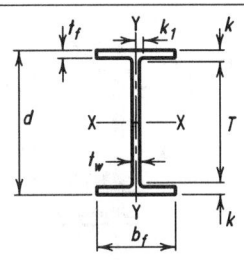

Nom-inal Wt. per ft	Compact Section Criteria			X_1	$X_2 \times 10^6$	Elastic Properties						Plastic Modulus	
						Axis X-X			Axis Y-Y				
	$\dfrac{b_f}{2t_f}$	$\dfrac{h}{t_w}$	F_y'''	X_1	$X_2 \times 10^6$	I	S	r	I	S	r	Z_x	Z_y
lb			ksi	ksi	$(1/ksi)^2$	in.^{4}	in.^{3}	in.	in.^{4}	in.^{3}	in.	in.^{3}	in.^{3}
132	7.1	17.7	—	4180	428	1530	209	6.28	548	74.5	3.76	234	113
120	7.8	19.3	—	3830	601	1380	190	6.24	495	67.5	3.74	212	102
109	8.5	21.7	—	3490	853	1240	173	6.22	447	61.2	3.73	192	92.7
99	9.3	23.5	—	3190	1220	1110	157	6.17	402	55.2	3.71	173	83.6
90	10.2	25.9	—	2900	1750	999	143	6.14	362	49.9	3.70	157	75.6
82	5.9	22.4	—	3600	846	882	123	6.05	148	29.3	2.48	139	44.8
74	6.4	25.3	—	3290	1190	796	112	6.04	134	26.6	2.48	126	40.6
68	7.0	27.5	—	3020	1650	723	103	6.01	121	24.2	2.46	115	36.9
61	7.7	30.4	—	2720	2460	640	92.2	5.98	107	21.5	2.45	102	32.8
53	6.1	30.8	—	2830	2250	541	77.8	5.89	57.7	14.3	1.92	87.1	22.0
48	6.7	33.5	57	2580	3220	485	70.3	5.85	51.4	12.8	1.91	78.4	19.6
43	7.5	37.4	46	2320	4900	428	62.7	5.82	45.2	11.3	1.89	69.6	17.3
38	6.6	39.6	41	2190	6850	385	54.6	5.87	26.7	7.88	1.55	61.5	12.1
34	7.4	43.1	35	1970	10600	340	48.6	5.83	23.3	6.91	1.53	54.6	10.6
30	8.7	45.4	31	1750	17600	291	42.0	5.73	19.6	5.82	1.49	47.3	8.99
26	6.0	48.1	28	1890	13900	245	35.3	5.65	8.91	3.54	1.08	40.2	5.54
22	7.5	53.3	22	1610	27300	199	29.0	5.54	7.00	2.80	1.04	33.2	4.39

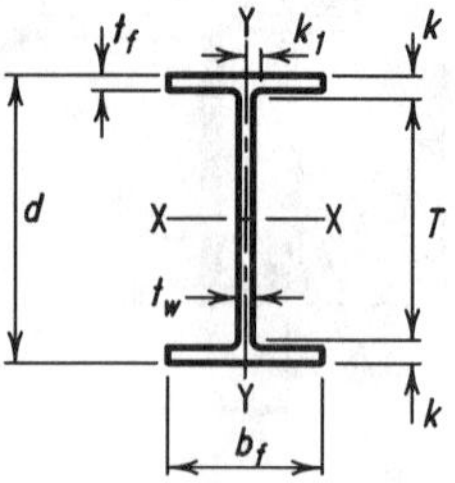

W SHAPES
Dimensions

Designation	Area A	Depth d		Web			Flange				Distance		
				Thickness t_w		$\frac{t_w}{2}$	Width b_f		Thickness t_f		T	k	k_1
	in.^{2}	in.		in.		in.	in.		in.		in.	in.	in.
W12×336*	98.8	16.82	16 7/8	1.775	1 3/4	7/8	13.385	13 3/8	2.955	2 15/16	9 1/2	3 11/16	1 1/2
×305*	89.6	16.32	16 3/8	1.625	1 5/8	13/16	13.235	13 1/4	2.705	2 11/16	9 1/2	3 7/16	1 7/16
×279*	81.9	15.85	15 7/8	1.530	1 1/2	3/4	13.140	13 1/8	2.470	2 1/2	9 1/2	3 3/16	1 3/8
×252*	74.1	15.41	15 3/8	1.395	1 3/8	11/16	13.005	13	2.250	2 1/4	9 1/2	2 15/16	1 5/16
×230*	67.7	15.05	15	1.285	1 5/16	11/16	12.895	12 7/8	2.070	2 1/16	9 1/2	2 3/4	1 1/4
×210*	61.8	14.71	14 3/4	1.180	1 3/16	5/8	12.790	12 3/4	1.900	1 7/8	9 1/2	2 5/8	1 1/4
×190	55.8	14.38	14 3/8	1.060	1 1/16	9/16	12.670	12 5/8	1.735	1 3/4	9 1/2	2 7/16	1 3/16
×170	50.0	14.03	14	0.960	15/16	1/2	12.570	12 5/8	1.560	1 9/16	9 1/2	2 1/4	1 1/8
×152	44.7	13.71	13 3/4	0.870	7/8	7/16	12.480	12 1/2	1.400	1 3/8	9 1/2	2 1/8	1 1/16
×136	39.9	13.41	13 3/8	0.790	13/16	7/16	12.400	12 3/8	1.250	1 1/4	9 1/2	1 15/16	1
×120	35.3	13.12	13 1/8	0.710	11/16	3/8	12.320	12 3/8	1.105	1 1/8	9 1/2	1 13/16	1
×106	31.2	12.89	12 7/8	0.610	5/8	5/16	12.220	12 1/4	0.990	1	9 1/2	1 11/16	15/16
×96	28.2	12.71	12 3/4	0.550	9/16	5/16	12.160	12 1/8	0.900	7/8	9 1/2	1 5/8	7/8
×87	25.6	12.53	12 1/2	0.515	1/2	1/4	12.125	12 1/8	0.810	13/16	9 1/2	1 1/2	7/8
×79	23.2	12.38	12 3/8	0.470	1/2	1/4	12.080	12 1/8	0.735	3/4	9 1/2	1 7/16	7/8
×72	21.1	12.25	12 1/4	0.430	7/16	1/4	12.040	12	0.670	11/16	9 1/2	1 3/8	7/8
×65	19.1	12.12	12 1/8	0.390	3/8	3/16	12.000	12	0.605	5/8	9 1/2	1 5/16	13/16
W12×58	17.0	12.19	12 1/4	0.360	3/8	3/16	10.010	10	0.640	5/8	9 1/2	1 3/8	13/16
×53	15.6	12.06	12	0.345	3/8	3/16	9.995	10	0.575	9/16	9 1/2	1 1/4	13/16
W12×50	14.7	12.19	12 1/4	0.370	3/8	3/16	8.080	8 1/8	0.640	5/8	9 1/2	1 3/8	13/16
×45	13.2	12.06	12	0.335	5/16	3/16	8.045	8	0.575	9/16	9 1/2	1 1/4	13/16
×40	11.8	11.94	12	0.295	5/16	3/16	8.005	8	0.515	1/2	9 1/2	1 1/4	3/4
W12×35	10.3	12.50	12 1/2	0.300	5/16	3/16	6.560	6 1/2	0.520	1/2	10 1/2	1	9/16
×30	8.79	12.34	12 3/8	0.260	1/4	1/8	6.520	6 1/2	0.440	7/16	10 1/2	15/16	1/2
×26	7.65	12.22	12 1/4	0.230	1/4	1/8	6.490	6 1/2	0.380	3/8	10 1/2	7/8	1/2
W12×22	6.48	12.31	12 1/4	0.260	1/4	1/8	4.030	4	0.425	7/16	10 1/2	7/8	1/2
×19	5.57	12.16	12 1/8	0.235	1/4	1/8	4.005	4	0.350	3/8	10 1/2	13/16	1/2
×16	4.71	11.99	12	0.220	1/4	1/8	3.990	4	0.265	1/4	10 1/2	3/4	1/2
×14	4.16	11.91	11 7/8	0.200	3/16	1/8	3.970	4	0.225	1/4	10 1/2	11/16	1/2

*Group 4 or Group 5 shape. See Notes in Table 1-2.

W SHAPES
Properties

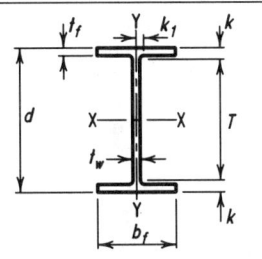

Nom-inal Wt. per ft	Compact Section Criteria			X₁	X₂ × 10⁶	Elastic Properties						Plastic Modulus	
						Axis X-X			Axis Y-Y				
	b_f	h	F_y'''	X_1	$X_2 \times 10^6$	I	S	r	I	S	r	Z_x	Z_y
lb	$\dfrac{b_f}{2t_f}$	$\dfrac{h}{t_w}$	ksi	ksi	(1/ksi)²	in.⁴	in.³	in.	in.⁴	in.³	in.	in.³	in.³
336	2.3	5.5	—	12800	6.05	4060	483	6.41	1190	177	3.47	603	274
305	2.4	6.0	—	11800	8.17	3550	435	6.29	1050	159	3.42	537	244
279	2.7	6.3	—	11000	10.8	3110	393	6.16	937	143	3.38	481	220
252	2.9	7.0	—	10100	14.7	2720	353	6.06	828	127	3.34	428	196
230	3.1	7.6	—	9390	19.7	2420	321	5.97	742	115	3.31	386	177
210	3.4	8.2	—	8670	26.6	2140	292	5.89	664	104	3.28	348	159
190	3.7	9.2	—	7940	37.0	1890	263	5.82	589	93.0	3.25	311	143
170	4.0	10.1	—	7190	54.0	1650	235	5.74	517	82.3	3.22	275	126
152	4.5	11.2	—	6510	79.3	1430	209	5.66	454	72.8	3.19	243	111
136	5.0	12.3	—	5850	119	1240	186	5.58	398	64.2	3.16	214	98.0
120	5.6	13.7	—	5240	184	1070	163	5.51	345	56.0	3.13	186	85.4
106	6.2	15.9	—	4660	285	933	145	5.47	301	49.3	3.11	164	75.1
96	6.8	17.7	—	4250	405	833	131	5.44	270	44.4	3.09	147	67.5
87	7.5	18.9	—	3880	586	740	118	5.38	241	39.7	3.07	132	60.4
79	8.2	20.7	—	3530	839	662	107	5.34	216	35.8	3.05	119	54.3
72	9.0	22.6	—	3230	1180	597	97.4	5.31	195	32.4	3.04	108	49.2
65	9.9	24.9	—	2940	1720	533	87.9	5.28	174	29.1	3.02	96.8	44.1
58	7.8	27.0	—	3070	1470	475	78.0	5.28	107	21.4	2.51	86.4	32.5
53	8.7	28.1	—	2820	2100	425	70.6	5.23	95.8	19.2	2.48	77.9	29.1
50	6.3	26.2	—	3170	1410	394	64.7	5.18	56.3	13.9	1.96	72.4	21.4
45	7.0	29.0	—	2870	2070	350	58.1	5.15	50.0	12.4	1.94	64.7	19.0
40	7.8	32.9	59	2580	3110	310	51.9	5.13	44.1	11.0	1.93	57.5	16.8
35	6.3	36.2	49	2420	4340	285	45.6	5.25	24.5	7.47	1.54	51.2	11.5
30	7.4	41.8	37	2090	7950	238	38.6	5.21	20.3	6.24	1.52	43.1	9.56
26	8.5	47.2	29	1820	13900	204	33.4	5.17	17.3	5.34	1.51	37.2	8.17
22	4.7	41.8	37	2160	8640	156	25.4	4.91	4.66	2.31	0.847	29.3	3.66
19	5.7	46.2	30	1880	15600	130	21.3	4.82	3.76	1.88	0.822	24.7	2.98
16	7.5	49.4	26	1610	32000	103	17.1	4.67	2.82	1.41	0.773	20.1	2.26
14	8.8	54.3	22	1450	49300	88.6	14.9	4.62	2.36	1.19	0.753	17.4	1.90

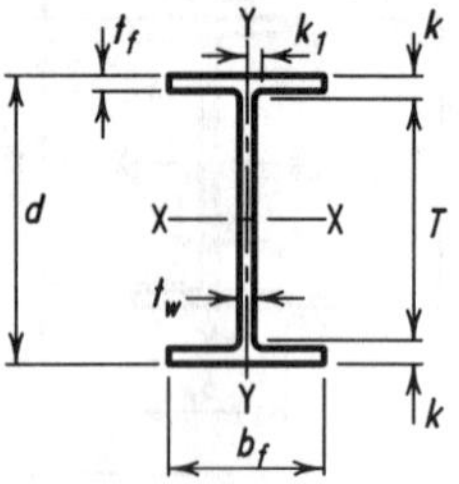

W SHAPES
Dimensions

Designation	Area A	Depth d		Web			Flange				Distance		
				Thickness t_w		$\frac{t_w}{2}$	Width b_f		Thickness t_f		T	k	k_1
	in.^{2}	in.		in.		in.	in.		in.		in.	in.	in.
W10×112	32.9	11.36	11 3/8	0.755	3/4	3/8	10.415	10 3/8	1.250	1 1/4	7 5/8	1 7/8	15/16
×100	29.4	11.10	11 1/8	0.680	11/16	3/8	10.340	10 3/8	1.120	1 1/8	7 5/8	1 3/4	7/8
×88	25.9	10.84	10 7/8	0.605	5/8	5/16	10.265	10 1/4	0.990	1	7 5/8	1 5/8	13/16
×77	22.6	10.60	10 5/8	0.530	1/2	1/4	10.190	10 1/4	0.870	7/8	7 5/8	1 1/2	13/16
×68	20.0	10.40	10 3/8	0.470	1/2	1/4	10.130	10 1/8	0.770	3/4	7 5/8	1 3/8	3/4
×60	17.6	10.22	10 1/4	0.420	7/16	1/4	10.080	10 1/8	0.680	11/16	7 5/8	1 5/16	3/4
×54	15.8	10.09	10 1/8	0.370	3/8	3/16	10.030	10	0.615	5/8	7 5/8	1 1/4	11/16
×49	14.4	9.98	10	0.340	5/16	3/16	10.000	10	0.560	9/16	7 5/8	1 3/16	11/16
W10×45	13.3	10.10	10 1/8	0.350	3/8	3/16	8.020	8	0.620	5/8	7 5/8	1 1/4	11/16
×39	11.5	9.92	9 7/8	0.315	5/16	3/16	7.985	8	0.530	1/2	7 5/8	1 1/8	11/16
×33	9.71	9.73	9 3/4	0.290	5/16	3/16	7.960	8	0.435	7/16	7 5/8	1 1/16	11/16
W10×30	8.84	10.47	10 1/2	0.300	5/16	3/16	5.810	5 3/4	0.510	1/2	8 5/8	15/16	1/2
×26	7.61	10.33	10 3/8	0.260	1/4	1/8	5.770	5 3/4	0.440	7/16	8 5/8	7/8	1/2
×22	6.49	10.17	10 1/8	0.240	1/4	1/8	5.750	5 3/4	0.360	3/8	8 5/8	3/4	1/2
W10×19	5.62	10.24	10 1/4	0.250	1/4	1/8	4.020	4	0.395	3/8	8 5/8	13/16	1/2
×17	4.99	10.11	10 1/8	0.240	1/4	1/8	4.010	4	0.330	5/16	8 5/8	3/4	1/2
×15	4.41	9.99	10	0.230	1/4	1/8	4.000	4	0.270	1/4	8 5/8	11/16	7/16
×12	3.54	9.87	9 7/8	0.190	3/16	1/8	3.960	4	0.210	3/16	8 5/8	5/8	7/16

W SHAPES
Properties

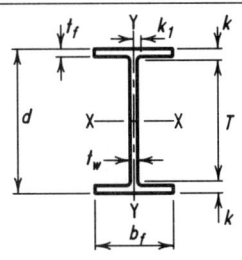

Nominal Wt. per ft	Compact Section Criteria					Elastic Properties						Plastic Modulus	
						Axis X-X			Axis Y-Y				
	b_f	h	F_y'''	X_1	$X_2 \times 10^6$	I	S	r	I	S	r	Z_x	Z_y
lb	$2t_f$	t_w	ksi	ksi	$(1/\text{ksi})^2$	in.^{4}	in.^{3}	in.	in.^{4}	in.^{3}	in.	in.^{3}	in.^{3}
112	4.2	10.4	—	7080	56.7	716	126	4.66	236	45.3	2.68	147	69.2
100	4.6	11.6	—	6400	83.8	623	112	4.60	207	40.0	2.65	130	61.0
88	5.2	13.0	—	5680	132	534	98.5	4.54	179	34.8	2.63	113	53.1
77	5.9	14.8	—	5010	213	455	85.9	4.49	154	30.1	2.60	97.6	45.9
68	6.6	16.7	—	4460	334	394	75.7	4.44	134	26.4	2.59	85.3	40.1
60	7.4	18.7	—	3970	525	341	66.7	4.39	116	23.0	2.57	74.6	35.0
54	8.2	21.2	—	3580	778	303	60.0	4.37	103	20.6	2.56	66.6	31.3
49	8.9	23.1	—	3280	1090	272	54.6	4.35	93.4	18.7	2.54	60.4	28.3
45	6.5	22.5	—	3650	758	248	49.1	4.32	53.4	13.3	2.01	54.9	20.3
39	7.5	25.0	—	3190	1300	209	42.1	4.27	45.0	11.3	1.98	46.8	17.2
33	9.1	27.1	—	2710	2510	170	35.0	4.19	36.6	9.20	1.94	38.8	14.0
30	5.7	29.5	—	2890	2160	170	32.4	4.38	16.7	5.75	1.37	36.6	8.84
26	6.6	34.0	55	2500	3790	144	27.9	4.35	14.1	4.89	1.36	31.3	7.50
22	8.0	36.9	47	2150	7170	118	23.2	4.27	11.4	3.97	1.33	26.0	6.10
19	5.1	35.4	51	2420	5160	96.3	18.8	4.14	4.29	2.14	0.874	21.6	3.35
17	6.1	36.9	47	2210	7820	81.9	16.2	4.05	3.56	1.78	0.844	18.7	2.80
15	7.4	38.5	43	1930	14300	68.9	13.8	3.95	2.89	1.45	0.810	16.0	2.30
12	9.4	46.6	30	1550	35400	53.8	10.9	3.90	2.18	1.10	0.785	12.6	1.74

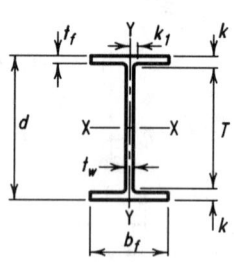

W SHAPES
Dimensions

Desig-nation	Area A in.^{2}	Depth d in.		Web Thickness t_w in.	$\frac{t_w}{2}$ in.	Flange Width b_f in.		Flange Thickness t_f in.		Distance T in.	k in.	k_1 in.
W8×67	19.7	9.00	9	0.570	$^9/_{16}$	$^5/_{16}$	8.280	$8^1/_4$	0.935	$^{15}/_{16}$	$6^1/_8$	$1^7/_{16}$	$1^1/_{16}$
×58	17.1	8.75	$8^3/_4$	0.510	$^1/_2$	$^1/_4$	8.220	$8^1/_4$	0.810	$^{13}/_{16}$	$6^1/_8$	$1^5/_{16}$	$1^1/_{16}$
×48	14.1	8.50	$8^1/_2$	0.400	$^3/_8$	$^3/_{16}$	8.110	$8^1/_8$	0.685	$^{11}/_{16}$	$6^1/_8$	$1^3/_{16}$	$^5/_8$
×40	11.7	8.25	$8^1/_4$	0.360	$^3/_8$	$^3/_{16}$	8.070	$8^1/_8$	0.560	$^9/_{16}$	$6^1/_8$	$1^1/_{16}$	$^5/_8$
×35	10.3	8.12	$8^1/_8$	0.310	$^5/_{16}$	$^3/_{16}$	8.020	8	0.495	$^1/_2$	$6^1/_8$	1	$^9/_{16}$
×31	9.13	8.00	8	0.285	$^5/_{16}$	$^3/_{16}$	7.995	8	0.435	$^7/_{16}$	$6^1/_8$	$^{15}/_{16}$	$^9/_{16}$
W8×28	8.25	8.06	8	0.285	$^5/_{16}$	$^3/_{16}$	6.535	$6^1/_2$	0.465	$^7/_{16}$	$6^1/_8$	$^{15}/_{16}$	$^9/_{16}$
×24	7.08	7.93	$7^7/_8$	0.245	$^1/_4$	$^1/_8$	6.495	$6^1/_2$	0.400	$^3/_8$	$6^1/_8$	$^7/_8$	$^9/_{16}$
W8×21	6.16	8.28	$8^1/_4$	0.250	$^1/_4$	$^1/_8$	5.270	$5^1/_4$	0.400	$^3/_8$	$6^5/_8$	$^{13}/_{16}$	$^1/_2$
×18	5.26	8.14	$8^1/_8$	0.230	$^1/_4$	$^1/_8$	5.250	$5^1/_4$	0.330	$^5/_{16}$	$6^5/_8$	$^3/_4$	$^7/_{16}$
W8×15	4.44	8.11	$8^1/_8$	0.245	$^1/_4$	$^1/_8$	4.015	4	0.315	$^5/_{16}$	$6^5/_8$	$^3/_4$	$^1/_2$
×13	3.84	7.99	8	0.230	$^1/_4$	$^1/_8$	4.000	4	0.255	$^1/_4$	$6^5/_8$	$^{11}/_{16}$	$^7/_{16}$
×10	2.96	7.89	$7^7/_8$	0.170	$^3/_{16}$	$^1/_8$	3.940	4	0.205	$^3/_{16}$	$6^5/_8$	$^5/_8$	$^7/_{16}$
W6×25	7.34	6.38	$6^3/_8$	0.320	$^5/_{16}$	$^3/_{16}$	6.080	$6^1/_8$	0.455	$^7/_{16}$	$4^3/_4$	$^{13}/_{16}$	$^7/_{16}$
×20	5.87	6.20	$6^1/_4$	0.260	$^1/_4$	$^1/_8$	6.020	6	0.365	$^3/_8$	$4^3/_4$	$^3/_4$	$^7/_{16}$
×15	4.43	5.99	6	0.230	$^1/_4$	$^1/_8$	5.990	6	0.260	$^1/_4$	$4^3/_4$	$^5/_8$	$^3/_8$
W6×16	4.74	6.28	$6^1/_4$	0.260	$^1/_4$	$^1/_8$	4.030	4	0.405	$^3/_8$	$4^3/_4$	$^3/_4$	$^7/_{16}$
×12	3.55	6.03	6	0.230	$^1/_4$	$^1/_8$	4.000	4	0.280	$^1/_4$	$4^3/_4$	$^5/_8$	$^3/_8$
×9	2.68	5.90	$5^7/_8$	0.170	$^3/_{16}$	$^1/_8$	3.940	4	0.215	$^3/_{16}$	$4^3/_4$	$^9/_{16}$	$^3/_8$
W5×19	5.54	5.15	$5^1/_8$	0.270	$^1/_4$	$^1/_8$	5.030	5	0.430	$^7/_{16}$	$3^1/_2$	$^{13}/_{16}$	$^7/_{16}$
×16	4.68	5.01	5	0.240	$^1/_4$	$^1/_8$	5.000	5	0.360	$^3/_8$	$3^1/_2$	$^3/_4$	$^7/_{16}$
W4×13	3.83	4.16	$4^1/_8$	0.280	$^1/_4$	$^1/_8$	4.060	4	0.345	$^3/_8$	$2^3/_4$	$^{11}/_{16}$	$^7/_{16}$

W SHAPES
Properties

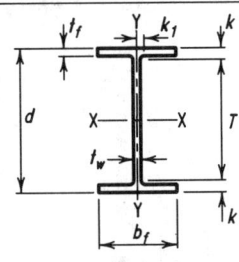

Nominal Wt. per ft	Compact Section Criteria			X_1	$X_2 \times 10^6$	Elastic Properties						Plastic Modulus	
						Axis X-X			Axis Y-Y				
	$\dfrac{b_f}{2t_f}$	$\dfrac{h}{t_w}$	F_y'''	X_1	$X_2 \times 10^6$	I	S	r	I	S	r	Z_x	Z_y
lb			ksi	ksi	$(1/\text{ksi})^2$	in.^{4}	in.^{3}	in.	in.^{4}	in.^{3}	in.	in.^{3}	in.^{3}
67	4.4	11.1	—	6620	73.9	272	60.4	3.72	88.6	21.4	2.12	70.2	32.7
58	5.1	12.4	—	5820	122	228	52.0	3.65	75.1	18.3	2.10	59.8	27.9
48	5.9	15.8	—	4860	238	184	43.3	3.61	60.9	15.0	2.08	49.0	22.9
40	7.2	17.6	—	4080	474	146	35.5	3.53	49.1	12.2	2.04	39.8	18.5
35	8.1	20.4	—	3610	761	127	31.2	3.51	42.6	10.6	2.03	34.7	16.1
31	9.2	22.2	—	3230	1180	110	27.5	3.47	37.1	9.27	2.02	30.4	14.1
28	7.0	22.2	—	3480	931	98.0	24.3	3.45	21.7	6.63	1.62	27.2	10.1
24	8.1	25.8	—	3020	1610	82.8	20.9	3.42	18.3	5.63	1.61	23.2	8.57
21	6.6	27.5	—	2890	2090	75.3	18.2	3.49	9.77	3.71	1.26	20.4	5.69
18	8.0	29.9	—	2490	3890	61.9	15.2	3.43	7.97	3.04	1.23	17.0	4.66
15	6.4	28.1	—	2670	3440	48.0	11.8	3.29	3.41	1.70	0.876	13.6	2.67
13	7.8	29.9	—	2370	5780	39.6	9.91	3.21	2.73	1.37	0.843	11.4	2.15
10	9.6	40.5	39	1760	17900	30.8	7.81	3.22	2.09	1.06	0.841	8.87	1.66
25	6.7	15.5	—	4410	369	53.4	16.7	2.70	17.1	5.61	1.52	18.9	8.56
20	8.2	19.1	—	3550	846	41.4	13.4	2.66	13.3	4.41	1.50	14.9	6.72
15	11.5	21.6	—	2740	2470	29.1	9.72	2.56	9.32	3.11	1.46	10.8	4.75
16	5.0	19.1	—	4010	591	32.1	10.2	2.60	4.43	2.20	0.966	11.7	3.39
12	7.1	21.6	—	3100	1740	22.1	7.31	2.49	2.99	1.50	0.918	8.30	2.32
9	9.2	29.2	—	2360	4980	16.4	5.56	2.47	2.19	1.11	0.905	6.23	1.72
19	5.8	14.0	—	5140	192	26.2	10.2	2.17	9.13	3.63	1.28	11.6	5.53
16	6.9	15.8	—	4440	346	21.3	8.51	2.13	7.51	3.00	1.27	9.59	4.57
13	5.9	10.6	—	5560	154	11.3	5.46	1.72	3.86	1.90	1.00	6.28	2.92

M SHAPES
Dimensions

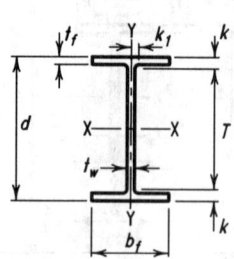

Desig-nation	Area A	Depth d		Web Thickness t_w		$\frac{t_w}{2}$	Flange Width b_f		Flange Thickness t_f		Distance T	k	k_1
	in.^{2}	in.		in.		in.	in.		in.		in.	in.	in.
M12×11.8	3.48	12.00	12	0.177	$^3/_{16}$	$^1/_{16}$	3.065	$3^1/_{16}$	0.225	$^1/_4$	$10^{15}/_{16}$	$^1/_2$	$^3/_8$
×10.8	3.20	11.97	12	0.162	$^3/_{16}$	$^1/_{16}$	3.065	$3^1/_{16}$	0.206	$^3/_{16}$	$10^7/_8$	$^1/_2$	$^3/_8$
M10×9	2.67	10.00	10	0.157	$^3/_{16}$	$^1/_{16}$	2.690	$2^{11}/_{16}$	0.206	$^3/_{16}$	$8^7/_8$	$^1/_2$	$^3/_8$
×8	2.38	9.95	10	0.139	$^1/_8$	$^1/_{16}$	2.690	$2^{11}/_{16}$	0.183	$^3/_{16}$	$8^{13}/_{16}$	$^1/_2$	$^3/_8$
M8×6.5	1.92	8.00	8	0.133	$^1/_8$	$^1/_{16}$	2.281	$2^1/_4$	0.189	$^3/_{16}$	$6^7/_8$	$^1/_2$	$^3/_8$
M5×18.9*	5.55	5.00	5	0.316	$^5/_{16}$	$^3/_{16}$	5.003	5	0.416	$^7/_{16}$	$3^1/_4$	$^7/_8$	$^1/_2$

*This shape has tapered flanges while all other M shapes have parallel flanges.

M SHAPES
Properties

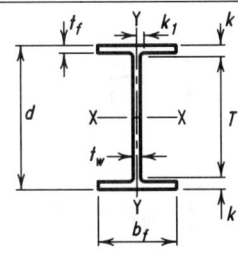

Nom-inal Wt. per ft	Compact Section Criteria					Elastic Properties						Plastic Modulus	
						Axis X-X			Axis Y-Y				
	$\dfrac{b_f}{2t_f}$	$\dfrac{h}{t_w}$	F_y'''	X_1	$X_2 \times 10^6$	I	S	r	I	S	r	Z_x	Z_y
lb			ksi	ksi	$(1/\text{ksi})^2$	in.^{4}	in.^{3}	in.	in.^{4}	in.^{3}	in.	in.^{3}	in.^{3}
11.8	6.8	61.4	17	1380	63700	71.7	12.1	4.54	1.09	0.709	0.559	14.3	1.16
10.8	7.4	67.0	14	1290	83400	65.8	11.1	4.54	0.995	0.649	0.558	13.1	1.05
9	6.5	56.4	20	1450	50500	38.5	7.82	3.80	0.673	0.501	0.502	9.21	0.815
8	7.3	63.7	16	1250	90000	34.3	6.99	3.80	0.597	0.444	0.502	8.20	0.718
6.5	6.1	51.7	24	1700	24800	18.1	4.62	3.07	0.371	0.325	0.439	5.40	0.527
18.9	6.0	11.2	—	5710	134	24.1	9.63	2.08	7.86	3.14	1.19	11.0	5.02

S SHAPES
Dimensions

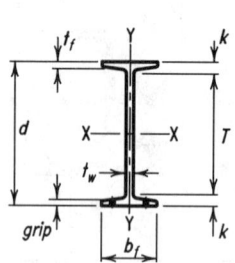

Desig-nation	Area A in.²	Depth d in.		Web Thickness t_w in.		$\dfrac{t_w}{2}$ in.	Flange Width b_f in.		Flange Thickness t_f in.		Distance T in.	Distance k in.	Grip in.	Max. Flge. Fastener in.
S24×121	35.6	24.50	24½	0.800	13/16	7/16	8.050	8	1.090	1 1/16	20½	2	1⅛	1
×106	31.2	24.50	24½	0.620	⅝	5/16	7.870	7⅞	1.090	1 1/16	20½	2	1⅛	1
S24×100	29.3	24.00	24	0.745	¾	⅜	7.245	7¼	0.870	⅞	20½	1¾	⅞	1
×90	26.5	24.00	24	0.625	⅝	5/16	7.125	7⅛	0.870	⅞	20½	1¾	⅞	1
×80	23.5	24.00	24	0.500	½	¼	7.000	7	0.870	⅞	20½	1¾	⅞	1
S20×96	28.2	20.30	20¼	0.800	13/16	7/16	7.200	7¼	0.920	15/16	16¾	1¾	15/16	1
×86	25.3	20.30	20¼	0.660	11/16	⅜	7.060	7	0.920	15/16	16¾	1¾	15/16	1
S20×75	22.0	20.00	20	0.635	⅝	5/16	6.385	6⅜	0.795	13/16	16¾	1⅝	13/16	⅞
×66	19.4	20.00	20	0.505	½	¼	6.255	6¼	0.795	13/16	16¾	1⅝	13/16	⅞
S18×70	20.6	18.00	18	0.711	11/16	⅜	6.251	6¼	0.691	11/16	15	1½	11/16	⅞
×54.7	16.1	18.00	18	0.461	7/16	¼	6.001	6	0.691	11/16	15	1½	11/16	⅞
S15×50	14.7	15.00	15	0.550	9/16	5/16	5.640	5⅝	0.622	⅝	12¼	1⅜	9/16	¾
×42.9	12.6	15.00	15	0.411	7/16	¼	5.501	5½	0.622	⅝	12¼	1⅜	9/16	¾
S12×50	14.7	12.00	12	0.687	11/16	⅜	5.477	5½	0.659	11/16	9⅛	1 7/16	11/16	¾
×40.8	12.0	12.00	12	0.462	7/16	¼	5.252	5¼	0.659	11/16	9⅛	1 7/16	⅝	¾
S12×35	10.3	12.00	12	0.428	7/16	¼	5.078	5⅛	0.544	9/16	9⅝	1 3/16	½	¾
×31.8	9.35	12.00	12	0.350	⅜	3/16	5.000	5	0.544	9/16	9⅝	1 3/16	½	¾
S10×35	10.3	10.00	10	0.594	⅝	5/16	4.944	5	0.491	½	7¾	1⅛	½	¾
×25.4	7.46	10.00	10	0.311	5/16	3/16	4.661	4⅝	0.491	½	7¾	1⅛	½	¾
S8×23	6.77	8.00	8	0.441	7/16	¼	4.171	4⅛	0.425	7/16	6	1	7/16	¾
×18.4	5.41	8.00	8	0.271	¼	⅛	4.001	4	0.425	7/16	6	1	7/16	¾
S6×17.25	5.07	6.00	6	0.465	7/16	¼	3.565	3⅝	0.359	⅜	4¼	⅞	⅜	⅝
×12.5	3.67	6.00	6	0.232	¼	⅛	3.332	3⅜	0.359	⅜	4¼	⅞	⅜	—
S5×10	2.94	5.00	5	0.214	3/16	⅛	3.004	3	0.326	5/16	3⅜	13/16	5/16	—
S4×9.5	2.79	4.00	4	0.326	5/16	3/16	2.796	2¾	0.293	5/16	2½	¾	5/16	—
×7.7	2.26	4.00	4	0.193	3/16	⅛	2.663	2⅝	0.293	5/16	2½	¾	5/16	—
S3×7.5	2.21	3.00	3	0.349	⅜	3/16	2.509	2½	0.260	¼	1⅝	11/16	¼	—
×5.7	1.67	3.00	3	0.170	3/16	⅛	2.330	2⅜	0.260	¼	1⅝	11/16	¼	—

S SHAPES
Properties

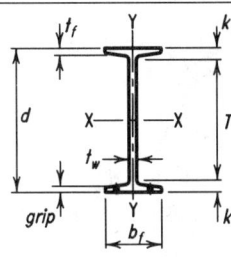

Nom-inal Wt. per ft	Compact Section Criteria					Elastic Properties						Plastic Modulus	
						Axis X-X			Axis Y-Y				
	b_f	h	F_y'''	X_1	$X_2 \times 10^6$	I	S	r	I	S	r	Z_x	Z_y
lb	$2t_f$	t_w	ksi	ksi	$(1/\text{ksi})^2$	in.^{4}	in.^{3}	in.	in.^{4}	in.^{3}	in.	in.^{3}	in.^{3}
121	3.7	26.4	—	3310	1770	3160	258	9.43	83.3	20.7	1.53	306	36.2
106	3.6	34.1	55	2960	2470	2940	240	9.71	77.1	19.6	1.57	279	33.2
100	4.2	28.3	—	3000	2940	2390	199	9.02	47.7	13.2	1.27	240	23.9
90	4.1	33.7	56	2710	4090	2250	187	9.21	44.9	12.6	1.30	222	22.3
80	4.0	42.1	36	2450	5480	2100	175	9.47	42.2	12.1	1.34	204	20.7
96	3.9	21.6	—	3730	1160	1670	165	7.71	50.2	13.9	1.33	198	24.9
86	3.8	26.2	—	3350	1630	1580	155	7.89	46.8	13.3	1.36	183	23.0
75	4.0	27.1	—	3140	2290	1280	128	7.62	29.8	9.32	1.16	153	16.7
66	3.9	34.1	55	2800	3250	1190	119	7.83	27.7	8.85	1.19	140	15.3
70	4.5	21.8	—	3590	1470	926	103	6.71	24.1	7.72	1.08	125	14.4
54.7	4.3	33.6	57	2770	3400	804	89.4	7.07	20.8	6.94	1.14	105	12.1
50	4.5	23.2	—	3450	1540	486	64.8	5.75	15.7	5.57	1.03	77.1	9.97
42.9	4.4	31.0	—	2960	2470	447	59.6	5.95	14.4	5.23	1.07	69.3	9.02
50	4.2	13.9	—	5070	333	305	50.8	4.55	15.7	5.74	1.03	61.2	10.3
40.8	4.0	20.7	—	4050	682	272	45.4	4.77	13.6	5.16	1.06	53.1	8.85
35	4.7	23.4	—	3500	1310	229	38.2	4.72	9.87	3.89	0.980	44.8	6.79
31.8	4.6	28.6	—	3190	1710	218	36.4	4.83	9.36	3.74	1.00	42.0	6.40
35	5.0	13.8	—	4960	374	147	29.4	3.78	8.36	3.38	0.901	35.4	6.22
25.4	4.7	26.4	—	3430	1220	124	24.7	4.07	6.79	2.91	0.954	28.4	4.96
23	4.9	14.5	—	4770	397	64.9	16.2	3.10	4.31	2.07	0.798	19.3	3.68
18.4	4.7	23.7	—	3770	821	57.6	14.4	3.26	3.73	1.86	0.831	16.5	3.16
17.25	5.0	9.9	—	6250	143	26.3	8.77	2.28	2.31	1.30	0.675	10.6	2.36
12.5	4.6	19.9	—	4290	477	22.1	7.37	2.45	1.82	1.09	0.705	8.47	1.85
10	4.6	17.4	—	4630	348	12.3	4.92	2.05	1.22	0.809	0.643	5.67	1.37
9.5	4.8	8.7	—	6830	87.4	6.79	3.39	1.56	0.903	0.646	0.569	4.04	1.13
7.7	4.5	14.7	—	5240	207	6.08	3.04	1.64	0.764	0.574	0.581	3.51	0.964
7.5	4.8	5.6	—	9160	28.1	2.93	1.95	1.15	0.586	0.468	0.516	2.36	0.826
5.7	4.5	11.4	—	6160	106	2.52	1.68	1.23	0.455	0.390	0.522	1.95	0.653

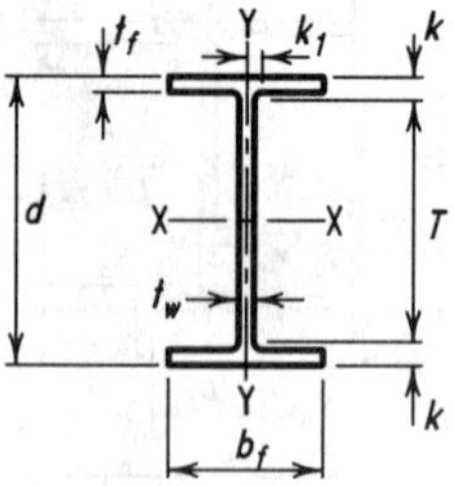

HP SHAPES
Dimensions

Designation	Area A	Depth d		Web Thickness t_w		Web $\frac{t_w}{2}$	Flange Width b_f		Flange Thickness t_f		Distance T	Distance k	Distance k_1
	in.²	in.		in.		in.	in.		in.		in.	in.	in.
HP14×117	34.4	14.21	14 1/4	0.805	13/16	7/16	14.885	14 7/8	0.805	13/16	11 1/4	1 1/2	1 1/16
×102	30.0	14.01	14	0.705	11/16	3/8	14.785	14 3/4	0.705	11/16	11 1/4	1 3/8	1
×89	26.1	13.83	13 7/8	0.615	5/8	5/16	14.695	14 3/4	0.615	5/8	11 1/4	1 5/16	15/16
×73	21.4	13.61	13 5/8	0.505	1/2	1/4	14.585	14 5/8	0.505	1/2	11 1/4	1 3/16	7/8
HP12×84	24.6	12.28	12 1/4	0.685	11/16	3/8	12.295	12 1/4	0.685	11/16	9 1/2	1 3/8	1
×74	21.8	12.13	12 1/8	0.605	5/8	5/16	12.215	12 1/4	0.610	5/8	9 1/2	1 5/16	15/16
×63	18.4	11.94	12	0.515	1/2	1/4	12.125	12 1/8	0.515	1/2	9 1/2	1 1/4	7/8
×53	15.5	11.78	11 3/4	0.435	7/16	1/4	12.045	12	0.435	7/16	9 1/2	1 1/8	7/8
HP10×57	16.8	9.99	10	0.565	9/16	5/16	10.225	10 1/4	0.565	9/16	7 5/8	1 3/16	13/16
×42	12.4	9.70	9 3/4	0.415	7/16	1/4	10.075	10 1/8	0.420	7/16	7 5/8	1 1/16	3/4
HP8×36	10.6	8.02	8	0.445	7/16	1/4	8.155	8 1/8	0.445	7/16	6 1/8	15/16	5/8

HP SHAPES
Properties

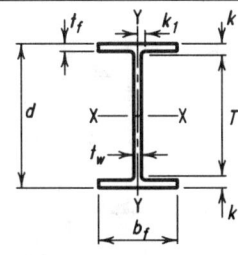

Nominal Wt. per ft	Compact Section Criteria			X_1	$X_2 \times 10^6$	Elastic Properties						Plastic Modulus	
						Axis X-X			Axis Y-Y				
	$\frac{b_f}{2t_f}$	$\frac{h}{t_w}$	F_y'''			I	S	r	I	S	r	Z_x	Z_y
lb			ksi	ksi	$(1/ksi)^2$	in.^{4}	in.^{3}	in.	in.^{4}	in.^{3}	in.	in.^{3}	in.^{3}
117	9.2	14.2	—	3870	659	1220	172	5.96	443	59.5	3.59	194	91.4
102	10.5	16.2	—	3400	1090	1050	150	5.92	380	51.4	3.56	169	78.8
89	11.9	18.5	—	2960	1840	904	131	5.88	326	44.3	3.53	146	67.7
73	14.4	22.6	—	2450	3880	729	107	5.84	261	35.8	3.49	118	54.6
84	9.0	14.2	—	3860	670	650	106	5.14	213	34.6	2.94	120	53.2
74	10.0	16.0	—	3440	1050	569	93.8	5.11	186	30.4	2.92	105	46.6
63	11.8	18.9	—	2940	1940	472	79.1	5.06	153	25.3	2.88	88.3	38.7
53	13.8	22.3	—	2500	3650	393	66.8	5.03	127	21.1	2.86	74.0	32.2
57	9.0	13.9	—	3920	631	294	58.8	4.18	101	19.7	2.45	66.5	30.3
42	12.0	18.9	—	2920	1970	210	43.4	4.13	71.7	14.2	2.41	48.3	21.8
36	9.2	14.2	—	3840	685	119	29.8	3.36	40.3	9.88	1.95	33.6	15.2

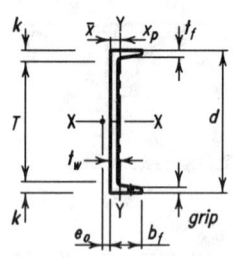

CHANNELS
AMERICAN STANDARD
Dimensions

Desig-nation	Area A in.^{2}	Depth d in.	Web Thickness t_w in.		$\dfrac{t_w}{2}$ in.	Flange Width b_f in.		Flange Thickness t_f in.		T in.	k in.	Grip in.	Max. Flge. Fastener in.
C15×50	14.7	15.00	0.716	$^{11}/_{16}$	$^3/_8$	3.716	$3^3/_4$	0.650	$^5/_8$	$12^1/_8$	$1^7/_{16}$	$^5/_8$	1
×40	11.8	15.00	0.520	$^1/_2$	$^1/_4$	3.520	$3^1/_2$	0.650	$^5/_8$	$12^1/_8$	$1^7/_{16}$	$^5/_8$	1
×33.9	9.96	15.00	0.400	$^3/_8$	$^3/_{16}$	3.400	$3^3/_8$	0.650	$^5/_8$	$12^1/_8$	$1^7/_{16}$	$^5/_8$	1
C12×30	8.82	12.00	0.510	$^1/_2$	$^1/_4$	3.170	$3^1/_8$	0.501	$^1/_2$	$9^3/_4$	$1^1/_8$	$^1/_2$	$^7/_8$
×25	7.35	12.00	0.387	$^3/_8$	$^3/_{16}$	3.047	3	0.501	$^1/_2$	$9^3/_4$	$1^1/_8$	$^1/_2$	$^7/_8$
×20.7	6.09	12.00	0.282	$^5/_{16}$	$^1/_8$	2.942	3	0.501	$^1/_2$	$9^3/_4$	$1^1/_8$	$^1/_2$	$^7/_8$
C10×30	8.82	10.00	0.673	$^{11}/_{16}$	$^5/_{16}$	3.033	3	0.436	$^7/_{16}$	8	1	$^7/_{16}$	$^3/_4$
×25	7.35	10.00	0.526	$^1/_2$	$^1/_4$	2.886	$2^7/_8$	0.436	$^7/_{16}$	8	1	$^7/_{16}$	$^3/_4$
×20	5.88	10.00	0.379	$^3/_8$	$^3/_{16}$	2.739	$2^3/_4$	0.436	$^7/_{16}$	8	1	$^7/_{16}$	$^3/_4$
×15.3	4.49	10.00	0.240	$^1/_4$	$^1/_8$	2.600	$2^5/_8$	0.436	$^7/_{16}$	8	1	$^7/_{16}$	$^3/_4$
C9×20	5.88	9.00	0.448	$^7/_{16}$	$^1/_4$	2.648	$2^5/_8$	0.413	$^7/_{16}$	$7^1/_8$	$^{15}/_{16}$	$^7/_{16}$	$^3/_4$
×15	4.41	9.00	0.285	$^5/_{16}$	$^1/_8$	2.485	$2^1/_2$	0.413	$^7/_{16}$	$7^1/_8$	$^{15}/_{16}$	$^7/_{16}$	$^3/_4$
×13.4	3.94	9.00	0.233	$^1/_4$	$^1/_8$	2.433	$2^3/_8$	0.413	$^7/_{16}$	$7^1/_8$	$^{15}/_{16}$	$^7/_{16}$	$^3/_4$
C8×18.75	5.51	8.00	0.487	$^1/_2$	$^1/_4$	2.527	$2^1/_2$	0.390	$^3/_8$	$6^1/_8$	$^{15}/_{16}$	$^3/_8$	$^3/_4$
×13.75	4.04	8.00	0.303	$^5/_{16}$	$^1/_8$	2.343	$2^3/_8$	0.390	$^3/_8$	$6^1/_8$	$^{15}/_{16}$	$^3/_8$	$^3/_4$
×11.5	3.38	8.00	0.220	$^1/_4$	$^1/_8$	2.260	$2^1/_4$	0.390	$^3/_8$	$6^1/_8$	$^{15}/_{16}$	$^3/_8$	$^3/_4$
C7×12.25	3.60	7.00	0.314	$^5/_{16}$	$^3/_{16}$	2.194	$2^1/_4$	0.366	$^3/_8$	$5^1/_4$	$^7/_8$	$^3/_8$	$^5/_8$
×9.8	2.87	7.00	0.210	$^3/_{16}$	$^1/_8$	2.090	$2^1/_8$	0.366	$^3/_8$	$5^1/_4$	$^7/_8$	$^3/_8$	$^5/_8$
C6×13	3.83	6.00	0.437	$^7/_{16}$	$^3/_{16}$	2.157	$2^1/_8$	0.343	$^5/_{16}$	$4^3/_8$	$^{13}/_{16}$	$^5/_{16}$	$^5/_8$
×10.5	3.09	6.00	0.314	$^5/_{16}$	$^3/_{16}$	2.034	2	0.343	$^5/_{16}$	$4^3/_8$	$^{13}/_{16}$	$^3/_8$	$^5/_8$
×8.2	2.40	6.00	0.200	$^3/_{16}$	$^1/_8$	1.920	$1^7/_8$	0.343	$^5/_{16}$	$4^3/_8$	$^{13}/_{16}$	$^5/_{16}$	$^5/_8$
C5×9	2.64	5.00	0.325	$^5/_{16}$	$^3/_{16}$	1.885	$1^7/_8$	0.320	$^5/_{16}$	$3^1/_2$	$^3/_4$	$^5/_{16}$	$^5/_8$
×6.7	1.97	5.00	0.190	$^3/_{16}$	$^1/_8$	1.750	$1^3/_4$	0.320	$^5/_{16}$	$3^1/_2$	$^3/_4$	—	—
C4×7.25	2.13	4.00	0.321	$^5/_{16}$	$^3/_{16}$	1.721	$1^3/_4$	0.296	$^5/_{16}$	$2^5/_8$	$^{11}/_{16}$	$^5/_{16}$	$^5/_8$
×5.4	1.59	4.00	0.184	$^3/_{16}$	$^1/_{16}$	1.584	$1^5/_8$	0.296	$^5/_{16}$	$2^5/_8$	$^{11}/_{16}$	—	—
C3×6	1.76	3.00	0.356	$^3/_8$	$^3/_{16}$	1.596	$1^5/_8$	0.273	$^1/_4$	$1^5/_8$	$^{11}/_{16}$	—	—
×5	1.47	3.00	0.258	$^1/_4$	$^1/_8$	1.498	$1^1/_2$	0.273	$^1/_4$	$1^5/_8$	$^{11}/_{16}$	—	—
×4.1	1.21	3.00	0.170	$^3/_{16}$	$^1/_{16}$	1.410	$1^3/_8$	0.273	$^1/_4$	$1^5/_8$	$^{11}/_{16}$	—	—

CHANNELS
AMERICAN STANDARD
Properties

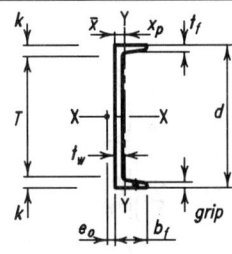

Nominal Wt. per ft	$\bar{x}$	Shear Center Location e_o	PNA Location x_p	Axis X-X				Axis Y-Y			
				I	Z	S	r	I	Z	S	r
lb	in.	in.	in.	in.^{4}	in.^{3}	in.^{3}	in.	in.^{4}	in.^{3}	in.^{3}	in.
50	0.798	0.583	0.488	404	68.2	53.8	5.24	11.0	8.17	3.78	0.867
40	0.777	0.767	0.390	349	57.2	46.5	5.44	9.23	6.87	3.37	0.886
33.9	0.787	0.896	0.330	315	50.4	42.0	5.62	8.13	6.23	3.11	0.904
30	0.674	0.618	0.366	162	33.6	27.0	4.29	5.14	4.33	2.06	0.763
25	0.674	0.746	0.305	144	29.2	24.1	4.43	4.47	3.84	1.88	0.780
20.7	0.698	0.870	0.252	129	25.4	21.5	4.61	3.88	3.49	1.73	0.799
30	0.649	0.369	0.439	103	26.6	20.7	3.42	3.94	3.78	1.65	0.669
25	0.617	0.494	0.366	91.2	23.0	18.2	3.52	3.36	3.19	1.48	0.676
20	0.606	0.637	0.292	78.9	19.3	15.8	3.66	2.81	2.71	1.32	0.692
15.3	0.634	0.796	0.223	67.4	15.8	13.5	3.87	2.28	2.35	1.16	0.713
20	0.583	0.515	0.325	60.9	16.8	13.5	3.22	2.42	2.47	1.17	0.642
15	0.586	0.682	0.243	51.0	13.5	11.3	3.40	1.93	2.05	1.01	0.661
13.4	0.601	0.743	0.217	47.9	12.5	10.6	3.48	1.76	1.95	0.962	0.669
18.75	0.565	0.431	0.343	44.0	13.8	11.0	2.82	1.98	2.17	1.01	0.599
13.75	0.553	0.604	0.251	36.1	10.9	9.03	2.99	1.53	1.73	0.854	0.615
11.5	0.571	0.697	0.209	32.6	9.55	8.14	3.11	1.32	1.58	0.781	0.625
12.25	0.525	0.538	0.255	24.2	8.40	6.93	2.60	1.17	1.43	0.703	0.571
9.8	0.540	0.647	0.203	21.3	7.12	6.08	2.72	0.968	1.26	0.625	0.581
13	0.514	0.380	0.317	17.4	7.26	5.80	2.13	1.05	1.36	0.642	0.525
10.5	0.499	0.486	0.255	15.2	6.15	5.06	2.22	0.866	1.15	0.564	0.529
8.2	0.511	0.599	0.198	13.1	5.13	4.38	2.34	0.693	0.993	0.492	0.537
9	0.478	0.427	0.262	8.90	4.36	3.56	1.83	0.632	0.918	0.450	0.489
6.7	0.484	0.552	0.217	7.49	3.51	3.00	1.95	0.479	0.763	0.378	0.493
7.25	0.459	0.386	0.264	4.59	2.81	2.29	1.47	0.433	0.697	0.343	0.450
5.4	0.457	0.502	0.241	3.85	2.26	1.93	1.56	0.319	0.569	0.283	0.449
6	0.455	0.322	0.291	2.07	1.72	1.38	1.08	0.305	0.544	0.268	0.416
5	0.438	0.392	0.242	1.85	1.50	1.24	1.12	0.247	0.466	0.233	0.410
4.1	0.436	0.461	0.284	1.66	1.30	1.10	1.17	0.197	0.401	0.202	0.404

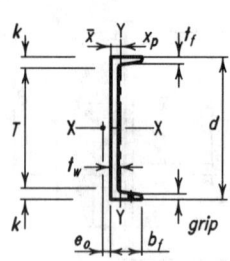

CHANNELS
MISCELLANEOUS
Dimensions

Desig-nation	Area A in.^{2}	Depth d in.	Web Thickness t_w in.		$\dfrac{t_w}{2}$ in.	Flange Width b_f in.		Flange Thickness t_f in.		Distance T in.	k in.	Grip in.	Max. Flge. Fastener in.
MC18×58	17.1	18.00	0.700	$^{11}/_{16}$	$^3/_8$	4.200	$4^1/_4$	0.625	$^5/_8$	$15^1/_4$	$1^3/_8$	$^5/_8$	1
×51.9	15.3	18.00	0.600	$^5/_8$	$^5/_{16}$	4.100	$4^1/_8$	0.625	$^5/_8$	$15^1/_4$	$1^3/_8$	$^5/_8$	1
×45.8	13.5	18.00	0.500	$^1/_2$	$^1/_4$	4.000	4	0.625	$^5/_8$	$15^1/_4$	$1^3/_8$	$^5/_8$	1
×42.7	12.6	18.00	0.450	$^7/_{16}$	$^1/_4$	3.950	4	0.625	$^5/_8$	$15^1/_4$	$1^3/_8$	$^5/_8$	1
MC13×50	14.7	13.00	0.787	$^3/_{16}$	$^3/_8$	4.412	$4^3/_8$	0.610	$^5/_8$	$10^1/_4$	$1^3/_8$	$^5/_8$	1
×40	11.8	13.00	0.560	$^9/_{16}$	$^1/_4$	4.185	$4^1/_8$	0.610	$^5/_8$	$10^1/_4$	$1^3/_8$	$^9/_{16}$	1
×35	10.3	13.00	0.447	$^7/_{16}$	$^1/_4$	4.072	$4^1/_8$	0.610	$^5/_8$	$10^1/_4$	$1^3/_8$	$^9/_{16}$	1
×31.8	9.35	13.00	0.375	$^3/_8$	$^3/_{16}$	4.000	4	0.610	$^5/_8$	$10^1/_4$	$1^3/_8$	$^9/_{16}$	1
MC12×50	14.7	12.00	0.835	$^{13}/_{16}$	$^7/_{16}$	4.135	$4^1/_8$	0.700	$^{11}/_{16}$	$9^3/_8$	$1^5/_{16}$	$^{11}/_{16}$	1
×45	13.2	12.00	0.712	$^{11}/_{16}$	$^3/_8$	4.012	4	0.700	$^{11}/_{16}$	$9^3/_8$	$1^5/_{16}$	$^{11}/_{16}$	1
×40	11.8	12.00	0.590	$^9/_{16}$	$^5/_{16}$	3.890	$3^7/_8$	0.700	$^{11}/_{16}$	$9^3/_8$	$1^5/_{16}$	$^{11}/_{16}$	1
×35	10.3	12.00	0.467	$^7/_{16}$	$^1/_4$	3.767	$3^3/_4$	0.700	$^{11}/_{16}$	$9^3/_8$	$1^5/_{16}$	$^{11}/_{16}$	1
×31	9.12	12.00	0.370	$^3/_8$	$^3/_{16}$	3.670	$3^5/_8$	0.700	$^{11}/_{16}$	$9^3/_8$	$1^5/_{16}$	$^{11}/_{16}$	1
MC12×10.6	3.10	12.00	0.190	$^3/_{16}$	$^1/_8$	1.500	$1^1/_2$	0.309	$^5/_{16}$	$10^5/_8$	$^{11}/_{16}$	—	—
MC10×41.1	12.1	10.00	0.796	$^{13}/_{16}$	$^3/_8$	4.321	$4^3/_8$	0.575	$^9/_{16}$	$7^1/_2$	$1^1/_4$	$^9/_{16}$	$^7/_8$
×33.6	9.87	10.00	0.575	$^9/_{16}$	$^5/_{16}$	4.100	$4^1/_8$	0.575	$^9/_{16}$	$7^1/_2$	$1^1/_4$	$^9/_{16}$	$^7/_8$
×28.5	8.37	10.00	0.425	$^7/_{16}$	$^3/_{16}$	3.950	4	0.575	$^9/_{16}$	$7^1/_2$	$1^1/_4$	$^9/_{16}$	$^7/_8$
MC10×25	7.35	10.00	0.380	$^3/_8$	$^3/_{16}$	3.405	$3^3/_8$	0.575	$^9/_{16}$	$7^1/_2$	$1^1/_4$	$^9/_{16}$	$^7/_8$
×22	6.45	10.00	0.290	$^5/_{16}$	$^1/_8$	3.315	$3^3/_8$	0.575	$^9/_{16}$	$7^1/_2$	$1^1/_4$	$^9/_{16}$	$^7/_8$
MC10×8.4	2.46	10.00	0.170	$^3/_{16}$	$^1/_{16}$	1.500	$1^1/_2$	0.280	$^1/_4$	$8^5/_8$	$^{11}/_{16}$	—	—

CHANNELS
MISCELLANEOUS
Properties

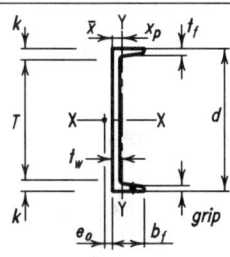

Nom-inal Wt. per ft	$\bar{x}$	Shear Center Location e_o	PNA Location x_p	Axis X-X				Axis Y-Y			
				I	Z	S	r	I	Z	S	r
lb	in.	in.	in.	in.^{4}	in.^{3}	in.^{3}	in.	in.^{4}	in.^{3}	in.^{3}	in.
58	0.862	0.695	0.472	676	94.6	75.1	6.29	17.8	9.94	5.32	1.02
51.9	0.858	0.797	0.422	627	86.5	69.7	6.41	16.4	9.13	5.07	1.04
45.8	0.866	0.909	0.372	578	78.4	64.3	6.56	15.1	8.42	4.82	1.06
42.7	0.877	0.969	0.347	554	74.4	61.6	6.64	14.4	8.10	4.69	1.07
50	0.974	0.815	0.564	314	60.5	48.4	4.62	16.5	10.1	4.79	1.06
40	0.963	1.03	0.450	273	50.9	42.0	4.82	13.7	8.57	4.26	1.08
35	0.980	1.16	0.394	252	46.2	38.8	4.95	12.3	7.95	3.99	1.10
31.8	1.00	1.24	0.358	239	43.1	36.8	5.06	11.4	7.60	3.81	1.11
50	1.05	0.741	0.610	269	56.1	44.9	4.28	17.4	10.2	5.65	1.09
45	1.04	0.844	0.549	252	51.7	42.0	4.36	15.8	9.35	5.33	1.09
40	1.04	0.952	0.488	234	47.3	39.0	4.46	14.3	8.59	5.00	1.10
35	1.05	1.07	0.426	216	42.8	36.1	4.59	12.7	7.91	4.67	1.11
31	1.08	1.18	0.416	203	39.3	33.8	4.71	11.3	7.44	4.39	1.12
10.6	0.269	0.284	0.129	55.4	11.6	9.23	4.22	0.382	0.639	0.310	0.351
41.1	1.09	0.864	0.601	158	38.9	31.5	3.61	15.8	8.71	4.88	1.14
33.6	1.08	1.06	0.490	139	33.4	27.8	3.75	13.2	7.51	4.38	1.16
28.5	1.12	1.21	0.415	127	29.6	25.3	3.89	11.4	6.83	4.02	1.17
25	0.953	1.03	0.364	110	25.8	22.0	3.87	7.35	5.21	3.00	1.00
22	0.990	1.13	0.468	103	23.6	20.5	3.99	6.50	4.86	2.80	1.00
8.4	0.284	0.332	0.122	32.0	7.86	6.40	3.61	0.328	0.552	0.270	0.365

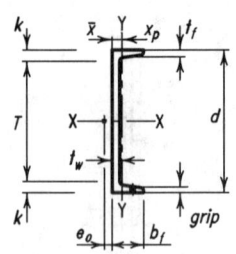

CHANNELS
MISCELLANEOUS
Dimensions

Desig- nation	Area A in.^{2}	Depth d in.	Web Thickness t_w in.		$\dfrac{t_w}{2}$ in.	Flange Width b_f in.		Thickness t_f in.		Distance T in.	k in.	Grip in.	Max. Flge. Fas- tener in.
MC9×25.4	7.47	9.00	0.450	$\frac{7}{16}$	$\frac{1}{4}$	3.500	$3\frac{1}{2}$	0.550	$\frac{9}{16}$	$6\frac{5}{8}$	$1\frac{3}{16}$	$\frac{9}{16}$	$\frac{7}{8}$
×23.9	7.02	9.00	0.400	$\frac{3}{8}$	$\frac{3}{16}$	3.450	$3\frac{1}{2}$	0.550	$\frac{9}{16}$	$6\frac{5}{8}$	$1\frac{3}{16}$	$\frac{9}{16}$	$\frac{7}{8}$
MC8×22.8	6.70	8.00	0.427	$\frac{7}{16}$	$\frac{3}{16}$	3.502	$3\frac{1}{2}$	0.525	$\frac{1}{2}$	$5\frac{5}{8}$	$1\frac{3}{16}$	$\frac{1}{2}$	$\frac{7}{8}$
×21.4	6.28	8.00	0.375	$\frac{3}{8}$	$\frac{3}{16}$	3.450	$3\frac{1}{2}$	0.525	$\frac{1}{2}$	$5\frac{5}{8}$	$1\frac{3}{16}$	$\frac{1}{2}$	$\frac{7}{8}$
MC8×20	5.88	8.00	0.400	$\frac{3}{8}$	$\frac{3}{16}$	3.025	3	0.500	$\frac{1}{2}$	$5\frac{3}{4}$	$1\frac{1}{8}$	$\frac{1}{2}$	$\frac{7}{8}$
×18.7	5.50	8.00	0.353	$\frac{3}{8}$	$\frac{3}{16}$	2.978	3	0.500	$\frac{1}{2}$	$5\frac{3}{4}$	$1\frac{1}{8}$	$\frac{1}{2}$	$\frac{7}{8}$
MC8×8.5	2.50	8.00	0.179	$\frac{3}{16}$	$\frac{1}{16}$	1.874	$1\frac{7}{8}$	0.311	$\frac{5}{16}$	$6\frac{1}{2}$	$\frac{3}{4}$	$\frac{5}{16}$	$\frac{5}{8}$
MC7×22.7	6.67	7.00	0.503	$\frac{1}{2}$	$\frac{1}{4}$	3.603	$3\frac{5}{8}$	0.500	$\frac{1}{2}$	$4\frac{3}{4}$	$1\frac{1}{8}$	$\frac{1}{2}$	$\frac{7}{8}$
×19.1	5.61	7.00	0.352	$\frac{3}{8}$	$\frac{3}{16}$	3.452	$3\frac{1}{2}$	0.500	$\frac{1}{2}$	$4\frac{3}{4}$	$1\frac{1}{8}$	$\frac{1}{2}$	$\frac{7}{8}$
MC6×18	5.29	6.00	0.379	$\frac{3}{8}$	$\frac{3}{16}$	3.504	$3\frac{1}{2}$	0.475	$\frac{1}{2}$	$3\frac{7}{8}$	$1\frac{1}{16}$	$\frac{1}{2}$	$\frac{7}{8}$
MC6×16.3	4.79	6.00	0.375	$\frac{3}{8}$	$\frac{3}{16}$	3.000	3	0.475	$\frac{1}{2}$	$3\frac{7}{8}$	$1\frac{1}{16}$	$\frac{1}{2}$	$\frac{3}{4}$
×15.1	4.44	6.00	0.316	$\frac{5}{16}$	$\frac{3}{16}$	2.941	3	0.475	$\frac{1}{2}$	$3\frac{7}{8}$	$1\frac{1}{16}$	$\frac{1}{2}$	$\frac{3}{4}$
MC6×12	3.53	6.00	0.310	$\frac{5}{16}$	$\frac{1}{8}$	2.497	$2\frac{1}{2}$	0.375	$\frac{3}{8}$	$4\frac{3}{8}$	$1\frac{3}{16}$	$\frac{3}{8}$	$\frac{5}{8}$

CHANNELS
MISCELLANEOUS
Properties

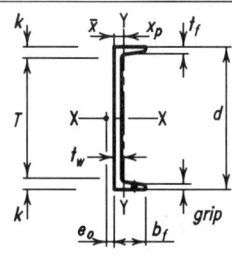

Nom-inal Wt. per ft	$\bar{x}$	Shear Center Loca-tion e_o	PNA Loca-tion x_p	Axis X-X				Axis Y-Y			
				I	Z	S	r	I	Z	S	r
lb	in.	in.	in.	in.^{4}	in.^{3}	in.^{3}	in.	in.^{4}	in.^{3}	in.^{3}	in.
25.4	0.970	0.986	0.411	88.0	23.2	19.6	3.43	7.65	5.23	3.02	1.01
23.9	0.981	1.04	0.386	85.0	22.2	18.9	3.48	7.22	5.05	2.93	1.01
22.8	1.01	1.04	0.415	63.8	18.8	16.0	3.09	7.07	4.88	2.84	1.03
21.4	1.02	1.09	0.449	61.6	18.0	15.4	3.13	6.64	4.71	2.74	1.03
20	0.840	0.843	0.364	54.5	16.2	13.6	3.05	4.47	3.57	2.05	0.872
18.7	0.849	0.889	0.341	52.5	15.4	13.1	3.09	4.20	3.44	1.97	0.874
8.5	0.428	0.542	0.155	23.3	6.91	5.83	3.05	0.628	0.882	0.434	0.501
22.7	1.04	1.01	0.473	47.5	16.2	13.6	2.67	7.29	4.86	2.85	1.05
19.1	1.08	1.15	0.567	43.2	14.3	12.3	2.77	6.11	4.34	2.57	1.04
18	1.12	1.17	0.622	29.7	11.5	9.91	2.37	5.93	4.14	2.48	1.06
16.3	0.927	0.930	0.464	26.0	10.2	8.68	2.33	3.82	3.18	1.84	0.892
15.1	0.940	0.982	0.537	25.0	9.69	8.32	2.37	3.51	3.00	1.75	0.889
12	0.704	0.725	0.292	18.7	7.38	6.24	2.30	1.87	1.79	1.04	0.728

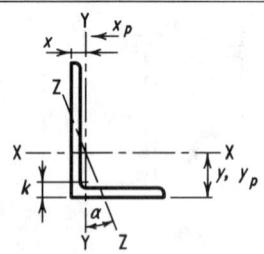

ANGLES
Equal legs and unequal legs
Properties for designing

Size and Thickness	k	Weight per ft	Area	Axis X-X					
				I	S	r	y	Z	y_p
in.	in.	lb	in.^{2}	in.^{4}	in.^{3}	in.	in.	in.^{3}	in.
L8×8×1⅛	1¾	56.9	16.7	98.0	17.5	2.42	2.41	31.6	1.05
1	1⅝	51.0	15.0	89.0	15.8	2.44	2.37	28.5	0.938
⅞	1½	45.0	13.2	79.6	14.0	2.45	2.33	25.3	0.827
¾	1⅜	38.9	11.4	69.7	12.2	2.47	2.28	22.0	0.715
⅝	1¼	32.7	9.61	59.4	10.3	2.49	2.23	18.6	0.601
⁹⁄₁₆	1³⁄₁₆	29.6	8.68	54.1	9.34	2.50	2.21	16.8	0.543
½	1⅛	26.4	7.75	48.6	8.36	2.50	2.19	15.1	0.484
L8×6×1	1½	44.2	13.0	80.8	15.1	2.49	2.65	27.3	1.50
⅞	1⅜	39.1	11.5	72.3	13.4	2.51	2.61	24.2	1.44
¾	1¼	33.8	9.94	63.4	11.7	2.53	2.56	21.1	1.38
⅝	1⅛	28.5	8.36	54.1	9.87	2.54	2.52	17.9	1.31
⁹⁄₁₆	1¹⁄₁₆	25.7	7.56	49.3	8.95	2.55	2.50	16.2	1.28
½	1	23.0	6.75	44.3	8.02	2.56	2.47	14.5	1.25
⁷⁄₁₆	¹⁵⁄₁₆	20.2	5.93	39.2	7.07	2.57	2.45	12.8	1.22
L8×4×1	1½	37.4	11.0	69.6	14.1	2.52	3.05	24.3	2.50
⅞	1⅜	33.1	9.73	62.5	12.5	2.53	3.00	21.6	2.44
¾	1¼	28.7	8.44	54.9	10.9	2.55	2.95	18.9	2.38
⅝	1⅛	24.2	7.11	46.9	9.21	2.57	2.91	16.0	2.31
⁹⁄₁₆	1¹⁄₁₆	21.9	6.43	42.8	8.35	2.58	2.88	14.5	2.28
½	1	19.6	5.75	38.5	7.49	2.59	2.86	13.0	2.25
⁷⁄₁₆	¹⁵⁄₁₆	17.2	5.06	34.1	6.60	2.60	2.83	11.5	2.22
L7×4×¾	1¼	26.2	7.69	37.8	8.42	2.22	2.51	14.8	1.88
⅝	1⅛	22.1	6.48	32.4	7.14	2.24	2.46	12.6	1.81
½	1	17.9	5.25	26.7	5.81	2.25	2.42	10.3	1.75
⁷⁄₁₆	¹⁵⁄₁₆	15.7	4.62	23.7	5.13	2.26	2.39	9.09	1.72
⅜	⅞	13.6	3.98	20.6	4.44	2.27	2.37	7.87	1.69

ANGLES
Equal legs and unequal legs
Properties for designing

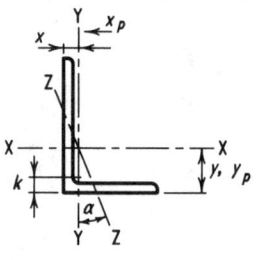

Size and Thickness	Axis Y-Y						Axis Z-Z	
	I	S	r	x	z	x_p	r	Tan
in.	in.^{4}	in.^{3}	in.	in.	in.^{3}	in.	in.	α
L8×8×1⅛	98.0	17.5	2.42	2.41	31.6	1.05	1.56	1.000
1	89.0	15.8	2.44	2.37	28.5	0.938	1.56	1.000
⅞	79.6	14.0	2.45	2.33	25.3	0.827	1.57	1.000
¾	69.7	12.2	2.47	2.28	22.0	0.715	1.58	1.000
⅝	59.4	10.3	2.49	2.23	18.6	0.601	1.58	1.000
9⁄16	54.1	9.34	2.50	2.21	16.8	0.543	1.59	1.000
½	48.6	8.36	2.50	2.19	15.1	0.484	1.59	1.000
L8×6×1	38.8	8.92	1.73	1.65	16.2	0.813	1.28	0.543
⅞	34.9	7.94	1.74	1.61	14.4	0.718	1.28	0.547
¾	30.7	6.92	1.76	1.56	12.5	0.621	1.29	0.551
⅝	26.3	5.88	1.77	1.52	10.5	0.522	1.29	0.554
9⁄16	24.0	5.34	1.78	1.50	9.52	0.472	1.30	0.556
½	21.7	4.79	1.79	1.47	8.51	0.422	1.30	0.558
7⁄16	19.3	4.23	1.80	1.45	7.50	0.371	1.31	0.560
L8×4×1	11.6	3.94	1.03	1.05	7.72	0.688	0.846	0.247
⅞	10.5	3.51	1.04	0.999	6.77	0.608	0.848	0.253
¾	9.36	3.07	1.05	0.953	5.81	0.527	0.852	0.258
⅝	8.10	2.62	1.07	0.905	4.86	0.444	0.857	0.262
9⁄16	7.43	2.38	1.07	0.882	4.38	0.402	0.861	0.265
½	6.74	2.15	1.08	0.859	3.90	0.359	0.865	0.267
7⁄16	6.02	1.90	1.09	0.835	3.42	0.316	0.869	0.269
L7×4×¾	9.05	3.03	1.09	1.01	5.65	0.549	0.860	0.324
⅝	7.84	2.58	1.10	0.963	4.74	0.463	0.865	0.329
½	6.53	2.12	1.11	0.917	3.83	0.375	0.872	0.335
7⁄16	5.83	1.88	1.12	0.893	3.37	0.330	0.875	0.337
⅜	5.10	1.63	1.13	0.870	2.90	0.285	0.880	0.340

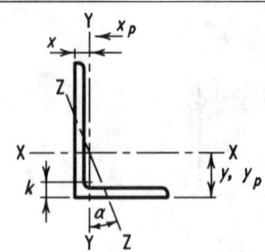

ANGLES
Equal legs and unequal legs
Properties for designing

Size and Thickness	k	Weight per ft	Area	Axis X-X					
				I	S	r	y	Z	y_p
in.	in.	lb	in.^{2}	in.^{4}	in.^{3}	in.	in.	in.^{3}	in.
L6×6×1	1½	37.4	11.0	35.5	8.57	1.80	1.86	15.5	0.917
⅞	1⅜	33.1	9.73	31.9	7.63	1.81	1.82	13.8	0.811
¾	1¼	28.7	8.44	28.2	6.66	1.83	1.78	12.0	0.703
⅝	1⅛	24.2	7.11	24.2	5.66	1.84	1.73	10.2	0.592
⁹⁄₁₆	1¹⁄₁₆	21.9	6.43	22.1	5.14	1.85	1.71	9.26	0.536
½	1	19.6	5.75	19.9	4.61	1.86	1.68	8.31	0.479
⁷⁄₁₆	¹⁵⁄₁₆	17.2	5.06	17.7	4.08	1.87	1.66	7.34	0.422
⅜	⅞	14.9	4.36	15.4	3.53	1.88	1.64	6.35	0.363
⁵⁄₁₆	¹³⁄₁₆	12.4	3.65	13.0	2.97	1.89	1.62	5.35	0.304
L6×4×⅞	1⅜	27.2	7.98	27.7	7.15	1.86	2.12	12.7	1.44
¾	1¼	23.6	6.94	24.5	6.25	1.88	2.08	11.2	1.38
⅝	1⅛	20.0	5.86	21.1	5.31	1.90	2.03	9.51	1.31
⁹⁄₁₆	1¹⁄₁₆	18.1	5.31	19.3	4.83	1.90	2.01	8.66	1.28
½	1	16.2	4.75	17.4	4.33	1.91	1.99	7.78	1.25
⁷⁄₁₆	¹⁵⁄₁₆	14.3	4.18	15.5	3.83	1.92	1.96	6.88	1.22
⅜	⅞	12.3	3.61	13.5	3.32	1.93	1.94	5.97	1.19
⁵⁄₁₆	¹³⁄₁₆	10.3	3.03	11.4	2.79	1.94	1.92	5.03	1.16
L6×3½×½	1	15.3	4.50	16.6	4.24	1.92	2.08	7.50	1.50
⅜	⅞	11.7	3.42	12.9	3.24	1.94	2.04	5.76	1.44
⁵⁄₁₆	¹³⁄₁₆	9.80	2.87	10.9	2.73	1.95	2.01	4.85	1.41
L5×5×⅞	1⅜	27.2	7.98	17.8	5.17	1.49	1.57	9.33	0.798
¾	1¼	23.6	6.94	15.7	4.53	1.51	1.52	8.16	0.694
⅝	1⅛	20.0	5.86	13.6	3.86	1.52	1.48	6.95	0.586
½	1	16.2	4.75	11.3	3.16	1.54	1.43	5.68	0.475
⁷⁄₁₆	¹⁵⁄₁₆	14.3	4.18	10.0	2.79	1.55	1.41	5.03	0.418
⅜	⅞	12.3	3.61	8.74	2.42	1.56	1.39	4.36	0.361
⁵⁄₁₆	¹³⁄₁₆	10.3	3.03	7.42	2.04	1.57	1.37	3.68	0.303

ANGLES
Equal legs and unequal legs
Properties for designing

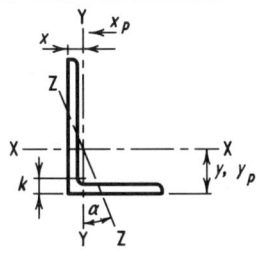

Size and Thickness	Axis Y-Y						Axis Z-Z	
	I	S	r	x	Z	x_p	r	Tan
in.	in.^{4}	in.^{3}	in.	in.	in.^{3}	in.	in.	α
L6×6×1	35.5	8.57	1.80	1.86	15.5	0.917	1.17	1.000
$\frac{7}{8}$	31.9	7.63	1.81	1.82	13.8	0.811	1.17	1.000
$\frac{3}{4}$	28.2	6.66	1.83	1.78	12.0	0.703	1.17	1.000
$\frac{5}{8}$	24.2	5.66	1.84	1.73	10.2	0.592	1.18	1.000
$\frac{9}{16}$	22.1	5.14	1.85	1.71	9.26	0.536	1.18	1.000
$\frac{1}{2}$	19.9	4.61	1.86	1.68	8.31	0.479	1.18	1.000
$\frac{7}{16}$	17.7	4.08	1.87	1.66	7.34	0.422	1.19	1.000
$\frac{3}{8}$	15.4	3.53	1.88	1.64	6.35	0.363	1.19	1.000
$\frac{5}{16}$	13.0	2.97	1.89	1.62	5.35	0.304	1.20	1.000
L6×4×$\frac{7}{8}$	9.75	3.39	1.11	1.12	6.31	0.665	0.857	0.421
$\frac{3}{4}$	8.68	2.97	1.12	1.08	5.47	0.578	0.860	0.428
$\frac{5}{8}$	7.52	2.54	1.13	1.03	4.62	0.488	0.864	0.435
$\frac{9}{16}$	6.91	2.31	1.14	1.01	4.19	0.442	0.866	0.438
$\frac{1}{2}$	6.27	2.08	1.15	0.987	3.75	0.396	0.870	0.440
$\frac{7}{16}$	5.60	1.85	1.16	0.964	3.30	0.349	0.873	0.443
$\frac{3}{8}$	4.90	1.60	1.17	0.941	2.85	0.301	0.877	0.446
$\frac{5}{16}$	4.18	1.35	1.17	0.918	2.40	0.252	0.882	0.448
L6×3$\frac{1}{2}$×$\frac{1}{2}$	4.25	1.59	0.972	0.833	2.91	0.375	0.759	0.344
$\frac{3}{8}$	3.34	1.23	0.988	0.787	2.20	0.285	0.767	0.350
$\frac{5}{16}$	2.85	1.04	0.996	0.763	1.85	0.239	0.772	0.352
L5×5×$\frac{7}{8}$	17.8	5.17	1.49	1.57	9.33	0.798	0.973	1.000
$\frac{3}{4}$	15.7	4.53	1.51	1.52	8.16	0.694	0.975	1.000
$\frac{5}{8}$	13.6	3.86	1.52	1.48	6.95	0.586	0.978	1.000
$\frac{1}{2}$	11.3	3.16	1.54	1.43	5.68	0.475	0.983	1.000
$\frac{7}{16}$	10.0	2.79	1.55	1.41	5.03	0.418	0.986	1.000
$\frac{3}{8}$	8.74	2.42	1.56	1.39	4.36	0.361	0.990	1.000
$\frac{5}{16}$	7.42	2.04	1.57	1.37	3.68	0.303	0.994	1.000

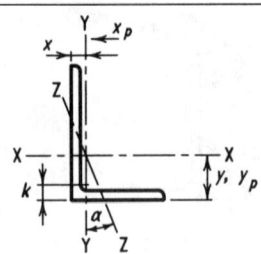

ANGLES
Equal legs and unequal legs
Properties for designing

Size and Thickness	k	Weight per ft	Area	Axis X-X					
				I	S	r	y	Z	y_p
in.	in.	lb	in.^{2}	in.^{4}	in.^{3}	in.	in.	in.^{3}	in.
L5×3½×¾	1¼	19.8	5.81	13.9	4.28	1.55	1.75	7.65	1.13
⅝	1⅛	16.8	4.92	12.0	3.65	1.56	1.70	6.55	1.06
½	1	13.6	4.00	9.99	2.99	1.58	1.66	5.38	1.00
⅜	⅞	10.4	3.05	7.78	2.29	1.60	1.61	4.14	0.938
5⁄16	13⁄16	8.70	2.56	6.60	1.94	1.61	1.59	3.49	0.906
¼	¾	7.00	2.06	5.39	1.57	1.62	1.56	2.83	0.875
L5×3×½	1	12.8	3.75	9.45	2.91	1.59	1.75	5.16	1.25
7⁄16	15⁄16	11.3	3.31	8.43	2.58	1.60	1.73	4.57	1.22
⅜	⅞	9.80	2.86	7.37	2.24	1.61	1.70	3.97	1.19
5⁄16	13⁄16	8.20	2.40	6.26	1.89	1.61	1.68	3.36	1.16
¼	¾	6.60	1.94	5.11	1.53	1.62	1.66	2.72	1.13
L4×4×¾	1⅛	18.5	5.44	7.67	2.81	1.19	1.27	5.07	0.680
⅝	1	15.7	4.61	6.66	2.40	1.20	1.23	4.33	0.576
½	⅞	12.8	3.75	5.56	1.97	1.22	1.18	3.56	0.469
7⁄16	13⁄16	11.3	3.31	4.97	1.75	1.23	1.16	3.16	0.414
⅜	¾	9.80	2.86	4.36	1.52	1.23	1.14	2.74	0.357
5⁄16	11⁄16	8.20	2.40	3.71	1.29	1.24	1.12	2.32	0.300
¼	⅝	6.60	1.94	3.04	1.05	1.25	1.09	1.88	0.242
L4×3½×½	15⁄16	11.9	3.50	5.32	1.94	1.23	1.25	3.50	0.500
⅜	13⁄16	9.10	2.67	4.18	1.49	1.25	1.21	2.71	0.438
5⁄16	¾	7.70	2.25	3.56	1.26	1.26	1.18	2.29	0.406
¼	11⁄16	6.20	1.81	2.91	1.03	1.27	1.16	1.86	0.375

ANGLES
Equal legs and unequal legs
Properties for designing

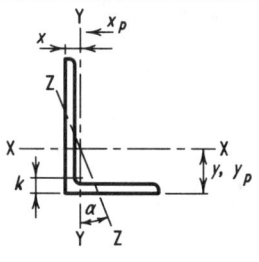

Size and Thickness	Axis Y-Y						Axis Z-Z	
	I	S	r	x	z	x_p	r	Tan
in.	in.^{4}	in.^{3}	in.	in.	in.^{3}	in.	in.	α
L5×3½×¾	5.55	2.22	0.977	0.996	4.10	0.581	0.748	0.464
⅝	4.83	1.90	0.991	0.951	3.47	0.492	0.751	0.472
½	4.05	1.56	1.01	0.906	2.83	0.400	0.755	0.479
⅜	3.18	1.21	1.02	0.861	2.16	0.305	0.762	0.486
⁵⁄₁₆	2.72	1.02	1.03	0.838	1.82	0.256	0.766	0.489
¼	2.23	0.830	1.04	0.814	1.47	0.206	0.770	0.492
L5×3×½	2.58	1.15	0.829	0.750	2.11	0.375	0.648	0.357
⁷⁄₁₆	2.32	1.02	0.837	0.727	1.86	0.331	0.651	0.361
⅜	2.04	0.888	0.845	0.704	1.60	0.286	0.654	0.364
⁵⁄₁₆	1.75	0.753	0.853	0.681	1.35	0.240	0.658	0.368
¼	1.44	0.614	0.861	0.657	1.09	0.194	0.663	0.371
L4×4×¾	7.67	2.81	1.19	1.27	5.07	0.680	0.778	1.000
⅝	6.66	2.40	1.20	1.23	4.33	0.576	0.779	1.000
½	5.56	1.97	1.22	1.18	3.56	0.469	0.782	1.000
⁷⁄₁₆	4.97	1.75	1.23	1.16	3.16	0.414	0.785	1.000
⅜	4.36	1.52	1.23	1.14	2.74	0.357	0.788	1.000
⁵⁄₁₆	3.71	1.29	1.24	1.12	2.32	0.300	0.791	1.000
¼	3.04	1.05	1.25	1.09	1.88	0.242	0.795	1.000
L4×3½×½	3.79	1.52	1.04	1.00	2.73	0.438	0.722	0.750
⅜	2.95	1.16	1.06	0.955	2.11	0.334	0.727	0.755
⁵⁄₁₆	2.55	0.994	1.07	0.932	1.78	0.281	0.730	0.757
¼	2.09	0.808	1.07	0.909	1.44	0.227	0.734	0.759

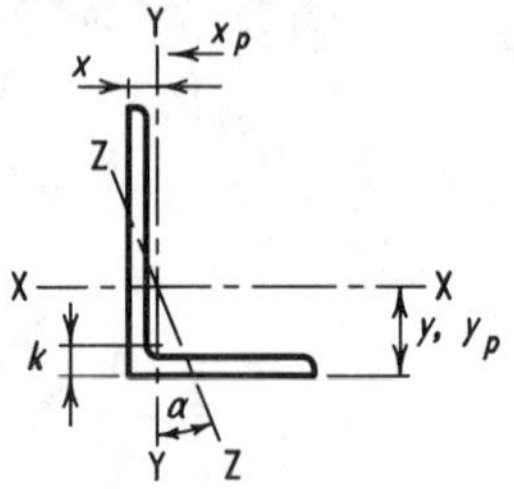

ANGLES
Equal legs and unequal legs
Properties for designing

Size and Thickness	k	Weight per ft	Area	Axis X-X					
				I	S	r	y	Z	y_p
in.	in.	lb	in.	in.^{4}	in.^{3}	in.	in.	in.^{3}	in.
L4×3×5/8	1 1/16	13.6	3.98	6.03	2.30	1.23	1.37	4.12	0.813
1/2	15/16	11.1	3.25	5.05	1.89	1.25	1.33	3.41	0.750
7/16	7/8	9.80	2.87	4.52	1.68	1.25	1.30	3.03	0.719
3/8	13/16	8.50	2.48	3.96	1.46	1.26	1.28	2.64	0.688
5/16	3/4	7.20	2.09	3.38	1.23	1.27	1.26	2.23	0.656
1/4	11/16	5.80	1.69	2.77	1.00	1.28	1.24	1.82	0.625
L3 1/2×3 1/2×1/2	7/8	11.1	3.25	3.64	1.49	1.06	1.06	2.68	0.464
7/16	13/16	9.80	2.87	3.26	1.32	1.07	1.04	2.38	0.410
3/8	3/4	8.50	2.48	2.87	1.15	1.07	1.01	2.08	0.355
5/16	11/16	7.20	2.09	2.45	0.976	1.08	0.990	1.76	0.299
1/4	5/8	5.80	1.69	2.01	0.794	1.09	0.968	1.43	0.241
L3 1/2×3×1/2	15/16	10.2	3.00	3.45	1.45	1.07	1.13	2.63	0.500
3/8	13/16	7.90	2.30	2.72	1.13	1.09	1.08	2.04	0.438
5/16	3/4	6.60	1.93	2.33	0.954	1.10	1.06	1.73	0.406
1/4	11/16	5.40	1.56	1.91	0.776	1.11	1.04	1.41	0.375
L3 1/2×2 1/2×1/2	15/16	9.40	2.75	3.24	1.41	1.09	1.20	2.53	0.750
3/8	13/16	7.20	2.11	2.56	1.09	1.10	1.16	1.97	0.688
1/4	11/16	4.90	1.44	1.80	0.755	1.12	1.11	1.36	0.625
L3×3×1/2	13/16	9.40	2.75	2.22	1.07	0.898	0.932	1.93	0.458
7/16	3/4	8.30	2.43	1.99	0.954	0.905	0.910	1.72	0.406
3/8	11/16	7.20	2.11	1.76	0.833	0.913	0.888	1.50	0.352
5/16	5/8	6.10	1.78	1.51	0.707	0.922	0.865	1.27	0.296
1/4	9/16	4.90	1.44	1.24	0.577	0.930	0.842	1.04	0.240
3/16	1/2	3.71	1.09	0.962	0.441	0.939	0.820	0.794	0.182

ANGLES
Equal legs and unequal legs
Properties for designing

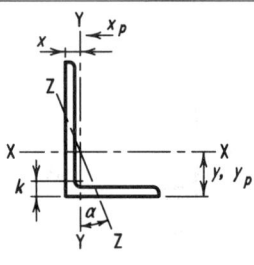

Size and Thickness	Axis Y-Y						Axis Z-Z	
	I	S	r	x	z	x_p	r	Tan
in.	in.^{4}	in.^{3}	in.	in.	in.^{3}	in.	in.	α
L4×3×⅝	2.87	1.35	0.849	0.871	2.48	0.498	0.637	0.534
½	2.42	1.12	0.864	0.827	2.03	0.406	0.639	0.543
⁷⁄₁₆	2.18	0.992	0.871	0.804	1.79	0.359	0.641	0.547
⅜	1.92	0.866	0.879	0.782	1.56	0.311	0.644	0.551
⁵⁄₁₆	1.65	0.734	0.887	0.759	1.31	0.261	0.647	0.554
¼	1.36	0.599	0.896	0.736	1.06	0.211	0.651	0.558
L3½×3½×½	3.64	1.49	1.06	1.06	2.68	0.464	0.683	1.000
⁷⁄₁₆	3.26	1.32	1.07	1.04	2.38	0.410	0.684	1.000
⅜	2.87	1.15	1.07	1.01	2.08	0.355	0.687	1.000
⁵⁄₁₆	2.45	0.976	1.08	0.990	1.76	0.299	0.690	1.000
¼	2.01	0.794	1.09	0.968	1.43	0.241	0.694	1.000
L3½×3×½	2.33	1.10	0.881	0.875	1.98	0.429	0.621	0.714
⅜	1.85	0.851	0.897	0.830	1.53	0.328	0.625	0.721
⁵⁄₁₆	1.58	0.722	0.905	0.808	1.30	0.276	0.627	0.724
¼	1.30	0.589	0.914	0.785	1.05	0.223	0.631	0.727
L3½×2½×½	1.36	0.760	0.704	0.705	1.40	0.393	0.534	0.486
⅜	1.09	0.592	0.719	0.660	1.07	0.301	0.537	0.496
¼	0.777	0.412	0.735	0.614	0.735	0.205	0.544	0.506
L3×3×½	2.22	1.07	0.898	0.932	1.93	0.458	0.584	1.000
⁷⁄₁₆	1.99	0.954	0.905	0.910	1.72	0.406	0.585	1.000
⅜	1.76	0.833	0.913	0.888	1.50	0.352	0.587	1.000
⁵⁄₁₆	1.51	0.707	0.922	0.865	1.27	0.296	0.589	1.000
¼	1.24	0.577	0.930	0.842	1.04	0.240	0.592	1.000
³⁄₁₆	0.962	0.441	0.939	0.820	0.794	0.182	0.596	1.000

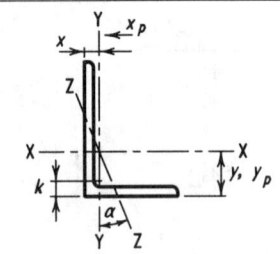

ANGLES
Equal legs and unequal legs
Properties for designing

Size and Thickness	k	Weight per ft	Area	Axis X-X					
				I	S	r	y	Z	y_p
in.	in.	lb	in.^{2}	in.^{4}	in.^{3}	in.	in.	in.^{3}	in.
L3×2½×½	⅞	8.50	2.50	2.08	1.04	0.913	1.000	1.88	0.500
⅜	¾	6.60	1.92	1.66	0.810	0.928	0.956	1.47	0.438
5⁄16	11⁄16	5.60	1.62	1.42	0.688	0.937	0.933	1.25	0.406
¼	⅝	4.50	1.31	1.17	0.561	0.945	0.911	1.02	0.375
3⁄16	9⁄16	3.39	0.996	0.907	0.430	0.954	0.888	0.781	0.344
L3×2×½	13⁄16	7.70	2.25	1.92	1.00	0.924	1.08	1.78	0.750
⅜	11⁄16	5.90	1.73	1.53	0.781	0.940	1.04	1.40	0.688
5⁄16	⅝	5.00	1.46	1.32	0.664	0.948	1.02	1.19	0.656
¼	9⁄16	4.10	1.19	1.09	0.542	0.957	0.993	0.973	0.625
3⁄16	½	3.07	0.902	0.842	0.415	0.966	0.970	0.746	0.594
L2½×2½×½	13⁄16	7.70	2.25	1.23	0.724	0.739	0.806	1.31	0.450
⅜	11⁄16	5.90	1.73	0.984	0.566	0.753	0.762	1.02	0.347
5⁄16	⅝	5.00	1.46	0.849	0.482	0.761	0.740	0.869	0.293
¼	9⁄16	4.10	1.19	0.703	0.394	0.769	0.717	0.711	0.238
3⁄16	½	3.07	0.902	0.547	0.303	0.778	0.694	0.545	0.180
L2½×2×⅜	11⁄16	5.30	1.55	0.912	0.547	0.768	0.831	0.986	0.438
5⁄16	⅝	4.50	1.31	0.788	0.466	0.776	0.809	0.843	0.406
¼	9⁄16	3.62	1.06	0.654	0.381	0.784	0.787	0.691	0.375
3⁄16	½	2.75	0.809	0.509	0.293	0.793	0.764	0.532	0.344
L2×2×⅜	11⁄16	4.70	1.36	0.479	0.351	0.594	0.636	0.633	0.340
5⁄16	⅝	3.92	1.15	0.416	0.300	0.601	0.614	0.541	0.288
¼	9⁄16	3.19	0.938	0.348	0.247	0.609	0.592	0.445	0.234
3⁄16	½	2.44	0.715	0.272	0.190	0.617	0.569	0.343	0.179
⅛	7⁄16	1.65	0.484	0.190	0.131	0.626	0.546	0.235	0.121

ANGLES
Equal legs and unequal legs
Properties for designing

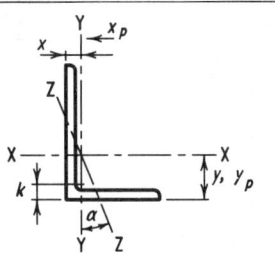

Size and Thickness	Axis Y-Y						Axis Z-Z	
	I	S	r	x	z	x_p	r	Tan
in.	in.^{4}	in.^{3}	in.	in.	in.^{3}	in.	in.	α
L3x2½x½	1.30	0.744	0.722	0.750	1.35	0.417	0.520	0.667
⅜	1.04	0.581	0.736	0.706	1.05	0.320	0.522	0.676
5⁄16	0.898	0.494	0.744	0.683	0.889	0.270	0.525	0.680
¼	0.743	0.404	0.753	0.661	0.724	0.219	0.528	0.684
3⁄16	0.577	0.310	0.761	0.638	0.553	0.166	0.533	0.688
L3x2x½	0.672	0.474	0.546	0.583	0.891	0.375	0.428	0.414
⅜	0.543	0.371	0.559	0.539	0.684	0.289	0.430	0.428
5⁄16	0.470	0.317	0.567	0.516	0.577	0.244	0.432	0.435
¼	0.392	0.260	0.574	0.493	0.468	0.198	0.435	0.440
3⁄16	0.307	0.200	0.583	0.470	0.357	0.150	0.439	0.446
L2½x2½x½	1.23	0.724	0.739	0.806	1.31	0.450	0.487	1.000
⅜	0.984	0.566	0.753	0.762	1.02	0.347	0.487	1.000
5⁄16	0.849	0.482	0.761	0.740	0.869	0.293	0.489	1.000
¼	0.703	0.394	0.769	0.717	0.711	0.238	0.491	1.000
3⁄16	0.547	0.303	0.778	0.694	0.545	0.180	0.495	1.000
L2½x2x⅜	0.514	0.363	0.577	0.581	0.660	0.309	0.420	0.614
5⁄16	0.446	0.310	0.584	0.559	0.561	0.262	0.422	0.620
¼	0.372	0.254	0.592	0.537	0.457	0.213	0.424	0.626
3⁄16	0.291	0.196	0.600	0.514	0.350	0.162	0.427	0.631
L2x2x⅜	0.479	0.351	0.594	0.636	0.633	0.340	0.389	1.000
5⁄16	0.416	0.300	0.601	0.614	0.541	0.288	0.390	1.000
¼	0.348	0.247	0.609	0.592	0.445	0.234	0.391	1.000
3⁄16	0.272	0.190	0.617	0.569	0.343	0.179	0.394	1.000
⅛	0.190	0.131	0.626	0.546	0.235	0.121	0.398	1.000

STRUCTURAL TEES (WT, MT, ST)

Structural tees are obtained by splitting the webs of various beams, generally with the aid of rotary shears, and straightening to meet established permissible variations listed in Standard Mill Practice in Part 1 of this Manual.

Although structural tees may be obtained by off-center splitting, or by splitting at two lines, as specified on order, the Dimensions and Properties are based on a depth of tee equal to one-half the published beam depth. Values of Q_s are given for $F_y = 36$ ksi and $F_y = 50$ ksi, for those tees having stems which exceed the limiting width-thickness ratio λ_r of LRFD Specification Section B5. Since the cross section is comprised entirely of unstiffened elements, $Q_a = 1.0$ and $Q = Q_s$ for all tee sections. The Flexural-Torsional Properties Table lists the dimensional values ($\bar{r}_o$ and H) and cross-section constants (J and C_w) needed for checking flexural-torsional buckling.

Use of Table

The table may be used as follows for checking the limit states of (1) flexural buckling about the x-axis and (2) flexural-torsional buckling. The lower of the two limit states must be used for design. See also Part 3 of this LRFD Manual.

(1) Flexural Buckling About the X-Axis

Where no value of Q_s is shown, the design compressive strength for this limit state is given by LRFD Specification Section E2. Where a value of Q_s is shown, the strength must be reduced in accordance with Appendix B5 of the LRFD Specification.

(2) Flexural-Torsional Buckling

The design compressive strength for this limit state is given by LRFD Specification Section E3. This involves calculations with J, $\bar{r}_o$, and H. Refer to the Flexural-Torsional Properties Tables, later in Part 1.

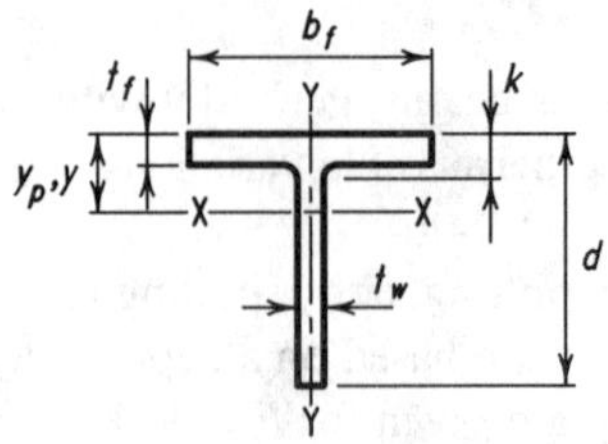

STRUCTURAL TEES
Cut from W shapes
Dimensions

Designation	Area	Depth of Tee d		Stem Thickness t_w		Stem $\frac{t_w}{2}$	Area of Stem	Flange Width b_f		Flange Thickness t_f		Distance k
	in.^{2}	in.		in.		in.	in.^{2}	in.		in.		in.
WT22×167.5	49.1	22.010	22	1.020	1	1/2	22.5	15.950	15 3/4	1.770	1 3/4	2 9/16
×145	42.9	21.810	21 13/16	0.870	7/8	7/16	19.0	15.830	15 7/8	1.580	1 9/16	2 3/8
×131	38.6	21.655	21 11/16	0.790	13/16	3/8	17.1	15.750	15 3/4	1.420	1 7/16	2 3/16
×115	33.8	21.455	21 7/16	0.710	11/16	3/8	15.2	15.750	15 3/4	1.220	1 1/4	2
WT20×296.5	87.0	21.495	21 1/2	1.790	1 13/16	1	38.5	16.690	16 3/4	3.230	3 1/4	4 7/16
×251.5	74.0	21.025	21	1.540	1 9/16	3/4	32.4	16.420	16 7/16	2.760	2 3/4	3 15/16
×215.5	63.4	20.630	20 5/8	1.340	1 5/16	11/16	27.6	16.220	16 1/4	2.360	2 3/8	3 9/16
×186	54.7	20.315	20 5/16	1.160	1 3/16	9/16	23.6	16.060	16 1/16	2.050	2 1/16	3 1/4
×160.5	47.0	20.040	20	1.000	1	1/2	20.0	15.910	15 7/8	1.770	1 3/4	2 15/16
×148.5	43.7	19.920	19 15/16	0.930	15/16	1/2	18.5	15.825	15 7/8	1.650	1 5/8	3 1/16
×138.5	40.7	19.845	19 7/8	0.830	13/16	7/16	16.5	15.830	15 7/8	1.575	1 9/16	2 3/4
×124.5	36.7	19.690	19 11/16	0.750	3/4	3/8	14.8	15.750	15 3/4	1.420	1 7/16	2 5/8
×107.5	31.7	19.490	19 1/2	0.650	5/8	5/16	12.7	15.750	15 3/4	1.220	1 1/4	2 3/8
×99.5	29.2	19.335	19 5/16	0.650	5/8	5/16	12.6	15.750	15 3/4	1.065	1 1/16	2 1/4
×87	25.5	19.100	19 1/8	0.650	5/8	5/16	12.4	15.750	15 3/4	0.830	13/16	2
WT20×233	68.4	21.220	21 3/16	1.67	1 11/16	13/16	35.4	12.640	12 5/8	2.950	2 15/16	4 1/8
×196	57.7	20.785	20 3/4	1.42	1 7/16	11/16	29.5	12.360	12 3/8	2.520	2 1/2	3 11/16
×165.5	48.8	20.395	20 3/8	1.22	1 1/4	5/8	24.9	12.170	12 3/16	2.130	2 1/8	3 5/16
×139	40.9	20.080	20 1/8	1.02	1	1/2	20.5	11.970	12	1.810	1 13/16	3
×132	38.8	20.000	20	0.960	1	1/2	19.2	11.930	12	1.730	1 3/4	2 15/16
×117.5	34.5	19.845	19 7/8	0.830	13/16	7/16	16.5	11.890	11 7/8	1.575	1 7/16	2 3/4
×105.5	31.0	19.685	19 11/16	0.750	3/4	3/8	14.8	11.810	11 3/4	1.415	1 9/16	2 5/8
×91.5	26.9	19.490	19 1/2	0.650	5/8	5/16	12.7	11.810	11 3/4	1.220	1 1/4	2 3/8
×83.5	24.6	19.295	19 5/16	0.650	5/8	5/16	12.5	11.810	11 3/4	1.025	1	2 3/16
×74.5	21.9	19.100	19 1/8	0.630	5/8	5/16	12.0	11.810	11 3/4	0.830	13/16	2
WT18×424	125	21.225	21 1/4	2.520	2 1/2	1 1/4	53.5	18.130	18 1/8	4.530	4 1/2	5 11/16
×399	117	20.985	21	2.380	2 3/8	1 3/16	49.9	17.990	18	4.290	4 5/16	5 7/16
×325	95.0	20.235	20 1/4	1.970	2	1	39.9	17.575	17 5/8	3.540	3 9/16	4 11/16
×263.5	77.0	19.605	19 5/8	1.610	1 5/8	13/16	31.6	17.220	17 1/4	2.910	2 15/16	4 1/16
×219.5	64.0	19.130	19 1/8	1.360	1 3/8	11/16	26.0	16.965	17	2.440	2 7/16	3 9/16
×196.5	57.5	18.900	18 7/8	1.220	1 1/4	5/8	23.1	16.830	16 7/8	2.200	2 3/16	3 5/16
×179.5	52.7	18.700	18 11/16	1.120	1 1/8	9/16	20.9	16.730	16 3/4	2.010	2	3 1/8
×164	48.2	18.545	18 9/16	1.020	1	1/2	18.9	16.630	16 5/8	1.850	1 7/8	3
×150	44.1	18.370	18 3/8	0.945	15/16	1/2	17.4	16.655	16 5/8	1.680	1 11/16	2 13/16
×140	41.2	18.260	18 1/4	0.885	7/8	7/16	16.2	16.595	16 5/8	1.570	1 9/16	2 11/16
×130	38.2	18.130	18 1/8	0.840	13/16	7/16	15.2	16.550	16 1/2	1.440	1 7/16	2 9/16
×122.5	36.0	18.040	18	0.800	13/16	7/16	14.4	16.510	16 1/2	1.350	1 3/8	2 1/2
×115	33.8	17.950	18	0.760	3/4	3/8	13.6	16.470	16 1/2	1.260	1 1/4	2 3/8

STRUCTURAL TEES
Cut from W shapes
Properties

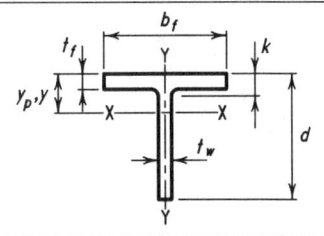

Nom-inal Wt. per ft		Axis X-X						Axis Y-Y				Q_s^*	
	h	I	S	r	y	Z	y_p	I	S	r	Z	F_y, ksi	
lb	t_w	in.^{4}	in.^{3}	in.	in.	in.^{3}	in.	in.^{4}	in.^{3}	in.	in.^{3}	36	50
167.5	19.1	2160	131	6.63	5.51	233	1.54	600	75.3	3.50	118	0.982	0.817
145	22.3	1840	111	6.55	5.27	197	1.35	524	66.1	3.49	103	0.833	0.636
131	24.6	1650	100	6.53	5.20	177	1.23	463	58.8	3.46	91.4	0.732	0.532
115	27.4	1440	88.6	6.53	5.17	157	1.07	398	50.5	3.43	78.4	0.608	0.438
296.5	9.5	3300	209	6.16	5.67	379	2.61	1260	151	3.81	240	—	—
251.5	11.1	2730	175	6.07	5.39	315	2.25	1020	125	3.72	197	—	—
215.5	12.8	2290	148	6.01	5.18	266	1.95	843	104	3.65	164	—	—
186	14.7	1930	126	5.95	4.97	225	1.70	710	88.5	3.60	139	—	—
160.5	17.1	1630	107	5.89	4.79	191	1.48	596	74.9	3.56	117	—	0.895
148.5	18.4	1500	98.9	5.87	4.71	176	1.38	546	69.1	3.54	108	0.989	0.825
138.5	20.6	1360	88.6	5.78	4.51	157	1.28	522	65.9	3.58	102	0.882	0.699
124.5	22.8	1210	79.3	5.75	4.41	140	1.16	463	58.8	3.56	91.0	0.782	0.580
107.5	26.3	1030	68.0	5.72	4.28	120	1.00	398	50.5	3.55	77.9	0.618	0.445
99.5	26.3	987	66.4	5.81	4.48	117	0.927	347	44.1	3.45	68.3	0.628	0.452
87	26.3	907	63.8	5.96	4.87	114	0.811	271	34.4	3.26	53.8	0.643	0.463
233	10.2	2770	185	6.36	6.22	333	2.71	504	79.8	2.72	131	—	—
196	12.0	2270	153	6.28	5.95	276	2.33	401	65.0	2.64	106	—	—
165.5	14.0	1880	128	6.21	5.74	231	2.01	323	53.1	2.57	86.2	—	—
139	16.8	1540	106	6.14	5.50	190	1.71	261	43.6	2.52	70.0	—	0.913
132	17.8	1450	99.3	6.11	5.40	178	1.63	246	41.3	2.52	66.2	—	0.855
117.5	20.6	1260	85.6	6.04	5.17	153	1.45	222	37.3	2.54	59.2	0.882	0.699
105.5	22.8	1120	76.7	6.01	5.08	137	1.31	195	33.0	2.51	52.3	0.782	0.581
91.5	26.3	957	65.8	5.97	4.94	117	1.14	168	28.5	2.50	44.8	0.618	0.445
83.5	26.3	898	63.7	6.05	5.20	115	1.04	141	23.9	2.40	38.0	0.630	0.454
74.5	27.1	815	59.7	6.10	5.45	108	1.82	115	19.4	2.29	31.1	0.604	0.435
424	6.3	4250	277	5.84	5.86	515	3.43	2270	251	4.27	399	—	—
399	6.6	3920	257	5.79	5.72	478	3.25	2100	234	4.24	371	—	—
325	8.0	3020	202	5.64	5.29	373	2.70	1610	184	4.12	290	—	—
263.5	9.8	2330	159	5.50	4.89	290	2.24	1240	145	4.02	227	—	—
219.5	11.6	1880	130	5.42	4.63	235	1.89	997	117	3.95	184	—	—
196.5	12.9	1660	115	5.37	4.46	207	1.71	877	104	3.90	162	—	—
179.5	14.1	1500	104	5.33	4.33	187	1.58	786	94.0	3.86	146	—	—
164	15.4	1350	94.1	5.29	4.21	168	1.45	711	85.5	3.84	132	—	—
150	16.7	1230	86.1	5.27	4.13	153	1.33	648	77.8	3.83	120	—	0.927
140	17.8	1140	80.0	5.25	4.07	142	1.24	599	72.2	3.81	112	—	0.867
130	18.7	1060	75.1	5.26	4.05	133	1.16	545	65.9	3.78	102	0.981	0.816
122.5	19.7	995	71.0	5.26	4.03	125	1.09	507	61.4	3.75	94.9	0.943	0.770
115	20.7	934	67.0	5.25	4.01	118	1.03	470	57.1	3.73	88.1	0.896	0.715

*Where no value of Q_s is shown, the Tee complies with LRFD Specification Section E2.

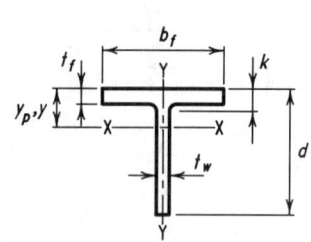

STRUCTURAL TEES
Cut from W shapes
Dimensions

Designation	Area in.²	Depth of Tee d in.		Stem Thickness t_w in.		$\frac{t_w}{2}$ in.	Area of Stem in.²	Flange Width b_f in.		Flange Thickness t_f in.		Distance k in.
WT18×128	37.7	18.715	18¹¹⁄₁₆	0.960	1	½	18.0	12.215	12¼	1.730	1¾	2⅝
×116	34.1	18.560	18⁹⁄₁₆	0.870	⅞	⁷⁄₁₆	16.1	12.120	12⅛	1.570	1⁹⁄₁₆	2½
×105	30.9	18.345	18⅜	0.830	¹³⁄₁₆	⁷⁄₁₆	15.2	12.180	12⅛	1.360	1⅜	2⁵⁄₁₆
×97	28.5	18.245	18¼	0.765	¾	⅜	14.0	12.115	12⅛	1.260	1¼	2³⁄₁₆
×91	26.8	18.165	18⅛	0.725	¾	⅜	13.2	12.075	12⅛	1.180	1³⁄₁₆	2⅛
×85	25.0	18.085	18⅛	0.680	¹¹⁄₁₆	⅜	12.3	12.030	12	1.100	1⅛	2
×80	23.5	18.005	18	0.650	⅝	⁵⁄₁₆	11.7	12.000	12	1.020	1	1¹⁵⁄₁₆
×75	22.1	17.925	17⅞	0.625	⅝	⁵⁄₁₆	11.2	11.975	12	0.940	¹⁵⁄₁₆	1⅞
×67.5	19.9	17.775	17¾	0.600	⅝	⁵⁄₁₆	10.7	11.950	12	0.790	¹³⁄₁₆	1¹¹⁄₁₆
WT16.5×177	52.1	17.775	17¾	1.160	1³⁄₁₆	⅝	20.6	16.100	16⅛	2.090	2¹⁄₁₆	2⅞
×159	46.7	17.580	17⁹⁄₁₆	1.040	1¹⁄₁₆	⁹⁄₁₆	18.3	15.985	16	1.890	1⅞	2¹¹⁄₁₆
×145.5	42.8	17.420	17⁷⁄₁₆	0.960	1	½	16.7	15.905	15⅞	1.730	1¾	2⁹⁄₁₆
×131.5	38.7	17.265	17¼	0.870	⅞	⁷⁄₁₆	15.0	15.805	15¾	1.570	1⁹⁄₁₆	2⅜
×120.5	35.4	17.090	17⅛	0.830	¹³⁄₁₆	⁷⁄₁₆	14.2	15.860	15⅞	1.400	1⅜	2³⁄₁₆
×110.5	32.5	16.965	17	0.775	¾	⅜	13.1	15.805	15¾	1.275	1¼	2¹⁄₁₆
×100.5	29.5	16.840	16⅞	0.715	¹¹⁄₁₆	⅜	12.0	15.745	15¾	1.150	1⅛	1¹⁵⁄₁₆
WT16.5×84.5	24.8	16.910	16¹⁵⁄₁₆	0.670	¹¹⁄₁₆	⅜	11.3	11.500	11½	1.220	1¼	2¹⁄₁₆
×76	22.4	16.745	16¾	0.635	⅝	⁵⁄₁₆	10.6	11.565	11⅝	1.055	1¹⁄₁₆	1⅞
×70.5	20.8	16.650	16⅝	0.605	⅝	⁵⁄₁₆	10.1	11.535	11½	0.960	¹⁵⁄₁₆	1¾
×65	19.2	16.545	16½	0.580	⁹⁄₁₆	⁵⁄₁₆	9.60	11.510	11½	0.855	⅞	1¹¹⁄₁₆
×59	17.3	16.430	16⅜	0.550	⁹⁄₁₆	⁵⁄₁₆	9.04	11.480	11½	0.740	¾	1⁹⁄₁₆
WT15×238.5	70.0	17.105	17⅛	1.630	1⅝	¹³⁄₁₆	27.9	15.865	15⅞	2.950	3	3¾
×195.5	57.0	16.595	16⅝	1.360	1⅜	¹¹⁄₁₆	22.6	15.590	15⅝	2.440	2⁷⁄₁₆	3¼
×163	47.9	16.200	16³⁄₁₆	1.140	1⅛	⁹⁄₁₆	18.5	15.370	15⅜	2.050	2¹⁄₁₆	2¹³⁄₁₆
×146	42.9	16.005	16	1.020	1	½	16.3	15.255	15½	1.850	1⅞	2⅝
×130.5	38.4	15.805	15¹³⁄₁₆	0.930	¹⁵⁄₁₆	½	14.7	15.155	15⅛	1.650	1⅝	2⁷⁄₁₆
×117.5	34.5	15.650	15⅝	0.830	¹³⁄₁₆	⁷⁄₁₆	13.0	15.055	15	1.500	1½	2¼
×105.5	31.0	15.470	15½	0.775	¾	⅜	12.0	15.105	15⅛	1.315	1⁵⁄₁₆	2⅛
×95.5	28.1	15.340	15⅜	0.710	¹¹⁄₁₆	⅜	10.9	15.040	15	1.185	1³⁄₁₆	1¹⁵⁄₁₆
×86.5	25.4	15.220	15¼	0.655	⅝	⁵⁄₁₆	9.97	14.985	15	1.065	1¹⁄₁₆	1⅞
WT15×74	21.7	15.335	15⁵⁄₁₆	0.650	⅝	⁵⁄₁₆	10.0	10.480	10½	1.180	1³⁄₁₆	2
×66	19.4	15.155	15⅛	0.615	⅝	⁵⁄₁₆	9.32	10.545	10½	1.000	1	1¾
×62	18.2	15.085	15⅛	0.585	⁹⁄₁₆	⁵⁄₁₆	8.82	10.515	10½	0.930	¹⁵⁄₁₆	1¹¹⁄₁₆
×58	17.1	15.005	15	0.565	⁹⁄₁₆	⁵⁄₁₆	8.48	10.495	10½	0.850	⅞	1⅝
×54	15.9	14.915	14⅞	0.545	⁹⁄₁₆	⁵⁄₁₆	8.13	10.475	10½	0.760	¾	1⁹⁄₁₆
×49.5	14.5	14.825	14⅞	0.520	½	¼	7.71	10.450	10½	0.670	¹¹⁄₁₆	1⁷⁄₁₆
×45	13.2	14.765	14¾	0.470	½	¼	6.94	10.400	10⅜	0.610	⁹⁄₁₆	1⁵⁄₁₆

STRUCTURAL TEES
Cut from W shapes
Properties

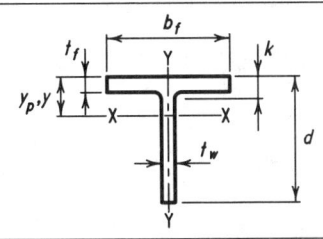

Nom-inal Wt. per ft		Axis X-X						Axis Y-Y				Q_s*	
	h	I	S	r	y	Z	y_p	I	S	r	Z	F_y, ksi	
lb	t_w	in.^{4}	in.^{3}	in.	in.	in.^{3}	in.	in.^{4}	in.^{3}	in.	in.^{3}	36	50
128	16.9	1200	87.4	5.66	4.92	156	1.54	264	43.2	2.65	68.6	—	0.927
116	18.7	1080	78.5	5.63	4.82	140	1.40	234	38.6	2.62	61.0	0.994	0.831
105	19.6	985	73.1	5.65	4.87	131	1.27	206	33.8	2.58	53.5	0.960	0.791
97	21.2	901	67.0	5.62	4.80	120	1.18	187	30.9	2.56	48.9	0.887	0.705
91	22.4	845	63.1	5.62	4.77	113	1.11	174	28.8	2.55	45.4	0.831	0.635
85	23.9	786	58.9	5.61	4.73	105	1.04	160	26.6	2.53	41.9	0.767	0.565
80	25.0	740	55.8	5.61	4.74	100	0.980	147	24.6	2.50	38.6	0.720	0.521
75	26.0	698	53.1	5.62	4.78	95.5	0.923	135	22.5	2.47	35.5	0.677	0.486
67.5	27.1	637	49.7	5.66	4.96	94.3	1.24	113	18.9	2.38	29.8	0.634	0.457
177	12.9	1320	96.8	5.03	4.16	174	1.62	729	90.6	3.74	141	—	—
159	14.4	1160	85.8	4.99	4.02	154	1.46	645	80.7	3.71	125	—	—
145.5	15.6	1050	78.3	4.97	3.94	140	1.34	581	73.1	3.69	113	—	0.993
131.5	17.2	943	70.2	4.94	3.84	125	1.22	517	65.5	3.66	101	—	0.907
120.5	18.1	871	65.8	4.96	3.85	116	1.12	466	58.8	3.63	90.9	—	0.867
110.5	19.3	799	60.8	4.96	3.81	107	1.03	420	53.2	3.59	82.1	0.968	0.801
100.5	21.0	725	55.5	4.95	3.78	97.7	0.938	375	47.6	3.56	73.4	0.896	0.715
84.5	22.4	649	51.1	5.12	4.21	90.8	1.08	155	27.0	2.50	42.2	0.827	0.630
76	23.6	592	47.4	5.14	4.26	84.5	0.967	136	23.6	2.47	37.0	0.775	0.574
70.5	24.8	552	44.7	5.15	4.29	79.8	0.901	123	21.3	2.43	33.5	0.728	0.529
65	25.8	513	42.1	5.18	4.36	75.6	0.832	109	18.9	2.39	29.7	0.685	0.492
59	27.3	469	39.2	5.20	4.47	74.8	0.862	93.6	16.3	2.32	25.7	0.621	0.447
238.5	8.3	1550	121	4.70	4.30	224	2.21	987	124	3.75	195	—	—
195.5	9.9	1210	96.6	4.61	4.04	177	1.83	774	99.2	3.68	155	—	—
163	11.8	981	78.9	4.53	3.76	143	1.56	622	81.0	3.61	126	—	—
146	13.2	861	69.6	4.48	3.63	125	1.40	549	71.9	3.58	111	—	—
130.5	14.5	764	62.3	4.46	3.54	112	1.27	480	63.3	3.54	97.9	—	—
117.5	16.2	674	55.1	4.42	3.42	98.2	1.15	427	56.8	3.52	87.5	—	0.952
105.5	17.4	610	50.5	4.43	3.40	89.5	1.03	378	50.1	3.49	77.2	—	0.897
95.5	19.0	549	45.7	4.42	3.35	80.8	0.933	336	44.7	3.46	68.9	0.981	0.816
86.5	20.6	497	41.7	4.42	3.31	73.4	0.848	299	39.9	3.43	61.4	0.913	0.735
74	20.8	466	40.6	4.63	3.84	72.2	1.04	113	21.7	2.28	34.0	0.896	0.715
66	22.0	421	37.4	4.66	3.90	66.8	0.921	98.0	18.6	2.25	29.2	0.853	0.664
62	23.1	396	35.3	4.66	3.90	63.1	0.867	90.4	17.2	2.23	27.0	0.801	0.601
58	23.9	373	33.7	4.67	3.94	60.4	0.815	82.1	15.7	2.19	24.6	0.767	0.565
54	24.8	349	32.0	4.69	4.01	57.7	0.757	73.0	13.9	2.15	22.0	0.733	0.533
49.5	26.0	322	30.0	4.71	4.09	57.4	0.912	63.9	12.2	2.10	19.3	0.685	0.492
45	28.7	291	27.1	4.69	4.03	49.4	0.445	57.3	11.0	2.08	17.3	0.563	0.405

*Where no value of Q_s is shown, the Tee complies with LRFD Specification Section E2.

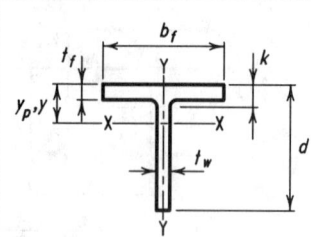

STRUCTURAL TEES
Cut from W shapes
Dimensions

Designation	Area in.^{2}	Depth of Tee d in.		Stem Thickness t_w in.		$\dfrac{t_w}{2}$ in.	Area of Stem in.^{2}	Flange Width b_f in.		Flange Thickness t_f in.		Distance k in.
WT13.5×269.5	79.0	16.260	16¼	1.970	2	1	32.0	15.255	15¼	3.540	3⁹⁄₁₆	4¼
×224	65.5	15.710	15¹¹⁄₁₆	1.650	1⅝	¹³⁄₁₆	25.9	14.940	15	2.990	3	3¹¹⁄₁₆
×184	54.0	15.195	15³⁄₁₆	1.380	1⅜	¹¹⁄₁₆	21.0	14.665	14⅝	2.480	2½	3³⁄₁₆
×153.5	45.1	14.805	14¹³⁄₁₆	1.160	1³⁄₁₆	⅝	17.2	14.445	14½	2.090	2¹⁄₁₆	2¹³⁄₁₆
×129	37.9	14.490	14½	0.980	1	½	14.2	14.270	14¼	1.770	1¾	2½
×117.5	34.6	14.330	14⁵⁄₁₆	0.910	¹⁵⁄₁₆	½	13.0	14.190	14¼	1.610	1⅝	2⁵⁄₁₆
×108.5	31.9	14.215	14³⁄₁₆	0.830	¹³⁄₁₆	⁷⁄₁₆	11.8	14.115	14⅛	1.500	1½	2³⁄₁₆
×97	28.5	14.055	14¹⁄₁₆	0.750	¾	⅜	10.5	14.035	14	1.340	1⁵⁄₁₆	2¹⁄₁₆
×89	26.1	13.905	13⅞	0.725	¾	⅜	10.1	14.085	14⅛	1.190	1³⁄₁₆	1⅞
×80.5	23.7	13.795	13¾	0.660	¹¹⁄₁₆	⅜	9.10	14.020	14	1.080	1¹⁄₁₆	1¹³⁄₁₆
×73	21.5	13.690	13¾	0.605	⅝	⁵⁄₁₆	8.28	13.965	14	0.975	1	1¹¹⁄₁₆
WT13.5×64.5	18.9	13.815	13¹³⁄₁₆	0.610	⅝	⁵⁄₁₆	8.43	10.010	10	1.100	1⅛	1¹³⁄₁₆
×57	16.8	13.645	13⅝	0.570	⁹⁄₁₆	⁵⁄₁₆	7.78	10.070	10⅛	0.930	¹⁵⁄₁₆	1⅝
×51	15.0	13.545	13½	0.515	½	¼	6.98	10.015	10	0.830	¹³⁄₁₆	1⁹⁄₁₆
×47	13.8	13.460	13½	0.490	½	¼	6.60	9.990	10	0.745	¾	1⁷⁄₁₆
×42	12.4	13.355	13⅜	0.460	⁷⁄₁₆	¼	6.14	9.960	10	0.640	⅝	1⅜
WT12×246	72.0	14.825	14¹³⁄₁₆	1.970	2	1	29.2	14.115	14⅛	3.540	3⁹⁄₁₆	4⁵⁄₁₆
×204	59.5	14.270	14¼	1.650	1⅝	¹³⁄₁₆	23.5	13.800	13¾	2.990	3	3¾
×167.5	49.2	13.760	13¾	1.380	1⅜	¹¹⁄₁₆	19.0	13.520	13½	2.480	2½	3¼
×139.5	41.0	13.365	13⅜	1.160	1³⁄₁₆	⅝	15.5	13.305	13¼	2.090	2¹⁄₁₆	2⅞
×125	36.8	13.170	13³⁄₁₆	1.040	1¹⁄₁₆	⁹⁄₁₆	13.7	13.185	13⅛	1.890	1⅞	2¹¹⁄₁₆
×114.5	33.6	13.010	13	0.960	1	½	12.5	13.110	13⅛	1.730	1¾	2½
×103.5	30.4	12.855	12⅞	0.870	⅞	⁷⁄₁₆	11.2	13.010	13	1.570	1⁹⁄₁₆	2⅜
×96	28.2	12.735	12¾	0.810	¹³⁄₁₆	⁷⁄₁₆	10.3	12.950	13	1.460	1⁷⁄₁₆	2¼
×88	25.8	12.620	12⅝	0.750	¾	⅜	9.47	12.890	12⅞	1.340	1⁵⁄₁₆	2⅛
×81	23.9	12.500	12½	0.705	¹¹⁄₁₆	⅜	8.81	12.955	13	1.220	1¼	2
×73	21.5	12.370	12⅜	0.650	⅝	⁵⁄₁₆	8.04	12.900	12⅞	1.090	1¹⁄₁₆	1⅞
×65.5	19.3	12.240	12¼	0.605	⅝	⁵⁄₁₆	7.41	12.855	12⅞	0.960	¹⁵⁄₁₆	1¾
×58.5	17.2	12.130	12⅛	0.550	⁹⁄₁₆	⁵⁄₁₆	6.67	12.800	12¾	0.850	⅞	1⅝
×52	15.3	12.030	12	0.500	½	¼	6.02	12.750	12¾	0.750	¾	1½
WT12×51.5	15.1	12.265	12¼	0.550	⁹⁄₁₆	⁵⁄₁₆	6.75	9.000	9	0.980	1	1¾
×47	13.8	12.155	12⅛	0.515	½	¼	6.26	9.065	9⅛	0.875	⅞	1⅝
×42	12.4	12.050	12	0.470	½	¼	5.66	9.020	9	0.770	¾	1⁹⁄₁₆
×38	11.2	11.960	12	0.440	⁷⁄₁₆	¼	5.26	8.990	9	0.680	¹¹⁄₁₆	1⁷⁄₁₆
×34	10.0	11.865	11⅞	0.415	⁷⁄₁₆	¼	4.92	8.965	9	0.585	⁹⁄₁₆	1⅜
WT12×31	9.11	11.870	11⅞	0.430	⁷⁄₁₆	¼	5.10	7.040	7	0.590	⁹⁄₁₆	1⅜
×27.5	8.10	11.785	11¾	0.395	⅜	³⁄₁₆	4.66	7.005	7	0.505	½	1⁵⁄₁₆

STRUCTURAL TEES
Cut from W shapes
Properties

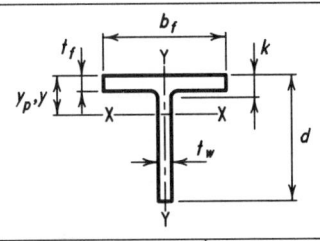

Nom-inal Wt. per ft		Axis X-X						Axis Y-Y				Q_s^*	
	h	I	S	r	y	Z	y_p	I	S	r	Z	F_y, ksi	
lb	$\dfrac{h}{t_w}$	in.^{4}	in.^{3}	in.	in.	in.^{3}	in.	in.^{4}	in.^{3}	in.	in.^{3}	36	50
269.5	6.2	1520	128	4.39	4.36	241	2.59	1060	138	3.66	218	—	—
224	7.4	1190	102	4.27	4.02	191	2.19	836	112	3.57	176	—	—
184	8.8	938	81.7	4.17	3.71	151	1.84	655	89.3	3.48	140	—	—
153.5	10.5	753	66.4	4.09	3.47	121	1.56	527	72.9	3.42	113	—	—
129	12.4	613	54.6	4.02	3.28	98.8	1.33	430	60.2	3.37	93.3	—	—
117.5	13.3	556	50.0	4.01	3.21	89.8	1.22	384	54.2	3.33	83.8	—	—
108.5	14.6	502	45.2	3.97	3.11	81.1	1.13	352	49.9	3.32	77.0	—	—
97	16.2	444	40.3	3.95	3.03	71.8	1.02	309	44.1	3.29	67.9	—	0.963
89	16.7	414	38.2	3.98	3.05	67.6	0.928	278	39.4	3.26	60.8	—	0.937
80.5	18.4	372	34.4	3.96	2.99	60.8	0.845	248	35.4	3.24	54.5	—	0.851
73	20.0	336	31.2	3.95	2.95	55.0	0.768	222	31.7	3.21	48.8	0.938	0.765
64.5	19.9	323	31.0	4.13	3.39	55.1	0.945	92.2	18.4	2.21	28.8	0.938	0.765
57	21.3	289	28.3	4.15	3.42	50.4	0.833	79.4	15.8	2.18	24.7	0.883	0.700
51	23.5	258	25.3	4.14	3.37	45.0	0.750	69.6	13.9	2.15	21.7	0.780	0.578
47	24.7	239	23.8	4.16	3.41	42.4	0.692	62.0	12.4	2.12	19.4	0.728	0.529
42	26.3	216	21.9	4.18	3.48	39.2	0.621	52.8	10.6	2.07	16.6	0.664	0.476
246	5.5	1130	105	3.96	4.07	200	2.55	837	119	3.41	187	—	—
204	6.5	874	83.1	3.83	3.74	157	2.16	659	95.5	3.33	150	—	—
167.5	7.8	685	66.3	3.73	3.42	123	1.82	513	75.9	3.23	119	—	—
139.5	9.3	546	53.6	3.65	3.18	98.8	1.54	412	61.9	3.17	96.4	—	—
125	10.4	478	47.2	3.61	3.05	86.5	1.39	362	54.9	3.14	85.3	—	—
114.5	11.2	431	42.9	3.58	2.97	78.1	1.28	326	49.7	3.11	77.0	—	—
103.5	12.4	382	38.3	3.55	2.87	69.3	1.17	289	44.4	3.08	68.6	—	—
96	13.3	350	35.2	3.53	2.80	63.5	1.09	265	40.9	3.07	63.1	—	—
88	14.4	319	32.2	3.51	2.74	57.8	1.00	240	37.2	3.04	57.3	—	—
81	15.3	293	29.9	3.50	2.70	53.3	0.921	221	34.2	3.05	52.6	—	—
73	16.6	264	27.2	3.50	2.66	48.2	0.833	195	30.3	3.01	46.6	—	0.947
65.5	17.8	238	24.8	3.52	2.65	43.9	0.750	170	26.5	2.97	40.7	—	0.887
58.5	19.6	212	22.3	3.51	2.62	39.2	0.672	149	23.2	2.94	35.7	0.960	0.791
52	21.6	189	20.0	3.51	2.59	35.1	0.600	130	20.3	2.91	31.2	0.874	0.690
51.5	19.6	204	22.0	3.67	3.01	39.2	0.841	59.7	13.3	1.99	20.7	0.951	0.781
47	20.9	186	20.3	3.67	2.99	36.1	0.764	54.5	12.0	1.98	18.8	0.896	0.715
42	22.9	166	18.3	3.67	2.97	32.5	0.685	47.2	10.5	1.95	16.3	0.810	0.610
38	24.5	151	16.9	3.68	3.00	30.1	0.622	41.3	9.18	1.92	14.3	0.741	0.541
34	26.0	137	15.6	3.70	3.06	27.9	0.560	35.2	7.85	1.87	12.3	0.681	0.489
31	25.1	131	15.6	3.79	3.46	28.4	1.28	17.2	4.90	1.38	7.87	0.724	0.525
27.5	27.3	117	14.1	3.80	3.50	25.6	1.53	14.5	4.15	1.34	6.67	0.626	0.450

*Where no value of Q_s is shown, the Tee complies with LRFD Specification Sect. E2.

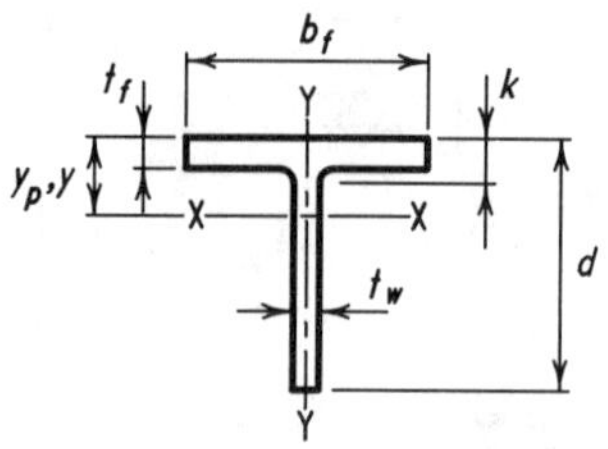

STRUCTURAL TEES
Cut from W shapes
Dimensions

Designation	Area	Depth of Tee d		Stem Thickness t_w		Stem $\frac{t_w}{2}$	Area of Stem	Flange Width b_f		Flange Thickness t_f		Distance k
	in.^{2}	in.		in.		in.	in.^{2}	in.		in.		in.
WT10.5×100.5	29.6	11.515	11 1/2	0.910	15/16	1/2	10.5	12.575	12 5/8	1.630	1 5/8	2 3/8
×91	26.8	11.360	11 3/8	0.830	13/16	7/16	9.43	12.500	12 1/2	1.480	1 1/2	2 1/4
×83	24.4	11.240	11 1/4	0.750	3/4	3/8	8.43	12.420	12 3/8	1.360	1 3/8	2 1/8
×73.5	21.6	11.030	11	0.720	3/4	3/8	7.94	12.510	12 1/2	1.150	1 1/8	1 7/8
×66	19.4	10.915	10 7/8	0.650	5/8	5/16	7.09	12.440	12 1/2	1.035	1 1/16	1 13/16
×61	17.9	10.840	10 7/8	0.600	5/8	5/16	6.50	12.390	12 3/8	0.960	15/16	1 11/16
×55.5	16.3	10.755	10 3/4	0.550	9/16	5/16	5.92	12.340	12 3/8	0.875	7/8	1 5/8
×50.5	14.9	10.680	10 5/8	0.500	1/2	1/4	5.34	12.290	12 1/4	0.800	13/16	1 9/16
WT10.5×46.5	13.7	10.810	10 3/4	0.580	9/16	5/16	6.27	8.420	8 3/8	0.930	15/16	1 11/16
×41.5	12.2	10.715	10 3/4	0.515	1/2	1/4	5.52	8.355	8 3/8	0.835	13/16	1 9/16
×36.5	10.7	10.620	10 5/8	0.455	7/16	1/4	4.83	8.295	8 1/4	0.740	3/4	1 1/2
×34	10.0	10.565	10 5/8	0.430	7/16	1/4	4.54	8.270	8 1/4	0.685	11/16	1 7/16
×31	9.13	10.495	10 1/2	0.400	3/8	3/16	4.20	8.240	8 1/4	0.615	5/8	1 3/8
WT10.5×28.5	8.37	10.530	10 1/2	0.405	3/8	3/16	4.26	6.555	6 1/2	0.650	5/8	1 3/8
×25	7.36	10.415	10 3/8	0.380	3/8	3/16	3.96	6.530	6 1/2	0.535	9/16	1 5/16
×22	6.49	10.330	10 3/8	0.350	3/8	3/16	3.62	6.500	6 1/2	0.450	7/16	1 3/16
WT9×155.5	45.8	11.160	11 3/16	1.520	1 1/2	3/4	17.0	12.005	12	2.740	2 3/4	3 7/16
×141.5	41.6	10.925	10 15/16	1.400	1 3/8	11/16	15.3	11.890	11 7/8	2.500	2 1/2	3 3/16
×129	38.0	10.730	10 3/4	1.280	1 1/4	5/8	13.7	11.770	11 3/4	2.300	2 5/16	3
×117	34.4	10.530	10 1/2	1.160	1 3/16	5/8	12.2	11.650	11 5/8	2.110	2 1/8	2 3/4
×105.5	31.1	10.335	10 5/16	1.060	1 1/16	9/16	11.0	11.555	11 1/2	1.910	1 15/16	2 9/16
×96	28.2	10.175	10 3/16	0.960	1	1/2	9.77	11.455	11 1/2	1.750	1 3/4	2 7/16
×87.5	25.7	10.020	10	0.890	7/8	7/16	8.92	11.375	11 3/8	1.590	1 9/16	2 1/4
×79	23.2	9.860	9 7/8	0.810	13/16	7/16	7.99	11.300	11 1/4	1.440	1 7/16	2 1/8
×71.5	21.0	9.745	9 3/4	0.730	3/4	3/8	7.11	11.220	11 1/4	1.320	1 5/16	2
×65	19.1	9.625	9 5/8	0.670	11/16	3/8	6.45	11.160	11 1/8	1.200	1 3/16	1 7/8
WT9×59.5	17.5	9.485	9 1/2	0.655	5/8	5/16	6.21	11.265	11 1/4	1.060	1 1/16	1 3/4
×53	15.6	9.365	9 3/8	0.590	9/16	5/16	5.53	11.200	11 1/4	0.940	15/16	1 5/8
×48.5	14.3	9.295	9 1/4	0.535	9/16	5/16	4.97	11.145	11 1/8	0.870	7/8	1 9/16
×43	12.7	9.195	9 1/4	0.480	1/2	1/4	4.41	11.090	11 1/8	0.770	3/4	1 7/16
×38	11.2	9.105	9 1/8	0.425	7/16	1/4	3.87	11.035	11	0.680	11/16	1 3/8
WT9×35.5	10.4	9.235	9 1/4	0.495	1/2	1/4	4.57	7.635	7 5/8	0.810	13/16	1 1/2
×32.5	9.55	9.175	9 1/8	0.450	7/16	1/4	4.13	7.590	7 5/8	0.750	3/4	1 7/16
×30	8.82	9.120	9 1/8	0.415	7/16	1/4	3.78	7.555	7 1/2	0.695	11/16	1 3/8
×27.5	8.10	9.055	9	0.390	3/8	3/16	3.53	7.530	7 1/2	0.630	5/8	1 5/16
×25	7.33	8.995	9	0.355	3/8	3/16	3.19	7.495	7 1/2	0.570	9/16	1 1/4
WT9×23	6.77	9.030	9	0.360	3/8	3/16	3.25	6.060	6	0.605	5/8	1 1/4
×20	5.88	8.950	9	0.315	5/16	3/16	2.82	6.015	6	0.525	1/2	1 3/16
×17.5	5.15	8.850	8 7/8	0.300	5/16	3/16	2.66	6.000	6	0.425	7/16	1 1/8

STRUCTURAL TEES
Cut from W shapes
Properties

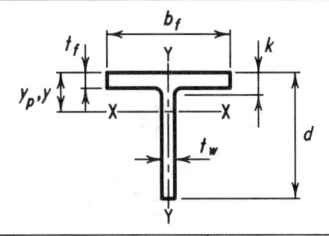

Nom-inal Wt. per ft		Axis X-X						Axis Y-Y				Q_s*	
	$\dfrac{h}{t_w}$	I	S	r	y	Z	y_p	I	S	r	Z	F_y, ksi	
lb		in.^{4}	in.^{3}	in.	in.	in.^{3}	in.	in.^{4}	in.^{3}	in.	in.^{3}	36	50
100.5	10.3	285	31.9	3.10	2.57	58.6	1.18	271	43.1	3.02	66.6	—	—
91	11.2	253	28.5	3.07	2.48	52.1	1.07	241	38.6	3.00	59.6	—	—
83	12.4	226	25.5	3.04	2.39	46.3	0.984	217	35.0	2.98	53.9	—	—
73.5	13.0	204	23.7	3.08	2.39	42.4	0.864	188	30.0	2.95	46.3	—	—
66	14.4	181	21.1	3.06	2.33	37.6	0.780	166	26.7	2.93	41.1	—	—
61	15.6	166	19.3	3.04	2.28	34.3	0.724	152	24.6	2.92	37.8	—	0.993
55.5	17.1	150	17.5	3.03	2.23	31.0	0.662	137	22.2	2.90	34.1	—	0.917
50.5	18.8	135	15.8	3.01	2.18	27.9	0.605	124	20.2	2.89	30.9	0.990	0.826
46.5	16.2	144	17.9	3.25	2.74	31.8	0.812	46.4	11.0	1.84	17.4	—	0.968
41.5	18.2	127	15.7	3.22	2.66	28.0	0.728	40.7	9.75	1.83	15.3	—	0.856
36.5	20.6	110	13.8	3.21	2.60	24.4	0.647	35.3	8.51	1.81	13.3	0.908	0.730
34	21.8	103	12.9	3.20	2.59	22.9	0.606	32.4	7.83	1.80	12.2	0.853	0.664
31	23.5	93.8	11.9	3.21	2.58	21.1	0.554	28.7	6.97	1.77	10.9	0.784	0.583
28.5	23.2	90.4	11.8	3.29	2.85	21.2	0.638	15.3	4.67	1.35	7.42	0.793	0.592
25	24.7	80.3	10.7	3.30	2.93	20.8	0.771	12.5	3.82	1.30	6.09	0.733	0.533
22	26.8	71.1	9.68	3.31	2.98	17.6	1.06	10.3	3.18	1.26	5.09	0.638	0.460
155.5	5.3	383	46.5	2.89	2.93	90.6	1.91	398	66.2	2.95	104	—	—
141.5	5.7	337	41.5	2.85	2.80	80.1	1.75	352	59.2	2.91	92.5	—	—
129	6.3	298	37.0	2.80	2.68	71.0	1.61	314	53.4	2.88	83.2	—	—
117	6.9	260	32.6	2.75	2.55	62.4	1.48	279	47.9	2.85	74.5	—	—
105.5	7.5	229	29.0	2.72	2.44	55.0	1.34	246	42.7	2.82	66.2	—	—
96	8.3	202	25.8	2.68	2.34	48.5	1.23	220	38.4	2.79	59.4	—	—
87.5	9.0	181	23.4	2.66	2.26	43.6	1.13	196	34.4	2.76	53.1	—	—
79	9.9	160	20.8	2.63	2.18	38.5	1.02	174	30.7	2.74	47.4	—	—
71.5	11.0	142	18.5	2.60	2.09	34.0	0.938	156	27.7	2.72	42.7	—	—
65	11.9	127	16.7	2.58	2.02	30.5	0.856	139	24.9	2.70	38.3	—	—
59.5	12.3	119	15.9	2.60	2.03	28.7	0.778	126	22.5	2.69	34.6	—	—
53	13.6	104	14.1	2.59	1.97	25.2	0.695	110	19.7	2.66	30.2	—	—
48.5	15.0	93.8	12.7	2.56	1.91	22.6	0.640	100	18.0	2.65	27.6	—	—
43	16.7	82.4	11.2	2.55	1.86	19.9	0.570	87.6	15.8	2.63	24.2	—	0.937
38	18.9	71.8	9.83	2.54	1.80	17.3	0.505	76.2	13.8	2.61	21.1	0.990	0.826
35.5	16.2	78.2	11.2	2.74	2.26	20.0	0.683	30.1	7.89	1.70	12.3	—	0.963
32.5	17.8	70.7	10.1	2.72	2.20	18.0	0.629	27.4	7.22	1.69	11.2	—	0.877
30	19.3	64.7	9.29	2.71	2.16	16.5	0.583	25.0	6.63	1.69	10.3	0.964	0.796
27.5	20.6	59.5	8.63	2.71	2.16	15.3	0.538	22.5	5.97	1.67	9.27	0.913	0.735
25	22.6	53.5	7.79	2.70	2.12	13.8	0.489	20.0	5.35	1.65	8.29	0.823	0.625
23	22.3	52.1	7.77	2.77	2.33	13.9	0.558	11.3	3.72	1.29	5.85	0.831	0.635
20	25.5	44.8	6.73	2.76	2.29	12.0	0.489	9.55	3.17	1.27	4.97	0.690	0.496
17.5	26.8	40.1	6.21	2.79	2.39	12.0	0.450	7.67	2.56	1.22	4.03	0.638	0.460

*Where no value of Q_s is shown, the Tee complies with LRFD Specification Section E2.

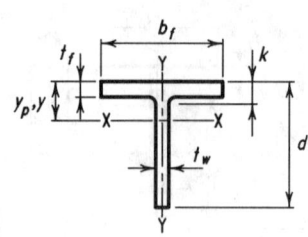

STRUCTURAL TEES
Cut from W shapes
Dimensions

Designation	Area in.²	Depth of Tee d in.		Stem Thickness t_w in.		$\dfrac{t_w}{2}$ in.	Area of Stem in.²	Flange Width b_f in.		Flange Thickness t_f in.		Distance k in.
WT8×50	14.7	8.485	8½	0.585	9/16	5/16	4.96	10.425	10⅜	0.985	1	1 11/16
×44.5	13.1	8.375	8⅜	0.525	½	¼	4.40	10.365	10⅜	0.875	⅞	1 9/16
×38.5	11.3	8.260	8¼	0.455	7/16	¼	3.76	10.295	10¼	0.760	¾	1 7/16
×33.5	9.84	8.165	8⅛	0.395	⅜	3/16	3.23	10.235	10¼	0.665	11/16	1⅜
WT8×28.5	8.38	8.215	8¼	0.430	7/16	¼	3.53	7.120	7⅛	0.715	11/16	1⅜
×25	7.37	8.130	8⅛	0.380	⅜	3/16	3.09	7.070	7⅛	0.630	⅝	1 5/16
×22.5	6.63	8.065	8⅛	0.345	⅜	3/16	2.78	7.035	7	0.565	9/16	1¼
×20	5.89	8.005	8	0.305	5/16	3/16	2.44	6.995	7	0.505	½	1 3/16
×18	5.28	7.930	7⅞	0.295	5/16	3/16	2.34	6.985	7	0.430	7/16	1⅛
WT8×15.5	4.56	7.940	8	0.275	¼	⅛	2.18	5.525	5½	0.440	7/16	1⅛
×13	3.84	7.845	7⅞	0.250	¼	⅛	1.96	5.500	5½	0.345	⅜	1 1/16
WT7×404	119	11.420	11 7/16	3.740	3¾	1⅞	42.7	18.560	18½	5.120	5⅛	5 13/16
×365	107	11.210	11¼	3.070	3 1/16	1 9/16	34.4	17.890	17⅞	4.910	4 15/16	5 9/16
×332.5	97.8	10.820	10⅞	2.830	2 13/16	1 7/16	30.6	17.650	17⅝	4.520	4½	5 3/16
×302.5	88.9	10.460	10½	2.595	2⅝	1 5/16	27.1	17.415	17⅜	4.160	4 3/16	4 13/16
×275	80.9	10.120	10⅛	2.380	2⅜	1 3/16	24.1	17.200	17¼	3.820	3 13/16	4½
×250	73.5	9.800	9¾	2.190	2 3/16	1⅛	21.5	17.010	17	3.500	3½	4 3/16
×227.5	66.9	9.510	9½	2.015	2	1	19.2	16.835	16⅞	3.210	3 3/16	3⅞
WT7×213	62.6	9.335	9⅜	1.875	1⅞	15/16	17.5	16.695	16¾	3.035	3 1/16	3 11/16
×199	58.5	9.145	9⅛	1.770	1¾	⅞	16.2	16.590	16⅝	2.845	2⅞	3½
×185	54.4	8.960	9	1.655	1⅝	13/16	14.8	16.475	16½	2.660	2 11/16	3 5/16
×171	50.3	8.770	8¾	1.540	1 9/16	13/16	13.5	16.360	16⅜	2.470	2½	3⅛
×155.5	45.7	8.560	8½	1.410	1 7/16	¾	12.1	16.230	16¼	2.260	2¼	2 15/16
×141.5	41.6	8.370	8⅜	1.290	1 5/16	11/16	10.8	16.110	16⅛	2.070	2 1/16	2¾
×128.5	37.8	8.190	8¼	1.175	1 3/16	⅝	9.62	15.995	16	1.890	1⅞	2 9/16
×116.5	34.2	8.020	8	1.070	1 1/16	9/16	8.58	15.890	15⅞	1.720	1¾	2⅜
×105.5	31.0	7.860	7⅞	0.980	1	½	7.70	15.800	15¾	1.560	1 9/16	2¼
×96.5	28.4	7.740	7¾	0.890	⅞	7/16	6.89	15.710	15¾	1.440	1 7/16	2⅛
×88	25.9	7.610	7⅝	0.830	13/16	7/16	6.32	15.650	15⅝	1.310	1 5/16	2
×79.5	23.4	7.490	7½	0.745	¾	⅜	5.58	15.565	15⅝	1.190	1 3/16	1⅞
×72.5	21.3	7.390	7⅜	0.680	11/16	⅜	5.03	15.500	15½	1.090	1 1/16	1¾

STRUCTURAL TEES
Cut from W shapes
Properties

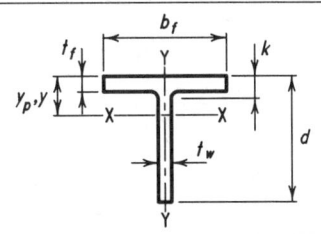

Nom-inal Wt. per ft		Axis X-X						Axis Y-Y				Q_s*	
	h	I	S	r	y	Z	y_p	I	S	r	Z	F_y, ksi	
lb	t_w	in.^{4}	in.^{3}	in.	in.	in.^{3}	in.	in.^{4}	in.^{3}	in.	in.^{3}	36	50
50	12.1	76.8	11.4	2.28	1.76	20.7	0.706	93.1	17.9	2.51	27.4	—	—
44.5	13.5	67.2	10.1	2.27	1.70	18.1	0.631	81.3	15.7	2.49	24.0	—	—
38.5	15.6	56.9	8.59	2.24	1.63	15.3	0.549	69.2	13.4	2.47	20.5	—	0.988
33.5	18.0	48.6	7.36	2.22	1.56	13.0	0.481	59.5	11.6	2.46	17.7	—	0.861
28.5	16.5	48.7	7.77	2.41	1.94	13.8	0.589	21.6	6.06	1.60	9.43	—	0.942
25	18.7	42.3	6.78	2.40	1.89	12.0	0.521	18.6	5.26	1.59	8.16	0.990	0.826
22.5	20.6	37.8	6.10	2.39	1.86	10.8	0.471	16.4	4.67	1.57	7.23	0.904	0.725
20	23.3	33.1	5.35	2.37	1.81	9.43	0.421	14.4	4.12	1.57	6.37	0.784	0.583
18	24.1	30.6	5.05	2.41	1.88	8.93	0.378	12.2	3.50	1.52	5.42	0.754	0.553
15.5	25.8	27.4	4.64	2.45	2.02	8.27	0.413	6.20	2.24	1.17	3.52	0.668	0.479
13	28.4	23.5	4.09	2.47	2.09	8.12	0.372	4.80	1.74	1.12	2.74	0.563	0.406
404	1.5	898	116	2.75	3.70	249	3.19	2760	297	4.82	463	—	—
365	1.9	739	95.4	2.62	3.47	211	3.00	2360	264	4.69	408	—	—
332.5	2.0	622	82.1	2.52	3.25	182	2.77	2080	236	4.62	365	—	—
302.5	2.2	524	70.6	2.43	3.05	157	2.55	1840	211	4.55	326	—	—
275	2.4	442	60.9	2.34	2.85	136	2.35	1630	189	4.49	292	—	—
250	2.6	375	52.7	2.26	2.67	117	2.16	1440	169	4.43	261	—	—
227.5	2.8	321	45.9	2.19	2.51	102	1.99	1280	152	4.38	234	—	—
213	3.0	287	41.4	2.14	2.40	91.7	1.88	1180	141	4.34	217	—	—
199	3.2	257	37.6	2.10	2.30	82.9	1.76	1090	131	4.31	201	—	—
185	3.4	229	33.9	2.05	2.19	74.4	1.65	994	121	4.27	185	—	—
171	3.7	203	30.4	2.01	2.09	66.2	1.54	903	110	4.24	169	—	—
155.5	4.0	176	26.7	1.96	1.97	57.7	1.41	807	99.4	4.20	152	—	—
141.5	4.4	153	23.5	1.92	1.86	50.4	1.29	722	89.7	4.17	137	—	—
128.5	4.9	133	20.7	1.88	1.75	43.9	1.18	645	80.7	4.13	123	—	—
116.5	5.3	116	18.2	1.84	1.65	38.2	1.08	576	72.5	4.10	110	—	—
105.5	5.8	102	16.2	1.81	1.57	33.4	0.980	513	65.0	4.07	99.0	—	—
96.5	6.4	89.8	14.4	1.78	1.49	29.4	0.903	466	59.3	4.05	90.2	—	—
88	6.9	80.5	13.0	1.76	1.43	26.3	0.827	419	53.5	4.02	81.4	—	—
79.5	7.7	70.2	11.4	1.73	1.35	22.8	0.751	374	48.1	4.00	73.0	—	—
72.5	8.4	62.5	10.2	1.71	1.29	20.2	0.688	338	43.7	3.98	66.3	—	—

*Where no value of Q_s is shown, the Tee complies with LRFD Specification Section E2.

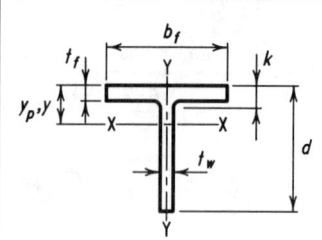

STRUCTURAL TEES
Cut from W shapes
Dimensions

Designation	Area in.²	Depth of Tee d in.		Stem Thickness t_w in.		$\dfrac{t_w}{2}$ in.	Area of Stem in.²	Flange Width b_f in.		Flange Thickness t_f in.		Distance k in.
WT7×66	19.4	7.330	7⅜	0.645	⅝	5/16	4.73	14.725	14¾	1.030	1	1 11/16
×60	17.7	7.240	7¼	0.590	9/16	5/16	4.27	14.670	14⅝	0.940	15/16	1⅝
×54.5	16.0	7.160	7⅛	0.525	½	¼	3.76	14.605	14⅝	0.860	⅞	1 9/16
×49.5	14.6	7.080	7⅛	0.485	½	¼	3.43	14.565	14⅝	0.780	¾	1 7/16
×45	13.2	7.010	7	0.440	7/16	¼	3.08	14.520	14½	0.710	11/16	1⅜
WT7×41	12.0	7.155	7⅛	0.510	½	¼	3.65	10.130	10⅛	0.855	⅞	1⅝
×37	10.9	7.085	7⅛	0.450	7/16	¼	3.19	10.070	10⅛	0.785	13/16	1 9/16
×34	9.99	7.020	7	0.415	7/16	¼	2.91	10.035	10	0.720	¾	1½
×30.5	8.96	6.945	7	0.375	⅜	3/16	2.60	9.995	10	0.645	⅝	1 7/16
WT7×26.5	7.81	6.960	7	0.370	⅜	3/16	2.58	8.060	8	0.660	11/16	1 7/16
×24	7.07	6.895	6⅞	0.340	5/16	3/16	2.34	8.030	8	0.595	⅝	1⅜
×21.5	6.31	6.830	6⅞	0.305	5/16	3/16	2.08	7.995	8	0.530	½	1 5/16
WT7×19	5.58	7.050	7	0.310	5/16	3/16	2.19	6.770	6¾	0.515	½	1 1/16
×17	5.00	6.990	7	0.285	5/16	3/16	1.99	6.745	6¾	0.455	7/16	1
×15	4.42	6.920	6⅞	0.270	¼	⅛	1.87	6.730	6¾	0.385	⅜	15/16
WT7×13	3.85	6.955	7	0.255	¼	⅛	1.77	5.025	5	0.420	7/16	15/16
×11	3.25	6.870	6⅞	0.230	¼	⅛	1.58	5.000	5	0.335	5/16	⅞

STRUCTURAL TEES
Cut from W shapes
Properties

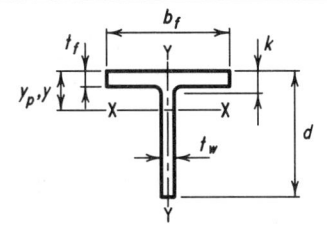

Nom- inal Wt. per ft	$\frac{h}{t_w}$	Axis X-X						Axis Y-Y				Q_s^*	
		I	S	r	y	Z	y_p	I	S	r	Z	F_y, ksi	
lb		in.^{4}	in.^{3}	in.	in.	in.^{3}	in.	in.^{4}	in.^{3}	in.	in.^{3}	36	50
66	8.8	57.8	9.57	1.73	1.29	18.6	0.658	274	37.2	3.76	56.6	—	—
60	9.7	51.7	8.61	1.71	1.24	16.5	0.602	247	33.7	3.74	51.2	—	—
54.5	10.9	45.3	7.56	1.68	1.17	14.4	0.549	223	30.6	3.73	46.4	—	—
49.5	11.8	40.9	6.88	1.67	1.14	12.9	0.500	201	27.6	3.71	41.8	—	—
45	13.0	36.4	6.16	1.66	1.09	11.5	0.456	181	25.0	3.70	37.8	—	—
41	11.2	41.2	7.14	1.85	1.39	13.2	0.594	74.2	14.6	2.48	22.4	—	—
37	12.7	36.0	6.25	1.82	1.32	11.5	0.541	66.9	13.3	2.48	20.3	—	—
34	13.7	32.6	5.69	1.81	1.29	10.4	0.498	60.7	12.1	2.46	18.5	—	—
30.5	15.2	28.9	5.07	1.80	1.25	9.16	0.448	53.7	10.7	2.45	16.4	—	0.973
26.5	15.4	27.6	4.94	1.88	1.38	8.87	0.484	28.8	7.16	1.92	11.0	—	0.958
24	16.8	24.9	4.48	1.87	1.35	8.00	0.440	25.7	6.40	1.91	9.82	—	0.882
21.5	18.7	21.9	3.98	1.86	1.31	7.05	0.395	22.6	5.65	1.89	8.66	0.947	0.775
19	19.8	23.3	4.22	2.04	1.54	7.45	0.412	13.3	3.94	1.55	6.07	0.934	0.760
17	21.5	20.9	3.83	2.04	1.53	6.74	0.371	11.7	3.45	1.53	5.32	0.857	0.669
15	22.7	19.0	3.55	2.07	1.58	6.25	0.329	9.79	2.91	1.49	4.49	0.810	0.610
13	24.1	17.3	3.31	2.12	1.72	5.89	0.383	4.45	1.77	1.08	2.77	0.737	0.537
11	26.7	14.8	2.91	2.14	1.76	5.20	0.325	3.50	1.40	1.04	2.19	0.621	0.447

*Where no value of Q_s is shown, the Tee complies with LRFD Specification Section E2.

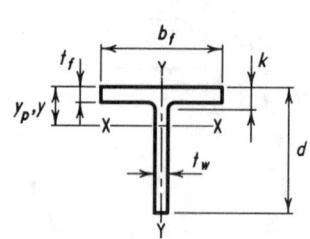

STRUCTURAL TEES
Cut from W shapes
Dimensions

Designation	Area in.^{2}	Depth of Tee d in.		Stem Thickness t_w in.		$\frac{t_w}{2}$ in.	Area of Stem in.^{2}	Flange Width b_f in.		Flange Thickness t_f in.		Distance k in.
WT6×168	49.4	8.410	$8\frac{3}{8}$	1.775	$1\frac{3}{4}$	$\frac{7}{8}$	14.9	13.385	$13\frac{3}{8}$	2.955	$2\frac{15}{16}$	$3\frac{11}{16}$
×152.5	44.8	8.160	$8\frac{1}{8}$	1.625	$1\frac{5}{8}$	$\frac{13}{16}$	13.3	13.235	$13\frac{1}{4}$	2.705	$2\frac{11}{16}$	$3\frac{7}{16}$
×139.5	41.0	7.925	$7\frac{7}{8}$	1.530	$1\frac{1}{2}$	$\frac{3}{4}$	12.1	13.140	$13\frac{1}{8}$	2.470	$2\frac{1}{2}$	$3\frac{3}{16}$
×126	37.0	7.705	$7\frac{3}{4}$	1.395	$1\frac{3}{8}$	$\frac{11}{16}$	10.7	13.005	13	2.250	$2\frac{1}{4}$	$2\frac{15}{16}$
×115	33.9	7.525	$7\frac{1}{2}$	1.285	$1\frac{5}{16}$	$\frac{11}{16}$	9.67	12.895	$12\frac{7}{8}$	2.070	$2\frac{1}{16}$	$2\frac{3}{4}$
×105	30.9	7.355	$7\frac{3}{8}$	1.180	$1\frac{3}{16}$	$\frac{5}{8}$	8.68	12.790	$12\frac{3}{4}$	1.900	$1\frac{7}{8}$	$2\frac{5}{8}$
×95	27.9	7.190	$7\frac{1}{4}$	1.060	$1\frac{1}{16}$	$\frac{9}{16}$	7.62	12.670	$12\frac{5}{8}$	1.735	$1\frac{3}{4}$	$2\frac{7}{16}$
×85	25.0	7.015	7	0.960	$\frac{15}{16}$	$\frac{1}{2}$	6.73	12.570	$12\frac{5}{8}$	1.560	$1\frac{9}{16}$	$2\frac{1}{4}$
×76	22.4	6.855	$6\frac{7}{8}$	0.870	$\frac{7}{8}$	$\frac{7}{16}$	5.96	12.480	$12\frac{1}{2}$	1.400	$1\frac{3}{8}$	$2\frac{1}{8}$
×68	20.0	6.705	$6\frac{3}{4}$	0.790	$\frac{13}{16}$	$\frac{7}{16}$	5.30	12.400	$12\frac{3}{8}$	1.250	$1\frac{1}{4}$	$1\frac{15}{16}$
×60	17.6	6.560	$6\frac{1}{2}$	0.710	$\frac{11}{16}$	$\frac{3}{8}$	4.66	12.320	$12\frac{3}{8}$	1.105	$1\frac{1}{8}$	$1\frac{13}{16}$
×53	15.6	6.445	$6\frac{1}{2}$	0.610	$\frac{5}{8}$	$\frac{5}{16}$	3.93	12.220	$12\frac{1}{4}$	0.990	1	$1\frac{11}{16}$
×48	14.1	6.355	$6\frac{3}{8}$	0.550	$\frac{9}{16}$	$\frac{5}{16}$	3.50	12.160	$12\frac{1}{8}$	0.900	$\frac{7}{8}$	$1\frac{5}{8}$
×43.5	12.8	6.265	$6\frac{1}{4}$	0.515	$\frac{1}{2}$	$\frac{1}{4}$	3.23	12.125	$12\frac{1}{8}$	0.810	$\frac{13}{16}$	$1\frac{1}{2}$
×39.5	11.6	6.190	$6\frac{1}{4}$	0.470	$\frac{1}{2}$	$\frac{1}{4}$	2.91	12.080	$12\frac{1}{8}$	0.735	$\frac{3}{4}$	$1\frac{7}{16}$
×36	10.6	6.125	$6\frac{1}{8}$	0.430	$\frac{7}{16}$	$\frac{1}{4}$	2.63	12.040	12	0.670	$\frac{11}{16}$	$1\frac{3}{8}$
×32.5	9.54	6.060	6	0.390	$\frac{3}{8}$	$\frac{3}{16}$	2.36	12.000	12	0.605	$\frac{5}{8}$	$1\frac{5}{16}$
WT6×29	8.52	6.095	$6\frac{1}{8}$	0.360	$\frac{3}{8}$	$\frac{3}{16}$	2.19	10.010	10	0.640	$\frac{5}{8}$	$1\frac{3}{8}$
×26.5	7.78	6.030	6	0.345	$\frac{3}{8}$	$\frac{3}{16}$	2.08	9.995	10	0.575	$\frac{9}{16}$	$1\frac{1}{4}$
WT6×25	7.34	6.095	$6\frac{1}{8}$	0.370	$\frac{3}{8}$	$\frac{3}{16}$	2.26	8.080	$8\frac{1}{8}$	0.640	$\frac{5}{8}$	$1\frac{3}{8}$
×22.5	6.61	6.030	6	0.335	$\frac{5}{16}$	$\frac{3}{16}$	2.02	8.045	8	0.575	$\frac{9}{16}$	$1\frac{1}{4}$
×20	5.89	5.970	6	0.295	$\frac{5}{16}$	$\frac{3}{16}$	1.76	8.005	8	0.515	$\frac{1}{2}$	$1\frac{1}{4}$
WT6×17.5	5.17	6.250	$6\frac{1}{4}$	0.300	$\frac{5}{16}$	$\frac{3}{16}$	1.88	6.560	$6\frac{1}{2}$	0.520	$\frac{1}{2}$	1
×15	4.40	6.170	$6\frac{1}{8}$	0.260	$\frac{1}{4}$	$\frac{1}{8}$	1.60	6.520	$6\frac{1}{2}$	0.440	$\frac{7}{16}$	$\frac{15}{16}$
×13	3.82	6.110	$6\frac{1}{8}$	0.230	$\frac{1}{4}$	$\frac{1}{8}$	1.41	6.490	$6\frac{1}{2}$	0.380	$\frac{3}{8}$	$\frac{7}{8}$
WT6×11	3.24	6.155	$6\frac{1}{8}$	0.260	$\frac{1}{4}$	$\frac{1}{8}$	1.60	4.030	4	0.425	$\frac{7}{16}$	$\frac{7}{8}$
×9.5	2.79	6.080	$6\frac{1}{8}$	0.235	$\frac{1}{4}$	$\frac{1}{8}$	1.43	4.005	4	0.350	$\frac{3}{8}$	$\frac{13}{16}$
×8	2.36	5.995	6	0.220	$\frac{1}{4}$	$\frac{1}{8}$	1.32	3.990	4	0.265	$\frac{1}{4}$	$\frac{3}{4}$
×7	2.08	5.955	6	0.200	$\frac{3}{16}$	$\frac{1}{8}$	1.19	3.970	4	0.225	$\frac{1}{4}$	$1\frac{1}{16}$

STRUCTURAL TEES
Cut from W shapes.
Properties

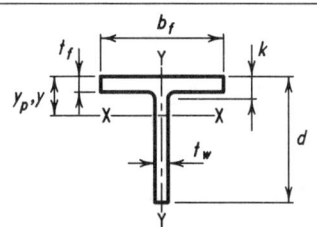

| Nom-inal Wt. per ft | h | Axis X-X | | | | | | Axis Y-Y | | | | $Q_s{}^*$ | |
| lb | t_w | I | S | r | y | Z | y_p | I | S | r | Z | F_y, ksi | |
		in.⁴	in.³	in.	in.	in.³	in.	in.⁴	in.³	in.	in.³	36	50
168	2.7	190	31.2	1.96	2.31	68.4	1.84	593	88.6	3.47	137	—	—
152.5	3.0	162	27.0	1.90	2.16	59.1	1.69	525	79.3	3.42	122	—	—
139.5	3.2	141	24.1	1.86	2.05	51.9	1.56	469	71.3	3.38	110	—	—
126	3.5	121	20.9	1.81	1.92	44.8	1.42	414	63.6	3.34	97.9	—	—
115	3.8	106	18.5	1.77	1.82	39.4	1.31	371	57.5	3.31	88.4	—	—
105	4.1	92.1	16.4	1.73	1.72	34.5	1.21	332	51.9	3.28	79.7	—	—
95	4.6	79.0	14.2	1.68	1.62	29.8	1.10	295	46.5	3.25	71.3	—	—
85	5.1	67.8	12.3	1.65	1.52	25.6	0.994	259	41.2	3.22	63.0	—	—
76	5.6	58.5	10.8	1.62	1.43	22.0	0.896	227	36.4	3.19	55.6	—	—
68	6.1	50.6	9.46	1.59	1.35	19.0	0.805	199	32.1	3.16	49.0	—	—
60	6.8	43.4	8.22	1.57	1.28	16.2	0.716	172	28.0	3.13	42.7	—	—
53	8.0	36.3	6.91	1.53	1.19	13.6	0.637	151	24.7	3.11	37.5	—	—
48	8.8	32.0	6.12	1.51	1.13	11.9	0.580	135	22.2	3.09	33.7	—	—
43.5	9.4	28.9	5.60	1.50	1.10	10.7	0.527	120	19.9	3.07	30.2	—	—
39.5	10.3	25.8	5.03	1.49	1.06	9.49	0.480	108	17.9	3.05	27.2	—	—
36	11.3	23.2	4.54	1.48	1.02	8.48	0.439	97.5	16.2	3.04	24.6	—	—
32.5	12.4	20.6	4.06	1.47	0.985	7.50	0.398	87.2	14.5	3.02	22.0	—	—
29	13.5	19.1	3.76	1.50	1.03	6.97	0.426	53.5	10.7	2.51	16.3	—	—
26.5	14.1	17.7	3.54	1.51	1.02	6.46	0.389	47.9	9.58	2.48	14.6	—	—
25	13.1	18.7	3.79	1.60	1.17	6.90	0.454	28.2	6.97	1.96	10.7	—	—
22.5	14.5	16.6	3.39	1.58	1.13	6.12	0.411	25.0	6.21	1.94	9.50	—	0.998
20	16.5	14.4	2.95	1.57	1.08	5.30	0.368	22.0	5.51	1.93	8.41	—	0.887
17.5	18.1	16.0	3.23	1.76	1.30	5.71	0.394	12.2	3.73	1.54	5.73	—	0.856
15	20.9	13.5	2.75	1.75	1.27	4.83	0.337	10.2	3.12	1.52	4.78	0.891	0.710
13	23.6	11.7	2.40	1.75	1.25	4.20	0.295	8.66	2.67	1.51	4.08	0.767	0.565
11	20.9	11.7	2.59	1.90	1.63	4.63	0.402	2.33	1.16	0.847	1.83	0.891	0.710
9.5	23.1	10.1	2.28	1.90	1.65	4.11	0.348	1.88	0.939	0.822	1.49	0.797	0.596
8	24.7	8.70	2.04	1.92	1.74	3.72	0.639	1.41	0.706	0.773	1.13	0.741	0.541
7	27.2	7.67	1.83	1.92	1.76	3.32	0.760	1.18	0.594	0.753	0.950	0.626	0.450

*Where no value of Q_s is shown, the Tee complies with LRFD Specification Section E2.

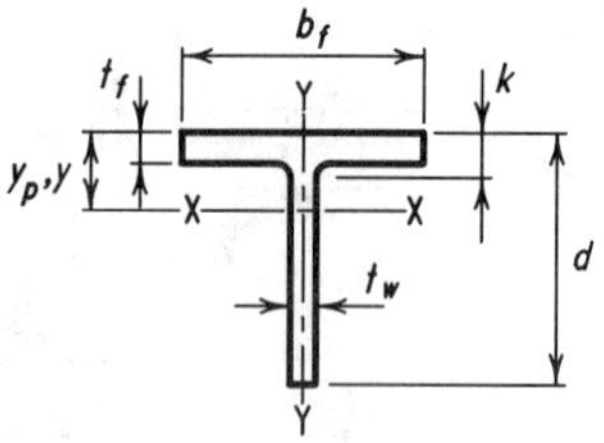

STRUCTURAL TEES
Cut from W shapes
Dimensions

Designation	Area	Depth of Tee d		Stem Thickness t_w		Stem $\frac{t_w}{2}$	Area of Stem	Flange Width b_f		Flange Thickness t_f		Distance k
	in.^{2}	in.		in.		in.	in.^{2}	in.		in.		in.
WT5×56	16.5	5.680	5 5/8	0.755	3/4	3/8	4.29	10.415	10 3/8	1.250	1 1/4	1 7/8
×50	14.7	5.550	5 1/2	0.680	11/16	3/8	3.77	10.340	10 3/8	1.120	1 1/8	1 3/4
×44	12.9	5.420	5 3/8	0.605	5/8	5/16	3.28	10.265	10 1/4	0.990	1	1 5/8
×38.5	11.3	5.300	5 1/4	0.530	1/2	1/4	2.81	10.190	10 1/4	0.870	7/8	1 1/2
×34	9.99	5.200	5 1/4	0.470	1/2	1/4	2.44	10.130	10 1/8	0.770	3/4	1 3/8
×30	8.82	5.110	5 1/8	0.420	7/16	1/4	2.15	10.080	10 1/8	0.680	11/16	1 5/16
×27	7.91	5.045	5	0.370	3/8	3/16	1.87	10.030	10	0.615	5/8	1 1/4
×24.5	7.21	4.990	5	0.340	5/16	3/16	1.70	10.000	10	0.560	9/16	1 3/16
WT5×22.5	6.63	5.050	5	0.350	3/8	3/16	1.77	8.020	8	0.620	5/8	1 1/4
×19.5	5.73	4.960	5	0.315	5/16	3/16	1.56	7.985	8	0.530	1/2	1 1/8
×16.5	4.85	4.865	4 7/8	0.290	5/16	3/16	1.41	7.960	8	0.435	7/16	1 1/16
WT5×15	4.42	5.235	5 1/4	0.300	5/16	3/16	1.57	5.810	5 3/4	0.510	1/2	15/16
×13	3.81	5.165	5 1/8	0.260	1/4	1/8	1.34	5.770	5 3/4	0.440	7/16	7/8
×11	3.24	5.085	5 1/8	0.240	1/4	1/8	1.22	5.750	5 3/4	0.360	3/8	3/4
WT5×9.5	2.81	5.120	5 1/8	0.250	1/4	1/8	1.28	4.020	4	0.395	3/8	13/16
×8.5	2.50	5.055	5	0.240	1/4	1/8	1.21	4.010	4	0.330	5/16	3/4
×7.5	2.21	4.995	5	0.230	1/4	1/8	1.15	4.000	4	0.270	1/4	11/16
×6	1.77	4.935	4 7/8	0.190	3/16	1/8	0.938	3.960	4	0.210	3/16	5/8

STRUCTURAL TEES
Cut from W shapes
Properties

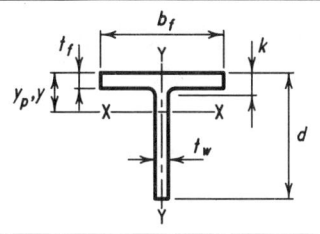

Nom-inal Wt. per ft	$\dfrac{h}{t_w}$	Axis X-X						Axis Y-Y				$Q_s{}^*$	
		I	S	r	y	Z	y_p	I	S	r	Z	F_y, ksi	
lb		in.^{4}	in.^{3}	in.	in.	in.^{3}	in.	in.^{4}	in.^{3}	in.	in.^{3}	36	50
56	5.2	28.6	6.40	1.32	1.21	13.4	0.791	118	22.6	2.68	34.6	—	—
50	5.8	24.5	5.56	1.29	1.13	11.4	0.711	103	20.0	2.65	30.5	—	—
44	6.5	20.8	4.77	1.27	1.06	9.65	0.631	89.3	17.4	2.63	26.5	—	—
38.5	7.4	17.4	4.04	1.24	0.990	8.06	0.555	76.8	15.1	2.60	22.9	—	—
34	8.4	14.9	3.49	1.22	0.932	6.85	0.493	66.8	13.2	2.59	20.0	—	—
30	9.4	12.9	3.04	1.21	0.884	5.87	0.438	58.1	11.5	2.57	17.5	—	—
27	10.6	11.1	2.64	1.19	0.836	5.05	0.395	51.7	10.3	2.56	15.7	—	—
24.5	11.6	10.0	2.39	1.18	0.807	4.52	0.361	46.7	9.34	2.54	14.2	—	—
22.5	11.2	10.2	2.47	1.24	0.907	4.65	0.413	26.7	6.65	2.01	10.1	—	—
19.5	12.5	8.84	2.16	1.24	0.876	3.99	0.359	22.5	5.64	1.98	8.59	—	—
16.5	13.6	7.71	1.93	1.26	0.869	3.48	0.305	18.3	4.60	1.94	7.01	—	—
15	14.8	9.28	2.24	1.45	1.10	4.01	0.380	8.35	2.87	1.37	4.42	—	—
13	17.0	7.86	1.91	1.44	1.06	3.39	0.330	7.05	2.44	1.36	3.75	—	0.902
11	18.4	6.88	1.72	1.46	1.07	3.02	0.282	5.71	1.99	1.33	3.05	0.999	0.836
9.5	17.7	6.68	1.74	1.54	1.28	3.10	0.349	2.15	1.07	0.874	1.68	—	0.872
8.5	18.4	6.06	1.62	1.56	1.32	2.90	0.311	1.78	0.888	0.844	1.40	—	0.841
7.5	19.2	5.45	1.50	1.57	1.37	3.03	0.306	1.45	0.723	0.810	1.15	0.977	0.811
6	23.3	4.35	1.22	1.57	1.36	2.50	0.323	1.09	0.551	0.785	0.872	0.793	0.592

*Where no value of Q_s is shown, the Tee complies with LRFD Specification Section E2.

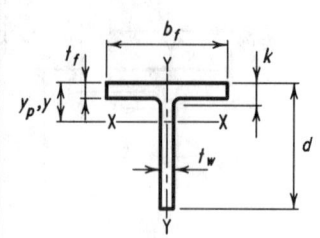

STRUCTURAL TEES
Cut from W shapes
Dimensions

Designation	Area in.²	Depth of Tee d (in.)		Stem Thickness t_w (in.)		$\dfrac{t_w}{2}$ in.	Area of Stem in.²	Flange Width b_f (in.)		Flange Thickness t_f (in.)		Distance k in.
WT4×33.5	9.84	4.500	4½	0.570	9/16	5/16	2.56	8.280	8¼	0.935	15/16	1 7/16
×29	8.55	4.375	4⅜	0.510	½	¼	2.23	8.220	8¼	0.810	13/16	1 5/16
×24	7.05	4.250	4¼	0.400	⅜	3/16	1.70	8.110	8⅛	0.685	11/16	1 3/16
×20	5.87	4.125	4⅛	0.360	⅜	3/16	1.48	8.070	8⅛	0.560	9/16	1 1/16
×17.5	5.14	4.060	4	0.310	5/16	3/16	1.26	8.020	8	0.495	½	1
×15.5	4.56	4.000	4	0.285	5/16	3/16	1.14	7.995	8	0.435	7/16	15/16
WT4×14	4.12	4.030	4	0.285	5/16	3/16	1.15	6.535	6½	0.465	7/16	15/16
×12	3.54	3.965	4	0.245	¼	⅛	0.971	6.495	6½	0.400	⅜	⅞
WT4×10.5	3.08	4.140	4⅛	0.250	¼	⅛	1.03	5.270	5¼	0.400	⅜	13/16
×9	2.63	4.070	4⅛	0.230	¼	⅛	0.936	5.250	5¼	0.330	5/16	¾
WT4×7.5	2.22	4.055	4	0.245	¼	⅛	0.993	4.015	4	0.315	5/16	¾
×6.5	1.92	3.995	4	0.230	¼	⅛	0.919	4.000	4	0.255	¼	11/16
×5	1.48	3.945	4	0.170	3/16	⅛	0.671	3.940	4	0.205	3/16	⅝
WT3×12.5	3.67	3.190	3¼	0.320	5/16	3/16	1.02	6.080	6⅛	0.455	7/16	13/16
×10	2.94	3.100	3⅛	0.260	¼	⅛	0.806	6.020	6	0.365	⅜	¾
×7.5	2.21	2.995	3	0.230	¼	⅛	0.689	5.990	6	0.260	¼	⅝
WT3×8	2.37	3.140	3⅛	0.260	¼	⅛	0.816	4.030	4	0.405	⅜	¾
×6	1.78	3.015	3	0.230	¼	⅛	0.693	4.000	4	0.280	¼	⅝
×4.5	1.34	2.950	3	0.170	3/16	⅛	0.502	3.940	4	0.215	3/16	9/16
WT2.5×9.5	2.77	2.575	2⅝	0.270	¼	⅛	0.695	5.030	5	0.430	7/16	13/16
×8	2.34	2.505	2½	0.240	¼	⅛	0.601	5.000	5	0.360	⅜	¾
WT2×6.5	1.91	2.080	2⅛	0.280	¼	⅛	0.582	4.060	4	0.345	⅜	11/16

STRUCTURAL TEES
Cut from W shapes
Properties

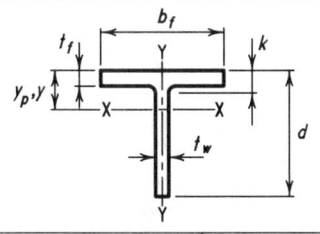

Nom-inal Wt. per ft	$\dfrac{h}{t_w}$	Axis X-X						Axis Y-Y				Q_s*	
		I	S	r	y	Z	y_p	I	S	r	Z	F_y, ksi	
lb		in.^{4}	in.^{3}	in.	in.	in.^{3}	in.	in.^{4}	in.^{3}	in.	in.^{3}	36	50
33.5	5.6	10.9	3.05	1.05	0.936	6.29	0.594	44.3	10.7	2.12	16.3	—	—
29	6.2	9.12	2.61	1.03	0.874	5.25	0.520	37.5	9.13	2.10	13.9	—	—
24	7.9	6.85	1.97	0.986	0.777	3.94	0.435	30.5	7.52	2.08	11.4	—	—
20	8.8	5.73	1.69	0.988	0.735	3.25	0.364	24.5	6.08	2.04	9.25	—	—
17.5	10.2	4.81	1.43	0.967	0.688	2.71	0.321	21.3	5.31	2.03	8.06	—	—
15.5	11.1	4.28	1.28	0.968	0.667	2.39	0.285	18.5	4.64	2.02	7.04	—	—
14	11.1	4.22	1.28	1.01	0.734	2.38	0.315	10.8	3.31	1.62	5.05	—	—
12	12.9	3.53	1.08	0.999	0.695	1.98	0.273	9.14	2.81	1.61	4.29	—	—
10.5	13.8	3.90	1.18	1.12	0.831	2.11	0.292	4.89	1.85	1.26	2.84	—	—
9	15.0	3.41	1.05	1.14	0.834	1.86	0.251	3.98	1.52	1.23	2.33	—	—
7.5	14.0	3.28	1.07	1.22	0.998	1.91	0.276	1.70	0.849	0.876	1.33	—	—
6.5	15.0	2.89	0.974	1.23	1.03	1.74	0.240	1.37	0.683	0.843	1.08	—	—
5	20.2	2.15	0.717	1.20	0.953	1.27	0.188	1.05	0.532	0.841	0.828	0.913	0.735
12.5	7.8	2.28	0.886	0.789	0.610	1.68	0.302	8.53	2.81	1.52	4.28	—	—
10	9.6	1.76	0.693	0.774	0.560	1.29	0.244	6.64	2.21	1.50	3.36	—	—
7.5	10.8	1.41	0.577	0.797	0.558	1.03	0.185	4.66	1.56	1.45	2.37	—	—
8	9.6	1.69	0.685	0.844	0.676	1.25	0.294	2.21	1.10	0.966	1.70	—	—
6	10.8	1.32	0.564	0.861	0.677	1.01	0.222	1.50	0.748	0.918	1.16	—	—
4.5	14.6	0.950	0.408	0.842	0.623	0.720	0.170	1.10	0.557	0.905	0.858	—	—
9.5	7.0	1.01	0.485	0.605	0.487	0.967	0.275	4.56	1.82	1.28	2.76	—	—
8	7.9	0.850	0.413	0.601	0.458	0.798	0.234	3.75	1.50	1.27	2.29	—	—
6.5	5.3	0.530	0.321	0.524	0.440	0.616	0.236	1.93	0.950	1.00	1.46	—	—

*Where no value of Q_s is shown, the Tee complies with LRFD Specification Section E2.

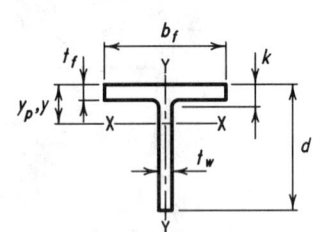

STRUCTURAL TEES
Cut from M shapes
Dimensions

Desig-nation	Area in.^{2}	Depth of Tee d in.		Stem Thickness t_w in.		$\dfrac{t_w}{2}$ in.	Area of Stem in.^{2}	Flange Width b_f in.		Flange Thickness t_f in.		Dis-tance k in.	Grip in.	Max. Flge. Fas-tener in.
MT6×5.9	1.73	6.000	6	0.177	3/16	1/8	1.06	3.065	3 1/8	0.225	1/4	1/2	1/4	—
×5.4	1.59	5.990	6	0.162	3/16	1/16	0.958	3.065	3 1/8	0.206	1/4	1/2	1/4	1/2
MT5×4.5	1.32	5.000	5	0.157	3/16	1/8	0.785	2.690	2 3/4	0.206	3/16	1/2	3/16	—
×4	1.18	4.980	5	0.139	3/16	1/16	0.702	2.690	2 3/4	0.183	3/16	1/2	3/16	3/8
MT4×3.25	0.958	4.000	4	0.133	1/8	1/16	0.540	2.281	2 1/4	0.189	3/16	1/2	3/16	—
MT2.5×9.45*	2.78	2.500	2 1/2	0.316	5/16	3/16	0.790	5.003	5	0.416	7/16	7/8	7/16	7/8

*This shape has tapered flanges, while all other MT shapes have parallel flanges.

STRUCTURAL TEES
Cut from M shapes
Properties

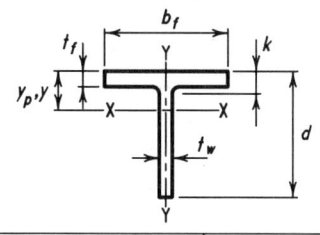

Nom-inal Wt. per ft	$\dfrac{h}{t_w}$	Axis X-X						Axis Y-Y				$Q_s{}^*$	
		I	S	r	y	Z	y_p	I	S	r	Z	F_y, ksi	
lb		in.^{4}	in.^{3}	in.	in.	in.^{3}	in.	in.^{4}	in.^{3}	in.	in.^{3}	36	50
5.9	31.3	6.60	1.60	1.95	1.89	2.89	1.09	0.490	0.320	0.532	0.577	0.483	0.348
5.4	31.8	6.03	1.46	1.95	1.85	2.63	1.01	0.453	0.295	0.533	0.525	0.397	0.286
4.5	29.2	3.46	0.997	1.62	1.53	1.81	0.778	0.305	0.227	0.480	0.405	0.549	0.396
4	29.7	3.09	0.893	1.62	1.52	1.62	0.778	0.269	0.200	0.477	0.333	0.446	0.321
3.25	26.9	1.57	0.556	1.28	1.17	1.01	0.446	0.172	0.150	0.423	0.265	0.634	0.457
9.45	5.6	1.05	0.527	0.615	0.511	1.03	0.278	3.93	1.57	1.19	2.66	—	—

*Where no value of Q_s is shown, the Tee complies with LRFD Specification Section E2.

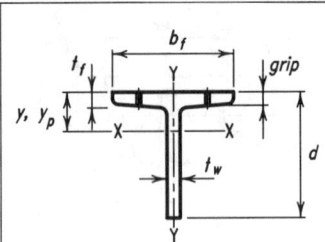

STRUCTURAL TEES
Cut from S shapes
Dimensions

Desig-nation	Area in.²	Depth of Tee d in.		Stem Thickness tw in.		$\frac{t_w}{2}$ in.	Area of Stem in.²	Flange Width bf in.		Flange Thickness tf in.	Dis-tance k in.	Grip in.	Max. Flge. Fas-tener in.
ST12×60.5	17.8	12.250	12¼	0.800	13/16	7/16	9.80	8.050	8	1.090	1 1/16	2	1⅛	1
×53	15.6	12.250	12¼	0.620	⅝	5/16	7.60	7.870	7⅞	1.090	1 1/16	2	1⅛	1
ST12×50	14.7	12.000	12	0.745	¾	⅜	8.94	7.245	7¼	0.870	⅞	1¾	⅞	1
×45	13.2	12.000	12	0.625	⅝	5/16	7.50	7.125	7⅛	0.870	⅞	1¾	⅞	1
×40	11.7	12.000	12	0.500	½	¼	6.00	7.000	7	0.870	⅞	1¾	⅞	1
ST10×48	14.1	10.150	10⅛	0.800	13/16	7/16	8.12	7.200	7¼	0.920	15/16	1¾	15/16	1
×43	12.7	10.150	10⅛	0.660	11/16	⅜	6.70	7.060	7	0.920	15/16	1¾	15/16	1
ST10×37.5	11.0	10.000	10	0.635	⅝	5/16	6.35	6.385	6⅜	0.795	13/16	1⅝	13/16	⅞
×33	9.70	10.000	10	0.505	½	¼	5.05	6.225	6¼	0.795	13/16	1⅝	13/16	⅞
ST9×35	10.3	9.000	9	0.711	11/16	⅜	6.40	6.251	6¼	0.691	11/16	1½	11/16	⅞
×27.35	8.04	9.000	9	0.461	7/16	¼	4.15	6.001	6	0.691	11/16	1½	11/16	⅞
ST7.5×25	7.35	7.500	7½	0.550	9/16	5/16	4.13	5.640	5⅝	0.622	⅝	1⅜	9/16	¾
×21.45	6.31	7.500	7½	0.411	7/16	¼	3.08	5.501	5½	0.622	⅝	1⅜	9/16	¾
ST6×25	7.35	6.000	6	0.687	11/16	⅜	4.12	5.477	5½	0.659	11/16	1 7/16	11/16	¾
×20.4	6.00	6.000	6	0.462	7/16	¼	2.77	5.252	5¼	0.659	11/16	1 7/16	⅝	¾
ST6×17.5	5.15	6.000	6	0.428	7/16	¼	2.57	5.078	5⅛	0.544	9/16	1 3/16	½	¾
×15.9	4.68	6.000	6	0.350	⅜	3/16	2.10	5.000	5	0.544	9/16	1 3/16	½	¾
ST5×17.5	5.15	5.000	5	0.594	⅝	5/16	2.97	4.944	5	0.491	½	1⅛	½	¾
×12.7	3.73	5.000	5	0.311	5/16	3/16	1.56	4.661	4⅝	0.491	½	1⅛	½	¾
ST4×11.5	3.38	4.000	4	0.441	7/16	¼	1.76	4.171	4⅛	0.425	7/16	1	7/16	¾
×9.2	2.70	4.000	4	0.271	¼	⅛	1.08	4.001	4	0.425	7/16	1	7/16	¾
ST3×8.625	2.53	3.000	3	0.465	7/16	¼	1.40	3.565	3⅝	0.359	⅜	⅞	⅜	⅝
×6.25	1.83	3.000	3	0.232	¼	⅛	0.70	3.332	3⅜	0.359	⅜	⅞	⅜	—
ST2.5×5	1.47	2.500	2½	0.214	3/16	⅛	0.535	3.004	3	0.326	5/16	13/16	5/16	—
ST2×4.75	1.40	2.000	2	0.326	5/16	3/16	0.652	2.796	2¾	0.293	5/16	¾	5/16	—
×3.85	1.13	2.000	2	0.193	3/16	⅛	0.386	2.663	2⅝	0.293	5/16	¾	5/16	—
ST1.5×3.75	1.10	1.500	1½	0.349	⅜	3/16	0.523	2.509	2½	0.260	¼	11/16	¼	—
×2.85	0.835	1.500	1½	0.170	3/16	⅛	0.255	2.330	2⅜	0.260	¼	11/16	¼	—

STRUCTURAL TEES
Cut from S shapes
Properties

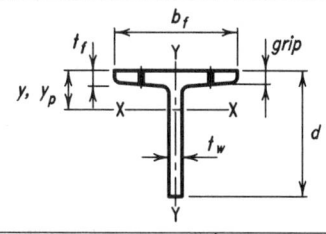

Nom-inal Wt. per ft		Axis X-X						Axis Y-Y				Q_s^*	
	h	I	S	r	y	Z	y_p	I	S	r	Z	F_y, ksi	
lb	$\dfrac{h}{t_w}$	in.^{4}	in.^{3}	in.	in.	in.^{3}	in.	in.^{4}	in.^{3}	in.	in.^{3}	36	50
60.5	13.2	259	30.1	3.82	3.63	54.5	1.28	41.7	10.4	1.53	18.1	—	—
53	17.0	216	24.1	3.72	3.28	43.3	1.03	38.5	9.80	1.57	16.6	—	0.907
50	14.1	215	26.3	3.83	3.84	47.5	2.20	23.8	6.58	1.27	12.0	—	—
45	16.8	190	22.6	3.79	3.60	41.1	1.48	22.5	6.31	1.30	11.2	—	0.937
40	21.1	162	18.7	3.72	3.29	33.6	0.922	21.1	6.04	1.34	10.4	0.878	0.695
48	10.8	143	20.3	3.18	3.13	36.9	1.40	25.1	6.97	1.33	12.5	—	—
43	13.1	125	17.2	3.14	2.91	31.1	0.985	23.4	6.63	1.36	11.6	—	—
37.5	13.6	109	15.8	3.15	3.07	28.6	1.40	14.9	4.66	1.16	8.37	—	—
33	17.0	93.1	12.9	3.10	2.81	23.4	0.855	13.8	4.43	1.19	7.70	—	0.907
35	10.9	84.7	14.0	2.87	2.94	25.1	1.81	12.1	3.86	1.08	7.21	—	—
27.35	16.8	62.4	9.61	2.79	2.50	17.3	0.747	10.4	3.47	1.14	6.07	—	0.922
25	11.6	40.6	7.73	2.35	2.25	14.0	0.872	7.85	2.78	1.03	5.01	—	—
21.45	15.5	33.0	6.00	2.29	2.01	10.8	0.613	7.19	2.61	1.07	4.54	—	0.988
25	7.00	25.2	6.05	1.85	1.84	11.0	0.770	7.85	2.87	1.03	5.19	—	—
20.4	10.3	18.9	4.28	1.78	1.58	7.71	0.581	6.78	2.58	1.06	4.45	—	—
17.5	11.7	17.2	3.95	1.83	1.64	7.12	0.548	4.94	1.95	0.980	3.41	—	—
15.9	14.3	14.9	3.31	1.78	1.51	5.94	0.485	4.68	1.87	1.00	3.22	—	—
17.5	6.90	12.5	3.63	1.56	1.56	6.58	0.702	4.18	1.69	0.901	3.11	—	—
12.7	13.2	7.83	2.06	1.45	1.20	3.70	0.408	3.39	1.46	0.954	2.49	—	—
11.5	7.30	5.03	1.77	1.22	1.15	3.19	0.447	2.15	1.03	0.798	1.84	—	—
9.2	11.8	3.51	1.15	1.14	0.941	2.07	0.341	1.86	0.932	0.831	1.59	—	—
8.63	5.00	2.13	1.02	0.917	0.914	1.85	0.401	1.15	0.648	0.675	1.18	—	—
6.25	10.0	1.27	0.552	0.833	0.691	1.01	0.275	0.911	0.547	0.705	0.929	—	—
5	8.70	0.681	0.353	0.681	0.569	0.650	0.243	0.608	0.405	0.643	0.685	—	—
4.75	4.30	0.470	0.325	0.580	0.553	0.592	0.255	0.451	0.323	0.569	0.566	—	—
3.85	7.30	0.316	0.203	0.528	0.448	0.381	0.209	0.382	0.287	0.581	0.483	—	—
3.75	2.80	0.204	0.191	0.430	0.432	0.351	0.223	0.293	0.234	0.516	0.412	—	—
2.85	5.70	0.118	0.101	0.376	0.329	0.196	0.175	0.227	0.195	0.522	0.327	—	—

*Where no value of Q_s is shown, the Tee complies with LRFD Specification Section E2.

AMERICAN INSTITUTE OF STEEL CONSTRUCTION

DOUBLE ANGLES

Properties of double angles in contact and separated are listed in the following tables. Each table shows properties of double angles in contact, and the radius of gyration about the Y-Y axis when the legs of the angles are separated. Values of Q_s are given for $F_y = 36$ ksi and $F_y = 50$ ksi for those angles exceeding the width-thickness ratio λ_r of LRFD Specification Section B5. Since the cross section is comprised entirely of unstiffened elements, $Q_a = 1.0$ and $Q = Q_s$, for all angle sections. The Flexural-Torsional Properties Table lists the dimensional values (J, $\bar{r}_o$, and H) needed for checking flexural-torsional buckling.

Use of Table

The table may be used as follows for checking the limit states of (1) flexural buckling and (2) flexural-torsional buckling. The lower of the two limit states must be used for design. See also Part 3 of this LRFD Manual.

(1) Flexural Buckling

Where no value of Q_s is shown, the design compressive strength for this limit state is given by LRFD Specification Section E2. Where a value of Q_s is shown, the strength must be reduced in accordance with Appendix B5 of the LRFD Specification.

(2) Flexural-Torsional Buckling

The design compressive strength for this limit state is given by LRFD Specification Sections E3 and E4. This involves calculations with J, $\bar{r}_o$, and H. The torsional constant J can be obtained by summing the respective values for single angles listed in the Flexural-Torsional Properties Tables in Part 1 of this Manual.

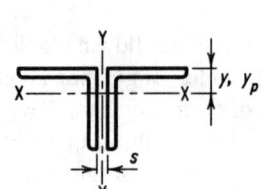

DOUBLE ANGLES
Two equal leg angles
Properties of sections

| Designation | Wt. per ft 2 Angles lb | Area of 2 Angles in.^{2} | Axis X-X | | | | | |
			I in.^{4}	S in.^{3}	r in.	y in.	Z in.^{3}	y_p in.
L8×8×1⅛	114	33.5	195	35.1	2.42	2.41	63.2	1.05
1	102	30.0	177	31.6	2.44	2.37	56.9	0.938
⅞	90.0	26.5	159	28.0	2.45	2.32	50.5	0.827
¾	77.8	22.9	139	24.4	2.47	2.28	43.9	0.715
⅝	65.4	19.2	118	20.6	2.49	2.23	37.1	0.601
½	52.8	15.5	97.3	16.7	2.50	2.19	30.1	0.484
L6×6×1	74.8	22.0	70.9	17.1	1.80	1.86	30.9	0.917
⅞	66.2	19.5	63.8	15.3	1.81	1.82	27.5	0.811
¾	57.4	16.9	56.3	13.3	1.83	1.78	24.0	0.703
⅝	48.4	14.2	48.3	11.3	1.84	1.73	20.4	0.592
½	39.2	11.5	39.8	9.23	1.86	1.68	16.6	0.479
⅜	29.8	8.72	30.8	7.06	1.88	1.64	12.7	0.363
L5×5×⅞	54.4	16.0	35.5	10.3	1.49	1.57	18.7	0.798
¾	47.2	13.9	31.5	9.06	1.51	1.52	16.3	0.694
½	32.4	9.50	22.5	6.31	1.54	1.43	11.4	0.475
⅜	24.6	7.22	17.5	4.84	1.56	1.39	8.72	0.361
⁵⁄₁₆	20.6	6.05	14.8	4.08	1.57	1.37	7.35	0.303
L4×4×¾	37.0	10.9	15.3	5.62	1.19	1.27	10.1	0.680
⅝	31.4	9.22	13.3	4.80	1.20	1.23	8.66	0.576
½	25.6	7.50	11.1	3.95	1.22	1.18	7.12	0.469
⅜	19.6	5.72	8.72	3.05	1.23	1.14	5.49	0.357
⁵⁄₁₆	16.4	4.80	7.43	2.58	1.24	1.12	4.64	0.300
¼	13.2	3.88	6.08	2.09	1.25	1.09	3.77	0.242

DOUBLE ANGLES
Two equal leg angles
Properties of sections

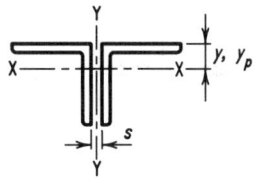

Designation	Axis Y-Y			Q_s*			
	Radii of Gyration			Angles in Contact		Angles Separated	
	Back to Back of Angles, in.			$F_y =$ 36 ksi	$F_y =$ 50 ksi	$F_y =$ 36 ksi	$F_y =$ 50 ksi
	0	$3/8$	$3/4$				
L8×8×1⅛	3.42	3.55	3.69	—	—	—	—
1	3.40	3.53	3.67	—	—	—	—
⅞	3.38	3.51	3.64	—	—	—	—
¾	3.36	3.49	3.62	—	—	—	—
⅝	3.34	3.47	3.60	—	—	0.997	0.935
½	3.32	3.45	3.58	0.995	0.921	0.911	0.834
L6×6×1	2.59	2.73	2.87	—	—	—	—
⅞	2.57	2.70	2.85	—	—	—	—
¾	2.55	2.68	2.82	—	—	—	—
⅝	2.53	2.66	2.80	—	—	—	—
½	2.51	2.64	2.78	—	—	—	0.961
⅜	2.49	2.62	2.75	0.995	0.921	0.911	0.834
L5×5×⅞	2.16	2.30	2.45	—	—	—	—
¾	2.14	2.28	2.42	—	—	—	—
½	2.10	2.24	2.38	—	—	—	—
⅜	2.09	2.22	2.35	—	—	0.982	0.919
⁵⁄₁₆	2.08	2.21	2.34	0.995	0.921	0.911	0.834
L4×4×¾	1.74	1.88	2.03	—	—	—	—
⅝	1.72	1.86	2.00	—	—	—	—
½	1.70	1.83	1.98	—	—	—	—
⅜	1.68	1.81	1.95	—	—	—	—
⁵⁄₁₆	1.67	1.80	1.94	—	—	0.997	0.935
¼	1.66	1.79	1.93	0.995	0.921	0.911	0.834

*Where no value of Q_s is shown, the angles comply with LRFD Specification Section E2.

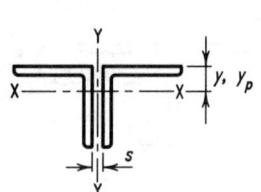

DOUBLE ANGLES
Two equal leg angles
Properties of sections

Designation	Wt. per ft 2 Angles	Area of 2 Angles	Axis X-X					
			I	S	r	y	Z	y_p
	lb	in.^{2}	in.^{4}	in.^{3}	in.	in.	in.^{3}	in.
L3½×3½×⅜	17.0	4.97	5.73	2.30	1.07	1.01	4.15	0.355
⁵⁄₁₆	14.4	4.18	4.90	1.95	1.08	0.990	3.52	0.299
¼	11.6	3.38	4.02	1.59	1.09	0.968	2.86	0.241
L3×3×½	18.8	5.50	4.43	2.14	0.898	0.932	3.87	0.458
⅜	14.4	4.22	3.52	1.67	0.913	0.888	3.00	0.352
⁵⁄₁₆	12.2	3.55	3.02	1.41	0.922	0.865	2.55	0.296
¼	9.80	2.88	2.49	1.15	0.930	0.842	2.08	0.240
³⁄₁₆	7.42	2.18	1.92	0.882	0.939	0.820	1.59	0.182
L2½×2½×⅜	11.8	3.47	1.97	1.13	0.753	0.762	2.04	0.347
⁵⁄₁₆	10.0	2.93	1.70	0.964	0.761	0.740	1.74	0.293
¼	8.20	2.38	1.41	0.789	0.769	0.717	1.42	0.238
³⁄₁₆	6.14	1.80	1.09	0.606	0.778	0.694	1.09	0.180
L2×2×⅜	9.40	2.72	0.958	0.702	0.594	0.636	1.27	0.340
⁵⁄₁₆	7.84	2.30	0.832	0.600	0.601	0.614	1.08	0.288
¼	6.38	1.88	0.695	0.494	0.609	0.592	0.890	0.234
³⁄₁₆	4.88	1.43	0.545	0.381	0.617	0.569	0.686	0.179
⅛	3.30	0.960	0.380	0.261	0.626	0.546	0.471	0.121

DOUBLE ANGLES
Two equal leg angles
Properties of sections

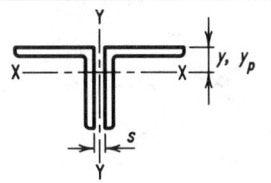

Designation	Axis Y-Y			Q_s*			
	Radii of Gyration			Angles in Contact		Angles Separated	
	Back to Back of Angles, in.			$F_y =$ 36 ksi	$F_y =$ 50 ksi	$F_y =$ 36 ksi	$F_y =$ 50 ksi
	0	³⁄₈	¾				
L3½x3½x³⁄₈	1.48	1.61	1.75	—	—	—	—
⁵⁄₁₆	1.47	1.60	1.74	—	—	—	0.986
¼	1.46	1.59	1.73	—	0.982	0.965	0.897
L3x3x½	1.29	1.43	1.59	—	—	—	—
³⁄₈	1.27	1.41	1.56	—	—	—	—
⁵⁄₁₆	1.26	1.40	1.55	—	—	—	—
¼	1.26	1.39	1.53	—	—	—	0.961
³⁄₁₆	1.25	1.38	1.52	0.995	0.921	0.911	0.834
L2½x2½x³⁄₈	1.07	1.21	1.36	—	—	—	—
⁵⁄₁₆	1.06	1.20	1.35	—	—	—	—
¼	1.05	1.19	1.34	—	—	—	—
³⁄₁₆	1.04	1.18	1.32	—	—	0.982	0.919
L2x2x³⁄₈	0.870	1.01	1.17	—	—	—	—
⁵⁄₁₆	0.859	1.00	1.16	—	—	—	—
¼	0.849	0.989	1.14	—	—	—	—
³⁄₁₆	0.840	0.977	1.13	—	—	—	—
⅛	0.831	0.965	1.11	0.995	0.921	0.911	0.834

*Where no value of Q_s is shown, the angles comply with LRFD Specification Section E2.

DOUBLE ANGLES
Two unequal leg angles
Properties of sections
Long legs back to back

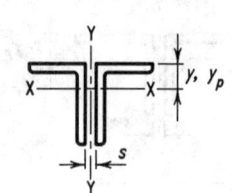

| Designation | Wt. per ft 2 Angles | Area of 2 Angles | Axis X-X | | | | | |
	lb	in.^{2}	I in.^{4}	S in.^{3}	r in.	y in.	Z in.^{3}	y_p in.
L8×6×1	88.4	26.0	161	30.2	2.49	2.65	54.5	1.50
¾	67.6	19.9	126	23.3	2.53	2.56	42.2	1.38
½	46.0	13.5	88.6	16.0	2.56	2.47	29.1	1.25
L8×4×1	74.8	22.0	139	28.1	2.52	3.05	48.5	2.50
¾	57.4	16.9	109	21.8	2.55	2.95	37.7	2.38
½	39.2	11.5	77.0	15.0	2.59	2.86	26.1	2.25
L7×4×¾	52.4	15.4	75.6	16.8	2.22	2.51	29.6	1.88
½	35.8	10.5	53.3	11.6	2.25	2.42	20.6	1.75
⅜	27.2	7.97	41.1	8.88	2.27	2.37	15.7	1.69
L6×4×¾	47.2	13.9	49.0	12.5	1.88	2.08	22.3	1.38
⅝	40.0	11.7	42.1	10.6	1.90	2.03	19.0	1.31
½	32.4	9.50	34.8	8.67	1.91	1.99	15.6	1.25
⅜	24.6	7.22	26.9	6.64	1.93	1.94	11.9	1.19
L6×3½×⅜	23.4	6.84	25.7	6.49	1.94	2.04	11.5	1.44
⁵⁄₁₆	19.6	5.74	21.8	5.47	1.95	2.01	9.70	1.41
L5×3½×¾	39.6	11.6	27.8	8.55	1.55	1.75	15.3	1.13
½	27.2	8.00	20.0	5.97	1.58	1.66	10.8	1.00
⅜	20.8	6.09	15.6	4.59	1.60	1.61	8.28	0.938
⁵⁄₁₆	17.4	5.12	13.2	3.87	1.61	1.59	6.99	0.906
L5×3×½	25.6	7.50	18.9	5.82	1.59	1.75	10.3	1.25
⅜	19.6	5.72	14.7	4.47	1.61	1.70	7.95	1.19
⁵⁄₁₆	16.4	4.80	12.5	3.77	1.61	1.68	6.71	1.16
¼	13.2	3.88	10.2	3.06	1.62	1.66	5.45	1.13

DOUBLE ANGLES
Two unequal leg angles
Properties of sections
Long legs back to back

Designation	Axis Y-Y			Q_s*			
	Radii of Gyration			Angles in Contact		Angles Separated	
	Back to Back of Angles, in.			$F_y =$ 36 ksi	$F_y =$ 50 ksi	$F_y =$ 36 ksi	$F_y =$ 50 ksi
	0	$\frac{3}{8}$	$\frac{3}{4}$				
L8×6×1	2.39	2.52	2.66	—	—	—	—
$\frac{3}{4}$	2.35	2.48	2.62	—	—	—	—
$\frac{1}{2}$	2.32	2.44	2.57	—	—	0.911	0.834
L8×4×1	1.47	1.61	1.75	—	—	—	—
$\frac{3}{4}$	1.42	1.55	1.69	—	—	—	—
$\frac{1}{2}$	1.38	1.51	1.64	—	—	0.911	0.834
L7×4×$\frac{3}{4}$	1.48	1.62	1.76	—	—	—	—
$\frac{1}{2}$	1.44	1.57	1.71	—	—	0.965	0.897
$\frac{3}{8}$	1.43	1.55	1.68	—	—	0.839	0.750
L6×4×$\frac{3}{4}$	1.55	1.69	1.83	—	—	—	—
$\frac{5}{8}$	1.53	1.67	1.81	—	—	—	—
$\frac{1}{2}$	1.51	1.64	1.78	—	—	—	0.961
$\frac{3}{8}$	1.50	1.62	1.76	—	—	0.911	0.834
L6×3$\frac{1}{2}$×$\frac{3}{8}$	1.26	1.39	1.53	—	—	0.911	0.834
$\frac{5}{16}$	1.26	1.38	1.51	—	—	0.825	0.733
L5×3$\frac{1}{2}$×$\frac{3}{4}$	1.40	1.53	1.68	—	—	—	—
$\frac{1}{2}$	1.35	1.49	1.63	—	—	—	—
$\frac{3}{8}$	1.34	1.46	1.60	—	—	0.982	0.919
$\frac{5}{16}$	1.33	1.45	1.59	—	—	0.911	0.834
L5×3×$\frac{1}{2}$	1.12	1.25	1.40	—	—	—	—
$\frac{3}{8}$	1.10	1.23	1.37	—	—	0.982	0.919
$\frac{5}{16}$	1.09	1.22	1.36	—	—	0.911	0.834
$\frac{1}{4}$	1.08	1.21	1.34	—	—	0.804	0.708

*Where no value of Q_s is shown the angles comply with LRFD Specification Section E2.

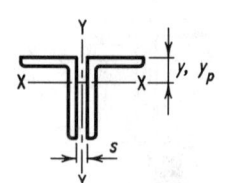

DOUBLE ANGLES
Two unequal leg angles
Properties of sections
Long legs back to back

| Designation | Wt. per ft 2 Angles | Area of 2 Angles | Axis X-X | | | | | |
| | | | I | S | r | y | Z | y_p |
	lb	in.^{2}	in.^{4}	in.^{3}	in.	in.	in.^{3}	in.
L4×3½×½	23.8	7.00	10.6	3.87	1.23	1.25	7.00	0.500
⅜	18.2	5.34	8.35	2.99	1.25	1.21	5.42	0.438
5⁄16	15.4	4.49	7.12	2.53	1.26	1.18	4.59	0.406
¼	12.4	3.63	5.83	2.05	1.27	1.16	3.73	0.375
L4×3×½	22.2	6.50	10.1	3.78	1.25	1.33	6.81	0.750
⅜	17.0	4.97	7.93	2.92	1.26	1.28	5.28	0.688
5⁄16	14.4	4.18	6.76	2.47	1.27	1.26	4.47	0.656
¼	11.6	3.38	5.54	2.00	1.28	1.24	3.63	0.625
L3½×3×⅜	15.8	4.59	5.45	2.25	1.09	1.08	4.08	0.438
5⁄16	13.2	3.87	4.66	1.91	1.10	1.06	3.46	0.406
¼	10.8	3.13	3.83	1.55	1.11	1.04	2.82	0.375
L3½×2½×⅜	14.4	4.22	5.12	2.19	1.10	1.16	3.94	0.688
¼	9.80	2.88	3.60	1.51	1.12	1.11	2.73	0.625
L3×2½×⅜	13.2	3.84	3.31	1.62	0.928	0.956	2.93	0.438
¼	9.00	2.63	2.35	1.12	0.945	0.911	2.04	0.375
3⁄16	6.77	1.99	1.81	0.859	0.954	0.888	1.56	0.344
L3×2×⅜	11.8	3.47	3.06	1.56	0.940	1.04	2.79	0.688
5⁄16	10.0	2.93	2.63	1.33	0.948	1.02	2.38	0.656
¼	8.20	2.38	2.17	1.08	0.957	0.993	1.95	0.625
3⁄16	6.14	1.80	1.68	0.830	0.966	0.970	1.49	0.594
L2½×2×⅜	10.6	3.09	1.82	1.09	0.768	0.831	1.97	0.438
5⁄16	9.00	2.62	1.58	0.932	0.776	0.809	1.69	0.406
¼	7.24	2.13	1.31	0.763	0.784	0.787	1.38	0.375
3⁄16	5.50	1.62	1.02	0.586	0.793	0.764	1.06	0.344

DOUBLE ANGLES
Two unequal leg angles
Properties of sections

Long legs back to back

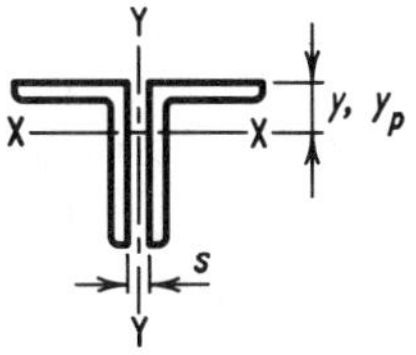

Designation	Axis Y-Y: Radii of Gyration, Back to Back of Angles, in.			Q_s*: Angles in Contact		Q_s*: Angles Separated	
	0	⅜	¾	F_y = 36 ksi	F_y = 50 ksi	F_y = 36 ksi	F_y = 50 ksi
L4×3½×½	1.44	1.58	1.72	—	—	—	—
⅜	1.42	1.56	1.70	—	—	—	—
5/16	1.42	1.55	1.69	—	—	0.997	0.935
¼	1.41	1.54	1.67	—	0.982	0.911	0.834
L4×3×½	1.20	1.33	1.48	—	—	—	—
⅜	1.18	1.31	1.45	—	—	—	—
5/16	1.17	1.30	1.44	—	—	0.997	0.935
¼	1.16	1.29	1.43	—	—	0.911	0.834
L3½×3×⅜	1.22	1.36	1.50	—	—	—	—
5/16	1.21	1.35	1.49	—	—	—	0.986
¼	1.20	1.33	1.48	—	—	0.965	0.897
L3½×2½×⅜	0.976	1.11	1.26	—	—	—	—
¼	0.958	1.09	1.23	—	—	0.965	0.897
L3×2½×⅜	1.02	1.16	1.31	—	—	—	—
¼	1.00	1.13	1.28	—	—	—	0.961
5/16	0.993	1.12	1.27	—	—	0.911	0.834
L3×2×⅜	0.777	0.917	1.07	—	—	—	—
5/16	0.767	0.903	1.06	—	—	—	—
¼	0.757	0.891	1.04	—	—	—	0.961
3/16	0.749	0.879	1.03	—	—	0.911	0.834
L2½×2×⅜	0.819	0.961	1.12	—	—	—	—
5/16	0.809	0.948	1.10	—	—	—	—
¼	0.799	0.935	1.09	—	—	—	—
3/16	0.790	0.923	1.07	—	—	0.982	0.919

*Where no value of Q_s is shown, the angles comply with LRFD Specification Section E2.

DOUBLE ANGLES
Two unequal leg angles
Properties of sections
Short legs back to back

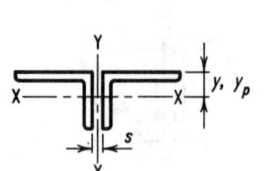

Designation	Wt. per ft 2 Angles	Area of 2 Angles	Axis X-X					
			I	S	r	y	Z	y_p
	lb	in.^{2}	in.^{4}	in.^{3}	in.	in.	in.^{3}	in.
L8×6×1	88.4	26.0	77.6	17.8	1.73	1.65	32.4	0.813
¾	67.6	19.9	61.4	13.8	1.76	1.56	24.9	0.621
½	46.0	13.5	43.4	9.58	1.79	1.47	17.0	0.422
L8×4×1	74.8	22.0	23.3	7.88	1.03	1.05	15.4	0.688
¾	57.4	16.9	18.7	6.14	1.05	0.953	11.6	0.527
½	39.2	11.5	13.5	4.29	1.08	0.859	7.80	0.359
L7×4×¾	52.4	15.4	18.1	6.05	1.09	1.01	11.3	0.549
½	35.8	10.5	13.1	4.23	1.11	0.917	7.66	0.375
⅜	27.2	7.97	10.2	3.26	1.13	0.870	5.80	0.285
L6×4×¾	47.2	13.9	17.4	5.94	1.12	1.08	10.9	0.578
⅝	40.0	11.7	15.0	5.07	1.13	1.03	9.24	0.488
½	32.4	9.50	12.5	4.16	1.15	0.987	7.50	0.396
⅜	24.6	7.22	9.81	3.21	1.17	0.941	5.71	0.301
L6×3½×⅜	23.4	6.84	6.68	2.46	0.988	0.787	4.41	0.285
⁵⁄₁₆	19.6	5.74	5.70	2.08	0.996	0.763	3.70	0.239
L5×3½×¾	39.6	11.6	11.1	4.43	0.977	0.996	8.20	0.581
½	27.2	8.00	8.10	3.12	1.01	0.906	5.65	0.400
⅜	20.8	6.09	6.37	2.41	1.02	0.861	4.32	0.305
⁵⁄₁₆	17.4	5.12	5.44	2.04	1.03	0.838	3.63	0.256
L5×3×½	25.6	7.50	5.16	2.29	0.829	0.750	4.22	0.375
⅜	19.6	5.72	4.08	1.78	0.845	0.704	3.21	0.286
⁵⁄₁₆	16.4	4.80	3.49	1.51	0.853	0.681	2.69	0.240
¼	13.2	3.88	2.88	1.23	0.861	0.657	2.17	0.194

mualc g##d DOUBLE ANGLES

1 - 101

DOUBLE ANGLES
Two unequal leg angles
Properties of sections
Short legs back to back

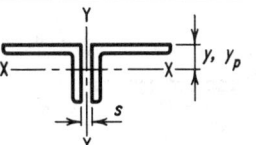

Designation	Axis Y-Y			Q_s*			
	Radii of Gyration			Angles in Contact		Angles Separated	
	Back to Back of Angles, in.			$F_y =$ 36 ksi	$F_y =$ 50 ksi	$F_y =$ 36 ksi	$F_y =$ 50 ksi
	0	3/8	3/4				
L8×6×1	3.64	3.78	3.92	—	—	—	—
3/4	3.60	3.74	3.88	—	—	—	—
1/2	3.56	3.69	3.83	0.995	0.921	0.911	0.834
L8×4×1	3.95	4.10	4.25	—	—	—	—
3/4	3.90	4.05	4.19	—	—	—	—
1/2	3.86	4.00	4.14	0.995	0.921	0.911	0.834
L7×4×3/4	3.35	3.49	3.64	—	—	—	—
1/2	3.30	3.44	3.59	—	0.982	0.965	0.897
3/8	3.28	3.42	3.56	0.926	0.838	0.839	0.750
L6×4×3/4	2.80	2.94	3.09	—	—	—	—
5/8	2.78	2.92	3.06	—	—	—	—
1/2	2.76	2.90	3.04	—	—	—	0.961
3/8	2.74	2.87	3.02	0.995	0.921	0.911	0.834
L6×3½×3/8	2.81	2.95	3.09	0.995	0.921	0.911	0.834
5/16	2.80	2.94	3.08	0.912	0.822	0.825	0.733
L5×3½×3/4	2.33	2.48	2.63	—	—	—	—
1/2	2.29	2.43	2.57	—	—	—	—
3/8	2.27	2.41	2.55	—	—	0.982	0.919
5/16	2.26	2.39	2.54	0.995	0.921	0.911	0.834
L5×3×1/2	2.36	2.50	2.65	—	—	—	—
3/8	2.34	2.48	2.63	—	—	0.982	0.919
5/16	2.33	2.47	2.61	0.995	0.921	0.911	0.834
1/4	2.32	2.46	2.60	0.891	0.797	0.804	0.708

*Where no value of Q_s is shown, the angles comply with LRFD Specification Section E2.

AMERICAN INSTITUTE OF STEEL CONSTRUCTION

DOUBLE ANGLES
Two unequal leg angles
Properties of sections
Short legs back to back

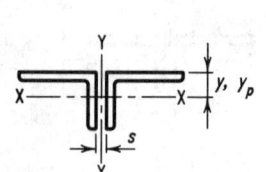

| Designation | Wt. per ft 2 Angles | Area of 2 Angles | Axis X-X | | | | | |
	lb	in.^{2}	I in.^{4}	S in.^{3}	r in.	y in.	Z in.^{3}	y_p in.
L4×3½×½	23.8	7.00	7.58	3.03	1.04	1.00	5.47	0.438
⅜	18.2	5.34	5.97	2.35	1.06	0.955	4.21	0.334
⁵⁄₁₆	15.4	4.49	5.10	1.99	1.07	0.932	3.56	0.281
¼	12.4	3.63	4.19	1.62	1.07	0.909	2.89	0.227
L4×3×½	22.2	6.50	4.85	2.23	0.864	0.827	4.06	0.406
⅜	17.0	4.97	3.84	1.73	0.879	0.782	3.11	0.311
⁵⁄₁₆	14.4	4.18	3.29	1.47	0.887	0.759	2.63	0.261
¼	11.6	3.38	2.71	1.20	0.896	0.736	2.13	0.211
L3½×3×⅜	15.8	4.59	3.69	1.70	0.897	0.830	3.06	0.328
⁵⁄₁₆	13.2	3.87	3.17	1.44	0.905	0.808	2.59	0.276
¼	10.8	3.13	2.61	1.18	0.914	0.785	2.10	0.223
L3½×2½×⅜	14.4	4.22	2.18	1.18	0.719	0.660	2.15	0.301
¼	9.80	2.88	1.55	0.824	0.735	0.614	1.47	0.205
L3×2½×⅜	13.2	3.84	2.08	1.16	0.736	0.706	2.10	0.320
¼	9.00	2.63	1.49	0.808	0.753	0.661	1.45	0.219
³⁄₁₆	6.77	1.99	1.15	0.620	0.761	0.638	1.11	0.166
L3×2×⅜	11.8	3.47	1.09	0.743	0.559	0.539	1.37	0.289
⁵⁄₁₆	10.0	2.93	0.941	0.634	0.567	0.516	1.16	0.244
¼	8.20	2.38	0.784	0.520	0.574	0.493	0.937	0.198
³⁄₁₆	6.14	1.80	0.613	0.401	0.583	0.470	0.713	0.150
L2½×2×⅜	10.6	3.09	1.03	0.725	0.577	0.581	1.32	0.309
⁵⁄₁₆	9.00	2.62	0.893	0.620	0.584	0.559	1.12	0.262
¼	7.24	2.13	0.745	0.509	0.592	0.537	0.915	0.213
³⁄₁₆	5.50	1.62	0.583	0.392	0.600	0.514	0.701	0.162

DOUBLE ANGLES
Two unequal leg angles
Properties of sections
Short legs back to back

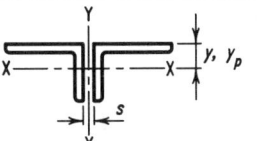

	Axis Y-Y			Q_s*			
	Radii of Gyration			Angles in Contact		Angles Separated	
	Back to Back of Angles, in.			$F_y =$ 36 ksi	$F_y =$ 50 ksi	$F_y =$ 36 ksi	$F_y =$ 50 ksi
Designation	0	3/8	3/4				
L4×3½×½	1.76	1.89	2.04	—	—	—	—
3/8	1.74	1.87	2.01	—	—	—	—
5/16	1.73	1.86	2.00	—	—	0.997	0.935
1/4	1.72	1.85	1.99	0.995	0.921	0.911	0.834
L4×3×½	1.82	1.96	2.11	—	—	—	—
3/8	1.80	1.94	2.08	—	—	—	—
5/16	1.79	1.93	2.07	—	—	0.997	0.935
1/4	1.78	1.92	2.06	0.995	0.921	0.911	0.834
L3½×3×3/8	1.53	1.67	1.82	—	—	—	—
5/16	1.52	1.66	1.80	—	—	—	0.986
1/4	1.52	1.65	1.79	—	0.982	0.965	0.897
L3½×2½×3/8	1.60	1.74	1.89	—	—	—	—
1/4	1.58	1.72	1.86	—	0.982	0.965	0.897
L3×2½×3/8	1.33	1.47	1.62	—	—	—	—
1/4	1.31	1.45	1.60	—	—	—	0.961
3/16	1.30	1.44	1.58	0.995	0.921	0.911	0.834
L3×2×3/8	1.40	1.55	1.70	—	—	—	—
5/16	1.39	1.53	1.68	—	—	—	—
1/4	1.38	1.52	1.67	—	—	—	0.961
3/16	1.37	1.51	1.66	0.995	0.921	0.911	0.834
L2½×2×3/8	1.13	1.28	1.43	—	—	—	—
5/16	1.12	1.26	1.42	—	—	—	—
1/4	1.11	1.25	1.40	—	—	—	—
3/16	1.10	1.24	1.39	—	—	0.982	0.911

*Where no value of Q_s is shown the angles comply with LRFD Specification Section E2.

COMBINATION SECTIONS

Standard rolled shapes are frequently combined to produce efficient and economical structural members for special applications. Experience has established a demand for certain combinations. When properly sized and connected to satisfy the design and specification criteria, these members may be used as struts, lintels, eave struts, and light crane and trolley runways. The W section with channel attached to the web is not recommended for use as a trolley or crane runway member. Properties of several combined sections are tabulated for those combinations that experience has proven to be in popular demand.

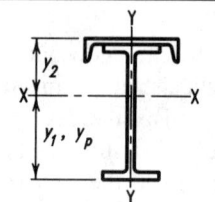

COMBINATION SECTIONS
W shapes and channels
Properties of sections

		Total Weight per ft	Total Area	Axis X-X			
				I	$S_1 = I/y_1$	$S_2 = I/y_2$	r
Beam	Channel	lb	in.^{2}	in.^{4}	in.^{3}	in.^{3}	in.
W12×26	C10×15.3	41.3	12.14	299	36.3	70.5	4.96
×26	C12×20.7	46.7	13.74	318	36.8	82.2	4.81
W14×30	C10×15.3	45.3	13.34	420	46.1	84.6	5.61
×30	C12×20.7	50.7	14.94	448	46.8	98.3	5.47
W16×36	C12×20.7	56.7	16.69	670	62.8	123	6.34
×36	C15×33.9	69.9	20.56	748	64.6	160	6.03
W18×50	C12×20.7	70.7	20.79	1120	97.4	166	7.34
×50	C15×33.9	83.9	24.66	1250	100	211	7.11
W21×62	C12×20.7	82.7	24.39	1800	138	218	8.59
×62	C15×33.9	95.9	28.26	2000	142	272	8.41
×68	C12×20.7	88.7	26.09	1960	152	232	8.68
×68	C15×33.9	101.9	29.96	2180	156	287	8.52
W24×68	C12×20.7	88.7	26.19	2450	168	258	9.67
×68	C15×33.9	101.9	30.06	2720	173	321	9.50
×84	C12×20.7	104.7	30.79	3040	212	303	9.93
×84	C15×33.9	117.9	34.66	3340	217	368	9.82
W27×84	C15×33.9	117.9	34.76	4050	237	404	10.8
×94	C15×33.9	127.9	37.66	4530	268	436	11.0
W30×99	C15×33.9	132.9	39.06	5540	300	480	11.9
×99	C18×42.7	141.7	41.70	5830	304	533	11.8
×116	C15×33.9	149.9	44.16	6590	360	544	12.2
×116	C18×42.7	158.7	46.80	6900	365	599	12.1
W33×118	C15×33.9	151.9	44.66	7900	395	596	13.3
×118	C18×42.7	160.7	47.30	8280	400	656	13.2
×141	C15×33.9	174.9	51.56	9580	484	689	13.6
×141	C18×42.7	183.7	54.20	10000	490	751	13.6
W36×150	C15×33.9	183.9	54.16	11500	546	765	14.6
×150	C18×42.7	192.7	56.80	12100	553	832	14.6

COMBINATION SECTIONS
W shapes and channels
Properties of sections

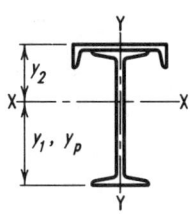

Beam	Channel	Axis X-X			Axis Y-Y			
		y_1	Z	y_p	I	S	r	Z
		in.	in.^{3}	in.	in.^{4}	in.^{3}	in.	in.^{3}
W12×26	C10×15.3	8.22	47.0	11.30	84.7	16.9	2.64	24.0
×26	C12×20.7	8.63	48.8	11.55	146	24.4	3.26	33.6
W14×30	C10×15.3	9.12	60.5	12.56	87.0	17.4	2.55	24.8
×30	C12×20.7	9.57	62.3	12.87	149	24.8	3.15	34.4
W16×36	C12×20.7	10.67	83.6	14.56	154	25.6	3.03	36.3
×36	C15×33.9	11.58	88.6	15.21	340	45.3	4.06	61.3
W18×50	C12×20.7	11.51	128	16.08	169	28.2	2.85	42.0
×50	C15×33.9	12.47	134	16.90	355	47.3	3.79	67.0
W21×62	C12×20.7	13.01	182	18.06	187	31.1	2.77	47.2
×62	C15×33.9	14.06	190	19.36	373	49.7	3.63	72.2
×68	C12×20.7	12.93	200	17.60	194	32.3	2.72	49.8
×68	C15×33.9	13.95	208	19.32	380	50.6	3.56	74.8
W24×68	C12×20.7	14.53	224	19.15	199	33.2	2.76	50.0
×68	C15×33.9	15.67	234	21.66	385	51.4	3.58	75.0
×84	C12×20.7	14.35	275	18.49	223	37.2	2.69	58.1
×84	C15×33.9	15.40	288	21.61	409	54.6	3.44	83.1
W27×84	C15×33.9	17.07	320	23.86	421	56.1	3.48	83.6
×94	C15×33.9	16.92	357	23.56	439	58.5	3.41	89.2
W30×99	C15×33.9	18.51	408	24.34	443	59.1	3.37	89.1
×99	C18×42.7	19.18	418	26.43	682	75.8	4.04	113
×116	C15×33.9	18.30	480	23.77	479	63.9	3.29	99.6
×116	C18×42.7	18.93	492	26.04	718	79.8	3.92	124
W33×118	C15×33.9	20.01	529	25.43	502	66.9	3.35	102
×118	C18×42.7	20.69	544	27.77	741	82.3	3.96	126
×141	C15×33.9	19.79	634	24.83	561	74.8	3.30	117
×141	C18×42.7	20.42	652	26.96	800	88.9	3.84	141
W36×150	C15×33.9	21.15	716	25.84	585	78.0	3.29	121
×150	C18×42.7	21.81	738	27.91	824	91.6	3.81	145

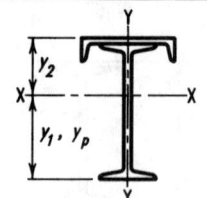

COMBINATION SECTIONS
S shapes and channels
Properties of sections

		Total Weight per ft	Total Area	Axis X-X			
				I	$S_1 = I/y_1$	$S_2 = I/y_2$	r
Beam	Channel	lb	in.^{2}	in.^{4}	in.^{3}	in.^{3}	in.
S10×25.4	C8×11.5	36.9	10.84	176	27.2	46.6	4.02
	C10×15.3	40.7	11.95	186	27.6	52.9	3.94
S12×31.8	C8×11.5	43.3	12.73	299	39.8	63.2	4.84
	C10×15.3	47.1	13.84	316	40.4	71.4	4.78
S15×42.9	C8×11.5	54.4	15.98	585	64.9	94.2	6.05
	C10×15.3	58.2	17.09	616	65.8	105	6.01
S20×66	C10×15.3	81.3	23.89	1530	130	181	8.00
	C12×20.7	86.7	25.49	1620	132	203	7.97
S24×80	C10×15.3	95.3	27.99	2610	188	252	9.66
	C12×20.7	100.7	29.59	2750	191	278	9.65

COMBINATION SECTIONS
S shapes and channels
Properties of sections

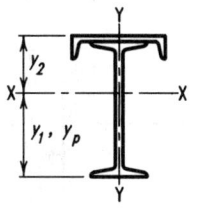

		Axis X-X			Axis Y-Y			
		y_1	Z	y_p	I	S	r	Z
Beam	Channel	in.	in.^{3}	in.	in.^{4}	in.^{3}	in.	in.^{3}
S10×25.4	C8×11.5	6.45	35.7	8.81	39.4	9.8	1.91	14.5
	C10×15.3	6.73	36.9	9.02	74.2	14.8	2.49	20.8
S12×31.8	C8×11.5	7.50	52.6	10.30	42.0	10.5	1.82	16.0
	C10×15.3	7.82	53.9	10.61	76.8	15.4	2.36	22.2
S15×42.9	C8×11.5	9.01	85.7	11.58	47.0	11.8	1.71	18.6
	C10×15.3	9.37	88.2	12.77	81.8	16.4	2.19	24.9
S20×66	C10×15.3	11.81	171	14.41	95.1	19.0	2.00	31.2
	C12×20.7	12.29	178	15.99	157	26.1	2.48	40.8
S24×80	C10×15.3	13.86	244	16.46	110	21.9	1.98	36.6
	C12×20.7	14.38	254	18.05	171	28.5	2.41	46.2

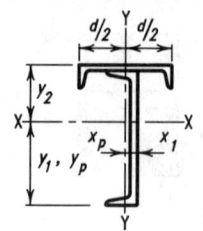

COMBINATION SECTIONS
Two channels
Properties of sections

Vertical Channel	Horizontal Channel	Total Weight per ft lb	Total Area in.^{2}	Axis X-X						
				I in.^{4}	$S_1 = I/y_1$ in.^{3}	$S_2 = I/y_2$ in.^{3}	r in.	y_1 in.	z in.^{3}	y_p in.
C3×4.1	C4×5.4	9.5	2.80	3.0	1.4	3.0	1.04	2.20	2.16	2.67
C4×5.4	C4×5.4	10.8	3.18	6.5	2.3	4.9	1.43	2.86	3.39	3.56
	C5×6.7	12.1	3.56	6.9	2.3	5.5	1.39	2.94	3.62	3.61
C5×6.7	C5×6.7	13.4	3.94	12.8	3.5	8.0	1.80	3.60	5.23	4.50
	C6×8.2	14.9	4.37	13.4	3.6	8.9	1.75	3.70	5.50	4.57
	C7×9.8	16.5	4.84	14.0	3.7	9.8	1.70	3.79	5.81	4.62
C6×8.2	C5×6.7	14.9	4.37	21.5	5.1	10.9	2.22	4.22	7.31	5.37
	C6×8.2	16.4	4.80	22.5	5.2	12.1	2.16	4.34	7.61	5.45
	C7×9.8	18.0	5.27	23.4	5.2	13.3	2.11	4.45	7.93	5.53
	C8×11.5	19.7	5.78	24.3	5.3	14.5	2.05	4.55	8.30	5.58
	C9×13.4	21.6	6.34	25.2	5.4	15.8	1.99	4.64	8.72	5.63
	C10×15.3	23.5	6.89	26.0	5.5	16.9	1.94	4.70	9.16	5.65
C7×9.8	C6×8.2	18.0	5.27	35.3	7.1	15.7	2.59	4.95	10.2	6.32
	C7×9.8	19.6	5.74	36.7	7.2	17.3	2.53	5.08	10.6	6.40
	C8×11.5	21.3	6.25	38.0	7.3	18.8	2.47	5.20	10.9	6.48
	C9×13.4	23.2	6.81	39.3	7.4	20.5	2.40	5.31	11.4	6.54
	C10×15.3	25.1	7.36	40.5	7.5	21.9	2.34	5.39	11.8	6.58
C8×11.5	C6×8.2	19.7	5.78	52.4	9.5	19.6	3.01	5.53	13.4	7.18
	C7×9.8	21.3	6.25	54.5	9.6	21.6	2.95	5.68	13.8	7.27
	C8×11.5	23.0	6.76	56.4	9.7	23.6	2.89	5.82	14.2	7.35
	C9×13.4	24.9	7.32	58.4	9.8	25.6	2.82	5.95	14.6	7.44
	C10×15.3	26.8	7.87	60.0	9.9	27.5	2.76	6.06	15.1	7.49
	C12×20.7	32.2	9.47	64.4	10.2	32.6	2.61	6.30	16.4	7.62

COMBINATION SECTIONS
Two channels
Properties of sections

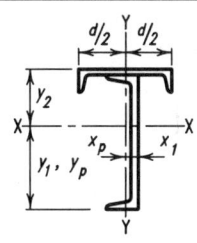

Vertical Channel	Horizontal Channel	Axis Y-Y					
		I in.^{4}	S in.^{3}	r in.	x_1 in.	Z in.^{3}	x_p in.
C3×4.1	C4×5.4	4.0	2.0	1.20	0.44	2.67	0.315
C4×5.4	C4×5.4	4.2	2.1	1.14	0.46	2.84	0.281
	C5×6.7	7.8	3.1	1.48	0.46	4.09	0.282
C5×6.7	C5×6.7	8.0	3.2	1.42	0.48	4.29	0.264
	C6×8.2	13.6	4.5	1.76	0.48	5.90	0.266
	C7×9.8	21.8	6.2	2.12	0.48	7.90	0.268
C6×8.2	C5×6.7	8.2	3.3	1.37	0.51	4.52	0.242
	C6×8.2	13.8	4.6	1.70	0.51	6.14	0.245
	C7×9.8	22.0	6.3	2.04	0.51	8.13	0.247
	C8×11.5	33.3	8.3	2.40	0.51	10.6	0.249
	C9×13.4	48.6	10.8	2.77	0.51	13.5	0.252
	C10×15.3	68.1	13.6	3.14	0.51	16.8	0.254
C7×9.8	C6×8.2	14.1	4.7	1.63	0.54	6.41	0.225
	C7×9.8	22.3	6.4	1.97	0.54	8.41	0.228
	C8×11.5	33.6	8.4	2.32	0.54	10.8	0.230
	C9×13.4	48.6	10.9	2.68	0.54	13.8	0.234
	C10×15.3	68.4	13.7	3.05	0.54	17.1	0.235
C8×11.5	C6×8.2	14.4	4.8	1.58	0.57	6.73	0.218
	C7×9.8	22.6	6.5	1.90	0.57	8.73	0.219
	C8×11.5	33.9	8.5	2.24	0.57	11.2	0.219
	C9×13.4	49.2	10.9	2.59	0.57	14.1	0.220
	C10×15.3	68.7	13.7	2.95	0.57	17.4	0.220
	C12×20.7	130	21.7	3.71	0.57	27.0	0.230

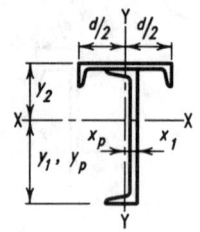

COMBINATION SECTIONS
Two channels
Properties of sections

Vertical Channel	Horizontal Channel	Total Weight per ft	Total Area	Axis X-X							
				I	$S_1 = I/y_1$	$S_2 = I/y_2$	r	y_1	Z	y_p
		lb	in.^{2}	in.^{4}	in.^{3}	in.^{3}	in.	in.	in.^{3}	in.
C9×13.4	C7×9.8	23.2	6.81	77.7	12.4	26.3	3.38	6.26	17.6	8.11
	C8×11.5	24.9	7.32	80.5	12.6	28.7	3.32	6.42	18.1	8.21
	C9×13.4	26.8	7.88	83.3	12.7	31.2	3.25	6.57	18.5	8.31
	C10×15.3	28.7	8.43	85.6	12.8	33.5	3.19	6.69	19.0	8.37
	C12×20.7	34.1	10.03	91.7	13.1	39.8	3.02	6.98	20.4	8.54
C10×15.3	C8×11.5	26.8	7.87	110	15.8	34.2	3.75	7.00	22.4	9.07
	C9×13.4	28.7	8.43	114	15.9	37.2	3.68	7.16	22.9	9.18
	C10×15.3	30.6	8.98	117	16.1	39.9	3.61	7.30	23.4	9.26
	C12×20.7	36.0	10.58	126	16.4	47.5	3.45	7.64	24.9	9.46
	C15×33.9	49.2	14.45	141	17.3	63.7	3.13	8.18	28.3	9.73
C12×20.7	C9×13.4	34.1	10.03	207	25.2	51.4	4.54	8.21	35.7	10.78
	C10×15.3	36.0	10.58	213	25.4	55.0	4.48	8.38	36.3	10.88
	C12×20.7	41.4	12.18	228	25.9	65.3	4.32	8.79	38.0	11.16
	C15×33.9	54.6	16.05	256	27.0	87.8	4.00	9.48	41.8	11.56
C15×33.9	C10×15.3	49.2	14.45	474	48.8	85.6	5.72	9.71	69.7	12.83
	C12×20.7	54.6	16.05	509	49.9	99.8	5.63	10.19	72.2	13.31
	C15×33.9	67.8	19.92	575	52.0	132	5.37	11.06	77.4	14.04
	MC18×42.7	76.6	22.56	608	53.1	152	5.19	11.45	80.7	14.37
MC18×42.7	MC12×20.7	63.4	18.69	860	72.9	133	6.78	11.80	106	15.51
	MC15×33.9	76.6	22.56	975	76.1	174	6.57	12.80	113	16.50
	MC18×42.7	85.4	25.20	1030	77.6	200	6.40	13.29	117	16.96

COMBINATION SECTIONS
Two channels
Properties of sections

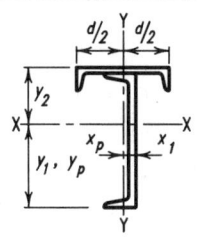

Vertical Channel	Horizontal Channel	Axis Y-Y					
		I in.^{4}	S in.^{3}	r in.	x_1 in.	Z in.^{3}	x_p in.
C9×13.4	C7×9.8	23.1	6.6	1.84	0.60	9.10	0.226
	C8×11.5	34.4	8.6	2.17	0.60	11.5	0.227
	C9×13.4	49.7	11.0	2.51	0.60	14.5	0.227
	C10×15.3	69.2	13.8	2.86	0.60	17.8	0.227
	C12×20.7	131	21.8	3.61	0.60	27.4	0.229
C10×15.3	C8×11.5	34.9	8.7	2.11	0.63	11.9	0.232
	C9×13.4	50.2	11.2	2.44	0.63	14.9	0.232
	C10×15.3	69.7	13.9	2.79	0.63	18.2	0.233
	C12×20.7	131	21.9	3.52	0.63	27.8	0.234
	C15×33.9	317	42.3	4.69	0.63	52.8	0.239
C12×20.7	C9×13.4	51.8	11.5	2.27	0.70	16.0	0.261
	C10×15.3	71.3	14.3	2.60	0.70	19.3	0.261
	C12×20.7	133	22.1	3.30	0.70	29.0	0.262
	C15×33.9	319	42.5	4.46	0.70	54.0	0.266
C15×33.9	C10×15.3	75.5	15.1	2.29	0.79	22.1	0.337
	C12×20.7	137	22.9	2.92	0.79	31.7	0.338
	C15×33.9	323	43.1	4.03	0.79	56.7	0.342
	MC18×42.7	562	62.5	4.99	0.79	80.7	0.343
MC18×42.7	MC12×20.7	143	23.9	2.77	0.88	33.6	0.355
	MC15×33.9	329	43.9	3.82	0.88	58.6	0.358
	MC18×42.7	568	63.2	4.75	0.88	82.6	0.359

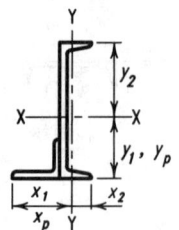

COMBINATION SECTIONS
Channels and angles
Properties of sections
Long leg of angle turned out

Channel	Angle	Total Weight per ft	Total Area	Axis X-X						
				I	$S_1 = I/y_1$	$S_2 = I/y_2$	r	y_1	z	y_p
		lb	in.^{2}	in.^{4}	in.^{3}	in.^{3}	in.	in.	in.^{3}	in.
C6×8.2	L2½×2½×¼	12.3	3.59	17.9	8.0	4.8	2.24	2.24	6.75	1.40
	L3 ×2½×¼	12.7	3.71	18.5	8.5	4.8	2.23	2.17	6.90	1.26
	L3½×3 ×¼	13.6	3.96	19.0	8.9	4.9	2.19	2.13	7.23	1.26
	×⁵⁄₁₆	14.8	4.33	19.8	9.8	5.0	2.14	2.02	7.54	1.11
	L4 ×3 ×¼	14.0	4.09	19.5	9.5	5.0	2.19	2.06	7.36	1.13
C7×9.8	L2½×2½×¼	13.9	4.06	28.5	10.6	6.6	2.65	2.68	9.13	1.67
	L3 ×2½×¼	14.3	4.18	29.3	11.2	6.7	2.65	2.61	9.31	1.53
	L3½×3 ×¼	15.2	4.43	30.0	11.8	6.7	2.60	2.54	9.64	1.53
	×⁵⁄₁₆	16.4	4.80	31.2	12.9	6.8	2.55	2.42	9.99	1.35
	L4 ×3 ×¼	15.6	4.56	30.8	12.4	6.8	2.60	2.48	9.81	1.39
	×⁵⁄₁₆	17.0	4.96	32.0	13.7	6.9	2.54	2.35	10.2	1.20
C8×11.5	L3 ×2½×¼	16.0	4.69	43.9	14.3	8.9	3.06	3.07	12.2	1.81
	L3½×3 ×¼	16.9	4.94	44.9	15.1	9.0	3.02	2.98	12.6	1.81
	×⁵⁄₁₆	18.1	5.31	46.7	16.4	9.0	2.97	2.84	13.0	1.60
	L4 ×3 ×¼	17.3	5.07	46.0	15.8	9.0	3.01	2.91	12.8	1.67
	×⁵⁄₁₆	18.7	5.47	47.8	17.3	9.1	2.96	2.76	13.2	1.45
	L5 ×3½×⁵⁄₁₆	20.2	5.94	49.9	18.9	9.3	2.90	2.64	13.9	1.30
C9×13.4	L3 ×2½×¼	17.9	5.25	63.1	17.8	11.6	3.47	3.54	15.8	2.11
	L3½×3 ×¼	18.8	5.50	64.6	18.8	11.6	3.43	3.45	16.1	2.11
	×⁵⁄₁₆	20.0	5.87	67.1	20.4	11.7	3.38	3.29	16.6	1.87
	L4 ×3 ×¼	19.2	5.63	66.0	19.6	11.7	3.42	3.37	16.3	1.98
	×⁵⁄₁₆	20.6	6.03	68.7	21.4	11.8	3.37	3.20	16.8	1.73
	L5 ×3½×⁵⁄₁₆	22.1	6.50	71.4	23.4	12.0	3.31	3.06	17.5	1.58
C10×15.3	L3½×3 ×¼	20.7	6.05	89.3	22.8	14.7	3.84	3.91	20.0	2.39
	×⁵⁄₁₆	21.9	6.42	92.7	24.8	14.8	3.80	3.74	20.6	2.12
	L4 ×3 ×¼	21.1	6.18	91.1	23.8	14.8	3.84	3.83	20.3	2.26
	×⁵⁄₁₆	22.5	6.58	94.7	25.9	14.9	3.79	3.65	20.9	1.98
	L5 ×3½×⁵⁄₁₆	24.0	7.05	98.4	28.2	15.1	3.74	3.49	21.6	1.84
	×³⁄₈	25.7	7.54	102	30.6	15.2	3.67	3.33	22.2	1.61

COMBINATION SECTIONS
Channels and angles
Properties of sections
Long leg of angle turned out

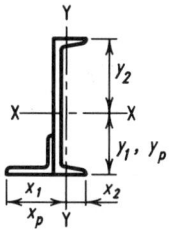

Channel	Angle	Axis Y-Y						
		I	$S_1 = I/x_1$	$S_2 = I/x_2$	r	x_1	z	x_p
		in.^{4}	in.^{3}	in.^{3}	in.	in.	in.^{3}	in.
C6×8.2	L2½×2½×¼	2.6	1.0	1.4	0.85	2.60	2.02	2.60
	L3 ×2½×¼	3.6	1.2	1.9	0.98	3.01	2.38	3.09
	L3½×3 ×¼	4.9	1.4	2.4	1.11	3.40	2.82	3.57
	×⁵⁄₁₆	5.7	1.7	2.7	1.14	3.31	3.27	3.54
	L4 ×3 ×¼	6.5	1.7	3.1	1.26	3.79	3.30	4.06
C7×9.8	L2½×2½×¼	3.0	1.1	1.6	0.86	2.67	2.31	2.62
	L3 ×2½×¼	4.0	1.3	2.0	0.98	3.09	2.66	3.11
	L3½×3 ×¼	5.4	1.6	2.6	1.10	3.48	3.11	3.59
	×⁵⁄₁₆	6.3	1.8	2.9	1.14	3.40	3.57	3.57
	L4 ×3 ×¼	7.1	1 8	3.2	1.25	3.88	3.59	4.08
	×⁵⁄₁₆	8.3	2.2	3.6	1.29	3.78	4.16	4.05
C8×11.5	L3 ×2½×¼	4.6	1 4	2.2	0.99	3.16	3.00	3.13
	L3½×3 ×¼	6.0	1.7	2.7	1.10	3.56	3.45	3.61
	×⁵⁄₁₆	6.9	2.0	3.0	1.14	3.48	3.91	3.59
	L4 ×3 ×¼	7.8	2.0	3.4	1.24	3.97	3.93	4.10
	×⁵⁄₁₆	9.0	2.3	3.8	1.28	3.87	4.51	4.08
	L5 ×3½×⁵⁄₁₆	14.7	3.2	5.6	1.57	4.64	5.97	5.05
C9×13.4	L3 ×2½×¼	5.2	1.6	2.3	0.99	3.22	3.38	3.14
	L3½×3 ×¼	6.7	1.8	2.9	1.10	3.64	3.83	3.63
	×⁵⁄₁₆	7.7	2.2	3.2	1.14	3.55	4.31	3.61
	L4 ×3 ×¼	8.5	2.1	3.6	1.23	4.05	4.32	4.12
	×⁵⁄₁₆	9.9	2.5	4.0	1.28	3.96	4.91	4.10
	L5 ×3½×⁵⁄₁₆	15.8	3.3	5.9	1.56	4.74	6.38	5.08
C10×15.3	L3½×3 ×¼	7.4	2.0	3.1	1.11	3.70	4.25	3.64
	×⁵⁄₁₆	8.5	2.3	3.4	1.15	3.62	4.73	3.63
	L4 ×3 ×¼	9.4	2.3	3.8	1.23	4.12	4.74	4.14
	×⁵⁄₁₆	10.8	2.7	4.2	1.28	4.03	5.34	4.12
	L5 ×3½×⁵⁄₁₆	16.9	3.5	6.1	1.55	4.83	6.82	5.09
	×³⁄₈	19.2	4.1	6.7	1.60	4.73	7.70	5.07

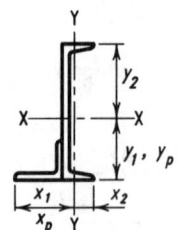

COMBINATION SECTIONS
Channels and angles
Properties of sections
Long leg of angle turned out

Channel	Angle	Total Weight per ft	Total Area	Axis X-X						
				I	$S_1 = I/y_1$	$S_2 = I/y_2$	r	y_1	z	y_p
		lb	in.^{2}	in.^{4}	in.^{3}	in.^{3}	in.	in.	in.^{3}	in.
C12×20.7	L3½×3 ×¼	26.1	7.65	164	33.2	23.2	4.63	4.94	31.4	3.23
	×⁵⁄₁₆	27.3	8.02	170	35.8	23.5	4.61	4.75	32.2	2.80
	L4 ×3 ×¼	26.5	7.78	167	34.4	23.4	4.63	4.86	31.8	3.01
	×⁵⁄₁₆	27.9	8.18	173	37.2	23.6	4.60	4.66	32.6	2.67
	L5 ×3½×⁵⁄₁₆	29.4	8.65	180	40.2	23.9	4.56	4.47	33.5	2.53
	×³⁄₈	31.1	9.14	186	43.4	24.1	4.51	4.29	34.2	2.25
	L6 ×4 ×³⁄₈	33.0	9.70	192	46.6	24.3	4.45	4.12	35.3	2.11
	×½	36.9	10.84	202	53.2	24.7	4.32	3.80	36.7	1.68
C12×25	L3½×3 ×¼	30.4	8.91	180	35.4	26.1	4.50	5.09	35.8	3.98
	×⁵⁄₁₆	31.6	9.28	187	38.0	26.4	4.49	4.92	36.8	3.50
	L4 ×3 ×¼	30.8	9.04	183	36.6	26.3	4.50	5.02	36.3	3.82
	×⁵⁄₁₆	32.2	9.44	190	39.3	26.6	4.49	4.84	37.3	3.30
	L5 ×3½×⁵⁄₁₆	33.7	9.91	197	42.3	26.9	4.46	4.67	38.3	3.05
	×³⁄₈	35.4	10.40	204	45.4	27.2	4.43	4.49	39.3	2.77
	L6 ×4 ×³⁄₈	37.3	10.96	211	48.7	27.5	4.39	4.33	40.4	2.65
	×½	41.2	12.10	223	55.3	28.0	4.29	4.03	42.2	2.20
C15×33.9	L4 ×3 ×¼	39.7	11.65	383	58.7	45.1	5.73	6.52	60.1	5.39
	×⁵⁄₁₆	41.1	12.05	395	62.4	45.6	5.73	6.33	61.8	4.89
	L5 ×3½×⁵⁄₁₆	42.6	12.52	408	66.5	46.1	5.71	6.14	63.4	4.30
	×³⁄₈	44.3	13.01	421	70.8	46.5	5.69	5.94	64.8	3.69
	L6 ×4 ×³⁄₈	46.2	13.57	434	75.4	46.9	5.65	5.76	66.2	3.48
	×½	50.1	14.71	458	84.8	47.7	5.58	5.40	68.6	2.92

COMBINATION SECTIONS
Channels and angles
Properties of sections
Long leg of angle turned out

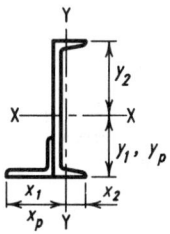

Channel	Angle	I in.^{4}	$S_1 = I/x_1$ in.^{3}	$S_2 = I/x_2$ in.^{3}	r in.	x_1 in.	Z in.^{3}	x_p in.
				Axis Y-Y				
C12×20.7	L3½×3 ×¼	9.5	2.5	3.7	1.12	3.84	5.45	3.69
	×⁵⁄₁₆	10.7	2.8	4.0	1.16	3.77	5.94	3.67
	L4 ×3 ×¼	11.6	2.7	4.4	1.22	4.28	5.94	4.18
	×⁵⁄₁₆	13.2	3.2	4.8	1.27	4.20	6.56	4.16
	L5 ×3½×⁵⁄₁₆	19.9	4 0	6.8	1.52	5.02	8.06	5.15
	×³⁄₈	22.5	4.6	7.5	1.57	4.93	8.97	5.13
	L6 ×4 ×³⁄₈	33.2	5.8	10.3	1.85	5.72	11.1	6.10
	×½	40.6	7.3	11.9	1.93	5.52	13.7	6.05
C12×25	L3½×3 ×¼	10.2	2.6	3.8	1.07	3.87	5.88	3.74
	×⁵⁄₁₆	11.4	3.0	4.2	1.11	3.81	6.40	3.72
	L4 ×3 ×¼	12.3	2.8	4.5	1.17	4.32	6.38	4.23
	×⁵⁄₁₆	13.9	3.3	5.0	1.22	4.25	7.02	4.22
	L5 ×3½×⁵⁄₁₆	20.8	4.1	7.0	1.45	5.09	8.54	5.20
	×³⁄₈	23.5	4.7	7.7	1.50	5.00	9.48	5.18
	L6 ×4 ×³⁄₈	34.5	5.9	10.7	1.77	5.81	11.7	6.15
	×½	42.3	7.5	12.4	1.87	5.63	14.3	6.11
C15×33.9	L4 ×3 ×¼	16.8	3.7	5.8	1.20	4.49	8.82	4.27
	×⁵⁄₁₆	18.7	4.2	6.3	1.25	4.43	9.47	4.26
	L5 ×3½×⁵⁄₁₆	26.2	4.9	8.5	1.45	5.30	11.0	5.24
	×³⁄₈	29.3	5.6	9.2	1.50	5.23	12.0	5.23
	L6 ×4 ×³⁄₈	41.3	6.8	12.4	1.75	6.06	14.2	6.21
	×½	50.3	8.5	14.3	1.85	5.89	16.9	6.17

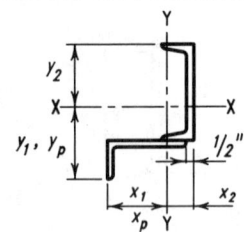

COMBINATION SECTIONS
Channels and angles
Properties of sections
Short leg of angle turned out

Channel	Angle	Total Weight per ft	Total Area	Axis X-X					
				$S_1 = I/y_1$	$S_2 = I/y_2$	r	y_1	z	y_p
		lb	in.^{2}	in.^{3}	in.^{3}	in.	in.	in.^{3}	in.
C 6×8.2	L3×2½×¼	12.7	3.71	6.0	5.9	2.61	4.21	7.86	2.79
	L3×3 ×¼	13.1	3.84	6.1	6.2	2.68	4.56	8.23	3.25
	L4×3 ×¼	14.0	4.09	6.4	6.2	2.63	4.46	8.32	3.18
	L5×3 ×⁵⁄₁₆	16.4	4.80	7.5	6.4	2.55	4.16	8.77	3.00
	L6×3½×⁵⁄₁₆	18.0	5.27	7.7	6.8	2.56	4.45	9.32	3.46
C 7×9.8	L3×2½×¼	14.3	4.18	8.0	7.8	3.00	4.70	10.5	2.95
	L3×3 ×¼	14.7	4.31	8.0	8.2	3.07	5.05	10.9	3.38
	L4×3 ×¼	15.6	4.56	8.5	8.2	3.03	4.93	11.0	3.29
	L5×3 ×⁵⁄₁₆	18.0	5.27	10.0	8.5	2.95	4.60	11.6	3.11
	L6×3½×⁵⁄₁₆	19.6	5.74	10.3	8.9	2.96	4.87	12.2	3.50
C 8×11.5	L3×2½×¼	16.0	4.69	10.4	10.2	3.39	5.20	13.7	3.52
	L3×3 ×¼	16.4	4.82	10.4	10.6	3.45	5.55	14.2	3.73
	L4×3 ×¼	17.3	5.07	10.9	10.6	3.42	5.42	14.3	3.42
	L5×3 ×⁵⁄₁₆	19.7	5.78	12.9	11.0	3.36	5.06	14.9	3.21
	L6×3½×⁵⁄₁₆	21.3	6.25	13.3	11.4	3.36	5.31	15.6	3.61
C 9×13.4	L3×2½×¼	17.9	5.25	13.1	12.9	3.78	5.71	17.4	4.18
	L3×3 ×¼	18.3	5.38	13.1	13.4	3.84	6.07	18.0	4.42
	L4×3 ×¼	19.2	5.63	13.8	13.5	3.81	5.93	18.3	3.88
	L5×3 ×⁵⁄₁₆	21.6	6.34	16.2	13.9	3.76	5.54	19.0	3.32
	L6×3½×⁵⁄₁₆	23.2	6.81	16.7	14.4	3.77	5.78	19.7	3.71
C10×15.3	L3×2½×¼	19.8	5.80	16.2	16.0	4.17	6.22	21.4	4.77
	L3×3 ×¼	20.2	5.93	16.1	16.5	4.22	6.58	22.1	5.01
	L4×3 ×¼	21.1	6.18	17.0	16.6	4.20	6.43	22.5	4.48
	L5×3 ×⁵⁄₁₆	23.5	6.89	19.9	17.1	4.17	6.02	23.5	3.44
	L6×3½×⁵⁄₁₆	25.1	7.36	20.5	17.7	4.18	6.25	24.2	3.81
C12×20.7	L3×2½×¼	25.2	7.40	24.3	24.7	4.90	7.32	32.6	6.17
	L3×3 ×¼	25.6	7.53	24.0	25.3	4.95	7.69	33.4	6.45
	L4×3 ×¼	26.5	7.78	25.3	25.5	4.95	7.54	34.3	6.01
	L5×3 ×⁵⁄₁₆	28.9	8.49	29.2	26.3	4.94	7.11	36.4	4.74
	L6×3½×⁵⁄₁₆	30.5	8.96	30.1	27.1	4.97	7.33	37.5	4.41
C15×33.9	L3×3 ×¼	38.8	11.40	42.7	47.2	5.95	9.45	61.1	8.70
	L4×3 ×¼	39.7	11.65	44.5	47.7	5.96	9.31	62.5	8.39
	L5×3 ×⁵⁄₁₆	42.1	12.36	50.1	49.1	6.01	8.91	66.5	7.50
	L6×3½×⁵⁄₁₆	43.7	12.83	51.4	50.3	6.05	9.15	69.0	7.41

COMBINATION SECTIONS
Channels and angles
Properties of sections
Short leg of angle turned out

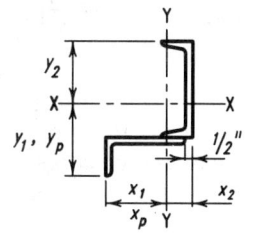

		Axis Y-Y					
Channel	Angle	$S_1 = I/x_1$ in.^{3}	$S_2 = I/x_2$ in.^{3}	r in.	x_1 in.	z in.^{3}	x_p in.
C 6×8.2	L3×2½×¼	2.4	4.4	1.22	2.25	3.70	2.65
	L3×3 ×¼	2.8	4.6	1.26	2.18	4.01	2.59
	L4×3 ×¼	3.8	6.6	1.64	2.85	5.46	3.46
	L5×3 ×⁵⁄₁₆	6.0	9.3	2.05	3.33	8.29	4.12
	L6×3½×⁵⁄₁₆	8.4	12.1	2.47	3.82	11.3	4.84
C 7×9.8	L3×2½×¼	2.6	5.0	1.19	2.32	4.03	2.72
	L3×3 ×¼	2.9	5.2	1.23	2.25	4.35	2.67
	L4×3 ×¼	3.9	7.5	1.60	2.95	5.82	3.55
	L5×3 ×⁵⁄₁₆	6.1	10.5	2.01	3.47	8.71	4.24
	L6×3½×⁵⁄₁₆	8.6	13.6	2.44	3.98	11.8	4.99
C 8×11.5	L3×2½×¼	2.7	5.6	1.16	2.37	4.40	2.78
	L3×3 ×¼	3.0	5.8	1.20	2.31	4.73	2.73
	L4×3 ×¼	4.0	8.3	1.55	3.03	6.21	3.62
	L5×3 ×⁵⁄₁₆	6.3	11.7	1.97	3.58	9.16	4.34
	L6×3½×⁵⁄₁₆	8.7	15.2	2.40	4 13	12.3	5.12
C 9×13.4	L3×2½×¼	2.8	6.2	1.14	2.40	4.81	2.84
	L3×3 ×¼	3.2	6.5	1.18	2.35	5.15	2.79
	L4×3 ×¼	4.2	9.2	1.51	3.10	6.65	3.70
	L5×3 ×⁵⁄₁₆	6.4	12.9	1.92	3.68	9.66	4.43
	L6×3½×⁵⁄₁₆	8.9	16.9	2.36	4.26	12.8	5.23
C10×15.3	L3×2½×¼	3.0	6.8	1.12	2.42	5.25	2.88
	L3×3 ×¼	3.4	7.1	1.16	2.37	5.59	2.84
	L4×3 ×¼	4.3	10.0	1.48	3.15	7.10	3.74
	L5×3 ×⁵⁄₁₆	6.5	14.0	1.88	3.76	10.1	4.49
	L6×3½×⁵⁄₁₆	9.0	18.3	2.31	4.36	13.3	5.31
C12×20.7	L3×2½×¼	3.6	8.6	1.10	2.47	6.47	3.01
	L3×3 ×¼	4.0	8.9	1.13	2.43	6.82	2.97
	L4×3 ×¼	4.7	12.2	1.40	3.25	8.37	3.89
	L5×3 ×⁵⁄₁₆	6.9	17.0	1.78	3.92	11.5	4.67
	L6×3½×⁵⁄₁₆	9.3	22.4	2 19	4.59	14.8	5.52
C15×33.9	L3×3 ×¼	5.6	13.5	1.10	2.48	9.54	3.12
	L4×3 ×¼	5.9	17.2	1.30	3.35	11.1	4.11
	L5×3 ×⁵⁄₁₆	7.8	23.4	1.61	4.12	14.5	5.02
	L6×3½×⁵⁄₁₆	10.2	30.7	1.97	4.88	18.0	5.90

STEEL PIPE AND STRUCTURAL TUBING

General

When designing and specifying steel pipe or tubing as compression members, refer to comments in the notes for Columns, Steel Pipe, and Structural Tubing, in Part 3. For standard mill practices and tolerances, refer to page 1-183. For material specifications and availability, see Tables 1-4 through 1-6, Part 1.

Steel Pipe

The Tables of Dimensions and Properties of Steel Pipe (unfilled) list a selected range of sizes of standard, extra strong, and double-extra strong pipe. For a complete range of sizes manufactured, refer to catalogs of the manufacturers or to the American Institute for Hollow Structural Sections (AIHSS).

Structural Tubing

The Tables of Dimensions and Properties of Square and Rectangular Structural Tubing (unfilled) list a selected range of frequently used sizes. For dimensions and properties of other sizes, refer to catalogs from the manufacturers or AIHSS.

The tables are based on an outside corner radius equal to two times the specified wall thickness. Material specifications stipulate that the outside corner radius may vary up to three times the specified wall thickness. This variation should be considered in those details where a close match or fit is important.

PIPE
Dimensions and properties

Dimensions				Properties						
Nominal Diameter	Outside Diameter	Inside Diameter	Wall Thickness	Weight per ft lbs Plain Ends	Area	I	S	r	J	Z
in.	in.	in.	in.		in.^{2}	in.^{4}	in.^{3}	in.	in.^{4}	in.^{3}
Standard Weight										
$\frac{1}{2}$	0.840	0.622	0.109	0.85	0.250	0.017	0.041	0.261	0.034	0.059
$\frac{3}{4}$	1.050	0.824	0.113	1.13	0.333	0.037	0.071	0.334	0.074	0.100
1	1.315	1.049	0.133	1.68	0.494	0.087	0.133	0.421	0.175	0.187
$1\frac{1}{4}$	1.660	1.380	0.140	2.27	0.669	0.195	0.235	0.540	0.389	0.324
$1\frac{1}{2}$	1.900	1.610	0.145	2.72	0.799	0.310	0.326	0.623	0.620	0.448
2	2.375	2.067	0.154	3.65	1.07	0.666	0.561	0.787	1.33	0.761
$2\frac{1}{2}$	2.875	2.469	0.203	5.79	1.70	1.53	1.06	0.947	3.06	1.45
3	3.500	3.068	0.216	7.58	2.23	3.02	1.72	1.16	6.03	2.33
$3\frac{1}{2}$	4.000	3.548	0.226	9.11	2.68	4.79	2.39	1.34	9.58	3.22
4	4.500	4.026	0.237	10.79	3.17	7.23	3.21	1.51	14.5	4.31
5	5.563	5.047	0.258	14.62	4.30	15.2	5.45	1.88	30.3	7.27
6	6.625	6.065	0.280	18.97	5.58	28.1	8.50	2.25	56.3	11.2
8	8.625	7.981	0.322	28.55	8.40	72.5	16.8	2.94	145	22.2
10	10.750	10.020	0.365	40.48	11.9	161	29.9	3.67	321	39.4
12	12.750	12.000	0.375	49.56	14.6	279	43.8	4.38	559	57.4
Extra Strong										
$\frac{1}{2}$	0.840	0.546	0.147	1.09	0.320	0.020	0.048	0.250	0.040	0.072
$\frac{3}{4}$	1.050	0.742	0.154	1.47	0.433	0.045	0.085	0.321	0.090	0.125
1	1.315	0.957	0.179	2.17	0.639	0.106	0.161	0.407	0.211	0.233
$1\frac{1}{4}$	1.660	1.278	0.191	3.00	0.881	0.242	0.291	0.524	0.484	0.414
$1\frac{1}{2}$	1.900	1.500	0.200	3.63	1.07	0.391	0.412	0.605	0.782	0.581
2	2.375	1.939	0.218	5.02	1.48	0.868	0.731	0.766	1.74	1.02
$2\frac{1}{2}$	2.875	2.323	0.276	7.66	2.25	1.92	1.34	0.924	3.85	1.87
3	3.500	2.900	0.300	10.25	3.02	3.89	2.23	1.14	8.13	3.08
$3\frac{1}{2}$	4.000	3.364	0.318	12.50	3.68	6.28	3.14	1.31	12.6	4.32
4	4.500	3.826	0.337	14.98	4.41	9.61	4.27	1.48	19.2	5.85
5	5.563	4.813	0.375	20.78	6.11	20.7	7.43	1.84	41.3	10.1
6	6.625	5.761	0.432	28.57	8.40	40.5	12.2	2.19	81.0	16.6
8	8.625	7.625	0.500	43.39	12.8	106	24.5	2.88	211	33.0
10	10.750	9.750	0.500	54.74	16.1	212	39.4	3.63	424	52.6
12	12.750	11.750	0.500	65.42	19.2	362	56.7	4.33	723	75.1
Double-Extra Strong										
2	2.375	1.503	0.436	9.03	2.66	1.31	1.10	0.703	2.62	1.67
$2\frac{1}{2}$	2.875	1.771	0.552	13.69	4.03	2.87	2.00	0.844	5.74	3.04
3	3.500	2.300	0.600	18.58	5.47	5.99	3.42	1.05	12.0	5.12
4	4.500	3.152	0.674	27.54	8.10	15.3	6.79	1.37	30.6	9.97
5	5.563	4.063	0.750	38.59	11.3	33.6	12.1	1.72	67.3	17.5
6	6.625	4.897	0.864	53.16	15.6	66.3	20.0	2.06	133	28.9
8	8.625	6.875	0.875	72.42	21.3	162	37.6	2.76	324	52.8

The listed sections are available in conformance with ASTM Specification A53 Grade B or A501. Other sections are made to these specifications. Consult with pipe manufacturers or distributors for availability.

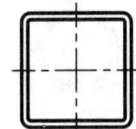

STRUCTURAL TUBING
Square
Dimensions and properties

Dimensions				Properties**					
Nominal* Size	Wall Thickness		Weight per ft	Area	I	S	r	J	Z
in.	in.		lb	in.^{2}	in.^{4}	in.^{3}	in.	in.^{4}	in.^{3}
30×30	0.6250	⅝	246.47	72.4	10300	690	12.0	16000	794
28×28	0.6250	⅝	229.45	67.4	8360	597	11.1	13000	689
26×26	0.6250	⅝	212.44	62.4	6650	511	10.3	10400	591
24×24	0.6250	⅝	195.43	57.4	5180	432	9.50	8100	500
	0.5000	½	157.74	46.4	4240	353	9.56	6570	407
	0.3750	⅜	119.35	35.1	3250	270	9.62	4990	310
22×22	0.6250	⅝	178.41	52.4	3950	359	8.68	6200	418
	0.5000	½	144.13	42.4	3240	294	8.74	5030	340
	0.3750	⅜	109.15	32.1	2490	226	8.80	3830	259
20×20	0.6250	⅝	161.40	47.4	2940	294	7.87	4620	342
	0.5000	½	130.52	38.4	2410	241	7.93	3760	279
	0.3750	⅜	98.94	29.1	1850	185	7.99	2870	213
18×18	0.6250	⅝	144.39	42.4	2110	234	7.05	3340	274
	0.5000	½	116.91	34.4	1740	193	7.11	2720	224
	0.3750	⅜	88.73	26.1	1340	149	7.17	2080	172

*Outside dimensions across flat sides.
**Properties are based upon a nominal outside corner radius equal to two times the wall thickness.

STRUCTURAL TUBING
Square
Dimensions and properties

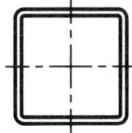

Dimensions			Properties**					
Nominal* **Size**	**Wall** **Thickness**	**Weight** **per ft**	**Area**	**I**	**S**	**r**	**J**	**Z**
in.	in.	lb	in.^{2}	in.^{4}	in.^{3}	in.	in.^{4}	in.^{3}
16×16	0.6250 ⅝	127.37	37.4	1450	182	6.23	2320	214
	0.5000 ½	103.30	30.4	1200	150	6.29	1890	175
	0.3750 ⅜	78.52	23.1	931	116	6.35	1450	134
	0.3125 ⁵⁄₁₆	65.87	19.4	789	98.6	6.38	1220	113
14×14	0.6250 ⅝	110.36	32.4	952	136	5.42	1530	161
	0.5000 ½	89.68	26.4	791	113	5.48	1250	132
	0.3750 ⅜	68.31	20.1	615	87.9	5.54	963	102
	0.3125 ⁵⁄₁₆	57.36	16.9	522	74.6	5.57	812	86.1
12×12	0.6250 ⅝	93.34	27.4	580	96.7	4.60	943	116
	0.5000 ½	76.07	22.4	485	80.9	4.66	777	95.4
	0.3750 ⅜	58.10	17.1	380	63.4	4.72	599	73.9
	0.3125 ⁵⁄₁₆	48.86	14.4	324	54.0	4.75	506	62.6
	0.2500 ¼	39.43	11.6	265	44.1	4.78	410	50.8
10×10	0.6250 ⅝	76.33	22.4	321	64.2	3.78	529	77.6
	0.5000 ½	62.46	18.4	271	54.2	3.84	439	64.6
	0.3750 ⅜	47.90	14.1	214	42.9	3.90	341	50.4
	0.3125 ⁵⁄₁₆	40.35	11.9	183	36.7	3.93	289	42.8
	0.2500 ¼	32.63	9.59	151	30.1	3.96	235	34.9
	0.1875 ³⁄₁₆	24.73	7.27	116	23.2	3.99	179	26.6

*Outside dimensions across flat sides.
**Properties are based upon a nominal outside corner radius equal to two times the wall thickness.

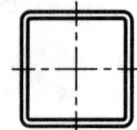

STRUCTURAL TUBING
Square
Dimensions and properties

Nominal* Size in.	Wall Thickness in.		Weight per ft lb	Area in.^{2}	I in.^{4}	S in.^{3}	r in.	J in.^{4}	Z in.^{3}
Dimensions				**Properties****					
8×8	0.6250	5/8	59.32	17.4	153	38.3	2.96	258	47.2
	0.5000	1/2	48.85	14.4	131	32.9	3.03	217	39.7
	0.3750	3/8	37.69	11.1	106	26.4	3.09	170	31.3
	0.3125	5/16	31.84	9.36	90.9	22.7	3.12	145	26.7
	0.2500	1/4	25.82	7.59	75.1	18.8	3.15	118	21.9
	0.1875	3/16	19.63	5.77	58.2	14.6	3.18	90.6	16.8
7×7	0.6250	5/8	50.81	14.9	97.5	27.9	2.56	166	34.8
	0.5000	1/2	42.05	12.4	84.6	24.2	2.62	141	29.6
	0.3750	3/8	32.58	9.58	68.7	19.6	2.68	112	23.5
	0.3125	5/16	27.59	8.11	59.5	17.0	2.71	95.6	20.1
	0.2500	1/4	22.42	6.59	49.4	14.1	2.74	78.3	16.5
	0.1875	3/16	17.08	5.02	38.5	11.0	2.77	60.2	12.7
6×6	0.6250	5/8	42.30	12.4	57.3	19.1	2.15	99.5	24.3
	0.5000	1/2	35.24	10.4	50.5	16.8	2.21	85.6	20.9
	0.3750	3/8	27.48	8.08	41.6	13.9	2.27	68.5	16.8
	0.3125	5/16	23.34	6.86	36.3	12.1	2.30	58.9	14.4
	0.2500	1/4	19.02	5.59	30.3	10.1	2.33	48.5	11.9
	0.1875	3/16	14.53	4.27	23.8	7.93	2.36	37.5	9.24
	0.1250	1/8	9.86	2.90	16.5	5.52	2.39	25.7	6.35
5½×5½	0.3750	3/8	24.93	7.33	31.2	11.4	2.07	51.9	13.8
	0.3125	5/16	21.21	6.23	27.4	9.95	2.10	44.8	12.0
	0.2500	1/4	17.32	5.09	23.0	8.36	2.13	37.0	9.91
	0.1875	3/16	13.25	3.89	18.1	6.58	2.16	28.6	7.70
	0.1250	1/8	9.01	2.65	12.6	4.60	2.19	19.7	5.31
5×5	0.5000	1/2	28.43	8.36	27.0	10.8	1.80	46.8	13.7
	0.3750	3/8	22.37	6.58	22.8	9.11	1.86	38.2	11.2
	0.3125	5/16	19.08	5.61	20.1	8.02	1.89	33.1	9.70
	0.2500	1/4	15.62	4.59	16.9	6.78	1.92	27.4	8.07
	0.1875	3/16	11.97	3.52	13.4	5.36	1.95	21.3	6.29
	0.1250	1/8	8.16	2.40	9.41	3.77	1.98	14.7	4.36

*Outside dimensions across flat sides.
**Properties are based upon a nominal outside corner radius equal to two times the wall thickness.

STRUCTURAL TUBING
Square
Dimensions and properties

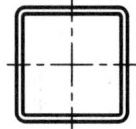

Nominal* Size in.	Wall Thickness in.		Weight per ft lb	Area in.^{2}	I in.^{4}	S in.^{3}	r in.	J in.^{4}	Z in.^{3}
Dimensions				**Properties**					
4½×4½	0.3750	⅜	19.82	5.83	16.0	7.10	1.66	27.1	8.81
	0.3125	⁵⁄₁₆	16.96	4.98	14.2	6.30	1.69	23.6	7.68
	0.2500	¼	13.91	4.09	12.1	5.36	1.72	19.7	6.43
	0.1875	³⁄₁₆	10.70	3.14	9.60	4.27	1.75	15.4	5.03
	0.1250	⅛	7.31	2.15	6.78	3.02	1.78	10.6	3.50
4×4	0.5000	½	21.63	6.36	12.3	6.13	1.39	21.8	8.02
	0.3750	⅜	17.27	5.08	10.7	5.35	1.45	18.4	6.72
	0.3125	⁵⁄₁₆	14.83	4.36	9.58	4.79	1.48	16.1	5.90
	0.2500	¼	12.21	3.59	8.22	4.11	1.51	13.5	4.97
	0.1875	³⁄₁₆	9.42	2.77	6.59	3.30	1.54	10.6	3.91
	0.1250	⅛	6.46	1.90	4.70	2.35	1.57	7.40	2.74
3½×3½	0.3125	⁵⁄₁₆	12.70	3.73	6.09	3.48	1.28	10.4	4.35
	0.2500	¼	10.51	3.09	5.29	3.02	1.31	8.82	3.69
	0.1875	³⁄₁₆	8.15	2.39	4.29	2.45	1.34	6.99	2.93
	0.1250	⅛	5.61	1.65	3.09	1.76	1.37	4.90	2.07
3×3	0.3125	⁵⁄₁₆	10.58	3.11	3.58	2.39	1.07	6.22	3.04
	0.2500	¼	8.81	2.59	3.16	2.10	1.10	5.35	2.61
	0.1875	³⁄₁₆	6.87	2.02	2.60	1.73	1.13	4.28	2.10
	0.1250	⅛	4.75	1.40	1.90	1.26	1.16	3.03	1.49
2½×2½	0.3125	⁵⁄₁₆	8.45	2.48	1.87	1.50	0.868	3.32	1.96
	0.2500	¼	7.11	2.09	1.69	1.35	0.899	2.92	1.71
	0.1875	³⁄₁₆	5.59	1.64	1.42	1.14	0.930	2.38	1.40
	0.1250	⅛	3.90	1.15	1.06	0.847	0.961	1.71	1.01
2×2	0.3125	⁵⁄₁₆	6.32	1.86	0.815	0.815	0.662	1.49	1.11
	0.2500	¼	5.41	1.59	0.766	0.766	0.694	1.36	1.00
	0.1875	³⁄₁₆	4.32	1.27	0.668	0.668	0.726	1.15	0.840
	0.1250	⅛	3.05	0.897	0.513	0.513	0.756	0.846	0.621
1½×1½	0.1875	³⁄₁₆	3.04	0.894	0.242	0.323	0.521	0.431	0.423

*Outside dimensions across flat sides.
**Properties are based upon a nominal outside corner radius equal to two times the wall thickness.

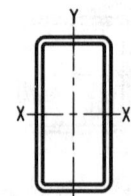

STRUCTURAL TUBING
Rectangular
Dimensions and properties

Dimensions			Properties**									
				X-X Axis				Y-Y Axis				
Nominal* Size	Wall Thickness	Weight per ft	Area	I	S	Z	r	I	S	Z	r	J
in.	in.	lb	in.^{2}	in.^{4}	in.^{3}	in.^{3}	in.	in.^{4}	in.^{3}	in.^{3}	in.	in.^{4}
30×24	0.5000 ½	178.16	52.4	7110	474	555	11.7	5070	422	477	9.84	9170
	0.3750 ⅜	134.67	39.6	5430	362	422	11.7	3870	323	363	9.89	6960
	0.3125 ⁵⁄₁₆	112.66	33.1	4570	305	354	11.7	3260	272	304	9.92	5830
28×24	0.5000 ½	171.35	50.4	6050	432	503	11.0	4790	399	454	9.75	8280
	0.3750 ⅜	129.56	38.1	4630	331	383	11.0	3660	305	345	9.81	6290
	0.3125 ⁵⁄₁₆	108.41	31.9	3890	278	321	11.1	3080	257	290	9.84	5270
26×24	0.5000 ½	164.55	48.4	5100	392	454	10.3	4510	376	430	9.66	7410
	0.3750 ⅜	124.46	36.6	3900	300	345	10.3	3460	288	327	9.72	5630
	0.3125 ⁵⁄₁₆	104.15	30.6	3280	253	290	10.4	2910	242	275	9.75	4720
24×22	0.5000 ½	150.93	44.4	3960	330	383	9.45	3470	315	361	8.84	5740
	0.3750 ⅜	114.25	33.6	3040	253	292	9.51	2660	242	275	8.90	4370
	0.3125 ⁵⁄₁₆	95.64	28.1	2560	213	245	9.54	2240	204	231	8.93	3660
22×20	0.5000 ½	137.32	40.4	3010	273	318	8.63	2600	260	298	8.03	4350
	0.3750 ⅜	104.04	30.6	2310	210	243	8.69	2000	200	228	8.09	3310
	0.3125 ⁵⁄₁₆	87.14	25.6	1950	177	204	8.72	1690	169	192	8.12	2780
20×18	0.5000 ½	123.71	36.4	2220	222	259	7.81	1890	210	242	7.21	3190
	0.3750 ⅜	93.83	27.6	1710	171	198	7.88	1460	162	185	7.27	2440
	0.3125 ⁵⁄₁₆	78.63	23.1	1440	144	167	7.91	1230	137	155	7.30	2050
20×12	0.5000 ½	103.30	30.4	1650	165	201	7.37	750	125	141	4.97	1650
	0.3750 ⅜	78.52	23.1	1280	128	154	7.45	583	97.2	109	5.03	1270
	0.3125 ⁵⁄₁₆	65.87	19.4	1080	108	130	7.47	495	82.5	91.8	5.06	1070
20×8	0.5000 ½	89.68	26.4	1270	127	162	6.94	300	75.1	84.7	3.38	806
	0.3750 ⅜	68.31	20.1	988	98.8	125	7.02	236	59.1	65.6	3.43	625
	0.3125 ⁵⁄₁₆	57.36	16.9	838	83.8	105	7.05	202	50.4	55.6	3.46	529
20×4	0.5000 ½	76.07	22.4	889	88.9	123	6.31	61.6	30.8	36.0	1.66	205
	0.3750 ⅜	58.10	17.1	699	69.9	95.3	6.40	50.3	25.1	28.5	1.72	165
	0.3125 ⁵⁄₁₆	48.86	14.4	596	59.6	80.8	6.44	43.7	21.8	24.3	1.74	143

*Outside dimensions across flat sides.
**Properties are based upon a nominal outside corner radius equal to two times the wall thickness.

STRUCTURAL TUBING
Rectangular
Dimensions and properties

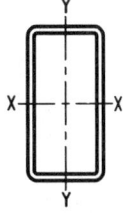

Dimensions				Properties**									
				X-X Axis				Y-Y Axis					
Nominal* Size	Wall Thickness		Weight per ft	Area	I	S	Z	r	I	S	Z	r	J
in.	in.		lb	in.^{2}	in.^{4}	in.^{3}	in.^{3}	in.	in.^{4}	in.^{3}	in.^{3}	in.	in.^{4}
18×12	0.5000	½	96.49	28.4	1280	142	172	6.71	684	114	130	4.91	1420
	0.3750	⅜	73.42	21.6	991	110	132	6.78	533	88.8	100	4.97	1090
	0.3125	⁵⁄₁₆	61.62	18.1	840	93.3	111	6.81	452	75.3	84.5	5.00	920
18×6	0.5000	½	76.07	22.4	818	90.9	119	6.05	141	47.2	53.9	2.52	410
	0.3750	⅜	58.10	17.1	641	71.3	92.2	6.13	113	37.6	42.1	2.57	322
	0.3125	⁵⁄₁₆	48.86	14.4	546	60.7	78.1	6.17	97.0	32.3	35.8	2.60	274
	0.2500	¼	39.43	11.6	447	49.6	63.5	6.21	80.0	26.7	29.2	2.63	224
16×12	0.5000	½	89.68	26.4	962	120	144	6.04	618	103	118	4.84	1200
	0.3750	⅜	68.31	20.1	748	93.5	111	6.11	482	80.3	91.3	4.90	922
	0.3125	⁵⁄₁₆	57.36	16.9	635	79.4	93.8	6.14	409	68.2	77.2	4.93	777
16×8	0.5000	½	76.07	22.4	722	90.2	113	5.68	244	61.0	69.7	3.30	599
	0.3750	⅜	58.10	17.1	565	70.6	87.6	5.75	193	48.2	54.2	3.36	465
	0.3125	⁵⁄₁₆	48.86	14.4	481	60.1	74.2	5.79	165	41.2	45.9	3.39	394
16×4	0.5000	½	62.46	18.4	481	60.2	82.2	5.12	49.3	24.6	29.0	1.64	157
	0.3750	⅜	47.90	14.1	382	47.8	64.2	5.21	40.4	20.2	23.0	1.69	127
	0.3125	⁵⁄₁₆	40.35	11.9	327	40.9	54.5	5.25	35.1	17.6	19.7	1.72	110
14×12	0.5000	½	82.88	24.4	699	99.9	119	5.36	552	91.9	107	4.76	983
	0.3750	⅜	63.21	18.6	546	78.0	91.7	5.42	431	71.9	82.6	4.82	757
14×10	0.5000	½	76.07	22.4	608	86.9	105	5.22	361	72.3	83.6	4.02	730
	0.3750	⅜	58.10	17.1	476	68.0	81.5	5.28	284	56.8	64.8	4.08	564
	0.3125	⁵⁄₁₆	48.86	14.4	405	57.9	69.0	5.31	242	48.4	54.9	4.11	477
14×6	0.6250	⅝	76.33	22.4	504	72.0	94.0	4.74	130	43.3	51.2	2.41	352
	0.5000	½	62.46	18.4	426	60.8	78.3	4.82	111	37.1	42.9	2.46	296
	0.3750	⅜	47.90	14.1	337	48.1	61.1	4.89	89.1	29.7	33.6	2.52	233
	0.3125	⁵⁄₁₆	40.35	11.9	288	41.2	51.9	4.93	76.7	25.6	28.7	2.54	199
	0.2500	¼	32.63	9.59	237	33.8	42.3	4.97	63.4	21.1	23.4	2.57	162

*Outside dimensions across flat sides.
**Properties are based upon a nominal outside corner radius equal to two times the wall thickness.

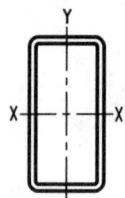

STRUCTURAL TUBING
Rectangular
Dimensions and properties

Nominal* Size	Wall Thickness		Weight per ft	Area	X-X Axis				Y-Y Axis				J
					I	S	Z	r	I	S	Z	r	
in.	in.		lb	in.^{2}	in.^{4}	in.^{3}	in.^{3}	in.	in.^{4}	in.^{3}	in.^{3}	in.	in.^{4}
14×4	0.6250	5/8	67.82	19.9	392	56.0	77.3	4.44	49.0	24.5	30.0	1.57	154
	0.5000	1/2	55.66	16.4	335	47.8	64.8	4.52	43.1	21.5	25.5	1.62	134
	0.3750	3/8	42.79	12.6	267	38.2	50.8	4.61	35.4	17.7	20.3	1.68	108
	0.3125	5/16	36.10	10.6	230	32.8	43.3	4.65	30.9	15.4	17.4	1.71	93.1
	0.2500	1/4	29.23	8.59	189	27.0	35.4	4.69	25.8	12.9	14.3	1.73	77.0
	0.1875	3/16	22.18	6.52	146	20.9	27.1	4.74	20.2	10.1	11.1	1.76	59.7
12×10	0.5000	1/2	69.27	20.4	419	69.9	83.9	4.54	316	63.3	74.1	3.94	581
	0.3750	3/8	53.00	15.6	330	55.0	65.2	4.60	249	49.8	57.6	4.00	450
	0.3125	5/16	44.60	13.1	281	46.9	55.2	4.63	213	42.6	48.8	4.03	381
	0.2500	1/4	36.03	10.6	230	38.4	44.9	4.66	174	34.9	39.7	4.06	309
12×8	0.6250	5/8	76.33	22.4	418	69.7	87.1	4.32	221	55.3	65.6	3.14	481
	0.5000	1/2	62.46	18.4	353	58.9	72.4	4.39	188	46.9	54.7	3.20	401
	0.3750	3/8	47.90	14.1	279	46.5	56.5	4.45	149	37.3	42.7	3.26	312
	0.3125	5/16	40.35	11.9	239	39.8	47.9	4.49	128	32.0	36.3	3.28	265
	0.2500	1/4	32.63	9.59	196	32.6	39.1	4.52	105	26.3	29.6	3.31	216
	0.1875	3/16	24.73	7.27	151	25.1	29.8	4.55	81.1	20.3	22.7	3.34	165
12×6	0.6250	5/8	67.82	19.9	337	56.2	72.9	4.11	112	37.2	44.5	2.37	286
	0.5000	1/2	55.66	16.4	287	47.8	60.9	4.19	96.0	32.0	37.4	2.42	241
	0.3750	3/8	42.79	12.6	228	38.1	47.7	4.26	77.2	25.7	29.4	2.48	190
	0.3125	5/16	36.10	10.6	196	32.6	40.6	4.30	66.6	22.2	25.1	2.51	162
	0.2500	1/4	29.23	8.59	161	26.9	33.2	4.33	55.2	18.4	20.6	2.53	132
	0.1875	3/16	22.18	6.52	124	20.7	25.4	4.37	42.8	14.3	15.8	2.56	101
12×4	0.6250	5/8	59.32	17.4	257	42.8	58.6	3.84	41.8	20.9	25.8	1.55	127
	0.5000	1/2	48.85	14.4	221	36.8	49.4	3.92	36.9	18.5	22.0	1.60	110
	0.3750	3/8	37.69	11.1	178	29.6	39.0	4.01	30.5	15.2	17.6	1.66	89.0
	0.3125	5/16	31.84	9.36	153	25.5	33.3	4.05	26.6	13.3	15.1	1.69	76.9
	0.2500	1/4	25.82	7.59	127	21.1	27.3	4.09	22.3	11.1	12.5	1.71	63.6
	0.1875	3/16	19.63	5.77	98.2	16.4	21.0	4.13	17.5	8.75	9.63	1.74	49.3
12×3	0.3125	5/16	29.72	8.73	132	22.0	29.7	3.89	13.8	9.19	10.6	1.26	43.6
	0.2500	1/4	24.12	7.09	109	18.2	24.4	3.93	11.7	7.79	8.80	1.28	36.5
	0.1875	3/16	18.35	5.39	85.1	14.2	18.8	3.97	9.28	6.19	6.84	1.31	28.7

*Outside dimensions across flat sides.
**Properties are based upon a nominal outside corner radius equal to two times the wall thickness.

STRUCTURAL TUBING
Rectangular
Dimensions and properties

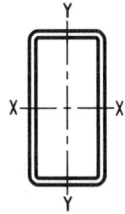

Dimensions			Properties**									
				X-X Axis				Y-Y Axis				
Nominal* Size	Wall Thickness	Weight per ft	Area	I	S	Z	r	I	S	Z	r	J
in.	in.	lb	in.^{2}	in.^{4}	in.^{3}	in.^{3}	in.	in.^{4}	in.^{3}	in.^{3}	in.	in.^{4}
12×2	0.2500 ¼	22.42	6.59	92.2	15.4	21.4	3.74	4.62	4.62	5.38	0.837	15.9
	0.1875 ³⁄₁₆	17.08	5.02	72.0	12.0	16.6	3.79	3.76	3.76	4.24	0.865	12.8
10×8	0.5000 ½	55.66	16.4	226	45.2	55.1	3.72	160	39.9	47.2	3.12	306
	0.3750 ⅜	42.79	12.6	180	35.9	43.1	3.78	127	31.8	37.0	3.18	239
	0.3125 ⁵⁄₁₆	36.10	10.6	154	30.8	36.7	3.81	109	27.3	31.5	3.21	203
	0.2500 ¼	29.23	8.59	127	25.4	30.0	3.84	90.2	22.5	25.8	3.24	166
	0.1875 ³⁄₁₆	22.18	6.52	97.9	19.6	23.0	3.87	69.7	17.4	19.7	3.27	127
10×6	0.5000 ½	48.85	14.4	181	36.2	45.6	3.55	80.8	26.9	31.9	2.37	187
	0.3750 ⅜	37.69	11.1	145	29.0	35.9	3.62	65.4	21.8	25.2	2.43	147
	0.3125 ⁵⁄₁₆	31.84	9.36	125	25.0	30.7	3.65	56.5	18.8	21.5	2.46	126
	0.2500 ¼	25.82	7.59	103	20.6	25.1	3.69	46.9	15.6	17.7	2.49	103
	0.1875 ³⁄₁₆	19.63	5.77	79.8	16.0	19.3	3.72	36.5	12.2	13.6	2.51	79.1
10×5	0.3750 ⅜	35.13	10.3	128	25.5	32.3	3.51	42.9	17.1	19.9	2.04	107
	0.3125 ⁵⁄₁₆	29.72	8.73	110	22.0	27.6	3.55	37.2	14.9	17.0	2.07	91.5
	0.2500 ¼	24.12	7.09	91.2	18.2	22.7	3.59	31.1	12.4	14.0	2.09	75.2
	0.1875 ³⁄₁₆	18.35	5.39	70.8	14.2	17.4	3.62	24.3	9.71	10.8	2.12	58.0
10×4	0.5000 ½	42.05	12.4	136	27.1	36.1	3.31	30.8	15.4	18.5	1.58	86.9
	0.3750 ⅜	32.58	9.58	110	22.0	28.7	3.39	25.5	12.8	14.9	1.63	70.4
	0.3125 ⁵⁄₁₆	27.59	8.11	95.5	19.1	24.6	3.43	22.4	11.2	12.8	1.66	60.8
	0.2500 ¼	22.42	6.59	79.3	15.9	20.2	3.47	18.8	9.39	10.6	1.69	50.4
	0.1875 ³⁄₁₆	17.08	5.02	61.7	12.3	15.6	3.51	14.8	7.39	8.20	1.72	39.1
10×3	0.3750 ⅜	30.03	8.83	92.8	18.6	25.1	3.24	13.0	8.66	10.3	1.21	39.8
	0.3125 ⁵⁄₁₆	25.46	7.48	80.8	16.2	21.6	3.29	11.5	7.68	8.92	1.24	34.9
	0.2500 ¼	20.72	6.09	67.4	13.5	17.8	3.33	9.79	6.53	7.42	1.27	29.3
	0.1875 ³⁄₁₆	15.80	4.64	52.7	10.5	13.8	3.37	7.80	5.20	5.79	1.30	23.0
10×2	0.3750 ⅜	27.48	8.08	75.4	15.1	21.5	3.06	4.85	4.85	6.05	0.775	16.5
	0.3125 ⁵⁄₁₆	23.34	6.86	66.1	13.2	18.5	3.10	4.42	4.42	5.33	0.802	14.9
	0.2500 ¼	19.02	5.59	55.5	11.1	15.4	3.15	3.85	3.85	4.50	0.830	12.8
	0.1875 ³⁄₁₆	14.53	4.27	43.7	8.74	11.9	3.20	3.14	3.14	3.56	0.858	10.3

*Outside dimensions across flat sides.
**Properties are based upon a nominal outside corner radius equal to two times the wall thickness.

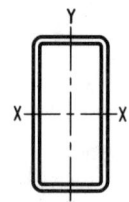

STRUCTURAL TUBING
Rectangular
Dimensions and properties

Nominal* Size in.	Wall Thickness in.		Weight per ft lb	Area in.^{2}	X-X Axis				Y-Y Axis				J in.^{4}
					I in.^{4}	S in.^{3}	Z in.^{3}	r in.	I in.^{4}	S in.^{3}	Z in.^{3}	r in.	
8×6	0.5000	½	42.05	12.4	103	25.8	32.2	2.89	65.7	21.9	26.4	2.31	135
	0.3750	⅜	32.58	9.58	83.7	20.9	25.6	2.96	53.5	17.8	21.0	2.36	107
	0.3125	⁵⁄₁₆	27.59	8.11	72.4	18.1	21.9	2.99	46.4	15.5	18.0	2.39	91.3
	0.2500	¼	22.42	6.59	60.1	15.0	18.0	3.02	38.6	12.9	14.8	2.42	74.9
	0.1875	³⁄₁₆	17.08	5.02	46.8	11.7	13.9	3.05	30.1	10.0	11.4	2.45	57.6
8×4	0.6250	⅝	42.30	12.4	85.1	21.3	28.8	2.62	27.4	13.7	17.3	1.49	73.2
	0.5000	½	35.24	10.4	75.1	18.8	24.7	2.69	24.6	12.3	15.0	1.54	64.1
	0.3750	⅜	27.48	8.08	61.9	15.5	19.9	2.77	20.6	10.3	12.2	1.60	52.2
	0.3125	⁵⁄₁₆	23.34	6.86	53.9	13.5	17.1	2.80	18.1	9.05	10.5	1.62	45.2
	0.2500	¼	19.02	5.59	45.1	11.3	14.1	2.84	15.3	7.63	8.72	1.65	37.5
	0.1875	³⁄₁₆	14.53	4.27	35.3	8.83	11.0	2.88	12.0	6.02	6.77	1.68	29.1
	0.1250	⅛	9.86	2.90	24.6	6.14	7.53	2.91	8.45	4.23	4.67	1.71	20.0
8×3	0.5000	½	31.84	9.36	61.0	15.3	21.0	2.55	12.1	8.05	10.1	1.14	35.7
	0.3750	⅜	24.93	7.33	51.0	12.7	17.0	2.64	10.4	6.92	8.31	1.19	29.9
	0.3125	⁵⁄₁₆	21.21	6.23	44.7	11.2	14.7	2.68	9.25	6.16	7.24	1.22	26.3
	0.2500	¼	17.32	5.09	37.6	9.40	12.2	2.72	7.90	5.26	6.05	1.25	22.1
	0.1875	³⁄₁₆	13.25	3.89	29.6	7.40	9.49	2.76	6.31	4.21	4.73	1.27	17.3
	0.1250	⅛	9.01	2.65	20.7	5.17	6.55	2.80	4.48	2.99	3.29	1.30	12.1
8×2	0.3750	⅜	22.37	6.58	40.1	10.0	14.2	2.47	3.85	3.85	4.83	0.765	12.6
	0.3125	⁵⁄₁₆	19.08	5.61	35.5	8.87	12.3	2.51	3.52	3.52	4.28	0.792	11.4
	0.2500	¼	15.62	4.59	30.1	7.52	10.3	2.56	3.08	3.08	3.63	0.819	9.84
	0.1875	³⁄₁₆	11.97	3.52	23.9	5.97	8.02	2.60	2.52	2.52	2.88	0.847	7.94
	0.1250	⅛	8.16	2.40	16.8	4.20	5.56	2.65	1.83	1.83	2.03	0.875	5.66
7×5	0.5000	½	35.24	10.4	63.5	18.1	23.1	2.48	37.2	14.9	18.2	1.90	79.9
	0.3750	⅜	27.48	8.08	52.2	14.9	18.5	2.54	30.8	12.3	14.6	1.95	64.2
	0.3125	⁵⁄₁₆	23.34	6.86	45.5	13.0	15.9	2.58	26.9	10.8	12.6	1.98	55.3
	0.2500	¼	19.02	5.59	38.0	10.9	13.2	2.61	22.6	9.04	10.4	2.01	45.6
	0.1875	³⁄₁₆	14.53	4.27	29.8	8.50	10.2	2.64	17.7	7.10	8.10	2.04	35.3
	0.1250	⅛	9.86	2.90	20.7	5.91	7.00	2.67	12.4	4.95	5.58	2.07	24.2
7×4	0.3750	⅜	24.93	7.33	44.0	12.6	16.0	2.45	18.1	9.06	10.8	1.57	43.3
	0.3125	⁵⁄₁₆	21.21	6.23	38.5	11.0	13.8	2.49	16.0	7.98	9.36	1.60	37.5
	0.2500	¼	17.32	5.09	32.3	9.23	11.5	2.52	13.5	6.75	7.78	1.63	31.2
	0.1875	³⁄₁₆	13.25	3.89	25.4	7.26	8.91	2.55	10.7	5.34	6.06	1.66	24.2
	0.1250	⅛	9.01	2.65	17.7	5.07	6.15	2.59	7.51	3.76	4.19	1.68	16.7

*Outside dimensions across flat sides.
**Properties are based upon a nominal outside corner radius equal to two times the wall thickness.

STRUCTURAL TUBING
Rectangular
Dimensions and properties

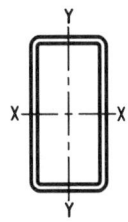

Dimensions			Properties**									
				X-X Axis				Y-Y Axis				
Nominal* Size	Wall Thickness	Weight per ft	Area	I	S	Z	r	I	S	Z	r	J
in.	in.	lb	in.^{2}	in.^{4}	in.^{3}	in.^{3}	in.	in.^{4}	in.^{3}	in.^{3}	in.	in.^{4}
7×3	0.3750 ³⁄₈	22.37	6.58	35.7	10.2	13.5	2.33	9.08	6.05	7.32	1.18	25.1
	0.3125 ⁵⁄₁₆	19.08	5.61	31.5	9.00	11.8	2.37	8.11	5.41	6.40	1.20	22.0
	0.2500 ¼	15.62	4.59	26.6	7.61	9.79	2.41	6.95	4.63	5.36	1.23	18.5
	0.1875 ³⁄₁₆	11.97	3.52	21.1	6.02	7.63	2.45	5.57	3.71	4.20	1.26	14.6
	0.1250 ⅛	8.16	2.40	14.8	4.22	5.29	2.48	3.96	2.64	2.93	1.29	10.2
6×4	0.5000 ½	28.43	8.36	35.3	11.8	15.4	2.06	18.4	9.21	11.5	1.48	42.1
	0.3750 ³⁄₈	22.37	6.58	29.7	9.90	12.5	2.13	15.6	7.82	9.44	1.54	34.6
	0.3125 ⁵⁄₁₆	19.08	5.61	26.2	8.72	10.9	2.16	13.8	6.92	8.21	1.57	30.1
	0.2500 ¼	15.62	4.59	22.1	7.36	9.06	2.19	11.7	5.87	6.84	1.60	25.0
	0.1875 ³⁄₁₆	11.97	3.52	17.4	5.81	7.06	2.23	9.32	4.66	5.34	1.63	19.5
	0.1250 ⅛	8.16	2.40	12.2	4.08	4.88	2.26	6.57	3.29	3.71	1.66	13.5
6×3	0.5000 ½	25.03	7.36	27.7	9.25	12.6	1.94	8.91	5.94	7.59	1.10	23.9
	0.3750 ³⁄₈	19.82	5.83	23.8	7.92	10.4	2.02	7.78	5.19	6.34	1.16	20.3
	0.3125 ⁵⁄₁₆	16.96	4.98	21.1	7.03	9.11	2.06	6.98	4.65	5.56	1.18	17.9
	0.2500 ¼	13.91	4.09	17.9	5.98	7.62	2.09	6.00	4.00	4.67	1.21	15.1
	0.1875 ³⁄₁₆	10.70	3.14	14.3	4.76	5.97	2.13	4.83	3.22	3.68	1.24	11.9
	0.1250 ⅛	7.31	2.15	10.1	3.36	4.15	2.17	3.45	2.30	2.57	1.27	8.27
6×2	0.3750 ³⁄₈	17.27	5.08	17.8	5.94	8.33	1.87	2.84	2.84	3.61	0.748	8.72
	0.3125 ⁵⁄₁₆	14.83	4.36	16.0	5.34	7.33	1.92	2.62	2.62	3.22	0.775	7.94
	0.2500 ¼	12.21	3.59	13.8	4.60	6.18	1.96	2.31	2.31	2.75	0.802	6.88
	0.1875 ³⁄₁₆	9.42	2.77	11.1	3.70	4.88	2.00	1.90	1.90	2.20	0.829	5.56
	0.1250 ⅛	6.46	1.90	7.92	2.64	3.42	2.04	1.39	1.39	1.56	0.857	3.98
5×4	0.3750 ³⁄₈	19.82	5.83	18.7	7.50	9.44	1.79	13.2	6.58	8.08	1.50	26.3
	0.3125 ⁵⁄₁₆	16.96	4.98	16.6	6.65	8.24	1.83	11.7	5.85	7.05	1.53	22.9
	0.2500 ¼	13.91	4.09	14.1	5.65	6.89	1.86	9.98	4.99	5.90	1.56	19.1
	0.1875 ³⁄₁₆	10.70	3.14	11.2	4.49	5.39	1.89	7.96	3.98	4.63	1.59	14.9
5×3	0.5000 ½	21.63	6.36	16.9	6.75	9.20	1.63	7.33	4.88	6.34	1.07	18.2
	0.3750 ³⁄₈	17.27	5.08	14.7	5.89	7.71	1.70	6.48	4.32	5.35	1.13	15.6
	0.3125 ⁵⁄₁₆	14.83	4.36	13.2	5.27	6.77	1.74	5.85	3.90	4.72	1.16	13.8
	0.2500 ¼	12.21	3.59	11.3	4.52	5.70	1.77	5.05	3.37	3.98	1.19	11.7
	0.1875 ³⁄₁₆	9.42	2.77	9.06	3.62	4.49	1.81	4.08	2.72	3.15	1.21	9.21
	0.1250 ⅛	6.46	1.90	6.44	2.58	3.14	1.84	2.93	1.95	2.21	1.24	6.44

*Outside dimensions across flatsides.
**Properties are based upon a nominal outside corner radius equal to two times the wall thickness.

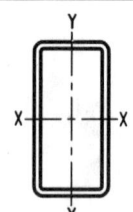

STRUCTURAL TUBING
Rectangular
Dimensions and properties

Dimensions			Properties**									
				X-X Axis				Y-Y Axis				
Nominal* Size	Wall Thickness	Weight per ft	Area	I	S	Z	r	I	S	Z	r	J
in.	in.	lb	in.^{2}	in.^{4}	in.^{3}	in.^{3}	in.	in.^{4}	in.^{3}	in.^{3}	in.	in.^{4}
5×2	0.3125 ⁵⁄₁₆	12.70	3.73	9.74	3.90	5.31	1.62	2.16	2.16	2.70	0.762	6.24
	0.2500 ¼	10.51	3.09	8.48	3.39	4.51	1.66	1.92	1.92	2.32	0.789	5.43
	0.1875 ³⁄₁₆	8.15	2.39	6.89	2.75	3.59	1.70	1.60	1.60	1.86	0.816	4.40
	0.1250 ⅛	5.61	1.65	4.96	1.98	2.53	1.73	1.17	1.17	1.32	0.844	3.15
4×3	0.3125 ⁵⁄₁₆	12.70	3.73	7.45	3.72	4.75	1.41	4.71	3.14	3.88	1.12	9.89
	0.2500 ¼	10.51	3.09	6.45	3.23	4.03	1.45	4.10	2.74	3.30	1.15	8.41
	0.1875 ³⁄₁₆	8.15	2.39	5.23	2.62	3.20	1.48	3.34	2.23	2.62	1.18	6.67
	0.1250 ⅛	5.61	1.65	3.76	1.88	2.25	1.51	2.41	1.61	1.85	1.21	4.68
4×2	0.3750 ⅜	12.17	3.58	5.75	2.87	4.00	1.27	1.83	1.83	2.39	0.715	4.97
	0.3125 ⁵⁄₁₆	10.58	3.11	5.32	2.66	3.60	1.31	1.71	1.71	2.17	0.743	4.58
	0.2500 ¼	8.81	2.59	4.69	2.35	3.09	1.35	1.54	1.54	1.88	0.770	4.01
	0.1875 ³⁄₁₆	6.87	2.02	3.87	1.93	2.48	1.38	1.29	1.29	1.52	0.798	3.26
	0.1250 ⅛	4.75	1.40	2.82	1.41	1.77	1.42	0.954	0.954	1.09	0.826	2.34
3×2	0.3125 ⁵⁄₁₆	8.45	2.48	2.44	1.63	2.20	0.992	1.26	1.26	1.64	0.714	2.97
	0.2500 ¼	7.11	2.09	2.21	1.47	1.92	1.03	1.15	1.15	1.44	0.742	2.63
	0.1875 ³⁄₁₆	5.59	1.64	1.86	1.24	1.57	1.06	0.977	0.977	1.18	0.771	2.16
	0.1250 ⅛	3.90	1.15	1.38	0.920	1.13	1.10	0.733	0.733	0.855	0.800	1.57
2½×1½	0.2500 ¼	5.41	1.59	1.05	0.844	1.15	0.815	0.458	0.610	0.793	0.537	1.14
	0.1875 ³⁄₁₆	4.32	1.27	0.920	0.736	0.964	0.852	0.405	0.540	0.669	0.565	0.976

*Outside dimensions across flat sides.
**Properties are based upon a nominal outside corner radius equal to two times the wall thickness.

BARS AND PLATES

Product Availability
Plates are readily available in seven of the structural steel specifications listed in Section A3.1 of the AISC LRFD Specification. These are: ASTM A36, A242, A529, A572, A588, A514, and A852. Bars are available in all of these steels except A514 and A852. Table 1-1 shows the availability of each steel in terms of plate thickness.

The Manual user is referred to the discussion on p. 1-5, Selection of the Appropriate Structural Steel, for guidance in selection of both plate and structural shapes.

Classification
Bars and plates are generally classified as follows:

Bars: 6 in. or less in width, .203 in. and over in thickness.
Over 6 in. to 8 in. in width, .230 in. and over in thickness.
Plates: Over 8 in. to 48 in. in width, .230 in. and over in thickness.
Over 48 in. in width, .180 in. and over in thickness.

Bars
Bars are available in various widths, thicknesses, diameters, and lengths. The preferred practice is to specify widths in ¼-in. increments and thickness and diameter in ⅛-in. increments.

Plates
Defined according to rolling procedure:

Sheared plates are rolled between horizontal rolls and trimmed (sheared or gas cut) on all edges.

Universal (UM) plates are rolled between horizontal and vertical rolls and trimmed (sheared or gas cut) on ends only.

Stripped plates are furnished to required widths by shearing or gas cutting from wider sheared plates.

Sizes
Plate mills are located in various districts, but the sizes of plates produced differ greatly and the catalogs of individual mills should be consulted for detail data. The extreme width of UM plates currently rolled is 60 inches and for sheared plates it is 200 inches, but their availability together with limiting thickness and lengths should be checked with the mills before specifying. The preferred increments for width and thickness are:

Widths: Various. The catalogs of individual mills should be consulted to determine the most economical widths.
Thickness: ¹⁄₃₂-in. increments up to ½-in.
¹⁄₁₆-in. increments over ½-in. to 1 in.
⅛-in. increments over 1 in. to 3 in.
¼-in. increments over 3 in.

Ordering
Plate thickness may be specified in inches or by weight per square foot, but no decimal edge thickness can be assured by the latter method. Separate tolerance tables apply to each method.

Table 1-7.
Theoretical Weights of Rolled Floor Plates

Gauge No.	Theoretical Weight per sq. ft lb	Nominal Thickness in.	Theoretical Weight per sq. ft, lb	Nominal Thickness in.	Theoretical Weight per sq. ft, lb
18	2.40	$1/8$	6.16	$1/2$	21.47
16	3.00	$3/16$	8.71	$9/16$	24.02
14	3.75	$1/4$	11.26	$5/8$	26.58
13	4.50	$5/16$	13.81	$3/4$	31.68
12	5.25	$3/8$	16.37	$7/8$	36.78
		$7/16$	18.92	1	41.89

Note:
Thickness is measured near the edge of the plate, exclusive of raised pattern.

Invoicing

Standard practice is to invoice plates to the fabricator at theoretical weight at point of shipment. Permissible variations in weight are in accordance with the tables of ASTM Specification A6.

All plates are invoiced at theoretical weight and, except as noted, are subject to the same weight variations which apply to rectangular plates. Odd shapes in most instances require gas cutting, for which gas cutting extras are applicable.

All plates ordered gas cut for whatever reason, or beyond published shearing limits, take extras for gas cutting in addition to all other extras. Rolled steel bearing plates are often gas cut to prevent distortion due to shearing but would also take the regular extra for the thickness involved.

Extras for thickness, width, length, cutting, quality and quantity, etc., which are added to the base price of plates, are subject to revision, and should be obtained by inquiry to the producer. The foregoing general statements are made as a guide toward economy in design.

Floor Plates

Floor plates having raised patterns are available from several mills, each offering its own style of surface projections and in a variety of widths, thicknesses, and lengths. A maximum width of 96 inches and a maximum thickness of one inch are available, but availability of matching widths, thicknesses, and lengths should be checked with the producer. Floor plates are generally not specified to chemical composition limits or mechanical property requirements; a commercial grade of carbon steel is furnished. However, when strength or corrosion resistance is a consideration, raised pattern floor plates are procurable in any of the regular steel specifications. As in the case of plain plates, the individual manufacturers should be consulted for precise information. The nominal or ordered thickness is that of the flat plate, exclusive of the height or raised pattern. The usual weights are as shown in Table 1-7.

SQUARE AND ROUND BARS
Weight and area

Size in.	Weight lb per ft ■	●	Area Sq in. □	○	Size in.	Weight lb per ft ■	●	Area Sq in. □	○
0					3	30.63	24.05	9.000	7.069
1/16	0.013	0.010	0.0039	0.0031	1/16	31.91	25.07	9.379	7.366
1/8	0.053	0.042	0.0156	0.0123	1/8	33.23	26.10	9.766	7.670
3/16	0.120	0.094	0.0352	0.0276	3/16	34.57	27.15	10.160	7.980
1/4	0.213	0.167	0.0625	0.0491	1/4	35.94	28.23	10.563	8.296
5/16	0.332	0.261	0.0977	0.0767	5/16	37.34	29.32	10.973	8.618
3/8	0.479	0.376	0.1406	0.1104	3/8	38.76	30.44	11.391	8.946
7/16	0.651	0.512	0.1914	0.1503	7/16	40.21	31.58	11.816	9.281
1/2	0.851	0.668	0.2500	0.1964	1/2	41.68	32.74	12.250	9.621
9/16	1.077	0.846	0.3164	0.2485	9/16	43.19	33.92	12.691	9.968
5/8	1.329	1.044	0.3906	0.3068	5/8	44.71	35.12	13.141	10.321
11/16	1.608	1.263	0.4727	0.3712	11/16	46.27	36.34	13.598	10.680
3/4	1.914	1.503	0.5625	0.4418	3/4	47.85	37.58	14.063	11.045
13/16	2.246	1.764	0.6602	0.5185	13/16	49.46	38.85	14.535	11.416
7/8	2.605	2.046	0.7656	0.6013	7/8	51.09	40.13	15.016	11.793
15/16	2.991	2.349	0.8789	0.6903	15/16	52.76	41.43	15.504	12.177
1	3.403	2.673	1.0000	0.7854	4	54.44	42.76	16.000	12.566
1/16	3.841	3.017	1.1289	0.8866	1/16	56.16	44.11	16.504	12.962
1/8	4.307	3.382	1.2656	0.9940	1/8	57.90	45.47	17.016	13.364
3/16	4.798	3.769	1.4102	1.1075	3/16	59.67	46.86	17.535	13.772
1/4	5.317	4.176	1.5625	1.2272	1/4	61.46	48.27	18.063	14.186
5/16	5.862	4.604	1.7227	1.3530	5/16	63.28	49.70	18.598	14.607
3/8	6.433	5.053	1.8906	1.4849	3/8	65.13	51.15	19.141	15.033
7/16	7.032	5.523	2.0664	1.6230	7/16	67.01	52.63	19.691	15.466
1/2	7.656	6.013	2.2500	1.7672	1/2	68.91	54.12	20.250	15.904
9/16	8.308	6.525	2.4414	1.9175	9/16	70.83	55.63	20.816	16.349
5/8	8.985	7.057	2.6406	2.0739	5/8	72.79	57.17	21.391	16.800
11/16	9.690	7.610	2.8477	2.2365	11/16	74.77	58.72	21.973	17.257
3/4	10.421	8.185	3.0625	2.4053	3/4	76.78	60.30	22.563	17.721
13/16	11.179	8.780	3.2852	2.5802	13/16	78.81	61.90	23.160	18.190
7/8	11.963	9.396	3.5156	2.7612	7/8	80.87	63.51	23.766	18.666
15/16	12.774	10.032	3.7539	2.9483	15/16	82.96	65.15	24.379	19.147
2	13.611	10.690	4.0000	3.1416	5	85.07	66.81	25.000	19.635
1/16	14.475	11.369	4.2539	3.3410	1/16	87.21	68.49	25.629	20.129
1/8	15.366	12.068	4.5156	3.5466	1/8	89.38	70.20	26.266	20.629
3/16	16.283	12.789	4.7852	3.7583	3/16	91.57	71.92	26.910	21.135
1/4	17.227	13.530	5.0625	3.9761	1/4	93.79	73.66	27.563	21.648
5/16	18.197	14.292	5.3477	4.2000	5/16	96.04	75.43	28.223	22.166
3/8	19.194	15.075	5.6406	4.4301	3/8	98.31	77.21	28.891	22.691
7/16	20.217	15.879	5.9414	4.6664	7/16	100.61	79.02	29.566	23.221
1/2	21.267	16.703	6.2500	4.9087	1/2	102.93	80.84	30.250	23.758
9/16	22.344	17.549	6.5664	5.1573	9/16	105.29	82.69	30.941	24.301
5/8	23.447	18.415	6.8906	5.4119	5/8	107.67	84.56	31.641	24.851
11/16	24.577	19.303	7.2227	5.6727	11/16	110.07	86.45	32.348	25.406
3/4	25.734	20.211	7.5625	5.9396	3/4	112.50	88.36	33.063	25.967
13/16	26.917	21.140	7.9102	6.2126	13/16	114.96	90.29	33.785	26.535
7/8	28.126	22.090	8.2656	6.4918	7/8	117.45	92.24	34.516	27.109
15/16	29.362	23.061	8.6289	6.7771	15/16	119.96	94.22	35.254	27.688

SQUARE AND ROUND BARS
Weight and area

Size in.	Weight lb per ft ■	Weight lb per ft ●	Area Sq in. ▫	Area Sq in. ○	Size in.	Weight lb per ft ■	Weight lb per ft ●	Area Sq in. ▫	Area Sq in. ○
6	122.50	96.21	36.000	28.274	9	275.63	216.48	81.000	63.617
1/16	125.07	98.23	36.754	28.867	1/16	279.47	219.49	82.129	64.504
1/8	127.66	100.26	37.516	29.465	1/8	283.33	222.53	83.266	65.397
3/16	130.28	102.32	38.285	30.069	3/16	287.23	225.59	84.410	66.296
1/4	132.92	104.40	39.063	30.680	1/4	291.15	228.67	85.563	67.201
5/16	135.59	106.49	39.848	31.296	5/16	295.10	231.77	86.723	68.112
3/8	138.29	108.61	40.641	31.919	3/8	299.07	234.89	87.891	69.029
7/16	141.02	110.75	41.441	32.548	7/16	303.07	238.03	89.066	69.953
1/2	143.77	112.91	42.250	33.183	1/2	307.10	241.20	90.250	70.882
9/16	146.55	115.10	43.066	33.824	9/16	311.15	244.38	91.441	71.818
5/8	149.35	117.30	43.891	34.472	5/8	315.24	247.59	92.641	72.760
11/16	152.18	119.52	44.723	35.125	11/16	319.34	250.81	93.848	73.708
3/4	155.04	121.77	45.563	35.785	3/4	323.48	254.06	95.063	74.662
13/16	157.92	124.03	46.410	36.451	13/16	327.64	257.33	96.285	75.622
7/8	160.83	126.32	47.266	37.122	7/8	331.82	260.61	97.516	76.589
15/16	163.77	128.63	48.129	37.800	15/16	336.04	263.92	98.754	77.561
7	166.74	130.95	49.000	38.485	10	340.28	267.25	100.000	78.540
1/16	169.73	133.30	49.879	39.175	1/16	344.54	270.61	101.254	79.525
1/8	172.74	135.67	50.766	39.871	1/8	348.84	273.98	102.516	80.516
3/16	175.79	138.06	51.660	40.574	3/16	353.16	277.37	103.785	81.513
1/4	178.86	140.48	52.563	41.283	1/4	357.50	280.78	105.063	82.516
5/16	181.96	142.91	53.473	41.997	5/16	361.88	284.22	106.348	83.525
3/8	185.08	145.36	54.391	42.718	3/8	366.28	287.67	107.641	84.541
7/16	188.23	147.84	55.316	43.446	7/16	370.70	291.15	108.941	85.563
1/2	191.41	150.33	56.250	44.179	1/2	375.16	294.65	110.250	86.590
9/16	194.61	152.85	57.191	44.918	9/16	379.64	298.17	111.566	87.624
5/8	197.84	155.38	58.141	45.664	5/8	384.14	301.70	112.891	88.664
11/16	201.10	157.94	59.098	46.415	11/16	388.67	305.26	114.223	89.710
3/4	204.38	160.52	60.063	47.173	3/4	393.23	308.85	115.563	90.763
13/16	207.69	163.12	61.035	47.937	13/16	397.82	312.45	116.910	91.821
7/8	211.03	165.74	62.016	48.707	7/8	402.43	316.07	118.266	92.886
15/16	214.39	168.38	63.004	49.483	15/16	407.07	319.71	119.629	93.957
8	217.78	171.04	64.000	50.266	11	411.74	323.38	121.000	95.033
1/16	221.19	173.73	65.004	51.054	1/16	416.43	327.06	122.379	96.116
1/8	224.64	176.43	66.016	51.849	1/8	421.15	330.77	123.766	97.206
3/16	228.11	179.15	67.035	52.649	3/16	425.89	334.50	125.160	98.301
1/4	231.60	181.90	68.063	53.456	1/4	430.66	338.24	126.563	99.402
5/16	235.12	184.67	69.098	54.269	5/16	435.46	342.01	127.973	100.510
3/8	238.67	187.45	70.141	55.088	3/8	440.29	345.80	129.391	101.623
7/16	242.25	190.26	71.191	55.914	7/16	445.14	349.61	130.816	102.743
1/2	245.85	193.09	72.250	56.745	1/2	450.02	353.44	132.250	103.869
9/16	249.48	195.94	73.316	57.583	9/16	454.92	357.30	133.691	105.001
5/8	253.13	198.81	74.391	58.426	5/8	459.85	361.17	135.141	106.139
11/16	256.82	201.70	75.473	59.276	11/16	464.81	365.06	136.598	107.284
3/4	260.53	204.62	76.563	60.132	3/4	469.80	368.98	138.063	108.434
13/16	264.26	207.55	77.660	60.994	13/16	474.81	372.91	139.535	109.591
7/8	268.02	210.50	78.766	61.863	7/8	479.84	376.87	141.016	110.754
15/16	271.81	213.48	79.879	62.737	15/16	484.91	380.85	142.504	111.923
					12	490.00	384.85	144.000	113.098

AREA OF RECTANGULAR SECTIONS
Square inches

Width in.	Thickness, inches														
	$3/16$	$1/4$	$5/16$	$3/8$	$7/16$	$1/2$	$9/16$	$5/8$	$11/16$	$3/4$	$13/16$	$7/8$	$15/16$	1
$1/4$	0.047	0.063	0.078	0.094	0.109	0.125	0.141	0.156	0.172	0.188	0.203	0.219	0.234	0.250
$1/2$	0.093	0.125	0.156	0.188	0.219	0.250	0.281	0.313	0.344	0.375	0.406	0.438	0.469	0.500
$3/4$	0.141	0.188	0.234	0.281	0.328	0.375	0.422	0.469	0.516	0.563	0.609	0.656	0.703	0.750
1	0.188	0.250	0.313	0.375	0.438	0.500	0.563	0.625	0.688	0.75	0.813	0.875	0.938	1.00
$1 1/4$	0.234	0.313	0.391	0.469	0.547	0.625	0.703	0.781	0.859	0.938	1.02	1.09	1.17	1.25
$1 1/2$	0.281	0.375	0.469	0.563	0.656	0.750	0.844	0.938	1.03	1.13	1.22	1.31	1.41	1.50
$1 3/4$	0.328	0.438	0.547	0.656	0.766	0.875	0.984	1.09	1.20	1.31	1.42	1.53	1.64	1.75
2	0.375	0.500	0.625	0.750	0.875	1.00	1.13	1.25	1.38	1.50	1.63	1.75	1.88	2.00
$2 1/4$	0.422	0.563	0.703	0.844	0.984	1.13	1.27	1.41	1.55	1.69	1.83	1.97	2.11	2.25
$2 1/2$	0.469	0.625	0.781	0.938	1.09	1.25	1.41	1.56	1.72	1.88	2.03	2.19	2.34	2.50
$2 3/4$	0.516	0.688	0.859	1.03	1.20	1.38	1.55	1.72	1.89	2.06	2.23	2.41	2.58	2.75
3	0.563	0.750	0.938	1.13	1.31	1.50	1.69	1.88	2.06	2.25	2.44	2.63	2.81	3.00
$3 1/4$	0.609	0.813	1.02	1.22	1.42	1.63	1.83	2.03	2.23	2.44	2.64	2.84	3.05	3.25
$3 1/2$	0.656	0.875	1.09	1.31	1.53	1.75	1.97	2.19	2.41	2.63	2.84	3.06	3.28	3.50
$3 3/4$	0.703	0.938	1.17	1.41	1.64	1.88	2.11	2.34	2.58	2.81	3.05	3.28	3.52	3.75
4	0.750	1.00	1.25	1.50	1.75	2.00	2.25	2.50	2.75	3.00	3.25	3.50	3.75	4.00
$4 1/4$	0.797	1.06	1.33	1.59	1.86	2.13	2.39	2.66	2.92	3.19	3.45	3.72	3.98	4.25
$4 1/2$	0.844	1.13	1.41	1.69	1.97	2.25	2.53	2.81	3.09	3.38	3.66	3.94	4.22	4.50
$4 3/4$	0.891	1.19	1.48	1.78	2.08	2.38	2.67	2.97	3.27	3.56	3.86	4.16	4.45	4.75
5	0.938	1.25	1.56	1.88	2.19	2.50	2.81	3.13	3.44	3.75	4.06	4.38	4.69	5.00
$5 1/4$	0.984	1.31	1.64	1.97	2.30	2.63	2.95	3.28	3.61	3.94	4.27	4.59	4.92	5.25
$5 1/2$	1.03	1.38	1.72	2.06	2.41	2.75	3.09	3.44	3.78	4.13	4.47	4.81	5.16	5.50
$5 3/4$	1.08	1.44	1.80	2.16	2.52	2.88	3.23	3.59	3.95	4.31	4.67	5.03	5.39	5.75
6	1.13	1.50	1.88	2.25	2.63	3.00	3.38	3.75	4.13	4.50	4.88	5.25	5.63	6.00
$6 1/4$	1.17	1.56	1.95	2.34	2.73	3.13	3.52	3.91	4.30	4.69	5.08	5.47	5.86	6.25
$6 1/2$	1.22	1.63	2.03	2.44	2.84	3.25	3.66	4.06	4.47	4.88	5.28	5.69	6.09	6.50
$6 3/4$	1.27	1.69	2.11	2.53	2.95	3.38	3.80	4.22	4.64	5.06	5.48	5.91	6.33	6.75
7	1.31	1.75	2.19	2.63	3.06	3.50	3.94	4.38	4.81	5.25	5.69	6.13	6.56	7.00
$7 1/4$	1.36	1.81	2.27	2.72	3.17	3.63	4.08	4.53	4.98	5.44	5.89	6.34	6.80	7.25
$7 1/2$	1.41	1.88	2.34	2.81	3.28	3.75	4.22	4.69	5.16	5.63	6.09	6.56	7.03	7.50
$7 3/4$	1.45	1.94	2.42	2.91	3.39	3.88	4.36	4.84	5.33	5.81	6.30	6.78	7.27	7.75
8	1.50	2.00	2.50	3.00	3.50	4.00	4.50	5.00	5.50	6.00	6.50	7.00	7.50	8.00
$8 1/2$	1.59	2.13	2.66	3.19	3.72	4.25	4.78	5.31	5.84	6.38	6.91	7.44	7.97	8.50
9	1.69	2.25	2.81	3.38	3.94	4.50	5.06	5.63	6.19	6.75	7.31	7.88	8.44	9.00
$9 1/2$	1.78	2.38	2.97	3.56	4.16	4.75	5.34	5.94	6.53	7.13	7.72	8.31	8.91	9.50
10	1.88	2.50	3.13	3.75	4.38	5.00	5.63	6.25	6.88	7.50	8.13	8.75	9.38	10.0
$10 1/2$	1.97	2.63	3.28	3.94	4.59	5.25	5.91	6.56	7.22	7.88	8.53	9.19	9.84	10.5
11	2.06	2.75	3.44	4.13	4.81	5.50	6.19	6.88	7.56	8.25	8.94	9.63	10.3	11.0
$11 1/2$	2.16	2.88	3.59	4.31	5.03	5.75	6.47	7.19	7.91	8.63	9.34	10.1	10.8	11.5
12	2.25	3.00	3.75	4.50	5.25	6.00	6.75	7.50	8.25	9.00	9.75	10.5	11.3	12.0

WEIGHT OF RECTANGULAR SECTIONS
Pounds per linear foot

Width in.	3/16	1/4	5/16	3/8	7/16	1/2	9/16	5/8	11/16	3/4	13/16	7/8	15/16	1
1/4	0.160	0.213	0.266	0.319	0.372	0.425	0.479	0.532	0.585	0.638	0.691	0.744	0.798	0.851
1/2	0.319	0.425	0.532	0.638	0.744	0.851	0.957	1.06	1.17	1.28	1.38	1.49	1.60	1.70
3/4	0.479	0.638	0.798	0.957	1.12	1.28	1.44	1.60	1.75	1.91	2.07	2.23	2.39	2.55
1	0.638	0.851	1.06	1.28	1.49	1.70	1.91	2.13	2.34	2.55	2.76	2.98	3.19	3.40
1 1/4	0.798	1.06	1.33	1.60	1.86	2.13	2.39	2.66	2.92	3.19	3.46	3.72	3.99	4.25
1 1/2	0.957	1.28	1.60	1.91	2.23	2.55	2.87	3.19	3.51	3.83	4.15	4.47	4.79	5.10
1 3/4	1.12	1.49	1.86	2.23	2.61	2.98	3.35	3.72	4.09	4.47	4.84	5.21	5.58	5.95
2	1.28	1.70	2.13	2.55	2.98	3.40	3.83	4.25	4.68	5.10	5.53	5.95	6.38	6.81
2 1/4	1.44	1.91	2.39	2.87	3.35	3.83	4.31	4.79	5.26	5.74	6.22	6.70	7.18	7.66
2 1/2	1.60	2.13	2.66	3.19	3.72	4.25	4.79	5.32	5.85	6.38	6.91	7.44	7.98	8.51
2 3/4	1.75	2.34	2.92	3.51	4.09	4.68	5.26	5.85	6.43	7.02	7.60	8.19	8.77	9.36
3	1.91	2.55	3.19	3.83	4.47	5.10	5.74	6.38	7.02	7.66	8.29	8.93	9.57	10.2
3 1/4	2.07	2.76	3.46	4.15	4.84	5.53	6.22	6.91	7.60	8.29	8.99	9.68	10.4	11.1
3 1/2	2.23	2.98	3.72	4.47	5.21	5.95	6.70	7.44	8.19	8.93	9.68	10.4	11.2	11.9
3 3/4	2.39	3.19	3.99	4.79	5.58	6.38	7.18	7.98	8.77	9.57	10.4	11.2	12.0	12.8
4	2.55	3.40	4.25	5.10	5.95	6.81	7.66	8.51	9.36	10.2	11.1	11.9	12.8	13.6
4 1/4	2.71	3.62	4.52	5.42	6.33	7.23	8.13	9.04	9.94	10.8	11.8	12.7	13.6	14.5
4 1/2	2.87	3.83	4.79	5.74	6.70	7.66	8.61	9.57	10.5	11.5	12.4	13.4	14.4	15.3
4 3/4	3.03	4.04	5.05	6.06	7.07	8.08	9.09	10.1	11.1	12.1	13.1	14.1	15.2	16.2
5	3.19	4.25	5.32	6.38	7.44	8.51	9.57	10.6	11.7	12.8	13.8	14.9	16.0	17.0
5 1/4	3.35	4.47	5.58	6.70	7.82	8.93	10.0	11.2	12.3	13.4	14.5	15.6	16.7	17.9
5 1/2	3.51	4.68	5.85	7.02	8.19	9.36	10.5	11.7	12.9	14.0	15.2	16.4	17.5	18.7
5 3/4	3.67	4.89	6.11	7.34	8.56	9.78	11.0	12.2	13.5	14.7	15.9	17.1	18.3	19.6
6	3.83	5.10	6.38	7.66	8.93	10.2	11.5	12.8	14.0	15.3	16.6	17.9	19.1	20.4
6 1/4	3.99	5.32	6.65	7.98	9.30	10.6	12.0	13.3	14.6	16.0	17.3	18.6	19.9	21.3
6 1/2	4.15	5.53	6.91	8.29	9.68	11.1	12.4	13.8	15.2	16.6	18.0	19.4	20.7	22.1
6 3/4	4.31	5.74	7.18	8.61	10.0	11.5	12.9	14.4	15.8	17.2	18.7	20.1	21.5	23.0
7	4.47	5.95	7.44	8.93	10.4	11.9	13.4	14.9	16.4	17.9	19.4	20.8	22.3	23.8
7 1/4	4.63	6.17	7.71	9.25	10.8	12.3	13.9	15.4	17.0	18.5	20.0	21.6	23.1	24.7
7 1/2	4.79	6.38	7.98	9.57	11.2	12.8	14.4	16.0	17.5	19.1	20.7	22.3	23.9	25.5
7 3/4	4.94	6.59	8.24	9.89	11.5	13.2	14.8	16.5	18.1	19.8	21.4	23.1	24.7	26.4
8	5.10	6.81	8.51	10.2	11.9	13.6	15.3	17.0	18.7	20.4	22.1	23.8	25.5	27.2
8 1/2	5.42	7.23	9.04	10.8	12.7	14.5	16.3	18.1	19.9	21.7	23.5	25.3	27.1	28.9
9	5.74	7.66	9.57	11.5	13.4	15.3	17.2	19.1	21.1	23.0	24.9	26.8	28.7	30.6
9 1/2	6.06	8.08	10.1	12.1	14.1	16.2	18.2	20.2	22.2	24.2	26.3	28.3	30.3	32.3
10	6.38	8.51	10.6	12.8	14.9	17.0	19.1	21.3	23.4	25.5	27.6	29.8	31.9	34.0
10 1/2	6.70	8.93	11.2	13.4	15.6	17.9	20.1	22.3	24.6	26.8	29.0	31.3	33.5	35.7
11	7.02	9.36	11.7	14.0	16.4	18.7	21.1	23.4	25.7	28.1	30.4	32.8	35.1	37.4
11 1/2	7.34	9.78	12.2	14.7	17.1	19.6	22.0	24.5	26.9	29.3	31.8	34.2	36.7	39.1
12	7.66	10.2	12.8	15.3	17.9	20.4	23.0	25.5	28.1	30.6	33.2	35.7	38.3	40.8

CRANE RAILS

General Notes

The ASCE rails and the 104- to 175-lb crane rails shown in Figure 1-2 are recommended for crane runway use. For complete details and for profiles and properties of rails not listed, consult manufacturers' catalogs.

Rails should be arranged so that joints on opposite sides of the crane runway will be staggered with respect to each other and with due consideration to the wheelbase of the crane. Rail joints should not occur at crane girder splices. Light 40-lb rails are available in 30-ft lengths, 60-lb rails in 30-, 33- or 39-ft lengths, standard rails in 33- or 39-ft lengths and crane rails up to 80 ft. Consult manufacturer for availability of other lengths. Odd lengths, which must be included to complete a run or obtain the necessary stagger, should be not less than 10 feet long. For crane rail service, 40-lb rails are furnished to manufacturers' specifications and tolerances. 60- and 85-lb rails are furnished to manufacturers' specifications and tolerances, or to ASTM A1. Crane rails are furnished to ASTM A759. Rails will be furnished with standard drilling in both standard and odd lengths unless stipulated otherwise on order. For controlled cooling, heat treatment, and rail end preparation, see manufacturers' catalogs. Purchase orders for crane rails should be noted "For crane service." (See Table 1-8.)

For maximum wheel loadings see manufacturers' catalogs.

Splices

Bolted Splices

It is often more desirable to use properly installed and maintained bolted splice bars in making up rail joints for crane service than welded splice bars.

Standard rail drilling and joint-bar punching, as furnished by manufacturers of light standard rails for track work, include round holes in rail ends and slotted holes in joint bars to receive standard oval-neck track bolts. Holes in rails are oversize and punching in joint bars is spaced to allow $\frac{1}{16}$- to $\frac{1}{8}$-in. clearance between rail ends (see manufacturers' catalogs for spacing and dimensions of holes and slots). Although this construction is satisfactory for track and light crane service, its use in general crane service may lead to joint failure.

For best service in bolted splices, it is recommended that tight joints be stipulated for all rails for crane service. This will require rail ends to be finished by milling or grinding, and the special rail drilling and joint-bar punching tabulated below. Special rail drilling is accepted by some mills, or rails may be ordered blank for shop drilling. End finishing of standard rails can be done at the mill; light rails must be end-finished in the fabricating shop or ground at the site prior to erection. In the crane rail range, from 104 to 175 lbs per yard, rails and joint bars are manufactured to obtain a tight fit and no further special end finishing, drilling, or punching is required. Because of cumulative tolerance variations in holes, bolt diameters, and rail ends, a slight gap may sometimes occur in the so-called tight joints. Conversely, it may sometimes be necessary to ream holes through joined bar and rail to permit entry of bolts.

Joint bars for crane service are provided in various sections to match the rails. Joint bars for light and standard rails may be purchased blank for special shop punching to obtain tight joints. See Bethlehem Steel Corp. Booklet 3351 for dimensions, material specifications, and the identification necessary to match the crane rail section.

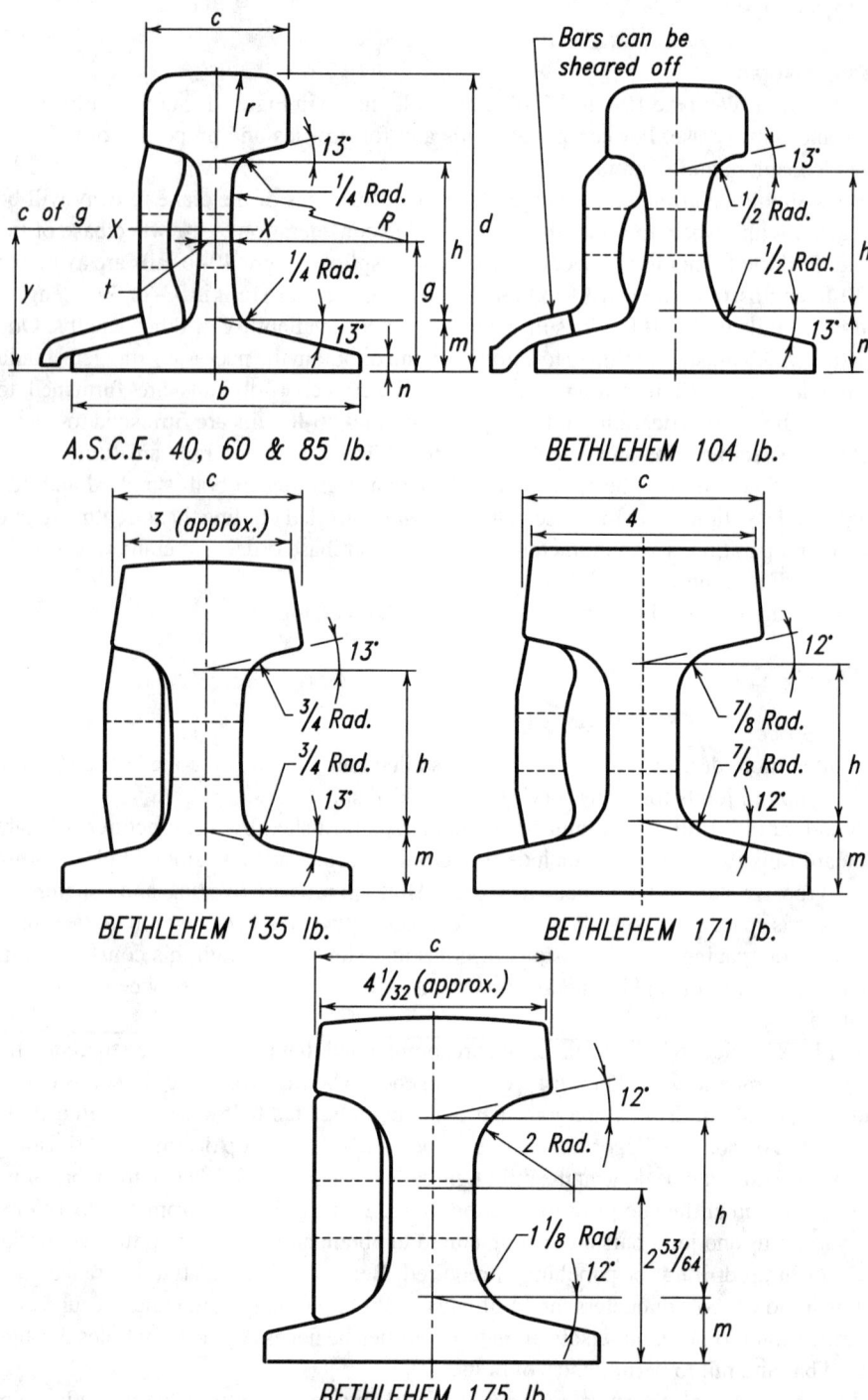

A.S.C.E. 40, 60 & 85 lb.

BETHLEHEM 104 lb.

BETHLEHEM 135 lb.

BETHLEHEM 171 lb.

BETHLEHEM 175 lb.

Nomenclature of sketch for A.S.C.E. rails also applies to the other sections.

Fig. 1-2. Crane rails.

Table 1-8.
Crane Rails

Dimensions and Properties

Type	Classi-fication	Nom-inal Wt. per Yd.	d	Gage g	b	m	n	c	r	t	h	R	Area	I_x	S_x		y
															Hd.	Base	
		lb	in.	in.	in.	in.	in.	in.	in.	in.	in.	in.	in.^{2}	in.^{4}	in.^{3}	in.^{3}	in.
ASCE	Light	30	3 1/8	1 25/64	3 1/8	17/32	11/64	1 11/16	12	21/64	1 23/32	12	3.00	4.10	2.55	—	—
ASCE	Light	40	3 1/2	1 71/128	3 1/2	5/8	7/32	1 7/8	12	25/64	1 55/64	12	3.94	6.54	3.59	3.89	1.68
ASCE	Light	50	3 7/8	1 23/32	3 7/8	11/16	1/4	2 1/8	12	7/16	2 1/16	12	4.90	10.1	5.10	—	1.88
ASCE	Light	60	4 1/4	1 115/128	4 1/4	48/64	9/32	2 3/8	12	31/64	2 17/64	·12	5.93	14.6	6.64	7.12	2.05
ASCE		70	4 5/8	2 3/64	4 5/8	13/16	9/32	2 7/16	12	33/64	2 15/32	12	6.81	19.7	8.19	8.87	2.22
ASCE		80	5	2 3/16	5	7/8	19/64	2 1/2	12	35/64	2 5/8	12	7.86	26.4	10.1	11.1	2.38
ASCE	Std.	85	5 3/16	2 17/64	5 3/16	57/64	19/64	2 9/16	12	9/16	2 3/4	12	8.33	30.1	11.1	12.2	2.47
ASCE	Std.	100	5 3/4	2 65/128	5 3/4	31/32	10/32	2 3/4	12	9/16	2 5/64	12	9.84	44.0	14.6	16.1	2.73
Bethlehem	Crane	104	5	2 7/16	5	1 1/16	1/2	2 1/2	12	1	2 7/16	3 1/2	10.3	29.8	10.7	13.5	2.21
Bethlehem	Crane	135	5 3/4	2 15/32	5 3/16	1 1/16	15/32	3 7/16	14	1 1/4	2 13/16	12	13.3	50.8	17.3	18.1	2.81
Bethlehem	Crane	171	6	2 5/8	6	1 1/4	5/8	4.3	Flat	1 1/4	2 3/4	Vert.	16.8	73.4	24.5	24.4	3.01
Bethlehem	Crane	175	6	2 21/32	6	1 9/64	1/2	4 1/4	18	1 1/2	3 7/64	Vert.	17.1	70.5	23.4	23.6	2.98

Joint-bar bolts, as distinguished from oval-neck track bolts, have straight shanks to the head and are manufactured to ASTM A449 specifications. Nuts are manufactured to ASTM A563 Gr. B specifications. ASTM A325 bolts and nuts may be used. Bolt assembly includes an alloy steel spring washer, furnished to AREA specifications.

After installation, bolts should be retightened within 30 days and every three months thereafter.

Welded Splices

When welded splices are specified, consult the manufacturer for recommended rail-end preparation, welding procedure, and method of ordering. Although joint continuity, made possible by this method of splicing, is desirable, it should be noted that the careful control required in all stages of the welding operation may be difficult to meet during crane rail installation.

Rails should not be attached to structural supports by welding. Rails with holes for joint bar bolts should not be used in making welded splices.

Fastenings

Hook Bolts

Hook bolts (Figure 1-3) are used primarily with light rails when attached to beams too narrow for clamps. Rail adjustment to $\pm\frac{1}{2}$-in. is inherent in the threaded shank. Hook bolts are paired alternately three to four inches apart, spaced at about 24-in. centers. The special rail drilling required must be done at the fabricator's shop. Hook bolts are not recommended for use with heavy duty cycle cranes (CMAA Classes, D, E, and F). It is generally recommended that hook bolts should not be used in runway systems which are longer than 500 feet because the bolts do not allow for longitudinal movement of the rail.

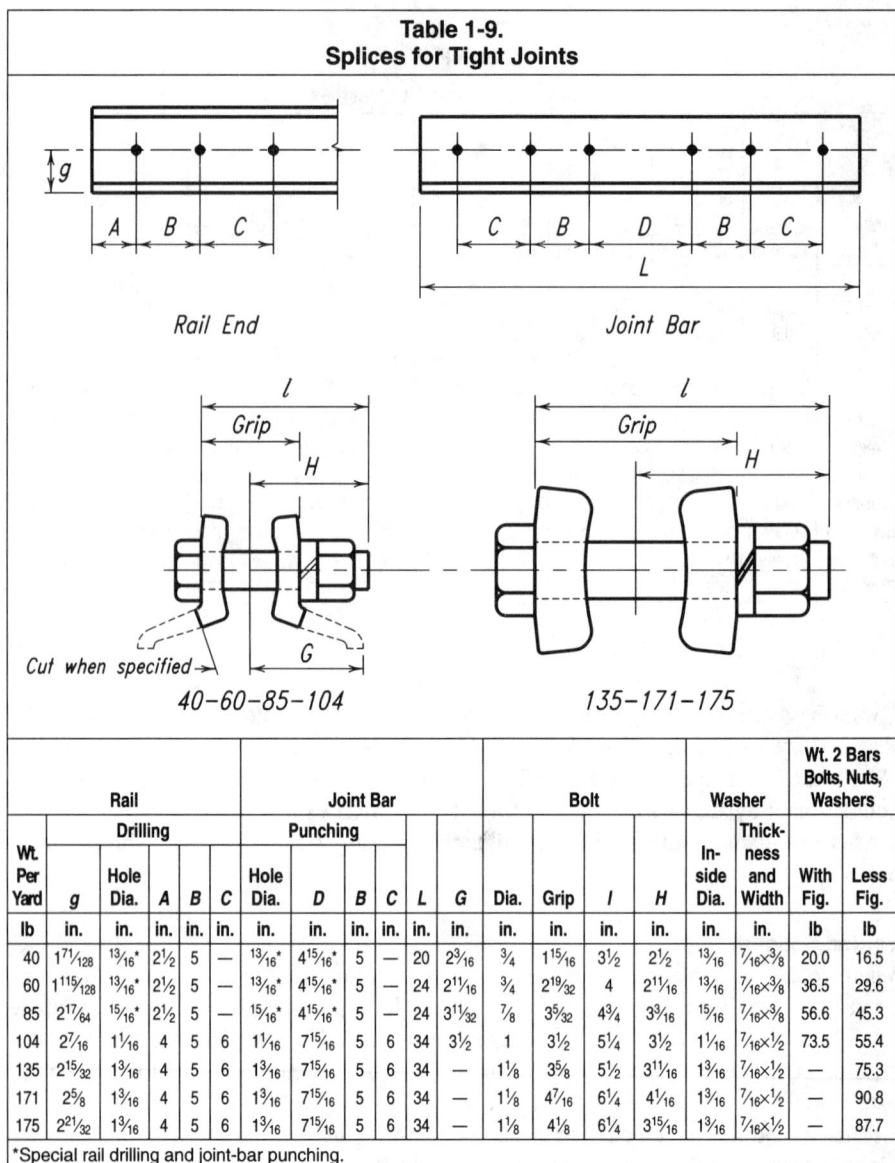

Table 1-9.
Splices for Tight Joints

Rail End　　　　　　　　　Joint Bar

Cut when specified →

40-60-85-104　　　　　　135-171-175

Wt. Per Yard	Rail					Joint Bar						Bolt				Washer		Wt. 2 Bars Bolts, Nuts, Washers	
		Drilling				Punching											Thick-		
	g	Hole Dia.	A	B	C	Hole Dia.	D	B	C	L	G	Dia.	Grip	I	H	In-side Dia.	ness and Width	With Fig.	Less Fig.
lb	in.	in.	in.	in.	in.	in.	in.	in.	in.	in.	in.	in.	in.	in.	in.	in.	in.	lb	lb
40	$1^{71}/_{128}$	$^{13}/_{16}$*	$2^{1}/_{2}$	5	—	$^{13}/_{16}$*	$4^{15}/_{16}$*	5	—	20	$2^{3}/_{16}$	$^{3}/_{4}$	$1^{15}/_{16}$	$3^{1}/_{2}$	$2^{1}/_{2}$	$^{13}/_{16}$	$^{7}/_{16}\times^{3}/_{8}$	20.0	16.5
60	$1^{115}/_{128}$	$^{13}/_{16}$*	$2^{1}/_{2}$	5	—	$^{13}/_{16}$*	$4^{15}/_{16}$*	5	—	24	$2^{11}/_{16}$	$^{3}/_{4}$	$2^{19}/_{32}$	4	$2^{11}/_{16}$	$^{13}/_{16}$	$^{7}/_{16}\times^{3}/_{8}$	36.5	29.6
85	$2^{17}/_{64}$	$^{15}/_{16}$*	$2^{1}/_{2}$	5	—	$^{15}/_{16}$*	$4^{15}/_{16}$*	5	—	24	$3^{11}/_{32}$	$^{7}/_{8}$	$3^{5}/_{32}$	$4^{3}/_{4}$	$3^{3}/_{16}$	$^{15}/_{16}$	$^{7}/_{16}\times^{3}/_{8}$	56.6	45.3
104	$2^{7}/_{16}$	$1^{1}/_{16}$	4	5	6	$1^{1}/_{16}$	$7^{15}/_{16}$	5	6	34	$3^{1}/_{2}$	1	$3^{1}/_{2}$	$5^{1}/_{4}$	$3^{1}/_{2}$	$1^{1}/_{16}$	$^{7}/_{16}\times^{1}/_{2}$	73.5	55.4
135	$2^{15}/_{32}$	$1^{3}/_{16}$	4	5	6	$1^{3}/_{16}$	$7^{15}/_{16}$	5	6	34	—	$1^{1}/_{8}$	$3^{5}/_{8}$	$5^{1}/_{2}$	$3^{11}/_{16}$	$1^{3}/_{16}$	$^{7}/_{16}\times^{1}/_{2}$	—	75.3
171	$2^{5}/_{8}$	$1^{3}/_{16}$	4	5	6	$1^{3}/_{16}$	$7^{15}/_{16}$	5	6	34	—	$1^{1}/_{8}$	$4^{7}/_{16}$	$6^{1}/_{4}$	$4^{1}/_{16}$	$1^{3}/_{16}$	$^{7}/_{16}\times^{1}/_{2}$	—	90.8
175	$2^{21}/_{32}$	$1^{3}/_{16}$	4	5	6	$1^{3}/_{16}$	$7^{15}/_{16}$	5	6	34	—	$1^{1}/_{8}$	$4^{1}/_{8}$	$6^{1}/_{4}$	$3^{15}/_{16}$	$1^{3}/_{16}$	$^{7}/_{16}\times^{1}/_{2}$	—	87.7

*Special rail drilling and joint-bar punching.

Rail Clips

Rail clips are forged or cast devices which are shaped to match specific rail profiles. They are usually bolted to the runway girder flange with one bolt or are sometimes welded. Rail clips have been used satisfactorily with all classes of cranes. However, one drawback is that when a single bolt is used the clip can rotate in response to rail longitudinal movement. This clip rotation can cause a camming action, thus forcing the rail out of alignment. Because of this limitation, rail clips should only be used in crane systems subject to infrequent use, and for runways less than 500 feet in length.

Rail Clamps
Rail clamps are a common method of attachment for heavy duty cycle cranes. Rail clamps are detailed to provide two types: tight and floating (Figure 1-4). Each clamp consists of two plates: an upper clamp plate and a lower filler plate.

The lower plate is flat and roughly matches the height of the toe of the rail flange. The upper plate covers the lower plate and extends over the top of the lower rail flange. In the tight clamp the upper plate is detailed to fit tightly to the lower tail flange top, thus "clamping" it tightly in place when the fasteners are tightened. In the past, the tight clamp had been illustrated with the filler plates fitted tightly against the rail flange toe. This tight fit-up was rarely achieved in practice and is not considered to be necessary to achieve a tight type clamp. In the floating type clamp, the pieces are detailed to provide a clearance both alongside the rail flange toe and below the upper plate. The floating type does not, in reality, clamp the rail but merely holds the rail within the limits of the clamp clearances.

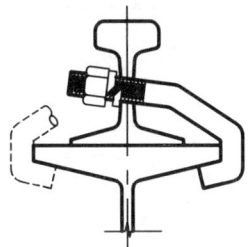

Fig. 1-3. Hook bolts.

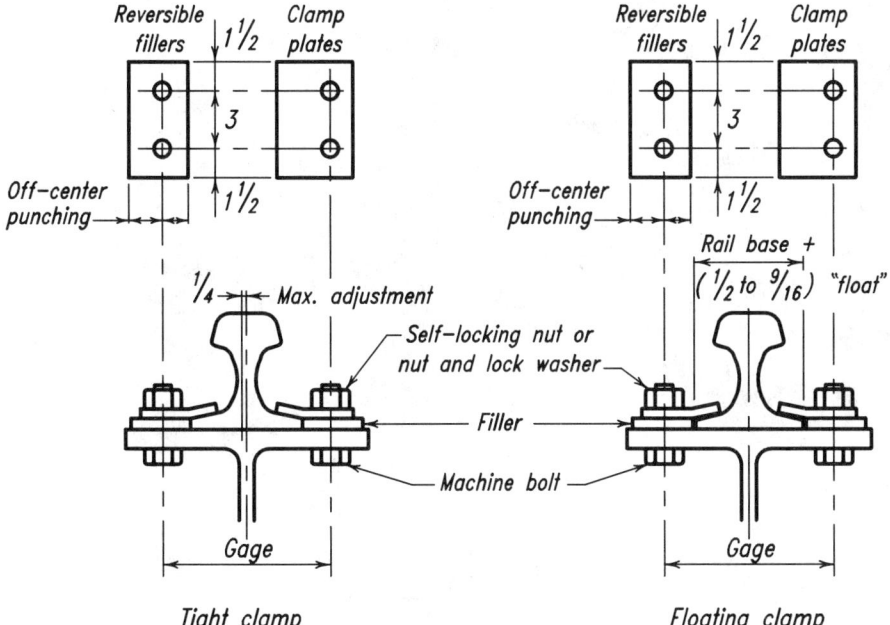

Fig. 1-4. Rail clamps.

High strength bolts are recommended for both clamp types. Both types should be spaced three feet or less apart.

Dimensions shown above are suggested. See manufacturers' catalogs for recommended gages, bolt sizes, and detail dimensions not shown.

Patented Rail Clips

Each manufacturer's literature presents in detail the desirable aspects of the various designs. In general patented rail clips are easy to install due to their range of adjustment while providing the proper limitations of lateral movement and allowance for longitudinal movement. Patented rail clips should be considered as a viable alternative to conventional hook bolts, clips, or clamps. Because of their desirable characteristics, patented rail clips can be used without restriction except as limited by the specific manufacturer's recommendations. Installations using patented rail clips sometimes incorporate pads beneath the rail. When this is done the lateral float of the rail should be limited as in the case of the tight rail clamps.

TORSION PROPERTIES

Torsional analysis is not required for the routine design of most structural steel members. When torsional analysis is required, the Table of Torsion Properties will be of assistance in utilizing current analysis methods. The reader is referred to the AISC publication *Torsional Analysis of Steel Members* (American Institute of Steel Construction, 1983) for additional information and appropriate design aids.

Torsion Properties are also required to determine the design compressive strength for torsional and flexural-torsional buckling as specified in the AISC LRFD Specification Appendix E3.

Nomenclature

C_w = warping constant for section, in.^{6}*

E = modulus of elasticity of steel (29,000 ksi)

G = shear modulus of elasticity of steel (11,200 ksi)

H = flexural constant in Equation E3-1, LRFD Specification

J = torsional constant for a section, in.^{4}

Q_f = statical moment for a point in the flange directly above the vertical edge of the web, in.^{3}

Q_w = statical moment at mid-depth of the section, in.^{3}

$\overline{r}_o$ = polar radius of gyration about the shear center, in.

S_w = warping statical moment at a point in the section, in.^{4}

W_{no} = normalized warping function at a point at the flange edge, in.^{2}

---

*Calculated values of C_w are given for all tabulated shapes. However, for many angles and T shapes, C_w is so small that for practical purposes it can be taken as zero.

TORSION PROPERTIES
W shapes

Designation	Torsional Constant J in.^{4}	Warping Constant C_w in.^{6}	$\sqrt{\dfrac{EC_w}{GJ}}$ in.	Normalized Warping Constant W_{no} in.^{2}	Warping Statical Moment S_w in.^{4}	Statical Moment Q_f in.^{3}	Statical Moment Q_w in.^{3}
W44×335	74.4	536000	137	168	1190	282	811
×290	51.5	463000	153	166	1040	251	709
×262	37.7	406000	167	165	922	225	636
×230	24.9	346000	190	164	789	194	551
W40×593	445	996000	76.1	166	2240	484	1380
×503	279	791000	85.7	158	1540	409	1160
×431	177	639000	96.7	156	1380	347	981
×372	116	528000	109	154	1240	300	836
×321	75.4	437000	123	152	1100	258	717
×297	61.2	397000	130	151	986	240	665
×277	38.1	378000	138	151	940	230	624
×249	37.7	333000	151	149	836	208	560
×215	24.4	283000	173	149	714	179	481
×199	18.1	245000	187	148	621	157	434
×174	11.2	189000	209	147	481	119	364
W40×466	277	393000	60.6	125	1160	322	1030
×392	172	306000	67.9	121	940	272	856
×331	106	242000	76.8	118	762	228	715
×278	64.7	192000	87.6	115	622	192	596
×264	56.1	181000	91.3	114	589	184	566
×235	41.3	161000	101	113	530	168	506
×211	30.4	140000	109	112	468	151	453
×183	19.6	119000	125	111	402	134	391
×167	14.0	99300	136	111	336	113	346
×149	9.60	79600	147	110	270	92.0	299
W36×848	1270	1620000	57.5	172	3530	674	1910
×798	1070	1480000	59.8	169	3270	634	1790
×650	600	1090000	68.6	162	2520	513	1420
×527	330	816000	80.0	156	1960	415	1130
×439	195	637000	92.0	152	1570	344	928
×393	143	554000	100	150	1390	309	830
×359	109	493000	108	148	1240	281	757
×328	84.5	441000	116	146	1130	258	691
×300	64.2	398000	127	146	1020	235	628
×280	52.6	366000	134	145	944	219	585
×260	41.5	330000	143	144	858	200	538
×245	34.6	306000	151	143	799	187	505
×230	28.6	282000	160	143	740	175	472
W36×256	53.3	168000	90.3	109	576	176	520
×232	39.8	148000	98.1	108	512	159	468
×210	28.0	128000	109	108	446	138	416
×194	22.2	116000	116	107	407	128	383
×182	18.4	107000	123	106	378	120	359
×170	15.1	98500	130	105	349	111	334
×160	12.4	90200	137	105	321	103	312
×150	10.1	82200	145	105	294	95.1	291
×135	6.99	68100	159	104	245	79.9	255
W33×354	115	408000	95.8	135	1130	263	709
×318	84.4	357000	105	133	1000	237	634
×291	65.0	319000	113	132	906	216	577
×263	48.5	281000	122	130	808	195	519
×241	35.8	250000	134	130	721	174	469
×221	27.5	224000	145	129	650	158	428
×201	20.5	198000	158	128	580	142	386
W33×169	17.7	82400	110	93.7	329	109	314
×152	12.4	71700	122	93.8	286	95.1	279
×141	9.70	64400	131	93.3	258	86.5	257
×130	7.37	56600	141	92.8	228	76.9	233
×118	5.30	48300	154	92.2	196	66.6	207

TORSION PROPERTIES
W shapes

Designation	Torsional Constant J in.^{4}	Warping Constant C_w in.^{6}	$\sqrt{\dfrac{EC_w}{GJ}}$ in.	Normalized Warping Constant W_{no} in.^{2}	Warping Statical Moment S_w in.^{4}	Statical Moment	
						Q_f in.^{3}	Q_w in.^{3}
W30×477	307	480000	63.6	124	1450	329	896
×391	174	364000	73.6	120	1140	268	716
×326	103	286000	84.8	117	919	223	595
×292	74.9	249000	92.8	115	812	200	530
×261	53.8	215000	102	114	710	177	470
×235	40.0	190000	111	112	633	160	422
×211	27.9	166000	124	112	556	141	374
×191	20.6	146000	135	111	494	126	337
×173	15.3	129000	148	110	439	113	303
W30×148	14.6	49400	93.6	77.3	239	86.8	250
×132	9.72	42100	106	77.3	204	74.0	219
×124	7.99	38600	112	76.9	188	68.8	204
×116	6.43	34900	119	76.5	171	62.8	189
×108	4.99	30900	127	76.1	152	56.1	173
×99	3.77	26800	136	75.7	133	49.5	156
×90	2.92	24000	146	75.0	119	45.0	142
W27×539	499	440000	47.8	111	1490	342	940
×448	297	336000	54.1	106	1190	283	766
×368	169	254000	62.4	102	930	231	620
×307	101	199000	71.4	99.4	750	192	511
×258	61.0	159000	82.2	98.2	613	161	424
×235	46.3	140000	88.5	96.0	548	146	384
×217	37.0	128000	94.6	95.0	503	135	354
×194	26.5	111000	104	93.9	442	120	314
×178	19.5	98300	114	93.7	393	107	284
×161	14.7	87300	124	92.9	352	96.6	256
×146	10.9	77200	135	92.2	314	87.0	231
W27×129	11.2	32500	86.7	66.4	183	69.5	197
×114	7.33	27600	98.7	66.4	155	59.2	171
×102	5.29	24000	108	65.7	137	52.7	153
×94	4.03	21300	117	65.4	122	47.3	139
×84	2.81	17900	128	64.9	103	40.6	122
W24×492	456	283000	40.1	92.1	1150	281	774
×408	271	214000	45.2	88.1	909	233	626
×335	154	160000	51.9	84.6	709	189	509
×279	91.7	125000	59.4	82.0	570	157	418
×250	67.3	108000	64.5	80.6	502	141	372
×229	51.8	95800	69.2	79.6	451	128	338
×207	38.6	83900	75.0	78.5	401	116	303
×192	31.0	76200	79.8	77.7	367	107	280
×176	24.1	68400	85.7	77.0	333	97.8	255
×162	18.5	62600	93.6	77.0	304	89.4	234
×146	13.4	54600	103	76.3	268	79.5	209
×131	9.50	47100	113	75.6	233	69.7	185
×117	6.72	40800	125	74.9	204	61.5	164
×104	4.72	35200	139	74.3	178	54.1	144
W24×103	7.10	16600	77.8	53.0	117	49.4	140
×94	5.26	15000	85.9	53.1	105	44.4	127
×84	3.70	12800	94.6	52.6	91.3	39.0	112
×76	2.68	11100	104	52.2	79.8	34.4	100
×68	1.87	9430	114	51.9	68.0	29.5	88.3
W24×62	1.71	4620	83.6	40.7	42.3	23.2	76.6
×55	1.18	3870	92.2	40.4	35.7	19.8	67.1

TORSION PROPERTIES
W shapes

Designation	Torsional Constant J in.^{4}	Warping Constant C_w in.^{6}	$\sqrt{\dfrac{EC_w}{GJ}}$ in.	Normalized Warping Constant W_{no} in.^{2}	Warping Statical Moment S_w in.^{4}	Statical Moment	
						Q_f in.^{3}	Q_w in.^{3}
W21×201	41.3	61800	62.2	67.0	345	102	265
×182	31.1	54300	67.2	66.0	307	92.3	238
×166	23.9	48500	72.5	65.6	277	84.4	216
×147	15.4	41100	83.1	65.4	235	71.4	187
×132	11.3	36000	90.8	64.7	208	64.0	167
×122	8.98	32700	97.1	64.2	191	59.2	154
×111	6.83	29200	105	63.7	172	53.7	139
×101	5.21	26200	114	63.2	155	49.0	127
W21×93	6.03	9940	65.3	43.6	85.3	38.2	110
×83	4.34	8630	71.8	43.0	75.0	34.2	98.0
×73	3.02	7410	79.7	42.5	65.2	30.3	86.2
×68	2.45	6760	84.5	42.3	59.9	28.0	79.9
×62	1.83	5960	91.8	42.0	53.2	25.1	72.2
W21×57	1.77	3190	68.3	33.4	35.6	20.9	64.3
×50	1.14	2570	76.4	33.1	28.9	17.2	55.0
×44	0.77	2110	84.2	32.8	24.0	14.5	47.7
W18×311	177	75700	33.3	58.8	483	141	376
×283	135	65600	35.5	57.5	427	127	338
×258	104	57400	37.8	56.4	382	116	306
×234	79.7	49900	40.3	55.2	339	105	274
×211	59.3	43200	43.4	54.2	299	94.3	245
×192	45.2	37900	46.6	53.3	267	85.7	221
×175	34.2	33200	50.1	52.5	237	77.2	199
×158	25.4	28900	54.3	51.6	210	69.4	178
×143	19.4	25700	58.6	51.0	189	63.2	161
×130	14.7	22700	63.2	50.4	169	57.1	145
W18×119	10.6	20300	70.4	50.4	151	50.6	131
×106	7.48	17400	77.6	49.8	131	44.6	115
×97	5.86	15800	83.6	49.4	120	41.2	105
×86	4.10	13600	92.7	48.9	104	36.3	92.8
×76	2.83	11700	103	48.4	90.7	31.9	81.4
W18×71	3.48	4700	59.1	33.7	52.1	25.8	72.7
×65	2.73	4240	63.4	33.4	47.5	23.8	66.6
×60	2.17	3850	67.8	33.1	43.5	22.1	61.4
×55	1.66	3430	73.1	32.9	39.0	19.9	55.9
×50	1.24	3040	79.7	32.6	34.9	18.0	50.4
W18×46	1.22	1710	60.2	26.4	24.2	15.3	45.3
×40	0.81	1440	67.8	26.1	20.6	13.3	39.2
×35	0.51	1140	76.1	25.9	16.5	10.7	33.2
W16×100	7.73	11900	63.1	41.7	107	39.0	99.0
×89	5.45	10200	69.6	41.1	93.3	34.4	87.3
×77	3.57	8590	78.9	40.6	79.3	29.7	75.0
×67	2.39	7300	88.9	40.1	68.2	25.9	64.9
W16×57	2.22	2660	55.7	28.0	35.6	19.0	52.6
×50	1.52	2270	62.2	27.6	30.8	16.7	46.0
×45	1.11	1990	68.1	27.4	27.2	15.0	41.1
×40	0.79	1730	75.3	27.1	23.9	13.4	36.5
×36	0.54	1460	83.7	26.9	20.2	11.4	32.0
W16×31	0.46	739	64.5	21.3	13.0	9.17	27.0
×26	0.26	565	75.0	21.1	10.0	7.20	22.1

TORSION PROPERTIES
W shapes

Designation	Torsional Constant J in.^{4}	Warping Constant C_w in.^{6}	$\sqrt{\dfrac{EC_w}{GJ}}$ in.	Normalized Warping Constant W_{no} in.^{2}	Warping Statical Moment S_w in.^{4}	Statical Moment	
						Q_f in.^{3}	Q_w in.^{3}
W14×808	1860	433000	24.6	82.2	1950	337	916
×730	1450	362000	25.4	78.3	1720	319	831
×665	1120	305000	26.6	75.5	1510	287	740
×605	870	258000	27.7	73.0	1320	259	660
×550	670	219000	29.1	70.6	1160	233	588
×500	514	187000	30.7	68.5	1020	209	524
×455	395	160000	32.4	66.5	899	189	468
W14×426	331	144000	33.6	65.3	827	176	434
×398	273	129000	35.0	64.1	756	163	401
×370	222	116000	36.8	62.9	689	151	368
×342	178	103000	38.7	61.6	623	138	336
×311	136	89100	41.2	60.3	553	125	301
×283	104	77700	44.0	59.1	493	113	271
×257	79.1	67800	47.1	57.9	438	102	243
×233	59.5	59000	50.7	56.9	389	91.7	218
×211	44.6	51500	54.7	55.9	345	82.3	195
×193	34.8	45900	58.4	55.1	312	75.4	177
×176	26.5	40500	62.9	54.4	279	68.0	160
×159	19.8	35600	68.2	53.7	248	61.3	143
×145	15.2	31700	73.5	53.0	224	55.8	130
W14×132	12.3	25500	73.3	50.2	190	49.9	117
×120	9.37	22700	79.2	49.7	171	45.3	106
×109	7.12	20200	85.7	49.1	154	41.2	95.9
×99	5.37	18000	93.2	48.7	138	37.2	86.6
×90	4.06	16000	101	48.3	125	33.7	78.3
W14×82	5.08	6710	58.5	34.1	73.8	28.1	69.3
×74	3.88	5990	63.2	33.7	66.6	25.7	62.8
×68	3.02	5380	67.9	33.4	60.4	23.5	57.3
×61	2.20	4710	74.5	33.1	53.3	21.0	51.1
W14×53	1.94	2540	58.2	26.7	35.5	17.3	43.6
×48	1.46	2240	63.0	26.5	31.6	15.6	39.2
×43	1.05	1950	69.3	26.2	27.8	13.9	34.8
W14×38	0.80	1230	63.1	23.0	20.0	11.5	30.7
×34	0.57	1070	69.7	22.8	17.5	10.2	27.3
×30	0.38	887	77.7	22.6	14.7	8.59	23.6
W14×26	0.36	405	54.0	16.9	8.94	6.98	20.1
×22	0.21	314	62.2	16.8	7.02	5.58	16.6
W12×336	243	57000	24.6	46.4	459	119	301
×305	185	48600	26.1	45.0	403	107	269
×279	143	42000	27.6	44.0	357	96.3	241
×252	108	35800	29.3	42.8	313	86.4	214
×230	83.8	31200	31.0	41.8	279	78.4	193
×210	64.7	27200	33.0	41.0	249	71.1	174
×190	48.8	23600	35.4	40.1	220	64.1	156
×170	35.6	20100	38.2	39.2	192	56.9	137
×152	25.8	17200	41.5	38.4	168	50.4	121
×136	18.5	14700	45.4	37.7	146	44.5	107
×120	12.9	12400	49.9	37.0	126	38.9	93.2
×106	9.13	10700	55.1	36.4	110	34.6	81.9
×96	6.86	9410	59.6	35.9	98.2	31.3	73.6
×87	5.10	8270	64.8	35.5	87.2	28.0	66.0
×79	3.84	7330	70.3	35.2	78.1	25.3	59.5
×72	2.93	6540	76.0	34.9	70.3	22.9	53.9
×65	2.18	5780	82.9	34.5	62.7	20.6	48.4

TORSION PROPERTIES
W shapes

Designation	Torsional Constant J in.^{4}	Warping Constant C_w in.^{6}	$\sqrt{\dfrac{EC_w}{GJ}}$ in.	Normalized Warping Constant W_{no} in.^{2}	Warping Statical Moment S_w in.^{4}	Statical Moment Q_f in.^{3}	Statical Moment Q_w in.^{3}
W12×58	2.10	3570	66.3	28.9	46.3	18.2	43.2
×53	1.58	3160	72.0	28.7	41.2	16.3	39.0
W12×50	1.78	1880	52.3	23.3	30.2	14.7	36.2
×45	1.31	1650	57.1	23.1	26.7	13.1	32.4
×40	0.95	1440	62.6	22.9	23.6	11.8	28.8
W12×35	0.74	879	55.5	19.6	16.8	9.86	25.6
×30	0.46	720	63.7	19.4	13.9	8.30	21.6
×26	0.30	607	72.4	19.2	11.8	7.15	18.6
W12×22	0.29	164	38.3	12.0	5.13	4.87	14.7
×19	0.18	131	43.4	11.8	4.14	4.01	12.4
×16	0.10	96.9	50.1	11.7	3.09	3.04	10.0
×14	0.07	80.4	54.5	11.6	2.59	2.59	8.72
W10×112	15.1	6020	32.1	26.3	85.7	30.8	73.7
×100	10.9	5150	35.0	25.8	74.7	27.2	64.9
×88	7.53	4330	38.6	25.3	64.2	23.8	56.4
×77	5.11	3630	42.9	24.8	54.9	20.7	48.8
×68	3.56	3100	47.5	24.4	47.6	18.1	42.6
×60	2.48	2640	52.5	24.0	41.2	15.9	37.3
×54	1.82	2320	57.5	23.8	36.6	14.3	33.3
×49	1.39	2070	62.1	23.6	33.0	13.0	30.2
W10×45	1.51	1200	45.4	19.0	23.6	11.5	27.5
×39	0.98	992	51.2	18.7	19.8	9.77	23.4
×33	0.58	790	59.4	18.5	16.0	7.98	19.4
W10×30	0.62	414	41.6	14.5	10.7	7.09	18.3
×26	0.40	345	47.3	14.3	9.05	6.08	15.6
×22	0.24	275	54.5	14.1	7.30	4.95	13.0
W10×19	0.23	104	34.2	9.89	3.93	3.76	10.8
×17	0.16	85.1	37.1	9.80	3.24	3.13	9.33
×15	0.10	68.3	42.1	9.72	2.62	2.56	8.00
×12	0.05	50.9	51.3	9.56	1.99	2.00	6.32
W8×67	5.06	1440	27.1	16.7	32.3	14.7	35.1
×58	3.34	1180	30.2	16.3	27.2	12.5	29.9
×48	1.96	931	35.1	15.8	22.0	10.4	24.5
×40	1.12	726	41.0	15.5	17.5	8.42	19.9
×35	0.77	619	45.6	15.3	15.2	7.39	17.3
×31	0.54	530	50.4	15.1	13.1	6.46	15.2
W8×28	0.54	312	38.7	12.4	9.43	5.64	13.6
×24	0.35	259	43.8	12.2	7.94	4.83	11.6
W8×21	0.28	152	37.5	10.4	5.47	4.03	10.2
×18	0.17	122	43.1	10.3	4.44	3.31	8.52
W8×15	0.14	51.8	31.0	7.82	2.47	2.39	6.78
×13	0.09	40.8	34.3	7.74	1.97	1.93	5.70
×10	0.04	30.9	44.7	7.57	1.53	1.56	4.43
W6×25	0.46	150	29.1	9.01	6.23	3.92	9.46
×20	0.24	113	34.9	8.78	4.82	3.10	7.45
×15	0.10	76.5	44.5	8.58	3.34	2.18	5.39
W6×16	0.22	38.2	21.2	5.92	2.42	2.28	5.84
×12	0.09	24.7	26.7	5.75	1.61	1.55	4.15
×9	0.04	17.7	33.8	5.60	1.19	1.19	3.12
W5×19	0.31	50.8	20.6	5.94	3.21	2.44	5.81
×16	0.19	40.6	23.5	5.81	2.62	2.02	4.82
W4×13	0.15	14.0	15.5	3.87	1.36	1.27	3.14

TORSION PROPERTIES
M shapes

Designation	Torsional Constant J in.^{4}	Warping Constant C_w in.^{6}	$\sqrt{\dfrac{EC_w}{GJ}}$ in.	Normalized Warping Constant W_{no} in.^{2}	Warping Statical Moment S_w in.^{4}	Statical Moment	
						Q_f in.^{3}	Q_w in.^{3}
M12×11.8	0.05	37.2	43.2	9.02	1.56	1.98	7.14
×10.8	0.04	33.8	46.6	9.01	1.45	1.86	6.58
M10×9	0.03	15.7	34.6	6.59	0.91	1.32	4.60
×8	0.02	13.8	38.4	6.57	0.80	1.18	4.06
M8×6.5	0.02	5.45	25.6	4.45	0.48	0.82	2.72
M5×18.9	0.34	41.3	17.7	5.73	2.98	2.28	5.53

TORSION PROPERTIES
S shapes

Designation	Torsional Constant J in.^{4}	Warping Constant C_w in.^{6}	$\sqrt{\dfrac{EC_w}{GJ}}$ in.	Normalized Warping Constant W_{no} in.^{2}	Warping Statical Moment S_w in.^{4}	Statical Moment	
						Q_f in.^{3}	Q_w in.^{3}
S24×121	12.8	11400	48.0	47.1	103	47.1	154
×106	10.1	10600	52.1	46.1	98.8	47.1	141
S24×100	7.58	6380	46.7	41.9	66.0	33.5	121
×90	6.04	6000	50.7	41.2	63.8	33.5	112
×80	4.88	5640	54.7	40.5	61.6	33.5	103
S20×96	8.39	4710	38.1	34.9	57.8	29.2	99.7
×86	6.64	4390	41.4	34.2	55.5	29.2	92.5
S20×75	4.59	2750	39.4	30.7	38.9	22.6	77.0
×66	3.58	2550	42.9	30.0	37.3	22.6	70.5
S18×70	4.15	1800	33.5	27.0	29.2	17.1	63.0
×54.7	2.37	1560	41.3	26.0	26.9	17.1	52.9
S15×50	2.12	811	31.5	20.3	17.8	11.8	39.0
×42.9	1.54	744	35.4	19.8	16.9	11.8	35.1
S12×50	2.82	505	21.5	15.5	14.0	9.30	31.0
×40.8	1.75	437	25.4	14.9	12.9	9.30	26.9
S12×35	1.08	324	27.9	14.5	10.0	7.48	22.7
×31.8	0.90	307	29.7	14.3	9.74	7.48	21.3
S10×35	1.29	189	19.5	11.8	7.13	5.24	17.9
×25.4	0.60	153	25.7	11.1	6.34	5.24	14.4
S8×23	0.55	61.8	17.1	7.90	3.50	3.10	9.74
×18.4	0.34	53.5	20.2	7.58	3.22	3.10	8.38
S6×17.25	0.37	18.4	11.3	5.03	1.61	1.63	5.35
×12.5	0.17	14.5	14.9	4.70	1.41	1.63	4.30
S5×10	0.11	6.66	12.3	3.51	0.86	1.11	2.88
S4×9.5	0.12	3.10	8.18	2.59	0.53	0.70	2.05
×7.7	0.07	2.62	9.84	2.47	0.48	0.70	1.79
S3×7.5	0.09	1.10	5.63	1.72	0.28	0.40	1.20
×5.7	0.04	0.85	7.42	1.60	0.24	0.40	1.00

TORSION PROPERTIES
HP shapes

Designation	Torsional Constant J in.^{4}	Warping Constant C_w in.^{6}	$\sqrt{\dfrac{EC_w}{GJ}}$ in.	Normalized Warping Constant W_{no} in.^{2}	Warping Statical Moment S_w in.^{4}	Statical Moment	
						Q_f in.^{3}	Q_w in.^{3}
HP14×117	8.02	19900	80.2	49.9	149	38.5	97.2
×102	5.40	16800	89.8	49.2	128	33.5	84.3
×89	3.60	14200	101	48.5	110	29.1	72.9
×73	2.01	11200	120	47.8	88.0	23.8	59.2
HP12×84	4.24	7160	66.1	35.6	75.0	23.5	59.8
×74	2.98	6170	73.2	35.2	65.5	20.8	52.7
×63	1.83	4990	84.0	34.6	54.1	17.5	44.2
×53	1.12	4090	97.2	34.2	44.7	14.7	37.0
HP10×57	1.97	2240	54.3	24.1	34.8	13.1	33.2
×42	0.81	1540	70.2	23.4	24.7	9.64	24.2
HP8×36	0.77	578	44.1	15.4	14.0	6.62	16.8

FLEXURAL-TORSIONAL PROPERTIES
Channels

[

Designation	Torsional Constant J in.^{4}	Warping Constant C_w in.^{6}	Polar Radius of Gyration $\bar{r}_o$* in.	Flexural Constant H* No Units
C15×50	2.67	492	5.49	.937
×40	1.46	411	5.72	.927
×33.9	1.02	358	5.94	.920
C12×30	0.87	151	4.55	.919
×25	0.54	130	4.72	.909
×20.7	0.37	112	4.93	.899
C10×30	1.23	79.3	3.63	.921
×25	0.69	68.4	3.75	.912
×20	0.37	56.9	3.93	.900
×15.3	0.21	45.6	4.19	.883
C9×20	0.43	39.4	3.46	.899
×15	0.21	31.0	3.69	.882
×13.4	0.17	28.2	3.79	.874
C8×18.75	0.44	25.1	3.06	.894
×13.75	0.19	19.2	3.27	.874
×11.5	0.13	16.5	3.42	.862
C7×12.25	0.16	11.2	2.87	.862
×9.8	0.10	9.18	3.02	.846
C6×13	0.24	7.22	2.37	.858
×10.5	0.13	5.95	2.49	.843
×8.2	0.08	4.72	2.65	.824
C5×9	0.11	2.93	2.10	.814
×6.7	0.06	2.22	2.26	.790
C4×7.25	0.08	1.24	1.75	.768
×5.4	0.04	0.92	1.89	.741
C3×6	0.07	0.46	1.39	.689
×5	0.04	0.38	1.45	.674
×4.1	0.03	0.31	1.53	.656

*See LRFD Specification Appendix E3.

FLEXURAL-TORSIONAL PROPERTIES
Channels

[

Designation	Torsional Constant J in.^{4}	Warping Constant C_w in.^{6}	Polar Radius of Gyration $\bar{r}_o{}^*$ in.	Flexural Constant H^* No Units
MC18×58	2.81	1070	6.56	.944
×51.9	2.03	986	6.70	.939
×45.8	1.45	897	6.88	.933
×42.7	1.23	852	6.97	.930
MC13×50	2.98	558	5.07	.875
×40	1.57	463	5.33	.860
×35	1.14	413	5.50	.849
×31.8	0.94	380	5.64	.842
MC12×50	3.24	411	4.77	.859
×45	2.35	374	4.87	.851
×40	1.70	336	5.01	.842
×35	1.25	297	5.18	.832
×31	1.01	268	5.34	.821
×10.6	0.06	11.7	4.27	.983
MC10×41.1	2.27	270	4.26	.790
×33.6	1.21	224	4.47	.771
×28.5	0.79	194	4.68	.752
MC10×25	0.64	125	4.46	.802
×22	0.51	111	4.63	.790
MC10×8.4	0.04	7.01	3.68	.972
MC9×25.4	0.69	104	4.08	.770
×23.9	0.60	98.2	4.15	.763
MC8×22.8	0.57	75.3	3.85	.716
×21.4	0.50	70.9	3.91	.709
×20	0.44	47.9	3.59	.780
×18.7	0.38	45.1	3.65	.773
×8.5	0.06	8.22	3.24	.910
MC7×22.7	0.63	58.5	3.53	.662
×19.1	0.41	49.4	3.71	.638
MC6×18	0.38	34.6	3.46	.562
MC6×16.3	0.34	22.1	3.11	.643
×15.1	0.29	20.6	3.18	.634
MC6×12	0.15	11.2	2.80	.740

*See LRFD Specification Appendix E3.

FLEXURAL-TORSIONAL PROPERTIES
Single Angles

Designation	Torsional Constant J in.^{4}	Warping Constant C_w in.^{6}	Polar Radius of Gyration $\bar{r}_o$* in.	Flexural Constant H* No Units
L8×8×1⅛	7.13	32.5	4.31	0.632
1	5.08	23.4	4.35	0.630
⅞	3.46	16.1	4.37	0.629
¾	2.21	10.4	4.41	0.627
⅝	1.30	6.16	4.45	0.627
⁹⁄₁₆	0.960	4.55	4.47	0.627
½	0.682	3.23	4.48	0.624
L8×6×1	4.35	16.3	3.89	—
⅞	2.96	11.3	3.93	—
¾	1.90	7.28	3.96	—
⅝	1.12	4.33	3.99	—
⁹⁄₁₆	0.822	3.20	4.01	—
½	0.584	2.28	4.02	—
⁷⁄₁₆	0.396	1.55	4.04	—
L8×4×1	3.68	12.9	3.77	—
⅞	2.48	8.89	3.79	—
¾	1.61	5.75	3.82	—
⅝	0.933	3.42	3.85	—
⁹⁄₁₆	0.704	2.53	3.86	—
½	0.501	1.80	3.88	—
⁷⁄₁₆	0.328	1.22	3.89	—
L7×4×¾	1.47	3.97	3.33	—
⅝	0.873	2.37	3.36	—
½	0.459	1.25	3.38	—
⁷⁄₁₆	0.300	0.851	3.40	—
⅜	0.200	0.544	3.42	—

*See LRFD Specification Appendix E3.

FLEXURAL-TORSIONAL PROPERTIES
Single Angles

Designation	Torsional Constant J in.^{4}	Warping Constant C_w in.^{6}	Polar Radius of Gyration $\bar{r}_o$* in.	Flexural Constant H* No Units
L6×6×1	3.68	9.24	3.19	0.637
7/8	2.51	6.41	3.22	0.632
3/4	1.61	4.17	3.26	0.629
5/8	0.954	2.50	3.29	0.628
9/16	0.704	1.85	3.31	0.627
1/2	0.501	1.32	3.32	0.627
7/16	0.340	0.899	3.34	0.627
3/8	0.218	0.575	3.36	0.626
5/16	0.129	0.338	3.38	0.625
L6×4×7/8	2.07	4.04	2.83	—
3/4	1.33	2.64	2.86	—
5/8	0.792	1.59	2.89	—
9/16	0.585	1.18	2.90	—
1/2	0.417	0.843	2.92	—
7/16	0.284	0.575	2.94	—
3/8	0.183	0.369	2.96	—
5/16	0.108	0.217	2.97	—
L6×3½×½	0.396	0.779	2.88	—
3/8	0.174	0.341	2.92	—
5/16	0.103	0.201	2.93	—
L5×5×7/8	2.07	3.53	2.65	0.634
3/4	1.33	2.32	2.68	0.634
5/8	0.792	1.40	2.71	0.630
1/2	0.417	0.744	2.74	0.630
7/16	0.284	0.508	2.77	0.629
3/8	0.183	0.327	2.79	0.627
5/16	0.108	0.193	2.81	0.626

*See LRFD Specification Appendix E3.

FLEXURAL-TORSIONAL PROPERTIES
Single Angles

Designation	Torsional Constant J in.^{4}	Warping Constant C_w in.^{6}	Polar Radius of Gyration $\bar{r}_o{}^*$ in.	Flexural Constant H^* No Units
L5×3½×¾	1.11	1.52	2.37	—
⅝	0.660	0.918	2.40	—
½	0.348	0.491	2.44	—
⅜	0.153	0.217	2.47	—
5⁄16	0.0905	0.128	2.49	—
¼	0.0479	0.0670	2.50	—
L5×3×½	0.322	0.444	2.39	—
7⁄16	0.219	0.304	2.41	—
⅜	0.141	0.196	2.42	—
5⁄16	0.0832	0.116	2.43	—
¼	0.0438	0.0606	2.45	—
L4×4×¾	1.02	1.12	2.11	0.639
⅝	0.610	0.680	2.14	0.631
½	0.322	0.366	2.17	0.632
7⁄16	0.219	0.252	2.19	0.631
⅜	0.141	0.162	2.20	0.625
5⁄16	0.0832	0.0963	2.22	0.623
¼	0.0438	0.0505	2.23	0.627
L4×3½×½	0.301	0.302	2.04	—
⅜	0.132	0.134	2.08	—
5⁄16	0.0782	0.0798	2.09	—
¼	0.0412	0.0419	2.11	—

*See LRFD Specification Appendix E3.

FLEXURAL-TORSIONAL PROPERTIES
Single Angles

Designation	Torsional Constant J in.^{4}	Warping Constant C_w in.^{6}	Polar Radius of Gyration $\bar{r}_o{}^*$ in.	Flexural Constant H^* No Units
L4×3×⅝	0.529	0.472	1.91	—
½	0.281	0.255	1.95	—
⁷⁄₁₆	0.192	0.176	1.96	—
⅜	0.123	0.114	1.98	—
⁵⁄₁₆	0.0731	0.0676	2.00	—
¼	0.0386	0.0356	2.01	—
L3½×3½×½	0.281	0.238	1.89	0.631
⁷⁄₁₆	0.192	0.164	1.91	0.629
⅜	0.123	0.106	1.91	0.628
⁵⁄₁₆	0.0731	0.0634	1.93	0.627
¼	0.0386	0.0334	1.95	0.626
L3½×3×½	0.260	0.191	1.76	—
⅜	0.114	0.0858	1.79	—
⁵⁄₁₆	0.0680	0.0512	1.81	—
¼	0.0360	0.0270	1.83	—
L3½×2½×½	0.234	0.159	1.67	—
⅜	0.103	0.0714	1.70	—
¼	0.0322	0.0225	1.73	—
L3×3×½	0.234	0.144	1.60	0.634
⁷⁄₁₆	0.160	0.100	1.61	0.632
⅜	0.103	0.0652	1.63	0.629
⁵⁄₁₆	0.0611	0.0390	1.65	0.628
¼	0.0322	0.0206	1.66	0.627
³⁄₁₆	0.0142	0.00899	1.68	0.626

*See LRFD Specification Appendix E3.

FLEXURAL-TORSIONAL PROPERTIES
Single Angles

Designation	Torsional Constant J in.^{4}	Warping Constant C_W in.^{6}	Polar Radius of Gyration $\bar{r}_o{}^*$ in.	Flexural Constant H^* No Units
L3×2½×½	0.213	0.112	1.47	—
⅜	0.0943	0.0507	1.50	—
⁵⁄₁₆	0.0560	0.0304	1.52	—
¼	0.0296	0.0161	1.54	—
³⁄₁₆	0.0131	0.00705	1.55	—
L3×2×½	0.192	0.0908	1.40	—
⅜	0.0855	0.0413	1.43	—
⁵⁄₁₆	0.0509	0.0248	1.45	—
¼	0.0270	0.0132	1.46	—
³⁄₁₆	0.0120	0.00576	1.48	—
L2½×2½×½	0.185	0.0791	1.31	0.639
⅜	0.0816	0.0362	1.34	0.632
⁵⁄₁₆	0.0483	0.0218	1.36	0.630
¼	0.0253	0.0116	1.37	0.628
³⁄₁₆	0.0110	0.00510	1.39	0.627
L2½×2×⅜	0.0728	0.0268	1.22	—
⁵⁄₁₆	0.0432	0.0162	1.24	—
¼	0.0227	0.00868	1.25	—
³⁄₁₆	0.00990	0.00382	1.27	—
L2×2×⅜	0.0640	0.0174	1.05	0.637
⁵⁄₁₆	0.0381	0.0106	1.07	0.633
¼	0.0201	0.00572	1.09	0.630
³⁄₁₆	0.00880	0.00254	1.10	0.628
⅛	0.00274	0.00079	1.12	0.626

*See LRFD Specification Appendix E3.

FLEXURAL-TORSIONAL PROPERTIES
Structural Tees

Designation	Torsional Constant J in.^{4}	Warping Constant C_w in.^{6}	Polar Radius of Gyration $\bar{r}_o{}^*$ in.	Flexural Constant H^* No Units
WT22×167.5	37.2	434	8.81	0.724
×145	25.7	279	8.67	0.733
×131	18.9	204	8.65	0.731
×115	12.4	139	8.67	0.723
WT20×296.5**	223	2340	8.30	0.761
×251.5**	140	1420	8.17	0.760
×215.5	88.5	881	8.09	0.756
×186	58.2	559	8.00	0.756
×160.5	37.7	350	7.92	0.756
×148.5	30.6	279	7.88	0.756
×138.5	25.8	218	7.75	0.770
×124.5	19.1	158	7.71	0.770
×107.5	12.4	101	7.66	0.770
×99.5	9.14	83.5	7.83	0.746
×87	5.60	65.3	8.12	0.699
WT20×233**	139	1360	8.39	0.680
×196**	86.1	802	8.27	0.678
×165.5	53.0	485	8.19	0.674
×139	32.4	278	8.07	0.676
×132	28.0	233	8.02	0.680
×117.5	20.6	156	7.88	0.690
×105.5	15.2	113	7.84	0.690
×91.5	10.0	72.1	7.79	0.691
×83.5	7.01	62.9	8.02	0.658
×74.5	4.68	51.9	8.24	0.626
WT18×424**	622	6880	8.08	0.802
×399**	527	5700	8.02	0.801
×325**	295	3010	7.82	0.797
×263.5**	163	1570	7.63	0.797
×219.5**	96.7	894	7.52	0.794
×196.5**	70.7	637	7.44	0.796
×179.5**	54.3	480	7.38	0.797
×164**	42.1	363	7.32	0.799
×150	32.0	278	7.30	0.797
×140	26.2	226	7.27	0.796
×130	20.7	181	7.28	0.791
×122.5	17.3	151	7.28	0.788
×115	14.3	125	7.27	0.784

*See LRFD Specification Section E3.
**Group 4 or Group 5 shape. See Notes in Table 1-2.

FLEXURAL-TORSIONAL PROPERTIES
Structural Tees

Designation	Torsional Constant J in.^{4}	Warping Constant C_w in.^{6}	Polar Radius of Gyration $\bar{r}_o{}^*$ in.	Flexural Constant H^* No Units
WT18×128	26.6	205	7.43	0.703
×116	19.8	151	7.40	0.703
×105	13.9	119	7.49	0.687
×97	11.1	92.7	7.45	0.687
×91	9.19	77.6	7.45	0.686
×85	7.51	63.2	7.44	0.684
×80	6.17	53.6	7.46	0.678
×75	5.04	46.0	7.50	0.670
×67.5	3.48	37.3	7.65	0.644
WT16.5×177**	57.2	468	7.00	0.802
×159**	42.1	335	6.94	0.803
×145.5**	32.4	256	6.90	0.801
×131.5**	24.2	188	6.86	0.802
×120.5	17.9	146	6.91	0.792
×110.5	13.7	113	6.90	0.788
×100.5	10.2	84.9	6.89	0.784
WT16.5×84.5	8.83	55.4	6.74	0.714
×76	6.16	43.0	6.82	0.700
×70.5	4.84	35.4	6.85	0.691
×65	3.67	29.3	6.93	0.678
×59	2.64	23.4	7.02	0.659
WT15×238.5**	151	1170	6.65	0.819
×195.5**	85.9	636	6.54	0.815
×163**	50.8	361	6.40	0.817
×146**	37.2	257	6.34	0.818
×130.5	26.7	184	6.31	0.815
×117.5	19.9	132	6.25	0.817
×105.5	13.9	96.4	6.27	0.809
×95.5	10.3	71.2	6.25	0.806
×86.5	7.61	53.0	6.25	0.802
WT15×74	7.27	37.6	6.10	0.716
×66	4.85	28.5	6.19	0.698
×62	3.98	23.9	6.20	0.693
×58	3.21	20.5	6.24	0.683
×54	2.49	17.3	6.31	0.669
×49.5	1.88	14.3	6.38	0.654
×45	1.42	10.5	6.34	0.655

*See LRFD Specification Section E3.
**Group 4 or Group 5 shape. See Notes in Table 1-2.

FLEXURAL-TORSIONAL PROPERTIES
Structural Tees

Designation	Torsional Constant J in.^{4}	Warping Constant C_w in.^{6}	Polar Radius of Gyration $\bar{r_o}^*$ in.	Flexural Constant H^* No Units
WT13.5×269.5**	245	1740	6.27	0.830
×224**	146	977	6.11	0.829
×184**	83.6	532	5.97	0.828
×153.5**	49.8	304	5.85	0.828
×129	30.2	178	5.77	0.828
×117.5	23.0	135	5.74	0.825
×108.5	18.5	105	5.72	0.830
×97	13.2	74.3	5.66	0.826
×89	9.74	57.7	5.70	0.815
×80.5	7.31	42.7	5.67	0.813
×73	5.44	31.7	5.65	0.810
WT13.5×64.5	5.60	24.0	5.48	0.731
×57	3.65	17.5	5.54	0.716
×51	2.64	12.6	5.52	0.714
×47	2.01	10.2	5.57	0.703
×42	1.40	7.79	5.63	0.685
WT12×246**	223	1340	5.71	0.838
×204**	133	748	5.55	0.836
×167.5**	76.0	405	5.40	0.837
×139.5**	45.3	230	5.28	0.837
×125**	33.3	165	5.22	0.838
×114.5	25.7	125	5.19	0.836
×103.5	19.1	91.3	5.14	0.836
×96	15.4	72.5	5.11	0.836
×88	12.0	55.8	5.09	0.835
×81	9.22	43.8	5.09	0.831
×73	6.70	31.9	5.08	0.827
×65.5	4.74	23.1	5.09	0.818
×58.5	3.35	16.4	5.08	0.813
×52	2.35	11.6	5.07	0.809
WT12×51.5	3.54	12.3	4.88	0.733
×47	2.62	9.57	4.89	0.727
×42	1.84	6.90	4.89	0.721
×38	1.34	5.30	4.93	0.709
×34	0.932	4.08	4.99	0.692
WT12×31	0.850	3.92	5.13	0.619
×27.5	0.588	2.93	5.18	0.606

*See LRFD Specification Section E3.
**Group 4 or Group 5 shape. See Notes in Table 1-2.

FLEXURAL-TORSIONAL PROPERTIES
Structural Tees

Designation	Torsional Constant J in.^{4}	Warping Constant C_w in.^{6}	Polar Radius of Gyration $\bar{r}_o{}^*$ in.	Flexural Constant H^* No Units
WT10.5×100.5	20.6	85.4	4.67	0.859
×91	15.4	63.0	4.64	0.859
×83	11.9	47.3	4.59	0.861
×73.5	7.69	32.5	4.64	0.847
×66	5.62	23.4	4.61	0.845
×61	4.47	18.4	4.58	0.846
×55.5	3.40	13.8	4.56	0.846
×50.5	2.60	10.4	4.54	0.846
WT10.5×46.5	3.01	9.33	4.37	0.729
×41.5	2.16	6.50	4.33	0.732
×36.5	1.51	4.42	4.31	0.732
×34	1.22	3.62	4.31	0.727
×31	0.513	2.78	4.31	0.722
WT10.5×28.5	0.884	2.50	4.36	0.665
×25	0.570	1.89	4.44	0.640
×22	0.383	1.40	4.49	0.623
WT9×155.5**	87.2	339	4.42	0.875
×141.5**	66.5	251	4.36	0.873
×129**	51.5	189	4.30	0.874
×117**	39.4	140	4.23	0.875
×105.5**	29.4	102	4.19	0.873
×96	22.4	75.7	4.14	0.875
×87.5	17.0	56.5	4.10	0.872
×79	12.6	41.2	4.06	0.872
×71.5	9.70	30.7	4.03	0.874
×65	7.30	22.8	3.99	0.874
WT9×59.5	5.30	17.4	4.03	0.862
×53	3.73	12.1	4.00	0.860
×48.5	2.92	9.29	3.97	0.862
×43	2.04	6.42	3.95	0.860
×38	1.41	4.37	3.92	0.862
WT9×35.5	1.74	3.96	3.72	0.751
×32.5	1.36	3.01	3.69	0.755
×30	1.08	2.35	3.67	0.756
×27.5	0.829	1.84	3.68	0.749
×25	0.613	1.36	3.66	0.748
WT9×23	0.609	1.20	3.67	0.694
×20	0.403	0.788	3.65	0.692
×17.5	0.252	0.598	3.74	0.662

*See LRFD Specification Section E3.
**Group 4 or Group 5 shape. See Notes in Table 1-2.

FLEXURAL-TORSIONAL PROPERTIES
Structural Tees

Designation	Torsional Constant J in.^{4}	Warping Constant C_W in.^{6}	Polar Radius of Gyration $\bar{r}_o$* in.	Flexural Constant H* No Units
WT8×50	3.85	10.4	3.62	0.877
×44.5	2.72	7.19	3.60	0.877
×38.5	1.78	4.61	3.56	0.877
×33.5	1.19	3.01	3.53	0.879
WT8×28.5	1.10	1.99	3.30	0.770
×25	0.760	1.34	3.28	0.770
×22.5	0.655	0.974	3.27	0.767
×20	0.396	0.673	3.24	0.769
×18	0.271	0.516	3.30	0.745
WT8×15.5	0.229	0.366	3.26	0.695
×13	0.130	0.243	3.32	0.667
WT7×404**	918	6970	5.67	0.959
×365**	714	5250	5.47	0.966
×332.5**	555	3920	5.36	0.966
×302.5**	430	2930	5.25	0.966
×275**	331	2180	5.15	0.967
×250**	255	1620	5.06	0.967
×227.5**	196	1210	4.98	0.967
WT7×213**	164	991	4.92	0.968
×199**	135	801	4.87	0.968
×185**	110	640	4.81	0.968
×171**	88.3	502	4.77	0.968
×155.5**	67.5	375	4.71	0.968
×141.5**	51.8	281	4.66	0.969
×128.5**	39.3	209	4.61	0.969
×116.5**	29.6	154	4.56	0.970
×105.5	22.2	113	4.52	0.970
×96.5	17.3	87.2	4.49	0.971
×88	13.2	65.2	4.46	0.971
×79.5	9.84	47.9	4.42	0.971
×72.5	7.56	36.3	4.40	0.971

*See LRFD Specification Section E3.
**Group 4 or Group 5 shape. See Notes in Table 1-2.

FLEXURAL-TORSIONAL PROPERTIES
Structural Tees

Designation	Torsional Constant J in.^{4}	Warping Constant C_w in.^{6}	Polar Radius of Gyration $\bar{r}_o{}^*$ in.	Flexural Constant H^* No Units
WT7×66	6.13	26.6	4.21	0.966
×60	4.67	20.0	4.18	0.966
×54.5	3.55	15.0	4.16	0.968
×49.5	2.68	11.1	4.14	0.968
×45	2.03	8.31	4.12	0.968
WT7×41	2.53	5.63	3.25	0.912
×37	1.94	4.19	3.21	0.917
×34	1.51	3.21	3.19	0.915
×30.5	1.10	2.29	3.18	0.915
WT7×26.5	0.970	1.46	2.89	0.868
×24	0.726	1.07	2.87	0.866
×21.5	0.524	0.751	2.85	0.866
WT7×19	0.398	0.554	2.87	0.800
×17	0.284	0.400	2.86	0.793
×15	0.190	0.287	2.90	0.772
WT7×13	0.179	0.207	2.82	0.713
×11	0.104	0.134	2.86	0.691

*See LRFD Specification Section E3.

FLEXURAL-TORSIONAL PROPERTIES
Structural Tees

Designation	Torsional Constant J in.^{4}	Warping Constant C_w in.^{6}	Polar Radius of Gyration $\bar{r}_o$* in.	Flexural Constant H* No Units
WT6×168**	120	481	4.07	0.958
×152.5**	92.0	356	4.00	0.959
×139.5**	70.9	267	3.94	0.957
×126**	53.5	195	3.88	0.958
×115**	41.6	148	3.84	0.958
×105**	32.2	112	3.79	0.958
×95	24.4	82.1	3.74	0.959
×85	17.7	58.3	3.69	0.960
×76	12.8	41.3	3.65	0.960
×68	9.22	28.9	3.61	0.960
×60	6.43	19.7	3.58	0.959
×53	4.55	13.6	3.54	0.961
×48	3.42	10.1	3.51	0.961
×43.5	2.54	7.34	3.49	0.960
×39.5	1.92	5.43	3.46	0.960
×36	1.46	4.07	3.45	0.961
×32.5	1.09	2.97	3.43	0.960
WT6×29	1.05	2.08	3.01	0.944
×26.5	0.788	1.53	3.00	0.940
WT6×25	0.889	1.23	2.67	0.899
×22.5	0.656	0.885	2.64	0.898
×20	0.476	0.620	2.62	0.901
WT6×17.5	0.369	0.437	2.56	0.835
×15	0.228	0.267	2.55	0.830
×13	0.150	0.174	2.54	0.826
WT6×11	0.146	0.137	2.52	0.683
×9.5	0.0899	0.0934	2.54	0.663
×8	0.0511	0.0678	2.62	0.624
×7	0.0350	0.0493	2.64	0.610

*See LRFD Specification Section E3.
**Group 4 or Group 5 shape. See Notes in Table 1-2.

FLEXURAL-TORSIONAL PROPERTIES
Structural Tees

Designation	Torsional Constant J in.^{4}	Warping Constant C_W in.^{6}	Polar Radius of Gyration $\bar{r}_o{}^*$ in.	Flexural Constant H^* No Units
WT5×56	7.50	16.9	3.04	0.963
×50	5.41	11.9	3.00	0.964
×44	3.75	8.02	2.98	0.964
×38.5	2.55	5.31	2.93	0.964
×34	1.78	3.62	2.92	0.965
×30	1.23	2.46	2.89	0.965
×27	0.909	1.78	2.87	0.966
×24.5	0.693	1.33	2.85	0.966
WT5×22.5	0.753	0.981	2.44	0.940
×19.5	0.487	0.616	2.42	0.936
×16.5	0.291	0.356	2.40	0.927
WT5×15	0.310	0.273	2.17	0.848
×13	0.201	0.173	2.15	0.848
×11	0.119	0.107	2.17	0.831
WT5×9.5	0.116	0.0796	2.08	0.728
×8.5	0.0776	0.061	2.12	0.702
×7.5	0.0518	0.0475	2.16	0.672
×6	0.0272	0.0255	2.16	0.662

*See LRFD Specification Section E3.

FLEXURAL-TORSIONAL PROPERTIES
Structural Tees

Designation	Torsional Constant J in.^{4}	Warping Constant C_w in.^{6}	Polar Radius of Gyration $\bar{r}_o{}^*$ in.	Flexural Constant H^* No Units
WT4×33.5	2.52	3.56	2.41	0.962
×29	1.66	2.28	2.39	0.961
×24	0.979	1.30	2.34	0.966
×20	0.559	0.715	2.31	0.961
×17.5	0.385	0.480	2.29	0.963
×15.5	0.268	0.327	2.29	0.961
WT4×14	0.268	0.230	1.97	0.935
×12	0.173	0.144	1.96	0.936
WT4×10.5	0.141	0.0916	1.80	0.877
×9	0.0855	0.0562	1.81	0.863
WT4×7.5	0.0679	0.0382	1.72	0.762
×6.5	0.0433	0.0269	1.74	0.732
×5	0.0212	0.0114	1.69	0.748
WT3×12.5	0.229	0.171	1.76	0.952
×10	0.120	0.0858	1.73	0.952
×7.5	0.0504	0.0342	1.71	0.937
WT3×8	0.111	0.0426	1.37	0.880
×6	0.0449	0.0178	1.37	0.846
×4.5	0.0202	0.0074	1.34	0.852
WT2.5×9.5	0.154	0.0775	1.44	0.964
×8	0.0930	0.0453	1.43	0.962
WT2×6.5	0.0750	0.0213	1.16	0.947

*See LRFD Specification Section E3.

FLEXURAL-TORSIONAL PROPERTIES
Structural Tees

Designation	Torsional Constant J in.^{4}	Warping Constant C_w in.^{6}	Polar Radius of Gyration $\bar{r}_o$* in.	Flexural Constant H* No Units
MT6×5.9	0.0307	0.0330	2.69	0.564
×5.4	0.0196	0.0252	2.67	0.572
MT5×4.5	0.0213	0.0133	2.21	0.584
×4	0.0116	0.00916	2.21	0.582
MT4×3.25	0.0146	0.00421	1.73	0.611
MT2.5×9.45**	0.165	0.0732	1.37	0.951

*See LRFD Specification Section E3.
**This shape has tapered flanges while other MT shapes have parallel flanges.

FLEXURAL-TORSIONAL PROPERTIES
Structural Tees

Designation	Torsional Constant J in.^{4}	Warping Constant C_w in.^{6}	Polar Radius of Gyration $\bar{r_o}^*$ in.	Flexural Constant H^* No Units
ST12×60.5	6.38	27.5	5.14	0.640
×53	5.04	15.0	4.87	0.685
ST12×50	3.76	19.5	5.27	0.584
×45	3.01	12.1	5.12	0.616
×40	2.43	6.94	4.89	0.657
ST10×48	4.15	15.0	4.36	0.625
×43	3.30	9.17	4.20	0.661
ST10×37.5	2.28	7.21	4.28	0.612
×33	1.78	4.02	4.10	0.655
ST9×35	2.05	7.03	4.01	0.583
×27.35	1.18	2.26	3.71	0.662
ST7.5×25	1.05	2.02	3.22	0.637
×21.45	0.767	0.995	3.04	0.689
ST6×25	1.39	1.97	2.60	0.663
×20.4	0.872	0.787	2.42	0.733
ST6×17.5	0.538	0.556	2.49	0.697
×15.9	0.449	0.364	2.39	0.731
ST5×17.5	0.633	0.725	2.23	0.653
×12.7	0.300	0.173	1.98	0.768
ST4×11.5	0.271	0.168	1.74	0.707
×9.2	0.167	0.0642	1.59	0.789
ST3×8.625	0.182	0.0772	1.36	0.706
×6.25	0.0838	0.0197	1.21	0.820
ST2.5×5	0.0568	0.0100	1.02	0.842
ST2×4.75	0.0589	0.00995	0.907	0.800
×3.85	0.0364	0.00457	0.841	0.872
ST1.5×3.75	0.0440	0.00496	0.737	0.832
×2.85	0.0220	0.00189	0.672	0.913

*See LRFD Specification Section E3.

FLEXURAL-TORSIONAL PROPERTIES
Double Angles

	Long Legs Vertical						Short Legs Vertical					
	Back to Back of Angles, in.						Back to Back of Angles, in.					
	0		3/8		3/4		0		3/8		3/4	
Designation	$\bar{r}_o{}^*$	H^*	$\bar{r}_o{}^*$	H^*	$\bar{r}_o{}^*$	H^*	$\bar{r}_o{}^*$	H^*	$\bar{r}_o{}^*$	H^*	$\bar{r}_o{}^*$	H^*
L8×8×1⅛	4.58	0.837	4.68	0.844	4.79	0.851	4.58	0.837	4.68	0.844	4.79	0.851
1	4.58	0.833	4.68	0.840	4.79	0.847	4.58	0.833	4.68	0.840	4.79	0.847
⅞	4.58	0.831	4.68	0.838	4.78	0.845	4.58	0.831	4.68	0.838	4.78	0.845
¾	4.58	0.828	4.68	0.835	4.78	0.842	4.58	0.828	4.68	0.835	4.78	0.842
⅝	4.58	0.825	4.68	0.832	4.78	0.839	4.58	0.825	4.68	0.832	4.78	0.839
½	4.59	0.822	4.69	0.829	4.78	0.836	4.59	0.822	4.69	0.829	4.78	0.836
L8×6×1	4.07	0.721	4.15	0.731	4.23	0.742	4.19	0.925	4.31	0.929	4.44	0.933
¾	4.08	0.714	4.16	0.724	4.24	0.735	4.17	0.919	4.29	0.924	4.41	0.928
½	4.11	0.708	4.18	0.718	4.26	0.728	4.17	0.914	4.28	0.919	4.40	0.923
L8×4×1	3.87	0.566	3.93	0.578	3.99	0.591	4.12	0.982	4.26	0.983	4.41	0.984
¾	3.89	0.562	3.94	0.573	4.00	0.586	4.08	0.980	4.22	0.981	4.36	0.982
½	3.93	0.558	3.97	0.568	4.03	0.580	4.05	0.977	4.19	0.979	4.33	0.980
L7×4×¾	3.42	0.609	3.48	0.623	3.55	0.637	3.58	0.968	3.71	0.971	3.85	0.973
½	3.45	0.604	3.50	0.616	3.57	0.629	3.55	0.965	3.68	0.967	3.82	0.969
⅜	3.46	0.602	3.51	0.614	3.57	0.627	3.54	0.963	3.67	0.965	3.80	0.968
L6×6×1	3.43	0.843	3.54	0.852	3.65	0.861	3.43	0.843	3.54	0.852	3.65	0.861
⅞	3.43	0.838	3.54	0.847	3.65	0.856	3.43	0.838	3.54	0.847	3.65	0.856
¾	3.44	0.833	3.54	0.842	3.65	0.852	3.44	0.833	3.54	0.842	3.65	0.852
⅝	3.44	0.830	3.54	0.839	3.64	0.848	3.44	0.830	3.54	0.839	3.64	0.848
½	3.44	0.827	3.54	0.836	3.64	0.845	3.44	0.827	3.54	0.836	3.64	0.845
⅜	3.44	0.822	3.54	0.831	3.64	0.841	3.44	0.822	3.54	0.831	3.64	0.841
L6×4×¾	2.98	0.672	3.05	0.687	3.13	0.704	3.10	0.948	3.23	0.952	3.36	0.956
⅝	2.98	0.668	3.05	0.683	3.13	0.699	3.09	0.946	3.21	0.950	3.34	0.954
½	3.00	0.663	3.06	0.678	3.14	0.693	3.08	0.943	3.20	0.947	3.34	0.951
⅜	3.01	0.661	3.07	0.675	3.15	0.690	3.07	0.940	3.19	0.944	3.32	0.948
L6×3½×⅜	2.97	0.610	3.02	0.624	3.09	0.640	3.05	0.961	3.17	0.964	3.31	0.967
5/16	2.97	0.610	3.02	0.624	3.09	0.639	3.03	0.960	3.16	0.963	3.29	0.966
L5×5×⅞	2.87	0.844	2.97	0.855	3.09	0.865	2.87	0.844	2.97	0.855	3.09	0.865
¾	2.85	0.839	2.96	0.850	3.07	0.861	2.85	0.839	2.96	0.850	3.07	0.861
½	2.86	0.830	2.96	0.841	3.07	0.852	2.86	0.830	2.96	0.841	3.07	0.852
⅜	2.87	0.824	2.96	0.835	3.07	0.846	2.87	0.824	2.96	0.835	3.07	0.846
5/16	2.87	0.821	2.97	0.833	3.07	0.844	2.87	0.821	2.97	0.833	3.07	0.844

*See LRFD Specification Section E3.

FLEXURAL-TORSIONAL PROPERTIES
Double Angles

	Long Legs Vertical						Short Legs Vertical					
	Back to Back of Angles, in.						Back to Back of Angles, in.					
	0		3/8		3/4		0		3/8		3/4	
Designation	$\bar{r}_o$*	H*	$\bar{r}_o$*	H*	$\bar{r}_o$*	H*	$\bar{r}_o$*	H*	$\bar{r}_o$*	H*	$\bar{r}_o$*	H*
L5×3½×¾	2.50	0.697	2.58	0.715	2.67	0.734	2.61	0.943	2.74	0.948	2.87	0.953
½	2.51	0.685	2.59	0.703	2.67	0.722	2.59	0.936	2.71	0.941	2.84	0.947
⅜	2.52	0.682	2.59	0.699	2.67	0.717	2.58	0.932	2.70	0.938	2.83	0.943
5/16	2.53	0.679	2.60	0.695	2.68	0.713	2.58	0.930	2.70	0.936	2.82	0.942
L5×3×½	2.45	0.626	2.52	0.645	2.59	0.665	2.55	0.962	2.69	0.965	2.82	0.969
⅜	2.46	0.623	2.52	0.641	2.60	0.661	2.54	0.959	2.67	0.963	2.80	0.966
5/16	2.47	0.621	2.53	0.638	2.60	0.657	2.54	0.957	2.67	0.961	2.80	0.965
¼	2.48	0.618	2.54	0.634	2.61	0.653	2.53	0.956	2.66	0.960	2.79	0.964
L4×4×¾	2.29	0.847	2.40	0.861	2.52	0.873	2.29	0.847	2.40	0.861	2.52	0.873
⅝	2.29	0.839	2.40	0.853	2.51	0.867	2.29	0.839	2.40	0.853	2.51	0.867
½	2.29	0.834	2.39	0.848	2.50	0.862	2.29	0.834	2.39	0.848	2.50	0.862
⅜	2.29	0.827	2.39	0.841	2.50	0.855	2.29	0.827	2.39	0.841	2.50	0.855
5/16	2.29	0.824	2.39	0.838	2.50	0.852	2.29	0.824	2.39	0.838	2.50	0.852
¼	2.29	0.823	2.39	0.837	2.49	0.850	2.29	0.823	2.39	0.837	2.49	0.850
L4×3½×½	2.15	0.783	2.24	0.801	2.34	0.818	2.17	0.881	2.29	0.892	2.41	0.903
⅜	2.15	0.774	2.24	0.792	2.34	0.809	2.17	0.875	2.28	0.887	2.40	0.898
5/16	2.15	0.774	2.24	0.791	2.34	0.808	2.17	0.872	2.28	0.884	2.40	0.895
¼	2.16	0.770	2.24	0.787	2.34	0.805	2.17	0.870	2.28	0.882	2.39	0.893
L4×3×½	2.04	0.719	2.12	0.740	2.22	0.762	2.10	0.924	2.22	0.933	2.35	0.940
⅜	2.04	0.714	2.12	0.735	2.21	0.757	2.09	0.919	2.21	0.928	2.34	0.935
5/16	2.05	0.710	2.13	0.731	2.22	0.752	2.09	0.917	2.21	0.925	2.33	0.933
¼	2.06	0.706	2.13	0.726	2.22	0.747	2.09	0.914	2.20	0.923	2.33	0.931
L3½×3½×⅜	2.00	0.831	2.10	0.847	2.22	0.862	2.00	0.831	2.10	0.847	2.22	0.862
5/16	2.01	0.827	2.10	0.843	2.21	0.858	2.01	0.827	2.10	0.843	2.21	0.858
¼	2.01	0.824	2.10	0.839	2.21	0.855	2.01	0.824	2.10	0.839	2.21	0.855
L3½×3×⅜	1.86	0.771	1.95	0.791	2.06	0.812	1.89	0.884	2.00	0.897	2.13	0.909
5/16	1.87	0.766	1.96	0.787	2.06	0.807	1.89	0.881	2.00	0.894	2.12	0.906
¼	1.87	0.762	1.96	0.782	2.06	0.803	1.89	0.878	2.00	0.891	2.12	0.903
L3½×2½×⅜	1.76	0.696	1.84	0.721	1.94	0.748	1.82	0.932	1.94	0.941	2.08	0.948
¼	1.77	0.691	1.85	0.715	1.93	0.740	1.81	0.927	1.93	0.936	2.06	0.944
L3×3×½	1.72	0.842	1.83	0.860	1.95	0.877	1.72	0.842	1.83	0.860	1.95	0.877
⅜	1.72	0.834	1.82	0.852	1.94	0.869	1.72	0.834	1.82	0.852	1.94	0.869
5/16	1.72	0.830	1.82	0.848	1.93	0.866	1.72	0.830	1.82	0.848	1.93	0.866
¼	1.72	0.825	1.82	0.844	1.93	0.862	1.72	0.825	1.82	0.844	1.93	0.862
3/16	1.72	0.822	1.82	0.841	1.93	0.858	1.72	0.822	1.82	0.841	1.93	0.858

*See LRFD Specification Section E3.

FLEXURAL-TORSIONAL PROPERTIES
Double Angles

	Long Legs Vertical						Short Legs Vertical					
	Back to Back of Angles, in.						Back to Back of Angles, in.					
	0		$\frac{3}{8}$		$\frac{3}{4}$		0		$\frac{3}{8}$		$\frac{3}{4}$	
Designation	$\bar{r}_o{}^*$	H^*	$\bar{r}_o{}^*$	H^*	$\bar{r}_o{}^*$	H^*	$\bar{r}_o{}^*$	H^*	$\bar{r}_o{}^*$	H^*	$\bar{r}_o{}^*$	H^*
L3×2½×⅜	1.58	0.763	1.67	0.789	1.78	0.813	1.61	0.896	1.73	0.910	1.86	0.922
¼	1.59	0.754	1.67	0.779	1.78	0.804	1.61	0.889	1.72	0.903	1.84	0.916
³⁄₁₆	1.59	0.750	1.67	0.775	1.77	0.800	1.61	0.885	1.72	0.899	1.84	0.912
L3×2×⅜	1.49	0.672	1.57	0.704	1.66	0.737	1.55	0.949	1.68	0.956	1.82	0.963
⁵⁄₁₆	1.50	0.667	1.57	0.698	1.66	0.730	1.55	0.946	1.68	0.954	1.82	0.961
¼	1.50	0.664	1.57	0.694	1.66	0.726	1.54	0.943	1.67	0.951	1.80	0.958
³⁄₁₆	1.50	0.661	1.57	0.690	1.66	0.721	1.54	0.940	1.66	0.949	1.80	0.956
L2½×2½×⅜	1.43	0.839	1.54	0.861	1.66	0.880	1.43	0.839	1.54	0.861	1.66	0.880
⁵⁄₁₆	1.43	0.834	1.54	0.856	1.66	0.876	1.43	0.834	1.54	0.856	1.66	0.876
¼	1.43	0.829	1.53	0.851	1.65	0.871	1.43	0.829	1.53	0.851	1.65	0.871
³⁄₁₆	1.43	0.825	1.53	0.847	1.65	0.867	1.43	0.825	1.53	0.847	1.65	0.867
L2½×2×⅜	1.29	0.752	1.39	0.785	1.50	0.816	1.33	0.912	1.45	0.927	1.59	0.939
⁵⁄₁₆	1.30	0.746	1.39	0.779	1.50	0.810	1.33	0.908	1.45	0.923	1.58	0.935
¼	1.30	0.741	1.39	0.773	1.50	0.804	1.33	0.903	1.45	0.919	1.58	0.932
³⁄₁₆	1.31	0.736	1.39	0.767	1.49	0.798	1.32	0.899	1.44	0.915	1.57	0.928
L2×2×⅜	1.15	0.846	1.26	0.873	1.39	0.896	1.15	0.846	1.26	0.873	1.39	0.896
⁵⁄₁₆	1.15	0.840	1.26	0.867	1.38	0.890	1.15	0.840	1.26	0.867	1.38	0.890
¼	1.15	0.834	1.25	0.861	1.38	0.885	1.15	0.834	1.25	0.861	1.38	0.885
³⁄₁₆	1.15	0.828	1.25	0.855	1.37	0.880	1.15	0.828	1.25	0.855	1.37	0.880
⅛	1.15	0.822	1.25	0.850	1.37	0.875	1.15	0.822	1.25	0.850	1.37	0.875

*See LRFD Specification Section E3.

SURFACE AREAS AND BOX AREAS
W shapes
Square feet per foot of length

Desig-nation	Case A	Case B	Case C	Case D	Desig-nation	Case A	Case B	Case C	Case D
W44×335	11.0	12.4	8.67	10.0	W36×256	9.02	10.0	7.26	8.27
×290	11.0	12.3	8.59	9.91	×232	8.96	9.97	7.20	8.21
×262	10.9	12.2	8.53	9.84	×210	8.91	9.93	7.13	8.15
×230	10.9	12.2	8.46	9.78	×194	8.88	9.89	7.09	8.10
					W36×182	8.85	9.85	7.06	8.07
W40×593	10.9	12.3	8.56	9.95	×170	8.82	9.82	7.03	8.03
×503	10.7	12.1	8.38	9.75	×160	8.79	9.79	7.00	8.00
×431	10.5	11.9	8.23	9.58	×150	8.76	9.76	6.97	7.97
×372	10.4	11.8	8.11	9.45	×135	8.71	9.70	6.92	7.92
×321	10.3	11.6	8.01	9.33					
×297	10.3	11.6	7.96	9.28	W33×354	9.66	11.0	7.27	8.61
×277	10.3	11.6	7.93	9.25	×318	9.58	10.9	7.19	8.52
×249	10.2	11.5	7.88	9.19	×291	9.52	10.8	7.13	8.46
×215	10.2	11.5	7.81	9.12	×263	9.46	10.8	7.07	8.39
×199	10.1	11.4	7.76	9.07	×241	9.42	10.7	7.02	8.34
×174	10.0	11.3	7.68	8.99	×221	9.38	10.7	6.97	8.29
					×201	9.33	10.6	6.93	8.24
W40×466	9.79	10.8	8.13	9.18					
×392	9.61	10.6	7.96	8.99	W33×169	8.30	9.26	6.60	7.55
×331	9.47	10.5	7.81	8.83	×152	8.27	9.23	6.55	7.51
×278	9.35	10.3	7.69	8.69	×141	8.23	9.19	6.51	7.47
×264	9.32	10.3	7.66	8.66	×130	8.20	9.15	6.47	7.43
×235	9.28	10.3	7.61	8.60	×118	8.15	9.11	6.43	7.39
×211	9.22	10.2	7.55	8.53					
×183	9.17	10.2	7.48	8.47	W30×477	9.30	10.6	7.02	8.35
×167	9.11	10.1	7.42	8.40	×391	9.11	10.4	6.83	8.13
×149	9.05	10.0	7.35	8.34	×326	8.96	10.2	6.68	7.96
					×292	8.88	10.2	6.61	7.88
W36×848	11.1	12.6	8.59	10.1	×261	8.81	10.1	6.53	7.79
×798	11.0	12.5	8.49	9.99	×235	8.75	10.0	6.47	7.73
×650	10.7	12.1	8.21	9.67	×211	8.71	9.97	6.42	7.67
×527	10.4	11.9	7.97	9.41	×191	8.66	9.92	6.37	7.62
×439	10.3	11.7	7.79	9.20	×173	8.62	9.87	6.32	7.57
×393	10.2	11.6	7.70	9.10					
×359	10.1	11.5	7.63	9.02	W30×148	7.53	8.40	5.99	6.86
×328	10.0	11.4	7.57	8.95	×132	7.49	8.37	5.93	6.81
×300	9.99	11.4	7.51	8.90	×124	7.47	8.34	5.90	6.78
×280	9.95	11.3	7.47	8.85	×116	7.44	8.31	5.88	6.75
×260	9.90	11.3	7.42	8.80	×108	7.41	8.28	5.84	6.72
×245	9.87	11.2	7.39	8.77	×99	7.37	8.25	5.81	6.68
×230	9.84	11.2	7.36	8.73	×90	7.35	8.22	5.79	6.66

Case A: Shape perimeter, minus one flange surface.
Case B: Shape perimeter.
Case C: Box perimeter, equal to one flange surface plus twice the depth.
Case D: Box perimeter, equal to two flange surfaces plus twice the depth.

SURFACE AREAS AND BOX AREAS
W shapes
Square feet per foot of length

Desig-nation	Case A	Case B	Case C	Case D	Desig-nation	Case A	Case B	Case C	Case D
W27×539	8.82	10.09	6.69	7.96	W21×201	6.75	7.80	4.89	5.93
×448	8.61	9.86	6.48	7.73	×182	6.69	7.74	4.83	5.87
×368	8.42	9.64	6.29	7.51	×166	6.65	7.68	4.78	5.82
×307	8.27	9.47	6.14	7.34	×147	6.61	7.66	4.72	5.76
×281	8.21	9.40	6.08	7.27	×132	6.57	7.61	4.68	5.71
×258	8.15	9.34	6.02	7.21	×122	6.54	7.57	4.65	5.68
×235	8.09	9.27	5.96	7.14	×111	6.51	7.54	4.61	5.64
×217	8.04	9.22	5.91	7.09	×101	6.48	7.50	4.58	5.61
×194	7.98	9.15	5.85	7.02					
×178	7.95	9.12	5.81	6.98	W21×93	5.54	6.24	4.31	5.01
×161	7.91	9.08	5.77	6.94	×83	5.50	6.20	4.27	4.96
×146	7.87	9.03	5.73	6.89	×73	5.47	6.16	4.23	4.92
					×68	5.45	6.14	4.21	4.90
W27×129	6.92	7.75	5.44	6.27	×62	5.42	6.11	4.19	4.87
×114	6.88	7.72	5.39	6.23					
×102	6.85	7.68	5.35	6.18	W21×57	5.01	5.56	4.06	4.60
×94	6.82	7.65	5.32	6.15	×50	4.97	5.51	4.02	4.56
×84	6.78	7.61	5.28	6.11	×44	4.94	5.48	3.99	4.53
W24×492	8.07	9.25	6.12	7.29	W18×311	6.41	7.41	4.72	5.72
×408	7.86	9.01	5.91	7.06	×283	6.32	7.31	4.63	5.62
×335	7.66	8.79	5.71	6.84	×258	6.24	7.23	4.56	5.54
×279	7.51	8.62	5.56	6.67	×234	6.17	7.14	4.48	5.45
×250	7.44	8.54	5.49	6.59	×211	6.10	7.06	4.41	5.37
×229	7.38	8.47	5.43	6.52	×192	6.03	6.99	4.35	5.30
×207	7.32	8.40	5.37	6.45	×175	5.97	6.92	4.29	5.24
×192	7.27	8.35	5.32	6.40	×158	5.92	6.86	4.23	5.17
×176	7.23	8.31	5.28	6.35	×143	5.87	6.81	4.18	5.12
×162	7.22	8.30	5.25	6.33	×130	5.83	6.76	4.14	5.07
×146	7.17	8.24	5.20	6.27					
×131	7.12	8.19	5.15	6.22	W18×119	5.81	6.75	4.10	5.04
×117	7.08	8.15	5.11	6.18	×106	5.77	6.70	4.06	4.99
×104	7.04	8.11	5.07	6.14	×97	5.74	6.67	4.03	4.96
					×86	5.70	6.62	3.99	4.91
W24×103	6.18	6.93	4.84	5.59	×76	5.67	6.59	3.95	4.87
×94	6.16	6.92	4.81	5.56					
×84	6.12	6.87	4.77	5.52	W18×71	4.85	5.48	3.71	4.35
×76	6.09	6.84	4.74	5.49	×65	4.82	5.46	3.69	4.32
×68	6.06	6.80	4.70	5.45	×60	4.80	5.43	3.67	4.30
					×55	4.78	5.41	3.65	4.27
W24×62	5.57	6.16	4.54	5.13	×50	4.76	5.38	3.62	4.25
×55	5.54	6.13	4.51	5.10					

Case A: Shape perimeter, minus one flange surface.
Case B: Shape perimeter.
Case C: Box perimeter, equal to one flange surface plus twice the depth.
Case D: Box perimeter, equal to two flange surfaces plus twice the depth.

SURFACE AREAS AND BOX AREAS
W shapes
Square feet per foot of length

Designation	Case A	Case B	Case C	Case D	Designation	Case A	Case B	Case C	Case D
W18×46	4.41	4.91	3.51	4.02	W14×82	4.75	5.59	3.23	4.07
×40	4.38	4.88	3.48	3.99	×74	4.72	5.56	3.20	4.04
×35	4.34	4.84	3.45	3.95	×68	4.69	5.53	3.18	4.01
					×61	4.67	5.50	3.15	3.98
W16×100	5.28	6.15	3.70	4.57					
×89	5.24	6.10	3.66	4.52	W14×53	4.19	4.86	2.99	3.66
×77	5.19	6.05	3.61	4.47	×48	4.16	4.83	2.97	3.64
×67	5.16	6.01	3.57	4.43	×43	4.14	4.80	2.94	3.61
W16×57	4.39	4.98	3.33	3.93	W14×38	3.93	4.50	2.91	3.48
×50	4.36	4.95	3.30	3.89	×34	3.91	4.47	2.89	3.45
×45	4.33	4.92	3.27	3.86	×30	3.89	4.45	2.87	3.43
×40	4.31	4.89	3.25	3.83					
×36	4.28	4.87	3.23	3.81	W14×26	3.47	3.89	2.74	3.16
					×22	3.44	3.86	2.71	3.12
W16×31	3.92	4.39	3.11	3.57					
×26	3.89	4.35	3.07	3.53	W12×336	5.77	6.88	3.92	5.03
					×305	5.67	6.77	3.82	4.93
W14×808	7.74	9.28	5.35	6.90	×279	5.59	6.68	3.74	4.83
×730	7.61	9.10	5.23	6.72	×252	5.50	6.58	3.65	4.74
×665	7.46	8.93	5.08	6.55	×230	5.43	6.51	3.58	4.66
×605	7.32	8.77	4.94	6.39	×210	5.37	6.43	3.52	4.58
×550	7.19	8.62	4.81	6.24	×190	5.30	6.36	3.45	4.51
×500	7.07	8.49	4.68	6.10	×170	5.23	6.28	3.39	4.43
×455	6.96	8.36	4.57	5.98	×152	5.17	6.21	3.33	4.37
					×136	5.12	6.15	3.27	4.30
W14×426	6.89	8.28	4.50	5.89	×120	5.06	6.09	3.21	4.24
×398	6.81	8.20	4.43	5.81	×106	5.02	6.03	3.17	4.19
×370	6.74	8.12	4.36	5.73	×96	4.98	5.99	3.13	4.15
×342	6.67	8.03	4.29	5.65	×87	4.95	5.96	3.10	4.11
×311	6.59	7.94	4.21	5.56	×79	4.92	5.93	3.07	4.08
×283	6.52	7.86	4.13	5.48	×72	4.89	5.90	3.05	4.05
×257	6.45	7.78	4.06	5.40	×65	4.87	5.87	3.02	4.02
×233	6.38	7.71	4.00	5.32					
×211	6.32	7.64	3.94	5.25	W12×58	4.39	5.22	2.87	3.70
×193	6.27	7.58	3.89	5.20	×53	4.37	5.20	2.84	3.68
×176	6.22	7.53	3.84	5.15					
×159	6.18	7.47	3.79	5.09	W12×50	3.90	4.58	2.71	3.38
×145	6.14	7.43	3.76	5.05	×45	3.88	4.55	2.68	3.35
					×40	3.86	4.52	2.66	3.32
W14×132	5.93	7.16	3.67	4.90					
×120	5.90	7.12	3.64	4.86					
×109	5.86	7.08	3.60	4.82					
×99	5.83	7.05	3.57	4.79					
×90	5.81	7.02	3.55	4.76					

Case A: Shape perimeter, minus one flange surface.
Case B: Shape perimeter.
Case C: Box perimeter, equal to one flange surface plus twice the depth.
Case D: Box perimeter, equal to two flange surfaces plus twice the depth.

SURFACE AREAS AND BOX AREAS
W shapes
Square feet per foot of length

Designation	Case A	Case B	Case C	Case D	Designation	Case A	Case B	Case C	Case D
W12×35	3.63	4.18	2.63	3.18	W8×21	2.61	3.05	1.82	2.26
×30	3.60	4.14	2.60	3.14	×18	2.59	3.03	1.79	2.23
×26	3.58	4.12	2.58	3.12					
					W8×15	2.27	2.61	1.69	2.02
W12×22	2.97	3.31	2.39	2.72	×13	2.25	2.58	1.67	2.00
×19	2.95	3.28	2.36	2.69	×10	2.23	2.56	1.64	1.97
×16	2.92	3.25	2.33	2.66					
×14	2.90	3.23	2.32	2.65	W6×25	2.49	3.00	1.57	2.08
					×20	2.46	2.96	1.54	2.04
W10×112	4.30	5.17	2.76	3.63	×15	2.42	2.92	1.50	2.00
×100	4.25	5.11	2.71	3.57					
×88	4.20	5.06	2.66	3.52	W6×16	1.98	2.31	1.38	1.72
×77	4.15	5.00	2.62	3.47	×12	1.93	2.26	1.34	1.67
×68	4.12	4.96	2.58	3.42	×9	1.90	2.23	1.31	1.64
×60	4.08	4.92	2.54	3.38					
×54	4.06	4.89	2.52	3.35	W5×19	2.04	2.45	1.28	1.70
×49	4.04	4.87	2.50	3.33	×16	2.01	2.43	1.25	1.67
W10×45	3.56	4.23	2.35	3.02	W4×13	1.63	1.96	1.03	1.37
×39	3.53	4.19	2.32	2.98					
×33	3.49	4.16	2.29	2.95					
W10×30	3.10	3.59	2.23	2.71					
×26	3.08	3.56	2.20	2.68					
×22	3.05	3.53	2.17	2.65					
W10×19	2.63	2.96	2.04	2.38					
×17	2.60	2.94	2.02	2.35					
×15	2.58	2.92	2.00	2.33					
×12	2.56	2.89	1.97	2.30					
W8×67	3.42	4.11	2.19	2.88					
×58	3.37	4.06	2.14	2.83					
×48	3.32	4.00	2.09	2.77					
×40	3.28	3.95	2.05	2.72					
×35	3.25	3.92	2.02	2.69					
×31	3.23	3.89	2.00	2.67					
W8×28	2.87	3.42	1.89	2.43					
×24	2.85	3.39	1.86	2.40					

Case A: Shape perimeter, minus one flange surface.
Case B: Shape perimeter.
Case C: Box perimeter, equal to one flange surface plus twice the depth.
Case D: Box perimeter, equal to two flange surfaces plus twice the depth.

CAMBER

Beams and Girders

Camber and sweep are used to form a desired curvature in either rolled beams or welded girders. Camber denotes a curve in the vertical plane. Beams and girders can be cambered to compensate for the anticipated deflection or for architectural reasons. Note that the required camber is determined at service (unfactored) load levels. Sweep denotes a curve in the horizontal plane. Camber and sweep may be induced through cold bending or through the application of heat.

The minimum radius for cold cambering in members up to a nominal depth of 30 inches is between 10 and 14 times the depth of the member; deeper members will require a larger minimum radius. Cold bending may be used to provide sweep in members to practically any radius desired. Note that a length limit of 40 to 50 feet is practical.

Heat cambering, sweeping, and straightening are provided through controlled heat application. The member is rapidly heated in selected areas which tend to expand, but are restrained by the adjacent cooler areas, causing plastic deformation of the heated areas and a change in the shape of the cooled member. The mechanical properties of steels are largely unaffected by such heating operations, provided the maximum temperature does not exceed 1,100°F for quenched and tempered alloy steels, and 1,300°F for other steels. The temperature should be carefully checked by temperature-indicating crayons or other suitable means during the heating process. Cambering and sweeping induces residual stresses similar to those that develop in rolled structural shapes as elements of the shape cool from the rolling temperature at different rates. In general, these residual stresses do not affect the ultimate strength of structural members. Additionally, the effect of residual stresses is incorporated in the provisions of the LRFD Specification.

Note that when a cambered beam bearing on a wall or other support is loaded, expansion of the unrestrained end must be considered. In Figure 1-5(a), the end will move a distance Δ, where

$$\Delta = \frac{4Cd}{L}$$

If instead the cambered beam is supported on a simple shear connection at both ends, the top and bottom flange will each move a distance of one-half Δ since end rotation will occur approximately about the neutral axis. The designer should be aware of the magnitude of these movements and make provisions to accommodate them. Figure 1-5(a) considers the geometry of a girder in the horizontal position, and Figure 1-5(b) illustrates the condition when the girder is not level.

Trusses

"Cambering" of trusses is accomplished by geometric relocation of panel points and adjustment of member lengths; it does not involve physical cold bending or the application of heat as with beams and girders.

The following discussion of cambering to compensate for the anticipated deflection of a truss is applicable for any parabolic condition; large-radius circular curves will be approximated very closely by the technique described. Cambering to compensate for the axial deformation of the members of a truss is beyond the scope of this Manual; refer to a textbook on mechanics of materials.

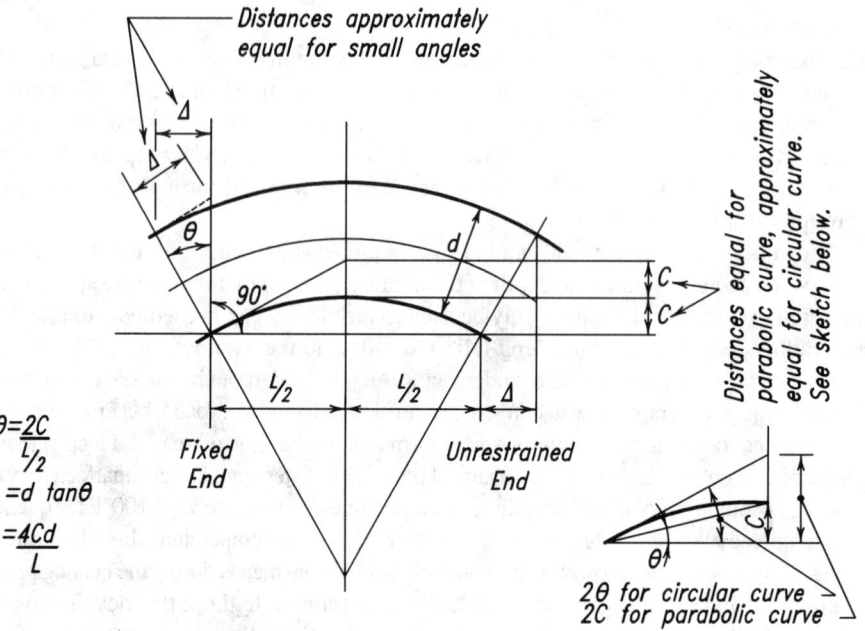

(a) Beam or Girder Ends at
Same Elevations

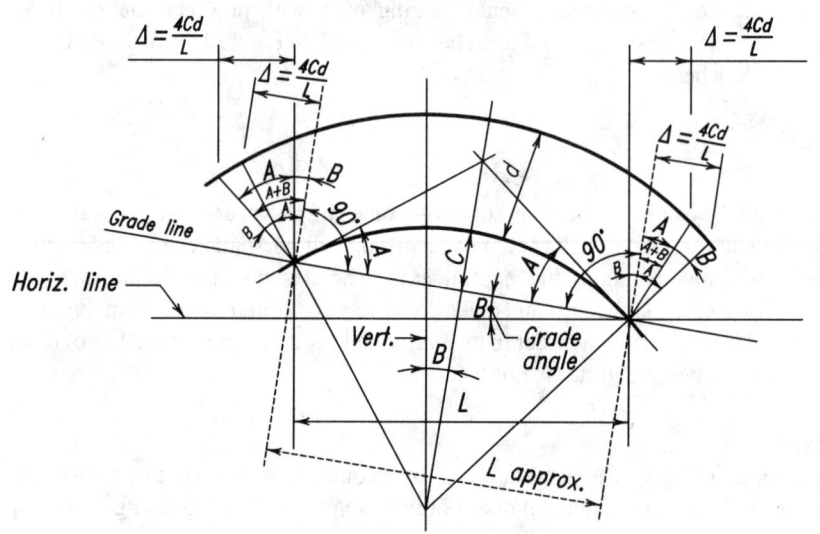

(b) Beam or Girder Ends at
Different Elevations

Fig. 1-5. Camber for beams and girders.

The usual method of providing camber in building trusses is to progressively raise each panel point. The lengths of the verticals are not changed, but the lengths of the diagonals are calculated on the basis of the adjusted elevation for the several panel points. For any simple-span truss, the offset above a straight base line, at the several panel points, can be computed from the following equations if the vertical curve forming the camber is taken as a parabola.

$$D = C - C\left(\frac{B}{A}\right)^2 = C\left[1 - \left(\frac{B}{A}\right)^2\right]$$

where

A = Horizontal distance from end panel point to mid-span of the truss (half the truss span).
B = Horizontal distance from mid-span of the truss to panel point for which offset is to be determined.
C = Required mid-span camber.
D = Offset from the base-line at panel point corresponding to distance B.

A and B must be expressed in the same units; similarly C and D must be expressed in the same units, but not necessarily the same units as A and B. When the truss is divided into any number of approximately equal panels, it may be convenient to express distances A and B in panel lengths.

For the truss of Figure 1-6(a) with eight equal panels, distance A is taken as four panel lengths. Assuming the camber at the midpoint is specified as 1½-in., the offset at panel point 1, where B equals three panel lengths, is:

$$D = 1\tfrac{1}{2}\text{-in.}\left[1 - \left(\frac{3}{4}\right)^2\right]$$
$$= {}^{21}\!/_{32}\text{-in.}$$

The offset at panel point 2, where B equals two panel lengths, is:

$$D = 1\tfrac{1}{2}\text{-in.}\left[1 - \left(\frac{2}{4}\right)^2\right]$$
$$= 1\tfrac{1}{8}\text{-in.}$$

The offset at panel point 3, where B equals one panel length, is:

$$D = 1\tfrac{1}{2}\text{-in.}\left[1 - \left(\frac{1}{4}\right)^2\right]$$
$$= 1^{13}\!/_{32}\text{-in.}$$

Finally, the offset at panel point 4, where B equals zero, is

$$D = C = 1\tfrac{1}{2}\text{-in.}$$

An alternative method of determining the amount of camber at intermediate panel points when all panel points are approximately the same distance apart is as follows. Using the truss in Figure 1-6(a) as an example, sketch the camber diagram and number the panel points, starting with the first panel point from the end of the truss, from 1 to 4, as shown in Figure 1-6(b) on line A. Next, on line B, reverse the numbering as shown. Finally, on line C, enter the product of the numbers on lines A and B.

The camber at any panel point is the amount of camber at the centerline of the truss multiplied by the fraction whose numerator is the figure on line C at the given panel point, and whose denominator is the figure on line C at the center line of the truss. Thus, at panel point 1, the camber is

$$\frac{7}{16} \times 1\frac{1}{2}\text{-in.} = \frac{21}{32}\text{-in.}$$

at panel point 2, the camber is

$$\frac{12}{16} \times 1\frac{1}{2}\text{-in.} = 1\frac{1}{8}\text{-in.}$$

at panel point 3, the camber is

$$\frac{15}{16} \times 1\frac{1}{2}\text{-in.} = 1\frac{13}{32}\text{-in.}$$

and at panel point 4, the camber is

$$\frac{16}{16} \times 1\frac{1}{2}\text{-in.} = 1\frac{1}{2}\text{-in.}$$

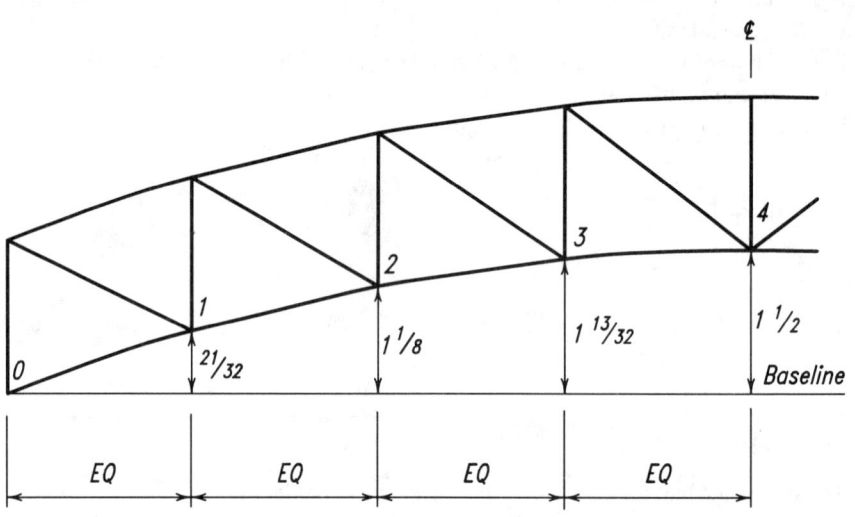

(a) Calculated camber ordinates by formula

Panel point	1	2	3	₵ 4
line A	1	2	3	4
line B	x7	x6	x5	x4
line C	7	12	15	16

(b) Alternative calculation method for approximately equal panels

Fig. 1-6. Camber for trusses.

STANDARD MILL PRACTICE

General Information

Rolling structural shapes and plates involves such factors as roll wear, subsequent roll dressing, temperature variations, etc., which cause the finished product to vary from published profiles. Such variations are limited by the provisions of the American Society for Testing and Materials Specification A6. Contained in this section is a summary of these provisions, not a reproduction of the complete specification. In its entirety, A6 covers a group of common requirements, which, unless otherwise specified in the purchase order or in an individual specification, apply to rolled steel plates, shapes, sheet piling, and bars.

As indicated in Table 1-1, carbon steel refers to ASTM designations A36 and A529; high-strength, low-alloy steel refers to designations A242, A572, and A588; alloy steel refers to designation A514; and low-alloy steel refers to A852.

For further information on mill practices, including permissible variations for rolled tees, zees, and bulb angles in structural and bar sizes, pipe, tubing, sheets, and strip, and for other grades of steel, see ASTM A6, A53, A500, A568, and A618; the Steel Products Manuals of the Iron and Steel Society (American Institute of Mining, Metallurgical, and Petroleum Engineers); and producers' catalogs.

The data on spreading rolls to increase areas and weights, and mill cambering of beams, is not a part of ASTM A6.

Additional material on mill practice is included in the descriptive material preceding the "Dimensions and Properties" tables for shapes and plates.

Letter symbols representing dimensions on sketches shown herein are in accordance with ASTM A6, AISI and mill catalogs and *not necessarily as defined by the general nomenclature of this manual.*

Methods of Increasing Areas and Weights by Spreading Rolls

W Shapes
To vary the area and weight within a given nominal size, the flange width, the flange thickness, and the web thickness are changed as shown in Figure 1-7(a).

S Shapes and American Standard Channels
To vary the area and weight within a given nominal size, the web thickness and the flange width are changed by an equal amount as shown in Figures 1-7(b) and (c).

Angles
To vary area and weight for a given leg length, the thickness of each leg is changed. Note that the leg length is changed slightly by this method (Figure 1-7(d)).

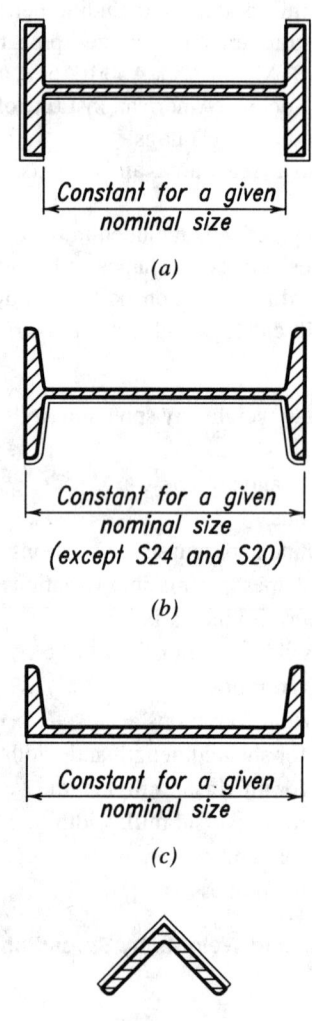

Fig. 1-7. Varying areas and weights.

Cambering of Rolled Beams

All beams are straightened after rolling to meet permissible variations for sweep and camber listed hereinafter for W shapes and S shapes. The following data refer to the subsequent cold cambering of beams to produce a predetermined dimension.

The maximum lengths that can be cambered depend on the length to which a given section can be rolled, with a maximum of 100 feet. Table 1-10 outlines the maximum and minimum induced camber of W shapes and S shapes.

Consult the producer for specific camber and/or lengths outside the above listed available lengths and sections.

Mill camber in beams of less depth than tabulated should not be specified.

A single minimum value for camber, within the ranges shown above for the length ordered, should be specified.

Camber is measured at the mill and will not necessarily be present in the same amount in the section of beam as received due to release of stress induced during the cambering operation. In general 75 percent of the specified camber is likely to remain.

Camber will approximate a simple regular curve nearly the full length of the beam, or between any two points specified.

Camber is ordinarily specified by the ordinate at the mid-length of the portion of the beam to be curved. Ordinates at the other points should not be specified.

Although mill cambering to achieve reverse or other compound curves is not considered practical, fabricating shop facilities for cambering by heat can accomplish such results as well as form regular curves in excess of the limits tabulated above. Refer to the earlier section Effect of Heat of Steel for further information.

Table 1-10.
Cambering of Rolled Beams

Maximum and Minimum Induced Camber

Sections, Nominal Depth, in.	Specified Length of Beam, ft				
	Over 30 to 42, incl.	Over 42 to 52, incl.	Over 52 to 65, incl.	Over 65 to 85, incl.	Over 85 to 100, incl.
	Max. and Min. Camber Acceptable, in.				
W shapes, 24 and over	1 to 2, incl.	1 to 3, incl.	2 to 4, incl.	3 to 5, incl.	3 to 6, incl.
W shapes, 14 to 21, incl. and S shapes, 12 in. and over	¾ to 2½, incl.	1 to 3, incl.	—	—	—

Permissible Variations for Camber Ordinate

Lengths	Plus Variation	Minus Variation
50 ft and less	½-in.	0
Over 50 ft	½-in. plus ⅛-in. for each 10 ft or fraction thereof in excess of 50 ft	0

Table 1-11.
Positions for Measuring Camber and Sweep

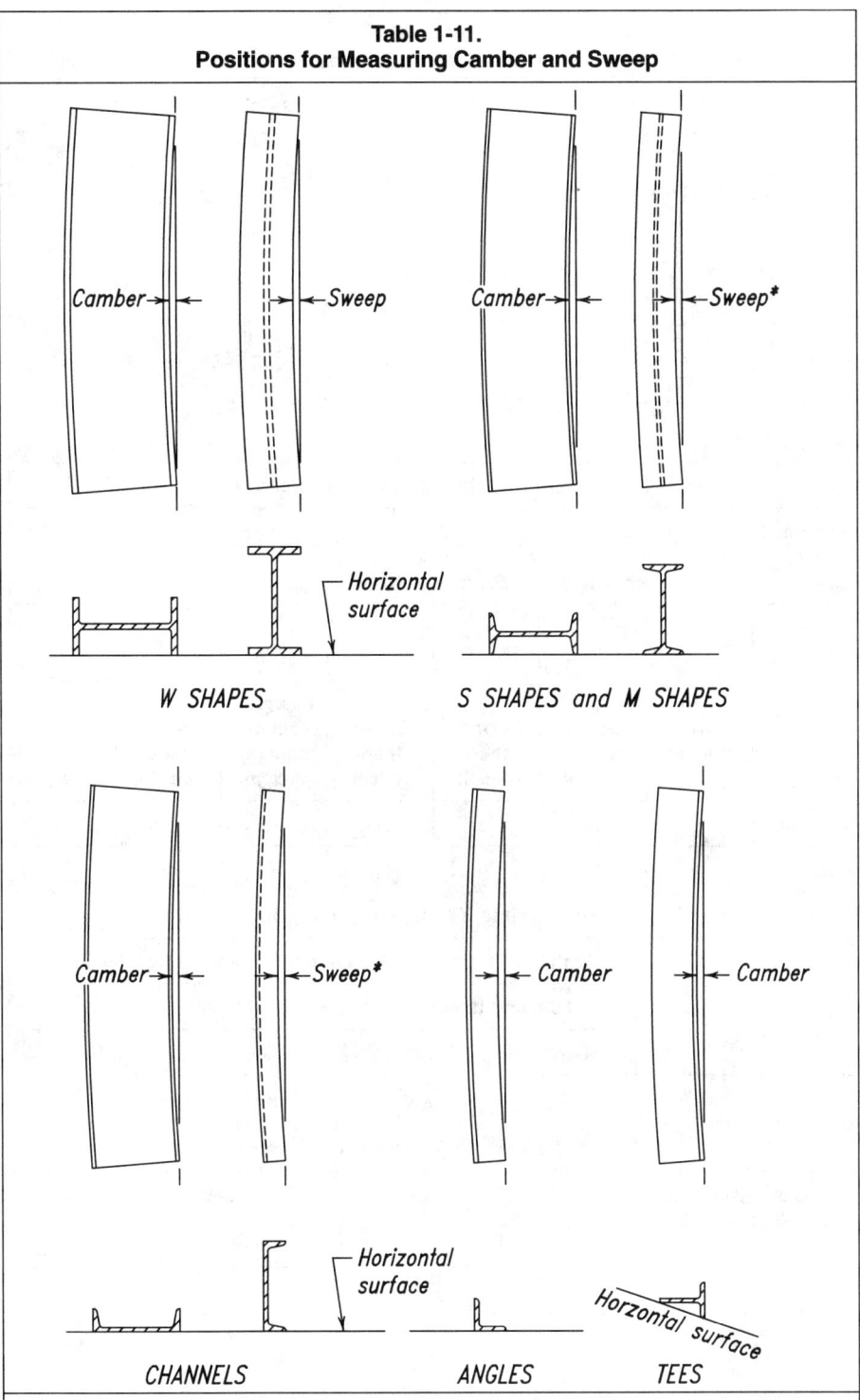

*Due to the extreme variations in flexibility of these shapes, straightness tolerances for sweep are subject to negotiations between manufacturer and purchaser for individual sections involved.

Table 1-12.
W Shapes, HP Shapes

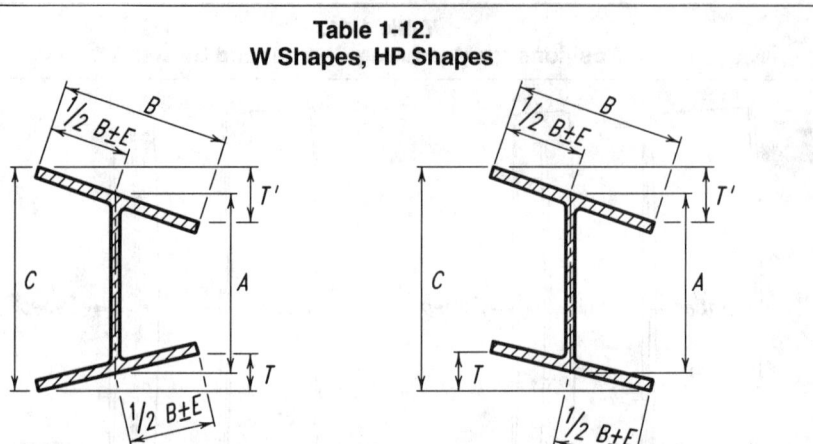

Permissible Variations in Cross Section

Section Nominal Size, in.	A, Depth, in.		B, Fig. Width, in.				C, Max. Depth at any Cross-section over Theoretical Depth, in.
	Over Theoretical	Under Theoretical	Over Theoretical	Under Theoretical	$T + T'$ Flanges, out of square, Max, in.	E^a, Web off Center, Max, in.	
To 12, inc.	$\frac{1}{8}$	$\frac{1}{8}$	$\frac{1}{4}$	$\frac{3}{16}$	$\frac{1}{4}$	$\frac{3}{16}$	$\frac{1}{4}$
Over 12	$\frac{1}{8}$	$\frac{1}{8}$	$\frac{1}{4}$	$\frac{3}{16}$	$\frac{5}{16}$	$\frac{3}{16}$	$\frac{1}{4}$

Permissible Variations in Length

W Shapes	Variations from Specified Length for Lengths for Given, in.			
	30 ft and Under		Over 30 ft	
	Over	Under	Over	Under
Beams 24 in. and under in nominal depth	$\frac{3}{8}$	$\frac{3}{8}$	$\frac{3}{8}$ plus $\frac{1}{16}$ for each additional 5 ft or fraction thereof	$\frac{3}{8}$
Beams over 24 in. nom. depth; all columns	$\frac{1}{2}$	$\frac{1}{2}$	$\frac{1}{2}$ plus $\frac{1}{16}$ for each additional 5 ft or fraction thereof	$\frac{1}{2}$

Notes:
[a]Variation of $\frac{5}{16}$ in. max. for sections over 426 lb / ft.

Table 1-12 (cont.).
W Shapes, HP Shapes

Other Permissible Variations

Area and weight variation: ±2.5 percent theoretical or specified amount.
Ends out-of-square: $\frac{1}{64}$-in. per in. of depth, or of flange width if it is greater than the depth.

Camber and Sweep

Sizes	Length	Permissible Variation, in.	
		Camber	Sweep
Sizes with flange width equal to or greater than 6 in.	All	$\frac{1}{8}$ in. $\times \dfrac{\text{(total length ft.)}}{10}$	
Sizes with flange width less than 6 in.	All	$\frac{1}{8}$ in. $\times \dfrac{\text{(total length ft.)}}{10}$	$\frac{1}{8}$ in. $\times \dfrac{\text{(total length ft.)}}{5}$
Certain sections with a flange width approx. equal to depth & specified on order as columns[b]	45 ft. and under	$\frac{1}{8}$ in. $\times \dfrac{\text{(total length ft.)}}{10}$ with $\frac{3}{8}$ in. max.	
	Over 45 ft.	$\frac{3}{8}$ in. $+ \left[\frac{1}{8} \text{ in.} \times \dfrac{\text{(total length ft.} - 45)}{10} \right]$	

[b]Applies only to W8×31 and heavier, W10×49 and heavier, W12×65 and heavier, W14×90 and heavier. If the other sections are specified on the order as columns, the tolerance will be subject to negotiation with the manufacturer.

Table 1-13.
S Shapes, M Shapes, and Channels

Permissible Variations in Cross Section

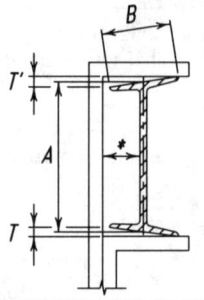 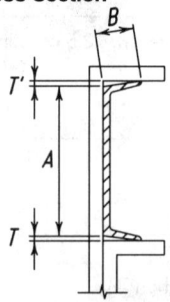

* Back of square and centerline of web to be parallel when measuring "out-of-square"

Section	Nominal Size in.	A, Depth in.[a]		B, Flange Width, in.		$T + T'$,[b] Out of Square per Inch of B, in.
		Over Theoretical	Under Theoretical	Over Theoretical	Under Theoretical	
S shapes and M shapes	3 to 7, incl.	½	1/16	⅛	⅛	1/32
	Over 7 to 14, incl.	⅛	3/32	5/32	5/32	1/32
	Over 14 to 24, incl.	3/16	⅛	3/16	3/16	1/32
Channels	3 to 4, incl.	3/32	1/16	⅛	⅛	1/32
	Over 7 to 14, incl.	⅛	3/32	⅛	5/32	1/32
	Over 14	3/16	⅛	⅛	3/16	1/32

Permissible Variations in Length

Section	Variations from Specified Length for Lengths Given, in.									
	to 30 ft., incl.		Over 30 to 40 ft., incl.		Over 40 to 50 ft., incl.		Over 50 to 65 ft., incl.		Over 65 ft.	
	Over	Under	Over	Under	Over	Under	Over	Under	Over	Under
S shapes, M shapes and Channels	½	¼	¾	¼	1	¼	1⅛	¼	1¼	¼

Other Permissible Variations

Area and weight variation: ±2.5 percent theoretical or specified amount.

Ends out-of-square: S shapes and channels 1/64-in. per in. of depth.

$$\text{Camber} = \tfrac{1}{8}\text{-in.} \times \frac{\text{total length, ft}}{5}$$

Notes:
[a] A is measured at center line of web for beams; and at back of web for channels.
[b] $T + T'$ applies when flanges of channels are toed in or out.

Table 1-14.
Tees Split from W, M, and S Shapes,
Angles Split from Channels

Permissible Variations in Depth

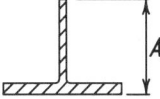 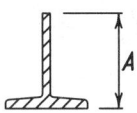 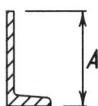

Dimension A may be approximately one-half beam or channel depth, or any dimension resulting from off-center splitting, or splitting on two lines as specified on the order.

Depth of Beam from which Tees or Angles are Split	Variations in Depth A Over and Under	
	Tees	Angles
To 6 in., excl.	$\frac{1}{8}$	$\frac{1}{8}$
6 to 16, excl.	$\frac{3}{16}$	$\frac{3}{16}$
16 to 20, excl.	$\frac{1}{4}$	$\frac{1}{4}$
20 to 24, excl.	$\frac{5}{16}$	—
24 and over	$\frac{3}{8}$	—

The above variations for depths to tees or angles include the permissible variations in depth for the beams and channels before splitting.

Other Permissible Variations

Other permissible variations in cross section as well as permissible variations in length, area, and weight variation, and ends out-of-square will correspond to those of the beam or channel before splitting, except

Camber = $\frac{1}{8}$-in. $\times \dfrac{\text{total length} ft}{5}$

Table 1-15.
Angles, Structural Size

Permissible Variations in Cross Section

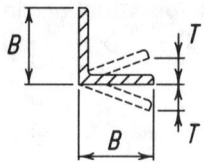

Section	Nominal Size, in.[a]	B Length of Leg, in.		T, Out of Square per in. of B, in.
		Over Theoretical	Under Theoretical	
Angles	3 to 4, incl.	$\frac{1}{8}$	$\frac{3}{32}$	$\frac{3}{128}$[b]
	Over 4 to 6, incl.	$\frac{1}{8}$	$\frac{1}{8}$	$\frac{3}{128}$[b]
	Over 6	$\frac{3}{16}$	$\frac{1}{8}$	$\frac{3}{128}$[b]

Permissible Variations in Length

Section	Variations from Specified Length for Lengths Given, in.									
	to 30 ft., incl.		Over 30 to 40 ft., incl.		Over 40 to 50 ft., incl.		Over 50 to 65 ft., incl.		Over 65 ft.	
	Over	Under	Over	Under	Over	Under	Over	Under	Over	Under
Angles	$\frac{1}{2}$	$\frac{1}{4}$	$\frac{3}{4}$	$\frac{1}{4}$	1	$\frac{1}{4}$	$1\frac{1}{8}$	$\frac{1}{4}$	$1\frac{1}{4}$	$\frac{1}{4}$

Other Permissible Variations

Area and weight variation: ±2.5 percent theoretical or specified amount.

Ends out-of square: $\frac{3}{128}$-in. per in. of leg length, or $1\frac{1}{2}$ degrees. Variations based on the longer leg of unequal angle.

Camber = $\frac{1}{8}$-in. $\times \dfrac{\text{total length, ft}}{5}$, applied to either leg

Notes;
[a]For unequal leg angles, longer leg determines classification.
[b]$\frac{1}{128}$ in. per in. = $1\frac{1}{2}$ deg.

Table 1-16.
Angles, Bar Size*

Permissible Variation in Cross Section

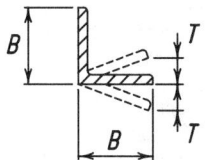

[a]Specified Length of Leg, in.	Variations from Thickness for Thicknesses Given, Over and Under, in.			B Length of Leg Over and Under, in.	T, Out of Square per Inch of B, in.
	3/16 and Under	Over 3/16 to 3/8 incl.	Over 3/8		
1 and Under	0.008	0.010		1/32	3/128[b]
Over 1 to 2, incl.	0.010	0.010	0.012	3/64	3/128[b]
Over 2 to 3, excl.	0.012	0.015	0.015	1/16	3/128[b]

Permissible Variations in Length

Section	Variations Over Specified Length for Lengths Given No Variation Under				
	50 to 10 ft. excl.	10 to 20 ft. excl.	20 to 30 ft. excl.	30 to 40 ft. excl.	40 to 65 ft. excl.
All sizes of bar-size angles	5/8	1	1½	2	2½

Other Permissible Variations

Camber: ¼-in. in any 5 feet, or ¼ in. $\times \dfrac{\text{total length,ft}}{5}$

Straightness: Because of warpage, permissible variations for straightness do not apply to bars if any subsequent heating operation has been performed.

Ends out-of-square: 3/128-in. per inch of leg length or 1½ degrees. Variation based on longer leg of an unequal angle.

Notes:
*A member is "bar size" when its greatest cross-sectional dimension is less than three inches.
[a]For unequal leg angles, longer leg determines classification.
[b]1/128 in. per in. = 1½ degrees.

Table 1-17.
Steel Pipe and Tubing

Dimensions and Weight Tolerances

Round Tubing and Pipe (see also Table 1-4)

ASTM A53

Weight—The weight of the pipe as specified in Table X2 and Table X3 (ASTM Specification A53) shall not vary by more than ±10 percent.

Note that the weight tolerance of ±10 percent is determined from the weights of the customary lifts of pipe as produced for shipment by the mill, divided by the number of feet of pipe in the lift. On pipe sizes over four inches where individual lengths may be weighed, the weight tolerance is applicable to the individual length.

Diameter—For pipe two inches and over in nominal diameter, the outside diameter shall not vary more than ±1 percent from the standard specified.

Thickness—The minimum wall thickness at any point shall not be more than 12.5 percent under the nominal wall thickness specified.

ASTM 500

Diameter—For pipe two inches and over in nominal diameter, the outside diameter shall not vary more than ±0.75 percent from the standard specified.

Thickness—The wall thickness at any point shall not be more than 10 percent under or over the nominal wall thickness specified.

ASTM A501 and ASTM 618

Outside dimensions—For round hot-formed structural tubing two inches and over in nominal size, the outside diameter shall not vary more than ±1 percent from the standard specified.

Weight (A501 only)—The weight of structural tubing shall be less than the specified value by more than 3.5 percent.

Mass (A618 only)—The mass of structural tubing shall not be less than the specified value by more than 3.5 percent.

Length—Structural tubing is commonly produced in random mill lengths and in definite cut lengths. When cut lengths are specified for structural tubing, the length tolerances shall be in accordance with the following table:

	22 ft and under		Over 22 to 44 ft, incl.	
	Over	Under	Over	Under
Length tolerance for specified cut lengths, in.	$\frac{1}{2}$	$\frac{1}{4}$	$\frac{3}{4}$	$\frac{1}{4}$

Straightness—The permissible variation for straightness of structural tubing shall be $\frac{1}{8}$-in. times the number of feet of total length divided by 5.

Table 1-17 (cont.).
Steel Pipe and Tubing

Dimensions and Weight Tolerances

Square and Rectangular Tubing (see also Table 1-4)

ASTM A500 and ASTM A618

Outside Dimensions—The specified dimensions, measured across the flats at positions at least two inches from either end-of-square or rectangular tubing and including an allowance for convexity or concavity, shall not exceed the plus and minus tolerance shown in the following table:

Largest Outside Dimension Across Flats, in.	Tolerance[a] Plus an Minus, in.
2½ and under	0.020
Over 2½ to 3½, incl.	0.025
Over 3½ to 5½, incl.	0.030
Over 5½	1 percent

[a]The respective outside dimension tolerances include the allowances for convexity and concavity.

Lengths—Structural tubing is commonly produced in random lengths, in multiple lengths, and in definite cut lengths. When cut lengths are specified for structural tubing, the length tolerances shall be in accordance with the following table:

	22 ft and under		Over 22 to 44 ft, incl.	
	Over	Under	Over	Under
Length tolerance for specified cut lengths, in.	½	¼	¾	¼

Mass (A618 only)—The mass of structural tubing shall not be less than the specified value by more than 3.5 percent.

Straightness—The permissible variation for straightness of structural tubing shall be ⅛-in. times the number of feet of total length divided by five.

Squareness of sides—For square or rectangular structural tubing, adjacent sides may deviate from 90 degrees by a tolerance of plus or minus two degrees maximum.

Radius of corners—For square or rectangular structural tubing, the radius of any outside corner of the section shall not exceed three times the specified wall thickness.

Twists—The tolerances for twist or variation with respect to axial alignment of the section, for square and rectangular structural tubing, shall be as shown in the following table:

Specified Dimension of Longest Side, in.	Maximum Twist per 3 ft of Length, in.
1½ and under	0.050
Over 1½ to 2½, incl.	0.062
Over 2½ to 4, incl.	0.075
Over 4 to 6 incl.	0.087
Over 6 to 8, incl.	0.100
Over 8	0.112

Twist is measured by holding down one end of a square or rectangular tube on a flat surface plate with the bottom side of the tube parallel to the surface plate and noting the height that either corner, at the opposite end of the bottom side of the tube, extends above the surface plate.

Wall thickness (A500 only)—The tolerance for wall thickness exclusive of the weld area shall be plus and minus 10 percent of the nominal wall thickness specified. The wall thickness is to be measured at the center of the flat.

Table 1-18.
Rectangular Sheared Plates and Universal Mill Plates

Permissible Variations in Width and Length for Sheared Plates
(1½-in. and under in thickness)
Permissible Variations in Length Only for Universal Mill Plates
(2½-in. and under in thickness)

Specified Dimensions, in.		Variations over Specified Width and Length for Thickness, in., and Equivalent Weights, lb per sq. ft., Given							
		To ⅜ excl.		⅜ to ⅝ excl.		⅝ to 1, excl.		1 to 2, incl.[a]	
		To 15.3, excl.		15.3 to 25.5, excl.		25.5 to 40.8, excl.		40.8 to 81.7, incl.	
Length	Width	Width	Length	Width	Length	Width	Length	Width	Length
To 120, excl.	To 60, excl.	⅜	½	⁷⁄₁₆	⅝	½	¾	⅝	1
	60 to 84, excl.	⁷⁄₁₆	⅝	½	¹¹⁄₁₆	⅝	⅞	¾	1
	84 to 108, excl	½	¾	⅝	⅞	¾	1	1	1⅛
	108 and over	⅝	⅞	¾	1	⅞	1⅛	1⅛	1¼
120 to 240, excl.	To 60, excl.	⅜	¾	½	⅞	⅝	1	¾	1⅛
	60 to 84, excl.	½	¾	⅝	⅞	¾	1	⅞	1¼
	84 to 108, excl.	⁹⁄₁₆	⅞	¹¹⁄₁₆	¹⁵⁄₁₆	¹³⁄₁₆	1⅛	1	1⅜
	108 and over	⅝	1	¾	1⅛	⅞	1¼	1⅛	1⅜
240 to 360, excl.	To 60, excl.	⅜	1	½	1⅛	⅝	1¼	¾	1½
	60 to 84, excl.	½	1	⅝	1⅛	¾	1¼	⅞	1½
	84 to 108, excl.	⁹⁄₁₆	1	¾	1⅛	⅞	1⅜	1	1½
	108 and over	¹¹⁄₁₆	1⅛	⅞	1¼	1	1⅜	1¼	1¾
360 to 480, excl.	To 60, excl.	⁷⁄₁₆	1⅛	½	1¼	⅝	1⅜	¾	1⅝
	60 to 84, excl.	½	1¼	⅝	1⅜	¾	1½	⅞	1⅝
	84 to 108, excl.	⁹⁄₁₆	1¼	¾	1⅜	⅞	1½	1	1⅞
	108 and over	¾	1⅜	⅞	1½	1	1⅝	1¼	1⅞
480 to 600, excl.	To 60, excl.	⁷⁄₁₆	1¼	½	1½	⅝	1⅝	¾	1⅞
	60 to 84, excl.	½	1⅜	⅝	1½	¾	1⅝	⅞	1⅞
	84 to 108, excl.	⅝	1⅜	¾	1½	⅞	1⅝	1	1⅞
	108 and over	¾	1½	⅞	1⅝	1	1¾	1¼	1⅞
600 to 720, excl.	To 60, excl.	½	1¼	⅝	1⅞	¾	1⅞	⅞	2¼
	60 to 84, excl.	⅝	1⅜	¾	1⅞	⅞	1⅞	1	2¼
	84 to 108, excl.	⅝	1⅜	¾	1⅞	⅞	1⅞	1⅛	2¼
	108 and over	⅞	1½	1	2	1⅛	2¼	1¼	2½
720 and over, excl.	To 60, excl.	⁹⁄₁₆	2	¾	2⅛	⅞	2¼	1	2¾
	60 to 84, excl.	¾	2	⅞	2⅛	1	2¼	1⅛	2¾
	84 to 108, excl.	¾	2	⅞	2⅛	1	2¼	1¼	2¾
	108 and over	1	2	1⅛	2⅜	1¼	2½	1⅜	3

Notes:
[a]Permissible variations in length apply also to Universal Mill plates up to 12 in. width for thicknesses over 2 to 2½-in., incl. except for alloy steels up to 1¾-in. thick.
Permissible variations under specified width and length, ¼-in. Table applies to all steels listed in ASTM A6.

Table 1-19.
Rectangular Sheared Plates and Universal Mill Plates

Permissible Variations from Flatness (Carbon Steel Only)

Specified Thickness, in.	Variations from Flatness for Specified Widths, in.							
	To 36 excl.	36 to 48, excl.	48 to 60, excl.	60 to 72, excl.	72 to 84, excl.	84 to 96, excl.	96 to 108, excl.	108 to 120, excl.
To 1/4, excl.	9/16	3/4	15/16	1 1/4	1 3/8	1 1/2	1 5/8	1 3/4
1/4 to 3/8, excl.	1/2	5/8	3/4	15/16	1 1/8	1 1/4	1 3/8	1 1/2
3/8 to 1/2, excl.	1/2	9/16	5/8	5/8	3/4	7/8	1	1 1/8
1/2 to 3/4, excl.	7/16	1/2	9/16	5/8	5/8	3/4	1	1
3/4 to 1, excl.	7/16	1/2	9/16	5/8	5/8	5/8	3/4	7/8
1 to 2, excl.	3/8	1/2	1/2	9/16	9/16	5/8	5/8	5/8
2 to 4, excl.	5/16	3/8	7/16	1/2	1/2	1/2	1/2	9/16
4 to 6, excl.	3/8	7/16	1/2	1/2	9/16	9/16	5/8	3/4
6 to 8, excl.	7/16	1/2	1/2	5/8	11/16	3/4	7/8	7/8

Permissible Variations in Camber for Carbon Steel Sheared and Gas Cut Rectangular Plates

Maximum permissible camber, in. (all thicknesses) $= \frac{1}{8}$-in. $\times \dfrac{\text{total length, ft}}{5}$

Permissible Variations in Camber for Carbon Steel Universal Mill Plates, High-Strength Low-Alloy Steel Sheared and Gas Cut Rectangular Plates, Universal Mill Plates, Special Cut Plates

Dimension, in.		Camber for Thicknesses and Widths Given
Thickness	Width	
To 2, incl.	All	$\frac{1}{8}$ in. $\times$ (total length, ft / 5)
Over 2 to 15, incl.	To 30, incl.	$\frac{3}{16}$ in. $\times$ (total length, ft / 5)
Over 2 to 15, incl.	Over 30 to 60, incl.	$\frac{1}{4}$ in. $\times$ (total length, ft / 5)

General Notes:
1. The longer dimension specified is considered the length, and permissible variations in flatness along the length should not exceed the tabular amount for the specified width in plates up to 12 feet in length.
2. The flatness variations across the width should not exceed the tabular amount for the specified width.
3. When the longer dimension is under 36 inches, the permissible variation should not exceed 1/4-in. When the longer dimension is from 36 to 72 inches, inclusive, the permissible variation should not exceed 75 percent of the tabular amount for the specified width, but in no case less than 1/4-in.
4. These variations apply to plates which have a specified minimum tensile strength of not more than 60 ksi or compatible chemistry or hardness. The limits in the table are increased 50 percent for plates specified to a higher minimum tensile strength or compatible chemistry or hardness.

Table 1-20.
Rectangular Sheared Plates and Universal Milled Plates

Permissible Variations from Flatness
(High-Strength Low-Alloy and Alloy Steel, Hot Rolled or Thermally Treated)

Specified Thickness, in.	Variations from Flatness for Specified Widths, in.							
	To 36 excl.	36 to 48, excl.	48 to 60, excl.	60 to 72, excl.	72 to 84, excl.	84 to 96, excl.	96 to 108, excl.	108 to 120, excl.
To $\frac{1}{4}$, excl.	$\frac{13}{16}$	$1\frac{1}{8}$	$1\frac{3}{8}$	$1\frac{7}{8}$	2	$2\frac{1}{4}$	$2\frac{3}{8}$	$2\frac{5}{8}$
$\frac{1}{4}$ to $\frac{3}{8}$, excl.	$\frac{3}{4}$	$\frac{15}{16}$	$1\frac{1}{8}$	$1\frac{3}{8}$	$1\frac{3}{4}$	$1\frac{7}{8}$	2	$2\frac{1}{4}$
$\frac{3}{8}$ to $\frac{1}{2}$, excl.	$\frac{3}{4}$	$\frac{7}{8}$	$\frac{15}{16}$	$\frac{15}{16}$	$1\frac{1}{8}$	$1\frac{5}{16}$	$1\frac{1}{2}$	$1\frac{5}{8}$
$\frac{1}{2}$ to $\frac{3}{4}$, excl.	$\frac{5}{8}$	$\frac{3}{4}$	$\frac{15}{16}$	$\frac{7}{8}$	1	$1\frac{1}{8}$	$1\frac{1}{4}$	$1\frac{3}{8}$
$\frac{3}{4}$ to 1, excl.	$\frac{5}{8}$	$\frac{3}{4}$	$\frac{7}{8}$	$\frac{7}{8}$	$\frac{15}{16}$	1	$1\frac{1}{8}$	$1\frac{5}{16}$
1 to 2, excl.	$\frac{9}{16}$	$\frac{5}{8}$	$\frac{3}{4}$	$\frac{13}{16}$	$\frac{7}{8}$	$\frac{15}{16}$	1	1
2 to 4, excl.	$\frac{1}{2}$	$\frac{9}{16}$	$\frac{9}{16}$	$\frac{3}{4}$	$\frac{3}{4}$	$\frac{3}{4}$	$\frac{3}{4}$	$\frac{7}{8}$
4 to 6, excl.	$\frac{9}{16}$	$\frac{11}{16}$	$\frac{3}{4}$	$\frac{3}{4}$	$\frac{7}{8}$	$\frac{7}{8}$	$\frac{15}{16}$	$1\frac{1}{8}$
6 to 8, excl.	$\frac{5}{8}$	$\frac{3}{4}$	$\frac{3}{4}$	$\frac{15}{16}$	1	$1\frac{1}{8}$	$1\frac{1}{4}$	$1\frac{5}{16}$

General Notes:
1. The longer dimension specified is considered the length, and variations from a flat surface along the length should not exceed the tabular amount for the specified width in plates up to 12 feet in length.
2. The flatness variation across the width should not exceed the tabular amount for the specified width.
3. When the longer dimension is under 36 inches, the variation should not exceed $\frac{3}{8}$-in. When the longer dimension is from 36 to 72 inches, inclusive the variation should not exceed 75 percent of the tabular amount for the specified width.

Permissible Variations in Width for Universal Mill Plates
(15 inches and under in thickness)

Specified Width, in.	Variations Over Specified Width for Thickness, in., and Equivalent Weights, lb per sq. ft., Given					
	To $\frac{3}{8}$, excl.	$\frac{3}{8}$ to $\frac{5}{8}$ excl.	$\frac{5}{8}$ to 1, excl.	1 to 2, excl.	Over 2 to 10, incl.	Over 10 to 15, incl.
	To 15.3, excl.	15.3 to 25.5, excl.	25.5 to 40.8, excl.	40.8 to 81.7, incl.	81.7 to 409.0, incl.	409.0 to 613.0, incl.
Over 8 to 20, excl.	$\frac{1}{8}$	$\frac{1}{8}$	$\frac{3}{16}$	$\frac{1}{4}$	$\frac{3}{8}$	$\frac{1}{2}$
20 to 36, excl.	$\frac{3}{16}$	$\frac{1}{4}$	$\frac{5}{16}$	$\frac{3}{8}$	$\frac{7}{16}$	$\frac{9}{16}$
36 and over	$\frac{5}{16}$	$\frac{3}{8}$	$\frac{7}{16}$	$\frac{1}{2}$	$\frac{9}{16}$	$\frac{5}{8}$

Notes:
Permissible variation under specified width, $\frac{1}{8}$-in.
Table applies to all steels listed in ASTM A6.

REFERENCES

American Institute of Steel Construction, 1973, "Commentary on Highly Restrained Welded Connections," *Engineering Journal,* 3rd Qtr., AISC, Chicago, IL.

American Iron and Steel Institute, 1979, *Fire Safe Structural Steel: A Design Guide,* AISI, Washington, DC.

AISI, 1980, *Designing Fire Protection for Steel Columns,* 3rd Edition.

AISI, 1981, *Designing Fire Protection for Steel Trusses,* 2nd Edition.

AISI, 1984, *Designing Fire Protection for Steel Beams.*

Brockenbrough, R. L. and B. G. Johnston, 1981, *USS Steel Design Manual,* R. L. Brockenbrough & Assoc. Inc., Pittsburgh, PA.

Dill, F. H., 1960, "Structural Steel After a Fire," *Proceedings of the 1960 National Engineering Conference,* AISC, New York, NY.

Fisher, J. W. and A. W. Pense, 1987, "Experience with Use of Heavy W Shapes in Tension," *Engineering Journal,* 2nd Qtr., AISC, Chicago.

Lightner, M. W. and R. W. Vanderbeck, 1956, "Factors Involved in Brittle Fracture," *Regional Technical Meetings,* AISI, Washington, DC.

Rolfe, S. T. and J. M. Barsom, 1986, *Fracture and Fatigue Control in Structures: Applications of Fracture Mechanics,* Prentice-Hall, Inc., Englewood Cliffs, NJ.

Rolfe, S. T., 1977, "Fracture and Fatigue Control in Steel Structures," *Engineering Journal,* 1st Qtr., AISC, Chicago.

Welding Research Council, 1957, *Control of Steel Construction to Avoid Brittle Failure,* New York.

PART 2

ESSENTIALS OF LRFD

OVERVIEW

The following LRFD topics are covered herein (with the letters A through I in the section headings referring to the corresponding chapters in the LRFD Specification):

INTRODUCTION TO LRFD

The intent of this part of the LRFD Manual is to provide a general introduction to the subject. It was written primarily for:

(1) engineers experienced in allowable stress design (ASD) who are unfamiliar with LRFD and

(2) students and novice engineers.

The emphasis is on understanding the most common cases, rather than on completeness and efficiency in design. Regular users of LRFD may also find it helpful to refer to the information provided herein. It should be noted, however, that the governing document is the LRFD Specification (in Part 6 of this volume of the Manual). For optimum design the use of the design aids elsewhere in this Manual is recommended. Among the topics not covered herein are:

(1) connections, the subject of Volume II, and

(2) noncompact beams and plate girders, for which the reader is referred to Appendices F and G of the LRFD Specification and Part 4 of this volume of the Manual.

LRFD Versus ASD

The primary objective of the LRFD Specification is to provide a uniform reliability for steel structures under various loading conditions. This uniformity cannot be obtained with the allowable stress design (ASD) format.

The ASD method can be represented by the inequality

$$\Sigma Q_i \leq R_n / F.S. \tag{2-1}$$

The left side is the summation of the load effects, Q_i (i.e., forces or moments). The right side is the nominal strength or resistance R_n divided by a factor of safety. When divided by the appropriate section property (e.g., area or section modulus), the two sides of the inequality become the calculated stress and allowable stress, respectively. The left side can be expanded as follows:

ΣQ_i = the maximum (absolute value) of the combinations

$D + L'$
$(D + L' + W) \times 0.75*$
$(D + L' + E) \times 0.75*$
$D - W$
$D - E$

where D, L', W, and E are, respectively, the effects of the dead, live, wind, and earthquake loads; total live load $L' = L + (L_r \text{ or } S \text{ or } R)$

L = Live load due to occupancy
L_r = Roof live load
S = Snow load
R = Nominal load due to initial rainwater or ice exclusive of the ponding contribution

---

*0.75 is the reciprocal of 1.33, which represents the $^1/_3$ increase in allowable stress permitted when wind or earthquake is taken simultaneously with live load.

ASD, then, is characterized by the use of unfactored service loads in conjunction with a single factor of safety applied to the resistance. Because of the greater variability and, hence, unpredictability of the live load and other loads in comparison with the dead load, a uniform reliability is not possible.

LRFD, as its name implies, uses separate factors for each load and for the resistance. Considerable research and experience were needed to establish the appropriate factors. Because the different factors reflect the degree of uncertainty of different loads and combinations of loads and the accuracy of predicted strength, a more uniform reliability is possible.

The LRFD method may be summarized by the formula

$$\Sigma \gamma_i Q_i \leq \phi R_n \tag{2-2}$$

On the left side of the inequality, the required strength is the summation of the various load effects Q_i multiplied by their respective load factors γ_i. The design strength, on the right side, is the nominal strength or resistance R_n multiplied by a resistance factor ϕ. Values of ϕ and R_n for columns, beams, etc. are provided throughout the LRFD Specification and will be covered here, as well.

According to the LRFD Specification (Section A4.1), $\Sigma \gamma_i Q_i$ = the maximum absolute value of the following combinations

$1.4D$	(A4-1)
$1.2D + 1.6L + 0.5(L_r$ or S or $R)$	(A4-2)
$1.2D + 1.6(L_r$ or S or $R) + (0.5L$ or $0.8W)$	(A4-3)
$1.2D + 1.3W + 0.5L + 0.5(L_r$ or S or $R)$	(A4-4)
$1.2D \pm 1.0E + 0.5L + 0.2S$	(A4-5)
$0.9D \pm (1.3W$ or $1.0E)$	(A4-6)

(Exception: The load factor on L in combinations A4-3, A4-4, A4-5 shall equal 1.0 for garages, areas occupied as places of public assembly, and all areas where the live load is greater than 100 psf).

The load effects D, L, L_r, S, R, W, and E are as defined above. The loads should be taken from the governing building code or from ASCE 7, *Minimum Design Loads in Buildings and Other Structures* (American Society of Civil Engineers, 1988). Where applicable, L should be determined from the reduced live load specified for the given member in the governing code. Earthquake loads should be from the AISC *Seismic Provisions for Structural Steel Buildings,* which appears in Part 6 of this Manual.

LRFD Fundamentals

The following is a brief discussion of the basic concepts of LRFD. A more complete treatment of the subject is available in the Commentary on the LRFD Specification (Section A4 and A5) and in the references cited therein.

LRFD is a method for proportioning structures so that no applicable limit state is exceeded when the structure is subjected to all appropriate factored load combinations. Strength limit states are related to safety and load carrying capacity (e.g., the limit states of plastic moment and buckling). Serviceability limit states (e.g., deflections) relate to performance under normal service conditions. In general, a structural member will have several limit states. For a beam, for example, they are flexural strength, shear strength, vertical deflection, etc. Each limit state has associated with it a value of R_n, which defines the boundary of structural usefulness.

Because the AISC Specification is concerned primarily with safety, strength limit states are emphasized. The load combinations for determining the required strength were given in expressions A4-1 through A4-6. (Other load combinations, with different values of γ_i, are appropriate for serviceability; see Chapter L in the LRFD Specification and Commentary.)

The AISC load factors (A4-1 through A4-6) are based on ASCE 7. They were originally developed by the A58 Load Factor Subcommittee of the American National Standards Institute, ANSI, (U.S. Department of Commerce, 1980) and are based strictly on load statistics. Being material-independent, they are applicable to all structural materials. Although others have written design codes similar in format to the LRFD Specification, the AISC was the first specification group to adopt the ANSI probability-based load factors.

The AISC load factors recognize that when several loads act in combination, only one assumes its maximum lifetime value at a time, while the others are at their "arbitrary-point-in-time" (APT) values. Each combination models the total design loading condition when a different load is at its maximum:

Load Combination	Load at its Lifetime (50-year) Maximum
A4-1	D (during construction; other loads not present)
A4-2	L
A4-3	L_r or S or R (a roof load)
A4-4	W (acting in direction of D)
A4-5	E (acting in direction of D)
A4-6	W or E (opposing D)

The other loads, which are APT loads, have mean values considerably lower than the lifetime maximums. To achieve a uniform reliability, every factored load (lifetime maximum or APT) is larger than its mean value by an amount depending on its variability.

The AISC resistance factors are based on research recommendations published by Washington University in St. Louis (Galambos et al., 1978) and reviewed by the AISC Specification Advisory Committee. Test data were analyzed to determine the variability of each resistance. In general, the resistance factors are less than one ($\phi < 1$). For uniform reliability, the greater the scatter in the data for a given resistance, the lower its ϕ factor.

Several representative LRFD ϕ factors for steel members (referenced to the corresponding chapters in the LRFD Specification) are:

$\phi_t = 0.90$ for tensile yielding (Chapter D)

$\phi_t = 0.75$ for tensile fracture (Chapter D)

$\phi_c = 0.85$ for compression (Chapter E)

$\phi_b = 0.90$ for flexure (Chapter F)

$\phi_v = 0.90$ for shear yielding (Chapter F)

Resistance factors for other member and connection limit states are given in the LRFD Specification.

The following sections (A through I) summarize and explain the corresponding chapters of the LRFD Specification.

A. GENERAL PROVISIONS

In the LRFD Specification, Sections A4 and A5 define Load and Resistance Factor Design. The remainder of Chapter A contains general provisions which are essentially the same as in the earlier ASD editions of the Specification.

Reference is made to the *Code of Standard Practice for Steel Buildings and Bridges* (adopted in 1992 by AISC), which appears with a Commentary in Part 6 of this LRFD Manual. The Code defines the practices and commonly accepted standards in the structural steel fabricating industry. In the absence of other instructions in the contract documents, these trade practices govern the fabrication and erection of structural steel.

The types of construction recognized by the AISC Specification have not changed, except that both "simple framing" (formerly Type 2) and "semi-rigid framing" (formerly Type 3) have been combined into one category, Type PR (partially restrained). "Rigid framing" (formerly Type 1) is now Type FR (fully restrained). Type FR construction is permitted unconditionally. Type PR is allowed only upon evidence that the connections to be used are capable of furnishing, as a minimum, a predictable portion of full end restraint. Type PR construction may necessitate some inelastic, but self-limiting, deformation of a structural steel part. When specifying Type PR construction, the designer should take into account the effects of reduced connection stiffness on the stability of the structure, lateral deflections, and second order bending moments.

Semi-rigid connections, once common, are again becoming popular. They offer economies in connection fabrication (compared with FR connections) and reduced member size (compared with simple framing). For information on connections, please refer to Volume II of this LRFD Manual.

The yield stresses of the grades of structural steel approved for use range from 36 ksi for the common A36 steel to 100 ksi for A514 steel. Not all rolled shapes and plate thicknesses are available for every yield stress. Availability tables for structural shapes, plates and bars are at the beginning of Part 1 of this LRFD Manual.

A36, for many years the dominant structural steel for buildings, is being replaced by the more economical 50 ksi steels. ASTM designations for structural steels with 50 ksi yield stress are: A572 for most applications, A529 for thin-plate members only, and A242 and A588 weathering steels for atmospheric corrosion resistance. A more complete explanation is provided by Table 1-1 in Part 1 of this Manual. However, A36 is still normally specified for connection material, where no appreciable savings can be realized from higher strength steels.

Complete and accurate drawings and specifications are necessary for all stages of steel construction. The requirements for design documents are set forth in Section A7 of the LRFD Specification and Section 3 of the AISC *Code of Standard Practice*. When beam end reactions are not shown on the drawings, the structural steel detailer will refer to the appropriate tables in Part 4 of the LRFD Manual. These tables, which are for uniform loads, may significantly underestimate the effects of the concentrated loads. The recording of beam end reactions on design drawings, which is recommended in all cases, is, therefore, absolutely essential when there are concentrated loads. Beam reactions, column loads, etc., shown on design drawings should be the required strengths calculated from the factored load combinations and should be so noted.

Loads and Load Combinations

LRFD Specification Sections A4 (Loads and Load Combinations) and A5 (Design Basis) describe the basic criteria of LRFD. This information was discussed above under

Introduction to LRFD. To illustrate the application of load factors, the AISC load combinations will be repeated here with design examples.

The required strength is the maximum absolute value of the combinations

$1.4D$	(A4-1)
$1.2D + 1.6L + 0.5(L_r \text{ or } S \text{ or } R)$	(A4-2)
$1.2D + 1.6(L_r \text{ or } S \text{ or } R) + (0.5L \text{ or } 0.8W)$	(A4-3)
$1.2D + 1.3W + 0.5L + 0.5(L_r \text{ or } S \text{ or } R)$	(A4-4)
$1.2D \pm 1.0E + 0.5L + 0.2S$	(A4-5)
$0.9D \pm (1.3W \text{ or } 1.0E)$	(A4-6)

(The load factor on L in combinations A4-3, A4-4 and A4-5 shall equal 1.0 for garages, areas occupied as placed of public assembly, and all areas where the live load is greater than 100 psf).

In the combinations the loads or load effects (i.e., forces or moments) are:

D = dead load due to the weight of the structural elements and the permanent features on the structure

L = live load due to occupancy and moveable equipment (reduced as permitted by the governing code)

L_r = roof live load

W = wind load

S = snow load

E = earthquake load

R = nominal load due to initial rainwater or ice exclusive of the ponding contribution

The loads are to be taken from the governing building code. In the absence of a code, one may use ASCE 7 *Minimum Design Loads for Buildings and Other Structures* (American Society of Civil Engineers, 1988). Earthquake loads should be determined from the AISC *Seismic Provisions for Structural Steel Buildings*, in Part 6 of this Manual.

Whether the loads themselves or the load effects are combined, the results are the same, provided the principle of superposition is valid. This is usually true because deflections are small and the stress-strain behavior is linear elastic; consequently, second order effects can usually be neglected. (The analysis of second order effects is covered in Chapter C of the LRFD Specification.) The linear elastic assumption, although not correct at the strength limit states, is valid under normal in-service loads and is permissible as a design assumption under the LRFD Specification. In fact, the Specification (in Section A.5.1) allows the designer the option of elastic or plastic analysis using the factored loads. However, to simplify this presentation, it is assumed that the more prevalent elastic analysis option has been selected.

EXAMPLE A-1

Given: Roof beams W16×31, spaced 7'-0 center-to-center, support a superimposed dead load of 40 psf. Code specified roof loads are 30 psf downward (due to roof live load, snow, or rain) and 20 psf upward or downward (due to wind). Determine the critical loading for LRFD.

Solution: D = 31 plf + 40 psf × 7.0 ft = 311 plf

$$L \quad = 0$$
$$(L_r \text{ or } S \text{ or } R) = 30 \text{ psf} \times 7.0 \text{ ft} = 210 \text{ plf}$$
$$W \quad = 20 \text{ psf} \times 7.0 \text{ ft} = 140 \text{ plf}$$
$$E \quad = 0$$

Load Combinations	Factored Loads	
A4-1	1.4(311 plf)	= 435 plf
A4-2	1.2(311 plf) + 0 + 0.5(210 plf)	= 478 plf
A4-3	1.2(311 plf) + 1.6 (210 plf) + 0.8(140 plf)	= 821 plf
A4-4	1.2(311 plf) + 1.3(140 plf) + 0 + 0.5(210 plf)	= 660 plf
A4-5	1.2(311 plf) + 0 + 0 + 0.2(210 plf)	= 415 plf
A4-6a	0.9 (311 plf) + 1.3 (140 plf)	= 462 plf
A4-6b	0.9(311 plf) − 1.3(140 plf)	= 98 plf

The critical factored load combination for design is the third, with a total factored load of 821 plf.

EXAMPLE A-2

Given: The axial loads on a building column resulting from the code-specified service loads have been calculated as: 100 kips from dead load, 150 kips from (reduced) floor live load, 30 kips from the roof (L_r or S or R), 60 kips due to wind, and 50 kips due to earthquake. Determine the required strength of this column.

Solution:

Load Combination	Factored Axial Load	
A4-1	1.4(100 kips)	= 140 kips
A4-2	1.2(100 kips) + 1.6(150 kips) + 0.5(30 kips)	= 375 kips
A4-3a	1.2(100 kips) + 1.6(30 kips) + 0.5(150 kips)	= 243 kips
A4-3b	1.2(100 kips) + 1.6(30 kips) + 0.8(60 kips)	= 216 kips
A4-4	1.2(100 kips) + 1.3(60 kips) + 0.5(150 kips) + 0.5(30 kips)	= 288 kips
A4-5a	1.2(100 kips) + 1.0(50 kips) + 0.5(150 kips) + 0.2(30 kips)	= 251 kips
A4-5b	1.2(100 kips) − 1.0(50 kips) + 0.5(150 kips) + 0.2(30 kips)	= 151 kips
A4-6a	0.9(100 kips) + 1.3(60 kips)	= 168 kips
A4-6b	0.9(100 kips) − 1.3(60 kips)	= 12 kips
A4-6c	0.9(100 kips) + 1.0(50 kips)	= 140 kips
A4-6d	0.9(100 kips) − 1.0(50 kips)	= 40 kips

The required strength of the column is 375 kips based on the second combination of factored axial loads. As none of the results above are negative, net tension need not be considered in the design of this column.

B. DESIGN REQUIREMENTS

Gross, Net, and Effective Net Areas for Tension Members

The concept of effective net area, which in earlier editions of the Specification was applied only to bolted members, has been extended to cover members connected by welding as well. As in the past, when tensile forces are transmitted directly to all elements of the member, the net area is used to determine stresses. However, when the tensile forces are transmitted through some, but not all, of the cross-sectional elements of the member, a reduced effective net area A_e is used instead. According to Section B3 of the LRFD Specification

$$A_e = AU \tag{B3-1}$$

where

A = area as defined below
U = reduction coefficient
  $= 1 - (\bar{x}/L) \le 0.9$, or as defined in (c) or (d) $\qquad$ (B3-2)
$\bar{x}$ = connection eccentricity. (See Commentary on the LRFD Specification, Section B3 and Figure C-B3.1.)
L = length of connection in the direction of loading

a. When the forces are transmitted only by bolts

$A = A_n$
  = net area of member, in.^{2}

b. When the forces are transmitted by longitudinal welds only or in combination with transverse welds

$A = A_g$
  = gross area of member, in.^{2}

c. When the forces are transmitted only by transverse welds

A = area of directly connected elements, in.^{2}
$U = 1.0$

d. When the forces are transmitted to a plate by longitudinal welds along both edges at the end of the plate

A = area of plate, in.^{2}
$l \ge w$

For $l \ge 2w$ $\qquad$ $U = 1.00$
For $2w > l \ge 1.5w$ $\qquad$ $U = 0.87$
For $1.5w > l \ge w$ $\qquad$ $U = 0.75$

where

l = weld length
w = plate width (distance between welds), in.

In computing the net area for tension and shear, the width of a bolt hole is taken as $\frac{1}{16}$-in. greater than the nominal dimension of the hole, which, for standard holes, is $\frac{1}{16}$-in. larger than the diameter of the bolt. Chains of holes, treated as in the past, are covered in Section B2 of the LRFD Specification.

Gross, Net, and Effective Net Areas for Flexural Members

Gross areas are used for elements in compression, in beams and columns. According to Section B10 of the LRFD Specification, the properties of beams and other flexural members are based on the gross section (with no deduction for holes in the tension flange) if

$$0.75F_u A_{fn} \geq 0.9F_y A_{fg} \tag{B10-1}$$

where

A_{fg} = gross flange area, in.^{2}
A_{fn} = net flange area (deducting bolt holes), in.^{2}
F_y = specified minimum yield stress, ksi
F_u = minimum tensile strength, ksi

Otherwise, an effective tension flange area A_{fe} is used to calculate flexural properties

$$A_{fe} = \frac{5}{6}\frac{F_u}{F_y} A_{fn} \tag{B10-3}$$

Local Buckling

Steel sections are classified as either compact, noncompact, or slender element sections:

- If the flanges are continuously connected to the web and the width-thickness ratios of all the compression elements do not exceed λ_p, then the section is compact.
- If the width-thickness ratio of at least one of its compression elements exceeds λ_p, but does not exceed λ_r, the section is noncompact.
- If the width-thickness ratio of any compression element exceeds λ_r, that element is called a slender compression element.

Columns with compact and noncompact cross sections are covered by Chapter E of the LRFD Specification. Column cross sections with slender elements require the special design procedure in Appendix B5.3 of the Specification.

Beams with compact sections are covered by Chapter F of the LRFD Specification. All other cross sections in bending must be designed in accordance with Appendices B5.3, F1 and/or G.

In general, reference to the appendices of the Specification is required for the design of members controlled by local buckling. In slender element sections, local buckling, occurring prior to initial yielding, will limit the strength of the member. Noncompact sections will yield first, but local buckling will precede the development of a fully plastic stress distribution. In actual practice, such cases are not common and can be easily avoided by designing so that:

Table B-1.
Limiting Width-Thickness Ratios for Compression Elements*

Beam Element	Width-Thickness Ratio	Limiting Width-Thickness Ratio, λ_p	
		General	For F_y = 50 ksi
Flanges of I shapes and channels	b/t	$65/\sqrt{F_y}$	9.2
Flanges of square and rectangular box beams	b/t	$190/\sqrt{F_y}$	26.9
Webs in flexural compression	h/t_w	$640/\sqrt{F_y}$	90.5
Webs in combined flexural and axial compression	h/t_w	$253/\sqrt{F_y}$**	35.8
Column Element	Width-Thickness Ratio	Limiting Width-Thickness Ratio, λ_r	
		General	For F_y = 50 ksi
Flanges of I shapes and channels and plates projecting from compression elements	b/t	$95/\sqrt{F_y}$	13.4
Webs in axial compression	h/t_w	$253/\sqrt{F_y}$	35.8

*For the complete table, see LRFD Specification, Section B5, Table B5.1.
**This is a simplified, conservative version of the corresponding entry in Table B5.1 of the LRFD Specification.

- for beams, the width-thickness ratios of all compression elements $\leq \lambda_p$;
- for columns, the width-thickness ratios of all elements $\leq \lambda_r$.

Table B-1, which is an abridged version of Table B5.1 in the LRFD Specification, should be useful for this purpose. The formulas for λ_p for beam elements and λ_r for column elements are tabulated, together with the corresponding numerical values for 50 ksi steel. The definitions of "width" for use in determining the width-thickness ratios of the elements of various structural shapes are stated in Section B5 of the LRFD Specification. They are shown graphically in Figure B-1. Compact section criteria for W shapes and other I-shaped cross sections are listed in the Properties Tables in Part 1 of LRFD Manual.

Limiting Slenderness Ratios

For members whose design is based on compressive force, the slenderness ratio Kl/r preferably should not exceed 200.

For members whose design is based on tensile force, the slenderness ratio l/r preferably should not exceed 300. The above limitation does not apply to rods in tension.

K = effective length factor, defined in Section C below

l = distance between points of lateral support (l_x or l_y), in.

r = radius of gyration (r_x or r_y), in.

C. FRAMES AND OTHER STRUCTURES

Second Order Effects

As stated in Section C1 of the LRFD Specification, an analysis of second order effects is required; i.e., the additional moments due to the axial loads acting on the deformed structure must be considered. In lieu of a second order analysis for M_u, the required flexural strength, the LRFD Specification (in Section C1) presents the following simplified method:

$$M_u = B_1 M_{nt} + B_2 M_{lt} \qquad (C1\text{-}1)$$

The components of the total factored moment, determined from a first order elastic analysis (neglecting second order effects) are divided into two groups, M_{nt} and M_{lt}. Each group is in turn multiplied by a magnification factor B_1 or B_2 and the results are added to approximate the actual second order factored moment M_u. (The method, as explained here, is valid where the moment connections are Type FR, fully restrained. The analysis for Type PR, or partially restrained, moment connections is beyond the scope of this section.)

Beam-columns are generally columns in frames, which are either braced ($M_{lt} = 0$) or unbraced ($M_{lt} \neq 0$). M_{nt} is the moment in the member assuming there is no lateral translation of the frame; M_{lt} is the moment due to lateral translation. M_{nt} includes the moments resulting from the gravity loads, as determined manually or by computer, using one of the customary (elastic, first order) methods. The moments from the lateral loads are classified as M_{lt}; i.e., due to lateral translation. If both the frame and its vertical loads are symmetric, M_{lt} from the vertical loads is zero. However, if either the vertical loads or the frame is asymmetric and the frame is not braced, lateral translation occurs and $M_{lt} \neq 0$. The procedure for obtaining M_{lt} in this case involves:

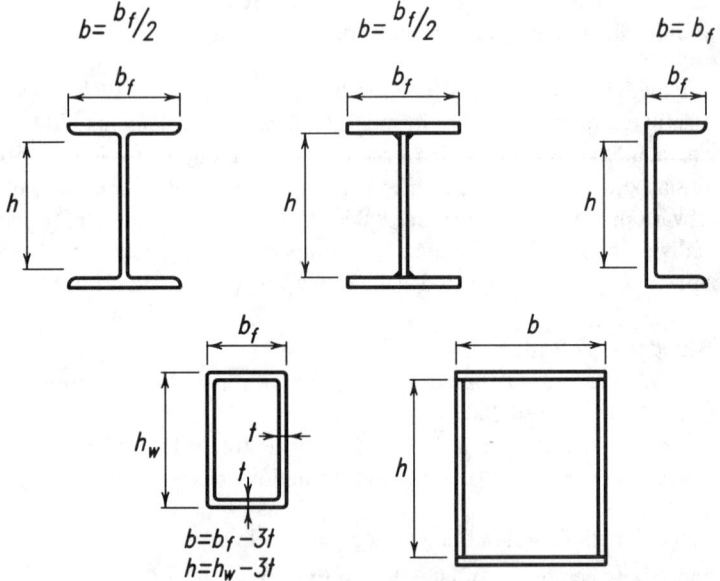

Fig. B-1. Definitions of widths (b and h) for use in Table B-1.

a. applying fictitious horizontal reactions at each floor level to prevent lateral translation, and

b. using the reverse of these reactions as the "sway forces" for determining M_{lt}.

In general, M_{lt} for an unbraced frame is the sum of the moments due to the lateral loads and these "sway forces," as illustrated in Figure C-1.

The magnification factors applied to M_{nt} and M_{lt} are, respectively, B_1 and B_2. As shown in Figure C-2, B_1 accounts for the secondary $P\delta$ member effect in all frames (including sway-inhibited) and B_2 covers the $P\Delta$ story effect in unbraced frames. The expressions for B_1 and B_2 follow:

$$B_1 = \frac{C_m}{(1 - P_u / P_{e1})} \ge 1.0 \tag{C1-2}$$

where

P_u = the factored axial compressive force on the member, kips

P_{e1} = P_e as listed in Table C-1 as a function of the slenderness ratio Kl / r, with effective length factor $K = 1.0$ and considering l / r in the plane of bending only

l = unbraced length of the member, in.

r = radius of gyration of its cross section, in.

C_m = a coefficient to be taken as follows:

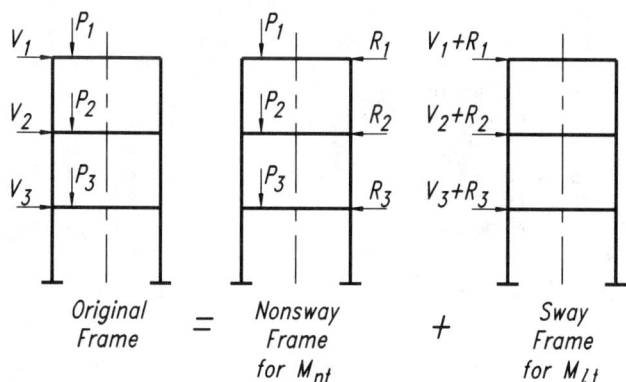

Fig. C-1. Frame models for M_{nt} and M_{lt}.

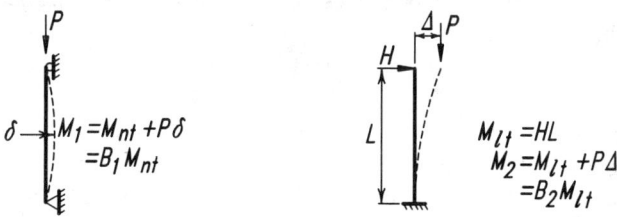

(a) Column in Braced Frame   (b) Column in Unbraced Frame

Fig. C-2. Illustrations of secondary effects.

Table C-1.
Values of P_e/A_g for Use in Equation C1-2 and C1-5 for Steel of Any Yield Stress

Note: Multiply tabulated values by A_g (the gross cross-sectional area of the member) to obtain P_e

Kl/r	P_e/A_g (ksi)	Kl/r	P_e/A_g (ksi)	Kl/r	P_e/A_g (ksi)	Kl/r	P_e/A_g (ksi)	Kl/r	P_e/A_g (ksi)	Kl/r	P_e/A_g (ksi)
21	649.02	51	110.04	81	43.62	111	23.23	141	14.40	171	9.79
22	591.36	52	105.85	82	42.57	112	22.82	142	14.19	172	9.67
23	541.06	53	101.89	83	41.55	113	22.42	143	14.00	173	9.56
24	496.91	54	98.15	84	40.56	114	22.02	144	13.80	174	9.45
25	457.95	55	94.62	85	39.62	115	21.64	145	13.61	175	9.35
26	423.40	56	91.27	86	38.70	116	21.27	146	13.43	176	9.24
27	392.62	57	88.08	87	37.81	117	20.91	147	13.25	177	9.14
28	365.07	58	85.08	88	36.96	118	20.56	148	13.07	178	9.03
29	340.33	59	82.22	89	36.13	119	20.21	149	12.89	179	8.93
30	318.02	60	79.51	90	35.34	120	19.88	150	12.72	180	8.83
31	297.83	61	76.92	91	34.56	121	19.55	151	12.55	181	8.74
32	279.51	62	74.46	92	33.82	122	19.23	152	12.39	182	8.64
33	262.83	63	72.11	93	33.09	123	18.92	153	12.23	183	8.55
34	247.59	64	69.88	94	32.39	124	18.61	154	12.07	184	8.45
35	233.65	65	67.74	95	31.71	125	18.32	155	11.91	185	8.36
36	220.85	66	65.71	96	31.06	126	18.03	156	11.76	186	8.27
37	209.07	67	63.76	97	30.42	127	17.75	157	11.61	187	8.18
38	198.21	68	61.90	98	29.80	128	17.47	158	11.47	188	8.10
39	188.18	69	60.12	99	29.20	129	17.20	159	11.32	189	8.01
40	178.89	70	58.41	100	28.62	130	16.94	160	11.18	190	7.93
41	170.27	71	56.78	101	28.06	131	16.68	161	11.04	191	7.85
42	162.26	72	55.21	102	27.51	132	16.43	162	10.91	192	7.76
43	154.80	73	53.71	103	26.98	133	16.18	163	10.77	193	7.68
44	147.84	74	52.57	104	26.46	134	15.94	164	10.64	194	7.60
45	141.34	75	50.88	105	25.96	135	15.70	165	10.51	195	7.53
46	135.26	76	49.55	106	25.47	136	15.47	166	10.39	196	7.45
47	129.57	77	48.27	107	25.00	137	15.25	167	10.26	197	7.38
48	124.23	78	47.04	108	24.54	138	15.03	168	10.14	198	7.30
49	119.21	79	45.86	109	24.09	139	14.81	169	10.02	199	7.23
50	114.49	80	44.72	110	23.65	140	14.60	170	9.90	200	7.16

Note: $P_e/A_g = \dfrac{\pi^2 E}{(Kl/r)^2}$

a. For compression members not subject to transverse loading between their supports in the plane of bending,

$$C_m = 0.6 - 0.4(M_1/M_2) \tag{C1-3}$$

where M_1/M_2 is the ratio of the smaller to larger moment at the ends of that portion of the member unbraced in the plane of bending under consideration. M_1/M_2 is positive when the member is bending in reverse curvature, negative when bending in single curvature.

b. For compression members subjected to transverse loading between their supports, the value of C_m can be determined by rational analysis, or the following values may be used:

for members with ends restrained against rotation . . . . $C_m = 0.85$
for members with ends unrestrained against rotation . . . . $C_m = 1.0$

Two alternative equations are given for B_2 in the LRFD Specification

$$B_2 = \cfrac{1}{1 - \cfrac{\Sigma P_u}{\Sigma H}\left(\cfrac{\Delta_{oh}}{L}\right)} \tag{C1-4}$$

$$B_2 = \cfrac{1}{1 - \cfrac{\Sigma P_u}{\Sigma P_{e2}}} \tag{C1-5}$$

where

ΣP_u = required axial strength of all columns in a story, i.e., the total factored gravity load above that level, kips

Δ_{oh} = translational deflection of the story under consideration, in.

ΣH = sum of all story horizontal forces producing Δ_{oh}, kips

L = story height, in.

ΣP_{e2} = the summation of P_{e2} for all rigid-frame columns in a story; P_{e2} is determined from Table C-1, considering the actual slenderness ratio Kl/r of each column in its plane of bending

K = effective length factor (see below)

Of the two expressions for B_2, the first (Equation C1-4) is better suited for design office practice. The quantity (Δ_{oh}/L) is the story drift index. For many structures, particularly tall buildings, a maximum drift index is one of the design criteria. Using this value in Equation C1-4 will facilitate the evaluation of B_2.

In general, two values of B_2 are obtained for each story of a building, one for each of the major directions. B_1 is evaluated separately for every column; two values of B_1 are needed for biaxial bending. Using Equations C1-1 through C1-5, the appropriate M_{ux} and M_{uy} are determined for each column.

Effective Length

As in previous editions of the AISC Specification, the effective length of Kl is used (instead of the actual unbraced length l) to account for the influence of end-conditions in the design of compression members. A number of acceptable methods have been utilized to evaluate K, the effective length factor. They are discussed in Section C2 of the Commentary on the LRFD Specification. One method will be shown here.

Table C-2, which is also Table C-C2.1 in the Commentary, is taken from the Structural Stability Research Council (SSRC) *Guide to Stability Design Criteria for Metal Structures*. It relates K to the rotational and translational restraints at the ends of the column. Theoretical values for K are given, as well as the recommendations of the SSRC. The basic case is d, the classical pin-ended column, for which $K = 1.0$. Theoretical K values for the other cases are determined by the distances between points of inflection. The more conservative SSRC recommendations reflect the fact that perfect fixity can never be attained in actual structures.

Table C-2. Effective Length Factors (K) for Columns						
Buckled shape of column is shown by dashed line	(a)	(b)	(c)	(d)	(e)	(f)
Theoretical K value	0.5	0.7	1.0	1.0	2.0	2.0
Recommended design value when ideal conditions are approximated	0.65	0.80	1.2	1.0	2.10	2.0
End condition code		Rotation fixed and translation fixed				
		Rotation free and translation fixed				
		Rotation fixed and translation free				
		Rotation free and translation free				

Like its predecessors, the LRFD Specification (in Section C2) distinguishes between columns in braced and unbraced frames. In braced frames, sidesway is inhibited by attachment to diagonal bracing or shear walls. Cases *a*, *b*, and *d* in Table C-2 represent columns in braced frames; $K \leq 1.0$. The LRFD Specification requires that for compression members in braced frames, K "shall be taken as unity, unless structural analysis shows that a smaller value may be used." Common practice is to assume conservatively $K = 1.0$ for columns in braced frames and compression members in trusses.

The other cases in Table C-2, *c*, *e*, and *f*, are in unbraced frames (sidesway uninhibited); $K \geq 1.0$. The SSRC recommendations given in Table C-2 are appropriate for design.

"Leaning" Columns

The concept of the "leaning" column, although not related exclusively to LRFD, is new to the 1993 LRFD Specification. A leaning column is one which is pin ended and does not participate in providing lateral stability to the structure. As a result it relies on the columns in other parts of the structure for stability. In analyzing and designing unbraced frames, the effects of the leaning columns must be considered (as required by Section C2.2 of the LRFD Specification). For further information the reader is referred to:

(1) Part 3 of this Manual.
(2) the Commentary on the LRFD Specification, Section C2, and
(3) a paper on this subject (Geschwindner, 1993).

D. TENSION MEMBERS

Design Tensile Strength

The design philosophy for tension members is the same in the LRFD and ASD Specifications:

a. The limit state of yielding in the gross section is intended to prevent excessive elongation of the member. Usually, the portion of the total member length occupied by fastener holes is small. The effect of early yielding at the reduced cross sections on the total member elongation is negligible. Use of the area of the gross section is appropriate.

b. The second limit state involves fracture at the section with the minimum effective net area.

The design strength of tension members, $\phi_t P_n$, as given in Section D1 of the LRFD Specification, is the lesser of the following:

a. For yielding in the gross section,

$$\phi_t = 0.90$$

$$P_n = F_y A_g \qquad \text{(D1-1)}$$

b. For fracture in the net section,

$$\phi_t = 0.75$$

$$P_n = F_u A_e \qquad \text{(D1-2)}$$

where

A_e = effective net area, in.^{2} (see Section B, above)
A_g = gross area of member, in.^{2}
F_y = specified minimum yield stress, ksi
F_u = specified minimum tensile strength, ksi
P_n = nominal axial strength, kips

For 50 ksi steels, $F_y = 50$ ksi and minimum $F_u = 65$ ksi. Accordingly

a. For yielding in the gross section,

$$\phi_t P_n = 0.9 \times 50 \text{ ksi} \times A_g = 45.0 \text{ ksi} \times A_g \qquad \text{(2-3)}$$

b. For fracture in the net section,

$$\phi_t P_n = 0.75 \times 65 \text{ ksi} \times A_e = 48.8 \text{ ksi} \times A_e \qquad \text{(2-4)}$$

The limit state of block shear rupture may govern the design tensile strength. For information on block shear, see Section J4.3 of the LRFD Specification and Part 8 (in Volume II) of this LRFD Manual.

EXAMPLE D-1

Given: Determine the design strength of a W8×24 as a tension member in 50 ksi steel. How much dead load can it support?

Solution: If there are no holes in the member, $A_e = A_g$ and Equation 2-3 governs

$$\phi_t P_n = 45.0 \text{ ksi} \times A_g = 45.0 \text{ ksi} \times 7.08 \text{ in.}^2 = 319 \text{ kips}$$

Assuming that dead load is the only load, the governing load combination from Section A is $1.4D$. Then, the required tensile strength

$$P_u = 1.4 P_D \le \phi_t P_n = 319 \text{ kips}$$

$P_D \le$ 319 kips/1.4 = 228 kips maximum dead load that can be supported by the member.

EXAMPLE D-2

Given: Repeat Example D-1 for a W8×24 in 50 ksi steel with four 1-in. diameter holes, two per flange, along the member (i.e., not at its ends) for miscellaneous attachments. See Figure D-1(a).

Solution: a. For yielding in the gross section

$$\phi_t P_n = 319 \text{ kips, as in Example D-1.}$$

b. For fracture in the net section

$$\begin{aligned} A_e &= A_n = A_g - 4 \times (d_{hole} + \tfrac{1}{16}\text{-in.}) \times t_f \\ &= 7.08 \text{ in.}^2 - 4 \times (1 + \tfrac{1}{16}\text{-in.}) \times 0.400 \text{ in.} \\ &= 5.38 \text{ in.}^2 \end{aligned}$$

$$\begin{aligned} \phi_t P_n &= 48.8 \text{ ksi} \times A_e \\ &= 48.8 \text{ ksi} \times 5.38 \text{ in.}^2 = 263 \text{ kips} < 319 \text{ kips} \end{aligned}$$

Fracture in the net section governs.

$$P_u = 1.4 P_D \le \phi_t P_n = 263 \text{ kips}$$
$$P_D \le 263 \text{ kips} / 1.4 = 188 \text{ kips}$$

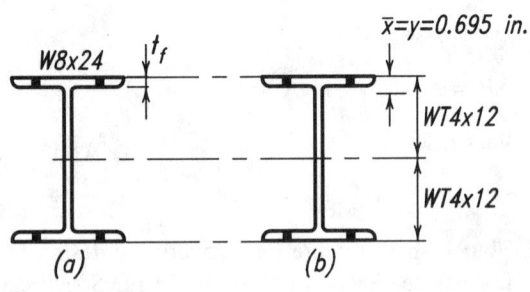

Fig. D-1

Note: If the holes had been at the end connection of the tension member, the U reduction coefficient would apply in the calculation of an effective net area.

EXAMPLE D-3

Given: Repeat Example D-2 for holes at a bolted end-connection. There are a total of eight 1-in. diameter holes, as shown in Figure D-1(a), on two planes, 4 in. center-to-center.

Solution: a. For yielding in the gross section $\phi_t P_n = 319$ kips, as in Example D-1.

b. For fracture in the net section, according to Equation B3-1 in Section B above, the effective net area

$$A_e = AU = A_n U$$

where

$$A_n = 5.38 \text{ in.}^2 \text{ as in Example D-2}$$

$$U = 1 - \frac{\bar{x}}{L}, L = 4 \text{ in.}^*$$

According to Commentary Figure C-B3.1(a), $\bar{x}$ for a W8×24 in this case is taken as that for a WT4×12. From the properties of a WT4×12 given in Part 1 of this Manual, $\bar{x} = y = 0.695$ in. See Figure D-1(b).

$$U = 1 - \frac{0.695 \text{ in.}}{4 \text{ in.}} = 0.826$$

Thus

$$A_e = 5.38 \text{ in.}^2 \times 0.826 = 4.45 \text{ in.}^2$$
$$\phi_t P_n = 48.8 \text{ ksi} \times A_e$$
$$= 48.8 \text{ ksi} \times 4.45 \text{ in.}^2 = 217 \text{ kips} < 319 \text{ kips}$$

Fracture in the net section governs. Again, assuming that dead load is the only load,

$$P_u = 1.4 P_D \le \phi_t P_n = 217 \text{ kips}$$

$P_D \le 217 \text{ kips} / 1.4 = 155$ kips maximum dead load that can be supported by the member.

Built-Up Members, Eyebars, and Pin-Connected Members
See Section D2 and D3 in the LRFD Specification.

---

*In lieu of calculating U, the Commentary on the LRFD Specification (Section B3) permits the use of more conservative values of U listed therein.

E. COLUMNS AND OTHER COMPRESSION MEMBERS

Effective Length
For a discussion of the effective length Kl for columns, refer to Section C above.

Design Compressive Strength
Although the column strength equations have been revised for compatibility with LRFD and recent research on column behavior, the philosophy and procedures of column design in LRFD are similar with those in ASD. The direct design of columns with W and other rolled shapes is facilitated by the column strength tables in Part 3 of this LRFD Manual, which show the design compressive strength $\phi_c P_n$ as a function of KL (the effective unbraced length in feet). Columns with cross sections not tabulated (e.g., built-up columns) can be designed iteratively, as in the past, with the aid of tables listing design stresses versus Kl/r, the slenderness ratio. Such tables are given in the Appendix of the LRFD Specification for 36 and 50 ksi structural steels, and below (Table E-1) for 50 ksi steel.

There are two equations governing column strength, based on the limit state of flexural buckling, one for inelastic buckling (Equation E2-2) and the other (Equation E2-3) for elastic, or Euler, buckling. Equation E2-2 is an empirical relationship for the inelastic range, while Equation E2-3 is the familiar Euler formula multiplied by 0.877. Both equations include the effects of residual stresses and initial out-of-straightness. The boundary between inelastic and elastic instability is $\lambda_c = 1.5$, where the parameter

$$\lambda_c = \frac{Kl}{r\pi}\sqrt{\frac{F_y}{E}} \qquad \text{(E2-4)}$$

For axially loaded columns with all elements having width-thickness ratios $< \lambda_r$ (in Section B5.1 of the LRFD Specification), the design compressive strength $= \phi_c P_n$

where

$$\phi_c = 0.85$$

$$P_n = A_g F_{cr} \qquad \text{(E2-1)}$$

$$A_g = \text{gross area of member, in.}^2$$

a. For $\lambda_c \leq 1.5$

$$F_{cr} = (0.658^{\lambda_c^2})F_y \qquad \text{(E2-2)}$$

As is done in the Commentary on Section E2, this equation can be expressed in exponential form

$$F_{cr} = [\exp(-0.419\lambda_c^2)]F_y \qquad \text{(C-E2-1)}$$

where $\exp(x) = e^x$

Table E-1.
Design Stress for Compression Members of 50 ksi Specified Minimum Yield Stress Steel, $\phi_c = 0.85$*

$\frac{Kl}{r}$	$\phi_c F_{cr}$ (ksi)	$\frac{Kl}{r}$	$\phi_c F_{cr}$ (ksi)	$\frac{Kl}{r}$	$\phi_c F_{cr}$ (ksi)	$\frac{Kl}{r}$	$\phi_c F_{cr}$ (ksi)	$\frac{Kl}{r}$	$\phi_c F_{cr}$ (ksi)
1	42.50	41	37.59	81	26.31	121	14.57	161	8.23
2	42.49	42	37.36	82	26.00	122	14.33	162	8.13
3	42.47	43	37.13	83	25.68	123	14.10	163	8.03
4	42.45	44	36.89	84	25.37	124	13.88	164	7.93
5	42.42	45	36.65	85	25.06	125	13.66	165	7.84
6	42.39	46	36.41	86	24.75	126	13.44	166	7.74
7	42.35	47	36.16	87	24.44	127	13.23	167	7.65
8	42.30	48	35.91	88	24.13	128	13.02	168	7.56
9	42.25	49	35.66	89	23.82	129	12.82	169	7.47
10	42.19	50	35.40	90	23.51	130	12.62	170	7.38
11	42.13	51	35.14	91	23.20	131	12.43	171	7.30
12	42.05	52	34.88	92	22.89	132	12.25	172	7.21
13	41.98	53	34.61	93	22.58	133	12.06	173	7.13
14	41.90	54	34.34	94	22.28	134	11.88	174	7.05
15	41.81	55	34.07	95	21.97	135	11.71	175	6.97
16	41.71	56	33.79	96	21.67	136	11.54	176	6.89
17	41.61	57	33.51	97	21.36	137	11.37	177	6.81
18	41.51	58	33.23	98	21.06	138	11.20	178	6.73
19	41.39	59	32.95	99	20.76	139	11.04	179	6.66
20	41.28	60	32.67	100	20.46	140	10.89	180	6.59
21	41.15	61	32.38	101	20.16	141	10.73	181	6.51
22	41.02	62	32.09	102	19.86	142	10.58	182	6.44
23	40.89	63	31.80	103	19.57	143	10.43	183	6.37
24	40.75	64	31.50	104	19.28	144	10.29	184	6.30
25	40.60	65	31.21	105	18.98	145	10.15	185	6.23
26	40.45	66	30.91	106	18.69	146	10.01	186	6.17
27	40.29	67	30.61	107	18.40	147	9.87	187	6.10
28	40.13	68	30.31	108	18.12	148	9.74	188	6.04
29	39.97	69	30.01	109	17.83	149	9.61	189	5.97
30	39.79	70	29.70	110	17.55	150	9.48	190	5.91
31	39.62	71	29.40	111	17.27	151	9.36	191	5.85
32	39.43	72	29.11	112	16.99	152	9.23	192	5.79
33	39.25	73	28.79	113	16.71	153	9.11	193	5.73
34	39.06	74	28.48	114	16.42	154	9.00	194	5.67
35	38.86	75	28.17	115	16.13	155	8.88	195	5.61
36	38.66	76	27.86	116	15.86	156	8.77	196	5.55
37	38.45	77	27.55	117	15.59	157	8.66	197	5.50
38	38.24	78	27.24	118	15.32	158	8.55	198	5.44
39	38.03	79	26.93	119	15.07	159	8.44	199	5.39
40	37.81	80	26.62	120	14.82	160	8.33	200	5.33

* When element width-to-thickness ratio exceeds λ_r, see Appendix B5.3 of LRFD Specification

b. For $\lambda_c > 1.5$

$$F_{cr} = \left[\frac{0.877}{\lambda_c^2}\right] F_y \qquad (E2\text{-}3)$$

where

F_y = specified minimum yield stress, ksi
E = modulus of elasticity, ksi
K = effective length factor
l = unbraced length of member, in.
r = governing radius of gyration about plane of buckling, in.

For 50 ksi steel

$$\lambda_c = \frac{Kl}{r}\frac{1}{\pi}\sqrt{\frac{50 \text{ ksi}}{29,000 \text{ ksi}}} = 0.0132\frac{Kl}{r} \text{ or } \frac{Kl}{r} = 75.7\lambda_c \qquad (2\text{-}5)$$

The boundary between inelastic and elastic buckling ($\lambda_c = 1.5$) for 50 ksi steel is

$$\frac{Kl}{r} = 75.7 \times 1.5 = 113.5$$

The column strength equations in terms of Kl/r for 50 ksi steel become

$$\phi_c P_n = (\phi_c F_{cr})A_g \qquad (2\text{-}6)$$

where $\phi_c = 0.85$

a. For $Kl/r \leq 113.5$

$$F_{cr} = \{\exp[-7.3 \times 10^{-5}(Kl/r)^2]\} \times 50 \text{ ksi} \qquad (2\text{-}7)$$

b. For $Kl/r \leq 113.5$

$$F_{cr} = \frac{2.51 \times 10^5}{(Kl/r)^2} \text{ ksi} \qquad (2\text{-}8)$$

Based on Equations 2-7 and 2-8, Table E-1 gives the design stresses for 50 ksi steel columns for the full range of slenderness ratios. Determining the design strength of a given 50 ksi steel column merely involves using Equation 2-6 in connection with Table E-1. The appropriate design stress ($\phi_c F_{cr}$) from Table E-1 is multiplied by the cross-sectional area to obtain the design strength $\phi_c P_n$.

EXAMPLE E-1

Given: Design a 25-ft high, free standing A618 ($F_y = 50$ ksi) steel pipe column to support a water tank with a weight of 75 kips at full capacity. See Figure E-1.

Solution: For a live load of 75 kips, the required column strength (from Section A) is $P_u = 1.6P_L = 1.6 \times 75$ kips $= 120$ kips.

From Table C-2, case *e*, recommended $K = 2.1$. $KL = 2.1 \times 25.0$ ft = 52.5 ft.

Try a standard 12-in. diameter pipe ($A = 14.6$ in.^{2}, $I = 279$ in.^{4}):

$$r = \sqrt{I/A} = \sqrt{279 \text{ in.} / 14.6 \text{ in.}^2} = 4.37 \text{ in.}$$

$$\frac{Kl}{r} = \frac{52.5 \text{ ft} \times 12 \text{ in./ft}}{4.37 \text{ in}} = 144.2$$

From Table E-1, $\phi_c F_{cr} = 10.3$ ksi

The design compressive strength

$$\phi_c P_n = (\phi_c F_{cr})A_g = 10.3 \text{ ksi} \times 14.6 \text{ in.}^2$$
$$= 150 \text{ kips} > 120 \text{ kips required} \quad \textbf{o.k.}$$

To complete the design, bending due to lateral loads (i.e., wind and earthquake) should also be considered. See Sections F and H.

EXAMPLE E-2   Determine the adequacy of a W14×120 building column.

Given: 50 ksi steel; $K = 1.0$; story height = 12.0 ft; required strength based on the maximum total factored load is 1,300 kips.

Solution: $K_x L_x = K_y L_y = 1.0 \times 12.0$ ft = 12.0 ft

Because $r_y < r_x$,

$$\left(\frac{Kl}{r}\right) \text{maximum} = \frac{K_y L_y}{r_y} = \frac{12.0 \text{ ft} \times 12 \text{ in./ft}}{3.74 \text{ in.}} = 38.5$$

From Table E-1, $\phi_c F_{cr} = 38.14$ ksi

Design compressive strength
$$\phi_c P_n = (\phi_c F_{cr})A_g = 38.14 \text{ ksi} \times 35.3 \text{ in.}^2$$
$$= 1,346 \text{ kips} > 1,300 \text{ kips required} \quad \textbf{o.k.}$$

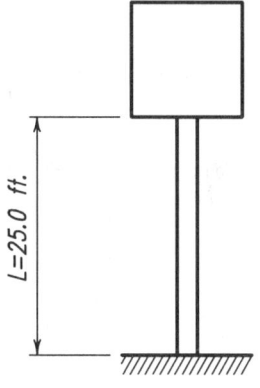

Fig. E-1

EXAMPLE E-3   Select the most economical W14 column for the case shown in Figures
E-2 and E-3.

Fig. E-2. Plan views.

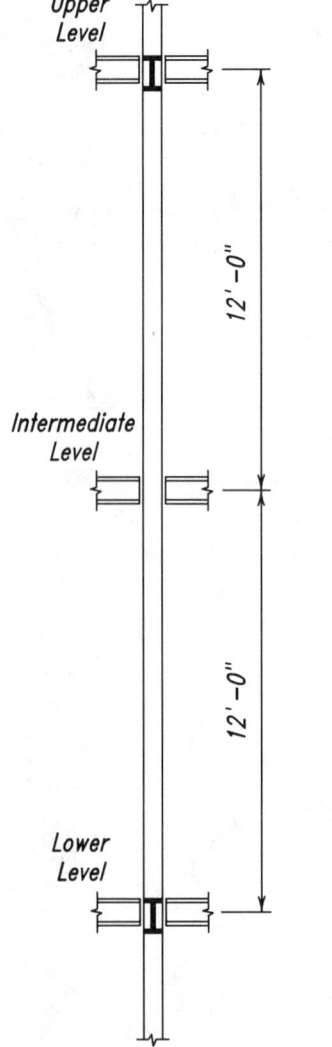

Fig. E-3. Elevation.

Given: 50 ksi steel; $K = 1.0$; required strength based on the maximum total factored load is 1,300 kips. The column is braced in both directions at the upper and lower levels, and in the weak direction at the intermediate level.

Solution: Try a W14×120 (as in Example E-2):

$$\frac{K_x\, l_x}{r_x} = \frac{1.0 \times 24.0 \text{ ft} \times 12 \text{ in./ft}}{6.24 \text{ in.}} = 46.2$$

$$\frac{K_y\, l_y}{r_y} = \frac{1.0 \times 12.0 \text{ ft} \times 12 \text{ in./ft}}{3.74 \text{ in.}} = 38.5$$

$$\frac{Kl}{r} \text{ max} = \frac{K_x\, l_x}{r_x} = 46.2$$

From Table E-1, $\phi_c F_{cr} = 36.35$ ksi

$$\text{Required } A_g = \frac{1{,}300 \text{ kips}}{36.35 \text{ ksi}} = 35.8 \text{ in.}^2 > 35.3 \text{ in.}^2 \text{ provided}$$

W14×120   **n.g.**

By inspection W14×132 is **o.k.**

Use W14×132

Flexural-Torsional Buckling

As stated in Section E3 of the LRFD Specification and Commentary, torsional and flexural-torsional buckling generally do not govern the design of doubly symmetric rolled shapes in compression. For other cross sections, see Section E3 and Appendix E3 of the LRFD Specification.

Built-Up and Pin-Connected Members

These members are covered, respectively, in Section E4 and E5 of the LRFD Specification.

F. BEAMS AND OTHER FLEXURAL MEMBERS

Chapter F of the LRFD Specification covers compact beams. Compactness criteria are given in Table B5.1 of the LRFD Specification and are summarized in Table B-1 above. To prevent torsion, wide-flange shapes must be loaded in either plane of symmetry, channels must be loaded through the shear center parallel to the web, or restraint against twisting must be provided at load points and points of support. Torsion combined with flexure and axial force combined with flexure are covered in Chapter H of the LRFD Specification.

This section explains the provisions of the LRFD Specification for compact rolled beams. For other compact and noncompact flexural members, refer to Appendix F of the Specification; plate girders are in Appendix G.

Flexure

To understand the provisions of the LRFD Specification regarding flexural design, it is helpful to review briefly some aspects of elementary beam theory.

Under working loads (and until initial yielding) the distributions of flexural strains and stresses over the cross-section of a beam are linear. As shown in Figure F-1, they vary from maximum compression at the extreme fibers on one side (the top) to zero at the neutral, or centroidal, axis to maximum tension at the extreme fibers on the other side (the bottom).

The relationship between moment and maximum bending stress (tension or compression) at a given cross section is

$$M = Sf_b \tag{2-9}$$

where

M = bending moment due to the applied loads, kip-in.
S = elastic section modulus, in the direction of bending, in.[3]

$$= \frac{I}{c}$$

f_b = maximum bending stress, ksi
I = moment of inertia of the cross section about its centroidal axis, in.[4]
c = distance from the elastic neutral axis to the extreme fiber, in.

Similarly, at initial yielding

$$M_r = SF_y \tag{2-10}$$

where

M_r = bending moment coinciding with first yielding, kip-in.

If additional load is applied, the strains continue to increase; the stresses, however, are limited to F_y. Yielding proceeds from the outer fibers inward until a plastic hinge is developed, as shown in Figure F-1. At full plastification of the cross section

$$M_p = ZF_y \tag{2-11}$$

where

M_p = plastic moment, kip-in
Z = plastic section modulus, in the direction of bending, in.[3]

Due to the presence of residual stresses (prior to loading, as a consequence of the rolling operation), yielding begins at an applied stress of $(F_y - F_r)$. Equation 2-10 should be modified to

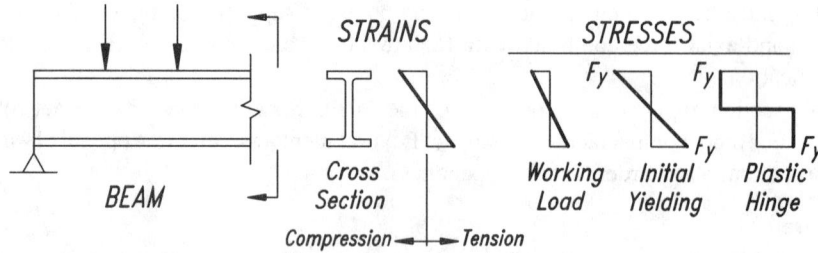

Fig. F-1. Flexural strains and stresses.

$$M_r = S(F_y - F_r) \tag{2-12}$$

where

F_r = the maximum compressive residual stress in either flange, ksi
  = 10 ksi for rolled shapes, 16.5 ksi for welded shapes

The definition of plastic moment in Equation 2-11 is still valid, because it is not affected by residual stresses.

Design for Flexure
a. Assuming $C_b = 1.0$

Compact sections will not experience local buckling before the formation of a plastic hinge. The occurrence of lateral-torsional buckling of the member depends on the unbraced length L_b. As implied by the term lateral-torsional buckling, overall instability of a beam requires that twisting of the member occur simultaneously with lateral buckling of the compression flange. L_b is the distance between points braced to prevent twist of the cross section. Many beams can be considered continuously braced; e.g., beams supporting a metal deck, if the deck is intermittently welded to the compression flange. Compact wide flange and channel members bending about their major (or x) axes can develop their full plastic moment M_p without buckling if $L_b \le L_p$. If $L_b = L_r$, the nominal flexural strength is M_r, the moment at first yielding adjusted for residual stresses. The nominal moment capacity (M_n) for $L_p < L_b < L_r$ is $M_r < M_n < M_p$. Compact shapes bent about their minor (or y) axes will not buckle before developing M_p, regardless of L_b.

Flexural design strength, governed by the limit state of lateral-torsional buckling, is $\phi_b M_n$, where $\phi_b = 0.90$ and M_n the nominal flexural strength is as follows:

$M_n = M_p = Z_x F_y$ for bending about the major axis if $L_b \le L_p$ (2-13)
$M_n = M_p = Z_y F_y$ for bending about the minor axis regardless of L_b (2-14)

$$L_p = \frac{300 r_y}{\sqrt{F_y}} = 42.4 r_y \quad \text{for 50 ksi steel} \tag{2-15}$$

$M_n = M_r = S_x(F_y - F_r)$ (2-16)
  = $S_x(F_y - 10 \text{ ksi})$ for rolled shapes bending about the major axis if $L_b = L_r$

M_n for bending about the major axis, if $L_p < L_b < L_r$, is determined by linear interpolation between Equations 2-13 and 2-16; i.e.,

$$M_n = M_p - (M_p - M_r)\left(\frac{L_b - L_p}{L_r - L_p}\right) \tag{2-17}$$

The definition for the limiting laterally unbraced length L_r is given in the LRFD Specification (in Equations F1-6, 8, and 9) and will not be repeated here. For bending about the major axis if $L_b > L_r$,

$$M_n = M_{cr} \le M_r \tag{2-18}$$

The case of $L_b > L_r$ is beyond the scope of this section. The reader is referred to Section F1.2b of LRFD Specification (specifically Equation F1-13, where the critical moment

Table F-1. Values of C_b for Simply Supported Beams Braced at Ends of Span		
Load	Lateral Bracing Along Span	C_b
Concentrated at center	None	1.32
	At centerline only	1.67
Uniform	None	1.14
	At centerline only	1.30

M_{cr} is controlled by lateral-torsional buckling). This case is also covered in the beam graphs in Part 4 of this LRFD Manual.

b. All values of C_b

C_b is the bending coefficient. A new expression for C_b is given in the LRFD Specification. (It is more accurate than the one previously shown.)

$$C_b = \frac{12.5M_{max}}{2.5M_{max} + 3M_A + 4M_B + 3M_c} \tag{F1-3}$$

where M is the absolute value of a moment in the unbraced beam segment as follows:

M_{max}, the maximum
M_A, at the quarter point
M_B, at the centerline
M_c, at the three-quarter point

The purpose of C_b is to account for the influence of moment gradient on lateral-torsional buckling. The flexural strength equations with $C_b = 1.0$ are based on a uniform moment along a laterally unsupported beam segment causing single curvature buckling of the member. Other loadings are less severe, resulting in higher flexural strengths; $C_b \geq 1.0$. Typical values of C_b are given in Table F-1. For unbraced cantilevers, $C_b = 1.0$. C_b can conservatively be taken as 1.0 for all cases.

For all values of C_b, the flexural design strength $\phi_b M_n$, where $\phi_b = 0.90$, is given in the LRFD Specification in terms of a nominal flexural strength M_n varying as follows:

$$M_n = M_p = Z_x F_y \tag{2-13}$$

for bending about the major axis if $L_b \leq L_m$

$$M_n = C_b M_r = C_b S_x (F_y - 10 \text{ ksi}) \leq M_p \tag{2-19}$$

for bending about the major axis if $L_b = L_r$.

For bending about the major axis if $L_m < L_b < L_r$, linear interpolation is used

$$M_n = C_b \left[M_p - (M_p - M_r) \left(\frac{L_b - L_p}{L_r - L_p} \right) \right] \leq M_p \tag{F1-2}$$

If $L_b > L_r$,

$$M_n = M_{cr} \leq C_b M_r \text{ and } M_p \tag{2-20}$$

The determination of M_n for a given L_b can best be done graphically, as illustrated in Figure F-2. The required parameters for each W shape are given in the beam design table in Part 4 of the LRFD Manual, an excerpt of which is shown herein as Table F-2. If $C_b = 1.0$, the coordinates for constructing the graph are (L_p, M_p), and (L_r, M_r). For $C_b > 1.0$, the key coordinates are $(L_p, C_b M_p)$ and $(L_r, C_b M_r)$. Note that M_n cannot exceed the plastic moment M_p. L_m, then, can be derived graphically as the upper limit of L_b for which $M_n = M_p$. If $L_b > L_r$, the beam graphs in Part 4 of the LRFD Manual can be used to determine M_{cr}.

EXAMPLE F-1

Given: Select the required W shape for a 30-foot simple floor beam with full lateral support carrying a dead load (including its own weight) of 1.5 kips per linear foot and a live load of 3.0 kips per linear foot. Assume 50 ksi steel and:

a. There is no member depth limitation

b. The deepest member is a W18

Solution: The governing load combination in Section A is A4-2:

$$1.2D + 1.6L + 0.5(L_r \text{ or } S \text{ or } R) = 1.2 \times 1.5 \text{ klf} + 1.6 \times 3.0 \text{ klf} + 0$$
$$= 6.6 \text{ klf}$$

$$\text{Required } M_u = \frac{wL^2}{8} = \frac{6.6 \text{ klf} \times (30.0 \text{ ft})^2}{8} = 743 \text{ kip-ft}$$

Flexural design strength $\phi_b M_n \geq 743$ kip-ft

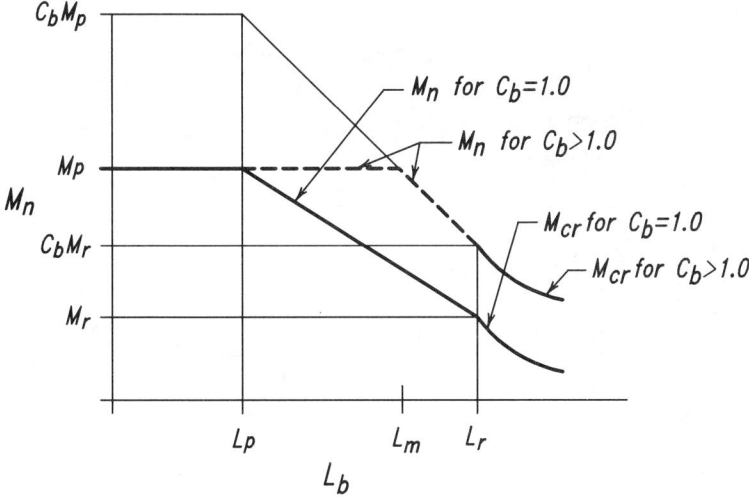

Fig. F-2. Determination of nominal flexural strength M_n.

Table F-2.
Excerpt from Load Factor Design Selection Table
(LRFD Manual, Part 4)

		For F_y = 50 ksi			
Z_x (in.^{3})	Shape	$\phi_b M_p$ (kip-ft)	$\phi_b M_r$ (kip-ft)	L_p (ft)	L_r (ft)
224	W24×84	840	588	6.9	18.6
221	W21×93	829	576	6.5	19.4
212	W14×120	795	570	13.2	46.2
211	W18×97	791	564	9.4	27.4
200	W24×76	750	528	6.8	18.0
198	W16×100	743	525	8.9	29.3
196	W21×83	735	513	6.5	18.5
192	W14×109	720	519	13.2	43.2
186	W18×86	698	498	9.3	26.1
186	W12×120	698	489	11.1	50.0
177	W24×68	664	462	6.6	17.4
175	W16×89	656	465	8.8	27.3

Note: Flexural design strength $\phi_b M_n = \phi_b M_p$, as tabulated is valid for $L_b \le L_m$. If $C_b = 1.0$, $L_m = L_p$; otherwise, $L_m > L_p$. $\phi_b = 0.90$.

a. In Table F-2, the most economical beams are in **boldface** print. Of the boldfaced beams, the lightest one with $\phi_b M_n = \phi_b M_p \ge 743$ kip-ft is a W24×76

b. By inspection of Table F-2, the lightest W18 with $\phi_b M_n = \phi_b M_p \ge 743$ kip-ft is a W18×97.

EXAMPLE F-2

Given: Determine the flexural design strength of a 30-ft long simply supported W24×76 girder (of 50 ksi steel) with a concentrated load and lateral support, both at midspan.

Solution: From Table F-1, $C_b = 1.67$

$$L_b = 30.0 \text{ ft}/2 = 15.0 \text{ ft}$$

From Equation F1-2:

$$\phi_b M_n = C_b \left[\phi_b M_p - (\phi_b M_p - \phi_b M_r) \left(\frac{L_b - L_p}{L_r - L_p} \right) \right] \le \phi_b M_p$$

From Table F-2 for a W24×76:

$\phi_b M_p = 750$ kip-ft
$\phi_b M_r = 528$ kip-ft
$L_p \quad = 6.8$ ft

$$L_r = 18.0 \text{ ft}$$

$$\phi_b M_n = 1.67 \left[750 \text{ kip-ft} - (750 - 528) \text{ kip-ft} \times \frac{15.0 \text{ ft} - 6.8 \text{ ft}}{18.0 \text{ ft} - 6.8 \text{ ft}} \right]$$

$$= 981 \text{ kip-ft} > 750 \text{ kip-ft}$$

Use $\phi_b M_n = \phi_b M_p = 750$ kip-ft

In this case, even though the unbraced length $L_b > L_p$, the design flexural strength is $\phi_b M_p$ because $C_b > 1.0$.

Design for Shear

The design shear strength is defined by the equations in Section F2 of the LRFD Specification. Shear in wide-flange and channel sections is resisted by the area of the web (A_w), which is taken as the overall depth d times the web thickness t_w. For webs of 50 ksi steel without transverse stiffeners, the design shear strength $\phi_v V_n$, where $\phi_v = 0.90$, and the nominal shear strength V_n are as follows:

For $\dfrac{h}{t_w} \leq 59$ (including all rolled W and channel shapes),

$$V_n = 30.0 \text{ ksi} \times dt_w$$
$$\phi_v V_n = 27.0 \text{ ksi} \times dt_w \tag{2-21}$$

For $59 < \dfrac{h}{t_w} \leq 74$,

$$V_n = 30.0 \text{ ksi} \times dt_w \times \frac{59}{h/t_w}$$

$$\phi_v V_n = 27.0 \text{ ksi} \times dt_w \times \frac{59}{h/t_w} \tag{2-22}$$

For $\dfrac{h}{t_w} > 74$,

$$V_n = \frac{132,000}{(h/t_w)^2} dt_w \text{ ksi}$$

$$\phi_v V_n = \frac{118,000}{(h/t_w)^2} dt_w \text{ ksi} \tag{2-23}$$

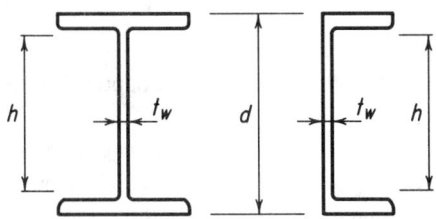

Fig. F-3. Definitions of d, h, and t_w for W and channel shapes.

Shear strength is governed by the following limit states; Equation 2-21 by yielding of the web; Equation 2-22, by inelastic buckling of the web; and Equation 2-23 by elastic buckling.

EXAMPLE F-3

Given: Check the adequacy of a W30×99 beam of 50 ksi steel to carry a load resulting in maximum shears of 100 kips due to dead load and 150 kips due to live load.

Solution: Required shear strength = V_u = 1.2D + 1.6L

$$= 1.2 \times 100 \text{ kips} + 1.6 \times 150 \text{ kips}$$
$$= 360 \text{ kips}$$

Design shear strength = $\phi_v V_n$ = 27.0 ksi × dt_w
$$= 27.0 \text{ ksi} \times 29.65 \text{ in.} \times 0.520 \text{ in.}$$
$$= 416 \text{ kips} > 360 \text{ kips required}   \textbf{o.k.}$$

Web Openings
See Section F4 of the LRFD Specification and Commentary, and the references given in the Commentary.

H. MEMBERS UNDER COMBINED FORCES AND TORSION

Symmetric Members Subject to Bending and Axial Tension
The interaction of flexure and tension in singly and doubly symmetric shapes is governed by Equations H1-1a and H1-1b, as follows:

For $\dfrac{P_u}{\phi P_n} \geq 0.2$,

$$\frac{P_u}{\phi P_n} + \frac{8}{9}\left(\frac{M_{ux}}{\phi_b M_{nx}} + \frac{M_{uy}}{\phi_b M_{ny}}\right) \leq 1.0 \qquad\qquad \text{(H1-1a)}$$

For $\dfrac{P_u}{\phi P_n} < 0.2$,

$$\frac{P_u}{2\phi P_n} + \left(\frac{M_{ux}}{\phi_b M_{nx}} + \frac{M_{uy}}{\phi_b M_{ny}}\right) \leq 1.0 \qquad\qquad \text{(H1-1b)}$$

where

P_u = required tensile strength; i.e., the total factored tensile force, kips

ϕP_n = design tensile strength, $\phi_t P_n$, kips

ϕ = resistance factor for tension, ϕ_t = 0.90

P_n = nominal tensile strength as defined in Chapter D of the LRFD Specification, kips

M_u = required flexural strength; i.e., the moment due to the total factored load, kip-in. or kip-ft. (Subscript x or y denotes the axis about which bending occurs.)

$\phi_b M_n$ = design flexural strength, kip-in. or kip-ft

ϕ_b = resistance factor for flexure = 0.90

M_n = nominal flexural strength determined in accordance with the appropriate equations in Chapter F of the LRFD Specification, kip-in. or kip-ft

Interaction Equations H1-1a and H1-1b cover the general case of biaxial bending combined with axial force. They are also valid for uniaxial bending (i.e., when $M_{ux} = 0$ or $M_{uy} = 0$). In this case, they reduce to the form plotted in Figure H-1. Pure biaxial bending (with $P_u = 0$) is covered by Equation H1-1b.

EXAMPLE H-1

Given: Check the adequacy of a W10×22 tension member of 50 ksi steel to carry loads resulting in the following factored load combination:

P_u = 55 kips
M_{uy} = 20 kip-ft
M_{ux} = 0

Solution: From Section D above for 50 ksi steel,

$$\phi P_n = \phi_t P_n = 45.0 \text{ ksi} \times A_g = 45.0 \text{ ksi} \times 6.49 \text{ in.}^2 = 292 \text{ kips}$$

$$\frac{P_u}{\phi P_n} = \frac{55 \text{ kips}}{292 \text{ kips}} = 0.188 < 0.20; \text{ therefore, Equation H1-1b governs.}$$

For bending about the y axis only, Equation H1-1b becomes:

$$\frac{P_u}{2\phi P_n} + \frac{M_{uy}}{\phi_b M_{ny}} \leq 1.0$$

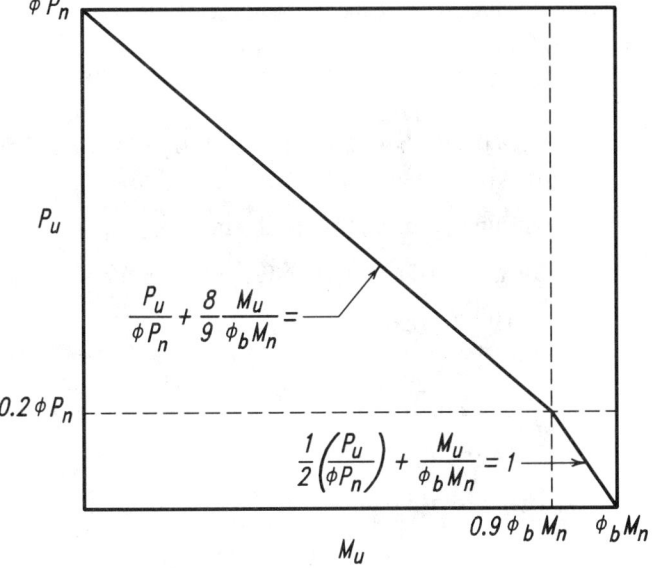

Fig. H-1. Interaction Equations H1-1a and H1-1b modified for axial load combined with bending about one axis only.

From Section F above for 50 ksi steel, $M_n = M_p = Z_y F_y = 50$ ksi $\times Z_y$ for minor-axis bending (regardless of the unbraced length).

$$\phi_b M_{ny} = 0.90 \times 50 \text{ ksi} \times Z_y = 45.0 \text{ ksi} \times Z_y$$

$$= 45.0 \text{ ksi} \times \frac{6.10 \text{ in.}^3}{12 \text{ in./ft}}$$

$$= 22.9 \text{ kip-ft for a W10}\times22 \text{ member}$$

$$\frac{P_u}{2\phi P_n} + \frac{M_{uy}}{\phi_b M_{ny}} = \frac{0.188}{2} + \frac{20 \text{ kip-ft}}{22.9 \text{ kip-ft}} = 0.094 + 0.873$$

$$= 0.967 < 1.0 \quad \textbf{o.k.}$$

EXAMPLE H-2

Given:

Check the same tension member, a W10×22 in 50 ksi steel, 4.0 ft long, subjected to the following combination of factored loads:

P_u = 140 kips
M_{ux} = 55 kip-ft
M_{uy} = 0
C_b = 1.0

Solution:

Again, $\phi P_n = 292$ kips

$$\frac{P_u}{\phi P_n} = \frac{140 \text{ kips}}{292 \text{ kips}} = 0.479 > 0.20; \text{ Equation H1-1a governs.}$$

For bending about the x axis only, Equation H1-1a becomes

$$\frac{P_u}{\phi P_n} + \frac{8}{9} \frac{M_{ux}}{\phi_b M_{nx}} \le 1.0$$

From Section F above for 50 ksi steel, $M_n = M_p = Z_x F_y = 50$ ksi $\times Z_x$ for major-axis bending if $L_b \le L_p$ for ($C_b = 1.0$).

Assume unbraced length, $L_b = 4.0$ ft.

By Equation F1-4a in Section F, $L_p = 42.4 r_y$ for 50 ksi steel.

For a W10×22, $r_y = 1.33$ in., $Z_x = 26.0$ in.^{3}

$$L_p = \frac{42.4 \times 1.33 \text{ in.}}{12 \text{ in./ft}} = 4.7 \text{ ft}$$

$$L_b = 4.0 \text{ ft} < L_p = 4.7 \text{ ft}$$

Then $M_{nx} = 50$ ksi $\times Z_x$

$$\phi_b M_{nx} = \frac{0.90 \times 50 \text{ ksi} \times 26.0 \text{ in.}^3}{12 \text{ in./ft}}$$

$$= 97.5 \text{ kip-ft for a W10}\times22 \text{ member}$$

$$\frac{P_u}{\phi P_n} + \frac{8M_{ux}}{9\phi_b\,M_{nx}} = 0.479 + \frac{8}{9} \times \frac{55 \text{ kip-ft}}{97.5 \text{ kip-ft}}$$

$$= 0.479 + 0.501 = 0.980 < 1.0 \quad \textbf{o.k.}$$

Symmetric Members Subject to Bending and Axial Compression

The interaction of compression and flexure in beam-columns with singly and doubly symmetric cross sections is governed by Equations H1-1a and H1-1b, repeated here for convenience:

For $\dfrac{P_u}{\phi P_n} \geq 0.2,$

$$\frac{P_u}{\phi P_n} + \frac{8}{9}\left(\frac{M_{ux}}{\phi_b\,M_{nx}} + \frac{M_{uy}}{\phi_b\,M_{ny}}\right) \leq 1.0 \qquad\qquad \text{(H1-1a)}$$

For $\dfrac{P_u}{\phi P_n} < 0.2,$

$$\frac{P_u}{2\phi P_n} + \left(\frac{M_{ux}}{\phi_b\,M_{nx}} + \frac{M_{uy}}{\phi_b\,M_{ny}}\right) \leq 1.0 \qquad\qquad \text{(H1-1b)}$$

The definitions of the terms in the formulas, which differ in some cases from those given above, are as follows:

P_u = required compressive strength; i.e., the total factored compressive force, kips

ϕP_n = design compressive strength, $\phi_c\,P_n$, kips

ϕ = resistance factor for compression, $\phi_c = 0.85$

P_n = nominal compressive strength as defined in Chapter E of the LRFD Specification, kips

M_u = required flexural strength including second-order effects, kip-in. or kip-ft

$\phi_b\,M_n$ = design flexural strength, kip-in. or kip-ft

ϕ_b = resistance factor for flexure = 0.90

M_n = nominal flexural strength from Chapter F of the LRFD Specification, kip-in. or kip-ft

The second-order analysis required for M_u involves the determination of the additional moment due to the action of the axial compressive forces on a deformed structure. In lieu of a second-order analysis, the simplified method given in Chapter C of the LRFD Specification (and in Section C above) may be used. However, in applying the simplified method, the additional moments obtained for beam-columns must also be distributed to connected members and connections (to satisfy equilibrium).

Bending and Axial Compression—Preliminary Design

The design of a beam-column is a trial and error process which can become tedious, particularly with the repeated solution of Interaction Equation H1-1a or H1-1b. A rapid method for the selection of a trial section is given in this LRFD Manual, Part 3, under the heading Combined Axial and Bending Loading (Interaction). As in earlier editions of the AISC Manual, the Interaction Equations are approximated by an equation which converts bending moments to equivalent axial loads:

$$P_{u\ eq} = P_u + M_{ux}m + M_{uy}mu$$

where

$P_{u\,eq}$ = equivalent axial load to be checked against the column load table, kips

P_u, M_{ux}, M_{uy} are defined in the Interaction Equations for compression and bending

m, u are factors tabulated in this LRFD Manual, Part 3

As soon as a satisfactory trial section has been found (i.e., one for which $P_{u\,eq} \leq$ tabulated $\phi_c\,P_n$), a final verification should be made with the appropriate Interaction Equation, H1-1a or H1-1b

---

EXAMPLE H-3

Given: Check the adequacy of a W14×176 beam-column, 14.0 ft in height floor-to-floor, in a braced symmetrical frame in 50 ksi steel. The member is subjected to the following factored forces due to symmetrical gravity loads: P_u = 1,400 kips; M_x = 200 kip-ft, M_y = 70 kip-ft (reverse curvature bending with equal end moments about both axes); and no loads along the member.

Solution: For a braced frame, K = 1.0 $K_x L_x = K_y L_y$ = 14.0 ft

For a W14×176:

A = 51.8 in.^{2}

Z_x = 320 in.^{3}

Z_y = 163 in.^{3}

r_x = 6.43 in.

r_y = 4.02 in.

Kl / r_x = (14.0 ft × 12 in./ft) / 6.43 in. = 26.1

Kl / r_y = (14.0 ft × 12 in./ft) / 4.02 in. = 41.8

From Table E-1, above, $\phi_c F_{cr}$ = 37.4 ksi for Kl / r = 41.8 in 50 ksi steel.

$\phi_c P_n = (\phi_c\,F_{cr})\,A$ = 37.4 ksi × 51.8 in.^{2} = 1,940 kips

Since $\dfrac{P_u}{\phi_c\,P_n} = \dfrac{1,400 \text{ kips}}{1,940 \text{ kips}}$ = 0.72 > 0.2, Interaction Equation H1-1a **governs.**

For a braced frame, M_{lt} = 0. From Equation C1-1:

$M_{ux} = B_{1x}\,M_{ntx}$, where M_{ntx} = 200 kip-ft; and

$M_{uy} = B_{1y}\,M_{nty}$, where M_{nty} = 70 kip-ft

From Equations C1-2 and C1-3:

$$B_1 = \frac{C_m}{(1 - P_u / P_{e1})} > 1.0$$

where in this case (a braced frame with no transverse loading),

$$C_m = 0.6 - 0.4(M_1 / M_2)$$

For reverse curvature bending and equal end moments:

$M_1 / M_2 = +1.0$

$C_m \quad = 0.6 - 0.4(1.0) = 0.2$

From Table C-1:

$P_{elx} = 420 \text{ ksi} \times A_g = 420 \text{ ksi} \times 51.8 \text{ in.}^2 = 21{,}756 \text{ kips}$

From Table C-1:

$P_{ely} = 164 \text{ ksi} \times A_g = 164 \text{ ksi} \times 51.8 \text{ in.}^2 = 8{,}495 \text{ kips}$

$$B_{1x} = \frac{C_{mx}}{(1 - P_u / P_{elx})} = \frac{0.2}{(1 - 1{,}400 \text{ kips} / 21{,}756 \text{ kips})} = 0.2$$

Use $B_{1x} = 1.0$, per Equation C1-2.

$$B_{1y} = \frac{C_{my}}{(1 - P_u / P_{ely})} = \frac{0.2}{(1 - 1{,}400 \text{ kips} / 8{,}495 \text{ kips})} = 0.2$$

Use $B_{1y} = 1.0$, per Equation C1-2.

$M_{ux} = 1.0 \times 200 \text{ kip-ft}$

$M_{uy} = 1.0 \times 70 \text{ kip-ft}$

From Equation 2-15 for 50 ksi steel,

$$L_p = 42.4 r_y = \frac{42.4 \times 4.02 \text{ in.}}{12 \text{ in./ft}} = 14.2 \text{ ft}$$

Since $L_b = 14.0 \text{ ft} < L_p = 14.2 \text{ ft}$, $M_{nx} = M_{px} = Z_x F_y$

$M_{ny} \quad = M_{py} = Z_y F_y$

$\phi_b F_y \quad = 0.90 \times 50 \text{ ksi} = 45.0 \text{ ksi}$

$$\phi_b M_{nx} = \phi_b F_y Z_x = \frac{45.0 \text{ ksi} \times 320 \text{ in.}^3}{12 \text{ in./ft}} = 1{,}200 \text{ kip-ft}$$

$$\phi_b M_{ny} = \phi_b F_y Z_y = \frac{45.0 \text{ ksi} \times 163 \text{ in.}^3}{12 \text{ in./ft}} = 611 \text{ kip-ft}$$

By Interaction Equation H1-1a

$$\frac{1{,}400 \text{ kips}}{1{,}940 \text{ kips}} + \frac{8}{9}\left(\frac{200 \text{ kip-ft}}{1{,}200 \text{ kip-ft}} + \frac{70 \text{ kip-ft}}{611 \text{ kip-ft}}\right) = 0.72 + \frac{8}{9}(0.17+0.11)$$

$$= 0.72 + 0.25$$

$$= 0.97 < 1.0$$

W14×176 is **o.k.**

EXAMPLE H-4

Given: Check the adequacy of a W14×176 beam-column (F_y = 50 ksi) in an unbraced symmetrical frame subjected to the following factored forces:

P_u = 1,400 kips (due to gravity plus wind)
M_{ux} = 300 kip-ft (due to wind only)
M_y = 0
$K_x L_x = K_y L_y$ = 14.0 ft

Drift index, $\Delta_{oh} / L \leq 0.0025$ (or $\frac{1}{400}$)

ΣP_u = 24,000 kips
ΣH = 800 kips

Solution: As in Example H-3, for a W14×176 with KL = 14.0 ft, $\phi_c P_n$ = 1,940 kips.

Since $\dfrac{P_u}{\phi_c P_n} = \dfrac{1,400 \text{ kips}}{1,940 \text{ kips}} = 0.72 > 0.2$, Interaction Equation H1-1a **governs**.

Because $M_{ntx} = M_{nty} = M_{lty} = 0$ and only $M_{ltx} \neq 0$, $M_{ux} = B_2 M_{ltx}$ and $M_{uy} = 0$.

M_{ltx} = 300 kip-ft

According to Equation C1-4,

$$B_2 = \cfrac{1}{1 - \dfrac{\Sigma P_u}{\Sigma H}\left(\dfrac{\Delta_{oh}}{L}\right)} = \cfrac{1}{1 - \dfrac{24,000 \text{ kips}}{800 \text{ kips}}(0.0025)} = 1.08$$

M_{ux} = 1.08 × 300 kip-ft = 324 kip-ft

Because $L_b < L_p$ = 14.2 ft, $M_{nx} = M_{px} = Z_x F_y$; $\phi_b M_{nx}$ = 1,200 kip-ft as in Example H-3.

By Interaction Equation H1-1a:

$$\frac{1,400 \text{ kips}}{1,940 \text{ kips}} + \frac{8}{9}\frac{324 \text{ kip-ft}}{1,200 \text{ kip-ft}} = 0.72 + \frac{8}{9}0.27 = 0.96 < 1.0$$

W14×176 is **o.k.**

Torsion and Combined Torsion, Flexure, and/or Axial Force

Criteria for members subjected to torsion and torsion combined with other forces are given in Section H2 of the LRFD Specification. They require the calculation of normal and shear stresses by elastic analysis of the member under the factored loads. The AISC book *Torsional Analysis of Steel Members* (American Institute of Steel Construction, 1983) provides design aids and examples for the determination of torsional stresses. Extensive coverage is given there to wide-flange shapes (W, S, and HP), channels (C and MC) and Z shapes. For these members, the charts and formulas simplify considerably the calculation of torsional rotations, torsional normal and shear stresses, and the combination of torsional with flexural stresses.

In the LRFD Specification,

f_{un} = the total normal stress under factored load (ksi) from torsion and all other causes
f_{uv} = the total shear stress under factored load (ksi) from torsion and all other causes

The criteria are as follows:

a. For the limit state of yielding under normal stress

$$f_{un} \leq \phi F_y, \text{ where } \phi = 0.90 \tag{H2-1}$$

For 50 ksi steel,

$$f_{un} \leq 0.90 \times 50 \text{ ksi} = 45.0 \text{ ksi} \tag{2-24}$$

b. For the limit state of yielding under shear stress,

$$f_{uv} \leq 0.60\phi F_y, \text{ where } \phi = 0.90 \tag{H2-2}$$

For 50 ksi steel,

$$f_{uv} \leq 0.60 \times 0.90 \times 50 \text{ ksi} = 27.0 \text{ ksi} \tag{2-25}$$

c. For the limit state of buckling,

$$f_{un} \leq \phi_c F_{cr} \text{ or } f_{uv} \leq \phi_c F_{cr}, \text{ as applicable, where } \phi_c = 0.85 \tag{H2-3}$$

For 50 ksi steel, values of $\phi_c F_{cr}$ are given in Table E-1, in Section E above.

Torsion will accompany flexure when the line of action of a lateral load does not pass through the shear center. For wide flange and other doubly symmetric shapes, the shear center is located at the centroid. Singly symmetric shapes have their shear centers on the axis of symmetry, but not at the centroid. (The location of the shear center of channel sections is given in the Properties tables in Part 1 of this LRFD Manual.)

Open sections, such as wide-flange and channel, are very inefficient in resisting torsion; i.e., torsional rotations can be large and torsional stresses relatively high. It is best to avoid torsion by detailing the loads and reactions to act through the shear center of the member. In the case of spandrel members supporting building facade elements, this may not be possible. Heavy exterior masonry walls and stone panels can impose severe torsional loads on spandrel beams. The following are suggestions for eliminating or reducing this kind of torsion:

1. Wall elements may span between floors. The moment due to the eccentricity of the wall with respect to the edge beams can be resisted by lateral forces acting through the floor diaphragms. Torsion would not be imposed on the spandrel beams.
2. If facade panels extend only a partial story height below the floor line, the use of diagonal steel "kickers" may be possible. These light members would provide lateral support to the wall panels. Torsion from the panels would be resisted by forces originating from structural elements other than the spandrel beams.
3. Even if torsion must be resisted by the edge members, providing intermediate torsional supports can be helpful. Reducing the span over which the torsion acts will reduce torsional stresses. If there are secondary beams framing into a spandrel girder,

the beams can act as intermediate torsional supports for the girder. By adding top and bottom moment plates to the connections of the beams with the girder, the bending resistances of the beams can be mobilized to provide the required torsional reactions along the girder.

4. Closed sections provide considerably better resistance to torsion than open sections; torsional rotations and stresses are much lower for box beams than for wide-flange members. For members subjected to torsion, it may be advisable to use box sections or to simulate a box shape by welding one or two side plates to a W shape.

I. COMPOSITE MEMBERS

Chapter I of the LRFD Specification covers composite members. Included are concrete-encased and concrete-filled steel columns and beam columns, as well as steel beams interactive with the concrete slabs they support and steel beams encased in concrete. Unlike traditional structural steel design, which considers only the strength of the steel, composite design assumes that the steel and concrete work together in resisting loads. This results in more economical designs, as the quantity of steel can be reduced.

Compression Members

Composite columns (concrete-encased and concrete-filled) must satisfy the limitations in Section I2 of the LRFD Specification. The design strength of axially loaded composite columns is $\phi_c P_n$, where $\phi_c = 0.85$ and the nominal axial compressive strength is determined from Equations E2-1 through E2-4 above with the following modifications:

A_s replaces A_g, r_m replaces r, F_{my} replaces F_y, and E_m replaces E.

$$F_{my} = F_y + c_1 F_{yr} \frac{A_r}{A_s} + c_2 f_c' \frac{A_c}{A_s} \qquad (I2\text{-}1)$$

$$E_m = E + c_3 E_c \frac{A_c}{A_s} \qquad (I2\text{-}2)$$

r_m = radius of gyration of the steel shape, pipe, or tubing, in. (For steel shapes it shall not be less than 0.3 times the overall thickness of the composite cross section in the plane of buckling.)

where

$$E_c = w^{1.5}\sqrt{f_c'}$$

and

F_{my}	= modified yield stress for the design of composite columns, ksi
F_y	= specified minimum yield stress of the structural steel shape, ksi
F_{yr}	= specified minimum yield stress of the longitudinal reinforcing bars, ksi
f_c'	= specified compressive strength of the concrete, ksi
E_m	= modified modulus of elasticity for the design of composite columns, ksi
E	= modulus of elasticity of steel = 29,000 ksi
E_c	= modulus of elasticity of concrete, ksi
w	= unit weight of concrete, lb/ft^3
A_c	= cross-sectional area of concrete, in.^{2}
A_r	= cross-sectional area of longitudinal reinforcing bars, in.^{2}

A_s = cross-sectional area of structural steel, in.^{2}

c_1, c_2, c_3 = numerical coefficients. For concrete-filled pipe and tubing: $c_1 = 1.0$, $c_2 = 0.85$, and $c_3 = 0.4$; for concrete-encased shapes $c_1 = 0.7$, $c_2 = 0.6$, and $c_3 = 0.2$

Composite columns can be designed by using the Composite Columns Tables in Part 5 of this LRFD Manual (or the numerous tables in AISC Steel Design Guide No. 6: *Load and Resistance Factor Design of W-Shapes Encased in Concrete*) for the cross sections tabulated therein, or the above equations for all cross sections.

Flexural Members

The most common case of a composite flexural member is a steel beam interacting with a concrete slab by means of stud or channel shear connectors. The slab can be a solid reinforced concrete slab, but is usually concrete on a corrugated metal deck.

The effective width of concrete slab acting compositely with a steel beam is determined by three criteria. On either side of the beam centerline, the effective width of concrete slab cannot exceed:

a. one-eighth of the beam span,
b. one-half the distance to the centerline of the adjacent beam, or
c. the distance to the edge of the slab.

The following pertains to rolled W shapes in regions of positive moment, the predominant use of composite beam design. Other cases (e.g., plate girders and negative moments) are covered in Chapter I of the LRFD Specification.

The horizontal shear force between the steel beam and concrete slab, to be transferred by the shear connectors between the points of zero and maximum positive moments, is the minimum of:

a. $0.85f_c'A_c$ (the maximum possible compressive force in the concrete),
b. A_sF_y (the maximum possible tensile force in the steel), and
c. ΣQ_n (the strength of the shear connectors).

For W shapes, the design flexural strength $\phi_b M_n$, with $\phi_b = 0.85$, is based on:

a. a uniform compressive stress of $0.85f_c'$ and zero tensile strength in the concrete
b. a uniform steel stress of F_y in the tension area and compression area (if any) of the steel section, and
c. equilibrium; i.e., the sum of the tensile forces equals the sum of the compressive forces.

The above is valid for shored and unshored construction. However, in the latter case, it is also necessary to check the bare steel beam for adequacy to support the wet concrete and other construction loads (properly factored).

The number of shear connectors required between a point of maximum moment and the nearest location of zero moment is

$$n = \frac{V_h}{Q_n} \tag{2-26}$$

where

V_h = the total horizontal shear force to be transferred, kips

= the minimum of $0.85f_c'A_c$, A_sF_y, and ΣQ_n

Q_n = the shear strength of one connector

The nominal strength of a single stud shear connector in a solid concrete slab is

$$Q_n = 0.5A_{sc}\sqrt{f_c'E_c} \leq A_{sc}F_u \qquad (I5\text{-}1)$$

where

A_{sc} = cross-sectional area of a stud shear connector, in.^{2}

f_c' = specified compressive strength of concrete, ksi

F_u = minimum specified tensile strength of a stud shear connector, ksi

E_c = modulus of elasticity of concrete, ksi

Special provisions for shear connectors embedded in concrete on formed steel deck are given in Section I3.5 of the LRFD Specification. Among them are reduction factors (given by Equation I3-1 and I3-2) to be applied to the middle term of Equation I5-1 above.

The design of composite beams and the selection of shear connectors can be accomplished with the tables in Part 5 of this LRFD Manual.

The design shear strength for composite beams is determined by the shear strength of the steel web, as for noncomposite beams; see Section F above.

Combined Compression and Flexure

Composite beam-columns are covered in Section I4 of the LRFD Specification.

COMPUTER SOFTWARE

ELRFD* (Electronic LRFD Specification)

ELRFD is a sophisticated computer program for interactively checking structural steel building components for compliance with the AISC Specification. All provisions of Chapters A through H and K of the LRFD Specification are included in the knowledge base of ELRFD.

The ELRFD program checks whether the member satisfies all limit states and limitation requirements set by the LRFD Specification and reports which sections of the specification are satisfied or violated. One can review in detail the formulas and rules used in the evaluation and interactively assess any mathematical expression appearing on the screen. Design data produced by the software can be viewed and/or printed in report form for permanent record. ELRFD has a fully interactive Windows-based user interface.

---

*ELRFD is copyright AISC and Visual Edge Software, Ltd.

REFERENCES

American Institute of Steel Construction, Inc., 1983, *Torsional Analysis of Steel Members,* AISC, Chicago, IL.

American Society of Civil Engineers, 1988, *Minimum Design Loads for Buildings and Other Structures,* ASCE 7-88, New York, NY.

Galambos, T. V., et al., 1978, Eight LRFD Papers, *Journal of the Structural Division,* ASCE, Vol. 104, No. ST9 (September 1978), New York.

Geschwindner, L., 1993, "The 'Leaning' Column in ASD and LRFD," *Proceedings of the 1993 National Steel Construction Conference,* AISC, Chicago.

U.S. Department of Commerce, 1980, *Development of a Probability Based Criterion for American National Standard A58,* NBS (National Bureau of Standards) Special Publication 577, Washington, DC.

PART 3

COLUMN DESIGN

COL.

OVERVIEW

Column tables with design compressive strengths, in kips, are located as follows:

Additional information related to column design is provided as follows:

DESIGN STRENGTH OF COLUMNS

General Notes

Column Load Tables

Column Load Tables are presented for W, WT, and HP shapes, pipe, structural tubing, double angles, and single angles. Tabular loads are computed in accordance with the AISC LRFD Specification, Sections E2 and E3 and Appendix E3, for axially loaded members having effective unsupported lengths indicated to the left of each table. The effective length KL is the actual unbraced length, in feet, multiplied by the factor K, which depends on the rotational restraint at the ends of the unbraced length and the means available to resist lateral movements.

Table C-C2.1 in the Commentary on the LRFD Specification is a guide in selecting the K-factor. Interpolation between the idealized cases is a matter of engineering judgment. Once sections have been selected for the several framing members, the alignment charts in Figure 3-1 [reproduced from the Structural Stability Research Council Guide (Galambos, 1988) here and in Figure C-C2.2 of the Commentary on the LRFD Specification] afford a means to obtain more precise values for K, if desired. For column behavior in the inelastic range, the values of G as defined in Figure 3-1 may be reduced by the values given in Table 3-1, as illustrated in Example 3-3.

Tables for W, WT, and HP shapes and for double and single angles are provided for 36 ksi and 50 ksi yield stress steels. Tables for steel pipe are provided for 36 ksi, and for structural tubing for 46 ksi yield stress steel. All design strengths are tabulated in kips. Values are not shown when Kl / r exceeds 200.

In all tables, except double angle and WT tables, design strengths are given for effective lengths with respect to the minor axis calculated by LRFD Specification Section E2. When the minor axis is braced at closer intervals than the major axis, the strength of the column must be investigated with reference to both major (X-X) and minor (Y-Y) axes. The ratio r_x / r_y included in these tables provides a convenient method for investigating the strength of a column with respect to its major axis. To obtain an effective length with respect to the minor axis equivalent in load carrying capacity to the actual effective length about the major axis, divide the major axis effective length by r_x / r_y ratio. Compare this length with the actual effective length about the minor axis. The longer of the two lengths will control the design, and the design strength may be taken from the table opposite the longer of the two effective lengths with respect to the minor axis. The double angle and WT tables show values for effective lengths about both axes.

Properties useful to the designer are listed at the bottom of the column design strength tables. Additional notes relating specifically to the W and HP shape tables, the steel pipe and structural tubing tables, and the double and single angle tables precede each of these groups of tables.

EXAMPLE 3-1

Given: Design the lightest W shape of $F_y = 50$ ksi steel to support a factored concentric load of 1,400 kips. The effective length with respect to its minor axis is 16 feet. The effective length with respect to its major axis is 31 feet.

Solution: Enter the appropriate Column Load Table for W shapes at effective length of KL = 16 ft. Since W14 columns are generally most efficient, begin with the W14 table and work downward, weightwise.

Select W14×145, good for 1,530 kips > 1,400 kips

r_x / r_y = 1.59. Equivalent L = 31 ft / 1.59 = 19.5 ft > 16 ft

Equivalent effective length for X-X axis controls.

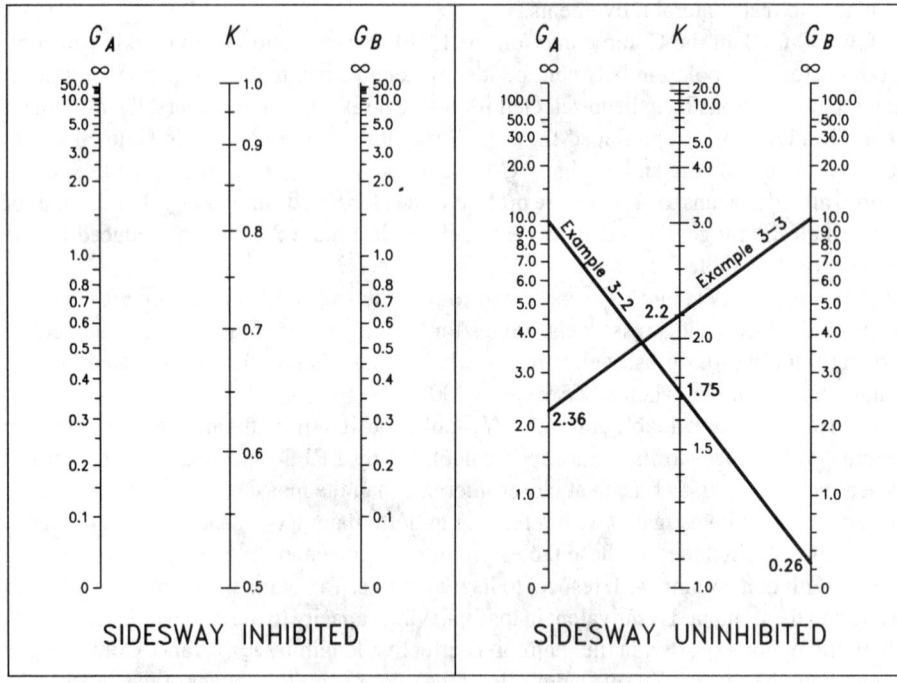

Fig. 3-1. Alignment charts for effective length of columns in continuous frames. The subscripts A and B refer to the joints at the two ends of the column section being considered. G is defined as

$$G = \frac{\Sigma(I_c / L_c)}{\Sigma(I_g / L_g)}$$

in which Σ indicates a summation of all members rigidly connected to that joint and lying on the plane in which buckling of the column is being considered. I_c is the moment of inertia and L_c the unsupported length of a column section, and I_g is the moment of inertia and L_g is the unsupported length of a girder or other restraining member. I_c and I_g are taken about axes perpendicular to the plane of buckling being considered.

For column ends supported but not rigidly connected to a footing or foundation, G is theoretically infinity, but, unless actually designed a true friction free pin, may be taken as 10 for practical designs. If the column end is rigidly attached to a properly designed footing, G may be taken as 1.0. Smaller values may be used if justified by analysis.

Table 3-1.
Stiffness Reduction Factors (SRF) for Columns

P_u/A ksi	F_y		P_u/A ksi	F_y	
	36 ksi	50 ksi		36 ksi	50 ksi
42	—	0.03	26	0.38	0.82
41	—	0.09	25	0.45	0.85
40	—	0.16	24	0.52	0.88
39	—	0.21	23	0.58	0.90
38	—	0.27	22	0.65	0.93
37	—	0.33	21	0.70	0.95
36	—	0.38	20	0.76	0.97
35	—	0.44	19	0.81	0.98
34	—	0.49	18	0.85	0.99
33	—	0.53	17	0.89	1.00
32	—	0.58	16	0.92	↓
31	—	0.63	15	0.95	
30	0.05	0.67	14	0.97	
29	0.14	0.71	13	0.99	
28	0.22	0.75	12	1.00	
27	0.30	0.79	11	↓	

— indicates not applicable.

Re-enter table for effective length of 19.5 ft to satisfy axial load of 1,400 kips, select W14×145.

By interpolation, the column is good for 1,410 kips.

Use W14×145 column

EXAMPLE 3-2

Given: Design an 11-ft long W12 interior bay column to support a factored concentric axial roof load of 1,100 kips. The column is rigidly framed at the top by 30-ft long W30×116 girders connected to each flange. Column moment is zero due to the assumption of equal and offsetting moments in the girders. The column is braced normal to its web at top and base so that sidesway is inhibited in this plane. Use F_y = 50 ksi steel.

Solution: a. Check Y-Y axis:

Assume the column is pin-connected at the top and bottom with sidesway inhibited.

From Table C-C2.1 in the Commentary for condition (d), K = 1.0:

Effective length = 11 ft

Enter Column Load Table:

W12×106 good for 1,160 kips > 1,100 kips   **o.k.**

b. Check X-X axis:

    1. Preliminary selection:

        Assume sidesway uninhibited and pin-connected at base.

        From Table C-C2.1 for condition (f):

        $K = 2.0$

Approximate effective length relative to X-X axis:

    $2.0 \times 11 = 22.0$ ft

From Properties section in tables, for **W**12 column:

    $r_x / r_y \approx 1.76$

Equivalent effective length relative to the Y-Y axis:

$$\frac{22.0}{1.76} \approx 12.5 \text{ ft} > 11.0 \text{ ft}$$

Therefore, effective length for X-X axis is critical.

Enter Column Load Table with an effective length of 12.5 ft:

    W12×106 column, by interpolation, good for 1,115 kips > 1,100 kips
o.k.

1. Final selection

    Try **W**12×106

    Using Figure 3-1 (sidesway uninhibited):

        I_x for **W**12×106 column = 933 in.[4]

        I_x for **W**30×116 girder = 4,930 in.[4]

        G (base)   = 10 (assume supported but not rigidly connected)

$$G \text{ (top)} \quad = \frac{933 / 11}{(4{,}930 \times 2) / 30} = 0.258, \text{ say } 0.26$$

Connect points $G_A = 10$ and $G_B = 0.26$, read $K = 1.75$

For **W**12×106, $r_x / r_y = 1.76$

Actual effective length relative to Y-Y axis:

$$\frac{1.75}{1.76} \times 11.0 = 10.9 \text{ ft} < 11.0 \text{ ft}$$

Since the effective length for Y-Y axis is not critical,

Use **W**12×106 column

EXAMPLE 3-3

Given: Using the alignment chart, Figure 3-1 (sidesway uninhibited) and Table 3-1 (Stiffness Reduction Factors), design columns for the bent shown, by the inelastic K-factor procedure. Let $F_y = 50$ ksi. Assume continuous support in the transverse direction.

Solution: The alignment charts in Figure 3-1 are applicable to elastic columns. By multiplying G-values times the stiffness reduction factor E_t / E, the charts may be used for inelastic columns.

Since $E_t / E \approx F_{cr,\,inelastic} / F_{cr,\,elastic}$, the relationship may be written as $G_{inelastic} = (F_{cr,\,inelastic} / F_{cr,\,elastic})G_{elastic}$.

By utilizing the calculated stress P_u / A a direct solution is possible, using the following steps:

1. For a known value of factored axial load, $P_u = 1,100$ kips, select a trial column size.

   Assume W12×120

   $A = 35.3$ in.^{2}, $I_x = 1,070$ in.^{4}, $r_x = 5.51$ in.

2. Calculate P_u / A:

   $P_u / A = 1,100$ kips $/ 35.3$ in.$^2 = 31.2$ ksi

3. From Table 3-1, determine the Stiffness Reduction Factor (SRF); SRF = 0.62. For values of P_u / A smaller than those with entries in Table 3-1, the column is elastic, and the reduction factor is 1.0.

4. Determine $G_{elastic}$:

   $G_{elastic}$ (bottom) = 10

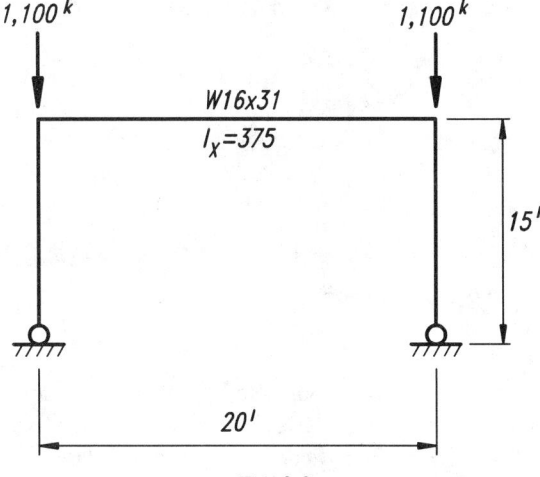

Fig. 3-2

$$G_{elastic} \text{ (top)} = \frac{1,070/15}{375/20} = 3.80$$

5. Calculate $G_{inelastic} = \text{SRF} \times G_{elastic}$:

$$G_{inelastic} \text{ (top)} = 0.62 \times 3.80 = 2.36$$

6. Determine K from Figure 3-1 using $G_{inelastic}$

For G (top) = 2.36 and G (bottom) = 10,

Read from Figure 3-1, $K = 2.2$

7. $KL_x = 2.2 \times 15$ ft = 33.0 ft

8. Calculate equivalent of KL_y:

$$\frac{KL_x}{r_x/r_y} = \frac{33.0 \text{ ft}}{1.76} = 18.75 \text{ ft}$$

9. From the column tables (for 50 ksi steel):

$\phi_c P_n = 1,030$ kips < 1,100 kips req'd.   **n.g.**

Try a stronger column.

1. Try a W12×136

$A = 39.9$ in.^{2}, $I_x = 1,240$ in.^{4}, $r_x = 5.58$ in.

2. $P_u / A = 1,100$ kips / 39.9 in.^{2} = 27.6 ksi

3. From Table 3-1: SRF = 0.77

4. $G_{elastic} \text{ (top)} = \dfrac{1,240/15}{375/20} = 4.41$

5. $G_{inelastic} \text{ (top)} = 0.77 \times 4.41 = 3.39$

6. $K = 2.3$

7. $KL_x = 2.3 \times 15$ ft = 34.5 ft

8. Equivalent KL_y: $\dfrac{KL_x}{r_x/r_y} = \dfrac{34.5 \text{ ft}}{1.77} = 19.5$ ft

9. $\phi_c P_n = 1,135$ kips > 1,100 kips req'd   **o.k.**

Use W12×136

"Leaning" Columns

A "leaning" column is one which is considered pin-ended and does not participate in providing lateral stability to the structure. As a result, it relies on other parts of the structure for stability. The LRFD Specification in Section C2.2 requires that for unbraced frames, "the destabilizing effects of gravity-loaded columns whose simple connections to the frame do not provide resistance to lateral loads shall be included in the design of the moment-frame columns."

Normal practice is to design leaning columns for their required strength with an effective length factor $K = 1$. To account for the effects of leaning columns on unbraced frames, one of the methods given in the Commentary on the LRFD Specification (Section C2) or in Geschwindner (1993) may be utilized. The simplest methods are:

1. The slightly conservative approach of adjusting the effective lengths of the rigid-frame columns,

$$K_i' = \sqrt{N} K_i$$

where

K_i' = the modified effective length factor of a column

K_i = the actual effective length factor of a column

N = ratio of the factored gravity load supported by all columns in the given story to that supported by the columns in the rigid frame

2. The more conservative approach of providing sufficient design compressive strength in the rigid-frame columns of a story to enable them to support the total factored gravity load of the story at their actual effective lengths.

Combined Axial and Bending Loading (Interaction)

Loads given in the Column Tables are for concentrically loaded columns. For columns subjected to both axial and bending stress, see Chapters C and H of the LRFD Specification.

The design of a beam-column is a trial and error process in which a trial section is checked for compliance with Equations H1-1a and H1-1b. A fast method for selecting an economical trial W section, using an equivalent axial load, is illustrated in the example problem, using Table 3-2 and the u values listed in the column properties at the bottom of the column load tables.

The procedure is as follows:

1. With the known value of KL (effective length), select a first approximate value of m from Table 3-2. Let u equal 2.

2. Solve for $P_{u\,eq} = P_u + M_{ux}\,m + M_{uy}\,mu$

where

P_u = actual factored axial load, kips

M_{ux} = factored bending moment about the strong axis, kip-ft

M_{uy} = factored bending moment about the weak axis, kip-ft

m = factor taken from Table 3-2

u = factor taken from column load table

3. From the appropriate Column Load Table, select a tentative section to support $P_{u\,eq}$.

4. Based on the section selected in Step 3, select a "subsequent approximate" value of m from Table 3-2 and a u value from the column load table.

5. With the values selected in Step 4, solve for $P_{u\,eq}$.

6. Repeat Steps 3 and 4 until the values of m and u stabilize.

7. Check section obtained in Step 6 per Equation H1-1a or H1-1b, as applicable.

**Table 3-2.
Preliminary Beam-Column Design
F_y = 36 ksi, F_y = 50 ksi**

Values of m

F_y	36 ksi							50 ksi						
KL (ft)	10	12	14	16	18	20	22 and over	10	12	14	16	18	20	22 and over
1st Approximation														
All Shapes	2.0	1.9	1.8	1.7	1.6	1.5	1.3	1.9	1.8	1.7	1.6	1.4	1.3	1.2
Subsequent Approximation														
W4	3.1	2.3	1.7	1.4	1.1	1.0	0.8	2.4	1.8	1.4	1.1	1.0	0.9	0.8
W5	3.2	2.7	2.1	1.7	1.4	1.2	1.0	2.8	2.2	1.7	1.4	1.1	1.0	0.9
W6	2.8	2.5	2.1	1.8	1.5	1.3	1.1	2.5	2.2	1.8	1.5	1.3	1.2	1.1
W8	2.5	2.3	2.2	2.0	1.8	1.6	1.4	2.4	2.2	2.0	1.7	1.5	1.3	1.2
W10	2.1	2.0	1.9	1.8	1.7	1.6	1.4	2.0	1.9	1.8	1.7	1.5	1.4	1.3
W12	1.7	1.7	1.6	1.5	1.5	1.4	1.3	1.7	1.6	1.5	1.5	1.4	1.3	1.2
W14	1.5	1.5	1.4	1.4	1.3	1.3	1.2	1.5	1.4	1.4	1.3	1.3	1.2	1.2

This table is from a paper in AISC *Engineering Journal* by Uang, Wattar, and Leet (1990).

EXAMPLE 3-4

Given: Design the following column:

P_u = 400 kips
M_{ntx} = 250 kip-ft
M_{ltx} = 0 (braced frame)
M_{nty} = 80 kip-ft
M_{lty} = 0 (braced frame)
$KL_x = KL_y$ = 14 ft
L_b = 14 ft
C_m = 0.85
F_y = 50 ksi

Solution: 1. For KL = 14 ft, from Table 3-2 select a first trial value of m = 1.7. Let u = 2

2. $P_{u\ eq} = P_u + M_{ux} m + M_{uy} mu = 400 + 250 \times 1.7 + 80 \times 1.7 \times 2 =$ 1,097 kips

3. From Column Load Tables select W14×109 ($\phi_c P_n$ = 1,170 kips) or W12×120 ($\phi_c P_n$ = 1,220 kips).

4. Select the W14 column, so the second trial value of m is 1.4. (Note: If a W14 column were required for architectural or other reasons,

the selection process could have started with $m = 1.4$). With $m = 1.4$ and $u = 1.97$ (for a W14×109) from Column Load Table,

$$P_{u\,eq} = 400 + 250 \times 1.4 + 80 \times 1.4 \times 1.97 = 971 \text{ kips}$$

5. From Column Load Tables select W14×90 ($\phi_c\,P_n = 969$ kips).

6. For W14×90, $m = 1.4$, $u = 1.94$. Repeat of Steps 3 and 4 not required.

7. Check W14×90 with the appropriate interaction formula.

A $= 26.5$ in.^{2}

r_y $= 3.70$ in., $\left(\dfrac{Kl}{r_y}\right) = \dfrac{14 \times 12}{3.70} = 45.4$

r_x $= 6.14$ in., $\left(\dfrac{Kl}{r_x}\right) = \dfrac{14 \times 12}{6.14} = 27.4$

The second-order moments, M_{ux} and M_{uy}, will be evaluated using the approximate method given in Section C1 of the LRFD Specification. Because $M_{ltx} = M_{lty} = 0$ (braced frames in both directions), Specification Equation C1-1 reduces to $M_u = B_1 M_{nt}$, where B_1 is a function of P_{e1} (Equation C1-2). The values of P_{e1} with respect to the x and y axes can be determined from LRFD Specification Table 8 as follows:

P_{ex} $= 382 \times 26.5 = 10{,}123$ kips
P_{ey} $= 139 \times 26.5 = 3{,}684$ kips

B_{1x} $= \dfrac{0.85}{1 - 400/10{,}123} < 1.0.$ Use $B_{1x} = 1.0$

B_{1y} $= \dfrac{0.85}{1 - 400/3{,}684} < 1.0.$ Use $B_{1y} = 1.0$

M_{ux} $= 1.0 \times 250 = 250$ kip-ft
M_{uy} $= 1.0 \times 80 = 80$ kip-ft
$\phi_b\,M_{ny} = \phi_b\,M_{py} = \dfrac{0.9 \times 50 \times 75.6}{12} = 284$ kip-ft

From the beam selection table in Part 4 of this Manual:

$\phi_b\,M_{nx} = 577$ kip-ft for $L_b < L_p = 15.0$ ft

$\dfrac{P_u}{\phi_c P_n}$ $= \dfrac{400}{969} = 0.412 > 0.2.$ Therefore, Equation H1-1a applies.

$\dfrac{400}{969} + \dfrac{8}{9}\left(\dfrac{250}{577} + \dfrac{80}{284}\right) = 0.412 + 0.636 = 1.05 < 1.0 \quad \textbf{n.g.}$

Use W14×99

Column Stiffening

Values of P_{wo}, P_{wi}, P_{wb}, and P_{fb}, listed in the Properties Section of the Column Load Tables for W and HP shapes, are useful in determining if a column requires stiffening because of forces transmitted into it from the flanges or connecting flange plates of a rigid beam connection to the column flange.

The parameters are defined as follows:

$P_{wo} = \phi 5 F_{yw}\, t_w\, k$ (kips), $\phi = 1.0$

$P_{wi} = \phi F_{yw}\, t_w$ (kips/in.), $\phi = 1.0$

$P_{wb} = \phi 4{,}100 t_w^3 \sqrt{F_{yw}} / h$ (kips), $\phi = 0.9$

$P_{fb} = \phi 6.25 t_f^2 F_{yf}$ (kips), $\phi = 0.9$

Column stiffening or a heavier column* is required if P_{bf}, the factored force transmitted into the column web, exceeds any one of the following three resisting forces:

P_{wb}

P_{fb}

$P_{wi}\, t_b + P_{wo}$, where t_b is the thickness of the beam flange delivering the concentrated force.

For a complete explanation of these design parameters, see the section Column Stiffening in Part 10 (Volume II) of this LRFD Manual.

---

*The designer should consider selecting a heavier column section to eliminate the need for stiffening. Although this will increase the material cost of the column, this heavier section may provide a more economical solution due to the reduction in labor cost associated with the elimination of stiffening.

W and HP Shapes

The design strengths in the tables that follow are tabulated for the effective lengths in feet KL (with respect to the minor axis), indicated at the left of each table. They are applicable to axially loaded members in accordance with Section E2 of the LRFD Specification. Two yield stresses are covered, 36 and 50 ksi.

The heavy horizontal lines appearing within the tables indicate $Kl / r = 200$. No values are listed beyond $Kl / r = 200$.

For discussion of effective length, range of l / r, strength about the major axis, combined axial and bending stress, and sample problems, see General Notes, above.

Properties and factors are listed at the bottom of the tables for checking strength about the strong axis, combined loading conditions, and column stiffener requirements.

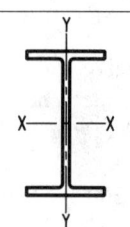

F_y = 36 ksi

F_y = 50 ksi

COLUMNS
W shapes
Design axial strength in kips (ϕ = 0.85)

Designation	W14									
Wt./ft	808		730		665		605		550	
F_y	36	50	36	50	36	50	36	50	36	50
Effective length KL (ft) with respect to least radius of gyration r_y: 0	7250	10100	6580	9140	6000	8330	5450	7570	4960	6890
11	6970	9540	6310	8620	5750	7850	5210	7110	4740	6460
12	6920	9440	6260	8530	5700	7760	5170	7030	4700	6390
13	6860	9330	6210	8430	5650	7660	5120	6940	4650	6300
14	6800	9220	6150	8320	5590	7560	5070	6850	4600	6220
15	6740	9100	6090	8200	5540	7450	5020	6750	4560	6120
16	6670	8970	6020	8080	5480	7340	4960	6640	4500	6020
17	6600	8840	5960	7960	5410	7220	4900	6530	4450	5920
18	6520	8700	5880	7820	5350	7100	4840	6420	4390	5810
19	6450	8550	5810	7690	5280	6970	4770	6300	4330	5700
20	6360	8400	5730	7550	5200	6840	4700	6170	4260	5590
22	6190	8090	5570	7250	5050	6560	4560	5910	4130	5350
24	6010	7760	5390	6940	4890	6270	4410	5640	3990	5100
26	5820	7410	5210	6610	4720	5970	4250	5360	3840	4840
28	5620	7060	5020	6280	4540	5660	4090	5080	3690	4570
30	5410	6700	4820	5940	4360	5340	3920	4790	3530	4300
32	5190	6330	4620	5600	4170	5030	3740	4490	3370	4030
34	4970	5970	4420	5250	3980	4710	3570	4200	3210	3760
36	4750	5600	4210	4910	3790	4400	3390	3910	3050	3500
38	4530	5240	4000	4580	3590	4090	3210	3630	2880	3240
40	4300	4880	3790	4250	3400	3780	3030	3350	2720	2990
42	4080	4530	3580	3930	3210	3490	2860	3080	2550	2740
44	3860	4190	3380	3620	3020	3200	2680	2820	2390	2500
46	3640	3860	3170	3310	2830	2930	2510	2580	2240	2290
48	3420	3550	2970	3040	2650	2690	2340	2370	2080	2100
50	3210	3240	2780	2800	2470	2480	2180	2180	1940	1940
Properties										
u	2.03	2.03	2.03	2.03	2.02	2.02	2.02	2.01	2.02	2.01
P_{wo} (kips)	3910	5430	3070	4270	2640	3670	2250	3120	1930	2680
P_{wi} (kips/in.)	135	187	111	154	102	142	93.4	130	85.7	119
P_{wb} (kips)	103000	122000	56400	66500	44300	52200	33900	39900	26100	30800
P_{fb} (kips)	5310	7370	4880	6780	4140	5750	3500	4870	2950	4100
L_p (ft)	20.1	17.0	19.5	16.6	19.3	16.3	19.0	16.1	18.7	15.9
L_r (ft)	415	270	372	241	342	222	313	203	288	188
A (in.^{2})	237		215		196		178		162	
I_x (in.^{4})	16000		14300		12400		10800		9430	
I_y (in.^{4})	5510		4720		4170		3680		3250	
r_y (in.)	4.82		4.69		4.62		4.55		4.49	
Ratio r_x / r_y	1.70		1.74		1.73		1.71		1.70	
$P_{ex}(KL)^2/10^4$	457000		411000		357000		310000		270000	
$P_{ey}(KL)^2/10^4$	158000		135000		120000		105000		93500	

| $F_y = 36$ ksi | | | | | | | | | | |

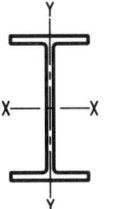

COLUMNS
W shapes
Design axial strength in kips ($\phi = 0.85$)

Designation		W14									
Wt./ft		**500**		**455**		**426**		**398**		**370**	
F_y		**36**	**50**	**36**	**50**	**36**	**50**	**36**	**50**	**36**	**50**
Effective length KL (ft) with respect to least radius of gyration r_y	0	4500	6250	4100	5700	3830	5310	3580	4970	3340	4630
	11	4290	5850	3910	5330	3640	4970	3410	4640	3170	4320
	12	4250	5780	3870	5260	3610	4900	3380	4580	3140	4260
	13	4210	5710	3840	5190	3570	4830	3340	4520	3110	4200
	14	4170	5620	3790	5110	3530	4760	3300	4450	3070	4140
	15	4120	5540	3750	5030	3490	4680	3270	4380	3040	4070
	16	4070	5450	3710	4950	3450	4600	3230	4300	3000	4000
	17	4020	5350	3660	4860	3400	4520	3180	4220	2960	3920
	18	3970	5250	3610	4770	3360	4430	3140	4140	2920	3840
	19	3910	5150	3560	4670	3310	4340	3090	4050	2870	3760
	20	3850	5040	3500	4570	3260	4250	3040	3960	2820	3680
	22	3730	4820	3390	4370	3150	4050	2940	3780	2730	3500
	24	3600	4590	3270	4150	3030	3850	2830	3590	2630	3320
	26	3460	4350	3140	3930	2910	3640	2720	3390	2520	3140
	28	3320	4100	3010	3700	2790	3430	2600	3190	2410	2950
	30	3180	3850	2870	3480	2660	3210	2480	2990	2290	2750
	32	3030	3610	2740	3250	2530	3000	2360	2780	2180	2560
	34	2880	3360	2600	3020	2400	2780	2230	2580	2060	2380
	36	2730	3120	2460	2800	2270	2570	2110	2390	1950	2190
	38	2580	2880	2320	2580	2140	2370	1990	2190	1830	2010
	40	2420	2650	2180	2370	2010	2170	1860	2010	1710	1840
	42	2280	2420	2040	2160	1880	1980	1740	1830	1600	1670
	44	2130	2210	1910	1970	1750	1800	1620	1660	1490	1520
	46	1990	2020	1780	1800	1630	1650	1510	1520	1380	1390
	48	1850	1860	1650	1650	1510	1510	1400	1400	1280	1280
	50	1710	1710	1520	1520	1400	1400	1290	1290	1180	1180

Properties											
u		2.01	2.00	1.99	1.99	2.00	1.99	1.99	1.98	1.98	1.97
P_{wo} (kips)		1650	2290	1410	1950	1240	1730	1120	1550	987	1370
P_{wi} (kips/in.)		78.8	110	72.5	101	67.5	93.8	63.7	88.5	59.6	82.8
P_{wb} (kips)		20400	24100	15800	18600	12800	15000	10800	12800	8790	10400
P_{fb} (kips)		2480	3450	2090	2900	1870	2590	1640	2280	1430	1990
L_p (ft)		18.5	15.7	18.3	15.5	18.1	15.3	18.0	15.2	17.8	15.1
L_r (ft)		264	172	242	157	227	148	213	139	199	129
A (in.^{2})		147		134		125		117		109	
I_x (in.^{4})		8210		7190		6600		6000		5440	
I_y (in.^{4})		2880		2560		2360		2170		1990	
r_y (in.)		4.43		4.38		4.34		4.31		4.27	
Ratio r_x / r_y		1.69		1.67		1.67		1.66		1.66	
$P_{ex}(KL)^2 / 10^4$		235000		206000		189000		172000		156000	
$P_{ey}(KL)^2 / 10^4$		82600		73600		67400		62200		56900	

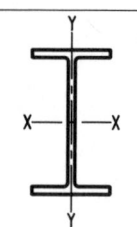

$F_y = 36$ ksi
$F_y = 50$ ksi

COLUMNS
W shapes
Design axial strength in kips ($\phi = 0.85$)

Designation		W14								
Wt./ft		342		311		283		257		233
F_y		36	50	36	50	36	50	36	50	36	50
	0	3090	4290	2800	3880	2550	3540	2310	3210	2100	2910
	6	3040	4200	2750	3800	2510	3460	2280	3140	2060	2850
	7	3030	4170	2740	3770	2500	3440	2260	3120	2050	2820
	8	3010	4130	2720	3740	2480	3410	2250	3090	2040	2800
	9	2990	4090	2700	3700	2460	3370	2230	3060	2020	2770
	10	2960	4050	2680	3660	2440	3330	2210	3020	2000	2730
	11	2940	4000	2660	3610	2420	3290	2190	2980	1980	2700
	12	2910	3950	2630	3560	2390	3240	2170	2940	1960	2660
	13	2880	3890	2600	3510	2370	3200	2150	2890	1940	2620
	14	2850	3830	2570	3460	2340	3140	2120	2850	1920	2570
	15	2810	3760	2540	3400	2310	3090	2090	2800	1890	2530
	16	2770	3690	2510	3330	2280	3030	2060	2740	1870	2480
	17	2740	3620	2470	3270	2250	2970	2030	2690	1840	2430
	18	2700	3550	2430	3200	2210	2910	2000	2630	1810	2380
	19	2650	3470	2390	3130	2180	2850	1970	2570	1780	2320
	20	2610	3400	2360	3060	2140	2780	1940	2510	1750	2270
	22	2520	3230	2270	2910	2060	2640	1870	2380	1690	2150
	24	2420	3060	2180	2750	1980	2500	1790	2250	1620	2030
	26	2320	2890	2090	2590	1900	2350	1710	2120	1550	1910
	28	2220	2710	2000	2430	1810	2200	1630	1980	1470	1780
	30	2110	2530	1900	2270	1720	2050	1550	1840	1400	1660
	32	2010	2360	1800	2110	1630	1900	1470	1710	1320	1530
	34	1900	2180	1700	1950	1540	1760	1380	1570	1240	1410
	36	1790	2010	1600	1790	1450	1620	1300	1440	1170	1290
	38	1680	1840	1500	1640	1360	1480	1220	1320	1090	1180
	40	1570	1680	1410	1490	1270	1340	1140	1190	1020	1070

Effective length KL (ft) with respect to least radius of gyration r_y

Properties										
u		1.98	1.97	1.97	1.96	1.97	1.95	1.96	1.94	1.95	1.93
P_{wo} (kips)		866	1200	746	1040	639	887	542	753	457	635
P_{wi} (kips/in.)		55.4	77.0	50.8	70.5	46.4	64.5	42.3	58.8	38.5	53.5
P_{wb} (kips)		7100	8360	5430	6400	4190	4930	3150	3710	2370	2790
P_{fb} (kips)		1240	1720	1030	1440	868	1210	723	1000	599	832
L_p (ft)		17.7	15.0	17.5	14.8	17.4	14.7	17.2	14.6	17.1	14.5
L_r (ft)		185	120	168	110	154	100	140	91.6	127	83.4
A (in.^{2})		101		91.4		83.3		75.6		68.5	
I_x (in.^{4})		4900		4330		3840		3400		3010	
I_y (in.^{4})		1810		1610		1440		1290		1150	
r_y (in.)		4.24		4.2		4.17		4.13		4.1	
Ratio r_x / r_y		1.65		1.64		1.63		1.62		1.62	
$P_{ex} (KL)^2 / 10^4$		141000		124000		110000		97400		86200	
$P_{ey} (KL)^2 / 10^4$		52000		46100		41500		36900		33000	

| $F_y = 36$ ksi |
| $F_y = 50$ ksi |

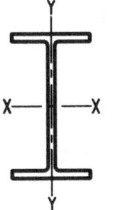

COLUMNS
W shapes
Design axial strength in kips ($\phi = 0.85$)

Designation		W14								
Wt./ft		211		193		176		159		145	
F_y		36	50	36	50	36	50	36	50	36	50
	0	1900	2640	1740	2410	1590	2200	1430	1980	1310	1810
	6	1870	2580	1710	2360	1560	2150	1400	1940	1280	1770
	7	1860	2550	1700	2340	1550	2130	1400	1920	1280	1760
	8	1840	2530	1690	2320	1540	2110	1390	1900	1270	1740
	9	1830	2500	1670	2290	1530	2090	1380	1880	1260	1720
	10	1810	2470	1660	2260	1510	2060	1360	1860	1250	1700
	11	1790	2440	1640	2230	1500	2030	1350	1830	1230	1670
	12	1780	2400	1630	2200	1480	2000	1330	1810	1220	1650
	13	1760	2370	1610	2170	1460	1970	1320	1780	1210	1620
	14	1730	2330	1590	2130	1450	1940	1300	1740	1190	1590
	15	1710	2280	1570	2090	1430	1900	1280	1710	1170	1560
	16	1690	2240	1540	2050	1410	1860	1270	1680	1160	1530
	17	1660	2190	1520	2010	1380	1820	1250	1640	1140	1500
	18	1640	2140	1500	1960	1360	1780	1230	1600	1120	1460
	19	1610	2090	1470	1910	1340	1740	1200	1570	1100	1430
	20	1580	2040	1440	1870	1310	1700	1180	1530	1080	1390
	22	1520	1940	1390	1770	1260	1610	1140	1440	1040	1320
	24	1460	1830	1330	1670	1210	1510	1090	1360	992	1240
	26	1390	1710	1270	1560	1150	1420	1040	1270	946	1160
	28	1330	1600	1210	1460	1100	1320	990	1180	898	1080
	30	1260	1490	1150	1350	1040	1220	930	1100	849	998
	32	1190	1370	1080	1250	980	1130	880	1010	800	919
	34	1120	1260	1020	1150	920	1040	830	928	752	842
	36	1050	1160	955	1050	863	946	773	846	703	767
	38	980	1050	892	956	805	859	721	767	655	694
	40	912	951	830	863	748	775	670	692	608	626

Effective length KL (ft) with respect to least radius of gyration r_y

Properties										
u		1.95	1.93	1.96	1.93	1.94	1.92	1.94	1.92	1.93	1.90
P_{wo} (kips)		397	551	340	473	299	415	251	349	214	298
P_{wi} (kips/in.)		35.3	49.0	32.0	44.5	29.9	41.5	26.8	37.3	24.5	34.0
P_{wb} (kips)		1830	2160	1370	1610	1110	1310	803	947	609	718
P_{fb} (kips)		493	684	420	583	348	483	287	398	241	334
L_p (ft)		17.0	14.4	16.9	14.3	16.8	14.2	16.7	14.1	16.6	14.1
L_r (ft)		116	76.0	106	70.1	97.5	64.5	88.6	59.0	81.5	54.7
A (in.^{2})		62.0		56.8		51.8		46.7		42.7	
I_x (in.^{4})		2660		2400		2140		1900		1710	
I_y (in.^{4})		1030		931		838		748		677	
r_y (in.)		4.07		4.05		4.02		4		3.98	
Ratio r_x / r_y		1.61		1.60		1.60		1.60		1.59	
$P_{ex} (KL)^2 / 10^4$		76100		68700		61300		54400		49000	
$P_{ey} (KL)^2 / 10^4$		29400		26700		24000		21400		19400	

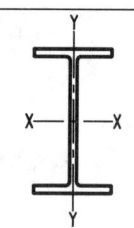

				F_y = 36 ksi
				F_y = 50 ksi

COLUMNS
W shapes
Design axial strength in kips (ϕ = 0.85)

Designation		W14								
Wt./ft		**132**		**120**		**109**		**99**		**90**
F_y		**36**	**50**	**36**	**50**	**36**	**50**	**36**	**50†**	**36**	**50†**
Effective length *KL* (ft) with respect to least radius of gyration r_y	0	1190	1650	1080	1500	979	1360	890	1240	811	1130
	6	1160	1610	1060	1460	960	1320	873	1200	795	1100
	7	1160	1590	1050	1450	953	1310	867	1190	789	1080
	8	1150	1570	1040	1430	946	1300	860	1180	783	1070
	9	1140	1550	1030	1410	937	1280	852	1160	775	1060
	10	1130	1530	1020	1390	927	1260	843	1150	767	1040
	11	1110	1510	1010	1370	917	1240	833	1130	758	1030
	12	1100	1480	999	1350	905	1220	823	1110	749	1010
	13	1080	1450	986	1320	893	1200	811	1090	738	989
	14	1070	1430	971	1290	880	1170	799	1060	727	969
	15	1050	1390	956	1270	866	1150	787	1040	716	947
	16	1030	1360	940	1240	852	1120	773	1020	704	925
	17	1020	1330	924	1210	837	1090	759	991	691	902
	18	997	1300	906	1180	821	1060	745	965	678	878
	19	978	1260	888	1140	804	1030	730	938	664	853
	20	958	1220	870	1110	787	1000	714	911	650	828
	22	916	1150	831	1040	752	943	682	854	620	776
	24	872	1070	791	972	715	879	648	796	589	723
	26	826	997	749	902	677	815	614	737	558	670
	28	780	920	706	832	639	751	578	679	525	616
	30	733	844	663	762	600	688	542	621	493	564
	32	686	769	620	694	560	627	507	565	460	512
	34	639	697	577	628	522	567	471	511	428	463
	36	593	627	535	565	483	509	436	458	396	415
	38	547	563	494	507	446	457	402	411	365	372
Properties										
u		2.03	1.99	2.04	1.99	2.02	1.97	2.02	1.95	2.02	1.94
P_{wo} (kips)		196	272	173	240	148	205	125	174	109	151
P_{wi} (kips/in.)		23.2	32.3	21.2	29.5	18.9	26.3	17.5	24.3	15.8	22.0
P_{wb} (kips)		520	613	399	471	281	331	222	261	165	195
P_{fb} (kips)		215	298	179	249	150	208	123	171	102	142
L_p (ft)		15.7	13.3	15.6	13.2	15.5	13.2	15.5	13.4	15.4	15.0
L_r (ft)		73.7	49.7	67.9	46.3	62.7	43.2	58.1	40.6	54.2	38.4
A (in.^{2})		38.8		35.3		32		29.1		26.5	
I_x (in.^{4})		1530		1380		1240		1110		999	
I_y (in.^{4})		548		495		447		402		362	
r_y (in.)		3.76		3.74		3.73		3.71		3.70	
Ratio r_x / r_y		1.67		1.67		1.67		1.66		1.66	
$P_{ex}(KL)^2/10^4$		43800		39300		35400		31700		28600	
$P_{ey}(KL)^2/10^4$		15700		14100		12700		11500		10400	

†Flange is noncompact; see discussion preceding column load tables.

F_y = 36 ksi
F_y = 50 ksi

COLUMNS
W shapes
Design axial strength in kips (ϕ = 0.85)

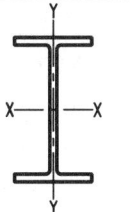

Designation															W14
Wt./ft	82		74		68		61		53		48		43	
F_y	36	50	36	50	36	50	36	50	36	50	36	50	36	50†
0	737	1020	667	927	612	850	548	761	477	663	431	599	386	536
6	705	963	638	871	585	798	523	714	443	598	400	540	357	482
7	694	942	628	852	576	781	515	698	432	576	390	520	347	463
8	682	918	616	830	565	760	505	680	418	552	378	498	337	443
9	667	892	604	807	553	738	494	660	404	526	365	474	325	422
10	652	863	590	781	540	714	483	638	389	498	351	449	312	399
11	635	833	575	753	526	689	470	615	372	469	336	423	298	375
12	618	800	559	724	511	662	457	591	355	439	320	395	284	350
14	579	732	524	662	479	604	428	539	319	379	287	340	254	301
16	538	661	487	598	444	544	396	486	282	319	253	286	224	252
18	495	588	447	532	408	484	364	431	245	263	220	235	194	206
20	450	516	407	467	371	424	331	377	210	213	188	191	165	167
22	406	447	367	405	334	366	297	325	176	176	157	157	138	138
24	363	381	328	345	297	311	265	276	148	148	132	132	116	116
26	321	325	290	294	262	265	233	236	126	126	113	113	99	99
28	280	280	253	253	229	229	203	203	109	109	97	97	85	85
30	244	244	221	221	199	199	177	177	95	95	85	85	74	74
31	229	229	207	207	187	187	166	166	89	89	79	79	69	69
32	214	214	194	194	175	175	155	155	83	83				
34	190	190	172	172	155	155	138	138						
36	169	169	153	153	138	138	123	123						
38	152	152	138	138	124	124	110	110						

The leftmost column label reads: Effective length KL (ft) with respect to least radius of gyration r_y

Properties														
u	2.85	2.68	2.82	2.62	2.80	2.56	2.74	2.44	3.2	2.7	3.12	2.56	2.97	2.37
P_{wo} (kips)	149	207	127	176	112	156	97.0	135	95.7	133	84.2	117	72.1	100
P_{wi} (kips/in.)	18.4	25.5	16.2	22.5	14.9	20.8	13.5	18.8	13.3	18.5	12.2	17.0	11.0	15.3
P_{wb} (kips)	257	303	177	209	139	163	102	121	98.4	116	76.4	90.0	55.1	64.9
P_{fb} (kips)	148	206	125	173	105	146	84.2	117	88.2	123	71.7	100	56.9	79.0
L_p (ft)	10.3	8.77	10.3	8.77	10.3	8.70	10.2	8.66	8.00	6.79	7.96	6.75	7.88	6.68
L_r (ft)	43.0	29.6	40.0	27.9	37.3	26.4	34.7	25.0	28.0	20.1	26.4	19.2	24.7	18.2
A (in.^{2})	24.1		21.8		20		17.9		15.6		14.1		12.6	
I_x (in.^{4})	882		796		723		640		541		485		428	
I_y (in.^{4})	148		134		121		107		57.7		51.4		45.2	
r_y (in.)	2.48		2.48		2.46		2.45		1.92		1.91		1.89	
Ratio r_x / r_y	2.44		2.44		2.44		2.44		3.07		3.06		3.08	
$P_{ex} (KL)^2 / 10^4$	25200		22800		20700		18300		15500		13800		12200	
$P_{ey} (KL)^2 / 10^4$	4240		3840		3460		3080		1650		1470		1290	

†Web may be noncompact for combined axial and bending stress; see AISC LRFD Specification Section B5.
Note: Heavy line indicates Kl / r of 200.

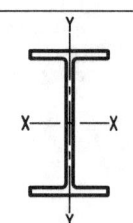

	F_y = 36 ksi
	F_y = 50 ksi

COLUMNS
W shapes
Design axial strength in kips (ϕ = 0.85)

Designation		W12								
Wt./ft		**336**		**305**		**279**		**252**		**230**	
F_y		**36**	**50**	**36**	**50**	**36**	**50**	**36**	**50**	**36**	**50**
	0	3020	4200	2740	3810	2510	3480	2270	3150	2070	2880
	6	2960	4070	2680	3690	2450	3370	2210	3040	2020	2780
	7	2930	4020	2660	3640	2430	3330	2190	3010	2000	2740
	8	2900	3970	2630	3590	2400	3280	2170	2960	1980	2710
	9	2870	3910	2600	3540	2370	3230	2150	2920	1960	2660
	10	2840	3850	2570	3480	2350	3170	2120	2870	1930	2610
	11	2800	3780	2530	3420	2310	3110	2090	2810	1910	2560
	12	2760	3700	2500	3350	2280	3050	2060	2750	1880	2500
	13	2720	3620	2460	3270	2240	2980	2020	2680	1840	2450
	14	2670	3540	2410	3190	2200	2910	1980	2620	1810	2380
	15	2620	3450	2370	3110	2160	2830	1950	2550	1770	2320
	16	2570	3360	2320	3020	2110	2750	1910	2470	1740	2250
	17	2520	3260	2270	2940	2070	2670	1860	2400	1700	2180
	18	2470	3160	2220	2840	2020	2580	1820	2320	1660	2110
	19	2410	3060	2170	2750	1970	2500	1770	2240	1610	2030
	20	2350	2960	2120	2660	1920	2410	1730	2160	1570	1960
	22	2230	2750	2000	2460	1820	2230	1630	1990	1480	1810
	24	2100	2540	1890	2270	1710	2050	1530	1830	1390	1650
	26	1980	2330	1770	2070	1600	1870	1430	1660	1300	1500
	28	1850	2120	1650	1880	1490	1690	1330	1500	1200	1350
	30	1720	1910	1530	1690	1380	1520	1230	1350	1110	1210
	32	1590	1720	1410	1510	1270	1350	1130	1200	1020	1070
	34	1460	1520	1300	1340	1160	1200	1030	1060	931	951
	36	1340	1360	1180	1200	1060	1070	940	945	845	848
	38	1220	1220	1080	1080	960	960	848	848	761	761
	40	1100	1100	971	971	866	866	766	766	687	687

Effective length KL (ft) with respect to least radius of gyration r_y

Properties										
u	2.18	2.17	2.18	2.16	2.16	2.15	2.16	2.14	2.15	2.13
P_{wo} (kips)	1180	1640	1010	1400	878	1220	738	1020	636	883
P_{wi} (kips/in.)	64	89	59	81	55	77	50	70	46	64
P_{wb} (kips)	12700	15000	9740	11500	8230	9700	6150	7250	4810	5670
P_{fb} (kips)	1770	2460	1480	2060	1240	1720	1030	1420	868	1210
L_p (ft)	14.5	12.3	14.3	12.1	14.1	12.0	13.9	11.8	13.8	11.7
L_r (ft)	202	131	184	120	169	110	154	100	141	92.0

A (in.^{2})	98.8	89.6	81.9	74.1	67.7
I_x (in.^{4})	4060	3550	3110	2720	2420
I_y (in.^{4})	1190	1050	937	828	742
r_y (in.)	3.47	3.42	3.38	3.34	3.31
Ratio r_x / r_y	1.85	1.84	1.82	1.81	1.80
$P_{ex} (KL)^2 / 10^4$	116000	101000	88900	77900	69100
$P_{ey} (KL)^2 / 10^4$	34000	30000	26800	23700	21200

F_y = 36 ksi

F_y = 50 ksi

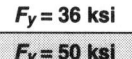

COLUMNS
W shapes
Design axial strength in kips (ϕ = 0.85)

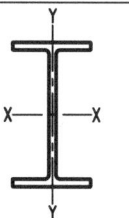

Designation	**W12**											
Wt./ft	**210**		**190**		**170**		**152**		**136**		**120**	
F_y	**36**	**50**	**36**	**50**	**36**	**50**	**36**	**50**	**36**	**50**	**36**	**50**
Effective length KL (ft) with respect to least radius of gyration r_y: 0	1890	2630	1710	2370	1530	2120	1370	1900	1220	1700	1080	1500
6	1840	2540	1660	2290	1490	2050	1330	1830	1190	1630	1050	1440
7	1830	2500	1650	2260	1480	2020	1320	1810	1180	1610	1040	1420
8	1810	2470	1630	2220	1460	1990	1300	1780	1160	1590	1030	1400
9	1790	2430	1610	2190	1440	1960	1290	1750	1150	1560	1010	1380
10	1760	2380	1590	2150	1420	1920	1270	1710	1130	1530	1000	1350
11	1740	2330	1570	2100	1400	1880	1250	1680	1110	1490	984	1320
12	1710	2280	1540	2050	1380	1840	1230	1640	1090	1460	966	1290
13	1680	2230	1510	2000	1350	1790	1210	1590	1070	1420	948	1250
14	1650	2170	1480	1950	1330	1740	1180	1550	1050	1380	928	1220
15	1610	2110	1450	1900	1300	1690	1160	1510	1030	1340	908	1180
16	1580	2040	1420	1840	1270	1640	1130	1460	1010	1290	886	1140
17	1540	1980	1390	1780	1240	1580	1100	1410	980	1250	864	1100
18	1510	1910	1350	1720	1210	1530	1070	1360	955	1210	841	1060
19	1470	1840	1320	1650	1180	1470	1050	1310	928	1160	817	1020
20	1430	1780	1280	1590	1140	1420	1020	1260	901	1110	793	976
22	1340	1640	1210	1460	1070	1300	954	1150	846	1020	743	892
24	1260	1490	1130	1340	1000	1180	891	1050	788	924	692	808
26	1170	1360	1050	1210	933	1070	827	944	731	831	640	726
28	1090	1220	973	1090	862	959	763	844	673	742	589	646
30	1000	1090	895	967	792	852	700	749	617	656	538	569
32	919	962	819	853	724	750	638	658	561	577	489	500
34	837	852	745	755	657	664	578	583	508	511	442	443
36	759	760	674	674	593	593	520	520	456	456	395	395
38	682	682	605	605	532	532	467	467	409	409	355	355
40	616	616	546	546	480	480	421	421	369	369	320	320
Properties												
u	2.16	2.13	2.14	2.11	2.14	2.11	2.15	2.11	2.13	2.09	2.12	2.07
P_{wo} (kips)	558	774	465	646	389	540	333	462	276	383	232	322
P_{wi} (kips/in.)	42	59	38	53	35	48	31	44	28	40	26	36
P_{wb} (kips)	3760	4430	2700	3190	2020	2380	1500	1760	1120	1320	815	960
P_{fb} (kips)	731	1020	610	847	493	684	397	551	316	439	247	343
L_p (ft)	13.7	11.6	13.5	11.5	13.4	11.4	13.3	11.3	13.2	11.2	13.0	11.1
L_r (ft)	129	84.2	117	76.6	105	68.9	94.7	62.1	84.6	55.7	75.5	50.0
A (in.^{2})	61.8		55.8		50.0		44.7		39.9		35.3	
I_x (in.^{4})	2140		1890		1650		1430		1240		1070	
I_y (in.^{4})	664		589		517		454		398		345	
r_y (in.)	3.28		3.25		3.22		3.19		3.16		3.13	
Ratio r_x / r_y	1.80		1.79		1.78		1.77		1.77		1.76	
$P_{ex}(KL)^2/10^4$	61400		54100		47100		41000		35600		30700	
$P_{ey}(KL)^2/10^4$	19000		16900		14800		13000		11400		9900	

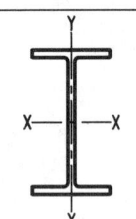

	F_y = 36 ksi
	F_y = 50 ksi

COLUMNS
W shapes
Design axial strength in kips (ϕ = 0.85)

Designation		W12										
Wt./ft		**106**		**96**		**87**		**79**		**72**		**65**	
F_y		36	50	36	50	36	50	36	50	36	50	36	50†
Effective length KL (ft) with respect to least radius of gyration r_y	0	955	1330	863	1200	783	1090	710	986	646	897	584	812
	6	928	1280	839	1150	761	1050	689	947	627	861	567	779
	7	919	1260	830	1140	753	1030	682	933	620	848	561	767
	8	908	1240	820	1120	744	1010	674	917	613	834	554	754
	9	896	1210	809	1100	734	994	665	900	604	818	546	739
	10	883	1190	797	1070	723	973	654	880	595	800	538	723
	11	868	1160	784	1050	711	950	643	860	585	781	529	706
	12	853	1130	770	1020	698	926	631	838	574	761	519	687
	13	836	1100	755	995	684	901	619	814	562	740	508	668
	14	819	1070	739	966	669	874	605	790	550	717	497	647
	15	800	1040	722	935	654	846	591	764	537	694	485	626
	16	781	1000	704	904	638	817	576	738	523	670	472	604
	17	761	968	686	871	621	788	561	711	509	645	460	581
	18	741	932	667	838	604	758	545	683	495	620	446	558
	19	719	895	648	805	586	727	529	655	480	594	433	535
	20	698	858	628	771	568	696	512	627	465	569	419	512
	22	653	783	588	703	531	634	479	570	434	517	391	464
	24	608	708	546	635	493	572	444	514	403	465	362	417
	26	562	635	505	569	455	511	409	459	371	415	333	372
	28	516	565	463	505	417	453	375	406	339	367	305	328
	30	472	497	422	443	380	397	341	355	309	321	277	287
	32	428	437	383	390	344	349	308	312	279	282	250	252
	34	386	387	345	345	309	309	277	277	250	250	223	223
	36	345	345	308	308	276	276	247	247	223	223	199	199
	38	310	310	276	276	248	248	221	221	200	200	179	179
	40	279	279	249	249	223	223	200	200	181	181	161	161

Properties												
u	2.12	2.06	2.10	2.04	2.10	2.02	2.09	2.01	2.08	1.98	2.06	1.95
P_{wo} (kips)	185	257	161	223	139	193	122	169	106	148	92.1	128
P_{wi} (kips/in.)	22	31	20	28	19	26	17	24	15	22	14	20
P_{wb} (kips)	518	611	378	446	311	366	236	278	181	213	135	159
P_{fb} (kips)	198	276	164	228	133	185	109	152	91	126	74	103
L_p (ft)	13.0	11.0	12.9	10.9	12.8	10.9	12.7	10.8	12.7	10.7	12.6	11.8
L_r (ft)	67.2	44.9	61.4	41.4	56.3	38.3	51.8	35.7	48.2	33.6	44.7	31.7

A (in.^{2})	31.2	28.2	25.6	23.2	21.1	19.1
I_x (in.^{4})	933	833	740	662	597	533
I_y (in.^{4})	301	270	241	216	195	174
r_y (in.)	3.11	3.09	3.07	3.05	3.04	3.02
Ratio r_x / r_y	1.76	1.76	1.75	1.75	1.75	1.75
$P_{ex}(KL)^2 / 10^4$	26700	23900	21200	18900	17000	15200
$P_{ey}(KL)^2 / 10^4$	8640	7710	6910	6180	5580	4990

†Flange is noncompact; see discussion preceding column load tables.

| F_y = 36 ksi |
| F_y = 50 ksi |

COLUMNS
W shapes
Design axial strength in kips (ϕ = 0.85)

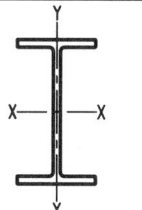

Designation						W12				
Wt./ft		58		53		50		45		40
F_y		36	50	36	50	36	50	36	50	36	50

	F_y	58-36	58-50	53-36	53-50	50-36	50-50	45-36	45-50	40-36	40-50
	0	520	723	477	663	450	625	404	561	361	502
	6	498	680	457	623	419	566	376	507	336	453
	7	490	666	449	610	408	546	366	489	327	437
	8	482	649	441	594	396	524	355	469	317	419
	9	472	631	432	577	383	500	343	447	306	399
	10	461	611	422	559	369	475	330	424	295	378
	11	450	590	411	539	354	448	317	400	282	356
	12	437	568	400	518	339	421	302	375	269	334
	13	424	545	388	496	322	393	287	350	256	311
	14	411	521	375	474	306	365	272	324	242	288
	15	397	496	362	451	289	337	257	299	228	266
	16	382	471	348	428	271	310	241	274	214	243
	18	352	420	320	381	237	257	210	227	187	201
	20	321	370	292	334	204	209	180	184	160	163
	22	291	322	263	290	173	173	152	152	135	135
	24	260	276	235	247	145	145	128	128	113	113
	26	231	235	207	210	124	124	109	109	96	96
	28	202	202	181	181	107	107	94	94	83	83
	30	176	176	158	158	93	93	82	82	72	72
	32	155	155	139	139	82	82	72	72	64	64
	34	137	137	123	123						
	38	110	110	98	98						
	41	94	94	85	85						

Left axis label: Effective length KL (ft) with respect to least radius of gyration r_y

Properties											
u		2.41	2.22	2.39	2.16	2.85	2.51	2.79	2.37	2.69	2.22
P_{wo} (kips)		89	124	78	108	92	127	75	105	66	92
P_{wi} (kips/in.)		13	18	12	17	13	19	12	17	11	15
P_{wb} (kips)		106	125	94	111	116	136	86	101	59	69
P_{fb} (kips)		83	115	67	93	83	115	67	93	54	75
L_p (ft)		10.5	8.9	10.3	8.8	8.2	6.9	8.1	6.9	8.0	6.8
L_r (ft)		38.3	27.0	35.8	25.6	30.8	21.6	28.4	20.3	26.5	19.3

A (in.^{2})		17.0		15.6		14.7		13.2		11.8	
I_x (in.^{4})		475		425		394		350		310	
I_y (in.^{4})		107		95.8		56.3		50		44.1	
r_y (in.)		2.51		2.48		1.96		1.94		1.93	
Ratio r_x / r_y		2.10		2.11		2.64		2.65		2.66	
$P_{ex} (KL)^2 / 10^4$		13600		12200		11300		10000		8890	
$P_{ey} (KL)^2 / 10^4$		3070		2750		1620		1420		1260	

Note: Heavy line indicates Kl / r of 200.

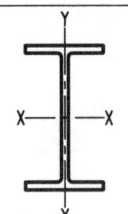

		F_y = 36 ksi
		F_y = 50 ksi

COLUMNS
W shapes
Design axial strength in kips (ϕ = 0.85)

Designation					W10					
Wt./ft		112		100		88		77		68
F_y		36	50	36	50	36	50	36	50	36	50
	0	1010	1400	900	1250	793	1100	692	961	612	850
	6	969	1330	865	1180	762	1040	664	908	588	803
	7	956	1300	853	1160	751	1020	655	890	579	787
	8	941	1270	840	1140	739	999	644	869	569	769
	9	924	1240	824	1110	725	973	632	847	558	749
	10	906	1210	808	1080	710	945	618	822	547	727
	11	886	1170	789	1040	694	916	604	796	534	703
	12	865	1130	770	1010	677	884	588	768	520	678
	13	842	1090	750	970	659	851	572	738	506	652
	14	819	1050	728	931	639	817	555	708	490	625
	15	794	1010	706	892	619	782	537	677	475	597
	16	768	961	682	851	599	746	519	645	458	569
	17	742	915	659	810	577	709	500	612	441	540
	18	715	870	634	769	556	672	481	580	424	511
	19	688	824	609	727	534	635	461	547	407	482
	20	660	778	584	686	511	599	442	515	389	454
	22	604	688	534	605	466	527	402	452	354	398
	24	548	601	483	527	422	458	362	392	319	344
	26	493	518	434	453	378	393	324	335	285	294
	28	440	447	386	390	336	339	287	289	252	254
	30	389	389	340	340	295	295	252	252	221	221
	32	342	342	299	299	259	259	221	221	194	194
	34	303	303	265	265	230	230	196	196	172	172
	36	270	270	236	236	205	205	175	175	153	153
	38	242	242	212	212	184	184	157	157	138	138
	40	219	219	191	191	166	166	141	141	124	124

Effective length KL (ft) with respect to least radius of gyration r_y

Properties										
u		2.06	2.02	2.06	2.01	2.04	1.99	2.03	1.96	2.01	1.93
P_{wo} (kips)		255	354	214	298	177	246	143	199	116	162
P_{wi} (kips/in.)		27	38	24	34	22	30	19	27	17	24
P_{wb} (kips)		1210	1430	883	1040	623	735	420	495	293	345
P_{fb} (kips)		316	439	254	353	198	276	153	213	120	167
L_p (ft)		11.2	9.5	11.0	9.4	11.0	9.3	10.8	9.2	10.8	9.2
L_r (ft)		86.4	56.5	77.4	50.8	68.4	45.1	60.0	39.8	53.8	36.0
A (in.^{2})		32.9		29.4		25.9		22.6		20.0	
I_x (in.^{4})		716		623		534		455		394	
I_y (in.^{4})		236		207		179		154		134	
r_y (in.)		2.68		2.65		2.63		2.60		2.59	
Ratio r_x / r_y		1.74		1.74		1.73		1.73		1.71	
$P_{ex} (KL)^2 / 10^4$		20400		17800		15300		13000		11300	
$P_{ey} (KL)^2 / 10^4$		6760		5910		5130		4370		3840	

| $F_y = 36$ ksi |
| $F_y = 50$ ksi |

COLUMNS
W shapes
Design axial strength in kips ($\phi = 0.85$)

Designation						W10						
Wt./ft	60		54		49		45		39		33	
F_y	36	50	36	50	36	50	36	50	36	50	36	50
0	539	748	483	672	441	612	407	565	352	489	297	413
6	517	706	464	634	422	577	380	515	328	444	276	373
7	509	692	457	621	416	565	371	497	320	428	269	360
8	500	675	449	606	409	551	361	478	311	412	261	345
9	491	657	440	590	401	536	350	458	301	393	252	329
10	480	638	431	572	392	520	337	436	290	374	243	312
11	469	617	420	553	382	502	324	412	278	353	233	294
12	457	595	409	533	372	484	311	388	266	332	222	276
13	444	571	398	512	361	464	296	364	254	310	211	257
14	430	547	385	490	350	444	282	339	241	289	200	238
15	416	523	373	468	338	424	267	314	228	267	189	220
16	401	497	360	445	326	403	252	290	215	246	177	202
17	387	472	346	422	314	382	237	266	201	225	166	184
18	371	446	332	399	301	361	222	243	188	205	155	167
19	356	421	318	376	288	340	207	221	175	185	144	150
20	340	395	304	353	275	319	192	199	162	167	133	135
22	309	346	276	309	250	278	164	164	138	138	112	112
24	278	299	248	266	224	239	138	138	116	116	94	94
26	248	255	221	227	199	204	118	118	99	99	80	80
28	219	220	195	196	175	176	102	102	85	85	69	69
30	191	191	170	170	153	153	88	88	74	74	60	60
32	168	168	150	150	134	134	78	78	65	65	53	53
33	158	158	141	141	126	126	73	73	61	61		
34	149	149	133	133	119	119						
36	133	133	118	118	106	106						

Left axis label: Effective length KL (ft) with respect to least radius of gyration r_y

Properties

u	2.00	1.90	1.97	1.87	1.96	1.83	2.37	2.17	2.31	2.04	2.23	1.87
P_{wo} (kips)	99	138	83	116	73	101	79	109	64	89	55	77
P_{wi} (kips/in.)	15	21	13	19	12	17	13	18	11	16	10	14
P_{wb} (kips)	209	246	143	168	111	131	121	142	88	104	69	81
P_{fb} (kips)	94	130	77	106	64	88	78	108	57	79	38	53
L_p (ft)	10.7	9.1	10.7	9.1	10.6	9.0	8.4	7.1	8.3	7.0	8.1	6.9
L_r (ft)	48.1	32.6	43.9	30.2	40.7	28.3	35.2	24.1	31.2	21.9	27.4	19.7

A (in.^{2})	17.6		15.8		14.4		13.3		11.5		9.71	
I_x (in.^{4})	341		303		272		248		209		170	
I_y (in.^{4})	116		103		93.4		53.4		45.0		36.6	
r_y (in.)	2.57		2.56		2.54		2.01		1.98		1.94	
Ratio r_x / r_y	1.71		1.71		1.71		2.15		2.16		2.16	
$P_{ex} (KL)^2 / 10^4$	9710		8640		7800		7100		6000		4880	
$P_{ey} (KL)^2 / 10^4$	3330		2960		2660		1540		1290		1050	

Note: Heavy line indicates Kl / r of 200.

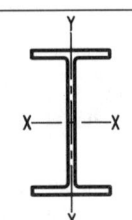

| | | F_y = 36 ksi |
| | | F_y = 50 ksi |

COLUMNS
W shapes
Design axial strength in kips (ϕ = 0.85)

Designation							W8						
Wt./ft		67		58		48		40		35		31	
F_y		36	50	36	50	36	50	36	50	36	50	36	50
Effective length KL (ft) with respect to least radius of gyration r_y	0	603	837	523	727	431	599	358	497	315	438	279	388
	6	567	770	492	667	405	549	335	454	295	399	261	354
	7	555	746	481	647	396	532	327	439	288	386	255	342
	8	541	721	469	624	386	513	319	423	280	372	248	329
	9	526	693	455	599	374	492	309	405	272	356	240	315
	10	509	662	441	572	362	470	298	386	262	339	232	300
	11	492	631	425	544	349	446	287	366	252	321	223	284
	12	473	598	409	515	335	422	275	345	242	303	214	268
	13	453	564	391	485	321	397	263	324	231	284	204	251
	14	433	529	374	455	306	372	251	303	220	265	194	234
	15	412	494	355	425	291	347	238	281	208	246	184	217
	16	391	460	337	394	276	321	225	260	197	228	174	200
	17	370	425	318	365	260	297	211	239	185	209	163	184
	18	349	392	300	335	245	272	198	219	174	191	153	168
	19	328	359	281	307	229	249	185	199	162	174	143	153
	20	307	328	263	279	214	226	173	180	151	157	133	138
	22	266	271	228	231	185	187	148	149	129	130	114	114
	24	228	228	194	194	157	157	125	125	109	109	96	96
	26	194	194	165	165	134	134	107	107	93	93	82	82
	28	167	167	143	143	115	115	92	92	80	80	70	70
	30	146	146	124	124	100	100	80	80	70	70	61	61
	32	128	128	109	109	88	88	70	70	61	61	54	54
	33	120	120	103	103	83	83	66	66	58	58	51	51
	34	113	113	97	97	78	78	62	62				
	35	107	107	91	91								

Properties													
u		2.03	1.96	2	1.93	1.97	1.87	1.93	1.8	1.89	1.74	1.85	1.65
P_{wo} (kips)		147	205	120	167	86	119	69	96	56	78	48	67
P_{wi} (kips/in.)		21	28	18	26	14	20	13	18	11	16	10	14
P_{wb} (kips)		648	764	464	547	224	264	163	192	104	123	81	95
P_{fb} (kips)		177	246	133	185	95	132	64	88	50	69	38	53
L_p (ft)		8.8	7.5	8.8	7.4	8.7	7.4	8.5	7.2	8.5	7.2	8.4	7.1
L_r (ft)		64.0	41.9	55.9	36.8	46.7	31.1	39.1	26.5	35.1	24.1	32.0	22.4
A (in.^{2})		19.7		17.1		14.1		11.7		10.3		9.13	
I_x (in.^{4})		272		228		184		146		127		110	
I_y (in.^{4})		88.6		75.1		60.9		49.1		42.6		37.1	
r_y (in.)		2.12		2.10		2.08		2.04		2.03		2.02	
Ratio r_x / r_y		1.75		1.74		1.74		1.73		1.73		1.72	
$P_{ex} (KL)^2 / 10^4$		7800		6520		5260		4170		3630		3150	
$P_{ey} (KL)^2 / 10^4$		2530		2160		1750		1390		1210		1070	

Note: Heavy line indicates Kl / r of 200.

F_y = 36 ksi
F_y = 50 ksi

COLUMNS
W shapes
Design axial strength in kips ($\phi = 0.85$)

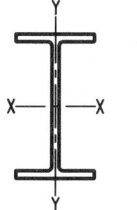

Designation		W8				W6				
Wt./ft		28		24		25		20		15
F_y		36	50	36	50	36	50	36	50	36†	50†
Effective length KL (ft) with respect to least radius of gyration r_y	0	252	351	217	301	225	312	180	249	136	188
	6	228	303	195	260	200	265	159	211	119	158
	7	219	288	188	247	191	250	152	198	114	148
	8	210	271	180	232	182	233	145	185	108	137
	9	200	253	171	217	172	216	137	171	102	126
	10	189	235	162	200	162	198	128	156	95	115
	11	178	216	152	184	151	180	119	142	88	104
	12	167	197	142	168	140	162	111	127	81	92
	13	155	178	132	151	129	144	102	113	74	82
	14	143	160	122	136	118	128	93	100	68	71
	15	132	142	112	121	107	112	84	87	61	62
	16	121	125	102	106	97	98	76	76	55	55
	17	110	111	93	94	87	87	68	68	48	48
	18	99	99	84	84	78	78	60	60	43	43
	19	89	89	75	75	70	70	54	54	39	39
	20	80	80	68	68	63	63	49	49	35	35
	22	66	66	56	56	52	52	40	40	29	29
	24	56	56	47	47	44	44	34	34	24	24
	25	51	51	44	44	40	40	31	31		
	26	47	47	40	40						
	27	44	44								

Properties

u	2.17	1.87	2.07	1.71	2.07	1.98	2.03	1.91	1.98	1.75
P_{wo} (kips)	48	67	39	54	47	65	35	49	26	36
P_{wi} (kips/in.)	10	14	9	12	12	16	9	13	8	12
P_{wb} (kips)	81	95	52	61	146	172	78	92	54	64
P_{fb} (kips)	44	61	32	45	42	58	27	37	14	19
L_p (ft)	6.8	5.7	6.7	5.7	6.3	5.4	6.3	5.3	6.7	6.8
L_r (ft)	27.2	18.8	24.3	17.2	31.2	21.0	25.6	17.6	20.8	15.0

A (in.^{2})	8.25	7.08	7.34	5.87	4.43
I_x (in.^{4})	98.0	82.8	53.4	41.4	29.1
I_y (in.^{4})	21.7	18.3	17.1	13.3	9.32
r_y (in.)	1.62	1.61	1.52	1.50	1.46
Ratio r_x / r_y	2.13	2.12	1.78	1.77	1.75
$P_{ex} (KL)^2 / 10^4$	2810	2370	1530	1190	831
$P_{ey} (KL)^2 / 10^4$	620	525	485	378	270

†Flange is noncompact; see discussion preceding column load tables.
Note: Heavy line indicates Kl / r of 200.

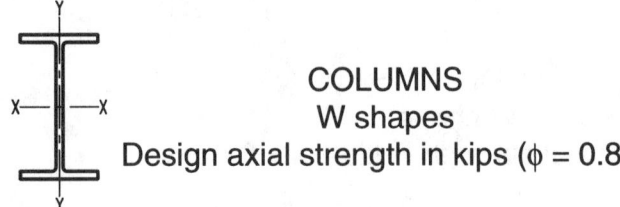

	F_y = 36 ksi
	F_y = 50 ksi

COLUMNS
W shapes
Design axial strength in kips (ϕ = 0.85)

Designation	W6						W5				W4	
Wt./ft	16		12		9		19		16		13	
F_y	36	50	36	50	36	50	36	50	36	50	36	50
0	145	201	109	151	82	114	170	235	143	199	117	163
2	140	193	105	144	79	108	166	229	141	194	114	156
3	135	182	100	135	75	101	163	222	137	188	109	148
4	127	168	94	124	71	93	157	212	133	179	104	138
5	118	152	87	110	65	83	151	201	127	169	97	125
6	108	134	79	96	59	72	144	187	121	157	89	111
7	97	116	70	82	52	61	135	172	114	144	81	97
8	86	98	61	68	45	50	126	156	106	131	72	83
9	75	81	52	55	39	40	117	140	98	117	63	69
10	64	66	44	44	32	33	107	124	90	104	55	57
11	54	54	37	37	27	27	97	108	81	90	47	47
12	46	46	31	31	23	23	87	93	73	78	39	39
13	39	39	26	26	19	19	78	80	65	66	34	34
14	33	33	23	23	17	17	68	69	57	57	29	29
15	29	29	20	20	14	14	60	60	50	50	25	25
16	26	26					53	53	44	44	22	22
17							47	47	39	39		
18							42	42	35	35		
19							37	37	31	31		
20							34	34	28	28		
21							30	30	25	25		

Effective length KL (ft) with respect to least radius of gyration r_y

Properties

u	2.84	2.5	2.62	2.13	2.24	1.72	1.84	1.72	1.79	1.63	1.89	1.77
P_{wo} (kips)	35	49	26	36	17	24	39	55	32	45	35	48
P_{wi} (kips/in.)	9	13	8	12	6	9	10	14	9	12	10	14
P_{wb} (kips)	78	92	54	64	22	26	115	136	81	95	164	193
P_{fb} (kips)	33	46	16	22	9	13	37	52	26	36	24	33
L_p (ft)	4.0	3.4	3.8	3.2	3.8	3.2	5.3	4.5	5.3	4.5	4.2	3.5
L_r (ft)	18.3	12.5	14.3	10.2	12.0	8.9	30.3	20.1	26.2	17.6	25.5	16.8
A (in.^{2})	4.74		3.55		2.68		5.54		4.68		3.83	
I_x (in.^{4})	32.1		22.1		16.4		26.2		21.3		11.3	
I_y (in.^{4})	4.43		2.99		2.19		9.13		7.51		3.86	
r_y (in.)	0.966		0.918		0.905		1.28		1.27		1.00	
Ratio r_x / r_y	2.69		2.71		2.73		1.70		1.68		1.72	
$P_{ex} (KL)^2 / 10^4$	917		630		468		747		608		324	
$P_{ey} (KL)^2 / 10^4$	127		85.6		62.8		260		216		110	

Note: Heavy line indicates Kl / r of 200.

| F_y = 36 ksi |
| F_y = 50 ksi |

COLUMNS
HP shapes
Design axial strength in kips (ϕ = 0.85)

Designation					HP14				HP13	
Wt./ft		117		102		89		73		100
F_y	36	50	36	50	36	50	36	50	36	50
0	1050	1460	918	1280	799	1110	655	909	900	1250
6	1030	1420	898	1240	781	1080	640	882	875	1200
7	1020	1400	891	1220	775	1060	635	872	867	1190
8	1010	1390	884	1210	768	1050	629	861	857	1170
9	1000	1370	875	1190	760	1040	623	848	846	1150
10	993	1350	865	1170	752	1020	615	834	834	1120
11	980	1320	854	1150	742	1000	607	819	821	1100
12	967	1300	842	1130	732	982	599	803	806	1070
13	953	1270	830	1110	721	962	589	786	791	1050
14	938	1250	816	1080	709	940	580	768	775	1020
15	922	1220	802	1060	696	917	569	749	758	986
16	905	1190	788	1030	683	893	558	729	741	954
17	888	1150	772	1000	670	869	547	708	722	921
18	870	1120	756	974	656	844	535	687	703	888
19	851	1090	740	945	641	818	523	666	684	854
20	832	1050	723	915	626	791	511	644	664	820
22	792	985	687	853	595	737	485	599	623	750
24	750	913	650	790	563	682	458	553	581	681
26	707	842	613	727	529	627	430	507	539	613
28	664	771	574	665	496	572	402	462	496	547
30	620	701	536	604	462	519	374	418	454	483
32	576	633	498	545	428	467	346	375	413	425
34	533	568	460	487	395	417	319	334	374	376
36	491	507	423	435	363	372	292	298	336	336
38	450	455	387	390	332	334	267	267	301	301
40	411	411	352	352	301	301	241	241	272	272

Effective length KL (ft) with respect to least radius of gyration r_y

Properties										
P_{wo} (kips)	217	302	174	242	145	202	108	150	198	275
P_{wi} (kips/in.)	29	40	25	35	22	31	18	25	28	38
P_{wb} (kips)	1010	1191	679	801	453	533	250	294	953	1123
P_{fb} (kips)	131	182	101	140	77	106	52	72	119	165
L_p (ft)	15.0	12.7	14.8	12.6	14.7	12.5	14.5	12.3	13.2	11.2
L_r (ft)	66.0	45.1	59.0	41.1	53.0	37.6	46.8	34.2	60.1	40.9
A (in.^{2})	34.4		30.0		26.1		21.4		29.4	
I_x (in.^{4})	1220		1050		904		729		886	
I_y (in.^{4})	443		380		326		261		294	
r_y (in.)	3.59		3.56		3.53		3.49		3.16	
Ratio r_x / r_y	1.66		1.66		1.67		1.67		1.74	
$P_{ex}(KL)^2 / 10^4$	35000		30100		25800		20900		25400	
$P_{ey}(KL)^2 / 10^4$	12700		10900		9310		7460		8400	

Note: Heavy line indicates Kl / r of 200.

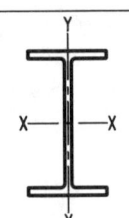

	F_y = 36 ksi
	F_y = 50 ksi

COLUMNS
HP shapes
Design axial strength in kips (ϕ = 0.85)

Designation		HP13					HP12			
Wt./ft		87		73		60		84		74	
F_y		36	50	36	50	36	50	36	50	36	50
Effective length KL (ft) with respect to least radius of gyration r_y	0	780	1080	661	918	536	744	753	1050	667	927
	6	759	1040	642	882	520	714	729	1000	646	886
	7	751	1030	636	870	515	704	721	985	639	872
	8	743	1010	628	856	509	692	712	967	630	856
	9	733	993	620	840	502	679	701	947	621	838
	10	722	973	611	823	494	665	690	926	610	819
	11	711	952	601	804	486	650	677	902	599	798
	12	698	928	590	784	477	633	663	877	587	776
	13	685	904	578	763	467	616	649	851	574	752
	14	670	878	566	741	457	597	634	823	560	727
	15	656	851	553	717	447	578	618	795	546	702
	16	640	823	540	693	436	559	601	765	531	675
	17	624	794	526	669	424	539	584	735	516	648
	18	607	765	512	644	413	518	567	705	500	621
	19	590	735	497	618	401	497	548	674	484	593
	20	573	705	482	592	388	476	530	642	467	565
	22	537	644	451	540	363	433	492	580	434	510
	24	500	584	420	488	337	391	454	518	400	455
	26	462	524	388	438	311	350	416	459	366	402
	28	425	467	356	389	285	310	378	402	332	351
	30	389	411	325	342	260	272	342	350	300	306
	32	353	361	295	300	235	239	307	308	268	269
	34	319	320	266	266	211	211	273	273	238	238
	36	286	286	237	237	189	189	243	243	213	213
	38	256	256	213	213	169	169	218	218	191	191
	40	231	231	192	192	153	153	197	197	172	172

Properties

P_{wo} (kips)	165	229	127	177	93	129	170	235	143	199
P_{wi} (kips/in.)	24	33	20	28	17	23	25	34	22	30
P_{wb} (kips)	624	735	384	453	206	243	732	862	506	597
P_{fb} (kips)	90	124	65	90	43	60	95	132	75	105
L_p (ft)	13.0	11.1	12.9	11.0	12.8	10.9	12.3	10.4	12.2	10.3
L_r (ft)	53.2	36.9	47.0	33.4	41.2	30.2	54.0	36.9	48.9	34.0
A (in.^{2})	25.5		21.6		17.5		24.6		21.8	
I_x (in.^{4})	755		630		503		650		569	
I_y (in.^{4})	250		207		165		213		186	
r_y (in.)	3.13		3.10		3.07		2.94		2.92	
Ratio r_x / r_y	1.74		1.74		1.75		1.75		1.75	
$P_{ex}(KL)^2 / 10^4$	21700		18000		14400		18600		16300	
$P_{ey}(KL)^2 / 10^4$	7150		5940		4720		6090		5320	

Note: Heavy line indicates Kl / r of 200.

F_y = 36 ksi
F_y = 50 ksi

COLUMNS
HP shapes
Design axial strength in kips (ϕ = 0.85)

Designation		HP12			HP10				HP8		
Wt./ft		63		53		57		42		36	
F_y		36	50	36	50	36	50	36	50	36	50
Effective length KL (ft) with respect to least radius of gyration r_y	0	563	782	474	659	514	714	379	527	324	451
	6	545	747	459	629	491	670	362	494	302	408
	7	538	735	453	618	483	655	356	482	294	393
	8	531	721	447	607	474	638	349	469	286	377
	9	523	706	440	594	464	619	341	455	276	360
	10	514	689	432	579	453	599	333	440	266	342
	11	504	671	424	564	441	577	324	423	255	322
	12	494	651	415	547	429	555	314	406	243	302
	13	482	631	406	530	415	531	304	388	232	282
	14	471	610	396	512	401	506	294	369	219	262
	15	458	588	385	493	387	481	283	350	207	242
	16	446	565	374	474	372	456	272	331	195	222
	17	432	542	363	454	357	430	260	312	182	202
	18	419	518	351	434	341	404	249	293	170	184
	19	405	495	339	414	326	379	237	274	158	165
	20	391	471	327	394	310	354	225	255	146	149
	22	362	423	303	353	279	305	202	219	123	123
	24	333	376	278	314	248	259	179	185	104	104
	26	304	332	253	276	219	221	157	158	88	88
	28	275	288	229	240	191	191	136	136	76	76
	30	247	251	206	209	166	166	119	119	66	66
	32	221	221	183	183	146	146	104	104	58	58
	34	196	196	163	163	129	129	92	92	52	52
	36	174	174	145	145	115	115	82	82	46	46
	38	157	157	130	130	103	103	74	74	41	41
	40	141	141	117	117	93	93	67	67	37	37

Properties

P_{wo} (kips)	116	161	88	122	121	168	79	110	75	104
P_{wi} (kips/in.)	19	26	16	22	20	28	15	21	16	22
P_{wb} (kips)	311	366	188	221	508	599	202	238	309	364
P_{fb} (kips)	54	75	38	53	65	90	36	50	40	56
L_p (ft)	12.0	10.2	11.9	10.1	10.2	8.7	10.0	8.5	8.1	6.9
L_r (ft)	43.0	30.7	38.7	28.3	45.6	31.1	35.9	25.6	35.7	24.4

A (in.^{2})	18.4		15.5		16.8		12.4		10.6	
I_x (in.^{4})	472		393		294		210		119	
I_y (in.^{4})	153		127		101		71.7		40.3	
r_y (in.)	2.88		2.86		2.45		2.41		1.95	
Ratio r_x / r_y	1.76		1.76		1.71		1.71		1.72	
$P_{ex} (KL)^2 / 10^4$	13500		11200		8400		6050		3420	
$P_{ey} (KL)^2 / 10^4$	4370		3630		2890		2060		1150	

Note: Heavy line indicates Kl / r of 200.

Steel Pipe and Structural Tubing

The design strengths in the tables that follow are tabulated for the effective lengths in feet KL (with respect to the least radius of gyration, r or r_y), indicated at the left of each table. They are applicable to axially loaded members in accordance with Section E2 of the LRFD Specification.

For discussion of effective length, range of l/r, strength about major axis, combined axial and bending stress, and sample problems, see General Notes. Properties and factors are listed at the bottom of the tables for checking strength about the strong axis.

Steel Pipe Columns

Design strengths for unfilled pipe columns are tabulated for F_y = 36 ksi. Steel pipe manufactured to ASTM A501 furnishes F_y = 36 ksi, and ASTM A53, Type E or S, Gr. B furnishes F_y = 35 ksi and may be designed for the strengths permitted for F_y = 36 ksi steel.

The heavy horizontal lines within the table indicate Kl/r = 200. No values are listed beyond Kl/r = 200.

Structural Tube Columns

Design strengths for square and rectangular structural tube columns are tabulated for F_y = 46 ksi. Structural tubing is manufactured to F_y = 46 ksi under ASTM A500, Gr. B.

All tubes listed in the column load tables satisfy Section B5 of the LRFD Specification. The heavy horizontal lines appearing within the tables indicate Kl/r = 200. No values are listed beyond Kl/r = 200.

F_y = 36 ksi

COLUMNS
Standard steel pipe
Design axial strength in kips (ϕ = 0.85)

Nominal Dia.	12	10	8	6	5	4	3½	3
Wall Thickness	0.375	0.365	0.322	0.280	0.258	0.237	0.226	0.216
Weight per ft	49.56	40.48	28.55	18.97	14.62	10.79	9.11	7.58
F_y	36 ksi							
0	447	364	257	171	132	97	82	68
6	440	357	249	162	122	86	70	56
7	438	354	246	159	118	82	67	52
8	436	351	243	155	115	78	63	48
9	433	348	239	151	111	74	58	43
10	429	344	235	147	106	70	54	39
11	426	340	231	142	102	65	49	35
12	422	336	227	138	97	60	45	30
13	418	331	222	133	92	55	40	26
14	413	326	216	127	86	51	36	23
15	409	321	211	122	81	46	32	20
16	404	315	205	116	76	41	28	17
17	399	309	199	111	71	37	25	15
18	393	303	193	105	66	33	22	14
19	387	297	187	99	61	30	20	12
20	381	291	181	94	56	27	18	
22	369	277	168	83	47	22	15	
24	356	263	155	72	39	19		
25	349	256	149	67	36	17		
26	342	249	142	62	33			
28	328	234	129	53	29			
30	313	219	117	47	25			
31	306	212	111	44	23			
32	298	205	105	41				
34	283	190	93	36				
36	268	176	83	32				
37	260	169	79	31				
38	253	162	75					
40	237	148	67					

Effective length KL (ft)

Properties

	12	10	8	6	5	4	3½	3
Area A (in.^{2})	14.6	11.9	8.40	5.58	4.30	3.17	2.68	2.23
I (in.^{4})	279	161	72.5	28.1	15.2	7.23	4.79	3.02
r (in.)	4.38	3.67	2.94	2.25	1.88	1.51	1.34	1.16

Note: Heavy line indicates Kl/r of 200.

F_y = 36 ksi

COLUMNS
Extra strong steel pipe
Design axial strength in kips (ϕ = 0.85)

Nominal Dia.	12	10	8	6	5	4	3½	3
Wall Thickness	0.500	0.500	0.500	0.432	0.375	0.337	0.318	0.300
Weight per ft	65.42	54.74	43.39	28.57	20.78	14.98	12.50	10.25
F_y					36 ksi			
0	588	493	392	257	187	135	113	92
6	579	483	379	243	172	119	96	75
7	576	479	375	238	168	114	91	69
8	573	475	369	232	162	108	85	64
9	569	470	364	226	156	102	79	58
10	564	465	357	219	149	95	72	52
11	559	460	351	212	143	89	66	46
12	554	453	343	205	135	82	60	40
13	549	447	336	197	128	75	53	34
14	543	440	327	189	121	68	47	30
15	536	433	319	180	113	62	42	26
16	530	425	310	172	105	56	37	23
18	515	409	291	154	91	44	29	18
19	508	400	282	145	83	40	26	16
20	500	391	272	137	76	36	23	
21	492	382	262	128	70	32	21	
22	483	373	252	120	63	30		
24	465	354	231	103	53	25		
26	447	334	211	88	45			
28	428	314	191	76	39			
30	408	294	172	66	34			
32	388	273	154	58				
34	368	253	136	52				
36	348	234	121	46				
38	328	215	109					
40	308	196	98					

Effective length KL (ft)

Properties								
Area A (in.^{2})	19.2	16.1	12.8	8.40	6.11	4.41	3.68	3.02
I (in.^{4})	362	212	106	40.5	20.7	9.61	6.28	3.89
r (in.)	4.33	3.63	2.88	2.19	1.84	1.48	1.31	1.14

Note: Heavy line indicates Kl/r of 200.

F_y = 36 ksi

COLUMNS
Double-extra strong steel pipe
Design axial strength in kips (ϕ = 0.85)

Nominal Dia.	8	6	5	4	3
Wall Thickness	0.875	0.864	0.750	0.674	0.600
Weight per ft	72.42	53.16	38.55	27.54	18.58
F_y	36 ksi				
0	652	477	346	248	167
6	629	448	315	214	131
7	621	437	305	203	120
8	612	426	293	191	108
9	601	413	281	179	96
10	590	399	268	165	84
11	578	385	254	152	73
12	565	369	239	139	62
13	551	353	224	125	53
14	536	336	209	112	46
15	521	319	194	100	40
16	505	302	179	88	35
17	489	285	165	78	31
18	472	268	151	70	
19	455	250	137	62	
20	438	234	124	56	
22	403	201	102	47	
24	367	170	86		
26	333	145	73		
28	299	125	63		
30	266	109			
32	235	96			
34	208	85			
36	186				
38	166				
40	150				

Effective length KL (ft)

Properties					
Area A (in.^{2})	21.3	15.6	11.3	8.10	5.47
I (in.^{4})	162	66.3	33.6	15.3	5.99
r (in.)	2.76	2.06	1.72	1.37	1.05

Note: Heavy line indicates Kl/r of 200.

F_y = 46 ksi

COLUMNS
Square structural tubing
Design axial strength in kips (ϕ = 0.85)

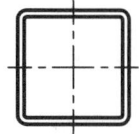

Nominal Size		16×16	14×14		12×12			
Thickness		½	½	⅜	⅝	½	⅜	⁵⁄₁₆
Wt./ft		103.30	89.68	68.31	93.34	76.07	58.10	48.86
F_y					46 ksi			
Effective length KL (ft)	0	1190	1030	786	1070	876	669	563
	6	1180	1020	777	1050	862	658	554
	7	1170	1020	774	1050	857	655	551
	8	1170	1010	770	1040	851	650	548
	9	1170	1010	766	1030	845	645	544
	10	1160	1000	761	1020	838	640	539
	11	1150	993	756	1010	830	634	535
	12	1150	985	751	1000	821	628	529
	13	1140	977	745	992	812	621	524
	14	1130	969	739	979	803	614	518
	15	1120	960	732	966	792	606	511
	16	1120	950	725	953	781	598	504
	17	1110	940	717	939	770	590	497
	18	1100	930	710	924	758	581	490
	19	1090	919	701	908	746	571	482
	20	1080	907	693	892	733	562	474
	21	1070	895	684	875	719	552	466
	22	1060	883	675	858	706	542	457
	23	1040	870	665	841	692	531	449
	24	1030	857	655	823	677	520	440
	25	1020	844	645	805	663	510	431
	26	1010	830	635	786	648	498	421
	27	994	816	624	767	633	487	412
	28	981	802	614	748	617	475	402
	29	967	787	603	729	602	464	392
	30	954	772	592	710	586	452	383
	32	925	742	569	670	555	428	363
	34	896	711	546	631	523	404	343
	36	865	680	522	592	491	381	323
	38	835	648	498	553	460	357	303
	40	803	616	474	515	429	333	283
Properties								
A (in².)		30.4	26.4	20.1	27.4	22.4	17.1	14.4
I (in.⁴)		1200	791	615	580	485	380	324
r (in.)		6.29	5.48	5.54	4.60	4.66	4.72	4.75

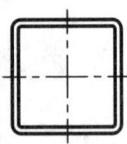

F_y = 46 ksi

COLUMNS
Square structural tubing
Design axial strength in kips (ϕ = 0.85)

Nominal Size		10×10				8×8			
Thickness	$\frac{5}{8}$	$\frac{1}{2}$	$\frac{3}{8}$	$\frac{5}{16}$	$\frac{5}{8}$	$\frac{1}{2}$	$\frac{3}{8}$	$\frac{5}{16}$	$\frac{1}{4}$
Wt./ft	76.33	62.46	47.90	40.35	59.32	48.85	37.69	31.84	25.82
F_y					46 ksi				
0	876	719	551	465	680	563	434	366	297
6	855	703	539	455	654	542	418	353	287
7	847	697	534	451	644	535	413	349	283
8	839	690	529	447	634	526	407	343	279
9	829	682	524	442	622	517	400	338	274
10	818	674	517	437	609	507	392	331	269
11	807	664	510	431	595	496	384	324	264
12	794	655	503	425	580	484	375	317	258
13	781	644	495	418	564	471	366	309	252
14	767	633	487	411	548	458	356	301	245
15	752	621	478	404	531	444	345	293	238
16	736	608	468	396	513	430	335	284	231
17	720	595	459	388	494	415	324	275	224
18	703	582	449	380	476	400	312	265	216
19	686	568	438	371	456	385	301	256	209
20	668	553	427	362	437	369	289	246	201
21	650	538	416	353	418	354	277	236	193
22	631	523	405	343	398	338	266	226	185
23	612	508	394	334	379	322	254	216	177
24	593	493	382	324	360	307	242	206	169
25	573	477	370	314	341	291	230	196	161
26	554	461	358	305	322	276	219	187	153
27	534	446	347	295	304	261	207	177	146
28	515	430	335	285	286	246	196	168	138
29	495	414	323	275	268	232	185	158	131
30	476	398	311	265	251	218	174	149	123
32	437	367	287	245	221	191	153	132	109
34	400	337	264	225	195	169	136	117	97
36	364	307	242	206	174	151	121	104	86
38	328	278	220	188	156	136	109	93	77
40	296	251	199	170	141	122	98	84	70
Properties									
A (in^2)	22.4	18.4	14.1	11.9	17.4	14.4	11.1	9.36	7.59
I (in.^{4})	321	271	214	183	153	131	106	90.9	75.1
r (in.)	3.78	3.84	3.90	3.93	2.96	3.03	3.09	3.12	3.15

Effective length KL (ft)

$F_y = 46$ ksi

COLUMNS
Square structural tubing
Design axial strength in kips ($\phi = 0.85$)

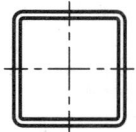

Nominal Size			7×7						6×6			
Thickness	$\frac{5}{8}$	$\frac{1}{2}$	$\frac{3}{8}$	$\frac{5}{16}$	$\frac{1}{4}$	$\frac{3}{16}$	$\frac{5}{8}$	$\frac{1}{2}$	$\frac{3}{8}$	$\frac{5}{16}$	$\frac{1}{4}$	$\frac{3}{16}$
Wt./ft	50.81	42.05	32.58	27.59	22.42	17.08	42.30	35.24	**27.48**	**23.34**	**19.02**	**14.53**
F_y						46 ksi						
Effective length KL (ft) 0	584	485	375	317	258	196	486	407	316	268	219	167
6	553	461	357	302	246	188	451	379	295	251	205	157
7	543	452	351	297	242	185	438	369	288	245	200	153
8	531	443	344	291	237	181	425	358	280	239	195	149
9	518	432	336	285	232	177	410	346	271	231	189	145
10	503	421	327	278	226	173	394	333	262	223	183	140
11	488	409	318	270	220	168	377	320	252	215	176	135
12	472	396	308	262	214	164	359	306	241	206	169	130
13	454	382	298	254	207	159	341	291	230	197	162	124
14	437	368	288	245	200	153	322	276	219	187	154	119
15	418	353	277	236	193	148	303	260	207	178	146	113
16	399	338	265	226	185	142	284	245	195	168	138	107
17	380	322	254	217	177	136	265	229	184	158	131	101
18	361	307	242	207	170	130	246	214	172	148	123	95
19	342	291	230	197	162	124	227	199	160	138	115	89
20	323	276	218	187	154	118	210	184	149	129	107	83
22	285	245	195	167	138	107	175	155	127	111	92	72
24	248	215	172	148	123	95	147	131	107	93	78	61
26	214	187	151	130	108	84	125	111	91	80	67	52
28	184	161	130	113	94	73	108	96	79	69	57	45
30	161	140	113	98	81	63	94	84	69	60	50	39
32	141	123	100	86	72	56	83	73	60	53	44	34
34	125	109	88	76	63	49	73	65	53	47	39	30
35	118	103	83	72	60	47	69	61	50	44	37	29
36	111	97	79	68	57	44		58	48	41	35	27
37	106	92	74	64	54	42			45	39	33	26
38	100	87	71	61	51	40				37	31	24
39	95	83	67	58	48	38						23
40	90	79	64	55	46	36						

Properties

A (in.^{2})	14.9	12.4	9.58	8.11	6.59	5.02	12.4	10.4	8.08	6.86	5.59	4.27
I (in.^{4})	97.5	84.6	68.7	59.5	49.4	38.5	57.3	50.5	41.6	36.3	30.3	23.8
r (in.)	2.56	2.62	2.68	2.71	2.74	2.77	2.15	2.21	2.27	2.30	2.33	2.36

Note: Heavy line indicates Kl / r of 200.

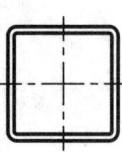

F

F_y = 46 ksi

COLUMNS
Square structural tubing
Design axial strength in kips ($\phi = 0.85$)

Nominal Size	5½×5½				5×5					
Thickness	$\frac{3}{8}$	$\frac{5}{16}$	$\frac{1}{4}$	$\frac{3}{16}$	$\frac{1}{2}$	$\frac{3}{8}$	$\frac{5}{16}$	$\frac{1}{4}$	$\frac{3}{16}$	$\frac{1}{8}$
Wt./ft	24.93	21.21	17.32	13.25	28.43	22.37	19.08	15.62	11.97	8.16
F_y	46 ksi									
0	286	244	199	152	327	257	219	179	138	94
6	264	225	184	141	294	233	199	163	126	86
7	256	219	179	137	282	224	192	158	121	83
8	248	212	173	133	270	215	184	152	117	80
9	238	204	167	129	257	205	176	145	112	77
10	228	195	161	124	242	194	167	138	107	73
11	218	187	154	118	228	183	158	131	101	70
12	207	177	146	113	213	172	148	123	95	66
13	195	168	139	107	197	160	139	115	89	62
14	184	158	131	101	182	149	129	107	84	58
15	172	148	123	95	167	137	119	99	78	54
16	160	139	115	89	152	126	110	92	72	50
17	149	129	107	83	138	115	100	84	66	46
18	137	119	99	77	124	104	91	77	60	42
19	126	110	92	72	111	93	82	69	55	38
20	115	101	84	66	100	84	74	63	50	35
22	96	84	70	55	83	70	61	52	41	29
24	80	70	59	47	70	59	52	44	34	24
26	68	60	50	40	59	50	44	37	29	21
28	59	52	43	34	51	43	38	32	25	18
30	51	45	38	30	45	37	33	28	22	15
31	48	42	35	28		35	31	26	21	15
32	45	40	33	26				24	19	14
33	43	37	31	25						13
34	40	35	29	23						
35			28	22						
Properties										
A (in^2)	7.33	6.23	5.09	3.89	8.36	6.58	5.61	4.59	3.52	2.40
I (in.^{4})	31.2	27.4	23.0	18.1	27.0	22.8	20.1	16.9	13.4	9.41
r (in.)	2.07	2.10	2.13	2.16	1.80	1.86	1.89	1.92	1.95	1.98

Effective length KL (ft)

Note: Heavy line indicates Kl / r of 200.

F_y = 46 ksi

COLUMNS
Square structural tubing
Design axial strength in kips (ϕ = 0.85)

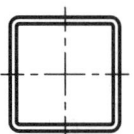

Nominal Size		$4\frac{1}{2}\times4\frac{1}{2}$					4×4					
Thickness		$\frac{3}{8}$	$\frac{5}{16}$	$\frac{1}{4}$	$\frac{3}{16}$	$\frac{1}{8}$	$\frac{1}{2}$	$\frac{3}{8}$	$\frac{5}{16}$	$\frac{1}{4}$	$\frac{3}{16}$	$\frac{1}{8}$
Wt./ft		19.82	16.96	13.91	10.70	7.31	21.63	17.27	14.83	12.21	9.42	6.46
F_y							46 ksi					
Effective length KL (ft)	0	228	195	160	123	84	249	199	170	140	108	74
	6	201	172	142	110	75	208	168	145	120	93	64
	7	192	165	136	105	72	195	158	137	114	89	61
	8	182	157	130	100	69	180	148	128	107	83	58
	9	171	148	123	95	65	166	137	119	100	78	54
	10	160	139	115	90	62	151	125	110	92	72	50
	11	149	129	107	84	58	136	114	100	84	66	46
	12	137	119	100	78	54	121	102	90	76	60	42
	13	125	110	92	72	50	107	91	81	68	54	38
	14	114	100	84	66	46	93	81	72	61	49	34
	15	103	91	76	60	42	81	70	63	54	43	31
	16	92	81	69	55	38	71	62	55	47	38	27
	17	82	73	62	49	35	63	55	49	42	34	24
	18	73	65	55	44	31	56	49	44	37	30	21
	19	66	58	49	39	28	50	44	39	34	27	19
	20	59	53	45	36	25	46	40	35	30	24	17
	21	54	48	41	32	23	41	36	32	28	22	16
	22	49	43	37	29	21	38	33	29	25	20	14
	23	45	40	34	27	19	34	30	27	23	18	13
	24	41	36	31	25	17		27	25	21	17	12
	25	38	34	29	23	16				19	16	11
	26	35	31	26	21	15						10
	27	32	29	25	20	14						
	28		27	23	18	13						
	29				17	12						
Properties												
A (in.^{2})		5.83	4.98	4.09	3.14	2.15	6.36	5.08	4.36	3.59	2.77	1.90
I (in.^{4})		16.0	14.2	12.1	9.60	6.78	12.3	10.7	9.58	8.22	6.59	4.70
r (in.)		1.66	1.69	1.72	1.75	1.78	1.39	1.45	1.48	1.51	1.54	1.57

Note: Heavy line indicates Kl / r of 200.

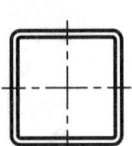

$F_y = 46$ ksi

COLUMNS
Square structural tubing
Design axial strength in kips ($\phi = 0.85$)

Nominal Size		3½×3½				3×3			
Thickness		$^5/_{16}$	$^1/_4$	$^3/_{16}$	$^1/_8$	$^5/_{16}$	$^1/_4$	$^3/_{16}$	$^1/_8$
Wt./ft		12.7	10.51	8.15	5.61	10.58	8.81	6.87	4.75
F_y					46 ksi				
	0	146	121	93	65	122	101	79	55
	6	118	99	77	54	90	76	60	42
	7	109	92	72	50	80	68	54	39
	8	100	84	66	46	71	61	49	35
	9	90	76	60	42	61	53	43	31
	10	81	69	54	39	52	45	37	27
	11	71	61	49	35	44	38	32	23
	12	62	54	43	31	37	32	27	20
	13	54	46	38	27	31	27	23	17
	14	46	40	32	23	27	24	19	14
	15	40	35	28	20	23	21	17	12
Effective length KL (ft)	16	35	31	25	18	21	18	15	11
	17	31	27	22	16	18	16	13	10
	18	28	24	20	14		14	12	9
	19	25	22	18	13				8
	20	23	20	16	11				
	21	21	18	14	10				
	22			13	9				
Properties									
A (in²)		3.73	3.09	2.39	1.65	3.11	2.59	2.02	1.40
I (in.⁴)		6.09	5.29	4.29	3.09	3.58	3.16	2.60	1.90
r (in.)		1.28	1.31	1.34	1.37	1.07	1.10	1.13	1.16

F_y = 46 ksi									

COLUMNS
Rectangular structural tubing
Design axial strength in kips (ϕ = 0.85)

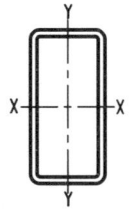

Nominal Size	16×12	16×8	14×12		14×10		12×10			
Thickness	½	½	½	⅜	½	⅜	½	⅜	⁵⁄₁₆	¼
Wt./ft	89.68	76.07	82.88	63.21	76.07	58.10	69.27	53.00	44.60	36.03
F_y	46 ksi									
0	1030	876	952	726	876	669	796	609	513	414
6	1020	848	938	715	857	655	778	596	502	405
7	1010	838	933	712	850	650	772	591	498	402
8	1010	827	927	707	843	644	765	586	493	399
9	998	815	920	702	834	638	757	580	488	395
10	990	801	912	697	825	631	748	573	483	390
11	982	786	904	691	815	623	738	566	477	386
12	973	771	895	684	803	615	728	558	470	380
13	963	754	886	677	791	606	716	550	463	375
14	952	736	876	669	779	597	704	541	456	369
15	941	717	865	661	765	587	692	531	448	363
16	929	697	854	653	751	576	679	522	440	356
17	916	677	842	644	737	565	665	511	431	349
18	903	657	829	634	721	554	650	501	422	342
19	889	635	816	625	705	542	636	489	413	335
20	875	614	803	615	689	530	620	478	404	327
22	845	569	774	593	655	504	589	454	384	311
24	813	525	744	571	620	478	556	430	363	295
26	781	480	713	548	584	451	522	404	342	278
28	746	436	681	524	547	424	488	379	321	261
30	711	393	648	499	511	396	454	353	300	244
32	676	352	615	474	474	368	420	328	278	227
34	640	313	581	448	438	341	387	302	257	210
36	604	279	547	423	403	315	355	278	237	193
38	568	250	514	398	369	289	324	254	217	177
40	533	226	480	372	335	264	293	231	197	162

Effective length KL (ft) with respect to least radius of gyration

Properties

A (in.^{2})	26.4	22.4	24.4	18.6	22.4	17.1	20.4	15.6	13.1	10.6
I_x (in.^{4})	962	722	699	546	608	476	419	330	281	230
I_y (in.^{4})	618	244	552	431	361	284	316	249	213	174
r_x / r_y	1.25	1.72	1.13	1.13	1.30	1.29	1.15	1.15	1.15	1.15
r_y (in.)	4.84	3.30	4.76	4.82	4.02	4.08	3.94	4.00	4.03	4.06

Note: Heavy line indicates Kl / r of 200.

	F_y = 46 ksi

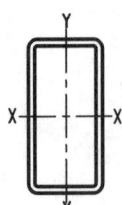

COLUMNS
Rectangular structural tubing
Design axial strength in kips (ϕ = 0.85)

Nominal Size		12×8				12×6		
Thickness	⅝	½	⅜	⁵⁄₁₆	⅝	½	⅜	⁵⁄₁₆
Wt./ft	76.33	62.46	47.90	40.35	67.82	55.66	42.79	36.10
F_y				46 ksi				
0	876	719	551	465	778	641	493	414
6	845	695	534	450	731	604	466	392
7	835	687	527	445	715	591	456	384
8	822	677	520	439	697	577	445	376
9	809	666	512	433	677	561	434	366
10	794	655	503	425	655	543	421	355
11	778	642	494	417	632	525	407	344
12	760	628	483	409	607	505	393	332
13	742	613	473	400	581	485	378	320
14	722	598	461	390	555	464	362	307
15	702	582	449	380	528	442	346	293
16	681	565	437	370	500	420	329	280
17	659	547	424	359	473	398	313	266
18	637	530	410	348	445	375	296	252
19	614	511	397	336	418	353	279	238
20	591	493	383	325	390	331	262	224
22	544	455	355	301	338	288	230	197
24	497	417	326	277	288	247	199	171
26	451	380	298	253	245	211	170	146
28	405	343	270	230	211	182	146	126
30	362	307	243	207	184	158	128	110
32	320	273	217	185	162	139	112	97
34	283	241	192	164	143	123	99	86
36	252	215	171	146	128	110	89	76
38	227	193	154	131	115	99	80	69
39	215	184	146	125	109	94	75	65
40	205	174	139	119		89	72	62

Effective length KL (ft) with respect to least radius of gyration

Properties								
A (in.^{2})	22.4	18.4	14.1	11.9	19.9	16.4	12.6	10.6
I_x (in.^{4})	418	353	279	239	337	287	228	196
I_y (in.^{4})	221	188	149	128	112	96.0	77.2	66.6
r_x / r_y	1.38	1.37	1.37	1.37	1.73	1.73	1.72	1.71
r_y (in.)	3.14	3.20	3.26	3.28	2.37	2.42	2.48	2.51

Note: Heavy line indicates Kl / r of 200.

| F_y = 46 ksi | |

COLUMNS
Rectangular structural tubing
Design axial strength in kips (ϕ = 0.85)

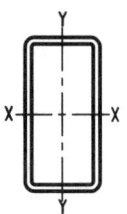

Nominal Size		10×8				10×6			
Thickness		½	⅜	5⁄16	¼	½	⅜	5⁄16	¼
Wt./ft		55.66	42.79	36.10	29.23	48.85	37.69	31.84	25.82
F_y		46 ksi							
Effective length KL (ft) with respect to least radius of gyration	0	641	493	414	336	563	434	366	297
	6	619	476	401	325	529	409	345	281
	7	611	470	396	321	517	400	338	275
	8	602	463	390	317	504	391	330	269
	9	592	456	384	312	490	380	321	261
	10	581	448	377	306	474	368	312	254
	11	568	439	370	300	457	356	302	246
	12	556	429	362	294	439	343	291	237
	13	542	419	354	287	421	329	279	228
	14	528	408	345	280	402	315	267	218
	15	513	397	335	273	382	300	255	209
	16	497	386	326	265	362	285	243	199
	17	481	374	316	257	342	270	230	189
	18	465	361	306	249	322	255	218	179
	19	448	349	295	241	302	240	205	169
	20	431	336	285	232	282	225	193	159
	22	396	310	263	215	244	196	169	139
	24	361	284	241	197	208	169	146	121
	26	327	258	220	180	177	144	124	103
	28	294	232	198	163	153	124	107	89
	30	262	208	178	146	133	108	93	77
	32	231	184	158	130	117	95	82	68
	34	205	163	140	116	104	84	73	60
	36	183	146	125	103	92	75	65	54
	38	164	131	112	93	83	67	58	48
	39	156	124	106	88	79	64	55	46
	40	148	118	101	84		61	52	44

Properties

		10×8				10×6			
A (in.^{2})		16.4	12.6	10.6	8.59	14.4	11.1	9.36	7.59
I_x (in.^{4})		226	180	154	127	181	145	125	103
I_y (in.^{4})		160	127	109	90.2	80.8	65.4	56.5	46.9
r_x / r_y		1.19	1.19	1.19	1.19	1.50	1.49	1.48	1.48
r_y (in.)		3.12	3.18	3.21	3.24	2.37	2.43	2.46	2.49

Note: Heavy line indicates Kl / r of 200.

$F_y = 46$ ksi

COLUMNS
Rectangular structural tubing
Design axial strength in kips ($\phi = 0.85$)

Nominal Size		10×5			8×6		
Thickness	$3/8$	$5/16$	$1/4$	$1/2$	$3/8$	$5/16$	$1/4$
Wt./ft	35.13	29.72	24.12	42.05	32.58	27.59	22.42
F_y				46 ksi			
0	403	341	277	485	375	317	258
6	370	315	256	454	352	298	243
7	359	306	249	444	344	292	238
8	347	295	241	432	335	284	232
9	334	284	232	419	325	276	225
10	319	272	222	404	315	268	218
11	304	260	212	389	303	258	211
12	288	246	201	373	292	248	203
13	272	233	191	357	279	238	195
14	255	219	179	340	266	227	186
15	239	205	168	322	253	217	178
16	222	191	157	305	240	205	169
17	206	178	146	287	227	194	160
18	189	164	135	269	213	183	151
19	174	151	124	252	200	172	142
20	159	138	114	235	187	161	133
22	131	115	95	201	161	140	116
24	110	96	80	170	137	119	99
26	94	82	68	145	117	102	85
28	81	71	59	125	101	88	73
30	71	62	51	109	88	76	64
32	62	54	45	96	77	67	56
34	55	48	40	85	68	59	49
36				76	61	53	44
38				68	55	48	40
39					52	45	38
40							36

Effective length KL (ft) with respect to least radius of gyration

Properties

A (in^2)	10.3	8.73	7.09	12.4	9.58	8.11	6.59
I_x (in.^{4})	128	110	91.2	103	83.7	72.4	60.1
I_y (in.^{4})	42.9	37.2	31.1	65.7	53.5	46.4	38.6
r_x / r_y	1.72	1.71	1.72	1.25	1.25	1.25	1.25
r_y (in.)	2.04	2.07	2.09	2.31	2.36	2.39	2.42

Note: Heavy line indicates Kl / r of 200.

F_y = 46 ksi										

COLUMNS
Rectangular structural tubing
Design axial strength in kips (ϕ = 0.85)

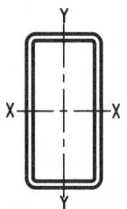

Nominal Size		8×4					7×5				
Thickness		5/8	1/2	3/8	5/16	1/4	1/2	3/8	5/16	1/4	3/16
Wt./ft		42.30	35.24	27.48	23.34	19.02	35.24	27.48	23.34	19.02	14.53
F_y		46 ksi									
	0	486	407	316	268	219	407	316	268	219	167
	6	415	351	276	235	192	369	288	245	200	154
	7	393	333	262	224	184	357	279	238	194	149
	8	368	313	248	212	174	342	268	229	187	144
	9	341	292	233	199	164	327	257	220	180	138
	10	314	270	216	185	153	311	245	210	172	132
	11	287	248	200	172	142	294	232	199	164	126
	12	259	226	183	158	131	276	219	188	155	119
	13	233	204	167	144	120	258	205	177	146	113
	14	207	183	150	130	109	240	192	165	137	106
	15	182	162	135	117	98	222	178	154	127	99
	16	160	143	120	104	88	205	165	142	118	92
	17	141	126	106	92	78	187	151	131	109	85
	18	126	113	95	82	70	170	138	120	101	79
	19	113	101	85	74	62	154	126	110	92	72
	20	102	91	77	67	56	139	114	100	84	66
	22	84	76	63	55	47	115	94	82	69	54
	24	71	63	53	46	39	97	79	69	58	46
	25		58	49	43	36	89	73	64	54	42
	26			45	39	33	82	67	59	50	39
	27				37	31	76	62	55	46	36
	28						71	58	51	43	34
	29						66	54	47	40	31
	30						62	51	44	37	29
	31						58	47	41	35	27
	32							44	39	33	26
	33								37	31	24
	34										23
Properties											
A (in.^{2})		12.4	10.4	8.08	6.86	5.59	10.40	8.08	6.86	5.59	4.27
I_x (in.^{4})		85.1	75.1	61.9	53.9	45.1	63.5	52.2	45.5	38.0	29.8
I_y (in.^{4})		27.4	24.6	20.6	18.1	15.3	37.2	30.8	26.9	22.6	17.7
r_x / r_y		1.76	1.75	1.73	1.73	1.72	1.31	1.30	1.30	1.30	1.29
r_y (in.)		1.49	1.54	1.60	1.62	1.65	1.90	1.95	1.98	2.01	2.04
Note: Heavy line indicates Kl / r of 200.											

(left axis label) Effective length KL (ft) with respect to least radius of gyration

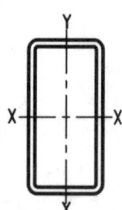

| | | | **Fy = 46 ksi** |

COLUMNS
Rectangular structural tubing
Design axial strength in kips ($\phi = 0.85$)

Nominal Size	7×4				6×4				
Thickness	³⁄₈	⁵⁄₁₆	¼	³⁄₁₆	½	³⁄₈	⁵⁄₁₆	¼	³⁄₁₆
Wt./ft	24.93	21.21	17.32	13.25	28.43	22.37	19.08	15.62	11.97
F_y	46 ksi								
0	287	244	199	152	327	257	219	179	138
6	249	213	175	134	279	222	190	157	121
7	236	202	166	128	263	211	181	149	115
8	223	191	158	121	246	198	171	141	109
9	208	179	148	114	228	185	160	132	102
10	193	167	138	107	210	171	148	123	96
11	178	154	128	99	191	157	136	114	89
12	163	141	118	92	173	143	125	104	81
13	148	129	107	84	155	129	113	95	74
14	133	116	97	76	137	116	102	85	67
15	118	104	88	69	121	103	91	77	61
16	105	92	78	62	106	90	80	68	54
17	93	82	69	55	94	80	71	60	48
18	83	73	62	49	84	71	63	54	43
19	74	65	56	44	75	64	57	48	38
20	67	59	50	40	68	58	51	44	35
21	61	54	45	36	62	52	46	39	31
22	55	49	41	33	56	48	42	36	29
23	51	45	38	30	51	44	39	33	26
24	46	41	35	28	47	40	36	30	24
25	43	38	32	25		37	33	28	22
26	40	35	30	23			30	26	20
27			27	22					19

Effective length KL (ft) with respect to least radius of gyration

Properties									
A (in²)	7.33	6.23	5.09	3.89	8.36	6.58	5.61	4.59	3.52
I_x (in.⁴)	44.0	38.5	32.3	25.4	35.3	29.7	26.2	22.1	17.4
I_y (in.⁴)	18.1	16.0	13.5	10.7	18.4	15.6	13.8	11.7	9.32
r_x / r_y	1.56	1.56	1.55	1.54	1.39	1.38	1.38	1.37	1.37
r_y (in.)	1.57	1.60	1.63	1.66	1.48	1.54	1.57	1.60	1.63

Note: Heavy line indicates Kl / r of 200.

F_y = 46 ksi

COLUMNS
Rectangular structural tubing
Design axial strength in kips (ϕ = 0.85)

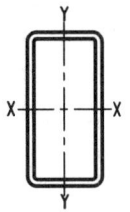

Nominal Size		6×3					5×4			
Thickness		$\frac{1}{2}$	$\frac{3}{8}$	$\frac{5}{16}$	$\frac{1}{4}$	$\frac{3}{16}$	$\frac{3}{8}$	$\frac{5}{16}$	$\frac{1}{4}$	$\frac{3}{16}$
Wt./ft		25.03	19.82	16.96	13.91	10.70	19.82	16.96	13.91	10.70
F_y						46 ksi				
	0	288	228	195	160	123	228	195	160	123
	6	216	176	152	126	98	195	168	139	107
	7	194	160	138	116	90	185	159	132	102
	8	172	144	125	105	82	173	149	124	96
	9	150	127	111	94	74	161	139	116	90
	10	129	111	97	83	65	148	129	107	84
	11	109	95	84	72	57	135	118	99	77
	12	92	81	71	62	50	123	107	90	71
	13	78	69	61	53	42	110	97	82	64
	14	67	59	52	45	36	98	87	73	58
	15	59	52	46	39	32	86	77	65	52
	16	52	45	40	35	28	76	67	58	46
	17	46	40	36	31	25	67	60	51	41
	18	41	36	32	27	22	60	53	46	36
	19		32	28	25	20	54	48	41	33
	20				22	18	49	43	37	29
	21						44	39	33	27
	22						40	36	30	24
	23						37	33	28	22
	24						34	30	26	20
	25						31	28	24	19
	26								22	17

Effective length KL (ft) with respect to least radius of gyration

Properties										
A (in.^{2})		7.36	5.83	4.98	4.09	3.14	5.83	4.98	4.09	3.14
I_x (in.^{4})		27.7	23.8	21.1	17.9	14.3	18.7	16.6	14.1	11.2
I_y (in.^{4})		8.91	7.78	6.98	6.00	4.83	13.2	11.7	9.98	7.96
r_x / r_y		1.76	1.74	1.75	1.73	1.72	1.19	1.20	1.19	1.19
r_y (in.)		1.10	1.16	1.18	1.21	1.24	1.50	1.53	1.56	1.59

Note: Heavy line indicates Kl / r of 200.

	F_y = 46 ksi

COLUMNS
Rectangular structural tubing
Design axial strength in kips ($\phi = 0.85$)

Nominal Size				5×3					4×3		
Thickness		½	⅜	⁵⁄₁₆	¼	³⁄₁₆	⅛	⁵⁄₁₆	¼	³⁄₁₆	⅛
Wt./ft		21.63	17.27	14.83	12.21	9.42	6.46	12.70	10.51	8.15	5.61
F_y							46 ksi				
	0	249	199	170	140	108	74	146	121	93	64
	6	183	151	132	110	85	59	110	93	73	51
	7	164	137	120	100	78	55	100	84	66	47
	8	145	122	108	91	71	50	89	76	60	42
	9	125	107	95	81	63	45	78	67	53	38
	10	107	93	83	71	56	40	67	58	47	33
	11	89	79	71	61	49	35	57	50	40	29
	12	75	67	60	52	42	30	48	42	34	25
	13	64	57	51	45	36	26	41	36	29	21
	14	55	49	44	38	31	22	35	31	25	18
	15	48	43	39	33	27	19	31	27	22	16
	16	42	38	34	29	23	17	27	24	19	14
	17	37	33	30	26	21	15	24	21	17	12
	18		30	27	23	19	13	21	19	15	11
	19			24	21	17	12		17	14	10
	20					15	11				9

Effective length KL (ft) with respect to least radius of gyration

Properties											
A (in.^{2})		6.36	5.08	4.36	3.59	2.77	1.90	3.73	3.09	2.39	1.65
I_x (in.^{4})		16.9	14.7	13.2	11.3	9.06	6.44	7.45	6.45	5.23	3.76
I_y (in.^{4})		7.33	6.48	5.85	5.05	4.08	2.93	4.71	4.10	3.34	2.41
r_x / r_y		1.52	1.50	1.50	1.49	1.50	1.48	1.26	1.26	1.25	1.25
r_y (in.)		1.07	1.13	1.16	1.19	1.21	1.24	1.12	1.15	1.18	1.21

Note: Heavy line indicates Kl/r of 200.

Double Angles and WT Shapes

Double Angles

Design strengths are tabulated for the effective length KL in feet with respect to both the X-X and Y-Y axes. Design strengths about the X-X axis are in accordance with LRFD Specification Section E2. For buckling about the Y-Y axis the shear deformation of the connectors may require the slenderness to be increased in accordance with the equations for $(Kl / r)_m$ in Section E4. Incorporating this slenderness ratio, the design strengths are determined from Section E2 or E3, whichever governs. In addition to the usual limit state of flexural buckling for columns, double angle and WT shapes in compression may also be governed by the limit state of flexural-torsional buckling, in accordance with Section E3 of the LRFD Specification. This has been included in the tables. Discussion under Section C2 of the LRFD Specification Commentary points out that for trusses it is usual practice to take $K = 1.0$. No values are listed beyond $KL / r = 200$.

For buckling about the X-X axis, both angles move parallel so that the design strength is not affected by the connectors. For buckling about the Y-Y axis, the design strengths are tabulated for the indicated number n of intermediate connectors. For connectors with snug-tight bolts or different spacings, the design strength must be recalculated using the corresponding modified slenderness and LRFD Specification Section E4. The number of intermediate connectors given in the table was selected so the design strength about the Y-Y axis is 90 percent or greater of that for buckling of the two angles acting as a unit. If fewer connectors are used, the strength must be reduced accordingly. According to Section E4 of the LRFD Specification, the connectors must be spaced so that the slenderness ratio a / r_z of the individual angle does not exceed 75 percent of the governing slenderness ratio of the built-up member.

In designing members fabricated of two angles connected to opposite faces of a gusset plate, Chapter J of the LRFD Specification states that eccentricity between the gage lines and gravity axis may be neglected. In the following tables, this eccentricity is neglected.

The tabulated loads for double angles referred to in the Y-Y axis assume a ⅜-in. spacing between angles. These values are conservative when a wider spacing is provided. Example 3-5 illustrates a method for determining the design strength when a ¾-in. gusset plate is used.

Examples 3-6 and 3-7 demonstrate how to determine the number of connectors when Kl_x / r_x governs and when the modified $(Kl_y / r_y)_m$ governs.

EXAMPLE 3-5

Given: Using 50 ksi steel, determine the design strength with respect to the Y-Y axis of a double angle member of 8×8×1 angles with an effective length equal to 12 ft, and connected to a ¾-in. thick gusset plate.

Solution: r_y = 3.53 in. (from Double Angle Column Design Strength Table for two L8×8×1 with ⅜-in. plate)

r_y' = 3.67 in. (from Part 1, Properties, Two Equal-Leg Angles, two L8×8×1 with ¾-in. plate)

$$\frac{r_y}{r_y'} = \frac{3.53}{3.67} = 0.962$$

Equivalent effective length = 0.962×12 ft = 11.5 ft

Enter Column Design Strength Table for two L8×8×1 with reference to Y-Y axis for effective lengths between 10 and 15 feet, read 1,120 and 1,000 kips, respectively.

$$\text{Equivalent design strength} = 1{,}120 - \left[(1{,}120 - 1{,}000) \times \frac{11.5 - 10}{15 - 10} \right]$$

$$= 1{,}084 \text{ kips}$$

EXAMPLE 3-6

Given:

Using a double angle member of 5×3×½ angles (short legs back to back) and 36 ksi steel, with $L_x = 10$ ft and $L_y = 20$ ft, and a factored axial load of 70 kips, determine the number of connectors required. Assume $K = 1.0$ and that the intermediate connectors are snug-tight bolted.

Solution:

$K_x L_x = 10$ ft, $K_x l_x / r_x = (10 \times 12) / 0.829 = 145$
$K_y L_y = 20$ ft, $K_y l_y / r_y = (20 \times 12) / 2.5 = 96$

The X-X axis governs. From the X-X axis portion of the table

$\phi P_n = 76$ kips > 70 kips   **o.k.**

Find number of connectors required based on Section E4:

$a / r_z \le 0.75 K L_x / r_x$
$a \quad \le 0.75 (K L_x / r_x) r_z = 0.75 \,(145)\, 0.648 = 70$ in.

Assume two connectors are required; $a = (10 \times 12) / 3 = 40$ in.

$a / r_z = 40 / r_z = 40.0 / 0.648 = 61.7$

Check that modified $(K_y l_y / r_y)_m$ **does not govern.**

According to Specification Equation E4-1,

$(K_y l_y / r_y)_m = \sqrt{96^2 + 61.7^2} = 114$

Modified $l_y' = 114 r_y / K_y = 114\,(2.50 \text{ in.}) / 1.0 = 285$ in. = 23.8 ft

Inspection of the tables indicates that $K_x l_x / r_x$ still governs, therefore one connector is required every 40 inches.

EXAMPLE 3-7

Given:

Using the same steel shape and bolts as Example 3-6, with $L_x = 10$ ft and $L_y = 30$ ft, determine the number of connectors required and the corresponding maximum design strength. Assume $K = 1.0$.

Solution:

$$K_x L_x = 10 \text{ ft}, K_x l_x / r_x = (10 \times 12) / 0.829 = 145$$
$$K_y L_y = 30 \text{ ft}, K_y l_y / r_y = (30 \times 12) / 2.5 = 144$$

$K_x l_x / r_x$ appears to govern, so try one connector in the 10-ft length. Check $(K_y l_y / r_y)_m$ with $a / r_z = 5 \times 12 / 0.648 = 93$

$$(K_y l_y / r_y)_m = \sqrt{144^2 + 93^2} = 171$$

Since $(K_y l_y / r_y)_m$ governs, the Y-Y portion of the table gives a design strength of 72 kips provided four connectors are used in the 30-ft length. This gives a spacing of 30 ft / 5 = 6.0 ft. Check if $(K_y l_y / r_y)_m$ governs with

$$a / r_z = (6.0 \times 12) / 0.648 = 111$$
$$(K_y l_y / r_y)_m = \sqrt{144^2 + 111^2} = 182$$

$(K_y l_y / r_y)_m$ still governs, so four connectors at 6.0 ft would be appropriate.

Verify that $a / r_z < 0.75$ governing Kl / r :

$$111 < (0.75 \times 182 = 137) \quad \textbf{o.k.}$$

Modified $l_y' = 182 r_y / K_y = 182(2.5 \text{ in.}) / 1.0 = 455 \text{ in.} = 37.9 \text{ ft}$

From the tables, the design strength is 45 kips.

The design strength can be increased by closer spacing of the connectors, which reduces $(K_y l_y / r_y)_m$.

| F_y = 36 ksi |
| F_y = 50 ksi |

COLUMNS
Double angles
Design axial strength in kips (ϕ = 0.85)
Equal legs
$3/8$ in. back to back of angles

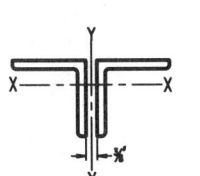

Size		8×8											No. of Connectors[a]
Thickness		$1\frac{1}{8}$		1		$7/8$		$3/4$		$5/8$		$1/2$	
Wt./ft		113.8		102.0		90.0		77.8		65.4		52.8	
F_y		36	50	36	50	36	50	36	50	36	50	36	50	

Effective length KL (ft) with respect to indicated axis

X-X AXIS

KL	36	50	36	50	36	50	36	50	36	50	36	50
0	1030	1420	918	1280	811	1130	701	973	586	763	432	550
10	901	1190	808	1070	715	945	619	819	518	651	387	478
14	795	1000	715	902	633	799	549	694	461	559	348	417
18	674	795	608	719	539	638	469	556	395	456	302	349
22	548	596	496	542	440	482	384	422	325	354	253	278
26	427	430	388	391	345	349	303	306	257	261	205	212
30	323	323	294	294	262	262	230	230	196	196	159	159
34	251	251	229	229	204	204	179	179	153	153	124	124
38	201	201	183	183	163	163	143	143	122	122	99	99
39	191	191	174	174	155	155	136	136	116	116	94	94
40	182	182	165	165	147	147	129	129	110	110	90	90
41							123	123	105	105	85	85

No. of Connectors: b

Y-Y AXIS

KL	36	50	36	50	36	50	36	50	36	50	36	50
0	1030	1420	918	1280	811	1130	701	973	586	763	432	550
10	935	1250	829	1100	720	951	607	791	484	587	331	382
15	865	1130	767	995	665	855	562	714	450	539	313	360
20	776	970	689	858	594	734	503	617	406	475	287	327
25	675	800	599	707	514	602	436	508	354	400	257	286
30	568	630	504	558	430	470	366	399	298	321	222	239
35	464	476	411	421	347	353	296	300	243	246	185	190
40	365	365	324	324	271	271	231	231	190	190	149	149
45	289	289	256	256	215	215	184	184	152	152	119	119
50	235	235	208	208	175	175	149	149	124	124	98	98
55	194	194	172	172	145	145	124	124	102	102	81	81
56	187	187	166	166	140	140	119	119	99	99	78	78
57	181	181	160	160	135	135	115	115	96	96	76	76
58	175	175	155	155	130	130	111	111				
59	169	169										

No. of Connectors: 2, 3

Properties of 2 angles—$3/8$ in. back to back						
A (in.²)	33.5	30.0	26.5	22.9	19.2	15.5
r_x (in.)	2.42	2.44	2.45	2.47	2.49	2.50
r_y (in.)	3.55	3.53	3.51	3.49	3.47	3.45

[a]For Y-Y axis, welded or fully tensioned bolted connectors only.
[b]For number of connectors, see double angle column discussion.

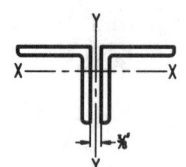

| | | | F_y = 36 ksi |
| | | | F_y = 50 ksi |

COLUMNS
Double angles
Design axial strength in kips (ϕ = 0.85)
Equal legs
$^3/_8$ in. back to back of angles

Size		6×6											No. of Connectors [a]
Thickness		1		$^7/_8$		$^3/_4$		$^5/_8$		$^1/_2$		$^3/_8$	
Wt./ft		74.8		66.2		57.4		48.4		39.2		29.8	
F_y		36	50	36	50	36	50	36	50	36	50	36	50
X-X AXIS	0	673	935	597	829	517	718	435	604	352	470	243	309
	8	580	759	515	675	447	587	377	495	306	389	214	264
	10	533	676	473	601	412	524	347	442	283	351	200	241
	12	481	586	428	522	373	457	315	386	257	308	183	216
	14	426	495	379	441	332	388	280	328	229	265	166	190
	16	370	407	330	364	290	321	245	272	201	222	147	164
	18	315	326	282	292	248	259	210	220	173	182	129	138
	22	218	218	196	196	173	173	147	147	122	122	94	94
	26	156	156	140	140	124	124	105	105	87	87	68	68
	30	117	117	105	105	93	93	79	79	65	65	51	51
	31									61	61	48	48
Y-Y AXIS	0	673	935	597	829	517	718	435	604	352	470	243	309
	10	593	784	521	686	446	584	367	476	283	349	180	207
	12	566	735	497	643	425	548	350	447	271	330	174	199
	14	535	681	470	596	402	508	331	415	256	308	166	190
	16	502	623	440	545	377	465	311	381	240	284	158	179
	18	467	564	409	492	350	420	289	344	223	258	149	167
	20	431	504	377	439	323	375	266	308	206	232	139	154
	22	394	445	344	387	294	330	243	271	187	205	129	140
	24	357	388	311	337	266	287	220	236	169	179	118	125
	26	320	334	279	289	239	247	197	203	151	154	107	111
	28	285	288	248	250	212	213	175	176	134	134	96	97
	30	251	251	218	218	186	186	154	154	117	117	85	85
	32	221	221	192	192	164	164	135	135	103	103	76	76
	34	196	196	170	170	145	145	120	120	92	92	68	68
	36	175	175	152	152	130	130	107	107	82	82	61	61
	38	157	157	136	136	117	117	96	96	74	74	55	55
	40	142	142	123	123	105	105	87	87	67	67	49	49
	42	129	129	112	112	96	96	79	79	61	61	45	45
	43	123	123	107	107	91	91	76	76	58	58	43	43
	44	117	117	102	102	87	87	72	72	55	55		
	45	112	112	97	97								

Effective length KL (ft) with respect to indicated axis

No. of Connectors [a]: b (X-X AXIS), 2, 3 (Y-Y AXIS)

Properties of 2 angles—$^3/_8$ in. back to back

	1	$^7/_8$	$^3/_4$	$^5/_8$	$^1/_2$	$^3/_8$
A (in^2)	22.0	19.5	16.9	14.2	11.5	8.72
r_x (in.)	1.80	1.81	1.83	1.84	1.86	1.88
r_y (in.)	2.73	2.70	2.68	2.66	2.64	2.62

[a]For Y-Y axis, welded or fully tensioned bolted connectors only.
[b]For number of connectors, see double angle column discussion.

| F_y = 36 ksi |
| F_y = 50 ksi |

COLUMNS
Double angles
Design axial strength in kips (ϕ = 0.85)
Equal legs
$3/8$ in. back to back of angles

Size											No. of Connectors [a]
		5×5									
Thickness		$7/8$		$3/4$		$1/2$		$3/8$		$5/16$	
Wt./ft		54.5		47.2		32.4		24.6		20.6	
F_y		36	50	36	50	36	50	36	50	36	50	
X-X AXIS	0	490	680	425	591	291	404	217	282	169	215	
	6	433	573	377	500	259	344	194	244	152	189	
	8	393	502	344	440	237	304	178	219	141	171	
	10	348	423	305	372	211	259	160	189	127	150	
	12	299	343	264	304	183	213	140	159	113	128	
	14	251	268	222	239	155	169	119	129	97	107	b
	16	204	206	182	183	128	130	99	102	82	86	
	18	162	162	145	145	103	103	80	80	68	68	
	20	132	132	117	117	83	83	65	65	55	55	
	22	109	109	97	97	69	69	54	54	46	46	
	24	91	91	82	82	58	58	45	45	38	38	
	25			75	75	53	53	42	42	35	35	
	26							39	39	33	33	
Y-Y AXIS	0	490	680	425	591	291	404	217	282	169	215	
	6	455	614	392	528	257	338	179	216	130	151	
	8	436	580	376	498	247	321	172	206	126	146	
	10	413	538	356	462	234	299	164	195	121	139	
	12	387	491	333	421	219	273	154	181	115	131	
	14	358	440	308	378	203	246	143	165	108	122	
	16	327	388	281	333	185	217	131	147	100	112	
	18	295	337	253	288	167	188	118	130	92	100	2
	20	262	287	225	245	149	160	105	112	83	88	
	22	231	240	198	205	130	134	93	94	74	76	
	24	201	202	172	173	113	113	80	80	65	65	
	26	172	172	147	147	97	97	69	69	56	56	
	28	149	149	127	127	84	84	60	60	49	49	
	30	130	130	111	111	73	73	52	52	43	43	
	32	114	114	98	98	64	64	46	46	38	38	
	34	101	101	86	86	57	57	41	41	34	34	
	36	90	90	77	77	51	51	37	37	30	30	
	37	85	85	73	73	48	48	35	35			
	38	81	81	69	69							3

(Effective length KL (ft) with respect to indicated axis)

Properties of 2 angles—$3/8$ in. back to back					
A (in^2)	16.0	13.9	9.50	7.22	6.05
r_x (in.)	1.49	1.51	1.54	1.56	1.57
r_y (in.)	2.30	2.28	2.24	2.22	2.21

[a] For Y-Y axis, welded or fully tensioned bolted connectors only.
[b] For number of connectors, see double angle column discussion.

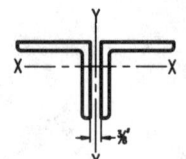

| | F_y = 36 ksi |
| | F_y = 50 ksi |

COLUMNS
Double angles
Design axial strength in kips (ϕ = 0.85)
Equal legs
$3/8$ in. back to back of angles

Size		4×4											No. of Connectors[a]
Thickness													

Size							4×4						No. of Connectors[a]
Thickness		$3/4$		$5/8$		$1/2$		$3/8$		$5/16$		$1/4$	
Wt./ft		37.0		31.4		25.6		19.6		16.4		13.2	
F_y		36	50	36	50	36	50	36	50	36	50	36	50	
	0	334	463	282	392	230	319	175	243	146	191	108	138	
X-X AXIS	4	306	411	259	349	212	285	162	217	135	172	101	126	
	6	275	354	233	301	191	247	146	189	123	152	92	112	
	8	237	288	201	245	166	203	127	156	107	127	82	96	
	10	195	220	167	189	138	157	106	121	90	101	70	78	
	12	154	159	132	137	110	115	85	89	72	76	57	61	b
	14	117	117	100	100	84	84	65	65	56	56	45	46	
	16	89	89	77	77	65	65	50	50	43	43	35	35	
	18	71	71	61	61	51	51	40	40	34	34	28	28	
	19	63	63	54	54	46	46	36	36	30	30	25	25	
	20			49	49	41	41	32	32	27	27	22	22	
	0	334	463	282	392	230	319	175	243	146	191	108	138	
	6	302	404	253	338	203	268	148	192	118	143	82	94	
	8	284	370	238	309	190	245	139	177	112	133	78	90	
	10	261	331	219	276	175	219	129	159	104	122	73	84	
	12	236	288	198	240	158	190	116	139	94	108	68	76	
	14	210	244	175	203	140	161	103	118	84	93	61	68	
Y-Y AXIS	16	183	202	153	168	121	132	90	97	73	78	55	59	
	18	157	162	130	135	103	106	77	78	63	64	48	49	3
	20	132	132	109	109	86	86	64	64	52	52	41	41	
	22	109	109	90	90	71	71	53	53	44	44	34	34	
	24	92	92	76	76	60	60	45	45	37	37	29	29	
	26	78	78	65	65	51	51	38	38	32	32	25	25	
	28	67	67	56	56	44	44	33	33	27	27	22	22	
	29	63	63	52	52	41	41	31	31	25	25	20	20	
	30	59	59	49	49	39	39	29	29	24	24			
	31	55	55	46	46									

Effective length *KL* (ft) with respect to indicated axis

Properties of 2 angles—$3/8$ in. back to back						
A (in.^{2})	10.9	9.22	7.50	5.72	4.80	3.88
r_x (in.)	1.19	1.20	1.22	1.23	1.24	1.25
r_y (in.)	1.88	1.86	1.83	1.81	1.80	1.79

[a]For Y-Y axis, welded or fully tensioned bolted connectors only.
[b]For number of connectors, see double angle column discussion.

F_y = 36 ksi
F_y = 50 ksi

COLUMNS
Double angles
Design axial strength in kips (ϕ = 0.85)
Equal legs
$3/8$ in. back to back of angles

Size		$3\frac{1}{2} \times 3\frac{1}{2}$						No. of Connectors [a]
Thickness		$3/8$		$5/16$		$1/4$		
Wt./ft		17.0		14.4		11.6		
F_y		36	50	36	50	36	50	
X-X AXIS	0	152	211	128	175	100	129	
	2	148	204	125	169	97	125	
	4	137	182	115	152	90	114	
	6	120	152	101	127	80	97	
	8	100	117	84	99	67	77	
	10	78	84	67	72	54	58	b
	12	59	59	50	50	41	41	
	14	43	43	37	37	30	30	
	16	33	33	28	28	23	23	
	17	29	29	25	25	21	21	
	18			22	22	18	18	
Y-Y axis	0	152	211	128	175	100	129	
	6	129	168	105	133	78	92	
	8	119	150	97	120	72	85	
	10	107	130	88	105	66	76	
	12	94	109	77	88	59	66	
	14	80	88	66	72	51	55	
	16	67	68	55	56	43	44	3
	18	54	54	45	45	35	35	
	20	44	44	37	37	29	29	
	22	37	37	30	30	24	24	
	24	31	31	26	26	20	20	
	26	26	26	22	22	17	17	

Effective length KL (ft) with respect to indicated axis

Properties of 2 angles—$3/8$ in. back to back			
A (in^2)	4.97	4.18	3.38
r_x (in.)	1.07	1.08	1.09
r_y (in.)	1.61	1.60	1.59

[a]For Y-Y axis, welded or fully tensioned bolted connectors only.
[b]For number of connectors, see double angle column discussion.

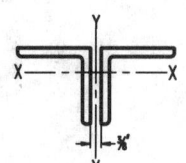

	F_y = 36 ksi
	F_y = 50 ksi

COLUMNS
Double angles
Design axial strength in kips (ϕ = 0.85)
Equal legs
$\frac{3}{8}$ in. back to back of angles

Size		\multicolumn{10}{c}{3×3}	No. of Connectors [a]
Thickness		\multicolumn{2}{c}{½}	\multicolumn{2}{c}{⅜}	\multicolumn{2}{c}{⁵⁄₁₆}	\multicolumn{2}{c}{¼}	\multicolumn{2}{c}{³⁄₁₆}	
Wt./ft		\multicolumn{2}{c}{18.8}	\multicolumn{2}{c}{14.4}	\multicolumn{2}{c}{12.2}	\multicolumn{2}{c}{9.8}	\multicolumn{2}{c}{7.42}	
F_y		36	50	36	50	36	50	36	50	36	50	

Effective length KL (ft) with respect to indicated axis	Axis	KL	36	50	36	50	36	50	36	50	36	50	No. of Connectors
	X-X AXIS	0	168	234	129	179	109	151	88	118	61	77	
		2	162	222	125	171	105	144	85	112	59	74	
		3	155	208	119	160	100	135	81	106	57	71	
		4	145	190	112	147	94	124	77	98	54	66	
		5	133	169	103	131	87	111	71	88	50	60	
		6	120	146	93	114	79	97	64	77	46	54	
		7	106	123	83	97	70	82	57	66	41	47	b
		8	92	101	72	80	61	68	50	56	37	41	
		9	79	81	62	64	53	55	43	46	32	34	
		10	66	66	52	52	45	45	37	37	28	28	
		11	54	54	43	43	37	37	31	31	24	24	
		12	46	46	36	36	31	31	26	26	20	20	
		13	39	39	31	31	26	26	22	22	17	17	
		14	34	34	27	27	23	23	19	19	15	15	
		15			23	23	20	20	16	16	13	13	
	Y-Y AXIS	0	168	234	129	179	109	151	88	118	61	77	
		2	162	222	122	165	100	134	77	96	48	55	
		4	155	208	116	155	96	126	74	92	46	53	
		6	143	187	108	139	89	114	69	84	44	50	
		8	128	160	96	120	79	98	62	74	40	46	
		10	111	132	84	98	69	81	54	62	36	40	
		12	93	104	70	77	58	64	46	50	31	34	3
		14	76	78	57	58	47	48	37	38	26	27	
		16	60	60	45	45	37	37	29	29	21	21	
		18	47	47	35	35	29	29	23	23	17	17	
		20	38	38	29	29	24	24	19	19	14	14	
		22	32	32	24	24	20	20	16	16	12	12	
		23	29	29	22	22	18	18	14	14	11	11	

Properties of 2 angles—⅜ in. back to back

A (in.^{2})	5.50	4.22	3.55	2.88	2.18
r_x (in.)	0.898	0.913	0.922	0.930	0.939
r_y (in.)	1.43	1.41	1.40	1.39	1.38

[a]For Y-Y axis, welded or fully tensioned bolted connectors only.
[b]For number of connectors, see double angle column discussion.

F_y = 36 ksi
F_y = 50 ksi

COLUMNS
Double angles
Design axial strength in kips (ϕ = 0.85)
Equal legs
⅜ in. back to back of angles

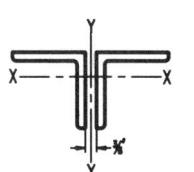

Size		\multicolumn{8}{c}{2½×2½}	No. of Connectors [a]
Thickness		\multicolumn{2}{c}{⅜}	\multicolumn{2}{c}{⁵⁄₁₆}	\multicolumn{2}{c}{¼}	\multicolumn{2}{c}{³⁄₁₆}	
Wt./ft		\multicolumn{2}{c}{11.8}	\multicolumn{2}{c}{10.0}	\multicolumn{2}{c}{8.2}	\multicolumn{2}{c}{6.14}	
F_y		36	50	36	50	36	50	36	50	
X-X AXIS	0	106	147	90	125	73	101	54	70	
	2	101	137	85	116	69	94	52	66	
	3	94	125	80	106	65	86	48	61	
	4	86	110	73	93	59	76	44	54	
	5	76	93	65	79	53	65	40	47	
	6	66	76	56	65	46	53	35	40	b
	7	55	59	47	51	39	42	30	32	
	8	45	46	39	39	32	33	25	25	
	9	36	36	31	31	26	26	20	20	
	10	29	29	25	25	21	21	16	16	
	11	24	24	21	21	17	17	13	13	
	12	20	20	17	17	14	14	11	11	
Y-Y AXIS	0	106	147	90	125	73	101	54	70	
	2	101	137	84	113	66	87	45	54	
	3	98	132	82	109	64	84	44	53	
	4	95	125	79	103	62	80	43	51	
	5	90	117	75	97	59	75	41	48	
	6	85	108	70	89	56	70	39	46	
	7	79	98	66	81	52	64	37	42	
	8	73	88	61	72	48	57	34	39	
	9	66	77	55	64	44	50	31	35	
	10	60	67	50	55	39	44	28	31	3
	11	53	57	44	47	35	37	25	27	
	12	47	48	39	40	31	31	22	23	
	13	41	41	34	34	27	27	20	20	
	14	35	35	29	29	23	23	17	17	
	15	31	31	26	26	20	20	15	15	
	16	27	27	22	22	18	18	13	13	
	17	24	24	20	20	16	16	12	12	
	18	21	21	18	18	14	14	11	11	
	19	19	19	16	16	13	13	9	9	
	20	17	17	14	14					

Effective length KL (ft) with respect to indicated axis

\multicolumn{9}{c}{**Properties of 2 angles—⅜ in. back to back**}
A (in²)	\multicolumn{2}{c}{3.47}	\multicolumn{2}{c}{2.93}	\multicolumn{2}{c}{2.38}	\multicolumn{2}{c}{1.80}
r_x (in.)	\multicolumn{2}{c}{0.753}	\multicolumn{2}{c}{0.761}	\multicolumn{2}{c}{0.769}	\multicolumn{2}{c}{0.778}
r_y (in.)	\multicolumn{2}{c}{1.21}	\multicolumn{2}{c}{1.20}	\multicolumn{2}{c}{1.19}	\multicolumn{2}{c}{1.18}

[a] For Y-Y axis, welded or fully tensioned bolted connectors only.
[b] For number of connectors, see double angle column discussion.

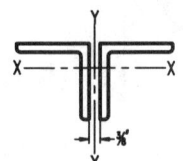

| | F_y = 36 ksi |
| | F_y = 50 ksi |

COLUMNS
Double angles
Design axial strength in kips (ϕ = 0.85)
Equal legs
⅜ in. back to back of angles

Size		2×2										No. of Connectors[a]
Thickness		⅜		⁵⁄₁₆		¼		³⁄₁₆		⅛		
Wt./ft		9.4		7.84		6.38		4.88		3.30		
F_y		36	50	36	50	36	50	36	50	36	50	
Effective length KL (ft) with respect to indicated axis	X-X AXIS											
	0	83	116	70	98	58	80	44	61	27	34	
	2	76	103	65	87	53	71	40	54	25	31	
	3	69	88	58	75	48	62	37	47	23	28	
	4	59	72	50	61	41	51	32	39	20	24	
	5	49	55	42	47	35	39	27	30	17	19	b
	6	38	39	33	34	28	29	21	22	14	15	
	7	29	29	25	25	21	21	16	16	11	11	
	8	22	22	19	19	16	16	13	13	9	9	
	9	18	18	15	15	13	13	10	10	7	7	
	10			12	12	10	10	8	8	6	6	
	Y-Y AXIS											
	0	83	116	70	98	58	80	44	61	27	34	
	2	79	108	66	90	53	71	39	51	21	24	
	3	76	102	64	85	51	68	37	48	20	24	
	4	72	94	60	79	48	63	35	45	20	23	
	5	67	85	56	71	45	57	33	41	19	21	
	6	61	76	51	63	41	50	30	36	17	20	
	7	55	65	46	54	37	44	27	32	16	18	
	8	49	55	41	46	33	37	24	27	14	16	3
	9	43	46	36	38	29	30	21	22	13	14	
	10	37	37	30	31	24	25	18	18	11	11	
	11	31	31	26	26	20	20	15	15	10	10	
	12	26	26	22	22	17	17	13	13	8	8	
	13	22	22	18	18	15	15	11	11	7	7	
	14	19	19	16	16	13	13	9	9	6	6	
	15	17	17	14	14	11	11	8	8	5	5	
	16	15	15	12	12	10	10	7	7	5	5	

Properties of 2 angles—⅜ in. back to back

A (in.^{2})	2.72	2.30	1.88	1.43	0.960
r_x (in.)	0.594	0.601	0.609	0.617	0.626
r_y (in.)	1.01	1.00	0.989	0.977	0.965

[a]For Y-Y axis, welded or fully tensioned bolted connectors only.
[b]For number of connectors, see double angle column discussion.

F_y = 36 ksi

F_y = 50 ksi

COLUMNS
Double angles
Design axial strength in kips (ϕ = 0.85)
Unequal legs
Long legs 3/8 in. back to back of angles

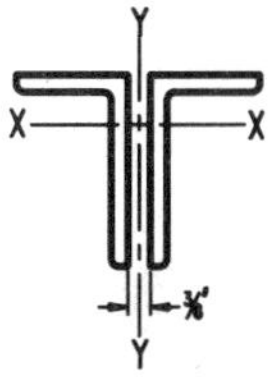

Size		8×6						No. of Connectors[a]		8×4						No. of Connectors[a]
Thickness		1		3/4		1/2				1		3/4		1/2		
Wt./ft		88.4		67.6		46.0				74.8		57.4		39.2		
F_y		36	50	36	50	36	50			36	50	36	50	36	50	
X-X AXIS	0	796	1110	609	846	376	479		0	673	935	517	718	321	408	
	10	704	932	541	717	339	419		10	597	792	460	611	289	358	
	12	667	865	513	667	323	395		12	567	736	437	569	276	338	
	14	626	792	483	613	306	368		14	533	676	412	523	262	315	
	16	582	715	450	555	287	340		16	496	612	384	475	246	292	
	18	535	637	415	496	267	310		18	457	546	354	425	230	267	
	20	488	560	379	438	247	280		20	418	482	324	376	212	242	
	22	440	486	343	381	226	250		22	378	419	294	328	195	216	
	24	393	415	308	328	205	221		24	338	359	264	283	177	192	
	26	348	353	273	279	185	193	b	26	300	306	235	241	160	168	b
	28	305	305	241	241	165	167		28	264	264	207	208	143	146	
	30	265	265	210	210	146	146		30	230	230	181	181	127	127	
	32	233	233	184	184	128	128		32	202	202	159	159	112	112	
	34	207	207	163	163	113	113		34	179	179	141	141	99	99	
	36	184	184	146	146	101	101		36	160	160	126	126	88	88	
	38	165	165	131	131	91	91		38	143	143	113	113	79	79	
	41	142	142	112	112	78	78		40	129	129	102	102	71	71	
	42			107	107	74	74		42	117	117	92	92	65	65	
									43					62	62	
Y-Y AXIS	0	796	1110	609	846	376	479		0	673	935	517	718	321	408	
	6	723	967	529	693	289	339		6	563	731	409	521	226	262	
	8	698	922	511	663	282	329		8	516	650	375	464	210	241	
	10	667	867	490	625	272	317		10	461	557	335	398	192	216	
	12	632	804	464	582	261	301		12	401	460	291	328	170	186	
	14	592	735	435	534	248	284		14	340	366	245	260	147	156	2
	16	548	662	404	483	234	264		16	281	298	201	202	124	125	
	18	503	588	371	430	218	242		18	237	237	161	161	101	101	
	20	457	515	337	377	201	220		20	193	193	132	132	84	84	
	22	410	444	303	326	184	196	2	22	160	160	109	109	70	70	
	24	364	376	269	277	166	173		24	135	135	92	92	59	59	
	26	320	322	237	237	148	150		25	124	124	85	85	55	55	
	28	278	278	206	206	131	131		26	115	115					3
	32	214	214	159	159	102	102									
	34	190	190	141	141	91	91									
	36	169	169	126	126	82	82									
	40	138	138	102	102	67	67									
	41	131	131	98	98											
	42	125	125													

Effective length *KL* (ft) with respect to indicated axis

Properties of 2 angles—3/8 in. back to back

	8×6, 1	8×6, 3/4	8×6, 1/2	8×4, 1	8×4, 3/4	8×4, 1/2
A (in²)	26.0	19.9	13.5	22.0	16.9	11.5
r_x (in.)	2.49	2.53	2.56	2.52	2.55	2.59
r_y (in.)	2.52	2.48	2.44	1.61	1.55	1.51

[a]For Y-Y axis, welded or fully tensioned bolted connectors only.
[b]For number of connectors, see double angle column discussion.

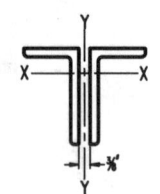

	F_y = 36 ksi
	F_y = 50 ksi

COLUMNS
Double angles
Design axial strength in kips (ϕ = 0.85)
Unequal legs
Long legs ³⁄₈ in. back to back of angles

Properties of 2 angles—³⁄₈ in. back to back

Size	7×4			No. of Connectors [a]	6×4				No. of Connectors [a]
Thickness	³⁄₄	½	³⁄₈		³⁄₄	⅝	½	³⁄₈	
Wt./ft	52.4	35.8	27.2		47.2	40.0	32.4	24.6	

Effective length KL (ft) with respect to indicated axis

X-X AXIS

F_y	36	50	36	50	36	50	Conn		36	50	36	50	36	50	36	50	Conn
0	471	655	310	401	205	254		0	425	591	358	497	291	388	201	256	
8	427	571	283	355	189	230		8	371	488	313	413	254	325	179	220	
10	404	529	268	332	181	218		10	343	439	290	371	236	294	167	202	
12	378	481	252	306	171	204		12	312	385	265	327	216	260	154	182	
14	349	431	233	278	161	188		14	279	329	237	281	193	225	140	161	
16	318	379	214	248	149	172		16	246	276	209	236	171	191	125	140	
18	286	328	194	219	137	155		18	212	225	181	193	148	158	110	119	b
20	255	278	174	190	125	138		20	180	182	155	156	127	128	96	100	
22	224	232	154	162	113	121	b	22	150	150	129	129	106	106	82	82	
24	194	195	135	137	101	105		24	126	126	109	109	89	89	69	69	
26	166	166	117	117	89	90		26	108	108	93	93	76	76	59	59	
28	143	143	100	100	78	78		28	93	93	80	80	65	65	51	51	
30	125	125	88	88	68	68		30	81	81	70	70	57	57	44	44	
32	110	110	77	77	59	59		31	76	76	65	65	53	53	41	41	
34	97	97	68	68	53	53		32							39	39	
36	87	87	61	61	47	47											
37	82	82	58	58	44	44											

Y-Y AXIS

F_y	36	50	36	50	36	50	Conn		36	50	36	50	36	50	36	50	Conn
0	471	655	310	401	205	254		0	425	591	358	497	291	388	201	256	
6	389	503	232	276	139	158		6	364	476	299	387	231	286	148	173	
8	358	450	215	252	132	148		8	336	428	276	349	214	259	139	161	
10	322	389	195	223	122	136		10	304	372	250	304	194	229	128	146	
12	282	324	173	192	111	121		12	268	313	221	256	172	195	116	129	
14	240	261	149	159	99	105		14	231	255	190	209	148	160	102	110	2
16	200	203	125	127	85	88	2	16	194	210	160	164	125	128	88	91	
18	162	162	102	102	72	72		18	159	167	131	131	102	102	74	74	
20	132	132	84	84	60	60		20	136	136	107	107	83	83	61	61	
22	109	109	70	70	50	50		22	112	112	89	89	69	69	51	51	
24	92	92	59	59	43	43		24	95	95	75	75	59	59	43	43	
25	85	85	54	54	40	40		26	81	81	64	64	50	50	37	37	
26	79	79	50	50				27	75	75	59	59	47	47	34	34	
27	73	73						28	70	70							3

Properties of 2 angles—³⁄₈ in. back to back

	³⁄₄	½	³⁄₈		³⁄₄	⅝	½	³⁄₈
A (in²)	15.4	10.5	7.97		13.9	11.7	9.50	7.22
r_x (in.)	2.22	2.25	2.27		1.88	1.90	1.91	1.93
r_y (in.)	1.62	1.57	1.55		1.69	1.67	1.64	1.62

[a] For Y-Y axis, welded or fully tensioned bolted connectors only.
[b] For number of connectors, see double angle column discussion.

Fy = 36 ksi
Fy = 50 ksi

COLUMNS
Double angles
Design axial strength in kips (φ = 0.85)
Unequal legs
Long legs 3/8 in. back to back of angles

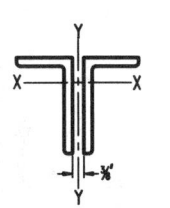

Size		6×3½			No. of Connectors a			5×3½							No. of Connectors a
Thickness		3/8		5/16				3/4		1/2		3/8		5/16	
Wt./ft		23.4		19.6				39.6		27.2		20.8		17.4	
Fy		36	50	36	50			36	50	36	50	36	50	36	50	
X-X AXIS	0	191	243	145	179		0	355	493	245	340	183	238	143	182	
	8	170	209	130	157		4	337	460	233	318	175	224	137	172	
	10	159	192	123	146		6	317	421	219	292	165	208	130	161	
	12	146	173	114	134		8	290	372	202	260	152	187	120	146	
	14	133	153	105	120		10	259	318	181	223	137	163	109	129	
	16	119	133	95	106		12	225	262	158	185	120	138	97	111	
	18	105	114	85	93	b	14	191	209	135	149	104	113	85	93	b
	20	92	95	75	79		16	158	161	113	116	87	90	72	76	
	22	78	79	65	67		18	127	127	91	91	71	71	60	61	
	24	66	66	56	56		20	103	103	74	74	58	58	49	49	
	26	56	56	48	48		22	85	85	61	61	48	48	41	41	
	28	49	49	41	41		24	72	72	51	51	40	40	34	34	
	30	42	42	36	36		25	66	66	47	47	37	37	31	31	
	32	37	37	32	32		26			44	44	34	34	29	29	
Y-Y AXIS	0	191	243	145	179		0	355	493	245	340	183	238	143	182	
	4	143	168	102	115		4	324	434	212	278	148	179	109	128	
	6	135	157	97	109		6	301	393	198	254	139	166	103	120	
	8	124	141	90	101		8	273	343	180	223	127	148	95	109	
	10	110	123	82	90		10	240	287	158	187	112	128	86	96	
	12	96	103	72	78	2	12	205	231	135	151	97	105	75	82	2
	14	80	82	62	65		14	170	187	112	116	81	83	64	67	
	16	65	65	52	52		16	144	144	90	90	65	65	53	53	
	18	52	52	42	42		18	114	114	72	72	52	52	42	42	
	20	43	43	35	35		20	93	93	58	58	42	42	35	35	
	22	36	36	29	29		22	77	77	48	48	35	35	29	29	
	23	33	33	27	27		24	65	65	41	41	30	30	25	25	
							25	60	60							3

Effective length KL (ft) with respect to indicated axis

Properties of 2 angles—3/8 in. back to back

A (in²)	6.84	5.74	11.6	8.00	6.09	5.12
rx (in.)	1.94	1.95	1.55	1.58	1.60	1.61
ry (in.)	1.39	1.38	1.53	1.49	1.46	1.45

a For Y-Y axis, welded or fully tensioned bolted connectors only.
b For number of connectors, see double angle column discussion.

	F_y = 36 ksi
	F_y = 50 ksi

COLUMNS
Double angles
Design axial strength in kips (ϕ = 0.85)
Unequal legs
Long legs ³⁄₈ in. back to back of angles

Size		5×3								No. of Connectors [a]
Thickness		½		³⁄₈		⁵⁄₁₆		¼		
Wt./ft		25.6		19.6		16.4		13.2		
F_y		36	50	36	50	36	50	36	50	
X-X AXIS	0	230	319	172	223	134	170	95	117	
	2	227	313	170	220	132	168	95	115	
	4	219	298	164	210	128	161	92	112	
	6	206	274	155	195	122	151	88	105	
	8	189	244	143	176	113	137	82	97	
	10	170	210	129	154	103	121	76	88	
	12	149	175	114	130	91	104	68	78	b
	14	128	141	98	107	79	88	61	67	
	16	107	110	82	86	68	71	53	56	
	18	87	87	68	68	56	57	45	47	
	20	70	70	55	55	46	46	38	38	
	22	58	58	45	45	38	38	31	31	
	24	49	49	38	38	32	32	26	26	
	26	42	42	32	32	27	27	22	22	
	27							21	21	
Y-Y AXIS	0	230	319	172	223	134	170	95	117	
	2	204	270	141	172	103	121	67	75	
	4	193	250	134	162	99	115	64	72	
	6	174	219	123	145	91	105	61	68	
	8	152	181	108	123	81	92	56	61	
	10	126	141	91	99	70	76	49	53	2
	12	101	103	73	75	58	60	42	44	
	14	80	80	56	56	45	45	35	35	
	16	62	62	44	44	36	36	27	27	
	18	49	49	35	35	28	28	22	22	
	20	40	40	28	28	23	23	18	18	
										3

Effective length *KL* (ft) with respect to indicated axis

Properties of 2 angles—³⁄₈ in. back to back

A (in²)	7.50	5.72	4.80	3.88
r_x (in.)	1.59	1.61	1.61	1.62
r_y (in.)	1.25	1.23	1.22	1.21

[a] For Y-Y axis, welded or fully tensioned bolted connectors only.
[b] For number of connectors, see double angle column discussion.

| F_y = 36 ksi |
| **F_y = 50 ksi** |

COLUMNS
Double angles
Design axial strength in kips (ϕ = 0.85)
Unequal legs
Long legs ³⁄₈ in. back to back of angles

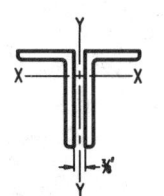

Size		4×3½								No. of Connectors [a]
Thickness		½		³⁄₈		⁵⁄₁₆		¼		
Wt./ft		23.8		18.2		15.4		12.4		
F_y		36	50	36	50	36	50	36	50	
X-X AXIS	0	214	298	163	227	137	179	101	129	
	2	210	289	160	221	134	174	99	126	
	4	198	266	151	204	127	162	94	118	
	6	179	232	137	178	115	143	87	106	
	8	155	191	120	147	101	120	77	91	
	10	130	148	101	116	85	96	66	75	b
	12	104	109	81	86	69	73	55	59	
	14	80	80	63	63	54	54	44	44	
	16	61	61	48	48	41	41	34	34	
	18	48	48	38	38	33	33	27	27	
	20	39	39	31	31	26	26	22	22	
	21					24	24	20	20	
Y-Y AXIS	0	214	298	163	227	137	179	101	129	
	2	202	273	147	195	118	144	80	94	
	4	194	259	142	186	114	139	78	91	
	6	182	237	133	171	107	129	74	86	
	8	166	210	122	152	99	117	70	80	
	10	148	178	109	130	88	102	63	72	
	12	128	146	95	107	77	85	56	62	2
	14	108	115	80	85	65	69	49	52	
	16	88	93	65	66	54	54	41	42	
	18	74	74	52	52	43	43	34	34	
	20	60	60	43	43	35	35	28	28	
	22	50	50	35	35	29	29	23	23	
	24	42	42	30	30	25	25	20	20	
	25	39	39	27	27	23	23	18	18	
	26	36	36	25	25					
										3

Effective length *KL* (ft) with respect to indicated axis

Properties of 2 angles—³⁄₈ in. back to back				
A (in²)	7.00	5.34	4.49	3.63
r_x (in.)	1.23	1.25	1.26	1.27
r_y (in.)	1.58	1.56	1.55	1.54

[a]For Y-Y axis, welded or fully tensioned bolted connectors only.
[b]For number of connectors, see double angle column discussion.

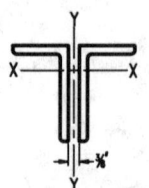

	F_y = 36 ksi
	F_y = 50 ksi

COLUMNS
Double angles
Design axial strength in kips ($\phi = 0.85$)
Unequal legs
Long legs ⅜ in. back to back of angles

Size		4×3								No. of Connectors[a]
Thickness		½		⅜		5⁄16		¼		
Wt./ft		22.2		17.0		14.4		11.6		
F_y		36	50	36	50	36	50	36	50	
X-X AXIS	0	199	276	152	211	127	166	94	120	
	2	195	269	149	206	125	162	93	117	
	4	184	248	141	190	118	151	88	110	
	6	167	217	128	166	108	133	81	99	
	8	146	179	112	138	94	112	72	85	
	10	122	141	94	109	80	90	62	70	b
	12	99	105	76	81	65	69	51	55	
	14	77	77	60	60	51	51	41	42	
	16	59	59	46	46	39	39	32	32	
	18	46	46	36	36	31	31	25	25	
	20	38	38	29	29	25	25	21	21	
	21			27	27	23	23	19	19	
Y-Y AXIS	0	199	276	152	211	127	166	94	120	
	2	185	250	136	179	108	133	74	88	
	4	176	232	129	167	103	125	71	84	
	6	160	205	118	148	95	113	67	77	
	8	141	172	104	125	84	97	60	68	
	10	119	136	88	100	72	79	52	58	
	12	97	107	72	75	59	61	44	47	2
	14	76	80	56	56	46	46	36	36	
	16	61	61	43	43	36	36	28	28	
	18	49	49	34	34	28	28	22	22	
	20	39	39	28	28	23	23	18	18	
	21	36	36	25	25	21	21	17	17	
	22	33	33							3

(left margin: Effective length KL (ft) with respect to indicated axis)

Properties of 2 angles—⅜ in. back to back				
A (in²)	6.50	4.97	4.18	3.38
r_x (in.)	1.25	1.26	1.27	1.28
r_y (in.)	1.33	1.31	1.30	1.29

[a]For Y-Y axis, welded or fully tensioned bolted connectors only.
[b]For number of connectors, see double angle column discussion.

| F_y = 36 ksi |
| F_y = 50 ksi |

COLUMNS
Double angles
Design axial strength in kips (ϕ = 0.85)
Unequal legs
Long legs ⅜ in. back to back of angles

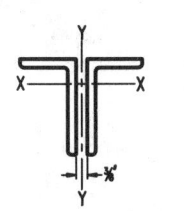

Size		3½×3						3½×2½				
Thickness		⅜		⁵⁄₁₆		¼		No. of Connectors[a]		⅜		¼	No. of Connectors[a]
Wt./ft		15.8		13.2		10.8				14.4		9.8	
F_y		36	50	36	50	36	50			36	50	36	50	
X-X AXIS	0	140	195	118	162	92	119	0		129	179	85	110	
	2	137	188	115	157	90	116	2		126	173	83	107	
	4	127	169	107	141	84	106	4		117	156	77	97	
	6	112	142	95	119	75	91	6		103	131	69	84	
	8	93	111	79	94	63	73	8		86	103	59	68	
	10	74	80	63	69	51	55	b	10	69	75	47	52	b
	12	56	56	48	48	39	40	12		52	53	37	37	
	14	41	41	35	35	29	29	14		39	39	27	27	
	16	32	32	27	27	22	22	16		30	30	21	21	
	17	28	28	24	24	20	20	18		23	23	17	17	
	18	25	25	21	21	18	18							
Y-Y AXIS	0	140	195	118	162	92	119	0		129	179	85	110	
	2	129	173	105	136	77	93	2		117	155	69	83	
	4	123	162	100	128	74	88	4		108	140	65	77	
	6	113	144	92	115	68	81	6		95	118	58	67	
	8	100	122	82	99	61	71	8		79	92	49	55	2
	10	85	99	70	80	53	59	10		62	66	39	42	
	12	70	75	58	62	44	47	2	12	49	49	30	30	
	14	56	59	46	46	36	36	14		36	36	22	22	
	16	45	45	36	36	28	28	16		28	28	17	17	
	18	36	36	28	28	22	22	18		22	22	14	14	
	20	29	29	23	23	18	18							3
	22	24	24	19	19	15	15	3						

Properties of 2 angles—⅜ in. back to back												
A (in²)		4.59		3.87		3.13				4.22		2.88	
r_x (in.)		1.09		1.10		1.11				1.10		1.12	
r_y (in.)		1.36		1.35		1.33				1.11		1.09	

[a]For Y-Y axis, welded or fully tensioned bolted connectors only.
[b]For number of connectors, see double angle column discussion.

Effective length KL (ft) with respect to indicated axis

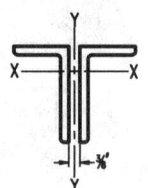

F_y = 36 ksi
F_y = 50 ksi

COLUMNS
Double angles
Design axial strength in kips (ϕ = 0.85)
Unequal legs
Long legs ⅜ in. back to back of angles

Size	3×2½						No. of Connectors [a]		3×2								No. of Connectors [a]
Thickness	⅜		¼		³⁄₁₆				⅜		⁵⁄₁₆		¼		³⁄₁₆		
Wt./ft	13.2		9.0		6.77				11.8		10.0		8.2		6.1		
F_y	36	50	36	50	36	50			36	50	36	50	36	50	36	50	
X-X AXIS																	
0	118	163	80	107	55	71		0	106	147	90	125	73	97	50	64	
2	113	155	78	103	54	68		2	103	141	87	119	70	93	49	61	
3	109	146	75	97	52	65		3	98	132	83	112	68	88	47	59	
4	102	134	70	90	49	60		4	93	122	78	103	64	81	45	55	
5	94	120	65	81	46	55		5	86	109	73	93	59	74	42	50	
6	86	105	59	71	42	50		6	78	96	66	82	54	65	38	45	
7	76	90	53	62	38	44		7	70	82	59	70	49	57	35	40	
8	67	75	47	52	34	38	b	8	61	69	52	59	43	48	31	35	b
9	58	60	40	43	30	32		9	53	56	45	48	37	40	28	30	
10	49	49	34	35	26	27		10	45	45	39	39	32	32	24	25	
11	40	40	29	29	22	22		11	38	38	32	32	27	27	20	21	
12	34	34	24	24	19	19		12	32	32	27	27	22	22	17	17	
13	29	29	21	21	16	16		13	27	27	23	23	19	19	15	15	
14	25	25	18	18	14	14		14	23	23	20	20	16	16	13	13	
15	22	22	15	15	12	12		15	20	20	17	17	14	14	11	11	
16								16							10	10	
Y-Y AXIS																	
0	118	163	80	107	55	71		0	106	147	90	125	73	97	50	64	
2	109	147	70	88	44	51		2	96	129	79	105	61	77	39	46	
3	106	141	68	85	43	50		3	92	121	76	98	58	72	37	44	2
4	102	133	65	80	41	48		4	86	110	71	90	55	67	35	41	
5	96	124	62	75	40	46		5	78	97	65	79	50	60	33	38	
6	90	113	58	69	38	43		6	70	84	58	68	45	52	30	34	
7	83	101	54	63	35	40		7	62	70	51	57	40	44	27	30	
8	76	89	49	56	33	37		8	53	60	44	49	34	36	24	25	
9	68	77	44	49	30	33	2	9	47	48	37	39	29	30	20	21	
10	61	65	39	42	27	29		10	39	39	32	32	25	25	17	17	
11	53	57	34	35	24	25		11	32	32	26	26	21	21	14	14	
12	48	48	30	30	21	21		12	27	27	22	22	18	18	12	12	
13	41	41	26	26	19	19		13	23	23	19	19	15	15	10	10	
14	36	36	22	22	16	16		14	20	20	17	17	13	13	10	10	3
15	31	31	20	20	14	14		15	18	18	14	14					
16	28	28	18	18	13	13											
17	24	24	16	16	11	11											
18	22	22	14	14	10	10											
19	20	20					3										

Effective length KL (ft) with respect to indicated axis

Properties of 2 angles—⅜ in. back to back								
A (in²)	3.84		2.63		1.99			
r_x (in.)	0.928		0.945		0.954			
r_y (in.)	1.16		1.13		1.12			

A (in²)	3.47	2.93	2.38	1.80
r_x (in.)	0.940	0.948	0.957	0.966
r_y (in.)	0.917	0.903	0.891	0.879

[a]For Y-Y axis, welded or fully tensioned bolted connectors only.
[b]For number of connectors, see double angle column discussion.

F_y = 36 ksi
F_y = 50 ksi

COLUMNS
Double angles
Design axial strength in kips (ϕ = 0.85)
Unequal legs
Long legs $3/8$ in. back to back of angles

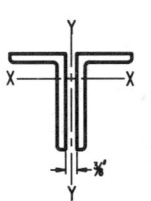

Size		\multicolumn{8}{c}{2½×2}	No. of Connectors a
Thickness		\multicolumn{2}{c}{$3/8$}	\multicolumn{2}{c}{$5/16$}	\multicolumn{2}{c}{$1/4$}	\multicolumn{2}{c}{$3/16$}	
Wt./ft		\multicolumn{2}{c}{10.6}	\multicolumn{2}{c}{9.0}	\multicolumn{2}{c}{7.2}	\multicolumn{2}{c}{5.5}	
F_y		36	50	36	50	36	50	36	50	

Effective length KL (ft) with respect to indicated axis

X-X AXIS

KL	3/8 36	3/8 50	5/16 36	5/16 50	1/4 36	1/4 50	3/16 36	3/16 50
0	95	131	80	111	65	91	49	63
2	90	122	76	104	62	85	46	59
3	84	112	72	95	58	78	44	55
4	77	99	66	84	54	69	40	49
5	69	84	59	72	48	59	36	43
6	60	69	51	59	42	49	32	36
7	50	55	43	47	36	39	27	30
8	42	42	36	37	30	30	23	24
9	33	33	29	29	24	24	19	19
10	27	27	23	23	19	19	15	15
11	22	22	19	19	16	16	12	12
12	19	19	16	16	13	13	10	10
13					11	11	9	9

(No. of Connectors: b)

Y-Y AXIS

KL	3/8 36	3/8 50	5/16 36	5/16 50	1/4 36	1/4 50	3/16 36	3/16 50
0	95	131	80	111	65	91	49	63
2	88	119	73	98	58	76	40	48
3	84	112	70	92	55	72	38	46
4	79	103	66	84	52	66	36	43
5	74	93	60	75	48	59	34	39
6	67	81	55	66	43	51	31	35
7	60	70	48	55	38	44	27	30
8	52	58	42	48	33	36	24	26
9	45	47	36	39	28	30	21	22
10	38	38	31	32	25	25	17	18
11	32	32	26	26	21	21	15	15
12	27	27	22	22	17	17	13	13
13	23	23	19	19	15	15	11	11
14	20	20	16	16	13	13	9	9
15	17	17	14	14	11	11	8	8
16	15	15						

(No. of Connectors: Y-Y axis shows 2 and 3)

Properties of 2 angles—$3/8$ in. back to back

	$3/8$	$5/16$	$1/4$	$3/16$
A (in²)	3.09	2.62	2.13	1.62
r_x (in.)	0.768	0.776	0.784	0.793
r_y (in.)	0.961	0.948	0.935	0.923

aFor Y-Y axis, welded or fully tensioned bolted connectors only.
bFor number of connectors, see double angle column discussion.

	F_y = 36 ksi
	F_y = 50 ksi

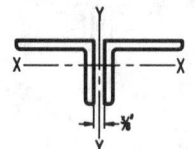

COLUMNS
Double angles
Design axial strength in kips (ϕ = 0.85)
Unequal legs
Short legs $\frac{3}{8}$ in. back to back of angles

X-X AXIS — Effective length KL (ft) with respect to indicated axis

	Size	8×6						No. of Connectors[a]		8×4						No. of Connectors[a]
	Thickness	1		3/4		1/2				1		3/4		1/2		
	Wt./ft	88.4		67.6		46.0				74.8		57.4		39.2		
	F_y	36	50	36	50	36	50			36	50	36	50	36	50	
X-X AXIS	0	796	1110	609	846	376	479		0	673	935	517	718	321	408	
	8	677	882	521	680	328	402		4	600	798	463	616	292	361	
	12	552	666	428	518	276	323		6	521	654	404	509	259	311	
	16	416	449	325	354	217	237		8	426	495	333	390	219	252	
	20	288	288	228	228	159	160	b	10	329	346	260	276	177	192	b
	24	200	200	159	159	111	111		12	240	240	192	192	137	138	
	28	147	147	116	116	82	82		14	176	176	141	141	101	101	
	29			109	109	76	76		16	135	135	108	108	78	78	
									17	120	120	96	96	69	69	
									18					61	61	

Y-Y AXIS — Effective length KL (ft) with respect to indicated axis

		8×6						No. of Connectors[a]		8×4						No. of Connectors[a]
	F_y	36	50	36	50	36	50			36	50	36	50	36	50	
Y-Y AXIS	0	796	1110	609	846	376	479		0	673	935	517	718	321	408	
	12	722	965	544	720	320	382		12	628	848	479	646	293	360	
	16	678	886	511	662	304	360		16	596	789	455	600	279	340	
	20	626	793	472	594	284	332		20	558	720	425	547	263	316	
	24	568	692	428	519	261	300		24	514	643	391	488	244	288	
	28	505	589	381	442	235	264		28	467	563	355	426	223	257	
	32	442	489	333	367	209	227		32	418	482	317	364	201	226	
	36	379	395	285	296	182	191		36	369	405	279	305	179	194	
	40	320	320	240	240	155	157	3	40	320	333	241	249	157	164	6
	44	265	265	199	199	130	130		44	274	275	206	206	135	137	
	48	223	223	168	168	110	110		48	231	231	173	173	115	115	
	52	190	190	143	143	94	94		52	197	197	148	148	98	98	
	56	164	164	123	123	81	81		56	170	170	128	128	85	85	
	60	143	143	108	108	71	71		60	148	148	111	111	74	74	
	61	138	138	104	104	69	69		64	130	130	98	98	65	65	
	62	134	134	101	101				66	122	122	92	92	61	61	
	63	130	130						67	119	119	89	89			
									68	115	115					

Properties of 2 angles—$\frac{3}{8}$ in. back to back

	8×6				8×4		
	1	3/4	1/2		1	3/4	1/2
A (in²)	26.0	19.9	13.5		22.0	16.9	11.5
r_x (in.)	1.73	1.76	1.79		1.03	1.05	1.08
r_y (in.)	3.78	3.74	3.69		4.10	4.05	4.00

[a] For Y-Y axis, welded or fully tensioned bolted connectors only.
[b] For number of connectors, see double angle column discussion.

| Fy = 36 ksi |
| Fy = 50 ksi |

COLUMNS
Double angles
Design axial strength in kips (ϕ = 0.85)
Unequal legs
Short legs ⅜ in. back to back of angles

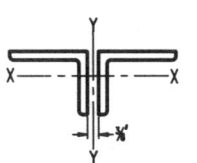

Size	7×4						No. of Connectors [a]	6×4								No. of Connectors [a]
Thickness	¾		½		⅜			¾		⅝		½		⅜		
Wt./ft	52.4		35.8		27.2			47.2		40.0		32.4		24.6		
Fy	36	50	36	50	36	50		36	50	36	50	36	50	36	50	

Effective length KL (ft) with respect to indicated axis

X-X AXIS

KL	¾-36	¾-50	½-36	½-50	⅜-36	⅜-50	Conn	KL	¾-36	¾-50	⅝-36	⅝-50	½-36	½-50	⅜-36	⅜-50	Conn
0	471	655	310	401	205	254		0	425	591	358	497	291	388	201	256	
4	426	568	282	354	189	230		4	386	517	326	436	265	343	186	231	
6	375	476	250	304	171	203		6	342	437	289	370	236	295	168	203	
8	313	371	212	245	149	171		8	289	345	245	293	201	238	146	170	
10	249	270	171	186	124	137	b	10	232	255	198	218	164	181	122	135	b
12	188	188	132	133	100	104		12	178	179	152	154	127	129	97	102	
14	138	138	98	98	77	77		14	132	132	113	113	95	95	75	75	
16	106	106	75	75	59	59		16	101	101	86	86	73	73	57	57	
18	84	84	59	59	47	47		18	80	80	68	68	57	57	45	45	
								19					52	52	41	41	

Y-Y AXIS

KL	¾-36	¾-50	½-36	½-50	⅜-36	⅜-50	Conn	KL	¾-36	¾-50	⅝-36	⅝-50	½-36	½-50	⅜-36	⅜-50	Conn
0	471	655	310	401	205	254		0	425	591	358	497	291	388	201	256	
8	448	610	289	364	185	217		8	397	537	332	447	266	343	179	216	
12	426	568	276	342	178	208		12	369	485	309	405	247	312	168	201	
16	397	515	257	314	168	196		16	333	422	279	351	223	273	154	181	
20	362	454	235	281	156	181		20	292	352	244	293	196	229	137	156	
24	323	388	211	244	143	162		24	249	281	208	234	167	185	118	130	
28	283	323	185	207	128	142		28	206	216	172	179	137	143	99	105	
32	243	261	160	170	113	121	5	32	165	165	137	137	110	110	81	81	4
36	204	207	135	137	97	101		36	131	131	109	109	87	87	64	64	
40	168	168	111	111	82	83		40	106	106	88	88	71	71	52	52	
44	139	139	92	92	69	69		44	88	88	73	73	58	58	43	43	
48	116	116	77	77	58	58		45	84	84	70	70	56	56	42	42	
52	99	99	66	66	49	49		46	80	80	67	67	53	53	40	40	
56	86	86	57	57	43	43		47	77	77	64	64	51	51	38	38	
57	83	83	55	55	41	41		48	74	74	61	61	49	49			
58	80	80						49	71	71							

Properties of 2 angles—⅜ in. back to back

	¾	½	⅜		¾	⅝	½	⅜
A (in²)	15.4	10.5	7.97		13.9	11.7	9.50	7.22
rx (in.)	1.09	1.11	1.13		1.12	1.13	1.15	1.17
ry (in.)	3.49	3.44	3.42		2.94	2.92	2.90	2.87

[a] For Y-Y axis, welded or fully tensioned bolted connectors only.
[b] For number of connectors, see double angle column discussion.

	Fy = 36 ksi
	Fy = 50 ksi

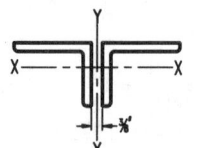

COLUMNS
Double angles
Design axial strength in kips (φ = 0.85)
Unequal legs
Short legs 3/8 in. back to back of angles

Size	6×3½				No. of Connectors[a]		5×3½								No. of Connectors[a]
Thickness	3/8		5/16				3/4		1/2		3/8		5/16		
Wt./ft	23.4		19.6				39.6		27.2		20.8		17.4		
F_y	36	50	36	50			36	50	36	50	36	50	36	50	

Effective length KL (ft) with respect to indicated axis

X-X AXIS

KL	36	50	36	50		KL	36	50	36	50	36	50	36	50	
0	191	243	145	179		0	355	493	245	340	183	238	143	182	
4	170	210	131	158		2	344	472	238	326	178	229	139	176	
6	148	175	115	135		4	313	413	217	288	163	205	129	159	
8	121	136	97	109		6	267	331	187	234	141	170	113	135	
10	94	99	77	82	b	8	214	243	152	176	116	131	94	107	b
						10	160	164	116	121	89	94	74	79	
12	69	69	58	59		12	114	114	84	84	65	65	56	56	
14	50	50	43	43		14	84	84	62	62	48	48	41	41	
16	39	39	33	33		16	64	64	47	47	37	37	31	31	
						17					32	32	28	28	

Y-Y AXIS

KL	36	50	36	50		KL	36	50	36	50	36	50	36	50	
0	191	243	145	179		0	355	493	245	340	183	238	143	182	
8	173	212	128	150		8	324	435	220	293	161	199	123	147	
12	163	197	122	142		10	310	408	210	275	154	188	118	140	
16	150	178	114	131		12	293	377	198	254	145	176	112	133	
20	134	155	103	117		14	274	344	185	231	136	162	106	123	
24	117	130	91	101		16	253	309	171	207	126	147	99	113	
28	99	105	79	84		18	232	274	156	183	116	131	91	103	
32	82	83	66	68		20	210	239	141	159	105	116	83	92	
36	65	65	54	54	5	22	189	205	127	136	94	100	75	81	4
40	53	53	44	44		24	168	174	112	115	83	86	67	70	
44	44	44	37	37		26	147	148	98	98	73	73	60	60	
48	37	37	31	31		28	128	128	85	85	63	63	52	52	
49	35	35	30	30		30	111	111	74	74	55	55	46	46	
						32	98	98	65	65	49	49	40	40	
						34	87	87	58	58	43	43	36	36	
						36	77	77	51	51	39	39	32	32	
						38	70	70	46	46	35	35	29	29	
						39	66	66	44	44	33	33	27	27	
						40	63	63	42	42	31	31			
						41	60	60							

Properties of 2 angles—3/8 in. back to back

A (in²)	6.84		5.74			11.6		8.00	6.09	5.12
r_x (in.)	0.988		0.996			0.977		1.01	1.02	1.03
r_y (in.)	2.95		2.94			2.48		2.43	2.41	2.39

[a]For Y-Y axis, welded or fully tensioned bolted connectors only.
[b]For number of connectors, see double angle column discussion.

F_y = 36 ksi
F_y = 50 ksi

COLUMNS
Double angles
Design axial strength in kips (ϕ = 0.85)
Unequal legs
Short legs ⅜ in. back to back of angles

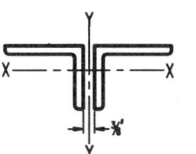

Size		\multicolumn{8}{c}{5×3}	No. of Connectors[a]
Thickness		\multicolumn{2}{c}{½}	\multicolumn{2}{c}{⅜}	\multicolumn{2}{c}{5⁄16}	\multicolumn{2}{c}{¼}	
Wt./ft		\multicolumn{2}{c}{25.6}	\multicolumn{2}{c}{19.6}	\multicolumn{2}{c}{16.4}	\multicolumn{2}{c}{13.2}	
F_y		36	50	36	50	36	50	36	50	
X-X AXIS	0	230	319	172	223	134	170	95	117	
	2	220	300	165	212	129	162	92	112	
	4	192	249	146	180	115	140	84	99	
	6	154	184	118	137	95	110	71	81	
	8	113	119	88	94	73	79	56	61	b
	10	76	76	61	61	52	52	42	43	
	12	53	53	42	42	36	36	30	30	
	13	45	45	36	36	31	31	25	25	
	14			31	31	26	26	22	22	
Y-Y AXIS	0	230	319	172	223	134	170	95	117	
	8	209	280	154	193	118	143	81	93	
	10	200	263	148	183	114	137	79	90	
	12	190	244	140	171	109	130	76	87	
	14	178	224	132	158	103	121	73	82	
	16	165	202	122	144	96	112	69	78	
	18	152	179	113	129	89	102	65	72	
	20	138	157	103	115	82	91	60	66	
	22	124	136	93	100	74	81	55	60	
	24	111	116	83	87	67	71	50	54	4
	26	98	99	73	74	60	61	46	48	
	28	85	85	64	64	53	53	41	42	
	30	74	74	56	56	46	46	36	36	
	32	65	65	49	49	41	41	32	32	
	34	58	58	43	43	36	36	28	28	
	36	52	52	39	39	32	32	25	25	
	38	46	46	35	35	29	29	23	23	
	40	42	42	31	31	26	26	21	21	
	41	40	40	30	30	25	25	20	20	
										5

Effective length KL (ft) with respect to indicated axis

Properties of 2 angles—⅜ in. back to back				
A (in.²)	7.50	5.72	4.80	3.88
r_x (in.)	0.829	0.845	0.853	0.861
r_y (in.)	2.50	2.48	2.47	2.46

[a]For Y-Y axis, welded or fully tensioned bolted connectors only.
[b]For number of connectors, see double angle column discussion.

	F_y = 36 ksi
	F_y = 50 ksi

COLUMNS
Double angles
Design axial strength in kips (ϕ = 0.85)
Unequal legs
Short legs ⅜ in. back to back of angles

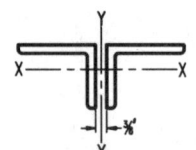

Size		4×3½								No. of Connectors [a]
Thickness		½		⅜		⁵⁄₁₆		¼		
Wt./ft		23.8		18.2		15.4		12.4		
F_y		36	50	36	50	36	50	36	50	
X-X AXIS	0	214	298	163	227	137	179	101	129	b
	2	208	286	159	219	133	172	99	125	
	4	191	255	147	195	123	156	92	114	
	6	166	210	128	162	108	131	81	98	
	8	137	160	106	125	90	103	69	79	
	10	106	112	83	89	71	76	55	60	
	12	78	78	62	62	53	53	42	43	
	14	57	57	45	45	39	39	31	31	
	16	44	44	35	35	30	30	24	24	
	17	39	39	31	31	26	26	21	21	
Y-Y AXIS	0	214	298	163	227	137	179	101	129	3
	4	201	272	149	198	121	149	84	99	
	6	192	256	143	187	116	142	81	95	
	8	181	235	135	173	109	132	78	90	
	10	167	211	124	156	101	121	73	84	
	12	152	184	113	136	92	107	67	77	
	14	135	157	101	117	83	93	61	68	
	16	118	131	88	97	73	79	55	59	
	18	101	106	76	78	62	65	48	50	
	20	86	86	64	64	53	53	41	41	
	22	71	71	53	53	44	44	35	35	
	24	60	60	45	45	37	37	29	29	
	26	51	51	38	38	32	32	25	25	
	28	44	44	33	33	27	27	22	22	
	30	38	38	29	29	24	24	19	19	
	31	36	36	27	27	22	22			

Effective length KL (ft) with respect to indicated axis

Properties of 2 angles—⅜ in. back to back

A (in.²)	7.00	5.34	4.49	3.63
r_x (in.)	1.04	1.06	1.07	1.07
r_y (in.)	1.89	1.87	1.86	1.85

[a]For Y-Y axis, welded or fully tensioned bolted connectors only.
[b]For number of connectors, see double angle column discussion.

F_y = 36 ksi
F_y = 50 ksi

COLUMNS
Double angles
Design axial strength in kips (ϕ = 0.85)
Unequal legs
Short legs $3/8$ in. back to back of angles

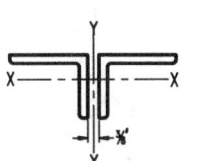

Size		4×3								No. of Connectors[a]
Thickness		$1/2$		$3/8$		$5/16$		$1/4$		
Wt./ft		22.2		17.0		14.4		11.6		
F_y		36	50	36	50	36	50	36	50	
X-X AXIS	0	199	276	152	211	127	166	94	120	
	2	191	261	146	200	123	158	91	115	
	4	169	220	130	170	109	136	82	101	
	6	138	166	107	129	90	106	69	81	b
	8	104	112	81	88	69	75	54	59	
	10	72	72	57	57	49	49	40	40	
	12	50	50	40	40	34	34	28	28	
	14	37	37	29	29	25	25	21	21	
Y-Y AXIS	0	199	276	152	211	127	166	94	120	
	4	189	258	142	191	116	146	83	99	
	6	182	244	137	181	112	139	80	96	
	8	172	225	129	168	106	130	76	91	
	10	160	204	120	152	98	118	72	84	
	12	146	180	110	134	90	106	66	77	
	14	131	155	99	116	81	92	60	68	
	16	116	131	87	98	71	78	54	59	3
	18	101	108	76	81	62	65	48	51	
	20	86	88	65	65	53	53	41	42	
	22	73	73	54	54	44	44	35	35	
	24	61	61	46	46	37	37	30	30	
	26	52	52	39	39	32	32	25	25	
	28	45	45	34	34	27	27	22	22	
	30	39	39	29	29	24	24	19	19	
	32	34	34	26	26	21	21	17	17	
										4

Effective length KL (ft) with respect to indicated axis

Properties of 2 angles—$3/8$ in. back to back					
A (in.^{2})	6.50		4.97	4.18	3.38
r_x (in.)	0.864		0.879	0.887	0.896
r_y (in.)	1.96		1.94	1.93	1.92

[a]For Y-Y axis, welded or fully tensioned bolted connectors only.
[b]For number of connectors, see double angle column discussion.

			F_y = 36 ksi
			F_y = 50 ksi

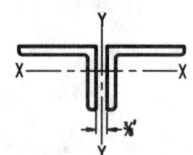

COLUMNS
Double angles
Design axial strength in kips (ϕ = 0.85)
Unequal legs
Short legs ⅜ in. back to back of angles

Size		3½×3						No. of Connectors[a]		3½×2½				No. of Connectors[a]
Thickness		⅜		⁵⁄₁₆		¼				⅜		¼		
Wt./ft		15.8		13.2		10.8				14.4		9.8		
F_y		36	50	36	50	36	50			36	50	36	50	
X-X AXIS	0	140	195	118	162	92	119		0	129	179	85	110	
	2	135	185	114	154	89	114		2	122	165	81	102	
	4	121	158	102	132	80	100		4	102	129	68	83	
	6	100	122	85	103	67	79		6	76	86	52	59	
	8	77	84	65	72	53	58	b	8	51	51	36	36	b
	10	55	55	47	47	38	39		10	32	32	23	23	
	12	38	38	33	33	27	27		11	27	27	19	19	
	14	28	28	24	24	20	20		12			16	16	
	15			21	21	17	17							
Y-Y AXIS	0	140	195	118	162	92	119		0	129	179	85	110	
	4	129	173	107	140	80	97		4	121	164	77	95	
	6	122	161	101	130	76	92		6	115	153	73	90	
	8	113	145	94	117	71	84		8	107	139	68	83	
	10	102	126	85	103	65	75		10	98	122	63	74	
	12	90	106	75	87	58	65		12	87	104	56	65	
	14	78	87	65	71	50	55	3	14	76	87	50	55	4
	16	66	68	55	57	43	45		16	65	70	43	45	
	18	54	54	45	45	36	36		18	55	55	36	37	
	20	44	44	37	37	29	29		20	45	45	30	30	
	22	36	36	30	30	24	24		22	37	37	25	25	
	24	31	31	25	25	20	20		24	31	31	21	21	
	26	26	26	22	22	17	17		26	27	27	18	18	
	27	24	24	20	20	16	16		28	23	23	15	15	
									29	21	21			

Effective length KL (ft) with respect to indicated axis

Properties of 2 angles—⅜ in. back to back

	3½×3			3½×2½	
A (in²)	4.59	3.87	3.13	4.22	2.88
r_x (in.)	0.897	0.905	0.914	0.719	0.735
r_y (in.)	1.67	1.66	1.65	1.74	1.72

[a]For Y-Y axis, welded or fully tensioned bolted connectors only.
[b]For number of connectors, see double angle column discussion.

| F_y = 36 ksi |
| F_y = 50 ksi |

COLUMNS
Double angles
Design axial strength in kips (ϕ = 0.85)
Unequal legs
Short legs $\frac{3}{8}$ in. back to back of angles

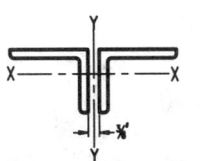

Size	3×2½						No. of Connectors [a]		3×2								No. of Connectors [a]
Thickness	$\frac{3}{8}$		$\frac{1}{4}$		$\frac{3}{16}$				$\frac{3}{8}$		$\frac{5}{16}$		$\frac{1}{4}$		$\frac{3}{16}$		
Wt./ft	13.2		9.0		6.77				11.8		10.0		8.2		6.1		
F_y	36	50	36	50	36	50			36	50	36	50	36	50	36	50	

Effective length KL (ft) with respect to indicated axis

X-X AXIS

KL	$\frac{3}{8}$ 36	50	$\frac{1}{4}$ 36	50	$\frac{3}{16}$ 36	50	KL	$\frac{3}{8}$ 36	50	$\frac{5}{16}$ 36	50	$\frac{1}{4}$ 36	50	$\frac{3}{16}$ 36	50
0	118	163	80	107	55	71	0	106	147	90	125	73	97	50	64
2	111	151	76	100	53	66	2	96	129	82	109	66	86	46	58
3	104	137	71	91	50	62	3	85	109	73	93	59	74	42	51
4	94	120	65	81	46	55	4	72	86	61	74	50	59	36	42
5	83	100	58	69	41	48	5	58	64	50	55	41	45	30	33
6	71	81	50	56	36	41	6	44	45	38	39	32	32	24	25
7	59	63	42	45	31	34	7	33	33	28	28	24	24	18	18
8	48	48	34	35	26	27	8	25	25	22	22	18	18	14	14
9	38	38	27	27	21	21	9	20	20	17	17	14	14	11	11
10	31	31	22	22	17	17									
11	25	25	18	18	14	14									
12	21	21	15	15	12	12									

No. of Connectors: b (left) / b (right)

Y-Y AXIS

KL	$\frac{3}{8}$ 36	50	$\frac{1}{4}$ 36	50	$\frac{3}{16}$ 36	50	KL	$\frac{3}{8}$ 36	50	$\frac{5}{16}$ 36	50	$\frac{1}{4}$ 36	50	$\frac{3}{16}$ 36	50
0	118	163	80	107	55	71	0	106	147	90	125	73	97	50	64
2	113	154	74	95	48	56	2	104	142	87	119	69	90	46	56
4	108	145	71	90	46	54	4	100	135	83	112	67	86	45	54
6	100	131	66	82	43	51	6	93	123	78	103	63	79	42	50
8	91	114	60	72	40	46	8	85	108	71	90	57	70	39	46
10	79	95	53	61	36	41	10	76	92	63	76	51	60	35	40
12	67	75	45	49	31	34	12	65	75	54	62	44	49	31	34
14	55	58	37	38	26	28	14	55	59	46	49	37	39	26	28
16	44	44	29	29	22	22	16	45	45	37	37	30	30	22	22
18	35	35	23	23	17	17	18	36	36	30	30	24	24	18	18
20	28	28	19	19	14	14	20	29	29	24	24	19	19	14	14
22	23	23	16	16	12	12	22	24	24	20	20	16	16	12	12
24	20	20	13	13	10	10	24	20	20	17	17	13	13	10	10
							25	19	19	15	15	12	12	9	9

No. of Connectors: 3 (left) / 4 (right)

Properties of 2 angles—$\frac{3}{8}$ in. back to back

	$\frac{3}{8}$	$\frac{1}{4}$	$\frac{3}{16}$		$\frac{3}{8}$	$\frac{5}{16}$	$\frac{1}{4}$	$\frac{3}{16}$
A (in^2)	3.84	2.63	1.99		3.47	2.93	2.38	1.80
r_x (in.)	0.736	0.753	0.761		0.559	0.567	0.574	0.583
r_y (in.)	1.47	1.45	1.44		1.55	1.53	1.52	1.51

[a] For Y-Y axis, welded or fully tensioned bolted connectors only.
[b] For number of connectors, see double angle column discussion.

	Fy = 36 ksi
	Fy = 50 ksi

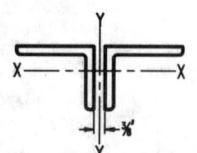

COLUMNS
Double angles
Design axial strength in kips (φ = 0.85)
Unequal legs
Short legs ⅜ in. back to back of angles

Size		2½×2							No. of Connectors[a]
Thickness		⅜		⁵⁄₁₆		¼		³⁄₁₆	
Wt./ft		10.6		9.0		7.2		5.5	
Fy		36	50	36	50	36	50	36	50	
X-X AXIS	0	95	131	80	111	65	91	49	63	
	2	86	116	73	98	60	80	45	57	
	3	77	99	66	84	54	69	40	50	
	4	66	79	56	68	46	56	35	41	
	5	54	60	46	51	38	43	29	32	b
	6	42	42	36	37	30	31	23	24	
	7	31	31	27	27	23	23	18	18	
	8	24	24	21	21	17	17	14	14	
	9	19	19	16	16	14	14	11	11	
	10							9	9	
Y-Y AXIS	0	95	131	80	111	65	91	49	63	
	2	91	125	77	105	61	83	44	54	
	4	86	116	72	96	58	76	41	51	
	6	78	101	65	84	52	67	38	45	
	8	68	84	57	69	45	55	33	38	
	10	57	66	47	53	38	43	28	31	3
	12	46	49	38	39	30	31	22	23	
	14	36	36	29	29	23	23	17	17	
	16	28	28	22	22	18	18	13	13	
	18	22	22	17	17	14	14	10	10	
	20	18	18	14	14	11	11	9	9	
	21	16	16	13	13					
										4

Effective length KL (ft) with respect to indicated axis

Properties of 2 angles—⅜ in. back to back			
A (in²)	3.09	2.62	2.13	1.62
rₓ (in.)	0.577	0.584	0.592	0.600
r_y (in.)	1.28	1.26	1.25	1.24

[a] For Y-Y axis, welded or fully tensioned bolted connectors only.
[b] For number of connectors, see double angle column discussion.

| F_y = 36 ksi |
| F_y = 50 ksi |

COLUMNS
Structural tees
cut from W shapes
Design axial strength in kips (ϕ = 0.85)

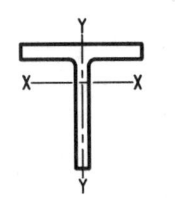

Designation		WT 18									
Wt./ft		150		140		130		122.5		115	
F_y		36	50	36	50	36	50	36	50	36	50
X-X AXIS	0	1350	1730	1260	1510	1150	1330	1040	1170	925	1030
	10	1310	1670	1230	1470	1120	1290	1010	1140	903	998
	12	1300	1650	1210	1440	1100	1270	998	1130	893	986
	14	1280	1620	1190	1420	1090	1250	985	1110	882	972
	16	1260	1590	1180	1390	1070	1220	970	1090	869	956
	18	1240	1550	1150	1360	1050	1200	953	1070	855	939
	20	1210	1510	1130	1330	1030	1170	935	1050	839	920
	22	1180	1460	1100	1290	1010	1140	915	1020	822	899
	24	1150	1420	1080	1250	983	1110	893	993	803	876
	26	1120	1370	1050	1210	957	1070	870	964	784	853
	28	1090	1320	1020	1170	930	1040	847	934	763	828
	30	1060	1260	984	1120	901	1000	822	903	742	802
	32	1020	1210	951	1080	871	965	796	871	719	776
	34	984	1160	917	1030	841	926	769	838	696	748
	36	947	1100	883	987	810	886	742	804	673	720
	38	910	1040	847	940	778	847	714	770	649	692
	40	872	989	812	893	746	806	686	736	624	663
Y-Y AXIS	0	1350	1730	1260	1510	1150	1330	1040	1170	925	1030
	10	1210	1510	1120	1320	1010	1140	906	1010	803	874
	12	1190	1470	1100	1290	990	1120	888	985	788	857
	14	1160	1430	1070	1250	966	1090	867	959	770	836
	16	1130	1370	1040	1210	938	1050	843	930	750	812
	18	1090	1320	1010	1160	908	1010	817	898	728	786
	20	1050	1260	972	1110	876	971	789	863	704	758
	22	1010	1190	932	1060	841	927	759	826	679	728
	24	962	1130	891	1000	804	880	727	787	651	696
	26	916	1060	848	943	766	833	693	747	623	662
	28	868	988	804	884	726	784	659	705	593	628
	30	820	918	758	825	686	734	623	662	563	592
	32	771	848	713	766	645	684	587	619	532	557
	34	721	780	667	708	604	634	551	577	501	521
	36	672	713	622	650	563	585	515	534	469	485
	38	624	647	577	595	523	537	480	493	438	450
	40	577	585	533	540	484	490	445	452	408	415

Effective length KL (ft) with respect to indicated axis

Properties					
A (in^2)	44.1	41.2	38.2	36.0	33.8
r_x (in.)	5.27	5.25	5.26	5.26	5.25
r_y (in.)	3.83	3.81	3.78	3.75	3.73

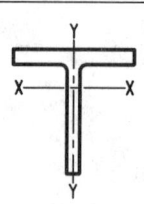

	F_y = 36 ksi
	F_y = 50 ksi

COLUMNS
Structural tees
cut from W shapes
Design axial strength in kips (ϕ = 0.85)

Designation			WT 18											
Wt./ft		105		97		91		85		80		75		67.5
F_y		36	50	36	50	36	50	36	50	36	50	36	50	36	50
X-X AXIS	0	908	1040	772	851	683	726	587	601	518	521	458	457	385	385
	10	887	1010	755	831	670	710	576	590	509	512	451	449	380	380
	12	878	1000	748	822	664	704	571	585	505	508	448	446	377	377
	14	868	986	740	812	657	696	566	579	500	503	444	442	374	374
	16	856	971	731	801	649	687	560	572	495	498	440	438	371	371
	18	843	954	720	788	640	677	553	565	489	492	435	433	367	367
	20	828	935	709	775	631	667	545	557	483	486	429	428	363	363
	22	813	915	696	759	620	655	537	548	476	479	424	422	359	359
	24	796	893	683	743	609	642	528	539	468	471	417	416	354	354
	26	778	870	668	726	597	629	518	529	460	463	411	409	348	348
	28	759	846	653	708	584	614	508	518	452	454	403	402	343	343
	30	739	821	637	689	571	599	497	507	443	445	396	395	337	337
	32	719	795	621	669	557	584	486	495	433	436	388	387	331	331
	36	675	740	586	628	527	551	462	470	413	415	371	370	317	317
	40	630	684	549	585	496	517	437	444	392	394	353	352	303	303
Y-Y AXIS	0	908	1040	772	851	683	726	587	601	518	521	458	457	385	385
	10	716	791	607	653	534	558	457	465	397	399	344	344	268	268
	12	687	756	585	627	515	538	442	450	385	387	335	334	261	261
	14	654	715	559	597	494	515	425	432	371	373	323	323	253	253
	16	617	670	530	563	470	488	406	412	356	357	311	310	244	244
	18	578	622	499	527	444	460	385	391	339	340	297	296	234	234
	20	537	572	465	489	416	430	363	368	320	321	281	281	223	223
	22	494	521	431	449	387	398	340	344	301	302	265	265	211	211
	24	450	469	395	409	358	366	315	319	280	281	248	248	198	198
	26	407	418	360	369	327	333	291	293	260	260	231	231	185	185
	28	364	367	325	330	297	301	266	268	239	239	213	213	172	172
	30	323	323	291	291	268	269	242	242	218	218	196	195	158	158
	32	285	285	258	258	239	239	218	218	197	197	178	178	144	144
	34	254	254	230	230	213	213	194	194	177	177	161	161	131	131
	36	228	228	206	206	191	191	174	174	159	159	144	144	118	118
	39	195	195	176	176	164	164	150	150	137	137	124	124	102	102
	41	177	177	160	160	149	149	136	136	124	124	113	113		
	42	169	169	153	153	142	142	130	130						
	43	161	161												

Effective length KL (ft) with respect to indicated axis

Properties							
A (in^2)		30.9	28.5	26.8	25.0	23.5	22.1	19.9
r_x (in.)		5.65	5.62	5.62	5.61	5.61	5.62	5.66
r_y (in.)		2.58	2.56	2.55	2.53	2.50	2.47	2.38

Note: Heavy line indicates Kl / r of 200.

$F_y = 36$ ksi
$F_y = 50$ ksi

COLUMNS
Structural tees
cut from W shapes
Design axial strength in kips ($\phi = 0.85$)

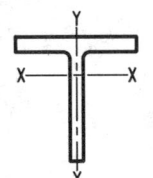

Designation		WT 16.5													
Wt./ft		120.5		110.5		100.5		76		70.5		65		59	
F_y		36	50	36	50	36	50	36	50	36	50	36	50		
X-X AXIS	0	1080	1300	964	1110	810	899	532	548	463	467	402	401	330	330
	10	1050	1260	935	1070	788	872	521	535	453	457	394	393	324	324
	12	1040	1240	923	1050	779	860	516	530	449	453	391	390	321	321
	14	1020	1210	909	1030	767	847	510	524	444	448	387	386	318	318
	16	1000	1190	893	1010	755	831	503	516	439	442	383	382	315	315
	18	980	1160	875	990	741	814	495	508	433	436	378	377	311	311
	20	958	1120	855	965	725	795	487	500	426	429	372	371	307	307
	22	933	1090	834	937	708	775	478	490	419	422	366	365	303	303
	24	907	1050	811	908	690	753	468	480	411	414	360	359	298	298
	26	880	1020	787	878	672	730	458	469	402	405	353	352	293	293
	28	851	975	762	846	652	706	447	458	393	396	345	345	287	287
	30	821	934	737	813	631	681	436	446	384	387	338	337	282	282
	32	790	892	710	779	610	656	424	433	374	377	330	329	276	276
	34	759	849	682	745	588	630	411	420	364	366	321	321	269	269
	36	727	806	654	710	565	603	399	407	354	356	313	312	263	263
	40	662	720	598	639	520	549	373	379	332	334	295	295	249	249
Y-Y AXIS	0	1080	1300	964	1110	810	899	532	548	463	467	402	401	330	330
	10	951	1110	833	934	692	753	420	429	358	360	300	299	234	234
	12	929	1080	815	911	678	736	406	414	346	348	291	290	228	228
	14	904	1050	793	884	661	717	389	396	333	335	280	280	220	220
	16	876	1010	769	854	643	695	370	377	318	319	268	268	212	212
	18	845	967	743	821	622	671	349	355	301	302	255	255	203	203
	20	811	922	714	785	600	644	327	332	283	284	241	241	192	192
	22	775	874	683	747	576	616	305	309	264	266	226	226	181	181
	24	738	824	651	707	551	587	281	284	245	246	211	210	170	170
	26	699	772	617	666	525	556	257	260	225	226	195	195	158	158
	28	659	720	583	624	497	525	234	235	206	206	179	178	146	146
	30	618	667	548	581	470	492	211	211	186	187	162	162	134	134
	34	537	564	477	496	413	427	167	167	149	149	131	131	109	109
	36	497	514	442	455	385	395	150	150	134	134	118	118	98	98
	38	458	464	408	414	357	363	135	135	121	121	107	107	89	89
	39	439	441	391	394	343	348	129	129	115	115	102	102		
	40	420	420	375	375	330	332	123	123	109	109				
	41	401	401	358	358	316	317	117	117						
Properties															
A (in^2)		35.4		32.5		29.5		22.4		20.8		19.2		17.3	
r_x (in.)		4.96		4.96		4.95		5.14		5.15		5.18		5.20	
r_y (in.)		3.63		3.59		3.56		2.47		2.43		2.39		2.32	

Effective length KL (ft) with respect to indicated axis

Note: Heavy line indicates Kl / r of 200.

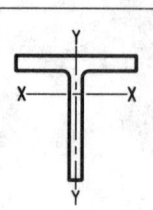

	Fy = 36 ksi
	Fy = 50 ksi

COLUMNS
Structural tees
cut from W shapes
Design axial strength in kips (ϕ = 0.85)

Designation						WT 15										
Wt./ft	105.5		95.5		86.5		66		62		58		54		49.5	
F_y	36	50	36	50	36	50	36	50	36	50	36	50	36	50	36	50
X-X AXIS																
0	949	1180	844	974	708	791	505	546	447	465	402	412	357	361	304	303
10	913	1130	812	932	684	761	490	529	434	452	392	401	348	352	297	296
12	897	1100	799	914	673	748	484	521	429	446	387	396	344	348	294	293
14	879	1080	783	894	661	733	477	513	423	439	382	391	340	343	290	290
16	859	1050	765	870	647	715	468	503	416	432	376	384	335	338	286	286
18	837	1010	746	845	632	696	459	492	408	423	369	377	329	332	282	281
20	813	976	724	817	615	676	449	480	399	414	361	369	323	326	277	276
22	787	938	702	787	597	654	437	467	390	404	353	361	316	319	271	271
24	759	897	678	756	578	630	426	454	380	393	345	352	309	311	266	265
26	731	855	652	724	558	606	413	439	370	382	336	343	301	304	259	259
28	701	811	626	690	537	581	400	424	359	370	326	333	293	295	253	253
30	670	767	599	656	515	555	387	409	347	358	316	322	284	287	246	246
32	639	723	571	621	493	528	373	393	335	345	306	311	276	278	239	239
34	607	678	543	586	471	501	358	377	323	332	295	300	267	269	232	232
36	575	634	515	551	448	474	344	360	311	319	284	289	257	259	224	224
40	511	547	459	482	402	421	314	327	285	292	262	266	238	240	209	209
Y-Y AXIS																
0	949	1180	844	974	708	791	505	546	447	465	402	412	357	361	304	303
10	838	1010	734	828	609	667	389	411	340	350	298	303	254	256	207	206
12	817	982	716	805	595	651	371	391	325	335	286	290	244	246	199	199
14	793	946	695	778	579	631	350	368	308	317	271	276	233	234	191	191
16	766	907	672	749	561	610	328	343	290	297	256	259	220	221	181	181
18	736	864	646	717	541	586	303	316	269	275	238	241	206	207	170	170
20	704	818	619	682	520	561	278	288	248	253	220	223	191	192	159	159
22	670	769	589	645	497	533	252	259	226	230	201	203	175	176	147	146
24	635	720	559	607	473	505	227	231	204	207	182	183	159	160	134	134
26	598	669	527	568	448	475	201	203	183	184	163	164	143	143	121	121
28	561	617	495	528	422	445	176	176	161	161	145	145	127	127	108	108
30	523	567	462	488	396	415	155	155	142	142	127	127	112	112	96	96
32	486	517	429	448	370	384	137	137	125	125	113	113	100	100	85	85
34	449	468	397	410	344	354	122	122	112	112	100	100	89	89	76	76
35	430	444	381	391	331	339	115	115	106	106	95	95	84	84		
36	412	420	365	372	318	325	109	109	100	100	90	90				
37	395	398	349	353	305	310	103	103	95	95						
38	377	378	334	335	293	296										
40	342	342	303	303	268	268										

Effective length KL (ft) with respect to indicated axis

Properties																
A (in²)	31.0		28.1		25.4		19.4		18.2		17.1		15.9		14.5	
r_x (in.)	4.43		4.42		4.42		4.66		4.66		4.67		4.69		4.71	
r_y (in.)	3.49		3.46		3.43		2.25		2.23		2.19		2.15		2.10	

Note: Heavy line indicates Kl / r of 200.

| F_y = 36 ksi |
| F_y = 50 ksi |

COLUMNS
Structural tees
cut from W shapes
Design axial strength in kips (ϕ = 0.85)

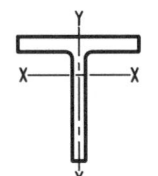

Designation								WT 13.5							
Wt./ft		89		80.5		73		57		51		47		42	
F_y		36	50	36	50	36	50	36	50	36	50	36	50	36	50
X-X AXIS	0	799	1040	725	857	617	698	453	498	358	369	308	311	251	250
	10	761	978	691	810	589	663	436	477	346	356	298	301	244	243
	12	745	951	676	790	577	648	428	468	341	350	294	297	241	240
	14	727	921	660	767	564	631	420	458	334	344	289	292	238	236
	16	707	887	641	741	549	612	410	447	328	337	284	286	234	232
	18	684	850	620	712	532	591	399	434	320	329	278	280	229	228
	20	660	811	598	682	514	568	388	420	312	320	271	273	224	223
	22	634	770	574	650	495	544	375	405	303	310	264	266	219	218
	24	606	727	549	617	474	519	362	390	293	300	256	258	213	212
	26	578	683	523	583	453	493	348	373	283	290	248	250	207	206
	28	549	638	496	548	431	466	334	357	273	279	240	241	201	200
	30	519	594	469	513	409	439	319	339	262	268	231	233	194	193
	32	489	550	442	478	387	412	304	322	251	256	222	223	187	187
	34	459	506	415	443	364	385	289	304	240	244	213	214	180	180
	36	430	464	388	409	342	358	274	287	229	233	204	205	173	173
	38	400	423	361	376	319	332	259	269	218	221	194	195	166	165
	40	371	383	335	344	298	306	244	252	206	209	185	186	159	158
Y-Y AXIS	0	799	1040	725	857	617	698	453	498	358	369	308	311	251	250
	10	701	877	627	720	527	583	349	374	275	281	231	233	181	180
	12	681	845	609	696	513	566	331	353	263	268	221	223	174	173
	14	658	808	588	669	497	546	311	330	248	253	210	211	166	165
	16	632	768	565	639	479	524	288	304	232	236	197	198	157	156
	18	604	725	540	606	459	500	264	277	215	218	184	185	147	147
	20	574	678	513	571	437	474	240	249	197	199	169	170	137	136
	22	542	631	485	534	415	446	215	221	179	180	154	155	126	125
	24	509	582	456	497	391	418	191	193	160	161	140	140	114	114
	26	476	533	426	458	367	389	167	167	142	143	125	125	103	103
	28	442	484	395	420	342	359	145	145	125	125	110	110	92	92
	30	408	436	365	382	317	330	127	127	109	109	97	97	81	81
	32	374	390	335	345	292	301	112	112	97	97	86	86	72	72
	34	342	347	305	309	268	273	100	100	86	86	76	76	64	64
	35	326	328	291	292	256	259	94	94	81	81	72	72		
	36	310	310	277	277	244	245	89	89						
	40	252	252	225	225	200	200								

Properties

	89	80.5	73	57	51	47	42
A (in²)	26.1	23.7	21.5	16.8	15.0	13.8	12.4
r_x (in.)	3.98	3.96	3.95	4.15	4.14	4.16	4.18
r_y (in.)	3.26	3.24	3.21	2.18	2.15	2.12	2.07

Note: Heavy line indicates Kl/r of 200.

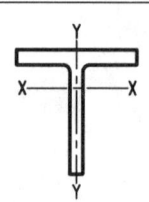

| | | | | | F_y = 36 ksi |
| | | | | | F_y = 50 ksi |

COLUMNS
Structural tees
cut from W shapes
Design axial strength in kips (ϕ = 0.85)

Designation		WT 12									
Wt./ft		81		73		65.5		58.5		52	
F_y		36	50	36	50	36	50	36	50	36	50
	0	731	1020	658	864	591	726	506	580	410	450
	10	687	932	618	797	556	673	477	542	389	424
	12	669	898	602	769	541	652	465	526	379	413
	14	648	858	583	737	524	627	451	508	369	401
	16	624	815	562	702	505	599	435	487	357	387
	18	598	769	538	664	484	569	418	465	344	371
X-X AXIS	20	571	720	514	624	462	537	400	442	331	355
	22	542	670	488	583	439	504	380	418	316	338
	24	512	619	461	541	415	471	360	392	301	320
	26	481	568	433	499	391	437	339	367	285	302
	28	450	518	405	457	366	403	318	341	269	283
	30	419	469	377	416	341	369	297	315	252	264
	32	388	421	349	376	316	336	276	290	236	246
	34	358	375	322	338	291	304	255	265	220	227
	36	328	335	295	301	267	273	235	241	204	209
	38	299	300	269	270	244	245	215	217	188	192
	40	271	271	244	244	221	221	196	196	173	175
	0	731	1020	658	864	591	726	506	580	410	450
	10	648	858	575	723	505	597	424	472	338	363
	12	626	819	555	691	487	573	410	455	328	352
	14	601	775	533	656	468	545	395	435	317	339
	16	574	727	508	617	446	515	377	414	304	324
	18	544	675	482	575	422	482	358	390	290	308
Y-Y AXIS	20	512	622	453	532	397	448	338	365	275	291
	22	480	568	424	487	371	412	316	339	260	273
	24	446	514	393	442	345	376	294	313	243	254
	26	412	460	363	398	317	340	272	286	226	235
	28	378	409	332	355	290	305	249	259	209	216
	30	345	359	303	313	264	271	227	233	192	196
	32	312	316	273	276	238	239	206	207	175	178
	34	281	281	245	245	213	213	184	184	159	159
	36	251	251	219	219	190	190	165	165	143	143
	38	225	225	197	197	171	171	149	149	129	129
	40	204	204	178	178	155	155	135	135	117	117

Effective length KL (ft) with respect to indicated axis

Properties					
A (in^2)	23.9	21.5	19.3	17.2	15.3
r_x (in.)	3.50	3.50	3.52	3.51	3.51
r_y (in.)	3.05	3.01	2.97	2.94	2.91

Note: Heavy line indicates Kl / r of 200.

F_y = 36 ksi
F_y = 50 ksi

COLUMNS
Structural tees
cut from W shapes
Design axial strength in kips (ϕ = 0.85)

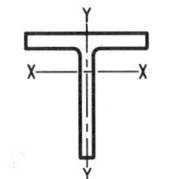

Designation		WT 12											
Wt./ft		47		42		38		34		31		27.5	
F_y		36	50	36	50	36	50	36	50	36	50	36	50
X-X AXIS	0	378	419	307	321	254	258	209	208	202	203	155	155
	10	360	396	293	306	244	247	201	200	194	196	150	150
	12	352	387	287	299	240	243	198	197	191	192	148	148
	14	343	376	280	292	234	237	194	193	187	188	145	145
	16	332	363	273	284	229	231	189	189	183	184	142	142
	18	321	350	265	275	222	225	185	184	178	179	139	139
	20	309	335	256	265	215	218	179	179	173	174	136	136
	22	296	320	246	255	208	210	174	173	168	169	132	132
	24	283	304	236	244	200	202	168	167	162	163	128	128
	26	269	287	225	233	192	194	162	161	156	157	124	124
	28	255	271	215	221	184	185	155	155	150	150	120	120
	30	240	254	204	209	175	176	148	148	143	144	115	115
	32	226	237	192	197	166	167	142	141	136	137	111	111
	34	211	220	181	185	157	158	135	135	130	130	106	106
	36	197	203	170	173	148	149	128	128	123	123	101	101
	38	183	187	159	161	140	140	121	121	116	117	96	96
	40	169	171	148	150	131	131	114	114	109	110	92	92
Y-Y AXIS	0	378	419	307	321	254	258	209	208	202	203	155	155
	10	288	310	231	239	188	190	148	147	121	122	91	91
	12	269	288	218	224	178	179	140	140	110	110	84	84
	14	248	263	202	208	166	167	132	132	97	98	75	75
	16	226	237	185	190	154	155	123	123	84	84	67	67
	18	202	211	168	171	140	141	113	113	71	71	57	57
	20	179	184	150	152	126	127	103	103	59	59	48	48
	22	156	158	132	134	113	113	92	92	49	49	41	41
	23	145	145	124	124	106	106	87	87	46	46		
	24	134	134	115	115	99	99	82	82				
	26	115	115	99	99	86	86	71	71				
	28	99	99	86	86	74	74	62	62				
	30	87	87	75	75	65	65	55	55				
	31	82	82	71	71	61	61	51	51				
	32	77	77	66	66	58	58						
	33	72	72										

Properties						
A (in.^{2})	13.8		12.4		11.2		10.0		9.11		8.10
r_x (in.)	3.67		3.67		3.68		3.70		3.79		3.80
r_y (in.)	1.98		1.95		1.92		1.87		1.38		1.34

Note: Heavy line indicates Kl / r of 200.

Effective length KL (ft) with respect to indicated axis

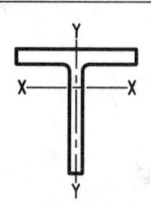

			F_y = 36 ksi
			F_y = 50 ksi

COLUMNS
Structural tees
cut from W shapes
Design axial strength in kips (ϕ = 0.85)

Designation										WT 10.5

Wt./ft			73.5		66		61		55.5		50.5	
F_y			36	50	36	50	36	50	36	50	36	50
		0	661	918	594	825	548	757	499	637	452	524
		10	610	822	547	737	505	676	459	573	416	476
		12	589	782	528	701	487	643	443	547	401	457
		14	565	739	507	661	466	606	424	518	384	434
		16	539	691	483	618	444	566	404	486	366	410
		18	510	641	457	573	420	524	382	453	346	384
	X-X AXIS	20	480	589	429	526	395	481	358	418	324	357
		22	449	536	401	478	368	437	334	382	303	329
		24	417	484	372	431	341	394	310	347	280	301
		26	385	434	343	386	315	352	285	312	258	274
		28	353	385	315	341	288	311	261	279	236	247
		30	322	337	286	299	262	272	237	246	214	221
		32	292	296	259	263	236	239	214	217	193	195
		34	262	263	233	233	212	212	192	192	173	173
		36	234	234	208	208	189	189	171	171	154	154
		38	210	210	186	186	170	170	154	154	139	139
		40	190	190	168	168	153	153	139	139	125	125
		0	661	918	594	825	548	757	499	637	452	524
		10	587	778	522	689	478	626	430	526	385	434
		12	566	740	503	655	461	596	415	503	372	417
		14	542	697	482	617	441	562	397	476	356	397
		16	515	650	458	576	419	524	377	447	339	375
		18	486	601	432	532	396	485	356	415	320	352
	Y-Y AXIS	20	456	550	405	487	371	444	334	383	300	327
		22	424	498	377	441	345	402	311	349	279	301
		24	392	447	348	395	319	361	287	316	258	275
		26	360	397	320	351	293	321	263	283	237	249
		28	328	349	291	308	267	281	239	251	216	223
		30	297	305	263	269	241	246	216	220	195	199
		32	267	268	237	237	216	217	194	194	175	175
		34	238	238	210	210	193	193	172	172	156	156
		36	213	213	188	188	172	172	154	154	139	139
		38	191	191	169	169	155	155	139	139	125	125
		40	173	173	153	153	140	140	125	125	113	113

Effective length KL (ft) with respect to indicated axis

Properties					
A (in^2)	21.6	19.4	17.9	16.3	14.9
r_x (in.)	3.08	3.06	3.04	3.03	3.01
r_y (in.)	2.95	2.93	2.92	2.90	2.89

Note: Heavy line indicates Kl/r of 200.

F_y = 36 ksi
F_y = 50 ksi

COLUMNS
Structural tees
cut from W shapes
Design axial strength in kips (ϕ = 0.85)

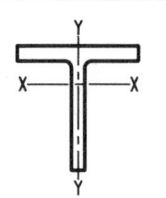

Designation						WT 10.5										
Wt./ft	46.5		41.5		36.5		34		31		28.5		25		22	
F_y	36	50	36	50	36	50	36	50	36	50	36	50	36	50	36	50
0 (X-X)	419	562	373	444	297	331	261	283	219	225	203	210	165	167	127	127
6	409	543	364	430	290	322	255	276	214	221	199	206	162	164	125	125
8	400	529	356	420	284	316	251	271	211	217	196	203	160	161	123	123
10	390	511	347	407	278	307	245	264	206	212	192	199	157	158	121	121
12	378	489	336	392	270	297	239	256	201	207	187	194	153	155	119	119
14	364	466	323	374	260	286	231	247	195	201	182	188	149	151	116	116
16	349	439	310	355	250	274	222	237	189	194	176	182	145	146	113	113
18	332	412	295	335	239	260	213	227	181	186	170	175	140	141	110	110
20	315	383	279	314	227	246	203	215	174	178	163	167	134	136	106	106
22	296	353	262	291	215	231	192	203	165	169	155	159	129	130	102	102
24	277	323	245	269	202	216	182	191	157	160	147	151	123	124	98	98
26	258	293	228	247	189	200	170	178	148	151	139	143	117	118	94	94
28	239	264	210	225	176	185	159	165	139	142	131	134	111	111	90	90
30	220	236	193	203	163	169	148	153	130	132	123	125	104	105	85	85
32	201	209	177	182	150	155	137	140	121	123	115	117	98	98	81	81
34	183	185	160	162	137	140	126	128	112	114	107	108	91	92	76	76
36	165	165	145	145	125	126	115	117	104	104	99	100	85	85	71	71
38	148	148	130	130	113	113	105	105	95	96	91	92	79	79	67	67
40	134	134	117	117	102	102	95	95	87	87	84	84	73	73	63	63
0 (Y-Y)	419	562	373	444	297	331	261	283	219	225	203	210	165	167	127	127
6	357	452	311	357	245	267	213	227	149	152	153	157	116	117	85	85
8	337	420	294	334	233	252	203	215	145	147	142	145	108	109	80	80
10	313	381	273	307	218	235	191	202	138	141	128	131	99	99	74	74
12	285	338	250	276	201	214	177	186	131	133	113	115	88	88	67	67
14	256	293	224	243	182	192	161	168	122	123	97	98	76	76	59	59
16	226	247	198	210	162	169	145	150	112	113	81	82	64	64	51	51
18	195	203	171	177	142	146	128	131	101	102	66	66	52	52	42	42
20	166	166	145	145	122	124	111	112	90	90	54	54	43	43	35	35
21	151	151	132	132	113	113	103	103	84	84	49	49	39	39	32	32
22	138	138	121	121	103	103	95	95	78	78	45	45				
24	117	117	102	102	87	87	80	80	67	67						
26	100	100	88	88	75	75	69	69	58	58						
28	86	86	76	76	65	65	60	60	51	51						
29	80	80	71	71	60	60	56	56	47	47						
30	75	75	66	66	57	57	52	52								

Effective length KL (ft) with respect to indicated axis — X-X AXIS / Y-Y AXIS

Properties

	46.5	41.5	36.5	34	31	28.5	25	22
A (in^2)	13.7	12.2	10.7	10.0	9.13	8.37	7.36	6.49
r_x (in.)	3.25	3.22	3.21	3.20	3.21	3.29	3.30	3.31
r_y (in.)	1.84	1.83	1.81	1.80	1.77	1.35	1.30	1.26

Note: Heavy line indicates Kl / r of 200.

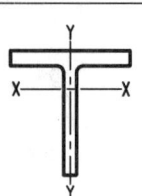

F_y = 36 ksi
F_y = 50 ksi

COLUMNS
Structural tees
cut from W shapes
Design axial strength in kips ($\phi = 0.85$)

Designation		WT 9									
Wt./ft		59.5		53		48.5		43		38	
F_y		36	50	36	50	36	50	36	50	36	50
X-X AXIS	0	536	744	477	663	438	608	389	507	339	393
	10	479	636	426	567	390	518	346	436	302	343
	12	456	594	406	529	370	482	329	407	287	323
	14	430	548	383	487	349	444	309	376	270	302
	16	402	499	357	444	325	403	288	344	252	278
	18	372	449	331	399	301	361	266	310	233	254
	20	342	399	304	354	275	320	244	276	213	229
	22	311	350	276	310	250	279	221	243	193	205
	24	281	303	249	268	225	241	199	211	174	181
	26	251	259	222	229	200	205	177	181	155	158
	28	222	224	197	198	177	177	156	156	136	137
	30	195	195	172	172	154	154	136	136	119	119
	34	152	152	134	134	120	120	106	106	93	93
	38	121	121	107	107	96	96	85	85	74	74
	42	99	99	88	88	79	79	69	69	61	61
	43	95	95	84	84						
Y-Y AXIS	0	536	744	477	663	438	608	389	507	339	393
	10	471	622	415	546	379	496	331	411	285	320
	12	450	584	397	513	362	467	317	388	272	304
	14	427	543	376	477	343	434	300	362	258	287
	16	401	499	353	438	322	398	282	334	243	267
	18	374	453	329	397	300	361	263	305	226	246
	20	346	407	304	356	277	324	242	275	209	225
	22	318	361	279	315	254	286	222	245	192	203
	24	289	317	253	276	230	251	201	216	174	182
	26	260	274	228	238	207	216	181	188	156	161
	28	233	237	203	206	185	187	161	163	139	140
	30	206	206	180	180	163	163	142	142	123	123
	34	161	161	140	140	127	127	111	111	96	96
	38	129	129	112	112	102	102	89	89	77	77
	42	106	106	92	92	84	84	73	73	63	63
	43	101	101	88	88	80	80	70	70	61	61
	44	96	96	84	84	76	76				

Effective length KL (ft) with respect to indicated axis

Properties					
A (in^2)	17.5	15.6	14.3	12.7	11.2
r_x (in.)	2.60	2.59	2.56	2.55	2.54
r_y (in.)	2.69	2.66	2.65	2.63	2.61

Note: Heavy line indicates Kl / r of 200.

F_y = 36 ksi
F_y = 50 ksi

COLUMNS
Structural tees
cut from W shapes
Design axial strength in kips (ϕ = 0.85)

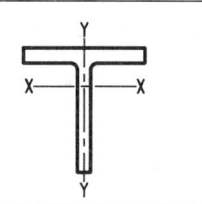

Designation		WT 9															
Wt./ft		35.5		32.5		30		27.5		25		23		20		17.5	
F_y		36	50	36	50	36	50	36	50	36	50	36	50	36	50	36	50
X-X AXIS	0	318	426	292	356	261	299	226	253	184	194	172	183	124	124	101	101
	10	288	372	264	314	236	266	206	227	169	177	159	168	116	116	95	95
	12	275	351	252	297	226	253	197	217	163	171	153	161	112	112	92	92
	14	261	327	239	279	214	239	188	206	156	163	147	154	108	108	89	89
	16	246	302	225	259	202	223	178	193	148	154	140	146	104	104	86	86
	18	229	275	210	238	189	206	167	180	140	145	132	138	99	99	82	82
	20	212	248	194	216	175	189	155	166	131	135	124	129	94	94	78	78
	22	195	222	178	195	161	172	143	152	122	126	116	120	89	89	74	74
	24	178	196	162	173	147	155	131	138	113	116	107	111	84	83	70	70
	26	161	171	146	153	133	138	120	124	103	106	99	101	78	78	66	66
	28	144	148	131	134	119	122	108	111	94	96	90	92	72	72	62	62
	30	128	129	116	116	106	107	97	98	85	86	82	83	67	67	57	57
	32	113	113	102	102	94	94	86	86	77	77	74	75	61	61	53	53
	34	100	100	91	91	83	83	76	76	68	68	67	67	56	56	49	49
	36	89	89	81	81	74	74	68	68	61	61	59	59	51	51	45	45
	38	80	80	72	72	66	66	61	61	55	55	53	53	46	46	41	41
	40	72	72	65	65	60	60	55	55	49	49	48	48	41	41	37	37
	42	66	66	59	59	54	54	50	50	45	45	44	44	38	38	34	34
	43	63	63	57	57	52	52	48	48	43	43	42	42	36	36	32	32
	45	57	57	52	52	47	47	44	44	39	39	38	38	33	33	29	29
	46											36	36	31	31	28	28
Y-Y AXIS	0	318	426	292	356	261	299	226	253	184	194	172	183	124	124	101	101
	10	231	279	210	238	187	204	161	173	132	137	107	111	80	80	61	61
	12	207	242	188	209	168	182	146	155	121	125	92	94	71	71	54	54
	14	182	204	165	179	149	158	130	136	109	111	77	78	61	61	47	47
	16	157	167	142	149	129	134	113	117	96	98	62	62	51	51	40	40
	18	132	133	119	120	109	111	96	98	83	84	50	50	41	41	32	32
	20	109	109	98	98	90	90	80	80	70	70	40	40	34	34	27	27
	21	99	99	89	89	82	82	73	73	64	64	37	37	31	31		
	22	90	90	82	82	75	75	67	67	59	59						
	24	76	76	69	69	63	63	57	57	50	50						
	26	65	65	59	59	54	54	48	48	43	43						
	27	60	60	55	55	50	50	45	45	40	40						
	28	56	56	51	51	47	47										

Effective length KL (ft) with respect to indicated axis

Properties								
A (in²)	10.4	9.55	8.82	8.10	7.33	6.77	5.88	5.15
r_x (in.)	2.74	2.72	2.71	2.71	2.70	2.77	2.76	2.79
r_y (in.)	1.70	1.69	1.69	1.67	1.65	1.29	1.27	1.22

Note: Heavy line indicates Kl / r of 200.

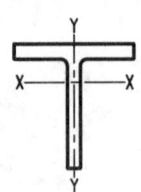

| | F_y = 36 ksi |
| | F_y = 50 ksi |

COLUMNS
Structural tees
cut from W shapes
Design axial strength in kips (ϕ = 0.85)

Designation		WT 8							
Wt./ft		50		44.5		38.5		33.5	
F_y		36	50	36	50	36	50	36	50
X-X AXIS	0	450	625	401	557	346	476	301	361
	10	389	510	346	454	297	386	258	300
	12	365	467	324	415	278	353	241	277
	14	338	420	300	373	257	316	223	251
	16	310	372	275	330	235	279	203	225
	18	280	324	249	287	212	243	183	199
	20	251	278	223	246	189	207	163	173
	22	222	234	197	207	166	174	143	148
	24	194	197	172	174	145	146	124	125
	26	167	167	148	148	124	124	106	106
	28	144	144	128	128	107	107	92	92
	30	126	126	111	111	93	93	80	80
	32	111	111	98	98	82	82	70	70
	34	98	98	87	87	73	73	62	62
	36	87	87	77	77	65	65	55	55
	37	83	83	73	73	61	61	52	52
	38	78	78						
Y-Y AXIS	0	450	625	401	557	346	476	301	361
	10	391	514	346	453	295	382	254	294
	12	371	479	328	422	280	356	241	276
	14	349	440	309	388	263	327	226	257
	16	325	399	287	351	245	297	211	236
	18	300	357	265	314	226	265	194	214
	20	274	315	242	277	206	234	177	192
	22	248	275	219	241	186	203	160	170
	24	223	236	196	206	166	174	143	148
	26	198	201	173	176	147	149	127	128
	28	174	174	152	152	129	129	111	111
	30	151	151	133	133	112	112	97	97
	32	133	133	117	117	99	99	85	85
	34	118	118	103	103	88	88	75	75
	36	105	105	92	92	78	78	67	67
	38	95	95	83	83	70	70	61	61
	41	81	81	71	71	60	60	52	52

Effective length KL (ft) with respect to indicated axis

Properties				
A (in^2)	14.7	13.1	11.3	9.84
r_x (in.)	2.28	2.27	2.24	2.22
r_y (in.)	2.51	2.49	2.47	2.46

Note: Heavy line indicates Kl/r of 200.

Fy = 36 ksi
Fy = 50 ksi

COLUMNS
Structural tees
cut from W shapes
Design axial strength in kips ($\phi = 0.85$)

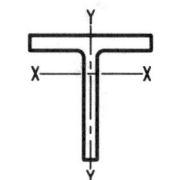

Designation		WT 8												
Wt./ft		28.5		25		22.5		20		18		15.5		13	
Fy		36	50	36	50	36	50	36	50	36	50	36	50	36	50
X-X AXIS	0	256	336	223	259	184	205	141	145	122	124	93	93	66	66
	6	245	316	213	245	176	195	136	140	118	120	91	90	65	65
	8	236	301	205	235	170	188	132	136	114	116	88	88	63	63
	10	225	283	196	223	163	179	127	130	111	112	86	85	62	62
	12	212	262	185	208	154	169	121	124	106	108	83	82	60	60
	14	199	240	173	193	145	157	115	117	101	102	79	79	58	58
	16	184	217	160	176	135	145	108	110	95	96	75	75	55	55
	18	168	193	146	159	124	133	100	102	89	90	71	71	53	53
	20	152	169	133	141	114	120	92	94	82	83	67	66	50	50
	22	136	147	119	125	103	107	85	86	76	76	62	62	47	47
	24	121	125	105	108	92	95	77	78	69	70	57	57	44	44
	26	106	107	93	93	82	83	69	70	63	63	53	53	41	41
	28	92	92	80	80	72	72	62	62	56	57	48	48	38	38
	30	80	80	70	70	62	62	54	54	50	50	44	44	35	35
	32	70	70	61	61	55	55	48	48	44	44	39	39	32	32
	34	62	62	54	54	49	49	42	42	39	39	35	35	29	29
	36	56	56	49	49	43	43	38	38	35	35	31	31	27	27
	38	50	50	44	44	39	39	34	34	31	31	28	28	24	24
	39	47	47	41	41	37	37	32	32	30	30	27	27	23	23
	40	45	45	39	39					28	28	25	25	22	22
	41													21	21
Y-Y AXIS	0	256	336	223	259	184	205	141	145	122	124	93	93	66	66
	6	216	268	185	207	153	167	116	119	95	97	70	70	47	47
	8	200	243	171	190	143	155	110	112	90	91	65	64	44	44
	10	181	214	155	170	130	140	102	104	84	85	58	57	40	40
	12	160	182	137	148	116	123	92	94	76	77	50	50	35	35
	14	138	151	119	125	101	106	82	84	68	69	42	42	30	30
	16	116	120	100	103	87	89	72	73	60	60	34	34	25	25
	18	96	96	83	83	72	72	62	62	51	51	27	27	21	21
	20	78	78	67	67	59	59	52	52	43	43				
	22	65	65	56	56	49	49	43	43	36	36				
	24	54	54	47	47	41	41	36	36	30	30				
	25	50	50	43	43	38	38	34	34	28	28				
	26	46	46	40	40	35	35	31	31						
Properties															
A (in²)		8.38		7.37		6.63		5.89		5.28		4.56		3.84	
rx (in.)		2.41		2.40		2.39		2.37		2.41		2.45		2.47	
ry (in.)		1.60		1.59		1.57		1.57		1.52		1.17		1.12	

Effective length KL (ft) with respect to indicated axis

Note: Heavy line indicates Kl / r of 200.

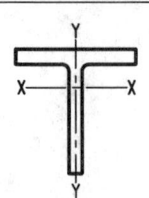

		F_y = 36 ksi
		F_y = 50 ksi

COLUMNS
Structural tees
cut from W shapes
Design axial strength in kips (ϕ = 0.85)

Designation		WT 7								
Wt./ft		66		60		54.5		49.5		45	
F_y		36	50	36	50	36	50	36	50	36	50
X-X AXIS	0	594	825	542	752	490	680	447	621	404	561
	2	588	813	536	741	484	670	442	611	399	552
	4	570	779	520	710	469	641	428	584	387	528
	6	542	726	493	661	444	595	405	542	366	489
	8	505	658	459	597	412	536	375	487	339	439
	10	461	580	418	525	374	468	340	425	307	383
	12	412	497	373	448	333	397	302	360	272	324
	14	361	414	326	371	289	327	262	296	236	265
	16	310	335	279	299	246	261	223	236	200	211
	18	261	266	234	237	205	207	185	186	166	166
	20	215	215	192	192	167	167	151	151	135	135
	22	178	178	158	158	138	138	125	125	111	111
	24	149	149	133	133	116	116	105	105	94	94
	26	127	127	113	113	99	99	89	89	80	80
	27	118	118	105	105	92	92	83	83	74	74
	28	110	110	98	98	85	85				
Y-Y AXIS	0	594	825	542	752	490	680	447	621	404	561
	6	578	794	526	722	475	651	432	591	389	531
	8	570	778	519	708	468	638	426	579	384	520
	10	559	758	509	689	459	621	418	564	376	506
	12	546	734	497	667	448	601	408	546	367	490
	14	531	706	483	642	436	578	396	525	357	472
	16	514	675	468	614	422	553	384	502	346	451
	18	496	642	451	584	407	526	370	477	333	429
	20	476	607	433	551	391	497	355	451	320	405
	22	455	571	414	518	373	467	339	423	305	381
	24	434	533	394	484	355	436	322	395	290	355
	26	411	495	373	449	337	404	305	366	275	329
	28	388	457	352	414	317	373	288	337	259	303
	30	365	420	331	380	298	342	270	309	243	278
	32	341	383	309	346	279	311	253	281	227	253
	34	318	347	288	314	260	282	235	255	211	229
	36	295	312	267	282	241	253	218	228	196	205
	38	273	281	247	253	222	227	201	205	180	184
	40	251	253	227	229	204	205	184	185	166	166

(Left margin label: Effective length KL (ft) with respect to indicated axis)

Properties

A (in²)	19.4		17.7	16.0	14.6	13.2
r_x (in.)	1.73		1.71	1.68	1.67	1.66
r_y (in.)	3.76		3.74	3.73	3.71	3.70

Note: Heavy line indicates Kl / r of 200.

F_y = 36 ksi
F_y = 50 ksi

COLUMNS
Structural tees
cut from W shapes
Design axial strength in kips (ϕ = 0.85)

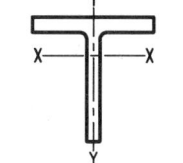

Designation		WT 7													
Wt./ft		41		37		34		30.5		26.5		24		21.5	
F_y		36	50	36	50	36	50	36	50	36	50	36	50	36	50
X-X AXIS	0	367	510	334	463	306	425	274	370	239	318	216	265	183	208
	4	354	486	322	440	295	403	264	352	231	303	209	254	177	200
	6	339	457	307	413	281	378	252	330	221	287	200	241	170	191
	8	319	419	288	378	264	346	236	302	208	265	188	224	160	179
	10	294	375	265	337	243	308	217	270	193	239	174	203	149	164
	12	267	327	240	293	219	267	196	235	175	211	158	181	136	148
	14	238	279	213	248	194	226	173	199	157	182	141	158	122	131
	16	208	232	186	205	169	186	151	165	138	153	124	134	108	114
	18	179	188	159	165	144	150	128	133	119	126	107	112	93	97
	20	151	152	134	134	121	121	108	108	101	102	91	92	80	81
	22	126	126	111	111	100	100	89	89	85	85	76	76	67	67
	24	106	106	93	93	84	84	75	75	71	71	64	64	56	56
	26	90	90	79	79	72	72	64	64	61	61	54	54	48	48
	28	78	78	68	68	62	62	55	55	52	52	47	47	41	41
	30	68	68	59	59	54	54	48	48	45	45	41	41	36	36
	31									43	43	38	38	34	34
Y-Y AXIS	0	367	510	334	463	306	425	274	370	239	318	216	265	183	208
	8	334	447	303	405	276	369	246	320	203	256	183	215	154	171
	10	320	421	290	382	265	347	236	302	189	233	170	197	144	158
	12	303	391	275	355	251	322	223	281	173	208	156	177	132	144
	14	285	359	258	325	235	295	210	258	156	181	140	156	119	128
	16	265	324	240	294	218	267	195	233	139	154	124	135	106	112
	18	244	289	221	262	201	238	179	209	121	129	108	114	93	97
	20	222	254	202	231	183	209	163	184	104	105	93	94	80	82
	22	201	221	182	200	165	181	147	160	87	87	78	78	68	68
	24	179	188	163	171	147	154	131	137	73	73	66	66	57	57
	26	159	161	144	146	130	131	116	117	63	63	56	56	49	49
	28	139	139	126	126	113	113	101	101	54	54	48	48	42	42
	30	121	121	110	110	99	99	88	88	47	47	42	42	37	37
	31	113	113	103	103	93	93	82	82	44	44	40	40	34	34
	32	106	106	97	97	87	87	77	77	41	41				
	34	94	94	86	86	77	77	69	69						
	36	84	84	76	76	69	69	61	61						
	40	68	68	62	62	56	56	50	50						
	41	65	65	59	59	53	53								

Effective length KL (ft) with respect to indicated axis

Properties

A (in²)	12.0	10.9	9.99	8.96	7.81	7.07	6.31
r_x (in.)	1.85	1.82	1.81	1.80	1.88	1.87	1.86
r_y (in.)	2.48	2.48	2.46	2.45	1.92	1.91	1.89

Note: Heavy line indicates Kl / r of 200.

AMERICAN INSTITUTE OF STEEL CONSTRUCTION

| | | F_y = 36 ksi |
| | | F_y = 50 ksi |

COLUMNS
Structural tees
cut from W shapes
Design axial strength in kips (ϕ = 0.85)

Designation		WT 7									
Wt./ft		19		17		15		13		11	
F_y		36	50	36	50	36	50	36	50	36	50
X-X AXIS	0	159	180	131	142	109	114	87	88	62	62
	2	158	178	130	141	109	114	87	88	62	62
	4	155	174	128	138	107	112	85	86	61	61
	6	150	168	124	134	104	108	83	84	60	60
	8	143	159	119	127	100	104	80	81	58	58
	10	134	148	112	120	95	98	77	78	56	56
	12	125	136	105	111	89	92	73	73	53	53
	14	114	123	96	102	83	85	68	69	51	51
	16	103	110	88	92	76	78	63	64	48	48
	18	92	97	79	82	69	70	58	58	44	44
	20	81	83	70	72	62	63	53	53	41	41
	22	70	71	62	63	55	55	48	48	38	38
	24	60	60	53	54	48	48	42	43	34	34
	26	51	51	46	46	42	42	37	38	31	31
	28	44	44	39	39	36	36	33	33	28	28
	30	38	38	34	34	31	31	28	28	24	24
	32	34	34	30	30	27	27	25	25	22	22
	34	30	30	27	27	24	24	22	22	19	19
	35							21	21	18	18
Y-Y AXIS	0	159	180	131	142	109	114	87	88	62	62
	6	132	146	108	115	86	89	65	66	45	45
	8	123	134	101	107	81	83	58	58	41	41
	10	111	120	92	97	74	76	50	50	35	35
	12	99	105	82	86	67	68	41	41	30	30
	14	86	89	72	74	59	60	32	32	24	24
	16	72	74	61	63	50	51	25	25	19	19
	17	66	66	56	57	46	47	22	22	17	17
	18	60	60	51	52	42	42	20	20		
	20	49	49	42	42	35	35				
	22	40	40	35	35	29	29				
	24	34	34	30	30	25	25				
	25	31	31	27	27						

Properties											
A (in^2)		5.58		5.00		4.42		3.85		3.25	
r_x (in.)		2.04		2.04		2.07		2.12		2.14	
r_y (in.)		1.55		1.53		1.49		1.08		1.04	

Note: Heavy line indicates Kl / r of 200.

Effective length KL (ft) with respect to indicated axis

| F_y = 36 ksi |
| F_y = 50 ksi |

COLUMNS
Structural tees
cut from W shapes
Design axial strength in kips (ϕ = 0.85)

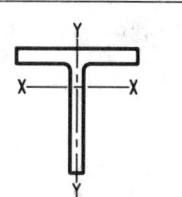

Designation		WT 6									
Wt./ft		29		26.6		25		22.5		20	
F_y		36	50	36	50	36	50	36	50	36	50
	0	261	362	238	331	225	312	202	280	180	221
X-X AXIS	2	257	355	235	325	222	307	200	276	178	218
	4	247	336	226	307	214	292	193	262	172	208
	6	231	306	211	280	202	269	181	241	161	193
	8	210	268	192	246	186	240	167	214	148	174
	10	186	227	171	208	167	207	149	184	133	152
	12	160	185	147	170	147	173	131	153	116	129
	14	135	145	124	134	126	139	112	123	99	106
	16	110	111	102	103	105	109	93	96	82	84
	18	88	88	81	81	86	86	75	75	66	66
	20	71	71	66	66	70	70	61	61	54	54
	22	59	59	54	54	58	58	51	51	44	44
	24	49	49	46	46	48	48	42	42	37	37
	25	45	45	42	42	45	45	39	39	34	34
	26					41	41	36	36	32	32
	0	261	362	238	331	225	312	202	280	180	221
	6	246	333	223	301	205	274	183	244	162	194
	8	238	318	216	287	194	255	174	227	154	182
	10	228	300	206	271	181	232	162	206	143	167
	12	216	279	196	252	166	206	148	183	131	150
	14	203	257	184	231	150	179	134	159	118	132
	16	189	233	171	209	134	152	119	135	105	114
	18	175	208	157	187	117	127	104	112	92	97
Y-Y AXIS	20	160	184	144	164	101	103	89	91	79	80
	22	144	160	130	143	86	86	75	75	66	66
	24	129	137	116	122	72	72	63	63	56	56
	26	115	117	103	104	61	61	54	54	48	48
	28	101	101	90	90	53	53	47	47	41	41
	30	88	88	78	78	46	46	41	41	36	36
	32	77	77	69	69	41	41	36	36	32	32
	34	69	69	61	61						
	36	61	61	54	54						
	38	55	55	49	49						
	41	47	47	42	42						
Properties											
A (in.^{2})		8.52		7.78		7.34		6.61		5.89	
r_x (in.)		1.50		1.51		1.60		1.58		1.57	
r_y (in.)		2.51		2.48		1.96		1.94		1.93	

Effective length KL (ft) with respect to indicated axis

Note: Heavy line indicates Kl/r of 200.

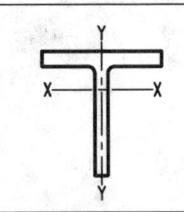

	F_y = 36 ksi
	F_y = 50 ksi

COLUMNS
Structural tees
cut from W shapes
Design axial strength in kips (ϕ = 0.85)

Designation		WT 6												
Wt./ft		17.5		15		13		11		9.5		8		7
F_y		36	50	36	50	36	50	36	50	36	50	36	50	36	50
X-X AXIS	0	158	188	120	132	90	92	88	98	68	71	53	54	40	40
	2	157	186	119	131	89	91	88	97	68	70	53	54	40	40
	4	152	179	116	127	87	89	86	95	66	69	52	53	39	39
	6	145	169	111	121	84	86	83	91	64	67	51	51	38	38
	8	135	156	104	113	80	81	78	86	61	63	48	49	37	37
	10	124	140	96	104	74	76	73	80	58	60	46	46	35	35
	12	111	124	87	93	68	70	68	73	54	55	43	43	33	33
	14	98	106	78	82	62	63	61	65	49	50	40	40	31	31
	16	85	89	68	71	55	56	55	58	44	45	36	36	29	29
	18	72	73	59	60	48	49	48	50	40	40	33	33	26	26
	20	59	59	50	50	42	42	42	43	35	35	29	29	24	24
	22	49	49	41	41	36	36	36	36	30	30	26	26	21	21
	24	41	41	35	35	30	30	30	30	26	26	22	22	19	19
	26	35	35	30	30	26	26	26	26	22	22	19	19	17	17
	28	30	30	25	25	22	22	22	22	19	19	16	16	14	14
	29	28	28	24	24	21	21	21	21	18	18	15	15	14	14
	30							19	19	17	17	14	14	13	13
	31							18	18	16	16	13	13	12	12
	32											13	13	11	11
Y-Y AXIS	0	158	188	120	132	90	92	88	98	68	71	53	54	40	40
	2	147	172	109	119	81	83	73	80	54	55	37	37	26	26
	4	142	165	106	115	79	80	67	72	49	51	34	34	25	25
	6	134	154	101	109	75	77	57	60	43	44	30	30	22	22
	8	123	139	93	100	71	72	45	46	34	35	24	25	19	19
	10	110	123	85	90	65	66	33	33	26	26	18	18	15	15
	12	96	105	75	79	59	59	23	23	18	18	13	13	11	11
	13	89	95	70	73	55	56	20	20	16	16				
	14	82	87	65	67	52	52	17	17						
	16	68	69	55	56	45	45								
	18	55	55	45	45	38	38								
	20	45	45	37	37	31	31								
	22	37	37	31	31	26	26								
	24	31	31	26	26	22	22								
	25	29	29	24	24	20	20								

Effective length *KL* (ft) with respect to indicated axis

Properties

	17.5	15	13	11	9.5	8	7
A (in²)	5.17	4.40	3.82	3.24	2.79	2.36	2.08
r_x (in.)	1.76	1.75	1.75	1.90	1.90	1.92	1.92
r_y (in.)	1.54	1.52	1.51	0.847	0.822	0.773	0.753

Note: Heavy line indicates *Kl* / *r* of 200.

| F_y = 36 ksi |
| F_y = 50 ksi |

COLUMNS
Structural tees
cut from W shapes
Design axial strength in kips ($\phi = 0.85$)

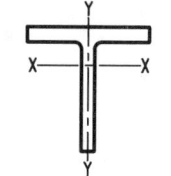

Designation		WT 5											
Wt./ft		22.5		19.5		16.5		15		13		11	
F_y		36	50	36	50	36	50	36	50	36	50	36	50
X-X AXIS	0	203	282	175	244	148	206	135	188	117	146	99	115
	2	199	274	172	237	146	201	133	184	115	144	98	113
	4	187	253	162	218	137	185	128	173	110	136	94	108
	6	170	220	147	190	125	162	119	157	102	124	87	99
	8	148	182	128	157	109	135	107	136	92	109	79	88
	10	124	142	107	123	92	106	94	114	81	92	69	76
	12	100	105	86	91	75	79	80	91	69	76	59	64
	14	77	77	67	67	58	58	67	70	57	60	49	51
	16	59	59	51	51	45	45	54	54	46	46	40	40
	18	47	47	40	40	35	35	42	42	36	36	32	32
	20	38	38	33	33	29	29	34	34	29	29	26	26
	21					26	26	31	31	27	27	23	23
	22							28	28	24	24	21	21
	24							24	24	20	20	18	18
Y-Y AXIS	0	203	282	175	244	148	206	135	188	117	146	99	115
	2	199	274	171	235	143	194	128	174	109	134	89	101
	4	195	266	167	227	140	188	122	163	104	126	85	96
	6	188	253	161	216	134	179	113	147	96	115	78	88
	8	178	235	153	201	127	166	101	126	86	100	70	78
	10	167	214	143	183	119	151	88	104	75	84	61	66
	12	153	191	131	163	109	134	74	82	63	68	51	54
	14	139	167	119	142	98	116	60	62	51	52	41	42
	16	125	143	106	121	87	99	47	47	40	40	32	32
	18	110	120	93	101	76	82	38	38	32	32	26	26
	20	95	99	80	83	66	67	30	30	26	26	21	21
	22	81	82	68	68	55	55	25	25	21	21	17	17
	24	69	69	57	57	47	47						
	26	59	59	49	49	40	40						
	28	50	50	42	42	34	34						
	30	44	44	37	37	30	30						
	32	39	39	32	32	26	26						
	33	36	36	30	30								

Effective length KL (ft) with respect to indicated axis

Properties

A (in²)	6.63	5.73	4.85	4.42	3.81	3.24
r_x (in.)	1.24	1.24	1.26	1.45	1.44	1.46
r_y (in.)	2.01	1.98	1.94	1.37	1.36	1.33

Note: Heavy line indicates Kl / r of 200.

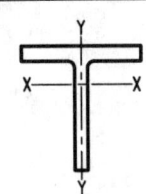

| | F_y = 36 ksi |
| | F_y = 50 ksi |

COLUMNS
Structural tees
cut from W shapes
Design axial strength in kips (ϕ = 0.85)

Designation		WT 5							
Wt./ft		9.5		8.5		7.5		6	
F_y		36	50	36	50	36	50	36	50
Effective length KL (ft) with respect to indicated axis	X-X AXIS								
	0	86	104	77	90	66	76	43	45
	2	85	103	76	88	65	75	43	44
	4	82	98	73	85	63	72	41	43
	6	77	91	68	79	59	67	39	41
	8	70	81	63	71	54	61	37	38
	10	62	71	56	62	49	54	34	35
	12	54	60	49	53	43	46	30	31
	14	46	49	42	44	37	39	27	27
	16	38	39	34	35	31	31	23	23
	18	30	30	28	28	25	25	19	20
	20	25	25	23	23	20	20	16	16
	22	20	20	19	19	17	17	13	13
	24	17	17	16	16	14	14	11	11
	25	16	16	14	14	13	13	10	10
	26			13	13	12	12	10	10
	Y-Y AXIS								
	0	86	104	77	90	66	76	43	45
	2	74	87	62	70	49	54	30	30
	4	67	77	56	62	45	48	28	28
	6	56	62	47	51	37	40	24	24
	8	43	46	36	37	29	29	19	20
	10	31	31	25	25	20	20	14	14
	12	22	22	18	18	14	14	10	10
	13	18	18	15	15	12	12	9	9
	14	16	16	13	13				

Properties

A (in^2)	2.81	2.50	2.21	1.77
r_x (in.)	1.54	1.56	1.57	1.57
r_y (in.)	0.874	0.844	0.810	0.785

Note: Heavy line indicates Kl / r of 200.

F_y = 36 ksi
F_y = 50 ksi

COLUMNS
Structural tees
cut from W shapes
Design axial strength in kips (ϕ = 0.85)

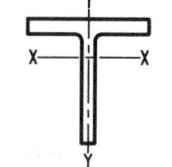

Designation									WT 4						
Wt./ft		14		12		10.5		9		7.5		6.5		5	
F_y		36	50	36	50	36	50	36	50	36	50	36	50	36	50
	0	126	175	108	150	94	131	80	112	68	94	59	82	41	46
	2	122	168	105	144	92	127	79	108	67	92	58	79	41	45
	3	118	160	101	137	89	121	76	104	65	89	56	77	40	44
	4	112	148	96	127	86	114	73	98	63	84	54	73	38	42
X-X AXIS	**5**	105	135	90	116	81	106	70	91	60	79	52	69	37	40
	6	96	121	82	103	76	97	65	83	57	73	49	64	35	38
	8	78	90	67	77	64	76	55	67	49	60	43	52	30	33
	10	60	62	51	52	52	57	45	50	41	47	36	41	26	27
	12	43	43	36	36	39	40	35	35	33	34	29	30	21	21
	14	32	32	27	27	29	29	26	26	25	25	22	22	16	16
	16	24	24	20	20	22	22	20	20	19	19	17	17	12	12
	18					18	18	16	16	15	15	13	13	10	10
	19							14	14	14	14	12	12	9	9
	20									12	12	11	11	8	8
	0	126	175	108	150	94	131	80	112	68	94	59	82	41	46
	2	123	169	105	143	89	121	75	100	60	79	49	63	33	35
	4	119	161	101	137	85	113	70	93	54	68	44	55	30	32
	6	112	149	96	127	77	99	64	82	45	53	36	43	25	27
	8	104	133	88	113	68	83	56	68	35	37	28	29	20	21
	10	93	116	80	98	57	66	47	54	25	25	19	19	15	15
	12	82	97	70	83	47	49	38	40	17	17	14	14	10	10
Y-Y AXIS	**14**	71	79	60	67	37	37	30	30	13	13	10	10	8	8
	16	60	62	51	53	28	28	23	23						
	18	49	49	42	42	22	22	18	18						
	20	40	40	34	34	18	18	15	15						
	21	36	36	31	31	16	16								
	22	33	33	28	28										
	24	28	28	24	24										
	26	24	24	20	20										
	27	22	22												

Effective length KL (ft) with respect to indicated axis

Properties							
A (in^2)		4.12	3.54	3.08	2.63	2.22	1.92	1.48
r_x (in.)		1.01	.999	1.12	1.14	1.22	1.23	1.20
r_y (in.)		1.62	1.61	1.26	1.23	0.876	0.843	0.841

Note: Heavy line indicates Kl/r of 200.

Single-Angle Struts

Design strengths of single-angle struts were formerly not tabulated in this Manual because of the difficulty in loading such struts concentrically. Concentric loading can be accomplished by milling the ends of an angle and loading it through bearing plates. However, in common practice, the eccentricity of loading is relatively large, and its neglect in design may lead to an under-designed member.

The design of single-angle struts is governed by the AISC *Specification for Load and Resistance Factor Design of Single-Angle Members,* which is reproduced in Part 6 of this Manual. The tables below conservatively include the effect of flexural-torsional buckling using the procedure discussed in Zureick (1993).

The following example illustrates the design procedure for an equal-leg angle loaded eccentrically. The design strengths for concentric loading, tabulated below, are useful in solving the interaction equations for combined axial force and bending.

EXAMPLE 3-8

Given:

An angle 2×2×¼ is loaded by a gusset plate attached to one leg with an eccentricity of 0.8 in. from the centroid, as shown in Figure 3-3. Determine the factored compressive load P_u which may be applied. The effective length KL is 4.0 ft.

$A = 0.938$ in.^{2}
$r_z = 0.391$ in.
$I_x = I_y = 0.348$ in.^{4}

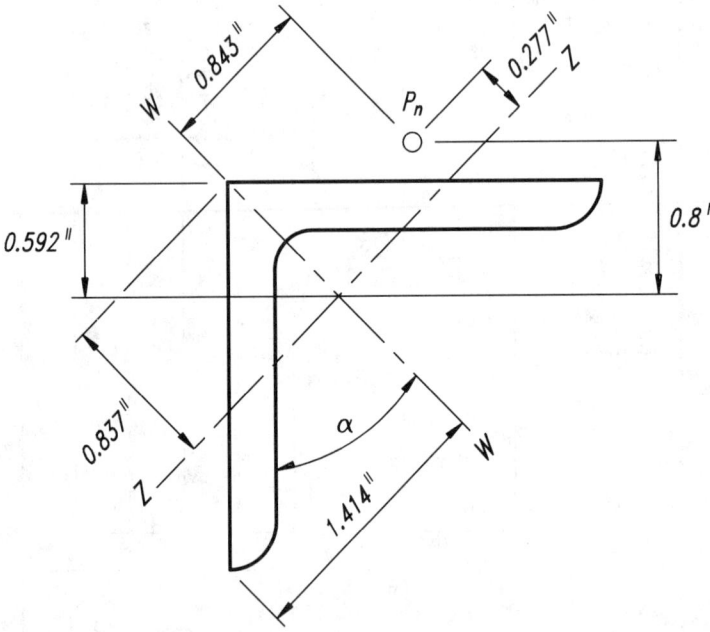

Fig. 3-3

$\alpha = 45°$

$F_y = 50$ ksi

Solution: Determine the properties for the principal axes Z-Z and W-W as follows:

$$I_z = Ar_z^2 = 0.938(0.391)^2 = 0.143 \text{ in.}^2$$
$$I_w + I_z = I_x + I_y$$
$$I_w = 0.348 + 0.348 - 0.143 = 0.552 \text{ in.}^4$$

$$r_w = \sqrt{\frac{I_w}{A}} = \sqrt{\frac{0.552}{0.938}} = 0.767 \text{ in.}$$

From the tables which follow, the design compressive strength $\phi_c P_n = 14$ kips for $KL = 4$ ft.

For combined axial compression and bending, the latter is taken about the principal axes in accordance with the Single-Angle LRFD Specification (Section 6).

For equal leg angles—

Major principal axis (W-W) bending (Section 5.3.1):

$$M_{ob} = C_b \frac{0.46Eb^2t^2}{l}$$
$$= 1.0 \times \frac{0.46(29,000 \text{ ksi})(2 \text{ in.})^2(0.25 \text{ in.})^2}{48 \text{ in.}}$$
$$= 69.5 \text{ k-in.}$$
$$M_y = F_y S_w = F_y \frac{I_w}{c_w} = 50 \text{ ksi} \times \frac{0.552 \text{ in.}^4}{1.414 \text{ in.}}$$
$$= 19.5 \text{ k-in.}$$

Since $M_{ob} > M_y$ (Section 5.1.3),

$$M_{nw} = [1.58 - 0.83 \sqrt{M_y/M_{ob}}] M_y \le 1.25M_y$$
$$= [1.58 - 0.83 \sqrt{19.5/69.5}] M_y = 1.14M_y$$
$$= 1.14 \times 19.5 \text{ k-in.}$$
$$= 22 \text{ k-in.}$$

According to Section 5.1.1,

for b/t (= 2 in. / 0.25 in. = 8) $< 0.382 \sqrt{E/F_y}$
(= $0.382 \sqrt{29,000/50} = 9.2$),

$$M_{nw} \le 1.25F_y S_c = 1.25F_y S_w = 1.25M_y$$

This is satisfied since $M_{nw} = 1.14M_y$.

Minor principal axis (Z-Z) bending (Section 5.3.1):

With the leg tips of the angle in tension and the angle corner in compression

$$M_{nz} = 1.25M_y = 1.25F_y S_z = 1.25F_y \frac{I_z}{c_z}$$

$$= 1.25 \times 50 \text{ ksi} \times \frac{0.143 \text{ in.}^4}{0.837 \text{ in.}}$$

$$= 11 \text{ k-in.}$$

Assuming $\frac{P_u}{\phi_c P_n} \geq 0.2$, Interaction Equation 6-1a governs.

$$\left| \frac{P_u}{\phi_c P_n} + \frac{8}{9} \left(\frac{M_{uw}}{\phi_b M_{nw}} + \frac{M_{uz}}{\phi_b M_{nz}} \right) \right| \leq 1.0$$

According to Section 6.1.1, for flexural compression M_u shall be multiplied by B_1 (Equation 6-2).

Major principal axis (W-W) bending:

$$Kl / r_w = 1.0 \times 48 / 0.767 = 62.2$$

From LRFD Specification Table 8, $P_e / A_g = 73.1$

$$P_{e1w} = 73.1(0.938) = 68.6 \text{ kips}$$

$$B_{1w} = \frac{C_m}{1 - P_u / P_{e1w}} = \frac{0.85}{1 - P_u / 68.6} < 1. \text{ Use } B_{1w} = 1.0.$$

Minor principal axis (Z-Z) bending:

$$Kl / r_z = 1.0 \times 48 / 0.391 = 122.8$$

From LRFD Specification Table 8, $P_e / A_g = 19.0$

$$P_{e1z} = 19.0(0.938) = 17.8 \text{ kips}$$

$$B_{1z} = \frac{C_m}{1 - P_u / P_{e1z}} = \frac{0.85}{1 - P_u / 17.8}$$

Conservatively adding the maximum axial and flexural terms, Equation 6-1a becomes

$$\frac{P_u}{14 \text{ kips}} + \frac{8}{9} \left(\frac{P_u \times 0.843 \text{ in.} \times 1.0}{0.9 \times 22 \text{ kip-in.}} + \frac{P_u \times 0.277 \text{ in.}}{0.9 \times 11 \text{ kip-in.}} \frac{0.85}{1 - P_u / 17.8} \right) \leq 1.0$$

$$P_u = 7 \text{ kips}$$

Checking $\dfrac{P_u}{\phi_c P_n} = \dfrac{7 \text{ kips}}{14 \text{ kips}} = 0.5 > 0.2$ **o.k.**

A less conservative approach would have involved applying the interaction equation separately at the corner and the two leg tips of the angle, with the proper signs (+ or −) for compression and tension.

| F_y = 36 ksi |
| F_y = 50 ksi |

COLUMNS
Single angles
Design axial strength in kips (ϕ = 0.90)

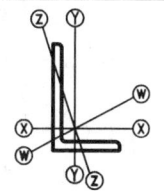

Size		8×6												
Thickness		1		$\frac{7}{8}$		$\frac{3}{4}$		$\frac{5}{8}$		$\frac{9}{16}$		$\frac{1}{2}$		$\frac{7}{16}$	
Wt./ft		44.2		39.1		33.8		28.5		25.7		23.0		20.2	
F_y		36	50	36	50	36	50	36	50	36	50	36	50	36	50
Effective length KL (ft)	0	421	585	373	518	322	447	270	352	235	303	199	253	163	203
	1	406	555	355	484	301	408	246	311	210	262	174	213	138	166
	2	398	541	347	468	292	391	235	294	199	245	163	197	128	151
	3	393	531	342	460	288	383	231	287	195	239	160	192	125	146
	4	384	515	336	448	284	375	228	282	193	235	158	188	123	144
	5	371	490	325	429	277	363	224	276	190	230	156	185	122	141
	6	353	458	311	402	266	343	217	265	186	224	153	181	120	139
	7	333	422	293	371	252	319	208	250	179	214	149	175	118	136
	8	311	383	274	338	236	291	195	231	170	200	143	166	115	131
	9	287	344	253	303	219	262	182	210	159	184	135	155	110	124
	10	263	305	232	268	201	233	167	189	148	167	126	142	104	116
	11	239	266	211	235	183	204	152	167	135	149	117	128	97	107
	12	215	230	190	203	165	177	137	146	123	131	107	114	90	97
	13	192	196	169	173	147	151	123	125	111	114	96	101	82	87
	14	169	169	149	149	130	130	109	109	99	99	87	88	75	77
	15	147	147	130	130	114	114	95	95	87	87	77	77	67	67
	16	130	130	115	115	100	100	84	84	77	77	68	68	60	60
	17	115	115	102	102	89	89	74	74	68	68	60	60	53	53
	18	103	103	91	91	79	79	66	66	61	61	54	54	48	48
	19	92	92	81	81	71	71	60	60	55	55	49	49	43	43
	20	83	83	74	74	64	64	54	54	49	49	44	44	39	39
	21	76	76	67	67	58	58	49	49	45	45	40	40	35	35

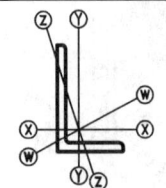

		F_y = 36 ksi
		F_y = 50 ksi

COLUMNS
Single angles
Design axial strength in kips (ϕ = 0.90)

Size														8×4
Thickness		1		7/8		3/4		5/8		9/16		1/2		7/16
Wt./ft		37.4		33.1		28.7		24.2		21.9		19.6		17.2
F_y		36	50	36	50	36	50	36	50	36	50	36	50	36	50

Effective length *KL* (ft)														
0	356	495	315	438	273	380	230	299	200	258	170	216	139	174
1	342	468	300	408	256	346	209	264	179	223	149	182	118	142
2	331	446	289	387	245	326	198	246	169	207	139	168	109	128
3	315	417	275	363	235	307	190	233	162	197	134	160	105	122
4	293	378	257	330	220	280	179	216	154	184	128	150	101	116
5	267	332	234	290	201	248	165	194	142	167	119	138	95	108
6	238	283	209	248	180	212	148	169	129	146	109	123	88	98
7	208	234	183	205	157	176	130	142	114	125	98	107	80	87
8	177	187	156	164	135	141	112	117	99	104	86	90	71	75
9	148	149	131	131	113	113	94	94	84	84	73	74	62	63
10	121	121	107	107	92	92	77	77	70	70	61	61	53	53
11	100	100	89	89	77	77	65	65	58	58	52	52	45	45
12	85	85	75	75	65	65	55	55	49	49	44	44	38	38
13	72	72	64	64	56	56	47	47	42	42	38	38	33	33
14	62	62	55	55	48	48	41	41	37	37	33	33	29	29

Size											7×4
Thickness		3/4		5/8		1/2		7/16		3/8	
Wt./ft		26.2		22.1		17.9		15.7		13.6	
F_y		36	50	36	50	36	50	36	50	36	50

Effective length *KL* (ft)											
0	249	346	210	288	164	212	136	173	108	134
1	236	320	194	258	146	182	118	144	90	107
2	228	306	187	245	139	171	111	133	85	99
3	219	289	180	233	134	164	107	128	82	95
4	205	264	170	215	128	154	103	122	79	91
5	188	233	156	191	119	140	97	113	76	86
6	168	200	140	165	108	124	89	101	70	79
7	147	166	123	138	96	106	80	88	64	70
8	126	134	106	113	84	89	71	75	57	61
9	106	107	89	90	71	72	61	62	50	52
10	87	87	73	73	59	59	51	51	43	43
11	72	72	61	61	49	49	43	43	37	37
12	61	61	52	52	42	42	37	37	31	31
13	52	52	44	44	36	36	31	31	27	27
14	45	45	38	38	31	31	27	27	23	23

$F_y = 36$ ksi
$F_y = 50$ ksi

COLUMNS
Single angles
Design axial strength in kips ($\phi = 0.90$)

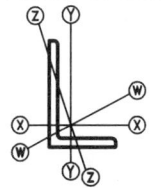

Size																6×4
Thickness		$^7/_8$		$^3/_4$		$^5/_8$		$^9/_{16}$		$^1/_2$		$^7/_{16}$		$^3/_8$		$^5/_{16}$
Wt./ft		27.2		23.6		20.0		18.1		16.2		14.3		12.3		10.3
F_y		36	50	36	50	36	50	36	50	36	50	36	50	36	50	36	50
Effective length *KL* (ft)	0	259	359	225	312	190	264	172	239	154	205	132	171	107	136	81	100
	1	250	343	215	293	178	241	159	214	139	180	116	145	91	111	66	78
	2	244	331	210	284	174	233	155	206	135	172	112	138	87	105	63	73
	3	233	310	201	268	168	222	150	197	131	166	109	133	85	102	61	70
	4	217	282	188	244	157	203	141	182	125	155	105	126	82	97	59	68
	5	198	248	172	215	144	180	130	162	115	139	98	116	78	91	57	65
	6	177	212	154	184	129	155	117	139	104	121	89	102	72	82	54	61
	7	155	176	135	153	113	129	102	116	91	102	79	87	65	72	50	55
	8	133	142	115	124	98	105	88	94	79	84	68	73	57	61	45	48
	9	111	112	97	98	82	83	74	75	67	67	58	59	49	50	39	41
	10	91	91	80	80	67	67	61	61	55	55	48	48	41	41	34	34
	11	75	75	66	66	56	56	51	51	46	46	40	40	34	34	29	29
	12	63	63	55	55	47	47	43	43	38	38	34	34	29	29	24	24
	13	54	54	47	47	40	40	37	37	33	33	29	29	25	25	21	21
	14	47	47	41	41	35	35	32	32	28	28	25	25	22	22	18	18

Size			6×3½			
Thickness		$^1/_2$		$^3/_8$		$^5/_{16}$
Wt./ft		15.3		11.7		9.8
F_y		36	50	36	50	36	50
Effective length *KL* (ft)	0	146	195	101	128	77	95
	1	132	171	86	105	63	74
	2	127	162	82	99	59	69
	3	121	152	79	94	57	66
	4	112	137	75	88	55	63
	5	100	118	69	79	51	58
	6	87	98	61	68	47	51
	7	74	78	53	56	41	44
	8	60	61	44	45	36	37
	9	48	48	36	36	30	30
	10	39	39	30	30	25	25
	11	33	33	25	25	21	21
	12	28	28	21	21	18	18

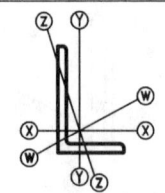

	F_y = 36 ksi
	F_y = 50 ksi

COLUMNS
Single angles
Design axial strength in kips ($\phi = 0.90$)

Size	5×3½											
Thickness	¾		⅝		½		⅜		5/16		¼	
Wt./ft	19.8		16.8		13.6		10.4		8.7		7.0	
F_y	36	50	36	50	36	50	36	50	36	50	36	50
Effective length KL (ft) 0	188	261	159	221	130	180	97	126	76	96	54	66
1	182	249	152	207	120	162	85	107	64	77	43	49
2	176	238	148	199	117	156	82	102	61	74	41	47
3	165	218	139	183	112	146	80	98	60	71	40	45
4	150	192	127	162	103	130	75	90	57	67	38	44
5	133	162	113	137	91	111	68	79	53	61	36	41
6	115	132	97	112	79	90	59	66	47	52	34	37
7	96	103	82	88	67	71	50	54	41	44	30	32
8	79	79	67	67	55	55	42	42	34	35	26	27
9	63	63	53	53	44	44	33	33	28	28	22	22
10	51	51	43	43	35	35	27	27	23	23	18	18
11	42	42	36	36	29	29	23	23	19	19	15	15
12	35	35	30	30	25	25	19	19	16	16	13	13

Size	5×3									
Thickness	½		7/16		⅜		5/16		¼	
Wt./ft	12.8		11.3		9.8		8.2		6.6	
F_y	36	50	36	50	36	50	36	50	36	50
Effective length KL (ft) 0	122	169	107	146	91	118	71	90	51	62
1	112	151	97	127	80	100	60	72	40	46
2	108	143	93	120	76	94	57	68	38	44
3	100	128	87	109	72	87	54	64	36	42
4	88	108	77	93	65	76	50	58	34	39
5	75	87	66	76	56	63	44	49	31	34
6	62	66	54	58	46	49	37	40	27	29
7	49	49	43	43	37	37	30	31	23	24
8	38	38	33	33	29	29	24	24	19	19
9	30	30	27	27	23	23	19	19	15	15
10	24	24	22	22	19	19	16	16	13	13

| F_y = 36 ksi |
| F_y = 50 ksi |

COLUMNS
Single angles
Design axial strength in kips ($\phi = 0.90$)

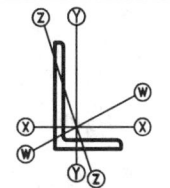

Size	4×3½							
Thickness	½		⅜		5/16		¼	
Wt./ft	11.9		9.1		7.7		6.2	
F_y	36	50	36	50	36	50	36	50
Effective length KL (ft) 0	113	158	87	120	73	95	53	68
1	107	146	78	104	63	79	43	52
2	105	142	77	102	61	76	42	50
3	99	131	75	98	60	74	42	49
4	90	114	68	87	57	69	41	48
5	79	95	60	73	51	59	38	44
6	67	76	51	58	43	48	33	37
7	56	58	43	45	36	38	28	30
8	45	45	35	34	29	29	23	24
9	35	35	27	27	23	23	19	19
10	29	29	22	22	19	19	15	15
11	24	24	18	18	15	15	13	13
12	20	20	15	15	13	13	11	11

Size	4×3											
Thickness	5/8		½		7/16		⅜		5/16		¼	
Wt./ft	13.6		11.1		9.8		8.5		7.2		5.8	
F_y	36	50	36	50	36	50	36	50	36	50	36	50
Effective length KL (ft) 0	129	179	105	146	93	129	80	112	67	88	50	63
1	124	170	100	136	87	117	73	98	59	74	41	50
2	119	160	96	129	84	112	71	94	57	71	40	48
3	108	141	88	114	78	101	66	86	55	67	39	46
4	95	118	78	96	68	84	59	73	49	58	36	42
5	81	93	66	76	58	67	50	58	42	47	32	36
6	66	70	54	57	47	51	41	44	35	37	27	29
7	52	51	42	42	37	37	32	32	27	27	22	22
8	39	39	32	32	29	29	25	25	21	21	17	17
9	31	31	26	26	23	23	20	20	17	17	14	14
10	25	25	21	21	18	18	16	16	14	14	11	11

Size	3½×3							
Thickness	½		⅜		5/16		¼	
Wt./ft	10.2		7.9		6.6		5.4	
F_y	36	50	36	50	36	50	36	50
Effective length KL (ft) 0	97	135	75	104	63	86	49	63
1	93	127	69	93	56	73	41	51
2	89	120	67	90	55	71	40	49
3	81	105	62	81	52	66	39	48
4	71	87	54	67	46	56	36	42
5	59	68	46	53	38	44	30	34
6	48	50	37	39	31	33	25	27
7	37	37	29	29	24	24	20	20
8	28	28	22	22	19	19	15	15
9	22	22	17	17	15	15	12	12
10	18	18	14	14	12	12	10	10

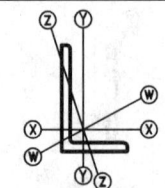

$F_y = 36$ ksi
$F_y = 50$ ksi

COLUMNS
Single angles
Design axial strength in kips ($\phi = 0.90$)

Size			$3\frac{1}{2}\times2\frac{1}{2}$			
Thickness	\multicolumn{2}{c}{$\frac{1}{2}$}		$\frac{3}{8}$		$\frac{1}{4}$	
Wt./ft	9.4		7.2		4.9	
F_y	36	50	36	50	36	50
Effective length KL (ft)						
0	89	124	68	95	45	58
1	85	116	63	85	38	47
2	79	105	60	79	37	45
3	70	88	53	67	34	41
4	58	68	44	52	29	33
5	46	49	35	38	24	25
6	34	34	26	26	18	18
7	25	25	19	19	13	13
8	19	19	15	15	10	10
9					8	8

Size					$3\times2\frac{1}{2}$					
Thickness	$\frac{1}{2}$		$\frac{3}{8}$		$\frac{5}{16}$		$\frac{1}{4}$		$\frac{3}{16}$	
Wt./ft	8.5		6.6		5.6		4.5		3.39	
F_y	36	50	36	50	36	50	36	50	36	50
Effective length KL (ft)										
0	81	113	62	86	52	73	42	57	29	37
1	78	107	58	79	48	65	37	48	24	29
2	72	96	55	73	46	61	36	46	23	28
3	63	79	48	61	41	51	33	40	22	26
4	52	60	40	46	34	39	27	31	19	22
5	40	42	31	33	26	28	21	23	16	17
6	29	29	23	23	19	19	16	16	12	12
7	22	22	17	17	14	14	12	12	9	9
8	17	17	13	13	11	11	9	9	7	7

Size					3×2					
Thickness	$\frac{1}{2}$		$\frac{3}{8}$		$\frac{5}{16}$		$\frac{1}{4}$		$\frac{3}{16}$	
Wt./ft	7.7		5.9		5.0		4.1		3.07	
F_y	36	50	36	50	36	50	36	50	36	50
Effective length KL (ft)										
0	73	101	56	78	47	66	39	51	27	34
1	69	94	52	71	43	58	34	43	22	27
2	61	80	47	61	39	51	31	39	21	25
3	50	60	38	46	32	39	26	30	18	21
4	37	40	29	31	24	26	20	21	14	15
5	26	26	20	20	17	17	14	14	10	10
6	18	18	14	14	12	12	10	10	7	7
7	13	13	10	10	9	9	7	7	5	5

Size				$2\frac{1}{2}\times2$				
Thickness	$\frac{3}{8}$		$\frac{5}{16}$		$\frac{1}{4}$		$\frac{3}{16}$	
Wt./ft	5.3		4.5		3.62		2.75	
F_y	36	50	36	50	36	50	36	50
Effective length KL (ft)								
0	50	70	42	59	34	48	26	33
1	48	65	40	54	31	42	22	27
2	42	55	36	46	29	37	21	25
3	34	41	29	34	23	28	17	20
4	25	27	21	21	17	18	13	14
5	17	17	15	15	12	12	9	9
6	12	12	10	10	8	8	6	6
7	9	9	7	7	6	6	5	5

| F_y = 36 ksi | | | | | | | | | | | | | |
| F_y = 50 ksi | | | | | | | | | | | | | |

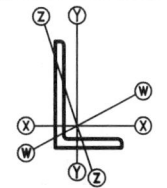

COLUMNS
Single angles
Design axial strength in kips (ϕ = 0.90)

Size								8×8						
Thickness		$1\frac{1}{8}$		1		$\frac{7}{8}$		$\frac{3}{4}$		$\frac{5}{8}$		$\frac{9}{16}$		$\frac{1}{2}$
Wt./ft		56.9		51.0		45.0		38.9		32.7		29.6		26.4
F_y		36	50	36	50	36	50	36	50	36	50	36	50	36	50
	0	541	752	486	675	428	594	369	513	310	405	270	348	229	291
	6	484	643	435	578	383	510	320	420	254	311	213	256	173	203
	7	465	608	417	546	368	482	318	417	252	309	212	255	172	202
	8	444	570	398	512	351	452	304	392	251	306	210	253	171	201
	9	421	530	378	476	334	421	289	365	243	294	209	250	170	199
	10	397	488	356	438	315	388	273	337	230	273	202	240	168	197
	11	372	446	334	401	295	355	256	308	215	251	191	222	165	191
	12	346	404	311	363	275	322	239	280	201	230	178	204	155	177
	13	320	363	288	326	255	289	221	252	186	208	166	186	144	162
	14	294	323	264	290	234	258	204	225	172	187	154	169	134	148
	15	269	283	242	255	215	227	187	198	157	167	141	151	124	133
	16	244	249	219	224	195	199	170	174	143	147	129	134	114	120
	17	221	221	198	198	176	177	154	154	130	130	118	119	104	106
	18	197	197	177	177	158	158	138	138	116	116	107	106	95	95
	19	177	177	159	159	141	141	124	124	104	104	95	95	85	85
	20	159	159	143	143	128	128	112	112	94	94	86	86	77	77
	21	145	145	130	130	116	116	101	101	85	85	78	78	70	70
	22	132	132	118	118	105	105	92	92	78	78	71	71	64	64
	23	121	121	108	108	96	96	84	84	71	71	65	65	58	58
	24	111	111	99	99	89	89	78	78	65	65	60	60	53	53
	25	102	102	92	92	82	82	71	71	60	60	55	55	49	49
	26	94	94	85	85	76	76	66	66	56	56	51	51	45	45

Effective length KL (ft)

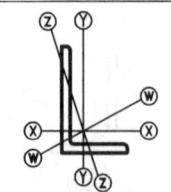

	Fy = 36 ksi
	Fy = 50 ksi

COLUMNS
Single angles
Design axial strength in kips (φ = 0.90)

Size	6×6																	
Thickness	1		7/8		3/4		5/8		9/16		1/2		7/16		3/8		5/16	
Wt./ft	37.4		33.1		28.7		24.2		21.9		19.6		17.2		14.9		12.4	
F_y	36	50	36	50	36	50	36	50	36	50	36	50	36	50	36	50	36	50
0	356	495	315	438	273	380	230	320	208	289	186	249	159	206	129	164	98	120
1	346	476	304	416	260	354	214	289	190	255	166	213	137	170	107	129	77	89
2	343	469	300	408	255	345	209	279	184	244	160	202	131	160	100	119	71	81
3	339	462	298	404	253	342	207	275	182	241	158	199	129	157	99	117	69	79
4	326	438	289	387	250	336	205	273	181	238	157	197	128	156	98	115	69	78
5	310	409	275	361	238	314	201	265	180	236	155	195	127	154	97	114	68	77
6	292	376	258	332	224	288	189	244	171	221	153	192	126	152	96	113	68	76
7	272	340	241	301	209	261	177	221	160	200	143	174	124	149	95	112	67	76
8	250	303	221	268	192	232	163	197	147	179	132	156	114	134	94	110	66	75
9	228	266	202	235	175	204	148	174	134	157	120	138	105	120	87	99	66	74
10	205	230	181	203	157	176	134	151	121	136	108	120	95	105	79	88	63	71
11	183	196	162	173	140	150	119	128	108	116	97	103	85	92	71	77	58	63
12	161	164	142	145	123	126	105	108	95	98	85	87	75	78	64	67	52	56
13	140	140	124	124	108	107	92	92	83	83	74	74	66	67	57	57	47	49
14	121	121	107	107	92	92	79	79	72	72	64	64	57	57	49	49	42	42
15	105	105	93	93	81	81	69	69	62	62	56	56	50	50	43	43	37	37
16	92	92	82	82	71	71	61	61	55	55	49	49	44	44	38	38	32	32
17	82	82	72	72	63	63	54	54	49	49	43	43	39	39	34	34	29	29
18	73	73	64	64	56	56	48	48	43	43	39	39	35	35	30	30	25	25
19	65	65	58	58	50	50	43	43	39	39	35	35	31	31	27	27	23	23

Effective length KL (ft)

Size	5×5														
Thickness	7/8		3/4		5/8		1/2		7/16		3/8		5/16	
Wt./ft	27.2		23.6		20.0		16.2		14.3		12.3		10.3	
F_y	36	50	36	50	36	50	36	50	36	50	36	50	36	50
0	259	359	225	312	190	264	154	214	135	184	115	149	89	114
1	251	345	216	296	179	244	141	189	121	157	99	122	73	88
2	249	341	214	291	177	239	138	183	117	152	95	117	70	83
3	241	325	209	283	176	236	137	181	116	150	94	115	69	81
4	228	301	198	262	167	221	135	179	115	148	93	114	68	80
5	212	272	184	237	156	200	127	163	112	141	92	112	67	79
6	194	241	169	210	143	178	116	145	102	126	87	105	67	78
7	175	209	152	182	129	154	105	126	93	110	79	92	64	74
8	155	177	135	154	114	131	93	107	82	94	71	80	57	65
9	135	146	118	128	100	108	82	89	72	78	62	67	51	55
10	116	119	102	104	86	88	70	72	62	64	54	56	45	47
11	98	98	86	86	73	73	60	60	53	53	46	46	38	39
12	82	82	72	72	61	61	50	50	44	44	39	39	33	33
13	70	70	61	61	52	52	43	43	38	38	33	33	28	28
14	60	60	53	53	45	45	37	37	33	33	28	28	24	24
15	53	53	46	46	39	39	32	32	28	28	25	25	21	21
16	46	46	40	40	34	34	28	28	25	25	22	22	18	18

Effective length KL (ft)

Fy = 36 ksi
Fy = 50 ksi

COLUMNS
Single angles
Design axial strength in kips (φ = 0.90)

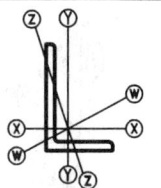

Size		4×4												
Thickness		3/4		5/8		1/2		7/16		3/8		5/16		1/4
Wt./ft		18.5		15.7		12.8		11.3		9.8		8.2		6.6
F_y		36	50	36	50	36	50	36	50	36	50	36	50	36	50
Effective length KL (ft)	0	176	245	149	207	122	169	107	149	93	129	78	101	57	73
	1	171	236	144	196	114	155	99	133	83	110	66	82	46	55
	2	168	228	142	194	113	152	97	130	81	107	64	79	44	52
	3	158	209	134	178	109	145	96	128	80	106	64	78	44	51
	4	144	185	122	157	100	128	88	113	76	98	63	77	43	51
	5	129	159	109	135	89	110	79	97	68	84	57	68	43	50
	6	112	131	95	111	78	91	69	81	60	70	50	57	39	44
	7	96	105	81	89	66	73	59	65	51	56	43	47	34	37
	8	79	81	67	69	55	56	49	50	43	44	36	37	29	30
	9	64	64	54	54	44	44	40	40	34	34	29	29	24	24
	10	52	52	44	44	36	36	32	32	28	28	24	24	19	19
	11	43	43	36	36	30	30	26	26	23	23	19	19	16	16
	12	36	36	30	30	25	25	22	22	19	19	16	16	13	13
	13					21	21	19	19	16	16	14	14	11	11

Size		3½×3½									
Thickness		1/2		7/16		3/8		5/16		1/4	
Wt./ft		11.1		9.8		8.5		7.2		5.8	
F_y		36	50	36	50	36	50	36	50	36	50
Effective length KL (ft)	0	105	146	93	129	80	112	68	93	53	68
	1	100	137	87	118	74	99	60	78	44	53
	2	99	134	86	116	73	97	59	76	43	52
	3	91	119	80	106	70	91	58	75	42	51
	4	81	102	72	90	62	78	53	65	41	50
	5	70	83	62	74	54	64	46	54	36	42
	6	59	65	52	58	45	50	38	42	31	34
	7	48	49	42	43	37	37	31	32	25	26
	8	37	37	33	33	29	29	24	24	20	20
	9	29	29	26	26	23	23	19	19	16	16
	10	24	24	21	21	18	18	16	16	13	13
	11	20	20	17	17	15	15	13	13	11	11

Size		3×3											
Thickness		1/2		7/16		3/8		5/16		1/4		3/16	
Wt./ft		9.4		8.3		7.2		6.1		4.9		3.71	
F_y		36	50	36	50	36	50	36	50	36	50	36	50
Effective length KL (ft)	0	89	124	79	109	68	95	58	80	47	62	32	41
	1	86	117	75	102	64	86	52	70	40	51	25	30
	2	82	109	72	97	63	84	52	69	39	50	25	29
	3	73	94	65	83	56	72	47	61	38	48	24	29
	4	62	76	55	67	48	58	41	49	33	39	24	28
	5	51	57	45	51	40	44	33	38	27	30	20	22
	6	40	41	36	36	31	32	26	27	21	22	16	17
	7	30	30	27	27	23	23	20	20	16	16	12	12
	8	23	23	20	20	18	18	15	15	12	12	9	9
	9	18	18	16	16	14	14	12	12	10	10	7	7

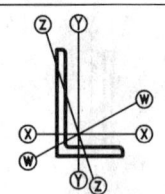

COLUMNS
Single angles
Design axial strength in kips ($\phi = 0.90$)

$F_y = 36$ ksi
$F_y = 50$ ksi

Size					$2\frac{1}{2} \times 2\frac{1}{2}$					
Thickness		$\frac{1}{2}$		$\frac{3}{8}$		$\frac{5}{16}$		$\frac{1}{4}$		$\frac{3}{16}$	
Wt./ft		7.7		5.9		5.0		4.1		3.07	
F_y		36	50	36	50	36	50	36	50	36	50
	0	73	101	56	78	47	66	39	54	29	37
Effective length KL (ft)	1	71	97	53	73	44	59	34	46	24	29
	2	64	85	49	65	42	55	34	45	23	28
	3	55	68	42	52	36	44	29	36	22	26
	4	44	50	34	38	29	33	23	27	18	20
	5	33	33	25	26	21	22	18	18	13	14
	6	23	23	18	18	15	15	13	13	10	10
	7	17	17	13	13	11	11	9	9	7	7
	8	13	13	10	10	9	9	7	7	5	5

Size					2×2					
Thickness		$\frac{3}{8}$		$\frac{5}{16}$		$\frac{1}{4}$		$\frac{3}{16}$		$\frac{1}{8}$	
Wt./ft		4.7		3.92		3.19		2.44		1.65	
F_y		36	50	36	50	36	50	36	50	36	50
	0	44	61	37	52	30	42	23	32	14	18
Effective length KL (ft)	1	42	57	35	48	28	38	20	27	11	13
	2	36	46	31	39	25	32	19	25	11	13
	3	28	33	24	28	19	23	15	18	10	11
	4	20	20	17	17	14	14	11	11	7	8
	5	13	13	11	11	9	9	7	7	5	5
	6	9	9	8	8	6	6	5	5	3	3

COLUMN BASE PLATES
The design of column base plates is covered in Part 11 (Volume II) of this LRFD Manual.

REFERENCES

Galambos, T. V. (ed.), 1988, *Guide to Stability Design Criteria for Metal Structures,* Fourth Edition, Structural Stability Research Council, John Wiley & Sons, New York, NY.

Geschwindner, L., 1993, "The 'Leaning' Column in ASD and LRFD," *Proceedings of the 1993 National Steel Construction Conference,* AISC, Chicago, IL.

Uang, C. M., S. W. Wattar, and K. M. Leet, 1990, "Proposed Revision of the Equivalent Axial Load Method for LRFD Steel and Composite Beam-Column Design," *Engineering Journal,* 4th Qtr., AISC, Chicago.

Zureick, A., 1993, "Design Strength of Concentrically Loaded Single-Angle Struts," *Engineering Journal,* 4th Qtr., AISC, Chicago.

PART 4

BEAM AND GIRDER DESIGN

OVERVIEW

Beam tables are located as follows:

Beam charts are located as follows:

Additional information related to beam design is provided as follows:

DESIGN STRENGTH OF BEAMS

General
Beams are proportioned so that no applicable strength limit state is exceeded when subjected to factored load combinations and that no serviceability limit state is exceeded when subjected to service loads. Strength limit states for beams include local buckling, lateral torsional buckling, and yielding. Serviceability limit states may include, but are not limited to, deflection and vibration.

The flexural design strength for beams must equal or exceed the required strength based on the factored loads. The design strength $\phi_b M_n$ for each applicable limit state shall equal or exceed the maximum moment M_u as determined from the applicable factored load combinations given in Section A4 of the LRFD Specification. Values of $\phi_b M_n$ are tabulated in the pages to follow. These values are based on beam behavior as shown in Figure 4-1 and explained in the following discussion.

It should be noted that the LRFD Specification expresses values for moments and lengths in kip-in. and inches. In this and other parts of the LRFD Manual, these values are tabulated in kip-ft and feet.

The required strength can be determined by either elastic or plastic analysis.

Design Strength If Elastic Analysis Is Used
The flexural design strength of rolled I and C shape beams designed using elastic analysis, according to LRFD Specification Section F1 is:

$$\phi_b M_n$$

where

$$\phi_b = 0.90$$

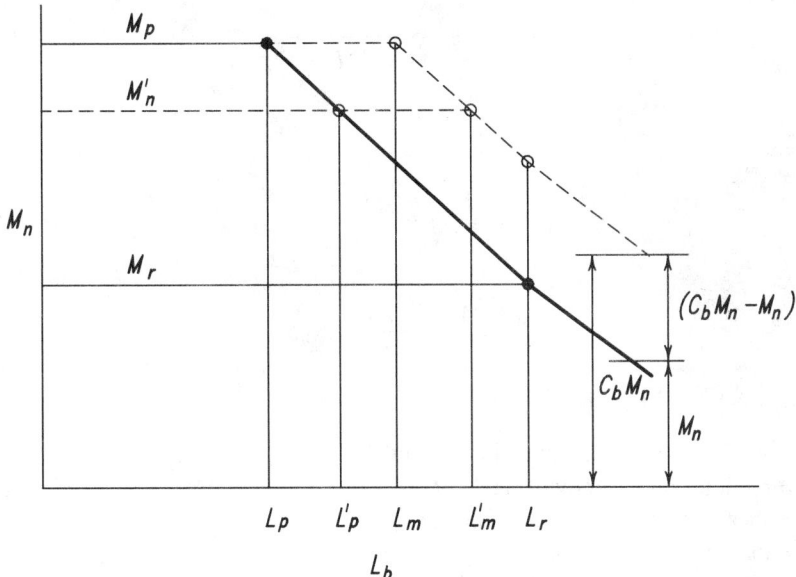

Fig. 4-1

M_n = nominal flexural strength as determined by the limit state of yielding, lateral-torsional buckling, or local buckling

Flexural Design Strength for $C_b = 1.0$

Compact Sections ($C_b = 1.0$)
When $L_b \leq L_p$

The flexural design strength of compact (flange and web local buckling $\lambda \leq \lambda_p$) I-shaped and C-shaped rolled beams (as defined in Section B5 of the LRFD Specification) bent about the major or minor axis is:

$$\phi_b M_n = \phi_b M_p = \phi_b Z F_y / 12$$

In minor axis flexure this is true for all unbraced lengths, but for bending about the major axis the distance L_b between points braced against lateral movement of the compression flange or between points braced to prevent twist of the cross-section shall not exceed the value L_p (see Figure 4-1).

$$L_p = \frac{300 r_y}{\sqrt{F_y}} \tag{F1-4}$$

When $L_p < L_b \leq L_r$

The flexural design strength of compact I or C rolled shapes bent about the major axis, from LRFD Specification Section F1.2, is:

$$\phi_b M_n = \phi_b M_p - \phi_b (M_p - M_r)\left(\frac{L_b - L_p}{L_r - L_p}\right) \leq \phi_b M_p$$

where the limiting length L_r and the corresponding buckling moment M_r (see Figure 4-1) are determined as follows:

$$L_r = \frac{r_y X_1}{(F_y - F_r)} \sqrt{1 + \sqrt{1 + X_2(F_y - F_r)^2}} \tag{F1-6}$$

where

$$X_1 = \frac{\pi}{S_x}\sqrt{\frac{EGJA}{2}} \tag{F1-8}$$

$$X_2 = \frac{4C_w}{I_y}\left(\frac{S_x}{GJ}\right)^2 \tag{F1-9}$$

$\phi_b M_r = \phi_b S_x (F_y - F_r) / 12$ kip-ft

S_x = section modulus about major axis, in.^{3}
E = modulus of elasticity of steel, 29,000 ksi
G = shear modulus of steel, 11,200 ksi
J = torsional constant, in.^{4}
A = cross-sectional area of beam, in.^{2}
C_w = warping constant, in.^{6}

F_r = compressive residual stress in flange: for rolled shapes F_r = 10 ksi; for welded shapes F_r = 16.5 ksi

Values of J and C_w are tabulated for some shapes in Part 1 of the LRFD Manual. For values not shown, see *Torsional Analysis of Steel Members* (AISC, 1983).

Compact and Noncompact Sections (C_b = 1.0)
When $L_b > L_r$

According to LRFD Specification Section F1.2b, the flexural design strength of compact and noncompact I or C rolled shapes bent about the major axis is:

$$\phi_b M_n = \phi_b M_{cr} = \phi_b \left(\frac{\pi}{L_b}\right)\sqrt{EI_yGJ + \left(\frac{\pi E}{L_b}\right)^2 I_y C_w}$$

$$= \phi_b \left(\frac{S_x X_1 \sqrt{2}}{(L_b/r_y)}\right)\sqrt{1 + \frac{X_1^2 X_2}{2(L_b/r_y)^2}} \le \phi_b M_r$$

Noncompact Sections (C_b = 1.0)
When $L_b \le L_p'$

All rolled W shapes are compact except the W40×174, W14×99, W14×90, W12×65, W10×12, W8×10, and W6×15 for 50 ksi and the W6×15 for 36 ksi. The flexural design strength $\phi_b M_n'$ (see Figure 4-1) for noncompact (flange or web local buckling $\lambda_p < \lambda \le \lambda_r$) I and C rolled shapes bent about the major or minor axis is the smaller value for either local flange buckling or local web buckling as determined by:

$$\phi_b M_n' = \phi_b M_p - \phi_b(M_p - M_r)\left(\frac{\lambda - \lambda_p}{\lambda_r - \lambda_p}\right)$$

For local flange buckling:

$\lambda = b_f/2t_f$ for I-shaped members
$\lambda = b_f/t_f$ for C-shaped members
$\lambda_p = 65/\sqrt{F_y}$
$\lambda_r = 141/\sqrt{F_y - 10}$

For local web buckling:

$\lambda = h/t_w$
$\lambda_p = 640/\sqrt{F_y}$
$\lambda_r = 970/\sqrt{F_y}$
$$L_p' = L_p + (L_r - L_p)\left(\frac{M_p - M_n'}{M_p - M_r}\right)$$

Sections with a width-to-thickness ratio exceeding the specified values for λ_r are slender shapes and must be analyzed using LRFD Specification Appendix B5.3.

When $L_p' < L_b \le L_r$

The flexural design strength of noncompact I or C rolled shapes bent about the major axis is determined by:

$$\phi_b M_n = \phi_b M_p - \phi_b (M_p - M_r) \left(\frac{L_b - L_p}{L_r - L_p} \right) \leq \phi_b M_n'$$

In the Load Factor Design Selection Table, in the case of the noncompact shapes, the values of $\phi_b M_n'$ and L_p' are tabulated as $\phi_b M_p$ and L_p. The formula above may be used with the tabulated values.

Flexural Design Strength for $C_b > 1.0$

C_b is a factor which varies with the moment gradient between bracing points (L_b). For C_b greater than 1.0, the design flexural strength is equal to the tabulated value of the design flexural strength (with $C_b = 1.0$) multiplied by the calculated C_b value. The maximum value is $\phi_b M_p$ for compact shapes or $\phi_b M_n'$ for noncompact shapes. The maximum unbraced lengths associated with the maximum flexural design strengths $\phi_b M_p$ and $\phi_b M_n'$ are L_m and L_m' (see Figure 4-1).

A new expression for C_b is given in the LRFD Specification. (It is more accurate than the one previously shown.)

$$C_b = \frac{12.5 M_{\max}}{2.5 M_{\max} + 3 M_A + 4 M_B + 3 M_c} \tag{F1-3}$$

where M is the absolute value of a moment in the unbraced beam segment as follows:

$M_{\max}$, the maximum
M_A , at the quarter point
M_B , at the centerline
M_c , at the three-quarter point

Values for C_b for some typical loading conditions are given in Table 4-1.

Compact Sections ($C_b > 1.0$)
When $L_b \leq L_m$

The flexural design strength for rolled I and C shapes is:

$$\phi_b M_n = \phi_b M_p$$

When $L_b > L_m$

The flexural design strength is:

$$\phi_b M_n = C_b [\phi_b M_n \text{ (for } C_b = 1.0)] \leq \phi_b M_p$$

For $L_m \leq L_r$

$$L_m = L_p + \frac{(C_b M_p - M_p)(L_r - L_p)}{C_b (M_p - M_r)}$$

Table 4-1.
Values of C_b for Simply Supported Beams

Load	Lateral Bracing Along Span	C_b
	None	1.32
	At load points	1.67 1.67
	None	1.14
	At load points	1.67 1.00 1.67
	None	1.14
	At load points	1.67 1.11 1.11 1.67
	None	1.14
	At centerline	1.30 1.30

For $L_m > L_r$

$$L_m = \frac{C_b \pi}{M_p} \sqrt{\frac{EI_y GJ}{2}} \sqrt{1 + \sqrt{1 + \frac{4C_w M_p^2}{I_y C_b^2 G^2 J^2}}}$$

The value of C_b for which L_m or $L_m{'}$ equals L_r for any rolled shape is:

$$C_b = \frac{F_y Z_x}{(F_y - 10)S_x}$$

Noncompact Sections ($C_b > 1.0$)
When $L_b \leq L_m{'}$

The flexural design strength for rolled I and C shapes is:

$$\phi_b M_n = \phi_b M_n{'} < \phi_b M_p$$

When $L_b > L_m{'}$

The flexural design strength is:

$$\phi_b M_n = C_b[\phi_b M_n \text{ (for } C_b = 1.0)] \leq \phi_b M_n{'}$$

For $L_m' \leq L_r$

$$L_m' = L_p' + \frac{(C_b M_n' - M_n')(L_r - L_p)}{C_b(M_p - M_r)}$$

For $L_m' > L_r$

$$L_m = \frac{C_b \pi}{M_p} \sqrt{\frac{EI_y GJ}{2}} \sqrt{1 + \sqrt{1 + \frac{4C_w M_p^2}{I_y C_b^2 G^2 J^2}}}$$

Design Strength If Plastic Analysis Is Used

The design flexural strength for plastic analysis is:

$$\phi_b M_n = \phi_b M_p$$

where

$\phi_b = 0.90$
$M_p = Z_x F_y / 12$ kip-ft

The yield strength of material that may be used with plastic analysis is limited to 65 ksi. Plastic analysis is limited to compact shapes as defined in Table B5.1 of the LRFD Specification as:

$\lambda_p = b_f / 2t_f \leq 65 / \sqrt{F_y}$ for the flanges of I shapes in flexure

$\lambda_p = b_f / t_f \leq 65 / \sqrt{F_y}$ for the flanges of C shapes in flexure

and

$\lambda_p = h / t_w \leq 640 / \sqrt{F_y}$ for beam webs in flexural compression

where

λ_p = limiting slenderness parameter for compact element
b_f = width of flange for I and C shapes, in.
t_f = flange thickness, in.
h = clear distance between flanges less the fillet at each flange, in.
t_w = beam web thickness, in

In addition, LRFD Specification Section F1.2d states: for a section bent about the major axis, the laterally unbraced length of the compression flange at plastic hinge locations associated with the failure mechanism shall not exceed:

$$L_{pd} = \frac{3,600 + 2,220(M_1 / M_2)}{F_y} r_y \tag{F1-17}$$

where

F_y	= specified yield strength of compression flange, ksi
M_1	= smaller moment at end of unbraced length of beam, kip-in.
M_2	= larger moment at end of unbraced length of beam, kip-in.
r_y	= radius of gyration about minor axis, in.

(M_1 / M_2) is positive when the moments cause reverse curvature

LOAD FACTOR DESIGN SELECTION TABLE FOR SHAPES USED AS BEAMS

This table facilitates the selection of beams designed on the basis of flexural strength in accordance with Section F of the LRFD Specification. It includes only W and M shapes designed as beams. A laterally supported beam can be selected by entering the table with the required plastic section modulus or factored bending moment, and comparing it with tabulated values of Z_x or $\phi_b M_p$ respectively.

The table is applicable to adequately braced beams with unbraced lengths not exceeding L_r, i.e., $L_b \leq L_r$. For beams with unbraced lengths greater than L_p, it may be convenient to use the unbraced beam charts. For most loading conditions, it is convenient to use this selection table. However, for adequately braced, simply supported beams with a uniform load over the entire length, or equivalent symmetrical loading, the tables of Factored Uniform Loads can also be used.

In this table, shapes are listed in groups by descending order plastic section modulus Z_x. Included also for steel of $F_y = 36$ ksi and 50 ksi are values for the maximum flexural design strength $\phi_b M_p$; the limiting buckling moment $\phi_b M_r$; the limiting laterally unbraced compression flange length for full plastic moment capacity and uniform moment ($C_b = 1.0$) L_p; limiting laterally unbraced length for inelastic lateral-torsional buckling L_r; and BF, a factor that can be used to calculate the resisting moment $\phi_b M_n$ for beams with unbraced lengths between the limiting bracing lengths L_p and L_r.

For noncompact shapes, as determined by Section B5 of the LRFD Specification, the maximum flexural design strength $\phi_b M_{n,\,max}$ as determined by LRFD Specification Formula A-F1-3 is tabulated as $\phi_b M_p$. The associated maximum unbraced length is tabulated as L_p. (See the previous discussion under Design Strength of Beams for further explanation.)

The symbols used in this table are:

Z_x = plastic section modulus, X-X axis, in.^{3}

$\phi_b M_p$ = design plastic bending moment, kip-ft

   = $\phi_b Z_x F_y / 12$ if shape is compact

   = $\phi_b M'_n = \phi_b M_p - \phi_b(M_p - M_r)\left(\dfrac{\lambda - \lambda_p}{\lambda_r - \lambda_p}\right)$ if shape is noncompact

$\phi_b M_r$ = limiting design buckling moment, kip-ft

   = $\phi_b S_x(F_y - F_r) / 12$

where

F_r = 10 ksi for rolled shapes

L_p = limiting laterally unbraced length for full plastic moment capacity, ft, uniform moment case ($C_b = 1$)

L_r = limiting laterally unbraced length for inelastic lateral-torsional buckling, ft

BF = a factor that can be used to calculate the design flexural strength for unbraced lengths L_b, between L_p and L_r, kip-ft

   $= \dfrac{\phi_b(M_p - M_r)}{L_r - L_p}$

where

   $\phi_b M_n = C_b[\phi_b M_p - BF(L_b - L_p)] \leq \phi_b M_p$

Use of the Table

Determine the required plastic section modulus Z_x from the maximum factored moment M_u (kip-ft) using the desired steel yield strength.

$$Z_x = \frac{12M_u}{\phi_b F_y}$$

Enter the column headed Z_x and find a value equal to or greater than the plastic section modulus required. Alternatively, enter the $\phi_b M_p$ column and find a value of $\phi_b M_p$ equal to or greater than the required factored load moment. The beam opposite these values (Z_x or $\phi_b M_p$) in the shapes column, and all beams above it, have sufficient flexural strength based only on these parameters. The first beam appearing in boldface type adjacent to or above the required Z_x or $\phi_b M_p$ is the lightest section that will serve for the steel yield stress used in the calculations. If the beam must not exceed a certain depth, proceed up the column headed "Shape" until a beam within the required depth is reached.

After a shape has been selected, the following checks should be made. If the lateral bracing of the compressive flange exceeds L_p, but is less than L_r, the design flexural strength may be calculated as follows:

$$\phi_b M_n = C_b [\phi_b M_p - BF(L_b - L_p)] \le \phi_b M_p$$

If the bracing length L_b is substantially greater than L_p, i.e., $L_b > L_r$, it is recommended the unbraced beam charts be used. A check should be made of the beam web shear strength by referring to the Factored Uniform Load Tables or by use of the formula:

$$\phi_v V_n = \phi_v 0.6 F_{yw} A_w \text{ (from LRFD Specification Section F2)}$$

where

$$\phi_v = 0.90$$

If a deflection limitation also exists, the adequacy of the selected beam should be checked accordingly.

EXAMPLE 4-1

Given: Select a beam of $F_y = 50$ ksi steel subjected to a factored uniform bending moment of 256 kip-ft, having its compression flange braced at 5.0 ft intervals. Assume $C_b = 1.0$.

Solution
(Z_x method): $$Z_x \text{ (req'd)} = \frac{M_u(12)}{\phi_b F_y} = \frac{256(12)}{0.9(50)} = 68.3 \text{ in.}^3$$

Enter the Load Factor Design Selection Table and find the nearest higher tabulated value of Z_x is 69.6 in., which corresponds to a W14×43. This beam, however, is not in boldface type. Proceed up the shape column and locate the first beam in boldface, W16×40. Note the values tabulated for $\phi_b M_p$ and L_p are 273 kip-ft and 5.6 ft, respectively.

Use W16x40

Alternatively, proceed up the shape column and select a W18×40. The tabulated values for $\phi_b M_p$ and L_p are 294 kip-ft and 4.5 ft, respectively. Since the bracing length L_b is larger than L_p and smaller than L_r, the maximum resisting moment may be calculated as follows:

$$\phi_b M_n = C_b[\phi_b M_p - BF(L_b - L_p)]$$
$$= 1.0[294 - (11.7)(5.0 - 4.5)]$$
$$= 288 \text{ kip-ft} > 256 \text{ kip-ft req'd} \quad \textbf{o.k.}$$

A W18×40 is satisfactory.

Alternate solution (M$_p$ method):

Enter the column of $\phi_b M_p$ values and note the tabulated value nearest and higher than the required factored moment (M_u) is 261 kip-ft, which corresponds to a W14×43. Scanning the $\phi_b M_p$ values for shapes listed higher in the column, a W16×40 is found to be the lightest suitable shape with $L_b < L_p$.

Use W16×40

EXAMPLE 4-2

Given:

Determine the design flexural strength of a W16×40 of F_y = 36 ksi and F_y = 50 ksi steel with the compression flange braced at intervals of 9.0 ft. Assume C_b = 1.1.

Solution:

Enter the Load Factor Design Table and note that for a W16×40, F_y = 36 ksi:

$\phi_b M_p$ = 197 kip-ft
L_p = 6.5 ft
L_r = 19.3 ft
BF = 5.54 kips
$\phi_b M_n = C_b[\phi_b M_p - BF(L_b - L_p)] \leq \phi_b M_p$
$\qquad = 1.1[197 - 5.54(9 - 6.5)] \leq 197 \text{ kip-ft}$
$\qquad = 197 \text{ kip-ft}$

Enter the Load Factor Design Selection Table and note that for a W16×40, F_y = 50 ksi:

$\phi_b M_p$ = 273 kip-ft
L_p = 5.6 ft
L_r = 14.7 ft
BF = 8.67 kips
$\phi_b M_n = C_b[\phi_b M_p - BF(L_b - L_p)] \leq \phi_b M_p$
$\qquad = 1.1[273 - 8.67(9 - 5.6)] \leq 273 \text{ kip-ft}$
$\qquad = 268 \text{ kip-ft}$

EXAMPLE 4-3

Given:

Select a beam of F_y = 50 ksi steel subjected to a factored uniform bending moment of 30 kip-ft having its compression flange braced at 4.0-ft intervals and a depth of eight inches or less. Assume C_b = 1.0.

*Solution
(Z_x method):*

Assume shape is compact and $L_b \leq L_p$.

$$Z_x \text{ req'd} = \frac{12M_u}{\phi_b F_y} = \frac{12(30)}{0.9(50)} = 8.0 \text{ in.}^3$$

Enter the Load Factor Design Selection Table and note that for a W8×10, F_y = 50 ksi, the shape is noncompact, however, the maximum resisting moment $\phi_b M_n$ listed in the $\phi_b M_p$ column is adequate. Further note:

$\phi_b M_n$ = 33.0 kip-ft
L_p = 3.1 ft
L_r = 7.8 ft
BF = 2.03 kips

Since $L_p < L_b \leq L_r$

$\phi_b M_n = C_b[\phi_b M_n - BF(L_b - L_p)]$
      $= 1.0[33.0 - 2.03(4.0 - 3.1)]$
      $= 33.0 - 1.8$
      $= 31.2$ kip-ft > 30 kip-ft req'd   **o.k.**

Use: W8×10

*Alternate
Solution
(M_p method):*

Enter the Selection Table and note that in the column of $\phi_b M_p$ values for W8×10, F_y = 50 ksi, the value of $\phi_b M_p$ is 33.0 kip-ft, which is adequate. Also note, however, L_p = 3.1 ft is less than the bracing interval L_b = 4.0 ft, and that BF is equal to 2.03 kips. Therefore:

$\phi_b M_n = 1.0[33.0 - 2.03(4 - 3.1)]$
      $= 31.2$ kip-ft > 30 kip-ft req'd   **o.k.**

Use: W8×10

LOAD FACTOR DESIGN SELECTION TABLE
For shapes used as beams
$$\phi_b = 0.90$$

Z_x

		$F_y = 36$ ksi							$F_y = 50$ ksi		
BF	L_r	L_p	$\phi_b M_r$	$\phi_b M_p$	Z_x	Shape	$\phi_b M_p$	$\phi_b M_r$	L_p	L_r	BF
Kips	Ft	Ft	Kip-ft	Kip-ft	in.^{3}		Kip-ft	Kip-ft	Ft	Ft	Kips
34.5	138.2	17.8	6180	10300	3830	W36×848[a]	14400	9510	15.1	90.5	64.3
34.1	130.1	17.7	5810	9639	3570	W36×798[a]	13400	8940	15.0	85.3	63.2
33.2	105.9	17.2	4720	7668	2840	W36×650[a]	10700	7260	14.6	70.0	61.1
41.9	84.8	15.9	4560	7450	2760	W40×593[a]	10400	7020	13.5	56.8	76.7
41.2	72.5	15.5	3860	6210	2300	W40×503[a]	8630	5940	13.2	49.5	73.9
33.2	86.8	16.8	3800	6129	2270	W36×527[a]	8510	5850	14.2	58.3	60.4
46.9	58.2	11.3	3330	5540	2050	W40×466[a]	7690	5130	9.6	39.4	85.9
41.0	63.3	15.2	3300	5270	1950	W40×431	7310	5070	12.9	44.1	72.0
19.3	119.4	15.3	3060	5080	1880	W27×539[a]	7050	4710	12.9	78.2	35.9
32.7	73.5	16.5	3160	5022	1860	W36×439[a]	6980	4860	14.0	50.3	58.2
23.7	93.6	15.6	2980	4830	1790	W30×477[a]	6710	4590	13.3	62.0	43.5
46.6	49.8	11.0	2810	4620	1710	W40×392[a]	6410	4320	9.6	34.3	83.7
39.9	56.6	15.0	2850	4510	1670	W40×372	6260	4380	12.7	40.3	68.4
32.5	67.2	16.3	2830	4482	1660	W36×393[a]	6230	4350	13.8	46.7	57.0
48.6	48.0	14.6	2750	4370	1620	W44×335	6080	4230	12.4	35.5	79.7
15.3	123.3	14.2	2520	4190	1550	W24×492[a]	5810	3870	12.1	80.5	28.4
18.9	99.2	14.9	2540	4130	1530	W27×448[a]	5740	3900	12.6	65.3	34.9
32.5	62.4	16.1	2570	4077	1510	W36×359[a]	5660	3960	13.7	44.0	56.2
46.2	43.2	10.7	2360	3860	1430	W40×331	5360	3630	9.1	30.5	80.9
22.9	77.5	15.3	2440	3860	1430	W30×391[a]	5360	3750	13.0	52.1	41.2
46.1	45.3	14.6	2420	3830	1420	W44×290	5330	3720	12.4	34.0	74.1
38.4	51.2	14.8	2440	3830	1420	W40×321	5330	3750	12.6	37.2	63.9
29.4	64.4	15.6	2400	3830	1420	W33×354[a]	5330	3690	13.2	44.7	51.9
32.2	58.5	16.0	2360	3726	1380	W36×328[a]	5180	3630	13.6	41.7	54.9
38.4	48.9	14.8	2280	3590	1330	W40×297	4990	3510	12.5	35.9	63.2
43.4	43.1	14.4	2180	3430	1270	W44×262	4760	3360	12.2	32.8	68.2
28.9	59.3	15.5	2160	3430	1270	W33×318[a]	4760	3330	13.1	41.8	49.9
31.6	55.1	16.0	2160	3402	1260	W36×300	4730	3330	13.5	39.9	52.9
14.7	103.0	13.9	2070	3380	1250	W24×408[a]	4690	3180	11.8	67.5	27.0
37.3	47.9	14.9	2150	3380	1250	W40×277	4690	3300	12.7	35.5	60.8
19.0	82.0	14.5	2070	3350	1240	W27×368[a]	4650	3180	12.3	54.6	34.8
44.2	38.2	10.5	1990	3210	1190	W40×278	4460	3060	8.9	27.6	74.9
23.4	66.6	15.0	2010	3210	1190	W30×326[a]	4460	3090	12.8	45.7	41.7
31.0	53.0	15.9	2010	3159	1170	W36×280	4390	3090	13.5	38.8	51.3
28.2	55.7	15.4	1970	3110	1150	W33×291[a]	4310	3030	13.0	39.8	48.0
43.7	37.0	10.5	1890	3050	1130	W40×264	4240	2910	8.9	27.0	73.3
35.6	45.4	14.8	1930	3020	1120	W40×249	4200	2980	12.6	34.1	56.9

[a]Group 4 or Group 5 shape. See Notes in Table 1-2 (Part 1).

| LOAD FACTOR DESIGN SELECTION TABLE |
| For shapes used as beams |
| $\phi_b = 0.90$ |

$F_y = 36$ ksi							$F_y = 50$ ksi				
BF	L_r	L_p	$\phi_b M_r$	$\phi_b M_p$	Z_x	Shape	$\phi_b M_p$	$\phi_b M_r$	L_p	L_r	BF
Kips	Ft	Ft	Kip-ft	Kip-ft	in.^{3}		Kip-ft	Kip-ft	Ft	Ft	Kips
40.1	41.2	14.3	1890	2970	1100	W44×230	4130	2910	12.1	31.7	62.0
30.3	50.6	15.8	1860	2916	1080	W36×260	4050	2860	13.4	37.5	49.4
23.1	60.5	14.9	1810	2860	1060	W30×292^a	3980	2780	12.7	42.1	40.4
37.0	39.7	11.0	1750	2810	1040	W36×256	3900	2690	9.4	28.8	62.7
27.7	52.0	15.3	1790	2810	1040	W33×263^a	3900	2750	12.9	37.8	46.3
15.0	84.5	13.5	1680	2750	1020	W24×335^a	3830	2590	11.4	55.8	27.8
18.7	69.4	14.3	1720	2750	1020	W27×307^a	3830	2650	12.1	46.9	33.7
41.8	35.1	10.6	1700	2730	1010	W40×235	3790	2620	9.0	26.0	68.6
29.6	48.8	15.6	1750	2727	1010	W36×245	3790	2690	13.3	36.4	47.7
33.1	42.7	14.8	1670	2600	963	W40×215	3610	2570	12.5	32.6	51.6
28.7	47.3	15.5	1630	2550	943	W36×230	3540	2510	13.2	35.6	45.8
22.8	55.4	14.8	1610	2540	941	W30×261	3530	2480	12.5	39.2	39.2
27.0	49.2	15.1	1620	2540	939	W33×241	3520	2490	12.8	36.2	44.2
36.1	37.2	10.9	1580	2530	936	W36×232	3510	2430	9.3	27.3	59.9
40.2	33.2	10.5	1530	2440	905	W40×211	3390	2360	8.9	24.9	64.7
31.6	41.1	14.4	1500	2340	868	W40×199	3260	2310	12.2	31.6	48.8
26.0	46.9	15.0	1480	2310	855	W33×221	3210	2270	12.7	35.0	41.9
18.6	59.6	14.0	1450	2300	850	W27×258	3190	2230	11.9	41.1	32.9
22.4	51.6	14.7	1450	2280	845	W30×235	3170	2240	12.4	37.1	37.7
14.7	71.2	13.2	1400	2250	835	W24×279^a	3130	2150	11.2	47.6	26.9
34.9	35.0	10.8	1400	2250	833	W36×210	3120	2160	9.1	26.1	56.8
37.4	31.2	10.4	1330	2110	781	W40×183	2930	2050	8.8	23.8	59.0
25.0	44.8	14.8	1330	2080	772	W33×201	2900	2050	12.6	33.8	39.7
18.5	55.0	13.9	1310	2080	769	W27×235	2880	2020	11.8	38.5	32.3
34.0	33.5	10.7	1290	2070	767	W36×194	2880	1990	9.1	25.2	54.6
8.40	109.5	12.3	1220	2030	753	W18×311^a	2820	1870	10.4	71.5	15.6
21.8	47.9	14.5	1290	2020	749	W30×211	2810	1990	12.3	35.1	36.0
14.7	64.3	13.1	1260	2010	744	W24×250^a	2790	1930	11.1	43.4	26.6
32.7	32.8	10.6	1210	1940	718	W36×182	2690	1870	9.0	24.9	52.0
27.5	38.4	13.6	1250	1930	715	W40×174^b	2660	1920	12.0	29.9	41.3
18.2	52.0	13.8	1220	1910	708	W27×217	2660	1870	11.7	36.8	31.3
35.6	29.7	10.0	1170	1870	692	W40×167	2600	1800	8.5	22.8	55.6
8.29	99.6	12.1	1100	1830	676	W18×283^a	2540	1690	10.3	65.1	15.4
14.7	59.3	13.0	1150	1830	676	W24×229	2540	1760	11.0	40.4	26.2
21.0	45.4	14.4	1170	1820	673	W30×191	2520	1790	12.2	33.7	33.9
31.5	31.9	10.5	1130	1800	668	W36×170	2510	1740	8.9	24.4	49.6
28.3	32.6	10.4	1070	1700	629	W33×169	2360	1650	8.8	24.5	45.4
17.8	48.0	13.7	1080	1700	628	W27×194	2360	1670	11.6	34.6	30.0
30.7	30.9	10.4	1060	1680	624	W36×160	2340	1630	8.8	23.7	48.0
8.21	90.9	12.0	1000	1650	611	W18×258^a	2290	1540	10.2	59.5	15.2
14.5	54.2	12.8	1040	1640	606	W24×207	2270	1590	10.9	37.4	25.6
20.2	43.2	14.3	1050	1630	605	W30×173	2270	1620	12.1	32.5	32.0

^{a}Group 4 or Group 5 shape. See Notes in Table 1-2 (Part 1).
^{b}Indicates noncompact shape; $F_y = 50$ ksi

LOAD FACTOR DESIGN SELECTION TABLE
For shapes used as beams
$$\phi_b = 0.90$$

Z_x

F_y = 36 ksi						Shape	F_y = 50 ksi				
BF	L_r	L_p	$\phi_b M_r$	$\phi_b M_p$	Z_x	Shape	$\phi_b M_p$	$\phi_b M_r$	L_p	L_r	BF
Kips	Ft	Ft	Kip-ft	Kip-ft	in.³		Kip-ft	Kip-ft	Ft	Ft	Kips
32.8	28.2	9.5	998	1610	597	W40×149	2240	1540	8.1	21.9	50.8
29.4	30.2	10.3	983	1570	581	W36×150	2180	1510	8.7	23.4	45.6
17.5	45.2	13.6	979	1530	567	W27×178	2130	1510	11.5	33.1	28.8
26.8	31.2	10.3	950	1510	559	W33×152	2100	1460	8.7	23.7	42.3
14.3	51.3	12.8	957	1510	559	W24×192	2100	1470	10.9	35.7	25.0
8.09	82.8	11.9	909	1480	549	W18×234[a]	2060	1400	10.1	54.4	14.9
11.0	60.8	12.6	899	1430	530	W21×201	1990	1380	10.7	41.1	19.9
25.7	30.1	10.1	874	1390	514	W33×141	1930	1340	8.6	23.1	40.2
16.9	42.8	13.5	887	1380	512	W27×161	1920	1370	11.5	31.7	27.4
14.3	47.8	12.7	878	1380	511	W24×176	1920	1350	10.7	33.8	24.6
27.5	28.8	9.9	856	1370	509	W36×135	1910	1320	8.4	22.4	42.2
23.7	30.6	9.5	850	1350	500	W30×148	1880	1310	8.1	22.8	38.6
8.00	75.0	11.8	817	1320	490	W18×211[a]	1840	1260	10.0	49.5	14.7
10.9	55.8	12.5	813	1290	476	W21×182	1790	1250	10.6	38.1	19.4
14.1	45.2	12.7	807	1260	468	W24×162	1760	1240	10.8	32.4	23.8
24.5	29.1	10.0	792	1260	467	W33×130	1750	1220	8.4	22.5	37.9
16.2	40.7	13.4	801	1240	461	W27×146	1730	1230	11.3	30.6	25.8
7.98	68.3	11.6	741	1190	442	W18×192	1660	1140	9.9	45.3	14.6
22.4	29.0	9.4	741	1180	437	W30×132	1640	1140	8.0	22.0	35.6
10.8	51.7	12.4	741	1170	432	W21×166	1620	1140	10.5	35.7	19.1
13.8	42.0	12.5	723	1130	418	W24×146	1570	1110	10.6	30.6	22.8
23.1	27.8	9.7	700	1120	415	W33×118	1560	1080	8.2	21.7	35.5
21.6	28.2	9.3	692	1100	408	W30×124	1530	1070	7.9	21.5	34.1
7.95	62.3	11.5	671	1070	398	W18×175	1490	1030	9.8	41.5	14.5
18.9	30.0	9.2	673	1070	395	W27×129	1480	1040	7.8	22.3	30.9
21.1	27.1	9.1	642	1020	378	W30×116	1420	987	7.7	20.8	33.0
10.7	46.4	12.3	642	1010	373	W21×147	1400	987	10.4	32.8	18.4
13.3	39.3	12.4	642	999	370	W24×131	1390	987	10.5	29.1	21.5
7.87	56.7	11.4	605	961	356	W18×158	1340	930	9.7	38.2	14.2
20.2	26.3	9.0	583	934	346	W30×108	1300	897	7.6	20.3	31.5
18.0	28.2	9.1	583	926	343	W27×114	1290	897	7.7	21.3	28.7
10.5	43.1	12.2	575	899	333	W21×132	1250	885	10.4	30.9	17.7
12.7	37.1	12.3	567	883	327	W24×117	1230	873	10.4	27.9	20.2
7.82	52.2	11.3	550	869	322	W18×143	1210	846	9.6	35.5	14.0
19.0	25.5	8.8	525	842	312	W30×99	1170	807	7.4	19.8	29.2
10.3	41.0	12.2	532	829	307	W21×122	1150	819	10.3	29.8	17.1
17.0	26.8	9.0	521	824	305	W27×102	1140	801	7.6	20.5	26.7
7.79	48.0	11.3	499	786	291	W18×130	1090	768	9.5	33.0	13.8
12.0	35.2	12.1	503	780	289	W24×104	1080	774	10.3	26.8	18.8

[a]Group 4 or Group 5 shape. See Notes in Table 1-2 (Part 1).

Z_X LOAD FACTOR DESIGN SELECTION TABLE
For shapes used as beams
$\phi_b = 0.90$

$F_y = 36$ ksi							$F_y = 50$ ksi				
BF	L_r	L_p	$\phi_b M_r$	$\phi_b M_p$	Z_x	Shape	$\phi_b M_p$	$\phi_b M_r$	L_p	L_r	BF
Kips	Ft	Ft	Kip-ft	Kip-ft	in.^{3}		Kip-ft	Kip-ft	Ft	Ft	Kips
17.8	24.8	8.7	478	764	283	W30×90	1060	735	7.4	19.4	27.1
14.8	27.1	8.3	478	756	280	W24×103	1050	735	7.0	20.1	24.1
10.1	38.7	12.1	486	753	279	W21×111	1050	747	10.3	28.5	16.4
16.2	25.9	8.8	474	751	278	W27×94	1040	729	7.5	19.9	25.2
7.72	44.1	11.2	450	705	261	W18×119	979	693	9.5	30.8	13.4
14.3	25.9	8.3	433	686	254	W24×94	953	666	7.0	19.4	23.0
9.61	37.1	12.0	443	683	253	W21×101	949	681	10.2	27.6	15.4
15.0	24.9	8.6	415	659	244	W27×84	915	639	7.3	19.3	23.0
3.87	73.6	15.7	408	632	234	W14×132	878	627	13.3	49.6	6.89
7.62	40.4	11.1	398	621	230	W18×106	863	612	9.4	28.7	13.0
13.6	24.5	8.1	382	605	224	W24×84	840	588	6.9	18.6	21.5
11.8	26.6	7.7	374	597	221	W21×93	829	576	6.5	19.4	19.6
3.86	67.9	15.6	371	572	212	W14×120	795	570	13.2	46.2	6.82
7.51	38.1	11.0	367	570	211	W18×97	791	564	9.4	27.4	12.6
12.7	23.4	8.0	343	540	200	W24×76	750	528	6.8	18.0	19.8
6.10	42.1	10.5	341	535	198	W16×100	743	525	8.9	29.3	10.7
11.3	24.9	7.6	333	529	196	W21×83	735	513	6.5	18.5	18.5
3.84	62.7	15.5	337	518	192	W14×109	720	519	13.2	43.2	6.70
2.95	75.5	13.0	318	502	186	W12×120	698	489	11.1	50.0	5.36
7.27	35.5	11.0	324	502	186	W18×86	698	498	9.3	26.1	11.9
12.1	22.4	7.8	300	478	177	W24×68	664	462	6.6	17.4	18.7
6.03	38.6	10.4	302	473	175	W16×89	656	465	8.8	27.3	10.3
3.77	58.2	15.5	306	467	173	W14×99^b	647	471	13.4	40.6	6.46
10.7	23.5	7.5	294	464	172	W21×73	645	453	6.4	17.7	17.0
2.95	67.2	13.0	283	443	164	W12×106	615	435	11.0	44.9	5.32
6.94	33.3	10.9	285	440	163	W18×76	611	438	9.2	24.8	11.1
10.4	22.8	7.5	273	432	160	W21×68	600	420	6.4	17.3	16.5
3.75	54.1	15.4	279	424	157	W14×90^b	577	429	15.0	38.4	6.31
13.8	17.2	5.8	255	413	153	W24×62	574	393	4.9	13.3	21.4
5.85	34.9	10.3	261	405	150	W16×77	563	402	8.7	25.2	9.75
2.01	86.4	11.2	246	397	147	W10×112	551	378	9.5	56.5	3.68
2.91	61.4	12.9	255	397	147	W12×96	551	393	10.9	41.3	5.20
8.29	24.4	7.1	248	392	145	W18×71	544	381	6.0	17.8	13.8
9.84	21.7	7.4	248	389	144	W21×62	540	381	6.3	16.6	15.3
4.15	43.0	10.3	240	375	139	W14×82	521	369	8.8	29.6	7.31

^{b}Indicates noncompact shape; $F_y = 50$ ksi.

LOAD FACTOR DESIGN SELECTION TABLE
For shapes used as beams
$\phi_b = 0.90$

$$Z_x$$

$F_y = 36$ ksi							$F_y = 50$ ksi				
BF	L_r	L_p	$\phi_b M_r$	$\phi_b M_p$	Z_x	Shape	$\phi_b M_p$	$\phi_b M_r$	L_p	L_r	BF
Kips	Ft	Ft	Kip-ft	Kip-ft	in.^{3}		Kip-ft	Kip-ft	Ft	Ft	Kips
12.7	16.6	5.6	222	362	134	W24×55	503	342	4.7	12.9	19.6
8.08	23.2	7.0	228	359	133	W18×65	499	351	6.0	17.1	13.3
2.90	56.4	12.8	230	356	132	W12×87	495	354	10.9	38.4	5.12
2.00	77.4	11.0	218	351	130	W10×100	488	336	9.4	50.8	3.66
5.57	32.3	10.3	228	351	130	W16×67	488	351	8.7	23.8	9.02
11.3	17.3	5.6	216	348	129	W21×57	484	333	4.8	13.1	18.0
4.10	40.0	10.3	218	340	126	W14×74	473	336	8.8	28.0	7.12
7.91	22.4	7.0	211	332	123	W18×60	461	324	6.0	16.7	12.8
2.88	51.8	12.7	209	321	119	W12×79	446	321	10.8	35.7	5.03
4.05	37.3	10.3	201	311	115	W14×68	431	309	8.7	26.4	6.91
1.97	68.4	11.0	192	305	113	W10×88	424	296	9.3	45.1	3.58
7.65	21.4	7.0	192	302	112	W18×55	420	295	5.9	16.1	12.2
10.5	16.2	5.4	184	297	110	W21×50	413	284	4.6	12.5	16.4
2.87	48.2	12.7	190	292	108	W12×72	405	292	10.7	33.6	4.93
6.43	22.8	6.7	180	284	105	W16×57	394	277	5.7	16.6	10.7
3.91	34.7	10.2	180	275	102	W14×61	383	277	8.7	24.9	6.51
7.31	20.5	6.9	173	273	101	W18×50	379	267	5.8	15.6	11.5
1.95	60.1	10.8	168	264	97.6	W10×77	366	258	9.2	39.9	3.53
2.80	44.7	12.6	171	261	96.8	W12×65^b	358	264	11.8	31.7	4.72
9.68	15.4	5.3	159	258	95.4	W21×44	358	245	4.5	12.0	14.9
6.18	21.3	6.6	158	248	92.0	W16×50	345	243	5.6	15.8	10.1
8.13	16.6	5.4	154	245	90.7	W18×46	340	236	4.6	12.6	13.0
4.17	28.0	8.0	152	235	87.1	W14×53	327	233	6.8	20.1	7.02
2.91	38.4	10.5	152	233	86.4	W12×58	324	234	8.9	27.0	4.96
1.93	53.7	10.8	148	230	85.3	W10×68	320	227	9.2	36.0	3.46
5.91	20.2	6.5	142	222	82.3	W16×45	309	218	5.6	15.2	9.43
7.51	15.7	5.3	133	212	78.4	W18×40	294	205	4.5	12.1	11.7
4.06	26.3	8.0	137	212	78.4	W14×48	294	211	6.8	19.2	6.70
2.85	35.8	10.3	138	210	77.9	W12×53	292	212	8.8	25.6	4.77
1.91	48.1	10.7	130	201	74.6	W10×60	280	200	9.1	32.6	3.38
5.54	19.3	6.5	126	197	72.9	W16×40	273	194	5.6	14.7	8.67
3.06	30.8	8.2	126	195	72.4	W12×50	272	194	6.9	21.7	5.25
1.30	64.0	8.8	118	190	70.2	W8×67	263	181	7.5	41.9	2.38
3.91	24.7	7.9	122	188	69.6	W14×43	261	188	6.7	18.2	6.32
1.89	43.9	10.7	117	180	66.6	W10×54	250	180	9.1	30.2	3.30
6.95	14.8	5.1	112	180	66.5	W18×35	249	173	4.3	11.5	10.7
3.01	28.5	8.1	113	175	64.7	W12×45	243	174	6.9	20.3	5.07
5.23	18.3	6.3	110	173	64.0	W16×36	240	170	5.4	14.1	8.08
4.41	20.0	6.5	106	166	61.5	W14×38	231	164	5.5	14.9	7.07
1.88	40.7	10.6	106	163	60.4	W10×49	227	164	9.0	28.3	3.25
1.27	56.0	8.8	101	161	59.8	W8×58	224	156	7.4	36.8	2.32
2.92	26.5	8.0	101	155	57.5	W12×40	216	156	6.8	19.3	4.82
1.96	35.1	8.4	95.7	148	54.9	W10×45	206	147	7.1	24.1	3.45

^{b}Indicates noncompact shape; $F_y = 50$ ksi.

LOAD FACTOR DESIGN SELECTION TABLE

Z_x For shapes used as beams

$\phi_b = 0.90$

		$F_y = 36$ ksi							$F_y = 50$ ksi		
BF	L_r	L_p	$\phi_b M_r$	$\phi_b M_p$	Z_x	Shape	$\phi_b M_p$	$\phi_b M_r$	L_p	L_r	BF
Kips	Ft	Ft	Kip-ft	Kip-ft	in.^{3}		Kip-ft	Kip-ft	Ft	Ft	Kips
4.18	19.0	6.4	94.8	147	54.6	W14×34	205	146	5.4	14.4	6.58
5.70	14.3	4.9	92.0	146	54.0	W16×31	203	142	4.1	11.0	8.85
3.47	20.6	6.4	88.9	138	51.2	W12×35	192	137	5.4	15.2	5.67
1.26	46.7	8.7	84.4	132	49.0	W8×48	184	130	7.4	31.1	2.27
3.92	17.9	6.2	81.9	128	47.3	W14×30	177	126	5.3	13.7	6.06
1.93	31.2	8.3	82.1	126	46.8	W10×39	176	126	7.0	21.8	3.32
5.15	13.3	4.7	74.9	119	44.2	W16×26	166	115	4.0	10.4	7.88
3.22	19.1	6.3	75.3	116	43.1	W12×30	162	116	5.4	14.4	5.10
4.44	13.4	4.5	68.8	109	40.2	W14×26	151	106	3.8	10.3	6.96
1.25	39.1	8.5	69.2	107	39.8	W8×40	149	107	7.2	26.4	2.22
1.89	27.4	8.1	68.3	105	38.8	W10×33	146	105	6.9	19.7	3.15
2.99	18.1	6.3	65.1	100	37.2	W12×26	140	100	5.3	13.8	4.64
2.44	20.3	5.7	63.2	98.8	36.6	W10×30	137	97.2	4.8	14.5	4.13
1.23	35.1	8.5	60.8	93.7	34.7	W8×35	130	93.6	7.2	24.1	2.16
4.06	12.5	4.3	56.6	89.6	33.2	W14×22	125	87.0	3.7	9.7	6.26
2.34	18.5	5.7	54.4	84.5	31.3	W10×26	117	83.7	4.8	13.5	3.85
1.21	32.0	8.4	53.6	82.1	30.4	W8×31	114	82.5	7.1	22.3	2.07
3.88	11.1	3.5	49.5	79.1	29.3	W12×22	110	76.2	3.0	8.4	6.24
1.27	27.3	6.8	47.4	73.4	27.2	W8×28	102	72.9	5.7	18.9	2.22
2.19	16.9	5.5	45.2	70.2	26.0	W10×22	97.5	69.6	4.7	12.7	3.50
3.61	10.4	3.4	41.5	66.7	24.7	W12×19	92.6	63.9	2.9	7.9	5.70
1.24	24.4	6.7	40.8	62.6	23.2	W8×24	87.0	62.7	5.7	17.2	2.11
2.60	12.0	3.6	36.7	58.3	21.6	W10×19	81.0	56.4	3.1	8.9	4.26
1.46	18.6	5.3	35.5	55.1	20.4	W8×21	76.5	54.6	4.5	13.3	2.47
3.30	9.6	3.2	33.3	54.3	20.1	W12×16	75.4	51.3	2.7	7.4	5.12
0.741	31.3	6.3	32.6	51.0	18.9	W6×25	70.9	50.1	5.4	21.0	1.33
2.46	11.2	3.5	31.6	50.5	18.7	W10×17	70.1	48.6	3.0	8.4	3.97
2.97	9.2	3.1	29.1	47.0	17.4	W12×14	65.3	44.7	2.7	7.2	4.56
1.40	16.7	5.1	29.6	45.9	17.0	W8×18	63.8	45.6	4.3	12.3	2.30
2.34	10.3	3.4	26.9	43.2	16.0	W10×15	60.0	41.4	2.9	7.9	3.69
0.728	25.6	6.3	26.1	40.2	14.9	W6×20	55.9	40.2	5.3	17.7	1.27
3.32	6.9	2.3	23.6	38.6	14.3	M12×11.8	53.7	36.3	2.0	5.4	5.10
1.53	12.6	3.7	23.0	36.7	13.6	W8×15	51.0	35.4	3.1	9.2	2.56

LOAD FACTOR DESIGN SELECTION TABLE
For shapes used as beams
$\phi_b = 0.90$

Z_x

$F_y = 36$ ksi							$F_y = 50$ ksi				
BF	L_r	L_p	$\phi_b M_r$	$\phi_b M_p$	Z_x	Shape	$\phi_b M_p$	$\phi_b M_r$	L_p	L_r	BF
Kips	Ft	Ft	Kip-ft	Kip-ft	in.^{3}		Kip-ft	Kip-ft	Ft	Ft	Kips
3.10	6.8	2.3	21.6	35.4	13.1	M12×10.8	49.2	33.3	2.0	5.3	4.74
2.03	9.5	3.3	21.3	34.0	12.6	W10×12[b]	47.0	32.7	2.9	7.4	3.13
0.817	18.3	4.0	19.9	31.6	11.7	W6×16	43.9	30.6	3.4	12.5	1.46
0.458	30.3	5.3	19.9	31.3	11.6	W5×19	43.5	30.6	4.5	20.1	0.830
1.44	11.5	3.5	19.3	30.8	11.4	W8×13	42.8	29.7	3.0	8.5	2.35
0.417	31.1	5.0	18.8	29.7	11.0	M5×18.9	41.3	28.9	4.2	20.5	0.758
0.693	20.8	6.7	19.0	28.8	10.8	W6×15[b,c]	38.6	29.2	6.8	15.0	1.16
0.444	26.3	5.3	16.6	25.9	9.59	W5×16	36.0	25.5	4.5	17.6	0.795
2.32	6.2	2.1	15.2	24.9	9.21	M10×9	34.5	23.5	1.8	4.9	3.59
1.30	10.2	3.5	15.2	23.9	8.87	W8×10[b]	33.0	23.4	3.1	7.8	2.03
0.775	14.4	3.8	14.3	22.4	8.30	W6×12	31.1	21.9	3.2	10.2	1.33
2.13	6.1	2.1	13.6	22.1	8.20	M10×8	30.8	21.0	1.8	4.8	3.26
0.295	25.5	4.2	10.6	17.0	6.28	W4×13	23.6	16.4	3.5	16.9	0.538
0.724	12.0	3.8	10.8	16.8	6.23	W6×9	23.4	16.7	3.2	8.9	1.17
1.50	5.5	1.8	9.01	14.6	5.40	M8×6.5	20.2	13.9	1.6	4.3	2.35

[b]Indicates noncompact shape; $F_y = 50$ ksi
[c]Indicates noncompact shape; $F_y = 36$ ksi

MOMENT OF INERTIA SELECTION TABLES FOR W AND M SHAPES

These two tables for moment of inertia (I_x and I_y) are provided to facilitate the selection of beams and columns on the basis of their stiffness properties with respect to the X-X axis or Y-Y axis, as applicable, where

I_x = moment of inertia, X-X axis, in.^{4}
I_y = moment of inertia, Y-Y axis, in.^{4}

In each table the shapes are listed in groups by descending order of moment of inertia for all W and M shapes. The boldface type identifies the shapes that are the lightest in weight in each group.

Enter the column headed I_x (or I_y) and find a value of I_x (or I_y) equal to or greater than the moment of inertia required. The shape opposite this value, and all shapes above it, have sufficient stiffness. Note that the member selected must also be checked for compliance with specification provisions governing its specific application.

I_x — MOMENT OF INERTIA SELECTION TABLE
For W and M shapes

Shape	I_x In.^{4}	Shape	I_x In.^{4}	Shape	I_x In.^{4}	Shape	I_x In.^{4}
W36×848*	**67400**	**W40×215**	**16700**	**W36×135**	**7800**	**W30×99**	**3990**
		W27×368*	16100	W24×229	7650	W18×192	3870
W36×798*	**62600**	W36×245	16100	W33×141	7450	W14×283*	3840
		W14×808*	16000	W14×455*	7190	W21×147	3630
W40×593*	**50400**	W33×263*	15800	W27×178	6990		
W36×650*	48900			W18×311*	6960		
		W40×211	**15500**	W24×207	6820		
W40×503*	**41700**	W24×408*	15100				
W36×527*	38300	W36×230	15000			**W30×90**	**3620**
		W36×232	15000			W27×102	3620
W40×466*	**36300**			**W33×130**	**6710**	W12×305*	3550
		W40×199	**14900**	W30×148	6680	W24×117	3540
W40×431	**34800**	W30×292*	14900	W14×426*	6600	W18×175	3450
		W14×730*	14300	W27×161	6280	W14×257*	3400
W44×335	**31100**	W33×241	14200	W24×192	6260	W27×94	3270
W36×439*	31000			W18×283*	6160	W21×132	3220
W40×392*	29900	**W40×183**	**13300**	W14×398*	6000	W12×279*	3110
W40×372	29600	W36×210	13200			W24×104	3100
W36×393*	27500	W27×307*	13100			W18×158	3060
		W30×261	13100			W14×233*	3010
W44×290	**27100**	W33×221	12800			W24×103	3000
W30×477*	26100	W14×665*	12400			W21×122	2960
W27×539*	25500			**W33×118**	**5900**		
W40×321	25100	**W40×174**	**12200**	W30×132	5770		
W36×359*	24800	W36×194	12100	W24×176	5680		
W40×331	24700	W24×335*	11900	W27×146	5630	**W27×84**	**2850**
		W30×235	11700	W18×258*	5510	W18×143	2750
W44×262	**24200**			W14×370*	5440	W12×252*	2720
W40×297	23200	**W40×167**	**11600**	W30×124	5360	W24×94	2700
W36×328*	22500	W33×201	11500	W21×201	5310	W21×111	2670
W40×277	21900	W36×182	11300	W24×162	5170	W14×211	2660
W33×354*	21900	W14×605*	10800			W18×130	2460
		W27×258	10800			W12×230*	2420
		W36×170	10500			W21×101	2420
		W30×211	10300	**W30×116**	**4930**	W14×193	2400
W44×230	**20800**			W18×234*	4900		
W30×391*	20700	**W40×149**	**9780**	W14×342*	4900		
W40×278	20500	W36×160	9750	W27×129	4760		
W27×448*	20400	W27×235	9660	W21×182	4730		
W36×300	20300	W24×279*	9600	W24×146	4580	**W24×84**	**2370**
W33×318*	19500	W14×550*	9430			W18×119	2190
W40×249	19500	W33×169	9290			W14×176	2140
W40×264	19400	W30×191	9170			W12×210*	2140
W24×492*	19100	W36×150	9040	**W30×108**	**4470**		
W36×280	18900	W27×217	8870	W18×211*	4330		
W33×291*	17700	W24×250*	8490	W14×311*	4330	**W24×76**	**2100**
W40×235	17400	W14×500*	8210	W21×166	4280	W21×93	2070
W36×260	17300	W30×173	8200	W27×114	4090	W18×106	1910
W30×326*	16800	W33×152	8160	W12×336*	4060	W14×159	1900
W36×256	16800	W27×194	7820	W24×131	4020	W12×190	1890

*Group 4 or 5 shape. See Notes in Table 1-2 (Part 1).

MOMENT OF INERTIA SELECTION TABLE
For W and M shapes I_x

Shape	I_x In.^{4}	Shape	I_x In.^{4}	Shape	I_x In.^{4}	Shape	I_x In.^{4}
W24×68	1830	W21×44	843	W16×26	301	M12×10.8	65.8
W21×83	1830	W12×96	833	W14×30	291	W8×18	61.9
W18×97	1750	W18×50	800	W12×35	285	W10×12	53.8
W14×145	1710	W14×74	796	W8×67	272	W6×25	53.4
W12×170	1650	W16×57	758	W10×49	272	W8×15	48.0
W21×73	1600	W12×87	740	W10×45	248	W6×20	41.4
		W14×68	723			W8×13	39.6
		W10×112	716	W14×26	245		
		W18×46	712	W12×30	238		
W24×62	1550	W12×79	662	W8×58	228		
W14×132	1530	W16×50	659	W10×39	209	M10×9	38.5
W18×86	1530	W14×61	640				
W16×100	1490	W10×100	623	W12×26	204		
W21×68	1480						
W12×152	1430	W18×40	612	W14×22	199		
W14×120	1380	W12×72	597	W8×48	184	M10×8	34.3
		W16×45	586	W10×30	170	W6×16	32.1
		W14×53	541	W10×33	170	W8×10	30.8
W24×55	1350	W10×88	534			W6×15	29.1
W18×76	1330	W12×65	533	W12×22	156	W5×19	26.2
W21×62	1330			W8×40	146	M5×18.9	24.1
W16×89	1300	W16×40	518	W10×26	144	W6×12	22.1
W14×109	1240					W5×16	21.3
W12×136	1240	W18×35	510	W12×19	130		
W18×71	1170	W14×48	485	W8×35	127		
W21×57	1170	W12×58	475	W10×22	118		
W14×99	1110	W10×77	455	W8×31	110		
W16×77	1110	W16×36	448				
W12×120	1070	W14×43	428	W12×16	103		
W18×65	1070	W12×53	425	W8×28	98.0	M8×6.5	18.1
W14×90	999	W12×50	394	W10×19	96.3	W6×9	16.4
W18×60	984	W10×68	394			W4×13	11.3
		W14×38	385				
				W12×14	88.6		
		W16×31	375	W8×24	82.8		
W21×50	984	W12×45	350	W10×17	81.9		
W16×67	954	W10×60	341	W8×21	75.3		
W12×106	933	W14×34	340				
W18×55	890	W12×40	310	M12×11.8	71.7		
W14×62	882	W10×54	303	W10×15	68.9		

MOMENT OF INERTIA SELECTION TABLE
I_y For W and M shapes

Shape	I_y In.^{4}	Shape	I_y In.^{4}	Shape	I_y In.^{4}	Shape	I_y In.^{4}
W14×808*	5510	W14×283*	1440	W14×176	838	W14×120	495
		W40×372	1420	W12×252*	828	W40×264	493
		W36×328*	1420	W24×279*	823	W18×211*	493
		W24×408*	1320	W40×392*	803	W21×182	483
W14×730*	4720	W27×368*	1310	W40×215	796	W24×176	479
W36×848*	4550	W36×300	1300	W44×230	796	W36×232	468
W36×798*	4200			W18×311*	795	W12×152	454
				W27×235	768		
		W14×257*	1290	W30×211	757	W14×109	447
		W33×318*	1290	W33×201	749	W40×235	444
W14×665*	4170	W30×326*	1240			W24×162	443
		W36×280	1200			W27×146	443
		W44×335	1200			W18×192	440
W14×605*	3680	W12×336*	1190			W21×166	435
		W40×321	1190	W14×159	748	W36×210	411
		W33×291*	1160	W12×230*	742		
W14×550*	3250			W24×250*	724	W14×99	402
W36×650*	3230			W18×283*	704	W12×136	398
		W14×233*	1150	W27×217	704	W18×175	391
		W30×292*	1100	W40×199	695	W24×146	391
W14×500*	2880	W40×297	1090			W40×211	390
		W36×260	1090			W21×147	376
W14×455*	2560	W44×290	1050			W36×194	375
W40×593*	2520	W27×307*	1050				
W36×527*	2490	W12×305*	1050	W14×145	677	W14×90	362
		W40×277	1040	W30×191	673	W36×182	347
W14×426*	2360	W33×263*	1030	W12×210*	664	W18×158	347
		W24×335*	1030	W24×229	651	W12×120	345
W14×398*	2170			W40×331	646	W24×131	340
W27×539*	2110			W18×258*	628	W40×183	336
W40×503*	2050			W27×194	618	W21×132	333
				W30×173	598	W36×170	320
W14×370*	1990	W14×211	1030	W12×190	589	W18×143	311
W36×439*	1990	W40×466*	1010	W24×207	578	W33×169	310
W30×477*	1970	W36×245	1010	W18×234*	558	W21×122	305
		W30×261	959	W27×178	555	W12×106	301
W14×342*	1810	W36×230	940			W24×117	297
W36×393*	1750	W12×279*	937			W36×160	295
W40×431	1690	W33×241	932			W40×167	283
W27×448*	1670					W18×130	278
W24×492*	1670			W14×132	548	W21×111	274
				W21×201	542	W33×152	273
		W14×193	931	W40×174	541	W36×150	270
		W44×262	927	W24×192	530	W12×96	270
W14×311*	1610	W40×249	926	W36×256	528	W24×104	259
W36×359*	1570	W27×258	859	W40×278	521	W18×119	253
W30×391*	1550	W30×235	855	W12×170	517	W21×101	248
W33×354*	1460	W33×221	840	W27×161	497	W33×141	246

*Group 4 or 5 shape. See Notes in Table 1-2 (Part 1).

MOMENT OF INERTIA SELECTION TABLE
For W and M shapes I_y

Shape	I_y In.^{4}	Shape	I_y In.^{4}	Shape	I_y In.^{4}	Shape	I_y In.^{4}
W12×87	241	W10×60	116	W10×39	45.0	W6×15	9.32
W10×112	236	W30×90	115	W18×55	44.9	W5×19	9.13
W40×149	229	W24×94	109	W12×40	44.1	W14×26	8.91
W30×148	227			W16×57	43.1	W8×18	7.97
W36×135	225	W12×58	107			M5×18.9	7.86
W18×106	220	W14×61	107	W8×35	42.6	W5×16	7.51
W33×130	218	W27×84	106	W18×50	40.1	W14×22	7.00
				W16×50	37.2	W12×22	4.66
		W10×54	103			W6×16	4.43
W12×79	216			W8×31	37.1	W10×19	4.29
W10×100	207	W12×53	95.8	W10×33	36.6		
W18×97	201	W24×84	94.4	W24×62	34.5		
W30×132	196			W16×45	32.8		
		W10×49	93.4	W21×57	30.6	W4×13	3.86
		W21×93	92.9	W24×55	29.1	W12×19	3.76
W12×72	195	W8×67	88.6	W16×40	28.9	W10×17	3.56
W33×118	187	W24×76	82.5	W14×38	26.7	W8×15	3.41
W16×100	186	W21×83	81.4	W21×50	24.9		
W27×129	184	W8×58	75.1	W12×35	24.5		
W30×124	181	W21×73	70.6	W16×36	24.5		
W10×88	179	W24×68	70.4	W14×34	23.3	W6×12	2.99
W18×86	175	W21×68	64.7	W18×46	22.5	W10×15	2.89
						W12×16	2.82
		W8×48	60.9	W8×28	21.7	W8×13	2.73
W12×65	174	W18×71	60.3	W21×44	20.7	W12×14	2.36
W30×116	164	W14×53	57.7	W12×30	20.3		
W16×89	163	W21×62	57.5	W14×30	19.6		
W27×114	159	W12×50	56.3	W18×40	19.1		
W10×77	154	W18×65	54.8			W6×9	2.19
W18×76	152			W8×24	18.3	W10×12	2.18
W14×82	148			W12×26	17.3	W8×10	2.09
W30×108	146			W6×25	17.1	M12×11.8	1.09
W27×102	139	W10×45	53.4	W10×30	16.7	M12×10.8	0.995
W16×77	138	W14×48	51.4	W18×35	15.3		
W10×68	134	W18×60	50.1	W10×26	14.1		
W14×74	134					M10×9	0.673
W30×99	128			W6×20	13.3		
W27×94	124	W12×45	50.0	W16×31	12.4		
W14×68	121			W10×22	11.4	M10×8	0.597
W24×103	119	W8×40	49.1	W8×21	9.77		
W16×67	119	W14×43	45.2	W16×26	9.59	M8×6.5	0.371

FACTORED UNIFORM LOAD TABLES

General Notes
The Tables of Factored Uniform Loads for W and S shapes and channels (C and MC) used as simple laterally supported beams give the maximum uniformly distributed factored loads in kips. The tables are based on the flexural design strengths specified in Section F1 of the LRFD Specification. Separate tables are presented for $F_y = 36$ ksi and $F_y = 50$ ksi. The tabulated loads include the weight of the beam, which should be deducted in the calculation to determine the net load that the beam will support.

The tables are also applicable to laterally supported simple beams for concentrated loading conditions. A method to determine the beam load capacity for several cases is shown in this discussion.

It is assumed, in all cases, that the loads are applied normal to the X-X axis (shown in the Tables of Properties of Shapes in Part 1 of this LRFD Manual) and that the beam deflects vertically in the plane of bending. If the conditions of loading involve forces outside this plane, design strengths must be determined from the general theory of flexure and torsion.

Lateral Support of Beams
The flexural design strength of a beam is dependent upon lateral support of its compression flange in addition to its section properties. In these tables the notation L_p is used to denote the maximum unbraced length of the compression flange, in feet, for the uniform moment case $(C_b = 1.0)$ and for which the design strengths for compact symmetrical shapes are calculated with a flexural design strength of:

$$\phi_b M_n = \phi_b M_p = \phi_b Z_x F_y / 12$$

Noncompact shapes are calculated with a flexural design strength of:

$$\phi_b M_n' = \phi_b M_p - \phi_b (M_p - M_r) \left(\frac{\lambda - \lambda_p}{\lambda_r - \lambda_p} \right)$$

as permitted in the LRFD Specification Appendix F1. The associated maximum unbraced length for $\phi_b M_n'$ is tabulated as L_p. The notation L_r is the unbraced length of the compression flange for which the flexural design strength for rolled shapes is:

$$\phi_b M_r = \phi_b S_x (F_y - 10) / 12$$

These tables are not applicable for beams with unbraced lengths greater than L_r. For such cases, the beam charts should be used.

Flexural Design Strength and Tabulated Factored Uniform Loads
For symmetrical rolled shapes designated W and S the flexural design strengths and resultant loads are based on the assumption that the compression flanges of the beams are laterally supported at intervals not greater than L_p.

The Uniform Load Constant $\phi_b W_c$ is obtained from the moment and stress relationship of a simply supported, uniformly loaded beam. The relationship results in the formula:

$$\phi_b W_c = \phi_b (2 Z_x F_y / 3), \text{ kip-ft for compact shapes}$$

The following expression may be used for calculating the tabulated uniformly distributed factored load W_u on a simply supported beam or girder:

$W_u = \phi_b W_c / L$, kips

For compact shapes, the tabulated constant is based on the yield stress $F_y = 36$ ksi or 50 ksi and the plastic section modulus Z_x. (See Section F1.1 of the LRFD Specification.) For noncompact sections, the tabulated constant is based on the nominal resisting moment as determined by Equation A-F1-3. (See LRFD Specification Appendix F1.)

Shear
For relatively short spans, the design strengths for beams and channels may be limited by the shear strength of the web instead of the bending strength. This limit is indicated in the tables by solid horizontal lines. Loads shown above these lines will produce the design shear strength in the beam web.

End and Interior Bearing
For a discussion of end and interior bearing and use of the tabulated values ϕR_1 through $\phi_r R_6$ and ϕR, see Part 9 in Volume II of this LRFD Manual.

Vertical Deflection
For rolled shapes designated W, M, S, C, and MC, the maximum vertical deflection may be calculated using the formula:

$\Delta = ML^2 / (C_1 I_x)$

where

M = maximum service load moment, kip-ft
L = span length, ft
I_x = moment of inertia, in.^{4}
C_1 = loading constant (see Figure 4-2)
Δ = maximum vertical deflection, in.

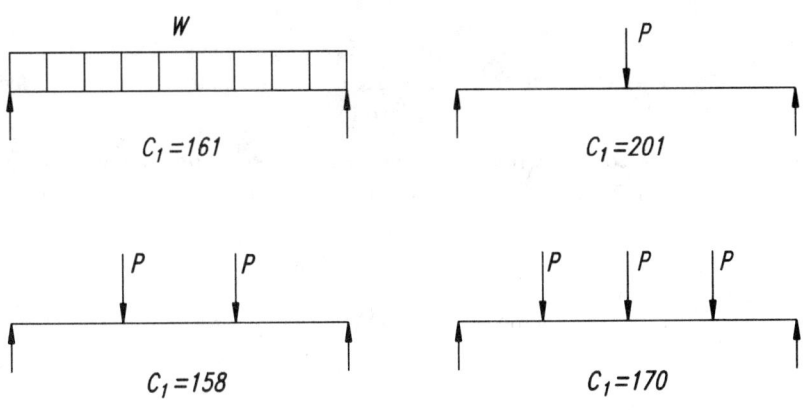

Fig. 4-2

Table 4-2. Recommended Span/Depth Ratios			
Service Load Ratios		Maximum Span/Depth Ratios	
Dead / Total	Dead / Live	F_y = 36 ksi	F_y = 50 ksi
0.2	0.25	20.0	14.0
0.3	0.43	22.2	16.0
0.4	0.67	25.0	18.0
0.5	1.00	29.0	21.0
0.6	1.50	—	26.0

Deflection can be controlled by limiting the span-depth ratio of a simply supported, uniformly loaded beam as shown in Table 4-2. A live-load deflection limit of $L / 360$ is assumed; i.e.,

$$\Delta_{LL} \leq \frac{\text{Span Length}}{360}$$

For large span/depth ratios, vibration may also be a consideration.

Use of Tables

Maximum factored uniform loads are tabulated for steels of F_y = 36 ksi and F_y = 50 ksi. They are based on the design flexural strength determined from the LRFD Specification: Equation F1-1 (in Section F1.1) for compact members, and Equation A-F1-3 (in Appendix F1) for noncompact members. The beams must be braced adequately and have an axis of symmetry in the plane of loading. Factored loads may be read directly from the tables when the distance between points of lateral support of the compression flange L_b does not exceed L_p (tabulated earlier in the Load Factor Design Selection Table for beams).

Loads above the heavy horizontal lines in the tables are governed by the design shear strength, determined from Section F2 of the LRFD Specification.

EXAMPLE 4-4

Given: A W16×45 floor beam of F_y = 50 ksi steel spans 20 feet. Determine the maximum uniform load, end reaction, and total service load deflection. The live load equals the dead load.

Solution: Based on Section A4 of the LRFD Specification, the governing load combination for a floor beam is 1.2 (dead load) + 1.6 (live load). As the two loads are equal,

factored load = 1.4 (total load)

Enter the Factored Uniform Loads Table for F_y = 50 ksi and note that:

Maximum factored uniform load = W_u

= 124 kips, or 124/20 = 6.2 kips/ft

Factored end reaction = W_u / 2 = 124 / 2 = 61.8 kips

$$\text{Service load moment} = \frac{W_u L}{8(LF)} = \frac{124(20)}{8(1.4)}$$

$$= 221 \text{ kip-ft}$$

Deflection:

$$\Delta = \frac{ML^2}{C_1 I_x} = \frac{221(20)^2}{161(586)} = 0.94 \text{ in.}$$

Live load deflection = 0.5×0.94 in. = 0.47 in. < ($L / 360 = 20 \times 12 / 360 = 0.66$ in.) **o.k.**

EXAMPLE 4-5

Given: A W10×45 beam of $F_y = 50$ ksi steel spans 6 feet. Determine the maximum load and corresponding end reaction.

Solution: Enter the Factored Uniform Loads Table for $F_y = 50$ ksi and note that:

Maximum factored uniform load = W_u
$$= 191 \text{ kips, or } 191/6 = 31.8 \text{ kips/ft}$$

As W_u appears above the horizontal line, it is limited by shear in the web.

Factored end reaction = $W_u / 2 = 191 / 2 = 96$ kips

EXAMPLE 4-6

Given: Using $F_y = 50$ ksi steel, select an 18-in. deep beam to span 30 feet and support two equal concentrated loads at the one-third and two-thirds points of the span. The service load intensities are 10 kips dead load and 24 kips live load. The beam is supported laterally at the points of load application and the ends. Determine the beam size and service live load deflection.

Solution: Refer to the Table of Concentrated Load Equivalents on page 4-189 and note that:

Equivalent uniform load = $2.67 P_u$

1. Required factored uniform load:

$$W_u = 2.67 P_u = 2.67[1.2(10) + 1.6(24)]$$
$$= 2.67(50.4)$$
$$= 135 \text{ kips}$$

2. Enter the Factored Uniform Loads Table for $F_y = 50$ ksi and $W_u \geq 135$ kips

For W18×71: $W_u = 145$ kips > 135 kips; however, $L_b = 10$ ft > $L_p = 6.0$ ft.

For W18×76: W_u = 163 kips > 135 kips; however, L_b = 10 ft > L_p = 9.2 ft.

3. Since $L_p < L_b < L_r$, use the Load Factor Design Selection Table.

$$\phi_b M_n = C_b[\phi_b M_p - BF(L_b - L_p)]$$

For the central third of the span (uniform moment), C_b = 1.0.

Required flexural strength: $M_u = P_u(L/3) = 50.4(30/3) = 504$ kip-ft

4. Try W18×71:

$\phi_b M_n$ = 1.0[544 − 13.8(10 − 6)]
      = 489 kip-ft < 504 kip-ft req'd.   **n.g.**

5. Try W18×76:

$\phi_b M_n$ = 1.0[611 − 11.1(10 − 9.2)]
      = 602 kip-ft > 504 kip-ft req'd.   **o.k.**

Use W18×76

6. Determine service live load deflection:

$M_{LL} = (P_{LL}/P_u)M_u = (24/50.4)504 = 240$ kip-ft

$$\text{Maximum } \Delta \text{ (at midspan)} = \frac{M_{LL}L^2}{C_1 I_x} = \frac{240(30)^2}{158(1,330)} = 1.03 \text{ in.}$$

EXAMPLE 4-7

Given: A W24×55 of 50 ksi steel spans 20 feet and is braced at 4-ft intervals. Determine the maximum factored load and end reaction.

Solution: 1. Enter the Factored Uniform Load Table for F_y = 50 ksi and note that:

Maximum factored uniform load = W_u = 201 kips, or 201 / 20 = 10.1 kips/ft

This is true for $L_b \le L_p$: 4.0 ft < 4.7 ft   **o.k.**

2. End reaction = $R = W_u/2 = 201/2 = 101$ kips

Reference Notes on Tables

1. Maximum factored uniform loads, in kips, are given for beams with adequate lateral support; i.e., $L_b \leq L_p$ for $C_b = 1.0$, $L_b \leq L_m$ for $C_b > 1.0$.
2. Loads below the heavy horizontal line are limited by design flexural strength, while loads above the line are limited by design shear strength.
3. Factored loads are given for span lengths up to the smaller of $L/d = 30$ or 72 ft.
4. The end bearing values at the bottom of the tables are for use in solving LRFD Specification Equations K1-3, K1-5a, and K1-5b. They are defined as follows:

$$\phi R_1 = \phi(2.5 k F_y t_w) \quad \text{kips}$$
$$\phi R_2 = \phi(F_y t_w) \quad \text{kips/in.}$$

Equation K1-3 becomes $\phi R_n = \phi R_1 + N(\phi R_2)$

$$\phi_r R_3 = \phi_r \left(68 t_w^2 \sqrt{\frac{F_y t_f}{t_w}} \right) \quad \text{kips}$$

$$\phi_r R_4 = \phi_r \left[68 t_w^2 \left(\frac{3}{d}\right) \left(\frac{t_w}{t_f}\right)^{1.5} \sqrt{\frac{F_y t_f}{t_w}} \right] \quad \text{kips/in.}$$

Equation K1-5a becomes $\phi_r R_n = \phi_r R_3 + N(\phi_r R_4)$

$$\phi_r R_5 = \phi_r \left\{ 68 t_w^2 \left[1 - 0.2 \left(\frac{t_w}{t_f}\right)^{1.5} \right] \sqrt{\frac{F_y t_f}{t_w}} \right\} \quad \text{kips}$$

$$\phi_r R_6 = \phi_r \left[68 t_w^2 \left(\frac{4}{d}\right) \left(\frac{t_w}{t_f}\right)^{1.5} \sqrt{\frac{F_y t_f}{t_w}} \right] \quad \text{kips/in.}$$

Equation K1-5b becomes $\phi_r R_n = \phi_r R_5 + N(\phi_r R_6)$

where $\phi = 1.00$, $\phi_r = 0.75$, $N =$ length of bearing (in.), and the other terms as defined in the LRFD Specification, Section K1.

ϕR ($N = 3\frac{1}{4}$) is defined as the design bearing strength for $N = 3\frac{1}{4}$-in.

For $N/d \leq 0.2$,
   ϕR is the minimum of
      $\phi R_1 + N(\phi R_2)$
      $\phi_r R_3 + N(\phi_r R_4)$

For $N/d > 0.2$,
   ϕR is the minimum of
      $\phi R_1 + N(\phi R_2)$
      $\phi_r R_5 + N(\phi_r R_6)$

For a complete explanation of end and interior bearing and use of the tabulated values, see Part 9 in Volume II of this LRFD Manual.

5. The other terms at the bottom of the tables are:

   Z_x = plastic section modulus for major axis bending, in.^{3}

$\phi_v V_n$ = design shear strength, kips

$\phi_b W_c$ = uniform load constant

        = $\phi_b(2Z_x F_y /3)$ kip-ft for compact shapes;

        per Equation A-F1-3 (LRFD Specification Appendix F1) for noncompact shapes

6. Tabulated maximum factored uniformly distributed load for the given beam and span is the minimum of

$$\frac{\phi_b W_c}{L} \text{ and } 2\phi_v V_n$$

See also Note 2 above.

| F_y = 36 ksi | | BEAMS
W Shapes
Maximum factored uniform loads in kips
for beams laterally supported
For beams laterally unsupported, see page 4-113 | | | W 44 |

Designation			W 44			
Wt./ft			335	290	262	230
F_y = 36 ksi	Span (ft)	20	1750	1480	1330	1180
		21	1670	1460	1310	1130
		22	1590	1390	1250	1080
		23	1520	1330	1190	1030
		24	1460	1280	1140	990
		25	1400	1230	1100	950
		26	1350	1180	1060	914
		27	1300	1140	1020	880
		28	1250	1100	980	849
		29	1210	1060	946	819
		30	1170	1020	914	792
		31	1130	989	885	766
		32	1090	959	857	743
		33	1060	929	831	720
		34	1030	902	807	699
		35	1000	876	784	679
		36	972	852	762	660
		38	921	807	722	625
		40	875	767	686	594
		42	833	730	653	566
		44	795	697	623	540
		46	761	667	596	517
		48	729	639	572	495
		50	700	613	549	475
		52	673	590	528	457
		54	648	568	508	440
		56	625	548	490	424
		58	603	529	473	410
		60	583	511	457	396
		62	564	495	442	383
		64	547	479	429	371
		66	530	465	416	360
		68	515	451	403	349
		70	500	438	392	339
		72	486	426	381	330

Properties and Reaction Values				
Z_x in.^{3}	1620	1420	1270	1100
$\phi_b W_c$ kip-ft	35000	30700	27400	23800
$\phi_v V_n$ kips	873	738	665	592
ϕR_1 kips	235	186	156	128
ϕR_2 kips/in.	36.7	31.3	28.4	25.6
$\phi_r R_3$ kips	419	312	256	202
$\phi_r R_4$ kips/in.	12.5	8.77	7.36	6.28
$\phi_r R_5$ kips	383	287	235	184
$\phi_r R_6$ kips/in.	16.7	11.7	9.81	8.37
ϕR (N = 3$\frac{1}{4}$) kips	355	288	248	211

Load above heavy line is limited by design shear strength.

W 40

BEAMS
W Shapes
Maximum factored uniform loads in kips
for beams laterally supported
For beams laterally unsupported, see page 4-113

$F_y = 36$ ksi

Designation						W 40				
Wt./ft		431	372	321	297	277	249	215	199	174
Span (ft)	15									965
	16									965
	17									908
	18									858
	19	2150	1830	1560	1440				977	813
	20	2110	1800	1530	1440				937	772
	21	2010	1720	1460	1370	1280	1150	985	893	735
	22	1910	1640	1390	1310	1230	1100	945	852	702
	23	1830	1570	1330	1250	1170	1050	904	815	671
	24	1760	1500	1280	1200	1130	1010	867	781	644
	25	1680	1440	1230	1150	1080	968	832	750	618
	26	1620	1390	1180	1100	1040	930	800	721	594
	27	1560	1340	1140	1060	1000	896	770	694	572
	28	1500	1290	1100	1030	964	864	743	670	552
	29	1450	1240	1060	991	931	834	717	647	533
	30	1400	1200	1020	958	900	806	693	625	515
	31	1360	1160	989	927	871	780	671	605	498
	32	1320	1130	959	898	844	756	650	586	483
	33	1280	1090	929	871	818	733	630	568	468
	34	1240	1060	902	845	794	712	612	551	454
	35	1200	1030	876	821	771	691	594	536	441
	36	1170	1000	852	798	750	672	578	521	429
	37	1140	975	829	776	730	654	562	507	417
	38	1110	949	807	756	711	637	547	493	406
	40	1050	902	767	718	675	605	520	469	386
	42	1000	859	730	684	643	576	495	446	368
	44	957	820	697	653	614	550	473	426	351
	46	916	784	667	625	587	526	452	408	336
	48	878	752	639	599	563	504	433	391	322
	50	842	721	613	575	540	484	416	375	309
	52	810	694	590	552	519	465	400	361	297
	54	780	668	568	532	500	448	385	347	286
	56	752	644	548	513	482	432	371	335	276
	58	726	622	529	495	466	417	359	323	266
	60	702	601	511	479	450	403	347	312	257
	62	679	582	495	463	435	390	335	302	249
	64	658	564	479	449	422	378	325	293	241
	66	638	547	465	435	409	367	315	284	234
	68	619	530	451	422	397	356	306	276	227
	70	602	515	438	410	386	346	297	268	221
	72	585	501	426	399	375	336	289	260	215

Properties and Reaction Values

	431	372	321	297	277	249	215	199	174
Z_x in.^{3}	1950	1670	1420	1330	1250	1120	963	868	715
$\phi_b W_c$ kip-ft	42100	36100	30700	28700	27000	24200	20800	18700	15400
$\phi_v V_n$ kips	1075	916	779	720	640	574	493	489	483
ϕR_1 kips	430	339	264	256	224	190	154	143	117
ϕR_2 kips/in.	48.2	41.8	36.0	33.5	29.9	27.0	23.4	23.4	23.4
$\phi_r R_3$ kips	729	547	407	353	290	237	177	165	146
$\phi_r R_4$ kips/in.	22.7	17.2	12.9	11.2	8.40	6.93	5.30	6.12	7.95
$\phi_r R_5$ kips	667	501	373	323	268	219	163	150	126
$\phi_r R_6$ kips/in.	30.2	22.9	17.3	15.0	11.2	9.23	7.07	8.16	10.6
ϕR ($N = 3\frac{1}{4}$) kips	586	475	381	365	318	259	194	185	172

Load above heavy line is limited by design shear strength.

F_y = 36 ksi				

BEAMS
W Shapes
Maximum factored uniform loads in kips
for beams laterally supported
For beams laterally unsupported, see page 4-113

W 40

Designation		W 40							
Wt./ft		331	278	264	235	211	183	167	149
Span (ft)	13								936
	14								921
	15	1930						975	860
	16	1930	1590	1490				934	806
	17	1820	1510	1440	1280	1150	985	879	759
	18	1720	1430	1360	1210	1090	937	830	716
	19	1630	1350	1280	1150	1030	888	787	679
	20	1540	1290	1220	1090	977	843	747	645
	21	1470	1220	1160	1040	931	803	712	614
	22	1400	1170	1110	992	889	767	679	586
	23	1340	1120	1060	949	850	733	650	561
	24	1290	1070	1020	909	815	703	623	537
	25	1240	1030	976	873	782	675	598	516
	26	1190	989	939	839	752	649	575	496
	27	1140	952	904	808	724	625	554	478
	28	1100	918	872	779	698	602	534	461
	29	1070	886	842	752	674	582	515	445
	30	1030	857	814	727	652	562	498	430
	31	996	829	787	704	631	544	482	416
	32	965	803	763	682	611	527	467	403
	33	936	779	740	661	592	511	453	391
	34	908	756	718	642	575	496	440	379
	35	883	734	697	623	559	482	427	368
	36	858	714	678	606	543	469	415	358
	37	835	695	660	590	528	456	404	349
	38	813	676	642	574	514	444	393	339
	40	772	643	610	545	489	422	374	322
	42	735	612	581	519	465	402	356	307
	44	702	584	555	496	444	383	340	293
	46	671	559	531	474	425	367	325	280
	48	644	536	509	455	407	351	311	269
	50	618	514	488	436	391	337	299	258
	52	594	494	469	420	376	324	287	248
	54	572	476	452	404	362	312	277	239
	56	552	459	436	390	349	301	267	230
	58	533	443	421	376	337	291	258	222
	60	515	428	407	364	326	281	249	215
	62	498	415	394	352	315	272	241	208
	64	483	402	381	341	305	264	234	201
	66	468	389	370	331	296	256	226	195
	68	454	378	359	321	287	248	220	190
	70	441	367	349	312	279	241	214	184
	72	429	357	339	303	272	234	208	179

Properties and Reaction Values								
Z_x in.^{3}	1430	1190	1130	1010	905	781	692	597
$\phi_b W_c$ kip-ft	30900	25700	24400	21800	19500	16900	14900	12900
$\phi_v V_n$ kips	967	796	746	640	574	493	488	468
ϕR_1 kips	364	275	254	205	177	154	143	128
ϕR_2 kips/in.	43.9	36.7	34.6	29.9	27.0	23.4	23.4	22.7
$\phi_r R_3$ kips	602	424	379	290	236	177	162	139
$\phi_r R_4$ kips/in.	19.2	13.4	11.7	8.40	6.95	5.30	6.37	7.24
$\phi_r R_5$ kips	550	388	347	268	218	163	146	121
$\phi_r R_6$ kips/in.	25.6	17.9	15.6	11.2	9.27	7.07	8.50	9.65
ϕR ($N = 3^1/_4$) kips	506	395	366	303	259	194	183	163

Load above heavy line is limited by design shear strength.

W 36	BEAMS			$F_y = 36$ ksi

I

BEAMS
W Shapes
Maximum factored uniform loads in kips
for beams laterally supported
For beams laterally unsupported, see page 4-113

Designation				W 36			
Wt./ft			300	280	260	245	230
		19			1180	1120	1060
		20	1350	1260	1170	1090	1020
		21	1300	1200	1110	1040	970
		22	1240	1150	1060	992	926
		23	1180	1100	1010	949	886
		24	1130	1050	972	909	849
		25	1090	1010	933	873	815
		26	1050	972	897	839	783
		27	1010	936	864	808	754
		28	972	903	833	779	727
		29	938	871	804	752	702
		30	907	842	778	727	679
		31	878	815	753	704	657
		32	851	790	729	682	637
		33	825	766	707	661	617
		34	800	743	686	642	599
		35	778	722	667	623	582
		36	756	702	648	606	566
$F_y = 36$ ksi	Span (ft)	37	736	683	630	590	551
		38	716	665	614	574	536
		39	698	648	598	559	522
		40	680	632	583	545	509
		41	664	616	569	532	497
		42	648	602	555	519	485
		43	633	588	543	507	474
		44	619	574	530	496	463
		46	592	549	507	474	443
		48	567	527	486	455	424
		50	544	505	467	436	407
		52	523	486	449	420	392
		54	504	468	432	404	377
		56	486	451	417	390	364
		58	469	436	402	376	351
		60	454	421	389	364	339
		62	439	408	376	352	329
		64	425	395	365	341	318
		66	412	383	353	331	309
		68	400	372	343	321	300
		70	389	361	333	312	291
		72	378	351	324	303	283

Properties and Reaction Values					
Z_x in.^{3}	1260	1170	1080	1010	943
$\phi_b W_c$ kip-ft	27200	25300	23300	21800	20400
$\phi_v V_n$ kips	675	628	592	561	530
ϕR_1 kips	239	214	194	180	162
ϕR_2 kips/in.	34.0	31.9	30.2	28.8	27.4
$\phi_r R_3$ kips	364	319	283	254	228
$\phi_r R_4$ kips/in.	12.6	11.1	10.4	9.65	8.91
$\phi_r R_5$ kips	334	292	258	231	206
$\phi_r R_6$ kips/in.	16.7	14.8	13.9	12.9	11.9
ϕR ($N = 3\frac{1}{4}$) kips	350	318	292	274	251

Load above heavy line is limited by design shear strength.

| F_y = 36 ksi | BEAMS
W Shapes
Maximum factored uniform loads in kips
for beams laterally supported
For beams laterally unsupported, see page 4-113 | W 36 |

Designation						W 36				
Wt./ft		256	232	210	194	182	170	160	150	135
	13									829
	14							910	871	785
	15			1180	1090	1020	956	899	837	733
	16	1400	1260	1120	1040	969	902	842	784	687
	17	1320	1190	1060	975	912	849	793	738	647
	18	1250	1120	1000	920	862	802	749	697	611
	19	1180	1060	947	872	816	759	709	661	579
	20	1120	1010	900	828	775	721	674	627	550
	21	1070	963	857	789	739	687	642	598	524
	22	1020	919	818	753	705	656	613	570	500
	23	977	879	782	720	674	627	586	546	478
	24	936	842	750	690	646	601	562	523	458
	25	899	809	720	663	620	577	539	502	440
	26	864	778	692	637	596	555	518	483	423
	27	832	749	666	614	574	534	499	465	407
	28	802	722	643	592	554	515	481	448	393
	29	775	697	620	571	535	498	465	433	379
	30	749	674	600	552	517	481	449	418	366
	31	725	652	580	534	500	465	435	405	355
	32	702	632	562	518	485	451	421	392	344
	33	681	613	545	502	470	437	408	380	333
	34	661	595	529	487	456	424	396	369	323
	35	642	578	514	473	443	412	385	359	314
	36	624	562	500	460	431	401	374	349	305
	38	591	532	473	436	408	380	355	330	289
	40	562	505	450	414	388	361	337	314	275
	42	535	481	428	394	369	344	321	299	262
	44	511	459	409	377	352	328	306	285	250
	46	488	440	391	360	337	314	293	273	239
	48	468	421	375	345	323	301	281	261	229
	50	449	404	360	331	310	289	270	251	220
	52	432	389	346	319	298	277	259	241	211
	54	416	374	333	307	287	267	250	232	204
	56	401	361	321	296	277	258	241	224	196
	58	387	349	310	286	267	249	232	216	190
	60	374	337	300	276	258	240	225	209	183
	62	362	326	290	267	250	233	217	202	177
	64	351	316	281	259	242	225	211	196	172
	66	340	306	273	251	235	219	204	190	167
	68	330	297	265	244	228	212	198	185	162
	70	321	289	257	237	222	206	193	179	157
	72	312	281	250	230	215	200	187	174	153

(Left margin: F_y = 36 ksi; Span (ft))

Properties and Reaction Values										
Z_x in.^{3}		1040	936	833	767	718	668	624	581	509
$\phi_b W_c$ kip-ft		22500	20200	18000	16600	15500	14400	13500	12500	11000
$\phi_v V_n$ kips		699	628	592	543	512	478	455	436	415
ϕR_1 kips		227	196	173	151	139	122	113	105	91.1
ϕR_2 kips/in.		34.6	31.3	29.9	27.5	26.1	24.5	23.4	22.5	21.6
$\phi_r R_3$ kips		379	311	270	230	205	180	162	147	126
$\phi_r R_4$ kips/in.		12.5	10.4	10.5	8.94	8.16	7.25	6.86	6.65	7.06
$\phi_r R_5$ kips		347	285	244	208	185	162	145	131	110
$\phi_r R_6$ kips/in.		16.7	13.8	14.0	11.9	10.9	9.67	9.15	8.87	9.41
ϕR ($N = 3\frac{1}{4}$) kips		339	298	270	240	223	202	184	168	149

Load above heavy line is limited by design shear strength.

| W 33 | | BEAMS | F_y = 36 ksi |

BEAMS
W Shapes
Maximum factored uniform loads in kips
for beams laterally supported
For beams laterally unsupported, see page 4-113

Designation		W 33			W 33				
Wt./ft		241	221	201	169	152	141	130	118
	16				849	755	694	630	560
	17			936	799	710	653	593	527
	18	1100	1020	926	755	671	617	560	498
	19	1070	972	878	715	635	584	531	472
	20	1010	923	834	679	604	555	504	448
	21	966	879	794	647	575	529	480	427
	22	922	839	758	618	549	505	459	407
	23	882	803	725	591	525	483	439	390
	24	845	770	695	566	503	463	420	374
	25	811	739	667	543	483	444	403	359
	26	780	710	641	523	464	427	388	345
	27	751	684	618	503	447	411	374	332
	28	724	660	596	485	431	397	360	320
	29	699	637	575	468	416	383	348	309
	30	676	616	556	453	402	370	336	299
	31	654	596	538	438	389	358	325	289
	32	634	577	521	425	377	347	315	280
	33	615	560	505	412	366	336	306	272
	34	597	543	490	400	355	327	297	264
	35	579	528	476	388	345	317	288	256
	36	563	513	463	377	335	308	280	249
	37	548	499	451	367	326	300	273	242
	38	534	486	439	358	318	292	265	236
	40	507	462	417	340	302	278	252	224
	42	483	440	397	323	287	264	240	213
	44	461	420	379	309	274	252	229	204
	46	441	401	363	295	262	241	219	195
	48	423	385	347	283	252	231	210	187
	50	406	369	334	272	241	222	202	179
	52	390	355	321	261	232	214	194	172
	54	376	342	309	252	224	206	187	166
	56	362	330	298	243	216	198	180	160
	58	350	318	288	234	208	191	174	155
	60	338	308	278	226	201	185	168	149
	62	327	298	269	219	195	179	163	145
	64	317	289	261	212	189	173	158	140
	66	307	280	253	206	183	168	153	136
	68	298	272	245	200	178	163	148	132
	70	290	264	238	194	172	159	144	128
	72	282	257	232	189	168	154	140	125

F_y = 36 ksi — Span (ft)

Properties and Reaction Values

	241	221	201	169	152	141	130	118
Z_x in.^{3}	939	855	772	629	559	514	467	415
$\phi_b W_c$ kip-ft	20300	18500	16700	13600	12100	11100	10100	8960
$\phi_v V_n$ kips	552	511	468	440	413	392	373	351
ϕR_1 kips	163	144	125	124	107	95.3	88.1	77.3
ϕR_2 kips/in.	29.9	27.9	25.7	24.1	22.9	21.8	20.9	19.8
$\phi_r R_3$ kips	274	236	198	185	159	141	125	107
$\phi_r R_4$ kips/in.	11.0	9.88	8.66	6.69	6.65	6.36	6.33	6.28
$\phi_r R_5$ kips	249	213	179	170	144	127	111	93.6
$\phi_r R_6$ kips/in.	14.6	13.2	11.6	8.92	8.87	8.48	8.44	8.37
ϕR (N = 3¼) kips	261	235	208	203	181	162	146	128

Load above heavy line is limited by design shear strength.

$F_y = 36$ ksi						

BEAMS
W Shapes
Maximum factored uniform loads in kips
for beams laterally supported
For beams laterally unsupported, see page 4-113

W 30

Designation		W 30				
Wt./ft		261	235	211	191	173
	16					775
	17	1140		932	847	769
	18	1130	1010	899	808	726
	19	1070	961	851	765	688
	20	1020	913	809	727	653
	21	968	869	770	692	622
	22	924	830	735	661	594
	23	884	794	703	632	568
	24	847	761	674	606	545
	25	813	730	647	581	523
	26	782	702	622	559	503
	27	753	676	599	538	484
	28	726	652	578	519	467
	29	701	629	558	501	451
	30	678	608	539	485	436
	31	656	589	522	469	422
	32	635	570	506	454	408
	33	616	553	490	441	396
	34	598	537	476	428	384
	36	565	507	449	404	363
	38	535	480	426	383	344
	40	508	456	404	363	327
	42	484	435	385	346	311
	44	462	415	368	330	297
	46	442	397	352	316	284
	48	423	380	337	303	272
	50	407	365	324	291	261
	52	391	351	311	280	251
	54	376	338	300	269	242
	56	363	326	289	260	233
	58	350	315	279	251	225
	60	339	304	270	242	218
	62	328	294	261	234	211
	64	318	285	253	227	204
	66	308	277	245	220	198
	68	299	268	238	214	192
	70	290	261	231	208	187
	72	282	254	225	202	182

$F_y = 36$ ksi · Span (ft)

Properties and Reaction Values

	261	235	211	191	173
Z_x in.^{3}	941	845	749	673	605
$\phi_b W_c$ kip-ft	20300	18300	16200	14500	13100
$\phi_v V_n$ kips	571	505	466	423	388
ϕR_1 kips	204	168	148	124	111
ϕR_2 kips/in.	33.5	29.9	27.9	25.6	23.6
$\phi_r R_3$ kips	353	283	239	199	167
$\phi_r R_4$ kips/in.	14.2	11.2	10.5	9.04	7.96
$\phi_r R_5$ kips	323	260	218	181	151
$\phi_r R_6$ kips/in.	18.9	14.9	14.0	12.0	10.6
ϕR ($N = 3\frac{1}{4}$) kips	313	265	239	207	187

Load above heavy line is limited by design shear strength.

W 30	**BEAMS**	$F_y = 36$ ksi

BEAMS
W Shapes
Maximum factored uniform loads in kips
for beams laterally supported
For beams laterally unsupported, see page 4-113

Designation			W 30						
Wt./ft		148	132	124	116	108	99	90
Span (ft)	11					632	599	540
	12			686	659	623	562	509
	13	775	725	678	628	575	518	470
	14	771	674	629	583	534	481	437
	15	720	629	588	544	498	449	408
	16	675	590	551	510	467	421	382
	17	635	555	518	480	440	396	360
	18	600	524	490	454	415	374	340
	19	568	497	464	430	393	355	322
	20	540	472	441	408	374	337	306
	21	514	449	420	389	356	321	291
	22	491	429	401	371	340	306	278
	23	470	410	383	355	325	293	266
	24	450	393	367	340	311	281	255
	25	432	378	353	327	299	270	245
	26	415	363	339	314	287	259	235
	27	400	350	326	302	277	250	226
	28	386	337	315	292	267	241	218
	29	372	325	304	282	258	232	211
	30	360	315	294	272	249	225	204
	31	348	304	284	263	241	217	197
	32	338	295	275	255	234	211	191
	33	327	286	267	247	226	204	185
	34	318	278	259	240	220	198	180
	36	300	262	245	227	208	187	170
	38	284	248	232	215	197	177	161
	40	270	236	220	204	187	168	153
	42	257	225	210	194	178	160	146
	44	245	215	200	186	170	153	139
	46	235	205	192	177	162	147	133
	48	225	197	184	170	156	140	127
	50	216	189	176	163	149	135	122
	52	208	182	169	157	144	130	118
	54	200	175	163	151	138	125	113
	56	193	169	157	146	133	120	109
	58	186	163	152	141	129	116	105
	60	180	157	147	136	125	112	102
	62	174	152	142	132	121	109	99
	64	169	147	138	128	117	105	96
	66	164	143	134	124	113	102	93
	68	159	139	130	120	110	99	90
	70	154	135	126	117	107	96	87
	72	150	131	122	113	104	94	85

Properties and Reaction Values

	148	132	124	116	108	99	90
Z_x in.^{3}	500	437	408	378	346	312	283
$\phi_b W_c$ kip-ft	10800	9440	8810	8160	7470	6740	6110
$\phi_v V_n$ kips	388	362	343	330	316	300	270
ϕR_1 kips	117	96.9	88.8	82.6	76.6	67.3	55.5
ϕR_2 kips/in.	23.4	22.1	21.1	20.3	19.6	18.7	16.9
$\phi_r R_3$ kips	174	148	132	120	107	93.9	77.0
$\phi_r R_4$ kips/in.	6.97	7.05	6.55	6.49	6.55	6.50	5.29
$\phi_r R_5$ kips	160	133	119	107	94.3	81.1	66.6
$\phi_r R_6$ kips/in.	9.29	9.39	8.73	8.65	8.74	8.66	7.05
ϕR ($N = 3\frac{1}{4}$) kips	193	169	153	141	129	115	94.2

Load above heavy line is limited by design shear strength.

AMERICAN INSTITUTE OF STEEL CONSTRUCTION

$F_y = 36$ ksi							

BEAMS
W Shapes
Maximum factored uniform loads in kips
for beams laterally supported
For beams laterally unsupported, see page 4-113

W 27

For beams laterally unsupported, see page 4-113

Designation		W 27						
Wt./ft		258	235	217	194	178	161	146
$F_y = 36$ ksi	Span (ft) 15					784	708	644
	16	1100	1010	917	820	765	691	622
	17	1080	977	900	798	720	651	586
	18	1020	923	850	754	680	614	553
	19	966	874	805	714	645	582	524
	20	918	831	765	678	612	553	498
	21	874	791	728	646	583	527	474
	22	835	755	695	617	557	503	453
	23	798	722	665	590	532	481	433
	24	765	692	637	565	510	461	415
	25	734	664	612	543	490	442	398
	26	706	639	588	522	471	425	383
	27	680	615	566	502	454	410	369
	28	656	593	546	484	437	395	356
	29	633	573	527	468	422	381	343
	30	612	554	510	452	408	369	332
	31	592	536	493	438	395	357	321
	32	574	519	478	424	383	346	311
	33	556	503	463	411	371	335	302
	34	540	489	450	399	360	325	293
	35	525	475	437	388	350	316	285
	36	510	461	425	377	340	307	277
	37	496	449	413	367	331	299	269
	38	483	437	402	357	322	291	262
	40	459	415	382	339	306	276	249
	42	437	395	364	323	292	263	237
	44	417	378	348	308	278	251	226
	46	399	361	332	295	266	240	216
	48	383	346	319	283	255	230	207
	50	367	332	306	271	245	221	199
	52	353	319	294	261	236	213	191
	54	340	308	283	251	227	205	184
	56	328	297	273	242	219	197	178
	58	317	286	264	234	211	191	172
	60	306	277	255	226	204	184	166
	62	296	268	247	219	198	178	161
	64	287	260	239	212	191	173	156
	66	278	252	232	206	186	168	151

Properties and Reaction Values

	258	235	217	194	178	161	146
Z_x in.^{3}	850	769	708	628	567	512	461
$\phi_b W_c$ kip-ft	18400	16600	15300	13600	12200	11100	9960
$\phi_v V_n$ kips	552	507	459	410	392	354	322
ϕR_1 kips	221	189	163	139	122	108	91.9
ϕR_2 kips/in.	35.3	32.8	29.9	27.0	26.1	23.8	21.8
$\phi_r R_3$ kips	395	337	283	230	206	171	142
$\phi_r R_4$ kips/in.	16.8	15.0	12.3	10.3	10.6	8.86	7.62
$\phi_r R_5$ kips	362	308	260	211	186	154	128
$\phi_r R_6$ kips/in.	22.5	20.0	16.4	13.7	14.1	11.8	10.2
ϕR ($N = 3\frac{1}{4}$) kips	335	296	261	227	207	185	163

Load above heavy line is limited by design shear strength.

| W 27 | BEAMS
W Shapes
Maximum factored uniform loads in kips
for beams laterally supported
For beams laterally unsupported, see page 4-113 | | | | $F_y = 36$ ksi |

Designation			W 27				
Wt./ft			129	114	102	94	84
		11				513	478
		12		605	542	500	439
		13	655	570	507	462	405
		14	609	529	471	429	376
		15	569	494	439	400	351
		16	533	463	412	375	329
		17	502	436	388	353	310
		18	474	412	366	334	293
		19	449	390	347	316	277
		20	427	370	329	300	264
		21	406	353	314	286	251
		22	388	337	299	273	240
		23	371	322	286	261	229
		24	356	309	275	250	220
		25	341	296	264	240	211
		26	328	285	253	231	203
		27	316	274	244	222	195
$F_y = 36$ ksi	Span (ft)	28	305	265	235	214	188
		29	294	255	227	207	182
		30	284	247	220	200	176
		31	275	239	213	194	170
		32	267	232	206	188	165
		33	259	225	200	182	160
		34	251	218	194	177	155
		36	237	206	183	167	146
		38	225	195	173	158	139
		40	213	185	165	150	132
		42	203	176	157	143	125
		44	194	168	150	136	120
		46	185	161	143	131	115
		48	178	154	137	125	110
		50	171	148	132	120	105
		52	164	142	127	115	101
		54	158	137	122	111	98
		56	152	132	118	107	94
		58	147	128	114	104	91
		60	142	123	110	100	88
		62	138	119	106	97	85
		64	133	116	103	94	82
		66	129	112	100	91	80

Properties and Reaction Values

	129	114	102	94	84
Z_x in.^{3}	395	343	305	278	244
$\phi_b W_c$ kip-ft	8530	7410	6590	6000	5270
$\phi_v V_n$ kips	328	302	271	256	239
ϕR_1 kips	99.5	83.4	72.4	63.4	56.9
ϕR_2 kips/in.	22.0	20.5	18.5	17.6	16.6
$\phi_r R_3$ kips	153	127	103	90.6	76.4
$\phi_r R_4$ kips/in.	6.86	6.70	5.58	5.39	5.23
$\phi_r R_5$ kips	140	115	93.0	80.9	67.1
$\phi_r R_6$ kips/in.	9.14	8.93	7.44	7.18	6.97
ϕR ($N = 3\frac{1}{4}$) kips	171	149	121	108	93.4

Load above heavy line is limited by design shear strength.

$F_y = 36$ ksi										

BEADS
W Shapes
Maximum factored uniform loads in kips
for beams laterally supported
For beams laterally unsupported, see page 4-113

W 24

Designation		W 24								
Wt./ft		229	207	192	176	162	146	131	117	104
Span (ft)	13							576	519	468
	14				736	685	625	571	505	446
	15	971	870	802	736	674	602	533	471	416
	16	913	818	755	690	632	564	500	441	390
	17	859	770	710	649	595	531	470	415	367
	18	811	727	671	613	562	502	444	392	347
	19	769	689	635	581	532	475	421	372	329
	20	730	654	604	552	505	451	400	353	312
	21	695	623	575	526	481	430	381	336	297
	22	664	595	549	502	459	410	363	321	284
	23	635	569	525	480	440	393	347	307	271
	24	608	545	503	460	421	376	333	294	260
	25	584	524	483	442	404	361	320	283	250
	26	562	503	464	425	389	347	307	272	240
	27	541	485	447	409	374	334	296	262	231
	28	521	467	431	394	361	322	285	252	223
	29	504	451	416	381	349	311	276	244	215
	30	487	436	402	368	337	301	266	235	208
	31	471	422	389	356	326	291	258	228	201
	32	456	409	377	345	316	282	250	221	195
	33	442	397	366	334	306	274	242	214	189
	34	429	385	355	325	297	266	235	208	184
	35	417	374	345	315	289	258	228	202	178
	36	406	364	335	307	281	251	222	196	173
	38	384	344	318	290	266	238	210	186	164
	40	365	327	302	276	253	226	200	177	156
	42	348	312	287	263	241	215	190	168	149
	44	332	297	274	251	230	205	182	161	142
	46	317	285	262	240	220	196	174	154	136
	48	304	273	252	230	211	188	167	147	130
	50	292	262	241	221	202	181	160	141	125
	52	281	252	232	212	194	174	154	136	120
	54	270	242	224	204	187	167	148	131	116
	56	261	234	216	197	181	161	143	126	111
	58	252	226	208	190	174	156	138	122	108
	60	243	218	201	184	168	150	133	118	104

Properties and Reaction Values

	229	207	192	176	162	146	131	117	104
Z_x in.^{3}	676	606	559	511	468	418	370	327	289
$\phi_b W_c$ kip-ft	14600	13100	12100	11000	10100	9030	7990	7060	6240
$\phi_v V_n$ kips	486	435	401	368	343	313	288	259	234
ϕR_1 kips	216	186	164	143	127	110	95.3	80.4	67.5
ϕR_2 kips/in.	34.6	31.3	29.2	27.0	25.4	23.4	21.8	19.8	18.0
$\phi_r R_3$ kips	379	311	270	230	200	167	141	115	93.7
$\phi_r R_4$ kips/in.	18.0	15.0	13.1	11.5	10.5	9.35	8.65	7.41	6.36
$\phi_r R_5$ kips	347	285	247	211	182	152	127	103	83.5
$\phi_r R_6$ kips/in.	24.1	20.0	17.5	15.3	14.1	12.5	11.5	9.88	8.48
ϕR ($N = 3^1/_4$) kips	328	288	259	231	209	186	166	139	114

Load above heavy line is limited by design shear strength.

			$F_y = 36$ ksi

W 24

BEAMS
W Shapes
Maximum factored uniform loads in kips
for beams laterally supported
For beams laterally unsupported, see page 4-113

Designation			W 24					W 24	
Wt./ft		103	94	84	76	68	62	55
Span (ft)	7							362
	8						397	362
	9					383	367	322
	10			440	409	382	330	289
	11	525	487	440	393	348	300	263
	12	504	457	403	360	319	275	241
	13	465	422	372	332	294	254	223
	14	432	392	346	309	273	236	207
	15	403	366	323	288	255	220	193
	16	378	343	302	270	239	207	181
	17	356	323	285	254	225	194	170
	18	336	305	269	240	212	184	161
	19	318	289	255	227	201	174	152
	20	302	274	242	216	191	165	145
	21	288	261	230	206	182	157	138
	22	275	249	220	196	174	150	132
	23	263	239	210	188	166	144	126
	24	252	229	202	180	159	138	121
	25	242	219	194	173	153	132	116
	26	233	211	186	166	147	127	111
	27	224	203	179	160	142	122	107
	28	216	196	173	154	137	118	103
	29	209	189	167	149	132	114	100
	30	202	183	161	144	127	110	96
	31	195	177	156	139	123	107	93
	32	189	171	151	135	119	103	90
	33	183	166	147	131	116	100	88
	34	178	161	142	127	112	97	85
	35	173	157	138	123	109	94	83
	36	168	152	134	120	106	92	80
	38	159	144	127	114	101	87	76
	40	151	137	121	108	96	83	72
	42	144	131	115	103	91	79	69
	44	137	125	110	98	87	75	66
	46	131	119	105	94	83	72	63
	48	126	114	101	90	80	69	60
	50	121	110	97	86	76	66	58
	52	116	106	93	83	74	64	56
	54	112	102	90	80	71	61	54
	56	108	98	86	77	68	59	52
	58	104	95	83	74	66	57	50
	60	101	91	81				

Properties and Reaction Values

	103	94	84	76	68	62	55
Z_x in.^{3}	280	254	224	200	177	153	134
$\phi_b W_c$ kip-ft	6050	5490	4840	4320	3820	3300	2890
$\phi_v V_n$ kips	262	243	220	205	191	198	181
ϕR_1 kips	86.6	75.3	66.1	56.9	51.4	53.2	46.7
ϕR_2 kips/in.	19.8	18.5	16.9	15.8	14.9	15.5	14.2
$\phi_r R_3$ kips	124	106	86.5	73.6	62.6	66.3	54.0
$\phi_r R_4$ kips/in.	6.35	5.89	5.14	4.81	4.73	5.21	4.75
$\phi_r R_5$ kips	113	96.2	78.3	66.0	55.1	58.0	46.5
$\phi_r R_6$ kips/in.	8.47	7.86	6.85	6.41	6.30	6.95	6.34
ϕR ($N = 3\frac{1}{4}$) kips	144	125	103	89.3	77.9	83.2	69.4

Load above heavy line is limited by design shear strength.

$F_y = 36$ ksi									

BEAMS
W Shapes
Maximum factored uniform loads in kips
for beams laterally supported

For beams laterally unsupported, see page 4-113

W 21

Designation					W 21					
Wt./ft		201	182	166	147	132	122	111	101	
	13				618	552	506	460	415	
	14	815	733	656	575	514	474	430	390	
	15	763	685	622	537	480	442	402	364	
	16	716	643	583	504	450	414	377	342	
	17	673	605	549	474	423	390	354	321	
	18	636	571	518	448	400	368	335	304	
	19	603	541	491	424	379	349	317	288	
	20	572	514	467	403	360	332	301	273	
	21	545	490	444	384	343	316	287	260	
	22	520	467	424	366	327	301	274	248	
	23	498	447	406	350	313	288	262	238	
	24	477	428	389	336	300	276	251	228	
	25	458	411	373	322	288	265	241	219	
	26	440	395	359	310	277	255	232	210	
	27	424	381	346	298	266	246	223	202	
	28	409	367	333	288	257	237	215	195	
	29	395	355	322	278	248	229	208	188	
	30	382	343	311	269	240	221	201	182	
	31	369	332	301	260	232	214	194	176	
	32	358	321	292	252	225	207	188	171	
	33	347	312	283	244	218	201	183	166	
	34	337	302	274	237	212	195	177	161	
	35	327	294	267	230	206	189	172	156	
	36	318	286	259	224	200	184	167	152	
	38	301	271	246	212	189	175	159	144	
	40	286	257	233	201	180	166	151	137	
	42	273	245	222	192	171	158	143	130	
	44	260	234	212	183	163	151	137	124	
	46	249	224	203	175	156	144	131	119	
	48	239	214	194	168	150	138	126	114	
	50	229	206	187	161	144	133	121	109	
	52	220	198	179	155	138	128	116	105	

($F_y = 36$ ksi) (Span (ft)) — left margin labels

Properties and Reaction Values

	201	182	166	147	132	122	111	101
Z_x in.^{3}	530	476	432	373	333	307	279	253
$\phi_b W_c$ kip-ft	11400	10300	9330	8060	7190	6630	6030	5460
$\phi_v V_n$ kips	407	367	328	309	276	253	230	208
ϕR_1 kips	195	168	143	122	106	91.1	80.4	70.3
ϕR_2 kips/in.	32.8	29.9	27.0	25.9	23.4	21.6	19.8	18.0
$\phi_r R_3$ kips	339	281	232	200	163	139	117	96.8
$\phi_r R_4$ kips/in.	18.4	15.6	12.7	13.5	11.2	9.53	8.11	6.72
$\phi_r R_5$ kips	311	258	213	181	147	126	105	87.2
$\phi_r R_6$ kips/in.	24.6	20.8	16.9	18.0	14.9	12.7	10.8	8.95
ϕR ($N = 3^{1}/_{4}$) kips	301	265	231	206	182	161	143	119

Load above heavy line is limited by design shear strength.

	F_y = 36 ksi

W 21

BEAMS
W Shapes
Maximum factored uniform loads in kips
for beams laterally supported
For beams laterally unsupported, see page 4-113

Designation		W 21					W 21		
Wt./ft		93	83	73	68	62	57	50	44
Span (ft)	7							308	281
	8						332	297	258
	9	488	429	376	353	326	310	264	229
	10	477	423	372	346	311	279	238	206
	11	434	385	338	314	283	253	216	187
	12	398	353	310	288	259	232	198	172
	13	367	326	286	266	239	214	183	159
	14	341	302	265	247	222	199	170	147
	15	318	282	248	230	207	186	158	137
	16	298	265	232	216	194	174	149	129
	17	281	249	219	203	183	164	140	121
	18	265	235	206	192	173	155	132	114
	19	251	223	196	182	164	147	125	108
	20	239	212	186	173	156	139	119	103
	21	227	202	177	165	148	133	113	98
	22	217	192	169	157	141	127	108	94
	23	208	184	162	150	135	121	103	90
	24	199	176	155	144	130	116	99	86
	25	191	169	149	138	124	111	95	82
	26	184	163	143	133	120	107	91	79
	27	177	157	138	128	115	103	88	76
	28	170	151	133	123	111	100	85	74
	29	165	146	128	119	107	96	82	71
	30	159	141	124	115	104	93	79	69
	31	154	137	120	111	100	90	77	66
	32	149	132	116	108	97	87	74	64
	33	145	128	113	105	94	84	72	62
	34	140	125	109	102	91	82	70	61
	35	136	121	106	99	89	80	68	59
	36	133	118	103	96	86	77	66	57
	38	126	111	98	91	82	73	63	54
	40	119	106	93	86	78	70	59	52
	42	114	101	88	82	74	66	57	49
	44	108	96	84	79	71	63	54	47
	46	104	92	81	75	68	61	52	45
	48	99	88	77	72	65	58	50	43
	50	95	85	74	69	62	56	48	41
	52	92	81	71	66	60	54	46	

Properties and Reaction Values									
Z_x in.^{3}		221	196	172	160	144	129	110	95.4
$\phi_b W_c$ kip-ft		4770	4230	3720	3460	3110	2790	2380	2060
$\phi_v V_n$ kips		244	215	188	177	163	166	154	141
ϕR_1 kips		88.1	72.4	61.4	55.6	49.5	50.1	44.9	37.4
ϕR_2 kips/in.		20.9	18.5	16.4	15.5	14.4	14.6	13.7	12.6
$\phi_r R_3$ kips		130	103	80.8	71.4	60.7	63.6	52.4	42.5
$\phi_r R_4$ kips/in.		8.91	7.01	5.50	5.04	4.55	4.45	4.52	4.23
$\phi_r R_5$ kips		118	93.3	73.0	64.3	54.3	57.3	46.2	36.7
$\phi_r R_6$ kips/in.		11.9	9.34	7.34	6.72	6.07	5.94	6.03	5.64
ϕR ($N = 3^1/_4$) kips		156	126	98.7	87.8	75.5	78.1	67.1	56.3

Load above heavy line is limited by design shear strength.

| F_y = 36 ksi | | **BEAMS** W Shapes Maximum factored uniform loads in kips for beams laterally supported For beams laterally unsupported, see page 4-113 | | | | | | | W 18 |

Designation				W 18					W 18		
Wt./ft		192	175	158	143	130	119	106	97	86	76
	11						483	430	387	343	301
	12	760	693	621	553	501	470	414	380	335	293
	13	734	661	592	535	484	434	382	351	309	271
	14	682	614	549	497	449	403	355	326	287	251
	15	636	573	513	464	419	376	331	304	268	235
	16	597	537	481	435	393	352	311	285	251	220
	17	562	506	452	409	370	332	292	268	236	207
	18	530	478	427	386	349	313	276	253	223	196
	19	502	452	405	366	331	297	261	240	211	185
	20	477	430	384	348	314	282	248	228	201	176
	21	455	409	366	331	299	268	237	217	191	168
	22	434	391	350	316	286	256	226	207	183	160
	23	415	374	334	302	273	245	216	198	175	153
	24	398	358	320	290	262	235	207	190	167	147
	25	382	344	308	278	251	226	199	182	161	141
	26	367	331	296	268	242	217	191	175	155	135
	27	354	318	285	258	233	209	184	169	149	130
	28	341	307	275	248	224	201	177	163	143	126
	29	329	296	265	240	217	194	171	157	139	121
	30	318	287	256	232	210	188	166	152	134	117
	31	308	277	248	224	203	182	160	147	130	114
	32	298	269	240	217	196	176	155	142	126	110
	33	289	261	233	211	190	171	151	138	122	107
	34	281	253	226	205	185	166	146	134	118	104
	35	273	246	220	199	180	161	142	130	115	101
	36	265	239	214	193	175	157	138	127	112	98
	37	258	232	208	188	170	152	134	123	109	95
	38	251	226	202	183	165	148	131	120	106	93
	39	245	220	197	178	161	145	127	117	103	90
	40	239	215	192	174	157	141	124	114	100	88
	42	227	205	183	166	150	134	118	109	96	84
	44	217	195	175	158	143	128	113	104	91	80

(Row labels at left: F_y = 36 ksi, Span (ft))

Properties and Reaction Values										
Z_x in.^{3}	442	398	356	322	291	261	230	211	186	163
$\phi_b W_c$ kip-ft	9550	8600	7690	6960	6290	5640	4970	4560	4020	3520
$\phi_v V_n$ kips	380	347	311	277	251	242	215	193	172	150
ϕR_1 kips	211	180	155	131	113	103	86.3	75.2	62.1	52.6
ϕR_2 kips/in.	34.6	32.0	29.2	26.3	24.1	23.6	21.2	19.3	17.3	15.3
$\phi_r R_3$ kips	381	324	268	219	184	167	134	112	89.3	69.9
$\phi_r R_4$ kips/in.	22.8	20.3	17.2	13.9	12.0	12.8	10.7	8.69	7.17	5.69
$\phi_r R_5$ kips	350	297	245	201	168	151	121	101	80.5	63.0
$\phi_r R_6$ kips/in.	30.4	27.1	22.9	18.5	15.9	17.1	14.3	11.6	9.56	7.59
ϕR (N = 3$^1/_4$) kips	323	284	250	217	191	180	155	138	113	88.4

Load above heavy line is limited by design shear strength.

F_y = 36 ksi

W 18

BEAMS
W Shapes
Maximum factored uniform loads in kips for beams laterally supported

For beams laterally unsupported, see page 4-113

Designation	W 18					W 18		
Wt./ft	71	65	60	55	50	46	40	35
F_y = 36 ksi, Span (ft)								
6								206
7						253	219	205
8	355	321		275	248	245	212	180
9	348	319	294	269	242	218	188	160
10	313	287	266	242	218	196	169	144
11	285	261	242	220	198	178	154	131
12	261	239	221	202	182	163	141	120
13	241	221	204	186	168	151	130	110
14	224	205	190	173	156	140	121	103
15	209	192	177	161	145	131	113	96
16	196	180	166	151	136	122	106	90
17	184	169	156	142	128	115	100	84
18	174	160	148	134	121	109	94	80
19	165	151	140	127	115	103	89	76
20	157	144	133	121	109	98	85	72
21	149	137	127	115	104	93	81	68
22	142	131	121	110	99	89	77	65
23	136	125	116	105	95	85	74	62
24	131	120	111	101	91	82	71	60
25	125	115	106	97	87	78	68	57
26	120	110	102	93	84	75	65	55
27	116	106	98	90	81	73	63	53
28	112	103	95	86	78	70	60	51
29	108	99	92	83	75	68	58	50
30	104	96	89	81	73	65	56	48
31	101	93	86	78	70	63	55	46
32	98	90	83	76	68	61	53	45
33	95	87	81	73	66	59	51	44
34	92	84	78	71	64	58	50	42
35	89	82	76	69	62	56	48	41
36	87	80	74	67	61	54	47	40
38	82	76	70	64	57	52	45	38
40	78	72	66	60	55	49	42	36
42	75	68	63	58	52	47	40	34
44	71	65	60	55	50	45	38	33

Properties and Reaction Values

	71	65	60	55	50	46	40	35
Z_x in.^{3}	145	133	123	112	101	90.7	78.4	66.5
$\phi_b W_c$ kip-ft	3130	2870	2660	2420	2180	1960	1690	1440
$\phi_v V_n$ kips	178	161	147	137	124	126	110	103
ϕR_1 kips	66.8	58.2	51.4	46.1	39.9	40.5	33.7	30.4
ϕR_2 kips/in.	17.8	16.2	14.9	14.0	12.8	13.0	11.3	10.8
$\phi_r R_3$ kips	95.9	80.0	68.2	59.2	48.9	51.4	39.2	32.8
$\phi_r R_4$ kips/in.	7.44	6.08	5.18	4.77	4.01	3.92	3.05	3.29
$\phi_r R_5$ kips	86.7	72.6	61.9	53.4	44.1	46.7	35.6	28.9
$\phi_r R_6$ kips/in.	9.92	8.10	6.90	6.36	5.34	5.23	4.07	4.39
ϕR ($N = 3\frac{1}{4}$) kips	120	99.8	85.0	74.7	61.9	64.2	49.1	43.5

Load above heavy line is limited by design shear strength.

F_y = 36 ksi

BEAMS
W Shapes
Maximum factored uniform loads in kips
for beams laterally supported
For beams laterally unsupported, see page 4-113

W 16

Designation		W 16				W 16					W 16	
Wt./ft		100	89	77	67	57	50	45	40	36	31	26
Span (ft)	6										170	153
	7									182	167	136
	8					275	240	216	190	173	146	119
	9					252	221	198	175	154	130	106
	10					227	199	178	157	138	117	95
	11	386	342	292	251	206	181	162	143	126	106	87
	12	356	315	270	234	189	166	148	131	115	97	80
	13	329	291	249	216	174	153	137	121	106	90	73
	14	305	270	231	201	162	142	127	112	99	83	68
	15	285	252	216	187	151	132	119	105	92	78	64
	16	267	236	203	176	142	124	111	98	86	73	60
	17	252	222	191	165	133	117	105	93	81	69	56
	18	238	210	180	156	126	110	99	87	77	65	53
	19	225	199	171	148	119	105	94	83	73	61	50
	20	214	189	162	140	113	99	89	79	69	58	48
	21	204	180	154	134	108	95	85	75	66	56	45
	22	194	172	147	128	103	90	81	72	63	53	43
	23	186	164	141	122	99	86	77	68	60	51	42
	24	178	158	135	117	95	83	74	66	58	49	40
	25	171	151	130	112	91	79	71	63	55	47	38
	26	164	145	125	108	87	76	68	61	53	45	37
	27	158	140	120	104	84	74	66	58	51	43	35
	28	153	135	116	100	81	71	63	56	49	42	34
	29	147	130	112	97	78	69	61	54	48	40	33
	30	143	126	108	94	76	66	59	52	46	39	32
	31	138	122	105	91	73	64	57	51	45	38	31
	32	134	118	101	88	71	62	56	49	43	36	30
	33	130	115	98	85	69	60	54	48	42	35	29
	34	126	111	95	83	67	58	52	46	41	34	28
	35	122	108	93	80	65	57	51	45	39	33	27
	36	119	105	90	78	63	55	49	44	38	32	27
	38	113	99	85	74	60	52	47	41	36	31	25
	40	107	95	81	70	57	50	44	39			

Properties and Reaction Values

	100	89	77	67	57	50	45	40	36	31	26
Z_x in.^{3}	198	175	150	130	105	92.0	82.3	72.9	64.0	54.0	44.2
$\phi_b W_c$ kip-ft	4280	3780	3240	2810	2270	1990	1780	1570	1380	1170	955
$\phi_v V_n$ kips	193	171	146	125	137	120	108	94.9	91.0	84.9	76.3
ϕR_1 kips	88.8	73.8	58.9	48.9	53.2	44.9	38.8	32.6	29.9	27.8	23.9
ϕR_2 kips/in.	21.1	18.9	16.4	14.2	15.5	13.7	12.4	11.0	10.6	9.90	9.00
$\phi_r R_3$ kips	136	109	81.9	61.9	73.0	56.9	46.6	36.6	32.2	29.3	22.5
$\phi_r R_4$ kips/in.	11.0	9.06	6.89	5.21	6.21	4.92	4.14	3.22	3.46	2.73	2.65
$\phi_r R_5$ kips	123	98.8	74.3	56.3	66.2	51.6	42.2	33.2	28.5	26.4	19.7
$\phi_r R_6$ kips/in.	14.7	12.1	9.18	6.95	8.28	6.56	5.52	4.30	4.61	3.64	3.53
ϕR ($N = 3\frac{1}{4}$) kips	157	135	104	78.9	93.2	72.9	60.1	47.1	43.5	38.2	31.2

Load above heavy line is limited by design shear strength.

| W 14 | | | | | | | | | $F_y = 36$ ksi |

BEAMS
W Shapes
Maximum factored uniform loads in kips
for beams laterally supported
For beams laterally unsupported, see page 4-113

Designation		W 14					W 14			
Wt./ft		132	120	109	99	90	82	74	68	61
Span (ft)	10						284	248	227	203
	11						273	247	226	200
	12						250	227	207	184
	13	368	332		267		231	209	191	169
	14	361	327	292	267	240	214	194	177	157
	15	337	305	276	249	226	200	181	166	147
	16	316	286	259	234	212	188	170	155	138
	17	297	269	244	220	199	177	160	146	130
	18	281	254	230	208	188	167	151	138	122
	19	266	241	218	197	178	158	143	131	116
	20	253	229	207	187	170	150	136	124	110
	21	241	218	197	178	161	143	130	118	105
	22	230	208	189	170	154	136	124	113	100
	23	220	199	180	162	147	131	118	108	96
	24	211	191	173	156	141	125	113	104	92
	25	202	183	166	149	136	120	109	99	88
	26	194	176	160	144	130	115	105	96	85
	27	187	170	154	138	126	111	101	92	82
	28	181	164	148	133	121	107	97	89	79
	29	174	158	143	129	117	104	94	86	76
	30	168	153	138	125	113	100	91	83	73
	31	163	148	134	121	109	97	88	80	71
	32	158	143	130	117	106	94	85	78	69
	33	153	139	126	113	103	91	82	75	67
	34	149	135	122	110	100	88	80	73	65

Properties and Reaction Values

	132	120	109	99	90	82	74	68	61
Z_x in.^{3}	234	212	192	173	157	139	126	115	102
$\phi_b W_c$ kip-ft	5050	4580	4150	3740	3390	3000	2720	2480	2200
$\phi_v V_n$ kips	184	166	146	134	120	142	124	113	101
ϕR_1 kips	98.0	86.3	73.8	62.7	54.5	74.6	63.3	56.0	48.5
ϕR_2 kips/in.	23.2	21.2	18.9	17.5	15.8	18.4	16.2	14.9	13.5
$\phi_r R_3$ kips	161	134	108	91.3	75.3	103	81.8	69.4	56.4
$\phi_r R_4$ kips/in.	16.3	13.9	10.8	9.48	7.86	9.95	7.52	6.49	5.40
$\phi_r R_5$ kips	145	121	97.6	82.3	67.9	93.6	74.7	63.3	51.4
$\phi_r R_6$ kips/in.	21.8	18.5	14.4	12.6	10.5	13.3	10.0	8.65	7.20
ϕR ($N = 3\frac{1}{4}$) kips	173	155	135	119	102	134	107	91.5	74.8

Load above heavy line is limited by design shear strength.

BEAMS
W Shapes
Maximum factored uniform loads in kips
for beams laterally supported
For beams laterally unsupported, see page 4-113

W 14

F_y = 36 ksi

Designation		W 14			W 14			W 14	
Wt./ft		53	48	43	38	34	30	26	22
Span (ft)	5								123
	6							138	120
	7				170	155	145	124	102
	8				166	147	128	109	90
	9	200	182	162	148	131	114	96	80
	10	188	169	150	133	118	102	87	72
	11	171	154	137	121	107	93	79	65
	12	157	141	125	111	98	85	72	60
	13	145	130	116	102	91	79	67	55
	14	134	121	107	95	84	73	62	51
	15	125	113	100	89	79	68	58	48
	16	118	106	94	83	74	64	54	45
	17	111	100	88	78	69	60	51	42
	18	105	94	84	74	66	57	48	40
	19	99	89	79	70	62	54	46	38
	20	94	85	75	66	59	51	43	36
	21	90	81	72	63	56	49	41	34
	22	86	77	68	60	54	46	39	33
	23	82	74	65	58	51	44	38	31
	24	78	71	63	55	49	43	36	30
	25	75	68	60	53	47	41	35	29
	26	72	65	58	51	45	39	33	28
	27	70	63	56	49	44	38	32	27
	28	67	60	54	47	42	36	31	26
	29	65	58	52	46	41	35	30	25
	30	63	56	50	44	39	34	29	24
	31	61	55	48	43	38	33	28	23
	32	59	53	47	42	37	32	27	22
	33	57	51	46	40	36	31	26	22
	34	55	50	44	39	35	30	26	21

Properties and Reaction Values

	53	48	43	38	34	30	26	22
Z_x in.[3]	87.1	78.4	69.6	61.5	54.6	47.3	40.2	33.2
$\phi_b W_c$ kip-ft	1880	1690	1500	1330	1180	1020	868	717
$\phi_v V_n$ kips	100	91.1	81.0	85.0	77.5	72.6	69.0	61.4
ϕR_1 kips	47.9	42.1	36.0	29.6	25.7	22.8	21.5	18.1
ϕR_2 kips/in.	13.3	12.2	11.0	11.2	10.3	9.72	9.18	8.28
$\phi_r R_3$ kips	55.9	46.8	37.5	37.9	31.4	26.6	25.5	19.5
$\phi_r R_4$ kips/in.	5.06	4.40	3.60	3.77	3.34	3.39	2.61	2.43
$\phi_r R_5$ kips	51.3	42.8	34.2	34.4	28.3	23.5	23.1	17.3
$\phi_r R_6$ kips/in.	6.75	5.86	4.80	5.02	4.45	4.52	3.47	3.24
ϕR ($N = 3\frac{1}{4}$) kips	73.2	61.8	49.8	50.7	42.8	38.2	34.4	27.8

Load above heavy line is limited by design shear strength.

| W 12 | | | BEAMS W Shapes Maximum factored uniform loads in kips for beams laterally supported For beams laterally unsupported, see page 4-113 | | | | | | F_y = 36 ksi | |

BEAMS
W Shapes
Maximum factored uniform loads in kips
for beams laterally supported
For beams laterally unsupported, see page 4-113

Designation						W 12					W 12	
Wt./ft			120	106	96	87	79	72	65	58	53
		10								171	162
		11	362	306	272	251	226	205	184	170	153
		12	335	295	265	238	214	194	174	156	140
		13	309	272	244	219	198	179	161	144	129
		14	287	253	227	204	184	167	149	133	120
		15	268	236	212	190	171	156	139	124	112
		16	251	221	198	178	161	146	131	117	105
		17	236	208	187	168	151	137	123	110	99
		18	223	197	176	158	143	130	116	104	93
F_y = 36 ksi	Span (ft)	19	211	186	167	150	135	123	110	98	89
		20	201	177	159	143	129	117	105	93	84
		21	191	169	151	136	122	111	100	89	80
		22	183	161	144	130	117	106	95	85	76
		23	175	154	138	124	112	101	91	81	73
		24	167	148	132	119	107	97	87	78	70
		25	161	142	127	114	103	93	84	75	67
		26	155	136	122	110	99	90	80	72	65
		27	149	131	118	106	95	86	77	69	62
		28	143	127	113	102	92	83	75	67	60
		29	139	122	109	98	89	80	72	64	58
		30	134	118	106	95	86	78	70	62	56

Properties and Reaction Values

	120	106	96	87	79	72	65	58	53
Z_x in.^{3}	186	164	147	132	119	108	96.8	86.4	77.9
$\phi_b W_c$ kip-ft	4020	3540	3180	2850	2570	2330	2090	1870	1680
$\phi_v V_n$ kips	181	153	136	125	113	102	91.9	85.3	80.9
ϕR_1 kips	116	92.6	80.4	69.5	60.8	53.2	46.1	44.6	38.8
ϕR_2 kips/in.	25.6	22.0	19.8	18.5	16.9	15.5	14.0	13.0	12.4
$\phi_r R_3$ kips	192	145	118	102	84.5	70.6	58.0	52.9	47.0
$\phi_r R_4$ kips/in.	22.7	16.3	13.4	12.4	10.5	8.89	7.43	5.49	5.44
$\phi_r R_5$ kips	173	131	107	91.5	75.9	63.4	52.0	48.4	42.6
$\phi_r R_6$ kips/in.	30.2	21.8	17.8	16.5	14.0	11.9	9.90	7.32	7.25
ϕR (N = 3¼) kips	**199**	**164**	**145**	**130**	**116**	102	84.1	72.2	66.2

Load above heavy line is limited by design shear strength.
Values of R in **bold face** exceed maximum design web shear $\phi_v V_n$.

FACTORED UNIFORM LOAD TABLES

$F_y = 36$ ksi

BEAMS
W Shapes
Maximum factored uniform loads in kips
for beams laterally supported

For beams laterally unsupported, see page 4-113

W 12

Designation		W 12			W 12			W 12		
Wt./ft		50	45	40	35	30	26	22	19	16	14
Span (ft)	4								111	103	93
	5							124	107	87	75
	6							105	89	72	63
	7				146	125	109	90	76	62	54
	8	175	157		138	116	100	79	67	54	47
	9	174	155	137	123	103	89	70	59	48	42
	10	156	140	124	111	93	80	63	53	43	38
	11	142	127	113	101	85	73	58	49	39	34
	12	130	116	104	92	78	67	53	44	36	31
	13	120	108	96	85	72	62	49	41	33	29
	14	112	100	89	79	66	57	45	38	31	27
	15	104	93	83	74	62	54	42	36	29	25
	16	98	87	78	69	58	50	40	33	27	23
	17	92	82	73	65	55	47	37	31	26	22
	18	87	78	69	61	52	45	35	30	24	21
	19	82	74	65	58	49	42	33	28	23	20
	20	78	70	62	55	47	40	32	27	22	19
	21	74	67	59	53	44	38	30	25	21	18
	22	71	64	56	50	42	37	29	24	20	17
	23	68	61	54	48	40	35	28	23	19	16
	24	65	58	52	46	39	33	26	22	18	16
	25	63	56	50	44	37	32	25	21	17	15
	26	60	54	48	43	36	31	24	21	17	14
	27	58	52	46	41	34	30	23	20	16	14
	28	56	50	44	39	33	29	23	19	16	13
	29	54	48	43	38	32	28	22	18	15	13
	30	52	47	37	31	27	21	18			

$F_y = 36$ ksi

Properties and Reaction Values

		50	45	40	35	30	26	22	19	16	14
Z_x in.^{3}		72.4	64.7	57.5	51.2	43.1	37.2	29.3	24.7	20.1	17.4
$\phi_b W_c$ kip-ft		1560	1400	1240	1110	931	804	633	534	434	376
$\phi_v V_n$ kips		87.7	78.5	68.5	72.9	62.4	54.6	62.2	55.6	51.3	46.3
ϕR_1 kips		45.8	37.7	33.2	27.0	21.9	18.1	20.5	17.2	14.9	12.4
ϕR_2 kips/in.		13.3	12.1	10.6	10.8	9.36	8.28	9.36	8.46	7.92	7.20
$\phi_r R_3$ kips		55.1	45.0	35.2	36.3	26.9	20.8	26.4	20.6	16.3	13.0
$\phi_r R_4$ kips/in.		5.96	4.98	3.83	3.81	2.97	2.41	3.08	2.80	3.08	2.74
$\phi_r R_5$ kips		50.3	41.0	32.1	33.1	24.5	18.8	23.9	18.4	13.8	10.8
$\phi_r R_6$ kips/in.		7.95	6.64	5.11	5.08	3.96	3.21	4.11	3.73	4.10	3.65
ϕR ($N = 3\frac{1}{4}$) kips		76.1	62.6	48.7	49.6	37.3	29.3	37.3	30.5	27.1	22.7

Load above heavy line is limited by design shear strength.

AMERICAN INSTITUTE OF STEEL CONSTRUCTION

W 10	BEAMS W Shapes Maximum factored uniform loads in kips for beams laterally supported	$F_y = 36$ ksi

For beams laterally unsupported, see page 4-113

Designation		W 10							
Wt./ft		112	100	88	77	68	60	54	49
Span (ft)	9	333	293	255	218	190	167	145	132
	10	318	281	244	211	184	161	144	130
	11	289	255	222	192	167	146	131	119
	12	265	234	203	176	154	134	120	109
	13	244	216	188	162	142	124	111	100
	14	227	201	174	151	132	115	103	93
	15	212	187	163	141	123	107	96	87
	16	198	176	153	132	115	101	90	82
	17	187	165	144	124	108	95	85	77
	18	176	156	136	117	102	90	80	72
	19	167	148	128	111	97	85	76	69
	20	159	140	122	105	92	81	72	65
	21	151	134	116	100	88	77	69	62
	22	144	128	111	96	84	73	65	59
	23	138	122	106	92	80	70	63	57
	24	132	117	102	88	77	67	60	54

$F_y = 36$ ksi

Properties and Reaction Values

	112	100	88	77	68	60	54	49
Z_x in.^{3}	147	130	113	97.6	85.3	74.6	66.6	60.4
$\phi_b W_c$ kip-ft	3180	2810	2440	2110	1840	1610	1440	1300
$\phi_v V_n$ kips	167	147	127	109	95.0	83.4	72.6	66.0
ϕR_1 kips	127	107	88.5	71.5	58.2	49.6	41.6	36.3
ϕR_2 kips/in.	27.2	24.5	21.8	19.1	16.9	15.1	13.3	12.2
$\phi_r R_3$ kips	224	182	143	110	86.5	68.7	54.0	45.4
$\phi_r R_4$ kips/in.	27.8	23.2	18.9	14.8	11.9	9.79	7.49	6.46
$\phi_r R_5$ kips	203	164	130	99.7	78.3	62.0	49.0	41.1
$\phi_r R_6$ kips/in.	37.1	31.0	25.3	19.8	15.9	13.0	9.99	8.61
ϕR ($N = 3\frac{1}{4}$) kips	**216**	**187**	**159**	**134**	**113**	**98.8**	**81.4**	**69.1**

Load above heavy line is limited by design shear strength.
Values of R in **bold face** exceed maximum design web shear $\phi_v V_n$.

F_y = 36 ksi	BEAMS	W 10

BEAMS
W Shapes
Maximum factored uniform loads in kips
for beams laterally supported
For beams laterally unsupported, see page 4-113

Designation			W 10			W 10			W 10		
Wt./ft		45	39	33	30	26	22	19	17	15	12
	3									89	73
	4							100	94	86	68
	5						95	93	81	69	54
	6					122	104	94	78	67	58	45
	7			110	113	97	80	67	58	49	39
	8	137	121	105	99	85	70	58	50	43	34
	9	132	112	93	88	75	62	52	45	38	30
	10	119	101	84	79	68	56	47	40	35	27
	11	108	92	76	72	61	51	42	37	31	25
	12	99	84	70	66	56	47	39	34	29	23
	13	91	78	64	61	52	43	36	31	27	21
	14	85	72	60	56	48	40	33	29	25	19
	15	79	67	56	53	45	37	31	27	23	18
	16	74	63	52	49	42	35	29	25	22	17
	17	70	59	49	47	40	33	27	24	20	16
	18	66	56	47	44	38	31	26	22	19	15
	19	62	53	44	42	36	30	25	21	18	14
	20	59	51	42	40	34	28	23	20	17	14
	21	56	48	40	38	32	27	22	19	16	13
	22	54	46	38	36	31	26	21	18	16	12
	23	52	44	36	34	29	24	20	18	15	12
	24	49	42	35	33	28	23	19	17	14	11

(Left margin: F_y = 36 ksi; Span (ft))

Properties and Reaction Values

	45	39	33	30	26	22	19	17	15	12
Z_x in.^{3}	54.9	46.8	38.8	36.6	31.3	26.0	21.6	18.7	16.0	12.6
$\phi_b W_c$ kip-ft	1190	1010	838	791	676	562	467	404	346	272
$\phi_v V_n$ kips	68.7	60.7	54.9	61.1	52.2	47.4	49.8	47.2	44.7	36.5
ϕR_1 kips	39.4	31.9	27.7	25.3	20.5	16.2	18.3	16.2	14.2	10.7
ϕR_2 kips/in.	12.6	11.3	10.4	10.8	9.36	8.64	9.00	8.64	8.28	6.84
$\phi_r R_3$ kips	49.9	39.4	31.5	35.9	26.9	21.6	24.0	20.7	17.5	11.6
$\phi_r R_4$ kips/in.	6.29	5.46	5.29	4.64	3.55	3.47	3.55	3.80	4.14	3.04
$\phi_r R_5$ kips	45.7	35.8	28.1	32.7	24.5	19.2	21.6	18.1	14.8	9.61
$\phi_r R_6$ kips/in.	8.38	7.28	7.05	6.19	4.73	4.62	4.73	5.07	5.52	4.05
ϕR ($N = 3\frac{1}{4}$) kips	**72.9**	**59.4**	51.0	52.8	39.8	34.3	37.0	34.6	32.7	22.8

Load above heavy line is limited by design shear strength.
Values of R in **bold face** exceed maximum design web shear $\phi_v V_n$.

		$F_y = 36$ ksi

W 8

BEAMS
W Shapes
Maximum factored uniform loads in kips
for beams laterally supported
For beams laterally unsupported, see page 4-113

Designation			W 8					
Wt./ft			67	58	48	40	35	31
		7	199	174		115	98	89
		8	190	161	132	107	94	82
		9	168	144	118	96	83	73
		10	152	129	106	86	75	66
		11	138	117	96	78	68	60
		12	126	108	88	72	62	55
		13	117	99	81	66	58	51
		14	108	92	76	61	54	47
		15	101	86	71	57	50	44
		16	95	81	66	54	47	41
		17	89	76	62	51	44	39
		18	84	72	59	48	42	36
		19	80	68	56	45	39	35
$F_y = 36$ ksi	Span (ft)	20	76	65	53	43	37	33

Properties and Reaction Values

	67	58	48	40	35	31
Z_x in.^{3}	70.2	59.8	49.0	39.8	34.7	30.4
$\phi_b W_c$ kip-ft	1520	1290	1060	860	750	657
$\phi_v V_n$ kips	99.7	86.8	66.1	57.7	48.9	44.3
ϕR_1 kips	73.7	60.2	42.8	34.4	27.9	24.0
ϕR_2 kips/in.	20.5	18.4	14.4	13.0	11.2	10.3
$\phi_r R_3$ kips	127	100	64.1	49.5	37.2	30.7
$\phi_r R_4$ kips/in.	20.2	17.2	10.1	9.27	6.80	6.11
$\phi_r R_5$ kips	115	90.3	58.4	44.4	33.5	27.4
$\phi_r R_6$ kips/in.	26.9	22.9	13.5	12.4	9.07	8.14
ϕR ($N = 3\frac{1}{4}$) kips	**140**	**120**	**89.6**	**76.5**	**63.0**	**53.9**

Load above heavy line is limited by design shear strength.
Values of R in **bold face** exceed maximum design web shear $\phi_v V_n$.

| $F_y = 36$ ksi | BEAMS W Shapes Maximum factored uniform loads in kips for beams laterally supported For beams laterally unsupported, see page 4-113 | | | | | | W 8 |

Designation		W 8		W 8		W 8		
Wt./ft		28	24	21	18	15	13	10
	3					77	71	52
	4					73	62	48
	5			80	73	59	49	38
	6	89	76	73	61	49	41	32
	7	84	72	63	52	42	35	27
	8	73	63	55	46	37	31	24
	9	65	56	49	41	33	27	21
	10	59	50	44	37	29	25	19
	11	53	46	40	33	27	22	17
	12	49	42	37	31	24	21	16
	13	45	39	34	28	23	19	15
	14	42	36	31	26	21	18	14
	15	39	33	29	24	20	16	13
	16	37	31	28	23	18	15	12
	17	35	29	26	22	17	14	11
	18	33	28	24	20	16	14	11
	19	31	26	23	19	15	13	10
	20	29	22	18	15			

$F_y = 36$ ksi   Span (ft)

Properties and Reaction Values

	28	24	21	18	15	13	10
Z_x in.^{3}	27.2	23.2	20.4	17.0	13.6	11.4	8.87
$\phi_b W_c$ kip-ft	588	501	441	367	294	246	192
$\phi_v V_n$ kips	44.7	37.8	40.2	36.4	38.6	35.7	26.1
ϕR_1 kips	24.0	19.3	18.3	15.5	16.5	14.2	9.56
ϕR_2 kips/in.	10.3	8.82	9.00	8.28	8.82	8.28	6.12
$\phi_r R_3$ kips	31.7	23.5	24.2	19.4	20.8	17.0	9.71
$\phi_r R_4$ kips/in.	5.67	4.26	4.33	4.16	5.28	5.48	2.79
$\phi_r R_5$ kips	28.7	21.2	21.8	17.1	18.0	14.1	8.24
$\phi_r R_6$ kips/in.	7.56	5.67	5.77	5.54	7.05	7.31	3.72
ϕR ($N = 3\frac{1}{4}$) kips	**53.3**	**39.7**	**40.6**	35.2	**40.9**	**37.9**	20.3

Load above heavy line is limited by design shear strength.
Values of R in **bold face** exceed maximum design web shear $\phi_v V_n$.

W 6–5–4

$F_y = 36$ ksi

BEAMS
W Shapes
Maximum factored uniform loads in kips
for beams laterally supported
For beams laterally unsupported, see page 4-113

Designation		W 6			W 6			W 5		W 4
Wt./ft		25	20	15*	16	12	9	19	16	13
	2									45
	3				63	54	39			45
	4			54	63	45	34	54	47	34
	5	79	63	46	51	36	27	50	41	27
	6	68	54	38	42	30	22	42	35	23
	7	58	46	33	36	26	19	36	30	19
	8	51	40	29	32	22	17	31	26	17
	9	45	36	26	28	20	15	28	23	15
	10	41	32	23	25	18	13	25	21	14
	11	37	29	21	23	16	12	23	19	
	12	34	27	19	21	15	11	21	17	
	13	31	25	18	19	14	10			
	14	29	23	16	18	13	9.6			

$F_y = 36$ ksi

Span (ft)

Properties and Reaction Values

Z_x in.^{3}	18.9	14.9	10.8	11.7	8.30	6.23	11.6	9.59	6.28
$\phi_b W_c$ kip-ft	408	322	230	253	179	135	251	207	136
$\phi_v V_n$ kips	39.7	31.3	26.8	31.7	27.0	19.5	27.0	23.4	22.6
ϕR_1 kips	23.4	17.5	12.9	17.5	12.9	8.61	19.7	16.2	17.3
ϕR_2 kips/in.	11.5	9.36	8.28	9.36	8.28	6.12	9.72	8.64	10.1
$\phi_r R_3$ kips	37.4	24.5	17.2	25.8	17.9	9.95	28.2	21.6	26.6
$\phi_r R_4$ kips/in.	10.4	7.13	7.17	6.34	6.62	3.56	8.16	7.04	14.0
$\phi_r R_5$ kips	33.0	21.6	14.3	23.2	15.2	8.55	25.4	19.2	22.7
$\phi_r R_6$ kips/in.	13.8	9.51	9.56	8.46	8.82	4.74	10.9	9.38	18.7
ϕR ($N = 3\frac{1}{4}$) kips	**60.8**	**48.0**	**39.8**	**48.0**	**39.8**	**24.0**	**51.3**	**44.3**	**50.1**

*Indicates noncompact shape.
Load above heavy line is limited by design shear strength.
Values of R in **bold face** exceed maximum design web shear $\phi_v V_n$.

F_y = 36 ksi

BEAMS
S Shapes
Maximum factored uniform loads in kips
for beams laterally supported

For beams laterally unsupported, see page 4-113

S 24–20

Designation		S 24		S 24			S 20		S 20	
Wt./ft		121	106	100	90	80	96	86	75	66
	6						631		494	
	7			695			611	521	472	393
	8	762		648	583		535	494	413	378
	9	734		576	533	467	475	439	367	336
	10	661	591	518	480	441	428	395	330	302
	11	601	548	471	436	401	389	359	300	275
	12	551	502	432	400	367	356	329	275	252
	13	508	464	399	369	339	329	304	254	233
	14	472	430	370	343	315	305	282	236	216
	15	441	402	346	320	294	285	264	220	202
	16	413	377	324	300	275	267	247	207	189
	17	389	354	305	282	259	252	233	194	178
	18	367	335	288	266	245	238	220	184	168
	19	348	317	273	252	232	225	208	174	159
	20	330	301	259	240	220	214	198	165	151
	21	315	287	247	228	210	204	188	157	144
	22	300	274	236	218	200	194	180	150	137
	23	287	262	225	208	192	186	172	144	131
	24	275	251	216	200	184	178	165	138	126
	25	264	241	207	192	176	171	158	132	121
	26	254	232	199	184	169	164	152	127	116
	27	245	223	192	178	163	158	146	122	112
	28	236	215	185	171	157	153	141	118	108
	29	228	208	179	165	152	147	136	114	104
	30	220	201	173	160	147	143	132	110	101
	32	207	188	162	150	138	134	124	103	95
	34	194	177	152	141	130	126	116	97	89
	36	184	167	144	133	122	119	110	92	84
	38	174	159	136	126	116	113	104	87	80
	40	165	151	130	120	110	107	99	83	76
	42	157	143	123	114	105	102	94	79	72
	44	150	137	118	109	100	97	90	75	69
	46	144	131	113	104	96	93	86	72	66
	48	138	126	108	100	92	89	82	69	63
	50	132	121	104	96	88	86	79	66	60
	52	127	116	100	92	85				
	54	122	112	96	89	82				
	56	118	108	93	86	79				
	58	114	104	89	83	76				
	60	110	100	86	80	73				

Span (ft) — F_y = 36 ksi

Properties and Reaction Values										
Z_x in.^{3}		306	279	240	222	204	198	183	153	140
$\phi_b W_c$ kip-ft		6610	6030	5180	4800	4410	4280	3950	3300	3020
$\phi_v V_n$ kips		381	295	348	292	233	316	260	247	196
ϕR_1 kips		144	112	117	98.4	78.8	126	104	92.9	73.9
ϕR_2 kips/in.		28.8	22.3	26.8	22.5	18.0	28.8	23.8	22.9	18.2
$\phi_r R_3$ kips		229	156	184	141	101	210	157	138	97.9
$\phi_r R_4$ kips/in.		17.6	8.19	18.2	10.7	5.50	25.2	14.1	14.8	7.44
$\phi_r R_5$ kips		200	143	154	124	92.1	176	138	118	88.0
$\phi_r R_6$ kips/in.		23.5	10.9	24.2	14.3	7.33	33.6	18.8	19.7	9.91
ϕR ($N = 3\frac{1}{4}$) kips		238	183	205	172	119	220	181	167	122

Load above heavy line is limited by design shear strength.

S 18–15–12–10	BEAMS — S Shapes — Maximum factored uniform loads in kips for beams laterally supported	F_y = 36 ksi

For beams laterally unsupported, see page 4-113

Designation		S 18		S 15		S 12		S 12		S 10	
Wt./ft		70	54.7	50	42.9	50	40.8	35	31.8	35	25.4
	3									231	
	4					321		200		191	
	5	498		321		264	216	194	163	153	121
	6	450		278	240	220	191	161	151	127	102
	7	386	323	238	214	189	164	138	130	109	88
	8	338	284	208	187	165	143	121	113	96	77
	9	300	252	185	166	147	127	108	101	85	68
	10	270	227	167	150	132	115	97	91	76	61
	11	245	206	151	136	120	104	88	82	70	56
	12	225	189	139	125	110	96	81	76	64	51
	13	208	174	128	115	102	88	74	70	59	47
	14	193	162	119	107	94	82	69	65	55	44
	15	180	151	111	100	88	76	65	60	51	41
	16	169	142	104	94	83	72	60	57	48	38
	17	159	133	98	88	78	67	57	53	45	36
	18	150	126	93	83	73	64	54	50	42	34
	19	142	119	88	79	70	60	51	48	40	32
	20	135	113	83	75	66	57	48	45	38	31
	21	129	108	79	71	63	55	46	43	36	29
	22	123	103	76	68	60	52	44	41	35	28
	23	117	99	72	65	57	50	42	39	33	27
	24	113	95	69	62	55	48	40	38	32	26
	25	108	91	67	60	53	46	39	36	31	25
	26	104	87	64	58	51	44	37	35		
	27	100	84	62	55	49	42	36	34		
	28	96	81	59	53	47	41	35	32		
	29	93	78	57	52	46	40	33	31		
	30	90	76	56	50	44	38	32	30		
	31	87	73	54	48						
	32	84	71	52	47						
	33	82	69	50	45						
	34	79	67	49	44						
	35	77	65	48	43						
	36	75	63	46	42						
	37	73	61	45	40						
	38	71	60								
	40	68	57								
	42	64	54								
	44	61	52								

F_y = 36 ksi — Span (ft)

Properties and Reaction Values

	S18 (70)	S18 (54.7)	S15 (50)	S15 (42.9)	S12 (50)	S12 (40.8)	S12 (35)	S12 (31.8)	S10 (35)	S10 (25.4)
Z_x in.^{3}	125	105	77.1	69.3	61.2	53.1	44.8	42.0	35.4	28.4
$\phi_b W_c$ kip-ft	2700	2270	1670	1500	1320	1150	968	907	765	613
$\phi_v V_n$ kips	249	161	160	120	160	108	99.8	81.6	115	60.5
ϕR_1 kips	96.0	62.2	68.1	50.9	88.9	59.8	45.7	37.4	60.1	31.5
ϕR_2 kips/in.	25.6	16.6	19.8	14.8	24.7	16.6	15.4	12.6	21.4	11.2
$\phi_r R_3$ kips	152	79.6	98.4	63.6	141	78.0	63.2	46.7	98.2	37.2
$\phi_r R_4$ kips/in.	26.5	7.23	16.4	6.83	37.6	11.4	11.0	6.03	39.2	5.62
$\phi_r R_5$ kips	121	70.9	82.1	56.8	111	68.8	54.4	41.9	72.0	33.4
$\phi_r R_6$ kips/in.	35.4	9.64	21.8	9.11	50.2	15.3	14.7	8.04	52.2	7.50
ϕR ($N = 3\frac{1}{4}$) kips	179	103	132	86.4	**169**	**114**	95.8	68.0	**130**	57.8

Load above heavy line is limited by design shear strength.
Values of R in **bold face** exceed maximum design web shear $\phi_v V_n$.

F_y = 36 ksi									

BEAMS
S Shapes
Maximum factored uniform loads in kips
for beams laterally supported

S 8–6–5–4–3

For beams laterally unsupported, see page 4-113

Designation		S 8		S 6		S 5	S 4		S 3	
Wt./ft		23	18.4	17.25	12.5	10	9.5	7.7	7.5	5.7
Span (ft)	1						51		41	
	2			108		42	44	30	25	20
	3	137		76	54	41	29	25	17	14
	4	104	84	57	46	31	22	19	13	11
	5	83	71	46	37	24	17	15	10	8.42
	6	69	59	38	30	20	15	13	8.50	7.02
	7	60	51	33	26	17	12	11	7.28	6.02
	8	52	45	29	23	15	11	9.48		
	9	46	40	25	20	14	9.70	8.42		
	10	42	36	23	18	12	8.73	7.58		
	11	38	32	21	17	11				
	12	35	30	19	15	10				
	13	32	27	18	14					

Properties and Reaction Values

Z_x in.^{3}		19.3	16.5	10.6	8.47	5.67	4.04	3.51	2.36	1.95
$\phi_b W_c$ kip-ft		417	356	229	183	122	87.3	75.8	51.0	42.1
$\phi_v V_n$ kips		68.6	42.1	54.2	27.1	20.8	25.3	15.0	20.4	9.91
ϕR_1 kips		39.7	24.4	36.6	18.3	15.6	22.0	13.0	21.6	10.5
ϕR_2 kips/in.		15.9	9.76	16.7	8.35	7.70	11.7	6.95	12.6	6.12
$\phi_r R_3$ kips		58.5	28.2	58.1	20.5	17.3	30.8	14.0	32.2	10.9
$\phi_r R_4$ kips/in.		23.1	5.36	42.9	5.32	5.52	27.1	5.63	50.0	5.78
$\phi_r R_5$ kips		46.2	25.3	41.0	18.4	15.5	23.6	12.5	22.2	9.78
$\phi_r R_6$ kips/in.		30.8	7.15	57.1	7.10	7.36	36.2	7.51	66.7	7.71
ϕR ($N = 3\frac{1}{4}$) kips		**91.3**	**48.5**	**91.0**	**41.4**	**39.4**	**60.1**	**35.6**	**62.4**	**30.4**

Load above heavy line is limited by design shear strength.
Values of R in **bold face** exceed maximum design web shear $\phi_v V_n$.

MC,C 18–15	BEAMS				$F_y = 36$ ksi	

Channels
Maximum factored uniform loads in kips
for beams laterally supported
For beams laterally unsupported, see page 4-113

Designation		MC 18				C 15		
Wt./ft		58	51.9	45.8	42.7	50	40	33.9
	3					418		
	4	490	420	350		368	303	233
	5	409	374	339	315	295	247	218
	6	341	311	282	268	246	206	181
	7	292	267	242	230	210	177	156
	8	255	234	212	201	184	154	136
	9	227	208	188	179	164	137	121
	10	204	187	169	161	147	124	109
	11	186	170	154	146	134	112	99
	12	170	156	141	134	123	103	91
	13	157	144	130	124	113	95	84
	14	146	133	121	115	105	88	78
	15	136	125	113	107	98	82	73
	16	128	117	106	100	92	77	68
	17	120	110	100	95	87	73	64
	18	114	104	94	89	82	69	60
	19	108	98	89	85	78	65	57
	20	102	93	85	80	74	62	54
	21	97	89	81	77	70	59	52
	22	93	85	77	73	67	56	49
	23	89	81	74	70	64	54	47
	24	85	78	71	67	61	51	45
	25	82	75	68	64	59	49	44
	26	79	72	65	62	57	48	42
	28	73	67	60	57	53	44	39
	30	68	62	56	54	49	41	36
	32	64	58	53	50	46	39	34
	34	60	55	50	47	43	36	32
	36	57	52	47	45	41	34	30
	38	54	49	45	42			
	40	51	47	42	40			
	42	49	44	40	38			
	44	46	42	38	37			

(Left margin: $F_y = 36$ ksi, Span (ft))

Properties and Reaction Values							
Z_x in.³	94.6	86.5	78.4	74.4	68.2	57.2	50.4
$\phi_b W_c$ kip-ft	2040	1870	1690	1610	1470	1240	1090
$\phi_v V_n$ kips	245	210	175	157	209	152	117
ϕR_1 kips	86.6	74.3	61.9	55.7	92.6	67.3	51.8
ϕR_2 kips/in.	25.2	21.6	18.0	16.2	25.8	18.7	14.4
$\phi_r R_3$ kips	142	112	85.5	73.0	149	92.5	62.4
$\phi_r R_4$ kips/in.	28.0	17.6	10.2	7.44	34.6	13.2	6.03
$\phi_r R_5$ kips	108	91.3	73.3	64.1	115	79.3	56.4
$\phi_r R_6$ kips/in.	37.3	23.5	13.6	9.91	46.1	17.7	8.03
ϕR ($N = 3\frac{1}{4}$) kips	169	144	119	97.2	176	128	82.5

Load above heavy line is limited by design shear strength.

$F_y = 36$ ksi		BEAMS			MC 13

BEAMS
Channels
Maximum factored uniform loads in kips
for beams laterally supported
For beams laterally unsupported, see page 4-113

Designation			MC 13			
Wt./ft			50	40	35	31.8
		3	398	283		
		4	327	275	226	190
		5	261	220	200	186
		6	218	183	166	155
		7	187	157	143	133
		8	163	137	125	116
		9	145	122	111	103
		10	131	110	100	93
		11	119	100	91	85
		12	109	92	83	78
		13	101	85	77	72
		14	93	79	71	66
		15	87	73	67	62
		16	82	69	62	58
		17	77	65	59	55
		18	73	61	55	52
		19	69	58	53	49
		20	65	55	50	47
$F_y = 36$ ksi	Span (ft)	21	62	52	48	44
		22	59	50	45	42
		23	57	48	43	40
		24	54	46	42	39
		25	52	44	40	37
		26	50	42	38	36
		27	48	41	37	34
		28	47	39	36	33
		29	45	38	34	32
		30	44	37	33	31
		31	42	35	32	30
		32	41	34	31	29

Properties and Reaction Values				
Z_x in.^{3}	60.5	50.9	46.2	43.1
$\phi_b W_c$ kip-ft	1310	1100	998	931
$\phi_v V_n$ kips	199	142	113	94.8
ϕR_1 kips	97.4	69.3	55.3	46.4
ϕR_2 kips/in.	28.3	20.2	16.1	13.5
$\phi_r R_3$ kips	167	100	71.4	54.9
$\phi_r R_4$ kips/in.	56.4	20.3	10.3	6.10
$\phi_r R_5$ kips	118	82.5	62.5	49.6
$\phi_r R_6$ kips/in.	75.2	27.1	13.8	8.14
ϕR ($N = 3\frac{1}{4}$) kips	189	135	107	76.0

Load above heavy line is limited by design shear strength.

| C, MC 12 | | BEAMS | | | | | F_y = 36 ksi | | |

BEAMS
Channels
Maximum factored uniform loads in kips
for beams laterally supported
For beams laterally unsupported, see page 4-113

F_y = 36 ksi

Designation		C 12			MC 12					MC 12
Wt./ft		30	25	20.7	50	45	40	35	31	10.6
	2									89
	3	238	181		390	332	275			84
	4	181	158	132	303	279	255	218	173	63
	5	145	126	110	242	223	204	185	170	50
	6	121	105	91	202	186	170	154	141	42
	7	104	90	78	173	160	146	132	121	36
	8	91	79	69	151	140	128	116	106	31
	9	81	70	61	135	124	114	103	94	28
	10	73	63	55	121	112	102	92	85	25
	11	66	57	50	110	102	93	84	77	23
	12	60	53	46	101	93	85	77	71	21
	13	56	49	42	93	86	79	71	65	19
	14	52	45	39	87	80	73	66	61	18
	15	48	42	37	81	74	68	62	57	17
	16	45	39	34	76	70	64	58	53	16
	17	43	37	32	71	66	60	54	50	15
	18	40	35	30	67	62	57	51	47	14
	19	38	33	29	64	59	54	49	45	13
	20	36	32	27	61	56	51	46	42	13
	21	35	30	26	58	53	49	44	40	12
	22	33	29	25	55	51	46	42	39	11
	23	32	27	24	53	49	44	40	37	11
	24	30	26	23	50	47	43	39	35	10
	25	29	25	22	48	45	41	37	34	10
	26	28	24	21	47	43	39	36	33	9.64
	27	27	23	20	45	41	38	34	31	9.28
	28	26	23	20	43	40	36	33	30	8.95
	29	25	22	19	42	39	35	32	29	8.64
	30	24	21	18	40	37	34	31	28	8.35

F_y = 36 ksi   Span (ft)

Properties and Reaction Values

	30	25	20.7	50	45	40	35	31	10.6
Z_x in.^{3}	33.6	29.2	25.4	56.1	51.7	47.3	42.8	39.3	11.6
$\phi_b W_c$ kip-ft	726	631	549	1210	1120	1020	924	849	251
$\phi_v V_n$ kips	119	90.3	65.8	195	166	138	109	86.3	44.3
ϕR_1 kips	51.6	39.2	28.6	98.6	84.1	69.7	55.2	43.7	11.8
ϕR_2 kips/in.	18.4	13.9	10.2	30.1	25.6	21.2	16.8	13.3	6.84
$\phi_r R_3$ kips	78.9	52.1	32.4	195	154	116	81.7	57.6	14.1
$\phi_r R_4$ kips/in.	20.3	8.85	3.42	63.6	39.4	22.4	11.1	5.54	1.70
$\phi_r R_5$ kips	62.7	45.1	29.7	144	122	98.1	72.8	53.2	12.7
$\phi_r R_6$ kips/in.	27.0	11.8	4.57	84.8	52.6	29.9	14.8	7.38	2.26
ϕR ($N = 3\frac{1}{4}$) kips	111	83.4	44.5	**196**	**167**	**139**	**110**	77.2	20.1

Load above heavy line is limited by design shear strength.
Values of R in **bold face** exceed maximum design web shear $\phi_v V_n$.

| $F_y = 36$ ksi | | | | | | | | | |

BEAMS
Channels
Maximum factored uniform loads in kips
for beams laterally supported
For beams laterally unsupported, see page 4-113

C, MC 10

[

Designation		C 10				MC 10			MC 10		MC 10
Wt./ft		30	25	20	15.3	41.1	33.6	28.5	25	22	8.4
	2	262	205	147		309					66
	3	192	166	139	93	280	224	165	148		57
	4	144	124	104	85	210	180	160	139	113	42
	5	115	99	83	68	168	144	128	111	102	34
	6	96	83	69	57	140	120	107	93	85	28
	7	82	71	60	49	120	103	91	80	73	24
	8	72	62	52	43	105	90	80	70	64	21
	9	64	55	46	38	93	80	71	62	57	19
	10	57	50	42	34	84	72	64	56	51	17
	11	52	45	38	31	76	66	58	51	46	15
	12	48	41	35	28	70	60	53	46	42	14
	13	44	38	32	26	65	55	49	43	39	13
	14	41	35	30	24	60	52	46	40	36	12
	15	38	33	28	23	56	48	43	37	34	11
	16	36	31	26	21	53	45	40	35	32	11
	17	34	29	25	20	49	42	38	33	30	10
	18	32	28	23	19	47	40	36	31	28	9.4
	19	30	26	22	18	44	38	34	29	27	8.9
	20	29	25	21	17	42	36	32	28	25	8.5
	21	27	24	20	16	40	34	30	27	24	8.1
	22	26	23	19	16	38	33	29	25	23	7.7
	23	25	22	18	15	37	31	28	24	22	7.4
	24	24	21	17	14	35	30	27	23	21	7.1

$F_y = 36$ ksi — Span (ft)

Properties and Reaction Values

Z_x in.^{3}	26.6	23.0	19.3	15.8	38.9	33.4	29.6	25.8	23.6	7.86
$\phi_b W_c$ kip-ft	575	497	417	341	840	721	639	557	510	170
$\phi_v V_n$ kips	131	102	73.7	46.7	155	112	82.6	73.9	56.4	33.0
ϕR_1 kips	60.6	47.3	34.1	21.6	89.6	64.7	47.8	42.8	32.6	10.5
ϕR_2 kips/in.	24.2	18.9	13.6	8.64	28.7	20.7	15.3	13.7	10.4	6.12
$\phi_r R_3$ kips	112	77.1	47.1	23.8	165	101	64.3	54.4	36.2	11.3
$\phi_r R_4$ kips/in.	64.2	30.6	11.5	2.91	80.5	30.4	12.3	8.76	3.89	1.61
$\phi_r R_5$ kips	68.8	56.7	39.5	21.8	111	80.9	56.1	48.5	33.6	10.3
$\phi_r R_6$ kips/in.	85.6	40.9	15.3	3.88	107	40.5	16.3	11.7	5.19	2.15
ϕR ($N = 3^1/_4$) kips	**139**	**109**	**78.5**	34.4	**183**	**132**	**97.5**	**86.5**	50.5	17.3

Load above heavy line is limited by design shear strength.
Values of R in **bold face** exceed maximum design web shear $\phi_v V_n$.

| C, MC 9 | **BEAMS** | | | F_y = 36 ksi |

BEAMS
Channels
Maximum factored uniform loads in kips
for beams laterally supported
For beams laterally unsupported, see page 4-113

Designation			C 9			MC 9	
Wt./ft		20	15	13.4	25.4	23.9
	2	157	100			
	3	121	97	82	157	140
	4	91	73	68	125	120
	5	73	58	54	100	96
	6	60	49	45	84	80
	7	52	42	39	72	69
	8	45	36	34	63	60
	9	40	32	30	56	53
	10	36	29	27	50	48
	11	33	27	25	46	44
	12	30	24	23	42	40
	13	28	22	21	39	37
	14	26	21	19	36	34
	15	24	19	18	33	32
	16	23	18	17	31	30
	17	21	17	16	29	28
	18	20	16	15	28	27
	19	19	15	14	26	25
	20	18	15	14	25	24
	21	17	14	13	24	23
	22	16	13	12	23	22

(Left margin: F_y = 36 ksi / Span (ft))

Properties and Reaction Values

	C 9 20	C 9 15	C 9 13.4	MC 9 25.4	MC 9 23.9
Z_x in.^{3}	16.8	13.5	12.5	23.2	22.2
$\phi_b W_c$ kip-ft	363	292	270	501	480
$\phi_v V_n$ kips	78.4	49.9	40.8	78.7	70.0
ϕR_1 kips	37.8	24.0	19.7	48.1	42.8
ϕR_2 kips/in.	16.1	10.3	8.39	16.2	14.4
$\phi_r R_3$ kips	59.0	29.9	22.1	68.5	57.4
$\phi_r R_4$ kips/in.	22.2	5.72	3.12	16.9	11.9
$\phi_r R_5$ kips	45.6	26.5	20.2	58.4	50.3
$\phi_r R_6$ kips/in.	29.6	7.62	4.17	22.5	15.8
ϕR (N = 3¼) kips	**90.2**	51.3	33.8	**101**	**89.6**

Load above heavy line is limited by design shear strength.
Values of R in **bold face** exceed maximum design web shear $\phi_v V_n$.

F_y = 36 ksi								

BEAMS
Channels
Maximum factored uniform loads in kips
for beams laterally supported
For beams laterally unsupported, see page 4-113

C, MC 8

Designation			C 8			MC 8		MC 8		MC 8
Wt./ft			18.75	13.75	11.5	22.8	21.4	20	18.7	8.5
		1	151							
		2	149	94				124		56
		3	99	78	68	133	117	117	110	50
		4	75	59	52	102	97	87	83	37
		5	60	47	41	81	78	70	67	30
		6	50	39	34	68	65	58	55	25
		7	43	34	29	58	56	50	48	21
		8	37	29	26	51	49	44	42	19
		9	33	26	23	45	43	39	37	17
		10	30	24	21	41	39	35	33	15
		11	27	21	19	37	35	32	30	14
		12	25	20	17	34	32	29	28	12
		13	23	18	16	31	30	27	26	11
		14	21	17	15	29	28	25	24	11
		15	20	16	14	27	26	23	22	10
F_y = 36 ksi	Span (ft)	16	19	15	13	25	24	22	21	9.3
		17	18	14	12	24	23	21	20	8.8
		18	17	13	11	23	22	19	18	8.3
		19	16	12	11	21	20	18	18	7.9
		20	15	12	10	20	19	17	17	7.5

Properties and Reaction Values

Z_x in.^{3}	13.8	10.9	9.55	18.8	18.0	16.2	15.4	6.91
$\phi_b W_c$ kip-ft	298	235	206	406	389	350	333	149
$\phi_v V_n$ kips	75.7	47.1	34.2	66.4	58.3	62.2	54.9	27.8
ϕR_1 kips	41.1	25.6	18.6	45.6	40.1	40.5	35.7	12.1
ϕR_2 kips/in.	17.5	10.9	7.92	15.4	13.5	14.4	12.7	6.44
$\phi_r R_3$ kips	64.9	31.9	19.7	61.9	50.9	54.7	45.4	12.9
$\phi_r R_4$ kips/in.	34.0	8.18	3.13	17.0	11.5	14.7	10.1	2.12
$\phi_r R_5$ kips	46.8	27.5	18.0	52.8	44.8	46.9	40.0	11.8
$\phi_r R_6$ kips/in.	45.3	10.9	4.18	22.7	15.4	19.6	13.5	2.82
ϕR (N = 3$\frac{1}{4}$) kips	**98.1**	**61.0**	31.6	**95.6**	**84.0**	**87.3,**	**77.0**	21.0

Load above heavy line is limited by design shear strength.
Values of R in **bold face** exceed maximum design web shear $\phi_v V_n$.

BEAMS
Channels
Maximum factored uniform loads in kips
for beams laterally supported
For beams laterally unsupported, see page 4-113

C, MC 7–6

[

$F_y = 36$ ksi

Designation			C 7		MC 7		C6			MC 6	MC 6		MC 6
Wt./ft			12.25	9.8	22.7	19.1	13	10.5	8.2	18	16.3	15.1	12
		1					102	73					
		2	85	57	137		78	66	47	88	87	74	72
		3	60	51	117	96	52	44	37	83	73	70	53
		4	45	38	87	77	39	33	28	62	55	52	40
		5	36	31	70	62	31	27	22	50	44	42	32
		6	30	26	58	51	26	22	18	41	37	35	27
		7	26	22	50	44	22	19	16	35	31	30	23
		8	23	19	44	39	20	17	14	31	28	26	20
		9	20	17	39	34	17	15	12	28	24	23	18
		10	18	15	35	31	16	13	11	25	22	21	16
		11	16	14	32	28	14	12	10	23	20	19	14
		12	15	13	29	26	13	11	9.23	21	18	17	13
		13	14	12	27	24	12	10	8.52	19	17	16	12
		14	13	11	25	22	11	9.49	7.91	18	16	15	11
		15	12	10	23	21	10	8.86	7.39	17	15	14	11
		16	11	9.61	22	19							

$F_y = 36$ ksi Span (ft)

Properties and Reaction Values

Z_x in.^{3}	8.40	7.12	16.2	14.3	7.26	6.15	5.13	11.5	10.2	9.69	7.38
$\phi_b W_c$ kip-ft	181	154	350	309	157	133	111	248	220	209	159
$\phi_v V_n$ kips	42.7	28.6	68.4	47.9	51.0	36.6	23.3	44.2	43.7	36.9	36.2
ϕR_1 kips	24.7	16.5	50.9	35.6	32.0	23.0	14.6	36.2	35.9	30.2	22.7
ϕR_2 kips/in.	11.3	7.56	18.1	12.7	15.7	11.3	7.20	13.6	13.5	11.4	11.2
$\phi_r R_3$ kips	32.6	17.8	77.2	45.2	51.8	31.5	16.0	49.2	48.4	37.5	32.3
$\phi_r R_4$ kips/in.	11.1	3.32	33.4	11.4	37.2	13.8	3.57	17.5	17.0	10.2	12.2
$\phi_r R_5$ kips	27.4	16.3	61.6	39.8	36.9	26.0	14.6	42.2	41.6	33.4	27.5
$\phi_r R_6$ kips/in.	14.8	4.42	44.5	15.3	49.6	18.4	4.76	23.4	22.6	13.6	16.2
ϕR ($N = 3\frac{1}{4}$) kips	**61.5**	**30.6**	**110**	**76.8**	**83.1**	**59.7**	**30.1**	**80.6**	**79.7**	**67.2**	**58.9**

Load above heavy line is limited by design shear strength.
Values of R in **bold face** exceed maximum design web shear $\phi_v V_n$.

| $F_y = 36$ ksi | BEAMS | | | | C 5–4–3 | | |

BEAMS
Channels
Maximum factored uniform loads in kips
for beams laterally supported
For beams laterally unsupported, see page 4-113

Designation		C 5		C 4		C 3		
Wt./ft		9	6.7	7.25	5.4	6	5	4.1
	1	63		50	29	37	30	20
	2	47	37	30	24	19	16	14
	3	31	25	20	16	12	11	9.4
	4	24	19	15	12	9.3	8.1	7.0
	5	19	15	12	9.8	7.4	6.5	5.6
	6	16	13	10	8.1	6.2	5.4	4.7
	7	13	11	8.7	7.0	5.3	4.6	4.0
	8	12	9.5	7.6	6.1			
	9	10	8.4	6.7	5.4			
	10	9.4	7.6	6.1	4.9			
	11	8.6	6.9					
	12	7.9	6.3					

$F_y = 36$ ksi   Span (ft)

Properties and Reaction Values

	C 5 9	C 5 6.7	C 4 7.25	C 4 5.4	C 3 6	C 3 5	C 3 4.1
Z_x in.^{3}	4.36	3.51	2.81	2.26	1.72	1.50	1.30
$\phi_b W_c$ kip-ft	94.2	75.8	60.7	48.8	37.2	32.4	28.1
$\phi_v V_n$ kips	31.6	18.5	25.0	14.3	20.8	15.0	9.91
ϕR_1 kips	21.9	12.8	19.9	11.4	22.0	16.0	10.5
ϕR_2 kips/in.	11.7	6.84	11.6	6.62	12.8	9.29	6.12
$\phi_r R_3$ kips	32.1	14.3	30.3	13.1	34.0	21.0	11.2
$\phi_r R_4$ kips/in.	19.7	3.94	25.6	4.83	50.6	19.2	5.51
$\phi_r R_5$ kips	25.5	13.0	23.4	11.9	23.8	17.1	10.1
$\phi_r R_6$ kips/in.	26.3	5.25	34.2	6.44	67.4	25.7	7.34
ϕR ($N = 3\frac{1}{4}$) kips	**60.0**	**30.1**	**57.4**	**32.8**	**63.7**	**46.1**	**30.4**

Load above heavy line is limited by design shear strength.
Values of R in **bold face** exceed maximum design web shear $\phi_v V_n$.

W 44

$$F_y = 50 \text{ ksi}$$

BEAMS
W Shapes
Maximum factored uniform loads in kips
for beams laterally supported
For beams laterally unsupported, see page 4-139

Designation	W 44			
Wt./ft	335	290	262	230
20	2420	2050	1850	1650
21	2310	2030	1810	1570
22	2210	1940	1730	1500
23	2110	1850	1660	1430
24	2030	1780	1590	1380
25	1940	1700	1520	1320
26	1870	1640	1470	1270
27	1800	1580	1410	1220
28	1740	1520	1360	1180
29	1680	1470	1310	1140
30	1620	1420	1270	1100
31	1570	1370	1230	1060
32	1520	1330	1190	1030
33	1470	1290	1150	1000
34	1430	1250	1120	971
35	1390	1220	1090	943
36	1350	1180	1060	917
38	1280	1120	1000	868
40	1220	1070	953	825
42	1160	1010	907	786
44	1100	968	866	750
46	1060	926	828	717
48	1010	888	794	688
50	972	852	762	660
52	935	819	733	635
54	900	789	706	611
56	868	761	680	589
58	838	734	657	569
60	810	710	635	550
62	784	687	615	532
64	759	666	595	516
66	736	645	577	500
68	715	626	560	485
70	694	609	544	471
72	675	592	529	458

$F_y = 50$ ksi — Span (ft)

Properties and Reaction Values

	335	290	262	230
Z_x in.^{3}	1620	1420	1270	1100
$\phi_b W_c$ kip-ft	48600	42600	38100	33000
$\phi_v V_n$ kips	1212	1025	924	823
ϕR_1 kips	327	258	216	178
ϕR_2 kips/in.	51.0	43.5	39.5	35.5
$\phi_r R_3$ kips	494	368	302	238
$\phi_r R_4$ kips/in.	14.7	10.3	8.67	7.40
$\phi_r R_5$ kips	451	338	277	217
$\phi_r R_6$ kips/in.	19.6	13.8	11.6	9.86
ϕR ($N = 3\frac{1}{4}$) kips	492	400	330	262

Load above heavy line is limited by design shear strength.

| $F_y = 50$ ksi | | BEAMS
W Shapes
Maximum factored uniform loads in kips
for beams laterally supported
For beams laterally unsupported, see page 4-139 | | | | | | W 40 | |

Designation		W 40								
Wt./ft		431	372	321	297	277	249	215	199	174*
	15									1340
	16									1330
	17									1250
	18									1180
	19	2990	2550	2160	2000				1360	1120
	20	2930	2510	2130	2000				1300	1070
	21	2790	2390	2030	1900	1780	1590	1370	1240	1010
	22	2660	2280	1940	1810	1700	1530	1310	1180	969
	23	2540	2180	1850	1730	1630	1460	1260	1130	927
	24	2440	2090	1780	1660	1560	1400	1200	1090	888
	25	2340	2000	1700	1600	1500	1340	1160	1040	852
	26	2250	1930	1640	1530	1440	1290	1110	1000	820
	27	2170	1860	1580	1480	1390	1240	1070	964	789
	28	2090	1790	1520	1430	1340	1200	1030	930	761
	29	2020	1730	1470	1380	1290	1160	996	898	735
	30	1950	1670	1420	1330	1250	1120	963	868	710
	31	1890	1620	1370	1290	1210	1080	932	840	687
	32	1830	1570	1330	1250	1170	1050	903	814	666
	33	1770	1520	1290	1210	1140	1020	875	789	646
	34	1720	1470	1250	1170	1100	988	850	766	627
	35	1670	1430	1220	1140	1070	960	825	744	609
	36	1630	1390	1180	1110	1040	933	803	723	592
	38	1540	1320	1120	1050	987	884	760	685	561
	40	1460	1250	1070	998	938	840	722	651	533
	42	1390	1190	1010	950	893	800	688	620	507
	44	1330	1140	968	907	852	764	657	592	484
	46	1270	1090	926	867	815	730	628	566	463
	48	1220	1040	888	831	781	700	602	543	444
	50	1170	1000	852	798	750	672	578	521	426
	52	1130	963	819	767	721	646	556	501	410
	54	1080	928	789	739	694	622	535	482	395
	56	1040	895	761	713	670	600	516	465	381
	58	1010	864	734	688	647	579	498	449	367
	60	975	835	710	665	625	560	482	434	355
	62	944	808	687	644	605	542	466	420	344
	64	914	783	666	623	586	525	451	407	333
	66	886	759	645	605	568	509	438	395	323
	68	860	737	626	587	551	494	425	383	313
	70	836	716	609	570	536	480	413	372	304
	72	813	696	592	554	521	467	401	362	296

The left axis labels read: $F_y = 50$ ksi / Span (ft)

Properties and Reaction Values										
Z_x in.^{3}		1950	1670	1420	1330	1250	1120	963	868	715
$\phi_b W_c$ kip-ft		58500	50100	42600	39900	37500	33600	28900	26000	21300
$\phi_v V_n$ kips		1493	1273	1082	1000	889	797	684	679	670
ϕR_1 kips		597	471	367	356	311	264	213	198	163
ϕR_2 kips/in.		67.0	58.0	50.0	46.5	41.5	37.5	32.5	32.5	32.5
$\phi_r R_3$ kips		859	645	480	415	342	279	209	195	172
$\phi_r R_4$ kips/in.		26.7	20.3	15.3	13.2	9.90	8.16	6.25	7.21	9.37
$\phi_r R_5$ kips		786	590	439	380	316	258	193	176	148
$\phi_r R_6$ kips/in.		35.6	27.0	20.3	17.7	13.2	10.9	8.33	9.62	12.5
ϕR ($N = 3\frac{1}{4}$) kips		814	660	529	458	374	306	229	218	203

*Noncompact shape; $F_y = 50$ ksi.
Load above heavy line is limited by design shear strength.

W 40	BEAMS W Shapes Maximum factored uniform loads in kips for beams laterally supported For beams laterally unsupported, see page 4-139							$F_y = 50$ ksi

Designation		W 40							
Wt./ft		331	278	264	235	211	183	167	149
	13								1300
	14								1280
	15	2690						1350	1190
	16	2680	2210	2070				1300	1120
	17	2520	2100	1990	1780	1590	1370	1220	1050
	18	2380	1980	1880	1680	1510	1300	1150	995
	19	2260	1880	1780	1590	1430	1230	1090	943
	20	2150	1790	1700	1520	1360	1170	1040	896
	21	2040	1700	1610	1440	1290	1120	989	853
	22	1950	1620	1540	1380	1230	1070	944	814
	23	1870	1550	1470	1320	1180	1020	903	779
	24	1790	1490	1410	1260	1130	976	865	746
	25	1720	1430	1360	1210	1090	937	830	716
	26	1650	1370	1300	1170	1040	901	798	689
	27	1590	1320	1260	1120	1010	868	769	663
	28	1530	1280	1210	1080	970	837	741	640
	29	1480	1230	1170	1040	936	808	716	618
	30	1430	1190	1130	1010	905	781	692	597
	31	1380	1150	1090	977	876	756	670	578
	32	1340	1120	1060	947	848	732	649	560
	33	1300	1080	1030	918	823	710	629	543
	34	1260	1050	997	891	799	689	611	527
	35	1230	1020	969	866	776	669	593	512
	36	1190	992	942	842	754	651	577	498
	38	1130	939	892	797	714	617	546	471
	40	1070	893	848	758	679	586	519	448
	42	1020	850	807	721	646	558	494	426
	44	975	811	770	689	617	533	472	407
	46	933	776	737	659	590	509	451	389
	48	894	744	706	631	566	488	433	373
	50	858	714	678	606	543	469	415	358
	52	825	687	652	583	522	451	399	344
	54	794	661	628	561	503	434	384	332
	56	766	638	605	541	485	418	371	320
	58	740	616	584	522	468	404	358	309
	60	715	595	565	505	453	391	346	299
	62	692	576	547	489	438	378	335	289
	64	670	558	530	473	424	366	324	280
	66	650	541	514	459	411	355	315	271
	68	631	525	499	446	399	345	305	263
	70	613	510	484	433	388	335	297	256
	72	596	496	471	421	377	325	288	249

$F_y = 50$ ksi, Span (ft)

Properties and Reaction Values

	331	278	264	235	211	183	167	149
Z_x in.^{3}	1430	1190	1130	1010	905	781	692	597
$\phi_b W_c$ kip-ft	42900	35700	33900	30300	27200	23400	20800	17900
$\phi_v V_n$ kips	1344	1106	1037	889	797	684	677	650
ϕR_1 kips	505	383	353	285	246	213	198	177
ϕR_2 kips/in.	61.0	51.0	48.0	41.5	37.5	32.5	32.5	31.5
$\phi_r R_3$ kips	709	500	446	342	279	209	191	164
$\phi_r R_4$ kips/in.	22.6	15.8	13.8	9.90	8.19	6.25	7.51	8.53
$\phi_r R_5$ kips	648	458	409	316	257	193	172	143
$\phi_r R_6$ kips/in.	30.1	21.1	18.4	13.2	10.9	8.33	10.0	11.4
ϕR ($N = 3\frac{1}{4}$) kips	703	548	491	374	305	229	216	192

Load above heavy line is limited by design shear strength.

| F_y = 50 ksi | BEAMS W Shapes Maximum factored uniform loads in kips for beams laterally supported | W 36 |

For beams laterally unsupported, see page 4-139

Designation			W 36				
Wt./ft			300	280	260	245	230
		19			1640	1560	1470
		20	1870	1750	1620	1520	1410
		21	1800	1670	1540	1440	1350
		22	1720	1600	1470	1380	1290
		23	1640	1530	1410	1320	1230
		24	1580	1460	1350	1260	1180
		25	1510	1400	1300	1210	1130
		26	1450	1350	1250	1170	1090
		27	1400	1300	1200	1120	1050
		28	1350	1250	1160	1080	1010
		29	1300	1210	1120	1040	976
		30	1260	1170	1080	1010	943
		31	1220	1130	1050	977	913
		32	1180	1100	1010	947	884
		33	1150	1060	982	918	857
		34	1110	1030	953	891	832
F_y = 50 ksi	Span (ft)	35	1080	1000	926	866	808
		36	1050	975	900	842	786
		38	995	924	853	797	744
		40	945	878	810	758	707
		42	900	836	771	721	674
		44	859	798	736	689	643
		46	822	763	704	659	615
		48	788	731	675	631	589
		50	756	702	648	606	566
		52	727	675	623	583	544
		54	700	650	600	561	524
		56	675	627	579	541	505
		58	652	605	559	522	488
		60	630	585	540	505	472
		62	610	566	523	489	456
		64	591	548	506	473	442
		66	573	532	491	459	429
		68	556	516	476	446	416
		70	540	501	463	433	404
		72	525	488	450	421	393

Properties and Reaction Values					
Z_x in.[3]	1260	1170	1080	1010	943
$\phi_b W_c$ kip-ft	37800	35100	32400	30300	28300
$\phi_v V_n$ kips	937	873	822	779	737
ϕR_1 kips	332	297	269	250	226
ϕR_2 kips/in.	47.3	44.3	42.0	40.0	38.0
$\phi_r R_3$ kips	429	376	333	300	268
$\phi_r R_4$ kips/in.	14.8	13.1	12.3	11.4	10.5
$\phi_r R_5$ kips	393	344	303	272	243
$\phi_r R_6$ kips/in.	19.7	17.4	16.4	15.2	14.0
ϕR (N = 3¼) kips	477	419	373	337	302

Load above heavy line is limited by design shear strength.

W 36			BEAMS					$F_y = 50$ ksi	

W Shapes

Maximum factored uniform loads in kips

for beams laterally supported

For beams laterally unsupported, see page 4-139

Designation		W 36								
Wt./ft		256	232	210	194	182	170	160	150	135
	13									1150
	14							1260	1210	1090
	15			1640	1510	1420	1330	1250	1160	1020
	16	1940	1740	1560	1440	1350	1250	1170	1090	954
	17	1840	1650	1470	1350	1270	1180	1100	1030	898
	18	1730	1560	1390	1280	1200	1110	1040	968	848
	19	1640	1480	1320	1210	1130	1050	985	917	804
	20	1560	1400	1250	1150	1080	1000	936	872	764
	21	1490	1340	1190	1100	1030	954	891	830	727
	22	1420	1280	1140	1050	979	911	851	792	694
	23	1360	1220	1090	1000	937	871	814	758	664
	24	1300	1170	1040	959	898	835	780	726	636
	25	1250	1120	1000	920	862	802	749	697	611
	26	1200	1080	961	885	828	771	720	670	587
	27	1160	1040	926	852	798	742	693	646	566
	28	1110	1000	893	822	769	716	669	623	545
	29	1080	968	862	793	743	691	646	601	527
	30	1040	936	833	767	718	668	624	581	509
	31	1010	906	806	742	695	646	604	562	493
	32	975	878	781	719	673	626	585	545	477
	33	945	851	757	697	653	607	567	528	463
	34	918	826	735	677	634	589	551	513	449
	35	891	802	714	657	615	573	535	498	436
	36	867	780	694	639	598	557	520	484	424
	38	821	739	658	606	567	527	493	459	402
	40	780	702	625	575	539	501	468	436	382
	42	743	669	595	548	513	477	446	415	364
	44	709	638	568	523	490	455	425	396	347
	46	678	610	543	500	468	436	407	379	332
	48	650	585	521	479	449	418	390	363	318
	50	624	562	500	460	431	401	374	349	305
	52	600	540	481	443	414	385	360	335	294
	54	578	520	463	426	399	371	347	323	283
	56	557	501	446	411	385	358	334	311	273
	58	538	484	431	397	371	346	323	301	263
	60	520	468	417	384	359	334	312	291	255
	62	503	453	403	371	347	323	302	281	246
	64	488	439	390	360	337	313	293	272	239
	66	473	425	379	349	326	304	284	264	231
	68	459	413	368	338	317	295	275	256	225
	70	446	401	357	329	308	286	267	249	218
	72	433	390	347	320	299	278	260	242	212

(Left margin: $F_y = 50$ ksi, Span (ft))

Properties and Reaction Values

	256	232	210	194	182	170	160	150	135
Z_x in.^{3}	1040	936	833	767	718	668	624	581	509
$\phi_b W_c$ kip-ft	31200	28100	25000	23000	21500	20000	18700	17400	15300
$\phi_v V_n$ kips	970	872	822	754	711	664	632	605	576
ϕR_1 kips	315	272	240	209	193	170	157	146	127
ϕR_2 kips/in.	48.0	43.5	41.5	38.3	36.3	34.0	32.5	31.3	30.0
$\phi_r R_3$ kips	446	367	318	271	242	212	191	173	149
$\phi_r R_4$ kips/in.	14.8	12.2	12.4	10.5	9.62	8.55	8.09	7.84	8.32
$\phi_r R_5$ kips	409	336	288	245	219	191	171	154	129
$\phi_r R_6$ kips/in.	19.7	16.3	16.5	14.0	12.8	11.4	10.8	10.5	11.1
$\phi R\ (N = 3\frac{1}{4})$ kips	471	406	358	305	273	240	217	198	176

Load above heavy line is limited by design shear strength.

$F_y = 50$ ksi

BEAMS
W Shapes
Maximum factored uniform loads in kips
for beams laterally supported
For beams laterally unsupported, see page 4-139

W 33

Designation		W 33			W 33				
Wt./ft		241	221	201	169	152	141	130	118
16					1180	1050	964	876	778
17				1300	1110	986	907	824	732
18		1530	1420	1290	1050	932	857	778	692
19		1480	1350	1220	993	883	812	737	655
20		1410	1280	1160	944	839	771	701	623
21		1340	1220	1100	899	799	734	667	593
22		1280	1170	1050	858	762	701	637	566
23		1220	1120	1010	820	729	670	609	541
24		1170	1070	965	786	699	643	584	519
25		1130	1030	926	755	671	617	560	498
26		1080	987	891	726	645	593	539	479
27		1040	950	858	699	621	571	519	461
28		1010	916	827	674	599	551	500	445
29		971	884	799	651	578	532	483	429
30		939	855	772	629	559	514	467	415
31		909	827	747	609	541	497	452	402
32		880	802	724	590	524	482	438	389
33		854	777	702	572	508	467	425	377
34		829	754	681	555	493	454	412	366
35		805	733	662	539	479	441	400	356
36		783	713	643	524	466	428	389	346
37		761	693	626	510	453	417	379	336
38		741	675	609	497	441	406	369	328
40		704	641	579	472	419	386	350	311
42		671	611	551	449	399	367	334	296
44		640	583	526	429	381	350	318	283
46		612	558	503	410	365	335	305	271
48		587	534	483	393	349	321	292	259
50		563	513	463	377	335	308	280	249
52		542	493	445	363	323	297	269	239
54		522	475	429	349	311	286	259	231
56		503	458	414	337	299	275	250	222
58		486	442	399	325	289	266	242	215
60		470	428	386	315	280	257	234	208
62		454	414	374	304	270	249	226	201
64		440	401	362	295	262	241	219	195
66		427	389	351	286	254	234	212	189
68		414	377	341	278	247	227	206	183
70		402	366	331	270	240	220	200	178
72		391	356	322	262	233	214	195	173

$F_y = 50$ ksi — Span (ft)

Properties and Reaction Values

Z_x in.^{3}		939	855	772	629	559	514	467	415
$\phi_b W_c$ kip-ft		28200	25700	23200	18900	16800	15400	14000	12500
$\phi_v V_n$ kips		766	710	650	612	574	544	518	488
ϕR_1 kips		227	200	173	173	149	132	122	107
ϕR_2 kips/in.		41.5	38.8	35.8	33.5	31.8	30.3	29.0	27.5
$\phi_r R_3$ kips		323	278	234	218	187	166	147	127
$\phi_r R_4$ kips/in.		12.9	11.6	10.2	7.89	7.84	7.49	7.46	7.40
$\phi_r R_5$ kips		293	251	211	201	170	150	131	110
$\phi_r R_6$ kips/in.		17.2	15.5	13.6	10.5	10.5	9.99	9.95	9.87
ϕR ($N = 3\frac{1}{4}$) kips		362	316	267	244	213	191	172	151

Load above heavy line is limited by design shear strength.

$F_y = 50$ ksi

W 30

BEAMS
W Shapes
Maximum factored uniform loads in kips
for beams laterally supported
For beams laterally unsupported, see page 4-139

Designation			W 30				
Wt./ft			261	235	211	191	173
$F_y = 50$ ksi	Span (ft)	16					1080
		17	1590		1290	1180	1070
		18	1570	1400	1250	1120	1010
		19	1490	1330	1180	1060	955
		20	1410	1270	1120	1010	908
		21	1340	1210	1070	961	864
		22	1280	1150	1020	918	825
		23	1230	1100	977	878	789
		24	1180	1060	936	841	756
		25	1130	1010	899	808	726
		26	1090	975	864	777	698
		27	1050	939	832	748	672
		28	1010	905	803	721	648
		29	973	874	775	696	626
		30	941	845	749	673	605
		31	911	818	725	651	585
		32	882	792	702	631	567
		33	855	768	681	612	550
		34	830	746	661	594	534
		36	784	704	624	561	504
		38	743	667	591	531	478
		40	706	634	562	505	454
		42	672	604	535	481	432
		44	642	576	511	459	413
		46	614	551	488	439	395
		48	588	528	468	421	378
		50	565	507	449	404	363
		52	543	488	432	388	349
		54	523	469	416	374	336
		56	504	453	401	361	324
		58	487	437	387	348	313
		60	471	423	375	337	303
		62	455	409	362	326	293
		64	441	396	351	315	284
		66	428	384	340	306	275
		68	415	373	330	297	267
		70	403	362	321	288	259
		72	392	352	312	280	252

Properties and Reaction Values

	261	235	211	191	173
Z_x in.^{3}	941	845	749	673	605
$\phi_b W_c$ kip-ft	28200	25400	22500	20200	18200
$\phi_v V_n$ kips	794	701	647	588	538
ϕR_1 kips	283	233	206	172	154
ϕR_2 kips/in.	46.5	41.5	38.8	35.5	32.8
$\phi_r R_3$ kips	415	334	282	235	197
$\phi_r R_4$ kips/in.	16.7	13.2	12.4	10.7	9.38
$\phi_r R_5$ kips	380	306	257	213	178
$\phi_r R_6$ kips/in.	22.2	17.6	16.5	14.2	12.5
ϕR ($N = 3\frac{1}{4}$) kips	434	368	322	269	228

Load above heavy line is limited by design shear strength.

F_y = 50 ksi		

BEAMS
W Shapes
Maximum factored uniform loads in kips
for beams laterally supported
For beams laterally unsupported, see page 4-139

W 30

Designation			W 30						
Wt./ft			148	132	124	116	108	99	90
F_y = 50 ksi	Span (ft)	11					878	833	749
		12			953	916	865	780	708
		13	1080	1010	942	872	798	720	653
		14	1070	936	874	810	741	669	606
		15	1000	874	816	756	692	624	566
		16	938	819	765	709	649	585	531
		17	882	771	720	667	611	551	499
		18	833	728	680	630	577	520	472
		19	789	690	644	597	546	493	447
		20	750	656	612	567	519	468	425
		21	714	624	583	540	494	446	404
		22	682	596	556	515	472	425	386
		23	652	570	532	493	451	407	369
		24	625	546	510	473	433	390	354
		25	600	524	490	454	415	374	340
		26	577	504	471	436	399	360	327
		27	556	486	453	420	384	347	314
		28	536	468	437	405	371	334	303
		29	517	452	422	391	358	323	293
		30	500	437	408	378	346	312	283
		31	484	423	395	366	335	302	274
		32	469	410	383	354	324	293	265
		33	455	397	371	344	315	284	257
		34	441	386	360	334	305	275	250
		36	417	364	340	315	288	260	236
		38	395	345	322	298	273	246	223
		40	375	328	306	284	260	234	212
		42	357	312	291	270	247	223	202
		44	341	298	278	258	236	213	193
		46	326	285	266	247	226	203	185
		48	313	273	255	236	216	195	177
		50	300	262	245	227	208	187	170
		52	288	252	235	218	200	180	163
		54	278	243	227	210	192	173	157
		56	268	234	219	203	185	167	152
		58	259	226	211	196	179	161	146
		60	250	219	204	189	173	156	142
		62	242	211	197	183	167	151	137
		64	234	205	191	177	162	146	133
		66	227	199	185	172	157	142	129
		68	221	193	180	167	153	138	125
		70	214	187	175	162	148	134	121
		72	208	182	170	158	144	130	118

Properties and Reaction Values							
Z_x in.^{3}	500	437	408	378	346	312	283
$\phi_b W_c$ kip-ft	15000	13100	12200	11300	10400	9360	8490
$\phi_v V_n$ kips	538	503	477	458	439	416	375
ϕR_1 kips	163	135	123	115	106	93.4	77.1
ϕR_2 kips/in.	32.5	30.8	29.2	28.3	27.3	26.0	23.5
$\phi_r R_3$ kips	205	174	156	141	126	111	90.8
$\phi_r R_4$ kips/in.	8.21	8.30	7.72	7.65	7.73	7.66	6.24
$\phi_r R_5$ kips	189	157	140	126	111	95.6	78.5
$\phi_r R_6$ kips/in.	10.9	11.1	10.3	10.2	10.3	10.2	8.31
ϕR ($N = 3\frac{1}{4}$) kips	232	201	181	166	152	136	111

Load above heavy line is limited by design shear strength.

W 27

BEAMS
W Shapes
Maximum factored uniform loads in kips
for beams laterally supported
For beams laterally unsupported, see page 4-139

$F_y = 50$ ksi

Designation		W 27						
Wt./ft		258	235	217	194	178	161	146
Span (ft)	15					1090	983	895
	16	1530	1410	1270	1140	1060	960	864
	17	1500	1360	1250	1110	1000	904	814
	18	1420	1280	1180	1050	945	853	768
	19	1340	1210	1120	992	895	808	728
	20	1280	1150	1060	942	851	768	692
	21	1210	1100	1010	897	810	731	659
	22	1160	1050	965	856	773	698	629
	23	1110	1000	923	819	740	668	601
	24	1060	961	885	785	709	640	576
	25	1020	923	850	754	680	614	553
	26	981	887	817	725	654	591	532
	27	944	854	787	698	630	569	512
	28	911	824	759	673	608	549	494
	29	879	796	732	650	587	530	477
	30	850	769	708	628	567	512	461
	31	823	744	685	608	549	495	446
	32	797	721	664	589	532	480	432
	33	773	699	644	571	515	465	419
	34	750	679	625	554	500	452	407
	35	729	659	607	538	486	439	395
	36	708	641	590	523	473	427	384
	37	689	624	574	509	460	415	374
	38	671	607	559	496	448	404	364
	40	638	577	531	471	425	384	346
	42	607	549	506	449	405	366	329
	44	580	524	483	428	387	349	314
	46	554	502	462	410	370	334	301
	48	531	481	443	393	354	320	288
	50	510	461	425	377	340	307	277
	52	490	444	408	362	327	295	266
	54	472	427	393	349	315	284	256
	56	455	412	379	336	304	274	247
	58	440	398	366	325	293	265	238
	60	425	385	354	314	284	256	231
	62	411	372	343	304	274	248	223
	64	398	360	332	294	266	240	216
	66	386	350	322	285	258	233	210

Properties and Reaction Values

	258	235	217	194	178	161	146
Z_x in.^{3}	850	769	708	628	567	512	461
$\phi_b W_c$ kip-ft	25500	23100	21200	18800	17000	15400	13800
$\phi_v V_n$ kips	767	704	637	569	544	492	447
ϕR_1 kips	306	263	227	193	170	150	128
ϕR_2 kips/in.	49.0	45.5	41.5	37.5	36.3	33.0	30.3
$\phi_r R_3$ kips	465	397	334	271	243	201	168
$\phi_r R_4$ kips/in.	19.9	17.7	14.5	12.1	12.5	10.4	8.97
$\phi_r R_5$ kips	427	363	306	248	220	182	151
$\phi_r R_6$ kips/in.	26.5	23.6	19.3	16.2	16.6	13.9	12.0
ϕR ($N = 3^1/_4$) kips	466	411	362	311	283	235	197

Load above heavy line is limited by design shear strength.

$F_y = 50$ ksi

$F_y = 50$ ksi

BEAMS
W Shapes
Maximum factored uniform loads in kips
for beams laterally supported
For beams laterally unsupported, see page 4-139

W 27

Designation			W 27				
Wt./ft			129	114	102	94	84
$F_y = 50$ ksi	Span (ft)	11				712	663
		12		840	753	695	610
		13	910	792	704	642	563
		14	846	735	654	596	523
		15	790	686	610	556	488
		16	741	643	572	521	458
		17	697	605	538	491	431
		18	658	572	508	463	407
		19	624	542	482	439	385
		20	593	515	458	417	366
		21	564	490	436	397	349
		22	539	468	416	379	333
		23	515	447	398	363	318
		24	494	429	381	348	305
		25	474	412	366	334	293
		26	456	396	352	321	282
		27	439	381	339	309	271
		28	423	368	327	298	261
		29	409	355	316	288	252
		30	395	343	305	278	244
		31	382	332	295	269	236
		32	370	322	286	261	229
		33	359	312	277	253	222
		34	349	303	269	245	215
		36	329	286	254	232	203
		38	312	271	241	219	193
		40	296	257	229	209	183
		42	282	245	218	199	174
		44	269	234	208	190	166
		46	258	224	199	181	159
		48	247	214	191	174	153
		50	237	206	183	167	146
		52	228	198	176	160	141
		54	219	191	169	154	136
		56	212	184	163	149	131
		58	204	177	158	144	126
		60	198	172	153	139	122
		62	191	166	148	135	118
		64	185	161	143	130	114
		66	180	156	139	126	111

Properties and Reaction Values

	129	114	102	94	84
Z_x in.^{3}	395	343	305	278	244
$\phi_b W_c$ kip-ft	11900	10300	9150	8340	7320
$\phi_v V_n$ kips	455	420	377	356	332
ϕR_1 kips	138	116	101	88.0	79.1
ϕR_2 kips/in.	30.5	28.5	25.8	24.5	23.0
$\phi_r R_3$ kips	180	150	121	107	90.0
$\phi_r R_4$ kips/in.	8.08	7.89	6.57	6.35	6.16
$\phi_r R_5$ kips	165	135	110	95.4	79.0
$\phi_r R_6$ kips/in.	10.8	10.5	8.76	8.46	8.21
ϕR ($N = 3\frac{1}{4}$) kips	206	175	143	127	110

Load above heavy line is limited by design shear strength.

AMERICAN INSTITUTE OF STEEL CONSTRUCTION

W 24

BEAMS
W Shapes
Maximum factored uniform loads in kips
for beams laterally supported
For beams laterally unsupported, see page 4-139

$F_y = 50$ ksi

Designation			W 24							
Wt./ft		229	207	192	176	162	146	131	117	104
Span (ft)	13							800	721	650
	14				1020	952	868	793	701	619
	15	1350	1210	1110	1020	936	836	740	654	578
	16	1270	1140	1050	958	878	784	694	613	542
	17	1190	1070	986	902	826	738	653	577	510
	18	1130	1010	932	852	780	697	617	545	482
	19	1070	957	883	807	739	660	584	516	456
	20	1010	909	839	767	702	627	555	491	434
	21	966	866	799	730	669	597	529	467	413
	22	922	826	762	697	638	570	505	446	394
	23	882	790	729	667	610	545	483	427	377
	24	845	758	699	639	585	523	463	409	361
	25	811	727	671	613	562	502	444	392	347
	26	780	699	645	590	540	482	427	377	333
	27	751	673	621	568	520	464	411	363	321
	28	724	649	599	548	501	448	396	350	310
	29	699	627	578	529	484	432	383	338	299
	30	676	606	559	511	468	418	370	327	289
	31	654	586	541	495	453	405	358	316	280
	32	634	568	524	479	439	392	347	307	271
	33	615	551	508	465	425	380	336	297	263
	34	596	535	493	451	413	369	326	289	255
	35	579	519	479	438	401	358	317	280	248
	36	563	505	466	426	390	348	308	273	241
	38	534	478	441	403	369	330	292	258	228
	40	507	455	419	383	351	314	278	245	217
	42	483	433	399	365	334	299	264	234	206
	44	461	413	381	348	319	285	252	223	197
	46	441	395	365	333	305	273	241	213	188
	48	423	379	349	319	293	261	231	204	181
	50	406	364	335	307	281	251	222	196	173
	52	390	350	323	295	270	241	213	189	167
	54	376	337	311	284	260	232	206	182	161
	56	362	325	299	274	251	224	198	175	155
	58	350	313	289	264	242	216	191	169	149
	60	338	303	280	256	234	209	185	164	145

Properties and Reaction Values

	229	207	192	176	162	146	131	117	104
Z_x in.^{3}	676	606	559	511	468	418	370	327	289
$\phi_b W_c$ kip-ft	20300	18200	16800	15300	14000	12500	11100	9810	8670
$\phi_v V_n$ kips	674	604	557	511	476	434	400	360	325
ϕR_1 kips	300	258	228	199	176	152	132	112	93.8
$\phi_r R_2$ kips/in.	48.0	43.5	40.5	37.5	35.3	32.5	30.3	27.5	25.0
$\phi_r R_3$ kips	446	367	318	271	236	197	166	136	110
$\phi_r R_4$ kips/in.	21.3	17.6	15.5	13.5	12.4	11.0	10.2	8.73	7.49
$\phi_r R_5$ kips	409	336	291	248	215	179	150	121	98.4
$\phi_r R_6$ kips/in.	28.4	23.5	20.6	18.0	16.6	14.7	13.6	11.6	9.99
ϕR ($N = 3\frac{1}{4}$) kips	456	400	359	315	276	233	199	164	135

Load above heavy line is limited by design shear strength.

$F_y = 50$ ksi

BEAMS
W Shapes
Maximum factored uniform loads in kips
for beams laterally supported
For beams laterally unsupported, see page 4-139

W 24

Designation			W 24					W 24	
Wt./ft		103	94	84	76	68	62	55
$F_y = 50$ ksi	Span (ft)							
	7							503
	8						551	503
	9					532	510	447
	10			612	568	531	459	402
	11	729	676	611	545	483	417	365
	12	700	635	560	500	443	383	335
	13	646	586	517	462	408	353	309
	14	600	544	480	429	379	328	287
	15	560	508	448	400	354	306	268
	16	525	476	420	375	332	287	251
	17	494	448	395	353	312	270	236
	18	467	423	373	333	295	255	223
	19	442	401	354	316	279	242	212
	20	420	381	336	300	266	230	201
	21	400	363	320	286	253	219	191
	22	382	346	305	273	241	209	183
	23	365	331	292	261	231	200	175
	24	350	318	280	250	221	191	168
	25	336	305	269	240	212	184	161
	26	323	293	258	231	204	177	155
	27	311	282	249	222	197	170	149
	28	300	272	240	214	190	164	144
	29	290	263	232	207	183	158	139
	30	280	254	224	200	177	153	134
	31	271	246	217	194	171	148	130
	32	263	238	210	188	166	143	126
	33	255	231	204	182	161	139	122
	34	247	224	198	176	156	135	118
	35	240	218	192	171	152	131	115
	36	233	212	187	167	148	128	112
	38	221	201	177	158	140	121	106
	40	210	191	168	150	133	115	101
	42	200	181	160	143	126	109	96
	44	191	173	153	136	121	104	91
	46	183	166	146	130	115	100	87
	48	175	159	140	125	111	96	84
	50	168	152	134	120	106	92	80
	52	162	147	129	115	102	88	77
	54	156	141	124	111	98	85	74
	56	150	136	120	107	95	82	72
	58	145	131	116	103	92	79	69
	60	140	127	112				

Properties and Reaction Values

	103	94	84	76	68	62	55
Z_x in.^{3}	280	254	224	200	177	153	134
$\phi_b W_c$ kip-ft	8400	7620	6720	6000	5310	4590	4020
$\phi_v V_n$ kips	364	338	306	284	266	276	251
ϕR_1 kips	120	105	91.8	79.1	71.3	73.9	64.8
ϕR_2 kips/in.	27.5	25.8	23.5	22.0	20.8	21.5	19.8
$\phi_r R_3$ kips	146	125	102	86.8	73.7	78.1	63.6
$\phi_r R_4$ kips/in.	7.49	6.95	6.05	5.67	5.57	6.14	5.60
$\phi_r R_5$ kips	133	113	92.2	77.8	64.9	68.4	54.8
$\phi_r R_6$ kips/in.	9.98	9.26	8.07	7.55	7.43	8.19	7.47
ϕR ($N = 3\frac{1}{4}$) kips	170	147	122	105	91.8	98.1	81.8

Load above heavy line is limited by design shear strength.

$F_y = 50$ ksi

W 21

BEAMS
W Shapes
Maximum factored uniform loads in kips for beams laterally supported

For beams laterally unsupported, see page 4-139

Designation	W 21							
Wt./ft	201	182	166	147	132	122	111	101
$F_y = 50$ ksi, Span (ft) 13				858	766	702	639	577
14	1130	1020	910	799	714	658	598	542
15	1060	952	864	746	666	614	558	506
16	994	893	810	699	624	576	523	474
17	935	840	762	658	588	542	492	446
18	883	793	720	622	555	512	465	422
19	837	752	682	589	526	485	441	399
20	795	714	648	560	500	461	419	380
21	757	680	617	533	476	439	399	361
22	723	649	589	509	454	419	380	345
23	691	621	563	487	434	400	364	330
24	663	595	540	466	416	384	349	316
25	636	571	518	448	400	368	335	304
26	612	549	498	430	384	354	322	292
27	589	529	480	414	370	341	310	281
28	568	510	463	400	357	329	299	271
29	548	492	447	386	344	318	289	262
30	530	476	432	373	333	307	279	253
31	513	461	418	361	322	297	270	245
32	497	446	405	350	312	288	262	237
33	482	433	393	339	303	279	254	230
34	468	420	381	329	294	271	246	223
35	454	408	370	320	285	263	239	217
36	442	397	360	311	278	256	233	211
38	418	376	341	294	263	242	220	200
40	398	357	324	280	250	230	209	190
42	379	340	309	266	238	219	199	181
44	361	325	295	254	227	209	190	173
46	346	310	282	243	217	200	182	165
48	331	298	270	233	208	192	174	158
50	318	286	259	224	200	184	167	152
52	306	275	249	215	192	177	161	146

Properties and Reaction Values

	201	182	166	147	132	122	111	101
Z_x in.^{3}	530	476	432	373	333	307	279	253
$\phi_b W_c$ kip-ft	15900	14300	13000	11200	9990	9210	8370	7590
$\phi_v V_n$ kips	566	509	455	429	383	351	319	288
ϕR_1 kips	270	233	199	169	147	127	112	97.7
ϕR_2 kips/in.	45.5	41.5	37.5	36.0	32.5	30.0	27.5	25.0
$\phi_r R_3$ kips	400	332	273	236	192	164	138	114
$\phi_r R_4$ kips/in.	21.7	18.4	14.9	15.9	13.1	11.2	9.56	7.91
$\phi_r R_5$ kips	366	304	251	213	173	148	124	103
$\phi_r R_6$ kips/in.	29.0	24.5	19.9	21.2	17.5	15.0	12.8	10.6
ϕR ($N = 3\frac{1}{4}$) kips	418	368	321	286	235	201	169	140

Load above heavy line is limited by design shear strength.

| $F_y = $ 50 ksi | BEAMS
W Shapes
Maximum factored uniform loads in kips
for beams laterally supported
For beams laterally unsupported, see page 4-139 | | | | | W 21 |

Designation		W 21					W 21		
Wt./ft		93	83	73	68	62	57	50	44
	7							427	390
	8						461	413	358
	9	677	596	522	491	453	430	367	318
	10	663	588	516	480	432	387	330	286
	11	603	535	469	436	393	352	300	260
	12	553	490	430	400	360	323	275	239
	13	510	452	397	369	332	298	254	220
	14	474	420	369	343	309	276	236	204
	15	442	392	344	320	288	258	220	191
	16	414	368	323	300	270	242	206	179
	17	390	346	304	282	254	228	194	168
	18	368	327	287	267	240	215	183	159
	19	349	309	272	253	227	204	174	151
	20	332	294	258	240	216	194	165	143
	21	316	280	246	229	206	184	157	136
	22	301	267	235	218	196	176	150	130
	23	288	256	224	209	188	168	143	124
	24	276	245	215	200	180	161	138	119
	25	265	235	206	192	173	155	132	114
	26	255	226	198	185	166	149	127	110
	27	246	218	191	178	160	143	122	106
	28	237	210	184	171	154	138	118	102
	29	229	203	178	166	149	133	114	99
	30	221	196	172	160	144	129	110	95
	31	214	190	166	155	139	125	106	92
	32	207	184	161	150	135	121	103	89
	33	201	178	156	145	131	117	100	87
	34	195	173	152	141	127	114	97	84
	35	189	168	147	137	123	111	94	82
	36	184	163	143	133	120	108	92	80
	38	174	155	136	126	114	102	87	75
	40	166	147	129	120	108	97	83	72
	42	158	140	123	114	103	92	79	68
	44	151	134	117	109	98	88	75	65
	46	144	128	112	104	94	84	72	62
	48	138	123	108	100	90	81	69	60
	50	133	118	103	96	86	77	66	57
	52	128	113	99	92	83	74	63	

Properties and Reaction Values									
Z_x in.^{3}		221	196	172	160	144	129	110	95.4
$\phi_b W_c$ kip-ft		6630	5880	5160	4800	4320	3870	3300	2860
$\phi_v V_n$ kips		339	298	261	245	227	230	214	195
ϕR_1 kips		122	101	85.3	77.3	68.8	69.6	62.3	52.0
ϕR_2 kips/in.		29.0	25.8	22.8	21.5	20.0	20.3	19.0	17.5
$\phi_r R_3$ kips		154	122	95.2	84.2	71.5	74.9	61.8	50.1
$\phi_r R_4$ kips/in.		10.5	8.26	6.48	5.94	5.36	5.25	5.33	4.99
$\phi_r R_5$ kips		138	110	86.0	75.8	64.0	67.6	54.4	43.2
$\phi_r R_6$ kips/in.		14.0	11.0	8.64	7.92	7.15	7.00	7.10	6.65
ϕR ($N = 3^1/_4$) kips		188	149	116	103	89.0	92.0	79.1	66.3

Load above heavy line is limited by design shear strength.

W 18

BEAMS
W Shapes
Maximum factored uniform loads in kips
for beams laterally supported
For beams laterally unsupported, see page 4-139

$F_y = 50$ ksi

Designation		W 18					W 18				
Wt./ft		192	175	158	143	130	119	106	97	86	76
	11						671	597	537	477	418
	12	1050	963	863	768	696	653	575	528	465	408
	13	1020	918	822	743	672	602	531	487	429	376
	14	947	853	763	690	624	559	493	452	399	349
	15	884	796	712	644	582	522	460	422	372	326
	16	829	746	668	604	546	489	431	396	349	306
	17	780	702	628	568	514	461	406	372	328	288
	18	737	663	593	537	485	435	383	352	310	272
	19	698	628	562	508	459	412	363	333	294	257
	20	663	597	534	483	437	392	345	317	279	245
	21	631	569	509	460	416	373	329	301	266	233
	22	603	543	485	439	397	356	314	288	254	222
	23	577	519	464	420	380	340	300	275	243	213
	24	553	498	445	403	364	326	288	264	233	204
	25	530	478	427	386	349	313	276	253	223	196
	26	510	459	411	372	336	301	265	243	215	188
	27	491	442	396	358	323	290	256	234	207	181
	28	474	426	381	345	312	280	246	226	199	175
	29	457	412	368	333	301	270	238	218	192	169
	30	442	398	356	322	291	261	230	211	186	163
	31	428	385	345	312	282	253	223	204	180	158
	32	414	373	334	302	273	245	216	198	174	153
	33	402	362	324	293	265	237	209	192	169	148
	34	390	351	314	284	257	230	203	186	164	144
	35	379	341	305	276	249	224	197	181	159	140
	36	368	332	297	268	243	218	192	176	155	136
	37	358	323	289	261	236	212	186	171	151	132
	38	349	314	281	254	230	206	182	167	147	129
	39	340	306	274	248	224	201	177	162	143	125
	40	332	299	267	242	218	196	173	158	140	122
	42	316	284	254	230	208	186	164	151	133	116
	44	301	271	243	220	198	178	157	144	127	111

$F_y = 50$ ksi   Span (ft)

Properties and Reaction Values

	192	175	158	143	130	119	106	97	86	76
Z_x in.^{3}	442	398	356	322	291	261	230	211	186	163
$\phi_b W_c$ kip-ft	13300	11900	10700	9660	8730	7830	6900	6330	5580	4890
$\phi_v V_n$ kips	527	482	431	384	348	335	298	269	238	209
ϕR_1 kips	293	250	215	183	157	143	120	104	86.3	73.0
ϕR_2 kips/in.	48.0	44.5	40.5	36.5	33.5	32.8	29.5	26.8	24.0	21.3
$\phi_r R_3$ kips	449	382	315	258	217	197	158	132	105	82.4
$\phi_r R_4$ kips/in.	26.9	23.9	20.2	16.4	14.1	15.1	12.6	10.2	8.45	6.71
$\phi_r R_5$ kips	412	350	289	237	199	178	143	119	94.9	74.3
$\phi_r R_6$ kips/in.	35.8	31.9	27.0	21.8	18.8	20.2	16.8	13.7	11.3	8.94
ϕR ($N = 3\frac{1}{4}$) kips	449	395	347	301	262	246	199	165	133	104

Load above heavy line is limited by design shear strength.

F_y = 50 ksi					

BEAMS
W Shapes
Maximum factored uniform loads in kips
for beams laterally supported
For beams laterally unsupported, see page 4-139

W 18

Designation		W 18					W 18		
Wt./ft		71	65	60	55	50	46	40	35
	6								287
	7						351	304	285
	8	494	446		381	345	340	294	249
	9	483	443	409	373	337	302	261	222
	10	435	399	369	336	303	272	235	200
	11	395	363	335	305	275	247	214	181
	12	363	333	308	280	253	227	196	166
	13	335	307	284	258	233	209	181	153
	14	311	285	264	240	216	194	168	143
	15	290	266	246	224	202	181	157	133
	16	272	249	231	210	189	170	147	125
	17	256	235	217	198	178	160	138	117
	18	242	222	205	187	168	151	131	111
	19	229	210	194	177	159	143	124	105
	20	218	200	185	168	152	136	118	100
	21	207	190	176	160	144	130	112	95
	22	198	181	168	153	138	124	107	91
	23	189	173	160	146	132	118	102	87
	24	181	166	154	140	126	113	98	83
	25	174	160	148	134	121	109	94	80
	26	167	153	142	129	117	105	90	77
	27	161	148	137	124	112	101	87	74
	28	155	143	132	120	108	97	84	71
	29	150	138	127	116	104	94	81	69
	30	145	133	123	112	101	91	78	67
	31	140	129	119	108	98	88	76	64
	32	136	125	115	105	95	85	74	62
	33	132	121	112	102	92	82	71	60
	34	128	117	109	99	89	80	69	59
	35	124	114	105	96	87	78	67	57
	36	121	111	103	93	84	76	65	55
	38	114	105	97	88	80	72	62	53
	40	109	100	92	84	76	68	59	50
	42	104	95	88	80	72	65	56	48
	44	99	91	84	76	69	62	53	45

(F_y = 50 ksi; Span (ft))

Properties and Reaction Values

	71	65	60	55	50	46	40	35
Z_x in.^{3}	145	133	123	112	101	90.7	78.4	66.5
$\phi_b W_c$ kip-ft	4350	3990	3690	3360	3030	2720	2350	2000
$\phi_v V_n$ kips	247	223	204	191	172	176	152	143
$\phi_r R_1$ kips	92.8	80.9	71.3	64.0	55.5	56.3	46.8	42.2
$\phi_r R_2$ kips/in.	24.8	22.5	20.8	19.5	17.8	18.0	15.8	15.0
$\phi_r R_3$ kips	113	94.3	80.4	69.7	57.6	60.6	46.2	38.6
$\phi_r R_4$ kips/in.	8.77	7.16	6.10	5.62	4.72	4.62	3.60	3.88
$\phi_r R_5$ kips	102	85.5	73.0	62.9	51.9	55.0	41.9	34.0
$\phi_r R_6$ kips/in.	11.7	9.55	8.13	7.50	6.29	6.16	4.80	5.18
ϕR ($N = 3\frac{1}{4}$) kips	142	118	100	88.0	72.9	75.6	57.9	51.3

Load above heavy line is limited by design shear strength.

W 16

BEAMS
W Shapes
Maximum factored uniform loads in kips
for beams laterally supported
For beams laterally unsupported, see page 4-139

$F_y = 50$ ksi

Designation		W 16				W 16				W 16	
Wt./ft		100	89	77	67	57	50	45	40	36	31	26
	6										236	212
	7									253	231	189
	8					382	334	301	264	240	203	166
	9					350	307	274	243	213	180	147
	10					315	276	247	219	192	162	133
	11	536	475	406	348	286	251	224	199	175	147	121
	12	495	438	375	325	263	230	206	182	160	135	111
	13	457	404	346	300	242	212	190	168	148	125	102
	14	424	375	321	279	225	197	176	156	137	116	95
	15	396	350	300	260	210	184	165	146	128	108	88
	16	371	328	281	244	197	173	154	137	120	101	83
	17	349	309	265	229	185	162	145	129	113	95	78
	18	330	292	250	217	175	153	137	122	107	90	74
	19	313	276	237	205	166	145	130	115	101	85	70
	20	297	263	225	195	158	138	123	109	96	81	66
	21	283	250	214	186	150	131	118	104	91	77	63
	22	270	239	205	177	143	125	112	99	87	74	60
	23	258	228	196	170	137	120	107	95	83	70	58
	24	248	219	188	163	131	115	103	91	80	68	55
	25	238	210	180	156	126	110	99	87	77	65	53
	26	228	202	173	150	121	106	95	84	74	62	51
	27	220	194	167	144	117	102	91	81	71	60	49
	28	212	188	161	139	113	99	88	78	69	58	47
	29	205	181	155	134	109	95	85	75	66	56	46
	30	198	175	150	130	105	92	82	73	64	54	44
	31	192	169	145	126	102	89	80	71	62	52	43
	32	186	164	141	122	98	86	77	68	60	51	41
	33	180	159	136	118	95	84	75	66	58	49	40
	34	175	154	132	115	93	81	73	64	56	48	39
	35	170	150	129	111	90	79	71	62	55	46	38
	36	165	146	125	108	88	77	69	61	53	45	37
	38	156	138	118	103	83	73	65	58	51	43	35
	40	149	131	113	98	79	69	62	55			

($F_y = 50$ ksi, Span (ft))

Properties and Reaction Values

	100	89	77	67	57	50	45	40	36	31	26
Z_x in.^{3}	198	175	150	130	105	92.0	82.3	72.9	64.0	54.0	44.2
$\phi_b W_c$ kip-ft	5940	5250	4500	3900	3150	2760	2470	2190	1920	1620	1330
$\phi_v V_n$ kips	268	237	203	174	191	167	150	132	126	118	106
ϕR_1 kips	123	103	81.8	67.9	73.9	62.3	53.9	45.3	41.5	38.7	33.2
ϕR_2 kips/in.	29.2	26.2	22.8	19.8	21.5	19.0	17.3	15.3	14.7	13.8	12.5
$\phi_r R_3$ kips	160	128	96.5	73.0	86.0	67.1	54.9	43.2	37.9	34.5	26.5
$\phi_r R_4$ kips/in.	13.0	10.7	8.12	6.14	7.32	5.80	4.87	3.80	4.07	3.22	3.12
$\phi_r R_5$ kips	145	116	87.5	66.3	78.0	60.8	49.7	39.1	33.6	31.1	23.2
$\phi_r R_6$ kips/in.	17.3	14.2	10.8	8.19	9.76	7.73	6.50	5.06	5.43	4.29	4.16
$\phi R\,(N = 3\frac{1}{4})$ kips	202	163	123	93.0	110	85.9	70.8	55.6	51.2	45.0	36.7

Load above heavy line is limited by design shear strength.

| F_y = 50 ksi | BEAMS
W Shapes
Maximum factored uniform loads in kips
for beams laterally supported
For beams laterally unsupported, see page 4-139 | W 14 |

Designation			W 14					W 14			
Wt./ft		132	120	109	99*	90*	82	74	68	61
Span (ft)	10						394	344	315	281
	11						379	344	314	278
	12						348	315	288	255
	13	511	461		371	333	321	291	265	235
	14	501	454	406	370	329	298	270	246	219
	15	468	424	384	345	307	278	252	230	204
	16	439	398	360	323	288	261	236	216	191
	17	413	374	339	304	271	245	222	203	180
	18	390	353	320	287	256	232	210	192	170
	19	369	335	303	272	243	219	199	182	161
	20	351	318	288	259	230	209	189	173	153
	21	334	303	274	246	219	199	180	164	146
	22	319	289	262	235	210	190	172	157	139
	23	305	277	250	225	200	181	164	150	133
	24	293	265	240	216	192	174	158	144	128
	25	281	254	230	207	184	167	151	138	122
	26	270	245	222	199	177	160	145	133	118
	27	260	236	213	192	171	154	140	128	113
	28	251	227	206	185	165	149	135	123	109
	29	242	219	199	178	159	144	130	119	106
	30	234	212	192	172	154	139	126	115	102
	31	226	205	186	167	149	135	122	111	99
	32	219	199	180	162	144	130	118	108	96
	33	213	193	175	157	140	126	115	105	93
	34	206	187	169	152	136	123	111	101	90

F_y = 50 ksi

Properties and Reaction Values									
Z_x in.^{3}	234	212	192	173	157	139	126	115	102
$\phi_b W_c$ kip-ft	7020	6360	5760	5170	4610	4170	3780	3450	3060
$\phi_v V_n$ kips	255	231	203	185	167	197	172	157	141
ϕR_1 kips	136	120	103	87.1	75.6	104	87.9	77.8	67.4
ϕR_2 kips/in.	32.3	29.5	26.2	24.3	22.0	25.5	22.5	20.8	18.8
$\phi_r R_3$ kips	190	158	127	108	88.7	121	96.5	81.8	66.5
$\phi_r R_4$ kips/in.	19.2	16.3	12.7	11.2	9.26	11.7	8.86	7.65	6.37
$\phi_r R_5$ kips	171	143	115	97.0	80.0	110	88.1	74.6	60.6
$\phi_r R_6$ kips/in.	25.6	21.8	16.9	14.9	12.3	15.6	11.8	10.2	8.49
ϕR (N = 3^1⁄$_4$) kips	241	213	170	145	120	161	126	108	88.2

*Noncompact shape; F_y = 50 ksi.
Load above heavy line is limited by design shear strength.

					F_y = 50 ksi		

W 14

BEAMS
W Shapes
Maximum factored uniform loads in kips
for beams laterally supported
For beams laterally unsupported, see page 4-139

Designation		W 14			W 14			W 14	
Wt./ft		53	48	43	38	34	30	26	22
	5								171
	6							192	166
	7				236	215	202	172	142
	8				231	205	177	151	125
	9	278	253	225	205	182	158	134	111
	10	261	235	209	185	164	142	121	100
	11	238	214	190	168	149	129	110	91
	12	218	196	174	154	137	118	101	83
	13	201	181	161	142	126	109	93	77
	14	187	168	149	132	117	101	86	71
	15	174	157	139	123	109	95	80	66
	16	163	147	131	115	102	89	75	62
	17	154	138	123	109	96	83	71	59
	18	145	131	116	103	91	79	67	55
	19	138	124	110	97	86	75	63	52
	20	131	118	104	92	82	71	60	50
	21	124	112	99	88	78	68	57	47
	22	119	107	95	84	74	65	55	45
	23	114	102	91	80	71	62	52	43
	24	109	98	87	77	68	59	50	42
	25	105	94	84	74	66	57	48	40
	26	101	90	80	71	63	55	46	38
	27	97	87	77	68	61	53	45	37
	28	93	84	75	66	59	51	43	36
	29	90	81	72	64	56	49	42	34
	30	87	78	70	62	55	47	40	33
	31	84	76	67	60	53	46	39	32
	32	82	74	65	58	51	44	38	31
	33	79	71	63	56	50	43	37	30
	34	77	69	61	54	48	42	35	29

F_y = 50 ksi   Span (ft)

Properties and Reaction Values

	53	48	43	38	34	30	26	22
Z_x in.^{3}	87.1	78.4	69.6	61.5	54.6	47.3	40.2	33.2
$\phi_b W_c$ kip-ft	2610	2350	2090	1850	1640	1420	1210	996
$\phi_v V_n$ kips	139	127	112	118	108	101	95.8	85.3
ϕR_1 kips	66.5	58.4	50.0	41.2	35.6	31.6	29.9	25.2
ϕR_2 kips/in.	18.5	17.0	15.3	15.5	14.3	13.5	12.8	11.5
$\phi_r R_3$ kips	65.9	55.1	44.2	44.7	37.0	31.4	30.1	23.0
$\phi_r R_4$ kips/in.	5.96	5.18	4.24	4.44	3.94	4.00	3.07	2.86
$\phi_r R_5$ kips	60.4	50.4	40.4	40.5	33.3	27.7	27.2	20.4
$\phi_r R_6$ kips/in.	7.95	6.91	5.65	5.92	5.25	5.33	4.09	3.81
ϕR ($N = 3\frac{1}{4}$) kips	86.2	72.8	58.7	59.7	50.4	45.0	40.6	32.8

Load above heavy line is limited by design shear strength.

| $F_y = 50$ ksi | | BEAMS
W Shapes
Maximum factored uniform loads in kips
for beams laterally supported
For beams laterally unsupported, see page 4-139 | | | | | | | W 12 | |

For beams laterally unsupported, see page 4-139

Designation			W 12							W 12	
Wt./ft			120	106	96	87	79	72	65*	58	53
		10								237	225
		11	503	425	377	348	314	284	255	236	212
		12	465	410	368	330	298	270	238	216	195
		13	429	378	339	305	275	249	220	199	180
		14	399	351	315	283	255	231	204	185	167
		15	372	328	294	264	238	216	191	173	156
		16	349	308	276	248	223	203	179	162	146
		17	328	289	259	233	210	191	168	152	137
		18	310	273	245	220	198	180	159	144	130
		19	294	259	232	208	188	171	151	136	123
		20	279	246	221	198	179	162	143	130	117
		21	266	234	210	189	170	154	136	123	111
		22	254	224	200	180	162	147	130	118	106
		23	243	214	192	172	155	141	124	113	102
		24	233	205	184	165	149	135	119	108	97
		25	223	197	176	158	143	130	114	104	93
$F_y = 50$ ksi	Span (ft)	26	215	189	170	152	137	125	110	100	90
		27	207	182	163	147	132	120	106	96	87
		28	199	176	158	141	128	116	102	93	83
		29	192	170	152	137	123	112	99	89	81
		30	186	164	147	132	119	108	95	86	78

Properties and Reaction Values										
Z_x in.³	186	164	147	132	119	108	96.8	86.4	77.9
$\phi_b W_c$ kip-ft	5580	4920	4410	3960	3570	3240	2860	2590	2340
$\phi_v V_n$ kips	252	212	189	174	157	142	128	118	112
ϕR_1 kips	161	129	112	96.6	84.5	73.9	64.0	61.9	53.9
ϕR_2 kips/in.	35.5	30.5	27.5	25.8	23.5	21.5	19.5	18.0	17.3
$\phi_r R_3$ kips	227	171	140	120	99.6	83.2	68.3	62.3	55.4
$\phi_r R_4$ kips/in.	26.7	19.2	15.7	14.6	12.3	10.5	8.75	6.47	6.41
$\phi_r R_5$ kips	203	154	126	108	89.4	74.7	61.2	57.1	50.3
$\phi_r R_6$ kips/in.	35.6	25.7	21.0	19.4	16.5	14.0	11.7	8.63	8.54
ϕR ($N = 3\frac{1}{4}$) kips	**276**	**228**	**194**	171	143	120	99.2	85.1	78.0

*Indicates noncompact shape; $F_y = 50$ ksi.
Load above heavy line is limited by design shear strength.
Values of R in **bold face** exceed maximum design web shear $\phi_v V_n$.

$F_y = 50$ ksi

W 12

BEAMS
W Shapes
Maximum factored uniform loads in kips
for beams laterally supported
For beams laterally unsupported, see page 4-139

Designation		W 12			W 12			W 12		
Wt./ft	50	45	40	35	30	26	22	19	16	14
4								154	142	129
5							173	148	121	104
6							147	124	101	87
7				203	173	152	126	106	86	75
8	244	218		192	162	140	110	93	75	65
9	241	216	190	171	144	124	98	82	67	58
10	217	194	173	154	129	112	88	74	60	52
11	197	176	157	140	118	101	80	67	55	47
12	181	162	144	128	108	93	73	62	50	44
13	167	149	133	118	99	86	68	57	46	40
14	155	139	123	110	92	80	63	53	43	37
15	145	129	115	102	86	74	59	49	40	35
16	136	121	108	96	81	70	55	46	38	33
17	128	114	101	90	76	66	52	44	35	31
18	121	108	96	85	72	62	49	41	34	29
19	114	102	91	81	68	59	46	39	32	27
20	109	97	86	77	65	56	44	37	30	26
21	103	92	82	73	62	53	42	35	29	25
22	99	88	78	70	59	51	40	34	27	24
23	94	84	75	67	56	49	38	32	26	23
24	91	81	72	64	54	47	37	31	25	22
25	87	78	69	61	52	45	35	30	24	21
26	84	75	66	59	50	43	34	29	23	20
27	80	72	64	57	48	41	33	27	22	19
28	78	69	62	55	46	40	31	26	22	19
29	75	67	59	53	45	38	30	26	21	18
30	72	65	51	43	37	29	25			

$F_y = 50$ ksi

Span (ft)

Properties and Reaction Values

	50	45	40	35	30	26	22	19	16	14
Z_x in.^{3}	72.4	64.7	57.5	51.2	43.1	37.2	29.3	24.7	20.1	17.4
$\phi_b W_c$ kip-ft	2170	1940	1730	1540	1290	1120	879	741	603	522
$\phi_v V_n$ kips	122	109	95.1	101	86.6	75.9	86.4	77.2	71.2	64.3
ϕR_1 kips	63.6	52.3	46.1	37.5	30.5	25.2	28.4	23.9	20.6	17.2
ϕR_2 kips/in.	18.5	16.8	14.7	15.0	13.0	11.5	13.0	11.8	11.0	10.0
$\phi_r R_3$ kips	64.9	53.0	41.5	42.7	31.7	24.5	31.2	24.3	19.2	15.3
$\phi_r R_4$ kips/in.	7.02	5.87	4.52	4.49	3.50	2.83	3.63	3.30	3.63	3.23
$\phi_r R_5$ kips	59.2	48.3	37.9	39.0	28.8	22.2	28.2	21.6	16.3	12.7
$\phi_r R_6$ kips/in.	9.37	7.82	6.02	5.99	4.67	3.78	4.85	4.40	4.83	4.31
ϕR ($N = 3\frac{1}{4}$) kips	89.7	73.7	57.4	58.5	44.0	34.5	43.9	35.9	32.0	26.7

Load above heavy line is limited by design shear strength.

$F_y = 50$ ksi									

BEAMS
W Shapes
Maximum factored uniform loads in kips
for beams laterally supported

For beams laterally unsupported, see page 4-139

W 10

Designation		W 10							
Wt./ft		112	100	88	77	68	60	54	49
	9	463	408	354	303	264	232	202	183
	10	441	390	339	293	256	224	200	181
	11	401	355	308	266	233	203	182	165
	12	368	325	283	244	213	187	167	151
	13	339	300	261	225	197	172	154	139
	14	315	279	242	209	183	160	143	129
	15	294	260	226	195	171	149	133	121
	16	276	244	212	183	160	140	125	113
	17	259	229	199	172	151	132	118	107
	18	245	217	188	163	142	124	111	101
	19	232	205	178	154	135	118	105	95
	20	221	195	170	146	128	112	100	91
	21	210	186	161	139	122	107	95	86
	22	200	177	154	133	116	102	91	82
	23	192	170	147	127	111	97	87	79
	24	184	163	141	122	107	93	83	76

$F_y = 50$ ksi   Span (ft)

Properties and Reaction Values

	112	100	88	77	68	60	54	49
Z_x in.^{3}	147	130	113	97.6	85.3	74.6	66.6	60.4
$\phi_b W_c$ kip-ft	4410	3900	3390	2930	2560	2240	2000	1810
$\phi_v V_n$ kips	232	204	177	152	132	116	101	91.6
ϕR_1 kips	177	149	123	99.4	80.8	68.9	57.8	50.5
ϕR_2 kips/in.	37.8	34.0	30.3	26.5	23.5	21.0	18.5	17.0
$\phi_r R_3$ kips	265	214	169	130	102	80.9	63.6	53.5
$\phi_r R_4$ kips/in.	32.8	27.4	22.3	17.5	14.0	11.5	8.83	7.61
$\phi_r R_5$ kips	240	194	153	117	92.2	73.1	57.7	48.4
$\phi_r R_6$ kips/in.	43.7	36.5	29.8	23.3	18.7	15.4	11.8	10.1
ϕR ($N = 3\frac{1}{4}$) kips	**300**	**259**	**221**	**185**	**153**	**123**	96.0	81.4

Load above heavy line is limited by design shear strength.
Values of ϕR ($N = 3\frac{1}{4}$) in **boldface** exceed maximum web shear $\phi_v V_n$.

	$F_y = 50$ ksi

W 10

BEAMS
W Shapes
Maximum factored uniform loads in kips
for beams laterally supported
For beams laterally unsupported, see page 4-139

Designation		W 10			W 10			W 10			
Wt./ft		45	39	33	30	26	22	19	17	15	12*
	3									124	101
	4							138	131	120	94
	5						132	130	112	96	75
	6				170	145	130	108	94	80	63
	7			152	157	134	111	93	80	69	54
	8	191	169	146	137	117	98	81	70	60	47
	9	183	156	129	122	104	87	72	62	53	42
	10	165	140	116	110	94	78	65	56	48	38
	11	150	128	106	100	85	71	59	51	44	34
	12	137	117	97	91	78	65	54	47	40	31
	13	127	108	90	84	72	60	50	43	37	29
	14	118	100	83	78	67	56	46	40	34	27
	15	110	94	78	73	63	52	43	37	32	25
	16	103	88	73	69	59	49	41	35	30	23
	17	97	83	68	65	55	46	38	33	28	22
	18	92	78	65	61	52	43	36	31	27	21
	19	87	74	61	58	49	41	34	30	25	20
	20	82	70	58	55	47	39	32	28	24	19
	21	78	67	55	52	45	37	31	27	23	18
	22	75	64	53	50	43	35	29	26	22	17
	23	72	61	51	48	41	34	28	24	21	16
	24	69	59	49	46	39	33	27	23	20	16

($F_y = 50$ ksi) · Span (ft)

Properties and Reaction Values

	45	39	33	30	26	22	19	17	15	12
Z_x in.^{3}	54.9	46.8	38.8	36.6	31.3	26.0	21.6	18.7	16.0	12.6
$\phi_b W_c$ kip-ft	1650	1400	1160	1100	939	780	648	561	480	376
$\phi_v V_n$ kips	95.4	84.4	76.2	84.8	72.5	65.9	69.1	65.5	62.0	50.6
ϕR_1 kips	54.7	44.3	38.5	35.2	28.4	22.5	25.4	22.5	19.8	14.8
ϕR_2 kips/in.	17.5	15.8	14.5	15.0	13.0	12.0	12.5	12.0	11.5	9.50
$\phi_r R_3$ kips	58.8	46.4	37.1	42.3	31.7	25.4	28.3	24.4	20.7	13.7
$\phi_r R_4$ kips/in.	7.41	6.43	6.23	5.47	4.18	4.08	4.18	4.48	4.88	3.58
$\phi_r R_5$ kips	53.8	42.2	33.1	38.5	28.8	22.7	25.5	21.3	17.4	11.3
$\phi_r R_6$ kips/in.	9.88	8.58	8.31	7.29	5.58	5.45	5.57	5.98	6.51	4.77
$\phi R (N = 3^1/_4)$ kips	85.9	70.0	60.1	62.2	47.0	40.4	43.6	40.8	38.6	26.8

*Indicates noncompact shape; $F_y = 50$ ksi.
Load above heavy line is limited by design shear strength.

| $F_y = 50$ ksi | BEAMS | W 8 |

BEAMS
W Shapes
Maximum factored uniform loads in kips
for beams laterally supported
For beams laterally unsupported, see page 4-139

Designation		W 8					
Wt./ft		67	58	48	40	35	31
Span (ft)	7	277	241		160	136	123
	8	263	224	184	149	130	114
	9	234	199	163	133	116	101
	10	211	179	147	119	104	91
	11	191	163	134	109	95	83
	12	175	150	123	100	87	76
	13	162	138	113	92	80	70
	14	150	128	105	85	74	65
	15	140	120	98	80	69	61
	16	132	112	92	75	65	57
	17	124	106	86	70	61	54
	18	117	100	82	66	58	51
	19	111	94	77	63	55	48
	20	105	90	74	60	52	46

$F_y = 50$ ksi

Properties and Reaction Values

		67	58	48	40	35	31
Z_x in.^{3}		70.2	59.8	49.0	39.8	34.7	30.4
$\phi_b W_c$ kip-ft		2110	1790	1470	1190	1040	912
$\phi_v V_n$ kips		139	120	91.8	80.2	68.0	61.6
ϕR_1 kips		102	83.7	59.4	47.8	38.8	33.4
ϕR_2 kips/in.		28.5	25.5	20.0	18.0	15.5	14.3
$\phi_r R_3$ kips		150	118	75.5	58.3	43.8	36.2
$\phi_r R_4$ kips/in.		23.8	20.2	11.9	10.9	8.02	7.20
$\phi_r R_5$ kips		136	106	68.8	52.3	39.5	32.3
$\phi_r R_6$ kips/in.		31.7	27.0	15.9	14.6	10.7	9.60
ϕR ($N = 3\frac{1}{4}$) kips		195	167	120	99.6	74.2	63.5

Load above heavy line is limited by design shear strength.
Values of R in **bold face** exceed maximum design web shear $\phi_v V_n$.

| W 8 | BEAMS W Shapes Maximum factored uniform loads in kips for beams laterally supported For beams laterally unsupported, see page 4-139 | $F_y = 50$ ksi |

For beams laterally unsupported, see page 4-139

Designation			W 8		W 8		W 8		
Wt./ft			28	24	21	18	15	13	10*
$F_y = 50$ ksi	Span (ft)	3					107	99	72
		4					102	86	66
		5			112	101	82	68	53
		6	124	105	102	85	68	57	44
		7	117	99	87	73	58	49	38
		8	102	87	77	64	51	43	33
		9	91	77	68	57	45	38	29
		10	82	70	61	51	41	34	26
		11	74	63	56	46	37	31	24
		12	68	58	51	43	34	29	22
		13	63	54	47	39	31	26	20
		14	58	50	44	36	29	24	19
		15	54	46	41	34	27	23	18
		16	51	44	38	32	26	21	16
		17	48	41	36	30	24	20	16
		18	45	39	34	28	23	19	15
		19	43	37	32	27	21	18	14
		20	41	31	26	20			

Properties and Reaction Values

	28	24	21	18	15	13	10*
Z_x in.^{3}	27.2	23.2	20.4	17.0	13.6	11.4	8.87
$\phi_b W_c$ kip-ft	816	696	612	510	408	342	264
$\phi_v V_n$ kips	62.0	52.5	55.9	50.5	53.6	49.6	36.2
ϕR_1 kips	33.4	26.8	25.4	21.6	23.0	19.8	13.3
ϕR_2 kips/in.	14.3	12.3	12.5	11.5	12.3	11.5	8.50
$\phi_r R_3$ kips	37.4	27.7	28.5	22.9	24.5	20.1	11.4
$\phi_r R_4$ kips/in.	6.68	5.02	5.10	4.90	6.23	6.46	3.29
$\phi_r R_5$ kips	33.8	25.0	25.7	20.2	21.2	16.6	9.72
$\phi_r R_6$ kips/in.	8.91	6.69	6.81	6.53	8.30	8.61	4.38
ϕR ($N = 3\frac{1}{4}$) kips	**62.8**	46.7	47.8	41.4	48.2	44.6	24.0

*Indicates noncompact shape; $F_y = 50$ ksi.
Load above heavy line is limited by design shear strength.
Values of R in **bold face** exceed maximum design web shear $\phi_v V_n$.

| $F_y = 50$ ksi | | | | | | | | W 6–5–4 |

BEAMS
W Shapes
Maximum factored uniform loads in kips
for beams laterally supported
For beams laterally unsupported, see page 4-139

Designation		W 6			W 6			W 5		W 4
Wt./ft		25	20	15*	16	12	9	19	16	13
	2									63
	3				88	75	54			63
	4			74	88	62	47	75	65	47
	5	110	87	62	70	50	37	70	58	38
	6	95	75	51	59	42	31	58	48	31
	7	81	64	44	50	36	27	50	41	27
	8	71	56	38	44	31	23	44	36	24
	9	63	50	34	39	28	21	39	32	21
	10	57	45	31	35	25	19	35	29	19
	11	52	41	28	32	23	17	32	26	
	12	47	37	26	29	21	16	29	24	
	13	44	34	24	27	19	14			
	14	41	32	22	25	18	13			

$F_y = 50$ ksi   Span (ft)

Properties and Reaction Values

	25	20	15*	16	12	9	19	16	13
Z_x in.^{3}	18.9	14.9	10.8	11.7	8.30	6.23	11.6	9.59	6.28
$\phi_b W_c$ kip-ft	567	447	308	351	249	187	348	288	188
$\phi_v V_n$ kips	55.1	43.5	37.2	44.1	37.4	27.1	37.5	32.5	31.4
ϕR_1 kips	32.5	24.4	18.0	24.4	18.0	12.0	27.4	22.5	24.1
ϕR_2 kips/in.	16.0	13.0	11.5	13.0	11.5	8.50	13.5	12.0	14.0
$\phi_r R_3$ kips	44.0	28.9	20.3	30.4	21.0	11.7	33.2	25.4	31.4
$\phi_r R_4$ kips/in.	12.2	8.40	8.45	7.48	7.80	4.19	9.62	8.29	16.5
$\phi_r R_5$ kips	38.8	25.4	16.9	27.3	17.9	10.1	29.9	22.7	26.8
$\phi_r R_6$ kips/in.	16.3	11.2	11.3	9.97	10.4	5.59	12.8	11.1	22.1
ϕR ($N = 3\frac{1}{4}$) kips	**84.5**	**61.8**	**53.5**	**59.7**	**51.7**	**28.2**	**71.3**	**58.6**	**69.6**

*Indicates noncompact shape; $F_y = 50$ ksi.
Load above heavy line is limited by design shear strength.
Values of R in **bold face** exceed maximum design web shear $\phi_v V_n$.

S 24–20		BEAMS			$F_y = 50$ ksi			

BEAMS
S Shapes
Maximum factored uniform loads in kips
for beams laterally supported
For beams laterally unsupported, see page 4-139

Designation		S 24		S 24			S 20		S 20	
Wt./ft		121	106	100	90	80	96	86	75	66
	6						877		686	
	7			966			849	723	656	545
	8	1060		900	810		743	686	574	525
	9	1020		800	740	648	660	610	510	467
	10	918	820	720	666	612	594	549	459	420
	11	835	761	655	605	556	540	499	417	382
	12	765	698	600	555	510	495	458	383	350
	13	706	644	554	512	471	457	422	353	323
	14	656	598	514	476	437	424	392	328	300
	15	612	558	480	444	408	396	366	306	280
	16	574	523	450	416	383	371	343	287	263
	17	540	492	424	392	360	349	323	270	247
	18	510	465	400	370	340	330	305	255	233
	19	483	441	379	351	322	313	289	242	221
	20	459	419	360	333	306	297	275	230	210
	21	437	399	343	317	291	283	261	219	200
	22	417	380	327	303	278	270	250	209	191
	23	399	364	313	290	266	258	239	200	183
	24	383	349	300	278	255	248	229	191	175
	25	367	335	288	266	245	238	220	184	168
	26	353	322	277	256	235	228	211	177	162
	27	340	310	267	247	227	220	203	170	156
	28	328	299	257	238	219	212	196	164	150
	29	317	289	248	230	211	205	189	158	145
	30	306	279	240	222	204	198	183	153	140
	32	287	262	225	208	191	186	172	143	131
	34	270	246	212	196	180	175	161	135	124
	36	255	233	200	185	170	165	153	128	117
	38	242	220	189	175	161	156	144	121	111
	40	230	209	180	167	153	149	137	115	105
	42	219	199	171	159	146	141	131	109	100
	44	209	190	164	151	139	135	125	104	95
	46	200	182	157	145	133	129	119	100	91
	48	191	174	150	139	128	124	114	96	88
	50	184	167	144	133	122	119	110	92	84
	52	177	161	138	128	118				
	54	170	155	133	123	113				
	56	164	149	129	119	109				
	58	158	144	124	115	106				
	60	153	140	120	111	102				

($F_y = 50$ ksi, Span (ft))

Properties and Reaction Values

	121	106	100	90	80	96	86	75	66
Z_x in.^{3}	306	279	240	222	204	198	183	153	140
$\phi_b W_c$ kip-ft	9180	8370	7200	6660	6120	5940	5490	4590	4200
$\phi_v V_n$ kips	529	410	483	405	324	438	362	343	273
ϕR_1 kips	200	155	163	137	109	175	144	129	103
ϕR_2 kips/in.	40.0	31.0	37.3	31.3	25.0	40.0	33.0	31.8	25.3
$\phi_r R_3$ kips	269	184	216	166	119	248	185	163	115
$\phi_r R_4$ kips/in.	20.7	9.66	21.4	12.6	6.48	29.7	16.7	17.4	8.76
$\phi_r R_5$ kips	236	168	182	146	109	207	163	139	104
$\phi_r R_6$ kips/in.	27.7	12.9	28.6	16.9	8.64	39.5	22.2	23.2	11.7
ϕR ($N = 3\frac{1}{4}$) kips	330	215	284	207	140	305	240	219	144

Load above heavy line is limited by design shear strength.

$F_y = 50$ ksi										

BEAMS
S Shapes
Maximum factored uniform loads in kips
for beams laterally supported
For beams laterally unsupported, see page 4-139

S 18–15–12–10

Designation		S 18		S 15		S 12		S 12		S 10	
Wt./ft		70	54.7	50	42.9	50	40.8	35	31.8	35	25.4
	3									321	
	4					445		277		266	
	5	691		446		367	299	269	227	212	168
	6	625		386	333	306	266	224	210	177	142
	7	536	448	330	297	262	228	192	180	152	122
	8	469	394	289	260	230	199	168	158	133	107
	9	417	350	257	231	204	177	149	140	118	95
	10	375	315	231	208	184	159	134	126	106	85
	11	341	286	210	189	167	145	122	115	97	77
	12	313	263	193	173	153	133	112	105	89	71
	13	288	242	178	160	141	123	103	97	82	66
	14	268	225	165	149	131	114	96	90	76	61
	15	250	210	154	139	122	106	90	84	71	57
	16	234	197	145	130	115	100	84	79	66	53
	17	221	185	136	122	108	94	79	74	62	50
	18	208	175	129	116	102	89	75	70	59	47
	19	197	166	122	109	97	84	71	66	56	45
	20	188	158	116	104	92	80	67	63	53	43
	21	179	150	110	99	87	76	64	60	51	41
	22	170	143	105	95	83	72	61	57	48	39
	23	163	137	101	90	80	69	58	55	46	37
	24	156	131	96	87	77	66	56	53	44	36
	25	150	126	93	83	73	64	54	50	42	34
	26	144	121	89	80	71	61	52	48		
	27	139	117	86	77	68	59	50	47		
	28	134	113	83	74	66	57	48	45		
	29	129	109	80	72	63	55	46	43		
	30	125	105	77	69	61	53	45	42		
	31	121	102	75	67						
	32	117	98	72	65						
	33	114	95	70	63						
	34	110	93	68	61						
	35	107	90	66	59						
	36	104	88	64	58						
	37	101	85	63	56						
	38	99	83								
	40	94	79								
	42	89	75								
	44	85	72								

Left margin label: $F_y = 50$ ksi; Span (ft)

Properties and Reaction Values

Z_x in.^{3}	125	105	77.1	69.3	61.2	53.1	44.8	42.0	35.4	28.4
$\phi_b W_c$ kip-ft	3750	3150	2310	2080	1840	1590	1340	1260	1060	852
$\phi_v V_n$ kips	346	224	223	166	223	150	139	113	160	84.0
ϕR_1 kips	133	86.4	94.5	70.6	123	83.0	63.5	52.0	83.5	43.7
ϕR_2 kips/in.	35.6	23.1	27.5	20.6	34.3	23.1	21.4	17.5	29.7	15.5
$\phi_r R_3$ kips	180	93.8	116	74.9	167	91.9	74.5	55.1	116	43.8
$\phi_r R_4$ kips/in.	31.3	8.52	19.3	8.05	44.4	13.5	13.0	7.11	46.2	6.63
$\phi_r R_5$ kips	142	83.6	96.7	66.9	131	81.1	64.1	49.4	84.9	39.4
$\phi_r R_6$ kips/in.	41.7	11.4	25.7	10.7	59.1	18.0	17.3	9.47	61.6	8.84
ϕR ($N = 3\frac{1}{4}$) kips	249	122	180	102	**235**	140	120	80.2	**180**	68.1

Load above heavy line is limited by design shear strength.
Values of R in **bold face** exceed maximum design web shear $\phi_v V_n$.

S 8–6–5–4–3

BEAMS
S Shapes
Maximum factored uniform loads in kips
for beams laterally supported
For beams laterally unsupported, see page 4-139

$F_y = 50$ ksi

$F_y = 50$ ksi

Span (ft)

Designation	S 8		S 6		S 5	S 4		S 3	
Wt./ft	23	18.4	17.25	12.5	10	9.5	7.7	7.5	5.7
1						70		57	
2			151		58	61	42	35	28
3	191		106	75	57	40	35	24	20
4	145	117	80	64	43	30	26	18	15
5	116	99	64	51	34	24	21	14	12
6	96	83	53	42	28	20	18	12	9.75
7	83	71	45	36	24	17	15	10	8.36
8	72	62	40	32	21	15	13		
9	64	55	35	28	19	13	12		
10	58	50	32	25	17	12	11		
11	53	45	29	23	15				
12	48	41	27	21	14				
13	45	38	24	20					

Properties and Reaction Values

Z_x in.^{3}	19.3	16.5	10.6	8.47	5.67	4.04	3.51	2.36	1.95
$\phi_b W_c$ kip-ft	579	495	318	254	170	121	105	70.8	58.5
$\phi_v V_n$ kips	95.3	58.5	75.3	37.6	28.9	35.2	20.8	28.3	13.8
ϕR_1 kips	55.1	33.9	50.9	25.4	21.7	30.6	18.1	30.0	14.6
ϕR_2 kips/in.	22.1	13.6	23.3	11.6	10.7	16.3	9.65	17.5	8.50
$\phi_r R_3$ kips	68.9	33.2	68.5	24.1	20.4	36.3	16.6	37.9	12.9
$\phi_r R_4$ kips/in.	27.2	6.32	50.5	6.27	6.50	32.0	6.64	59.0	6.81
$\phi_r R_5$ kips	54.4	29.8	48.3	21.6	18.2	27.8	14.8	26.1	11.5
$\phi_r R_6$ kips/in.	36.3	8.42	67.3	8.36	8.67	42.6	8.85	78.6	9.09
ϕR ($N = 3\frac{1}{4}$) kips	**127**	57.2	**126**	48.8	46.4	83.5	43.5	86.7	41.1

Load above heavy line is limited by design shear strength.
Values of R in **bold face** exceed maximum design web shear $\phi_v V_n$.

$F_y = 50$ ksi

BEAMS
Channels
Maximum factored uniform loads in kips
for beams laterally supported
For beams laterally unsupported, see page 4-139

MC,C 18–15

[

Designation			MC 18				C 15		
Wt./ft		58	51.9	45.8	42.7	50	40	33.9
	3					580		
	4	680	583	486		511	421	324
	5	568	519	470	437	409	343	302
	6	473	433	392	372	341	286	252
	7	405	371	336	319	292	245	216
	8	355	324	294	279	256	215	189
	9	315	288	261	248	227	191	168
	10	284	260	235	223	205	172	151
	11	258	236	214	203	186	156	137
	12	237	216	196	186	170	143	126
	13	218	200	181	172	157	132	116
	14	203	185	168	159	146	123	108
	15	189	173	157	149	136	114	101
	16	177	162	147	140	128	107	95
	17	167	153	138	131	120	101	89
	18	158	144	131	124	114	95	84
	19	149	137	124	117	108	90	80
	20	142	130	118	112	102	86	76
	21	135	124	112	106	97	82	72
	22	129	118	107	101	93	78	69
	23	123	113	102	97	89	75	66
	24	118	108	98	93	85	72	63
	25	114	104	94	89	82	69	60
	26	109	100	90	86	79	66	58
	28	101	93	84	80	73	61	54
	30	95	87	78	74	68	57	50
	32	89	81	74	70	64	54	47
	34	83	76	69	66	60	50	44
	36	79	72	65	62	57	48	42
	38	75	68	62	59			
	40	71	65	59	56			
	42	68	62	56	53			
	44	65	59	53	51			

$F_y = 50$ ksi — Span (ft)

Properties and Reaction Values							
Z_x in.^{3}	94.6	86.5	78.4	74.4	68.2	57.2	50.4
$\phi_b W_c$ kip-ft	2840	2600	2350	2230	2050	1720	1510
$\phi_v V_n$ kips	340	292	243	219	290	211	162
ϕR_1 kips	120	103	85.9	77.3	129	93.4	71.9
ϕR_2 kips/in.	35.0	30.0	25.0	22.5	35.8	26.0	20.0
$\phi_r R_3$ kips	167	133	101	86.1	176	109	73.6
$\phi_r R_4$ kips/in.	33.0	20.8	12.0	8.76	40.7	15.6	7.10
$\phi_r R_5$ kips	127	108	86.4	75.5	135	93.4	66.5
$\phi_r R_6$ kips/in.	44.0	27.7	16.0	11.7	54.3	20.8	9.47
ϕR ($N = 3\frac{1}{4}$) kips	234	200	140	115	245	161	97.2

Load above heavy line is limited by design shear strength.

MC 13					$F_y = 50$ ksi

BEAMS
Channels
Maximum factored uniform loads in kips
for beams laterally supported
For beams laterally unsupported, see page 4-139

Designation			MC 13			
Wt./ft		50	40	35	31.8
	3	552	393		
	4	454	382	314	263
	5	363	305	277	259
	6	303	255	231	216
	7	259	218	198	185
	8	227	191	173	162
	9	202	170	154	144
	10	182	153	139	129
	11	165	139	126	118
	12	151	127	116	108
	13	140	117	107	99
	14	130	109	99	92
	15	121	102	92	86
	16	113	95	87	81
	17	107	90	82	76
	18	101	85	77	72
	19	96	80	73	68
	20	91	76	69	65
	21	86	73	66	62
	22	83	69	63	59
	23	79	66	60	56
	24	76	64	58	54
	25	73	61	55	52
	26	70	59	53	50
	27	67	57	51	48
	28	65	55	50	46
	29	63	53	48	45
	30	61	51	46	43
	31	59	49	45	42
	32	57	48	43	40

$F_y = 50$ ksi   Span (ft)

Properties and Reaction Values

	50	40	35	31.8
Z_x in.^{3}	60.5	50.9	46.2	43.1
$\phi_b W_c$ kip-ft	1820	1530	1390	1290
$\phi_v V_n$ kips	276	197	157	132
ϕR_1 kips	135	96.3	76.8	64.5
ϕR_2 kips/in.	39.3	28.0	22.4	18.8
$\phi_r R_3$ kips	197	118	84.2	64.7
$\phi_r R_4$ kips/in.	66.5	24.0	12.2	7.19
$\phi_r R_5$ kips	139	97.3	73.6	58.4
$\phi_r R_6$ kips/in.	88.7	31.9	16.2	9.59
ϕR ($N = 3\frac{1}{4}$) kips	263	187	126	89.6

Load above heavy line is limited by design shear strength.

$F_y = 50$ ksi

BEAMS
Channels
Maximum factored uniform loads in kips
for beams laterally supported
For beams laterally unsupported, see page 4-139

C,MC 12

[

Designation		C 12			MC 12					MC 12
Wt./ft		30	25	20.7	50	45	40	35	31	10.6
	2									123
	3	330	251		541	461	382			116
	4	252	219	183	421	388	355	303	240	87
	5	202	175	152	337	310	284	257	236	70
	6	168	146	127	281	259	237	214	197	58
	7	144	125	109	240	222	203	183	168	50
	8	126	110	95	210	194	177	161	147	44
	9	112	97	85	187	172	158	143	131	39
	10	101	88	76	168	155	142	128	118	35
	11	92	80	69	153	141	129	117	107	32
	12	84	73	64	140	129	118	107	98	29
	13	78	67	59	129	119	109	99	91	27
	14	72	63	54	120	111	101	92	84	25
	15	67	58	51	112	103	95	86	79	23
	16	63	55	48	105	97	89	80	74	22
	17	59	52	45	99	91	83	76	69	20
	18	56	49	42	94	86	79	71	66	19
	19	53	46	40	89	82	75	68	62	18
	20	50	44	38	84	78	71	64	59	17
	21	48	42	36	80	74	68	61	56	17
	22	46	40	35	77	71	65	58	54	16
	23	44	38	33	73	67	62	56	51	15
	24	42	37	32	70	65	59	54	49	15
	25	40	35	30	67	62	57	51	47	14
	26	39	34	29	65	60	55	49	45	13
	27	37	32	28	62	57	53	48	44	13
	28	36	31	27	60	55	51	46	42	12
	29	35	30	26	58	53	49	44	41	12
	30	34	29	25	56	52	47	43	39	12

$F_y = 50$ ksi   Span (ft)

Properties and Reaction Values

	30	25	20.7	50	45	40	35	31	10.6
Z_x in.³	33.6	29.2	25.4	56.1	51.7	47.3	42.8	39.3	11.6
$\phi_b W_c$ kip-ft	1010	876	762	1680	1550	1420	1280	1180	348
$\phi_v V_n$ kips	165	125	91.4	271	231	191	151	120	61.6
ϕR_1 kips	71.7	54.4	39.7	137	117	96.8	76.6	60.7	16.3
ϕR_2 kips/in.	25.5	19.4	14.1	41.8	35.6	29.5	23.4	18.5	9.50
$\phi_r R_3$ kips	93.0	61.5	38.2	230	181	137	96.3	67.9	16.6
$\phi_r R_4$ kips/in.	23.9	10.4	4.04	75.0	46.5	26.5	13.1	6.52	2.00
$\phi_r R_5$ kips	73.9	53.1	35.0	170	144	116	85.8	62.7	15.0
$\phi_r R_6$ kips/in.	31.8	13.9	5.38	100.0	62.0	35.3	17.5	8.70	2.67
ϕR ($N = 3\frac{1}{4}$) kips	155	98.3	52.5	**273**	**233**	**193**	143	91.0	23.7

Load above heavy line is limited by design shear strength.
Values of R in **bold face** exceed maximum design web shear $\phi_v V_n$.

| C,MC 10 | | | | | BEAMS Channels Maximum factored uniform loads in kips for beams laterally supported | | | | | F_y = 50 ksi |

For beams laterally unsupported, see page 4-139

Designation				C 10				MC 10			MC 10		MC 10
Wt./ft			30	25	20	15.3	41.1	33.6	28.5	25	22	8.4
		2	363	284	205		430					92
		3	266	230	193	130	389	311	230	205		79
		4	200	173	145	119	292	251	222	194	157	59
		5	160	138	116	95	233	200	178	155	142	47
		6	133	115	96	79	195	167	148	129	118	39
		7	114	99	83	68	167	143	127	111	101	34
		8	100	86	72	59	146	125	111	97	89	29
		9	89	77	64	53	130	111	99	86	79	26
		10	80	69	58	47	117	100	89	77	71	24
		11	73	63	53	43	106	91	81	70	64	21
		12	67	58	48	40	97	84	74	65	59	20
		13	61	53	45	36	90	77	68	60	54	18
		14	57	49	41	34	83	72	63	55	51	17
		15	53	46	39	32	78	67	59	52	47	16
		16	50	43	36	30	73	63	56	48	44	15
		17	47	41	34	28	69	59	52	46	42	14
		18	44	38	32	26	65	56	49	43	39	13
F_y = 50 ksi	Span (ft)	19	42	36	30	25	61	53	47	41	37	12
		20	40	35	29	24	58	50	44	39	35	12
		21	38	33	28	23	56	48	42	37	34	11
		22	36	31	26	22	53	46	40	35	32	11
		23	35	30	25	21	51	44	39	34	31	10
		24	33	29	24	20	49	42	37	32	30	9.83

Properties and Reaction Values

	30	25	20	15.3	41.1	33.6	28.5	25	22	8.4
Z_x in.^{3}	26.6	23.0	19.3	15.8	38.9	33.4	29.6	25.8	23.6	7.86
$\phi_b W_c$ kip-ft	798	690	579	474	1170	1000	888	774	708	236
$\phi_v V_n$ kips	182	142	102	64.8	215	155	115	103	78.3	45.9
ϕR_1 kips	84.1	65.8	47.4	30.0	124	89.8	66.4	59.4	45.3	14.6
ϕR_2 kips/in.	33.6	26.3	19.0	12.0	39.8	28.8	21.3	19.0	14.5	8.50
$\phi_r R_3$ kips	131	90.8	55.6	28.0	194	119	75.8	64.1	42.7	13.4
$\phi_r R_4$ kips/in.	75.6	36.1	13.5	3.43	94.9	35.8	14.4	10.3	4.59	1.90
$\phi_r R_5$ kips	81.0	66.8	46.6	25.7	131	95.4	66.1	57.2	39.6	12.1
$\phi_r R_6$ kips/in.	101	48.1	18.0	4.57	127	47.7	19.3	13.8	6.12	2.53
ϕR (N = 3¼) kips	193	151	105	40.6	254	183	129	102	59.5	20.3

Load above heavy line is limited by design shear strength.
Values of R in **bold face** exceed maximum design web shear $\phi_v V_n$.

BEAMS
Channels

C,MC 9

Maximum factored uniform loads in kips
for beams laterally supported

For beams laterally unsupported, see page 4-139

$F_y = 50$ ksi

Designation			C 9			MC 9	
Wt./ft			20	15	13.4	25.4	23.9
		2	218	139			
		3	168	135	113	219	194
		4	126	101	94	174	167
		5	101	81	75	139	133
		6	84	68	63	116	111
		7	72	58	54	99	95
		8	63	51	47	87	83
		9	56	45	42	77	74
		10	50	41	38	70	67
		11	46	37	34	63	61
		12	42	34	31	58	56
		13	39	31	29	54	51
		14	36	29	27	50	48
		15	34	27	25	46	44
$F_y = 50$ ksi	Span (ft)	16	31	25	23	44	42
		17	30	24	22	41	39
		18	28	23	21	39	37
		19	27	21	20	37	35
		20	25	20	19	35	33
		21	24	19	18	33	32
		22	23	18	17	32	30

Properties and Reaction Values

	C 9			MC 9	
Z_x in.^{3}	16.8	13.5	12.5	23.2	22.2
$\phi_b W_c$ kip-ft	504	405	375	696	666
$\phi_v V_n$ kips	109	69.3	56.6	109	97.2
ϕR_1 kips	52.5	33.4	27.3	66.8	59.4
ϕR_2 kips/in.	22.4	14.3	11.7	22.5	20.0
$\phi_r R_3$ kips	69.5	35.3	26.1	80.7	67.7
$\phi_r R_4$ kips/in.	26.2	6.74	3.68	19.9	14.0
$\phi_r R_5$ kips	53.8	31.2	23.9	68.8	59.3
$\phi_r R_6$ kips/in.	34.9	8.98	4.91	26.6	18.7
ϕR ($N = 3\frac{1}{4}$) kips	125	60.4	39.8	140	120

Load above heavy line is limited by design shear strength.
Values of R in **bold face** exceed maximum design web shear $\phi_v V_n$.

| C,MC 8 | | | | BEAMS
Channels
Maximum factored uniform loads in kips
for beams laterally supported
For beams laterally unsupported, see page 4-139 | | | | $F_y = 50$ ksi |

Designation			C 8			MC 8		MC 8		MC 8
Wt./ft			18.75	13.75	11.5	22.8	21.4	20	18.7	8.5
		1	210							
		2	207	131				173		77
		3	138	109	95	184	162	162	152	69
		4	104	82	72	141	135	122	116	52
		5	83	65	57	113	108	97	92	41
		6	69	55	48	94	90	81	77	35
		7	59	47	41	81	77	69	66	30
		8	52	41	36	70	68	61	58	26
		9	46	36	32	63	60	54	51	23
		10	41	33	29	56	54	49	46	21
		11	38	30	26	51	49	44	42	19
		12	35	27	24	47	45	41	39	17
		13	32	25	22	43	42	37	36	16
		14	30	23	20	40	39	35	33	15
$F_y = 50$ ksi	Span (ft)	15	28	22	19	38	36	32	31	14
		16	26	20	18	35	34	30	29	13
		17	24	19	17	33	32	29	27	12
		18	23	18	16	31	30	27	26	12
		19	22	17	15	30	28	26	24	11
		20	21	16	14	28	27	24	23	10

Properties and Reaction Values									
Z_x in.^{3}	13.8	10.9	9.55	18.8	18.0	16.2	15.4	6.91
$\phi_b W_c$ kip-ft	414	327	287	564	540	486	462	207
$\phi_v V_n$ kips	105	65.4	47.5	92.2	81.0	86.4	76.2	38.7
ϕR_1 kips	57.1	35.5	25.8	63.4	55.7	56.3	49.6	16.8
ϕR_2 kips/in.	24.3	15.2	11.0	21.3	18.8	20.0	17.7	8.95
$\phi_r R_3$ kips	76.5	37.6	23.2	72.9	60.0	64.5	53.5	15.2
$\phi_r R_4$ kips/in.	40.1	9.65	3.69	20.1	13.6	17.3	11.9	2.49
$\phi_r R_5$ kips	55.2	32.4	21.3	62.2	52.8	55.3	47.1	13.9
$\phi_r R_6$ kips/in.	53.4	12.9	4.92	26.7	18.1	23.1	15.9	3.33
ϕR ($N = 3\frac{1}{4}$) kips	**136**	**74.2**	37.3	**133**	**112**	**121**	**98.7**	24.7

Load above heavy line is limited by design shear strength.
Values of R in **bold face** exceed maximum design web shear $\phi_v V_n$.

$F_y = 50$ ksi

BEAMS
Channels
Maximum factored uniform loads in kips
for beams laterally supported

C,MC 7–6

For beams laterally unsupported, see page 4-139

Designation		C 7		MC 7		C 6			MC 6	MC 6		MC 6
Wt./ft		12.25	9.8	22.7	19.1	13	10.5	8.2	18	16.3	15.1	12
	1					142	102					
	2	119	79	190		109	92	65	123	122	102	100
	3	84	71	162	133	73	62	51	115	102	97	74
	4	63	53	122	107	54	46	38	86	77	73	55
	5	50	43	97	86	44	37	31	69	61	58	44
	6	42	36	81	72	36	31	26	58	51	48	37
	7	36	31	69	61	31	26	22	49	44	42	32
	8	31	27	61	54	27	23	19	43	38	36	28
	9	28	24	54	48	24	21	17	38	34	32	25
	10	25	21	49	43	22	18	15	35	31	29	22
	11	23	19	44	39	20	17	14	31	28	26	20
	12	21	18	41	36	18	15	13	29	26	24	18
	13	19	16	37	33	17	14	12	27	24	22	17
	14	18	15	35	31	16	13	11	25	22	21	16
	15	17	14	32	29	15	12	10	23	20	19	15
	16	16	13	30	27							

$F_y = 50$ ksi

Span (ft)

Properties and Reaction Values

Z_x in.^{3}	8.40	7.12	16.2	14.3	7.26	6.15	5.13	11.5	10.2	9.69	7.38
$\phi_b W_c$ kip-ft	252	214	486	429	218	185	154	345	306	291	221
$\phi_v V_n$ kips	59.3	39.7	95.1	66.5	70.8	50.9	32.4	61.4	60.8	51.2	50.2
ϕR_1 kips	34.3	23.0	70.7	49.5	44.4	31.9	20.3	50.3	49.8	42.0	31.5
ϕR_2 kips/in.	15.7	10.5	25.2	17.6	21.9	15.7	10.0	19.0	18.8	15.8	15.5
$\phi_r R_3$ kips	38.4	21.0	91.0	53.3	61.0	37.2	18.9	58.0	57.1	44.2	38.1
$\phi_r R_4$ kips/in.	13.1	3.91	39.3	13.5	43.9	16.3	4.21	20.7	20.0	12.0	14.3
$\phi_r R_5$ kips	32.3	19.2	72.6	47.0	43.5	30.7	17.2	49.7	49.1	39.4	32.4
$\phi_r R_6$ kips/in.	17.4	5.21	52.5	18.0	58.5	21.7	5.61	27.6	26.7	16.0	19.1
ϕR ($N = 3\frac{1}{4}$) kips	**85.4**	36.1	**152**	**105**	**115**	**82.9**	35.4	**112**	**111**	**91.3**	**81.9**

Load above heavy line is limited by design shear strength.
Values of R in **bold face** exceed maximum design web shear $\phi_v V_n$.

C 5–4–3	$F_y = 50$ ksi

BEAMS
Channels
Maximum factored uniform loads in kips
for beams laterally supported
For beams laterally unsupported, see page 4-139

Designation			C 5		C 4		C 3		
Wt./ft			9	6.7	7.25	5.4	6	5	4.1
		1	88		69	40	52	42	28
		2	65	51	42	34	26	23	20
		3	44	35	28	23	17	15	13
		4	33	26	21	17	13	11	9.8
		5	26	21	17	14	10	9.0	7.8
		6	22	18	14	11	8.6	7.5	6.5
		7	19	15	12	9.7	7.4	6.4	5.6
		8	16	13	11	8.5			
		9	15	12	9.4	7.5			
		10	13	11	8.4	6.8			
		11	12	9.6					
		12	11	8.8					

$F_y = 50$ ksi Span (ft)

Properties and Reaction Values

Z_x in.^{3}	4.36	3.51	2.81	2.26	1.72	1.50	1.30
$\phi_b W_c$ kip-ft	131	105	84.3	67.8	51.6	45.0	39.0
$\phi_v V_n$ kips	43.9	25.7	34.7	19.9	28.8	20.9	13.8
ϕR_1 kips	30.5	17.8	27.6	15.8	30.6	22.2	14.6
ϕR_2 kips/in.	16.3	9.50	16.1	9.20	17.8	12.9	8.50
$\phi_r R_3$ kips	37.8	16.9	35.7	15.5	40.0	24.7	13.2
$\phi_r R_4$ kips/in.	23.2	4.64	30.2	5.69	59.6	22.7	6.49
$\phi_r R_5$ kips	30.1	15.3	27.6	14.0	28.1	20.2	11.9
$\phi_r R_6$ kips/in.	30.9	6.18	40.3	7.59	79.5	30.2	8.65
ϕR ($N = 3\frac{1}{4}$) kips	**83.3**	**35.4**	**79.7**	**38.6**	**88.4**	**64.1**	**40.0**

Load above heavy line is limited by design shear strength.
Values of R in **bold face** exceed maximum design web shear $\phi_v V_n$.

DESIGN FLEXURAL STRENGTH OF BEAMS WITH UNBRACED LENGTH GREATER THAN L_p

General Notes

Spacing of lateral bracing at distances greater than L_p creates a problem in which the designer is confronted with a given laterally unbraced length (usually less than the total span) along the compression flange, and a calculated required bending moment. The beam cannot be selected from its plastic section modulus alone, since depth, flange proportions, and other properties have an influence on its bending strength.

The following charts show the design moment $\phi_b M_n$ for W and M shapes of $F_y = 36$ ksi and $F_y = 50$ ksi steels, used as beams, with respect to the maximum unbraced length for which this moment is permissible. In bending, ϕ_b of 0.9 is given in Section F1.2 of the LRFD Specification. The charts extend over varying unbraced lengths, depending upon the flexural strengths of the beams represented. In general, they extend beyond most unbraced lengths frequently encountered in design practice. The design moment $\phi_b M_n$, kip-ft, is plotted with respect to the unbraced length with no consideration of the moment due to weight of the beam. Design moments are shown for unbraced lengths in feet, starting at spans less than L_p, for spans between L_p and L_r and for spans beyond L_r.

The unbraced length L_p, in feet, with the limit indicated by a solid symbol, ●, is the maximum unbraced length of the compression flange, with $C_b = 1.0$, for which the design moment is given by $\phi_b M_p$,

where

$$L_p = 300 r_y / \sqrt{F_y} \tag{F1-4}$$
$$M_p = Z_x F_y$$

For those noncompact rolled shapes, which meet the requirements of compact sections except that $b_f / 2t_f$ exceeds $65 / \sqrt{F_y}$, but is less than $141 / \sqrt{F_y - F_r}$, the design moment is obtained from Equation A-F1-3 in Appendix F1 of the LRFD Specification. This criterion applies to one W shape when F_y is equal to 36 ksi and to seven W shapes when F_y is equal to 50 ksi. (Noncompact W shapes are given on p. 4-7.)

For the case $C_b = 1.0$ and noncompact shapes:

$$M_n' = M_p - (M_p - M_r)\left(\frac{\lambda - \lambda_p}{\lambda_r - \lambda_p}\right) \tag{A-F1-3}$$

$$L_p' = L_p + (L_r - L_p)\left(\frac{M_p - M_n'}{M_p - M_r}\right)$$

$$\lambda = b_f / 2t_f$$

$$\lambda_p = 65 / \sqrt{F_y}$$

$$\lambda_r = 141 / \sqrt{F_y - F_r}$$

$$L_r = \frac{r_y X_1}{F_y - F_r}\sqrt{1 + \sqrt{1 + X_2(F_y - F_r)^2}} \tag{F1-6}$$

$$X_1 = \frac{\pi}{S_x}\sqrt{\frac{EGJA}{2}} \tag{F1-8}$$

$$X_2 = \frac{4C_w}{I_y}\left(\frac{S_x}{GJ}\right)^2 \qquad\qquad (F1\text{-}9)$$

$$M_r = (F_y - F_r)S_x \qquad\qquad (F1\text{-}7)$$
$$M_p = Z_xF_y$$
$$F_r = 10 \text{ ksi for rolled shapes}$$

The unbraced length in the charts may be either the total span or any part of the total span between braced points. The plots shown in these charts were computed for beams for which $C_b = 1.0$. When a moment gradient exists between points of bracing, C_b may be larger than unity. (See Table 4-1.) Using this larger value of C_b may provide a more liberal flexural strength for the section chosen if the unbraced length is greater than L_p. In these cases, the design moment can be determined using the provisions of Section F1.2a of the LRFD Specification.

$$\phi_b M_n = \phi_b C_b \left[M_p - (M_p - M_r)\left(\frac{L_b - L_p}{L_r - L_p}\right)\right] \le \phi_b M_p$$

The unbraced length L_r, ft, with the limit indicated by an open symbol $\bigcirc$, is the maximum unbraced length of the compression flange beyond which the design moment is governed by Specification Section F1.2b. For unbraced lengths greater than L_r:

$$\phi_b M_n = \phi_b M_{cr} = \phi_b C_b \frac{\pi}{L_b}\sqrt{EI_yGJ + \left(\frac{\pi E}{L_b}\right)^2 I_y C_w} \le \phi_b C_b M_r \text{ and } \phi_b M_p$$

In computing the points for the curves, C_b in the above formulas was taken as unity, $E = 29,000$ ksi and $G = 11,200$ ksi. The properties of the beams are taken from the Tables of Dimensions and Properties in Part 1 of this LRFD Manual. The beam strengths have been reduced by multiplying the nominal flexural strength M_n by 0.9, the resistance factor ϕ_b for flexure.

Over a limited range of length, a given beam is the lightest available for various combinations of unbraced length and design moment. The charts are designed to assist in selection of the lightest available beam for the given combination.

The solid portion of each curve indicates the most economical section by weight. The dashed portion of each curve indicates ranges in which a lighter weight beam will satisfy the loading conditions.

The curves are plotted without regard to shear strength and deflection criteria, therefore due care must be exercised in their use. The curves do not extend beyond an arbitrary span/depth limit of 30.

The following examples illustrate the use of the charts.

EXAMPLE 4-8

Given: Using $F_y = 50$ ksi steel, determine the size of a "simple" framed girder with a span of 35 feet, which supports two equal concentrated loads. The factored loads produce a required moment of 440 kip-ft in the center 15-ft section between the loads. The load points are laterally braced.

Solution: For this loading condition, $C_b = 1.0$ due to nearly uniform moment across the central portion of the span.

Center section of 15 feet is longest unbraced length.

With total span equal to 35 feet and $M_n = 440$ kip-ft, assume approximate weight of beam at 70 lbs/ft (equal to 0.07 kips/ft).

$$\text{Total } M_u = 440 + \left[\frac{0.07 \times (35)^2}{8} \times 1.2 \right] = 453 \text{ kip-ft}$$

Entering chart, with unbraced length equal to 15 feet on the bottom scale (abscissa), proceed upward to meet the horizontal line corresponding to a design moment equal to 453 kip-ft on the left hand scale (ordinate). Any beam listed above and to the right of the point so located satisfies the design moment requirement. In this case, the lightest section satisfying this criterion is a W21×68, for which the total design moment with an unbraced length of 15 feet is 457 kip-ft.

Use: W21×68

Note: If depth is limited, a W14×82 could be selected, provided deflection conditions are not critical.

EXAMPLE 4-9

Given: A "fixed end" girder with a span of 60 feet supports a concentrated load at the center. The compression flange is laterally supported at the concentrated load point and at the inflection points. The factored load produces a maximum calculated moment of 440 kip-ft at the load point and the supports. Determine the size of the beam using $F_y = 50$ ksi steel.

Solution: For this loading condition, $C_b = 1.67$ (by comparison with Table 4-1), with an unbraced length of 15 feet. With the total span equal to 60 feet and $M_u = 440$ kip-ft, assume approximate weight of beam at 60 lbs/ft (0.06 kips/ft).

$$\text{Total } M_u = 440 + \left[\frac{0.06 \times (60)^2}{24} \times 1.2 \right]$$
$$= 451 \text{ kip-ft at the centerline and 462 at the supports}$$

Compute M_{equiv} by dividing the required design moment by C_b

$$M_{equiv} = 462 / 1.67 = 277 \text{ kip-ft}$$

Enter charts with unbraced length equal to 15 feet and proceed upward to 277 kip-ft. Any beam listed above and to the right of the point satisfies the design moment.

The lightest section satisfying the criteria of a design moment of 277 kip-ft at an unbraced length of 15 feet and $\phi_b M_p$ greater than 462 kip-ft is a W21×62. The design moment for a W21×62 with an unbraced length of 15 feet is 406 kip-ft and $\phi_b M_p$ is 540 kip-ft.

Since $(\phi_b M_n = 406 \text{ kip-ft}) > (M_{equiv} = 277 \text{ kip-ft})$ and $(\phi_b M_p = 540 \text{ kip-ft}) > (M_u = 462 \text{ kip-ft})$, a W21×62 is **o.k.**

A 21-in. nominal depth beam spanning 60 feet should be checked for deflection since the span/depth ratio exceeds 30.

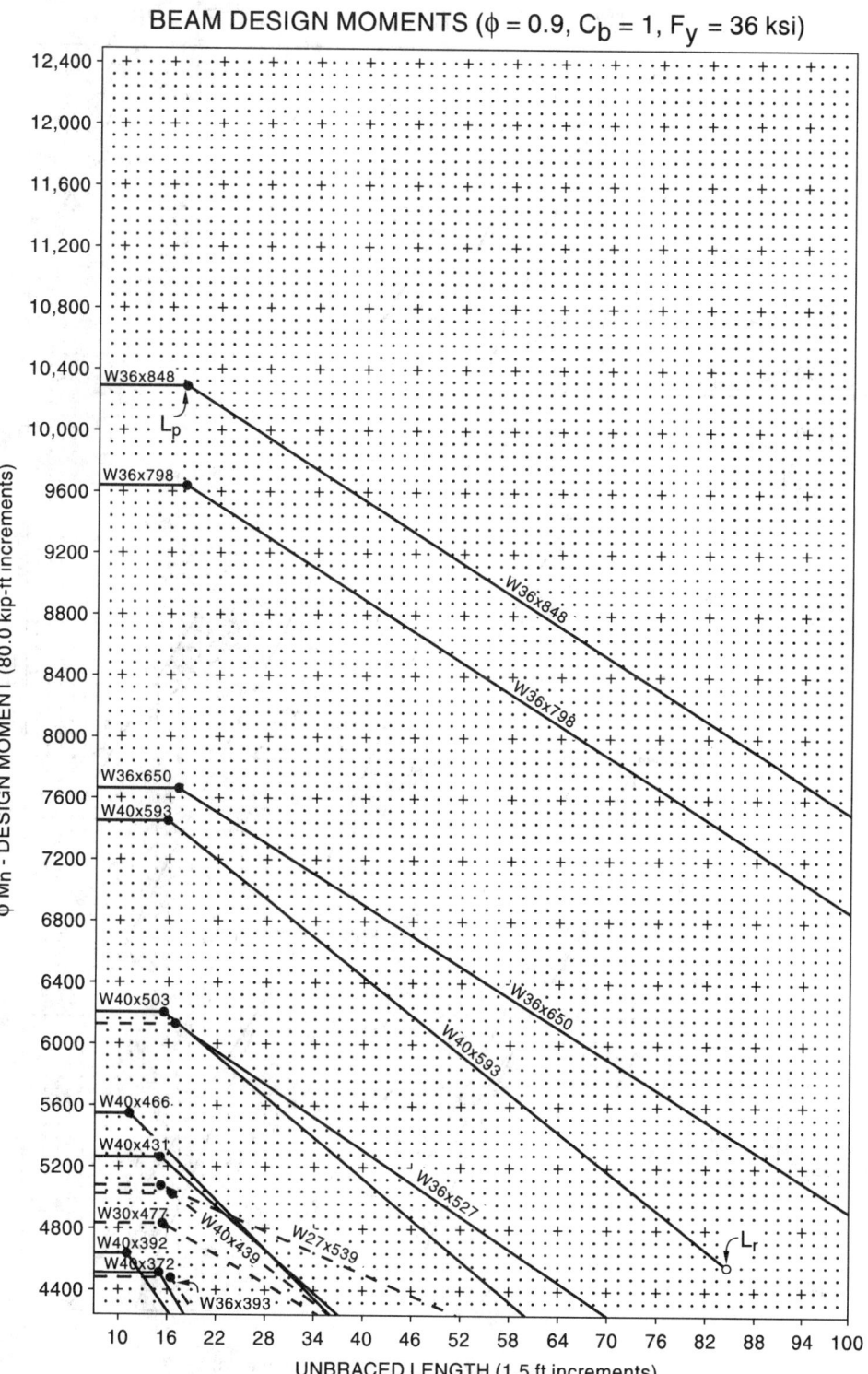

BEAM DESIGN MOMENTS (ϕ = 0.9, C_b = 1, F_y = 36 ksi)

ϕM_n - DESIGN MOMENT (80.0 kip-ft increments)

UNBRACED LENGTH (1.5 ft increments)

BEAM DESIGN MOMENTS ($\phi = 0.9$, $C_b = 1$, $F_y = 36$ ksi)

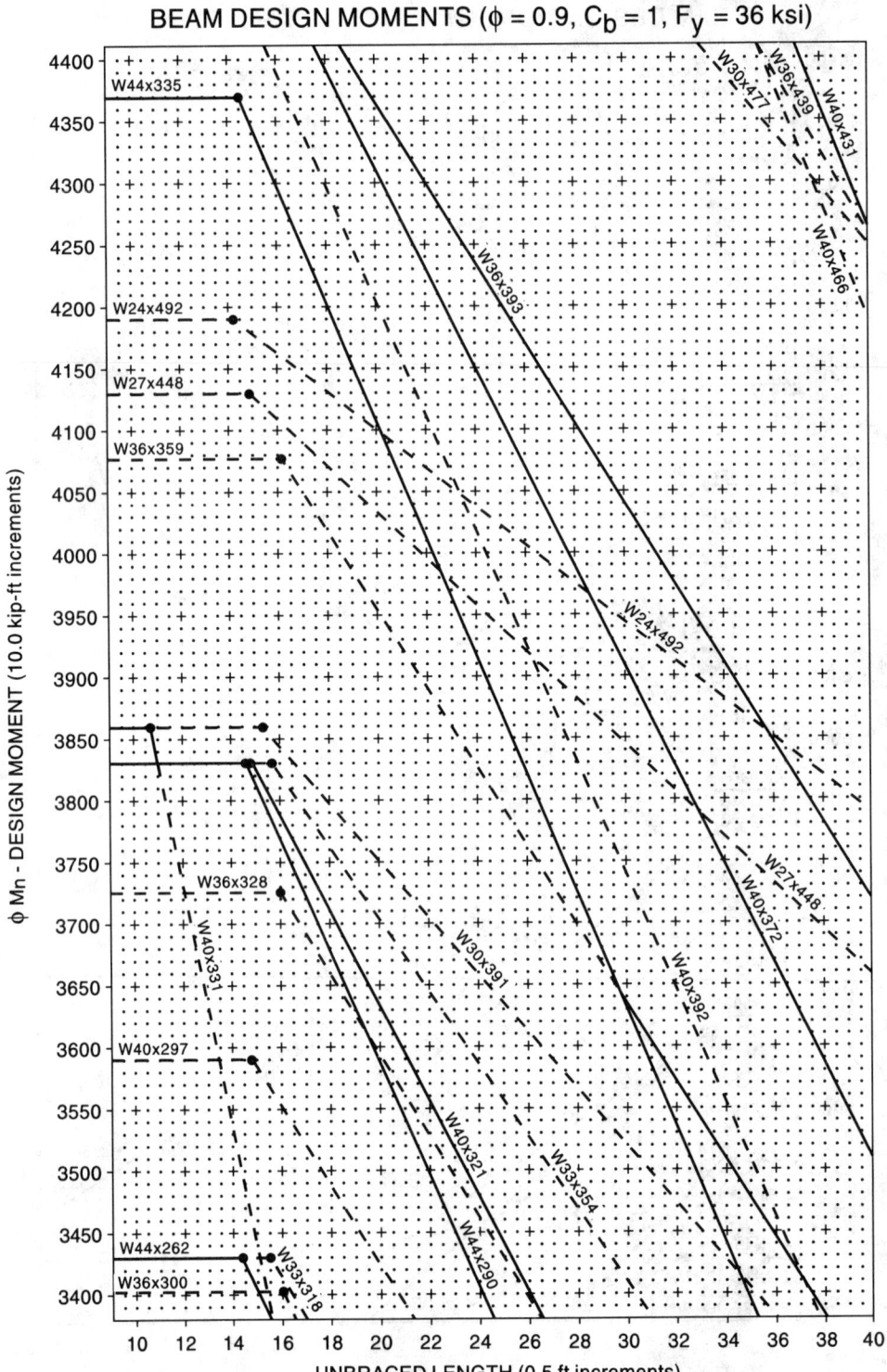

ϕM_n - DESIGN MOMENT (10.0 kip-ft increments)

UNBRACED LENGTH (0.5 ft increments)

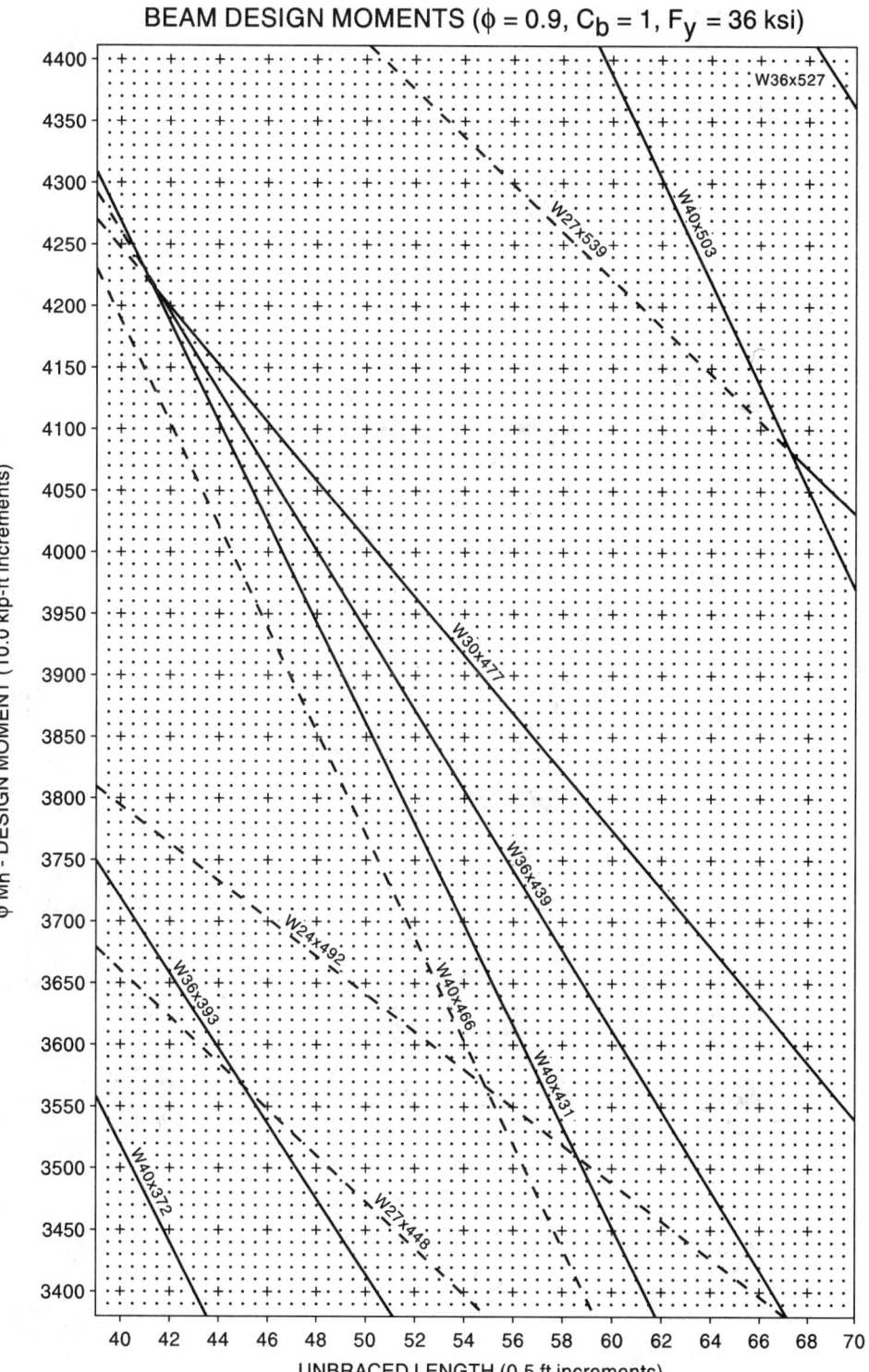

BEAM DESIGN MOMENTS ($\phi = 0.9$, $C_b = 1$, $F_y = 36$ ksi)

BEAM DESIGN MOMENTS ($\phi = 0.9$, $C_b = 1$, $F_y = 36$ ksi)

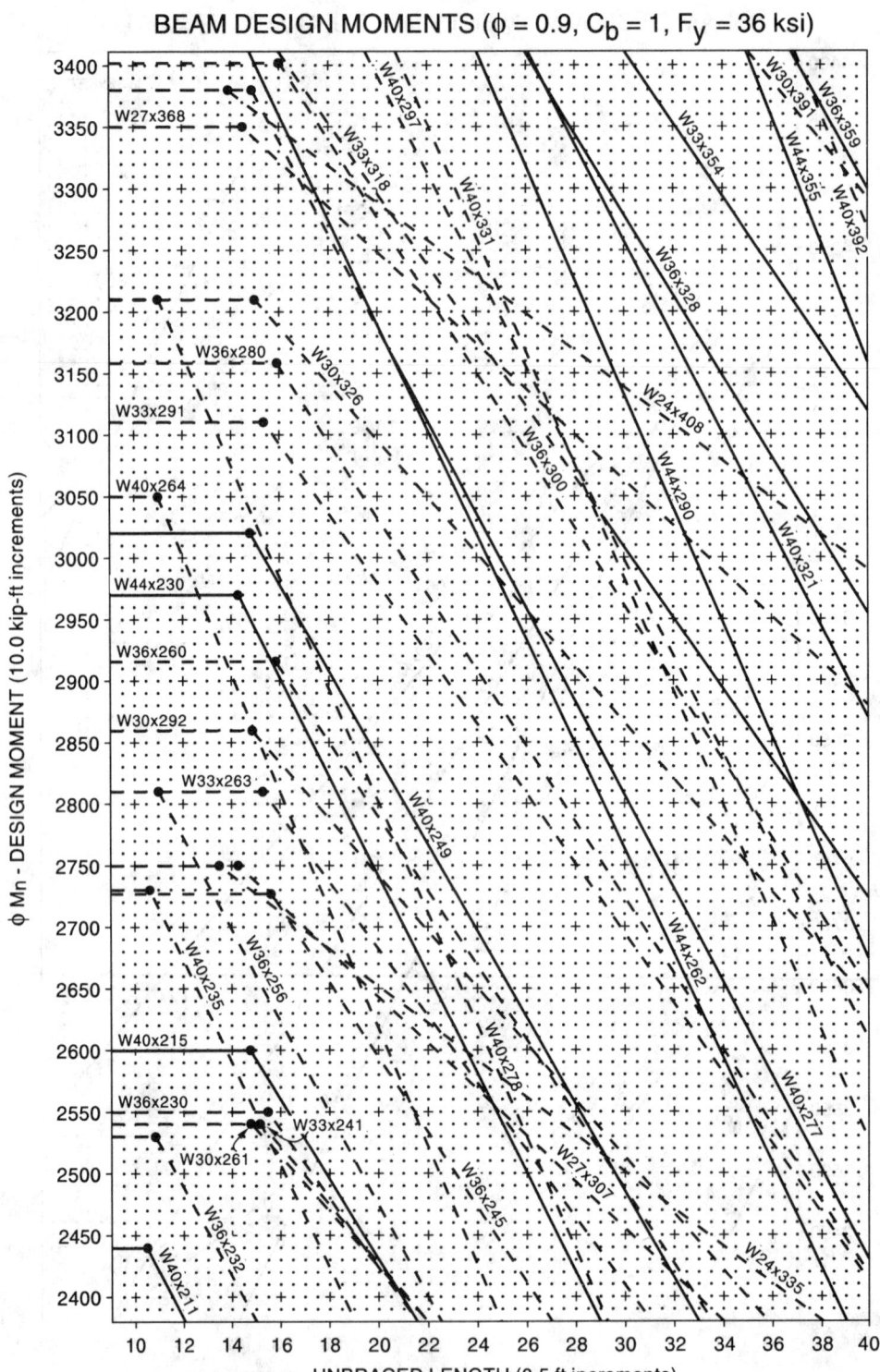

ϕM_n - DESIGN MOMENT (10.0 kip-ft increments)

UNBRACED LENGTH (0.5 ft increments)

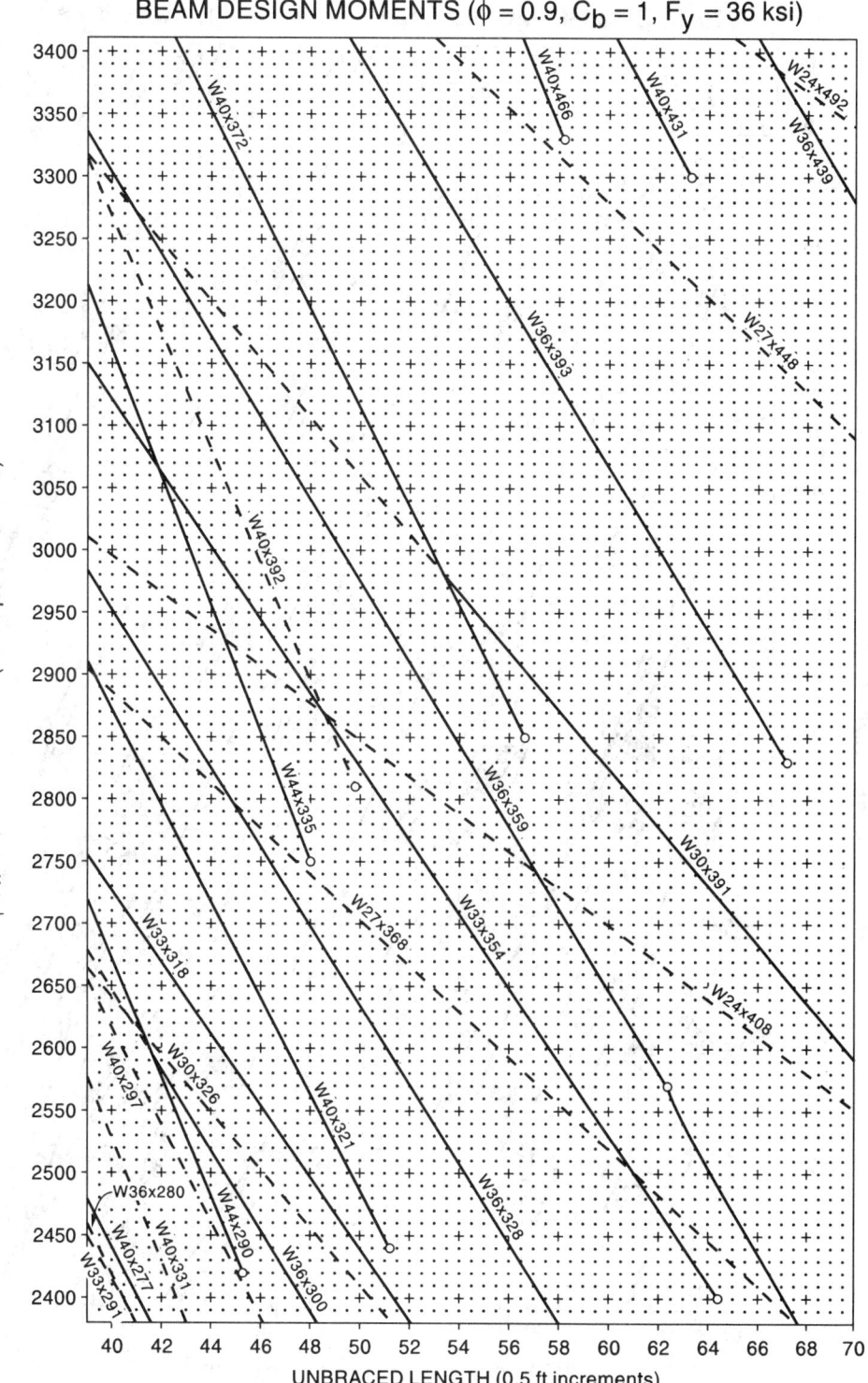

BEAM DESIGN MOMENTS (ϕ = 0.9, C_b = 1, F_y = 36 ksi)

BEAM DESIGN MOMENTS ($\phi = 0.9$, $C_b = 1$, $F_y = 36$ ksi)

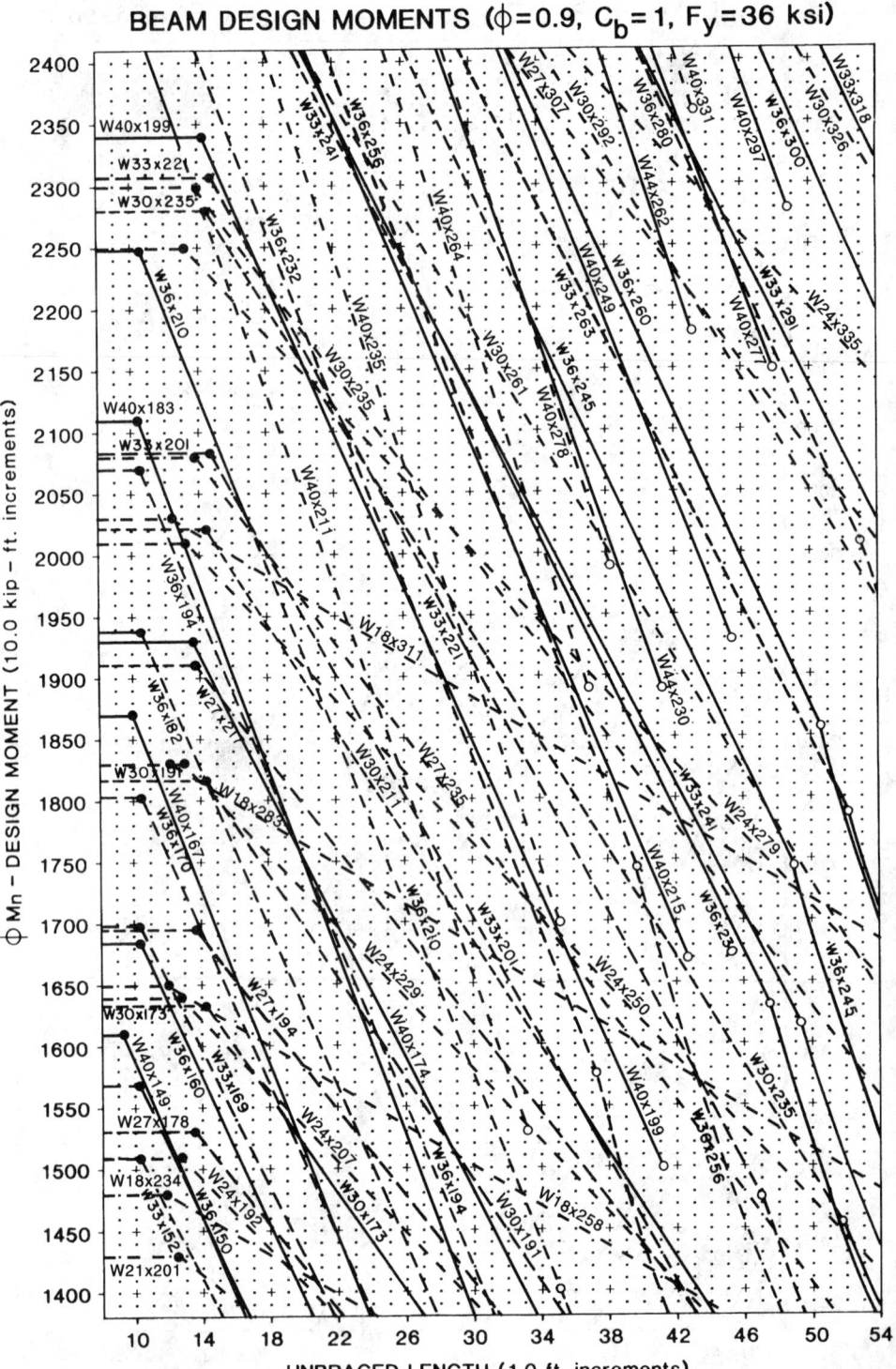

ϕM_n – DESIGN MOMENT (10.0 kip – ft. increments)

UNBRACED LENGTH (1.0 ft. increments)

BEAM DESIGN MOMENTS ($\phi = 0.9$, $C_b = 1$, $F_y = 36$ ksi)

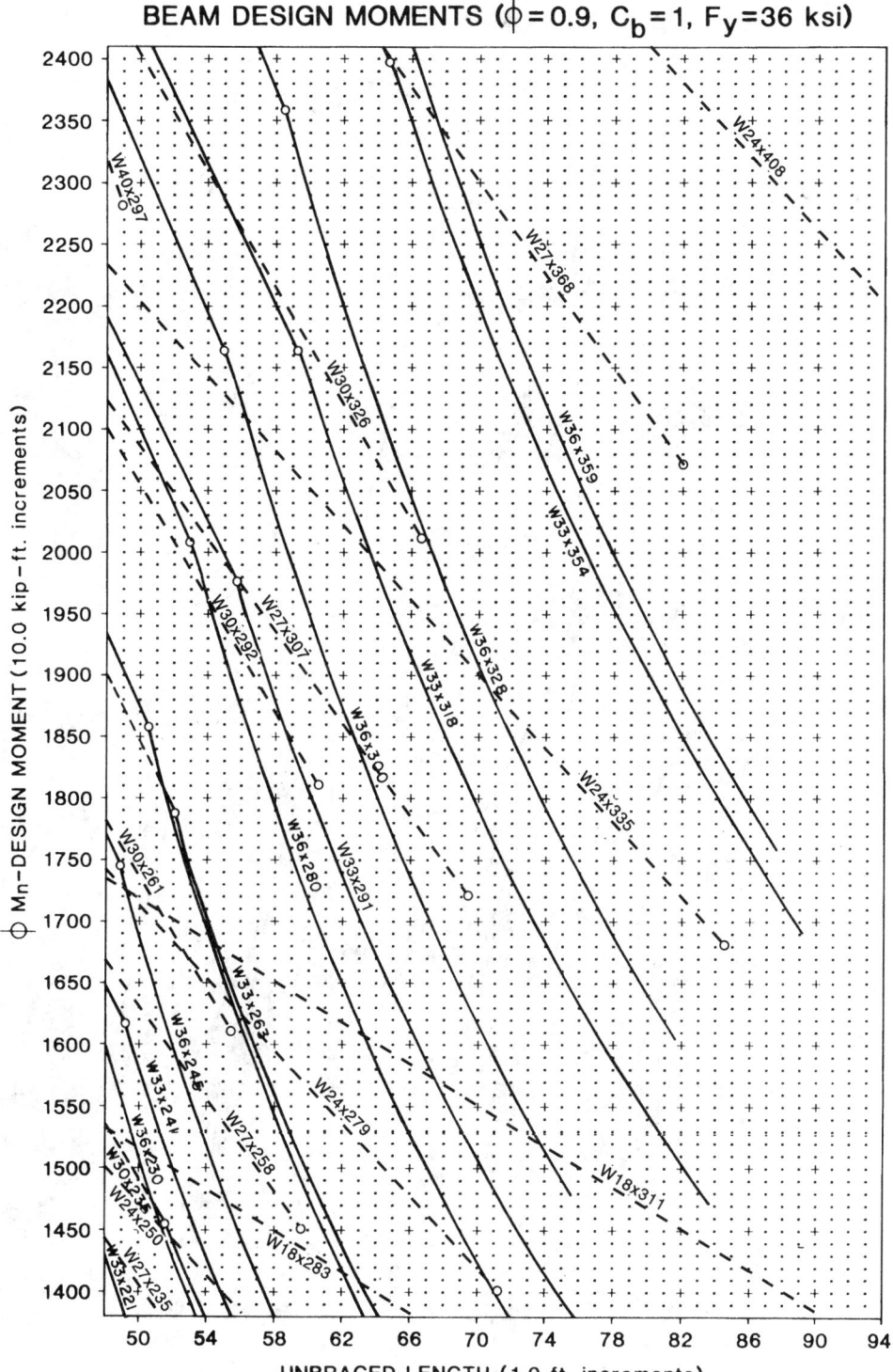

ϕM_n–DESIGN MOMENT (10.0 kip–ft. increments)

UNBRACED LENGTH (1.0 ft. increments)

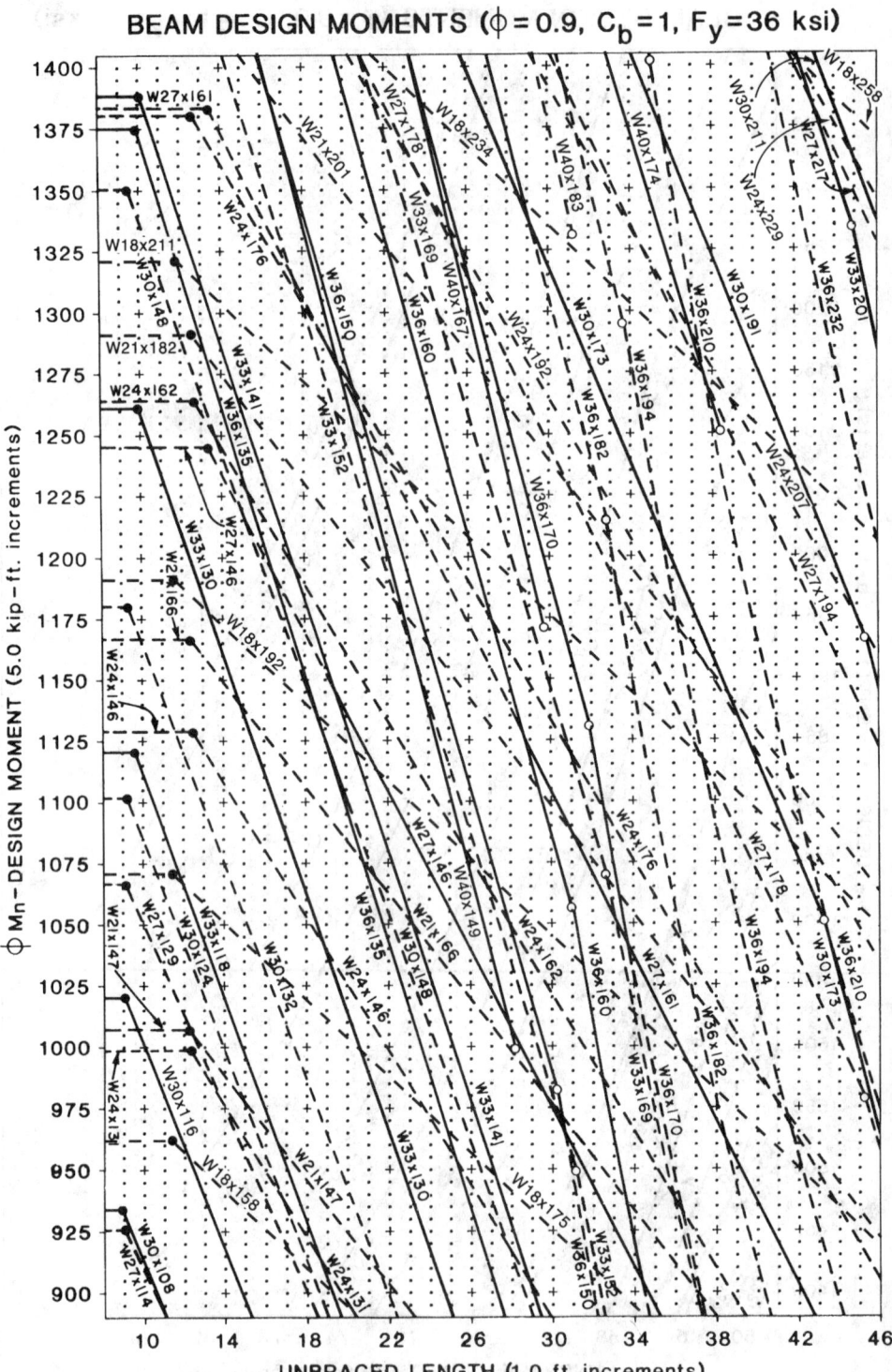

BEAM DESIGN MOMENTS ($\phi = 0.9$, $C_b = 1$, $F_y = 36$ ksi)

ϕM_n – DESIGN MOMENT (5.0 kip – ft. increments)

UNBRACED LENGTH (1.0 ft. increments)

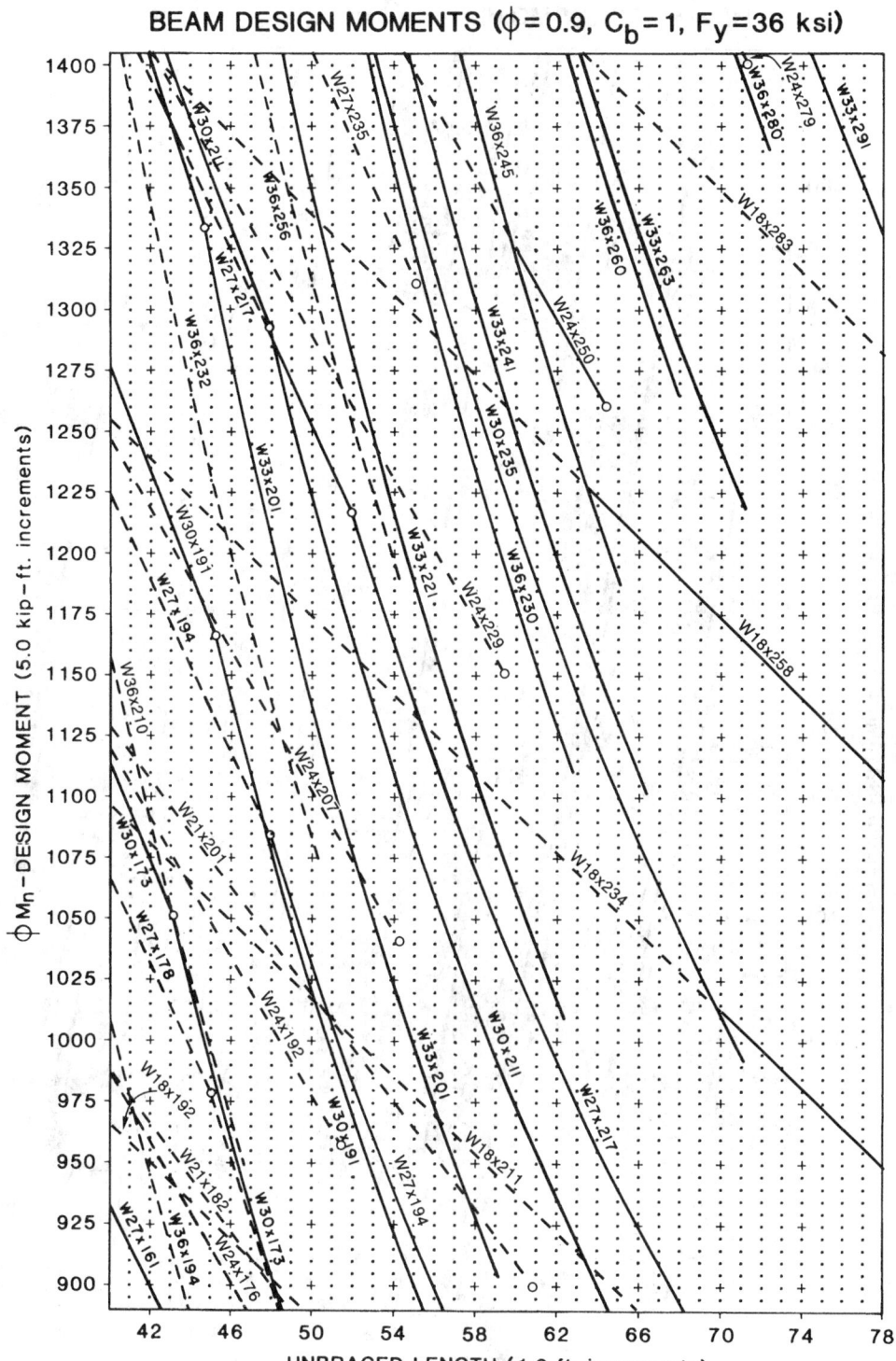

BEAM DESIGN MOMENTS ($\phi = 0.9$, $C_b = 1$, $F_y = 36$ ksi)

ϕM_n—DESIGN MOMENT (5.0 kip-ft. increments)

UNBRACED LENGTH (1.0 ft. increments)

BEAM DESIGN MOMENTS ($\phi = 0.9$, $C_b = 1$, $F_y = 36$ ksi)

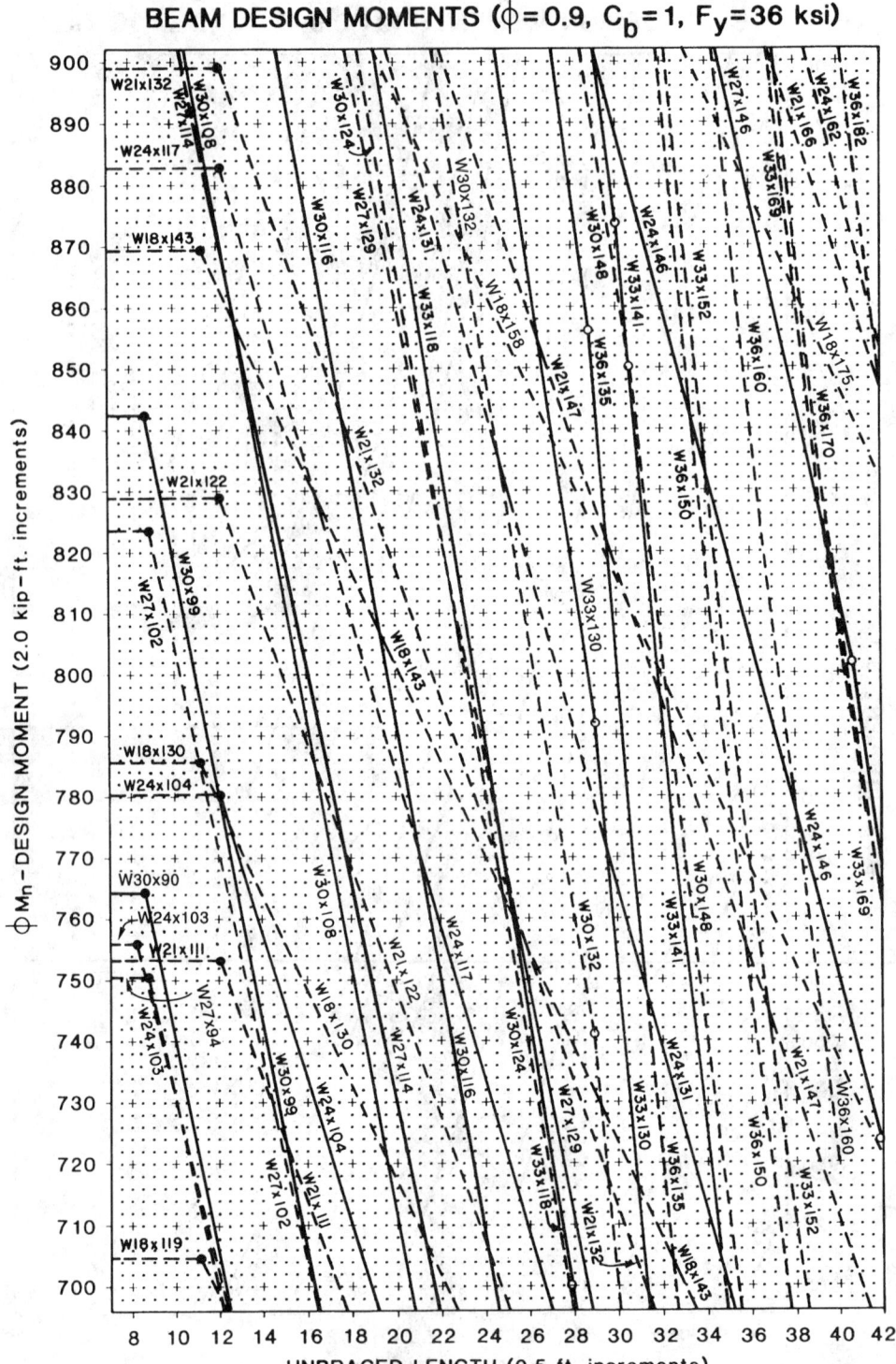

ϕM_n – DESIGN MOMENT (2.0 kip-ft. increments)

UNBRACED LENGTH (0.5 ft. increments)

BEAM DESIGN MOMENTS ($\phi = 0.9$, $C_b = 1$, $F_y = 36$ ksi)

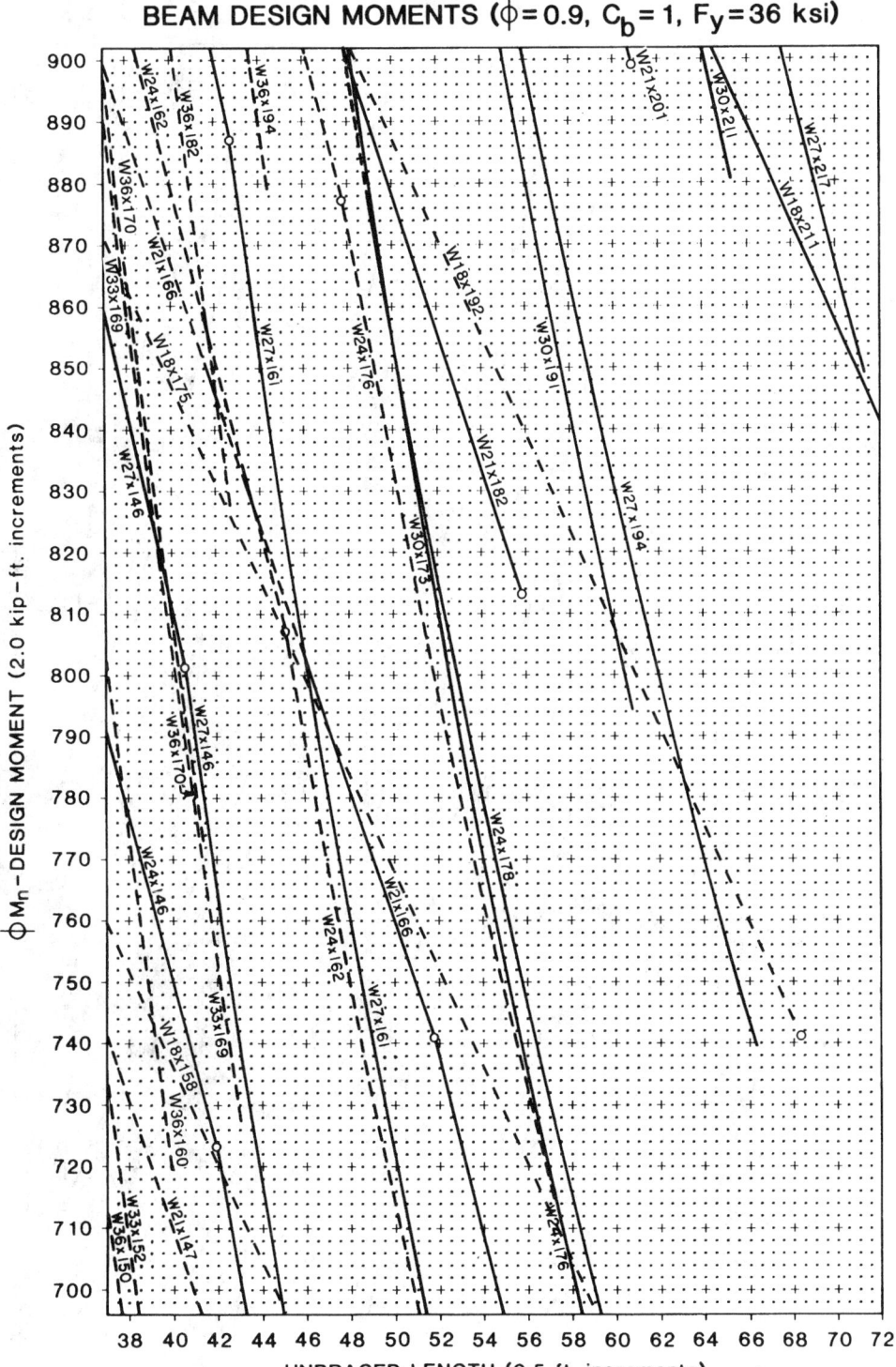

BEAM DESIGN MOMENTS (ϕ=0.9, C_b=1, F_y=36 ksi)

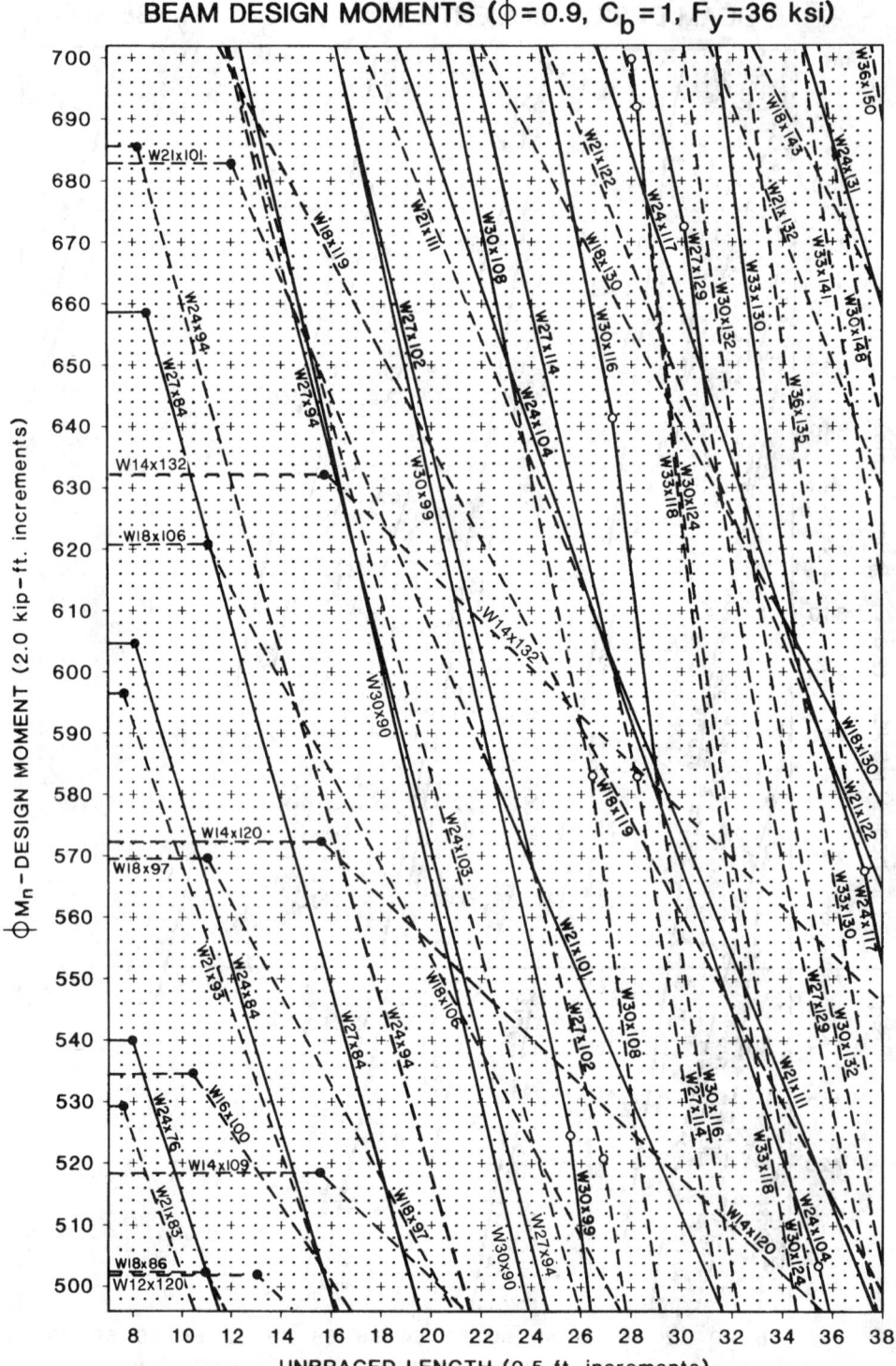

BEAM DESIGN MOMENTS ($\phi = 0.9$, $C_b = 1$, $F_y = 36$ ksi)

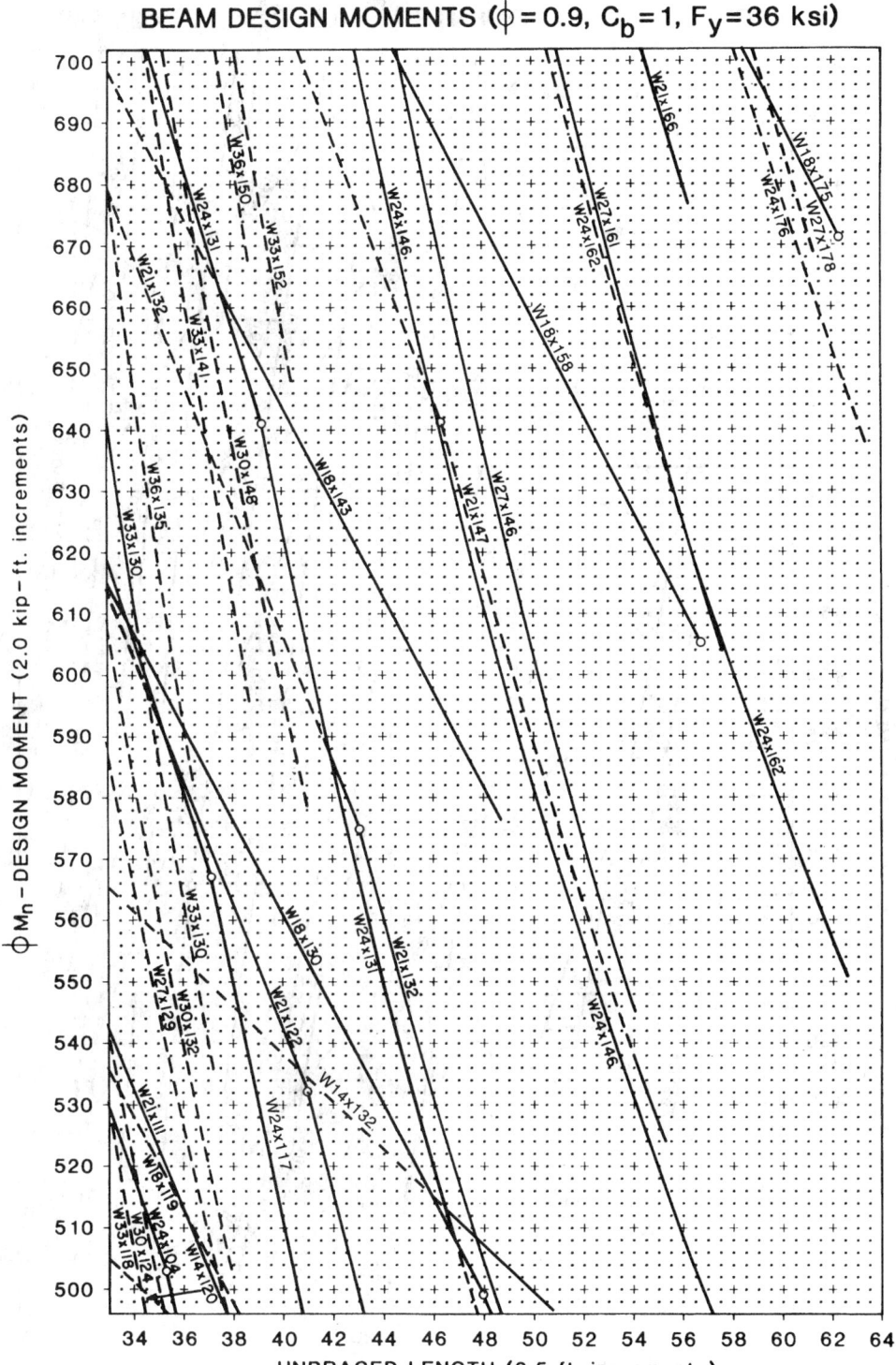

ϕM_n – DESIGN MOMENT (2.0 kip – ft. increments)

UNBRACED LENGTH (0.5 ft. increments)

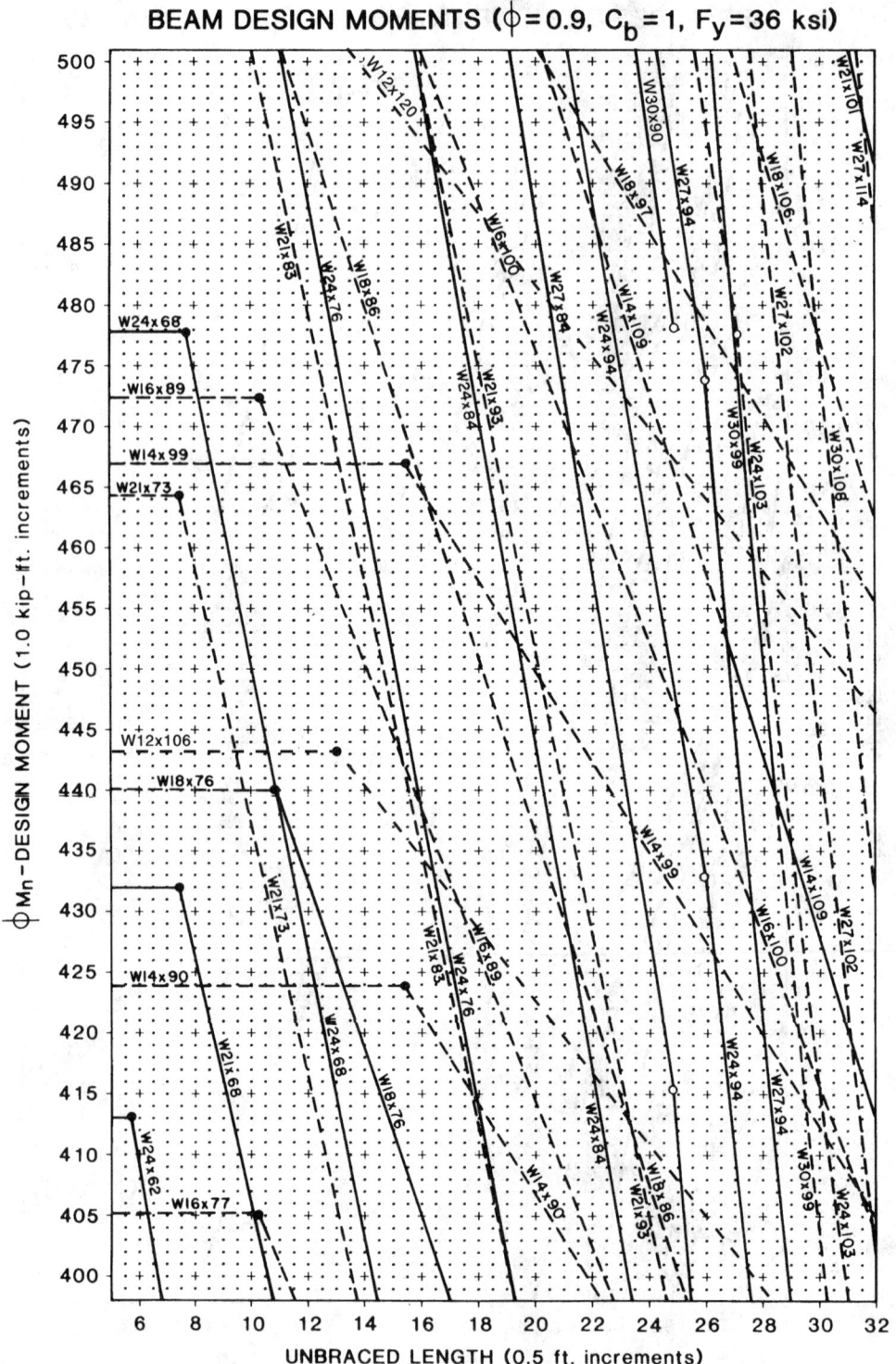

BEAM DESIGN MOMENTS (ϕ = 0.9, C_b = 1, F_y = 36 ksi)

ϕM_n – DESIGN MOMENT (1.0 kip–ft. increments)

UNBRACED LENGTH (0.5 ft. increments)

BEAM DESIGN MOMENTS ($\phi = 0.9$, $C_b = 1$, $F_y = 36$ ksi)

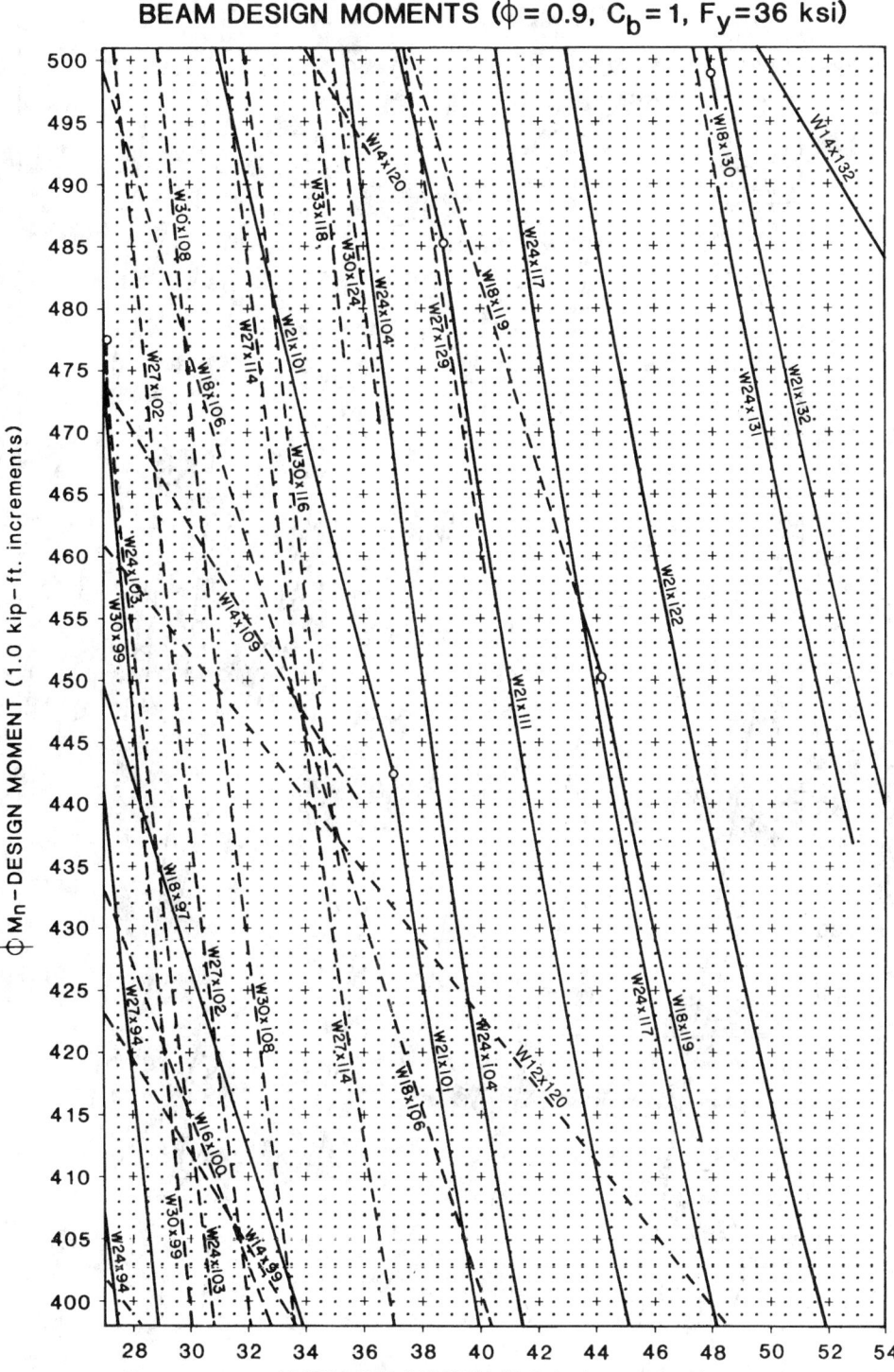

ϕM_n – DESIGN MOMENT (1.0 kip – ft. increments)

UNBRACED LENGTH (0.5 ft. increments)

BEAM DESIGN MOMENTS ($\phi = 0.9$, $C_b = 1$, $F_y = 36$ ksi)

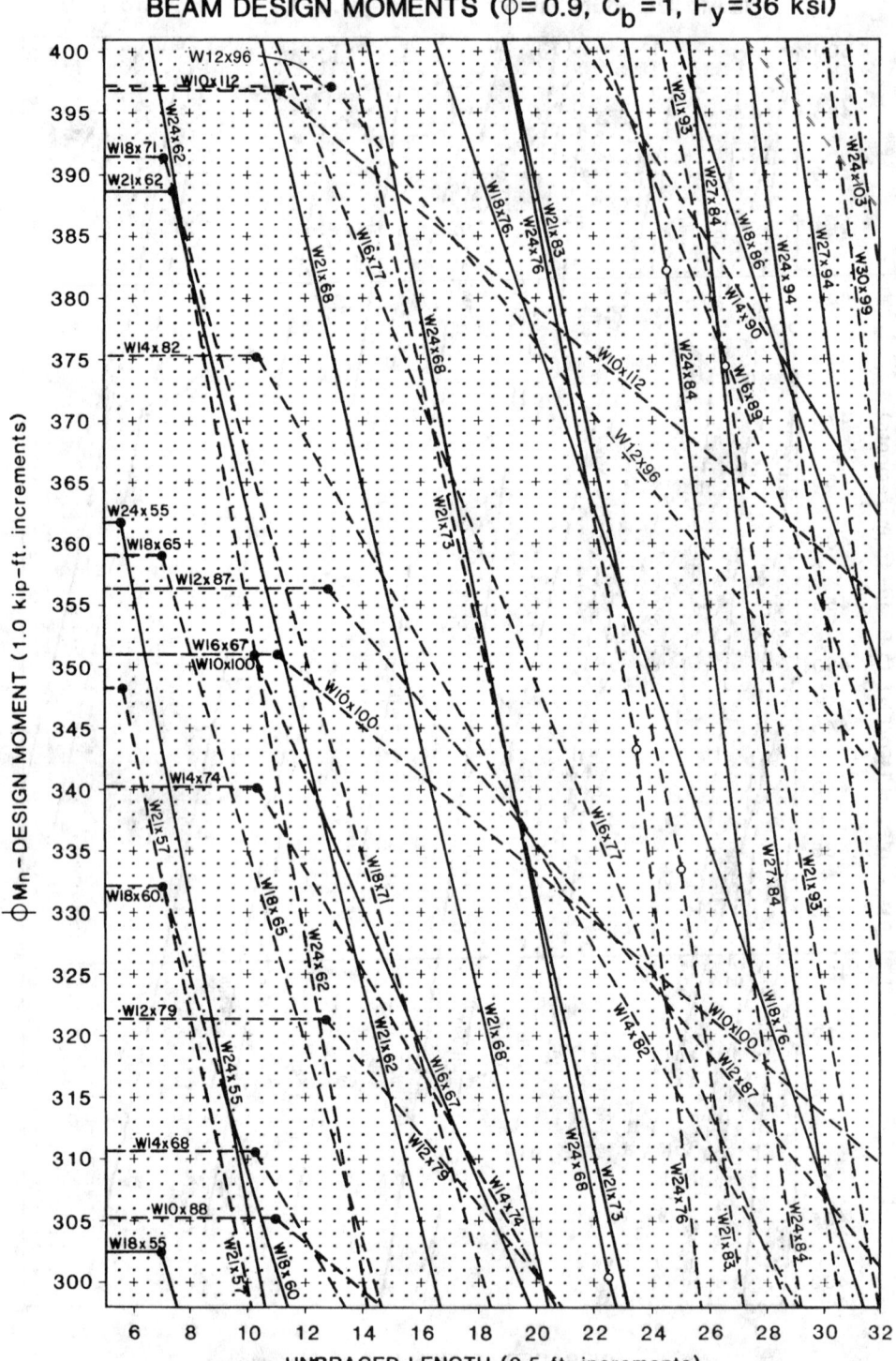

ϕM_n – DESIGN MOMENT (1.0 kip-ft. increments)

UNBRACED LENGTH (0.5 ft. increments)

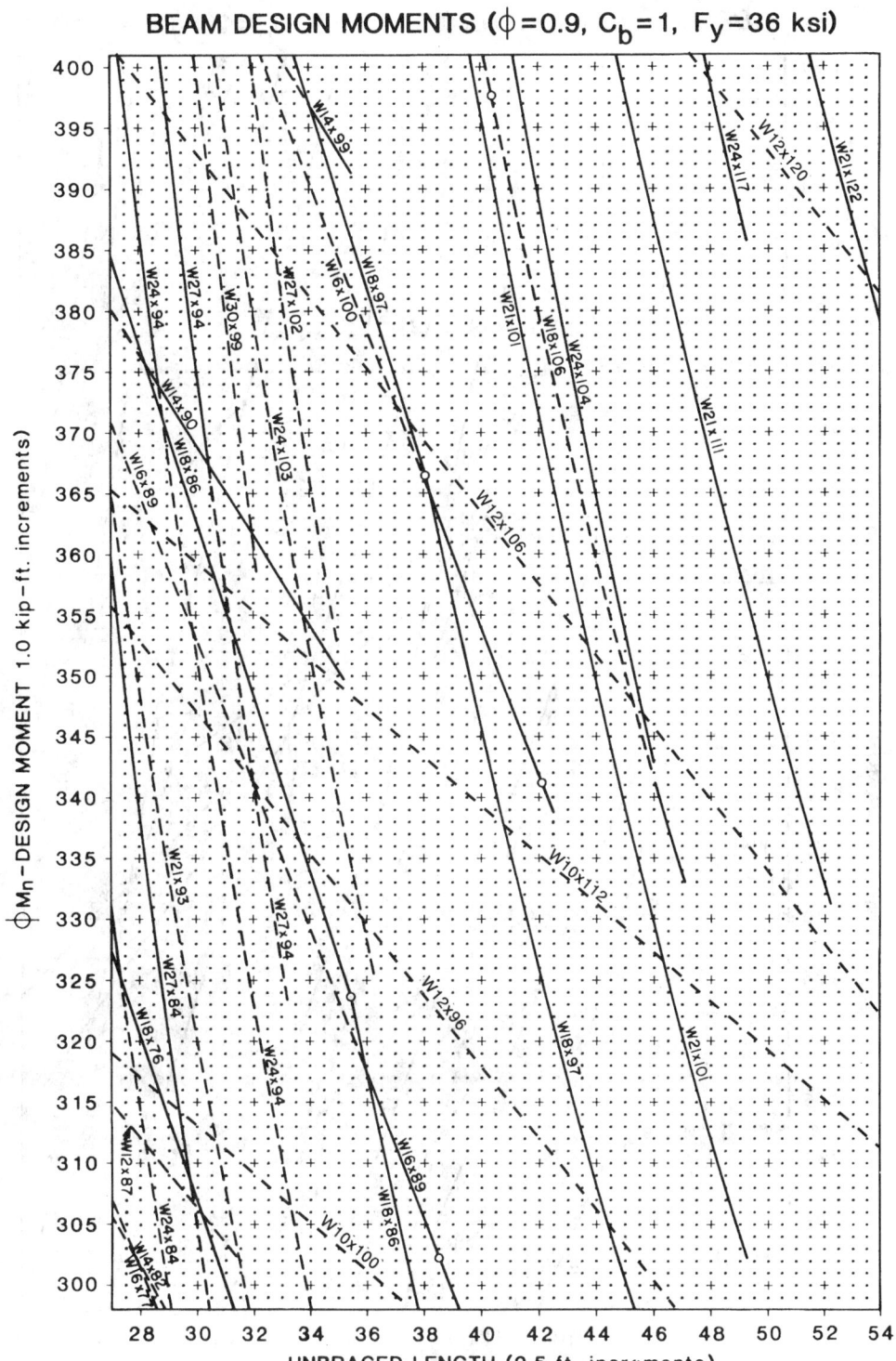

BEAM DESIGN MOMENTS (ϕ=0.9, C_b=1, F_y=36 ksi)

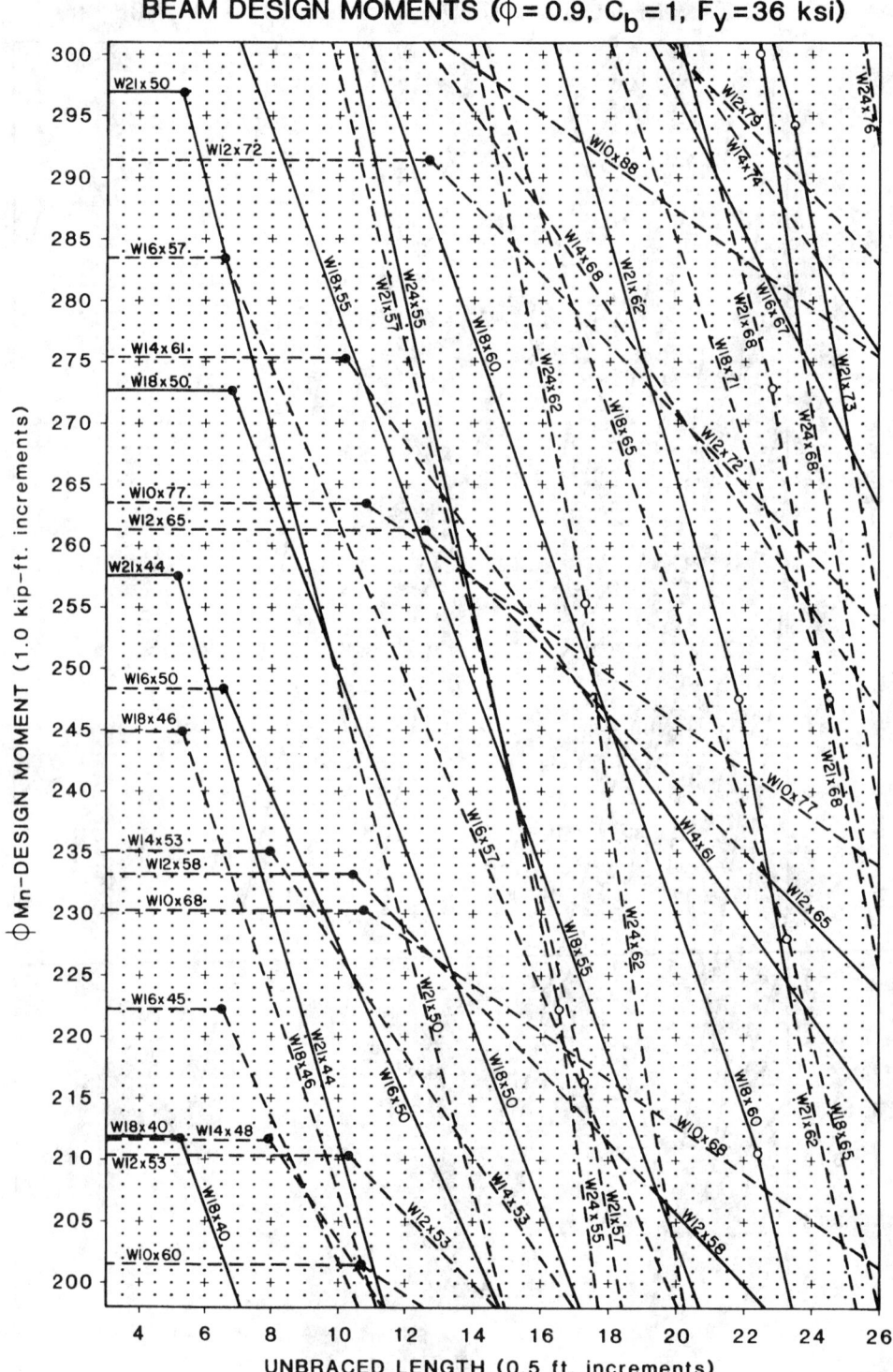

BEAM DESIGN MOMENTS ($\phi = 0.9$, $C_b = 1$, $F_y = 36$ ksi)

ϕM_n–DESIGN MOMENT (1.0 kip-ft. increments)

UNBRACED LENGTH (0.5 ft. increments)

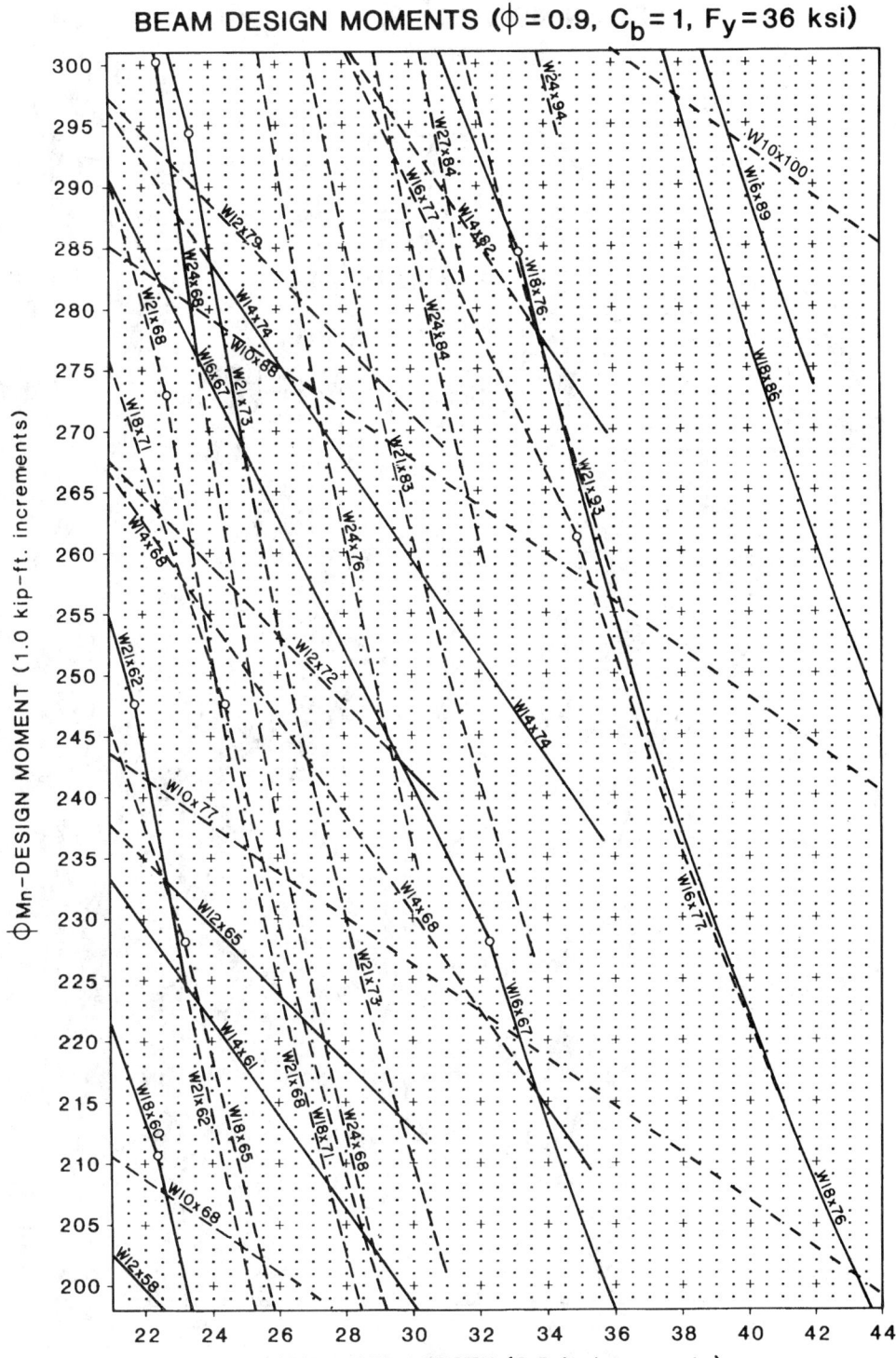

BEAM DESIGN MOMENTS ($\phi = 0.9$, $C_b = 1$, $F_y = 36$ ksi)

ϕM_n – DESIGN MOMENT (1.0 kip-ft. increments)

UNBRACED LENGTH (0.5 ft. increments)

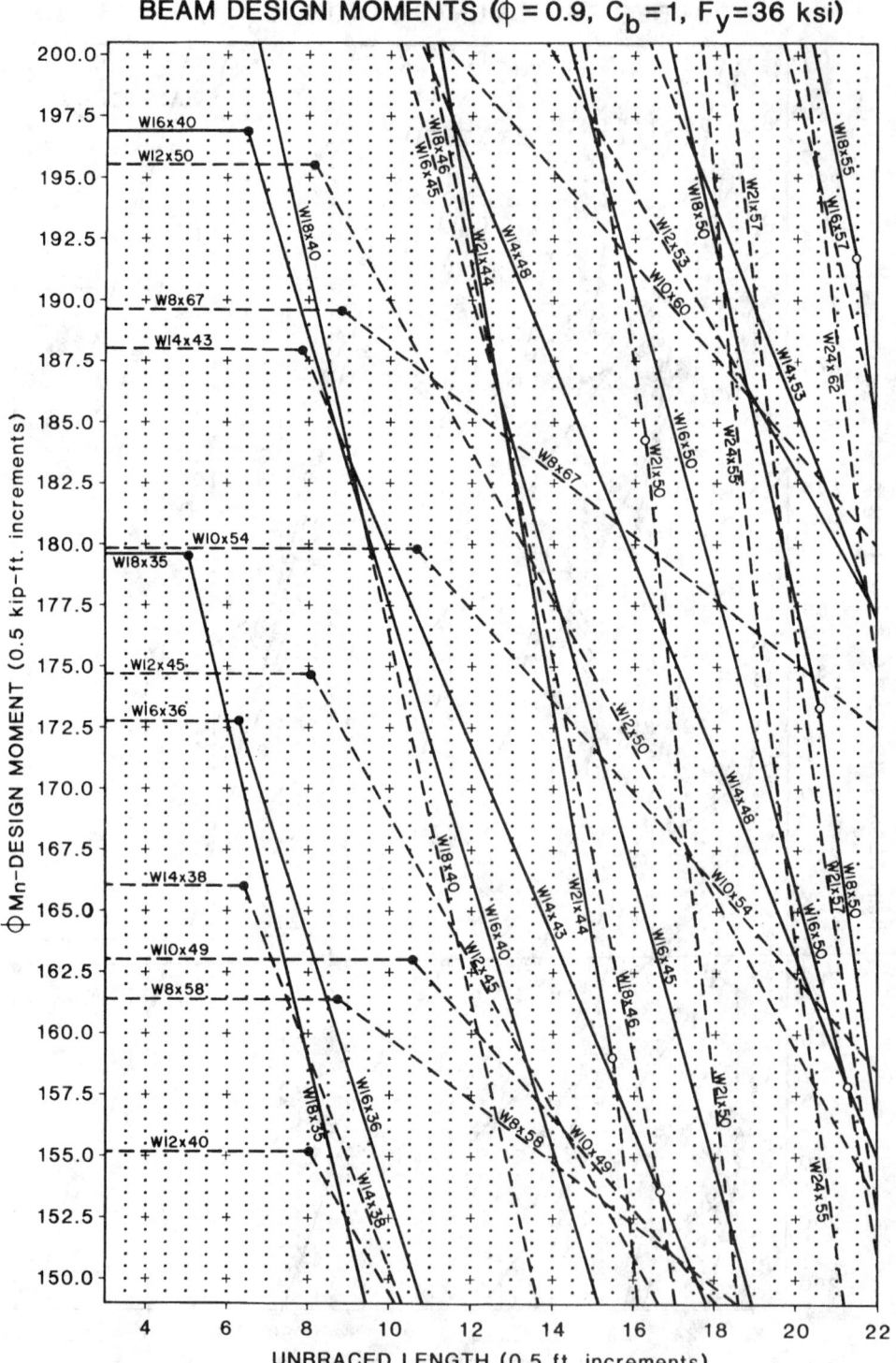

BEAM DESIGN MOMENTS (ϕ = 0.9, C_b = 1, F_y = 36 ksi)

ϕM_n –DESIGN MOMENT (0.5 kip-ft. increments)

UNBRACED LENGTH (0.5 ft. increments)

BEAM DESIGN MOMENTS (ϕ = 0.9, C_b = 1, F_y = 36 ksi)

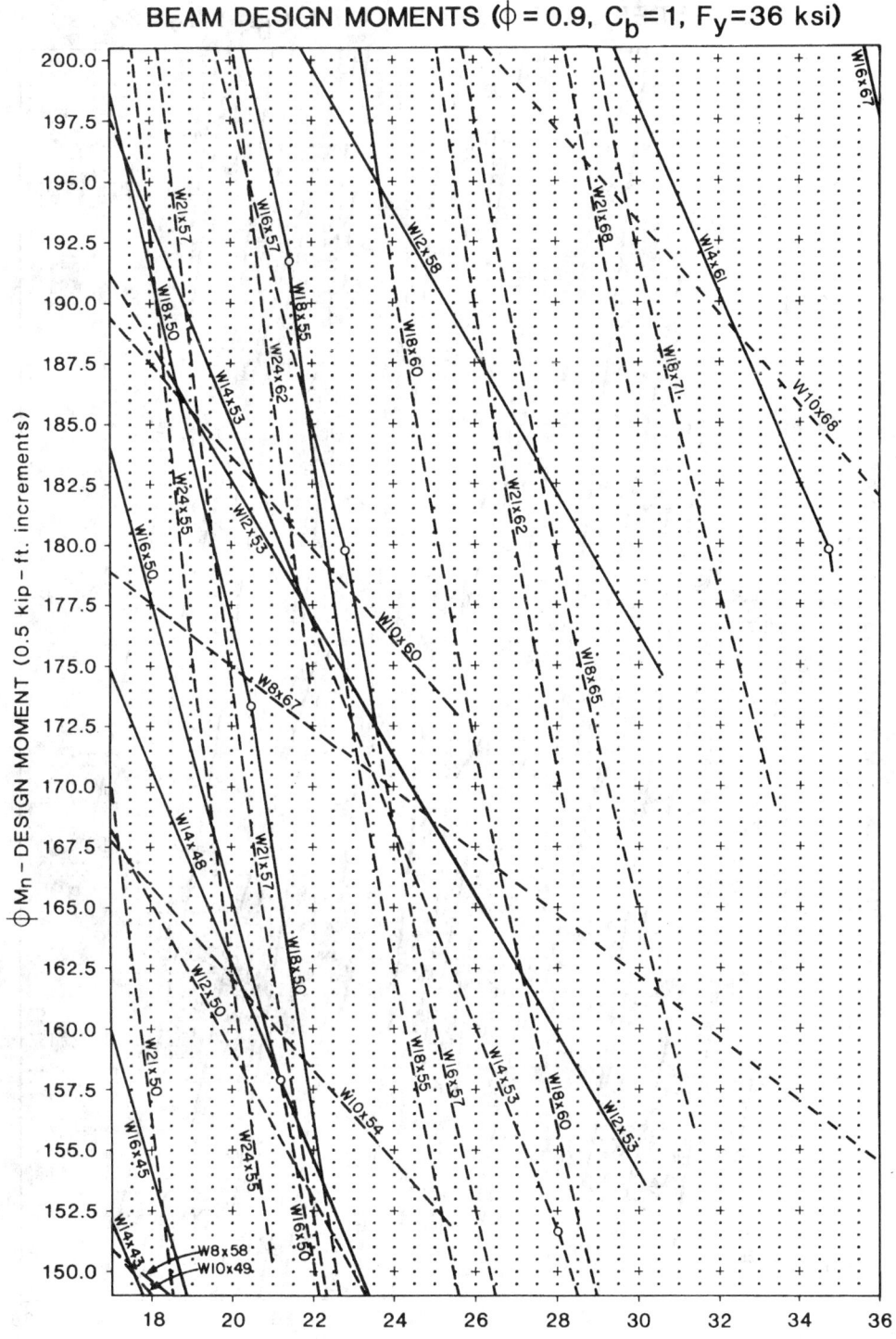

ϕM_n – DESIGN MOMENT (0.5 kip – ft. increments)

UNBRACED LENGTH (0.5 ft. increments)

BEAM DESIGN MOMENTS ($\phi = 0.9$, $C_b = 1$, $F_y = 36$ ksi)

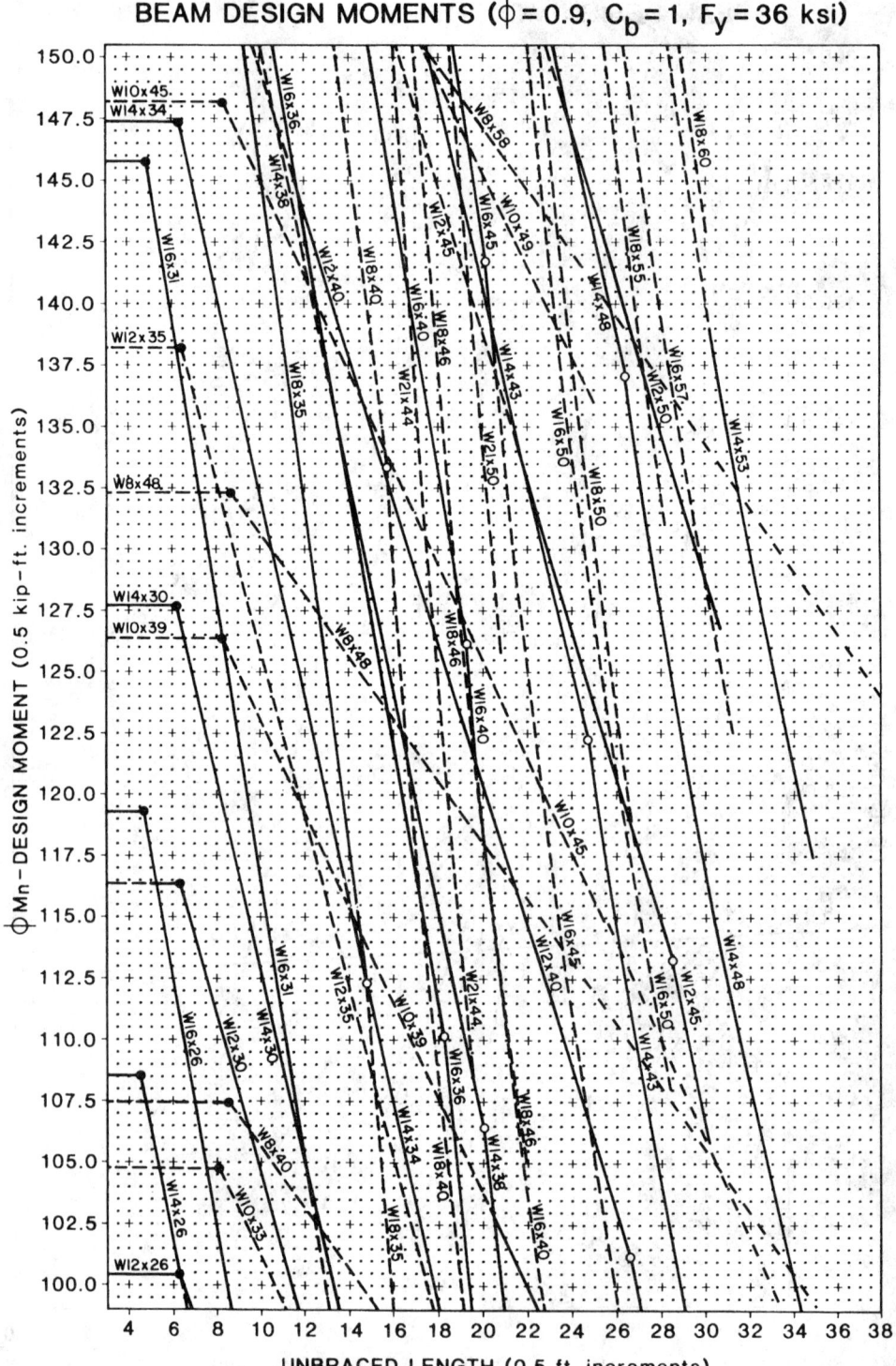

ϕM_n—DESIGN MOMENT (0.5 kip–ft. increments)

UNBRACED LENGTH (0.5 ft. increments)

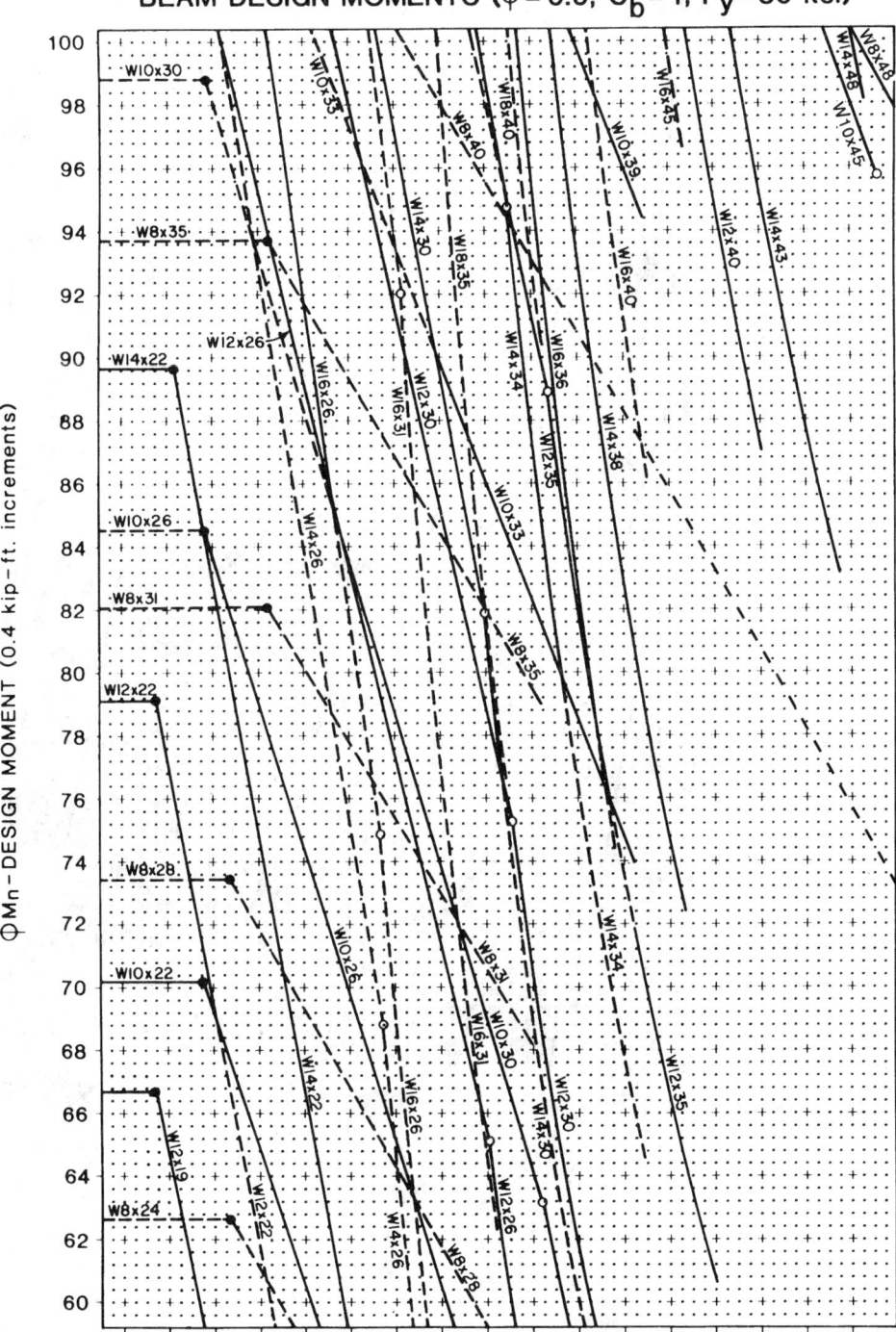

BEAM DESIGN MOMENTS (ϕ = 0.9, C_b = 1, F_y = 36 ksi)

ϕM_n - DESIGN MOMENT (0.4 kip-ft. increments)

UNBRACED LENGTH (0.5 ft. increments)

BEAM DESIGN MOMENTS ($\phi = 0.9$, $C_b = 1$, $F_y = 36$ ksi)

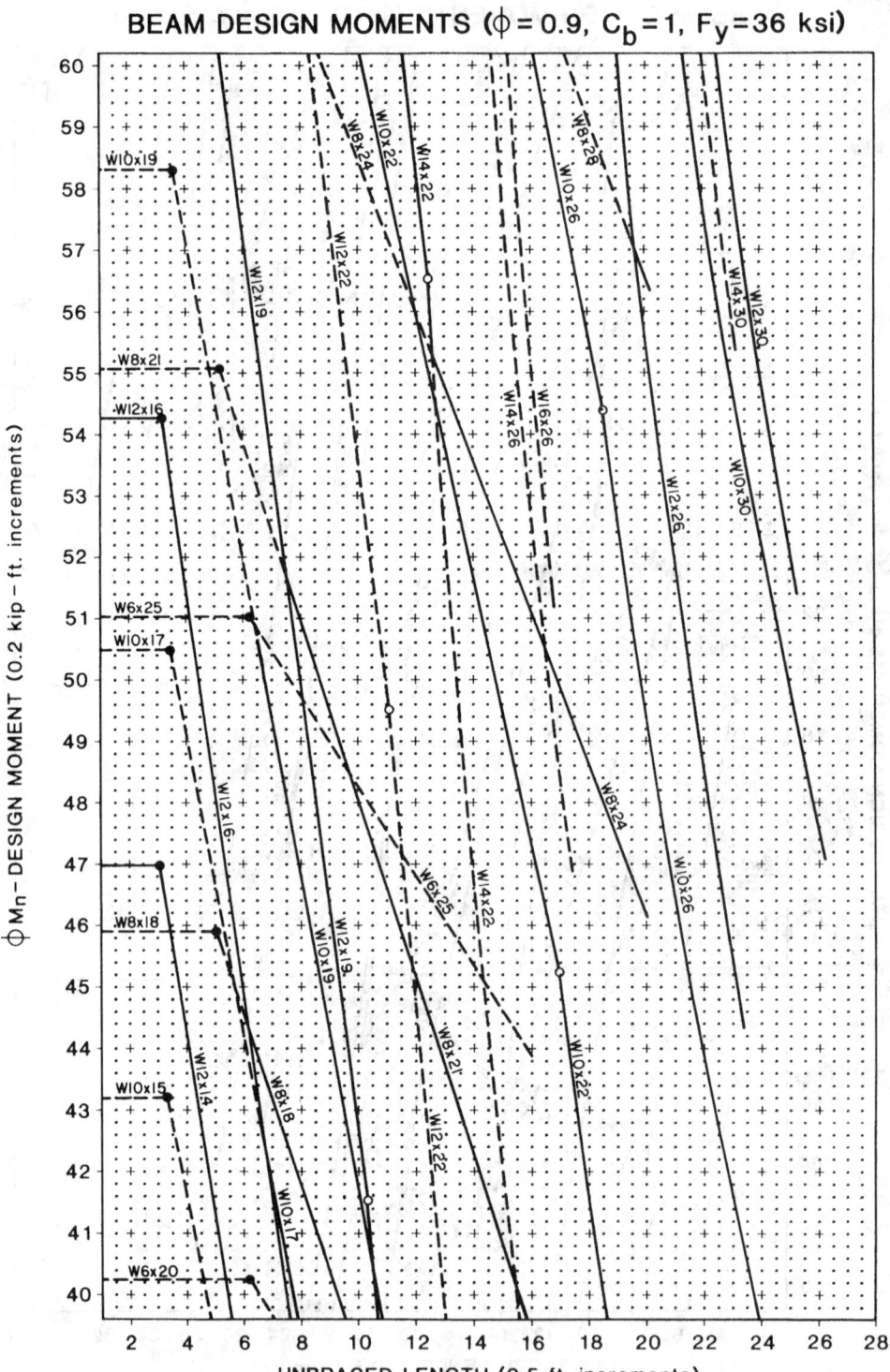

ϕM_n – DESIGN MOMENT (0.2 kip – ft. increments)

UNBRACED LENGTH (0.5 ft. increments)

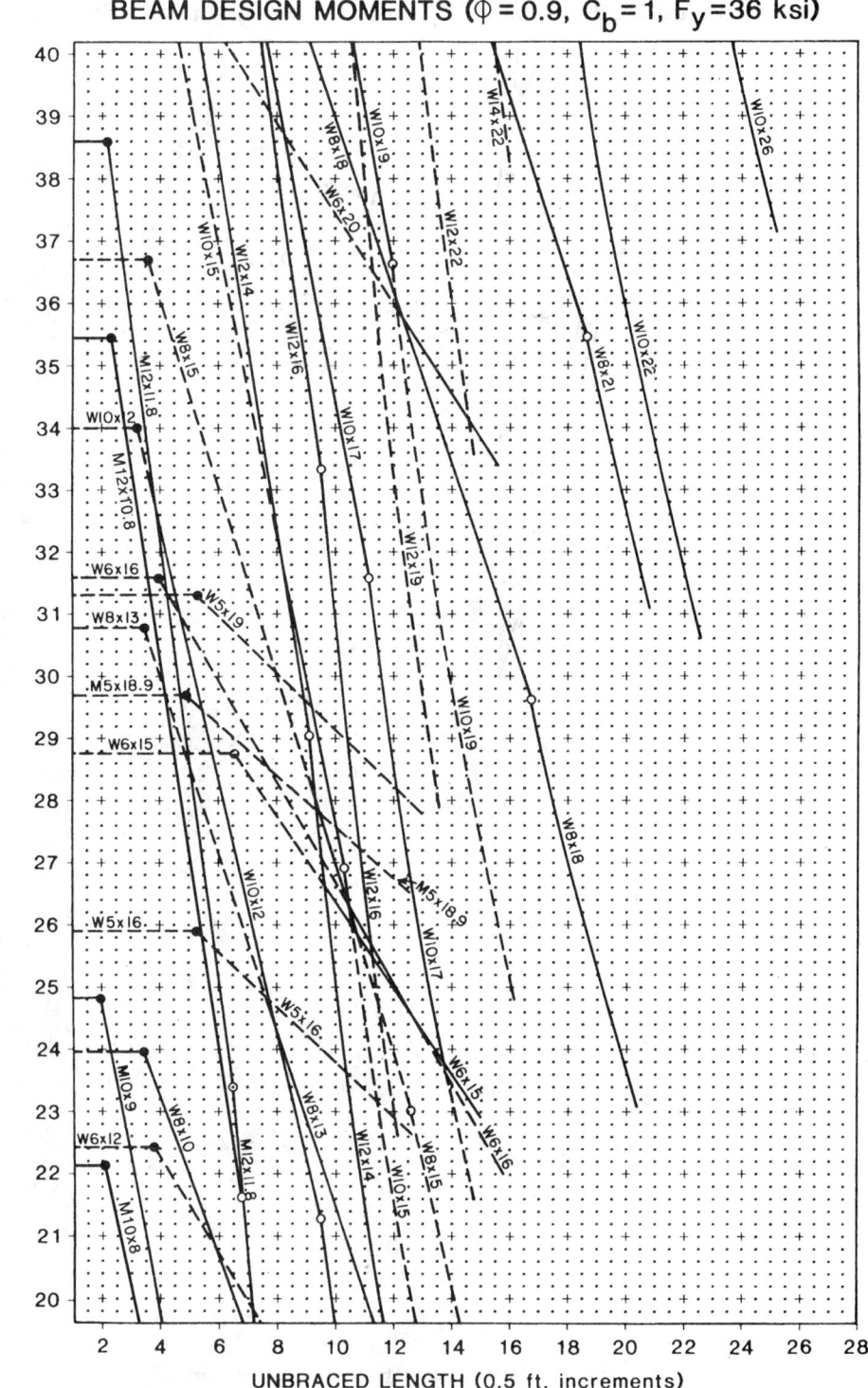

BEAM DESIGN MOMENTS ($\phi = 0.9$, $C_b = 1$, $F_y = 36$ ksi)

BEAM DESIGN MOMENTS ($\phi = 0.9$, $C_b = 1$, $F_y = 36$ ksi)

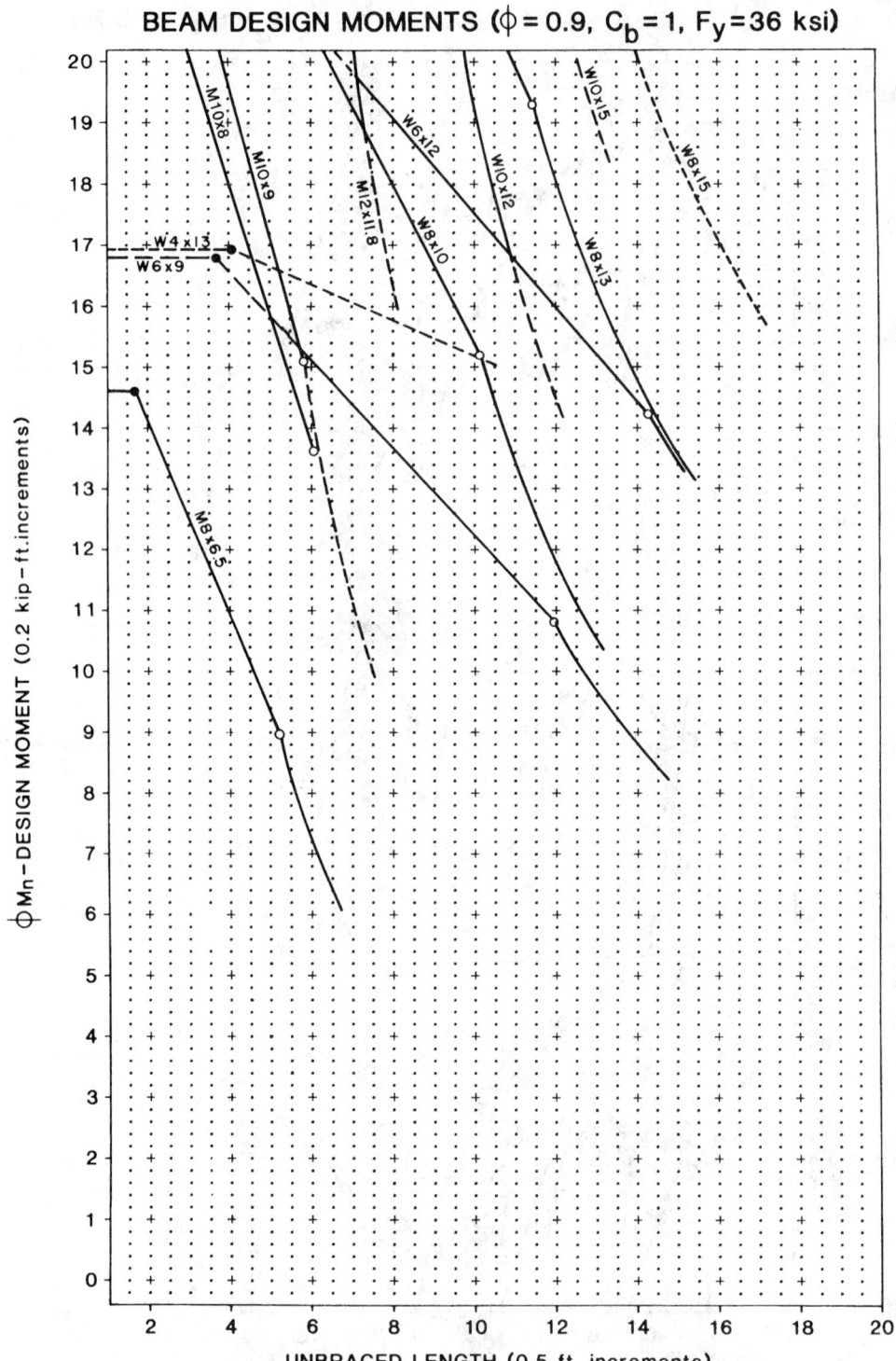

ϕM_n – DESIGN MOMENT (0.2 kip – ft. increments)

UNBRACED LENGTH (0.5 ft. increments)

BEAM DESIGN MOMENTS ($\phi = 0.9$, $C_b = 1$, $F_y = 50$ ksi)

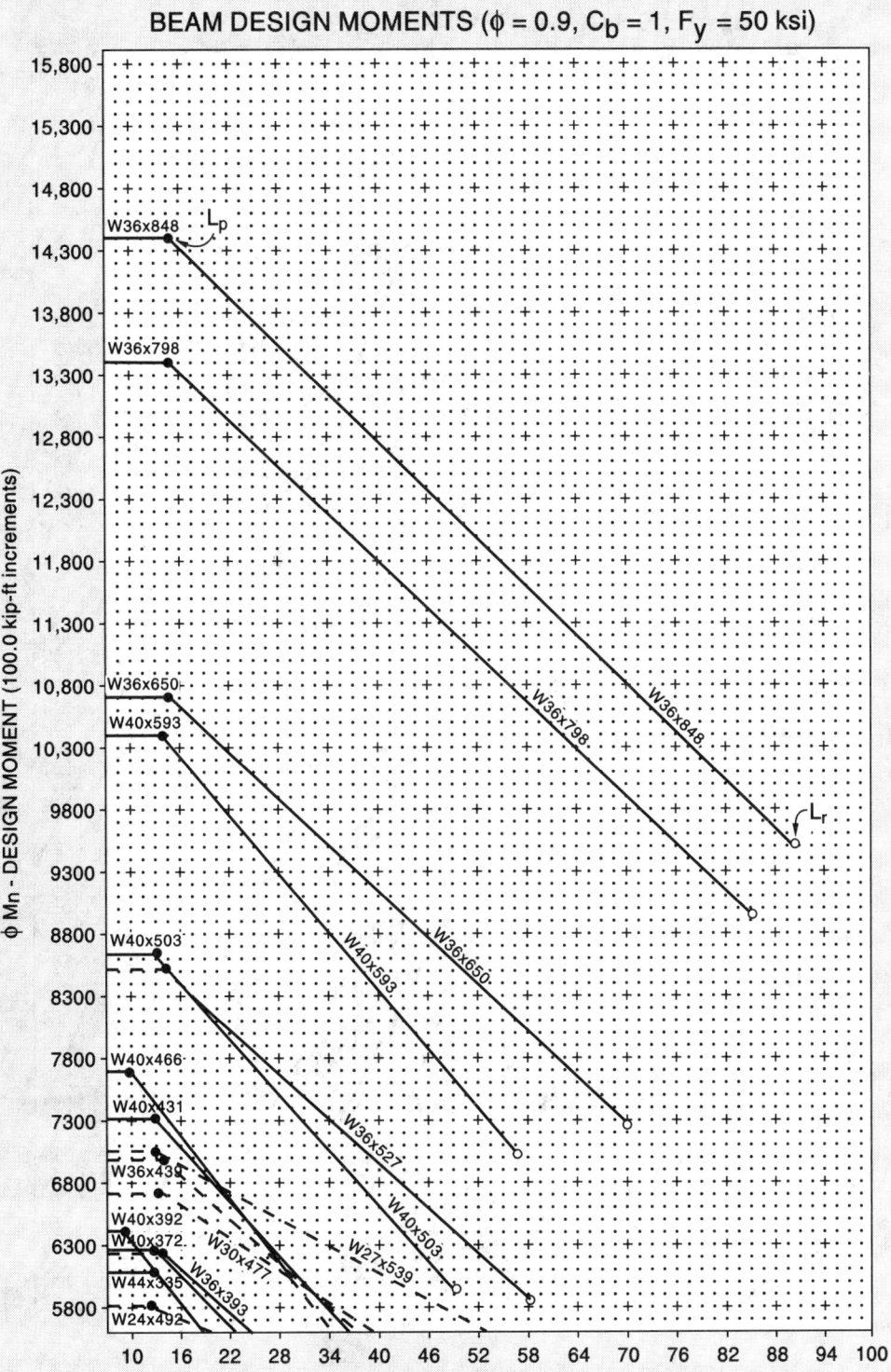

BEAM DESIGN MOMENTS ($\phi = 0.9$, $C_b = 1$, $F_y = 50$ ksi)

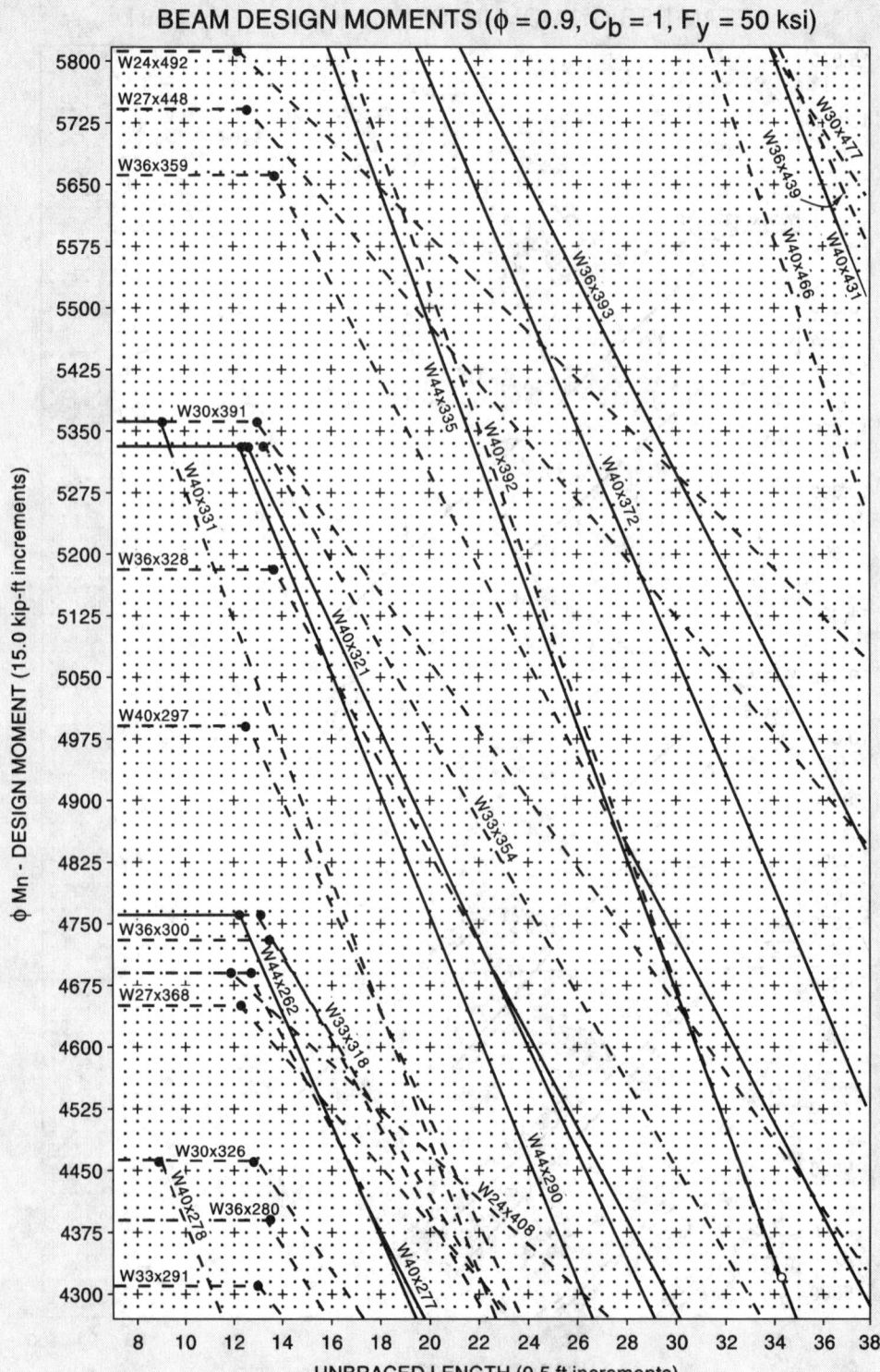

ϕM_n - DESIGN MOMENT (15.0 kip-ft increments)

UNBRACED LENGTH (0.5 ft increments)

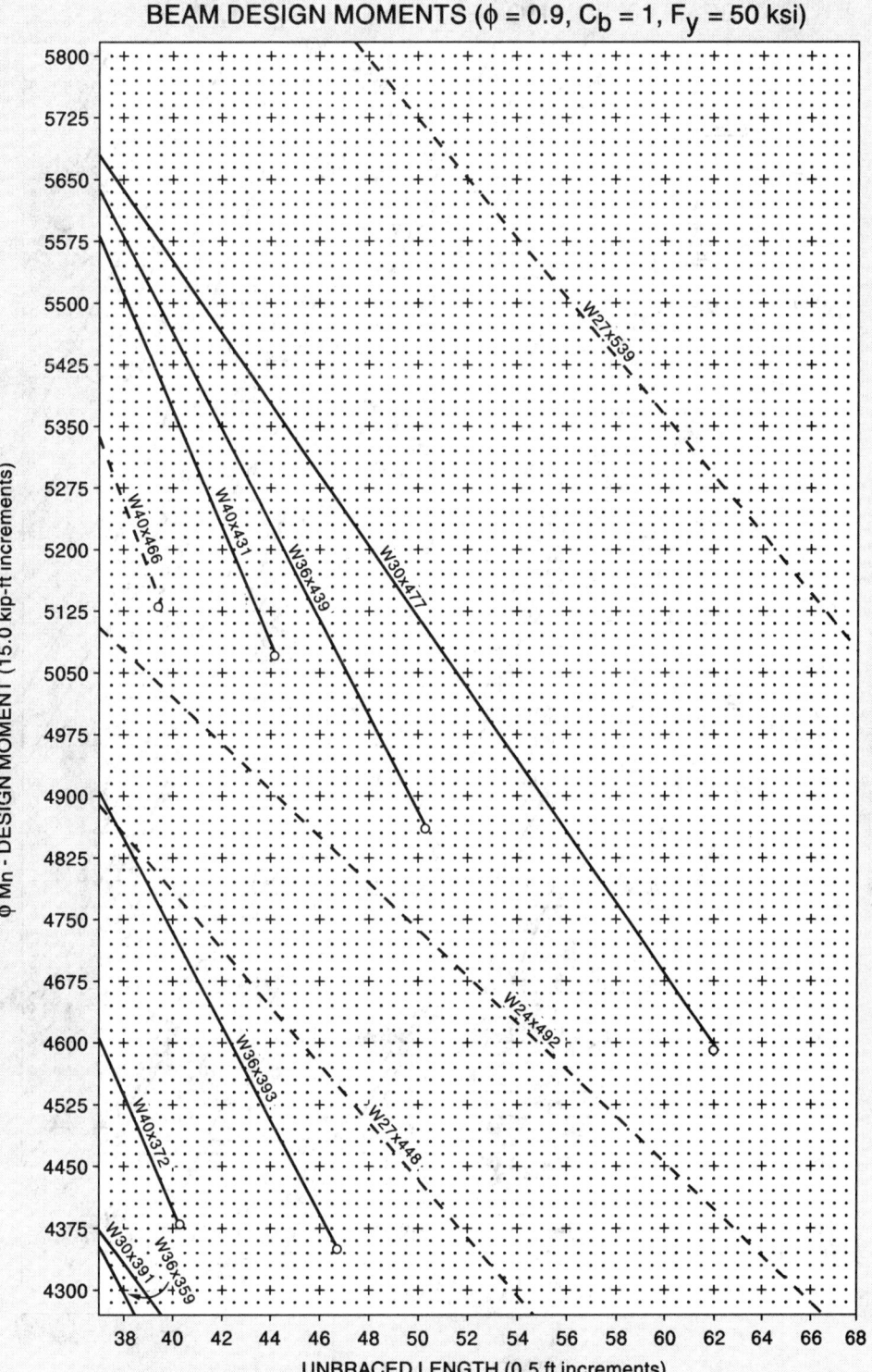

BEAM DESIGN MOMENTS (ϕ = 0.9, C_b = 1, F_y = 50 ksi)

BEAM DESIGN MOMENTS ($\phi = 0.9$, $C_b = 1$, $F_y = 50$ ksi)

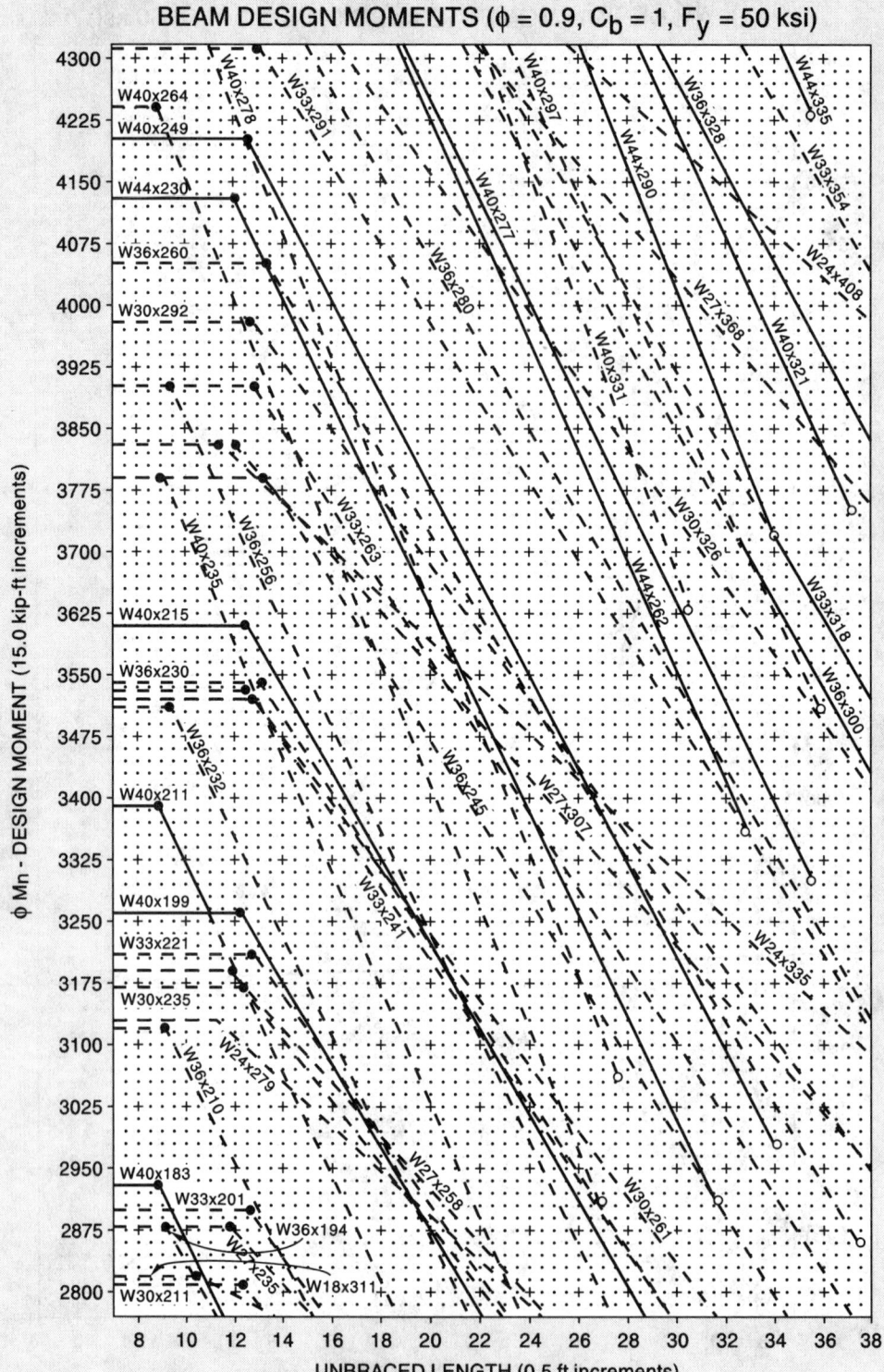

φ Mn - DESIGN MOMENT (15.0 kip-ft increments)

UNBRACED LENGTH (0.5 ft increments)

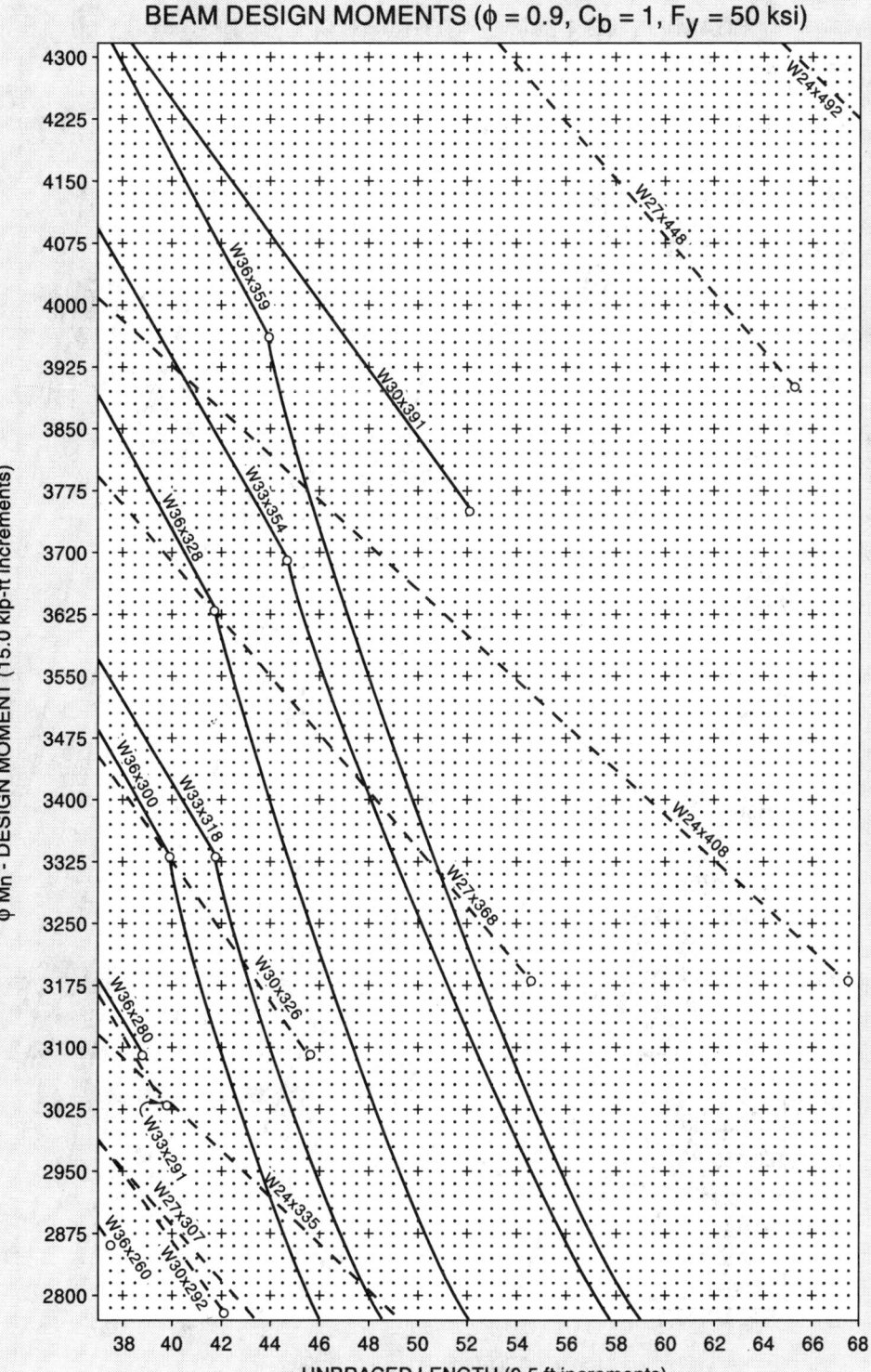

BEAM DESIGN MOMENTS ($\phi = 0.9$, $C_b = 1$, $F_y = 50$ ksi)

BEAM DESIGN MOMENTS ($\phi = 0.9$, $C_b = 1$, $F_y = 50$ ksi)

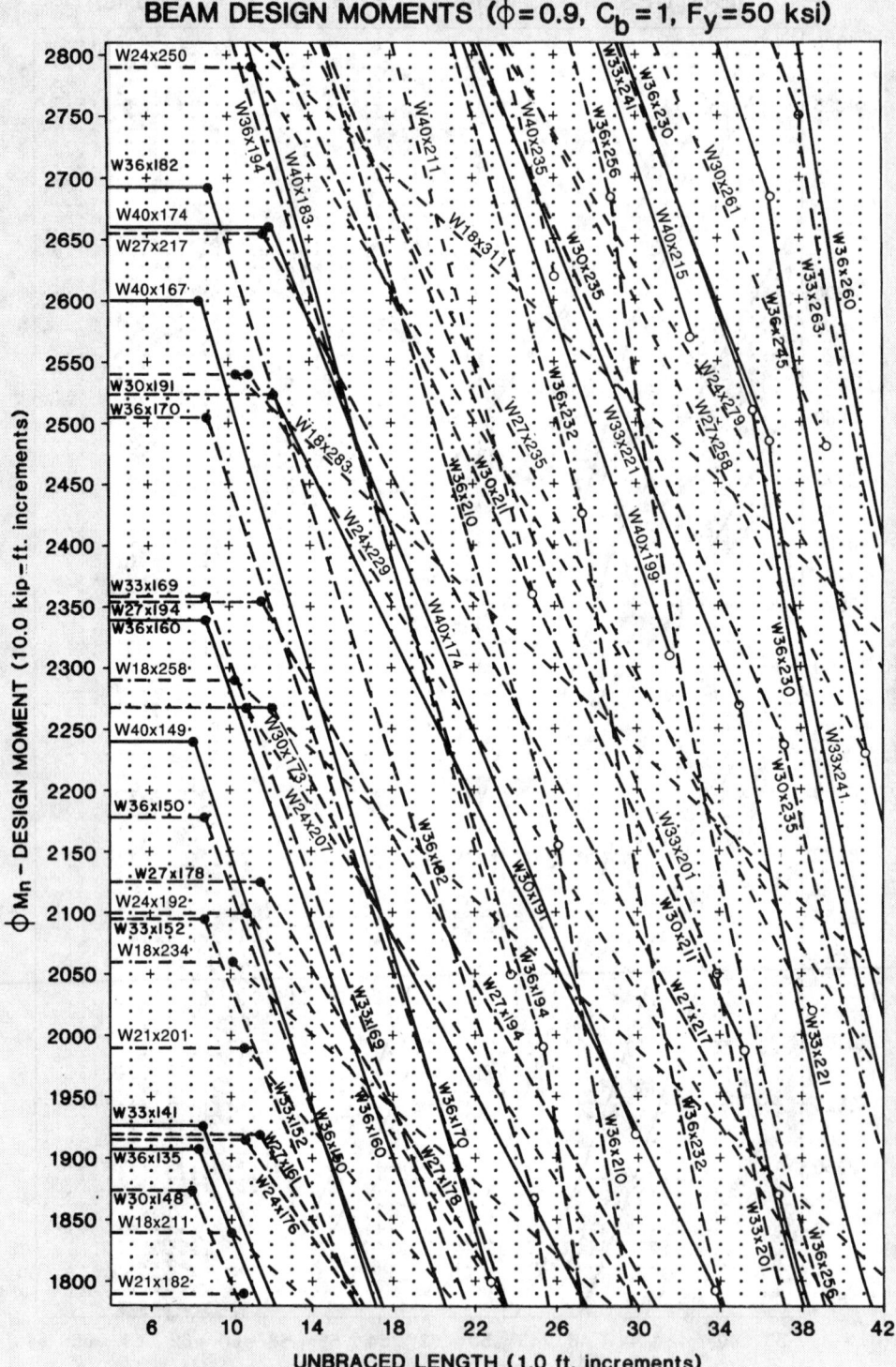

ϕM_n – DESIGN MOMENT (10.0 kip–ft. increments)

UNBRACED LENGTH (1.0 ft. increments)

BEAM DESIGN MOMENTS (ϕ=0.9, C_b=1, F_y= 50 ksi)

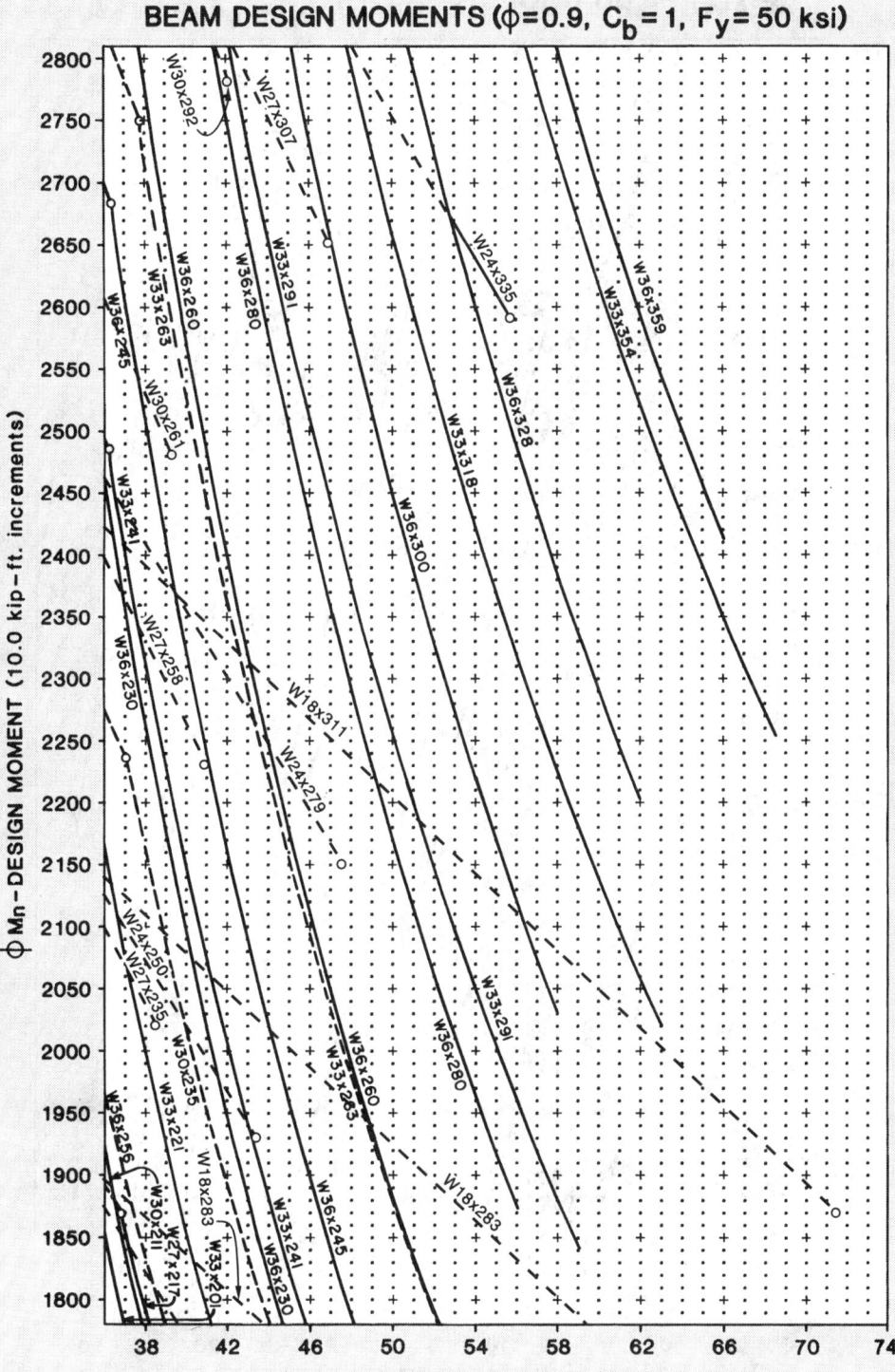

ϕM_n - DESIGN MOMENT (10.0 kip-ft. increments)

UNBRACED LENGTH (1.0 ft. increments)

BEAM DESIGN MOMENTS (ϕ = 0.9, C_b = 1, F_y = 50 ksi)

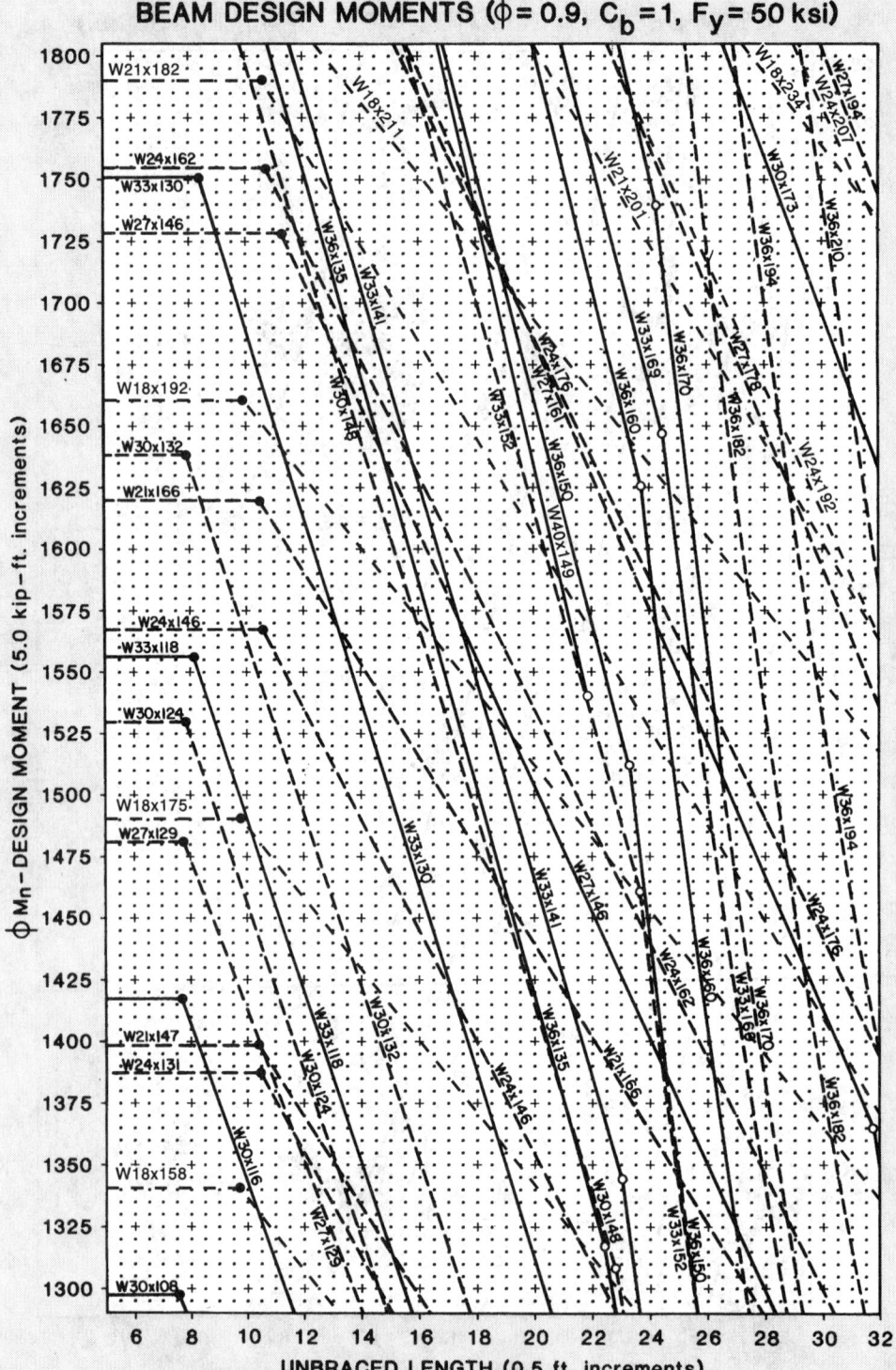

ϕM_n – DESIGN MOMENT (5.0 kip–ft. increments)

UNBRACED LENGTH (0.5 ft. increments)

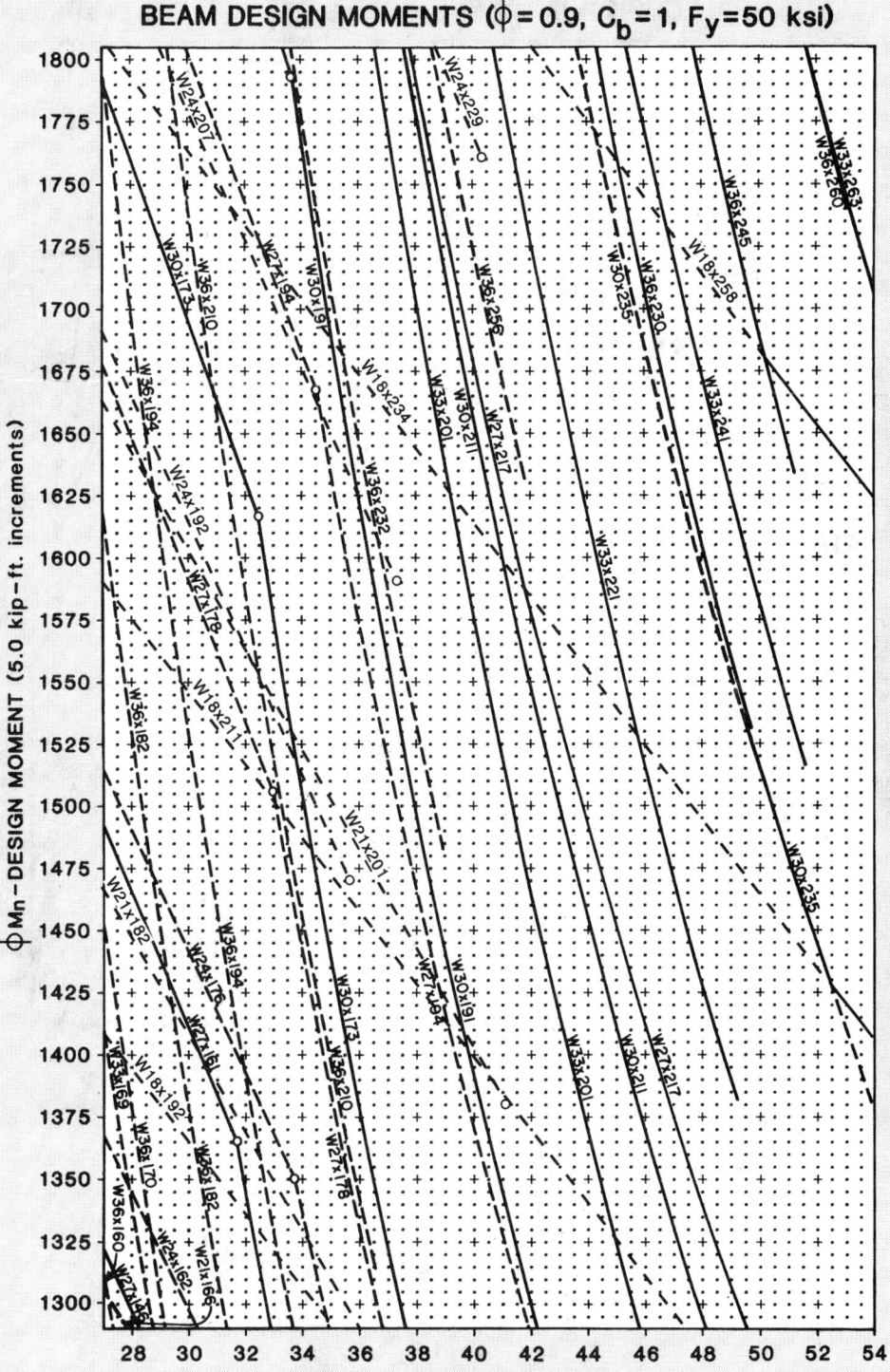

BEAM DESIGN MOMENTS (ϕ = 0.9, C_b = 1, F_y = 50 ksi)

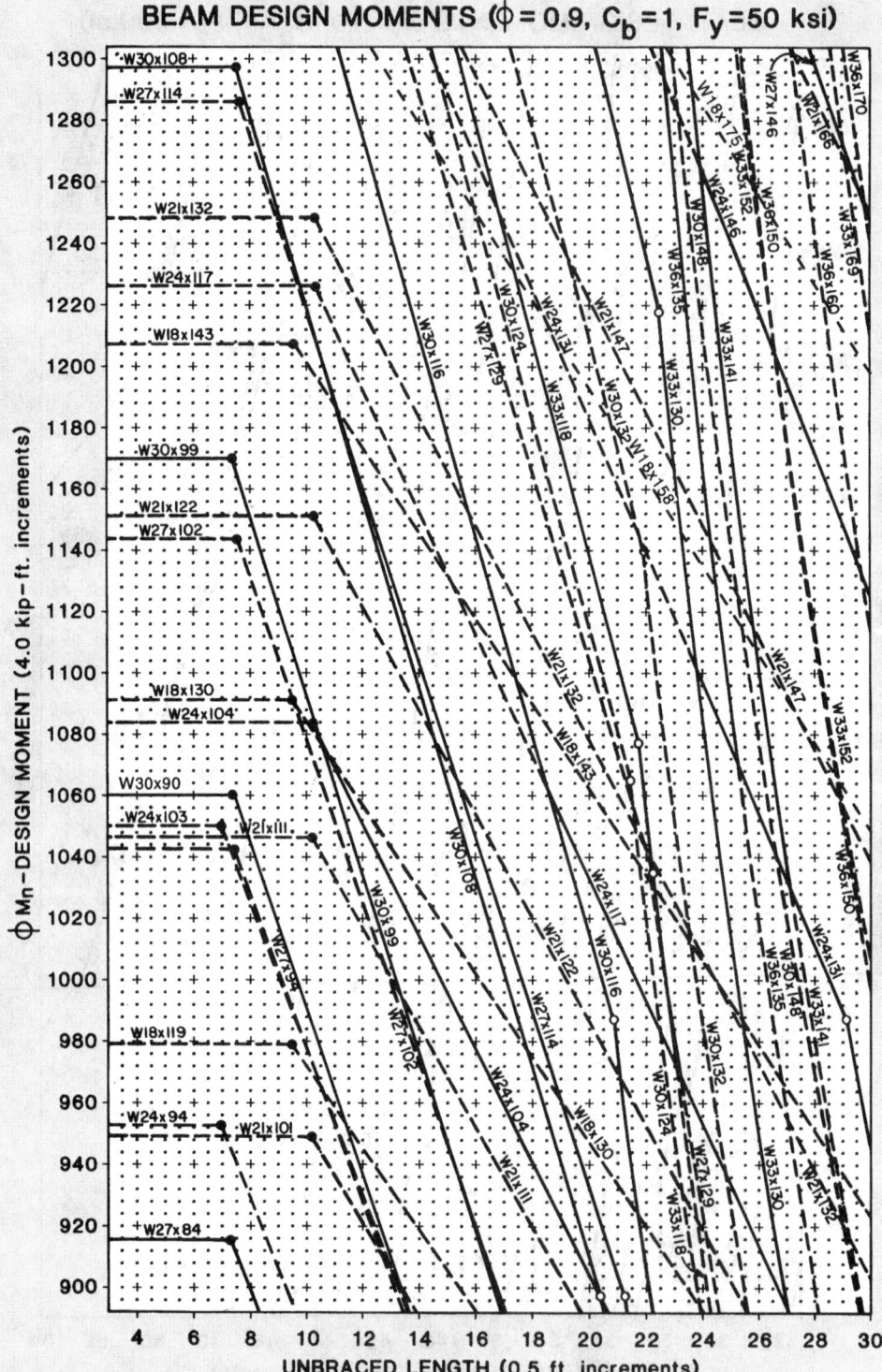

ϕM_n – DESIGN MOMENT (4.0 kip–ft. increments)

UNBRACED LENGTH (0.5 ft. increments)

BEAM DESIGN MOMENTS (ϕ = 0.9, C_b = 1, F_y = 50 ksi)

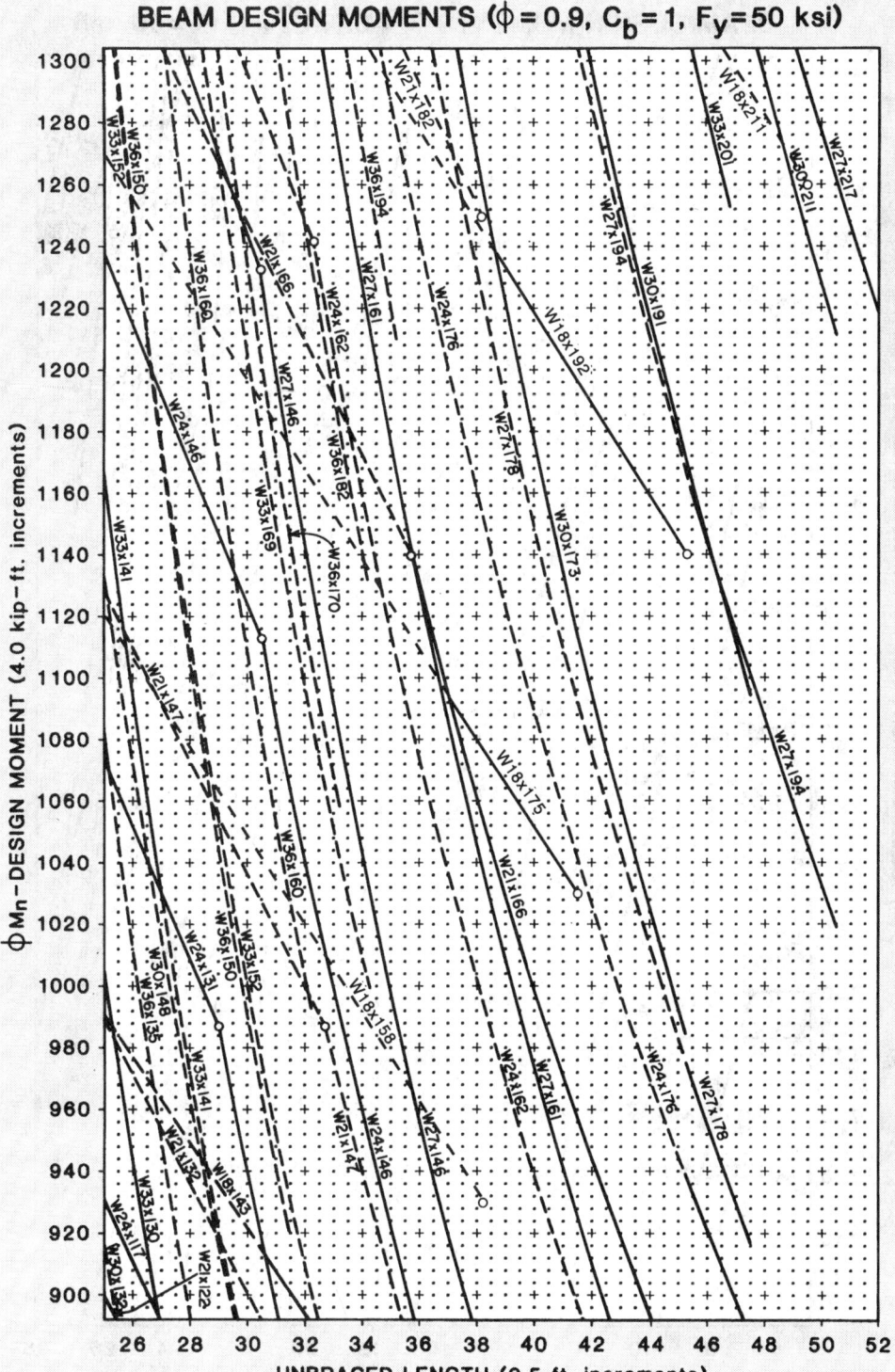

BEAM DESIGN MOMENTS ($\phi = 0.9$, $C_b = 1$, $F_y = 50$ ksi)

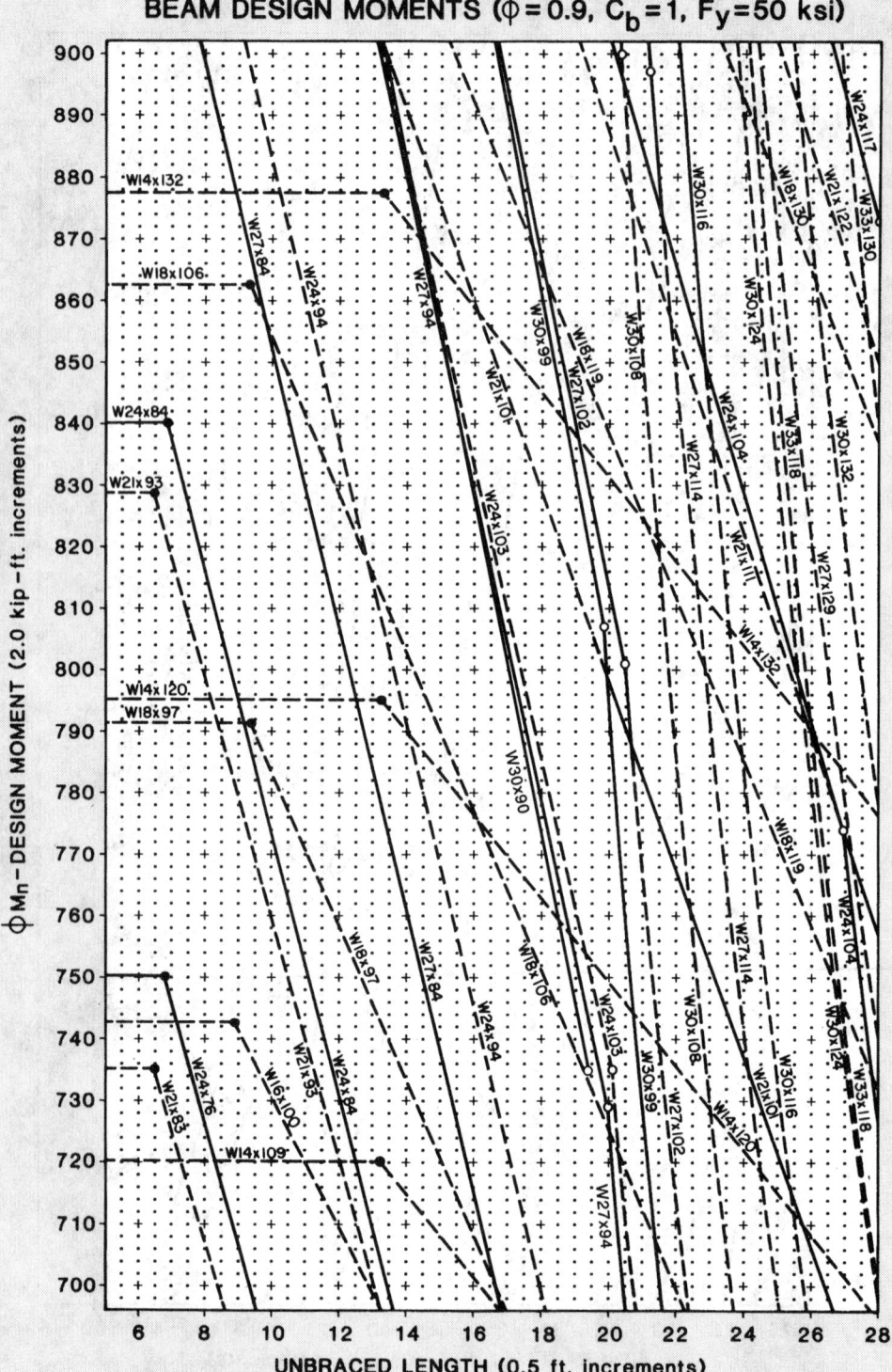

ϕM_n – DESIGN MOMENT (2.0 kip – ft. increments)

UNBRACED LENGTH (0.5 ft. increments)

BEAM DESIGN MOMENTS ($\phi = 0.9$, $C_b = 1$, $F_y = 50$ ksi)

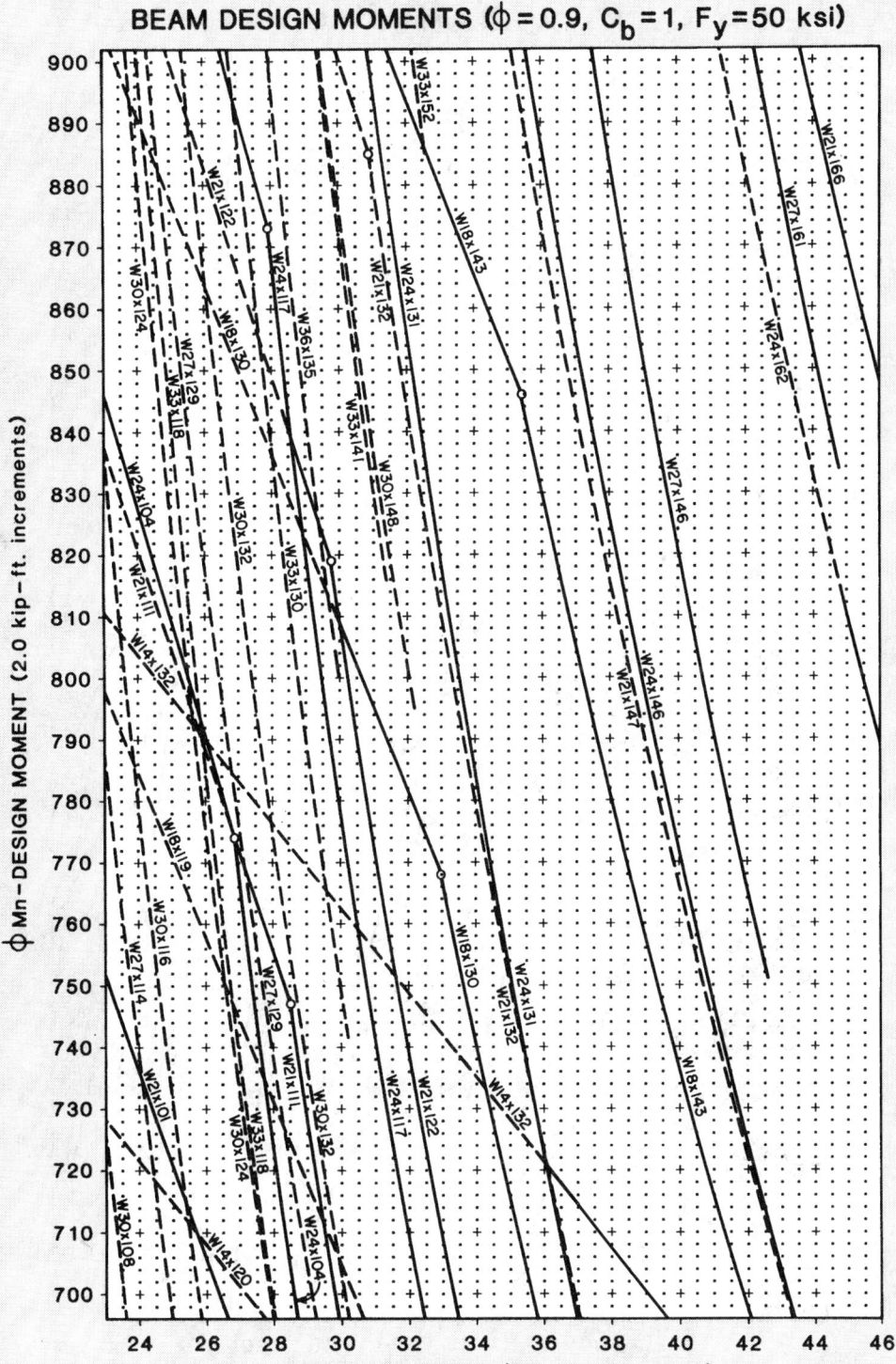

ϕ Mn - DESIGN MOMENT (2.0 kip - ft. increments)

UNBRACED LENGTH (0.5 ft. increments)

BEAM DESIGN MOMENTS ($\phi = 0.9$, $C_b = 1$, $F_y = 50$ ksi)

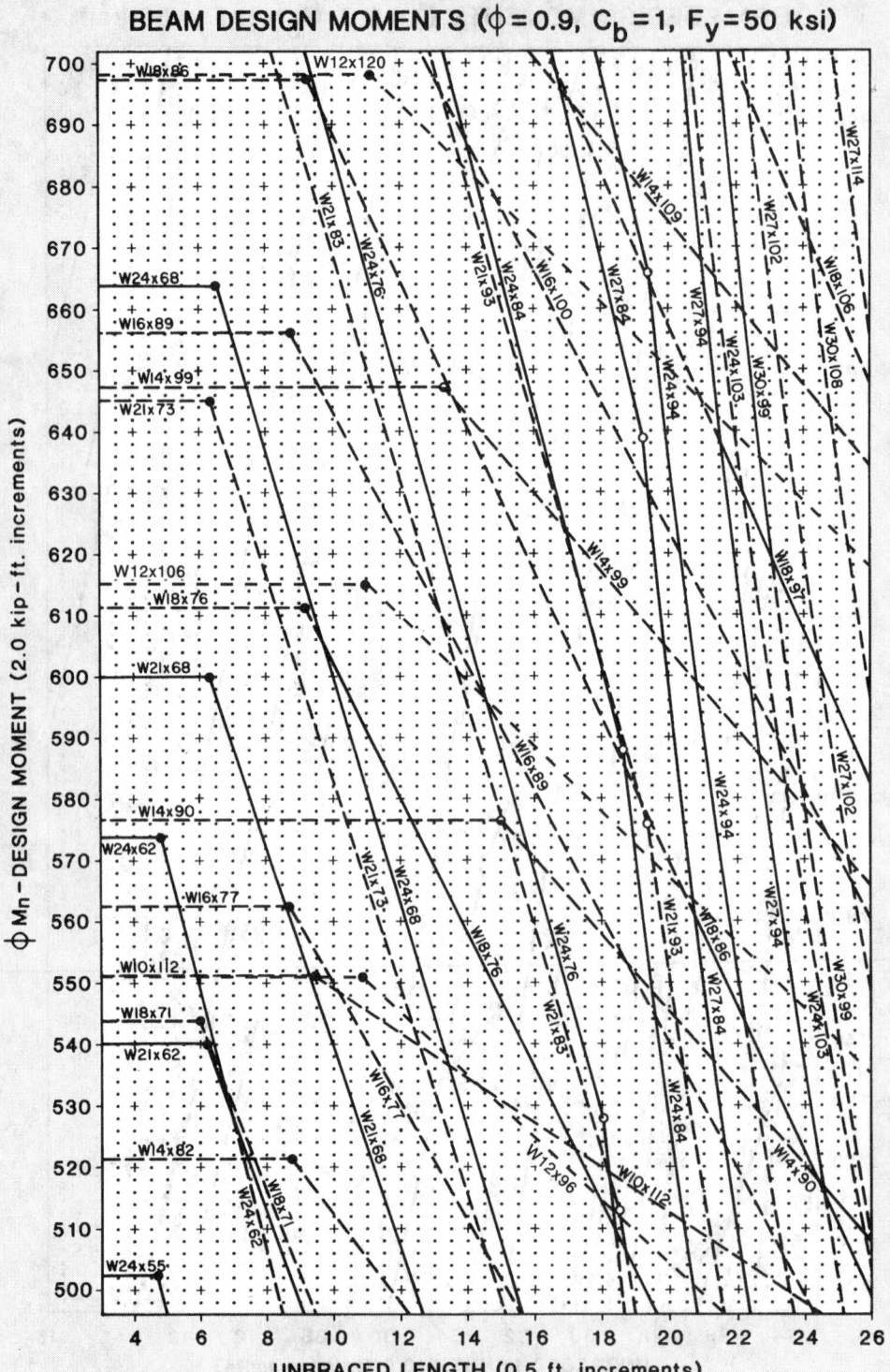

ϕM_n – DESIGN MOMENT (2.0 kip-ft. increments)

UNBRACED LENGTH (0.5 ft. increments)

BEAM DESIGN MOMENTS ($\phi = 0.9$, $C_b = 1$, $F_y = 50$ ksi)

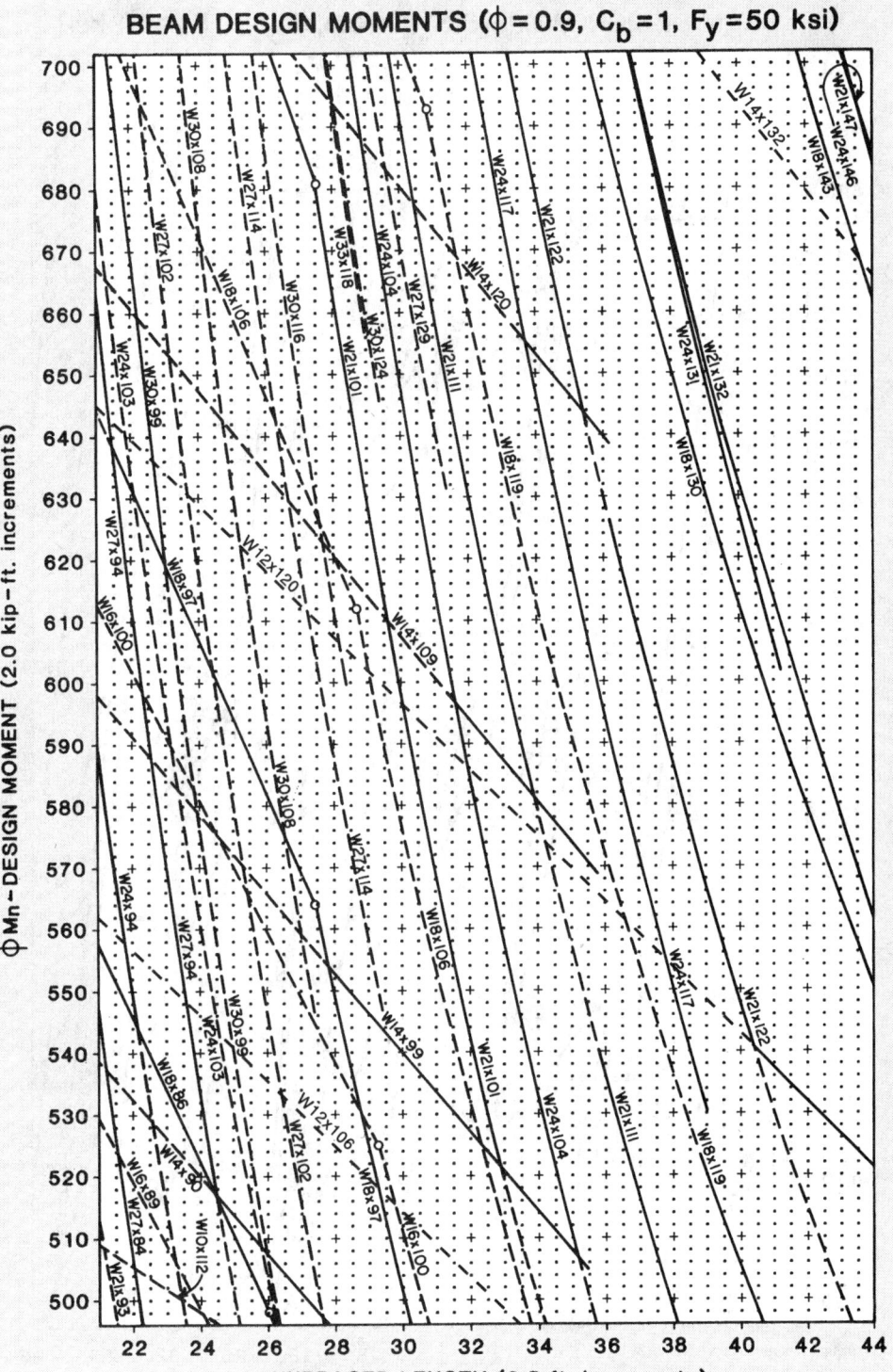

ϕM_n – DESIGN MOMENT (2.0 kip–ft. increments)

UNBRACED LENGTH (0.5 ft. increments)

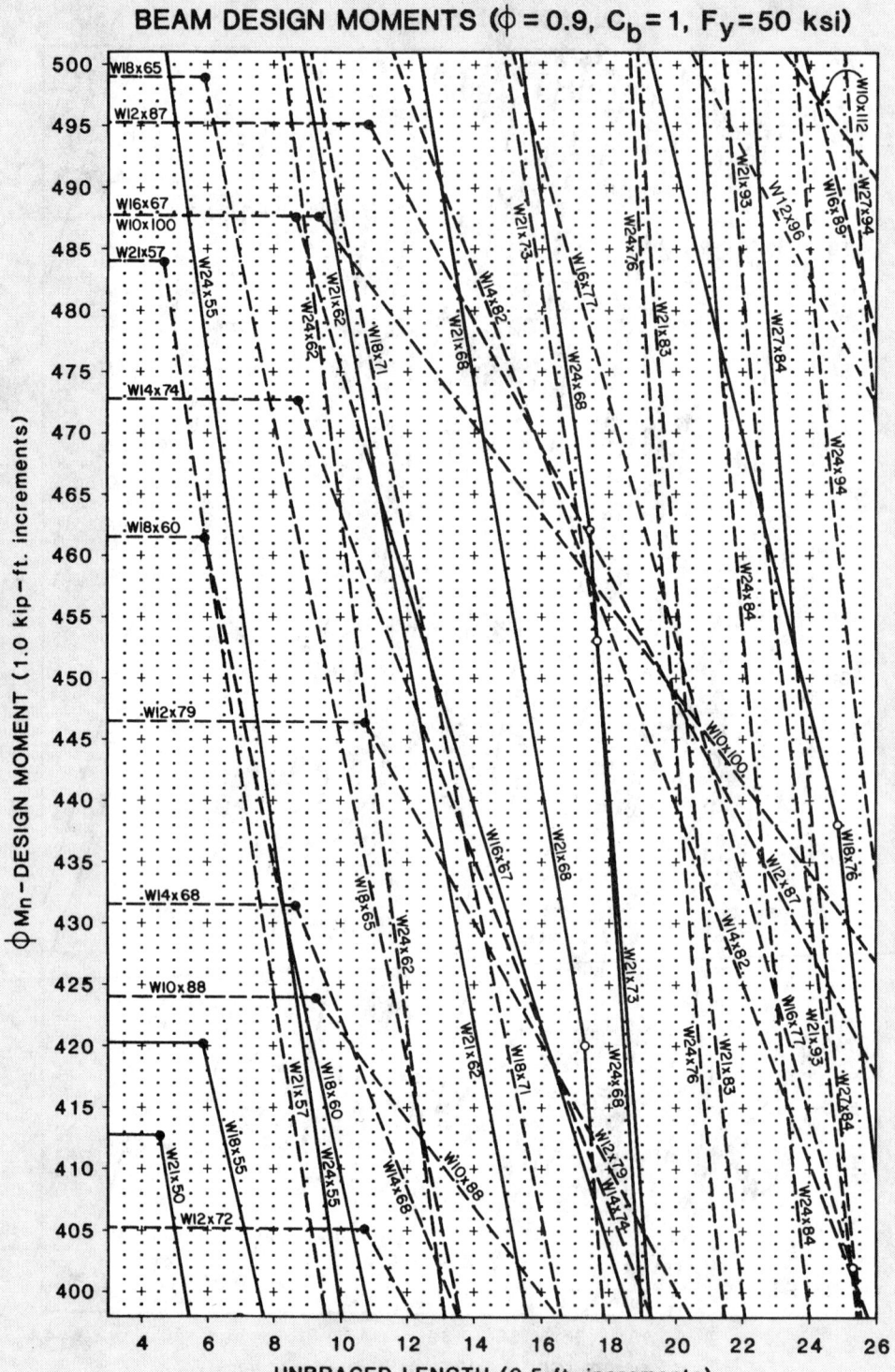

BEAM DESIGN MOMENTS (ϕ = 0.9, C_b = 1, F_y = 50 ksi)

ϕM_n - DESIGN MOMENT (1.0 kip–ft. increments)

UNBRACED LENGTH (0.5 ft. increments)

BEAM DESIGN MOMENTS ($\phi = 0.9$, $C_b = 1$, $F_y = 50$ ksi)

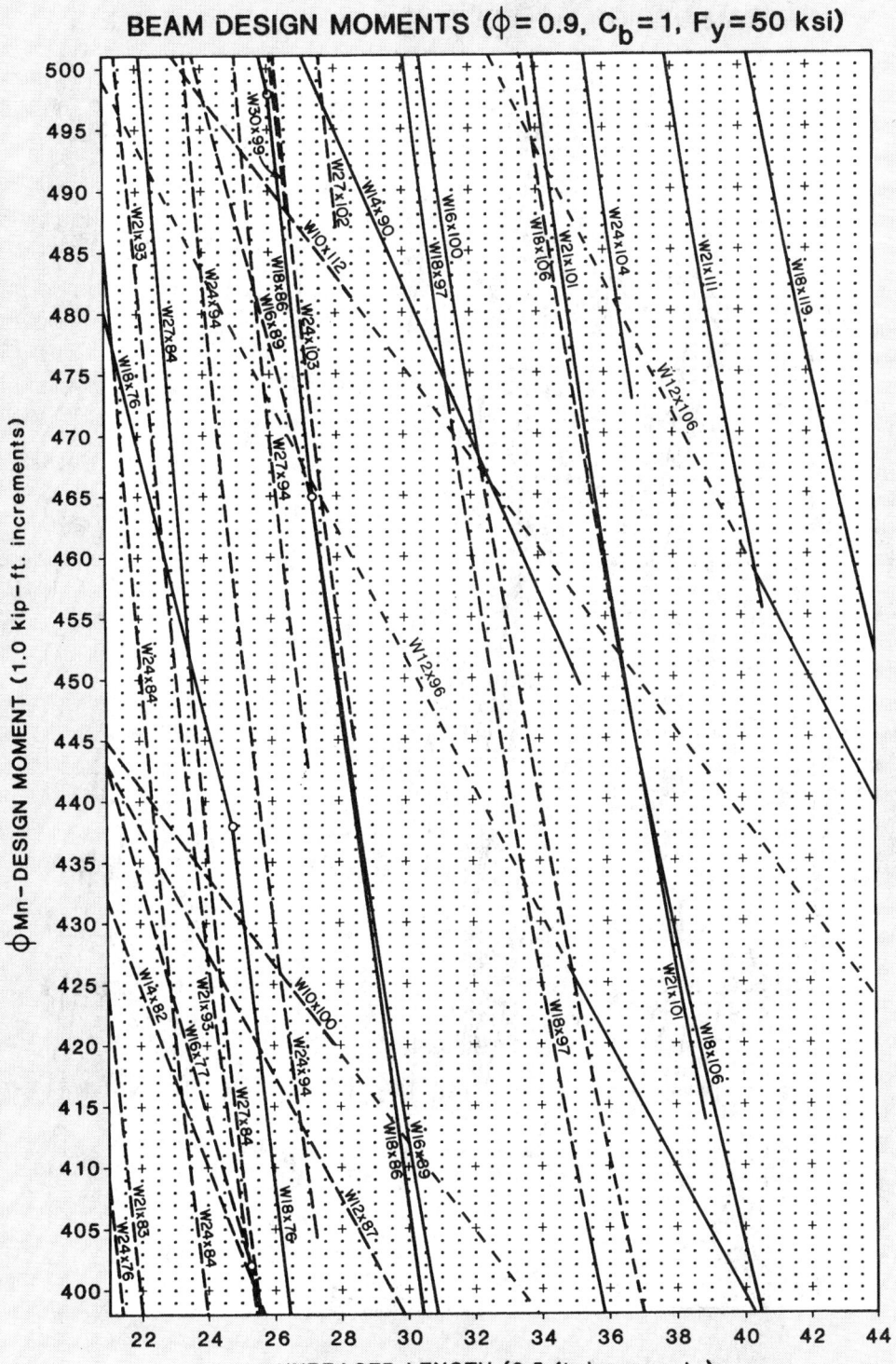

ϕMn – DESIGN MOMENT (1.0 kip–ft. increments)

UNBRACED LENGTH (0.5 ft. increments)

BEAM DESIGN MOMENTS ($\phi = 0.9$, $C_b = 1$, $F_y = 50$ ksi)

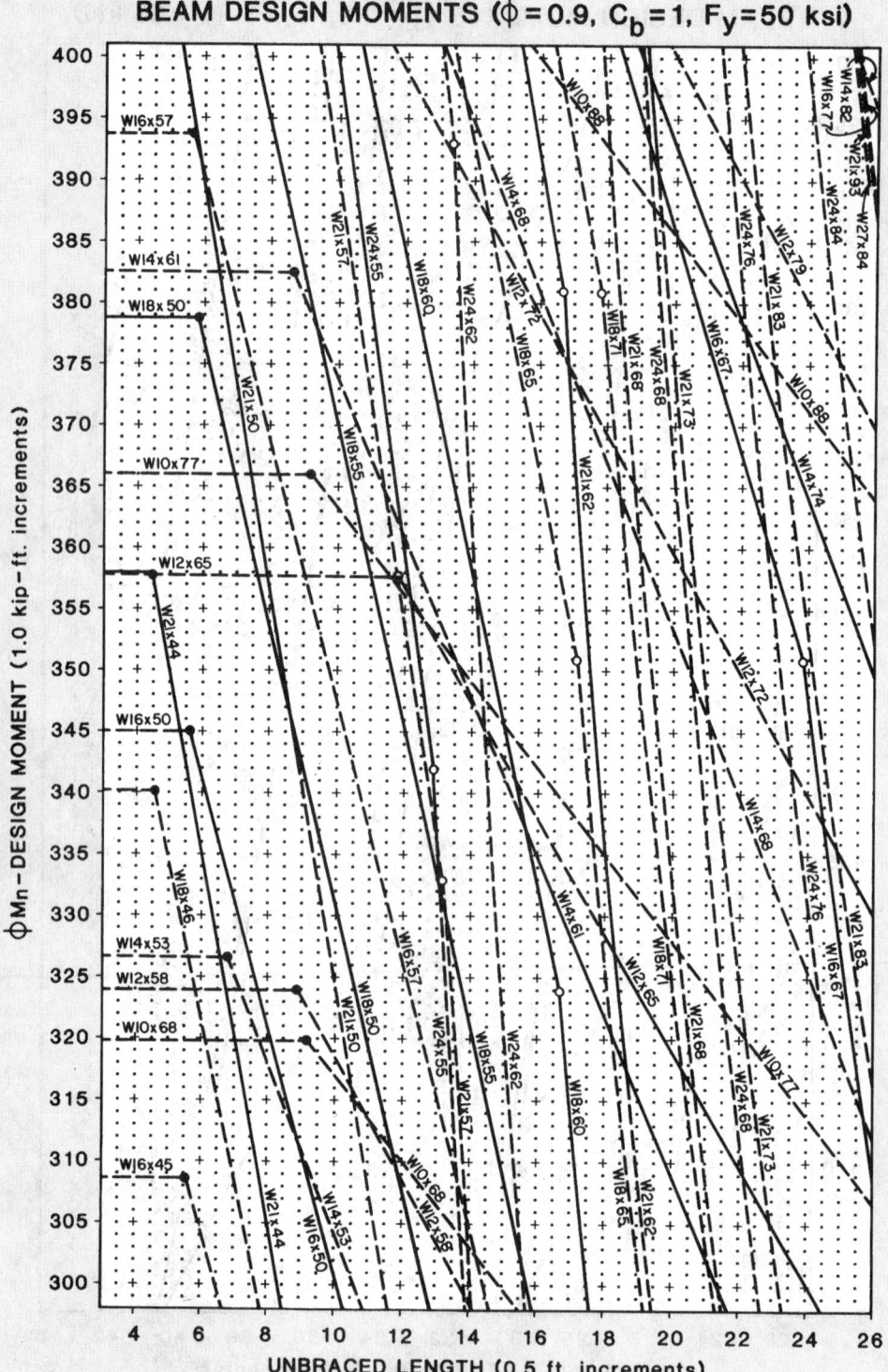

φMn – DESIGN MOMENT (1.0 kip – ft. increments)

UNBRACED LENGTH (0.5 ft. increments)

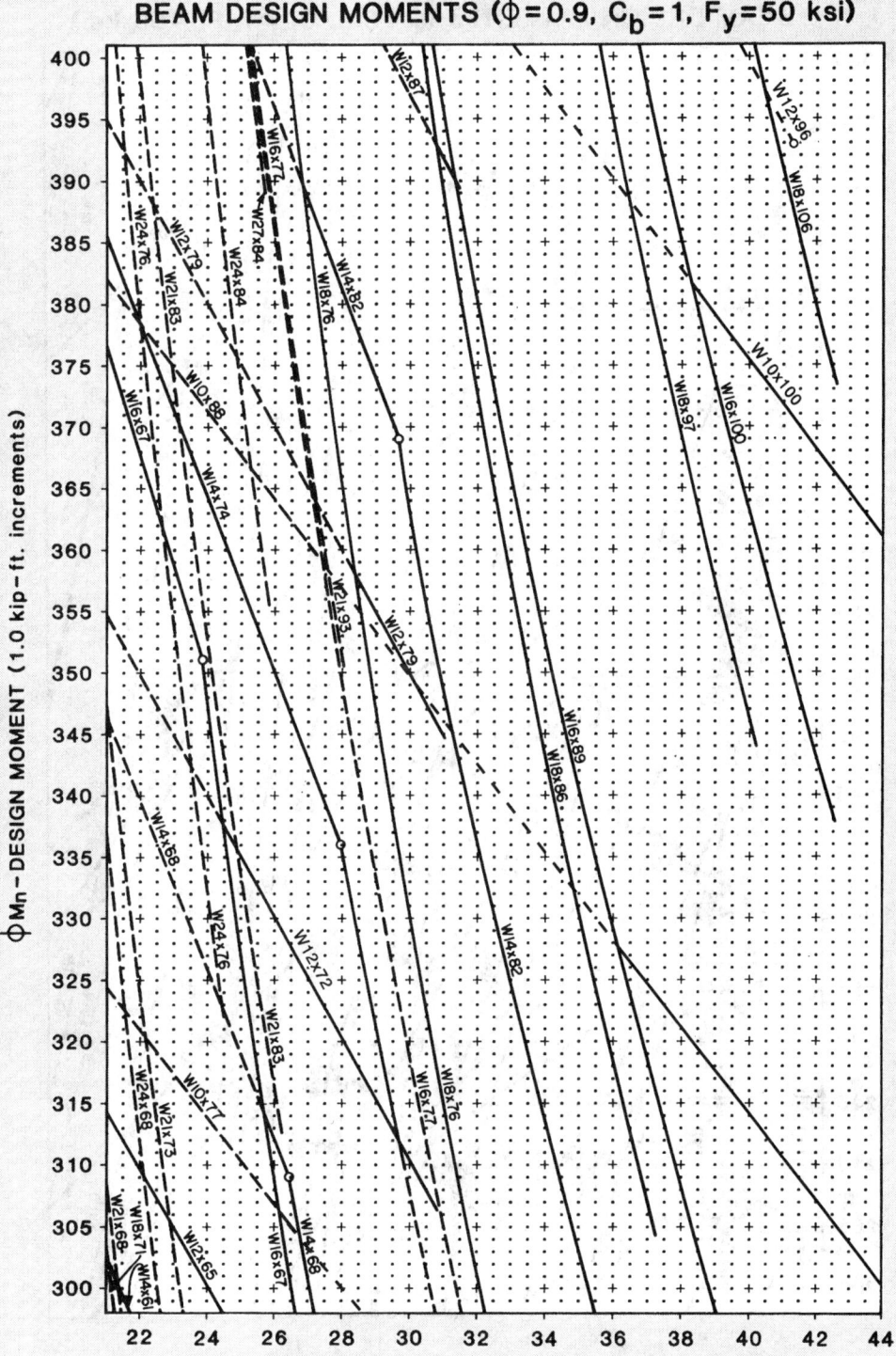

BEAM DESIGN MOMENTS ($\phi = 0.9$, $C_b = 1$, $F_y = 50$ ksi)

BEAM DESIGN MOMENTS (ϕ = 0.9, C_b = 1, F_y = 50 ksi)

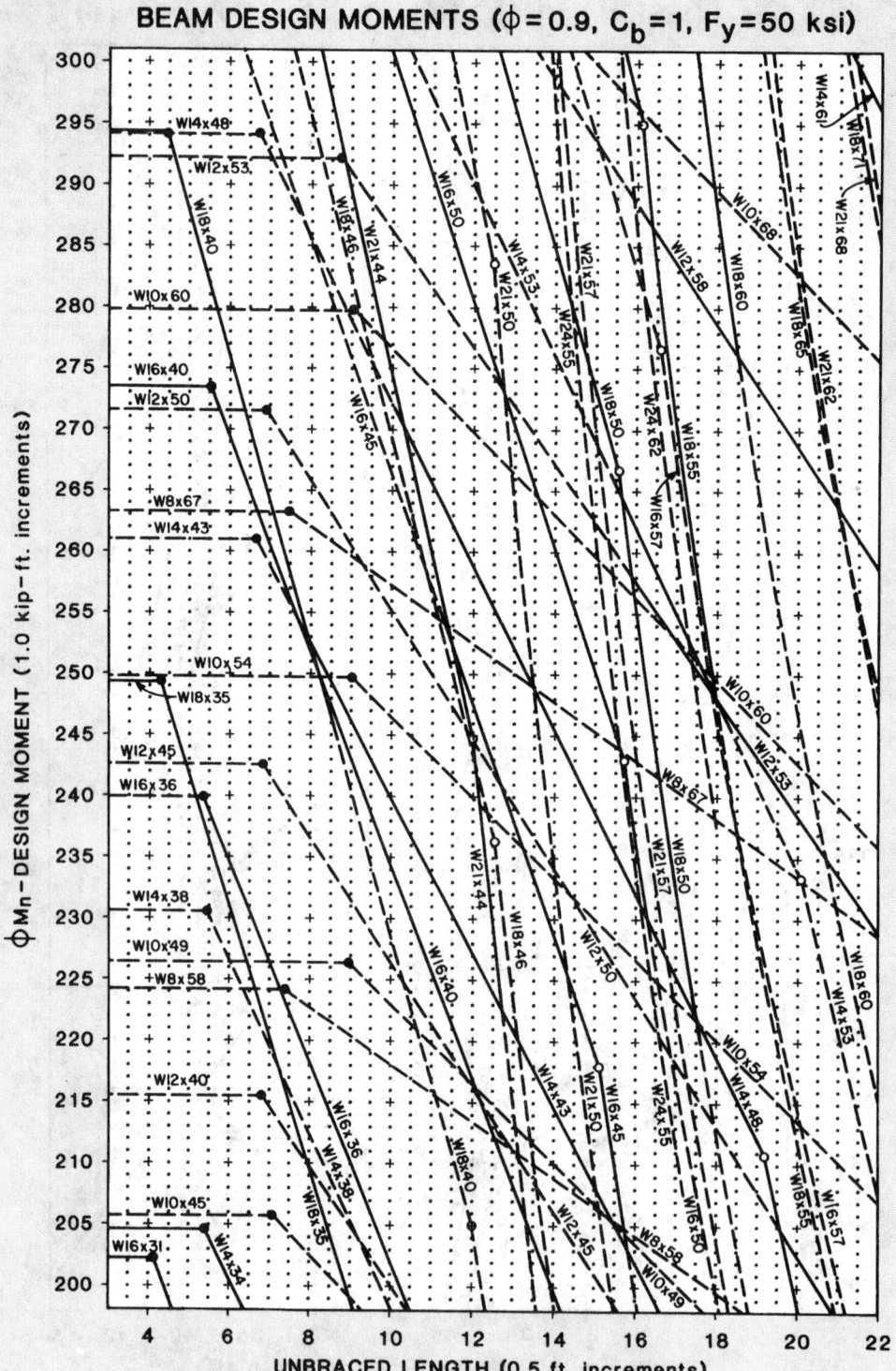

φMn – DESIGN MOMENT (1.0 kip – ft. increments)

UNBRACED LENGTH (0.5 ft. increments)

BEAM DESIGN MOMENTS ($\phi = 0.9$, $C_b = 1$, $F_y = 50$ ksi)

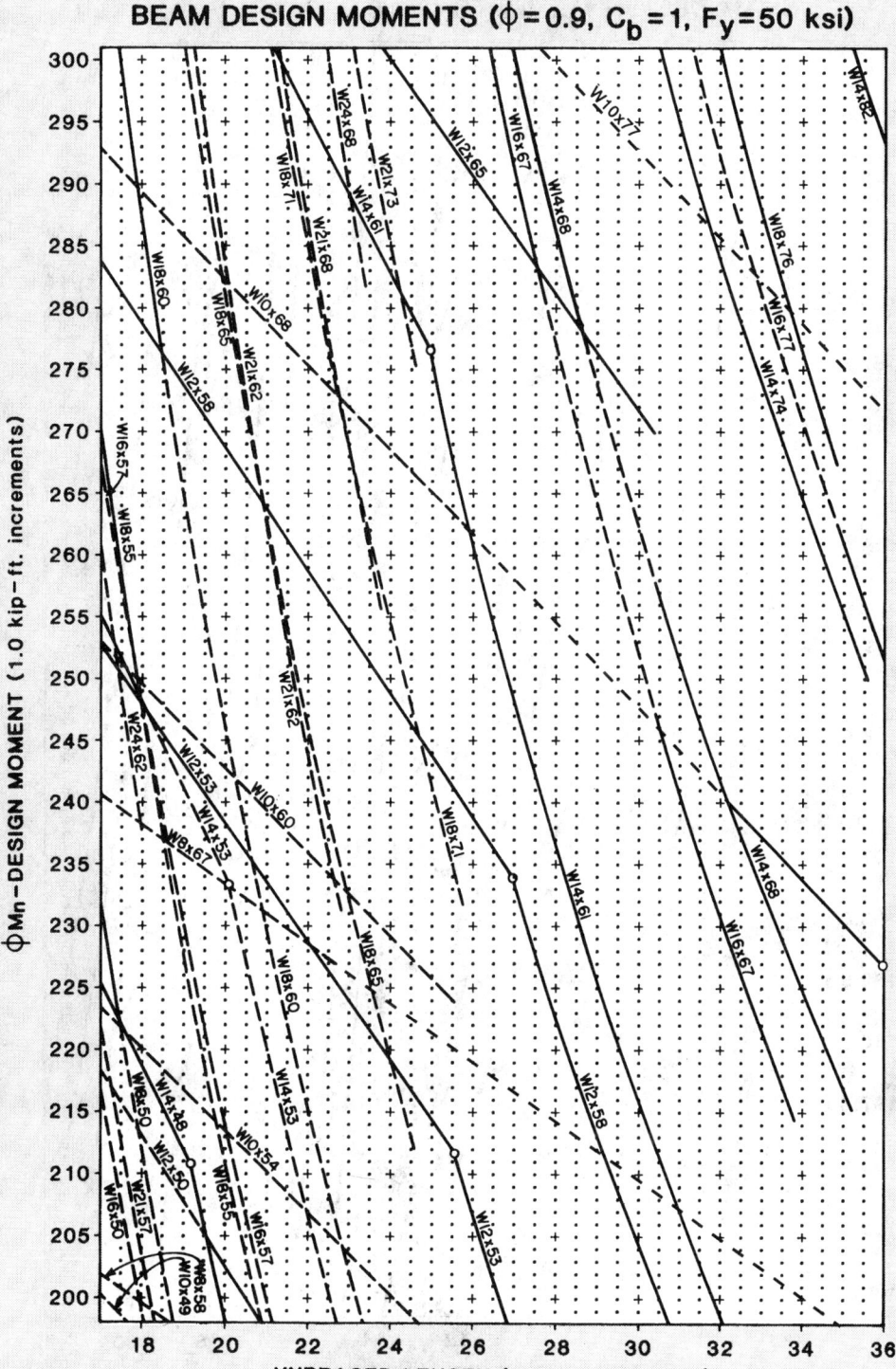

BEAM DESIGN MOMENTS ($\phi = 0.9$, $C_b = 1$, $F_y = 50$ ksi)

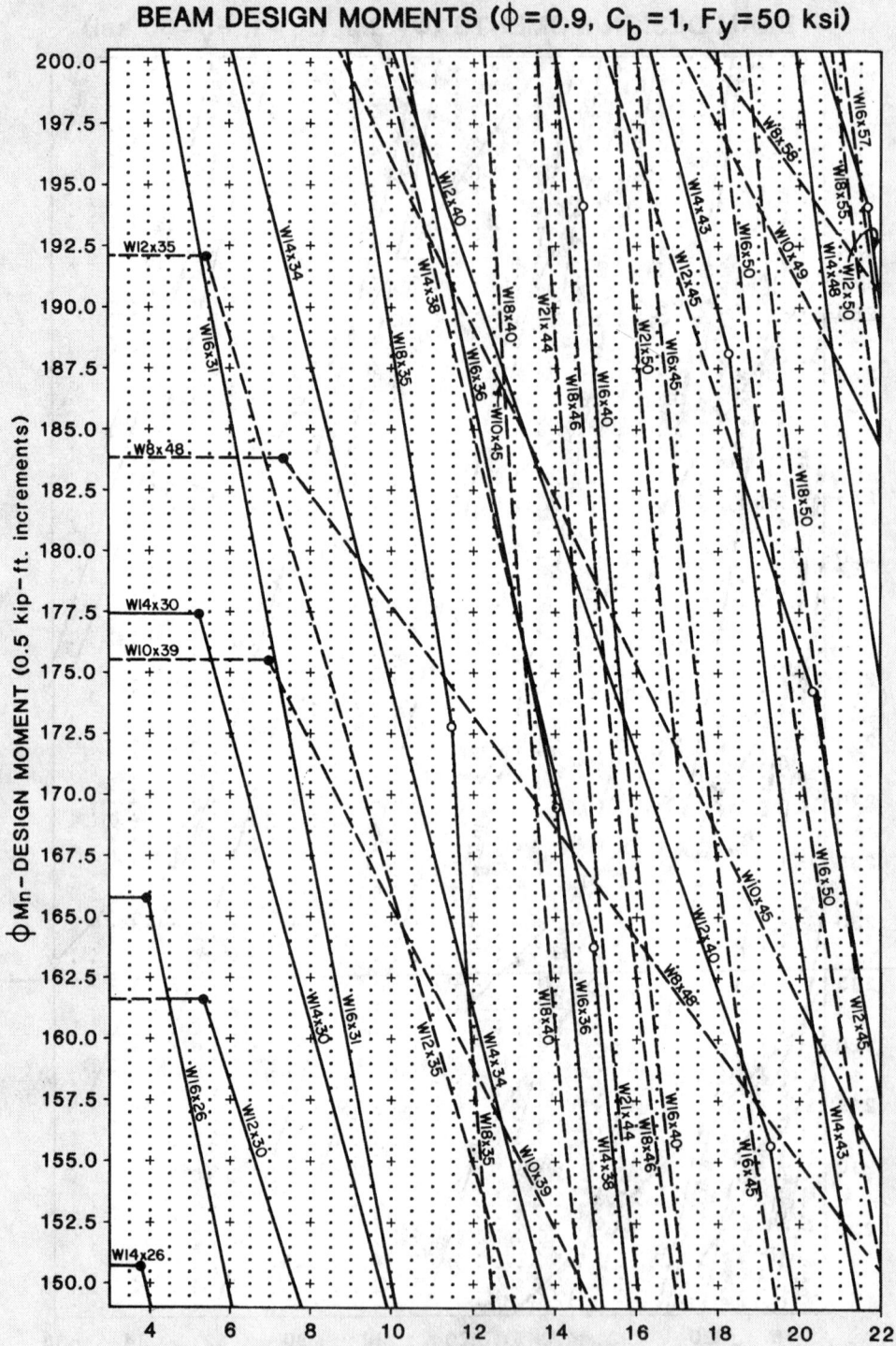

ϕM_n – DESIGN MOMENT (0.5 kip–ft. increments)

UNBRACED LENGTH (0.5 ft. increments)

BEAM DESIGN MOMENTS (ϕ = 0.9, C$_b$ = 1, F$_y$ = 50 ksi)

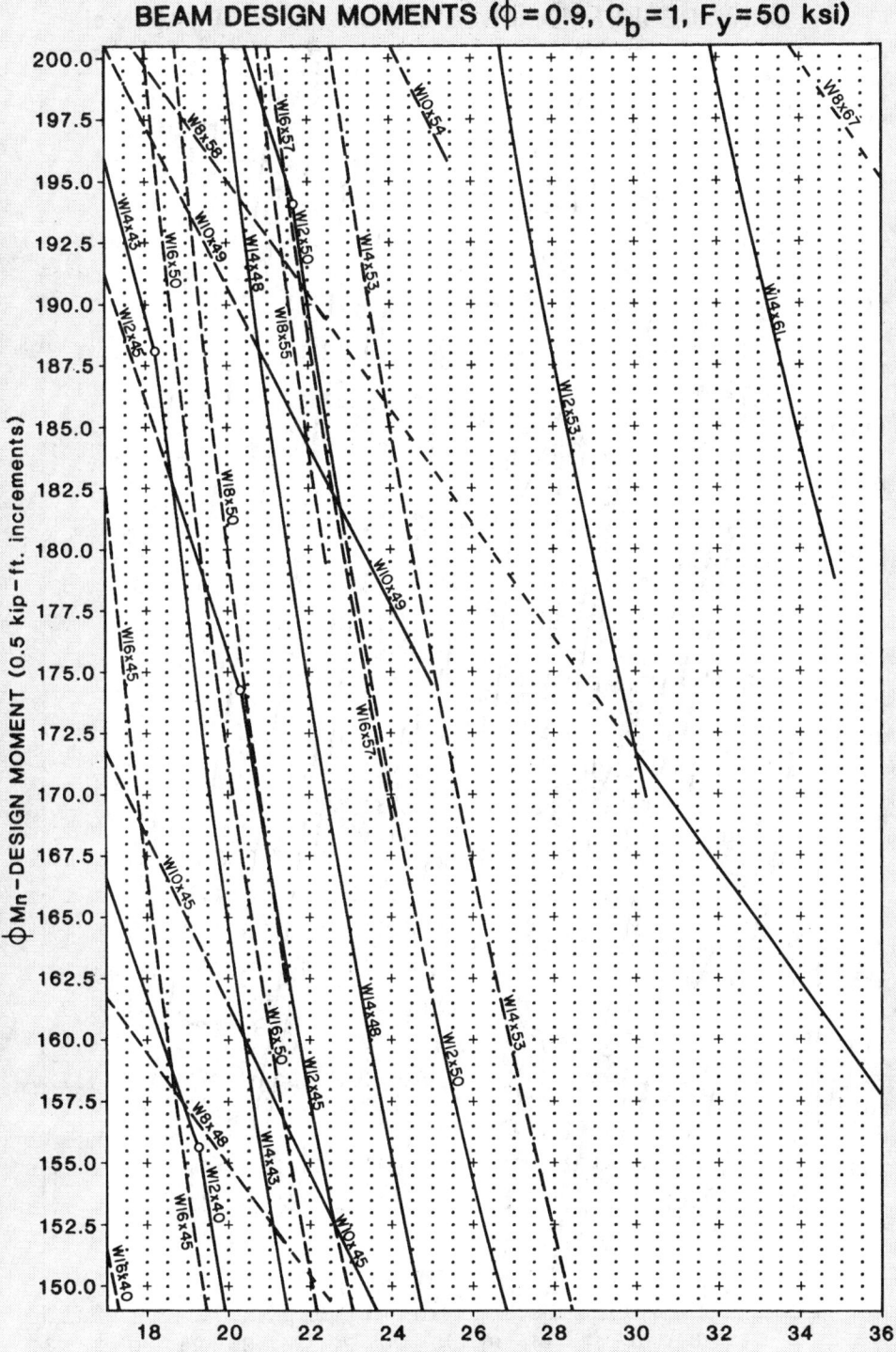

BEAM DESIGN MOMENTS ($\phi = 0.9$, $C_b = 1$, $F_y = 50$ ksi)

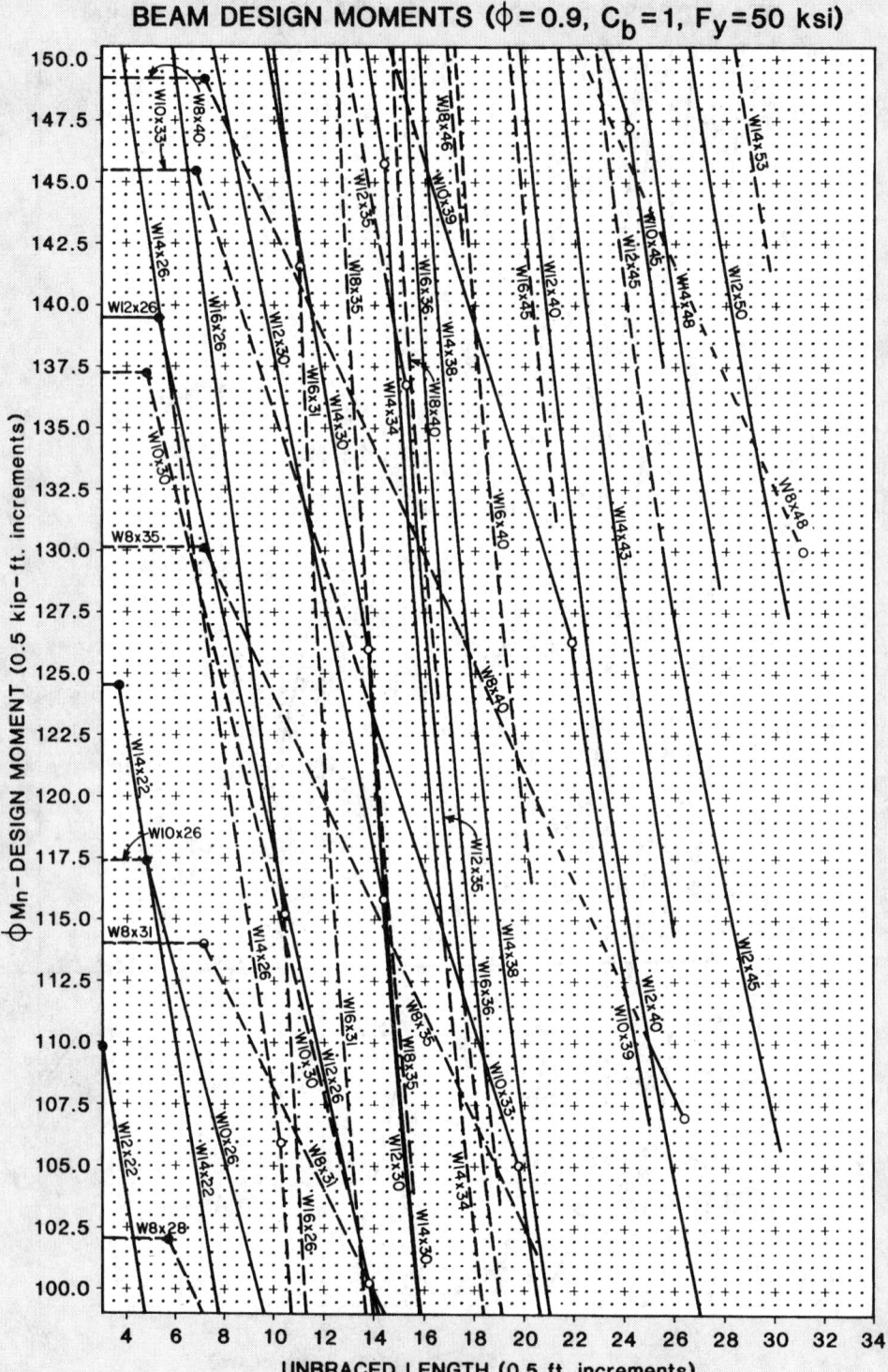

ϕM_n — DESIGN MOMENT (0.5 kip-ft. increments)

UNBRACED LENGTH (0.5 ft. increments)

BEAM DESIGN MOMENTS (ϕ=0.9, C_b=1, F_y=50 ksi)

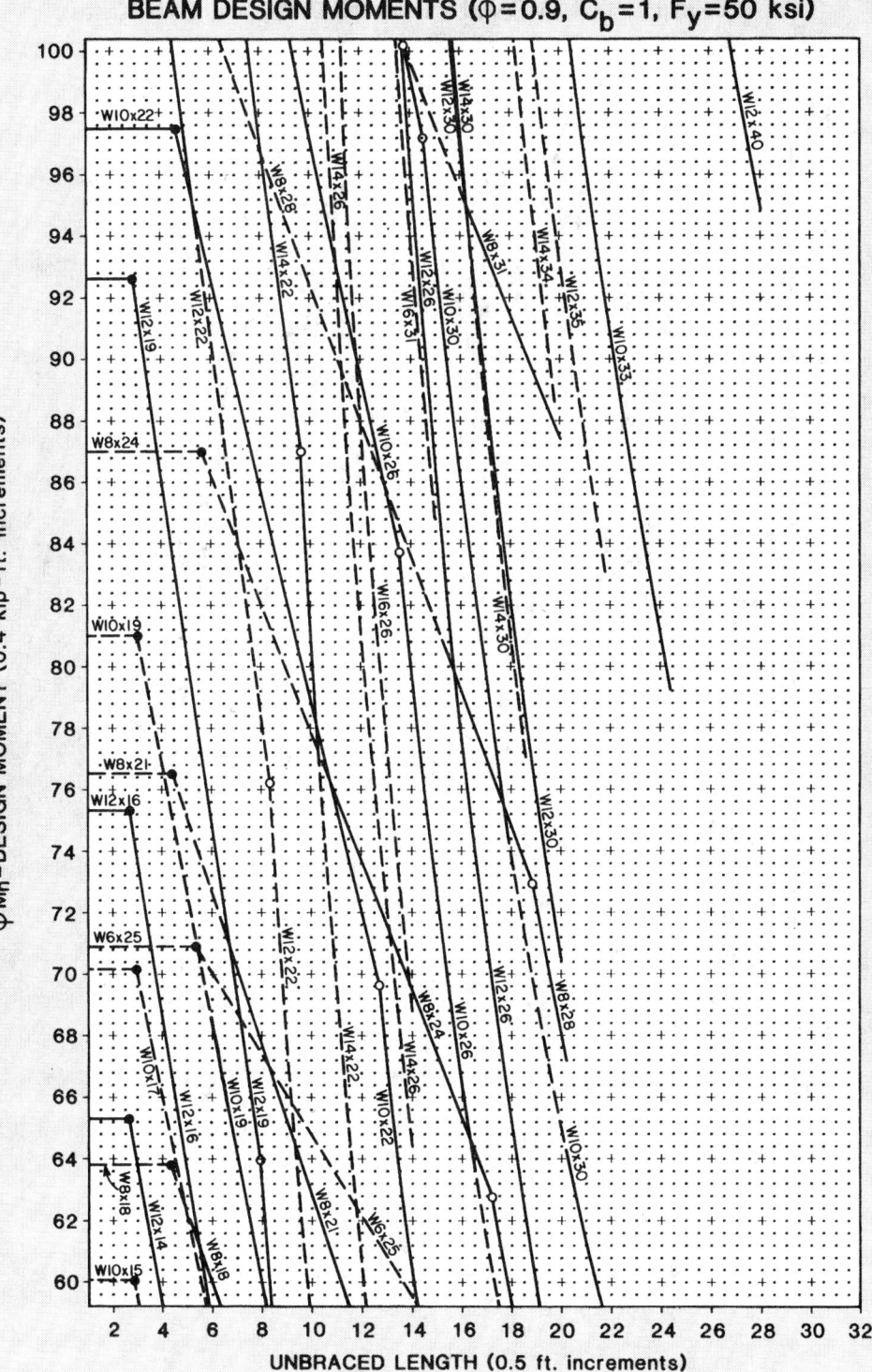

ϕM_n–DESIGN MOMENT (0.4 kip–ft. increments)

UNBRACED LENGTH (0.5 ft. increments)

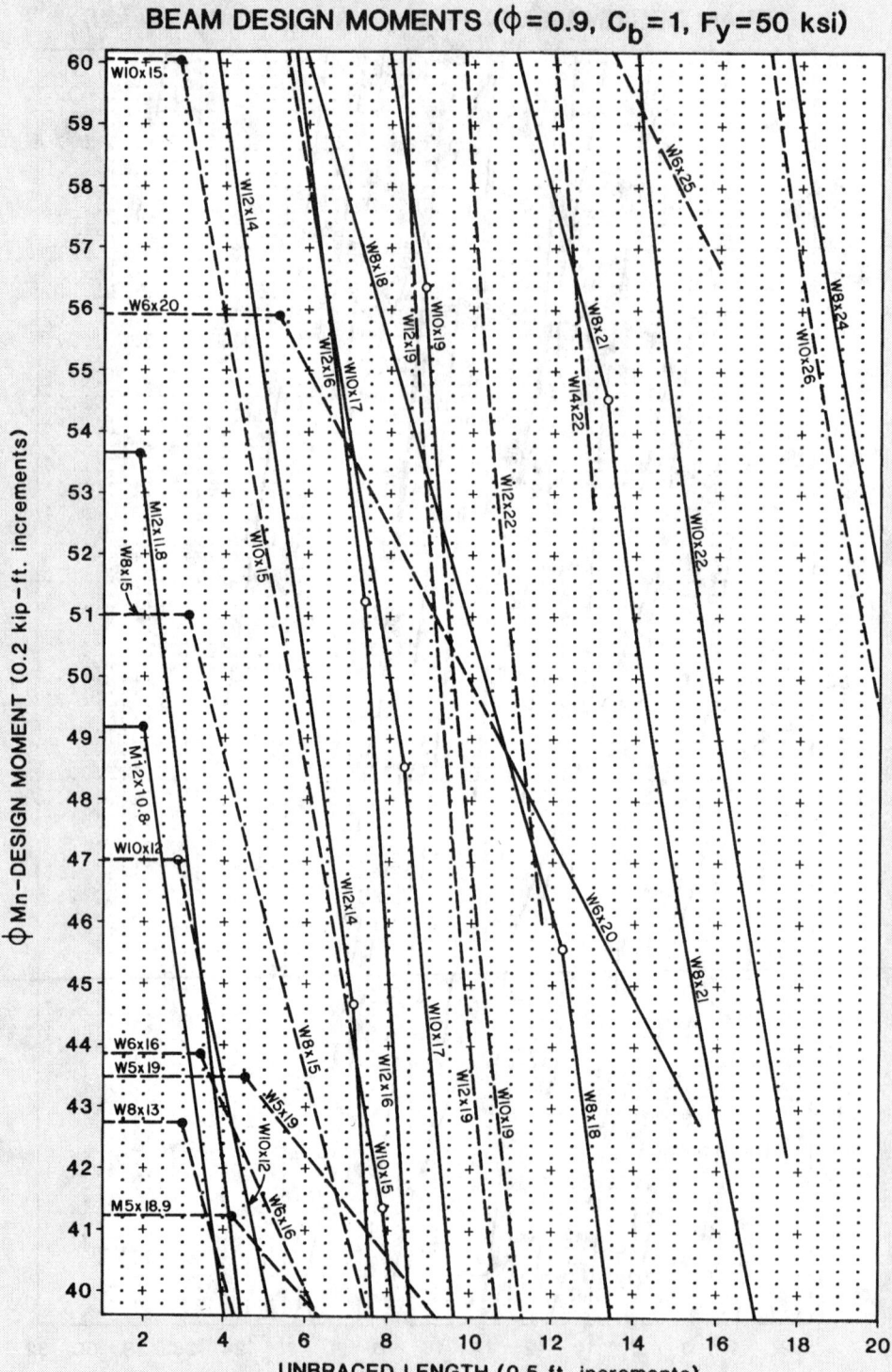

BEAM DESIGN MOMENTS ($\phi = 0.9$, $C_b = 1$, $F_y = 50$ ksi)

BEAM DESIGN MOMENTS (ϕ=0.9, C_b=1, F_y=50 ksi)

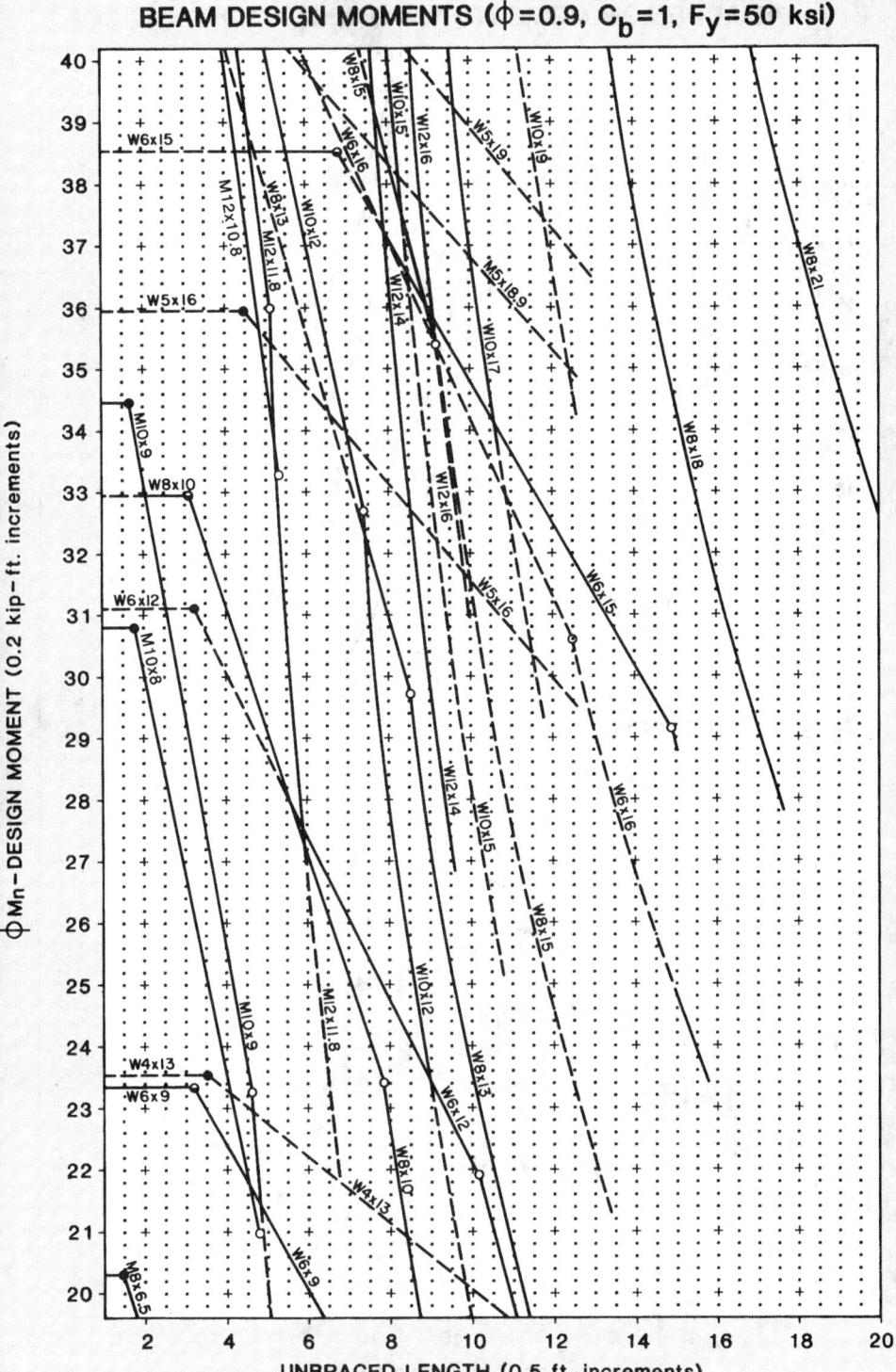

ϕM_n – DESIGN MOMENT (0.2 kip–ft. increments)

UNBRACED LENGTH (0.5 ft. increments)

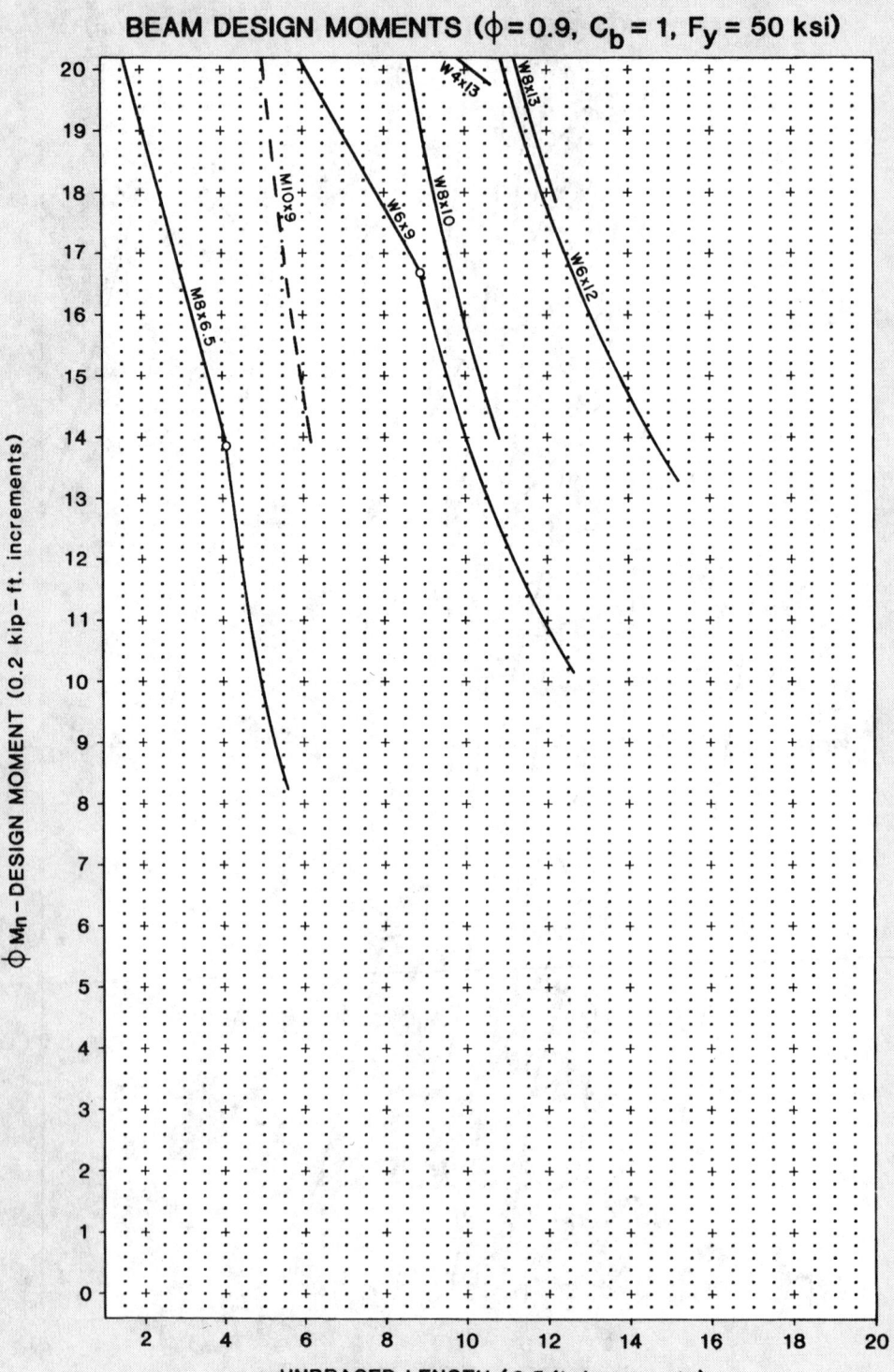

BEAM DESIGN MOMENTS ($\phi = 0.9$, $C_b = 1$, $F_y = 50$ ksi)

ϕM_n – DESIGN MOMENT (0.2 kip – ft. increments)

UNBRACED LENGTH (0.5 ft. increments)

PLATE GIRDER DESIGN

General Notes

The distinction between a beam and a plate girder, according to Chapter G of the LRFD Specification, must be made before the design can be undertaken. A beam can be a rolled or welded shape, but its web width-thickness ratio h / t_w must be less than or equal to $970 / \sqrt{F_{yf}}$. For doubly symmetric plate girders h / t_w is greater than $970 / \sqrt{F_{yf}}$.

The limit states that must be considered in plate girder design include: flexural strength, bearing under concentrated loads, shear strength, and flexure-shear interaction (for tension field action only). From these checks, the adequacy of the design and the need for stiffeners can be determined. This section contains design examples to explain these items from the LRFD Specification. A flowchart covering plate girder design has been published (Zahn, 1987).

Flexural and Shear Strength

General

In the design of welded girders, the flexural strength of the trial section must be determined to ensure that an adequate section modulus is provided. Although there are preliminary steps, flexural strength, using elastic design, is determined from LRFD Specification Section F1 if the section is compact. For sections with more slender webs, either LRFD Specification Appendix F1 or Appendix G2 is used, depending on the section's classification as a beam or plate girder.

A shear strength calculation is required to ascertain if there is a need for intermediate stiffeners. The applicable formulas are found in LRFD Specification Section F2, or Appendix G3 if tension field action is implemented. Note, however, that Appendix G cannot be used if h / t_w exceeds the limits given in Appendix G1.

Table of Dimensions and Properties of Built-up Wide Flange sections

This table serves as a guide for selecting welded built-up I sections of economical proportions. It provides dimensions and properties for a wide range of sections with nominal depths from 45 to 92 inches. No preference is intended for the tabulated flange plate dimensions, as compared to other flange plates having the same area. Substitution of wider but thinner flange plates, without a change in flange area, will result in a slight reduction in section modulus.

In analyzing overall economy, weight savings must be balanced against higher fabrication costs incurred in splicing the flanges. In some cases, it may prove economical to reduce the size of flange plates at one or more points near the girder ends, where the bending moment is substantially less. Economy through reduction of flange plate sizes is most likely to be realized with long girders, where flanges must be spliced in any case.

Only one thickness of web plate is given for each depth of girder. When the design is dominated by shear in the web, rather than flexural strength, overall economy may dictate selection of a thicker web plate. The resultant increase in elastic section modulus can be obtained by multiplying the value S', given in the table, by the number of sixteenths of an inch increase in web thickness, and adding the value obtained to the section modulus value S for the girder profile shown in the table. The increase in plastic section modulus Z can be calculated in the same way with Z'.

Overall economy may often be obtained by using a web plate of such thickness that intermediate stiffeners are not required. This is not always the case, however. The girder section listed in the table will provide a "balanced" design with respect to bending moment and web shear without excessive use of intermediate stiffeners.

The maximum design end shear strength without transverse stiffeners is given in the table column labeled $\phi_v V_n$. These values come from the equation,

$$\phi_v V_n = 0.6\phi_v A_w F_{yw} C_v$$

where

$$C_v = \frac{44,000k}{(h/t_w)^2 F_{yw}} \text{ with } k = 5.0$$

It is evident from this formula that a thicker web plate increases the design shear strength.

Design Examples

Design of a plate girder should begin with a preliminary design or selection of a trial section. The initial choice may require one or more adjustments before arriving at a final cross section that satisfies all the provisions of the LRFD Specification with maximum economy. In the following design examples, applicable provisions of the LRFD Specification are indicated at the left of each page.

In addition, references to Tables 9 and 10 in the LRFD Specification are listed. These tables may be used in place of the equations for $\phi_v V_n$. Values for $\phi_v V_n / A_w$ are given in ksi for plate girders. Tables 9-36 and 9-50 do not include the tension field action equation and, therefore, are based on LRFD Specification Section F2. For design with tension field action, Tables 10-36 and 10-50, based on Appendix G3, are applicable. Table 10 also includes the required gross area of pairs of stiffeners, as a percent of $(h \times t_w)$, from LRFD Specification Formula A-G4-1.

Example 4-10 illustrates a recommended procedure for designing a welded plate girder of constant depth. The selection of a suitable trial cross section is obtained by the flange area method, and then checked by the moment of inertia method.

Example 4-11 shows a recommended procedure for designing a welded hybrid girder of constant depth.

Example 4-12 illustrates use of the Table of Dimensions and Properties of Built-up Wide-Flange Sections to obtain an efficient trial profile. The 52-in. depth specified for this example demonstrates how tabular data may be used for girder depths intermediate to those listed. Another design requirement in this example is the omission of intermediate web stiffeners.

EXAMPLE 4-10 Design a welded plate girder to support a factored uniform load of 7 kips per foot and two concentrated factored loads of 150 kips located 17 feet from each end. The compression flange of the girder will be laterally supported only at points of concentrated load. (See Figure 4-3.)

Given: Maximum bending moment: 4,566 kip-ft

Maximum vertical shear: 318 kips

Span: 48 feet

Maximum depth: 72 inches

Steel: $F_y = 50$ ksi

Solution: A. Preliminary web design:

LRFD
Specification
Reference   1. Assume web depth, $h = 70$ inches. For noncompact web,

Section B5 & $640/\sqrt{F_y} < h/t_w \le 970/\sqrt{F_y} = 137$
Table B5.1

                     Corresponding thickness of web = 70 / 137 = 0.51 in.

(A-G1-2)   2. Assuming $a/h > 1.5$, minimum thickness of web = 70 / 243 = 0.29 in.

                     Choose thinnest web.

                     Try web plate $\frac{5}{16} \times 70$:

                     $A_w\ \ = 21.9$ in.^{2}
                     $h/t_w = 70/0.313 = 224$

                     Since $0.31 < 0.51$ in. as calculated above, expect R_{PG} to be less than 1.0

                     B. Preliminary flange design:

                     1. Required flange area:

                     An approximate formula for the area of one flange is:

                     $$A_f \approx \frac{M_u}{F_y h} = \frac{4,566(12)}{50(70)} = 15.7 \text{ in.}^2$$

                     Try 1×16 plate. $A_f = 16$ in.^{2}

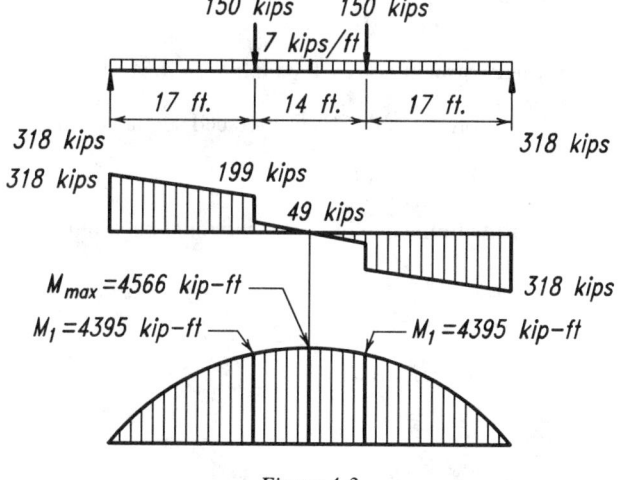

Figure 4-3

AMERICAN INSTITUTE OF STEEL CONSTRUCTION

Table B5.1   2. Check for compactness for no reduction in critical stress:

$$\frac{b_f}{2t_f} = \frac{16}{2(1)} = 8 \le 65/\sqrt{50} = 9.2 \quad \textbf{o.k.}$$

C. Trial girder section:

Web $\frac{5}{16}\times 70$; two flange plates 1×16

1. Find section modulus by moment of inertia method:

Section	A in.^{2}	y in.	ΣA_y^2 in.^{4}	I_o in.^{4}	I_{gr} in.^{4}
1 web $\frac{5}{16}\times 70$	21.9			8,932	8,932
1 flange 1×16 1 flange 1×16	16 16	35.5	40,328	3	40,331
Moment of inertia					49,263

Section modulus furnished: $49,263 / 36.0 = 1,368$ in.^{3}

Section F1 &   2. Check flexural strength using elastic design:
Appendix G2

Since $h/t_w > 970/\sqrt{F_{yf}}$, Appendix G2 applies.

Moment of inertia of flange plus $\frac{1}{6}$ web about Y-Y axis:

$$I_{oy} = 1 \times (16)^3/12 = 341 \text{ in.}^4$$
$$A_f + \frac{1}{6}A_w = 16.0 + \frac{1}{6}(21.9) = 19.65 \text{ in.}^2$$
$$r_T = \sqrt{341/19.65} = 4.17 \text{ in.}$$

a. Check limitations of Appendix G:

Assume $a/h \le 1.5$

(A-G1-1) $$(h/t_w) \max = \frac{2,000}{\sqrt{F_{yf}}} = 283 > 224 \quad \textbf{o.k.}$$

b. Check strength of 14-ft panel: $M_u = 4,566$ kip-ft

The moment in the 14-ft unbraced segment is nearly constant.

Section F1.2a   Therefore, $C_b \approx 1.0$

Appendix G2   For the limit state of lateral-torsional buckling:

(A-G2-7) $$\lambda = \frac{L_b}{r_T} = \frac{14 \times 12}{4.67} = 36.0$$

(A-G2-8) $$\lambda_p = 300/\sqrt{F_{yf}} = 42.4$$

(A-G2-9) $$\lambda_r = 756/\sqrt{F_{yf}} = 106.9$$

(A-G2-4)   Since $\lambda \le \lambda_p$, $F_{cr} = F_{yf} = 50$ ksi

For the limit state of flange local buckling:

(A-G2-11) $\quad\quad\lambda = b_f / 2t_f = 16 / (2 \times 1.0) = 8.0$

(A-G2-12) $\quad\quad\lambda_p = 65 / \sqrt{F_{yf}} = 9.2$

(A-G2-13) $\quad\quad\lambda_r = \dfrac{230}{\sqrt{F_{yf} / k_c}}$

(A-G2-4) $\quad\quad$ Since $\lambda \le \lambda_p$, $F_{cr} = F_{yf} = 50$ ksi

Design flexural strength:

$\quad\quad a_r \quad = 21.9 / 16 = 1.37$

(A-G2-3) $\quad\quad R_{PG} = 1 - \dfrac{1.37}{1,200 + 300(1.37)} \left(\dfrac{70}{0.313} - \dfrac{970}{\sqrt{50}} \right) = 0.927$

With $F_{cr} = 50$ ksi use Equation A-G2-1 or A-G2-2 as applicable:

(A-G2-1) or $\quad M_n = 1,368(1 / 12)(0.927)(1.0)(50) = 5,284$ kip-ft
(A-G2-2)

Therefore, $\phi M_n = 0.90(5,284) = 4,756$ kip-ft > 4,566 kip-ft req'd
o.k.

c. Check strength of 17-ft panels:

$\quad\quad M_u = 4,395$ kip-ft

Appendix G2 $\quad$ For moment increasing approximately linearly from zero at one end
and (F1-3) $\quad$ of the unbraced segment to a maximum value at the other end,
$\quad\quad C_b \approx 1.67$.

For the limit state of lateral-torsional buckling:

(A-G2-7) $\quad\quad\lambda \; = \dfrac{L_b}{r_T} = \dfrac{17(12)}{4.17} = 48.9$

(A-G2-8) $\quad\quad\lambda_p = 300 / \sqrt{F_{yf}} = 42.4$

(A-G2-9) $\quad\quad\lambda_r = 756 / \sqrt{F_{yf}} = 106.9$

(A-G2-5) $\quad\quad$ Since $\lambda_p \le \lambda \le \lambda_r$, $F_{cr} = C_b F_{yf} \left[1 - \dfrac{1}{2} \left(\dfrac{\lambda - \lambda_p}{\lambda_r - \lambda_p} \right) \right] \le F_{yf}$

As the middle term exceeds F_{yf}, $F_{cr} = F_{yf} = 50.0$ ksi.

For the limit state of flange local buckling:

(A-G2-4) $\quad\quad F_{cr} \; = F_{yf} = 50$ ksi (as for the 14-ft panel)

(A-G2-3) $\quad\quad R_{PG} = 0.927$ (as for the 14-ft panel)

Again, with $F_{cr} = 50$ ksi use Equation A-G2-1 or A-G2-2 as
applicable:

$\quad\quad M_n \quad = 5,284$ kip-ft (see Step C.2a)

(A-G2-1) or $\phi_b M_n = 0.90 \times 5{,}284 = 4{,}756$ kip-ft $> 4{,}395$ kip-ft req'd.   **o.k.**
(A-G2-2)

Use: Web: One plate $\frac{5}{16} \times 70$

Flanges: Two plates $\frac{7}{8} \times 18$

D. Stiffener requirements:

1. Bearing stiffeners:

Section K1   a. Check bearing at end reactions:

Assume point bearing ($N = 0$) and $\frac{5}{16}$-in. web-to-flange welds.

Check local web yielding:

(K1-2) $R_n = (5k + N)F_{yw}t_w$; $k = \frac{7}{8} + \frac{5}{16} = 1.188$ in.
$\phi R_n = 1.0[5(1.188) + 0](50)(\frac{5}{16})$
     $= 92.8$ kips < 318 **n.g.**

Therefore, provide bearing stiffeners at unframed girder ends.

(*Note:* If local web yielding criteria are satisfied, criteria set forth in Section K1.4 and K1.5 should also be checked.)

b. Bearing stiffeners are also required at concentrated load points since $92.8 < 150$ **n.g.**

2. Intermediate stiffeners:

Appendix G3   a. Check shear strength in unstiffened end panel:

$h / t_w = 224 > 418 / \sqrt{F_{yw}} = 59.1$
$a / h = 17 \times 12 / 70 = 2.9$
$V_u / A_w = 318 / 21.9 = 14.5$ ksi

Appendix G3   Tension field action is not permitted for end panels, or when $a / h > 3.0$ or $[260 / (h / t_w)]^2$. Here, $2.9 > (260 / 224)^2 = 1.35$. In either of these cases, Equations A-G3-3 and F2-3 are both applicable, as they are equivalent formulas.

Section F2.2   Using Equation F2-3,

(F2-3) or $\dfrac{\phi_v V_n}{A_w} = \dfrac{0.9(132{,}000)}{(224)^2} = 2.4 < 14.5$ ksi
(A-G3-3) or
Table 9-50

Therefore, provide intermediate stiffeners.

b. End panel stiffener spacing

(F2-3) or   Let $\dfrac{\phi_v V_n}{A_w} = 14.5$ ksi and solve for a / h.
(A-G3-3) or
Table 9-50

Result: $a / h = 0.45$

$$a \leq (0.45)(70) = 31.5 \text{ in.}$$

Use: 30 in.

c. Check for additional stiffeners:

Shear at first intermediate stiffener:

$$V_u = 318 - [7(30 / 12)] = 301 \text{ kips}$$

$$\frac{V_u}{A_w} = \frac{301}{21.9} = 13.7 \text{ ksi}$$

Distance between first intermediate stiffener and concentrated load:

$$a = (17)(12) - 30 = 174 \text{ in.}$$

(A-G3-4) $a / h = 174 / 70 = 2.5$

Then $k = 5.8$, and the shear strength is inadequate.

Therefore, provide intermediate stiffeners spaced at $174 / 2 = 87$ in.

$$a / h = 87 / 70 = 1.24$$

Appendix G3   Maximum a / h for tension field action:

$$\left[\frac{260}{(h / t_w)} \right]^2 = \left(\frac{260}{224} \right)^2 = 1.35 > 1.24$$

Design for tension field action:

For $a / h = 1.24$ and $h / t_w = 224$,

(A-G3-4) $$k = 5 + \frac{5}{(1.24)^2} = 8.2$$

Appendix G3 $$h / t_w = 224 > 234 \sqrt{8.2 / 50} = 95$$

(A-G3-6) $$C_v = \frac{44,000(8.2)}{(224)^2(50)} = 0.14$$

(A-G3-2) or
Table 10-50 $$\frac{\phi_v V_n}{A_w} = (0.9)(0.6)(50) \left[0.14 + \frac{1 - 0.14}{1.15\sqrt{1 + (1.24)^2}} \right]$$

$$= 16.5 \text{ ksi} > 13.7 \text{ ksi} \quad \textbf{o.k.}$$

d. Check center 14-ft panel:

$$h / t_w = 224$$
$$a / h = (14)(12) / 70 = 2.4 > 1.35$$
$$k = 5.0$$
$$C_v = 0.12$$

(A-G3-3) or
Table 9-50 $$\frac{\phi_v V_n}{A_w} = 2.4 \text{ ksi}$$

$$\frac{V_u}{A_w} = \frac{49}{21.9} = 2.2 \text{ ksi} < 2.4 \text{ ksi} \quad \textbf{o.k.}$$

3. Flexure-shear interaction:

Appendix G5 Check $V_u / \phi V_n$ and $M_u / \phi M_n$ at intermediate stiffener and concentrated load locations in tension field panel:

Location	V_u	ϕV_n	$V_u/\phi V_n$	M_u	ϕM_n	$M_u/\phi M_n$
2.5 ft	301	318	0.95	744	4756	0.16
9.75 ft	250	361	0.69	2769	4756	0.58
17.0 ft	199	361	0.55	4395	4756	0.92

Since $0.6 \le \dfrac{V_u}{\phi V_n} \le 1.0$ and $0.75 \le \dfrac{M_u}{\phi M_n} \le 1.0$ (with $\phi = 0.9$ for both shear and bending) do not occur simultaneously at 2.5 ft, 9.75 ft, and 17.0 ft, Interaction Equation A-G5-1 need not be checked.

Summary: space stiffeners as shown in Figure 4-4:

E. Stiffener design: Let stiffener $F_{yst} = 36$ ksi.

1. For intermediate stiffeners:

a. Area required (single plate stiffener):

For a single plate stiffener, or when $\dfrac{V_u}{\phi V_n} < 1$, use Equation A-G4-2 instead of Table 10.

(A-G4-1)
$$A_{st} = \frac{F_{yw}}{F_{yst}}\left[0.15Dht_w(1 - C_v)\frac{V_u}{\phi V_n} - 18t_w^2\right] \ge 0$$

where

h = 70 in.
t_w = 0.3125 in.
D = 2.4
C_v = 0.14
V_u = 250 kips
ϕV_n = 361 kips

$$A_{st} = \frac{50}{36}\left[0.15(2.4)(70)(0.3125)(10.14)\left(\frac{250}{361}\right) - 18(0.3125)^2\right]$$
$$= 4.07 \text{ in.}^2$$

Try one bar $\tfrac{5}{8}\times 7$

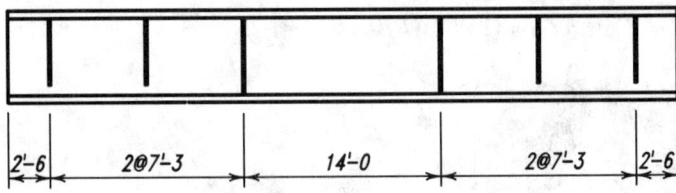

Figure 4-4

$$A_{st} = 4.38 \text{ in.}^2 > 4.07 \text{ in.}^2 \text{ req'd.} \quad \textbf{o.k.}$$

b. Check width-thickness ratio:

Table B5.1 $7 / 0.625 = 11.2 < 95 / \sqrt{F_y} = 15.8 \quad \textbf{o.k.}$

c. Check moment of inertia:

Appendix F2.3 $I_{req'd} = a t_w^3 j$

(A-F2-4)

$$j = \frac{2.5}{(1.24)^2} - 2 = -0.4 < 0.5; \text{ take } j = 0.5$$

$$I_{req'd} = 87(\tfrac{5}{16})^3 (0.5) = 1.33 \text{ in.}^4$$

$$I_{furn} = \tfrac{1}{3}(0.625)(7)^3 = 71.5 \text{ in.}^4$$

$$71.5 \text{ in.}^4 > 1.33 \text{ in.}^4 \quad \textbf{o.k.}$$

d. Minimum length required:

Section F3   It is suggested that intermediate stiffeners be stopped short of the tension flange and the weld by which they are attached to the web not closer than four times nor more than six times the web thickness from the near toe of the web-to-flange weld.*

$$70 - \tfrac{5}{16} - (4)(\tfrac{5}{16}) = 68.4 \text{ in.}$$

$$70 - \tfrac{5}{16} - (6)(\tfrac{5}{16}) = 67.8 \text{ in.}$$

Use for intermediate stiffeners:

One plate $\tfrac{5}{8} \times 7 \times 5'$-8, fillet-welded to the compression flange and web.

2. For bearing stiffeners:

At end of girder, design for end reaction.

Try two $\tfrac{5}{8} \times 8$-in. bars (see Figure 4-5).

a. Check width-thickness ratio (local buckling check):

Table B5.1 $8 / 0.625 = 12.8 < 95 / \sqrt{F_y} = 15.8 \quad \textbf{o.k.}$

b. Check compressive strength:

$$I = (\tfrac{5}{8}) \frac{(16.31)^3}{12} = 226 \text{ in.}^4$$

$$A_{eff} = (2)(8)(\tfrac{5}{8}) + [(12)(\tfrac{5}{16})^2] = 11.17 \text{ in.}^2$$

$$r = \sqrt{\frac{226}{11.17}} = 4.50 \text{ in.}$$

---

*When single stiffeners are used, they shall be attached to the compression flange, if it consists of a rectangular plate, to resist any uplift tendency due to torsion in the plate. When lateral bracing is attached to a stiffener, or a pair of stiffeners, these, in turn, shall be connected to the compression flange to transmit one percent of the total flange stress, unless the flange is composed only of angles.

Section K1.9 $KL = 0.75h = (0.75)70 = 52.5$ in.

$$\frac{Kl}{r} = \frac{52.5}{4.50} = 11.7$$

$$\lambda_c = 0.13$$

(E2-2) or   Design stress: $\phi F_{cr} = 30.38$
Table 3-36

Design strength:

$$\phi P_n = \phi F_{cr} A_g = (30.38)11.17 = 339 \text{ kips}$$

339 kips > 318 kips req'd   **o.k.**

Section J8   c. Check bearing criterion

Design strength:

$$\phi R_n = (0.75)1.8 F_y A_{pb}$$
$$A_{pb} = 2(16 - 0.5)(\tfrac{5}{8}) = 19.4 \text{ in.}^2$$

(The 0.5 accounts for cutout for welds.)

$\phi R_n = 943$ kips > 318 kips req'd.   **o.k.**

Use for bearing stiffeners: Two plates $\tfrac{5}{8}\times8\times5'\text{-}9\tfrac{3}{4}$ with close bearing on flange receiving reaction or concentrated loads.

Use same size stiffeners for bearing under concentrated loads.*

EXAMPLE 4-11   Design a hybrid girder to support a factored uniform load of three kips per foot and three concentrated factored loads of 300 kips located at the quarter points. The girder depth must be limited to five feet. The compression flange will be laterally supported throughout its length. (See Figure 4-6.)

Given: Maximum bending moment: 14,400 kip-ft

---

*In this example, bearing stiffeners were designed for end bearing; however, $25t_w$ may be used in determining effective area of web for bearing stiffeners under concentrated loads at interior panels (Section K1-9).

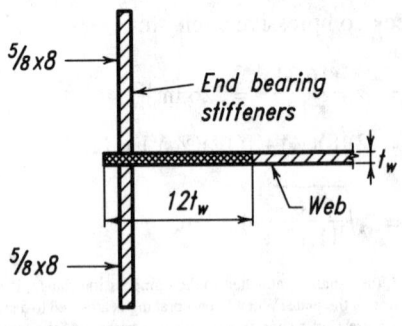

Figure 4-5

Maximum vertical shear: 570 kips
Span: 80 ft
Maximum depth: 60 in.
Steel: Flanges: $F_y = 50$ ksi
Web: $F_y = 36$ ksi

Solution: A. Preliminary web design:

LRFD
Specification   Assume web depth, $h = 54$ in.
Reference

(A-G1-2)   For $a/h > 1.5$ minimum thickness of web: $54/243 = 0.22$ in.

(A-G2-3)   For $R_{PG} = 1.0$, $h/t_w \leq 970/\sqrt{F_{yf}} = 137$

Corresponding web thickness $= 54/137 = 0.39$

(F2-1) or   Minimum t_w required for maximum $\dfrac{\phi_v V_n}{A_w}$ of 19.4 ksi:
Table 9-36

$$t_w = \frac{V_n}{19.4h} = \frac{570}{19.4 \times 54} = 0.54 \text{ in.}$$

Try web plate $\frac{5}{8} \times 54$; $A_w = 33.75$ in.^{2}

$$\frac{V_u}{A_w} = 570/33.75 = 16.9 \text{ ksi} < 19.4 \text{ ksi} \quad \textbf{o.k.}$$

Table B5.1 $h/t_w = 54/0.625 = 86.4 < (640/\sqrt{F_{yf}} = 90.5) \quad \textbf{o.k.}$

Web is compact.

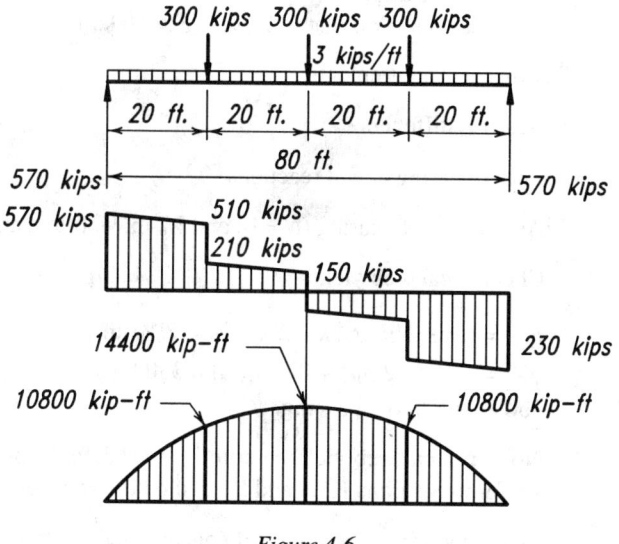

Figure 4-6

B. Preliminary flange design:

1. An approximate formula for the area of one flange is:

$$A_f \approx \frac{M_u}{F_{yf}h} = \frac{14,400(12)}{(50)(54)} = 64.0 \text{ in.}^2$$

2. Check adequacy against local buckling:

Table B5.1 $b_f / 2t_f = 24 / (2)(2.625) = 4.6 < (65 / \sqrt{F_y} = 9.2)$ **o.k.**

Flange is compact.

C. Trial girder section:

Web ⅝×54; two flange plates 2⅝×24

1. Determine plastic section moduli:

$$Z_f = 2\left[(2.625)(24)\left(\frac{54}{2} + \frac{2.625}{2}\right)\right] = 3.567 \text{ in.}^3$$

$$Z_w = 2\left[(⅝)\left(\frac{54}{2}\right)\left(\frac{54}{4}\right)\right] = 456 \text{ in.}^3$$

2. Check flexural strength:

Compression flange is supported laterally for its full length and the section is compact.

Appendix F1 $M_n = M_p = F_{yf}Z_f + F_{yw}Z_w$
$M_n = [(50)(3,567) + (36)(456)] \, 1 / 12 = 16,230 \text{ kip-ft}$
$\phi_b M_n = (0.90)16,230 = 14,610 \text{ kip-ft} > 14,400 \text{ kip-ft}$ **o.k.**

Use: Web: One plate ⅝×54 ($F_y = 36$ ksi)
       Flanges: Two plates 2⅝×24 ($F_y = 50$ ksi)

D. Stiffener requirements:

1. Bearing stiffeners

Section K1   a. Check bearing at end reactions:

Assume point bearing ($N = 0$) and ⁵⁄₁₆-in. web-to-flange welds.

Check local web yielding:

(K1-2) $R_n = (5k + N)F_{yw}t_w; \; k = 2⅝ + ⁵⁄₁₆ = 2¹⁵⁄₁₆$-in.
$\phi R_n = 1.0[(5)(2¹⁵⁄₁₆) + 0](36)(⅝) = 330$ kips
330 kips < 570 kips   **n.g.**

Note: If local web yielding criteria are satisfied, applicable criteria set for in Sections K1.4 and K1.5 should also be checked.

b. Bearing stiffeners at points of concentrated loads are also required.

2. Intermediate stiffeners:

The LRFD Specification does not permit design of hybrid girders on the basis of tension field action. Therefore, determine the need for intermediate stiffeners by use of Equations F2-1, F2-2, F2-3, Table 9-36.

a. Check shear strength without intermediate stiffeners:

Section F2 $h/t_w = 86.4$, a/h exceeds 3.0

Therefore, $k = 5.0$

$$\frac{V_u}{A_w} = \frac{570}{33.75} = 16.9 \text{ ksi}$$

Since $h/t_w = 86.4 > 523/\sqrt{36} = 87$, but $h/t_w = 86.4 < 418/\sqrt{36} = 70$, use Equation F2-2:

(F2-2) or
Table 9-36
$$\frac{\phi_v V_n}{A_w} = 15.5 \text{ ksi} < \frac{V_u}{A_w} = 16.9 \text{ ksi}$$

Therefore, intermediate stiffeners required.

b. End panel stiffener spacing

(F2-2) or
Table 9-36
$$\frac{\phi_v V_n}{A_w} = 16.9 \text{ ksi}$$

Therefore, $a/h = 2.5$ for $h/t_w = 86.4$.

Max. $a_1 = 2.5(54) = 135$ in. (Use 10 ft = 120 in.)

c. Check need for stiffeners between concentrated loads:

$h/t_w = 86.4$, a/h is over 3, $k = 5$

$$\frac{V_u}{A_w} = \frac{210}{33.75} = 6.2 \text{ ksi}$$

(F2-3) or
Table 9-36
$$\frac{\phi_v V_n}{A_w} = 15.5 \text{ ksi} > 6.2 \text{ ksi} \quad \textbf{o.k.}$$

Therefore, intermediate stiffeners not required between the concentrated loads.

Summary (see Figure 4-7):

E. Stiffener design:

1. Bearing stiffeners:

See Step E.2, Example1, for design procedure.

Use for bearing stiffeners: Two plates ¾×11×4′-5¾ with close bearing on flange receiving reaction or concentrated loads.

2. Intermediate stiffeners:

Assume $\frac{5}{16} \times 4$ in., $F_y = 36$ ksi, on each side of web.

a. Check width-thickness ratio:

Table B5.1 $4 / 0.313 = 12.8 < 95 / \sqrt{F_y} = 15.8$ **o.k.**

b. Check moment of inertia:

Appendix F2.3 $I_{req'd} = 120(\frac{5}{8})^3(0.5) = 14.6$ in.^{4}
$I_{fum} = \frac{1}{12}(0.313)(8.63)^3 = 16.8$ in.^{4}
16.8 in.$^4 > 14.6$ in.^{4} **o.k.**

c. Length required (see Step E.1.d. Example 1):

$54 - \frac{5}{16} - (6)(\frac{9}{16}) = 50\frac{5}{16}$

$54 - \frac{5}{16} - (4)(\frac{9}{16}) = 51\frac{7}{16}$ (use 51 in.)

Use for intermediate stiffeners: Two plates $\frac{5}{16} \times 4 \times 4'$-3, fillet-welded to the compression flange and web, one on each side of the web.

EXAMPLE 4-12 Design the section of a nominal 52-in. deep welded girder with no intermediate stiffeners to support a factored uniform load of 5.0 kips per linear foot on an 85-ft span. The girder will be framed between columns and its compression flange will be laterally supported for its entire length.

Given: Required bending moment: 4,516 kip-ft
Required vertical shear: 213 kips
Span: 85 ft
Nominal depth: 52 in.
Steel: $F_y = 50$ ksi

Solution: For compact web and flange, $M_n = F_y Z$

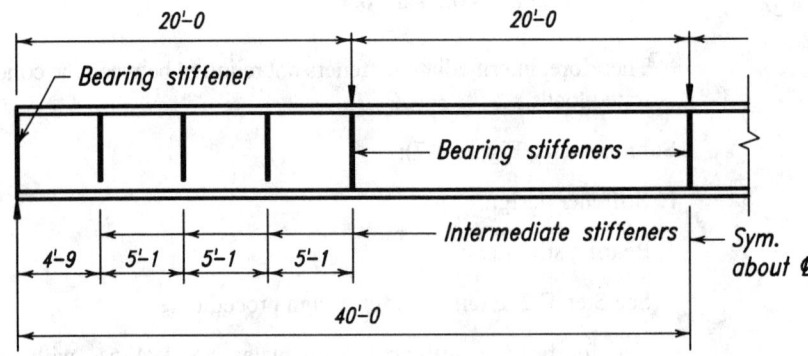

Figure 4-7

LRFD
Specification
Reference

Design for a compact web and flange:

Design for plastic moment, $F_y Z$:

Required plastic section modulus:

$$Z_{req'd} = \frac{M_u}{\phi F_y} = \frac{4,516 \times 12}{0.90 \times 50} = 1,204 \text{ in.}^3$$

Enter Table of Built-up Wide-Flange Sections, Dimensions and Properties:

For girder having $\frac{3}{8} \times 48$ web with $1\frac{1}{4} \times 16$ flange plates:

$Z = 1,200 \text{ in.}^3 < 1,204 \text{ in.}^3$

For girder having $\frac{3}{8} \times 52$ web with $1\frac{1}{4} \times 18$ flange plates:

$Z = 1,450 \text{ in.}^3 > 1,204 \text{ in.}^3$

A. Determine web required:

Table B5.1

For compact web, $h/t_w \leq 640/\sqrt{F_{yf}} = 91$

Assume $h = 50$ in.

Minimum $t_w = 50/91 = 0.55$ in.

Try: web $= \frac{9}{16} \times 50$; $A_w = 28.1 \text{ in.}^2$

$h/t_w = 50/0.56 = 89 < 91$

The web is compact.

Intermediate stiffeners can be avoided if the design shear strength of the web is adequate. (For plate girders with $h/t_w > 970/\sqrt{F_{yf}}$, refer to Appendix G4 of the LRFD Specification.)

(F2-3)

$\phi_v V_n = \phi_v (132,000 A_w)/(h/t_w)^2$
$= 0.9(132,000 \times 28.1)/(89)^2$
$= 421 \text{ kips} > 213 \text{ kips req'd.}$ **o.k.**

Therefore, no intermediate stiffeners are necessary.

B. Determine flange required.

$$A_f \approx \frac{(4,516)(12)}{50 \times 50} = 21.7 \text{ in.}^2$$

Try 1×18 plate: $A_f = 18.0 \text{ in.}^2$

Table B5.1

$b_f/2t_f = 18/(2)(1.0) = 9.0 < 65/\sqrt{F_y} = 9.2$ **o.k.**

Flange is compact.

C. Check plastic section modulus:

$$Z_x = 2\left[(18)(1.0)\left(\frac{51.0}{2}\right) + \left(\frac{9}{16}\right)\left(\frac{50}{2}\right)\left(\frac{50}{4}\right)\right] = 1{,}270 \text{ in.}^3$$

$1{,}270 \text{ in.}^3 > 1{,}204 \text{ in.}^3$ req'd.   **o.k.**

Use: Web: One plate $\frac{9}{16}{\times}50$

      Flanges:* Two plates $1{\times}18$

Section K1.8   *Note:* Because this girder will be framed between columns, the usual end bearing stiffeners are not required.

---

*Because this girder is longer than 60 feet, some economy may be gained by decreasing the flange size in areas of smaller moment, i.e., near ends of girder.

BUILT-UP WIDE-FLANGE SECTIONS
Dimensions and properties

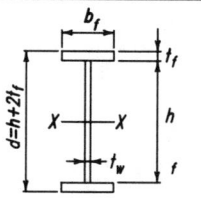

Nominal Size (h/tw) In.	Wt. per Ft Lb.	Area In.²	Depth d In.	Flange Width bf In.	Flange Thick tf In.	Web Depth h In.	Web Thick tw In.	I In.⁴	S In.³	ªS' In.³	Z In.³	ᵇZ In.³	ᶜφᵥVn Kips
92×30	823	242	96.00	30	3	90	11/16	431000	8980	79.1	9760	127	428.9
(131)	721	212	95.00	30	2½	90	11/16	363000	7640	79.9	8330	127	428.9
	619	182	94.00	30	2	90	11/16	296000	6290	80.8	6910	127	428.9
	568	167	93.50	30	1¾	90	11/16	263000	5620	81.2	6210	127	428.9
	517	152	93.00	30	1½	90	11/16	230000	4950	81.7	5510	127	428.9
	466	137	92.50	30	1¼	90	11/16	198000	4280	82.1	4810	127	428.9
	415	122	92.00	30	1	90	11/16	166000	3610	82.5	4120	127	428.9
86×28	750	220	90.00	28	3	84	5/8	349000	7750	68.6	8410	110	345.3
(134)	654	192	89.00	28	2½	84	5/8	293000	6580	69.4	7160	110	345.3
	559	164	88.00	28	2	84	5/8	238000	5410	70.2	5920	110	345.3
	512	150	87.50	28	1¾	84	5/8	211000	4820	70.6	5300	110	345.3
	464	136	87.00	28	1½	84	5/8	184000	4240	71.0	4690	110	345.3
	416	122	86.50	28	1¼	84	5/8	158000	3650	71.4	4090	110	345.3
	369	108	86.00	28	1	84	5/8	132000	3070	71.8	3480	110	345.3
80×26	696	205	84.00	26	3	78	5/8	281000	6680	58.8	7270	95.1	371.8
(125)	609	179	83.00	26	2½	78	5/8	235000	5670	59.6	6180	95.1	371.8
	519	153	82.00	26	2	78	5/8	191000	4660	60.3	5110	95.1	371.8
	475	140	81.50	26	1¾	78	5/8	169000	4160	60.7	4580	95.1	371.8
	431	127	81.00	26	1½	78	5/8	148000	3650	61.0	4050	95.1	371.8
	387	114	80.50	26	1¼	78	5/8	127000	3150	61.4	3530	95.1	371.8
	343	101	80.00	26	1	78	5/8	106000	2650	61.8	3000	95.1	371.8
	320	94.2	79.75	26	7/8	78	5/8	95500	2390	62.0	2750	95.1	371.8
74×24	627	184	78.00	24	3	72	9/16	220000	5640	49.8	6130	81.0	293.7
(128)	546	160	77.00	24	2½	72	9/16	184000	4780	50.5	5200	81.0	293.7
	464	136	76.00	24	2	72	9/16	149000	3920	51.2	4280	81.0	293.7
	423	124	75.50	24	1¾	72	9/16	132000	3490	51.5	3830	81.0	293.7
	382	112	75.00	24	1½	72	9/16	115000	3060	51.8	3380	81.0	293.7
	342	100	74.50	24	1¼	72	9/16	98000	2630	52.2	2930	81.0	293.7
	301	88.5	74.00	24	1	72	9/16	81400	2200	52.5	2480	81.0	293.7
	280	82.5	73.75	24	7/8	72	9/16	73300	1990	52.7	2260	81.0	293.7
68×22	561	165	72.00	22	3	66	½	169000	4700	41.6	5100	68.1	225.0
(132)	486	143	71.00	22	2½	66	½	141000	3970	42.2	4310	68.1	225.0
	411	121	70.00	22	2	66	½	114000	3250	42.8	3540	68.1	225.0
	374	110	69.50	22	1¾	66	½	100000	2890	43.1	3150	68.1	225.0
	337	99.0	69.00	22	1½	66	½	87000	2530	43.4	2770	68.1	225.0
	299	88.0	68.50	22	1¼	66	½	74000	2170	43.7	2390	68.1	225.0
	262	77.0	68.00	22	1	66	½	61000	1800	44.0	2020	68.1	225.0
	243	71.5	67.75	22	7/8	66	½	55000	1620	44.2	1830	68.1	225.0
	224	66.0	67.50	22	¾	66	½	49000	1440	44.4	1650	68.1	225.0

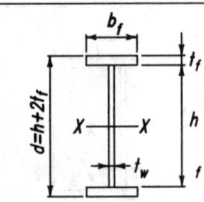

BUILT-UP WIDE-FLANGE SECTIONS
Dimensions and properties

Nominal Size (h/t_w)	Wt. per Ft	Area	Depth	Flange		Web		Axis X-X					
			d	Width b_f	Thick t_f	Depth h	Thick t_w	I	S	$^aS'$	Z	^bZ	$^c\phi_v V_n$
In.	Lb.	In.^{2}	In.	In.	In.	In.	In.	In.^{4}	In.^{3}	In.^{3}	In.^{3}	In.^{3}	Kips
61×20	429	126	65.00	20	$2\frac{1}{2}$	60	$\frac{7}{16}$	106000	3250	34.6	3520	56.3	165.8
(137)	361	106	64.00	20	2	60	$\frac{7}{16}$	84800	2650	35.2	2870	56.3	165.8
	327	96.2	63.50	20	$1\frac{3}{4}$	60	$\frac{7}{16}$	74600	2350	35.4	2560	56.3	165.8
	293	86.2	63.00	20	$1\frac{1}{2}$	60	$\frac{7}{16}$	64600	2050	35.7	2240	56.3	165.8
	259	76.2	62.50	20	$1\frac{1}{4}$	60	$\frac{7}{16}$	54800	1750	36.0	1930	56.3	165.8
	225	66.2	62.00	20	1	60	$\frac{7}{16}$	45100	1450	36.3	1610	56.3	165.8
	208	61.2	61.75	20	$\frac{7}{8}$	60	$\frac{7}{16}$	40300	1310	36.4	1460	56.3	165.8
	191	56.2	61.50	20	$\frac{3}{4}$	60	$\frac{7}{16}$	35600	1160	36.6	1310	56.3	165.8
57×18	389	115	61.00	18	$2\frac{1}{2}$	56	$\frac{7}{16}$	83500	2740	30.0	2980	49.0	177.6
(128)	328	96.5	60.00	18	2	56	$\frac{7}{16}$	67000	2230	30.5	2430	49.0	177.6
	298	87.5	59.50	18	$1\frac{3}{4}$	56	$\frac{7}{16}$	58900	1980	30.7	2160	49.0	177.6
	267	78.5	59.00	18	$1\frac{1}{2}$	56	$\frac{7}{16}$	51000	1730	31.0	1900	49.0	177.6
	236	69.5	58.50	18	$1\frac{1}{4}$	56	$\frac{7}{16}$	43300	1480	31.3	1630	49.0	177.6
	206	60.5	58.00	18	1	56	$\frac{7}{16}$	35600	1230	31.5	1370	49.0	177.6
	190	56.0	57.75	18	$\frac{7}{8}$	56	$\frac{7}{16}$	31900	1100	31.7	1240	49.0	177.6
	175	51.5	57.50	18	$\frac{3}{4}$	56	$\frac{7}{16}$	28100	979	31.8	1110	49.0	177.6
	160	47.0	57.25	18	$\frac{5}{8}$	56	$\frac{7}{16}$	24400	854	32.0	980	49.0	177.6
53×18	342	100	56.50	18	$2\frac{1}{4}$	52	$\frac{3}{8}$	64000	2270	25.9	2450	42.3	120.5
(138)	311	91.5	56.00	18	2	52	$\frac{3}{8}$	56900	2030	26.2	2200	42.3	120.5
	280	82.5	55.50	18	$1\frac{3}{4}$	52	$\frac{3}{8}$	49900	1800	26.4	1950	42.3	120.5
	250	73.5	55.00	18	$1\frac{1}{2}$	52	$\frac{3}{8}$	43000	1570	26.6	1700	42.3	120.5
	219	64.5	54.50	18	$1\frac{1}{4}$	52	$\frac{3}{8}$	36300	1330	26.9	1450	42.3	120.5
	189	55.5	54.00	18	1	52	$\frac{3}{8}$	29700	1100	27.1	1210	42.3	120.5
	173	51.0	53.75	18	$\frac{7}{8}$	52	$\frac{3}{8}$	26400	983	27.2	1090	42.3	120.5
	158	46.5	53.50	18	$\frac{3}{4}$	52	$\frac{3}{8}$	23200	866	27.4	966	42.3	120.5
	143	42.0	53.25	18	$\frac{5}{8}$	52	$\frac{3}{8}$	20000	750	27.5	846	42.3	120.5
49×16	306	90.0	52.50	16	$2\frac{1}{4}$	48	$\frac{3}{8}$	48900	1860	21.9	2030	36.0	130.5
(128)	279	82.0	52.00	16	2	48	$\frac{3}{8}$	43500	1670	22.2	1820	36.0	130.5
	252	74.0	51.50	16	$1\frac{3}{4}$	48	$\frac{3}{8}$	38100	1480	22.4	1610	36.0	130.5
	224	66.0	51.00	16	$1\frac{1}{2}$	48	$\frac{3}{8}$	32900	1290	22.6	1400	36.0	130.5
	197	58.0	50.50	16	$1\frac{1}{4}$	48	$\frac{3}{8}$	27700	1100	22.8	1200	36.0	130.5
	170	50.0	50.00	16	1	48	$\frac{3}{8}$	22700	910	23.0	1000	36.0	130.5
	156	46.0	49.75	16	$\frac{7}{8}$	48	$\frac{3}{8}$	20200	811	23.2	900	36.0	130.5
	143	42.0	49.50	16	$\frac{3}{4}$	48	$\frac{3}{8}$	17700	716	23.3	801	36.0	130.5
	129	38.0	49.25	16	$\frac{5}{8}$	48	$\frac{3}{8}$	15300	620	23.4	702	36.0	130.5
45×16	237	69.8	47.50	16	$1\frac{3}{4}$	44	$\frac{5}{16}$	31500	1330	18.7	1430	30.3	82.4
(141)	210	61.8	47.00	16	$1\frac{1}{2}$	44	$\frac{5}{16}$	27100	1150	18.9	1240	30.3	82.4
	183	53.8	46.50	16	$1\frac{1}{4}$	44	$\frac{5}{16}$	22700	976	19.1	1060	30.3	82.4

BUILT-UP WIDE-FLANGE SECTIONS
Dimensions and properties

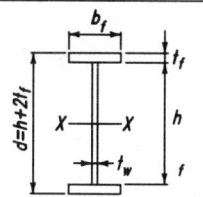

Nominal Size (h/t_w)	Wt. per Ft	Area	Depth	Flange		Web		Axis X-X					
			d	Width b_f	Thick t_f	Depth h	Thick t_w	I	S	$^aS'$	Z	$^bZ'$	$^c\phi_v V_n$
In.	Lb.	In.^{2}	In.	In.	In.	In.	In.	In.^{4}	In.^{3}	In.^{3}	In.^{3}	In.^{3}	Kips
45×16	156	45.8	46.00	16	1	44	$^5/_{16}$	18400	801	19.3	871	30.3	82.4
(141)	142	41.8	45.75	16	$^7/_8$	44	$^5/_{16}$	16300	713	19.4	780	30.3	82.4
	129	37.8	45.50	16	$^3/_4$	44	$^5/_{16}$	14200	626	19.5	688	30.3	82.4
	116	33.8	45.25	16	$^5/_8$	44	$^5/_{16}$	12200	538	19.6	598	30.3	82.4

$^a S'$ = Additional section modulus corresponding to $^1/_{16}$" increase in web thickness.
$^b Z'$ = Additional plastic section modulus corresponding to $^1/_{16}$" increase in web thickness.
$^c \phi_v V_n$ = Maximum design end shear strength permissible without transverse stiffeners for tabulated web plate (LRFD Specification Section F2). $\phi_v = 0.90$.

Notes:
Based on their width-thickness ratios the girders in this table are noncompact shapes in accordance with LRFD Specification Section B5 for $F_y = 36$ ksi steel. For steels of higher yield strengths, check flanges for compliance with this section.
This table does not consider local effects on the web due to concentrated loads. (See LRFD Specification Section K1.)
See LRFD Specification Appendix G4 for design of stiffeners.
Welds are not included in the tabulated weight per foot.

BEAM DIAGRAMS AND FORMULAS

Nomenclature

E = modulus of elasticity of steel at 29,000 ksi

I = moment of inertia of beam (in.^{4})

L = total length of beam between reaction points (ft)

M_{max} = maximum moment (kip-in.)

M_1 = maximum moment in left section of beam (kip-in.)

M_2 = maximum moment in right section of beam (kip-in.)

M_3 = maximum positive moment in beam with combined end moment conditions (kip-in.)

M_x = moment at distance x from end of beam (kip-in.)

P = concentrated load (kips)

P_1 = concentrated load nearest left reaction (kips)

P_2 = concentrated load nearest right reaction, and of different magnitude than P_1 (kips)

R = end beam reaction for any condition of symmetrical loading (kips)

R_1 = left end beam reaction (kips)

R_2 = right end or intermediate beam reaction (kips)

R_3 = right end beam reaction (kips)

V = maximum vertical shear for any condition of symmetrical loading (kips)

V_1 = maximum vertical shear in left section of beam (kips)

V_2 = vertical shear at right reaction point, or to left of intermediate reaction point of beam (kips)

V_3 = vertical shear at right reaction point, or to right of intermediate reaction point of beam (kips)

V_x = vertical shear at distance x from end of beam (kips)

W = total load on beam (kips)

a = measured distance along beam (in.)

b = measured distance along beam which may be greater or less than a (in.)

l = total length of beam between reaction points (in.)

w = uniformly distributed load per unit of length (kips per in.)

w_1 = uniformly distributed load per unit of length nearest left reaction (kips per in.)

w_2 = uniformly distributed load per unit of length nearest right reaction, and of different magnitude than w_1 (kips per in.)

x = any distance measured along beam from left reaction (in.)

x_1 = any distance measured along overhang section of beam from nearest reaction point (in.)

Δ_{max} = maximum deflection (in.)

Δ_a = deflection at point of load (in.)

Δ_x = deflection at any point x distance from left reaction (in.)

Δ_{x1} = deflection of overhang section of beam at any distance from nearest reaction point (in.)

BEAM DIAGRAMS AND FORMULAS
Frequently Used Formulas

The formulas given below are frequently required in structural designing. They are included herein for the convenience of those engineers who have infrequent use for such formulas and hence may find reference necessary. Variation from the standard nomenclature on page 4-187 is noted.

BEAMS

Flexural stress at extreme fiber:

$$f = Mc / I = M / S$$

Flexural stress at any fiber:

$$f = My / I \qquad y = \text{distance from neutral axis to fiber}$$

Average vertical shear (for maximum see below):

$$v = V / A = V / dt \text{ (for beams and girders)}$$

Horizontal shearing stress at any section A-A:

$$v = VQ / Ib \qquad Q = \text{statical moment about the neutral axis of that portion of the cross section lying outside of section A-A}$$

$$b = \text{ width at section A-A}$$

(Intensity of vertical shear is equal to that of horizontal shear acting normal to it at the same point and both are usually a maximum at mid-height of beam.)

Slope and deflection at any point:

$$EI\frac{d^2y}{dx^2} = M \qquad x \text{ and } y \text{ are abscissa and ordinate respectively of a point on the neutral axis, referred to axes of rectangular coordinates through a selected point of support.}$$

(First integration gives slopes; second integration gives deflections. Constants of integration must be determined.)

CONTINUOUS BEAMS (the theorem of three moments)

Uniform load:

$$M_a\frac{l_1}{I_1} + 2M_b\left(\frac{l_1}{I_1} + \frac{l_2}{I_2}\right) + M_c\frac{l_2}{I_2} = -\tfrac{1}{4}\left(\frac{w_1 l_1^3}{I_1} + \frac{w_2 l_2^3}{I_2}\right)$$

Concentrated loads:

$$M_a\frac{l_1}{I_1} + 2M_b\left(\frac{l_1}{I_1} + \frac{l_2}{I_2}\right) + M_c\frac{l_2}{I_2} = -\frac{P_1 a_1 b_1}{I_1}\left(1 + \frac{a_1}{l_1}\right) - \frac{P_2 a_2 b_2}{I_2}\left(1 + \frac{b_2}{l_2}\right)$$

Considering any two consecutive spans in any continuous structure:

M_a, M_b, M_c = moments at left, center, and right supports respectively, of any pair of adjacent spans

l_1 and l_2 = length of left and right spans, respectively, of the pair

I_1 and I_2 = moment of inertia of left and right spans, respectively

w_1 and w_2 = load per unit of length on left and right spans, respectively

P_1 and P_2 = concentrated loads on left and right spans, respectively

a_1 and a_2 = distance of concentrated loads from left support, in left and right spans, respectively

b_1 and b_2 = distance of concentrated loads from right support, in left and right spans, respectively

The above equations are for beam with moment of inertia constant in each span but differing in different spans, continuous over three or more supports. By writing such an equation for each successive pair of spans and introducing the known values (usually zero) of end moments, all other moments can be found.

BEAM DIAGRAMS AND FORMULAS
Table of Concentrated Load Equivalents

n	Loading	Coeff.	Simple Beam	Beam Fixed One End, Supported at Other	Beam Fixed Both Ends
∞		a	0.125	0.070	0.042
		b	—	0.125	0.083
		c	0.500	0.375	—
		d	—	0.625	0.500
		e	0.013	0.005	0.003
		f	1.000	1.000	0.667
		g	1.000	0.415	0.300
2		a	0.250	0.156	0.125
		b	—	0.188	0.125
		c	0.500	0.313	—
		d	—	0.688	0.500
		e	0.021	0.009	0.005
		f	2.000	1.500	1.000
		g	0.800	0.477	0.400
3		a	0.333	0.222	0.111
		b	—	0.333	0.222
		c	1.000	0.667	—
		d	—	1.333	1.000
		e	0.036	0.015	0.008
		f	2.667	2.667	1.778
		g	1.022	0.438	0.333
4		a	0.500	0.266	0.188
		b	—	0.469	0.313
		c	1.500	1.031	—
		d	—	1.969	1.500
		e	0.050	0.021	0.010
		f	4.000	3.750	2.500
		g	0.950	0.428	0.320
5		a	0.600	0.360	0.200
		b	—	0.600	0.400
		c	2.000	1.400	—
		d	—	2.600	2.000
		e	0.063	0.027	0.013
		f	4.800	4.800	3.200
		g	1.008	0.424	0.312

Maximum positive moment (kip-ft): aPL
Maximum negative moment (kip-ft): bPL
Pinned end reaction (kips): cP
Fixed end reaction (kips): dP
Maximum deflection (in): ePl^3 / EI

Equivalent simple span uniform load (kips): fP
Deflection coefficient for equivalent simple span uniform load: g
Number of equal load spaces: n
Span of beam (ft): L
Span of beam (in): l

BEAM DIAGRAMS AND FORMULAS
For Various Static Loading Conditions

For meaning of symbols, see page 4-187

---

1. SIMPLE BEAM—UNIFORMLY DISTRIBUTED LOAD

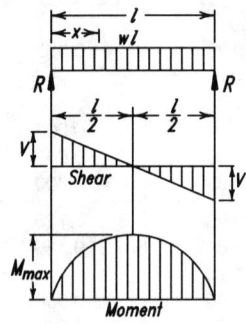

Total Equiv. Uniform Load $\ldots\ldots = wl$

$R = V \quad\ldots\ldots\ldots\ldots = \dfrac{wl}{2}$

$V_x \quad\quad\quad\ldots\ldots\ldots = w\left(\dfrac{l}{2} - x\right)$

M_{max} (at center) $\ldots\ldots\ldots = \dfrac{wl^2}{8}$

$M_x \quad\quad\ldots\ldots\ldots\ldots = \dfrac{wx}{2}(l - x)$

Δ_{max} (at center) $\ldots\ldots\ldots = \dfrac{5wl^4}{384EI}$

$\Delta_x \quad\quad\ldots\ldots\ldots\ldots = \dfrac{wx}{24EI}(l^3 - 2lx^2 + x^3)$

---

2. SIMPLE BEAM—LOAD INCREASING UNIFORMLY TO ONE END

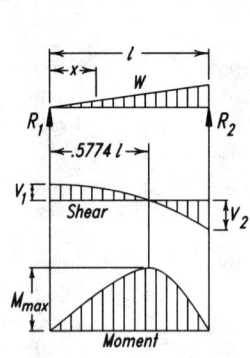

Total Equiv. Uniform Load $\ldots\ldots = \dfrac{16W}{9\sqrt{3}} = 1.0264W$

$R_1 = V_1 \ldots\ldots\ldots\ldots = \dfrac{W}{3}$

$R_2 = V_2 = V_{max} \ldots\ldots\ldots = \dfrac{2W}{3}$

$V_x \quad\quad\quad\ldots\ldots\ldots = \dfrac{W}{3} - \dfrac{Wx^2}{l^2}$

M_{max} (at $x = \dfrac{l}{\sqrt{3}} = .5774\,l$) $\ldots\ldots = \dfrac{2Wl}{9\sqrt{3}} = .1283\,Wl$

$M_x \quad\quad\ldots\ldots\ldots\ldots = \dfrac{Wx}{3l^2}(l^2 - x^2)$

Δ_{max} (at $x = l\sqrt{1 - \sqrt{\dfrac{8}{15}}} = .5193\,l$) $\ldots = 0.01304\dfrac{Wl^3}{EI}$

$\Delta_x \quad\ldots\ldots\ldots\ldots = \dfrac{Wx}{180EIl^2}(3x^4 - 10l^2x^2 + 7l^4)$

---

3. SIMPLE BEAM—LOAD INCREASING UNIFORMLY TO CENTER

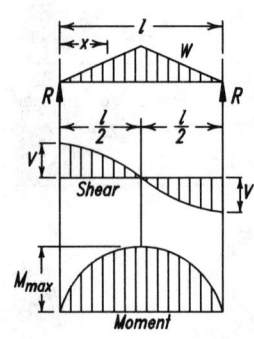

Total Equiv. Uniform Load $\ldots\ldots = \dfrac{4W}{3}$

$R = V \quad\ldots\ldots\ldots\ldots = \dfrac{W}{2}$

$V_x \quad$ (when $x < \dfrac{l}{2}$) $\ldots\ldots\ldots = \dfrac{W}{2l^2}(l^2 - 4x^2)$

M_{max} (at center) $\ldots\ldots\ldots = \dfrac{Wl}{6}$

$M_x \quad$ (when $x < \dfrac{l}{2}$) $\ldots\ldots\ldots = Wx\left(\dfrac{1}{2} - \dfrac{2x^2}{3l^2}\right)$

Δ_{max} (at center) $\ldots\ldots\ldots = \dfrac{Wl^3}{60EI}$

$\Delta_x \quad$ (when $x < \dfrac{l}{2}$) $\ldots\ldots\ldots = \dfrac{Wx}{480EIl^2}(5l^2 - 4x^2)^2$

BEAM DIAGRAMS AND FORMULAS
For Various Static Loading Conditions

For meaning of symbols, see page 4-187

---

4. SIMPLE BEAM—UNIFORM LOAD PARTIALLY DISTRIBUTED

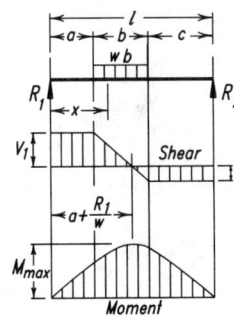

$R_1 = V_1$ (max. when $a < c$) . . . . $= \dfrac{wb}{2l}(2c + b)$

$R_2 = V_2$ (max. when $a > c$) . . . . $= \dfrac{wb}{2l}(2a + b)$

V_x (when $x > a$ and $< (a + b)$) . . . $= R_1 - w(x - a)$

M_{max} $\left(\text{at } x = a + \dfrac{R_1}{w}\right)$ . . . . $= R_1\left(a + \dfrac{R_1}{2w}\right)$

M_x (when $x < a$) . . . . $= R_1 x$

M_x (when $x > a$ and $< (a + b)$) . . . $= R_1 x - \dfrac{w}{2}(x - a)^2$

M_x (when $x > (a + b)$) . . . . $= R_2(l - x)$

---

5. SIMPLE BEAM—UNIFORM LOAD PARTIALLY DISTRIBUTED AT ONE END

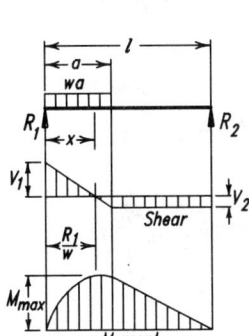

$R_1 = V_1 = V_{max}$ . . . . $= \dfrac{wa}{2l}(2l - a)$

$R_2 = V_2$ . . . . $= \dfrac{wa^2}{2l}$

V_x (when $x < a$) . . . . $= R_1 - wx$

M_{max} $\left(\text{at } x = \dfrac{R_1}{w}\right)$ . . . . $= \dfrac{R_1^2}{2w}$

M_x (when $x < a$) . . . . $= R_1 x - \dfrac{wx^2}{2}$

M_x (when $x > a$) . . . . $= R_2(l - x)$

Δ_x (when $x < a$) . . . . $= \dfrac{wx}{24EIl}(a^2(2l - a)^2 - 2ax^2(2l - a) + lx^3)$

Δ_x (when $x > a$) . . . . $= \dfrac{wa^2(l - x)}{24EIl}(4xl - 2x^2 - a^2)$

---

6. SIMPLE BEAM—UNIFORM LOAD PARTIALLY DISTRIBUTED AT EACH END

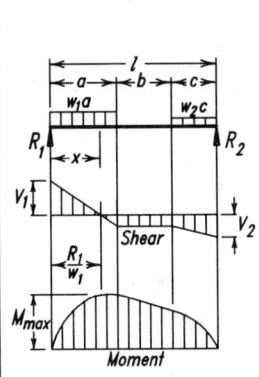

$R_1 = V_1$ . . . . $= \dfrac{w_1 a(2l - a) + w_2 c^2}{2l}$

$R_2 = V_2$ . . . . $= \dfrac{w_2 c(2l - c) + w_1 a^2}{2l}$

V_x (when $x < a$) . . . . $= R_1 - w_1 x$

V_x (when $x > a$ and $< (a + b)$) $= R_1 - w_1 a$

V_x (when $x > (a + b)$) . . . . . . $= R_2 - w_2(l - x)$

M_{max} $\left(\text{at } x = \dfrac{R_1}{w_1} \text{ when } R_1 < w_1 a\right)$ $= \dfrac{R_1^2}{2w_1}$

M_{max} $\left(\text{at } x = l - \dfrac{R_2}{w_2} \text{ when } R_2 < w_2 c\right)$ $= \dfrac{R_2^2}{2w_2}$

M_x (when $x < a$) . . . . $= R_1 x - \dfrac{w_1 x^2}{2}$

M_x (when $x > a$ and $< (a + b)$) . . . $= R_1 x - \dfrac{w_1 a}{2}(2x - a)$

M_x (when $x > (a + b)$) . . . . $= R_2(l - x) - \dfrac{w_2(l - x)^2}{2}$

BEAM DIAGRAMS AND FORMULAS
For Various Static Loading Conditions
For meaning of symbols, see page 4-187

7. SIMPLE BEAM—CONCENTRATED LOAD AT CENTER

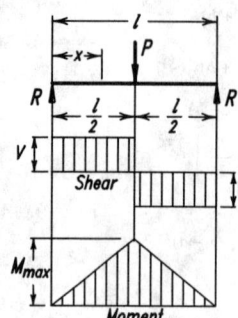

Total Equiv. Uniform Load $\dots\dots\dots = 2P$

$R = V \quad \dots\dots\dots\dots\dots\dots = \dfrac{P}{2}$

$M_{max} \quad$ (at point of load) $\dots\dots\dots = \dfrac{Pl}{4}$

$M_x \quad \left(\text{when } x < \dfrac{l}{2}\right) \dots\dots\dots = \dfrac{Px}{2}$

$\Delta_{max} \quad$ (at point of load) $\dots\dots\dots = \dfrac{Pl^3}{48EI}$

$\Delta_x \quad \left(\text{when } x < \dfrac{l}{2}\right) \dots\dots\dots = \dfrac{Px}{48EI}(3l^2 - 4x^2)$

8. SIMPLE BEAM—CONCENTRATED LOAD AT ANY POINT

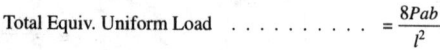

Total Equiv. Uniform Load $\dots\dots\dots = \dfrac{8Pab}{l^2}$

$R_1 = V_1 \ (\text{max when } a < b) \dots\dots\dots = \dfrac{Pb}{l}$

$R_2 = V_2 \ (\text{max when } a > b) \dots\dots\dots = \dfrac{Pa}{l}$

$M_{max} \quad$ (at point of load) $\dots\dots\dots = \dfrac{Pab}{l}$

$M_x \quad$ (when $x < a$) $\dots\dots\dots\dots = \dfrac{Pbx}{l}$

$\Delta_{max} \quad \left(\text{at } x = \sqrt{\dfrac{a(a+2b)}{3}} \text{ when } a > b\right) \dots = \dfrac{Pab(a+2b)\sqrt{3a(a+2b)}}{27EIl}$

$\Delta_a \quad$ (at point of load) $\dots\dots\dots = \dfrac{Pa^2b^2}{3EIl}$

$\Delta_x \quad$ (when $x < a$) $\dots\dots\dots\dots = \dfrac{Pbx}{6EIl}(l^2 - b^2 - x^2)$

9. SIMPLE BEAM—TWO EQUAL CONCENTRATED LOADS SYMMETRICALLY PLACED

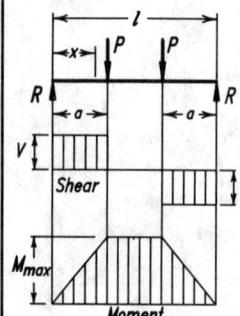

Total Equiv. Uniform Load $\dots\dots\dots = \dfrac{8Pa}{l}$

$R = V \quad \dots\dots\dots\dots\dots\dots = P$

$M_{max} \quad$ (between loads) $\dots\dots\dots = Pa$

$M_x \quad$ (when $x < a$) $\dots\dots\dots\dots = Px$

$\Delta_{max} \quad$ (at center) $\dots\dots\dots\dots = \dfrac{Pa}{24EI}(3l^2 - 4a^2)$

$\Delta_x \quad$ (when $x < a$) $\dots\dots\dots\dots = \dfrac{Px}{6EI}(3la - 3a^2 - x^2)$

$\Delta_x \quad$ (when $x > a$ and $< (l-a)$) $\dots\dots = \dfrac{Pa}{6EI}(3lx - 3x^2 - a^2)$

BEAM DIAGRAMS AND FORMULAS
For Various Static Loading Conditions
For meaning of symbols, see page 4-187

---

10. SIMPLE BEAM—TWO EQUAL CONCENTRATED LOADS UNSYMMETRICALLY PLACED

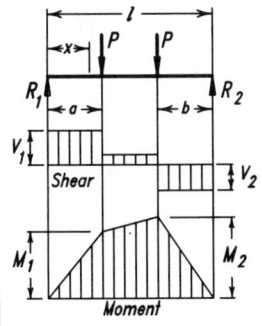

$R_1 = V_1$ (max. when $a < b$) . . . . $= \dfrac{P}{l}(l - a + b)$

$R_2 = V_2$ (max. when $a > b$) . . . . $= \dfrac{P}{l}(l - b + a)$

V_x (when $x > a$ and $< (l - b)$) . . . . . $= \dfrac{P}{l}(b - a)$

M_1 (max. when $a > b$) . . . . $= R_1 a$

M_2 (max. when $a < b$) . . . . $= R_2 b$

M_x (when $x < a$) . . . . $= R_1 x$

M_x (when $x > a$ and $< (l - b)$) . . . . . $= R_1 x - P_1 (x - a)$

---

11. SIMPLE BEAM—TWO UNEQUAL CONCENTRATED LOADS UNSYMMETRICALLY PLACED

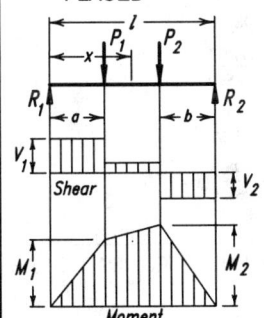

$R_1 = V_1$ . . . . $= \dfrac{P_1(l - a) + P_2 b}{l}$

$R_2 = V_2$ . . . . $= \dfrac{P_1 a + P_2(l - b)}{l}$

V_x (when $x > a$ and $< (l - b)$) . . . . . $= R_1 - P_1$

M_1 (max. when $R_1 < P_1$) . . . . $= R_1 a$

M_2 (max. when $R_2 < P_2$) . . . . $= R_2 b$

M_x (when $x < a$) . . . . $= R_1 x$

M_x (when $x > a$ and $< (l - b)$) . . . . . $= R_1 x - P_1(x - a)$

---

12. BEAM FIXED AT ONE END, SUPPORTED AT OTHER—UNIFORMLY DISTRIBUTED LOAD

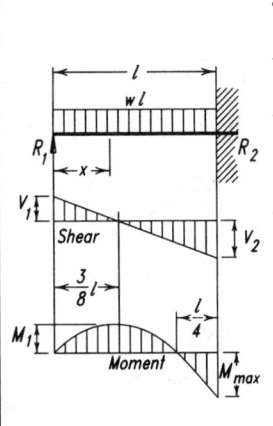

Total Equiv. Uniform Load . . . . $= wl$

$R_1 = V_1$ . . . . $= \dfrac{3wl}{8}$

$R_2 = V_2 = V_{max}$ . . . . $= \dfrac{5wl}{8}$

V_x . . . . $= R_1 - wx$

M_{max} . . . . $= \dfrac{wl^2}{8}$

M_1 $\left(\text{at } x = \dfrac{3}{8}l\right)$ . . . . $= \dfrac{9}{128}wl^2$

M_x . . . . $= R_1 x - \dfrac{wx^2}{2}$

Δ_{max} $\left(\text{at } x = \dfrac{l}{16}(1 + \sqrt{33}) = .4215\,l\right)$ . . . $= \dfrac{wl^4}{185EI}$

Δ_x . . . . $= \dfrac{wx}{48EI}(l^3 - 3lx^2 + 2x^3)$

BEAM DIAGRAMS AND FORMULAS
For Various Static Loading Conditions

For meaning of symbols, see page 4-187

13. BEAM FIXED AT ONE END, SUPPORTED AT OTHER—CONCENTRATED LOAD AT CENTER

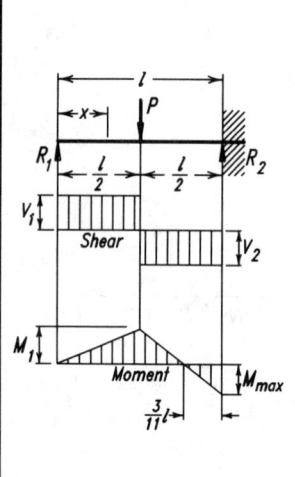

Total Equiv. Uniform Load $\ldots\ldots = \dfrac{3P}{2}$

$R_1 = V_1 \ldots\ldots = \dfrac{5P}{16}$

$R_2 = V_2 = V_{\max} \ldots\ldots = \dfrac{11P}{16}$

$M_{\max}$ (at fixed end) $\ldots\ldots = \dfrac{3Pl}{16}$

M_1 (at point of load) $\ldots\ldots = \dfrac{5Pl}{32}$

$M_x \left(\text{when } x < \dfrac{l}{2}\right) \ldots\ldots = \dfrac{5Px}{16}$

$M_x \left(\text{when } x > \dfrac{l}{2}\right) \ldots\ldots = P\left(\dfrac{l}{2} - \dfrac{11x}{16}\right)$

$\Delta_{\max}\left(\text{at } x = l\sqrt{\dfrac{1}{5}} = .4472l\right) \ldots\ldots = \dfrac{Pl^3}{48EI\sqrt{5}} = .009317\dfrac{Pl^3}{EI}$

Δ_x (at point of load) $\ldots\ldots = \dfrac{7PL^3}{768EI}$

$\Delta_x \left(\text{when } x < \dfrac{l}{2}\right) \ldots\ldots = \dfrac{Px}{96EI}(3l^2 - 5x^2)$

$\Delta_x \left(\text{when } x > \dfrac{l}{2}\right) \ldots\ldots = \dfrac{P}{96EI}(x - l)^2(11x - 2l)$

14. BEAM FIXED AT ONE END, SUPPORTED AT OTHER—CONCENTRATED LOAD AT ANY POINT

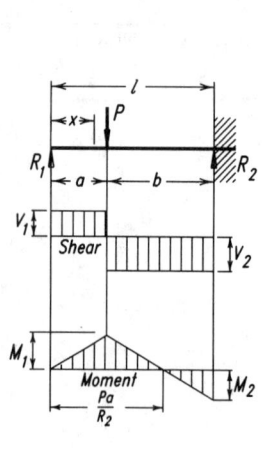

$R_1 = V_1 \ldots\ldots = \dfrac{Pb^2}{2l^3}(a + 2l)$

$R_2 = V_2 \ldots\ldots = \dfrac{Pa}{2l^3}(3l^2 - a^2)$

M_1 (at point of load) $\ldots\ldots = R_1a$

M_2 (at fixed end) $\ldots\ldots = \dfrac{Pab}{2l^2}(a + l)$

M_x (when $x < a$) $\ldots\ldots = R_1x$

M_x (when $x > a$) $\ldots\ldots = R_1x - P(x - a)$

$\Delta_{\max}\left(\text{when } a < .414l \text{ at } x = l\dfrac{(l^2 + a^2)}{(3l^2 - a^2)}\right) = \dfrac{Pa(l^2 - a^2)^3}{3EI(3l^2 - a^2)^2}$

$\Delta_{\max}\left(\text{when } a > .414l \text{ at } x = l\sqrt{\dfrac{a}{2l + a}}\right) = \dfrac{Pab^2}{6EI}\sqrt{\dfrac{a}{2l + a}}$

Δ_a (at point of load) $\ldots\ldots = \dfrac{Pa^2b^3}{12EIl^3}(3l + a)$

Δ_x (when $x < a$) $\ldots\ldots = \dfrac{Pb^2x}{12EIl^3}(3al^2 - 2lx^2 - ax^2)$

Δ_x (when $x > a$) $\ldots\ldots = \dfrac{Pa}{12EIl^3}(l - x)^2(3l^2x - a^2x - 2a^2l)$

BEAM DIAGRAMS AND FORMULAS
For Various Static Loading Conditions
For meaning of symbols, see page 4-187

15. BEAM FIXED AT BOTH ENDS—UNIFORMLY DISTRIBUTED LOADS

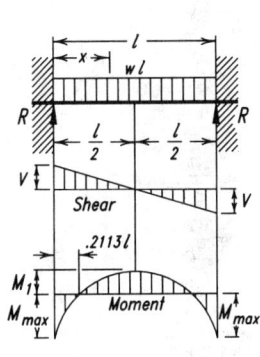

Total Equiv. Uniform Load $\ldots\ldots\ldots = \dfrac{2wl}{3}$

$R = V \quad \ldots\ldots\ldots\ldots\ldots = \dfrac{wl}{2}$

$V_x \quad \ldots\ldots\ldots\ldots\ldots = w\left(\dfrac{l}{2} - x\right)$

$M_{max} \quad \text{(at ends)} \quad \ldots\ldots\ldots\ldots = \dfrac{wl^2}{12}$

$M_1 \quad \text{(at center)} \ldots\ldots\ldots = \dfrac{wl^2}{24}$

$M_x \quad \ldots\ldots\ldots\ldots = \dfrac{w}{12}(6lx - l^2 - 6x^2)$

$\Delta_{max} \quad \text{(at center)} \ldots\ldots\ldots = \dfrac{wl^4}{384EI}$

$\Delta_x \quad \ldots\ldots\ldots\ldots = \dfrac{wx^2}{24EI}(l - x)^2$

16. BEAM FIXED AT BOTH ENDS—CONCENTRATED LOAD AT CENTER

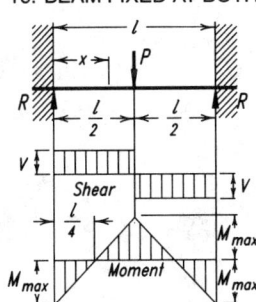

Total Equiv. Uniform Load $\ldots\ldots\ldots = P$

$R = V \quad \ldots\ldots\ldots\ldots\ldots = \dfrac{P}{2}$

$M_{max} \quad \text{(at center and ends)} \ldots\ldots\ldots = \dfrac{Pl}{8}$

$M_x \quad \left(\text{when } x < \dfrac{l}{2}\right) \ldots\ldots\ldots = \dfrac{P}{8}(4x - l)$

$\Delta_{max} \quad \text{(at center)} \ldots\ldots\ldots\ldots = \dfrac{Pl^3}{192EI}$

$\Delta_x \quad \left(\text{when } x < \dfrac{l}{2}\right) \ldots\ldots\ldots = \dfrac{Px^2}{48EI}(3l - 4x)$

17. BEAM FIXED AT BOTH ENDS—CONCENTRATED LOAD AT ANY POINT

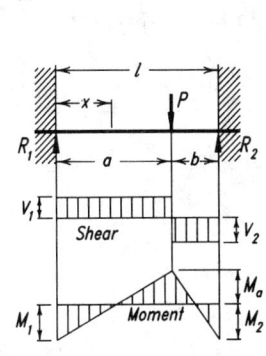

$R_1 = V_1 \quad \text{(max. when } a < b) \ldots\ldots\ldots = \dfrac{Pb^2}{l^3}(3a + b)$

$R_2 = V_2 \quad \text{(max. when } a > b) \ldots\ldots\ldots = \dfrac{Pa^2}{l^3}(a + 3b)$

$M_1 \quad \text{(max. when } a < b) \ldots\ldots\ldots = \dfrac{Pab^2}{l^2}$

$M_2 \quad \text{(max. when } a > b) \ldots\ldots\ldots = \dfrac{Pa^2b}{l^2}$

$M_a \quad \text{(at point of load)} \ldots\ldots\ldots = \dfrac{2Pa^2b^2}{l^3}$

$M_x \quad \text{(when } x < a) \ldots\ldots\ldots = R_1x - \dfrac{Pab^2}{l^2}$

$\Delta_{max} \quad \left(\text{when } a > b \text{ at } x = \dfrac{2al}{3a + b}\right) \ldots = \dfrac{2Pa^3b^2}{3EI(3a + b)^2}$

$\Delta_a \quad \text{(at point of load)} \ldots\ldots\ldots = \dfrac{Pa^3b^3}{3EIl^3}$

$\Delta_x \quad \text{(when } x < a) \ldots\ldots\ldots = \dfrac{Pb^2x^2}{6EIl^3}(3al - 3ax - bx)$

BEAM DIAGRAMS AND FORMULAS
For Various Static Loading Conditions

For meaning of symbols, see page 4-187

---

18. CANTILEVER BEAM—LOAD INCREASING UNIFORMLY TO FIXED END

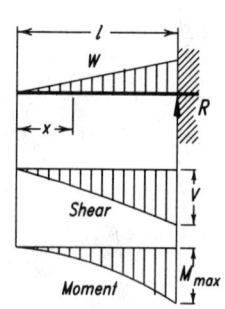

Total Equiv. Uniform Load $\ldots\ldots\ldots = \dfrac{8}{3}W$

$R = V \ldots\ldots\ldots\ldots\ldots\ldots = W$

$V_x \ldots\ldots\ldots\ldots\ldots\ldots = W\dfrac{x^2}{l^2}$

M_{max} (at fixed end) $\ldots\ldots\ldots = \dfrac{Wl}{3}$

$M_x \ldots\ldots\ldots\ldots\ldots\ldots = \dfrac{Wx^3}{3l^2}$

Δ_{max} (at free end) $\ldots\ldots\ldots\ldots = \dfrac{Wl^3}{15EI}$

$\Delta_x \ldots\ldots\ldots\ldots\ldots = \dfrac{W}{60EIl^2}(x^5 - 5l^4x + 4l^5)$

---

19. CANTILEVER BEAM—UNIFORMLY DISTRIBUTED LOAD

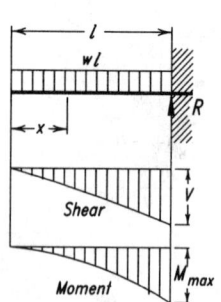

Total Equiv. Uniform Load $\ldots\ldots\ldots = 4wl$

$R = V \ldots\ldots\ldots\ldots\ldots\ldots = wl$

$V_x \ldots\ldots\ldots\ldots\ldots\ldots = wx$

M_{max} (at fixed end) $\ldots\ldots\ldots\ldots = \dfrac{wl^2}{2}$

$M_x \ldots\ldots\ldots\ldots\ldots\ldots = \dfrac{wx^2}{2}$

Δ_{max} (at free end) $\ldots\ldots\ldots\ldots = \dfrac{wl^4}{8EI}$

$\Delta_x \ldots\ldots\ldots\ldots\ldots = \dfrac{w}{24EI}(x^4 - 4l^3x + 3l^4)$

---

20. BEAM FIXED AT ONE END, FREE TO DEFLECT VERTICALLY BUT NOT ROTATE AT OTHER—UNIFORMLY DISTRIBUTED LOAD

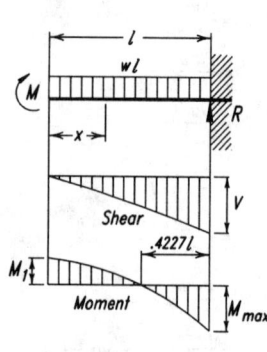

Total Equiv. Uniform Load $\ldots\ldots\ldots = \dfrac{8}{3}wl$

$R = V \ldots\ldots\ldots\ldots\ldots = wl$

$V_x \ldots\ldots\ldots\ldots\ldots\ldots = wx$

M_{max} (at fixed end) $\ldots\ldots\ldots\ldots = \dfrac{wl^2}{3}$

M_1 (at deflected end) $\ldots\ldots\ldots\ldots = \dfrac{wl^2}{6}$

$M_x \ldots\ldots\ldots\ldots\ldots\ldots = \dfrac{w}{6}(l^2 - 3x^2)$

Δ_{max} (at deflected end) $\ldots\ldots\ldots\ldots = \dfrac{wl^4}{24EI}$

$\Delta_x \ldots\ldots\ldots\ldots\ldots = \dfrac{w(l^2 - x^2)^2}{24EI}$

BEAM DIAGRAMS AND FORMULAS
For Various Static Loading Conditions

For meaning of symbols, see page 4-187

---

21. CANTILEVER BEAM—CONCENTRATED LOAD AT ANY POINT

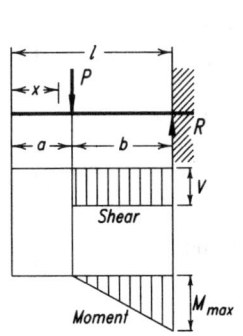

Total Equiv. Uniform Load $\ldots\ldots\ldots = \dfrac{8Pb}{l}$

$R = V \ldots\ldots\ldots\ldots\ldots\ldots = P$

$M_{\max}$ (at fixed end) $\ldots\ldots\ldots\ldots = Pb$

M_x (when $x > a$) $\ldots\ldots\ldots\ldots = P(x - a)$

$\Delta_{\max}$ (at free end) $\ldots\ldots\ldots\ldots = \dfrac{Pb^2}{6EI}(3l - b)$

Δ_a (at point of load) $\ldots\ldots\ldots\ldots = \dfrac{Pb^3}{3EI}$

Δ_x (when $x < a$) $\ldots\ldots\ldots\ldots = \dfrac{Pb^2}{6EI}(3l - 3x - b)$

Δ_x (when $x > a$) $\ldots\ldots\ldots\ldots = \dfrac{P(l - x)^2}{6EI}(3b - l + x)$

---

22. CANTILEVER BEAM—CONCENTRATED LOAD AT FREE END

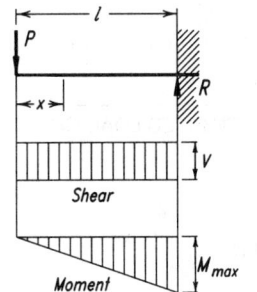

Total Equiv. Uniform Load $\ldots\ldots\ldots = 8P$

$R = V \ldots\ldots\ldots\ldots\ldots\ldots = P$

$M_{\max}$ (at fixed end) $\ldots\ldots\ldots\ldots = Pl$

$M_x \ldots\ldots\ldots\ldots\ldots\ldots = Px$

$\Delta_{\max}$ (at free end) $\ldots\ldots\ldots\ldots = \dfrac{Pl^3}{3EI}$

$\Delta_x \ldots\ldots\ldots\ldots\ldots\ldots = \dfrac{P}{6EI}(2l^3 - 3l^2x + x^3)$

---

23. BEAM FIXED AT ONE END, FREE TO DEFLECT VERTICALLY BUT NOT ROTATE AT OTHER—CONCENTRATED LOAD AT DEFLECTED END

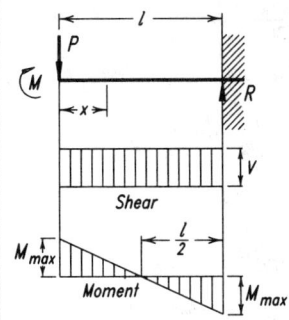

Total Equiv. Uniform Load $\ldots\ldots\ldots = 4P$

$R = V \ldots\ldots\ldots\ldots\ldots\ldots = P$

$M_{\max}$ (at both ends) $\ldots\ldots\ldots\ldots = \dfrac{Pl}{2}$

$M_x \ldots\ldots\ldots\ldots\ldots\ldots = P\left(\dfrac{l}{2} - x\right)$

$\Delta_{\max}$ (at deflected end) $\ldots\ldots\ldots\ldots = \dfrac{Pl^3}{12EI}$

$\Delta_x \ldots\ldots\ldots\ldots\ldots\ldots = \dfrac{P(l - x)^2}{12EI}(l + 2x)$

BEAM DIAGRAMS AND FORMULAS
For Various Static Loading Conditions

For meaning of symbols, see page 4-187

24. BEAM OVERHANGING ONE SUPPORT—UNIFORMLY DISTRIBUTED LOAD

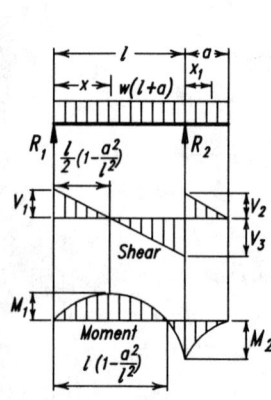

$$R_1 = V_1 \quad \ldots\ldots\ldots\ldots\ldots = \frac{w}{2l}(l^2 - a^2)$$

$$R_2 = V_2 + V_3 \quad \ldots\ldots\ldots\ldots = \frac{w}{2l}(l + a)^2$$

$$V_2 \quad \ldots\ldots\ldots\ldots\ldots = wa$$

$$V_3 \quad \ldots\ldots\ldots\ldots\ldots = \frac{w}{2l}(l^2 + a^2)$$

$$V_x \quad \text{(between supports)} \ldots\ldots = R_1 - wx$$

$$V_{x_1} \quad \text{(for overhang)} \ldots\ldots\ldots = w(a - x_1)$$

$$M_1 \left(\text{at } x = \frac{l}{2}\left[1 - \frac{a^2}{l^2} \right] \right) \ldots\ldots = \frac{w}{8l^2}(l + a)^2(l - a)^2$$

$$M_2 \quad \text{(at } R_2) \ldots\ldots\ldots\ldots = \frac{wa^2}{2}$$

$$M_x \quad \text{(between supports)} \ldots\ldots = \frac{wx}{2l}(l^2 - a^2 - xl)$$

$$M_{x_1} \quad \text{(for overhang)} \ldots\ldots\ldots = \frac{w}{2}(a - x_1)^2$$

$$\Delta_x \quad \text{(between supports)} \ldots\ldots = \frac{wx}{24EIl}(l^4 - 2l^2x^2 + lx^3 - 2a^2l^2 + 2a^2x^2)$$

$$\Delta_{x_1} \quad \text{(for overhang)} \ldots\ldots\ldots = \frac{wx_1}{24EI}(4a^2l - l^3 + 6a^2x_1 - 4ax_1^2 + x_1^3)$$

25. BEAM OVERHANGING ONE SUPPORT—UNIFORMLY DISTRIBUTED LOAD ON OVERHANG

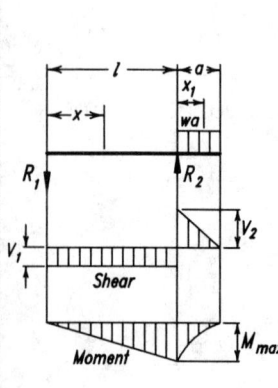

$$R_1 = V_1 \quad \ldots\ldots\ldots\ldots\ldots = \frac{wa^2}{2l}$$

$$R_2 = V_1 + V_2 \quad \ldots\ldots\ldots\ldots = \frac{wa}{2l}(2l + a)$$

$$V_2 \quad \ldots\ldots\ldots\ldots\ldots = wa$$

$$V_{x_1} \quad \text{(for overhang)} \ldots\ldots\ldots = w(a - x_1)$$

$$M_{\max} \quad \text{(at } R_2) \ldots\ldots\ldots\ldots = \frac{wa^2}{2}$$

$$M_x \quad \text{(between supports)} \ldots\ldots = \frac{wa^2x}{2l}$$

$$M_{x_1} \quad \text{(for overhang)} \ldots\ldots\ldots = \frac{w}{2}(a - x_1)^2$$

$$\Delta_{\max}\left(\text{between supports at } x = \frac{l}{\sqrt{3}} \right) = \frac{wa^2l^2}{18\sqrt{3}EI} = 0.03208 \frac{wa^2l^2}{EI}$$

$$\Delta_{\max} \text{ (for overhang at } x_1 = a) \ldots = \frac{wa^3}{24EI}(4l + 3a)$$

$$\Delta_x \quad \text{(between supports)} \ldots\ldots = \frac{wa^2x}{12EIl}(l^2 - x^2)$$

$$\Delta_{x_1} \quad \text{(for overhang)} \ldots\ldots\ldots = \frac{wx_1}{24EI}(4a^2l + 6a^2x_1 - 4ax_1^2 + x_1^3)$$

BEAM DIAGRAMS AND FORMULAS
For Various Static Loading Conditions

For meaning of symbols, see page 4-187

---

26. BEAM OVERHANGING ONE SUPPORT—CONCENTRATED LOAD AT END OF OVERHANG

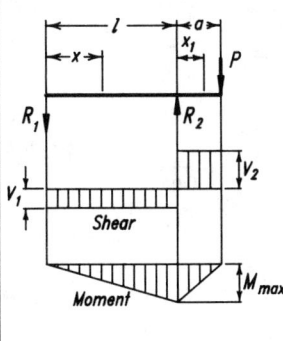

$R_1 = V_1$ . . . . $= \dfrac{Pa}{l}$

$R_2 = V_1 + V_2$ . . . . $= \dfrac{P}{l}(l + a)$

V_2 . . . . $= P$

M_{max} (at R_2) . . . . $= Pa$

M_x (between supports) . . . . $= \dfrac{Pax}{l}$

M_{x_1} (for overhang) . . . . $= P(a - x_1)$

Δ_{max} $\left(\text{between supports at } x = \dfrac{l}{\sqrt{3}}\right)$ . . . . $= \dfrac{Pal^2}{9\sqrt{3}\,EI} = .06415\dfrac{Pal^2}{EI}$

Δ_{max} (for overhang at $x_1 = a$) . . . . $= \dfrac{Pa^2}{3EI}(l + a)$

Δ_x (between supports) . . . . $= \dfrac{Pax}{6EIl}(l^2 - x^2)$

Δ_{x_1} (for overhang) . . . . $= \dfrac{Px_1}{6EI}(2al + 3ax_1 - x_1^2)$

---

27. BEAM OVERHANGING ONE SUPPORT—UNIFORMLY DISTRIBUTED LOAD BETWEEN SUPPORTS

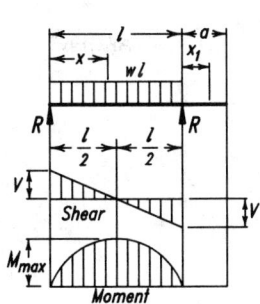

Total Equiv. Uniform Load . . . . $= wl$

$R = V$ . . . . $= \dfrac{wl}{2}$

V_x . . . . $= w\left(\dfrac{l}{2} - x\right)$

M_{max} (at center) . . . . $= \dfrac{wl^2}{8}$

M_x . . . . $= \dfrac{wx}{2}(l - x)$

Δ_{max} (at center) . . . . $= \dfrac{5wl^4}{384EI}$

Δ_x . . . . $= \dfrac{wx}{24EI}(l^3 - 2lx^2 + x^3)$

Δ_{x_1} . . . . $= \dfrac{wl^3 x_1}{24EI}$

---

28. BEAM OVERHANGING ONE SUPPORT—CONCENTRATED LOAD AT ANY POINT BETWEEN SUPPORTS

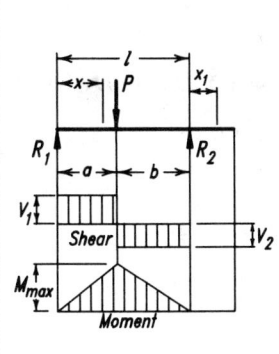

Total Equiv. Uniform Load . . . . $= \dfrac{8Pab}{l^2}$

$R_1 = V_1$ (max. when $a < b$) . . . . $= \dfrac{Pb}{l}$

$R_2 = V_2$ (max. when $a > b$) . . . . $= \dfrac{Pa}{l}$

M_{max} (at point of load) . . . . $= \dfrac{Pab}{l}$

M_x (when $x < a$) . . . . $= \dfrac{Pbx}{l}$

Δ_{max} $\left(\text{at } x = \sqrt{\dfrac{a(a + 2b)}{3}} \text{ when } a > b\right)$ . . . $= \dfrac{Pab(a + 2b)\sqrt{3a(a + 2b)}}{27EIl}$

Δ_a (at point of load) . . . . $= \dfrac{Pa^2 b^2}{3EIl}$

Δ_x (when $x < a$) . . . . $= \dfrac{Pbx}{6EIl}(l^2 - b^2 - x^2)$

Δ_x (when $x > a$) . . . . $= \dfrac{Pa(l - x)}{6EIl}(2lx - x^2 - a^2)$

Δ_{x_1} . . . . $= \dfrac{Pabx_1}{6EIl}(l + a)$

BEAM DIAGRAMS AND FORMULAS
For Various Static Loading Conditions
For meaning of symbols, see page 4-187

---

29. CONTINUOUS BEAM—TWO EQUAL SPANS—UNIFORM LOAD ON ONE SPAN

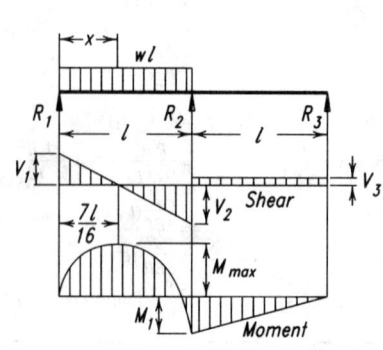

Total Equiv. Uniform Load $= \dfrac{49}{64}\, wl$

$R_1 = V_1 \ldots \ldots \ldots = \dfrac{7}{16}\, wl$

$R_2 = V_2 + V_3 \ldots \ldots = \dfrac{5}{8}\, wl$

$R_3 = V_3 \ldots \ldots \ldots = -\dfrac{1}{16}\, wl$

$V_2 \qquad \ldots \ldots \ldots = \dfrac{9}{16}\, wl$

$M_{max} \left(\text{at } x = \dfrac{7}{16}l \right) \ldots = \dfrac{49}{512}\, wl^2$

$M_1 \quad (\text{at support } R_2) \ldots = \dfrac{1}{16}\, wl^2$

$M_x \quad (\text{when } x < l) \ldots = \dfrac{wx}{16}(7l - 8x)$

$\Delta_{max} \ (\text{at } 0.472\, l \text{ from } R_1) \ldots = 0.0092\, wl^4 / EI$

---

30. CONTINUOUS BEAM—TWO EQUAL SPANS—CONCENTRATED LOAD AT CENTER OF ONE SPAN

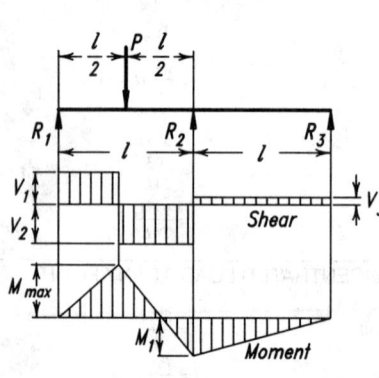

Total Equiv. Uniform Load $= \dfrac{13}{8}\, P$

$R_1 = V_1 \ldots \ldots \ldots = \dfrac{13}{32}\, P$

$R_2 = V_2 + V_3 \ldots \ldots = \dfrac{11}{16}\, P$

$R_3 = V_3 \ldots \ldots \ldots = -\dfrac{3}{32}\, P$

$V_2 \qquad \ldots \ldots \ldots = \dfrac{19}{32}\, P$

$M_{max} \ (\text{at point of load}) \ldots = \dfrac{13}{64}\, Pl$

$M_1 \quad (\text{at support } R_2) \ldots = \dfrac{3}{32}\, Pl$

$\Delta_{max} \ (\text{at } 0.480\, l \text{ from } R_1) \ldots = 0.015\, Pl^3 / EI$

---

31. CONTINUOUS BEAM—TWO EQUAL SPANS—CONCENTRATED LOAD AT ANY POINT

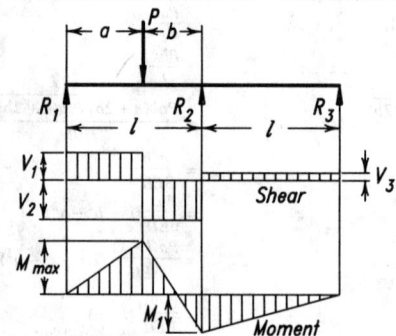

$R_1 = V_1 \ldots \ldots \ldots = \dfrac{Pb}{4l^3}(4l^2 - a(l + a))$

$R_2 = V_2 + V_3 \ldots \ldots = \dfrac{Pa}{2l^3}(2l^2 + b(l + a))$

$R_3 = V_3 \ldots \ldots \ldots = -\dfrac{Pab}{4l^3}(l + a)$

$V_2 \qquad \ldots \ldots \ldots = \dfrac{Pa}{4l^3}(4l^2 + b(l + a))$

$M_{max} \ (\text{at point of load}) \ldots = \dfrac{Pab}{4l^3}(4l^2 - a(l + a))$

$M_1 \quad (\text{at support } R_2) \ldots = \dfrac{Pab}{4l^2}(l + a)$

BEAM DIAGRAMS AND FORMULAS
For Various Static Loading Conditions

For meaning of symbols, see page 4-187

32. BEAM—UNIFORMLY DISTRIBUTED LOAD AND VARIABLE END MOMENTS

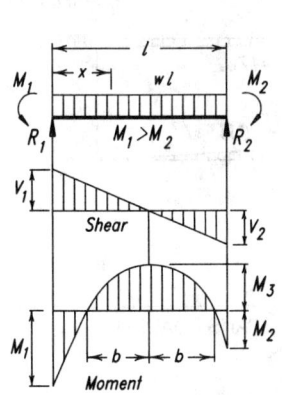

$$R_1 = V_1 \quad \ldots \ldots \ldots \ldots \quad = \frac{wl}{2} + \frac{M_1 - M_2}{l}$$

$$R_2 = V_2 \quad \ldots \ldots \ldots \ldots \quad = \frac{wl}{2} - \frac{M_1 - M_2}{l}$$

$$V_x \quad \ldots \ldots \ldots \ldots \quad = w\left(\frac{l}{2} - x\right) + \frac{M_1 - M_2}{l}$$

$$M_3 \left(\text{at } x = \frac{l}{2} + \frac{M_1 - M_2}{wl}\right) \ldots \quad = \frac{wl^2}{8} - \frac{M_1 + M_2}{2} + \frac{(M_1 - M_2)^2}{2wl^2}$$

$$M_x \quad \ldots \ldots \ldots \ldots \quad = \frac{wx}{2}(l - x) + \left(\frac{M_1 - M_2}{l}\right)x - M_1$$

$$b \quad \text{(to locate inflection points)} = \sqrt{\frac{l^2}{4} - \left(\frac{M_1 + M_2}{w}\right) + \left(\frac{M_1 - M_2}{wl}\right)^2}$$

$$\Delta_x = \frac{wx}{24EI}\left[x^3 - \left(2l + \frac{4M_1}{wl} - \frac{4M_2}{wl}\right)x^2 + \frac{12M_1}{w}x + l^3 - \frac{8M_1 l}{w} - \frac{4M_2 l}{w}\right]$$

33. BEAM—CONCENTRATED LOAD AT CENTER AND VARIABLE END MOMENTS

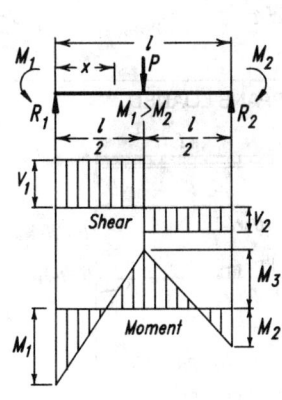

$$R_1 = V_1 \quad \ldots \ldots \ldots \ldots \quad = \frac{P}{2} + \frac{M_1 - M_2}{l}$$

$$R_2 = V_2 \quad \ldots \ldots \ldots \ldots \quad = \frac{P}{2} - \frac{M_1 - M_2}{l}$$

$$M_3 \text{ (at center)} \quad \ldots \ldots \ldots \quad = \frac{Pl}{4} - \frac{M_1 + M_2}{2}$$

$$M_x \left(\text{when } x < \frac{l}{2}\right) \ldots \ldots \quad = \left(\frac{P}{2} + \frac{M_1 - M_2}{l}\right)x - M_1$$

$$M_x \left(\text{when } x > \frac{l}{2}\right) \ldots \ldots \quad = \frac{P}{2}(l - x) + \frac{(M_1 - M_2)x}{l} - M_1$$

$$\Delta_x \left(\text{when } x < \frac{l}{2}\right) = \frac{Px}{48EI}\left(3l^2 - 4x^2 - \frac{8(l - x)}{Pl}[M_1(2l - x) + M_2(l + x)]\right)$$

BEAM DIAGRAMS AND FORMULAS
For Various Static Loading Conditions

For meaning of symbols, see page 4-187

34. CONTINUOUS BEAM—THREE EQUAL SPANS—ONE END SPAN UNLOADED

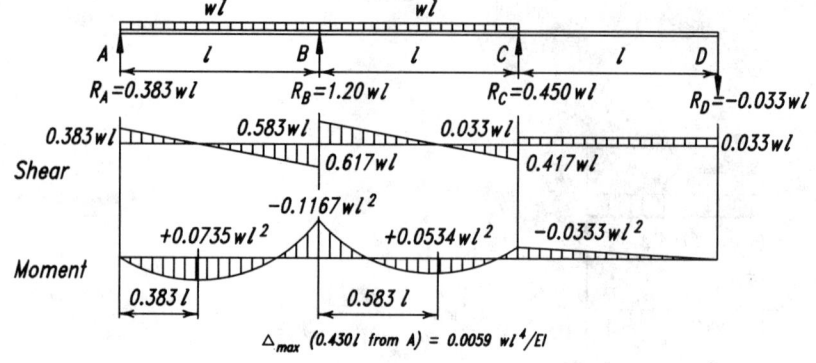

$\triangle_{max}$ (0.430 l from A) = 0.0059 wl^4/EI

35. CONTINUOUS BEAM—THREE EQUAL SPANS—END SPANS LOADED

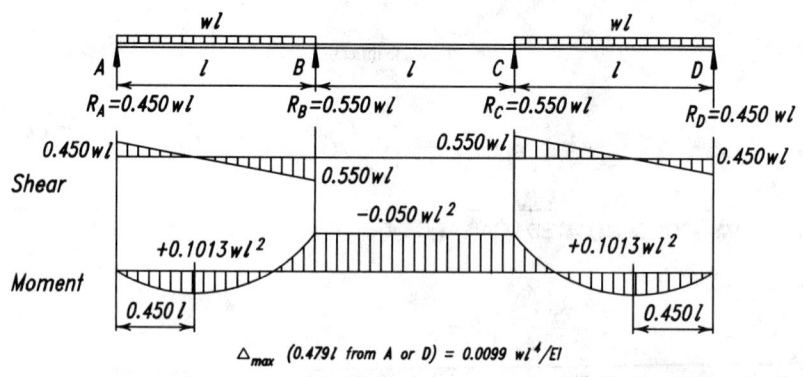

$\triangle_{max}$ (0.479 l from A or D) = 0.0099 wl^4/EI

36. CONTINUOUS BEAM—THREE EQUAL SPANS—ALL SPANS LOADED

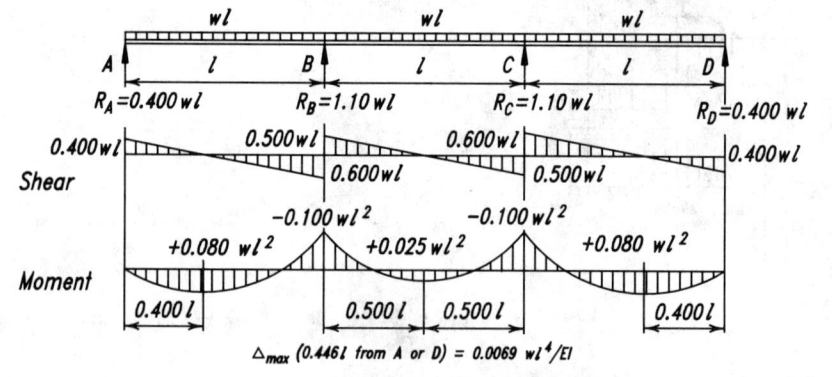

$\triangle_{max}$ (0.446 l from A or D) = 0.0069 wl^4/EI

BEAM DIAGRAMS AND FORMULAS
For Various Static Loading Conditions
For meaning of symbols, see page 4-187

37. CONTINUOUS BEAM—FOUR EQUAL SPANS—THIRD SPAN UNLOADED

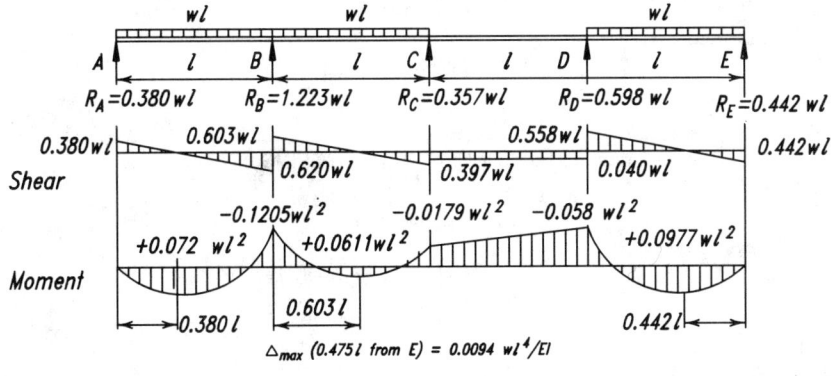

$R_A=0.380\,wl$ $R_B=1.223\,wl$ $R_C=0.357\,wl$ $R_D=0.598\,wl$ $R_E=0.442\,wl$

Shear

$\triangle_{max}$ (0.475 l from E) = 0.0094 wl^4/EI

38. CONTINUOUS BEAM—FOUR EQUAL SPANS—LOAD FIRST AND THIRD SPANS

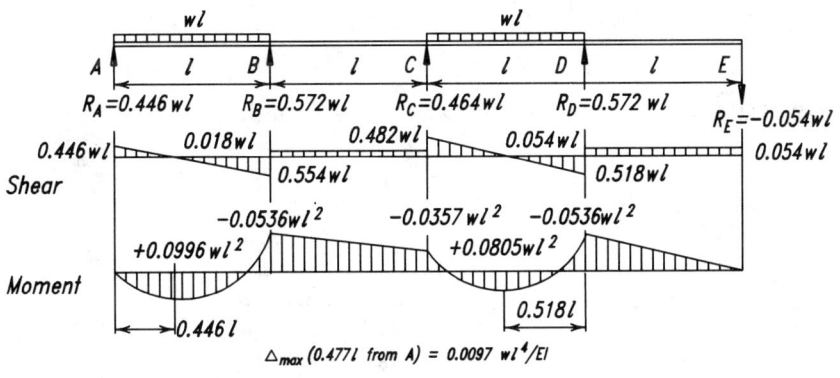

$R_A=0.446\,wl$ $R_B=0.572\,wl$ $R_C=0.464\,wl$ $R_D=0.572\,wl$ $R_E=-0.054\,wl$

Shear

$\triangle_{max}$ (0.477 l from A) = 0.0097 wl^4/EI

39. CONTINUOUS BEAM—FOUR EQUAL SPANS—ALL SPANS LOADED

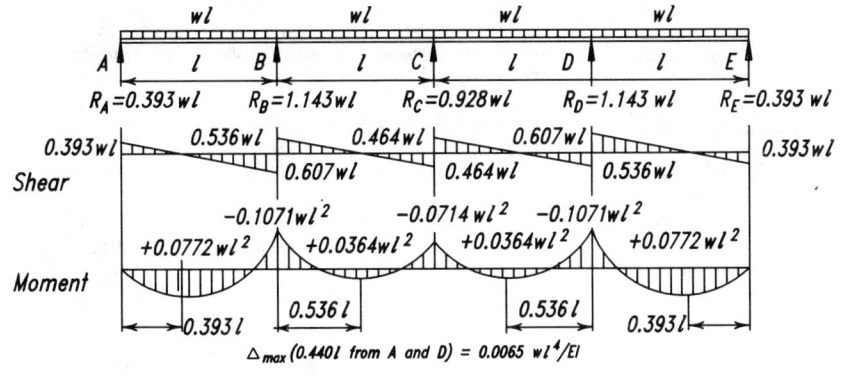

$R_A=0.393\,wl$ $R_B=1.143\,wl$ $R_C=0.928\,wl$ $R_D=1.143\,wl$ $R_E=0.393\,wl$

Shear

$\triangle_{max}$ (0.440 l from A and D) = 0.0065 wl^4/EI

BEAM DIAGRAMS AND FORMULAS
For Various Static Loading Conditions

For meaning of symbols, see page 4-187

---

40. SIMPLE BEAM—ONE CONCENTRATED MOVING LOAD

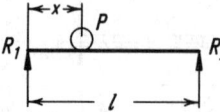

$R_{1\,max} = V_{1\,max}$ (at $x = 0$) . . . . $= P$

M_{max} $\left(\text{at point of load, when } x = \dfrac{l}{2}\right)$ . . . . . $= \dfrac{Pl}{4}$

---

41. SIMPLE BEAM—TWO EQUAL CONCENTRATED MOVING LOADS

$R_{1\,max} = V_{1\,max}$ (at $x = 0$) . . . . $= P\left(2 - \dfrac{a}{l}\right)$

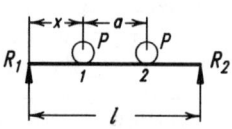

M_{max}
$\begin{cases}
\text{when } a < (2 - \sqrt{2})l = 0.586\,l \\[2mm]
\text{under load 1 at } x\,\dfrac{1}{2}\left(l - \dfrac{a}{2}\right) . . \; = \dfrac{P}{2l}\left(l - \dfrac{a}{2}\right)^2 \\[2mm]
\text{when } a > (2 - \sqrt{2})l = 0.586\,l \\[2mm]
\text{with one load at center of span} \\[2mm]
\text{(Case 40)} \quad . . . . \; = \dfrac{Pl}{4}
\end{cases}$

---

42. SIMPLE BEAM—TWO UNEQUAL CONCENTRATED MOVING LOADS

$R_{1\,max} = V_{1\,max}$ (at $x = 0$) . . . . $= P_1 + P_2\,\dfrac{l - a}{l}$

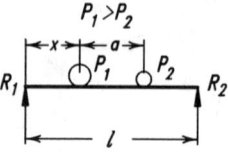

M_{max}
$\begin{cases}
\text{under } P_1, \text{ at } x = \dfrac{1}{2}\left(l - \dfrac{P_2 a}{P_1 + P_2}\right) \; = (P_1 + P_2)\dfrac{x^2}{l} \\[2mm]
M_{max} \text{ may occur with larger} \\[2mm]
\text{load at center of span and other} \\[2mm]
\text{load off span (Case 40)} \; . . . \; = \dfrac{P_1 l}{4}
\end{cases}$

GENERAL RULES FOR SIMPLE BEAMS CARRYING MOVING CONCENTRATED LOADS

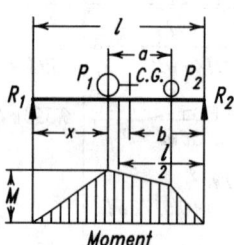

Moment

The maximum shear due to moving concentrated loads occurs at one support when one of the loads is at that support. With several moving loads, the location that will produce maximum shear must be determined by trial.

The maximum bending moment produced by moving concentrated loads occurs under one of the loads when that load is as far from one support as the center of gravity of all the moving loads on the beam is from the other support.

In the accompanying diagram, the maximum bending moment occurs under load P_1 when $x = b$. It should also be noted that this condition occurs when the centerline of the span is midway between the center of gravity of loads and the nearest concentrated load.

BEAM DIAGRAMS AND FORMULAS
Design properties of cantilevered beams

Equal loads, equally spaced

No. Spans	System
2	
3	
4	
5	
≥6 (even)	
≥7 (odd)	

n	∞	2	3	4	5
Typical Span Loading					

		∞	2	3	4	5
Moments	M_1	0.086PL	0.167PL	0.250PL	0.333PL	0.429PL
	M_2	0.096PL	0.188PL	0.278PL	0.375PL	0.480PL
	M_3	0.063PL	0.125PL	0.167PL	0.250PL	0.300PL
	M_4	0.039PL	0.083PL	0.083PL	0.167PL	0.171PL
	M_5	0.051PL	0.104PL	0.139PL	0.208PL	0.249PL
Reactions	A	0.414P	0.833P	1.250P	1.667P	2.071P
	B	1.172P	2.333P	3.500P	4.667P	5.857P
	C	0.438P	0.875P	1.333P	1.750P	2.200P
	D	1.063P	2.125P	3.167P	4.250P	5.300P
	E	1.086P	2.167P	3.250P	4.333P	5.429P
	F	1.109P	2.208P	3.333P	4.417P	5.557P
	G	0.977P	1.958P	2.917P	3.917P	4.871P
	H	1.000P	2.000P	3.000P	4.000P	5.000P
Cantilever Dimensions	a	0.172L	0.250L	0.200L	0.182L	0.176L
	b	0.125L	0.200L	0.143L	0.143L	0.130L
	c	0.220L	0.333L	0.250L	0.222L	0.229L
	d	0.204L	0.308L	0.231L	0.211L	0.203L
	e	0.157L	0.273L	0.182L	0.176L	0.160L
	f	0.147L	0.250L	0.167L	0.167L	0.150L

BEAM DIAGRAMS AND FORMULAS
CONTINUOUS BEAMS

MOMENT AND SHEAR COEFFICIENTS
EQUAL SPANS, EQUALLY LOADED

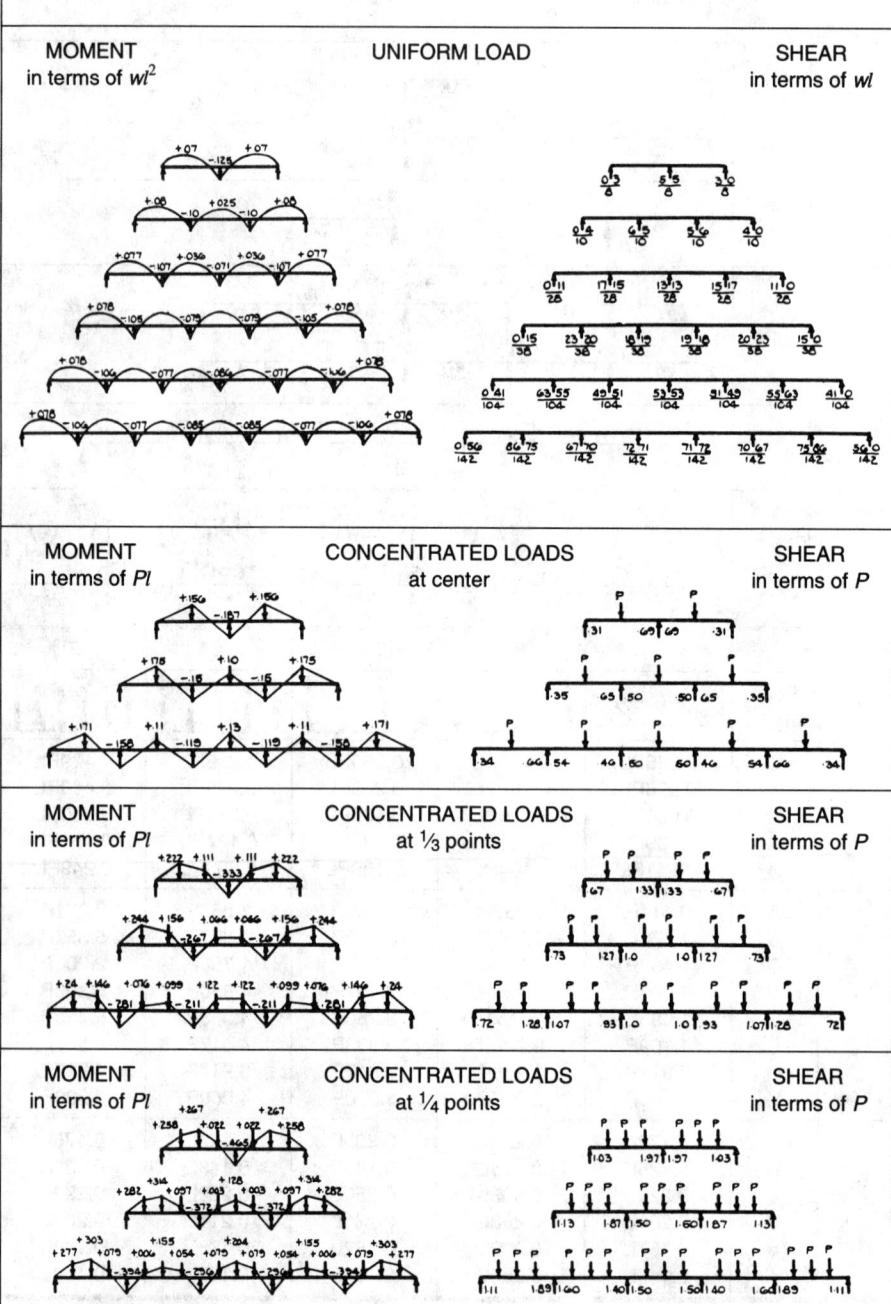

MOMENT
in terms of wl^2

UNIFORM LOAD

SHEAR
in terms of wl

MOMENT
in terms of Pl

CONCENTRATED LOADS
at center

SHEAR
in terms of P

MOMENT
in terms of Pl

CONCENTRATED LOADS
at ⅓ points

SHEAR
in terms of P

MOMENT
in terms of Pl

CONCENTRATED LOADS
at ¼ points

SHEAR
in terms of P

FLOOR DEFLECTIONS AND VIBRATIONS

Serviceability
Serviceability checks are necessary in design to provide for the satisfactory performance of structures. Chapter L of the LRFD Specification and Commentary contains general guidelines on serviceability. In contrast with the factored forces used to determine the required strength, the (unfactored) working loads are used in serviceability calculations.

The primary concern regarding the serviceability of floor beams is the prevention of excessive deflections and vibrations. The use of higher strength steels and composite construction has resulted in shallower and lighter beams. Serviceability has become a more important consideration than in the past, as the design of more beams is governed by deflection and vibration criteria.

Deflections and Camber
Criteria for acceptable vertical deflections have traditionally been set by the design engineer, based on the intended use of the given structure. What is appropriate for an office building, for example, may not be satisfactory for a hospital. An illustration of deflection criteria is the following:

1. Live load deflections shall not exceed a specified fraction of the span (e.g., $\frac{1}{360}$) nor a specific quantity (e.g., one inch). A deeper and/or heavier beam shall be selected, if necessary, to meet these requirements.
2. Under dead load, plus a given portion of the design live load (say, 10 psf), the floor shall be theoretically level. Where feasible and necessary, upward camber of the beam shall be specified.

Regarding camber, the engineer is cautioned that:

1. It is unrealistic to expect precision in cambering. The limits and tolerances given in Part 1 of of this Manual for cambering of rolled beams are typical for mill camber. Kloiber (1989) states that camber tolerances are dependent on the method used (hot or cold cambering) and whether done at the mill or the fabrication shop. According to the AISC Code of Standard Practice, Section 6.4.5: "When members are specified on the contract documents as requiring camber, the shop fabrication tolerance shall be $-0/+\frac{1}{2}$ in. for members 50 ft and less in length, or $-0/+(\frac{1}{2}$ in. $+\frac{1}{8}$ in. for each 10 ft or fraction thereof in excess of 50 ft in length) for members over 50 ft. Members received from the rolling mill with 75 percent of the specified camber require no further cambering. For purposes of inspection, camber must be measured in the fabricator's shop in the unstressed condition." Some of the camber may be lost in transportation prior to placement of the beam, due to vibration.
2. There are two methods for erection of floors: uniform slab thickness and level floor. As a consequence of possible overcamber, the latter may result in a thinner concrete slab for composite action and fire protection at midspan, and may cause the shear studs to protrude above the slab.
3. Due to end restraint at the connections, actual beam deflections are often less than the calculated values.
4. The deflections of a composite beam (under live load for shored construction, and under dead and live loads for unshored) cannot be determined as easily and accurately as the deflections of a bare steel beam. Equation C-I3-6 in Section I3.2 of the

Commentary on the LRFD Specification provides an approximate effective moment of inertia for partially composite beams.

5. Cambers of less than ¾-in. should not be specified, and beams less than 24 ft in length should not be cambered (Kloiber, 1989).

Vibrations

Annoying floor motion may be caused by the normal activities of the occupants. Remedial action is usually very difficult and expensive and not always effective. The prevention of excessive and objectionable floor vibration should be part of the design process.

Several researchers have developed procedures to enable structural engineers to predict occupant acceptability of proposed floor systems. Based on field measurement of approximately 100 floor systems, Murray (1991) developed the following acceptability criterion:

$$D > 35A_o f + 2.5 \tag{4-1}$$

where

D = damping in percent of critical
A_o = maximum initial amplitude of the floor system due to a heel-drop excitation, in.
f = first natural frequency of the floor system, hz

Damping in a completed floor system can be estimated from the following ranges:

Bare Floor: 1–3 percent
Lower limit for thin slab of lightweight concrete; upper limit for thick slab of normal weight concrete.

Ceiling: 1–3 percent
Lower limit for hung ceiling; upper limit for sheetrock on furring attached to beams or joists.

Ductwork and Mechanical: 1–10 percent
Depends on amount and attachment.

Partitions: 10–20 percent
If attached to the floor system and not spaced more than every five floor beams or the effective joist floor width.

Note: The above values are based on observation only.

Beam or girder frequency can be estimated from

$$f = K \left[\frac{gEI_t}{WL^3} \right]^{1/2} \tag{4-2}$$

where

f = first natural frequency, hz
K = 1.57 for simply supported beams
  = 0.56 for cantilevered beams
  = from Figure 4-8 for overhanging beams
g = acceleration of gravity = 386 in./sec^2

E = modulus of elasticity, psi

I_t = transformed moment of inertia of the tee-beam model, Figure 4-9, in.[4] (to be used for both composite and noncomposite construction)

W = total weight supported by the tee beam, dead load plus 10–25 percent of design live load, lbs

L = tee-beam span, in.

System frequency is estimated using

$$\frac{1}{f_s^2} = \frac{1}{f_b^2} + \frac{1}{f_g^2}$$

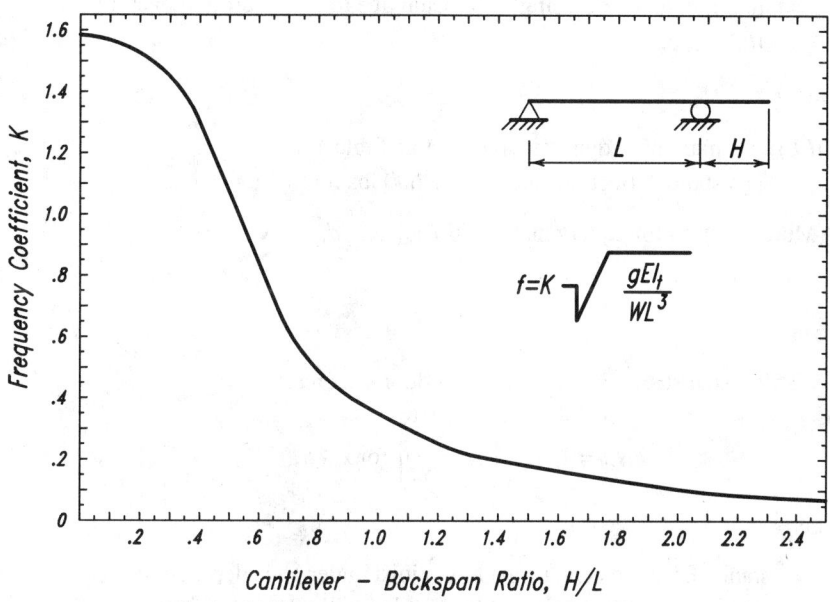

Fig. 4-8. Frequency coefficients for overhanging beams.

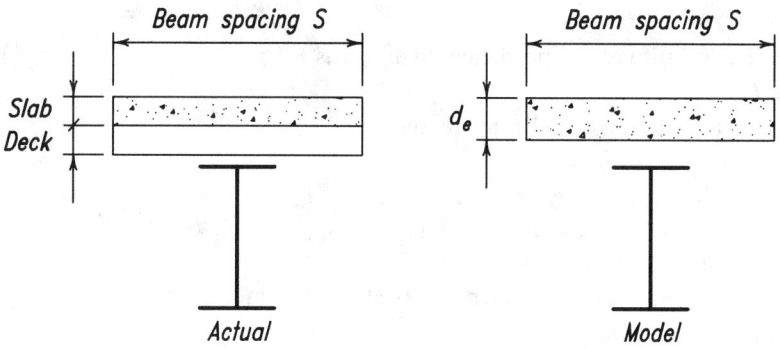

Fig. 4-9. Tee-beam model for computing transformed moment of inertia.

where

f_s = system frequency, hz
f_b = beam or joist frequency, hz
f_g = girder frequency, hz

Amplitude from a heel-drop impact can be estimated from

$$A_o = \frac{A_{ot}}{N_{eff}}$$

(4-3)

where

A_o = initial amplitude of the floor system due to a heel-drop impact, in.
N_{eff} = number of effective tee beams
A_{ot} = initial amplitude of a single tee beam due to a heel-drop impact, in.
   = $(DLF)_{max}d_s$

(4-4)

where

$(DLF)_{max}$ = maximum dynamic load factor, Table 4-2
d_s = static deflection caused by a 600 lbs force, in.

See (Murray, 1975) for equations for $(DLF)_{max}$ and d_s

For girders, N_{eff} = 1.0.

For beams:

1. $S < 2.5$ft, usual steel joist-concrete slab floor systems.

$$N_{eff} = 1 + 2\Sigma\left(\cos\frac{\pi x}{2x_o}\right) \text{ for } x \le x_o$$

(4-5)

where

x = distance from the center joist to the joist under consideration, in.
x_o = distance from the center joist to the edge of the effective floor, in.
   = $1.06\varepsilon L$
L = joist span, in.
ε = $(D_x/D_y)^{0.25}$
D_x = flexural stiffness perpendicular to the joists
   = $E_c t^3 / 12$
D_y = flexural stiffness parallel to the joists
   = EI_t / S
E_c = modulus of elasticity of concrete, psi
E = modulus of elasticity of steel, psi
t = slab thickness, in.
I_t = transformed moment of inertia of the tee beam, in.[4]
S = joist spacing, in.

2. $S > 2.5$ ft, usual steel beam-concrete slab floor systems.

$$N_{\it eff} = 2.97 - \frac{S}{17.3 d_e} + \frac{L^4}{1.35 EI_T} \tag{4-6}$$

where E is defined above and

S = beam spacing, in.
d_e = effective slab depth, in.
L = beam span, in.

Limitations:

$15 \le (S / d_e) < 40; \; 1 \times 10^6 \le (L^4 / I_T) \le 50 \times 10^6$

The amplitude of a two-way system can be estimated from

$A_{os} = A_{ob} + A_{og} / 2$

where

A_{os} = system amplitude
$A_{ob} = A_{ot}$ for beam
$A_{og} = A_{ot}$ for girder

Additional information on building floor vibrations can be obtained from the above-referenced paper by Murray (1991) and the references cited therein.

BEAMS: OTHER SUBJECTS

Other topics related to the design of flexural members covered elsewhere in this Manual include:

Beam Bearing Plates, in Part 11 (Volume II);
Beam Web Penetrations, in Part 12 (Volume II).

\textbf{\textit{Table 4-2.}}
\textbf{Dynamic Load Factors for Heel-Drop Impact}

f, hz	DLF	F, hz	DLF	F, hz	DLF
1.00	0.1541	5.50	0.7819	10.00	1.1770
1.10	0.1695	5.60	0.7937	10.10	1.1831
1.20	0.1847	5.70	0.8053	10.20	1.1891
1.30	0.2000	5.80	0.8168	10.30	1.1949
1.40	0.2152	5.90	0.8282	10.40	1.2007
1.50	0.2304	6.00	0.8394	10.50	1.2065
1.60	0.2456	6.10	0.8505	10.60	1.2121
1.70	0.2607	6.20	0.8615	10.70	1.2177
1.80	0.2758	6.30	0.8723	10.80	1.2231
1.90	0.2908	6.40	0.8830	10.90	1.2285
2.00	0.3058	6.50	0.8936	11.00	1.2339
2.10	0.3207	6.60	0.9040	11.10	1.2391
2.20	0.3356	6.70	0.9143	11.20	1.2443
2.30	0.3504	6.80	0.9244	11.30	1.2494
2.40	0.3651	6.90	0.9344	11.40	1.2545
2.50	0.3798	7.00	0.9443	11.50	1.2594
2.60	0.3945	7.10	0.9540	11.60	1.2643
2.70	0.4091	7.20	0.9635	11.70	1.2692
2.80	0.4236	7.30	0.9729	11.80	1.2740
2.90	0.4380	7.40	0.9821	11.90	1.2787
3.00	0.4524	7.50	0.9912	12.00	1.2834
3.10	0.4667	7.60	1.0002	12.10	1.2879
3.20	0.4809	7.70	1.0090	12.20	1.2925
3.30	0.4950	7.80	1.0176	12.30	1.2970
3.40	0.5091	7.90	1.0261	12.40	1.3014
3.50	0.5231	8.00	1.0345	12.50	1.3058
3.60	0.5369	8.10	1.0428	12.60	1.3101
3.70	0.5507	8.20	1.0509	12.70	1.3143
3.80	0.5645	8.30	1.0588	12.80	1.3185
3.90	0.5781	8.40	1.0667	12.90	1.3227
4.00	0.5916	8.50	1.0744	13.00	1.3268
4.10	0.6050	8.60	1.0820	13.10	1.3308
4.20	0.6184	8.70	1.0895	13.20	1.3348
4.30	0.6316	8.80	1.0969	13.30	1.3388
4.40	0.6448	8.90	1.1041	13.40	1.3427
4.50	0.6578	9.00	1.1113	13.50	1.3466
4.60	0.6707	9.10	1.1183	13.60	1.3504
4.70	0.6835	9.20	1.1252	13.70	1.3541
4.80	0.6962	9.30	1.1321	13.80	1.3579
4.90	0.7088	9.40	1.1388	13.90	1.3615
5.00	0.7213	9.50	1.1434	14.00	1.3652
5.10	0.7337	9.60	1.1519	14.10	1.3688
5.20	0.7459	9.70	1.1583	14.20	1.3723
5.30	0.7580	9.80	1.1647	14.30	1.3758
5.40	0.7700	9.90	1.1709	14.40	1.3793

REFERENCES

Allison, H., 1991, *Low- and Medium-Rise Steel Buildings,* AISC Steel Design Guide Series No. 5, American Institute of Steel Construction, Chicago, IL.

American Institute of Steel Construction, 1983, *Torsional Analysis of Steel Members,* AISC, Chicago.

Fisher, J. M. and M. A. West, 1990, *Serviceability Design Considerations for Low-Rise Buildings,* AISC Steel Design Guide Series No. 3, AISC, Chicago.

Kloiber, L. A., 1989, "Cambering of Steel Beams," *Steel Structures: Proceedings of Structures Congress '89,* American Society of Civil Engineers (ASCE), New York.

Murray, T. M., 1991, "Building Floor Vibrations," *Proceedings of the 1991 National Steel Construction Conference,* AISC, Chicago.

Zahn, C. J., 1987, "Plate Girder Design Using LRFD," *Engineering Journal,* 1st Qtr., AISC, Chicago.

PART 5

COMPOSITE DESIGN

OVERVIEW

Tables are given for the design of composite beams and columns.

Composite Beam tables are located as follows:

Composite Column tables are located as follows:

COMPOSITE BEAMS

General Notes
The Composite Beam Tables can be used for the design and analysis of simple composite steel beams. Values for the design flexural strength ϕM_n for rolled I-shaped beams with yield strengths of 36 ksi and 50 ksi are tabulated, as well as lower bound moments of inertia. The values tabulated are independent of the concrete flange properties. The strength evaluation of the concrete flange portion of the composite section is left to the design engineer. The preparation of these tables is based upon the fact that the location of the plastic neutral axis (PNA) is uniquely determined by the horizontal shear force ΣQ_n at the interface between the steel section and the concrete slab. With the knowledge of the location of the PNA and the distance to the centroid of the concrete flange force ΣQ_n, the design flexural strengths ϕM_n for the rolled section can be computed.

Design Flexural Strength (Positive)
The design flexural strength of simple steel beams with composite concrete flanges is computed from the equilibrium of internal forces using the plastic stress distribution as shown in Figure 5-1:

$$\phi M_n = \phi T_{Tot} y = \phi C_{Tot} y$$

where

ϕ = 0.85

T_{Tot} = sum of tensile forces = $F_y \times$ (tensile force beam area)

C_{Tot} = sum of compressive forces = concrete flange force + $F_y \times$ (compressive force beam area)

y = moment arm between centroid of tensile force and the resultant compressive force

The model used in the calculation of the design strengths tabulated herein is given in Figure 5-2. A summary of the model properties follows:

A_s = area of steel cross section, in.^{2}

A_f = flange area = $b_f \times t_f$, in.^{2}

A_w = web area = $(d - 2k)t_w$, in.^{2}

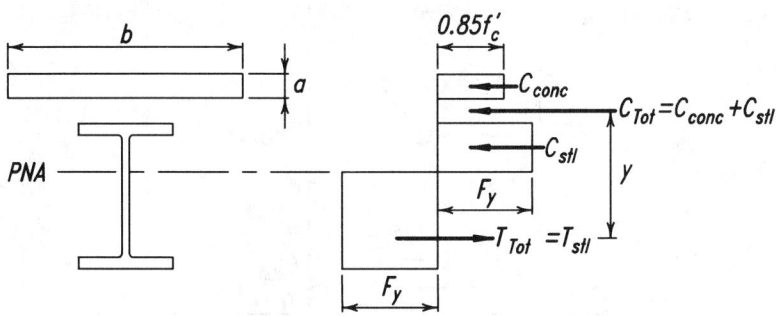

Fig. 5-1. Plastic stress distribution.

$K_{dep} = k - t_f$, in.

$K_{area} = (A_s - 2A_f - A_w)/2$, in.^{2}

Limitations for the tabulated values include the following:

$(d - 2k)/t_w \leq 640/\sqrt{F_y}$

and

$\Sigma Q_n \text{ (min.)} = 0.25 A_s F_y$

The limitation of ΣQ_n (min.) is not required by the Specification, but is deemed to be a practical minimum value. Design strength moment values are tabulated for plastic neutral axis (PNA) locations at the top and intermediate quarter points through the thickness of the steel beam top flange. In addition, PNA locations are computed at the point where ΣQ_n equals $0.25 A_s F_y$, and the point where ΣQ_n is the average of the minimum value of $(0.25 A_s F_y)$ and the value of ΣQ_n when the PNA is at the bottom of the top flange (see Figure 5-3).

To use the tables, select a valid value of ΣQ_n, determine the appropriate value of $Y2$ and read the design flexural strength moment ϕM_n directly. Values for $Y1$ are also tabulated for convenience. The parameters $Y1$ and $Y2$ are defined as follows:

$Y1$ = distance from PNA to beam top flange
$Y2$ = distance from concrete flange force to beam top flange

Valid values for ΣQ_n are the smaller of the following three expressions (LRFD Specification Section I5):

$0.85 f_c' A_c$
$A_s F_y$
$n Q_n$

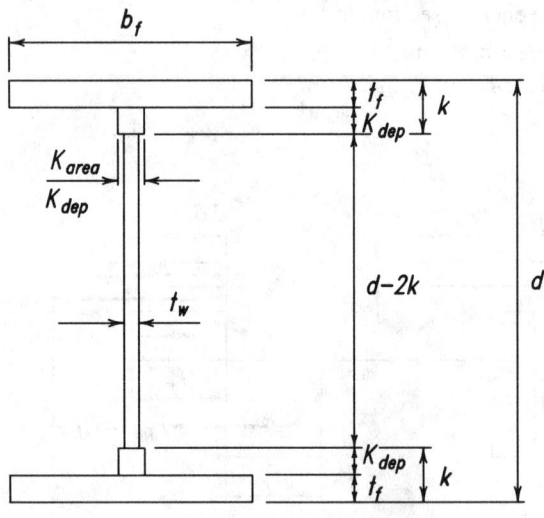

Fig. 5-2. Composite beam model.

where

f'_c = specified compressive strength of concrete, ksi
A_c = area of concrete slab within effective width, in.^{2}
A_s = area of steel cross section, in.^{2}
F_y = specified minimum yield stress, ksi
n = number of shear connectors between the point of maximum positive moment and the point of zero moment to either side
Q_n = shear capacity of single shear connector, kips

Concrete Flange

According to LRFD Specification Section I3.1 the effective width of the concrete slab on each side of the beam centerline shall not exceed:

    a. one-eighth of the beam span, center to center, of supports;
    b. one-half the distance to the centerline of the adjacent beams; or
    c. the distance to the edge of the slab.

The maximum concrete flange force is equal to $0.85 f'_c A_c$ where A_c is based on the actual slab thickness, t_c. However, often the maximum concrete flange force exceeds the maximum capacity of the specified steel beam. In that case, the effective concrete flange force is determined from a value of ΣQ_n, which will be the smaller of $A_s F_y$ or $n Q_n$. The effective concrete flange force is:

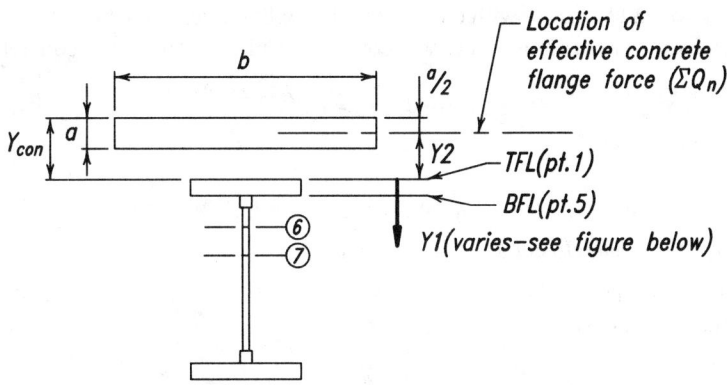

$Y1$ = Distance from top of steel flange to any of the seven tabulated PNA locations

$$\Sigma Q_n(@ \text{ point } ⑥) = \frac{\Sigma Q_n(@ \text{ pt. } 5) + \Sigma Q_n(@ \text{ pt. } 7)}{2}$$

$$\Sigma Q_n(@ \text{ point } ⑦) = .25 A_s F_y$$

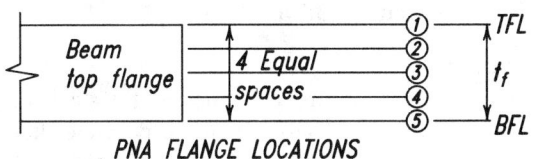

PNA FLANGE LOCATIONS

Fig. 5-3. Composite beam table parameters.

$$\Sigma Q_n = C_{conc} = 0.85 f_c' b a$$

where

C_{conc} = effective concrete flange force, kips

b = effective concrete flange width, in.

a = effective concrete flange thickness, in.

The basis of the design of most composite beams will be the relationship:

$$a = \frac{\Sigma Q_n}{0.85 f_c' b}$$

From this relationship, the value of Y2 can be computed as:

$$Y2 = Y_{con} - a/2$$

where

Y_{con} = distance from top of steel beam to top of concrete, in.

Shear Connectors

Shear connectors must be headed steel studs, not less than four stud diameters in length after installation, or hot-rolled steel channels. Shear connectors must be embedded in concrete slabs made with ASTM C33 aggregate or with rotary kiln produced aggregates conforming to ASTM C330, with concrete unit weight not less than 90 pcf.

The nominal strength of one stud shear connector embedded in a solid concrete slab is:

$$Q_n = 0.5 A_{sc} \sqrt{f_c' E_c} \leq A_{sc} F_u \qquad (\text{I5-1})$$

where

A_{sc} = cross-sectional area of a stud shear connector, in.^{2}

f_c' = specified concrete compressive strength, ksi

F_u = minimum specified tensile strength of stud, ksi

E_c = modulus of elasticity of concrete, ksi

    = $(w^{1.5})\sqrt{f_c'}$

w = unit weight of concrete, lb/cu ft

The nominal shear strengths of ¾-in. headed studs embedded in concrete slabs are listed in Table 5-1.

Note the effective shear strengths of studs used in conjunction with composite or non-composite metal forms may be affected by the shape of the deck and spacing of the studs. See LRFD Specification Sections I3.5 and I5.6.

Strength During Construction

When temporary shores are not used during construction, the steel section must have sufficient strength to support the applied loads prior to the concrete attaining 75 percent of the specified concrete strength f_c' (LRFD Specification Section I3.4). The effect of deflection on unshored steel beams during construction should be considered.

Table 5-1.
Nominal Stud Shear Strength Q_n (kips) for ¾-in Headed Studs

f_c' (ksi)	w (lbs/cu. ft)	Q_n (kips)
3.0	115	17.7
3.0	145	21.0
3.5	115	19.8
3.5	145	23.6
4.0	115	21.9
4.0	145	26.1

Lateral Support

Adequate lateral support for the compression flange of the steel section will be provided by the concrete slab after hardening. During construction, however, lateral support must be provided, or the design strength must be reduced in accordance with Section F1 of the LRFD Specification. Steel deck with adequate attachment to the compression flange will usually provide the necessary lateral support. For construction using fully encased beams, particular attention should be given to lateral support during construction.

Design Shear Strength

The design shear strength of composite beams is determined by the strength of the steel web, in accordance with the requirements of Section F2 of the LRFD Specification.

Lower Bound Moment of Inertia

With regard to serviceability, a table of lower bound moments of inertia of composite sections is included to assist in the evaluation of deflection. If calculated deflections using the lower bound moment of inertia are acceptable, a complete elastic analysis of the composite section can be avoided.

The lower bound moment of inertia is based on the area of the beam and an equivalent concrete area of $\Sigma Q_n / F_y$. The analysis includes only the horizontal shear force transferred by the shear connectors supplied; and, thus, neglects the contribution of the concrete flange not considered in the plastic distribution of forces (see Figure 5-4). The lower bound moment of inertia, therefore, is the moment of inertia of the section at the factored (ultimate) load. This is smaller than the moment of inertia at service loads where deflection is calculated. The value for the lower bound moment of inertia can be calculated as follows:

$$I_{LB} = I_x + A_s \left(Y_{ENA} - \frac{d}{2}\right)^2 + \left(\frac{\Sigma Q_n}{F_y}\right)(d + Y2 - Y_{ENA})^2$$

where

Y_{ENA} = distance from bottom of beam to elastic neutral axis (ENA)

$$= \frac{\left[\frac{A_s d}{2} + \left(\frac{\Sigma Q_n}{F_y}\right)(d + Y2)\right]}{\left[A_s + \left(\frac{\Sigma Q_n}{F_y}\right)\right]}$$

Composite Beam Reactions

Design reactions for symmetrically loaded composite beams may be computed using the Composite Beam Tables. Two situations will be considered. First, an upper bound value for a beam reaction may be computed neglecting the composite concrete flange properties other than concrete strength. Second, a more refined value for a beam reaction can be computed if the properties of the composite concrete flange are determined initially.

When the properties of the composite concrete flange have not been computed, a conservative value for the maximum horizontal shear between the composite concrete slab and the steel section (ΣQ_n) may be taken as the smaller of $A_s F_s$ or $n Q_n$. Here, n is the number of headed studs between the reaction point and point of maximum moment. The value of Q_n may be taken from Table 5-1 or determined from LRFD Specification Section I5. A value for ϕM_n of the composite section may be obtained from the Composite Beam Tables using the sum of horizontal shear ΣQ_n as described above. In this case, $Y2$ is defined as the distance from the top of the steel beam to the top of the concrete slab. The design reaction may be determined from the value of ϕM_n as discussed in the following paragraph.

When the properties of the concrete flange have been computed (effective width and depth), a slightly different method is used to find ϕM_n. The stud efficiency can be determined in accordance with Section I5 of the LRFD Specification, or Table 5-1 can be used for $\frac{3}{4}$-in. diameter stud shear connectors. The value for the sum of the horizontal shear force ΣQ_n can be taken as the smaller of $n Q_n$, $A_s F_y$, or $0.85 f_c' A_c$, where f_c' is the concrete cylinder strength (ksi) and A_c is the maximum permitted concrete flange area (LRFD Specification Section I5.2). The distance $Y2$ is the distance from the top of the steel beam to the top of the concrete slab less $[\Sigma Q_n / (0.85 f_c' b)] / 2$. Using these values for ΣQ_n and $Y2$, the value for ϕM_n can be selected from the Composite Beam Tables.

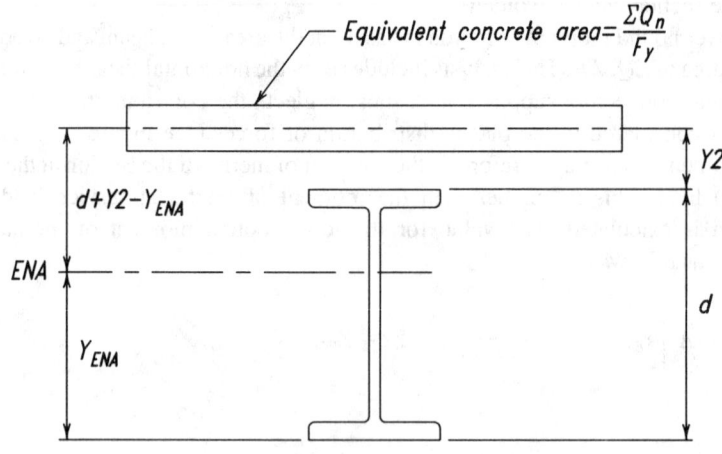

Fig. 5-4. Moment of inertia.

The design beam reaction for a symmetrically loaded composite beam may be computed from known values of ϕM_n and the span length as:

$$R = C_c \phi M_n / L$$

where

R = design beam reaction, kips
C_c = coefficient from Figure 5-5
ϕM_n = composite beam flexural design strength, kip-ft
L = span length, ft

Preliminary Section Selection

When using the Composite Beam Tables, the approximate beam weight per unit length required for several different beam depths may be calculated as follows:

$$\text{Beam weight (lb/ft)} = \left[\frac{M_u(12)}{(d/2 + Y_{con} - a/2)\phi F_y} \right] 3.4$$

where

M_u = required flexural strength, kip-ft
d = nominal beam depth, in.
Y_{con} = distance from top of steel beam to top of concrete slab, in.
a = effective concrete slab thickness, in.
F_y = steel yield stress, ksi
ϕ = 0.85
3.4 = ratio of the weight of a beam to its area, lb/in.^{2}

For convenience in the preliminary selection phase the nominal depth may be used. A value for $a/2$ must also be selected. For relatively light sections and loads, this value can be assumed to be one inch. With the PNA at the top of the steel beam, i.e., $\Sigma Q_n = A_s F_y$, the flexural design strength is:

$$\phi M_n = \phi A_s F_y (d/2 + Y_{con} - a/2) / 12$$

where

ϕM_n = flexural design strength, kip-ft
A_s = steel beam cross-sectional area, in.^{2}

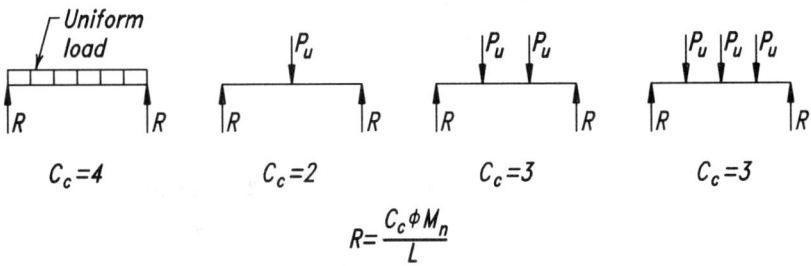

Fig. 5-5. Beam reaction coefficients.

Floor Deflections and Vibrations

Refer to the discussion of Floor Deflections and Vibrations at the end of Part 4 of this LRFD Manual.

EXAMPLE 5-1

Given: Determine the beam, with $F_y = 50$ ksi, required to support a service live load of 1.3 kips/ft and a service dead load of 0.9 kips/ft. The beam span is 30 ft and the beam spacing 10 ft. The slab is 3¼-in. light weight concrete ($f_c' = 3.5$ ksi, 115 pcf) supported by a 3-in. deep composite metal deck with an average rib width of six inches. The ribs are oriented perpendicular to the beam. Shored construction is specified. Also, determine the number of ¾-in. diameter headed studs required and the service live load deflection.

Solution: A. Load tabulation:

	Service load (kips/ft)	(L.F.)	Factored load (kips/ft)
LL	1.3	(1.6)	2.1
DL	0.9	(1.2)	1.1
Total	2.2		3.2

B. Flexural design strength:

Beam moments

$$M_u = 3.2(30)^2/8 = 360 \text{ kip-ft}$$
$$M_{LL} = 1.3(30)^2/8 = 146 \text{ kip-ft}$$

C. Select section and determine properties:

At this point, go directly to the Composite Beam Tables and select a section or compute a preliminary trial section size using the formula:

$$\text{Beam weight} = \left[\frac{M_u(12)}{(d/2 + Y_{con} - a/2)\phi F_y} \right] 3.4$$

where

$Y_{con} = 3 + 3.25 = 6.25$ in.
$a/2 = 1$ in. (estimate)
$\phi = 0.85$
$F_y = 50$ ksi

d	$\dfrac{M_u(12)(3.4)}{\phi F_y}$	$d/2$	$(Y_{con} - a/2)$	Beam Weight
16	346	8	5.25	26
18	346	9	5.25	24

From the results above, a W16×26 would be the most appropriate selection.

Let $\Sigma Q_n = A_s F_y = 7.68(50) = 384$ kips.

The effective width of the concrete flange is

$$b \leq \begin{cases} 2 \times L/8 = 2 \times 30 \text{ ft}/8 = 7.5 \text{ ft} = 90 \text{ in. (governs)} \\ 10 \text{ ft spacing} \end{cases}$$

$$a_{req'd} = \frac{\Sigma Q_n}{0.85 f_c' b} = \frac{384}{0.85(3.5)(90)} = 1.43 \text{ in.}$$

$$Y2 = 6.25 - 1.43/2 = 5.53 \text{ in.}$$

By interpolation from the Composite Beam Tables for a W16×26 and a value of $Y2$ equal to 5.53 in.,

$$\phi M_n = 363 + (0.03/0.50)(377 - 363)$$
$$= 364 \text{ kip-ft} > 360 \text{ kip-ft req'd} \quad \textbf{o.k.}$$

The selected section is adequate for $Y2 = 5.5$ in. and $Y1 = 0.0$ in., for which $\phi M_n = 363$ kip-ft

D. Compute number of studs required:

The stud reduction is calculated to be:

$$\text{Reduction factor} = \frac{0.85}{\sqrt{N_r}} (w_r/h_r)(H_s/h_r - 1.0) \leq 1.0$$

$$= \frac{0.85}{\sqrt{2}} (6/3)(5.5/3 - 1.0) = 1.0 \qquad \text{(I3-1)}$$

where

N_r = number of stud connectors in one rib; not to exceed three in computations, although more than three may be installed

w_r = average width of rib, in.

h_r = nominal rib height, in.

H_s = length of stud connector after welding, in.; not to exceed the value ($h_r + 3$) in computations, although actual length may be greater. Also must not be less than four stud diameters

The value for $H_s = 5.5$ was selected to ensure the stud capacity reduction factor is 1.0.

The number of studs required is:

with $Q_n = 19.8$ kips (Table 5-1)

$2(\Sigma Q_n)/Q_n = 2(384)/19.8 = 38.8$, say 40 studs

E. Check deflection:

For the selected section, a W16×26, F_y = 50 ksi, $Y2$ = 5.5 in. and $Y1$ = 0.0 in.; from the Elastic Moment of Inertia Tables one can find the lower bound moment of inertia is 985 in.[4] Thus, the service live load deflection can be calculated as follows (see LRFD Manual Part 4):

$$\Delta_{LL} = \frac{M_{LL}L^2}{161 I_{LB}} = \frac{146(30)^2}{161(985)} = 0.83 \text{ in.} = \frac{L}{434} < \frac{L}{360} \quad \textbf{o.k.}$$

F. Shear check:

V_u = 3.2(15) = 48 kips

$\phi V_n = \phi 0.6 F_{yw} A_w$

= (0.9)(0.6)(50)(15.69 × 0.250)

= 106 kips > 48 kips req'd   **o.k.**

EXAMPLE 5-2

Given:

Determine the beam, with F_y = 50 ksi, required to support a service live load of 250 psf and a service dead load of 90 psf. The beam span is 40 ft and the beam spacing is 10 ft. Assume 3 in. metal deck is used with a 4.5 in. slab of 4 ksi normal weight concrete (145 pcf). The stud reduction factor is 1.0. Unshored construction is specified. Determine the beam size and service dead and live load deflections. Also select a non-composite section (no shear connectors).

Solution:

A. Load tabulation:

	Service load (kips/ft)	(L.F.)	Factored load (kips/ft)
LL	2.5	(1.6)	4.0
DL	0.9	(1.2)	1.1
Total	3.4		5.1

B. Beam moments:

M_u = 5.1(40)² / 8 = 1,020 kip-ft

M_{LL} = 2.5(40)² / 8 = 500 kip-ft

M_{DL} = 0.9(40)² / 8 = 180 kip-ft

C. Select section and determine properties:

Assume a = 2 in.; therefore, take $Y2$ = 7.5 – 2 / 2 = 6.5 in. From the Composite Beam Tables, for F_y = 50 ksi and $Y2$ = 6.5 in., W21×62, W24×55, and W24×62 are possible sizes.

Try a W24×55:

F_y = 50 ksi

$Y2$ = 6.5 in.

$Y1$ = 0.0 in.

Q_n = 810 kips

ϕM_n = 1,050 kip-ft

Compute $Y2$ for ΣQ_n = 810 kips:

$$b \leq \begin{cases} 2 \times L/8 = 2 \times 40 \text{ ft}/8 = 10 \text{ ft} \\ 10 \text{ ft spacing} \end{cases}$$
$$= 120 \text{ in.}$$

$$a = \frac{\Sigma Q_n}{0.85 f'_c b} = \frac{810}{0.85(4)(120)} = 1.99$$

$Y2 = 7.5 - 1.99/2 = 6.5$ in.

D. Compute the number of studs required:

Q_n = 26.1 kips (Table 5-1)

Number of studs = $(2)\Sigma Q_n / Q_n = 2(810) / 26.1 = 62.1$, say 64 studs

E. Construction phase strength check:

A construction live load of 20 psf will be assumed. From the LRFD Specification (Section A4.1), the relevant load combinations are

$1.4D$ $= 1.4 \times 0.9 = 1.26$ k/ft

$1.2D + 1.6L = 1.2 \times 0.9 + 1.6 \times 0.2 = 1.40$ k/ft

M_u $= 1.40 \times (40)^2 / 8 = 280$ kip-ft

From the Composite Beam Tables for a W24×55 with F_y = 50 ksi, and assuming adequate lateral support is provided by the attachment of the steel deck to the compression flange,

$\phi M_n = \phi M_p = 503$ kip-ft > 280 kip-ft

F. Service load deflections:

Assume that the wet concrete load moment is equal to the service dead load moment. With I_x = 1,350 in.[4] for a W24×55,

$$\Delta_{DL} = \frac{180(40)^2}{161(1,350)} = 1.33 \text{ in.}$$

For the W24×55 with $Y2$ = 6.5 in. and $Y1$ = 0.0 in., the lower bound moment of inertia can be found in the Lower Bound Elastic Moment of Inertia Tables; I_{LB} = 4,060 in.[4]

$$\Delta_{LL} = \frac{500(40)^2}{161(4,060)} = 1.22 \text{ in.}$$

$$= \frac{L}{393} < \frac{L}{360} \quad \textbf{o.k.}$$

Specify a beam camber of 1¼-in. to overcome the dead load deflection.

G. Check shear:

$$V_u = 5.1(40) / 2 = 102 \text{ kips}$$
$$\phi V = \phi(0.6)F_{yw}A_w$$
$$= (0.9)(0.6)(50)(23.57 \times 0.395)$$
$$= 251 \text{ kips} > 102 \text{ kips} \quad \textbf{o.k.}$$

H. Final section selection:

Use: W24×55, $F_y = 50$ ksi, camber 1¼-in., 64 studs, ¾-in. diameter (32 each side of midspan)

I. Noncomposite section:

Considering the given problem without shear connectors (i.e., non-composite), a steel section can be selected from the ϕM_p values tabulated under each section in either the Composite Beam Tables or the Load Factor Design Selection Tables.

For $M_u = 1,020$ kip-ft, select a W27×94, $F_y = 50$ ksi, with a ϕM_p flexural design strength equal to 1,040 kip-ft.

$$\Delta_{DL} = \frac{180(40)^2}{161(3,270)} = 0.55 \text{ in.}$$

$$\Delta_{LL} = \frac{500(40)^2}{161(3,270)} = 1.52 \text{ in.} > L / 360$$

For $\Delta = L / 360 = 1.33$ in.

$$I_{req'd} = \frac{500(40)^2}{161(1.33)} = 3,736 \text{ in.}^4$$

Use: W30×99, $F_y = 50$ ksi, $\phi M_n = \phi_b M_p = 1,170$ kip-ft

---

EXAMPLE 5-3

Given:

A W21×44, $F_y = 50$ ksi, steel girder spans 30 feet and supports intermediate beams at the third points. A total of fifty ¾-in. diameter headed studs are applied to the beam as follows: 24 between each support and the beams at the one-third points, and two between the intermediate beams. The slab consists of 3¼-in. light-weight concrete (115 pcf) with a specified design strength of 3.5 ksi over a 3-in. deep composite metal deck with an average rib width of six inches. The ribs are oriented parallel to the beam centerline. Determine the design beam reactions.

Solutions:

For studs in a single row the spacing between the support and first intermediate beam would be $10(12) / 24 = 5.0$ in. which is greater than the specified minimum of six stud diameters (LRFD Specification Section I5.6). Since $w_r / h_r = 6 / 3 = 2$ is greater than 1.5, the stud reduction factor is not necessary (LRFD Specification I3.5c). Therefore, from Table 5.1, the stud shear strength is:

$$\Sigma Q_n = nQ_n = 24(19.8) = 475 \text{ kips}$$

For $\Sigma Q_n = 475$ kips, the required effective concrete flange thickness can be calculated to be:

$$a = \frac{475}{0.85(3.5)(7.5)(12)} = 1.77 \text{ in.}$$

$$Y2 = 3 + 3.25 - 1.77 / 2 = 5.36 \text{ in.}$$

Beam reaction:

From the Composite Beam Selection Table, for a W21×44, $F_y = 50$ ksi,

$\Sigma Q_n = 475$ kips places the PNA at $Y1 = 0.27$ in.

For $Y2 = 5.36$ in. and $Y1 = 0.27$ in., $\phi M_n = 655$ kip-ft

$$R = C_c \phi M_n / L$$
$$= 3(655) / 30$$
$$= 65.5 \text{ kips}$$

where

R = design reaction, kips
C_c = coefficient from Figure 5-5
ϕM_n = flexural design strength of beam, kip-ft
L = span length, ft

Note: The beam weight was neglected in this example.

					$F_y = 36$ ksi

COMPOSITE DESIGN
COMPOSITE BEAM SELECTION TABLE
W Shapes
$\phi = 0.85$ $\phi_b = 0.90$

Shape	$\phi_b M_p$	PNA[c]	Y1[a]	ΣQ_n	ϕM_n (kip-ft)										
					Y2[b] (in.)										
	Kip-ft		In.	Kips	2	2.5	3	3.5	4	4.5	5	5.5	6	6.5	7
W40×297	3590	TFL	0.00	3150	4890	5000	5110	5220	5330	5440	5550	5670	5780	5890	6000
		2	0.41	2680	4810	4910	5000	5100	5190	5290	5380	5480	5570	5660	5760
		3	0.83	2210	4720	4800	4880	4960	5040	5120	5190	5270	5350	5430	5510
		4	1.24	1740	4620	4690	4750	4810	4870	4930	4990	5050	5120	5180	5240
		BFL	1.65	1270	4510	4550	4600	4640	4690	4730	4780	4820	4870	4910	4960
		6	4.59	1030	4420	4460	4500	4540	4570	4610	4640	4680	4720	4750	4790
		7	8.17	787	4280	4310	4340	4370	4390	4420	4450	4480	4500	4530	4560
W 40 X 278	3210	TFL	0.00	2940	4610	4710	4810	4920	5020	5130	5230	5340	5440	5540	5650
		2	0.45	2550	4540	4630	4730	4820	4910	5000	5090	5180	5270	5360	5450
		3	0.91	2160	4470	4550	4620	4700	4780	4850	4930	5010	5080	5160	5240
		4	1.36	1770	4380	4450	4510	4570	4640	4700	4760	4820	4890	4950	5010
		BFL	1.81	1380	4280	4330	4380	4430	4480	4530	4580	4630	4680	4730	4780
		6	5.64	1060	4160	4190	4230	4270	4310	4340	4380	4420	4460	4490	4530
		7	10.06	736	3930	3960	3980	4010	4040	4060	4090	4110	4140	4170	4190
W40×277	3380	TFL	0.00	2930	4530	4630	4740	4840	4940	5050	5150	5250	5360	5460	5570
		2	0.39	2480	4460	4550	4630	4720	4810	4900	4990	5070	5160	5250	5340
		3	0.79	2030	4380	4450	4520	4590	4660	4740	4810	4880	4950	5020	5100
		4	1.18	1580	4280	4340	4390	4450	4510	4560	4620	4670	4730	4790	4840
		BFL	1.58	1130	4170	4210	4250	4290	4330	4370	4410	4450	4490	4540	4580
		6	4.25	932	4110	4140	4170	4210	4240	4270	4300	4340	4370	4400	4440
		7	7.60	732	3990	4020	4050	4070	4100	4120	4150	4180	4200	4230	4250
W40×264	3050	TFL	0.00	2790	4350	4450	4550	4650	4750	4850	4950	5050	5140	5240	5340
		2	0.43	2420	4300	4380	4470	4550	4640	4720	4810	4900	4980	5070	5150
		3	0.87	2050	4230	4300	4370	4440	4520	4590	4660	4730	4810	4880	4950
		4	1.30	1680	4140	4200	4260	4320	4380	4440	4500	4560	4620	4680	4740
		BFL	1.73	1310	4050	4100	4140	4190	4240	4280	4330	4380	4420	4470	4510
		6	5.49	1000	3930	3970	4000	4040	4070	4110	4150	4180	4220	4250	4290
		7	9.90	698	3730	3750	3780	3800	3820	3850	3870	3900	3920	3950	3970

[a] Y1 = distance from top of the steel beam to plastic neutral axis.
[b] Y2 = distance from top of the steel beam to concrete flange force.
[c] See Figure 5-3 for PNA locations.

F_y = 36 ksi															

COMPOSITE DESIGN
COMPOSITE BEAM SELECTION TABLE
W Shapes

$$\phi = 0.85 \qquad\qquad \phi_b = 0.90$$

Shape	$\phi_b M_p$	PNA[c]	Y1[a]	ΣQ_n	ϕM_n (kip-ft)										
					Y2[b] (in.)										
	Kip-ft		In.	Kips	2	2.5	3	3.5	4	4.5	5	5.5	6	6.5	7
W40×249	3020	TFL	0.00	2640	4050	4150	4240	4330	4430	4520	4610	4710	4800	4900	4990
		2	0.36	2240	3990	4070	4150	4230	4310	4390	4470	4550	4630	4700	4780
		3	0.71	1830	3920	3980	4050	4110	4180	4240	4310	4370	4440	4500	4570
		4	1.07	1430	3840	3890	3940	3990	4040	4090	4140	4190	4240	4290	4340
		BFL	1.42	1030	3750	3780	3820	3850	3890	3930	3960	4000	4040	4070	4110
		6	4.06	844	3690	3720	3750	3770	3800	3830	3860	3890	3920	3950	3980
		7	7.47	660	3580	3610	3630	3650	3680	3700	3720	3750	3770	3790	3820
W40×235	2730	TFL	0.00	2480	3840	3930	4010	4100	4190	4280	4370	4450	4540	4630	4720
		2	0.39	2140	3790	3860	3940	4010	4090	4170	4240	4320	4390	4470	4540
		3	0.79	1810	3720	3790	3850	3920	3980	4040	4110	4170	4240	4300	4360
		4	1.18	1470	3650	3700	3760	3810	3860	3910	3960	4020	4070	4120	4170
		BFL	1.58	1130	3570	3610	3650	3690	3730	3770	3810	3850	3890	3930	3970
		6	5.18	876	3480	3510	3540	3570	3600	3630	3660	3690	3730	3760	3790
		7	9.47	620	3310	3330	3350	3370	3400	3420	3440	3460	3480	3510	3530
W40×215	2600	TFL	0.00	2280	3470	3550	3630	3710	3790	3870	3950	4030	4110	4200	4280
		2	0.31	1930	3420	3480	3550	3620	3690	3760	3830	3900	3960	4030	4100
		3	0.61	1590	3360	3410	3470	3520	3580	3640	3690	3750	3810	3860	3920
		4	0.92	1240	3290	3330	3380	3420	3460	3510	3550	3600	3640	3680	3730
		BFL	1.22	895	3210	3240	3280	3310	3340	3370	3400	3440	3470	3500	3530
		6	3.84	733	3160	3190	3210	3240	3270	3290	3320	3340	3370	3400	3420
		7	7.32	570	3080	3100	3120	3140	3160	3180	3200	3220	3240	3260	3280
W40×211	2440	TFL	0.00	2230	3430	3510	3590	3670	3740	3820	3900	3980	4060	4140	4220
		2	0.35	1930	3380	3450	3520	3590	3660	3720	3790	3860	3930	4000	4070
		3	0.71	1630	3330	3390	3440	3500	3560	3620	3670	3730	3790	3850	3910
		4	1.06	1330	3270	3310	3360	3410	3460	3500	3550	3600	3640	3690	3740
		BFL	1.42	1030	3200	3230	3270	3310	3340	3380	3420	3450	3490	3530	3560
		6	4.99	793	3110	3140	3170	3200	3230	3250	3280	3310	3340	3370	3400
		7	9.35	558	2960	2980	3000	3020	3040	3060	3080	3100	3120	3140	3160

[a] Y1 = distance from top of the steel beam to plastic neutral axis.
[b] Y2 = distance from top of the steel beam to concrete flange force.
[c] See Figure 5-3 for PNA locations.

										F_y = 36 ksi				

COMPOSITE DESIGN
COMPOSITE BEAM SELECTION TABLE
W Shapes
$$\phi = 0.85 \qquad\qquad \phi_b = 0.90$$

Shape	$\phi_b M_p$	PNA[c]	Y1[a]	ΣQ_n	ϕM_n (kip-ft)										
					Y2[b] (in.)										
	Kip-ft		In.	Kips	2	2.5	3	3.5	4	4.5	5	5.5	6	6.5	7
W40×199	2340	TFL	0.00	2100	3180	3250	3330	3400	3480	3550	3620	3700	3770	3850	3920
		2	0.27	1800	3130	3200	3260	3320	3390	3450	3510	3580	3640	3710	3770
		3	0.53	1500	3080	3130	3190	3240	3290	3350	3400	3450	3500	3560	3610
		4	0.80	1200	3020	3070	3110	3150	3190	3240	3280	3320	3360	3400	3450
		BFL	1.07	895	2960	2990	3020	3060	3090	3120	3150	3180	3210	3250	3280
		6	4.16	710	2900	2930	2950	2980	3000	3030	3050	3080	3100	3130	3150
		7	8.10	526	2800	2820	2830	2850	2870	2890	2910	2930	2950	2960	2980
W40×183	2110	TFL	0.00	1930	2940	3010	3080	3150	3220	3290	3350	3420	3490	3560	3630
		2	0.31	1670	2900	2960	3020	3080	3140	3200	3260	3320	3380	3440	3500
		3	0.61	1410	2860	2910	2960	3010	3060	3110	3160	3210	3260	3310	3360
		4	0.92	1160	2810	2850	2890	2930	2970	3010	3050	3090	3130	3180	3220
		BFL	1.22	896	2750	2780	2810	2850	2880	2910	2940	2970	3000	3040	3070
		6	4.76	690	2680	2710	2730	2750	2780	2800	2830	2850	2880	2900	2930
		7	9.16	483	2550	2570	2580	2600	2620	2640	2650	2670	2690	2700	2720
W40×174	1930	TFL	0.00	1840	2750	2810	2880	2940	3010	3080	3140	3210	3270	3340	3400
		2	0.21	1600	2710	2770	2830	2880	2940	3000	3060	3110	3170	3230	3280
		3	0.42	1370	2680	2720	2770	2820	2870	2920	2970	3020	3060	3110	3160
		4	0.62	1130	2630	2670	2710	2750	2790	2830	2870	2910	2960	3000	3040
		BFL	0.83	898	2590	2620	2650	2680	2720	2750	2780	2810	2840	2870	2910
		6	4.59	679	2520	2540	2570	2590	2620	2640	2660	2690	2710	2740	2760
		7	9.27	460	2380	2400	2410	2430	2450	2460	2480	2490	2510	2530	2540
W40×167	1870	TFL	0.00	1770	2670	2730	2790	2850	2920	2980	3040	3100	3170	3230	3290
		2	0.26	1550	2630	2690	2740	2800	2850	2910	2960	3020	3070	3130	3180
		3	0.51	1330	2600	2640	2690	2740	2790	2830	2880	2930	2970	3020	3070
		4	0.77	1110	2560	2600	2630	2670	2710	2750	2790	2830	2870	2910	2950
		BFL	1.03	896	2510	2540	2570	2610	2640	2670	2700	2730	2760	2800	2830
		6	5.00	669	2430	2460	2480	2510	2530	2550	2580	2600	2620	2650	2670
		7	9.85	442	2280	2300	2310	2330	2350	2360	2380	2390	2410	2420	2440
W40×149	1610	TFL	0.00	1580	2360	2410	2470	2520	2580	2640	2690	2750	2800	2860	2920
		2	0.21	1400	2330	2380	2430	2480	2530	2580	2630	2680	2730	2780	2830
		3	0.42	1220	2300	2340	2390	2430	2470	2520	2560	2600	2650	2690	2730
		4	0.62	1050	2270	2310	2340	2380	2420	2460	2490	2530	2570	2600	2640
		BFL	0.83	871	2240	2270	2300	2330	2360	2390	2420	2450	2480	2510	2540
		6	5.15	633	2160	2180	2200	2220	2250	2270	2290	2310	2340	2360	2380
		7	10.41	394	1990	2010	2020	2030	2050	2060	2070	2090	2100	2120	2130

[a] Y1 = distance from top of the steel beam to plastic neutral axis.
[b] Y2 = distance from top of the steel beam to concrete flange force.
[c] See Figure 5-3 for PNA locations.

| F_y = 36 ksi | | COMPOSITE DESIGN COMPOSITE BEAM SELECTION TABLE W Shapes $\phi = 0.85$ $\phi_b = 0.90$ | | | | | | | | | | | | |

Shape	$\phi_b M_p$	PNA[c]	Y1[a]	ΣQ_n	ϕM_n (kip-ft)										
					Y2[b] (in.)										
	Kip-ft		In.	Kips	2	2.5	3	3.5	4	4.5	5	5.5	6	6.5	7
W 36×300	3400	TFL	0.00	3180	4590	4700	4810	4920	5040	5150	5260	5370	5490	5600	5710
		2	0.42	2680	4510	4600	4700	4790	4890	4980	5080	5170	5270	5360	5460
		3	0.84	2170	4410	4490	4570	4640	4720	4800	4880	4950	5030	5110	5180
		4	1.26	1670	4310	4360	4420	4480	4540	4600	4660	4720	4780	4840	4900
		BFL	1.68	1160	4180	4220	4260	4310	4350	4390	4430	4470	4510	4550	4590
		6	3.97	979	4120	4150	4190	4220	4260	4290	4330	4360	4400	4430	4470
		7	6.69	795	4020	4050	4080	4110	4140	4160	4190	4220	4250	4280	4300
W 36×280	3160	TFL	0.00	2970	4260	4360	4470	4570	4680	4780	4890	4990	5100	5200	5310
		2	0.39	2500	4180	4270	4360	4450	4540	4630	4710	4800	4890	4980	5070
		3	0.79	2030	4100	4170	4240	4310	4390	4460	4530	4600	4670	4740	4820
		4	1.18	1560	4000	4050	4110	4160	4220	4280	4330	4390	4440	4500	4550
		BFL	1.57	1090	3890	3930	3960	4000	4040	4080	4120	4160	4200	4230	4270
		6	3.88	916	3830	3860	3890	3930	3960	3990	4020	4060	4090	4120	4150
		7	6.62	742	3740	3770	3790	3820	3850	3870	3900	3920	3950	3980	4000
W 36×260	2920	TFL	0.00	2750	3930	4020	4120	4220	4320	4410	4510	4610	4710	4800	4900
		2	0.36	2330	3860	3940	4030	4110	4190	4270	4350	4440	4520	4600	4680
		3	0.72	1900	3780	3850	3920	3980	4050	4120	4190	4250	4320	4390	4450
		4	1.08	1470	3700	3750	3800	3850	3900	3960	4010	4060	4110	4160	4210
		BFL	1.44	1040	3600	3630	3670	3710	3740	3780	3820	3850	3890	3930	3960
		6	3.86	863	3540	3570	3600	3630	3660	3690	3720	3750	3780	3820	3850
		7	6.75	689	3450	3470	3500	3520	3550	3570	3600	3620	3640	3670	3690
W 36×245	2730	TFL	0.00	2600	3680	3780	3870	3960	4050	4140	4240	4330	4420	4510	4600
		2	0.34	2190	3620	3700	3780	3860	3930	4010	4090	4170	4240	4320	4400
		3	0.68	1790	3550	3620	3680	3740	3810	3870	3930	4000	4060	4120	4190
		4	1.01	1390	3470	3520	3570	3620	3670	3720	3770	3820	3870	3910	3960
		BFL	1.35	991	3380	3420	3450	3490	3520	3560	3590	3630	3660	3700	3730
		6	3.81	820	3330	3360	3380	3410	3440	3470	3500	3530	3560	3590	3620
		7	6.77	649	3240	3260	3280	3310	3330	3350	3380	3400	3420	3440	3470
W 36×230	2550	TFL	0.00	2430	3440	3530	3610	3700	3780	3870	3960	4040	4130	4210	4300
		2	0.32	2060	3380	3450	3530	3600	3670	3750	3820	3890	3970	4040	4110
		3	0.63	1690	3320	3380	3440	3500	3560	3620	3670	3730	3790	3850	3910
		4	0.95	1310	3240	3290	3340	3380	3430	3480	3520	3570	3610	3660	3710
		BFL	1.26	939	3160	3190	3230	3260	3290	3330	3360	3390	3430	3460	3490
		6	3.81	774	3110	3140	3160	3190	3220	3250	3270	3300	3330	3360	3380
		7	6.83	608	3020	3040	3070	3090	3110	3130	3150	3170	3200	3220	3240

[a] Y1 = distance from top of the steel beam to plastic neutral axis.
[b] Y2 = distance from top of the steel beam to concrete flange force.
[c] See Figure 5-3 for PNA locations.

$$F_y = 36 \text{ ksi}$$

COMPOSITE DESIGN
COMPOSITE BEAM SELECTION TABLE
W Shapes

$$\phi = 0.85 \qquad\qquad \phi_b = 0.90$$

Shape	$\phi_b M_p$	PNA[c]	Y1[a]	ΣQ_n	ϕM_n (kip-ft)										
					Y2[b] (in.)										
	Kip-ft		In.	Kips	2	2.5	3	3.5	4	4.5	5	5.5	6	6.5	7
W 36×210	2250	TFL	0.00	2220	3210	3280	3360	3440	3520	3600	3680	3760	3840	3920	3990
		2	0.34	1930	3160	3230	3300	3370	3430	3500	3570	3640	3710	3770	3840
		3	0.68	1630	3110	3170	3220	3280	3340	3400	3450	3510	3570	3630	3680
		4	1.02	1330	3050	3090	3140	3190	3240	3280	3330	3380	3420	3470	3520
		BFL	1.36	1030	2980	3020	3050	3090	3130	3160	3200	3240	3270	3310	3350
		6	5.06	794	2890	2920	2950	2980	3010	3030	3060	3090	3120	3150	3170
		7	9.04	556	2740	2760	2780	2800	2820	2840	2860	2880	2900	2920	2940
W 36×194	2070	TFL	0.00	2050	2940	3020	3090	3160	3230	3310	3380	3450	3520	3600	3670
		2	0.32	1780	2900	2960	3030	3090	3150	3220	3280	3340	3400	3470	3530
		3	0.63	1500	2850	2910	2960	3010	3070	3120	3170	3220	3280	3330	3380
		4	0.95	1230	2800	2840	2890	2930	2970	3020	3060	3100	3150	3190	3230
		BFL	1.26	953	2740	2770	2810	2840	2870	2910	2940	2970	3010	3040	3080
		6	4.94	733	2660	2690	2710	2740	2760	2790	2820	2840	2870	2890	2920
		7	8.93	513	2520	2540	2560	2580	2590	2610	2630	2650	2670	2680	2700
W 36×182	1940	TFL	0.00	1930	2760	2820	2890	2960	3030	3100	3170	3230	3300	3370	3440
		2	0.29	1670	2720	2780	2840	2890	2950	3010	3070	3130	3190	3250	3310
		3	0.59	1420	2670	2720	2770	2820	2870	2920	2970	3020	3070	3120	3170
		4	0.89	1160	2620	2660	2710	2750	2790	2830	2870	2910	2950	2990	3030
		BFL	1.18	904	2570	2600	2630	2660	2700	2730	2760	2790	2820	2860	2890
		6	4.89	693	2490	2520	2540	2570	2590	2620	2640	2670	2690	2720	2740
		7	8.92	482	2360	2380	2400	2410	2430	2450	2460	2480	2500	2520	2530
W 36×170	1800	TFL	0.00	1800	2560	2620	2690	2750	2820	2880	2940	3010	3070	3130	3200
		2	0.28	1560	2520	2580	2640	2690	2750	2800	2860	2910	2970	3020	3080
		3	0.55	1320	2480	2530	2580	2620	2670	2720	2770	2810	2860	2910	2950
		4	0.83	1090	2440	2480	2520	2550	2590	2630	2670	2710	2750	2780	2820
		BFL	1.10	847	2390	2420	2450	2480	2510	2540	2570	2600	2630	2660	2690
		6	4.84	649	2320	2340	2370	2390	2410	2440	2460	2480	2500	2530	2550
		7	8.89	450	2200	2210	2230	2240	2260	2280	2290	2310	2320	2340	2360
W 36×160	1680	TFL	0.00	1690	2400	2460	2520	2580	2640	2700	2760	2820	2880	2940	3000
		2	0.26	1470	2360	2420	2470	2520	2570	2630	2680	2730	2780	2830	2890
		3	0.51	1250	2330	2370	2420	2460	2500	2550	2590	2640	2680	2730	2770
		4	0.77	1030	2290	2320	2360	2400	2430	2470	2510	2540	2580	2610	2650
		BFL	1.02	811	2240	2270	2300	2330	2360	2380	2410	2440	2470	2500	2530
		6	4.82	617	2170	2200	2220	2240	2260	2280	2310	2330	2350	2370	2390
		7	8.97	423	2050	2070	2080	2100	2110	2130	2140	2160	2170	2190	2200

[a] Y1 = distance from top of the steel beam to plastic neutral axis.
[b] Y2 = distance from top of the steel beam to concrete flange force.
[c] See Figure 5-3 for PNA locations.

F_y = 36 ksi														

COMPOSITE DESIGN
COMPOSITE BEAM SELECTION TABLE
W Shapes
$\phi = 0.85$ $\phi_b = 0.90$

Shape	$\phi_b M_p$	PNA[c]	Y1[a]	ΣQ_n	ϕM_n (kip-ft) Y2[b] (in.)										
	Kip-ft		In.	Kips	2	2.5	3	3.5	4	4.5	5	5.5	6	6.5	7
W 36×150	1570	TFL	0.00	1590	2250	2300	2360	2410	2470	2530	2580	2640	2700	2750	2810
		2	0.24	1390	2220	2260	2310	2360	2410	2460	2510	2560	2610	2660	2710
		3	0.47	1190	2180	2220	2270	2310	2350	2390	2430	2480	2520	2560	2600
		4	0.71	983	2140	2180	2210	2250	2280	2320	2350	2390	2420	2460	2490
		BFL	0.94	781	2100	2130	2160	2190	2210	2240	2270	2300	2330	2350	2380
		6	4.83	589	2040	2060	2080	2100	2120	2140	2160	2190	2210	2230	2250
		7	9.08	398	1920	1930	1950	1960	1970	1990	2000	2020	2030	2040	2060
W 36×135	1370	TFL	0.00	1430	2000	2050	2100	2150	2200	2260	2310	2360	2410	2460	2510
		2	0.20	1260	1980	2020	2070	2110	2160	2200	2240	2290	2330	2380	2420
		3	0.40	1090	1950	1990	2030	2060	2100	2140	2180	2220	2260	2300	2330
		4	0.59	919	1920	1950	1980	2020	2050	2080	2110	2150	2180	2210	2240
		BFL	0.79	749	1890	1910	1940	1970	1990	2020	2050	2070	2100	2130	2150
		6	4.96	553	1820	1840	1860	1880	1900	1920	1940	1960	1980	2000	2020
		7	9.50	357	1690	1710	1720	1730	1740	1760	1770	1780	1790	1810	1820
W 33×221	2310	TFL	0.00	2340	3140	3230	3310	3390	3470	3560	3640	3720	3810	3890	3970
		2	0.32	1980	3090	3160	3230	3300	3370	3440	3510	3580	3650	3720	3790
		3	0.64	1610	3020	3080	3140	3200	3250	3310	3370	3420	3480	3540	3600
		4	0.96	1250	2950	3000	3040	3090	3130	3170	3220	3260	3310	3350	3400
		BFL	1.28	889	2870	2900	2940	2970	3000	3030	3060	3090	3120	3160	3190
		6	3.76	737	2820	2850	2880	2900	2930	2960	2980	3010	3030	3060	3090
		7	6.48	585	2750	2770	2790	2810	2830	2850	2870	2890	2910	2930	2960
W 33×201	2080	TFL	0.00	2130	2840	2910	2990	3070	3140	3220	3290	3370	3440	3520	3590
		2	0.29	1800	2790	2850	2920	2980	3050	3110	3170	3240	3300	3360	3430
		3	0.58	1480	2730	2790	2840	2890	2940	2990	3050	3100	3150	3200	3260
		4	0.86	1150	2670	2710	2750	2790	2830	2870	2920	2960	3000	3040	3080
		BFL	1.15	824	2600	2630	2660	2690	2720	2750	2780	2810	2830	2860	2890
		6	3.67	678	2560	2580	2600	2630	2650	2680	2700	2720	2750	2770	2800
		7	6.51	532	2480	2500	2520	2540	2560	2580	2600	2620	2630	2650	2670
W 33×141	1390	TFL	0.00	1500	1980	2030	2080	2140	2190	2240	2300	2350	2400	2460	2510
		2	0.24	1300	1950	1990	2040	2090	2130	2180	2220	2270	2320	2360	2410
		3	0.48	1100	1920	1950	1990	2030	2070	2110	2150	2190	2230	2270	2300
		4	0.72	900	1880	1910	1940	1970	2010	2040	2070	2100	2130	2170	2200
		BFL	0.96	700	1840	1860	1890	1910	1940	1960	1990	2010	2040	2060	2090
		6	4.31	537	1790	1810	1820	1840	1860	1880	1900	1920	1940	1960	1980
		7	8.05	374	1690	1710	1720	1730	1740	1760	1770	1780	1800	1810	1820

[a] Y1 = distance from top of the steel beam to plastic neutral axis.
[b] Y2 = distance from top of the steel beam to concrete flange force.
[c] See Figure 5-3 for PNA locations.

COMPOSITE DESIGN
COMPOSITE BEAM SELECTION TABLE
W Shapes

$F_y = 36$ ksi

$\phi = 0.85$ $\phi_b = 0.90$

Shape	$\phi_b M_p$	PNA[c]	Y1[a]	ΣQ_n	ϕM_n (kip-ft) Y2[b] (in.)										
	Kip-ft		In.	Kips	2	2.5	3	3.5	4	4.5	5	5.5	6	6.5	7
W 33×130	1260	TFL	0.00	1380	1810	1860	1910	1960	2010	2060	2100	2150	2200	2250	2300
		2	0.21	1200	1780	1830	1870	1910	1950	2000	2040	2080	2130	2170	2210
		3	0.43	1020	1760	1790	1830	1860	1900	1940	1970	2010	2050	2080	2120
		4	0.64	847	1720	1750	1780	1810	1840	1870	1900	1930	1960	1990	2020
		BFL	0.86	670	1690	1710	1740	1760	1780	1810	1830	1860	1880	1900	1930
		6	4.39	507	1640	1660	1670	1690	1710	1730	1750	1760	1780	1800	1820
		7	8.29	345	1540	1550	1570	1580	1590	1600	1610	1630	1640	1650	1660
W 33×118	1120	TFL	0.00	1250	1630	1680	1720	1760	1810	1850	1900	1940	1980	2030	2070
		2	0.19	1100	1610	1650	1690	1720	1760	1800	1840	1880	1920	1960	2000
		3	0.37	943	1580	1620	1650	1680	1720	1750	1780	1820	1850	1880	1920
		4	0.56	790	1560	1580	1610	1640	1670	1700	1720	1750	1780	1810	1840
		BFL	0.74	638	1530	1550	1570	1600	1620	1640	1660	1690	1710	1730	1750
		6	4.44	475	1480	1490	1510	1530	1540	1560	1580	1590	1610	1630	1650
		7	8.54	312	1380	1390	1400	1410	1420	1430	1450	1460	1470	1480	1490
W 30×116	1020	TFL	0.00	1230	1480	1530	1570	1610	1660	1700	1740	1790	1830	1880	1920
		2	0.21	1070	1460	1500	1530	1570	1610	1650	1690	1720	1760	1800	1840
		3	0.43	910	1430	1460	1500	1530	1560	1590	1630	1660	1690	1720	1750
		4	0.64	749	1400	1430	1460	1480	1510	1540	1560	1590	1620	1640	1670
		BFL	0.85	589	1370	1390	1410	1440	1460	1480	1500	1520	1540	1560	1580
		6	3.98	448	1330	1350	1360	1380	1390	1410	1430	1440	1460	1470	1490
		7	7.44	308	1250	1260	1270	1290	1300	1310	1320	1330	1340	1350	1360
W 30×108	934	TFL	0.00	1140	1370	1410	1450	1490	1530	1570	1610	1650	1690	1730	1770
		2	0.19	998	1350	1380	1420	1450	1490	1520	1560	1590	1630	1660	1700
		3	0.38	855	1320	1350	1380	1410	1440	1470	1500	1530	1570	1600	1630
		4	0.57	711	1300	1320	1350	1370	1400	1420	1450	1470	1500	1520	1550
		BFL	0.76	568	1270	1290	1310	1330	1350	1370	1390	1410	1430	1450	1470
		6	4.04	427	1230	1240	1260	1270	1290	1300	1320	1330	1350	1360	1380
		7	7.64	285	1150	1160	1170	1180	1190	1200	1210	1220	1230	1240	1250
W 30×99	842	TFL	0.00	1050	1250	1290	1320	1360	1400	1430	1470	1510	1550	1580	1620
		2	0.17	922	1230	1260	1300	1330	1360	1390	1430	1460	1490	1520	1560
		3	0.34	796	1210	1240	1270	1290	1320	1350	1380	1410	1440	1460	1490
		4	0.50	670	1190	1210	1240	1260	1280	1310	1330	1350	1380	1400	1430
		BFL	0.67	543	1170	1180	1200	1220	1240	1260	1280	1300	1320	1340	1360
		6	4.07	403	1120	1140	1150	1170	1180	1190	1210	1220	1240	1250	1270
		7	7.83	262	1040	1050	1060	1070	1080	1090	1100	1110	1120	1130	1140

[a] Y1 = distance from top of the steel beam to plastic neutral axis.
[b] Y2 = distance from top of the steel beam to concrete flange force.
[c] See Figure 5-3 for PNA locations.

| $F_y = 36$ ksi |

COMPOSITE DESIGN
COMPOSITE BEAM SELECTION TABLE
W Shapes
$\phi = 0.85$ $\phi_b = 0.90$

Shape	$\phi_b M_p$	PNA[c]	$Y1$[a]	ΣQ_n	ϕM_n (kip-ft)										
					$Y2$[b] (in.)										
	Kip-ft		In.	Kips	2	2.5	3	3.5	4	4.5	5	5.5	6	6.5	7
W 27×102	824	TFL	0.00	1080	1190	1230	1270	1300	1340	1380	1420	1460	1500	1530	1570
		2	0.21	930	1170	1200	1230	1270	1300	1330	1360	1400	1430	1460	1500
		3	0.42	781	1140	1170	1200	1230	1250	1280	1310	1340	1360	1390	1420
		4	0.62	631	1120	1140	1160	1180	1210	1230	1250	1270	1290	1320	1340
		BFL	0.83	482	1090	1100	1120	1140	1160	1170	1190	1210	1220	1240	1260
		6	3.41	376	1060	1070	1080	1100	1110	1120	1140	1150	1160	1180	1190
		7	6.26	270	1010	1010	1020	1030	1040	1050	1060	1070	1080	1090	1100
W 27×94	751	TFL	0.00	997	1090	1130	1160	1200	1230	1270	1300	1340	1370	1410	1450
		2	0.19	863	1070	1100	1130	1160	1190	1230	1260	1290	1320	1350	1380
		3	0.37	729	1050	1080	1100	1130	1150	1180	1210	1230	1260	1280	1310
		4	0.56	595	1030	1050	1070	1090	1110	1130	1150	1170	1200	1220	1240
		BFL	0.75	461	1000	1020	1030	1050	1070	1080	1100	1120	1130	1150	1170
		6	3.39	355	972	985	997	1010	1020	1040	1050	1060	1070	1090	1100
		7	6.39	249	921	929	938	947	956	965	974	982	991	1000	1010
W 27×84	659	TFL	0.00	893	971	1000	1030	1070	1100	1130	1160	1190	1220	1260	1290
		2	0.16	778	954	982	1010	1040	1060	1090	1120	1150	1170	1200	1230
		3	0.32	663	936	959	983	1010	1030	1050	1080	1100	1120	1150	1170
		4	0.48	549	916	936	955	975	994	1010	1030	1050	1070	1090	1110
		BFL	0.64	434	896	911	926	942	957	972	988	1000	1020	1030	1050
		6	3.44	329	866	878	890	901	913	925	936	948	960	971	983
		7	6.62	223	814	822	830	838	846	854	861	869	877	885	893
W 24×76	540	TFL	0.00	806	797	826	855	883	912	940	969	997	1030	1050	1080
		2	0.17	696	781	806	830	855	880	904	929	954	978	1000	1030
		3	0.34	586	764	784	805	826	847	867	888	909	930	950	971
		4	0.51	476	745	762	778	795	812	829	846	863	880	896	913
		BFL	0.68	366	724	737	750	763	776	789	802	815	828	841	854
		6	3.00	284	703	713	723	733	743	753	763	773	783	793	803
		7	5.60	202	666	673	680	687	694	702	709	716	723	730	737
W 24×68	478	TFL	0.00	724	711	736	762	788	813	839	864	890	916	941	967
		2	0.15	629	697	719	741	764	786	808	830	853	875	897	920
		3	0.29	535	682	701	720	739	758	777	796	815	833	852	871
		4	0.44	440	666	682	697	713	729	744	760	775	791	807	822
		BFL	0.59	346	649	662	674	686	698	711	723	735	747	760	772
		6	3.05	263	628	637	646	656	665	674	684	693	702	712	721
		7	5.81	181	590	596	603	609	616	622	628	635	641	648	654

[a] $Y1$ = distance from top of the steel beam to plastic neutral axis.
[b] $Y2$ = distance from top of the steel beam to concrete flange force.
[c] See Figure 5-3 for PNA locations.

COMPOSITE DESIGN
COMPOSITE BEAM SELECTION TABLE
W Shapes

$\phi = 0.85$ $\phi_b = 0.90$

$F_y = 36$ ksi

Shape	$\phi_b M_p$	PNA[c]	Y1[a]	ΣQ_n	ϕM_n (kip-ft)										
					Y2[b] (in.)										
	Kip-ft		In.	Kips	2	2.5	3	3.5	4	4.5	5	5.5	6	6.5	7
W 24×62	413	TFL	0.00	655	644	667	690	713	737	760	783	806	829	853	876
		2	0.15	580	633	653	674	694	715	736	756	777	797	818	838
		3	0.29	506	621	639	657	675	693	711	728	746	764	782	800
		4	0.44	431	608	624	639	654	669	685	700	715	731	746	761
		BFL	0.59	356	595	608	620	633	646	658	671	683	696	709	721
		6	3.47	260	568	577	587	596	605	614	623	633	642	651	660
		7	6.58	164	520	526	532	538	543	549	555	561	567	572	578
W 24×55	362	TFL	0.00	583	569	590	611	631	652	673	693	714	735	755	776
		2	0.13	520	560	579	597	615	634	652	671	689	707	726	744
		3	0.25	456	550	566	583	599	615	631	647	663	679	696	712
		4	0.38	392	540	554	568	582	595	609	623	637	651	665	679
		BFL	0.51	328	529	540	552	564	575	587	599	610	622	634	645
		6	3.45	237	504	512	520	529	537	546	554	562	571	579	588
		7	6.66	146	458	463	468	474	479	484	489	494	499	505	510
W 21×62	389	TFL	0.00	659	583	606	630	653	676	700	723	746	770	793	816
		2	0.15	568	570	590	610	630	650	670	690	710	730	751	771
		3	0.31	476	555	572	589	606	623	640	656	673	690	707	724
		4	0.46	385	540	553	567	581	594	608	622	635	649	663	676
		BFL	0.62	294	523	534	544	555	565	575	586	596	607	617	628
		6	2.53	229	507	516	524	532	540	548	556	564	572	581	589
		7	4.78	165	482	487	493	499	505	511	517	522	528	534	540
W 21×57	348	TFL	0.00	601	534	555	576	597	619	640	661	683	704	725	747
		2	0.16	525	522	541	559	578	597	615	634	652	671	689	708
		3	0.33	448	510	526	542	558	574	589	605	621	637	653	669
		4	0.49	371	497	510	523	536	550	563	576	589	602	615	628
		BFL	0.65	294	483	493	504	514	525	535	546	556	566	577	587
		6	2.90	222	464	472	480	488	496	504	511	519	527	535	543
		7	5.38	150	433	438	443	449	454	459	465	470	475	481	486
W 21×50	297	TFL	0.00	529	465	484	503	522	540	559	578	597	615	634	653
		2	0.13	466	456	473	489	506	522	539	555	572	588	605	621
		3	0.27	403	446	461	475	489	504	518	532	546	561	575	589
		4	0.40	341	436	448	460	472	484	496	508	520	532	545	557
		BFL	0.54	278	425	435	445	454	464	474	484	494	504	513	523
		6	2.92	205	406	413	421	428	435	442	450	457	464	472	479
		7	5.58	132	374	379	383	388	393	397	402	407	411	416	421

[a] Y1 = distance from top of the steel beam to plastic neutral axis.
[b] Y2 = distance from top of the steel beam to concrete flange force.
[c] See Figure 5-3 for PNA locations.

F_y = 36 ksi															

COMPOSITE DESIGN
COMPOSITE BEAM SELECTION TABLE
W Shapes

$$\phi = 0.85 \qquad \phi_b = 0.90$$

Shape	$\phi_b M_p$	PNA[c]	Y1[a]	ΣQ_n	ϕM_n (kip-ft)										
					Y2[b] (in.)										
	Kip-ft		In.	Kips	2	2.5	3	3.5	4	4.5	5	5.5	6	6.5	7
W 21×44	258	TFL	0.00	468	409	425	442	458	475	492	508	525	541	558	574
		2	0.11	415	401	416	430	445	460	475	489	504	519	533	548
		3	0.23	363	393	406	419	432	444	457	470	483	496	509	521
		4	0.34	310	384	395	406	417	428	439	450	461	472	483	494
		BFL	0.45	257	376	385	394	403	412	421	430	439	448	458	467
		6	2.90	187	358	364	371	378	384	391	398	404	411	417	424
		7	5.69	117	326	331	335	339	343	347	351	355	360	364	368
W 18×60	332	TFL	0.00	634	499	522	544	566	589	611	634	656	679	701	723
		2	0.17	539	485	504	523	542	561	581	600	619	638	657	676
		3	0.35	445	470	486	501	517	533	549	564	580	596	612	627
		4	0.52	350	454	466	478	491	503	516	528	540	553	565	578
		BFL	0.70	256	436	445	454	463	472	481	491	500	509	518	527
		6	2.19	207	424	432	439	446	454	461	468	476	483	490	498
		7	3.82	158	407	413	418	424	430	435	441	447	452	458	463
W 18×55	302	TFL	0.00	583	457	477	498	519	539	560	581	601	622	643	663
		2	0.16	498	444	462	479	497	515	532	550	568	585	603	620
		3	0.32	412	431	445	460	474	489	504	518	533	547	562	577
		4	0.47	327	416	428	439	451	462	474	486	497	509	520	532
		BFL	0.63	242	401	409	418	426	435	443	452	461	469	478	486
		6	2.16	194	389	396	403	410	417	424	430	437	444	451	458
		7	3.86	146	372	377	383	388	393	398	403	408	414	419	424
W 18×50	273	TFL	0.00	529	412	431	450	468	487	506	525	543	562	581	600
		2	0.14	452	401	417	433	449	465	481	497	513	529	545	561
		3	0.29	375	389	402	415	429	442	455	469	482	495	508	522
		4	0.43	299	376	387	397	408	418	429	439	450	461	471	482
		BFL	0.57	222	362	370	378	386	394	402	409	417	425	433	441
		6	2.07	177	352	358	365	371	377	383	390	396	402	408	415
		7	3.82	132	336	341	346	350	355	360	365	369	374	379	383
W 18×46	245	TFL	0.00	486	380	397	414	431	449	466	483	500	517	535	552
		2	0.15	420	370	385	400	415	430	444	459	474	489	504	519
		3	0.30	354	360	372	385	397	410	422	435	447	460	472	485
		4	0.45	288	348	359	369	379	389	399	410	420	430	440	450
		BFL	0.61	222	337	345	352	360	368	376	384	392	400	407	415
		6	2.40	172	324	330	336	343	349	355	361	367	373	379	385
		7	4.34	122	305	310	314	318	322	327	331	335	340	344	348

[a] Y1 = distance from top of the steel beam to plastic neutral axis.
[b] Y2 = distance from top of the steel beam to concrete flange force.
[c] See Figure 5-3 for PNA locations.

COMPOSITE DESIGN

Shape	$\phi_b M_p$	PNA[c]	$Y1$[a]	ΣQ_n	ϕM_n (kip-ft)										
					$Y2$[b] (in.)										
	Kip-ft		In.	Kips	2	2.5	3	3.5	4	4.5	5	5.5	6	6.5	7
W 18×40	212	TFL	0.00	425	329	345	360	375	390	405	420	435	450	465	480
		2	0.13	368	321	334	347	360	373	386	399	412	425	438	451
		3	0.26	311	312	323	334	345	356	367	378	389	400	411	423
		4	0.39	254	303	312	321	330	339	348	357	366	375	384	393
		BFL	0.53	197	293	300	307	314	321	328	335	342	349	356	363
		6	2.26	152	282	288	293	298	304	309	315	320	325	331	336
		7	4.27	106	265	269	273	277	280	284	288	292	295	299	303
W 18×35	180	TFL	0.00	371	285	298	311	324	338	351	364	377	390	403	416
		2	0.11	325	278	290	301	313	324	336	347	359	370	382	393
		3	0.21	279	271	281	291	301	311	321	331	340	350	360	370
		4	0.32	233	264	272	280	289	297	305	313	322	330	338	346
		BFL	0.43	187	256	263	269	276	283	289	296	303	309	316	323
		6	2.37	140	245	250	255	260	265	270	275	280	285	290	295
		7	4.56	92.7	227	230	233	237	240	243	246	250	253	256	260
W 16×36	173	TFL	0.00	382	268	282	295	309	322	336	349	363	377	390	404
		2	0.11	328	261	272	284	295	307	319	330	342	353	365	377
		3	0.22	273	252	262	272	281	291	301	310	320	330	339	349
		4	0.32	219	244	251	259	267	275	282	290	298	306	314	321
		BFL	0.43	165	234	240	246	252	258	264	270	275	281	287	293
		6	1.79	130	227	232	236	241	245	250	255	259	264	268	273
		7	3.44	95.4	215	219	222	226	229	232	236	239	243	246	249
W 16×31	146	TFL	0.00	328	231	243	254	266	278	289	301	313	324	336	347
		2	0.11	285	225	235	245	255	265	275	285	295	305	315	326
		3	0.22	241	218	227	235	244	252	261	269	278	286	295	303
		4	0.33	197	211	218	225	232	239	246	253	260	267	274	281
		BFL	0.44	153	204	209	214	220	225	231	236	242	247	253	258
		6	2.00	118	196	200	204	208	212	217	221	225	229	233	237
		7	3.79	82.1	183	186	189	192	195	198	201	204	207	209	212
W 16×26	119	TFL	0.00	276	193	203	212	222	232	242	252	261	271	281	291
		2	0.09	242	188	196	205	214	222	231	239	248	257	265	274
		3	0.17	208	183	190	197	205	212	220	227	234	242	249	256
		4	0.26	174	177	184	190	196	202	208	214	220	227	233	239
		BFL	0.35	140	172	177	182	187	192	197	202	206	211	216	221
		6	2.04	104	164	168	171	175	179	182	186	190	194	197	201
		7	4.01	69.1	151	154	156	159	161	164	166	169	171	173	176

COMPOSITE DESIGN
COMPOSITE BEAM SELECTION TABLE
W Shapes

$\phi = 0.85 \qquad \phi_b = 0.90$

$F_y = 36$ ksi

[a] $Y1$ = distance from top of the steel beam to plastic neutral axis.
[b] $Y2$ = distance from top of the steel beam to concrete flange force.
[c] See Figure 5-3 for PNA locations.

AMERICAN INSTITUTE OF STEEL CONSTRUCTION

F_y = 36 ksi														

COMPOSITE DESIGN
COMPOSITE BEAM SELECTION TABLE
W Shapes
$$\phi = 0.85 \qquad \phi_b = 0.90$$

Shape	$\phi_b M_p$	PNA[c]	$Y1^a$	ΣQ_n	ϕM_n (kip-ft)										
					$Y2^b$ (in.)										
	Kip-ft		In.	Kips	2	2.5	3	3.5	4	4.5	5	5.5	6	6.5	7
W 14×38	166	TFL	0.00	403	258	273	287	301	316	330	344	358	373	387	401
		2	0.13	340	249	261	273	285	298	310	322	334	346	358	370
		3	0.26	278	240	249	259	269	279	289	299	308	318	328	338
		4	0.39	215	229	237	244	252	260	267	275	283	290	298	305
		BFL	0.52	152	218	224	229	234	240	245	251	256	261	267	272
		6	1.38	126	213	218	222	226	231	235	240	244	249	253	258
		7	2.53	101	206	209	213	217	220	224	227	231	234	238	242
W 14×34	147	TFL	0.00	360	229	242	255	267	280	293	306	318	331	344	357
		2	0.11	305	221	232	243	254	264	275	286	297	308	318	329
		3	0.23	250	213	222	230	239	248	257	266	275	283	292	301
		4	0.34	194	204	211	218	224	231	238	245	252	259	266	273
		BFL	0.46	139	194	199	204	209	214	219	224	229	234	239	244
		6	1.41	115	189	193	197	202	206	210	214	218	222	226	230
		7	2.60	90.0	182	186	189	192	195	198	202	205	208	211	214
W 14×30	128	TFL	0.00	319	201	213	224	235	246	258	269	280	292	303	314
		2	0.10	272	195	204	214	223	233	243	252	262	272	281	291
		3	0.19	225	187	195	203	211	219	227	235	243	251	259	267
		4	0.29	179	180	186	193	199	205	212	218	224	231	237	243
		BFL	0.39	132	172	177	182	186	191	196	200	205	210	214	219
		6	1.48	106	167	171	174	178	182	186	189	193	197	201	204
		7	2.82	79.7	159	162	165	168	171	173	176	179	182	185	187
W 14×26	109	TFL	0.00	277	176	185	195	205	215	225	234	244	254	264	274
		2	0.11	239	170	179	187	195	204	212	221	229	238	246	255
		3	0.21	201	164	171	179	186	193	200	207	214	221	228	235
		4	0.32	163	158	164	170	175	181	187	193	199	204	210	216
		BFL	0.42	125	152	156	161	165	170	174	178	183	187	192	196
		6	1.67	97.0	146	149	153	156	160	163	167	170	173	177	180
		7	3.19	69.2	137	140	142	145	147	149	152	154	157	159	162
W 14×22	89.6	TFL	0.00	234	147	155	163	172	180	188	196	205	213	221	230
		2	0.08	203	142	150	157	164	171	178	186	193	200	207	215
		3	0.17	173	138	144	150	156	162	169	175	181	187	193	199
		4	0.25	143	133	138	143	148	153	159	164	169	174	179	184
		BFL	0.34	113	128	132	136	140	144	148	152	156	160	164	168
		6	1.69	85.7	123	126	129	132	135	138	141	144	147	150	153
		7	3.34	58.4	114	116	118	120	122	124	126	128	130	132	135

[a] $Y1$ = distance from top of the steel beam to plastic neutral axis.
[b] $Y2$ = distance from top of the steel beam to concrete flange force.
[c] See Figure 5-3 for PNA locations.

$$F_y = 36 \text{ ksi}$$

COMPOSITE DESIGN
COMPOSITE BEAM SELECTION TABLE
W Shapes

$$\phi = 0.85 \qquad \phi_b = 0.90$$

Shape	$\phi_b M_p$	PNA[c]	Y1[a]	ΣQ_n	ϕM_n (kip-ft)										
					Y2[b] (in.)										
	Kip-ft		In.	Kips	2	2.5	3	3.5	4	4.5	5	5.5	6	6.5	7
W 12×30	116	TFL	0.00	316	183	194	206	217	228	239	250	262	273	284	295
		2	0.11	265	176	185	194	204	213	223	232	241	251	260	269
		3	0.22	213	168	175	183	190	198	205	213	221	228	236	243
		4	0.33	162	159	165	171	177	182	188	194	199	205	211	217
		BFL	0.44	110	151	155	158	162	166	170	174	178	182	186	190
		6	1.12	94.5	148	151	154	158	161	164	168	171	174	178	181
		7	1.94	79.1	144	147	149	152	155	158	161	163	166	169	172
W 12×26	100	TFL	0.00	275	158	168	178	187	197	207	217	226	236	246	256
		2	0.10	231	152	160	168	176	184	193	201	209	217	225	234
		3	0.19	187	145	152	158	165	171	178	185	191	198	205	211
		4	0.29	142	138	143	148	153	158	163	168	173	178	183	188
		BFL	0.38	97.8	131	134	138	141	145	148	151	155	158	162	165
		6	1.08	83.3	128	131	134	137	140	143	146	149	151	154	157
		7	1.95	68.9	124	127	129	132	134	136	139	141	144	146	149
W 12×22	79.1	TFL	0.00	233	135	143	151	160	168	176	184	193	201	209	217
		2	0.11	202	130	137	145	152	159	166	173	180	188	195	202
		3	0.21	172	126	132	138	144	150	156	162	168	174	180	186
		4	0.32	141	121	126	131	136	141	146	151	156	160	165	170
		BFL	0.43	110	115	119	123	127	131	135	139	143	147	150	154
		6	1.66	84.1	110	113	116	119	122	125	128	131	134	137	140
		7	3.04	58.3	102	104	106	108	110	112	114	116	119	121	123
W 12×19	66.7	TFL	0.00	201	115	122	129	136	143	150	157	164	172	179	186
		2	0.09	175	111	117	124	130	136	142	148	155	161	167	173
		3	0.18	150	107	113	118	123	129	134	139	145	150	155	160
		4	0.26	125	103	108	112	117	121	125	130	134	139	143	148
		BFL	0.35	99.6	99.2	103	106	110	113	117	120	124	127	131	134
		6	1.66	74.9	94.0	96.7	99.3	102	105	107	110	113	115	118	121
		7	3.12	50.1	86.3	88.1	89.9	91.7	93.5	95.2	97.0	98.8	101	102	104
W 12×16	54.3	TFL	0.00	170	96.0	102	108	114	120	126	132	138	144	150	156
		2	0.07	151	93.3	98.6	104	109	115	120	125	131	136	141	147
		3	0.13	131	90.5	95.1	99.8	104	109	114	118	123	128	132	137
		4	0.20	112	87.5	91.5	95.5	99.5	103	107	111	115	119	123	127
		BFL	0.26	93.4	84.5	87.8	91.1	94.5	97.8	101	104	108	111	114	118
		6	1.71	67.9	79.2	81.6	84.0	86.4	88.8	91.2	93.6	96.1	98.5	101	103
		7	3.32	42.4	71.1	72.6	74.1	75.6	77.1	78.6	80.1	81.6	83.1	84.6	86.1

[a] Y1 = distance from top of the steel beam to plastic neutral axis.
[b] Y2 = distance from top of the steel beam to concrete flange force.
[c] See Figure 5-3 for PNA locations.

| F_y = 36 ksi |

COMPOSITE DESIGN
COMPOSITE BEAM SELECTION TABLE
W Shapes
$$\phi = 0.85 \qquad \phi_b = 0.90$$

Shape	$\phi_b M_p$	PNA[c]	Y1[a]	ΣQ_n	ϕM_n (kip-ft)											
					Y2[b] (in.)											
	Kip-ft		In.	Kips	2	2.5	3	3.5	4	4.5	5	5.5	6	6.5	7
W 12×14	47.0	TFL	0.00	150	84.4	89.7	95.0	100	106	111	116	122	127	132	137
		2	0.06	134	82.1	86.8	91.5	96.3	101	106	110	115	120	125	129
		3	0.11	118	79.7	83.9	88.0	92.2	96.4	101	105	109	113	117	121
		4	0.17	102	77.3	80.9	84.5	88.1	91.6	95.2	98.8	102	106	110	113
		BFL	0.23	85.4	74.8	77.8	80.8	83.8	86.9	89.9	92.9	95.9	99.0	102	105
		6	1.69	61.4	69.8	72.0	74.2	76.4	78.5	80.7	82.9	85.1	87.2	89.4	91.6
		7	3.36	37.4	62.2	63.5	64.8	66.1	67.5	68.8	70.1	71.4	72.8	74.1	75.4
W 10×26	84.5	TFL	0.00	274	139	149	158	168	178	188	197	207	217	226	236
		2	0.11	228	132	140	149	157	165	173	181	189	197	205	213
		3	0.22	183	125	132	138	145	151	158	164	171	177	184	190
		4	0.33	137	118	123	128	133	137	142	147	152	157	162	166
		BFL	0.44	91.2	110	114	117	120	123	126	130	133	136	139	143
		6	0.90	79.8	108	111	114	117	119	122	125	128	131	134	136
		7	1.51	68.5	106	108	110	113	115	118	120	123	125	127	130
W 10×22	70.2	TFL	0.00	234	117	126	134	142	150	159	167	175	183	192	200
		2	0.09	196	112	119	126	133	140	147	154	161	167	174	181
		3	0.18	159	106	112	117	123	129	134	140	146	151	157	163
		4	0.27	122	100	105	109	113	118	122	126	131	135	139	144
		BFL	0.36	84.6	94.2	97.2	100	103	106	109	112	115	118	121	124
		6	0.95	71.5	91.8	94.3	96.9	99.4	102	104	107	110	112	115	117
		7	1.70	58.4	88.7	90.8	92.9	94.9	97.0	99.1	101	103	105	107	109
W 10×19	58.3	TFL	0.00	202	102	109	116	124	131	138	145	152	159	167	174
		2	0.10	174	97.9	104	110	116	123	129	135	141	147	153	159
		3	0.20	145	93.5	98.7	104	109	114	119	124	130	135	140	145
		4	0.30	117	89.0	93.1	97.2	101	106	110	114	118	122	126	130
		BFL	0.40	88.0	84.2	87.4	90.5	93.6	96.7	99.8	103	106	109	112	115
		6	1.27	69.3	80.5	83.0	85.4	87.9	90.4	92.8	95.3	97.7	100	103	105
		7	2.31	50.6	75.5	77.3	79.1	80.9	82.7	84.5	86.3	88.1	89.8	91.6	93.4
W 10×17	50.5	TFL	0.00	180	89.8	96.1	102	109	115	122	128	134	141	147	153
		2	0.08	156	86.3	91.8	97.4	103	108	114	119	125	130	136	142
		3	0.17	132	82.7	87.4	92.1	96.8	101	106	111	115	120	125	129
		4	0.25	108	79.0	82.9	86.7	90.5	94.3	98.2	102	106	110	114	117
		BFL	0.33	84.4	75.2	78.1	81.1	84.1	87.1	90.1	93.1	96.1	99.1	102	105
		6	1.31	64.6	71.3	73.6	75.8	78.1	80.4	82.7	85.0	87.3	89.6	91.9	94.2
		7	2.46	44.9	65.8	67.4	69.0	70.6	72.2	73.8	75.4	77.0	78.6	80.2	81.7

[a] Y1 = distance from top of the steel beam to plastic neutral axis.
[b] Y2 = distance from top of the steel beam to concrete flange force.
[c] See Figure 5-3 for PNA locations.

| | | | | | | | | | | $F_y = 36$ ksi | | | |

COMPOSITE DESIGN
COMPOSITE BEAM SELECTION TABLE
W Shapes
$\phi = 0.85$ $\qquad\qquad$ $\phi_b = 0.90$

Shape	$\phi_b M_p$	PNA[c]	Y1[a]	ΣQ_n	ϕM_n (kip-ft)										
					Y2[b] (in.)										
	Kip-ft		In.	Kips	2	2.5	3	3.5	4	4.5	5	5.5	6	6.5	7
W 10×15	43.2	TFL	0.00	159	78.7	84.3	89.9	95.5	101	107	112	118	124	129	135
		2	0.07	139	75.9	80.8	85.7	90.7	95.6	101	105	110	115	120	125
		3	0.14	120	73.0	77.2	81.5	85.7	90.0	94.2	98.4	103	107	111	115
		4	0.20	100	70.0	73.5	77.1	80.7	84.2	87.8	91.3	94.9	98.4	102	106
		BFL	0.27	81.0	66.9	69.8	72.6	75.5	78.4	81.2	84.1	87.0	89.9	92.7	95.6
		6	1.35	60.3	62.9	65.0	67.1	69.3	71.4	73.5	75.7	77.8	80.0	82.1	84.2
		7	2.60	39.7	57.0	58.4	59.9	61.3	62.7	64.1	65.5	66.9	68.3	69.7	71.1
W 10×12	34.0	TFL	0.00	127	62.6	67.1	71.6	76.1	80.7	85.2	89.7	94.2	98.7	103	108
		2	0.05	112	60.5	64.4	68.4	72.4	76.4	80.4	84.4	88.3	92.3	96.3	100
		3	0.11	97.5	58.2	61.7	65.2	68.6	72.1	75.5	79.0	82.4	85.9	89.3	92.8
		4	0.16	82.5	56.0	58.9	61.8	64.8	67.7	70.6	73.5	76.5	79.4	82.3	85.2
		BFL	0.21	67.6	53.7	56.1	58.5	60.9	63.2	65.6	68.0	70.4	72.8	75.2	77.6
		6	1.30	49.7	50.3	52.0	53.8	55.5	57.3	59.1	60.8	62.6	64.3	66.1	67.9
		7	2.61	31.9	45.3	46.4	47.5	48.6	49.8	50.9	52.0	53.2	54.3	55.4	56.5
W 8×28	73.4	TFL	0.00	297	127	137	148	158	169	179	190	200	211	222	232
		2	0.12	242	119	127	136	145	153	162	170	179	188	196	205
		3	0.23	188	110	117	124	130	137	144	150	157	164	170	177
		4	0.35	133	102	106	111	116	120	125	130	135	139	144	149
		BFL	0.47	78.2	92.3	95.0	97.8	101	103	106	109	112	114	117	120
		6	0.53	76.2	91.9	94.6	97.3	100	103	105	108	111	114	116	119
		7	0.59	74.3	91.5	94.2	96.8	99.4	102	105	107	110	113	115	118
W 8×24	62.6	TFL	0.00	255	108	117	126	135	144	153	162	171	180	189	198
		2	0.10	208	101	108	116	123	130	138	145	152	160	167	175
		3	0.20	161	93.8	99.5	105	111	117	122	128	134	139	145	151
		4	0.30	115	86.3	90.4	94.4	98.5	103	107	111	115	119	123	127
		BFL	0.40	67.8	78.5	80.9	83.3	85.7	88.2	90.6	93.0	95.4	97.8	100	103
		6	0.47	65.8	78.2	80.5	82.8	85.2	87.5	89.8	92.2	94.5	96.8	99.2	101
		7	0.55	63.7	77.8	80.1	82.3	84.6	86.9	89.1	91.4	93.6	95.9	98.1	100
W 8×21	55.1	TFL	0.00	222	96.4	104	112	120	128	136	144	151	159	167	175
		2	0.10	184	90.9	97.4	104	110	117	123	130	137	143	150	156
		3	0.20	146	85.2	90.3	95.5	101	106	111	116	121	126	132	137
		4	0.30	108	79.1	82.9	86.8	90.6	94.4	98.2	102	106	110	114	117
		BFL	0.40	70.0	72.8	75.3	77.8	80.2	82.7	85.2	87.7	90.1	92.6	95.1	97.6
		6	0.70	62.7	71.5	73.7	75.9	78.1	80.4	82.6	84.8	87.0	89.3	91.5	93.7
		7	1.06	55.4	70.0	72.0	73.9	75.9	77.9	79.8	81.8	83.8	85.7	87.7	89.6

[a] Y1 = distance from top of the steel beam to plastic neutral axis.
[b] Y2 = distance from top of the steel beam to concrete flange force.
[c] See Figure 5-3 for PNA locations.

| F_y = 36 ksi | | | | COMPOSITE DESIGN COMPOSITE BEAM SELECTION TABLE W Shapes ϕ = 0.85 ϕ_b = 0.90 | | | | | | | | | | | |

Shape	$\phi_b M_p$	PNA[c]	Y1[a]	ΣQ_n	ϕM_n (kip-ft) Y2[b] (in.)										
	Kip-ft		In.	Kips	2	2.5	3	3.5	4	4.5	5	5.5	6	6.5	7
W 8×18	45.9	TFL	0.00	189	81.4	88.1	94.8	102	108	115	122	128	135	142	148
		2	0.08	158	76.9	82.5	88.1	93.7	99.3	105	111	116	122	127	133
		3	0.17	127	72.2	76.7	81.2	85.7	90.2	94.7	99.2	104	108	113	117
		4	0.25	95.8	67.3	70.7	74.1	77.5	80.9	84.3	87.7	91.1	94.5	97.9	101
		BFL	0.33	64.6	62.3	64.6	66.9	69.2	71.4	73.7	76.0	78.3	80.6	82.9	85.2
		6	0.71	56.0	60.7	62.7	64.7	66.7	68.7	70.7	72.6	74.6	76.6	78.6	80.6
		7	1.21	47.3	58.9	60.6	62.3	64.0	65.6	67.3	69.0	70.7	72.4	74.0	75.7
W 8×15	36.7	TFL	0.00	160	68.6	74.2	79.9	85.5	91.2	96.9	103	108	114	120	125
		2	0.08	137	65.3	70.1	75.0	79.8	84.7	89.5	94.4	99.2	104	109	114
		3	0.16	114	61.9	65.9	69.9	74.0	78.0	82.1	86.1	90.2	94.2	98.3	102
		4	0.24	91.5	58.3	61.6	64.8	68.0	71.3	74.5	77.8	81.0	84.2	87.5	90.7
		BFL	0.32	68.8	54.6	57.1	59.5	61.9	64.4	66.8	69.3	71.7	74.1	76.6	79.0
		6	0.97	54.4	52.0	53.9	55.8	57.7	59.7	61.6	63.5	65.4	67.4	69.3	71.2
		7	1.79	40.0	48.5	49.9	51.3	52.8	54.2	55.6	57.0	58.4	59.8	61.2	62.7
W 8×13	30.8	TFL	0.00	138	58.7	63.6	68.5	73.4	78.3	83.2	88.1	93.0	97.9	103	108
		2	0.06	120	56.1	60.3	64.6	68.8	73.0	77.3	81.5	85.8	90.0	94.3	98.5
		3	0.13	102	53.3	56.9	60.5	64.1	67.7	71.3	74.9	78.5	82.1	85.7	89.3
		4	0.19	83.2	50.5	53.5	56.4	59.4	62.3	65.3	68.2	71.1	74.1	77.0	80.0
		BFL	0.26	64.8	47.6	49.9	52.2	54.5	56.8	59.1	61.4	63.7	66.0	68.3	70.6
		6	1.00	49.7	44.9	46.6	48.4	50.1	51.9	53.7	55.4	57.2	58.9	60.7	62.5
		7	1.91	34.6	41.2	42.4	43.6	44.8	46.1	47.3	48.5	49.7	51.0	52.2	53.4
W 8×10	23.9	TFL	0.00	107	44.9	48.6	52.4	56.2	60.0	63.7	67.5	71.3	75.1	78.8	82.6
		2	0.05	92.0	42.8	46.0	49.3	52.6	55.8	59.1	62.3	65.6	68.9	72.1	75.4
		3	0.10	77.5	40.6	43.4	46.1	48.9	51.6	54.4	57.1	59.9	62.6	65.3	68.1
		4	0.15	62.9	38.5	40.7	42.9	45.1	47.4	49.6	51.8	54.1	56.3	58.5	60.7
		BFL	0.21	48.4	36.2	37.9	39.6	41.4	43.1	44.8	46.5	48.2	49.9	51.6	53.4
		6	0.88	37.5	34.3	35.6	36.9	38.3	39.6	40.9	42.2	43.6	44.9	46.2	47.6
		7	1.77	26.6	31.7	32.7	33.6	34.5	35.5	36.4	37.4	38.3	39.3	40.2	41.1

[a] Y1 = distance from top of the steel beam to plastic neutral axis.
[b] Y2 = distance from top of the steel beam to concrete flange force.
[c] See Figure 5-3 for PNA locations.

F_y = 50 ksi

COMPOSITE DESIGN
COMPOSITE BEAM SELECTION TABLE
W Shapes

$\phi = 0.85$ $\qquad\qquad$ $\phi_b = 0.90$

Shape	$\phi_b M_p$	PNA[c]	Y1[a]	ΣQ_n	ϕM_n (kip-ft) Y2[b] (in.)										
	Kip-ft		In.	Kips	2	2.5	3	3.5	4	4.5	5	5.5	6	6.5	7
W40×297	4990	TFL	0.00	4370	6790	6940	7090	7250	7400	7560	7710	7870	8020	8180	8330
		2	0.41	3720	6680	6810	6950	7080	7210	7340	7470	7600	7740	7870	8000
		3	0.83	3060	6560	6670	6780	6890	7000	7100	7210	7320	7430	7540	7650
		4	1.24	2410	6420	6510	6590	6680	6760	6850	6930	7020	7110	7190	7280
		BFL	1.65	1760	6260	6320	6390	6450	6510	6570	6640	6700	6760	6820	6890
		6	4.59	1430	6150	6200	6250	6300	6350	6400	6450	6500	6550	6600	6650
		7	8.17	1090	5950	5990	6020	6060	6100	6140	6180	6220	6260	6300	6330
W40×278	4460	TFL	0.00	4090	6400	6540	6690	6830	6980	7120	7270	7410	7560	7700	7850
		2	0.45	3550	6310	6440	6560	6690	6810	6940	7070	7190	7320	7440	7570
		3	0.91	3010	6210	6320	6420	6530	6630	6740	6850	6950	7060	7170	7270
		4	1.36	2470	6090	6180	6260	6350	6440	6520	6610	6700	6790	6870	6960
		BFL	1.81	1920	5950	6020	6090	6160	6220	6290	6360	6430	6500	6560	6630
		6	5.64	1470	5770	5820	5880	5930	5980	6030	6080	6140	6190	6240	6290
		7	10.06	1020	5460	5500	5530	5570	5600	5640	5680	5710	5750	5790	5820
W40×277	4690	TFL	0.00	4070	6290	6430	6580	6720	6870	7010	7150	7300	7440	7590	7730
		2	0.39	3440	6190	6310	6440	6560	6680	6800	6920	7050	7170	7290	7410
		3	0.79	2820	6080	6180	6280	6380	6480	6580	6680	6780	6880	6980	7080
		4	1.18	2200	5950	6020	6100	6180	6260	6340	6410	6490	6570	6650	6720
		BFL	1.58	1570	5800	5850	5910	5960	6020	6080	6130	6190	6240	6300	6350
		6	4.25	1290	5700	5750	5790	5840	5880	5930	5980	6020	6070	6110	6160
		7	7.60	1020	5550	5580	5620	5660	5690	5730	5760	5800	5840	5870	5910
W40×264	4240	TFL	0.00	3880	6050	6180	6320	6460	6600	6730	6870	7010	7150	7280	7420
		2	0.43	3360	5970	6080	6200	6320	6440	6560	6680	6800	6920	7040	7160
		3	0.87	2850	5870	5970	6070	6170	6270	6370	6470	6570	6680	6780	6880
		4	1.30	2330	5760	5840	5920	6000	6090	6170	6250	6330	6420	6500	6580
		BFL	1.73	1820	5630	5690	5760	5820	5880	5950	6010	6080	6140	6210	6270
		6	5.49	1390	5470	5510	5560	5610	5660	5710	5760	5810	5860	5910	5960
		7	9.90	970	5170	5210	5240	5280	5310	5350	5380	5420	5450	5480	5520

[a] Y1 = distance from top of the steel beam to plastic neutral axis.
[b] Y2 = distance from top of the steel beam to concrete flange force.
[c] See Figure 5-3 for PNA locations.

$F_y = 50$ ksi														

COMPOSITE DESIGN
COMPOSITE BEAM SELECTION TABLE
W Shapes
$\phi = 0.85$ $\phi_b = 0.90$

Shape	$\phi_b M_p$	PNA[c]	$Y1$[a]	ΣQ_n	ϕM_n (kip-ft)									
					$Y2$[b] (in.)									
	Kip-ft		In.	Kips	2	2.5	3	3.5	4	4.5	5	5.5	6	6.5	7
W40×249	4200	TFL	0.00	3670	5630	5760	5890	6020	6150	6280	6410	6540	6670	6800	6930
		2	0.36	3110	5540	5650	5760	5870	5980	6090	6200	6310	6420	6530	6640
		3	0.71	2550	5440	5530	5620	5710	5810	5900	5990	6080	6170	6260	6350
		4	1.07	1990	5330	5400	5470	5540	5610	5680	5750	5820	5890	5960	6030
		BFL	1.42	1430	5200	5250	5300	5350	5400	5450	5510	5560	5610	5660	5710
		6	4.06	1170	5120	5160	5200	5240	5280	5320	5370	5410	5450	5490	5530
		7	7.47	916	4980	5010	5040	5070	5110	5140	5170	5200	5240	5270	5300
W40×235	3790	TFL	0.00	3450	5330	5450	5570	5700	5820	5940	6060	6180	6310	6430	6550
		2	0.39	2980	5260	5360	5470	5570	5680	5780	5890	6000	6100	6210	6310
		3	0.79	2510	5170	5260	5350	5440	5530	5620	5700	5790	5880	5970	6060
		4	1.18	2040	5070	5150	5220	5290	5360	5430	5510	5580	5650	5720	5800
		BFL	1.58	1570	4960	5020	5070	5130	5180	5240	5290	5350	5410	5460	5520
		6	5.18	1220	4830	4880	4920	4960	5000	5050	5090	5130	5180	5220	5260
		7	9.47	861	4600	4630	4660	4690	4720	4750	4780	4810	4840	4870	4900
W40×215	3610	TFL	0.00	3170	4820	4930	5040	5150	5270	5380	5490	5600	5710	5830	5940
		2	0.31	2680	4740	4840	4930	5030	5120	5220	5320	5410	5510	5600	5700
		3	0.61	2200	4660	4740	4820	4900	4970	5050	5130	5210	5290	5360	5440
		4	0.92	1720	4570	4630	4690	4750	4810	4870	4930	4990	5060	5120	5180
		BFL	1.22	1240	4460	4510	4550	4590	4640	4680	4730	4770	4810	4860	4900
		6	3.84	1020	4390	4430	4470	4500	4540	4570	4610	4650	4680	4720	4760
		7	7.32	791	4270	4300	4330	4360	4380	4410	4440	4470	4500	4520	4550
W40×211	3390	TFL	0.00	3100	4760	4870	4980	5090	5200	5310	5420	5530	5640	5750	5860
		2	0.35	2680	4700	4790	4890	4980	5080	5170	5270	5360	5460	5550	5650
		3	0.71	2260	4620	4700	4780	4860	4940	5020	5100	5180	5260	5340	5420
		4	1.06	1850	4540	4600	4670	4730	4800	4860	4930	4990	5060	5130	5190
		BFL	1.42	1430	4440	4490	4540	4590	4640	4690	4740	4800	4850	4900	4950
		6	4.99	1100	4330	4360	4400	4440	4480	4520	4560	4600	4640	4680	4710
		7	9.35	775	4110	4140	4170	4200	4220	4250	4280	4310	4330	4360	4390

[a] $Y1$ = distance from top of the steel beam to plastic neutral axis.
[b] $Y2$ = distance from top of the steel beam to concrete flange force.
[c] See Figure 5-3 for PNA locations.

					$F_y = 50$ ksi

COMPOSITE DESIGN
COMPOSITE BEAM SELECTION TABLE
W Shapes
$\phi = 0.85$ $\phi_b = 0.90$

Shape	$\phi_b M_p$	PNA[c]	$Y1$[a]	ΣQ_n	\multicolumn{13}{c}{ϕM_n (kip-ft)}												
					\multicolumn{13}{c}{$Y2$[b] (in.)}												
	Kip-ft		In.	Kips	2	2.5	3	3.5	4	4.5	5	5.5	6	6.5	7
W40×199	3250	TFL	0.00	2920	4410	4520	4620	4720	4830	4930	5030	5140	5240	5340	5450
		2	0.27	2500	4350	4440	4530	4620	4700	4790	4880	4970	5060	5150	5240
		3	0.53	2080	4280	4350	4430	4500	4570	4650	4720	4790	4870	4940	5020
		4	0.80	1660	4200	4260	4320	4380	4430	4490	4550	4610	4670	4730	4790
		BFL	1.07	1240	4110	4160	4200	4240	4290	4330	4380	4420	4460	4510	4550
		6	4.16	986	4030	4070	4100	4140	4170	4210	4240	4280	4310	4350	4380
		7	8.10	730	3880	3910	3940	3960	3990	4010	4040	4070	4090	4120	4140
W40×183	2930	TFL	0.00	2690	4090	4180	4280	4370	4470	4560	4660	4750	4850	4940	5040
		2	0.31	2320	4030	4110	4200	4280	4360	4440	4530	4610	4690	4770	4860
		3	0.61	1960	3970	4040	4110	4180	4250	4320	4390	4460	4530	4600	4670
		4	0.92	1600	3900	3960	4010	4070	4130	4180	4240	4300	4350	4410	4470
		BFL	1.22	1240	3820	3860	3910	3950	4000	4040	4090	4130	4170	4220	4260
		6	4.76	958	3720	3760	3790	3830	3860	3890	3930	3960	4000	4030	4060
		7	9.16	671	3540	3570	3590	3610	3640	3660	3690	3710	3730	3760	3780
W40×174	2680	TFL	0.00	2560	3820	3910	4000	4090	4180	4270	4360	4450	4540	4630	4720
		2	0.21	2230	3770	3850	3930	4010	4090	4160	4240	4320	4400	4480	4560
		3	0.42	1900	3720	3780	3850	3920	3990	4050	4120	4190	4260	4320	4390
		4	0.62	1570	3660	3710	3770	3830	3880	3940	3990	4050	4100	4160	4220
		BFL	0.83	1250	3600	3640	3680	3730	3770	3820	3860	3900	3950	3990	4040
		6	4.59	943	3500	3530	3570	3600	3630	3670	3700	3730	3770	3800	3830
		7	9.27	639	3310	3330	3350	3370	3400	3420	3440	3460	3490	3510	3530
W40×167	2600	TFL	0.00	2460	3700	3790	3880	3960	4050	4140	4220	4310	4400	4490	4570
		2	0.26	2150	3660	3730	3810	3890	3960	4040	4110	4190	4270	4340	4420
		3	0.51	1850	3610	3670	3740	3800	3870	3930	4000	4060	4130	4200	4260
		4	0.77	1550	3550	3600	3660	3710	3770	3820	3880	3930	3990	4040	4100
		BFL	1.03	1240	3490	3530	3580	3620	3660	3710	3750	3800	3840	3880	3930
		6	5.00	929	3380	3410	3450	3480	3510	3550	3580	3610	3640	3680	3710
		7	9.85	614	3170	3190	3210	3240	3260	3280	3300	3320	3340	3370	3390
W40×149	2240	TFL	0.00	2190	3270	3350	3430	3510	3580	3660	3740	3820	3890	3970	4050
		2	0.21	1940	3240	3310	3370	3440	3510	3580	3650	3720	3790	3860	3930
		3	0.42	1700	3200	3260	3320	3380	3440	3500	3560	3620	3680	3740	3800
		4	0.62	1450	3150	3200	3260	3310	3360	3410	3460	3510	3560	3620	3670
		BFL	0.83	1210	3110	3150	3190	3230	3280	3320	3360	3410	3450	3490	3530
		6	5.15	879	2990	3030	3060	3090	3120	3150	3180	3210	3240	3280	3310
		7	10.41	548	2770	2780	2800	2820	2840	2860	2880	2900	2920	2940	2960

[a] $Y1$ = distance from top of the steel beam to plastic neutral axis.
[b] $Y2$ = distance from top of the steel beam to concrete flange force.
[c] See Figure 5-3 for PNA locations.

F_y = 50 ksi															
COMPOSITE DESIGN COMPOSITE BEAM SELECTION TABLE W Shapes ϕ = 0.85 ϕ_b = 0.90															

Shape	$\phi_b M_p$	PNA[c]	$Y1^a$	ΣQ_n	ϕM_n (kip-ft)										
					$Y2^b$ (in.)										
	Kip-ft		In.	Kips	2	2.5	3	3.5	4	4.5	5	5.5	6	6.5	7
W 36×300	4730	TFL	0.00	4420	6370	6530	6680	6840	7000	7150	7310	7460	7620	7780	7930
		2	0.42	3720	6260	6390	6520	6660	6790	6920	7050	7180	7310	7450	7580
		3	0.84	3020	6130	6240	6340	6450	6560	6660	6770	6880	6990	7090	7200
		4	1.26	2320	5980	6060	6140	6230	6310	6390	6470	6550	6640	6720	6800
		BFL	1.68	1620	5810	5860	5920	5980	6040	6090	6150	6210	6270	6320	6380
		6	3.97	1360	5720	5770	5820	5870	5910	5960	6010	6060	6110	6150	6200
		7	6.69	1100	5590	5630	5670	5710	5740	5780	5820	5860	5900	5940	5980
W 36×280	4390	TFL	0.00	4120	5910	6060	6200	6350	6500	6640	6790	6930	7080	7230	7370
		2	0.39	3470	5810	5930	6060	6180	6300	6430	6550	6670	6790	6920	7040
		3	0.79	2820	5690	5790	5890	5990	6090	6190	6290	6390	6490	6590	6690
		4	1.18	2170	5550	5630	5710	5780	5860	5940	6010	6090	6170	6240	6320
		BFL	1.57	1510	5400	5450	5510	5560	5610	5670	5720	5770	5830	5880	5930
		6	3.88	1270	5320	5360	5410	5450	5500	5540	5590	5630	5680	5720	5770
		7	6.62	1030	5190	5230	5270	5300	5340	5380	5410	5450	5490	5520	5560
W 36×260	4050	TFL	0.00	3830	5450	5590	5720	5860	6000	6130	6270	6400	6540	6670	6810
		2	0.36	3230	5360	5480	5590	5710	5820	5930	6050	6160	6280	6390	6510
		3	0.72	2630	5250	5350	5440	5530	5630	5720	5810	5910	6000	6090	6190
		4	1.08	2040	5130	5200	5280	5350	5420	5490	5570	5640	5710	5780	5850
		BFL	1.44	1440	4990	5050	5100	5150	5200	5250	5300	5350	5400	5450	5510
		6	3.86	1200	4920	4960	5000	5040	5090	5130	5170	5210	5260	5300	5340
		7	6.75	956	4790	4830	4860	4890	4930	4960	4990	5030	5060	5100	5130
W 36×245	3790	TFL	0.00	3610	5120	5240	5370	5500	5630	5760	5880	6010	6140	6270	6390
		2	0.34	3050	5030	5140	5250	5360	5460	5570	5680	5790	5900	6000	6110
		3	0.68	2490	4930	5020	5110	5200	5290	5370	5460	5550	5640	5730	5810
		4	1.01	1930	4820	4890	4960	5030	5090	5160	5230	5300	5370	5440	5510
		BFL	1.35	1380	4690	4740	4790	4840	4890	4940	4990	5040	5080	5130	5180
		6	3.81	1140	4620	4660	4700	4740	4780	4820	4860	4900	4940	4980	5020
		7	6.77	901	4500	4530	4560	4590	4620	4660	4690	4720	4750	4780	4820
W 36×230	3540	TFL	0.00	3380	4780	4900	5020	5140	5260	5370	5490	5610	5730	5850	5970
		2	0.32	2860	4700	4800	4900	5000	5100	5200	5310	5410	5510	5610	5710
		3	0.63	2340	4610	4690	4770	4860	4940	5020	5100	5190	5270	5350	5440
		4	0.95	1820	4500	4570	4630	4700	4760	4830	4890	4960	5020	5090	5150
		BFL	1.26	1300	4390	4440	4480	4530	4570	4620	4670	4710	4760	4810	4850
		6	3.81	1070	4320	4350	4390	4430	4470	4510	4540	4580	4620	4660	4690
		7	6.83	845	4200	4230	4260	4290	4320	4350	4380	4410	4440	4470	4500

[a] $Y1$ = distance from top of the steel beam to plastic neutral axis.
[b] $Y2$ = distance from top of the steel beam to concrete flange force.
[c] See Figure 5-3 for PNA locations.

					$F_y = 50$ ksi								

COMPOSITE DESIGN
COMPOSITE BEAM SELECTION TABLE
W Shapes
$\phi = 0.85$ $\phi_b = 0.90$

Shape	$\phi_b M_p$	PNA[c]	$Y1$[a]	ΣQ_n	ϕM_n (kip-ft) $Y2$[b] (in.)										
	Kip-ft		In.	Kips	2	2.5	3	3.5	4	4.5	5	5.5	6	6.5	7
W 36×210	3120	TFL	0.00	3090	4450	4560	4670	4780	4890	5000	5110	5220	5330	5440	5550
		2	0.34	2680	4390	4480	4580	4670	4770	4860	4960	5050	5150	5240	5340
		3	0.68	2260	4320	4400	4480	4560	4640	4720	4800	4880	4960	5040	5120
		4	1.02	1850	4230	4300	4360	4430	4490	4560	4620	4690	4760	4820	4890
		BFL	1.36	1430	4140	4190	4240	4290	4340	4390	4440	4490	4540	4600	4650
		6	5.06	1100	4020	4060	4100	4130	4170	4210	4250	4290	4330	4370	4410
		7	9.04	773	3810	3830	3860	3890	3920	3940	3970	4000	4030	4050	4080
W 36×194	2880	TFL	0.00	2850	4090	4190	4290	4390	4490	4590	4690	4790	4890	5000	5100
		2	0.32	2470	4030	4120	4200	4290	4380	4470	4550	4640	4730	4820	4900
		3	0.63	2090	3960	4040	4110	4180	4260	4330	4410	4480	4550	4630	4700
		4	0.95	1710	3890	3950	4010	4070	4130	4190	4250	4310	4370	4430	4490
		BFL	1.26	1320	3800	3850	3900	3940	3990	4040	4080	4130	4180	4220	4270
		6	4.94	1020	3690	3730	3770	3800	3840	3880	3910	3950	3980	4020	4060
		7	8.93	713	3500	3530	3550	3580	3600	3630	3650	3680	3700	3730	3750
W 36×182	2690	TFL	0.00	2680	3830	3920	4020	4110	4210	4300	4400	4490	4590	4680	4780
		2	0.29	2320	3770	3860	3940	4020	4100	4190	4270	4350	4430	4510	4600
		3	0.59	1970	3710	3780	3850	3920	3990	4060	4130	4200	4270	4340	4410
		4	0.89	1610	3640	3700	3760	3810	3870	3930	3990	4040	4100	4160	4210
		BFL	1.18	1260	3570	3610	3660	3700	3740	3790	3830	3880	3920	3970	4010
		6	4.89	963	3460	3500	3530	3570	3600	3640	3670	3700	3740	3770	3810
		7	8.92	670	3280	3300	3330	3350	3370	3400	3420	3450	3470	3490	3520
W 36×170	2500	TFL	0.00	2500	3560	3650	3730	3820	3910	4000	4090	4180	4270	4350	4440
		2	0.28	2170	3510	3580	3660	3740	3810	3890	3970	4040	4120	4200	4270
		3	0.55	1840	3450	3520	3580	3650	3710	3780	3840	3910	3970	4040	4100
		4	0.83	1510	3390	3440	3490	3550	3600	3650	3710	3760	3810	3870	3920
		BFL	1.10	1180	3320	3360	3400	3440	3480	3530	3570	3610	3650	3690	3730
		6	4.84	901	3220	3250	3290	3320	3350	3380	3410	3450	3480	3510	3540
		7	8.89	625	3050	3070	3090	3120	3140	3160	3180	3200	3230	3250	3270
W 36×160	2340	TFL	0.00	2350	3330	3410	3500	3580	3660	3750	3830	3910	4000	4080	4160
		2	0.26	2040	3280	3360	3430	3500	3570	3650	3720	3790	3860	3940	4010
		3	0.51	1740	3230	3290	3360	3420	3480	3540	3600	3660	3720	3790	3850
		4	0.77	1430	3180	3230	3280	3330	3380	3430	3480	3530	3580	3630	3680
		BFL	1.02	1130	3110	3150	3190	3230	3270	3310	3350	3390	3430	3470	3510
		6	4.82	857	3020	3050	3080	3110	3140	3170	3200	3230	3260	3290	3320
		7	8.97	588	2850	2870	2890	2910	2930	2960	2980	3000	3020	3040	3060

[a] $Y1$ = distance from top of the steel beam to plastic neutral axis.
[b] $Y2$ = distance from top of the steel beam to concrete flange force.
[c] See Figure 5-3 for PNA locations.

$F_y = 50$ ksi															

COMPOSITE DESIGN
COMPOSITE BEAM SELECTION TABLE
W Shapes
$$\phi = 0.85 \qquad\qquad \phi_b = 0.90$$

Shape	$\phi_b M_p$	PNA[c]	Y1[a]	ΣQ_n	ϕM_n (kip-ft)										
					Y2[b] (in.)										
	Kip-ft		In.	Kips	2	2.5	3	3.5	4	4.5	5	5.5	6	6.5	7
W 36×150	2180	TFL	0.00	2210	3120	3200	3280	3350	3430	3510	3590	3670	3750	3820	3900
		2	0.24	1930	3080	3150	3210	3280	3350	3420	3490	3560	3620	3690	3760
		3	0.47	1650	3030	3090	3150	3210	3260	3320	3380	3440	3500	3560	3610
		4	0.71	1370	2980	3030	3080	3120	3170	3220	3270	3320	3370	3410	3460
		BFL	0.94	1080	2920	2960	3000	3040	3080	3110	3150	3190	3230	3270	3310
		6	4.83	818	2830	2860	2890	2920	2950	2980	3010	3040	3060	3090	3120
		7	9.08	553	2660	2680	2700	2720	2740	2760	2780	2800	2820	2840	2860
W 36×135	1910	TFL	0.00	1990	2780	2850	2920	2990	3060	3130	3200	3270	3340	3410	3480
		2	0.20	1750	2750	2810	2870	2930	2990	3060	3120	3180	3240	3300	3360
		3	0.40	1510	2710	2760	2810	2870	2920	2970	3030	3080	3140	3190	3240
		4	0.59	1280	2670	2710	2760	2800	2850	2890	2940	2980	3030	3070	3120
		BFL	0.79	1040	2620	2660	2690	2730	2770	2800	2840	2880	2920	2950	2990
		6	4.97	769	2530	2560	2580	2610	2640	2660	2690	2720	2750	2770	2800
		7	9.50	496	2350	2370	2390	2400	2420	2440	2460	2470	2490	2510	2530
W 33×221	3210	TFL	0.00	3250	4370	4480	4600	4710	4830	4940	5060	5170	5290	5400	5520
		2	0.32	2750	4290	4390	4480	4580	4680	4780	4870	4970	5070	5160	5260
		3	0.64	2240	4200	4280	4360	4440	4520	4600	4680	4760	4840	4920	4990
		4	0.96	1740	4100	4160	4220	4290	4350	4410	4470	4530	4590	4650	4720
		BFL	1.28	1230	3990	4030	4080	4120	4160	4210	4250	4300	4340	4380	4430
		6	3.76	1020	3920	3960	3990	4030	4070	4100	4140	4170	4210	4250	4280
		7	6.48	813	3820	3850	3870	3900	3930	3960	3990	4020	4050	4080	4100
W 33×201	2900	TFL	0.00	2960	3940	4050	4150	4260	4360	4470	4570	4680	4780	4890	4990
		2	0.29	2500	3870	3960	4050	4140	4230	4320	4410	4500	4580	4670	4760
		3	0.58	2050	3800	3870	3940	4010	4090	4160	4230	4300	4380	4450	4520
		4	0.86	1600	3710	3770	3820	3880	3940	3990	4050	4110	4160	4220	4280
		BFL	1.15	1140	3610	3650	3690	3730	3780	3820	3860	3900	3940	3980	4020
		6	3.67	942	3550	3580	3620	3650	3680	3720	3750	3780	3820	3850	3880
		7	6.51	739	3450	3480	3500	3530	3550	3580	3610	3630	3660	3680	3710
W 33×141	1930	TFL	0.00	2080	2750	2820	2900	2970	3040	3120	3190	3260	3340	3410	3480
		2	0.24	1800	2710	2770	2830	2900	2960	3030	3090	3150	3220	3280	3340
		3	0.48	1530	2660	2710	2770	2820	2880	2930	2980	3040	3090	3150	3200
		4	0.72	1250	2610	2650	2700	2740	2790	2830	2870	2920	2960	3010	3050
		BFL	0.96	973	2550	2590	2620	2660	2690	2730	2760	2790	2830	2860	2900
		6	4.31	746	2480	2510	2530	2560	2590	2610	2640	2670	2690	2720	2750
		7	8.05	520	2350	2370	2390	2410	2420	2440	2460	2480	2500	2520	2530

[a] Y1 = distance from top of the steel beam to plastic neutral axis.
[b] Y2 = distance from top of the steel beam to concrete flange force.
[c] See Figure 5-3 for PNA locations.

					F_y = 50 ksi								

COMPOSITE DESIGN
COMPOSITE BEAM SELECTION TABLE
W Shapes
ϕ = 0.85 ϕ_b = 0.90

Shape	$\phi_b M_p$	PNA[c]	Y1[a]	ΣQ_n	ϕM_n (kip-ft)										
					Y2[b] (in.)										
	Kip-ft		In.	Kips	2	2.5	3	3.5	4	4.5	5	5.5	6	6.5	7
W 33×130	1750	TFL	0.00	1920	2520	2580	2650	2720	2790	2850	2920	2990	3060	3130	3190
		2	0.21	1670	2480	2540	2600	2660	2720	2770	2830	2890	2950	3010	3070
		3	0.43	1420	2440	2490	2540	2590	2640	2690	2740	2790	2840	2890	2940
		4	0.64	1180	2390	2440	2480	2520	2560	2600	2640	2690	2730	2770	2810
		BFL	0.86	931	2350	2380	2410	2450	2480	2510	2540	2580	2610	2640	2680
		6	4.39	705	2270	2300	2320	2350	2370	2400	2420	2450	2470	2500	2520
		7	8.29	479	2140	2160	2170	2190	2210	2230	2240	2260	2280	2290	2310
W 33×118	1560	TFL	0.00	1740	2260	2330	2390	2450	2510	2570	2630	2700	2760	2820	2880
		2	0.19	1520	2230	2290	2340	2400	2450	2500	2560	2610	2660	2720	2770
		3	0.37	1310	2200	2250	2290	2340	2380	2430	2480	2520	2570	2620	2660
		4	0.56	1100	2160	2200	2240	2280	2320	2360	2400	2430	2470	2510	2550
		BFL	0.74	885	2120	2150	2190	2220	2250	2280	2310	2340	2370	2400	2440
		6	4.44	660	2050	2070	2100	2120	2140	2170	2190	2210	2240	2260	2280
		7	8.54	434	1920	1930	1950	1960	1980	1990	2010	2020	2040	2050	2070
W 30×116	1420	TFL	0.00	1710	2060	2120	2180	2240	2300	2360	2420	2480	2540	2600	2670
		2	0.21	1490	2030	2080	2130	2180	2240	2290	2340	2400	2450	2500	2550
		3	0.43	1260	1990	2030	2080	2120	2170	2210	2260	2300	2350	2390	2440
		4	0.64	1040	1950	1990	2020	2060	2100	2130	2170	2210	2240	2280	2320
		BFL	0.85	818	1910	1940	1960	1990	2020	2050	2080	2110	2140	2170	2200
		6	3.98	623	1850	1870	1890	1910	1940	1960	1980	2000	2020	2050	2070
		7	7.44	428	1740	1760	1770	1790	1800	1820	1830	1850	1860	1880	1890
W 30×108	1300	TFL	0.00	1590	1900	1960	2010	2070	2120	2180	2240	2290	2350	2400	2460
		2	0.19	1390	1870	1920	1970	2020	2070	2110	2160	2210	2260	2310	2360
		3	0.38	1190	1840	1880	1920	1960	2010	2050	2090	2130	2170	2220	2260
		4	0.57	988	1800	1840	1870	1910	1940	1980	2010	2050	2080	2120	2150
		BFL	0.76	789	1760	1790	1820	1850	1880	1900	1930	1960	1990	2020	2040
		6	4.04	593	1710	1730	1750	1770	1790	1810	1830	1850	1870	1890	1920
		7	7.64	396	1600	1610	1620	1640	1650	1670	1680	1690	1710	1720	1740
W 30×99	1170	TFL	0.00	1460	1730	1790	1840	1890	1940	1990	2040	2090	2150	2200	2250
		2	0.17	1280	1710	1750	1800	1840	1890	1930	1980	2030	2070	2120	2160
		3	0.34	1100	1680	1720	1760	1800	1840	1880	1920	1950	1990	2030	2070
		4	0.50	930	1650	1680	1720	1750	1780	1810	1850	1880	1910	1950	1980
		BFL	0.67	755	1620	1640	1670	1700	1730	1750	1780	1810	1830	1860	1890
		6	4.07	559	1560	1580	1600	1620	1640	1660	1680	1700	1720	1740	1760
		7	7.83	364	1450	1460	1480	1490	1500	1510	1530	1540	1550	1570	1580

[a] Y1 = distance from top of the steel beam to plastic neutral axis.
[b] Y2 = distance from top of the steel beam to concrete flange force.
[c] See Figure 5-3 for PNA locations.

F_y = 50 ksi																	

COMPOSITE DESIGN
COMPOSITE BEAM SELECTION TABLE
W Shapes

$$\phi = 0.85 \qquad \qquad \phi_b = 0.90$$

Shape	$\phi_b M_p$	PNA[c]	Y1[a]	ΣQ_n	ϕM_n (kip-ft)										
					Y2[b] (in.)										
	Kip-ft		In.	Kips	2	2.5	3	3.5	4	4.5	5	5.5	6	6.5	7
W 27×102	1140	TFL	0.00	1500	1650	1700	1760	1810	1860	1920	1970	2020	2080	2130	2180
		2	0.21	1290	1620	1670	1710	1760	1800	1850	1900	1940	1990	2030	2080
		3	0.42	1080	1590	1630	1660	1700	1740	1780	1820	1860	1890	1930	1970
		4	0.62	877	1550	1580	1610	1640	1670	1700	1740	1770	1800	1830	1860
		BFL	0.83	669	1510	1530	1560	1580	1600	1630	1650	1680	1700	1720	1750
		6	3.41	522	1470	1490	1500	1520	1540	1560	1580	1600	1620	1630	1650
		7	6.26	375	1400	1410	1420	1440	1450	1460	1480	1490	1500	1520	1530
W 27×94	1040	TFL	0.00	1390	1520	1570	1610	1660	1710	1760	1810	1860	1910	1960	2010
		2	0.19	1200	1490	1530	1570	1620	1660	1700	1740	1790	1830	1870	1910
		3	0.37	1010	1460	1490	1530	1570	1600	1640	1670	1710	1750	1780	1820
		4	0.56	827	1430	1460	1490	1510	1540	1570	1600	1630	1660	1690	1720
		BFL	0.75	641	1390	1410	1440	1460	1480	1510	1530	1550	1570	1600	1620
		6	3.39	493	1350	1370	1390	1400	1420	1440	1460	1470	1490	1510	1530
		7	6.39	346	1280	1290	1300	1320	1330	1340	1350	1360	1380	1390	1400
W 27×84	915	TFL	0.00	1240	1350	1390	1440	1480	1520	1570	1610	1660	1700	1740	1790
		2	0.16	1080	1330	1360	1400	1440	1480	1520	1550	1590	1630	1670	1710
		3	0.32	921	1300	1330	1370	1400	1430	1460	1500	1530	1560	1590	1630
		4	0.48	762	1270	1300	1330	1350	1380	1410	1430	1460	1490	1520	1540
		BFL	0.64	603	1240	1270	1290	1310	1330	1350	1370	1390	1410	1440	1460
		6	3.44	456	1200	1220	1240	1250	1270	1280	1300	1320	1330	1350	1360
		7	6.62	310	1130	1140	1150	1160	1170	1190	1200	1210	1220	1230	1240
W 24×76	750	TFL	0.00	1120	1110	1150	1190	1230	1270	1310	1350	1390	1420	1460	1500
		2	0.17	967	1080	1120	1150	1190	1220	1260	1290	1320	1360	1390	1430
		3	0.34	814	1060	1090	1120	1150	1180	1200	1230	1260	1290	1320	1350
		4	0.51	662	1030	1060	1080	1100	1130	1150	1170	1200	1220	1250	1270
		BFL	0.68	509	1010	1020	1040	1060	1080	1100	1110	1130	1150	1170	1190
		6	3.00	394	976	990	1000	1020	1030	1050	1060	1070	1090	1100	1120
		7	5.60	280	925	935	945	955	964	974	984	994	1000	1010	1020
W 24×68	664	TFL	0.00	1010	987	1020	1060	1090	1130	1160	1200	1240	1270	1310	1340
		2	0.15	874	968	999	1030	1060	1090	1120	1150	1180	1220	1250	1280
		3	0.29	743	947	973	1000	1030	1050	1080	1100	1130	1160	1180	1210
		4	0.44	612	925	947	969	990	1010	1030	1060	1080	1100	1120	1140
		BFL	0.59	481	902	919	936	953	970	987	1000	1020	1040	1060	1070
		6	3.05	366	872	885	898	910	923	936	949	962	975	988	1000
		7	5.81	251	819	828	837	846	855	864	873	882	891	899	908

[a] Y1 = distance from top of the steel beam to plastic neutral axis.
[b] Y2 = distance from top of the steel beam to concrete flange force.
[c] See Figure 5-3 for PNA locations.

								F_y = 50 ksi				

COMPOSITE DESIGN
COMPOSITE BEAM SELECTION TABLE
W Shapes
$$\phi = 0.85 \qquad\qquad \phi_b = 0.90$$

Shape	$\phi_b M_p$	PNA[c]	Y1[a]	ΣQ_n	ϕM_n (kip-ft)										
					Y2[b] (in.)										
	Kip-ft		In.	Kips	2	2.5	3	3.5	4	4.5	5	5.5	6	6.5	7
W 24×62	574	TFL	0.00	910	894	926	958	991	1020	1060	1090	1120	1150	1180	1220
		2	0.15	806	879	907	936	964	993	1020	1050	1080	1110	1140	1160
		3	0.29	702	862	887	912	937	962	987	1010	1040	1060	1090	1110
		4	0.44	598	845	866	887	909	930	951	972	993	1010	1040	1060
		BFL	0.59	495	827	844	862	879	897	914	932	949	967	984	1000
		6	3.47	361	789	802	815	827	840	853	866	879	891	904	917
		7	6.58	228	723	731	739	747	755	763	771	779	787	795	803
W 24×55	503	TFL	0.00	810	791	820	848	877	906	934	963	992	1020	1050	1080
		2	0.13	722	778	804	829	855	880	906	931	957	982	1010	1030
		3	0.25	633	764	787	809	832	854	876	899	921	944	966	989
		4	0.38	545	750	769	788	808	827	846	866	885	904	923	943
		BFL	0.51	456	734	751	767	783	799	815	831	848	864	880	896
		6	3.45	329	700	711	723	735	746	758	770	781	793	805	816
		7	6.66	203	636	643	651	658	665	672	679	686	694	701	708
W 21×62	540	TFL	0.00	915	810	842	875	907	939	972	1000	1040	1070	1100	1130
		2	0.15	788	791	819	847	875	903	931	959	987	1010	1040	1070
		3	0.31	662	771	795	818	841	865	888	912	935	959	982	1010
		4	0.46	535	750	769	788	807	826	845	863	882	901	920	939
		BFL	0.62	408	727	741	756	770	785	799	814	828	843	857	872
		6	2.53	318	705	716	727	739	750	761	772	784	795	806	818
		7	4.78	229	669	677	685	693	701	709	717	726	734	742	750
W 21×57	484	TFL	0.00	835	741	771	800	830	859	889	919	948	978	1010	1040
		2	0.16	728	725	751	777	803	829	854	880	906	932	958	983
		3	0.33	622	708	730	753	775	797	819	841	863	885	907	929
		4	0.49	515	690	709	727	745	763	782	800	818	836	855	873
		BFL	0.65	409	671	685	700	714	729	743	758	772	787	801	816
		6	2.90	309	645	656	667	677	688	699	710	721	732	743	754
		7	5.38	209	601	608	616	623	631	638	645	653	660	668	675
W 21×50	413	TFL	0.00	735	646	672	698	724	750	777	803	829	855	881	907
		2	0.13	648	634	657	679	702	725	748	771	794	817	840	863
		3	0.27	560	620	640	660	679	699	719	739	759	779	799	818
		4	0.40	473	606	622	639	656	673	689	706	723	740	756	773
		BFL	0.54	386	590	604	618	631	645	659	672	686	700	713	727
		6	2.92	285	564	574	584	594	604	615	625	635	645	655	665
		7	5.58	184	519	526	532	539	545	552	559	565	572	578	585

[a] Y1 = distance from top of the steel beam to plastic neutral axis.
[b] Y2 = distance from top of the steel beam to concrete flange force.
[c] See Figure 5-3 for PNA locations.

$F_y = 50$ ksi														

COMPOSITE DESIGN
COMPOSITE BEAM SELECTION TABLE
W Shapes

$$\phi = 0.85 \qquad \phi_b = 0.90$$

Shape	$\phi_b M_p$	PNA[c]	Y1[a]	ΣQ_n	ϕM_n (kip-ft) Y2[b] (in.)										
	Kip-ft		In.	Kips	2	2.5	3	3.5	4	4.5	5	5.5	6	6.5	7
W 21×44	358	TFL	0.00	650	568	591	614	637	660	683	706	729	752	775	798
		2	0.11	577	557	577	598	618	639	659	680	700	720	741	761
		3	0.23	504	546	564	581	599	617	635	653	671	689	706	724
		4	0.34	431	534	549	564	580	595	610	626	641	656	671	687
		BFL	0.45	358	522	534	547	560	572	585	598	610	623	636	648
		6	2.90	260	497	506	515	525	534	543	552	561	571	580	589
		7	5.69	163	453	459	465	471	476	482	488	494	499	505	511
W 18×60	461	TFL	0.00	880	693	724	755	787	818	849	880	911	942	974	1000
		2	0.17	749	674	700	727	753	780	806	833	859	886	912	939
		3	0.35	617	653	675	696	718	740	762	784	806	828	850	871
		4	0.52	486	630	647	665	682	699	716	733	751	768	785	802
		BFL	0.70	355	606	618	631	644	656	669	681	694	706	719	732
		6	2.19	287	590	600	610	620	630	640	651	661	671	681	691
		7	3.82	220	566	573	581	589	597	605	612	620	628	636	644
W 18×55	420	TFL	0.00	810	634	663	692	720	749	778	806	835	864	892	921
		2	0.16	691	617	641	666	690	715	739	764	788	813	837	862
		3	0.32	573	598	618	639	659	679	699	720	740	760	781	801
		4	0.47	454	578	594	610	626	642	658	674	691	707	723	739
		BFL	0.63	336	556	568	580	592	604	616	628	640	652	663	675
		6	2.16	269	541	550	560	569	579	588	598	607	617	626	636
		7	3.86	203	517	524	531	539	546	553	560	567	574	582	589
W 18×50	379	TFL	0.00	735	572	598	624	651	677	703	729	755	781	807	833
		2	0.14	628	557	579	601	624	646	668	690	712	735	757	779
		3	0.29	521	540	558	577	595	614	632	651	669	688	706	725
		4	0.43	415	522	537	552	566	581	596	610	625	640	654	669
		BFL	0.57	308	503	514	525	536	547	558	569	580	590	601	612
		6	2.07	246	489	498	506	515	524	532	541	550	559	567	576
		7	3.82	184	467	474	480	487	493	500	506	513	519	526	532
W 18×46	340	TFL	0.00	675	527	551	575	599	623	647	671	695	719	743	766
		2	0.15	583	514	535	555	576	597	617	638	659	679	700	720
		3	0.30	492	499	517	534	552	569	587	604	621	639	656	674
		4	0.45	400	484	498	512	527	541	555	569	583	597	612	626
		BFL	0.61	308	468	478	489	500	511	522	533	544	555	566	577
		6	2.40	239	450	459	467	476	484	493	501	510	518	526	535
		7	4.34	169	424	430	436	442	448	454	460	466	472	478	484

[a] Y1 = distance from top of the steel beam to plastic neutral axis.
[b] Y2 = distance from top of the steel beam to concrete flange force.
[c] See Figure 5-3 for PNA locations.

Fy = 50 ksi

COMPOSITE DESIGN
COMPOSITE BEAM SELECTION TABLE
W Shapes

$$\phi = 0.85 \qquad\qquad \phi_b = 0.90$$

Shape	$\phi_b M_p$	PNA[c]	$Y1$[a]	ΣQ_n	ϕM_n (kip-ft)										
					$Y2$[b] (in.)										
	Kip-ft		In.	Kips	2	2.5	3	3.5	4	4.5	5	5.5	6	6.5	7
W 18×40	294	TFL	0.00	590	458	479	499	520	541	562	583	604	625	646	667
		2	0.13	511	446	464	482	500	518	537	555	573	591	609	627
		3	0.26	432	434	449	464	480	495	510	526	541	556	572	587
		4	0.39	353	421	433	446	458	471	483	496	508	521	533	546
		BFL	0.53	274	407	417	426	436	446	456	465	475	485	494	504
		6	2.26	211	392	400	407	415	422	429	437	444	452	459	467
		7	4.27	148	369	374	379	384	389	395	400	405	410	416	421
W 18×35	249	TFL	0.00	515	396	414	432	451	469	487	505	523	542	560	578
		2	0.11	451	387	403	418	434	450	466	482	498	514	530	546
		3	0.21	388	377	391	404	418	432	445	459	473	487	500	514
		4	0.32	324	367	378	389	401	412	424	435	447	458	470	481
		BFL	0.43	260	356	365	374	383	393	402	411	420	430	439	448
		6	2.37	194	340	347	354	361	368	375	382	389	395	402	409
		7	4.56	129	315	320	324	329	333	338	342	347	351	356	361
W 16×36	240	TFL	0.00	530	373	392	410	429	448	467	485	504	523	542	560
		2	0.11	455	362	378	394	410	426	442	459	475	491	507	523
		3	0.22	380	350	364	377	391	404	418	431	445	458	471	485
		4	0.32	305	338	349	360	371	381	392	403	414	425	435	446
		BFL	0.43	230	326	334	342	350	358	366	374	383	391	399	407
		6	1.79	181	315	322	328	334	341	347	354	360	366	373	379
		7	3.44	133	299	304	309	313	318	323	327	332	337	342	346
W 16×31	203	TFL	0.00	456	321	337	353	370	386	402	418	434	450	466	483
		2	0.11	395	312	326	340	354	368	382	396	410	424	438	452
		3	0.22	334	303	315	327	338	350	362	374	386	398	410	421
		4	0.33	274	293	303	312	322	332	342	351	361	371	380	390
		BFL	0.44	213	283	290	298	305	313	321	328	336	343	351	358
		6	2.00	163	272	278	283	289	295	301	307	312	318	324	330
		7	3.79	114	255	259	263	267	271	275	279	283	287	291	295
W 16×26	166	TFL	0.00	384	268	281	295	309	322	336	349	363	377	390	404
		2	0.09	337	261	273	285	297	309	321	332	344	356	368	380
		3	0.17	289	254	264	274	284	295	305	315	325	336	346	356
		4	0.26	242	246	255	263	272	281	289	298	306	315	323	332
		BFL	0.35	194	239	245	252	259	266	273	280	287	294	301	307
		6	2.04	145	228	233	238	243	248	253	259	264	269	274	279
		7	4.01	96.0	210	214	217	220	224	227	231	234	237	241	244

[a] $Y1$ = distance from top of the steel beam to plastic neutral axis.
[b] $Y2$ = distance from top of the steel beam to concrete flange force.
[c] See Figure 5-3 for PNA locations.

$F_y = 50$ ksi															

COMPOSITE DESIGN
COMPOSITE BEAM SELECTION TABLE
W Shapes

$$\phi = 0.85 \qquad\qquad \phi_b = 0.90$$

| Shape | $\phi_b M_p$ | PNA[c] | $Y1$[a] | ΣQ_n | ϕM_n (kip-ft) | | | | | | | | | | |
| | | | | | $Y2$[b] (in.) | | | | | | | | | | |
	Kip-ft		In.	Kips	2	2.5	3	3.5	4	4.5	5	5.5	6	6.5	7
W 14×38	231	TFL	0.00	560	359	379	399	418	438	458	478	498	518	537	557
		2	0.13	473	346	363	380	396	413	430	447	463	480	497	514
		3	0.26	386	333	346	360	374	387	401	415	428	442	456	469
		4	0.39	299	318	329	340	350	361	371	382	392	403	414	424
		BFL	0.52	211	303	311	318	326	333	341	348	356	363	371	378
		6	1.38	176	296	302	308	315	321	327	333	339	346	352	358
		7	2.53	140	286	291	296	301	306	311	316	321	326	331	335
W 14×34	205	TFL	0.00	500	318	336	354	372	389	407	425	442	460	478	495
		2	0.11	423	307	322	337	352	367	382	397	412	427	442	457
		3	0.23	347	295	308	320	332	345	357	369	381	394	406	418
		4	0.34	270	283	293	302	312	321	331	340	350	359	369	379
		BFL	0.46	193	270	277	284	290	297	304	311	318	325	332	338
		6	1.41	159	263	269	274	280	286	291	297	302	308	314	319
		7	2.60	125	253	258	262	267	271	275	280	284	289	293	298
W 14×30	177	TFL	0.00	443	280	295	311	327	342	358	374	389	405	421	436
		2	0.10	378	270	284	297	310	324	337	350	364	377	391	404
		3	0.19	313	260	271	283	294	305	316	327	338	349	360	371
		4	0.29	248	250	259	268	276	285	294	303	312	320	329	338
		BFL	0.39	183	239	246	252	259	265	272	278	285	291	298	304
		6	1.48	147	232	237	242	248	253	258	263	268	274	279	284
		7	2.82	111	221	225	229	233	237	241	245	249	253	256	260
W 14×26	151	TFL	0.00	385	244	258	271	285	298	312	326	339	353	366	380
		2	0.11	332	236	248	260	271	283	295	307	318	330	342	354
		3	0.21	279	228	238	248	258	268	278	287	297	307	317	327
		4	0.32	226	220	228	236	244	252	260	268	276	284	292	300
		BFL	0.42	173	211	217	223	229	235	242	248	254	260	266	272
		6	1.67	135	203	207	212	217	222	227	231	236	241	246	250
		7	3.19	96.1	191	194	197	201	204	208	211	214	218	221	225
W 14×22	125	TFL	0.00	325	204	215	227	238	250	261	273	284	296	307	319
		2	0.08	283	198	208	218	228	238	248	258	268	278	288	298
		3	0.17	241	192	200	209	217	226	234	243	251	260	268	277
		4	0.25	199	185	192	199	206	213	220	227	234	241	248	255
		BFL	0.34	157	178	184	189	195	200	206	212	217	223	228	234
		6	1.69	119	170	174	179	183	187	191	196	200	204	208	212
		7	3.34	81.1	158	161	164	167	170	172	175	178	181	184	187

[a] $Y1$ = distance from top of the steel beam to plastic neutral axis.
[b] $Y2$ = distance from top of the steel beam to concrete flange force.
[c] See Figure 5-3 for PNA locations.

					F_y = 50 ksi

COMPOSITE DESIGN
COMPOSITE BEAM SELECTION TABLE
W Shapes

$$\phi = 0.85 \qquad\qquad \phi_b = 0.90$$

Shape	$\phi_b M_p$	PNA[c]	Y1[a]	ΣQ_n	ϕM_n (kip-ft)										
					Y2[b] (in.)										
	Kip-ft		In.	Kips	2	2.5	3	3.5	4	4.5	5	5.5	6	6.5	7
W 12×30	162	TFL	0.00	440	254	270	285	301	317	332	348	363	379	394	410
		2	0.11	368	244	257	270	283	296	309	322	335	348	361	374
		3	0.22	296	233	243	254	264	275	285	296	306	317	327	338
		4	0.33	224	221	229	237	245	253	261	269	277	285	293	301
		BFL	0.44	153	209	215	220	225	231	236	242	247	252	258	263
		6	1.12	131	205	210	214	219	224	228	233	238	242	247	252
		7	1.94	110	200	204	207	211	215	219	223	227	231	235	239
W 12×26	140	TFL	0.00	383	220	233	247	260	274	287	301	315	328	342	355
		2	0.10	321	211	222	234	245	256	268	279	290	302	313	324
		3	0.19	259	201	211	220	229	238	247	257	266	275	284	293
		4	0.29	198	192	199	206	213	220	227	234	241	248	255	262
		BFL	0.38	136	181	186	191	196	201	206	210	215	220	225	230
		6	1.08	116	178	182	186	190	194	198	202	206	210	215	219
		7	1.95	95.6	173	176	179	183	186	190	193	196	200	203	206
W 12×22	110	TFL	0.00	324	187	199	210	222	233	245	256	267	279	290	302
		2	0.11	281	181	191	201	211	221	231	241	251	261	271	281
		3	0.21	238	174	183	191	200	208	217	225	233	242	250	259
		4	0.32	196	168	174	181	188	195	202	209	216	223	230	237
		BFL	0.43	153	160	166	171	177	182	187	193	198	204	209	214
		6	1.66	117	153	157	161	165	169	173	178	182	186	190	194
		7	3.04	81.0	142	145	147	150	153	156	159	162	165	167	170
W 12×19	92.6	TFL	0.00	279	159	169	179	189	199	209	219	228	238	248	258
		2	0.09	243	154	163	172	180	189	197	206	215	223	232	241
		3	0.18	208	149	156	164	171	179	186	193	201	208	215	223
		4	0.26	173	144	150	156	162	168	174	180	187	193	199	205
		BFL	0.35	138	138	143	148	152	157	162	167	172	177	182	187
		6	1.66	104	131	134	138	142	145	149	153	156	160	164	167
		7	3.12	69.6	120	122	125	127	130	132	135	137	140	142	145
W 12×16	75.4	TFL	0.00	236	133	142	150	158	167	175	183	192	200	208	217
		2	0.07	209	130	137	144	152	159	167	174	181	189	196	204
		3	0.13	183	126	132	139	145	152	158	164	171	177	184	190
		4	0.20	156	122	127	133	138	144	149	155	160	166	171	177
		BFL	0.26	130	117	122	127	131	136	140	145	150	154	159	163
		6	1.71	94.3	110	113	117	120	123	127	130	133	137	140	143
		7	3.32	58.9	98.7	101	103	105	107	109	111	113	115	117	120

[a] Y1 = distance from top of the steel beam to plastic neutral axis.
[b] Y2 = distance from top of the steel beam to concrete flange force.
[c] See Figure 5-3 for PNA locations.

F_y = 50 ksi															

COMPOSITE DESIGN
COMPOSITE BEAM SELECTION TABLE
W Shapes
$\phi = 0.85$ $\phi_b = 0.90$

Shape	$\phi_b M_p$	PNA[c]	$Y1$[a]	ΣQ_n	ϕM_n (kip-ft)										
					$Y2$[b] (in.)										
	Kip-ft		In.	Kips	2	2.5	3	3.5	4	4.5	5	5.5	6	6.5	7
W 12×14	65.2	TFL	0.00	208	117	125	132	139	147	154	161	169	176	184	191
		2	0.06	186	114	121	127	134	140	147	153	160	167	173	180
		3	0.11	163	111	116	122	128	134	140	145	151	157	163	169
		4	0.17	141	107	112	117	122	127	132	137	142	147	152	157
		BFL	0.23	119	104	108	112	116	121	125	129	133	137	142	146
		6	1.69	85.3	97.0	100	103	106	109	112	115	118	121	124	127
		7	3.36	52.0	86.3	88.2	90.0	91.8	93.7	95.5	97.4	99.2	101	103	105
W 10×26	117	TFL	0.00	381	193	207	220	234	247	260	274	287	301	314	328
		2	0.11	317	184	195	206	218	229	240	251	262	274	285	296
		3	0.22	254	174	183	192	201	210	219	228	237	246	255	264
		4	0.33	190	164	171	177	184	191	198	204	211	218	225	231
		BFL	0.44	127	153	158	162	167	171	176	180	185	189	194	198
		6	0.90	111	150	154	158	162	166	170	174	178	182	186	189
		7	1.51	95.1	147	150	153	157	160	163	167	170	174	177	180
W 10×22	97.5	TFL	0.00	325	163	174	186	197	209	220	232	243	255	266	278
		2	0.09	273	155	165	175	184	194	204	213	223	233	242	252
		3	0.18	221	148	155	163	171	179	187	194	202	210	218	226
		4	0.27	169	139	145	151	157	163	169	175	181	187	193	199
		BFL	0.36	117	131	135	139	143	148	152	156	160	164	168	173
		6	0.95	99.3	127	131	135	138	142	145	149	152	156	159	163
		7	1.70	81.1	123	126	129	132	135	138	140	143	146	149	152
W 10×19	81.0	TFL	0.00	281	142	152	162	172	182	191	201	211	221	231	241
		2	0.10	241	136	145	153	162	170	179	187	196	204	213	221
		3	0.20	202	130	137	144	151	158	166	173	180	187	194	201
		4	0.30	162	124	129	135	141	147	152	158	164	169	175	181
		BFL	0.40	122	117	121	126	130	134	139	143	147	152	156	160
		6	1.27	96.2	112	115	119	122	125	129	132	136	139	143	146
		7	2.31	70.3	105	107	110	112	115	117	120	122	125	127	130
W 10×17	70.1	TFL	0.00	249	125	134	142	151	160	169	178	187	195	204	213
		2	0.08	216	120	128	135	143	151	158	166	174	181	189	197
		3	0.17	183	115	121	128	134	141	147	154	160	167	173	180
		4	0.25	150	110	115	120	126	131	136	142	147	152	158	163
		BFL	0.33	117	104	109	113	117	121	125	129	133	138	142	146
		6	1.31	89.8	99.0	102	105	109	112	115	118	121	124	128	131
		7	2.46	62.4	91.4	93.7	95.9	98.1	100	102	105	107	109	111	114

[a] $Y1$ = distance from top of the steel beam to plastic neutral axis.
[b] $Y2$ = distance from top of the steel beam to concrete flange force.
[c] See Figure 5-3 for PNA locations.

$F_y = 50$ ksi

COMPOSITE DESIGN
COMPOSITE BEAM SELECTION TABLE
W Shapes
$\phi = 0.85$ $\phi_b = 0.90$

Shape	$\phi_b M_p$	PNA[c]	$Y1^a$	ΣQ_n	ϕM_n (kip-ft)										
					$Y2^b$ (in.)										
	Kip-ft		In.	Kips	2	2.5	3	3.5	4	4.5	5	5.5	6	6.5	7
W 10×15	60.0	TFL	0.00	221	109	117	125	133	140	148	156	164	172	180	187
		2	0.07	194	105	112	119	126	133	140	146	153	160	167	174
		3	0.14	167	101	107	113	119	125	131	137	143	149	154	160
		4	0.20	139	97.2	102	107	112	117	122	127	132	137	142	147
		BFL	0.27	112	92.9	96.9	101	105	109	113	117	121	125	129	133
		6	1.35	83.8	87.3	90.3	93.2	96.2	99.2	102	105	108	111	114	117
		7	2.60	55.1	79.2	81.2	83.1	85.1	87.0	89.0	90.9	92.9	94.8	96.8	98.7
W 10×12	47.3	TFL	0.00	177	86.9	93.2	99.5	106	112	118	125	131	137	143	150
		2	0.05	156	84.0	89.5	95.0	101	106	112	117	123	128	134	139
		3	0.11	135	80.9	85.7	90.5	95.3	100	105	110	114	119	124	129
		4	0.16	115	77.8	81.8	85.9	89.9	94.0	98.1	102	106	110	114	118
		BFL	0.21	93.8	74.5	77.9	81.2	84.5	87.8	91.2	94.5	97.8	101	104	108
		6	1.30	69.0	69.8	72.3	74.7	77.1	79.6	82.0	84.5	86.9	89.4	91.8	94.3
		7	2.61	44.3	62.9	64.4	66.0	67.6	69.1	70.7	72.3	73.8	75.4	77.0	78.5
W 8×28	102	TFL	0.00	413	176	191	205	220	235	249	264	278	293	308	322
		2	0.12	337	165	177	189	201	213	225	237	249	260	272	284
		3	0.23	261	153	163	172	181	190	200	209	218	227	236	246
		4	0.35	185	141	148	154	161	167	174	180	187	193	200	206
		BFL	0.47	109	128	132	136	140	144	147	151	155	159	163	167
		6	0.53	106	128	131	135	139	143	146	150	154	158	161	165
		7	0.59	103	127	131	134	138	142	145	149	153	156	160	164
W 8×24	87.0	TFL	0.00	354	150	162	175	187	200	212	225	237	250	262	275
		2	0.10	289	140	150	161	171	181	191	202	212	222	232	243
		3	0.20	224	130	138	146	154	162	170	178	186	194	202	210
		4	0.30	159	120	126	131	137	142	148	154	159	165	171	176
		BFL	0.40	94.2	109	112	116	119	122	126	129	132	136	139	142
		6	0.47	91.3	109	112	115	118	122	125	128	131	134	138	141
		7	0.55	88.5	108	111	114	117	121	124	127	130	133	136	139
W 8×21	76.5	TFL	0.00	308	134	145	156	167	178	188	199	210	221	232	243
		2	0.10	255	126	135	144	153	162	172	181	190	199	208	217
		3	0.20	203	118	125	133	140	147	154	161	169	176	183	190
		4	0.30	150	110	115	120	126	131	136	142	147	152	158	163
		BFL	0.40	97.2	101	105	108	111	115	118	122	125	129	132	136
		6	0.70	87.1	99.3	102	105	109	112	115	118	121	124	127	130
		7	1.06	77.0	97.2	100	103	105	108	111	114	116	119	122	125

[a] $Y1$ = distance from top of the steel beam to plastic neutral axis.
[b] $Y2$ = distance from top of the steel beam to concrete flange force.
[c] See Figure 5-3 for PNA locations.

F_y = 50 ksi

COMPOSITE DESIGN
COMPOSITE BEAM SELECTION TABLE
W Shapes

$\phi = 0.85$ $\phi_b = 0.90$

Shape	$\phi_b M_p$	PNA[c]	Y1[a]	ΣQ_n	ϕM_n (kip-ft)										
					Y2[b] (in.)										
	Kip-ft		In.	Kips	2	2.5	3	3.5	4	4.5	5	5.5	6	6.5	7
W 8×18	63.7	TFL	0.00	263	113	122	132	141	150	160	169	178	188	197	206
		2	0.08	220	107	115	122	130	138	146	154	161	169	177	185
		3	0.17	176	100	107	113	119	125	132	138	144	150	157	163
		4	0.25	133	93.5	98.2	103	108	112	117	122	127	131	136	141
		BFL	0.33	89.8	86.5	89.7	92.9	96.0	99.2	102	106	109	112	115	118
		6	0.71	77.8	84.4	87.1	89.9	92.6	95.4	98.1	101	104	106	109	112
		7	1.21	65.8	81.9	84.2	86.5	88.8	91.2	93.5	95.8	98.2	100	103	105
W 8×15	51.0	TFL	0.00	222	95.2	103	111	119	127	135	142	150	158	166	174
		2	0.08	190	90.6	97.4	104	111	118	124	131	138	145	151	158
		3	0.16	159	85.9	91.5	97.1	103	108	114	120	125	131	137	142
		4	0.24	127	81.0	85.5	90.0	94.5	99.0	103	108	113	117	122	126
		BFL	0.32	95.5	75.9	79.3	82.7	86.0	89.4	92.8	96.2	99.6	103	106	110
		6	0.97	75.5	72.2	74.8	77.5	80.2	82.9	85.5	88.2	90.9	93.6	96.2	98.9
		7	1.79	55.5	67.4	69.3	71.3	73.3	75.2	77.2	79.2	81.1	83.1	85.1	87.0
W 8×13	42.7	TFL	0.00	192	81.5	88.3	95.1	102	109	116	122	129	136	143	150
		2	0.06	167	77.9	83.8	89.7	95.6	101	107	113	119	125	131	137
		3	0.13	141	74.1	79.1	84.1	89.1	94.1	99.0	104	109	114	119	124
		4	0.19	116	70.2	74.3	78.4	82.4	86.5	90.6	94.7	98.8	103	107	111
		BFL	0.26	90.0	66.2	69.3	72.5	75.7	78.9	82.1	85.3	88.5	91.7	94.8	98.0
		6	0.99	69.0	62.3	64.7	67.2	69.6	72.1	74.5	77.0	79.4	81.8	84.3	86.7
		7	1.91	48.0	57.2	58.9	60.6	62.3	64.0	65.7	67.4	69.1	70.8	72.5	74.2
W 8×10	33.3	TFL	0.00	148	62.3	67.6	72.8	78.0	83.3	88.5	93.8	99.0	104	109	115
		2	0.05	128	59.4	64.0	68.5	73.0	77.5	82.1	86.6	91.1	95.6	100	105
		3	0.10	108	56.5	60.3	64.1	67.9	71.7	75.5	79.3	83.1	86.9	90.8	94.6
		4	0.15	87.4	53.4	56.5	59.6	62.7	65.8	68.9	72.0	75.1	78.2	81.3	84.4
		BFL	0.21	67.2	50.3	52.7	55.1	57.4	59.8	62.2	64.6	67.0	69.3	71.7	74.1
		6	0.88	52.1	47.6	49.5	51.3	53.1	55.0	56.8	58.7	60.5	62.4	64.2	66.1
		7	1.77	37.0	44.0	45.4	46.7	48.0	49.3	50.6	51.9	53.2	54.5	55.8	57.2

[a] Y1 = distance from top of the steel beam to plastic neutral axis.
[b] Y2 = distance from top of the steel beam to concrete flange force.
[c] See Figure 5-3 for PNA locations.

LOWER BOUND
ELASTIC MOMENT OF INERTIA
FOR PLASTIC COMPOSITE SECTIONS

I_{LB}

Shape[d]	PNA[c]	Y1[a]	I_{LB} (in.[4])										
			$Y2$[b] (in.)										
		In.	2	2.5	3	3.5	4	4.5	5	5.5	6	6.5	7
W40×297	TFL	0.00	44200	45200	46200	47200	48200	49300	50300	51400	52600	53700	54900
(23200)	2	0.41	42500	43400	44300	45200	46200	47200	48100	49200	50200	51200	52300
	3	0.83	40500	41300	42100	43000	43800	44700	45600	46500	47400	48300	49300
	4	1.24	38100	38800	39500	40200	41000	41700	42500	43300	44100	44900	45700
	BFL	1.65	35300	35800	36400	37000	37600	38200	38800	39400	40100	40700	41400
	6	4.59	33500	34000	34500	35000	35500	36000	36600	37100	37600	38200	38800
	7	8.17	31600	32000	32400	32800	33200	33600	34100	34500	34900	35400	35900
W40×278	TFL	0.00	40400	41400	42300	43200	44200	45200	46200	47300	48300	49400	50500
(20500)	2	0.45	39000	39900	40700	41600	42500	43500	44400	45400	46300	47300	48400
	3	0.91	37400	38200	39000	39800	40600	41400	42300	43200	44100	45000	45900
	4	1.36	35500	36200	36900	37600	38300	39100	39800	40600	41400	42200	43100
	BFL	1.81	33300	33800	34400	35000	35700	36300	37000	37600	38300	39000	39700
	6	5.64	31100	31500	32000	32500	33100	33600	34100	34700	35200	35800	36400
	7	10.06	28500	28800	29200	29600	30000	30400	30800	31200	31600	32100	32500
W40×277	TFL	0.00	41300	42200	43100	44100	45000	46000	47000	48000	49100	50100	51200
(21900)	2	0.39	39700	40500	41400	42200	43100	44000	44900	45800	46800	47800	48800
	3	0.79	37800	38500	39300	40000	40800	41600	42400	43300	44100	45000	45900
	4	1.18	35500	36100	36800	37400	38100	38800	39500	40200	40900	41700	42400
	BFL	1.58	32700	33200	33700	34300	34800	35300	35900	36500	37000	37600	38200
	6	4.25	31300	31700	32100	32600	33100	33500	34000	34500	35000	35500	36000
	7	7.60	29700	30000	30400	30800	31100	31500	31900	32300	32800	33200	33600
W40×264	TFL	0.00	38200	39000	39900	40800	41700	42700	43700	44600	45600	46600	47700
(19400)	2	0.43	36800	37600	38500	39300	40200	41000	41900	42800	43800	44700	45700
	3	0.87	35300	36000	36800	37500	38300	39100	39900	40800	41600	42500	43300
	4	1.30	33500	34100	34800	35500	36200	36900	37600	38300	39100	39900	40600
	BFL	1.73	31400	31900	32500	33100	33700	34300	34900	35500	36100	36800	37400
	6	5.49	29300	29800	30200	30700	31200	31700	32200	32700	33300	33800	34300
	7	9.90	26900	27300	27600	28000	28300	28700	29100	29500	29900	30300	30700

[a] Y1 = distance from top of the steel beam to plastic neutral axis.
[b] Y2 = distance from top of the steel beam to concrete flange force.
[c] See Figure 5-3 for PNA locations.
[d] Value in parentheses is I_x (in.[4]) of non-composite steel shape.

LOWER BOUND
ELASTIC MOMENT OF INERTIA
FOR PLASTIC COMPOSITE SECTIONS

 I_{LB}

Shape[d]	PNA[c]	Y1[a]	I_{LB} (in.[4])										
			$Y2^{b}$ (in.)										
		In.	2	2.5	3	3.5	4	4.5	5	5.5	6	6.5	7
W40×249	TFL	0.00	36700	37500	38400	39200	40100	40900	41800	42800	43700	44600	45600
(19500)	2	0.36	35300	36100	36800	37600	38400	39200	40000	40800	41700	42600	43500
	3	0.71	33600	34300	35000	35700	36400	37100	37800	38600	39300	40100	40900
	4	1.07	31600	32200	32800	33400	34000	34600	35200	35900	36500	37200	37900
	BFL	1.42	29200	29600	30100	30600	31000	31500	32000	32500	33100	33600	34100
	6	4.06	27900	28200	28600	29100	29500	29900	30300	30800	31200	31700	32200
	7	7.47	26400	26700	27000	27400	27700	28100	28400	28800	29200	29600	29900
W40×235	TFL	0.00	33800	34600	35400	36200	37000	37800	38700	39500	40400	41300	42200
(17400)	2	0.39	32600	33300	34100	34800	35600	36300	37100	37900	38700	39600	40400
	3	0.79	31300	31900	32600	33200	33900	34600	35300	36000	36800	37500	38300
	4	1.18	29600	30200	30800	31400	32000	32600	33200	33900	34500	35200	35900
	BFL	1.58	27700	28200	28700	29200	29700	30200	30700	31300	31800	32400	33000
	6	5.18	26000	26400	26800	27200	27600	28100	28500	29000	29400	29900	30400
	7	9.47	24000	24300	24600	24900	25200	25600	25900	26300	26600	27000	27300
W40×215	TFL	0.00	31300	32000	32700	33400	34200	34900	35700	36500	37300	38100	38900
(16700)	2	0.31	30100	30700	31400	32100	32700	33400	34100	34800	35600	36300	37100
	3	0.61	28700	29300	29800	30400	31000	31700	32300	32900	33600	34300	34900
	4	0.92	27000	27500	28000	28500	29000	29500	30100	30600	31200	31800	32400
	BFL	1.22	24900	25300	25700	26100	26600	27000	27400	27900	28300	28800	29200
	6	3.84	23800	24100	24500	24800	25200	25600	25900	26300	26700	27100	27500
	7	7.32	22500	22800	23100	23400	23700	24000	24300	24600	24900	25300	25600
W40×211	TFL	0.00	30100	30800	31500	32200	32900	33600	34400	35200	36000	36800	37600
(15500)	2	0.35	29000	29700	30300	31000	31600	32300	33000	33700	34500	35200	36000
	3	0.71	27800	28400	29000	29600	30200	30800	31400	32100	32800	33400	34100
	4	1.06	26400	26900	27400	27900	28500	29000	29600	30200	30800	31400	32000
	BFL	1.42	24700	25100	25600	26000	26500	26900	27400	27900	28400	28900	29400
	6	4.99	23100	23500	23900	24200	24600	25000	25400	25800	26200	26600	27100
	7	9.35	21300	21600	21900	22200	22500	22800	23100	23400	23700	24000	24300

[a] $Y1$ = distance from top of the steel beam to plastic neutral axis.
[b] $Y2$ = distance from top of the steel beam to concrete flange force.
[c] See Figure 5-3 for PNA locations.
[d] Value in parentheses is I_x (in.[4]) of non-composite steel shape.

LOWER BOUND
ELASTIC MOMENT OF INERTIA
FOR PLASTIC COMPOSITE SECTIONS

Shape[d]	PNA[c]	Y1[a]	I_{LB} (in.^{4})										
			Y2[b] (in.)										
		in.	2	2.5	3	3.5	4	4.5	5	5.5	6	6.5	7
W40×199	TFL	0.00	28200	28800	29500	30100	30800	31500	32200	32900	33600	34400	35200
(14900)	2	0.27	27200	27700	28300	28900	29600	30200	30900	31500	32200	32900	33600
	3	0.53	26000	26500	27000	27600	28100	28700	29300	29900	30500	31100	31800
	4	0.80	24500	25000	25500	25900	26400	26900	27400	28000	28500	29000	29600
	BFL	1.07	22800	23200	23600	24000	24400	24800	25200	25700	26100	26500	27000
	6	4.16	21600	21900	22300	22600	22900	23300	23600	24000	24400	24700	25100
	7	8.10	20200	20500	20700	21000	21300	21500	21800	22100	22400	22700	23000
W40×183	TFL	0.00	25700	26300	26900	27500	28100	28800	29400	30100	30700	31400	32100
(13300)	2	0.31	24800	25400	25900	26500	27100	27600	28200	28900	29500	30100	30800
	3	0.61	23800	24300	24800	25300	25800	26400	26900	27500	28000	28600	29200
	4	0.92	22600	23000	23500	23900	24400	24900	25300	25800	26400	26900	27400
	BFL	1.22	21200	21500	21900	22300	22700	23100	23500	23900	24300	24800	25200
	6	4.76	19800	20100	20400	20800	21100	21400	21800	22100	22500	22800	23200
	7	9.16	18300	18500	18700	19000	19200	19500	19700	20000	20300	20600	20800
W40×174	TFL	0.00	23600	24100	24700	25200	25800	26400	27000	27700	28300	28900	29600
(12200)	2	0.21	22800	23300	23800	24400	24900	25500	26000	26600	27200	27800	28400
	3	0.42	21900	22400	22800	23300	23800	24300	24900	25400	25900	26500	27100
	4	0.62	20900	21300	21700	22200	22600	23100	23500	24000	24500	25000	25500
	BFL	0.83	19700	20000	20400	20800	21100	21500	21900	22300	22800	23200	23600
	6	4.59	18300	18600	18900	19200	19600	19900	20200	20500	20900	21200	21600
	7	9.27	16800	17000	17200	17400	17700	17900	18100	18400	18600	18900	19200
W40×167	TFL	0.00	22700	23300	23800	24400	24900	25500	26100	26700	27300	27900	28600
(11600)	2	0.26	22000	22500	23000	23500	24000	24600	25100	25700	26300	26900	27500
	3	0.51	21200	21600	22100	22600	23000	23500	24100	24600	25100	25600	26200
	4	0.77	20200	20600	21000	21500	21900	22300	22800	23300	23700	24200	24700
	BFL	1.03	19100	19400	19800	20200	20600	21000	21300	21800	22200	22600	23000
	6	5.00	17700	18000	18300	18600	18900	19200	19600	19900	20200	20600	20900
	7	9.85	16100	16300	16500	16700	16900	17200	17400	17600	17900	18100	18400
W40×149	TFL	0.00	19500	20000	20500	21000	21500	22000	22500	23000	23600	24100	24700
(9780)	2	0.21	19000	19400	19800	20300	20800	21300	21700	22200	22800	23300	23800
	3	0.42	18300	18700	19100	19600	20000	20400	20900	21400	21800	22300	22800
	4	0.62	17600	17900	18300	18700	19100	19500	19900	20400	20800	21200	21700
	BFL	0.83	16700	17100	17400	17700	18100	18500	18800	19200	19600	20000	20400
	6	5.15	15400	15600	15900	16200	16500	16800	17100	17400	17700	18000	18300
	7	10.41	13700	13900	14100	14300	14500	14700	14900	15100	15300	15500	15700

[a] Y1 = distance from top of the steel beam to plastic neutral axis.
[b] Y2 = distance from top of the steel beam to concrete flange force.
[c] See Figure 5-3 for PNA locations.
[d] Value in parentheses is I_x (in.^{4}) of non-composite steel shape.

LOWER BOUND
ELASTIC MOMENT OF INERTIA
FOR PLASTIC COMPOSITE SECTIONS

I_{LB}

Shape[d]	PNA[c]	Y1[a]	I_{LB} (in.[4])										
			$Y2^b$ (in.)										
		In.	2	2.5	3	3.5	4	4.5	5	5.5	6	6.5	7
W 36×300	TFL	0.00	38600	39500	40500	41400	42400	43400	44400	45500	46500	47600	48700
(20300)	2	0.42	37000	37900	38700	39600	40500	41400	42300	43300	44300	45300	46300
	3	0.84	35200	35900	36700	37400	38200	39000	39900	40700	41600	42500	43400
	4	1.26	32900	33500	34200	34800	35500	36200	36900	37600	38300	39100	39900
	BFL	1.68	30100	30600	31100	31600	32100	32700	33200	33800	34400	34900	35500
	6	3.97	28900	29400	29800	30200	30700	31200	31700	32200	32700	33200	33700
	7	6.69	27600	28000	28400	28700	29100	29500	29900	30400	30800	31200	31700
W 36×280	TFL	0.00	35800	36700	37500	38400	39300	40200	41200	42200	43100	44200	45200
(18900)	2	0.39	34400	35100	35900	36700	37600	38400	39300	40200	41100	42000	42900
	3	0.79	32600	33300	34000	34700	35500	36200	37000	37800	38600	39400	40300
	4	1.18	30600	31100	31700	32300	33000	33600	34300	34900	35600	36300	37000
	BFL	1.57	28000	28400	28900	29400	29900	30400	30900	31400	31900	32500	33000
	6	3.88	26900	27300	27700	28100	28500	29000	29400	29900	30300	30800	31300
	7	6.62	25700	26000	26300	26700	27100	27400	27800	28200	28600	29000	29400
W 36×260	TFL	0.00	32800	33600	34400	35200	36000	36900	37800	38700	39600	40500	41500
(17300)	2	0.36	31500	32200	32900	33700	34500	35200	36000	36900	37700	38500	39400
	3	0.72	29900	30600	31200	31900	32600	33300	34000	34700	35500	36200	37000
	4	1.08	28100	28600	29200	29700	30300	30900	31500	32100	32800	33400	34100
	BFL	1.44	25800	26200	26700	27100	27600	28000	28500	29000	29500	30000	30500
	6	3.86	24700	25100	25500	25800	26200	26600	27100	27500	27900	28400	28800
	7	6.75	23500	23800	24100	24500	24800	25100	25500	25800	26200	26600	27000
W 36×245	TFL	0.00	30600	31300	32100	32800	33600	34400	35200	36100	36900	37800	38700
(16100)	2	0.34	29400	30000	30700	31400	32100	32900	33600	34400	35200	36000	36800
	3	0.68	27900	28500	29100	29800	30400	31100	31700	32400	33100	33800	34600
	4	1.01	26200	26700	27200	27800	28300	28900	29500	30000	30600	31300	31900
	BFL	1.35	24100	24500	24900	25300	25800	26200	26700	27100	27600	28100	28600
	6	3.81	23100	23400	23800	24100	24500	24900	25300	25700	26100	26500	27000
	7	6.77	21900	22200	22500	22800	23100	23400	23800	24100	24400	24800	25100
W 36×230	TFL	0.00	28500	29100	29800	30600	31300	32000	32800	33600	34400	35200	36000
(15000)	2	0.32	27300	28000	28600	29300	29900	30600	31300	32000	32800	33500	34300
	3	0.63	26000	26600	27100	27700	28300	28900	29600	30200	30900	31500	32200
	4	0.95	24400	24900	25400	25900	26400	26900	27500	28000	28600	29200	29700
	BFL	1.26	22500	22900	23300	23700	24100	24500	24900	25400	25800	26300	26700
	6	3.81	21500	21800	22200	22500	22900	23200	23600	24000	24400	24800	25200
	7	6.83	20400	20700	20900	21200	21500	21800	22100	22400	22800	23100	23400

[a] Y1 = distance from top of the steel beam to plastic neutral axis.
[b] Y2 = distance from top of the steel beam to concrete flange force.
[c] See Figure 5-3 for PNA locations.
[d] Value in parentheses is I_x (in.[4]) of non-composite steel shape.

I $_{LB}$

LOWER BOUND
ELASTIC MOMENT OF INERTIA
FOR PLASTIC COMPOSITE SECTIONS

Shape[d]	PNA[c]	Y1[a]	I_{LB} (in.^{4})										
			Y2[b] (in.)										
		In.	2	2.5	3	3.5	4	4.5	5	5.5	6	6.5	7
W 36×210	TFL	0.00	26000	26600	27300	27900	28600	29300	30000	30800	31500	32300	33000
(13200)	2	0.34	25100	25700	26300	26900	27500	28200	28800	29500	30200	30900	31600
	3	0.68	24000	24500	25100	25700	26200	26800	27400	28100	28700	29300	30000
	4	1.02	22800	23200	23700	24200	24700	25300	25800	26300	26900	27500	28100
	BFL	1.36	21300	21700	22100	22500	23000	23400	23900	24300	24800	25300	25800
	6	5.06	19900	20300	20600	21000	21300	21700	22100	22400	22800	23200	23600
	7	9.04	18300	18600	18800	19100	19400	19700	19900	20200	20500	20800	21100
W 36×194	TFL	0.00	23800	24400	25000	25600	26200	26800	27500	28200	28900	29600	30300
(12100)	2	0.32	22900	23500	24000	24600	25200	25800	26400	27000	27700	28300	29000
	3	0.63	22000	22500	23000	23500	24000	24600	25100	25700	26300	26900	27500
	4	0.95	20800	21300	21700	22200	22700	23100	23600	24100	24600	25200	25700
	BFL	1.26	19500	19900	20300	20600	21000	21500	21900	22300	22700	23200	23600
	6	4.94	18200	18600	18900	19200	19500	19900	20200	20600	20900	21300	21700
	7	8.93	16800	17000	17200	17500	17700	18000	18300	18500	18800	19100	19400
W 36×182	TFL	0.00	22200	22700	23300	23900	24500	25100	25700	26300	26900	27600	28300
(11300)	2	0.29	21400	21900	22500	23000	23500	24100	24700	25200	25800	26400	27100
	3	0.59	20500	21000	21500	22000	22400	23000	23500	24000	24600	25100	25700
	4	0.89	19500	19900	20300	20700	21200	21600	22100	22600	23100	23500	24000
	BFL	1.18	18300	18600	19000	19300	19700	20100	20500	20900	21300	21700	22100
	6	4.89	17100	17300	17600	17900	18300	18600	18900	19200	19600	19900	20300
	7	8.92	15700	15900	16100	16300	16600	16800	17100	17300	17600	17800	18100
W 36×170	TFL	0.00	20600	21100	21600	22100	22700	23300	23800	24400	25000	25600	26200
(10500)	2	0.28	19900	20300	20800	21300	21800	22300	22900	23400	24000	24500	25100
	3	0.55	19000	19500	19900	20400	20800	21300	21800	22300	22800	23300	23800
	4	0.83	18100	18500	18900	19300	19700	20100	20500	21000	21400	21900	22300
	BFL	1.10	17000	17300	17600	18000	18300	18700	19000	19400	19800	20200	20600
	6	4.84	15800	16100	16400	16700	17000	17300	17600	17900	18200	18500	18800
	7	8.89	14500	14700	14900	15200	15400	15600	15800	16100	16300	16500	16800
W 36×160	TFL	0.00	19200	19600	20100	20600	21100	21700	22200	22700	23300	23900	24400
(9750)	2	0.26	18500	18900	19400	19900	20300	20800	21300	21800	22300	22900	23400
	3	0.51	17700	18200	18600	19000	19400	19900	20300	20800	21300	21700	22200
	4	0.77	16900	17200	17600	18000	18400	18800	19200	19600	20000	20400	20900
	BFL	1.02	15800	16200	16500	16800	17100	17500	17800	18200	18500	18900	19300
	6	4.82	14800	15000	15300	15600	15800	16100	16400	16700	17000	17300	17600
	7	8.97	13500	13700	13900	14100	14300	14500	14700	14900	15200	15400	15600

[a] Y1 = distance from top of the steel beam to plastic neutral axis.
[b] Y2 = distance from top of the steel beam to concrete flange force.
[c] See Figure 5-3 for PNA locations.
[d] Value in parentheses is I_x (in.^{4}) of non-composite steel shape.

LOWER BOUND
ELASTIC MOMENT OF INERTIA
FOR PLASTIC COMPOSITE SECTIONS

I $_{LB}$

Shape[d]	PNA[c]	Y1[a]	I_{LB} (in.[4])										
			Y2[b] (in.)										
		In.	2	2.5	3	3.5	4	4.5	5	5.5	6	6.5	7
W 36×150	TFL	0.00	17800	18300	18700	19200	19700	20200	20700	21200	21700	22200	22800
(9040)	2	0.24	17200	17600	18100	18500	18900	19400	19900	20300	20800	21300	21800
	3	0.47	16500	16900	17300	17700	18100	18500	19000	19400	19800	20300	20800
	4	0.71	15700	16100	16400	16800	17200	17500	17900	18300	18700	19100	19500
	BFL	0.94	14800	15100	15400	15700	16000	16400	16700	17000	17400	17700	18100
	6	4.83	13800	14000	14300	14500	14800	15000	15300	15600	15900	16200	16500
	7	9.08	12500	12700	12900	13100	13300	13500	13700	13900	14100	14300	14500
W 36×135	TFL	0.00	15600	16000	16400	16800	17200	17600	18100	18600	19000	19500	20000
(7800)	2	0.20	15100	15400	15800	16200	16600	17000	17400	17900	18300	18800	19200
	3	0.40	14500	14900	15200	15600	15900	16300	16700	17100	17500	17900	18300
	4	0.59	13900	14200	14500	14800	15200	15500	15900	16200	16600	17000	17300
	BFL	0.79	13100	13400	13700	14000	14300	14600	14900	15200	15500	15800	16200
	6	4.96	12100	12400	12600	12800	13100	13300	13500	13800	14100	14300	14600
	7	9.50	10900	11100	11200	11400	11600	11700	11900	12100	12300	12500	12700
W 33×221	TFL	0.00	24500	25100	25800	26400	27100	27800	28500	29200	29900	30700	31500
(12800)	2	0.32	23500	24100	24700	25300	25900	26500	27200	27800	28500	29200	29900
	3	0.64	22300	22900	23400	23900	24500	25000	25600	26200	26800	27400	28000
	4	0.96	20900	21400	21800	22300	22800	23200	23700	24200	24700	25300	25800
	BFL	1.28	19200	19600	19900	20300	20700	21000	21400	21800	22200	22700	23100
	6	3.76	18400	18700	19000	19300	19600	20000	20300	20700	21000	21400	21700
	7	6.48	17500	17700	18000	18200	18500	18800	19100	19400	19700	20000	20300
W 33×201	TFL	0.00	22000	22600	23100	23700	24300	25000	25600	26200	26900	27600	28300
(11500)	2	0.29	21100	21600	22200	22700	23300	23800	24400	25000	25600	26300	26900
	3	0.58	20100	20600	21000	21500	22000	22500	23000	23600	24100	24700	25300
	4	0.86	18900	19300	19700	20100	20500	20900	21400	21800	22300	22800	23300
	BFL	1.15	17400	17700	18000	18300	18700	19000	19400	19700	20100	20500	20900
	6	3.67	16600	16800	17100	17400	17700	18000	18300	18600	18900	19300	19600
	7	6.51	15700	15900	16200	16400	16600	16900	17100	17400	17700	17900	18200
W 33×141	TFL	0.00	14700	15100	15500	15900	16300	16800	17200	17700	18100	18600	19100
(7450)	2	0.24	14200	14500	14900	15300	15700	16100	16500	16900	17400	17800	18300
	3	0.48	13600	13900	14200	14600	15000	15300	15700	16100	16500	16900	17300
	4	0.72	12900	13200	13500	13800	14100	14400	14800	15100	15500	15800	16200
	BFL	0.96	12100	12300	12600	12800	13100	13400	13700	14000	14200	14600	14900
	6	4.31	11300	11500	11700	11900	12100	12400	12600	12800	13100	13300	13600
	7	8.05	10300	10500	10700	10800	11000	11200	11300	11500	11700	11900	12100

[a] Y1 = distance from top of the steel beam to plastic neutral axis.
[b] Y2 = distance from top of the steel beam to concrete flange force.
[c] See Figure 5-3 for PNA locations.
[d] Value in parentheses is I_x (in.[4]) of non-composite steel shape.

LOWER BOUND
ELASTIC MOMENT OF INERTIA
FOR PLASTIC COMPOSITE SECTIONS

I_{LB}

Shape[d]	PNA[c]	Y1[a]	I_{LB} (in.^{4})										
			$Y2^b$ (in.)										
		In.	2	2.5	3	3.5	4	4.5	5	5.5	6	6.5	7
W 33×130	TFL	0.00	13300	13700	14000	14400	14800	15200	15600	16000	16400	16900	17300
(6710)	2	0.21	12800	13200	13500	13900	14200	14600	15000	15400	15800	16200	16600
	3	0.43	12300	12600	12900	13300	13600	13900	14300	14600	15000	15400	15800
	4	0.64	11700	12000	12300	12600	12900	13200	13500	13800	14100	14500	14800
	BFL	0.86	11000	11300	11500	11700	12000	12300	12500	12800	13100	13400	13700
	6	4.39	10300	10400	10600	10900	11100	11300	11500	11700	11900	12200	12400
	7	8.29	9340	9490	9640	9790	9940	10100	10300	10400	10600	10800	11000
W 33×118	TFL	0.00	11800	12100	12500	12800	13100	13500	13900	14200	14600	15000	15400
(5900)	2	0.19	11400	11700	12000	12300	12700	13000	13300	13700	14100	14400	14800
	3	0.37	11000	11300	11500	11800	12100	12400	12800	13100	13400	13700	14100
	4	0.56	10500	10700	11000	11200	11500	11800	12100	12400	12700	13000	13300
	BFL	0.74	9880	10100	10300	10600	10800	11000	11300	11500	11800	12100	12300
	6	4.44	9150	9330	9510	9700	9890	10100	10300	10500	10700	10900	11100
	7	8.54	8260	8390	8520	8660	8800	8940	9090	9240	9390	9550	9710
W 30×116	TFL	0.00	9870	10200	10500	10800	11100	11400	11800	12100	12500	12800	13200
(4930)	2	0.21	9530	9800	10100	10400	10700	11000	11300	11600	11900	12300	12600
	3	0.43	9130	9380	9640	9910	10200	10500	10700	11000	11300	11700	12000
	4	0.64	8670	8900	9130	9360	9600	9850	10100	10400	10600	10900	11200
	BFL	0.85	8130	8320	8520	8720	8930	9140	9360	9580	9810	10000	10300
	6	3.98	7570	7730	7890	8060	8230	8400	8580	8770	8960	9150	9350
	7	7.44	6910	7030	7150	7270	7400	7530	7670	7810	7950	8090	8240
W 30×108	TFL	0.00	9000	9280	9560	9840	10100	10400	10800	11100	11400	11700	12100
(4470)	2	0.19	8700	8960	9220	9480	9760	10000	10300	10600	10900	11300	11600
	3	0.38	8350	8590	8830	9070	9330	9590	9850	10100	10400	10700	11000
	4	0.57	7950	8160	8380	8600	8820	9060	9300	9540	9790	10100	10300
	BFL	0.76	7480	7660	7850	8040	8240	8440	8650	8860	9080	9300	9530
	6	4.04	6940	7090	7240	7400	7560	7720	7890	8070	8240	8430	8610
	7	7.64	6280	6390	6500	6620	6740	6860	6980	7110	7240	7380	7510
W 30×99	TFL	0.00	8110	8360	8610	8880	9150	9420	9710	10000	10300	10600	10900
(3990)	2	0.17	7850	8080	8320	8560	8820	9080	9340	9620	9900	10200	10500
	3	0.34	7550	7760	7980	8210	8440	8680	8930	9180	9440	9700	9970
	4	0.50	7200	7400	7600	7800	8010	8230	8450	8680	8910	9150	9390
	BFL	0.67	6800	6970	7150	7330	7510	7700	7900	8100	8300	8510	8720
	6	4.07	6280	6420	6560	6700	6850	7010	7170	7330	7490	7660	7840
	7	7.83	5640	5740	5840	5940	6050	6160	6280	6390	6510	6640	6760

[a] Y1 = distance from top of the steel beam to plastic neutral axis.
[b] Y2 = distance from top of the steel beam to concrete flange force.
[c] See Figure 5-3 for PNA locations.
[d] Value in parentheses is I_x (in.^{4}) of non-composite steel shape.

LOWER BOUND
ELASTIC MOMENT OF INERTIA
FOR PLASTIC COMPOSITE SECTIONS

I LB

Shape[d]	PNA[c]	Y1[a]	I_{LB} (in.[4])										
			Y2[b] (in.)										
		In.	2	2.5	3	3.5	4	4.5	5	5.5	6	6.5	7
W 27×102	TFL	0.00	7240	7480	7730	7980	8240	8500	8780	9060	9350	9650	9950
(3620)	2	0.21	6970	7190	7420	7650	7890	8140	8390	8660	8920	9200	9480
	3	0.42	6660	6860	7070	7280	7490	7720	7950	8190	8430	8680	8930
	4	0.62	6290	6470	6650	6830	7030	7220	7430	7630	7850	8070	8290
	BFL	0.83	5860	6000	6150	6310	6470	6630	6800	6980	7150	7340	7520
	6	3.41	5490	5610	5740	5870	6000	6140	6280	6430	6580	6730	6890
	7	6.26	5070	5160	5260	5360	5470	5570	5680	5800	5910	6030	6150
W 27×94	TFL	0.00	6580	6800	7020	7250	7490	7740	7990	8250	8510	8790	9070
(3270)	2	0.19	6340	6540	6750	6970	7190	7420	7650	7890	8140	8390	8650
	3	0.37	6070	6250	6440	6640	6840	7040	7260	7480	7700	7930	8170
	4	0.56	5740	5910	6080	6250	6430	6610	6800	6990	7190	7400	7600
	BFL	0.75	5360	5500	5640	5790	5940	6100	6260	6420	6590	6760	6940
	6	3.39	5010	5120	5240	5360	5490	5620	5750	5890	6030	6170	6320
	7	6.39	4590	4680	4770	4860	4960	5060	5160	5260	5370	5480	5590
W 27×84	TFL	0.00	5770	5970	6170	6370	6580	6800	7030	7260	7500	7740	7990
(2850)	2	0.16	5570	5750	5940	6130	6330	6530	6740	6960	7180	7400	7630
	3	0.32	5340	5510	5680	5850	6030	6220	6410	6610	6810	7020	7230
	4	0.48	5080	5220	5370	5530	5690	5860	6030	6210	6390	6570	6760
	BFL	0.64	4760	4890	5020	5150	5290	5440	5580	5730	5890	6050	6210
	6	3.44	4420	4530	4630	4750	4860	4980	5100	5220	5350	5480	5610
	7	6.62	4020	4100	4180	4260	4340	4430	4520	4610	4710	4810	4910
W 24×76	TFL	0.00	4280	4440	4610	4780	4950	5130	5320	5510	5710	5920	6130
(2100)	2	0.17	4120	4270	4420	4580	4740	4910	5090	5260	5450	5640	5830
	3	0.34	3940	4070	4210	4350	4500	4650	4810	4970	5140	5310	5490
	4	0.51	3720	3840	3960	4090	4220	4350	4490	4640	4780	4930	5090
	BFL	0.68	3460	3560	3670	3770	3880	4000	4110	4230	4360	4480	4610
	6	3.00	3240	3320	3410	3490	3590	3680	3780	3880	3980	4090	4200
	7	5.60	2970	3040	3100	3170	3240	3310	3390	3470	3550	3630	3710
W 24×68	TFL	0.00	3760	3900	4050	4200	4360	4520	4690	4860	5040	5220	5410
(1830)	2	0.15	3630	3760	3900	4040	4180	4330	4490	4650	4810	4980	5160
	3	0.29	3470	3590	3720	3850	3980	4120	4260	4410	4560	4710	4870
	4	0.44	3290	3400	3510	3630	3740	3870	3990	4120	4260	4390	4540
	BFL	0.59	3080	3170	3270	3370	3470	3570	3680	3790	3910	4020	4140
	6	3.05	2860	2940	3020	3100	3180	3270	3360	3450	3540	3640	3740
	7	5.81	2600	2660	2720	2780	2840	2910	2970	3040	3110	3190	3260

[a] Y1 = distance from top of the steel beam to plastic neutral axis.
[b] Y2 = distance from top of the steel beam to concrete flange force.
[c] See Figure 5-3 for PNA locations.
[d] Value in parentheses is I_x (in.[4]) of non-composite steel shape.

LOWER BOUND
ELASTIC MOMENT OF INERTIA
FOR PLASTIC COMPOSITE SECTIONS

Shape[d]	PNA[c]	Y1[a]	I_{LB} (in.^{4})										
			Y2[b] (in.)										
		In.	2	2.5	3	3.5	4	4.5	5	5.5	6	6.5	7
W 24×62	TFL	0.00	3300	3430	3560	3700	3840	3990	4140	4300	4460	4620	4790
(1550)	2	0.15	3190	3320	3440	3570	3700	3840	3980	4130	4280	4440	4590
	3	0.29	3080	3190	3300	3420	3550	3670	3810	3940	4080	4230	4370
	4	0.44	2940	3040	3150	3260	3370	3480	3600	3730	3860	3990	4120
	BFL	0.59	2780	2870	2970	3060	3160	3270	3370	3480	3600	3710	3830
	6	3.47	2540	2620	2690	2770	2850	2940	3020	3110	3200	3290	3390
	7	6.58	2250	2300	2350	2410	2470	2530	2590	2650	2710	2780	2850
W 24×55	TFL	0.00	2890	3000	3120	3240	3370	3500	3630	3770	3910	4060	4210
(1350)	2	0.13	2800	2910	3020	3130	3250	3370	3500	3630	3760	3900	4040
	3	0.25	2700	2800	2900	3010	3120	3230	3350	3470	3600	3730	3860
	4	0.38	2590	2680	2770	2870	2970	3080	3190	3300	3410	3530	3650
	BFL	0.51	2460	2540	2630	2710	2800	2900	2990	3090	3200	3300	3410
	6	3.45	2240	2310	2370	2440	2520	2590	2670	2750	2830	2920	3000
	7	6.66	1970	2010	2060	2110	2160	2210	2260	2320	2370	2430	2490
W 21×62	TFL	0.00	2760	2880	3000	3120	3250	3390	3530	3670	3820	3970	4130
(1330)	2	0.15	2650	2760	2870	2990	3110	3230	3360	3500	3630	3780	3920
	3	0.31	2530	2630	2730	2830	2940	3060	3170	3290	3420	3550	3680
	4	0.46	2380	2470	2560	2650	2750	2850	2950	3060	3170	3280	3400
	BFL	0.62	2210	2280	2360	2440	2520	2600	2690	2770	2870	2960	3060
	6	2.53	2070	2130	2190	2260	2320	2390	2460	2540	2620	2690	2780
	7	4.78	1900	1950	2000	2050	2100	2150	2210	2270	2330	2390	2450
W 21×57	TFL	0.00	2480	2590	2700	2810	2930	3060	3180	3320	3450	3590	3740
(1170)	2	0.16	2390	2490	2590	2700	2810	2930	3050	3170	3300	3430	3560
	3	0.33	2290	2380	2480	2570	2680	2780	2890	3000	3120	3240	3360
	4	0.49	2170	2250	2340	2420	2520	2610	2710	2810	2910	3020	3130
	BFL	0.65	2030	2100	2170	2250	2330	2410	2490	2580	2670	2760	2860
	6	2.90	1880	1940	2000	2060	2120	2190	2260	2330	2400	2480	2560
	7	5.38	1690	1740	1780	1830	1880	1920	1980	2030	2080	2140	2200
W 21×50	TFL	0.00	2120	2210	2310	2410	2510	2620	2730	2850	2960	3090	3210
(984)	2	0.13	2050	2130	2220	2320	2410	2520	2620	2730	2840	2950	3070
	3	0.27	1960	2040	2130	2220	2310	2400	2500	2590	2700	2800	2910
	4	0.40	1870	1940	2020	2100	2180	2260	2350	2440	2530	2630	2730
	BFL	0.54	1760	1830	1890	1960	2040	2110	2190	2270	2350	2430	2520
	6	2.92	1620	1670	1720	1780	1840	1900	1960	2020	2090	2160	2230
	7	5.58	1440	1470	1510	1550	1590	1640	1680	1730	1780	1830	1880

[a] Y1 = distance from top of the steel beam to plastic neutral axis.
[b] Y2 = distance from top of the steel beam to concrete flange force.
[c] See Figure 5-3 for PNA locations.
[d] Value in parentheses is I_x (in.^{4}) of non-composite steel shape.

LOWER BOUND
ELASTIC MOMENT OF INERTIA
FOR PLASTIC COMPOSITE SECTIONS

I LB

Shape[d]	PNA[c]	Y1[a]	I_{LB} (in.⁴)										
			Y2[b] (in.)										
		In.	2	2.5	3	3.5	4	4.5	5	5.5	6	6.5	7
W 21×44	TFL	0.00	1830	1910	2000	2090	2180	2270	2370	2470	2580	2680	2800
(843)	2	0.11	1770	1850	1930	2010	2100	2190	2280	2370	2470	2570	2680
	3	0.23	1710	1780	1850	1930	2010	2090	2180	2270	2360	2450	2550
	4	0.34	1630	1700	1760	1830	1910	1980	2060	2140	2220	2310	2400
	BFL	0.45	1540	1600	1660	1730	1790	1860	1930	2000	2070	2150	2230
	6	2.90	1410	1450	1500	1550	1610	1660	1720	1770	1830	1900	1960
	7	5.69	1240	1270	1300	1340	1380	1410	1450	1490	1540	1580	1620
W 18×60	TFL	0.00	2070	2170	2280	2390	2500	2620	2740	2860	3000	3130	3270
(984)	2	0.17	1980	2080	2170	2270	2380	2480	2600	2710	2830	2960	3090
	3	0.35	1880	1960	2050	2140	2230	2330	2430	2540	2640	2750	2870
	4	0.52	1760	1830	1900	1980	2060	2150	2230	2320	2420	2510	2610
	BFL	0.70	1610	1670	1730	1790	1850	1920	1990	2070	2140	2220	2300
	6	2.19	1520	1570	1620	1670	1730	1790	1850	1910	1970	2040	2110
	7	3.82	1420	1460	1500	1540	1590	1640	1690	1740	1790	1840	1900
W 18×55	TFL	0.00	1880	1970	2070	2170	2270	2380	2490	2610	2730	2850	2980
(890)	2	0.16	1800	1890	1970	2070	2160	2260	2360	2470	2580	2700	2810
	3	0.32	1710	1790	1870	1950	2030	2120	2220	2310	2410	2510	2620
	4	0.47	1600	1670	1740	1810	1880	1960	2040	2120	2210	2300	2390
	BFL	0.63	1470	1520	1580	1640	1700	1760	1830	1900	1970	2040	2110
	6	2.16	1380	1430	1480	1530	1580	1630	1690	1750	1810	1870	1930
	7	3.86	1290	1320	1360	1400	1440	1490	1530	1580	1620	1670	1730
W 18×50	TFL	0.00	1690	1770	1860	1950	2040	2140	2240	2340	2450	2560	2680
(800)	2	0.14	1620	1700	1770	1860	1940	2030	2130	2220	2320	2430	2530
	3	0.29	1540	1610	1680	1750	1830	1910	1990	2080	2170	2260	2360
	4	0.43	1440	1500	1560	1630	1700	1770	1840	1910	1990	2070	2160
	BFL	0.57	1320	1370	1420	1480	1530	1590	1650	1710	1780	1840	1910
	6	2.07	1250	1290	1330	1380	1420	1470	1520	1570	1630	1680	1740
	7	3.82	1160	1190	1220	1260	1300	1340	1380	1420	1460	1510	1550
W 18×46	TFL	0.00	1530	1610	1690	1770	1860	1950	2040	2140	2240	2340	2450
(712)	2	0.15	1470	1540	1620	1690	1770	1860	1940	2030	2130	2220	2320
	3	0.30	1400	1470	1540	1610	1680	1750	1830	1910	2000	2080	2170
	4	0.45	1320	1380	1440	1500	1560	1630	1700	1770	1850	1920	2000
	BFL	0.61	1230	1270	1320	1380	1430	1490	1550	1610	1670	1730	1800
	6	2.40	1140	1180	1220	1270	1310	1360	1410	1460	1510	1560	1620
	7	4.34	1040	1070	1100	1140	1170	1210	1240	1280	1320	1360	1410

[a] Y1 = distance from top of the steel beam to plastic neutral axis.
[b] Y2 = distance from top of the steel beam to concrete flange force.
[c] See Figure 5-3 for PNA locations.
[d] Value in parentheses is I_x (in.⁴) of non-composite steel shape.

LOWER BOUND
ELASTIC MOMENT OF INERTIA
FOR PLASTIC COMPOSITE SECTIONS

I_{LB}

Shape[d]	PNA[c]	Y1[a]	I_{LB} (in.[4])										
			$Y2^{[b]}$ (in.)										
		In.	2	2.5	3	3.5	4	4.5	5	5.5	6	6.5	7
W 18×40	TFL	0.00	1320	1390	1450	1530	1600	1680	1760	1840	1930	2020	2110
(612)	2	0.13	1270	1330	1390	1460	1530	1600	1680	1760	1840	1920	2010
	3	0.26	1210	1270	1320	1390	1450	1510	1580	1650	1730	1800	1880
	4	0.39	1140	1190	1240	1300	1350	1410	1470	1530	1600	1670	1740
	BFL	0.53	1060	1100	1150	1190	1240	1290	1340	1390	1450	1510	1560
	6	2.26	985	1020	1060	1090	1130	1170	1220	1260	1310	1350	1400
	7	4.27	895	921	949	978	1010	1040	1070	1100	1140	1180	1210
W 18×35	TFL	0.00	1120	1170	1230	1300	1360	1430	1500	1570	1650	1720	1800
(510)	2	0.11	1080	1130	1190	1240	1300	1370	1430	1500	1570	1640	1720
	3	0.21	1030	1080	1130	1180	1240	1300	1360	1420	1490	1550	1620
	4	0.32	978	1020	1070	1120	1170	1220	1270	1330	1390	1450	1510
	BFL	0.43	917	955	995	1040	1080	1130	1170	1220	1270	1320	1380
	6	2.37	842	874	906	940	976	1010	1050	1090	1130	1170	1220
	7	4.56	753	775	799	824	850	877	905	934	964	995	1030
W 16×36	TFL	0.00	971	1020	1080	1140	1200	1270	1330	1400	1480	1550	1630
(448)	2	0.11	931	981	1030	1090	1140	1200	1270	1330	1400	1470	1540
	3	0.22	884	929	977	1030	1080	1130	1190	1250	1310	1370	1430
	4	0.32	830	869	910	954	999	1050	1090	1150	1200	1250	1310
	BFL	0.43	764	797	831	867	904	943	984	1030	1070	1120	1160
	6	1.79	714	742	770	801	832	865	899	935	972	1010	1050
	7	3.44	657	679	701	725	750	776	802	830	859	889	921
W 16×31	TFL	0.00	826	872	921	972	1030	1080	1140	1200	1260	1330	1390
(375)	2	0.11	793	837	882	929	979	1030	1080	1140	1200	1260	1320
	3	0.22	756	796	837	880	925	972	1020	1070	1120	1180	1240
	4	0.33	713	748	784	823	863	904	948	993	1040	1090	1140
	BFL	0.44	662	691	722	755	789	824	861	899	939	980	1020
	6	2.00	613	637	663	690	718	747	778	810	843	877	912
	7	3.79	555	574	593	614	635	657	680	704	729	755	782
W 16×26	TFL	0.00	673	712	753	795	840	886	935	985	1040	1090	1150
(301)	2	0.09	649	685	723	763	804	848	893	940	989	1040	1090
	3	0.17	621	654	689	726	764	804	845	888	933	980	1030
	4	0.26	589	618	650	683	717	753	790	829	870	911	955
	BFL	0.35	551	577	604	633	663	694	727	760	796	832	870
	6	2.04	505	526	549	572	597	622	649	676	705	734	765
	7	4.01	450	465	482	499	517	535	554	575	595	617	639

[a] Y1 = distance from top of the steel beam to plastic neutral axis.
[b] Y2 = distance from top of the steel beam to concrete flange force.
[c] See Figure 5-3 for PNA locations.
[d] Value in parentheses is I_x (in.[4]) of non-composite steel shape.

LOWER BOUND
ELASTIC MOMENT OF INERTIA
FOR PLASTIC COMPOSITE SECTIONS

I $_{LB}$

Shape[d]	PNA[c]	Y1[a]	I_{LB} (in.^{4})										
			$Y2^b$ (in.)										
		In.	2	2.5	3	3.5	4	4.5	5	5.5	6	6.5	7
W 14×38	TFL	0.00	844	896	951	1010	1070	1130	1200	1270	1340	1410	1490
(385)	2	0.13	805	853	903	956	1010	1070	1130	1190	1260	1330	1400
	3	0.26	759	802	846	893	943	994	1050	1100	1160	1220	1290
	4	0.39	704	740	778	818	861	905	950	998	1050	1100	1150
	BFL	0.52	636	665	695	727	760	794	831	868	908	948	991
	6	1.38	604	629	655	683	712	742	773	806	840	876	913
	7	2.53	568	589	611	634	659	684	710	738	766	796	827
W 14×34	TFL	0.00	744	790	839	890	944	1000	1060	1120	1180	1250	1320
(340)	2	0.11	711	753	798	844	894	945	999	1060	1110	1170	1240
	3	0.23	671	709	749	790	834	880	929	979	1030	1080	1140
	4	0.34	623	656	690	726	763	803	844	887	931	978	1030
	BFL	0.46	565	591	618	647	677	708	741	775	810	847	885
	6	1.41	535	557	581	606	631	659	687	716	747	779	812
	7	2.60	502	520	540	560	582	604	628	652	677	704	731
W 14×30	TFL	0.00	643	684	726	771	819	868	920	974	1030	1090	1150
(291)	2	0.10	615	653	692	734	777	823	870	920	971	1020	1080
	3	0.19	583	616	652	689	728	769	812	857	903	951	1000
	4	0.29	544	573	604	636	670	706	743	782	822	864	907
	BFL	0.39	497	521	546	573	600	629	659	691	724	758	793
	6	1.48	467	487	508	531	554	579	605	631	659	688	719
	7	2.82	432	448	465	483	502	522	542	564	586	610	634
W 14×26	TFL	0.00	553	589	626	665	706	750	795	841	890	941	994
(245)	2	0.11	531	563	598	634	672	712	754	798	843	890	939
	3	0.21	504	534	565	598	633	669	707	747	788	830	875
	4	0.32	473	500	527	556	587	619	652	687	723	761	800
	BFL	0.42	437	459	482	506	532	559	587	616	646	678	711
	6	1.67	405	423	443	463	485	507	530	555	580	606	634
	7	3.19	368	382	397	413	430	447	465	484	503	523	545
W 14×22	TFL	0.00	454	484	515	548	582	619	656	696	736	779	823
(199)	2	0.08	437	464	493	524	556	590	625	661	699	739	780
	3	0.17	416	442	468	496	526	556	588	622	657	693	731
	4	0.25	393	416	439	464	490	518	546	576	607	640	673
	BFL	0.34	365	385	405	427	449	473	497	523	550	577	606
	6	1.69	336	352	369	386	405	424	444	466	488	510	534
	7	3.34	301	313	325	339	352	367	382	398	414	431	449

[a] Y1 = distance from top of the steel beam to plastic neutral axis.
[b] Y2 = distance from top of the steel beam to concrete flange force.
[c] See Figure 5-3 for PNA locations.
[d] Value in parentheses is I_x (in.^{4}) of non-composite steel shape.

I$_{LB}$

LOWER BOUND
ELASTIC MOMENT OF INERTIA
FOR PLASTIC COMPOSITE SECTIONS

Shape[d]	PNA[c]	Y1[a]	I_{LB} (in.⁴)										
			Y2[b] (in.)										
		In.	2	2.5	3	3.5	4	4.5	5	5.5	6	6.5	7
W 12×30	TFL	0.00	531	568	608	649	693	738	786	837	889	944	1000
(238)	2	0.11	505	539	575	612	652	694	738	783	831	881	933
	3	0.22	474	504	536	569	604	641	679	720	762	806	852
	4	0.33	436	461	488	516	545	576	609	643	678	715	753
	BFL	0.44	389	408	429	450	472	496	521	547	574	602	631
	6	1.12	373	390	408	427	447	468	490	513	537	562	589
	7	1.94	355	370	386	402	420	438	457	477	498	520	543
W 12×26	TFL	0.00	456	488	521	557	595	635	676	720	765	812	861
(204)	2	0.10	434	463	494	526	561	597	635	674	716	759	804
	3	0.19	407	433	460	489	520	552	585	621	657	695	735
	4	0.29	375	397	420	445	470	497	526	555	586	618	652
	BFL	0.38	336	353	370	389	409	430	452	474	498	523	549
	6	1.08	321	336	351	368	386	404	423	444	465	487	509
	7	1.95	305	317	331	345	360	376	393	410	428	447	467
W 12×22	TFL	0.00	371	399	428	458	490	524	559	596	635	675	717
(156)	2	0.11	356	382	408	437	466	498	531	565	601	638	677
	3	0.21	339	362	386	412	439	468	498	529	562	596	631
	4	0.32	318	339	360	383	408	433	459	487	516	547	578
	BFL	0.43	294	312	330	350	370	392	414	438	463	488	515
	6	1.66	270	285	300	316	333	351	370	389	410	431	453
	7	3.04	242	253	265	277	290	303	317	332	347	364	380
W 12×19	TFL	0.00	312	335	360	386	413	442	472	503	536	571	606
(130)	2	0.09	300	321	344	368	394	421	449	478	509	541	574
	3	0.18	286	306	327	349	372	397	423	450	478	507	538
	4	0.26	270	287	306	326	347	369	392	417	442	468	496
	BFL	0.35	251	266	282	300	318	337	357	378	400	423	446
	6	1.66	229	241	255	269	284	299	316	333	351	370	389
	7	3.12	203	212	222	232	243	255	267	279	293	306	321
W 12×16	TFL	0.00	254	273	294	315	338	362	388	414	442	471	501
(103)	2	0.07	245	263	282	303	324	347	371	396	422	449	477
	3	0.13	234	251	269	288	309	330	352	375	399	424	450
	4	0.20	223	239	255	272	291	310	330	351	373	396	420
	BFL	0.26	210	224	238	254	270	287	305	324	344	364	386
	6	1.71	189	200	212	224	238	251	266	281	297	313	330
	7	3.32	163	171	179	188	197	207	217	227	239	250	262

[a] Y1 = distance from top of the steel beam to plastic neutral axis.
[b] Y2 = distance from top of the steel beam to concrete flange force.
[c] See Figure 5-3 for PNA locations.
[d] Value in parentheses is I_x (in.⁴) of non-composite steel shape.

LOWER BOUND ELASTIC MOMENT OF INERTIA FOR PLASTIC COMPOSITE SECTIONS

LB

| Shape[d] | PNA[c] | Y1[a] | I_{LB} (in.^{4}) | | | | | | | | | | |
| | | | Y2[b] (in.) | | | | | | | | | | |
		In.	2	2.5	3	3.5	4	4.5	5	5.5	6	6.5	7
W 12×14	TFL	0.00	220	237	255	275	295	316	338	362	386	411	438
(88.6)	2	0.06	213	229	246	264	283	303	324	346	369	393	418
	3	0.11	204	219	235	252	270	289	308	329	350	372	396
	4	0.17	195	209	223	239	255	272	290	309	329	349	371
	BFL	0.23	184	197	210	224	238	254	270	287	305	323	342
	6	1.69	165	175	186	197	209	221	234	247	262	276	292
	7	3.36	141	148	155	163	171	180	188	198	208	218	228
W 10×26	TFL	0.00	339	368	398	430	464	499	537	577	618	662	707
(144)	2	0.11	322	347	375	404	435	467	501	537	575	615	656
	3	0.22	300	323	347	372	400	428	458	490	523	558	594
	4	0.33	274	293	313	334	357	381	406	432	460	489	519
	BFL	0.44	242	256	271	287	304	321	340	360	381	403	425
	6	0.90	232	245	258	273	288	304	321	339	358	378	398
	7	1.51	222	233	245	258	272	286	301	317	334	351	369
W 10×22	TFL	0.00	281	305	330	357	386	416	448	482	517	554	592
(118)	2	0.09	267	289	312	336	363	390	419	450	482	516	551
	3	0.18	250	269	290	312	335	360	385	413	441	471	502
	4	0.27	230	246	263	282	302	322	344	367	391	417	443
	BFL	0.36	205	217	231	245	260	277	293	311	330	350	370
	6	0.95	194	205	217	230	244	258	273	288	305	322	340
	7	1.70	183	193	203	214	225	237	250	263	277	292	308
W 10×19	TFL	0.00	239	259	282	305	330	356	384	413	444	476	509
(96.3)	2	0.10	228	247	267	289	312	337	362	389	417	447	478
	3	0.20	215	233	251	271	292	314	337	361	387	413	441
	4	0.30	200	216	232	249	267	286	307	328	350	374	398
	BFL	0.40	183	195	209	223	238	254	271	288	307	326	347
	6	1.27	169	180	191	203	216	229	243	258	274	290	307
	7	2.31	153	162	170	180	190	200	211	223	235	248	261
W 10×17	TFL	0.00	206	224	244	265	286	310	334	360	387	415	444
(81.9)	2	0.08	197	214	232	252	272	294	316	340	365	391	419
	3	0.17	187	203	219	237	255	275	296	317	340	364	389
	4	0.25	175	189	204	219	236	253	272	291	311	332	354
	BFL	0.33	161	173	185	199	213	227	243	260	277	295	314
	6	1.31	148	157	168	179	190	202	215	229	243	258	274
	7	2.46	132	139	147	155	164	173	183	193	204	215	227

[a] Y1 = distance from top of the steel beam to plastic neutral axis.
[b] Y2 = distance from top of the steel beam to concrete flange force.
[c] See Figure 5-3 for PNA locations.
[d] Value in parentheses is I_x (in.^{4}) of non-composite steel shape.

LOWER BOUND
ELASTIC MOMENT OF INERTIA
FOR PLASTIC COMPOSITE SECTIONS

Shape[d]	PNA[c]	Y1[a]	I_{LB} (in.^{4})										
			$Y2^b$ (in.)										
		In.	2	2.5	3	3.5	4	4.5	5	5.5	6	6.5	7
W 10×15	TFL	0.00	177	193	210	228	247	268	289	312	335	360	386
(68.9)	2	0.07	170	185	201	218	236	255	275	296	318	341	365
	3	0.14	162	175	190	206	222	240	258	278	298	320	342
	4	0.20	153	165	178	192	207	223	240	257	275	295	315
	BFL	0.27	142	153	164	176	189	203	218	233	249	266	283
	6	1.35	128	137	147	157	167	178	190	203	216	229	244
	7	2.60	112	118	125	133	140	148	157	166	176	185	196
W 10×12	TFL	0.00	139	152	165	180	195	211	229	247	265	285	306
(53.8)	2	0.05	134	146	158	172	186	202	218	235	252	271	290
	3	0.11	128	139	150	163	176	190	205	221	237	254	272
	4	0.16	121	131	141	153	165	178	191	205	220	236	252
	BFL	0.21	113	122	131	141	152	163	175	187	200	214	229
	6	1.30	102	109	116	124	133	142	152	162	173	184	195
	7	2.61	87.9	92.9	98.4	104	110	117	124	131	138	146	155
W 8×28	TFL	0.00	248	274	302	332	364	398	434	473	513	555	600
(98)	2	0.12	233	256	281	308	337	368	400	435	471	509	549
	3	0.23	214	234	256	279	304	330	358	388	419	452	487
	4	0.35	191	207	224	243	262	284	306	330	355	381	408
	BFL	0.47	161	171	183	196	209	223	238	254	271	289	307
	6	0.53	159	170	181	194	207	221	235	251	268	285	303
	7	0.59	158	168	180	192	204	218	233	248	264	281	299
W 8×24	TFL	0.00	209	231	255	280	307	336	367	400	434	470	508
(82.8)	2	0.10	196	216	237	260	285	311	339	368	399	431	465
	3	0.20	180	198	216	236	257	279	303	329	355	383	413
	4	0.30	161	175	189	205	222	240	259	280	301	323	347
	BFL	0.40	136	145	155	166	177	189	202	216	231	246	262
	6	0.47	134	143	153	164	175	187	200	213	227	242	257
	7	0.55	133	142	151	162	173	184	197	210	223	238	253
W 8×21	TFL	0.00	191	211	232	255	279	305	333	362	392	424	458
(75.3)	2	0.10	181	198	218	238	260	284	309	335	362	391	422
	3	0.20	167	183	200	218	237	258	279	302	327	352	379
	4	0.30	151	164	178	193	209	226	244	263	283	304	326
	BFL	0.40	131	140	151	162	173	186	199	213	227	243	259
	6	0.70	126	135	145	155	165	177	189	201	215	229	244
	7	1.06	122	130	138	147	157	167	178	190	202	215	228

[a] Y1 = distance from top of the steel beam to plastic neutral axis.
[b] Y2 = distance from top of the steel beam to concrete flange force.
[c] See Figure 5-3 for PNA locations.
[d] Value in parentheses is I_x (in.^{4}) of non-composite steel shape.

LOWER BOUND
ELASTIC MOMENT OF INERTIA
FOR PLASTIC COMPOSITE SECTIONS

 LB

Shape[d]	PNA[c]	Y1[a]	I_{LB} (in.[4])										
			$Y2^b$ (in.)										
		In.	2	2.5	3	3.5	4	4.5	5	5.5	6	6.5	7
W 8×18	TFL	0.00	159	175	193	213	233	255	278	303	329	356	384
(61.9)	2	0.08	150	165	182	199	218	238	259	281	305	329	355
	3	0.17	140	153	167	183	199	217	236	255	276	298	321
	4	0.25	127	138	150	163	177	192	207	224	241	259	278
	BFL	0.33	111	120	129	139	149	160	172	184	198	211	226
	6	0.71	106	114	122	131	140	150	161	172	184	196	209
	7	1.21	101	107	114	122	130	139	148	158	169	179	191
W 8×15	TFL	0.00	129	143	158	175	192	210	230	251	272	295	319
(48)	2	0.08	123	136	150	165	181	198	216	235	255	276	299
	3	0.16	116	128	140	154	168	183	200	217	235	254	274
	4	0.24	107	117	128	140	153	166	181	196	211	228	246
	BFL	0.32	97.0	105	114	124	135	146	158	170	183	197	211
	6	0.97	89.3	96.4	104	112	121	130	140	151	162	174	186
	7	1.79	80.6	86.2	92.2	98.7	106	113	121	129	138	147	157
W 8×13	TFL	0.00	109	121	134	147	162	178	195	213	231	251	272
(39.6)	2	0.06	104	115	127	140	154	168	184	200	218	236	255
	3	0.13	98.0	108	119	131	144	157	171	186	202	219	236
	4	0.19	91.4	100	110	121	132	144	156	170	184	198	214
	BFL	0.26	83.6	91.3	99.6	108	118	128	139	150	162	175	188
	6	1.00	76.1	82.4	89.3	96.6	104	113	122	131	141	151	162
	7	1.91	67.2	72.0	77.2	82.7	88.7	95.0	102	109	116	124	132
W 8×10	TFL	0.00	83.1	92.3	102	113	124	136	149	163	177	192	208
(30.8)	2	0.05	79.3	87.8	97.0	107	117	129	141	153	166	180	195
	3	0.10	74.8	82.6	90.9	99.9	109	120	131	142	154	167	180
	4	0.15	69.6	76.5	83.8	91.7	100	109	119	129	140	151	162
	BFL	0.21	63.5	69.2	75.4	82.0	89.2	96.7	105	113	122	132	142
	6	0.88	58.0	62.8	68.0	73.5	79.5	85.8	92.5	99.6	107	115	123
	7	1.77	51.7	55.4	59.4	63.6	68.2	73.0	78.2	83.6	89.4	95.4	102

[a] Y1 = distance from top of the steel beam to plastic neutral axis.
[b] Y2 = distance from top of the steel beam to concrete flange force.
[c] See Figure 5-3 for PNA locations.
[d] Value in parentheses is I_x (in.[4]) of non-composite steel shape.

COMPOSITE COLUMNS

General Notes

Load tables for composite columns are presented for a variety of W shapes, pipe, and structural tubing. Tabular loads have been computed in accordance with Section I2.2 of the LRFD Specification for axially loaded members having effective unsupported lengths indicated at the left of each table. The effective length KL is the actual unbraced length, in feet, multiplied by factor K, which depends on the rotational restraint at the ends of the unbraced length and the means available to resist lateral movements. The K factor may be selected using as a guide Table C-C2.1 in the Commentary of the LRFD Specification. Interpolation between the idealized cases is a matter of engineering judgment. More precise values for K may be obtained, if desired, from the alignment chart in Figure C-C2.2 in the LRFD Commentary (also shown in Part 3 of this Manual under Design Strength of Columns) once sections have been selected for several framing members.

Load tables are provided for W shapes encased in reinforced normal-weight concrete of square or rectangular cross section, and for steel pipe and structural tubing filled with normal-weight concrete. The following W shapes are included:

Nominal depth, in.	Weight, lb per ft
14	426-61
12	336-58
10	112-45
8	67-35

Two values of yield stress, F_y equal to 36 and 50 ksi, and three values of concrete cylinder strength, f_c' equal to 3.5, 5, and 8 ksi, are included for W shapes. All reinforcing steel is Grade 60. The tables for steel pipe columns include nominal pipe diameters of 4 to 12 in., yield stress $F_y = 36$ ksi, and concrete cylinder strength f_c' equal to 3.5 and 5 ksi. The tables for tubular columns include tubes of nominal side dimensions 4 to 16 in., yield stress F_y equal to 46 ksi, and concrete cylinder strength f_c' equal to 3.5 and 5 ksi.

All axial design strengths are tabulated in kips. Strength values are omitted when Kl/r exceeds 200. Resistance factor $\phi = 0.85$ was used in computing the axial design strengths of all composite columns.

In all tables, the design strengths are given for effective lengths with respect to the minor axis. When the minor axis is braced at closer intervals than the major axis, the strength of the column must be investigated with reference to both the major (X-X) and minor (Y-Y) axes. The ratio r_{mx}/r_{my} included in the tables provides a convenient method for investigating the strength of a column with respect to its major axis.

Properties useful to the designer are listed at the bottom of the Column Load Tables. They are helpful in considering buckling about the major axis as discussed above and in the design of members under combined axial compression and bending as discussed subsequently. Both of these cases are illustrated with design examples. The properties have the following units:

modified radius of gyration r_m, in.
nominal flexural strength M_n, kip-ft
Euler buckling term $P_e (KL)^2$, kip-ft^2

Subscripts x and y refer to the major and minor axes. Resistance factor $\phi_b = 0.9$ was used in computing the flexural design strength $\phi_b M_n$.

Additional notes relating specifically to the W shapes, steel pipe, and structural tube column tables precede each of these groups of tables.

EXAMPLE 5-4

Given: Using tables on the pages that follow, design the smallest composite column with a W shape of $F_y = 50$ ksi to support a factored concentric load of 1,000 kips. The effective length with respect to its minor axis is 16 feet and that with respect to its major axis is 31 feet.

Solution: Use composite column tables for W shapes of $f_c' = 8$ ksi since the strongest concrete requires the smallest size column. An inspection of the tables reveals that the tabulated values of r_{mx} / r_{my} do not exceed 1.22 and in most cases are equal to 1.0 (square columns).

Assuming that r_{mx} / r_{my} is equal to 1.0, enter the tables with equivalent effective column length of $KL = 31$ ft. Select an 18 in. × 18 in. column with a W10×49 and $r_{mx} / r_{my} = 1.0$. By interpolation, the column has an axial design strength ϕP_n of 1,029 kips.

Use: 18 in. × 18 in. column with W10×49 of $F_y = 50$ ksi, $f_c' = 8$ ksi, four #8 Gr. 60 longitudinal bars and #3 Gr. 60 ties spaced 12 inches on center.

EXAMPLE 5-5

Given: Redesign the column from Example 5-4 using (a) rectangular and (b) square structural tubing, both filled with structural concrete.

Solution: a. Enter the Composite Column Tables for rectangular structural tubing filled with 5 ksi concrete at an effective column length of $KL = 16$ ft.

Select 14×10 tubing with ½-in. thick walls; $\phi P_n = 1,090$ kips.

$r_{mx} / r_{my} = 1.30$

Equivalent effective length for X-X axis:

$31 / 1.30 = 23.8$ ft

Since 23.8 ft > 16 ft, X-X axis controls.

Re-enter the table at an effective length of $KL = 23.8$ ft. It is apparent that 14×10 tube will not satisfy the axial load of 1,000 kips. Select 16×12 steel tube with ½-in. walls:

$r_{mx} / r_{my} = 1.25$
$KL \qquad = 31 / 1.25 = 24.8$ ft

By interpolation, the column provides an axial design strength of 1,206 kips. The same tubing filled with 3.5 ksi concrete is good for 1,094 kips > 1,000 kips.

Use: 16-in. × 12-in. × ½-in. tubing filled with 3.5-ksi concrete.

b. Enter the Composite Column Tables for square structural tubing filled with 5 ksi concrete at effective column length $KL = 31$ ft.

Select 14×14 tubing with ½-in. thick walls good for 1,135 kips. The same tubing filled with 3.5 ksi concrete is good for 1,035 kips > 1,000 kips.

Use: 14-in. × 14-in. × ½-in. tubing filled with 3.5-ksi concrete.

Note: The weight of both the 16×12 and 14×14 steel tubing with ½-in. thick walls is 89.68 lb. per ft.

Combined Axial Compression and Bending (Interaction)
Loads given in the Composite Column Tables are for concentrically loaded columns. For columns subjected to combined compression and flexure, the nominal flexural design strength $\phi_b M_n$ determined from Equation C-I4-1 of the Commentary on the LRFD Specification (only valid for $P_u / \phi_c P_n > 0.3$) and the elastic buckling load P_e times the square of the effective column length are given at the bottom of the tables. With these quantities and the loads tabulated for concentric loading, the column may be designed by successive approximations based on LRFD Specification Section I4. The procedure is illustrated in Example 5-6 for a column subjected to a large bending moment combined with a moderate axial load and in Example 5-7 for a column subjected to a large axial load combined with a moderate bending moment.

EXAMPLE 5-6

Given: Design a composite encased W shape column to resist a factored axial load of 350 kips and a factored moment about the X-X axis of 240 kip-ft. The unsupported length of the column is 12 feet, $F_y = 50$ ksi, $f_c' = 3.5$ ksi and $C_m = 1.0$. The loads were obtained by first order elastic analysis and there is no lateral translation of column ends.

Solution: Since the moment is large in relation to the axial load, assume that $P_u / \phi P_n = \frac{1}{2}$ and $B_1 = 1.0$. From Equation H1-1a

$$\phi_b M_{nx} = (8 / 9) \times 240 \times 2 = 427 \text{ kip-ft}$$

From the Composite Column Tables, find a column with $\phi_b M_{nx}$ close to 427 kip-ft.

Try a W10×60 with 18-in. encasement: $\phi_b M_{nx} = 439$ kip-ft, $\phi P_n = 1,300$ kips and $P_{ex} = 142 \times 10^4 / 12^2 = 9,861$ kips. Therefore,

$$P_u / \phi P_n = 350 / 1,300 = 0.2692 \geq 0.2$$

From Equation C1-1 and C1-2, with $M_{lt} = 0$ since there is no lateral translation,

$$M_{ux} = \frac{240}{1 - \frac{350}{9,861}} = 248.8 \text{ kip-ft}$$

and from Equation H1-1a

$$\phi_b M_{nx} = \frac{8(248.8)}{9\left(1 - \frac{350}{1,300}\right)} = 302.6 \text{ kip-ft}$$

Since 302.6 kip-ft is much less than the 439 kip-ft provided, select a smaller size. Try a W8×58 with 16-in. × 16-in. encasement: $\phi_b M_{nx}$ = 345 kip-ft, ϕP_n = 1,130 kips and P_{ex} = 103 × 10⁴ / 12² = 7,153 kips.

$$M_{ux} = \frac{240}{1 - \frac{350}{7,153}} = 252.4 \text{ kip-ft}$$

From Equation H1-1a, with $\phi_b M_{nx}$ = 345 kip-ft since $\phi P_n / P_u > 0.3$,

$$\frac{350}{1,130} + \frac{8(252.3)}{9(345)} = 0.960 < 1.0 \quad \textbf{o.k.}$$

Use: 16-in. × 16-in. column with W8×58 of F_y = 50 ksi, f_c' = 3.5 ksi, four #7 Gr. 60 longitudinal bars and #3 Gr. 60 ties spaced at 10 inches.

EXAMPLE 5-7

Given:

Design a composite encased W shape column to resist a factored axial load of 1,100 kips and factored moment of 200 kip-ft. Use 50 ksi structural steel and 5 ksi concrete. The unsupported column length is 11 feet and C_m = 0.85. Assume that sidesway is prevented.

Solution:

Since the axial load is large in relation to the moment, assume that:

$$\frac{8}{9}\frac{M_u}{\phi_b M_{nx}} = 0.5$$

From Equation H1-1a

ϕP_n = 1,100 / 0.5 = 2,200 kips

From the Composite Column Tables, find a column with ϕP_n close to 2,200 kips at KL = 11 ft.

Try a W12×106 with 20-in. encasement: ϕP_n = 2,270 kips, $\phi_b M_{nx}$ = 899 kip-ft and P_{ex} = 294 × 10⁴ / 11² = 24,300 kips.

From Equation C1-2

$$B_1 = \frac{0.85}{1 - \dfrac{1,100}{24,300}} = 0.890 < 1.0$$

Therefore, $B_1 = 1.0$. From Equation H1-1a

$$\frac{P_u}{\phi P_n} = 1 - \frac{8}{9} \times \frac{200}{899} = 0.802$$

and

$$\phi P_n = 1,100 / 0.802 = 1,372 \text{ kips}$$

Try a W10×45 with 18-in. × 18-in. encasement: $\phi P_n = 1,360$ kips, $\phi_b M_{nx} = 356$ kip-ft and $P_{ex} = 125 \times 10^4 / 11^2 = 10,330$ kips.

From Equation H1-3

$$B_1 = \frac{0.85}{1 - \dfrac{1,100}{10,300}} = 0.951 < 1.0$$

Therefore, $B_1 = 1.0$ and

$$\frac{P_u}{\phi P_n} = 1 - \frac{8}{9} \times \frac{200}{356} = 0.501$$

and

$$\phi P_n = 1,100 / 0.501 = 2,196 \text{ kips}$$

Since the convergence is slow, estimate a new ϕP_n as:

$$\phi P_n = \frac{2,196 + 1,372}{2} = 1,784 \text{ kips}$$

Try a W10×77 with 18-in. × 18-in. encasement: $\phi P_n = 1,720$ kips, $\phi_b M_{nx} = 555$ kip-ft and $P_{ex} = 178 \times 10^4 / 11^2 = 14,710$ kips.

$$B_1 = \frac{0.85}{1 - \dfrac{1,100}{14,710}} = 0.919 < 1.0$$

and

$$\frac{1,100}{1,720} + \frac{8}{9} \times \frac{200}{555} = 0.960 < 1.0 \quad \textbf{o.k.}$$

Use: 18-in. × 18-in. column with W10×77 of $F_y = 50$ ksi, $f_c' = 5$ ksi, four #8 Gr. 60 longitudinal bars and #3 Gr. 60 ties spaced at 12 inches.

COMPOSITE COLUMNS—W SHAPES ENCASED IN CONCRETE

General Notes

Concentric load design strengths in the tables that follow are tabulated for the effective length KL in feet, listed at the left of each table. They are applicable to axially loaded members with respect to their minor axis in accordance with Section I2.2 of the LRFD Specification. Two steel yield stresses, $F_y = 36$ ksi and $F_y = 50$ ksi, and three concrete strengths, $f_c' = 3.5$ ksi, $f_c' = 5$ ksi, and $f_c' = 8$ ksi, are covered. The tables apply to normal-weight concrete. All reinforcing steel is Gr. 60; however, $F_{yr} = 55$ ksi is used in the calculation of ϕP_n in accordance with LRFD Specification Section I2.1.

Each W shape is embedded in concrete of a square or rectangular cross section reinforced with four longitudinal reinforcing bars placed in the four corners and with lateral ties spaced as required in Section I2.1. For the design of additional confinement reinforcement, see LRFD Specification Section I2.1. The size of the concrete section was selected so as to provide at least the minimum required cover over the reinforcing bars in the column.

For discussion of the effective length, range of Kl / r, strength about the major axis, combined axial and bending strength, and for sample problems, see Composite Columns, General Notes.

The properties listed at the bottom of each table are for use in checking strength about the strong axis and in design for combined axial load and bending.

Additional information on W shapes encased in concrete, including numerous tables for columns and beam-columns, is available in the AISC Steel Design Guide No. 6, *Load and Resistance Factor Design of W-Shapes Encased in Concrete* (Griffis, 1992).

	F_y = 36 ksi
	F_y = 50 ksi

COMPOSITE COLUMNS
W Shapes
$f_c' = 3.5$ ksi
All reinforcing steel is Grade 60
Axial design strength in kips

Size $b \times h$		24 in.×26 in.								
Reinf. bars		4-#11 bars								
Ties		#4 bars spaced 16 in. c. to c.								
Steel Shape Designation		**W 14**								
Wt./ft		426		398		370		342		311	
F_y		36	50	36	50	36	50	36	50	36	50
	0	4910	6400	4680	6070	4450	5740	4220	5420	3940	5030
	6	4880	6340	4650	6020	4420	5700	4190	5370	3910	4990
	7	4870	6330	4640	6000	4410	5680	4180	5360	3910	4970
	8	4850	6300	4630	5980	4400	5660	4170	5340	3890	4950
	9	4840	6280	4610	5960	4380	5640	4160	5320	3880	4930
	10	4820	6250	4600	5930	4370	5620	4140	5300	3870	4910
	11	4810	6220	4580	5910	4350	5590	4130	5270	3850	4890
	12	4790	6190	4560	5880	4340	5560	4110	5240	3840	4860
	13	4770	6160	4540	5840	4320	5530	4090	5210	3820	4830
	14	4750	6120	4520	5810	4300	5490	4070	5180	3800	4800
	15	4720	6080	4500	5770	4270	5460	4050	5150	3780	4770
	16	4700	6040	4470	5730	4250	5420	4030	5110	3760	4740
	17	4670	5990	4450	5690	4230	5380	4000	5070	3740	4700
	18	4640	5950	4420	5640	4200	5340	3980	5030	3710	4660
	19	4610	5900	4390	5590	4170	5290	3950	4990	3690	4620
	20	4580	5850	4360	5540	4140	5240	3930	4940	3660	4580
	22	4520	5740	4300	5440	4080	5140	3870	4850	3610	4490
	24	4440	5620	4230	5330	4020	5040	3800	4750	3550	4400
	26	4370	5490	4160	5210	3950	4920	3740	4640	3480	4300
	28	4290	5360	4080	5080	3870	4800	3660	4520	3420	4190
	30	4200	5220	4000	4950	3790	4680	3590	4410	3340	4080
	32	4110	5080	3910	4810	3710	4550	3510	4280	3270	3960
	34	4020	4930	3820	4670	3630	4410	3430	4150	3190	3840
	36	3930	4780	3730	4530	3540	4270	3340	4020	3110	3720
	38	3830	4620	3640	4380	3450	4130	3260	3890	3030	3590
	40	3720	4460	3540	4230	3350	3990	3170	3750	2940	3460

Effective length KL (ft) with respect to least radius of gyration r_{my}

Properties										
$\phi_b M_{nx}$ (kip-ft)		2970	3850	2780	3610	2600	3380	2420	3150	2210	2890
$\phi_b M_{ny}$ (kip-ft)		1750	2170	1660	2070	1560	1960	1460	1840	1360	1720
$P_{ex}(K_x L_x)^2/10^4$ (kip-ft^2)		1650	1650	1550	1550	1460	1460	1360	1360	1250	1250
$P_{ey}(K_y L_y)^2/10^4$ (kip-ft^2)		1400	1400	1320	1320	1240	1240	1160	1160	1060	1060
r_{my} (in.)		7.20		7.20		7.20		7.20		7.20	
r_{mx} / r_{my} (in./in.)		1.08		1.08		1.08		1.08		1.08	

| F_y = 36 ksi |
| F_y = 50 ksi |

COMPOSITE COLUMNS
W Shapes
$f_c' = 3.5$ ksi
All reinforcing steel is Grade 60
Axial design strength in kips

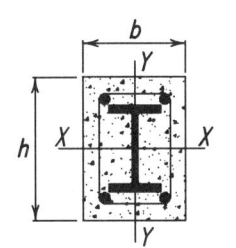

Size $b \times h$	24 in.×24 in.									
Reinf. bars	4-#11 bars									
Ties	#4 bars spaced 16 in. c. to c.									
	Designation	**W 14**								
Steel Shape	Wt./ft	283		257		233		211		193	
	F_y	36	50	36	50	36	50	36	50	36	50

Effective length KL (ft) with respect to least radius of gyration r_{my}		36	50	36	50	36	50	36	50	36	50
	0	3620	4610	3400	4300	3200	4010	3010	3750	2860	3530
	6	3600	4570	3380	4260	3170	3980	2990	3710	2840	3500
	7	3590	4560	3370	4250	3170	3960	2980	3700	2830	3490
	8	3580	4540	3360	4230	3160	3950	2970	3690	2820	3480
	9	3570	4530	3350	4220	3150	3930	2960	3670	2810	3460
	10	3560	4510	3340	4200	3130	3910	2950	3660	2800	3450
	11	3540	4480	3320	4180	3120	3900	2940	3640	2790	3430
	12	3530	4460	3310	4150	3110	3870	2920	3620	2780	3410
	13	3510	4430	3290	4130	3090	3850	2910	3590	2760	3390
	14	3490	4410	3280	4100	3080	3830	2890	3570	2750	3370
	15	3470	4380	3260	4080	3060	3800	2880	3550	2730	3340
	16	3460	4340	3240	4050	3040	3770	2860	3520	2720	3320
	17	3430	4310	3220	4010	3020	3740	2840	3490	2700	3290
	18	3410	4280	3200	3980	3000	3710	2820	3460	2680	3260
	19	3390	4240	3180	3950	2980	3680	2800	3430	2660	3230
	20	3360	4200	3150	3910	2960	3640	2780	3400	2640	3200
	22	3310	4120	3100	3830	2910	3570	2740	3330	2590	3140
	24	3260	4030	3050	3750	2860	3490	2690	3250	2550	3060
	26	3200	3940	2990	3660	2810	3410	2630	3180	2500	2990
	28	3140	3840	2930	3570	2750	3320	2580	3090	2440	2910
	30	3070	3740	2870	3470	2690	3230	2520	3010	2390	2830
	32	3000	3630	2810	3370	2630	3140	2460	2920	2330	2740
	34	2930	3520	2740	3270	2560	3040	2400	2820	2270	2660
	36	2850	3410	2670	3160	2490	2940	2330	2730	2210	2560
	38	2780	3290	2590	3050	2420	2830	2270	2630	2140	2470
	40	2700	3170	2520	2940	2350	2730	2200	2530	2080	2380

Properties										
$\phi_b M_{nx}$ (kip-ft)	1970	2580	1810	2360	1650	2160	1510	1980	1400	1830
$\phi_b M_{ny}$ (kip-ft)	1250	1570	1150	1460	1070	1360	991	1260	924	1170
$P_{ex}(K_x L_x)^2/10^4$ (kip-ft²)	971	971	894	894	822	822	757	757	704	704
$P_{ey}(K_y L_y)^2/10^4$ (kip-ft²)	971	971	894	894	822	822	757	757	704	704
r_{my} (in.)	7.20		7.20		7.20		7.20		7.20	
r_{mx}/r_{my} (in./in.)	1.00		1.00		1.00		1.00		1.00	

| | | | | | F_y = 36 ksi |
| | | | | | F_y = 50 ksi |

COMPOSITE COLUMNS
W Shapes
f_c' = 3.5 ksi
All reinforcing steel is Grade 60
Axial design strength in kips

Size $b \times h$		24 in.×24 in.							
Reinf. bars		4-#10 bars							
Ties		#3 bars spaced 16 in. c. to c.							

	Designation		**W 14**						
	Wt./ft		176		159		145		132
	F_y		36	50	36	50	36	50	36	50
	0	2680	3290	2530	3090	2420	2920	2300	2770
	6	2660	3270	2510	3060	2400	2900	2290	2740
	7	2650	3250	2510	3050	2390	2890	2280	2730
	8	2640	3240	2500	3040	2380	2880	2270	2720
	9	2630	3230	2490	3020	2370	2860	2260	2710
	10	2620	3210	2480	3010	2370	2850	2250	2690
	11	2610	3200	2470	2990	2350	2830	2240	2680
	12	2600	3180	2460	2980	2340	2820	2230	2660
	13	2590	3160	2440	2960	2330	2800	2220	2650
	14	2570	3140	2430	2940	2320	2780	2210	2630
	15	2560	3120	2420	2920	2300	2760	2190	2610
	16	2540	3090	2400	2890	2290	2740	2180	2590
	17	2530	3070	2380	2870	2270	2720	2160	2560
	18	2510	3040	2370	2840	2260	2690	2150	2540
	19	2490	3010	2350	2820	2240	2670	2130	2520
	20	2470	2980	2330	2790	2220	2640	2110	2490
	22	2430	2920	2290	2730	2180	2580	2070	2440
	24	2380	2860	2250	2670	2140	2520	2030	2380
	26	2340	2790	2200	2600	2090	2460	1990	2320
	28	2290	2710	2150	2530	2050	2390	1940	2250
	30	2230	2630	2100	2460	2000	2320	1890	2190
	32	2180	2550	2050	2380	1940	2250	1840	2120
	34	2120	2470	1990	2310	1890	2170	1790	2050
	36	2060	2390	1940	2230	1840	2100	1740	1970
	38	2000	2300	1880	2140	1780	2020	1680	1900
	40	1940	2210	1820	2060	1720	1940	1630	1820

Steel Shape — Effective length KL (ft) with respect to least radius of gyration r_{my}

Properties									
$\phi_b M_{nx}$ (kip-ft)		1260	1660	1150	1520	1060	1400	983	1290
$\phi_b M_{ny}$ (kip-ft)		837	1070	771	987	719	921	656	835
$P_{ex}(K_x L_x)^2/10^4$ (kip-ft^2)		654	654	603	603	563	563	523	523
$P_{ey}(K_y L_y)^2/10^4$ (kip-ft^2)		654	654	603	603	563	563	523	523
r_{my} (in.)		7.20		7.20		7.20		7.20	
r_{mx}/r_{my} (in./in.)		1.00		1.00		1.00		1.00	

| F_y = 36 ksi |
| F_y = 50 ksi |

COMPOSITE COLUMNS
W Shapes
$f_c' = 3.5$ ksi
All reinforcing steel is Grade 60
Axial design strength in kips

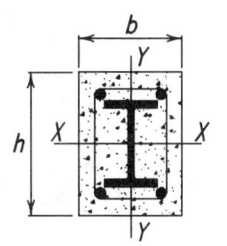

Size $b \times h$		22 in.×22 in.						
Reinf. bars		4-#10 bars						
Ties		#3 bars spaced 14 in. c. to c.						

Steel Shape	Designation	W 14							
	Wt./ft	120		109		99		90	
	F_y	36	50	36	50	36	50	36	50
Effective length KL (ft) with respect to least radius of gyration r_{my}	0	2040	2460	1940	2320	1860	2210	1780	2100
	6	2020	2430	1920	2300	1840	2180	1770	2080
	7	2010	2420	1920	2290	1840	2170	1760	2070
	8	2010	2410	1910	2280	1830	2160	1750	2060
	9	2000	2400	1900	2270	1820	2150	1750	2050
	10	1990	2380	1890	2250	1810	2140	1740	2030
	11	1980	2370	1880	2240	1800	2120	1730	2020
	12	1960	2350	1870	2220	1790	2110	1710	2000
	13	1950	2330	1860	2200	1780	2090	1700	1990
	14	1940	2310	1850	2190	1760	2070	1690	1970
	15	1920	2290	1830	2170	1750	2050	1680	1950
	16	1910	2270	1820	2140	1740	2030	1660	1930
	17	1890	2250	1800	2120	1720	2010	1650	1910
	18	1880	2220	1780	2100	1700	1990	1630	1890
	19	1860	2200	1770	2070	1690	1970	1610	1870
	20	1840	2170	1750	2050	1670	1940	1600	1840
	22	1800	2120	1710	2000	1630	1890	1560	1790
	24	1760	2060	1670	1940	1590	1830	1520	1740
	26	1710	1990	1630	1880	1550	1780	1480	1680
	28	1670	1930	1580	1820	1510	1720	1440	1630
	30	1620	1860	1530	1750	1460	1650	1390	1570
	32	1570	1790	1480	1680	1410	1590	1340	1500
	34	1520	1720	1430	1620	1360	1520	1300	1440
	36	1460	1650	1380	1550	1310	1460	1250	1380
	38	1410	1570	1330	1480	1260	1390	1200	1310
	40	1350	1500	1280	1410	1210	1320	1150	1250

Properties								
$\phi_b M_{nx}$ (kip-ft)	871	1140	801	1050	740	966	684	892
$\phi_b M_{ny}$ (kip-ft)	574	729	533	677	498	631	465	587
$P_{ex}(K_x L_x)^2/10^4$ (kip-ft^2)	392	392	364	364	340	340	318	318
$P_{ey}(K_y L_y)^2/10^4$ (kip-ft^2)	392	392	364	364	340	340	318	318
r_{my} (in.)	6.60		6.60		6.60		6.60	
r_{mx}/r_{my} (in./in.)	1.00		1.00		1.00		1.00	

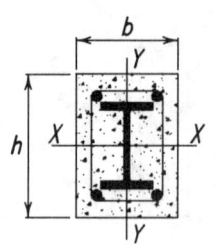

	F_y = 36 ksi
	F_y = 50 ksi

COMPOSITE COLUMNS
W Shapes
f_c' = 3.5 ksi
All reinforcing steel is Grade 60
Axial design strength in kips

Size $b \times h$		18 in.×22 in.						
Reinf. bars		4-#9 bars						
Ties		#3 bars spaced 12 in. c. to c.						
Designation		**W 14**						
Wt./ft		82		74		68		61
F_y		36	50	36	50	36	50	36	50

Steel Shape

Effective length KL (ft) with respect to least radius of gyration r_{my}

KL	82, 36	82, 50	74, 36	74, 50	68, 36	68, 50	61, 36	61, 50
0	1530	1810	1460	1720	1410	1640	1350	1560
6	1500	1780	1440	1690	1390	1620	1330	1530
7	1500	1770	1430	1680	1380	1610	1320	1520
8	1490	1760	1420	1670	1370	1590	1310	1510
9	1480	1740	1410	1650	1360	1580	1300	1500
10	1470	1730	1400	1640	1350	1570	1290	1480
11	1450	1710	1390	1620	1340	1550	1280	1470
12	1440	1690	1370	1600	1320	1530	1260	1450
13	1430	1670	1360	1580	1310	1510	1250	1430
14	1410	1650	1350	1560	1300	1490	1240	1410
15	1390	1630	1330	1540	1280	1470	1220	1390
16	1380	1600	1310	1520	1260	1450	1200	1370
17	1360	1580	1300	1490	1250	1430	1190	1350
18	1340	1550	1280	1470	1230	1400	1170	1320
19	1320	1530	1260	1440	1210	1380	1150	1300
20	1300	1500	1240	1420	1190	1350	1130	1270
22	1260	1440	1200	1360	1150	1300	1090	1220
24	1210	1380	1150	1300	1100	1240	1050	1170
26	1160	1310	1100	1240	1060	1180	1000	1110
28	1110	1250	1060	1170	1010	1120	957	1050
30	1060	1180	1010	1110	963	1060	910	991
32	1010	1110	957	1050	914	993	862	931
34	960	1050	906	981	864	930	814	871
36	908	978	856	917	815	868	766	812
38	855	911	805	853	765	807	718	753
40	804	846	755	791	717	747	671	696

Properties								
$\phi_b M_{nx}$ (kip-ft)	623	808	574	744	535	693	490	633
$\phi_b M_{ny}$ (kip-ft)	319	398	296	370	279	348	259	323
$P_{ex}(K_x L_x)^2/10^4$ (kip-ft^2)	280	280	261	261	246	246	228	228
$P_{ey}(K_y L_y)^2/10^4$ (kip-ft^2)	188	188	175	175	164	164	153	153
r_{my} (in.)	5.40		5.40		5.40		5.40	
r_{mx}/r_{my} (in./in.)	1.22		1.22		1.22		1.22	

| $F_y = 36$ ksi |
| $F_y = 50$ ksi |

COMPOSITE COLUMNS
W Shapes
$f'_c = 3.5$ ksi
All reinforcing steel is Grade 60
Axial design strength in kips

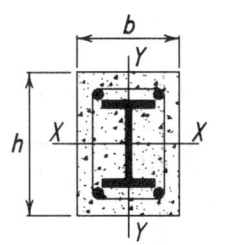

Size $b \times h$										
		22 in.×24 in.								
Reinf. bars		4-#10 bars								
Ties		#3 bars spaced 14 in. c. to c.								
Designation		W 12								
Wt./ft		336		305		279		252		230	
F_y		36	50	36	50	36	50	36	50	36	50
Effective length KL (ft) with respect to least radius of gyration r_{my}	0	3950	5120	3680	4750	3460	4430	3230	4120	3050	3860
	6	3920	5070	3650	4700	3430	4390	3210	4080	3030	3820
	7	3910	5060	3640	4680	3420	4370	3200	4060	3020	3800
	8	3890	5030	3630	4670	3410	4360	3190	4040	3010	3790
	9	3880	5010	3620	4640	3400	4340	3180	4020	3000	3770
	10	3870	4990	3600	4620	3390	4310	3160	4000	2980	3750
	11	3850	4960	3590	4590	3370	4290	3150	3980	2970	3730
	12	3830	4930	3570	4570	3350	4260	3130	3950	2950	3700
	13	3810	4890	3550	4530	3340	4230	3120	3930	2940	3680
	14	3790	4860	3530	4500	3320	4200	3100	3900	2920	3650
	15	3770	4820	3510	4470	3300	4170	3080	3870	2900	3620
	16	3740	4780	3490	4430	3270	4130	3060	3830	2880	3590
	17	3720	4740	3460	4390	3250	4100	3040	3800	2860	3550
	18	3690	4690	3440	4350	3230	4060	3010	3760	2840	3520
	19	3660	4650	3410	4300	3200	4020	2990	3720	2810	3480
	20	3630	4600	3380	4260	3180	3970	2960	3680	2790	3450
	22	3570	4500	3320	4160	3120	3880	2910	3600	2740	3360
	24	3500	4390	3260	4060	3060	3780	2850	3510	2680	3280
	26	3430	4270	3190	3950	2990	3680	2790	3410	2620	3190
	28	3350	4150	3120	3840	2920	3570	2730	3310	2560	3090
	30	3270	4020	3040	3720	2850	3460	2660	3200	2500	2990
	32	3190	3890	2970	3590	2780	3350	2590	3100	2430	2890
	34	3110	3750	2880	3470	2700	3230	2510	2980	2360	2780
	36	3020	3610	2800	3340	2620	3110	2440	2870	2290	2680
	38	2930	3470	2710	3210	2540	2980	2360	2750	2210	2570
	40	2830	3330	2630	3070	2450	2860	2280	2640	2140	2460
Properties											
$\phi_b M_{nx}$ (kip-ft)		2110	2730	1920	2490	1770	2290	1610	2090	1480	1930
$\phi_b M_{ny}$ (kip-ft)		1180	1450	1090	1350	1020	1270	942	1180	879	1100
$P_{ex}(K_x L_x)^2 / 10^4$ (kip-ft^2)		1120	1120	1020	1020	946	946	868	868	803	803
$P_{ey}(K_y L_y)^2 / 10^4$ (kip-ft^2)		938	938	860	860	795	795	729	729	675	675
r_{my} (in.)		6.60		6.60		6.60		6.60		6.60	
r_{mx} / r_{my} (in./in.)		1.09		1.09		1.09		1.09		1.09	

	F_y = 36 ksi
	F_y = 50 ksi

COMPOSITE COLUMNS
W Shapes
f_c' = 3.5 ksi
All reinforcing steel is Grade 60
Axial design strength in kips

Size $b \times h$		20 in.×22 in.								
Reinf. bars		4-#10 bars								
Ties		#3 bars spaced 13 in. c. to c.								

Steel Shape Designation		W 12								
Wt./ft		210		190		170		152		136	
F_y		36	50	36	50	36	50	36	50	36	50
Effective length KL (ft) with respect to least radius of gyration r_{my}	0	2720	3460	2550	3210	2380	2980	2230	2760	2090	2570
	6	2700	3420	2530	3170	2360	2940	2210	2730	2070	2530
	7	2690	3400	2520	3160	2350	2930	2200	2710	2060	2520
	8	2680	3380	2510	3140	2340	2910	2190	2700	2050	2510
	9	2670	3360	2490	3130	2330	2900	2180	2680	2040	2490
	10	2650	3340	2480	3110	2320	2880	2170	2670	2030	2480
	11	2640	3320	2470	3080	2300	2850	2150	2650	2020	2460
	12	2620	3290	2450	3060	2290	2830	2140	2620	2010	2440
	13	2600	3270	2440	3030	2270	2810	2120	2600	1990	2410
	14	2580	3240	2420	3000	2260	2780	2110	2580	1970	2390
	15	2560	3200	2400	2970	2240	2750	2090	2550	1960	2370
	16	2540	3170	2380	2940	2220	2720	2070	2520	1940	2340
	17	2520	3140	2360	2910	2200	2690	2050	2490	1920	2310
	18	2500	3100	2330	2880	2180	2660	2030	2460	1900	2280
	19	2470	3060	2310	2840	2150	2630	2010	2430	1880	2250
	20	2450	3020	2290	2800	2130	2590	1990	2400	1860	2220
	22	2390	2930	2230	2720	2080	2510	1940	2330	1810	2150
	24	2330	2840	2180	2640	2030	2430	1890	2250	1760	2080
	26	2270	2750	2120	2550	1970	2350	1840	2170	1710	2010
	28	2210	2650	2060	2450	1910	2260	1780	2090	1660	1930
	30	2140	2550	1990	2360	1850	2170	1720	2010	1600	1850
	32	2070	2440	1930	2260	1790	2080	1660	1920	1550	1770
	34	2000	2340	1860	2160	1720	1990	1600	1830	1490	1690
	36	1930	2230	1790	2060	1660	1890	1540	1740	1430	1600
	38	1850	2120	1720	1960	1590	1800	1470	1650	1370	1520
	40	1780	2010	1650	1850	1520	1700	1410	1560	1300	1440

Properties										
$\phi_b M_{nx}$ (kip-ft)	1310	1690	1190	1550	1080	1410	983	1280	892	1160
$\phi_b M_{ny}$ (kip-ft)	759	946	704	883	646	813	594	748	547	689
$P_{ex}(K_x L_x)^2/10^4$ (kip-ft^2)	608	608	557	557	508	508	463	463	423	423
$P_{ey}(K_y L_y)^2/10^4$ (kip-ft^2)	502	502	460	460	420	420	383	383	349	349
r_{my} (in.)	6.00		6.00		6.00		6.00		6.00	
r_{mx} / r_{my} (in./in.)	1.10		1.10		1.10		1.10		1.10	

AMERICAN INSTITUTE OF STEEL CONSTRUCTION

| F_y = 36 ksi |
| F_y = 50 ksi |

COMPOSITE COLUMNS
W Shapes
$f'_c = 3.5$ ksi
All reinforcing steel is Grade 60
Axial design strength in kips

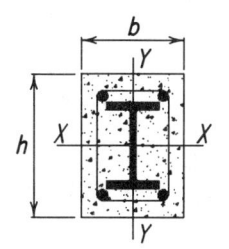

Size $b \times h$		20 in.×20 in.								
Reinf. bars		4-#9 bars								
Ties		#3 bars spaced 13 in. c. to c.								
Designation		W 12								
Wt./ft		120		106		96		87		79	
F_y		36	50	36	50	36	50	36	50	36	50
	0	1850	2270	1740	2110	1650	1990	1580	1880	1510	1780
	6	1840	2250	1720	2080	1630	1960	1560	1850	1490	1760
	7	1830	2240	1710	2070	1630	1950	1550	1850	1480	1750
	8	1820	2220	1700	2060	1620	1940	1540	1830	1470	1740
	9	1810	2210	1690	2050	1610	1930	1540	1820	1470	1730
	10	1800	2190	1680	2030	1600	1910	1530	1810	1460	1710
	11	1790	2180	1670	2020	1590	1900	1520	1800	1450	1700
	12	1780	2160	1660	2000	1580	1880	1500	1780	1440	1690
	13	1760	2140	1650	1980	1570	1860	1490	1760	1420	1670
	14	1750	2120	1640	1960	1550	1850	1480	1740	1410	1650
	15	1740	2100	1620	1940	1540	1830	1470	1730	1400	1630
	16	1720	2070	1610	1920	1520	1800	1450	1710	1380	1610
	17	1700	2050	1590	1900	1510	1780	1440	1680	1370	1590
	18	1690	2020	1570	1870	1490	1760	1420	1660	1350	1570
	19	1670	2000	1560	1850	1470	1730	1400	1640	1340	1550
	20	1650	1970	1540	1820	1460	1710	1390	1610	1320	1530
	22	1610	1910	1500	1760	1420	1660	1350	1560	1280	1480
	24	1570	1850	1460	1700	1380	1600	1310	1510	1250	1420
	26	1520	1780	1410	1640	1340	1540	1270	1450	1210	1370
	28	1470	1710	1370	1580	1290	1480	1230	1390	1160	1310
	30	1420	1640	1320	1510	1250	1420	1180	1330	1120	1260
	32	1370	1570	1270	1450	1200	1350	1140	1270	1080	1200
	34	1320	1500	1220	1380	1150	1290	1090	1210	1030	1140
	36	1270	1430	1170	1310	1100	1220	1040	1150	984	1080
	38	1210	1350	1120	1240	1050	1160	993	1080	937	1020
	40	1160	1280	1070	1170	1000	1090	945	1020	891	957

Effective length KL (ft) with respect to least radius of gyration r_{my}

Steel Shape

Properties										
$\phi_b M_{nx}$ (kip-ft)		746	977	669	878	613	803	566	739	521	680
$\phi_b M_{ny}$ (kip-ft)		474	599	429	545	398	505	372	471	347	438
$P_{ex}(K_xL_x)^2/10^4$ (kip-ft²)		311	311	282	282	261	261	243	243	226	226
$P_{ey}(K_yL_y)^2/10^4$ (kip-ft²)		311	311	282	282	261	261	243	243	226	226
r_{my} (in.)		6.00		6.00		6.00		6.00		6.00	
r_{mx} / r_{my} (in./in.)		1.00		1.00		1.00		1.00		1.00	

	F_y = 36 ksi
	F_y = 50 ksi

COMPOSITE COLUMNS
W Shapes
$f_c' = 3.5$ ksi
All reinforcing steel is Grade 60
Axial design strength in kips

Size $b \times h$			20 in.×20 in.			
Reinf. bars			4-#9 bars			
Ties			#3 bars spaced 13 in. c. to c.			

Steel Shape	Designation		**W 12**					
	Wt./ft		72		65		58	
	F_y		36	50	36	50	36	50
	Effective length KL (ft) with respect to least radius of gyration r_{my}	0	1450	1700	1390	1620	1330	1530
		6	1430	1670	1370	1590	1310	1510
		7	1420	1660	1360	1580	1300	1500
		8	1410	1650	1360	1570	1300	1490
		9	1410	1640	1350	1560	1290	1480
		10	1400	1630	1340	1550	1280	1470
		11	1390	1620	1330	1540	1270	1460
		12	1380	1600	1320	1520	1260	1440
		13	1370	1590	1310	1510	1250	1430
		14	1350	1570	1300	1490	1240	1410
		15	1340	1550	1280	1480	1220	1390
		16	1330	1530	1270	1460	1210	1380
		17	1310	1510	1260	1440	1200	1360
		18	1300	1490	1240	1420	1180	1340
		19	1280	1470	1220	1400	1170	1320
		20	1260	1450	1210	1370	1150	1300
		22	1230	1400	1170	1330	1120	1250
		24	1190	1350	1140	1280	1080	1210
		26	1150	1300	1100	1230	1040	1160
		28	1110	1240	1060	1180	1000	1110
		30	1070	1190	1020	1120	962	1060
		32	1020	1130	973	1070	920	1000
		34	979	1070	930	1010	878	949
		36	934	1020	886	958	835	896
		38	889	958	842	902	792	843
		40	843	901	797	847	749	790
Properties								
$\phi_b M_{nx}$ (kip-ft)			483	629	445	577	410	530
$\phi_b M_{ny}$ (kip-ft)			325	409	302	380	264	328
$P_{ex}(K_x L_x)^2/10^4$ (kip-ft^2)			211	211	197	197	183	183
$P_{ey}(K_y L_y)^2/10^4$ (kip-ft^2)			211	211	197	197	183	183
r_{my} (in.)			6.00		6.00		6.00	
r_{mx}/r_{my} (in./in.)			1.00		1.00		1.00	

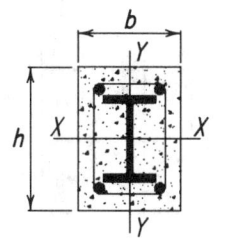

$F_y = 36$ ksi
$F_y = 50$ ksi

COMPOSITE COLUMNS
W Shapes
$f'_c = 3.5$ ksi
All reinforcing steel is Grade 60
Axial design strength in kips

Size $b \times h$		18 in.×18 in.								
Reinf. bars		4-#8 bars								
Ties		#3 bars spaced 12 in. c. to c.								
	Designation	W 10								
	Wt./ft	112		100		88		77		68	
	F_y	36	50	36	50	36	50	36	50	36	50
	0	1620	2020	1520	1870	1420	1730	1330	1600	1250	1490
	6	1600	1980	1500	1840	1400	1700	1310	1570	1230	1470
	7	1600	1970	1500	1830	1400	1690	1300	1560	1230	1460
	8	1590	1960	1490	1820	1390	1680	1300	1550	1220	1450
	9	1580	1950	1480	1810	1380	1670	1290	1540	1210	1430
	10	1570	1930	1470	1790	1370	1650	1280	1520	1200	1420
	11	1560	1910	1460	1780	1360	1640	1270	1510	1190	1410
	12	1540	1890	1450	1760	1350	1620	1260	1490	1180	1390
	13	1530	1870	1430	1740	1340	1600	1240	1480	1170	1380
	14	1520	1850	1420	1720	1320	1580	1230	1460	1160	1360
	15	1500	1830	1400	1700	1310	1560	1220	1440	1150	1340
	16	1480	1800	1390	1670	1290	1540	1200	1420	1130	1320
	17	1470	1780	1370	1650	1280	1520	1190	1400	1120	1300
	18	1450	1750	1360	1620	1260	1490	1170	1370	1100	1280
	19	1430	1720	1340	1600	1240	1470	1160	1350	1080	1260
	20	1410	1690	1320	1570	1230	1440	1140	1330	1070	1230
	22	1370	1630	1280	1510	1190	1390	1100	1280	1030	1190
	24	1330	1570	1240	1450	1150	1330	1060	1220	996	1130
	26	1280	1500	1190	1390	1110	1270	1020	1170	957	1080
	28	1230	1430	1150	1320	1060	1210	981	1110	917	1030
	30	1180	1360	1100	1260	1020	1150	939	1050	876	973
	32	1130	1290	1050	1190	972	1090	895	994	833	918
	34	1080	1220	1000	1120	926	1030	850	935	791	862
	36	1030	1150	955	1060	879	963	806	876	748	807
	38	979	1080	905	988	832	901	761	818	705	752
	40	927	1000	856	922	785	839	717	761	663	699
	Properties									
$\phi_b M_{nx}$ (kip-ft)		576	756	522	685	467	612	415	543	373	488
$\phi_b M_{ny}$ (kip-ft)		366	464	336	426	305	387	276	350	251	318
$P_{ex}(K_x L_x)^2/10^4$ (kip-ft²)		228	228	208	208	189	189	170	170	155	155
$P_{ey}(K_y L_y)^2/10^4$ (kip-ft²)		228	228	208	208	189	189	170	170	155	155
r_{my} (in.)		5.40		5.40		5.40		5.40		5.40	
r_{mx} / r_{my} (in./in.)		1.00		1.00		1.00		1.00		1.00	

Steel Shape

Effective length KL (ft) with respect to least radius of gyration r_{my}

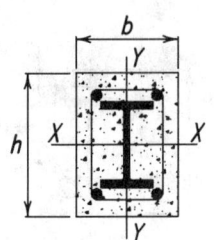

	F_y = 36 ksi
	F_y = 50 ksi

COMPOSITE COLUMNS
W Shapes
f_c' = 3.5 ksi
All reinforcing steel is Grade 60
Axial design strength in kips

Size $b \times h$		18 in.×18 in.						
Reinf. bars		4-#8 bars						
Ties		#3 bars spaced 12 in. c. to c.						

	Designation	**W 10**						
	Wt./ft	60		54		49		45	
	F_y	36	50	36	50	36	50	36	50
Effective length *KL* (ft) with respect to least radius of gyration r_{my}	0	1180	1390	1130	1320	1090	1260	1060	1220
	6	1170	1370	1110	1300	1070	1240	1040	1200
	7	1160	1360	1110	1290	1070	1230	1040	1190
	8	1150	1350	1100	1280	1060	1220	1030	1180
	9	1140	1340	1090	1270	1050	1210	1020	1170
	10	1140	1330	1080	1260	1040	1200	1010	1160
	11	1130	1310	1070	1240	1040	1190	1000	1140
	12	1120	1300	1060	1230	1020	1170	994	1130
	13	1100	1280	1050	1210	1010	1160	983	1120
	14	1090	1270	1040	1200	1000	1140	971	1100
	15	1080	1250	1030	1180	989	1130	958	1090
	16	1060	1230	1010	1160	976	1110	945	1070
	17	1050	1210	1000	1140	962	1090	931	1050
	18	1040	1190	986	1120	947	1070	917	1030
	19	1020	1170	971	1100	932	1050	902	1010
	20	1000	1150	955	1080	917	1030	886	992
	22	969	1100	921	1040	884	989	854	950
	24	933	1050	886	992	849	944	820	907
	26	896	1000	849	944	813	898	784	861
	28	857	952	811	895	775	850	747	815
	30	817	900	772	845	737	802	709	768
	32	776	848	733	795	698	753	671	721
	34	735	795	693	745	659	705	633	674
	36	694	743	653	695	620	657	595	627
	38	653	692	613	646	582	610	557	581
	40	612	641	574	598	543	564	519	537
Properties									
$\phi_b M_{nx}$ (kip-ft)		337	439	307	399	286	370	272	351
$\phi_b M_{ny}$ (kip-ft)		230	290	212	267	199	249	179	222
$P_{ex}(K_x L_x)^2 / 10^4$ (kip-ft²)		142	142	131	131	123	123	117	117
$P_{ey}(K_y L_y)^2 / 10^4$ (kip-ft²)		142	142	131	131	123	123	117	117
r_{my} (in.)		5.40		5.40		5.40		5.40	
r_{mx} / r_{my} (in./in.)		1.00		1.00		1.00		1.00	

| F_y = 36 ksi |
| F_y = 50 ksi |

COMPOSITE COLUMNS
W Shapes
f_c' = 3.5 ksi
All reinforcing steel is Grade 60
Axial design strength in kips

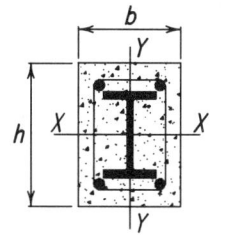

Size $b \times h$										
		16 in.×16 in.								

Reinf. bars										
		4-#7 bars								

Ties										
		#3 bars spaced 10 in. c. to c.								

Steel Shape	Designation									

	Designation	W 8								
	Wt./ft	67		58		48		40		35
	F_y	36	50	36	50	36	50	36	50	36	50
	0	1100	1330	1020	1230	938	1110	868	1010	828	951
	6	1080	1310	1010	1200	920	1080	851	985	811	929
	7	1070	1300	1000	1190	914	1070	845	977	805	921
	8	1070	1290	992	1180	907	1060	839	968	799	912
	9	1060	1270	984	1170	899	1050	831	957	791	902
	10	1050	1260	975	1160	890	1040	822	946	783	891
	11	1040	1240	965	1140	881	1030	813	933	773	879
	12	1030	1230	954	1130	870	1010	803	920	763	866
	13	1010	1210	942	1110	859	997	792	905	753	852
	14	1000	1190	930	1090	847	981	780	890	741	837
	15	988	1170	917	1080	834	963	768	874	729	822
	16	973	1150	903	1060	821	945	755	857	717	805
	17	958	1130	888	1040	807	927	742	839	703	788
	18	943	1110	873	1010	793	907	728	821	690	770
	19	926	1080	857	992	778	887	713	802	675	752
	20	909	1060	841	970	762	866	698	782	661	733
	22	874	1010	807	923	730	823	667	742	630	695
	24	836	958	772	874	696	778	634	700	598	654
	26	798	904	735	824	661	732	601	657	565	613
	28	758	850	697	773	625	685	566	614	532	572
	30	717	795	658	722	588	638	532	570	498	530
	32	677	740	619	671	552	592	497	527	464	489
	34	636	686	580	621	515	546	462	485	431	449
	36	595	633	542	572	479	501	428	444	398	410
	38	554	581	504	524	444	458	395	404	366	373
	40	515	532	466	478	410	416	363	366	335	337

Left label for rows: Effective length KL (ft) with respect to least radius of gyration r_{my}

Properties										
$\phi_b M_{nx}$ (kip-ft)	298	390	264	345	223	291	194	251	174	224
$\phi_b M_{ny}$ (kip-ft)	197	250	178	225	153	193	136	171	124	155
$P_{ex}(K_x L_x)^2/10^4$ (kip-ft²)	114	114	103	103	89.3	89.3	78.5	78.5	72.3	72.3
$P_{ey}(K_y L_y)^2/10^4$ (kip-ft²)	114	114	103	103	89.3	89.3	78.5	78.5	72.3	72.3
r_{my} (in.)	4.80		4.80		4.80		4.80		4.80	
r_{mx}/r_{my} (in./in.)	1.00		1.00		1.00		1.00		1.00	

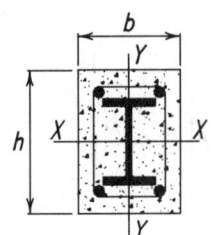

		F_y = 36 ksi
		F_y = 50 ksi

COMPOSITE COLUMNS
W Shapes
f_c' = 5 ksi
All reinforcing steel is Grade 60
Axial design strength in kips

Size $b \times h$		24 in.×26 in.								
Reinf. bars		4-#11 bars								
Ties		#4 bars spaced 16 in. c. to c.								

	Designation	**W 14**								
	Wt./ft	426		398		370		342		311	
	F_y	36	50	36	50	36	50	36	50	36	50
	0	5290	6770	5060	6450	4840	6130	4610	5810	4340	5430
	6	5250	6720	5030	6400	4800	6080	4580	5760	4310	5380
	7	5240	6700	5020	6380	4790	6060	4570	5750	4300	5370
	8	5220	6670	5000	6360	4780	6040	4560	5730	4290	5350
	9	5210	6650	4990	6330	4760	6020	4540	5700	4270	5320
	10	5190	6620	4970	6300	4750	5990	4520	5680	4260	5300
	11	5170	6580	4950	6270	4730	5960	4510	5650	4240	5270
	12	5150	6550	4930	6240	4710	5930	4490	5620	4220	5240
	13	5120	6510	4910	6200	4690	5890	4470	5580	4200	5210
	14	5100	6470	4880	6160	4660	5850	4440	5550	4180	5180
	15	5070	6430	4850	6120	4640	5810	4420	5510	4160	5140
	16	5040	6380	4830	6070	4610	5770	4390	5470	4130	5100
	17	5010	6330	4800	6030	4580	5720	4360	5420	4100	5060
	18	4980	6280	4770	5980	4550	5680	4330	5380	4070	5010
	19	4950	6220	4730	5930	4520	5630	4300	5330	4050	4970
	20	4910	6170	4700	5870	4490	5570	4270	5280	4010	4920
	22	4840	6050	4630	5760	4410	5460	4200	5170	3950	4820
	24	4760	5920	4550	5630	4340	5340	4130	5060	3880	4710
	26	4670	5780	4460	5500	4260	5220	4050	4940	3800	4600
	28	4580	5640	4380	5360	4170	5090	3970	4810	3720	4480
	30	4480	5490	4280	5220	4080	4950	3880	4680	3640	4350
	32	4380	5330	4190	5060	3990	4800	3790	4540	3550	4220
	34	4280	5170	4080	4910	3890	4650	3690	4400	3460	4090
	36	4170	5000	3980	4750	3790	4500	3600	4250	3370	3950
	38	4060	4830	3870	4590	3680	4340	3500	4100	3270	3810
	40	3940	4660	3760	4420	3580	4180	3390	3950	3170	3660

Left axis label: Effective length KL (ft) with respect to least radius of gyration r_{my}

Properties										
$\phi_b M_{nx}$ (kip-ft)	3090	4070	2890	3810	2690	3550	2490	3300	2280	3010
$\phi_b M_{ny}$ (kip-ft)	1850	2380	1750	2250	1640	2120	1540	1980	1420	1830
$P_{ex}(K_xL_x)^2/10^4$ (kip-ft^2)	1670	1670	1580	1580	1480	1480	1390	1390	1280	1280
$P_{ey}(K_yL_y)^2/10^4$ (kip-ft^2)	1420	1420	1340	1340	1260	1260	1180	1180	1090	1090
r_{my} (in.)	7.20		7.20		7.20		7.20		7.20	
r_{mx} / r_{my} (in./in.)	1.08		1.08		1.08		1.08		1.08	

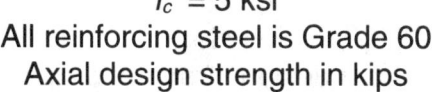

| $F_y = 36$ ksi | | |
| $F_y = 50$ ksi | | |

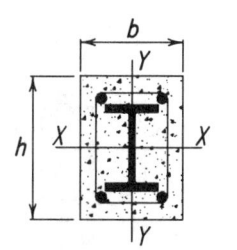

COMPOSITE COLUMNS
W Shapes
$f_c' = 5$ ksi
All reinforcing steel is Grade 60
Axial design strength in kips

Size $b \times h$		24 in.×24 in.									
Reinf. bars		4-#11 bars									
Ties		#4 bars spaced 16 in. c. to c.									
	Designation	**W 14**									
	Wt./ft	283		257		233		211		193	
	F_y	36	50	36	50	36	50	36	50	36	50
	0	3990	4980	3780	4680	3580	4390	3400	4130	3250	3930
	6	3970	4940	3750	4640	3550	4350	3370	4100	3220	3890
	7	3960	4920	3740	4620	3540	4340	3360	4080	3220	3880
	8	3940	4910	3730	4600	3530	4320	3350	4070	3200	3860
	9	3930	4890	3720	4580	3520	4300	3340	4050	3190	3840
	10	3920	4860	3700	4560	3500	4280	3320	4030	3180	3820
	11	3900	4840	3690	4540	3490	4260	3310	4010	3160	3800
	12	3880	4810	3670	4510	3470	4230	3290	3980	3150	3780
	13	3860	4780	3650	4480	3450	4210	3280	3960	3130	3750
	14	3840	4750	3630	4450	3440	4180	3260	3930	3110	3730
	15	3820	4720	3610	4420	3410	4150	3240	3900	3090	3700
	16	3800	4680	3590	4390	3390	4110	3210	3870	3070	3670
	17	3770	4640	3560	4350	3370	4080	3190	3830	3050	3640
	18	3750	4600	3540	4310	3340	4040	3170	3800	3030	3600
	19	3720	4560	3510	4270	3320	4010	3140	3760	3000	3570
	20	3690	4520	3480	4230	3290	3970	3120	3720	2980	3530
	22	3630	4420	3420	4140	3230	3880	3060	3640	2920	3450
	24	3560	4320	3360	4050	3170	3790	3000	3560	2860	3370
	26	3490	4220	3290	3950	3110	3690	2940	3460	2800	3280
	28	3420	4110	3220	3840	3040	3590	2870	3370	2730	3190
	30	3340	3990	3150	3730	2970	3490	2800	3270	2670	3090
	32	3260	3870	3070	3610	2890	3380	2730	3160	2590	2990
	34	3180	3750	2990	3500	2810	3270	2650	3060	2520	2890
	36	3090	3620	2900	3380	2730	3150	2570	2950	2440	2780
	38	3000	3490	2820	3250	2650	3030	2490	2830	2360	2670
	40	2910	3360	2730	3130	2560	2920	2410	2720	2290	2570
	Properties										
$\phi_b M_{nx}$ (kip-ft)		2020	2680	1850	2450	1690	2230	1540	2040	1420	1880
$\phi_b M_{ny}$ (kip-ft)		1300	1680	1200	1550	1110	1430	1020	1320	949	1220
$P_{ex}(K_x L_x)^2 / 10^4$ (kip-ft²)		993	993	916	916	845	845	780	780	728	728
$P_{ey}(K_y L_y)^2 / 10^4$ (kip-ft²)		993	993	916	916	845	845	780	780	728	728
r_{my} (in.)		7.20		7.20		7.20		7.20		7.20	
r_{mx} / r_{my} (in./in.)		1.00		1.00		1.00		1.00		1.00	

Steel Shape

Effective length KL (ft) with respect to least radius of gyration r_{my}

	F_y = 36 ksi
	F_y = 50 ksi

COMPOSITE COLUMNS
W Shapes
$f_c' = 5$ ksi
All reinforcing steel is Grade 60
Axial design strength in kips

Size $b \times h$					24 in.×26 in.				
Reinf. bars					4-#10 bars				
Ties					#3 bars spaced 16 in. c. to c.				
	Designation				**W 14**				
	Wt./ft	176		159		145		132	
	F_y	36	50	36	50	36	50	36	50
	0	3080	3690	2930	3490	2820	3330	2710	3170
	6	3050	3660	2910	3450	2800	3290	2690	3140
	7	3040	3640	2900	3440	2790	3280	2680	3130
	8	3030	3630	2890	3430	2780	3270	2670	3110
	9	3020	3610	2880	3410	2770	3250	2660	3100
	10	3010	3590	2870	3390	2750	3240	2650	3080
	11	2990	3570	2850	3370	2740	3220	2630	3060
	12	2980	3550	2840	3350	2730	3200	2620	3040
	13	2960	3530	2820	3330	2710	3170	2600	3020
	14	2940	3500	2800	3310	2690	3150	2580	3000
	15	2920	3480	2780	3280	2670	3120	2570	2970
	16	2900	3450	2760	3250	2650	3100	2550	2950
	17	2880	3420	2740	3220	2630	3070	2530	2920
	18	2860	3380	2720	3190	2610	3040	2500	2890
	19	2840	3350	2700	3160	2590	3010	2480	2860
	20	2810	3320	2670	3130	2560	2980	2460	2830
	22	2760	3240	2620	3050	2510	2910	2410	2760
	24	2700	3160	2570	2980	2460	2830	2350	2690
	26	2640	3080	2510	2900	2400	2750	2300	2620
	28	2580	2990	2450	2810	2340	2670	2240	2540
	30	2510	2900	2380	2720	2280	2590	2180	2450
	32	2450	2800	2320	2630	2210	2500	2110	2370
	34	2370	2710	2250	2540	2140	2410	2040	2280
	36	2300	2610	2170	2440	2070	2320	1980	2190
	38	2230	2510	2100	2350	2000	2220	1910	2100
	40	2150	2400	2030	2250	1930	2130	1830	2010

Effective length KL (ft) with respect to least radius of gyration r_{my}

Steel Shape

Properties									
$\phi_b M_{nx}$ (kip-ft)		1280	1700	1170	1550	1080	1430	996	1320
$\phi_b M_{ny}$ (kip-ft)		860	1110	789	1020	734	950	670	861
$P_{ex}(K_x L_x)^2/10^4$ (kip-ft^2)		678	678	627	627	587	587	547	547
$P_{ey}(K_y L_y)^2/10^4$ (kip-ft^2)		678	678	627	627	587	587	547	547
r_{my} (in.)		7.20		7.20		7.20		7.20	
r_{mx} / r_{my} (in./in.)		1.00		1.00		1.00		1.00	

| F_y = 36 ksi |
| F_y = 50 ksi |

COMPOSITE COLUMNS
W Shapes
f_c' = 5 ksi
All reinforcing steel is Grade 60
Axial design strength in kips

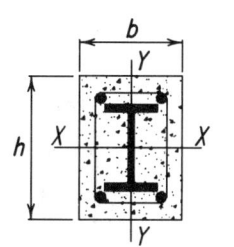

Size $b \times h$		22 in.×22 in.						
Reinf. bars		4-#10 bars						
Ties		#3 bars spaced 14 in. c. to c.						
	Designation	**W 14**						
	Wt./ft	120		109		99		90	
	F_y	36	50	36	50	36	50	36	50
	0	2380	2800	2290	2670	2200	2550	2130	2450
	6	2350	2760	2260	2630	2180	2520	2110	2410
	7	2340	2750	2250	2620	2170	2510	2100	2400
	8	2330	2740	2240	2610	2160	2490	2090	2390
	9	2320	2720	2230	2590	2150	2480	2080	2380
	10	2310	2700	2220	2580	2140	2460	2070	2360
	11	2300	2690	2200	2560	2120	2440	2050	2340
	12	2280	2670	2190	2540	2110	2420	2040	2320
	13	2270	2640	2170	2520	2090	2400	2020	2300
	14	2250	2620	2160	2490	2080	2380	2000	2280
	15	2230	2590	2140	2470	2060	2360	1990	2260
	16	2210	2570	2120	2440	2040	2330	1970	2230
	17	2190	2540	2100	2410	2020	2300	1950	2200
	18	2170	2510	2080	2380	2000	2270	1920	2180
	19	2140	2480	2050	2350	1970	2250	1900	2150
	20	2120	2440	2030	2320	1950	2210	1880	2120
	22	2070	2380	1980	2260	1900	2150	1830	2050
	24	2020	2300	1930	2190	1850	2080	1780	1990
	26	1960	2230	1870	2110	1790	2010	1720	1920
	28	1900	2150	1810	2030	1740	1930	1670	1840
	30	1840	2070	1750	1950	1680	1860	1610	1770
	32	1770	1980	1690	1870	1610	1780	1550	1690
	34	1710	1890	1620	1790	1550	1700	1480	1610
	36	1640	1810	1560	1710	1490	1620	1420	1530
	38	1570	1720	1490	1620	1420	1530	1350	1450
	40	1500	1630	1420	1540	1350	1450	1290	1380
Properties								
$\phi_b M_{nx}$ (kip-ft)		883	1160	811	1070	748	982	691	905
$\phi_b M_{ny}$ (kip-ft)		586	752	543	695	507	647	472	600
$P_{ex}(K_x L_x)^2/10^4$ (kip-ft²)		409	409	381	381	357	357	335	335
$P_{ey}(K_y L_y)^2/10^4$ (kip-ft²)		409	409	381	381	357	357	335	335
r_{my} (in.)		6.60		6.60		6.60		6.60	
r_{mx}/r_{my} (in./in.)		1.00		1.00		1.00		1.00	

Column left side: Steel Shape / Effective length *KL* (ft) with respect to least radius of gyration r_{my}

			F_y = 36 ksi
			F_y = 50 ksi

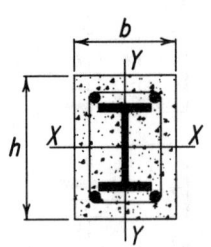

COMPOSITE COLUMNS
W Shapes
f_c' = 5 ksi
All reinforcing steel is Grade 60
Axial design strength in kips

Size $b \times h$			18 in.×22 in.						
Reinf. bars			4-#9 bars						
Ties			#3 bars spaced 12 in. c. to c.						
	Designation		**W 14**						
Steel Shape	Wt./ft		82		74		68		61
	F_y		36	50	36	50	36	50	36	50
Effective length KL (ft) with respect to least radius of gyration r_{my}	0	1810	2090	1740	2000	1690	1930	1630	1850
	6	1780	2050	1710	1960	1660	1890	1600	1810
	7	1770	2040	1700	1950	1650	1880	1590	1800
	8	1760	2020	1690	1930	1640	1860	1580	1780
	9	1740	2010	1680	1920	1630	1850	1570	1760
	10	1730	1990	1660	1900	1610	1830	1550	1740
	11	1710	1960	1650	1880	1600	1810	1540	1720
	12	1690	1940	1630	1850	1580	1780	1520	1700
	13	1670	1920	1610	1830	1560	1760	1500	1680
	14	1650	1890	1590	1800	1540	1730	1480	1650
	15	1630	1860	1570	1770	1520	1710	1460	1630
	16	1610	1830	1550	1740	1500	1680	1440	1600
	17	1590	1800	1520	1710	1470	1650	1410	1570
	18	1560	1770	1500	1680	1450	1620	1390	1540
	19	1530	1730	1470	1650	1420	1580	1360	1510
	20	1510	1700	1450	1620	1400	1550	1340	1470
	22	1450	1620	1390	1540	1340	1480	1280	1410
	24	1390	1550	1330	1470	1280	1410	1230	1340
	26	1330	1470	1270	1390	1220	1330	1170	1260
	28	1270	1390	1210	1310	1160	1260	1110	1190
	30	1200	1310	1140	1240	1100	1180	1040	1110
	32	1140	1220	1080	1160	1040	1100	982	1040
	34	1070	1140	1020	1080	972	1030	920	965
	36	1010	1060	952	999	909	950	858	892
	38	941	983	888	923	847	876	797	821
	40	877	906	826	849	786	805	738	752
Properties									
$\phi_b M_{nx}$ (kip-ft)		634	830	582	761	543	708	496	644
$\phi_b M_{ny}$ (kip-ft)		328	416	303	384	285	360	264	332
$P_{ex}(K_x L_x)^2/10^4$ (kip-ft^2)		294	294	275	275	260	260	242	242
$P_{ey}(K_y L_y)^2/10^4$ (kip-ft^2)		197	197	184	184	174	174	162	162
r_{my} (in.)		5.40		5.40		5.40		5.40	
r_{mx} / r_{my} (in./in.)		1.22		1.22		1.22		1.22	

F_y = 36 ksi
F_y = 50 ksi

COMPOSITE COLUMNS
W Shapes
f_c' = 5 ksi
All reinforcing steel is Grade 60
Axial design strength in kips

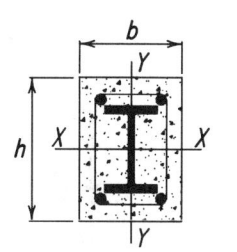

Size $b \times h$										
		22 in.×24 in.								
Reinf. bars		4-#10 bars								
Ties		#3 bars spaced 14 in. c. to c.								

Steel Shape		Designation	**W 12**								
		Wt./ft	336		305		279		252		230	
		F_y	36	50	36	50	36	50	36	50	36	50
Effective length KL (ft) with respect to least radius of gyration r_{my}		0	4270	5450	4010	5080	3800	4770	3580	4460	3400	4200
		6	4240	5390	3980	5030	3770	4720	3550	4410	3370	4160
		7	4230	5370	3970	5010	3750	4700	3540	4400	3360	4140
		8	4210	5350	3960	4990	3740	4680	3520	4380	3350	4120
		9	4200	5320	3940	4960	3730	4660	3510	4350	3330	4100
		10	4180	5300	3920	4940	3710	4640	3490	4330	3320	4080
		11	4160	5260	3900	4910	3690	4610	3480	4300	3300	4050
		12	4140	5230	3880	4870	3670	4580	3460	4270	3280	4030
		13	4110	5190	3860	4840	3650	4540	3440	4240	3260	4000
		14	4090	5150	3840	4800	3630	4510	3420	4210	3240	3960
		15	4060	5110	3810	4760	3610	4470	3390	4170	3220	3930
		16	4040	5070	3790	4720	3580	4430	3370	4140	3190	3890
		17	4010	5020	3760	4680	3550	4390	3340	4100	3170	3860
		18	3980	4970	3730	4630	3520	4340	3310	4050	3140	3820
		19	3940	4920	3700	4580	3490	4300	3290	4010	3110	3770
		20	3910	4870	3670	4530	3460	4250	3260	3960	3090	3730
		22	3840	4750	3600	4420	3400	4150	3190	3870	3020	3640
		24	3760	4630	3520	4310	3320	4040	3120	3760	2960	3540
		26	3680	4500	3450	4190	3250	3920	3050	3650	2890	3430
		28	3590	4370	3360	4060	3170	3800	2970	3540	2810	3320
		30	3500	4230	3280	3930	3090	3680	2890	3420	2730	3210
		32	3410	4080	3190	3790	3000	3550	2810	3300	2650	3090
		34	3310	3930	3090	3650	2910	3410	2720	3170	2570	2970
		36	3210	3780	3000	3510	2820	3280	2630	3040	2480	2850
		38	3110	3630	2900	3360	2720	3140	2540	2910	2400	2730
		40	3000	3470	2800	3220	2630	3000	2450	2780	2310	2600

Properties										
$\phi_b M_{nx}$ (kip-ft)	2190	2890	1990	2630	1830	2410	1660	2190	1530	2010
$\phi_b M_{ny}$ (kip-ft)	1260	1600	1150	1480	1080	1380	990	1270	919	1180
$P_{ex}(K_x L_x)^2 / 10^4$ (kip-ft^2)	1140	1140	1040	1040	966	966	888	888	824	824
$P_{ey}(K_y L_y)^2 / 10^4$ (kip-ft^2)	954	954	877	877	812	812	746	746	692	692

r_{my} (in.)	6.60		6.60		6.60		6.60		6.60	
r_{mx} / r_{my} (in./in.)	1.09		1.09		1.09		1.09		1.09	

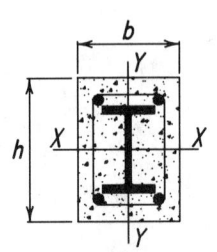

		F_y = 36 ksi
		F_y = 50 ksi

COMPOSITE COLUMNS
W Shapes
$f_c' = 5$ ksi
All reinforcing steel is Grade 60
Axial design strength in kips

Size $b \times h$		20 in.×22 in.								
Reinf. bars		4-#10 bars								
Ties		#3 bars spaced 13 in. c. to c.								

Steel Shape	Designation	W 12									
	Wt./ft	210		190		170		152		136	
	F_y	36	50	36	50	36	50	36	50	36	50
Effective length *KL* (ft) with respect to least radius of gyration r_{my}	0	3010	3740	2840	3500	2680	3270	2530	3060	2390	2870
	6	2980	3700	2810	3460	2650	3230	2500	3020	2370	2830
	7	2970	3680	2800	3440	2640	3210	2490	3000	2360	2810
	8	2950	3660	2790	3420	2630	3200	2480	2990	2340	2800
	9	2940	3640	2770	3400	2610	3180	2470	2970	2330	2780
	10	2920	3610	2760	3380	2600	3150	2450	2950	2320	2760
	11	2910	3590	2740	3350	2580	3130	2430	2920	2300	2740
	12	2890	3560	2720	3320	2560	3100	2420	2900	2280	2710
	13	2870	3520	2700	3290	2540	3070	2400	2870	2270	2690
	14	2840	3490	2680	3260	2520	3040	2380	2840	2250	2660
	15	2820	3450	2660	3230	2500	3010	2360	2810	2220	2630
	16	2790	3420	2630	3190	2480	2970	2330	2780	2200	2600
	17	2770	3380	2610	3150	2450	2940	2310	2740	2180	2560
	18	2740	3330	2580	3110	2430	2900	2280	2700	2150	2530
	19	2710	3290	2550	3070	2400	2860	2260	2670	2130	2490
	20	2680	3240	2520	3030	2370	2820	2230	2630	2100	2450
	22	2620	3150	2460	2940	2310	2730	2170	2540	2040	2370
	24	2550	3040	2390	2840	2250	2640	2110	2460	1980	2290
	26	2480	2940	2320	2740	2180	2540	2040	2360	1920	2200
	28	2400	2830	2250	2630	2110	2440	1970	2270	1850	2110
	30	2320	2710	2180	2520	2030	2340	1900	2170	1780	2020
	32	2240	2590	2100	2410	1960	2230	1830	2070	1710	1920
	34	2160	2470	2020	2300	1880	2130	1760	1970	1640	1830
	36	2070	2350	1930	2180	1800	2020	1680	1870	1570	1730
	38	1980	2230	1850	2070	1720	1910	1600	1760	1490	1630
	40	1900	2110	1770	1950	1640	1800	1530	1660	1420	1530

Properties										
$\phi_b M_{nx}$ (kip-ft)	1350	1770	1230	1610	1110	1460	1010	1320	910	1190
$\phi_b M_{ny}$ (kip-ft)	796	1020	734	941	671	860	614	787	563	721
$P_{ex}(K_xL_x)^2/10^4$ (kip-ft^2)	622	622	572	572	523	523	478	478	438	438
$P_{ey}(K_yL_y)^2/10^4$ (kip-ft^2)	514	514	472	472	432	432	395	395	362	362

r_{my} (in.)	6.00		6.00		6.00		6.00		6.00	
r_{mx} / r_{my} (in./in.)	1.10		1.10		1.10		1.10		1.10	

F_y = 36 ksi
F_y = 50 ksi

COMPOSITE COLUMNS
W Shapes
f_c' = 5 ksi
All reinforcing steel is Grade 60
Axial design strength in kips

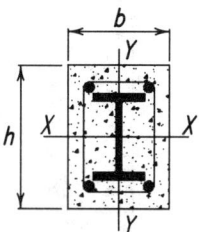

Size $b \times h$		20 in.×20 in.								
Reinf. bars		4-#9 bars								
Ties		#3 bars spaced 13 in. c. to c.								
Steel Shape	Designation	**W 12**								
	Wt./ft	120		106		96		87		79	
	F_y	36	50	36	50	36	50	36	50	36	50
Effective length KL (ft) with respect to least radius of gyration r_{my}	0	2130	2550	2020	2390	1930	2270	1860	2160	1790	2070
	6	2110	2520	1990	2350	1910	2230	1830	2130	1770	2040
	7	2100	2500	1980	2340	1900	2220	1830	2120	1760	2020
	8	2090	2490	1970	2330	1890	2210	1820	2110	1750	2010
	9	2080	2470	1960	2310	1880	2190	1810	2090	1740	2000
	10	2060	2450	1950	2290	1870	2180	1790	2070	1730	1980
	11	2050	2430	1930	2270	1850	2160	1780	2060	1710	1960
	12	2030	2410	1920	2250	1840	2140	1760	2040	1700	1940
	13	2020	2390	1900	2230	1820	2120	1750	2020	1680	1920
	14	2000	2360	1890	2210	1800	2090	1730	1990	1670	1900
	15	1980	2340	1870	2180	1790	2070	1710	1970	1650	1880
	16	1960	2310	1850	2150	1770	2040	1700	1940	1630	1850
	17	1940	2280	1830	2130	1750	2010	1680	1920	1610	1830
	18	1920	2250	1810	2100	1730	1990	1650	1890	1590	1800
	19	1890	2220	1780	2070	1700	1960	1630	1860	1570	1770
	20	1870	2180	1760	2030	1680	1920	1610	1830	1540	1740
	22	1820	2110	1710	1970	1630	1860	1560	1770	1500	1680
	24	1770	2040	1660	1890	1580	1790	1510	1700	1450	1620
	26	1710	1960	1600	1820	1530	1720	1460	1630	1390	1550
	28	1650	1880	1550	1740	1470	1640	1400	1560	1340	1480
	30	1590	1800	1490	1660	1410	1570	1340	1480	1280	1410
	32	1530	1710	1430	1580	1350	1490	1290	1410	1220	1330
	34	1460	1620	1360	1500	1290	1410	1230	1330	1170	1260
	36	1400	1540	1300	1420	1230	1330	1170	1260	1110	1190
	38	1330	1450	1240	1340	1170	1250	1110	1180	1050	1110
	40	1260	1370	1170	1260	1100	1180	1040	1110	989	1040

Properties										
$\phi_b M_{nx}$ (kip-ft)	760	1000	680	899	622	820	573	754	528	693
$\phi_b M_{ny}$ (kip-ft)	489	628	440	566	407	522	380	486	353	450
$P_{ex}(K_x L_x)^2/10^4$ (kip-ft²)	322	322	294	294	273	273	255	255	238	238
$P_{ey}(K_y L_y)^2/10^4$ (kip-ft²)	322	322	294	294	273	273	255	255	238	238
r_{my} (in.)	6.00		6.00		6.00		6.00		6.00	
r_{mx}/r_{my} (in./in.)	1.00		1.00		1.00		1.00		1.00	

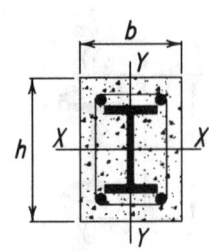

	$F_y = 36$ ksi
	$F_y = 50$ ksi

COMPOSITE COLUMNS
W Shapes
$f_c' = 5$ ksi
All reinforcing steel is Grade 60
Axial design strength in kips

Size $b \times h$		20 in.×20 in.					
Reinf. bars		4-#9 bars					
Ties		#3 bars spaced 13 in. c. to c.					
	Designation	**W 12**					
	Wt./ft	72		65		58	
	F_y	36	50	36	50	36	50
Effective length KL (ft) with respect to least radius of gyration r_{my}	0	1730	1980	1680	1900	1620	1820
	6	1710	1950	1650	1870	1590	1790
	7	1700	1940	1640	1860	1590	1780
	8	1690	1930	1630	1850	1580	1770
	9	1680	1910	1620	1840	1560	1750
	10	1670	1900	1610	1820	1550	1740
	11	1650	1880	1600	1800	1540	1720
	12	1640	1860	1580	1780	1520	1700
	13	1620	1840	1570	1760	1510	1680
	14	1610	1820	1550	1740	1490	1660
	15	1590	1800	1530	1720	1480	1640
	16	1570	1770	1520	1700	1460	1620
	17	1550	1750	1500	1670	1440	1590
	18	1530	1720	1480	1650	1420	1570
	19	1510	1690	1450	1620	1400	1540
	20	1490	1660	1430	1590	1370	1510
	22	1440	1600	1380	1530	1330	1460
	24	1390	1540	1340	1470	1280	1400
	26	1340	1480	1280	1410	1230	1330
	28	1280	1410	1230	1340	1170	1270
	30	1230	1340	1180	1270	1120	1200
	32	1170	1270	1120	1200	1060	1140
	34	1110	1200	1060	1130	1010	1070
	36	1060	1120	1000	1060	951	1000
	38	997	1050	948	996	895	935
	40	939	984	891	929	840	871
Properties							
$\phi_b M_{nx}$ (kip-ft)		489	640	449	586	414	537
$\phi_b M_{ny}$ (kip-ft)		330	419	307	388	268	335
$P_{ex}(K_x L_x)^2/10^4$ (kip-ft^2)		223	223	209	209	195	195
$P_{ey}(K_y L_y)^2/10^4$ (kip-ft^2)		223	223	209	209	195	195
r_{my} (in.)		6.00		6.00		6.00	
r_{mx} / r_{my} (in./in.)		1.00		1.00		1.00	

| F_y = 36 ksi |
| F_y = 50 ksi |

COMPOSITE COLUMNS
W Shapes
f_c' = 5 ksi
All reinforcing steel is Grade 60
Axial design strength in kips

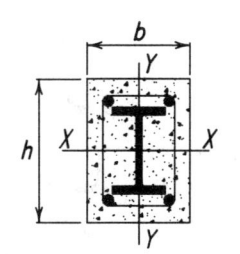

Size $b \times h$	18 in.×18 in.									
Reinf. bars	4-#8 bars									
Ties	#3 bars spaced 12 in. c. to c.									

Steel Shape	Designation		W 10								
	Wt./ft	112		100		88		77		68	
	F_y	36	50	36	50	36	50	36	50	36	50
Effective length KL (ft) with respect to least radius of gyration r_{my}	0	1840	2240	1750	2100	1650	1960	1560	1820	1480	1720
	6	1820	2200	1720	2060	1620	1920	1530	1790	1460	1690
	7	1810	2190	1710	2050	1610	1910	1520	1780	1450	1680
	8	1800	2170	1700	2030	1600	1900	1510	1770	1440	1660
	9	1790	2150	1690	2020	1590	1880	1500	1750	1430	1650
	10	1770	2130	1680	2000	1580	1860	1490	1730	1420	1630
	11	1760	2110	1660	1980	1570	1840	1480	1720	1400	1620
	12	1740	2090	1650	1960	1550	1820	1460	1700	1390	1600
	13	1730	2070	1630	1930	1540	1800	1450	1670	1370	1580
	14	1710	2040	1610	1910	1520	1780	1430	1650	1360	1550
	15	1690	2010	1600	1880	1500	1750	1410	1630	1340	1530
	16	1670	1980	1580	1850	1480	1720	1390	1600	1320	1510
	17	1650	1950	1560	1830	1460	1700	1370	1580	1300	1480
	18	1630	1920	1530	1790	1440	1670	1350	1550	1280	1450
	19	1600	1890	1510	1760	1420	1640	1330	1520	1260	1430
	20	1580	1850	1490	1730	1400	1610	1310	1490	1240	1400
	22	1530	1780	1440	1660	1350	1540	1260	1430	1190	1340
	24	1480	1710	1390	1590	1300	1470	1210	1360	1140	1270
	26	1420	1630	1330	1520	1250	1400	1160	1300	1090	1210
	28	1360	1550	1280	1440	1190	1330	1110	1230	1040	1140
	30	1300	1470	1220	1360	1140	1260	1050	1160	990	1080
	32	1240	1380	1160	1280	1080	1180	1000	1090	936	1010
	34	1180	1300	1100	1210	1020	1110	944	1020	883	942
	36	1120	1220	1040	1130	963	1030	889	946	829	876
	38	1060	1140	982	1050	906	962	834	878	776	811
	40	995	1060	923	975	850	891	779	811	723	748

Properties										
$\phi_b M_{nx}$ (kip-ft)	589	780	532	704	475	627	421	555	378	497
$\phi_b M_{ny}$ (kip-ft)	379	488	346	445	313	402	282	361	256	327
$P_{ex}(K_x L_x)^2/10^4$ (kip-ft^2)	236	236	216	216	196	196	178	178	163	163
$P_{ey}(K_y L_y)^2/10^4$ (kip-ft^2)	236	236	216	216	196	196	178	178	163	163

r_{my} (in.)	5.40		5.40		5.40		5.40		5.40	
r_{mx} / r_{my} (in./in.)	1.00		1.00		1.00		1.00		1.00	

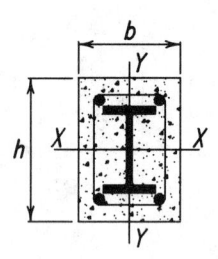

			F_y = 36 ksi
			F_y = 50 ksi

COMPOSITE COLUMNS
W Shapes
$f_c' = 5$ ksi
All reinforcing steel is Grade 60
Axial design strength in kips

Size $b \times h$		18 in.×18 in.						
Reinf. bars		4-#8 bars						
Ties		#3 bars spaced 12 in. c. to c.						
Designation		**W 10**						
Wt./ft		60		54		49		45
F_y		36	50	36	50	36	50	36	50
	0	1420	1620	1360	1550	1330	1500	1290	1450
	6	1390	1590	1340	1520	1300	1470	1270	1420
	7	1380	1580	1330	1510	1290	1460	1260	1410
	8	1370	1570	1320	1500	1280	1440	1250	1400
	9	1360	1560	1310	1490	1270	1430	1240	1390
	10	1350	1540	1300	1470	1260	1420	1230	1370
	11	1340	1520	1290	1450	1250	1400	1220	1360
	12	1320	1500	1270	1430	1230	1380	1200	1340
	13	1310	1480	1260	1410	1220	1360	1190	1320
	14	1290	1460	1240	1390	1200	1340	1170	1300
	15	1270	1440	1220	1370	1190	1320	1150	1280
	16	1260	1420	1210	1350	1170	1300	1140	1250
	17	1240	1390	1190	1320	1150	1270	1120	1230
	18	1220	1370	1170	1300	1130	1250	1100	1210
	19	1200	1340	1150	1270	1110	1220	1080	1180
	20	1170	1310	1120	1250	1090	1200	1060	1160
	22	1130	1250	1080	1190	1040	1140	1010	1100
	24	1080	1190	1030	1130	995	1080	965	1040
	26	1030	1130	984	1070	947	1020	917	987
	28	981	1070	934	1010	897	963	868	927
	30	929	1000	883	947	847	903	818	868
	32	877	938	832	884	796	842	768	808
	34	825	874	781	822	746	782	718	749
	36	772	811	729	761	695	723	668	692
	38	721	749	679	702	646	665	620	636
	40	670	689	630	644	598	609	572	581

Steel Shape / Effective length KL (ft) with respect to least radius of gyration r_{my}

Properties								
$\phi_b M_{nx}$ (kip-ft)	341	446	310	405	288	375	275	356
$\phi_b M_{ny}$ (kip-ft)	234	297	215	272	201	254	182	227
$P_{ex}(K_x L_x)^2 / 10^4$ (kip-ft²)	149	149	139	139	131	131	125	125
$P_{ey}(K_y L_y)^2 / 10^4$ (kip-ft²)	149	149	139	139	131	131	125	125
r_{my} (in.)	5.40		5.40		5.40		5.40	
r_{mx} / r_{my} (in./in.)	1.00		1.00		1.00		1.00	

F_y = 36 ksi
F_y = 50 ksi

COMPOSITE COLUMNS
W Shapes
$f_c' = 5$ ksi
All reinforcing steel is Grade 60
Axial design strength in kips

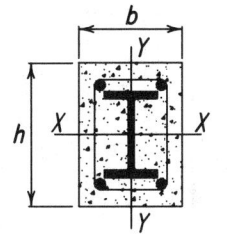

Size $b \times h$						16 in.×16 in.					
Reinf. bars						4-#7 bars					
Ties						#3 bars spaced 10 in. c. to c.					
Steel Shape	Designation						**W 18**				
	Wt./ft		67		58		48		40		35
	F_y		36	50	36	50	36	50	36	50	36	50
Effective length KL (ft) with respect to least radius of gyration r_{my}	0	1280	1510	1200	1410	1120	1290	1050	1190	1010	1140
	6	1250	1480	1180	1380	1100	1260	1030	1160	991	1110
	7	1250	1470	1170	1360	1090	1250	1020	1150	982	1100
	8	1240	1450	1160	1350	1080	1230	1010	1140	973	1090
	9	1220	1440	1150	1340	1070	1220	1000	1130	962	1070
	10	1210	1420	1140	1320	1060	1200	990	1110	951	1060
	11	1200	1400	1130	1300	1040	1190	977	1100	938	1040
	12	1180	1380	1110	1280	1030	1170	963	1080	924	1020
	13	1170	1360	1100	1260	1010	1150	948	1060	909	1010
	14	1150	1340	1080	1240	999	1130	932	1040	893	986
	15	1130	1310	1060	1220	982	1110	916	1020	877	965
	16	1120	1290	1050	1190	964	1080	898	996	859	944
	17	1100	1260	1030	1170	946	1060	880	973	841	922
	18	1080	1230	1010	1140	927	1040	861	949	822	898
	19	1060	1210	987	1120	907	1010	841	925	803	875
	20	1030	1180	966	1090	886	984	821	900	783	850
	22	989	1120	922	1030	844	929	779	848	741	800
	24	942	1050	876	971	799	873	736	795	698	748
	26	894	990	829	910	754	816	692	741	655	696
	28	844	926	781	849	707	759	647	686	610	643
	30	794	861	733	787	661	702	602	632	566	591
	32	743	796	684	727	614	646	557	579	522	540
	34	693	733	636	667	569	591	513	528	480	491
	36	644	672	589	610	524	537	470	478	438	444
	38	595	612	542	554	480	486	429	431	398	398
	40	548	555	498	501	438	438	389	389	360	360
Properties											
$\phi_b M_{nx}$ (kip-ft)		303	400	268	353	226	296	196	255	175	227
$\phi_b M_{ny}$ (kip-ft)		202	259	182	233	155	198	138	175	125	157
$P_{ex}(K_x L_x)^2/10^4$ (kip-ft^2)		119	119	107	107	94.1	94.1	83.4	83.4	77.2	77.2
$P_{ey}(K_y L_y)^2/10^4$ (kip-ft^2)		119	119	107	107	94.1	94.1	83.4	83.4	77.2	77.2
r_{my} (in.)		4.80		4.80		4.80		4.80		4.80	
r_{mx}/r_{my} (in./in.)		1.00		1.00		1.00		1.00		1.00	

COMPOSITE DESIGN

| | F_y = 36 ksi |
| | F_y = 50 ksi |

COMPOSITE COLUMNS
W Shapes
$f_c' = 8$ ksi
All reinforcing steel is Grade 60
Axial design strength in kips

Size $b \times h$										
					24 in.×26 in.					

Reinf. bars	4-#11 bars

Ties	#4 bars spaced 16 in. c. to c.

	Designation	W 14									
Steel Shape	Wt./ft	426		398		370		342		311	
	F_y	36	50	36	50	36	50	36	50	36	50
Effective length KL (ft) with respect to least radius of gyration r_{my}	0	6040	7530	5830	7220	5620	6910	5400	6610	5150	6240
	6	6000	7460	5780	7150	5570	6850	5360	6540	5110	6180
	7	5980	7430	5770	7130	5560	6820	5350	6520	5090	6150
	8	5960	7410	5750	7100	5540	6800	5330	6490	5070	6130
	9	5940	7370	5730	7070	5520	6770	5310	6460	5060	6100
	10	5920	7340	5710	7040	5500	6730	5290	6430	5030	6070
	11	5890	7300	5680	7000	5470	6700	5260	6400	5010	6030
	12	5870	7260	5660	6960	5450	6660	5240	6360	4980	6000
	13	5840	7210	5630	6910	5420	6610	5210	6310	4960	5960
	14	5800	7160	5590	6860	5390	6570	5180	6270	4930	5910
	15	5770	7110	5560	6810	5350	6520	5140	6220	4890	5860
	16	5730	7050	5530	6760	5320	6460	5110	6170	4860	5820
	17	5690	6990	5490	6700	5280	6410	5070	6120	4820	5760
	18	5650	6930	5450	6640	5240	6350	5030	6060	4790	5710
	19	5610	6870	5410	6580	5200	6290	4990	6000	4750	5650
	20	5570	6800	5360	6510	5160	6230	4950	5940	4700	5590
	22	5470	6660	5270	6370	5070	6090	4860	5810	4620	5460
	24	5370	6500	5170	6220	4970	5940	4770	5670	4520	5330
	26	5260	6340	5060	6070	4860	5790	4660	5520	4420	5190
	28	5150	6170	4950	5900	4750	5630	4550	5360	4310	5040
	30	5030	5990	4830	5730	4640	5460	4440	5200	4200	4880
	32	4900	5800	4710	5550	4520	5290	4320	5030	4090	4720
	34	4770	5610	4580	5360	4390	5110	4200	4850	3970	4550
	36	4640	5420	4450	5170	4260	4920	4070	4680	3840	4380
	38	4500	5220	4320	4980	4130	4740	3940	4500	3720	4210
	40	4360	5010	4180	4780	4000	4550	3810	4310	3590	4030

Properties											
$\phi_b M_{nx}$ (kip-ft)		3190	4270	2970	3980	2770	3700	2560	3430	2330	3120
$\phi_b M_{ny}$ (kip-ft)		1940	2550	1830	2410	1710	2260	1600	2100	1470	1930
$P_{ex}(K_x L_x)^2 / 10^4$ (kip-ft^2)		1710	1710	1620	1620	1530	1530	1430	1430	1320	1320
$P_{ey}(K_y L_y)^2 / 10^4$ (kip-ft^2)		1460	1460	1380	1380	1300	1300	1220	1220	1130	1130
r_{my} (in.)		7.20		7.20		7.20		7.20		7.20	
r_{mx} / r_{my} (in./in.)		1.08		1.08		1.08		1.08		1.08	

AMERICAN INSTITUTE OF STEEL CONSTRUCTION

F_y = 36 ksi
F_y = 50 ksi

COMPOSITE COLUMNS
W Shapes
f'_c = 8 ksi
All reinforcing steel is Grade 60
Axial design strength in kips

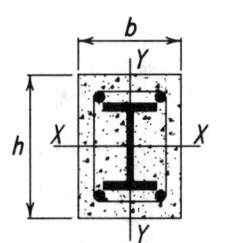

Size $b \times h$		24 in.×24 in.								
Reinf. bars		4-#11 bars								
Ties		#4 bars spaced 16 in. c. to c.								
Designation		**W 14**								
Wt./ft		283		257		233		211		193	
F_y		36	50	36	50	36	50	36	50	36	50
Effective length KL (ft) with respect to least radius of gyration r_{my}	0	4740	5730	4530	5430	4350	5160	4170	4910	4040	4710
	6	4700	5670	4500	5380	4310	5110	4140	4860	4000	4660
	7	4690	5650	4480	5360	4290	5090	4120	4840	3980	4640
	8	4670	5630	4470	5340	4280	5070	4110	4820	3970	4620
	9	4650	5600	4450	5310	4260	5040	4090	4790	3950	4600
	10	4630	5570	4430	5280	4240	5010	4070	4770	3930	4570
	11	4610	5540	4410	5250	4220	4980	4050	4740	3910	4540
	12	4590	5510	4380	5220	4200	4950	4020	4710	3890	4510
	13	4560	5470	4360	5180	4170	4920	4000	4670	3860	4480
	14	4530	5430	4330	5140	4140	4880	3970	4630	3830	4440
	15	4500	5390	4300	5100	4110	4840	3940	4590	3810	4400
	16	4470	5340	4270	5060	4080	4790	3910	4550	3780	4360
	17	4440	5290	4240	5010	4050	4750	3880	4510	3740	4320
	18	4400	5240	4200	4960	4020	4700	3850	4460	3710	4270
	19	4370	5190	4170	4910	3980	4650	3810	4410	3670	4220
	20	4330	5130	4130	4860	3940	4600	3770	4360	3640	4170
	22	4250	5020	4050	4740	3860	4490	3690	4250	3560	4070
	24	4160	4890	3960	4620	3780	4370	3610	4140	3470	3960
	26	4060	4760	3870	4490	3690	4250	3520	4020	3380	3840
	28	3970	4620	3770	4360	3590	4120	3430	3890	3290	3710
	30	3860	4480	3670	4220	3490	3980	3330	3760	3190	3590
	32	3760	4330	3570	4070	3390	3840	3220	3630	3090	3450
	34	3640	4170	3460	3930	3280	3700	3120	3490	2990	3320
	36	3530	4010	3340	3770	3170	3550	3010	3340	2880	3180
	38	3410	3850	3230	3620	3060	3400	2900	3200	2770	3040
	40	3300	3690	3110	3460	2950	3250	2790	3060	2660	2900
Properties										
$\phi_b M_{nx}$ (kip-ft)		2070	2770	1890	2530	1720	2300	1570	2090	1440	1920
$\phi_b M_{ny}$ (kip-ft)		1350	1770	1240	1620	1140	1490	1050	1370	972	1270
$P_{ex}(K_xL_x)^2/10^4$ (kip-ft^2)		1030	1030	952	952	882	882	817	817	765	765
$P_{ey}(K_yL_y)^2/10^4$ (kip-ft^2)		1030	1030	952	952	882	882	817	817	765	765
r_{my} (in.)		7.20		7.20		7.20		7.20		7.20	
r_{mx} / r_{my} (in./in.)		1.00		1.00		1.00		1.00		1.00	

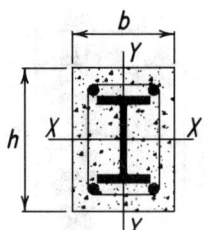

	Fy = 36 ksi
	Fy = 50 ksi

COMPOSITE COLUMNS
W Shapes
$f_c' = 8$ ksi
All reinforcing steel is Grade 60
Axial design strength in kips

Size $b \times h$		24 in.×24 in.						
Reinf. bars		4-#10 bars						
Ties		#3 bars spaced 16 in. c. to c.						

	Designation	**W 14**						
	Wt./ft	176		159		145		132
	F_y	36	50	36	50	36	50	36	50
Effective length KL (ft) with respect to least radius of gyration r_{my}	0	3870	4490	3730	4290	3630	4140	3520	3990
	6	3830	4440	3700	4240	3590	4090	3490	3940
	7	3820	4420	3680	4220	3580	4070	3470	3920
	8	3800	4400	3670	4200	3560	4050	3460	3900
	9	3790	4370	3650	4180	3540	4030	3440	3880
	10	3770	4350	3630	4160	3530	4000	3420	3850
	11	3750	4320	3610	4130	3500	3980	3400	3830
	12	3720	4290	3590	4100	3480	3950	3380	3800
	13	3700	4260	3560	4070	3460	3910	3350	3770
	14	3670	4220	3540	4030	3430	3880	3330	3730
	15	3640	4180	3510	3990	3400	3840	3300	3700
	16	3610	4140	3480	3950	3370	3800	3270	3660
	17	3580	4100	3450	3910	3340	3760	3240	3620
	18	3550	4060	3410	3870	3310	3720	3200	3580
	19	3510	4010	3380	3820	3270	3680	3170	3530
	20	3480	3960	3340	3780	3240	3630	3130	3490
	22	3400	3860	3270	3680	3160	3530	3050	3390
	24	3320	3750	3180	3570	3080	3430	2970	3290
	26	3230	3640	3100	3460	2990	3320	2880	3180
	28	3140	3520	3010	3340	2900	3200	2790	3070
	30	3040	3400	2910	3220	2800	3080	2700	2950
	32	2950	3270	2810	3100	2710	2960	2600	2830
	34	2840	3140	2710	2970	2610	2840	2500	2710
	36	2740	3010	2610	2840	2500	2710	2400	2580
	38	2630	2870	2500	2710	2400	2580	2300	2460
	40	2530	2740	2400	2580	2300	2450	2190	2330

Properties								
$\phi_b M_{nx}$ (kip-ft)	1300	1740	1190	1580	1090	1450	1010	1340
$\phi_b M_{ny}$ (kip-ft)	879	1150	805	1050	748	975	681	884
$P_{ex}(K_x L_x)^2/10^4$ (kip-ft²)	716	716	665	665	625	625	587	587
$P_{ey}(K_y L_y)^2/10^4$ (kip-ft²)	716	716	665	665	625	625	587	587
r_{my} (in.)	7.20		7.20		7.20		7.20	
r_{mx} / r_{my} (in./in.)	1.00		1.00		1.00		1.00	

F_y = 36 ksi
F_y = 50 ksi

COMPOSITE COLUMNS
W Shapes
$f_c' = 8$ ksi
All reinforcing steel is Grade 60
Axial design strength in kips

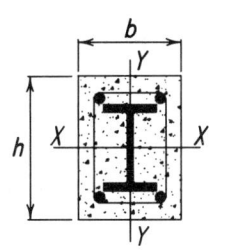

Size $b \times h$	22 in.×22 in.
Reinf. bars	4-#10 bars
Ties	#3 bars spaced 14 in. c. to c.

Steel Shape		Designation	W 14							
		Wt./ft	120		109		99		90	
		F_y	36	50	36	50	36	50	36	50
Effective length KL (ft) with respect to least radius of gyration r_{my}		0	3060	3480	2970	3350	2890	3240	2820	3140
		6	3020	3430	2930	3300	2850	3190	2780	3090
		7	3010	3410	2920	3280	2840	3170	2770	3070
		8	2990	3390	2900	3260	2820	3150	2750	3050
		9	2970	3370	2880	3240	2810	3130	2740	3030
		10	2950	3340	2860	3220	2790	3110	2720	3010
		11	2930	3320	2840	3190	2770	3080	2700	2980
		12	2910	3290	2820	3160	2740	3050	2670	2950
		13	2880	3250	2790	3130	2720	3020	2650	2920
		14	2860	3220	2770	3090	2690	2990	2620	2890
		15	2830	3180	2740	3060	2660	2950	2590	2850
		16	2800	3140	2710	3020	2630	2910	2560	2810
		17	2770	3100	2680	2980	2600	2870	2530	2770
		18	2730	3060	2640	2940	2560	2830	2490	2730
		19	2700	3020	2610	2900	2530	2790	2460	2690
		20	2660	2970	2570	2850	2490	2740	2420	2650
		22	2590	2870	2500	2750	2420	2650	2340	2550
		24	2510	2770	2420	2650	2340	2550	2260	2450
		26	2420	2670	2330	2550	2250	2450	2180	2350
		28	2330	2560	2240	2440	2160	2340	2090	2250
		30	2240	2440	2150	2330	2070	2230	2000	2140
		32	2150	2330	2060	2220	1980	2120	1910	2030
		34	2050	2210	1960	2100	1880	2000	1810	1920
		36	1950	2090	1870	1980	1790	1890	1720	1810
		38	1860	1970	1770	1870	1690	1780	1620	1700
		40	1760	1850	1670	1760	1600	1670	1530	1590

Properties								
$\phi_b M_{nx}$ (kip-ft)	894	1190	819	1080	755	996	697	917
$\phi_b M_{ny}$ (kip-ft)	597	773	551	712	514	661	478	612
$P_{ex}(K_x L_x)^2 / 10^4$ (kip-ft^2)	436	436	409	409	385	385	363	363
$P_{ey}(K_y L_y)^2 / 10^4$ (kip-ft^2)	436	436	409	409	385	385	363	363

r_{my} (in.)	6.60		6.60		6.60		6.60	
r_{mx} / r_{my} (in./in.)	1.00		1.00		1.00		1.00	

| | F_y = 36 ksi |
| | F_y = 50 ksi |

COMPOSITE COLUMNS
W Shapes
$f_c' = 8$ ksi
All reinforcing steel is Grade 60
Axial design strength in kips

Size $b \times h$		18 in.×22 in.						
Reinf. bars		4-#9 bars						
Ties		#3 bars spaced 12 in. c. to c.						

Steel Shape	Designation		W 14						
	Wt./ft		82		74		68		61
	F_y		36	50	36	50	36	50	36	50
Effective length KL (ft) with respect to least radius of gyration r_{my}	0	2370	2660	2310	2570	2260	2500	2200	2420
	6	2320	2600	2260	2510	2210	2440	2160	2360
	7	2310	2580	2240	2490	2200	2420	2140	2340
	8	2290	2550	2230	2470	2180	2400	2120	2320
	9	2270	2530	2200	2440	2160	2370	2100	2290
	10	2240	2500	2180	2410	2130	2340	2070	2260
	11	2220	2470	2150	2380	2110	2310	2050	2230
	12	2190	2430	2130	2340	2080	2280	2020	2200
	13	2160	2390	2100	2310	2050	2240	1990	2160
	14	2130	2350	2060	2270	2010	2200	1960	2120
	15	2090	2310	2030	2230	1980	2160	1920	2080
	16	2060	2270	1990	2180	1940	2120	1890	2040
	17	2020	2220	1960	2140	1910	2070	1850	1990
	18	1980	2180	1920	2090	1870	2020	1810	1950
	19	1940	2130	1880	2040	1830	1980	1770	1900
	20	1900	2080	1840	1990	1790	1930	1730	1850
	22	1820	1970	1750	1890	1700	1820	1640	1750
	24	1730	1860	1660	1780	1610	1720	1550	1640
	26	1630	1750	1570	1670	1520	1610	1460	1540
	28	1540	1640	1480	1560	1430	1500	1360	1430
	30	1440	1520	1380	1450	1330	1390	1270	1320
	32	1350	1410	1290	1340	1240	1280	1180	1220
	34	1250	1300	1190	1230	1140	1180	1090	1110
	36	1160	1190	1100	1130	1050	1080	998	1010
	38	1070	1090	1010	1030	966	977	911	917
	40	983	989	926	929	882	882	827	827
Properties									
$\phi_b M_{nx}$ (kip-ft)		644	849	590	776	549	720	501	654
$\phi_b M_{ny}$ (kip-ft)		336	431	309	396	290	370	268	341
$P_{ex}(K_x L_x)^2 / 10^4$ (kip-ft^2)		317	317	298	298	283	283	265	265
$P_{ey}(K_y L_y)^2 / 10^4$ (kip-ft^2)		212	212	199	199	189	189	178	178
r_{my} (in.)		5.40		5.40		5.40		5.40	
r_{mx} / r_{my} (in./in.)		1.22		1.22		1.22		1.22	

F_y = 36 ksi										

F_y = 50 ksi

COMPOSITE COLUMNS
W Shapes
f_c' = 8 ksi
All reinforcing steel is Grade 60
Axial design strength in kips

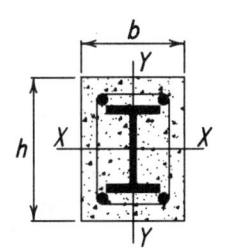

Size $b \times h$		22 in.×24 in.								
Reinf. bars		4-#10 bars								
Ties		#3 bars spaced 14 in. c. to c.								

		Designation		W 12						
Steel Shape		Wt./ft	336		305		279		252		230	
		F_y	36	50	36	50	36	50	36	50	36	50

Effective length KL (ft) with respect to least radius of gyration r_{my}		336 F_y=36	336 F_y=50	305 F_y=36	305 F_y=50	279 F_y=36	279 F_y=50	252 F_y=36	252 F_y=50	230 F_y=36	230 F_y=50
	0	4920	6100	4680	5740	4470	5450	4260	5150	4100	4900
	6	4880	6030	4630	5680	4430	5380	4220	5090	4050	4840
	7	4860	6000	4620	5650	4410	5360	4210	5060	4040	4820
	8	4840	5980	4600	5630	4400	5340	4190	5040	4020	4800
	9	4820	5950	4580	5600	4380	5310	4170	5010	4000	4770
	10	4800	5910	4560	5570	4360	5270	4150	4980	3980	4740
	11	4770	5870	4530	5530	4330	5240	4130	4950	3960	4710
	12	4750	5830	4510	5490	4310	5200	4100	4910	3930	4670
	13	4720	5790	4480	5450	4280	5160	4070	4870	3910	4630
	14	4690	5740	4450	5400	4250	5120	4040	4830	3880	4590
	15	4650	5690	4420	5350	4220	5070	4010	4780	3840	4540
	16	4620	5640	4380	5300	4180	5020	3980	4730	3810	4500
	17	4580	5580	4340	5250	4140	4970	3940	4680	3780	4450
	18	4540	5520	4310	5190	4110	4910	3910	4630	3740	4400
	19	4500	5460	4260	5130	4070	4850	3870	4570	3700	4340
	20	4460	5390	4220	5070	4030	4790	3830	4510	3660	4290
	22	4360	5260	4130	4930	3940	4670	3740	4390	3580	4170
	24	4270	5110	4040	4790	3840	4530	3650	4260	3480	4040
	26	4160	4960	3940	4650	3740	4390	3550	4120	3390	3910
	28	4050	4790	3830	4490	3640	4240	3450	3980	3290	3770
	30	3940	4630	3720	4330	3530	4080	3340	3830	3180	3620
	32	3820	4450	3600	4170	3420	3920	3230	3680	3070	3480
	34	3700	4280	3480	4000	3300	3760	3120	3520	2960	3330
	36	3570	4100	3360	3830	3180	3600	3000	3370	2850	3170
	38	3440	3920	3240	3650	3060	3430	2880	3210	2730	3020
	40	3310	3730	3110	3480	2940	3260	2760	3050	2620	2870

Properties											
$\phi_b M_{nx}$ (kip-ft)		2270	3030	2050	2750	1880	2510	1710	2280	1570	2090
$\phi_b M_{ny}$ (kip-ft)		1320	1730	1210	1580	1130	1470	1030	1350	954	1250
$P_{ex}(K_x L_x)^2 / 10^4$ (kip-ft^2)		1170	1170	1080	1080	999	999	921	921	857	857
$P_{ey}(K_y L_y)^2 / 10^4$ (kip-ft^2)		980	980	904	904	839	839	774	774	720	720
r_{my} (in.)		6.60		6.60		6.60		6.60		6.60	
r_{mx} / r_{my} (in./in.)		1.09		1.09		1.09		1.09		1.09	

	F_y = 36 ksi
	F_y = 50 ksi

COMPOSITE COLUMNS
W Shapes
f_c' = 8 ksi
All reinforcing steel is Grade 60
Axial design strength in kips

Size $b \times h$			20 in.×22 in.								
Reinf. bars			4-#10 bars								
Ties			#3 bars spaced 13 in. c. to c.								
	Designation		**W 12**								
	Wt./ft		210		190		170		152		136
	F_y		36	50	36	50	36	50	36	50	36	50
	0		3580	4320	3420	4080	3270	3860	3130	3660	3000	3470
	6		3540	4250	3380	4020	3230	3800	3080	3600	2960	3420
	7		3520	4230	3360	4000	3210	3780	3070	3580	2940	3400
	8		3500	4210	3350	3980	3190	3760	3050	3560	2930	3380
	9		3480	4180	3330	3950	3170	3730	3030	3530	2910	3350
	10		3460	4150	3310	3920	3150	3700	3010	3500	2880	3320
	11		3440	4110	3280	3890	3130	3670	2990	3470	2860	3290
	12		3410	4070	3260	3850	3100	3630	2960	3440	2840	3260
	13		3380	4030	3230	3810	3080	3600	2940	3400	2810	3220
	14		3350	3990	3200	3770	3050	3560	2910	3360	2780	3180
	15		3320	3940	3170	3730	3010	3510	2880	3320	2750	3140
	16		3290	3900	3130	3680	2980	3470	2840	3270	2720	3100
	17		3250	3850	3100	3630	2950	3420	2810	3230	2680	3050
	18		3220	3790	3060	3580	2910	3370	2770	3180	2650	3000
	19		3180	3740	3020	3520	2870	3320	2730	3130	2610	2950
	20		3140	3680	2980	3470	2830	3260	2690	3070	2570	2900
	22		3050	3560	2900	3350	2750	3150	2610	2960	2490	2800
	24		2960	3430	2810	3230	2660	3030	2520	2850	2400	2680
	26		2860	3300	2710	3100	2570	2900	2430	2730	2310	2570
	28		2760	3160	2610	2960	2470	2770	2340	2600	2210	2440
	30		2660	3010	2510	2830	2370	2640	2240	2470	2120	2320
	32		2550	2870	2410	2690	2270	2510	2140	2340	2020	2190
	34		2440	2720	2300	2540	2160	2370	2030	2210	1920	2070
	36		2330	2570	2190	2400	2060	2240	1930	2080	1820	1940
	38		2220	2430	2080	2260	1950	2100	1830	1950	1710	1820
	40		2110	2280	1980	2120	1850	1970	1730	1830	1610	1690
Properties											
$\phi_b M_{nx}$ (kip-ft)			1380	1840	1250	1670	1130	1510	1020	1360	926	1230
$\phi_b M_{ny}$ (kip-ft)			828	1080	760	991	692	901	631	821	578	749
$P_{ex}(K_xL_x)^2/10^4$ (kip-ft²)			645	645	595	595	546	546	502	502	462	462
$P_{ey}(K_yL_y)^2/10^4$ (kip-ft²)			533	533	492	492	452	452	415	415	382	382
r_{my} (in.)			6.00		6.00		6.00		6.00		6.00	
r_{mx}/r_{my} (in./in.)			1.10		1.10		1.10		1.10		1.10	

Steel Shape

Effective length KL (ft) with respect to least radius of gyration r_{my}

$F_y = 36$ ksi										

$F_y = 50$ ksi

COMPOSITE COLUMNS
W Shapes
$f_c' = 8$ ksi
All reinforcing steel is Grade 60
Axial design strength in kips

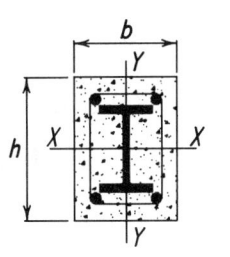

Size $b \times h$			20 in.×20 in.								
Reinf. bars			4-#9 bars								
Ties			#3 bars spaced 13 in. c. to c.								

		Designation	W 12								
Steel Shape		Wt./ft	120		106		96		87		79
		F_y	36	50	36	50	36	50	36	50	36	50
Effective length KL (ft) with respect to least radius of gyration r_{my}		0	2680	3100	2570	2950	2490	2830	2430	2730	2360	2640
		6	2650	3050	2540	2900	2460	2780	2390	2680	2320	2590
		7	2630	3040	2520	2880	2440	2760	2370	2670	2310	2570
		8	2620	3010	2510	2860	2430	2740	2360	2650	2290	2550
		9	2600	2990	2490	2840	2410	2720	2340	2620	2280	2530
		10	2580	2970	2470	2810	2390	2700	2320	2600	2260	2510
		11	2560	2940	2450	2780	2370	2670	2300	2570	2240	2480
		12	2540	2910	2430	2750	2350	2640	2280	2540	2210	2450
		13	2510	2880	2400	2720	2320	2610	2250	2510	2190	2420
		14	2490	2840	2380	2690	2300	2580	2230	2480	2160	2390
		15	2460	2800	2350	2650	2270	2540	2200	2440	2130	2350
		16	2430	2770	2320	2610	2240	2500	2170	2410	2100	2320
		17	2400	2730	2290	2570	2210	2460	2140	2370	2070	2280
		18	2370	2680	2260	2530	2180	2420	2110	2330	2040	2240
		19	2330	2640	2220	2490	2140	2380	2070	2290	2010	2200
		20	2300	2590	2190	2450	2110	2340	2040	2240	1970	2150
		22	2220	2500	2110	2350	2030	2240	1960	2150	1900	2060
		24	2140	2400	2040	2250	1960	2150	1890	2060	1820	1970
		26	2060	2290	1960	2150	1880	2050	1800	1960	1740	1870
		28	1980	2180	1870	2050	1790	1940	1720	1860	1660	1770
		30	1890	2070	1790	1940	1710	1840	1640	1750	1570	1670
		32	1800	1960	1700	1830	1620	1730	1550	1650	1480	1570
		34	1710	1850	1610	1720	1530	1630	1460	1550	1400	1470
		36	1620	1730	1520	1610	1440	1520	1380	1440	1310	1370
		38	1530	1620	1430	1500	1360	1420	1290	1340	1230	1270
		40	1440	1510	1340	1400	1270	1320	1200	1240	1140	1170

Properties											
$\phi_b M_{nx}$ (kip-ft)			773	1030	690	917	629	835	580	767	534	704
$\phi_b M_{ny}$ (kip-ft)			501	653	450	584	415	537	387	499	359	461
$P_{ex}(K_xL_x)^2/10^4$ (kip-ft^2)			340	340	312	312	291	291	273	273	257	257
$P_{ey}(K_yL_y)^2/10^4$ (kip-ft^2)			340	340	312	312	291	291	273	273	257	257
r_{my} (in.)			6.00		6.00		6.00		6.00		6.00	
r_{mx} / r_{my} (in./in.)			1.00		1.00		1.00		1.00		1.00	

F_y = 36 ksi
F_y = 50 ksi

COMPOSITE COLUMNS
W Shapes
$$f'_c = 8 \text{ ksi}$$
All reinforcing steel is Grade 60
Axial design strength in kips

Size $b \times h$			20 in.×20 in.					
Reinf. bars			4-#9 bars					
Ties			#3 bars spaced 13 in. c. to c.					
		Designation	**W 12**					
		Wt./ft	72		65		58	
		F_y	36	50	36	50	36	50
Steel Shape	Effective length KL (ft) with respect to least radius of gyration r_{my}	0	2310	2560	2250	2480	2200	2400
		6	2270	2510	2210	2430	2160	2350
		7	2250	2490	2200	2420	2140	2340
		8	2240	2470	2180	2400	2130	2320
		9	2220	2450	2170	2380	2110	2290
		10	2200	2430	2150	2350	2090	2270
		11	2180	2400	2120	2320	2070	2240
		12	2160	2370	2100	2300	2040	2220
		13	2130	2340	2080	2270	2020	2190
		14	2100	2310	2050	2230	1990	2150
		15	2080	2270	2020	2200	1960	2120
		16	2050	2240	1990	2160	1930	2080
		17	2010	2200	1960	2130	1900	2050
		18	1980	2160	1920	2090	1870	2010
		19	1950	2120	1890	2040	1830	1970
		20	1910	2080	1860	2000	1790	1920
		22	1840	1990	1780	1910	1720	1840
		24	1760	1900	1700	1820	1640	1750
		26	1680	1800	1620	1730	1560	1650
		28	1600	1700	1540	1630	1480	1560
		30	1510	1600	1450	1530	1390	1460
		32	1430	1500	1370	1430	1310	1360
		34	1340	1400	1280	1340	1220	1270
		36	1260	1300	1200	1240	1140	1170
		38	1170	1210	1120	1150	1060	1080
		40	1090	1110	1040	1050	978	991
Properties								
$\phi_b M_{nx}$ (kip-ft)			494	649	453	593	417	544
$\phi_b M_{ny}$ (kip-ft)			335	428	311	396	271	342
$P_{ex}(K_x L_x)^2/10^4$ (kip-ft^2)			242	242	229	229	214	214
$P_{ey}(K_y L_y)^2/10^4$ (kip-ft^2)			242	242	229	229	214	214
r_{my} (in.)			6.00		6.00		6.00	
r_{mx}/r_{my} (in./in.)			1.00		1.00		1.00	

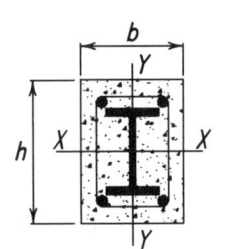

COMPOSITE COLUMNS
W Shapes
$f'_c = 8$ ksi
All reinforcing steel is Grade 60
Axial design strength in kips

| Fy = 36 ksi |
| Fy = 50 ksi |

Size $b \times h$		18 in.×18 in.								
Reinf. bars		4-#8 bars								
Ties		#3 bars spaced 12 in. c. to c.								
Designation		**W 10**								
Wt./ft		112		100		88		77		68	
F_y		36	50	36	50	36	50	36	50	36	50
	0	2280	2680	2190	2540	2100	2410	2010	2280	1940	2180
	6	2250	2630	2160	2490	2060	2360	1970	2230	1910	2130
	7	2230	2610	2140	2470	2050	2340	1960	2220	1890	2120
	8	2220	2590	2130	2450	2030	2320	1950	2200	1880	2100
	9	2200	2560	2110	2430	2020	2300	1930	2170	1860	2080
	10	2180	2540	2090	2410	2000	2270	1910	2150	1840	2050
	11	2160	2510	2070	2380	1980	2250	1890	2120	1820	2020
	12	2140	2480	2050	2350	1950	2220	1870	2090	1800	2000
	13	2120	2450	2020	2320	1930	2190	1840	2060	1770	1970
	14	2090	2410	2000	2280	1900	2150	1820	2030	1750	1930
	15	2060	2370	1970	2250	1880	2120	1790	2000	1720	1900
	16	2030	2340	1940	2210	1850	2080	1760	1960	1690	1860
	17	2000	2290	1910	2170	1820	2040	1730	1920	1660	1830
	18	1970	2250	1880	2130	1790	2000	1700	1880	1630	1790
	19	1940	2210	1850	2080	1750	1960	1670	1840	1600	1750
	20	1900	2160	1810	2040	1720	1920	1630	1800	1560	1710
	22	1830	2070	1740	1950	1650	1830	1560	1710	1490	1620
	24	1760	1970	1670	1850	1580	1730	1490	1620	1420	1530
	26	1680	1870	1590	1750	1500	1640	1410	1530	1340	1440
	28	1600	1760	1510	1650	1420	1540	1340	1430	1270	1350
	30	1520	1660	1430	1550	1340	1440	1260	1340	1190	1260
	32	1430	1550	1350	1450	1260	1340	1180	1240	1110	1160
	34	1350	1450	1270	1350	1180	1250	1100	1150	1030	1070
	36	1270	1340	1190	1250	1100	1150	1020	1060	957	985
	38	1180	1240	1110	1150	1020	1060	946	970	883	900
	40	1100	1140	1030	1050	948	968	872	885	811	817
Properties										
$\phi_b M_{nx}$ (kip-ft)		599	800	541	721	482	640	426	565	382	505
$\phi_b M_{ny}$ (kip-ft)		389	509	354	462	320	416	287	371	260	335
$P_{ex}(K_x L_x)^2 / 10^4$ (kip-ft²)		248	248	228	228	208	208	190	190	175	175
$P_{ey}(K_y L_y)^2 / 10^4$ (kip-ft²)		248	248	228	228	208	208	190	190	175	175
r_{my} (in.)		5.40		5.40		5.40		5.40		5.40	
r_{mx} / r_{my} (in./in.)		1.00		1.00		1.00		1.00		1.00	

Row labels at left: Steel Shape | Effective length KL (ft) with respect to least radius of gyration r_{my}

	F_y = 36 ksi
	F_y = 50 ksi

COMPOSITE COLUMNS
W Shapes
f_c' = 8 ksi
All reinforcing steel is Grade 60
Axial design strength in kips

Size $b \times h$	18 in.×18 in.							
Reinf. bars	4-#8 bars							
Ties	#3 bars spaced 12 in. c. to c.							

	Designation	W 10							
	Wt./ft	60		54		49		45	
	F_y	36	50	36	50	36	50	36	50
	0	1880	2090	1830	2020	1790	1970	1770	1920
	6	1840	2040	1790	1970	1760	1920	1730	1880
	7	1830	2020	1780	1960	1740	1900	1710	1860
	8	1810	2010	1760	1940	1730	1880	1700	1840
	9	1790	1980	1750	1910	1710	1860	1680	1820
	10	1770	1960	1730	1890	1690	1840	1660	1800
	11	1750	1930	1700	1870	1670	1810	1640	1770
	12	1730	1910	1680	1840	1640	1780	1610	1740
	13	1710	1880	1660	1810	1620	1750	1590	1710
	14	1680	1840	1630	1780	1590	1720	1560	1680
	15	1650	1810	1600	1740	1560	1690	1530	1650
	16	1620	1770	1570	1710	1530	1650	1500	1610
	17	1590	1740	1540	1670	1500	1620	1470	1580
	18	1560	1700	1510	1630	1470	1580	1440	1540
	19	1530	1660	1480	1590	1440	1540	1410	1500
	20	1490	1620	1440	1550	1400	1500	1370	1460
	22	1420	1540	1370	1470	1330	1420	1300	1380
	24	1350	1450	1300	1380	1260	1330	1230	1290
	26	1280	1360	1230	1300	1180	1250	1150	1210
	28	1200	1270	1150	1210	1110	1160	1080	1120
	30	1120	1180	1070	1120	1030	1070	1000	1040
	32	1050	1090	996	1030	956	987	925	951
	34	970	1000	921	946	882	903	851	868
	36	895	916	847	863	809	822	779	788
	38	822	834	776	784	739	743	709	711
	40	752	754	707	707	671	671	642	642

Steel Shape — Effective length KL (ft) with respect to least radius of gyration r_{my}

Properties								
$\phi_b M_{nx}$ (kip-ft)	344	452	313	410	290	379	277	361
$\phi_b M_{ny}$ (kip-ft)	237	304	217	277	204	258	184	231
$P_{ex}(K_x L_x)^2/10^4$ (kip-ft^2)	162	162	152	152	144	144	138	138
$P_{ey}(K_y L_y)^2/10^4$ (kip-ft^2)	162	162	152	152	144	144	138	138
r_{my} (in.)	5.40		5.40		5.40		5.40	
r_{mx}/r_{my} (in./in.)	1.00		1.00		1.00		1.00	

F_y = 36 ksi										

F_y = 50 ksi

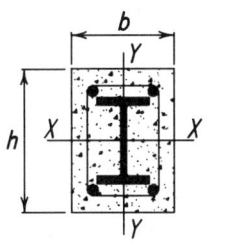

COMPOSITE COLUMNS
W Shapes
f'_c = 8 ksi
All reinforcing steel is Grade 60
Axial design strength in kips

Size $b \times h$		16 in.×16 in.								
Reinf. bars		4-#7 bars								
Ties		#3 bars spaced 10 in. c. to c.								

	Designation	**W 8**								
Steel Shape	Wt./ft	67		58		48		40		35	
	F_y	36	50	36	50	36	50	36	50	36	50
Effective length KL (ft) with respect to least radius of gyration r_{my}	0	1640	1870	1570	1770	1490	1650	1420	1560	1390	1510
	6	1600	1820	1530	1720	1450	1610	1380	1520	1350	1460
	7	1590	1800	1520	1710	1440	1590	1370	1500	1330	1450
	8	1570	1790	1500	1690	1420	1570	1360	1480	1320	1430
	9	1550	1760	1480	1660	1400	1550	1340	1460	1300	1410
	10	1530	1740	1470	1640	1380	1530	1320	1440	1280	1380
	11	1510	1710	1440	1620	1360	1500	1300	1410	1260	1360
	12	1490	1680	1420	1590	1340	1470	1270	1380	1240	1330
	13	1470	1650	1400	1560	1320	1450	1250	1360	1210	1300
	14	1440	1620	1370	1530	1290	1410	1220	1320	1180	1270
	15	1420	1590	1350	1490	1260	1380	1200	1290	1160	1240
	16	1390	1550	1320	1460	1240	1350	1170	1260	1130	1210
	17	1360	1520	1290	1420	1210	1310	1140	1220	1100	1170
	18	1330	1480	1260	1380	1180	1280	1110	1190	1070	1140
	19	1300	1440	1230	1350	1150	1240	1080	1150	1040	1100
	20	1270	1400	1200	1310	1120	1200	1050	1120	1010	1060
	22	1200	1310	1130	1230	1050	1120	981	1040	940	989
	24	1130	1230	1060	1140	983	1040	914	961	873	912
	26	1060	1140	996	1060	915	963	847	883	806	836
	28	993	1060	926	977	846	884	779	807	739	761
	30	922	972	857	895	778	805	713	731	673	687
	32	852	888	789	815	712	729	648	659	609	617
	34	784	806	722	737	648	656	586	589	548	549
	36	717	729	657	663	586	586	525	525	490	490
	38	653	654	595	595	526	526	471	471	439	439
	40	590	590	537	537	475	475	425	425	397	397
Properties											
$\phi_b M_{nx}$ (kip-ft)		307	408	272	359	228	300	197	258	177	230
$\phi_b M_{ny}$ (kip-ft)		206	268	186	240	157	202	140	178	127	160
$P_{ex}(K_x L_x)^2 / 10^4$ (kip-ft^2)		127	127	115	115	102	102	91.3	91.3	85.1	85.1
$P_{ey}(K_y L_y)^2 / 10^4$ (kip-ft^2)		127	127	115	115	102	102	91.3	91.3	85.1	85.1
r_{my} (in.)		4.80		4.80		4.80		4.80		4.80	
r_{mx} / r_{my} (in./in.)		1.00		1.00		1.00		1.00		1.00	

COMPOSITE COLUMNS—CONCRETE-FILLED STEEL PIPE AND STRUCTURAL TUBING

General Notes

Concentric load design strengths in the tables that follow are tabulated for the effective lengths KL in feet, shown at the left of each table. They are applicable to axially loaded members with respect to their minor axis in accordance with Section I2.2 of the LRFD Specification. The tables apply to normal-weight concrete.

For discussion of the effective length, range of Kl/r strength about the major axis, combined axial and bending strength, and for sample problems, see Composite Columns, General Notes.

The properties listed at the bottom of each table are for use in checking strength about the strong axis and in design for combined axial compression and bending.

The heavy horizontal lines within the tables indicate $Kl/r = 200$. No values are listed beyond these lines.

Steel Pipe Filled with Concrete

Design strengths for filled pipe are tabulated for $F_y = 36$ ksi and f_c' equal to 3.5 and 5 ksi. Steel pipe is manufactured to $F_y = 36$ ksi under ASTM A501 and to $F_y = 35$ ksi under ASTM A53 Types E or S, Grade B. Both are designed for 36 ksi yield stress.

Structural Tubing Filled with Concrete

Design strengths for square and rectangular structural tubing filled with concrete are tabulated for $F_y = 46$ ksi and f_c' equal to 3.5 and 5 ksi. Structural tubing is manufactured to $F_y = 46$ ksi under ASTM A500, Grade B.

F_y = 36 ksi	COMPOSITE COLUMNS Steel pipe f_c' = 3.5 ksi Axial design strength in kips	

Steel Pipe	Nominal Diameter (in.)		12		10		8		
	Wall Thickness (in.)		0.500	0.375	0.500	0.365	0.875	0.500	0.322
	Wt./ft		65.42	49.56	54.74	40.48	72.42	43.39	28.55
	F_y					36 ksi			
Effective length KL (ft) with respect to radius of gyration		0	862	733	681	564	746	507	384
		6	847	720	665	550	718	489	370
		7	842	716	660	545	708	482	365
		8	836	711	653	540	697	475	359
		9	829	705	646	534	684	467	353
		10	822	698	638	527	671	458	346
		11	814	691	629	520	656	448	338
		12	805	684	620	512	640	438	330
		13	796	675	610	503	623	427	322
		14	786	667	599	494	606	415	313
		15	775	658	587	485	588	403	304
		16	764	648	576	475	569	390	294
		17	752	638	563	464	549	377	284
		18	739	627	550	454	529	364	274
		19	727	616	537	442	509	351	264
		20	713	604	523	431	488	337	254
		22	686	580	495	407	447	309	232
		24	656	555	466	383	405	281	211
		26	626	529	436	358	365	254	191
		28	595	502	406	333	325	228	170
		30	563	475	376	308	288	202	151
		32	531	448	347	284	253	178	133
		34	499	420	318	260	224	158	118
		36	467	393	290	236	200	141	105
		38	436	366	263	214	179	126	94
		40	405	339	237	193	162	114	85

Properties									
r_m (in.)			4.33	4.38	3.63	3.67	2.76	2.88	2.94
$\phi_b M_n$ (kip-ft)			203	155	142	106	143	89.2	60.0
P_e $(KL)^2 / 10^4$ (kip-ft^2)			89.8	75.1	51.0	41.4	34.8	24.5	18.3

Note: Heavy line indicates Kl / r of 200.

| | | | | | F_y = 36 ksi |

COMPOSITE COLUMNS
Steel pipe
f'_c = 3.5 ksi
Axial design strength in kips

Steel Pipe	Nominal Diameter (in.)		6			5			4		
	Wall Thickness (in.)		0.864	0.432	0.280	0.750	0.375	0.258	0.674	0.337	0.237
	Wt./ft		53.16	28.57	18.97	38.55	20.78	14.62	27.54	14.98	10.79
	F_y						36 ksi				
Effective length KL (ft) with respect to radius of gyration		0	525	323	244	379	233	182	268	164	129
		6	491	303	229	344	213	167	230	143	113
		7	479	297	224	332	206	161	218	136	107
		8	466	289	218	319	199	156	205	129	102
		9	451	281	212	305	191	149	191	121	95
		10	436	271	205	290	182	142	176	112	89
		11	419	262	197	274	173	135	162	104	82
		12	401	252	190	258	163	128	147	95	75
		13	383	241	182	241	154	120	132	87	69
		14	364	230	173	225	144	112	118	78	62
		15	345	219	165	208	134	105	105	70	56
		16	326	207	156	191	124	97	92	62	49
		17	306	196	147	175	114	89	82	55	44
		18	287	184	139	160	105	82	73	49	39
		19	268	173	130	145	96	75	65	44	35
		20	249	161	121	131	87	68	59	40	32
		22	213	139	105	108	72	56	49	33	26
		24	180	119	89	91	60	47		28	22
		26	153	101	76	77	51	40			
		28	132	87	66	67	44	35			
		30	115	76	57		39	30			
		32	101	67	50						
		34	90	59	45						
		36		53	40						

Properties											
r_m (in.)			2.06	2.19	2.25	1.72	1.84	1.88	1.37	1.48	1.51
$\phi_b M_n$ (kip-ft)			78.0	44.8	30.5	47.3	27.3	19.6	26.9	15.8	11.6
$P_e\,(KL)^2 / 10^4$ (kip-ft^2)			13.9	9.13	6.92	6.99	4.66	3.65	3.15	2.15	1.70

Note: Heavy line indicates $Kl\,/\,r$ of 200.

F_y = 36 ksi

COMPOSITE COLUMNS
Steel pipe
f'_c = 5 ksi
Axial design strength in kips

	Nominal Diameter (in.)		12		10		8		
Steel Pipe	Wall Thickness (in.)		0.500	0.375	0.500	0.365	0.875	0.500	0.322
	Wt./ft		65.42	49.56	54.74	40.48	72.42	43.39	28.55
	F_y					**36 ksi**			
Effective length KL (ft) with respect to radius of gyration		0	979	855	762	649	786	557	438
		6	961	839	743	632	755	535	420
		7	955	833	736	626	745	528	414
		8	947	827	728	619	732	519	407
		9	939	819	720	611	719	510	399
		10	930	811	710	603	704	499	391
		11	920	802	699	594	688	488	382
		12	909	792	688	584	671	476	372
		13	897	781	676	573	653	463	361
		14	885	770	663	562	633	449	351
		15	872	758	649	550	614	435	339
		16	858	746	635	537	593	421	327
		17	844	733	620	524	572	406	315
		18	828	719	605	511	550	391	303
		19	813	705	589	497	528	375	291
		20	797	691	573	483	506	360	278
		22	763	660	540	454	461	328	253
		24	727	628	505	424	417	297	228
		26	691	596	471	394	374	266	203
		28	653	562	436	364	332	236	180
		30	615	528	401	334	292	208	158
		32	577	495	367	305	256	183	139
		34	539	461	334	276	227	162	123
		36	502	428	302	249	203	145	109
		38	465	395	272	224	182	130	98.2
		40	429	363	245	202	164	117	88.7
Properties									
r_m (in.)			4.33	4.38	3.63	3.67	2.76	2.88	2.94
$\phi_b M_n$ (kip-ft)			203	155	142	106	143	89.2	60.0
P_e $(KL)^2 / 10^4$ (kip-ft^2)			93.3	78.9	52.7	43.2	35.3	25.2	19.1

Note: Heavy line indicates Kl / r of 200.

COMPOSITE COLUMNS
Steel pipe
$f_c' = 5$ ksi
Axial design strength in kips

$F_y = 36$ ksi

	Nominal Diameter (in.)		6			5			4		
Steel Pipe	Wall Thickness (in.)		0.864	0.432	0.280	0.750	0.375	0.258	0.674	0.337	0.237
	Wt./ft		53.16	28.57	18.97	38.55	20.78	14.62	27.54	14.98	10.79
	F_y						36 ksi				
Effective length KL (ft) with respect to radius of gyration		0	545	351	275	393	253	204	276	176	143
		6	509	329	257	356	230	185	237	153	124
		7	496	321	251	343	222	179	224	145	117
		8	482	312	244	330	214	172	210	137	110
		9	467	302	236	315	204	164	196	128	103
		10	450	292	228	299	195	156	180	118	96
		11	432	281	219	282	184	148	165	109	88
		12	414	269	209	265	173	139	149	99	80
		13	394	257	200	247	162	130	134	90	72
		14	374	245	190	230	151	121	120	81	65
		15	354	232	180	212	140	112	106	72	58
		16	334	219	169	195	129	103	93	64	51
		17	313	206	159	178	119	94	82	56	45
		18	293	193	149	162	108	86	74	50	40
		19	273	180	139	146	98	78	66	45	36
		20	253	168	129	132	89	70	60	41	33
		22	216	144	110	109	73	58	49	34	27
		24	181	121	93	92	62	49		28	23
		26	155	104	79	78	53	42			
		28	133	89	68	67	45	36			
		30	116	78	59		40	31			
		32	102	68	52						
		34	90	61	46						
		36		54	41						
Properties											
r_m (in.)			2.06	2.19	2.25	1.72	1.84	1.88	1.37	1.48	1.51
$\phi_b M_n$ (kip-ft)			78.0	44.8	30.5	47.3	27.3	19.6	26.9	15.8	11.6
$P_e (KL)^2 / 10^4$ (kip-ft^2)			14.0	9.35	7.18	7.06	4.77	3.78	3.18	2.19	1.75

Note: Heavy line indicates Kl / r of 200.

F_y = 46 ksi

COMPOSITE COLUMNS
Square structural tubing
$f_c' = 3.5$ ksi
Axial design strength in kips

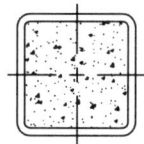

Steel Tube	Nominal Size	16×16	14×14		12×12			
	Thickness (in.)	½	½	⅜	⅝	½	⅜	⁵⁄₁₆
	Wt./ft	103.30	89.68	68.31	93.34	76.07	58.10	48.86
	F_y				46 ksi			

Effective length KL (ft) with respect to radius of gyration		16×16	14×14		12×12			
	0	1760	1460	1230	1360	1180	988	890
	6	1740	1440	1210	1340	1160	971	875
	7	1730	1430	1210	1330	1150	965	869
	8	1730	1430	1200	1320	1150	958	863
	9	1720	1420	1190	1310	1140	950	856
	10	1710	1410	1190	1300	1130	942	848
	11	1700	1400	1180	1280	1110	932	839
	12	1690	1390	1170	1270	1100	922	830
	13	1680	1370	1160	1250	1090	911	820
	14	1670	1360	1150	1240	1070	899	810
	15	1650	1350	1140	1220	1060	887	798
	16	1640	1330	1120	1200	1040	874	786
	17	1620	1320	1110	1180	1030	860	774
	18	1610	1300	1100	1160	1010	845	761
	19	1590	1280	1080	1140	992	830	747
	20	1580	1270	1070	1120	973	815	734
	22	1540	1230	1040	1080	934	783	704
	24	1500	1190	1000	1030	894	749	674
	26	1460	1150	968	979	852	713	642
	28	1420	1110	931	928	808	677	609
	30	1380	1060	894	877	764	640	576
	32	1330	1020	855	826	720	603	543
	34	1280	969	816	774	675	566	509
	36	1240	922	777	723	631	529	476
	38	1190	875	737	672	587	493	443
	40	1140	828	697	623	544	457	411

Properties								
r_m (in.)		6.29	5.48	5.54	4.60	4.66	4.72	4.75
$\phi_b M_n$ (kip-ft)		604	455	352	400	329	255	216
P_e $(KL)^2 / 10^4$ (kip-ft²)		319	203	171	137	120	101	90.7

Note: Heavy line indicates Kl / r of 200.

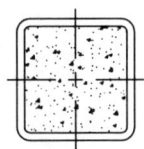

		F_y = 46 ksi

COMPOSITE COLUMNS
Square structural tubing
$f_c' = 3.5$ ksi
Axial design strength in kips

Steel Tube	Nominal Size		10×10					8×8				
	Thickness (in.)		⅝	½	⅜	⁵⁄₁₆	¼	⅝	½	⅜	⁵⁄₁₆	¼
	Wt./ft		76.33	62.46	47.90	40.35	32.63	59.32	48.85	37.69	31.84	25.82
	F_y							46 ksi				
Effective length KL (ft) with respect to radius of gyration		0	1070	924	767	687	603	795	686	567	503	439
		6	1040	900	748	670	588	762	659	545	484	422
		7	1030	892	742	664	583	751	650	537	477	416
		8	1020	883	734	657	577	738	639	528	470	409
		9	1010	872	725	650	570	724	627	519	461	402
		10	995	861	716	641	562	708	614	508	452	394
		11	980	848	705	632	554	691	600	496	441	385
		12	964	835	694	622	545	673	584	484	431	376
		13	947	820	682	611	536	654	568	471	419	365
		14	929	805	669	600	526	634	551	457	407	355
		15	909	788	656	588	515	613	534	443	394	344
		16	889	771	642	575	504	591	516	428	381	333
		17	869	754	627	562	493	569	497	413	368	321
		18	847	735	612	549	481	546	478	398	354	309
		19	825	716	597	535	469	524	459	382	340	297
		20	802	697	581	520	456	501	439	366	326	285
		22	755	657	548	491	430	454	400	334	298	260
		24	707	616	514	460	403	408	361	302	269	235
		26	658	574	479	430	376	364	323	271	242	211
		28	609	532	445	398	349	321	286	241	215	188
		30	560	490	410	368	322	280	251	212	189	165
		32	513	449	376	337	295	246	220	186	166	145
		34	466	409	343	308	269	218	195	165	147	129
		36	422	371	311	279	244	195	174	147	131	115
		38	379	333	280	251	220	175	156	132	118	103
		40	342	301	253	227	198	158	141	119	106	93
	Properties											
r_m (in.)			3.78	3.84	3.90	3.93	3.96	2.96	3.03	3.09	3.12	3.15
$\phi_b M_n$ (kip-ft)			268	223	174	148	120	163	137	108	92.1	75.6
$P_e (KL)^2 / 10^4$ (kip-ft²)			73.4	64.6	54.3	48.7	42.6	33.9	30.3	25.5	22.8	20.0

Note: Heavy line indicates Kl/r of 200.

F_y = 46 ksi

COMPOSITE COLUMNS
Square structural tubing
$f_c' = 3.5$ ksi
Axial design strength in kips

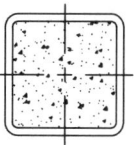

Steel Tube	Nominal Size		7×7					
	Thickness (in.)		$5/8$	$1/2$	$3/8$	$5/16$	$1/4$	$3/16$
	Wt./ft		50.81	42.05	32.58	27.59	22.42	17.08
	F_y		46 ksi					
Effective length KL (ft) with respect to radius of gyration		0	667	575	473	420	364	307
		6	631	545	449	399	346	292
		7	618	535	441	391	340	287
		8	604	523	431	383	333	281
		9	588	510	421	374	325	274
		10	571	496	410	364	316	267
		11	553	481	397	353	307	259
		12	534	465	384	342	297	250
		13	514	448	371	330	287	242
		14	493	430	357	317	276	233
		15	471	412	342	305	265	223
		16	449	394	327	291	254	214
		17	427	375	312	278	242	204
		18	405	356	297	265	230	194
		19	382	337	281	251	219	184
		20	360	318	266	237	207	174
		22	316	281	236	211	184	155
		24	274	245	206	185	161	136
		26	235	211	178	160	140	118
		28	202	182	154	138	120	101
		30	176	158	134	120	105	88
		32	155	139	118	106	92	78
		34	137	123	104	94	82	69
		36	122	110	93	83	73	61
		38	110	99	84	75	65	55
		40	99	89	75	68	59	50
Properties								
r_m (in.)			2.56	2.62	2.68	2.71	2.74	2.77
$\phi_b M_n$ (kip-ft)			120	102	81.1	69.3	56.9	43.8
$P_e (KL)^2 / 10^4$ (kip-ft^2)			21.3	19.1	16.2	14.5	12.7	10.7

Note: Heavy line indicates Kl / r of 200.

	F_y = 46 ksi

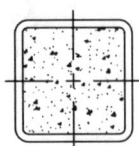

COMPOSITE COLUMNS
Square structural tubing
$$f_c' = 3.5 \text{ ksi}$$
Axial design strength in kips

Steel Tube			6×6					
	Nominal Size				6×6			
	Thickness (in.)		$\frac{5}{8}$	$\frac{1}{2}$	$\frac{3}{8}$	$\frac{5}{16}$	$\frac{1}{4}$	$\frac{3}{16}$
	Wt./ft		42.30	35.24	27.48	23.34	19.02	14.53
	F_y		46 ksi					

Effective length KL (ft) with respect to radius of gyration							
0	542	469	385	341	295	247
6	502	436	359	318	275	230
7	488	424	350	310	268	225
8	472	411	339	301	261	218
9	455	397	328	291	252	211
10	436	382	316	280	243	204
11	417	365	303	269	233	196
12	397	348	289	257	223	187
13	376	331	275	245	213	178
14	354	313	261	232	202	169
15	333	295	246	220	191	160
16	311	276	232	207	180	151
17	290	258	217	194	169	141
18	268	240	202	181	157	132
19	248	222	188	168	147	123
20	228	205	174	156	136	114
22	189	172	147	132	116	97
24	159	145	123	111	97	82
26	136	123	105	95	83	70
28	117	106	91	82	71	60
30	102	93	79	71	62	52
32	90	81	69	62	55	46
34	79	72	62	55	48	41
36		64	55	49	43	36
38				44	39	33

Properties						
r_m (in.)	2.15	2.21	2.27	2.30	2.33	2.36
$\phi_b M_n$ (kip-ft)	83.9	72.1	58.0	49.7	41.1	31.9
P_e $(KL)^2 / 10^4$ (kip-ft^2)	12.3	11.2	9.54	8.58	7.50	6.30

Note: Heavy line indicates Kl / r of 200.

F_y = 46 ksi

COMPOSITE COLUMNS
Square structural tubing
$f_c' = 3.5$ ksi
Axial design strength in kips

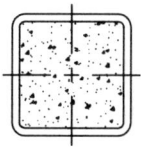

	Nominal Size		5½×5½					5×5			
	Thickness (in.)	⅜	⁵⁄₁₆	¼	³⁄₁₆	⅛	½	⅜	⁵⁄₁₆	¼	³⁄₁₆
	Wt./ft	24.93	21.21	17.32	13.25	9.01	28.43	22.37	19.08	15.62	11.97
	F_y					46 ksi					
	0	343	304	262	219	173	367	303	268	231	192
	6	315	279	241	201	159	328	272	241	208	173
	7	305	271	234	195	155	315	262	232	201	167
	8	294	261	226	189	149	301	251	223	193	160
	9	283	251	217	182	144	286	239	212	184	153
	10	270	240	208	174	137	269	226	201	174	145
	11	257	229	198	166	131	252	212	189	164	137
	12	243	217	188	157	124	235	198	177	154	128
	13	229	204	177	148	117	218	184	165	143	120
	14	215	192	167	139	110	200	170	152	133	111
	15	200	179	156	130	103	183	156	140	122	102
	16	186	167	145	121	96	166	143	128	112	94
	17	172	154	134	113	89	150	130	117	102	86
	18	158	142	124	104	82	134	117	106	93	78
	19	145	130	114	95	75	121	105	95	83	70
	20	132	119	104	87	69	109	95	86	75	63
	22	109	98	86	72	57	90	78	71	62	52
	24	91	82	72	61	48	76	66	59	52	44
	26	78	70	62	52	41	64	56	51	44	37
	28	67	61	53	45	35	56	48	44	38	32
	30	59	53	46	39	31	48	42	38	33	28
	32	51	46	41	34	27				29	25
	34	46	41	36	30	24					
	36					21					
	Properties										
r_m (in.)		2.07	2.10	2.13	2.16	2.19	1.80	1.86	1.89	1.92	1.95
$\phi_b M_n$ (kip-ft)		47.7	41.2	34.2	26.5	18.3	47.3	38.6	33.5	27.8	21.7
$P_e (KL)^2 / 10^4$ (kip-ft²)		7.07	6.37	5.58	4.69	3.69	5.84	5.08	4.59	4.03	3.39

Column labels along left side: **Steel Tube** / **Effective length KL (ft) with respect to radius of gyration**

Note: Heavy line indicates Kl/r of 200.

| | | | | $F_y = 46$ ksi |

COMPOSITE COLUMNS
Square structural tubing
$f'_c = 3.5$ ksi
Axial design strength in kips

Steel Tube	Nominal Size		4½×4½				
	Thickness (in.)		⅜	⁵⁄₁₆	¼	³⁄₁₆	⅛
	Wt./ft		19.82	16.96	13.91	10.70	7.31
	F_y		46 ksi				
Effective length KL (ft) with respect to radius of gyration		0	263	233	200	166	130
		6	230	204	177	147	115
		7	220	195	169	140	110
		8	208	185	160	133	104
		9	195	174	151	126	98
		10	182	163	141	118	92
		11	168	151	131	109	86
		12	155	139	121	101	79
		13	141	127	111	93	73
		14	128	115	101	84	66
		15	115	104	91	76	60
		16	102	93	82	69	54
		17	91	83	73	61	48
		18	81	74	65	55	43
		19	73	66	58	49	38
		20	66	60	53	44	35
		22	54	49	43	37	29
		24	46	41	36	31	24
		26	39	35	31	26	21
		28		30	27	23	18

Properties

r_m (in.)			1.66	1.69	1.72	1.75	1.78
$\phi_b M_n$ (kip-ft)			30.4	26.5	22.2	17.4	12.1
$P_e (KL)^2 / 10^4$ (kip-ft²)			3.52	3.20	2.82	2.37	1.86

Note: Heavy line indicates Kl / r of 200.

$F_y = 46$ ksi

COMPOSITE COLUMNS
Square structural tubing
$f_c' = 3.5$ ksi
Axial design strength in kips

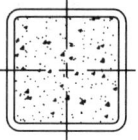

Steel Tube	Nominal Size		4×4					
	Thickness (in.)		½	⅜	⁵⁄₁₆	¼	³⁄₁₆	⅛
	Wt./ft		21.63	17.27	14.83	12.21	9.42	6.46
	F_y		46 ksi					

Effective length KL (ft) with respect to radius of gyration							
	0	271	225	199	171	141	110
	6	225	189	169	146	121	94
	7	211	178	159	137	114	89
	8	195	166	148	128	107	83
	9	179	153	137	119	99	77
	10	162	140	125	109	91	71
	11	145	126	114	99	83	65
	12	129	113	102	90	75	59
	13	114	100	91	80	67	53
	14	99	88	80	71	60	47
	15	86	77	70	62	52	41
	16	76	68	62	55	46	36
	17	67	60	55	48	41	32
	18	60	53	49	43	36	29
	19	54	48	44	39	33	26
	20	48	43	40	35	30	23
	22	40	36	33	29	24	19
	24		30	27	24	21	16
	26						14

Properties						
r_m (in.)	1.39	1.45	1.48	1.51	1.54	1.57
$\phi_b M_n$ (kip-ft)	27.7	23.2	20.4	17.1	13.5	9.5
$P_e (KL)^2 / 10^4$ (kip-ft²)	2.59	2.33	2.12	1.88	1.58	1.24

Note: Heavy line indicates Kl / r of 200.

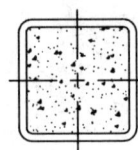

			F_y = 46 ksi

COMPOSITE COLUMNS
Square structural tubing
$f_c' = 5$ ksi
Axial design strength in kips

	Nominal Size		16×16	14×14		12×12			
Steel Tube	Thickness (in.)		½	½	⅜	⅝	½	⅜	⁵⁄₁₆
	Wt./ft		103.30	89.68	68.31	93.34	76.07	58.10	48.86
	F_y		46 ksi						

Effective length KL (ft) with respect to radius of gyration		16×16 ½	14×14 ½	14×14 ⅜	12×12 ⅝	12×12 ½	12×12 ⅜	12×12 ⁵⁄₁₆
	0	2000	1640	1420	1490	1310	1130	1030
	6	1980	1620	1400	1460	1290	1100	1010
	7	1970	1610	1390	1450	1280	1100	1000
	8	1960	1600	1380	1440	1270	1090	996
	9	1950	1590	1380	1430	1260	1080	987
	10	1940	1580	1370	1410	1250	1070	977
	11	1930	1570	1350	1400	1230	1060	966
	12	1920	1550	1340	1380	1220	1040	954
	13	1900	1540	1330	1360	1200	1030	942
	14	1890	1520	1320	1340	1190	1020	928
	15	1870	1510	1300	1320	1170	1000	914
	16	1860	1490	1290	1300	1150	984	899
	17	1840	1470	1270	1280	1130	967	884
	18	1820	1450	1250	1260	1110	950	868
	19	1800	1430	1230	1230	1090	932	851
	20	1780	1410	1220	1210	1070	913	833
	22	1730	1360	1180	1160	1020	873	797
	24	1690	1320	1140	1100	974	832	759
	26	1640	1270	1090	1050	924	790	720
	28	1590	1220	1050	991	874	747	680
	30	1530	1160	1000	933	823	703	640
	32	1480	1110	954	875	772	658	599
	34	1420	1060	907	817	721	614	559
	36	1370	1000	859	760	671	571	519
	38	1310	946	811	704	621	528	480
	40	1250	891	763	649	573	487	442

Properties

	16×16 ½	14×14 ½	14×14 ⅜	12×12 ⅝	12×12 ½	12×12 ⅜	12×12 ⁵⁄₁₆
r_m (in.)	6.29	5.48	5.54	4.60	4.66	4.72	4.75
$\phi_b M_n$ (kip-ft)	604	455	352	400	329	255	216
$P_e (KL)^2 / 10^4$ (kip-ft²)	334	212	180	141	125	106	95.8

Note: Heavy line indicates Kl/r of 200.

$F_y = 46$ ksi

COMPOSITE COLUMNS
Square structural tubing
$f_c' = 5$ ksi
Axial design strength in kips

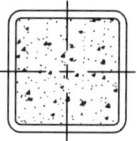

Steel Tube	Nominal Size		10×10					8×8				
	Thickness (in.)	$5/8$	$1/2$	$3/8$	$5/16$	$1/4$	$5/8$	$1/2$	$3/8$	$5/16$	$1/4$
	Wt./ft	76.33	62.46	47.90	40.35	32.63	59.32	48.85	37.69	31.84	25.82
	F_y					46 ksi					

Effective length KL (ft) with respect to radius of gyration											
0	1150	1010	860	782	701	844	739	623	562	500
6	1120	984	837	761	682	808	709	598	539	479
7	1110	975	829	754	675	796	698	589	531	472
8	1100	964	820	745	667	781	686	579	522	464
9	1080	952	809	736	659	766	672	567	512	454
10	1070	938	798	725	649	748	657	555	500	444
11	1050	924	785	714	639	730	641	542	488	434
12	1030	908	772	702	628	710	624	527	475	422
13	1010	891	758	689	616	689	606	512	462	410
14	993	873	742	675	603	667	587	496	447	397
15	972	855	727	660	590	644	568	480	433	384
16	949	835	710	645	576	620	548	463	417	370
17	926	815	693	629	562	596	527	446	401	356
18	902	794	675	613	547	571	506	428	385	341
19	877	772	656	596	531	547	484	410	369	327
20	852	750	637	578	516	522	463	392	353	312
22	800	704	598	543	484	471	419	355	320	283
24	746	658	559	507	451	422	376	319	287	254
26	692	610	518	470	417	374	335	284	256	226
28	638	563	478	433	384	328	295	251	226	199
30	585	516	438	397	352	286	257	219	197	173
32	532	471	399	361	320	251	226	192	173	152
34	482	426	362	327	289	223	200	170	153	135
36	433	383	325	294	259	199	179	152	137	120
38	389	344	292	263	232	178	160	136	123	108
40	351	311	263	238	210	161	145	123	111	98

Properties										
r_m (in.)	3.78	3.84	3.90	3.93	3.96	2.96	3.03	3.09	3.12	3.15
$\phi_b M_n$ (kip-ft)	268	223	174	148	120	163	137	108	92.1	75.6
$P_e (KL)^2 / 10^4$ (kip-ft^2)	75.3	66.7	56.5	51.0	45.0	34.5	31.1	26.4	23.8	20.9

Note: Heavy line indicates Kl/r of 200.

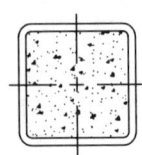

			F_y = 46 ksi

COMPOSITE COLUMNS
Square structural tubing
f'_c = 5 ksi
Axial design strength in kips

Steel Tube	Nominal Size		7×7					
	Thickness (in.)		$\frac{5}{8}$	$\frac{1}{2}$	$\frac{3}{8}$	$\frac{5}{16}$	$\frac{1}{4}$	$\frac{3}{16}$
	Wt./ft		50.81	42.05	32.58	27.59	22.42	17.08
	F_y		46 ksi					
Effective length KL (ft) with respect to radius of gyration		0	702	614	515	464	410	355
		6	663	581	488	439	388	335
		7	649	569	478	430	381	329
		8	634	556	467	421	372	321
		9	617	542	455	410	362	313
		10	599	526	442	398	352	304
		11	579	509	429	386	341	294
		12	558	491	414	373	329	284
		13	536	473	398	359	317	273
		14	514	453	382	344	304	262
		15	490	433	366	329	291	250
		16	467	413	349	314	277	238
		17	443	393	332	299	264	227
		18	419	372	315	283	250	215
		19	395	351	297	268	236	203
		20	371	331	280	252	223	191
		22	324	290	247	222	196	167
		24	280	252	214	193	170	145
		26	239	216	184	166	146	124
		28	206	186	159	143	126	107
		30	179	162	138	124	110	93
		32	158	142	121	109	96	82
		34	140	126	107	97	85	73
		36	125	112	96	86	76	65
		38	112	101	86	78	68	58
		40	101	91	78	70	62	52
	Properties							
	r_m (in.)		2.56	2.62	2.68	2.71	2.74	2.77
	$\phi_b M_n$ (kip-ft)		120	102	81.1	69.3	56.9	43.8
	P_e $(KL)^2 / 10^4$ (kip-ft^2)		21.7	19.5	16.7	15.0	13.2	11.3

Note: Heavy line indicates Kl / r of 200.

$F_y = 46$ ksi

COMPOSITE COLUMNS
Square structural tubing
$$f_c' = 5 \text{ ksi}$$
Axial design strength in kips

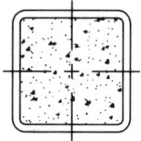

Steel Tube	Nominal Size		6×6					
	Thickness (in.)		⅝	½	⅜	⁵⁄₁₆	¼	³⁄₁₆
	Wt./ft		42.30	35.24	27.48	23.34	19.02	14.53
	F_y				46 ksi			
Effective length KL (ft) with respect to radius of gyration		0	566	496	415	372	328	281
		6	523	459	385	346	304	261
		7	508	447	375	336	296	254
		8	491	432	363	326	287	246
		9	473	417	350	315	277	237
		10	453	400	337	303	266	228
		11	432	383	322	290	255	218
		12	410	364	307	276	243	208
		13	388	345	292	262	231	197
		14	365	326	276	248	218	187
		15	343	306	259	234	206	176
		16	320	286	243	219	193	164
		17	297	267	227	205	180	154
		18	275	248	211	190	167	143
		19	253	229	195	176	155	132
		20	232	210	180	162	143	122
		22	192	175	151	136	120	102
		24	161	147	127	114	101	86
		26	138	126	108	98	86	73
		28	119	108	93	84	74	63
		30	103	94	81	73	64	55
		32	91	83	71	64	57	48
		34	80	73	63	57	50	43
		36		66	56	51	45	38
		38				46	40	34
Properties								
r_m (in.)			2.15	2.21	2.27	2.30	2.33	2.36
$\phi_b M_n$ (kip-ft)			83.9	72.1	58.0	49.7	41.1	31.9
P_e $(KL)^2 / 10^4$ (kip-ft²)			12.5	11.4	9.79	8.84	7.79	6.61

Note: Heavy line indicates Kl / r of 200.

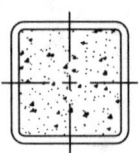

COMPOSITE COLUMNS
Square structural tubing
$f_c' = 5$ ksi
Axial design strength in kips

$F_y = 46$ ksi

Steel Tube	Nominal Size		5½×5½					5×5			
	Thickness (in.)	3/8	5/16	1/4	3/16	1/8	1/2	3/8	5/16	1/4	3/16
	Wt./ft	24.93	21.21	17.32	13.25	9.01	28.43	22.37	19.08	15.62	11.97
	F_y	46 ksi									

Effective length KL (ft) with respect to radius of gyration:

KL	3/8	5/16	1/4	3/16	1/8	1/2	3/8	5/16	1/4	3/16
0	367	329	289	247	203	384	322	288	252	215
6	336	301	265	226	185	342	289	259	227	193
7	325	292	256	219	179	328	277	249	218	185
8	313	281	247	211	172	313	265	238	208	177
9	300	269	237	202	165	297	251	226	198	169
10	286	257	226	193	157	279	237	213	187	159
11	272	244	215	183	149	261	223	200	176	149
12	256	231	203	173	141	243	207	187	164	140
13	241	217	191	162	132	224	192	173	152	129
14	225	203	178	152	123	206	177	160	141	119
15	209	189	166	141	114	187	162	146	129	109
16	194	175	154	131	106	170	147	133	117	100
17	178	161	142	121	97	153	133	121	106	90
18	163	148	130	111	89	136	119	108	96	81
19	149	135	119	101	81	122	107	97	86	73
20	135	122	108	91	73	111	97	88	78	66
22	111	101	89	76	60	91	80	73	64	54
24	94	85	75	63	51	77	67	61	54	46
26	80	72	64	54	43	65	57	52	46	39
28	69	62	55	47	37	56	49	45	40	34
30	60	54	48	41	32	49	43	39	34	29
32	53	48	42	36	29				30	26
34	47	42	37	32	25					
36					23					

Properties

	3/8	5/16	1/4	3/16	1/8	1/2	3/8	5/16	1/4	3/16
r_m (in.)	2.07	2.10	2.13	2.16	2.19	1.80	1.86	1.89	1.92	1.95
$\phi_b M_n$ (kip-ft)	47.7	41.2	34.2	26.5	18.3	47.3	38.6	33.5	27.8	21.7
$P_e (KL)^2 / 10^4$ (kip-ft^2)	7.23	6.55	5.78	4.90	3.92	5.93	5.19	4.71	4.16	3.53

Note: Heavy line indicates Kl/r of 200.

F_y = 46 ksi	

COMPOSITE COLUMNS
Square structural tubing
$f_c' = 5$ ksi
Axial design strength in kips

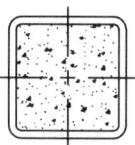

Steel Tube	Nominal Size		$4\frac{1}{2}\times4\frac{1}{2}$				
	Thickness (in.)		$\frac{3}{8}$	$\frac{5}{16}$	$\frac{1}{4}$	$\frac{3}{16}$	$\frac{1}{8}$
	Wt./ft		19.82	16.96	13.91	10.70	7.31
	F_y		46 ksi				
Effective length KL (ft) with respect to radius of gyration		0	278	249	217	184	149
		6	242	217	190	161	130
		7	231	207	181	154	124
		8	218	196	172	146	117
		9	204	184	161	137	110
		10	190	171	150	127	103
		11	175	158	139	118	95
		12	160	145	128	108	87
		13	146	132	117	99	79
		14	132	119	105	90	72
		15	118	107	95	80	64
		16	104	95	84	72	57
		17	92	84	75	64	51
		18	82	75	67	57	45
		19	74	68	60	51	40
		20	67	61	54	46	37
		22	55	50	45	38	30
		24	46	42	38	32	25
		26	40	36	32	27	22
		28		31	28	23	19

Properties						
r_m (in.)			1.66	1.69	1.72	1.75	1.78
$\phi_b M_n$ (kip-ft)			30.4	26.5	22.2	17.4	12.1
P_e $(KL)^2/10^4$ (kip-ft²)			3.58	3.27	2.90	2.46	1.96

Note: Heavy line indicates Kl/r of 200.

| | | | F_y = 46 ksi |

COMPOSITE COLUMNS
Square structural tubing
f'_c = 5 ksi
Axial design strength in kips

Steel Tube	Nominal Size		4×4					
	Thickness (in.)		½	⅜	5⁄16	¼	3⁄16	⅛
	Wt./ft		21.63	17.27	14.83	12.21	9.42	6.46
	F_y		46 ksi					
Effective length KL (ft) with respect to radius of gyration		0	280	236	211	184	156	125
		6	232	198	178	156	132	105
		7	217	186	167	146	124	99
		8	200	172	155	136	115	92
		9	183	158	143	126	107	85
		10	166	144	131	115	98	78
		11	148	130	118	104	88	71
		12	131	116	106	93	79	63
		13	115	103	94	83	71	56
		14	100	90	82	73	62	49
		15	87	78	72	64	54	43
		16	76	69	63	56	48	38
		17	68	61	56	50	42	34
		18	60	54	50	44	38	30
		19	54	49	45	40	34	27
		20	49	44	40	36	31	24
		22	40	36	33	30	25	20
		24		31	28	25	21	17
		26						14

Properties								
r_m (in.)			1.39	1.45	1.48	1.51	1.54	1.57
$\phi_b M_n$ (kip-ft)			27.7	23.2	20.4	17.1	13.5	9.45
$P_e\,(KL)^2/10^4$ (kip-ft^2)			2.62	2.36	2.16	1.92	1.64	1.30

Note: Heavy line indicates Kl / r of 200.

F_y = 46 ksi

COMPOSITE COLUMNS
Rectangular structural tubing
$f_c' = 3.5$ ksi
Axial design strength in kips

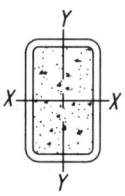

	Nominal Size	16×12	16×8	14×10		12×10		
Steel Tube	Thickness (in.)	½	½	½	⅜	½	⅜	⁵⁄₁₆
	Wt./ft	89.68	76.07	76.07	58.10	69.27	53.00	44.60
	F_y				46 ksi			

Effective length KL (ft) with respect to least radius of gyration, r_{my}								
	0	1450	1140	1170	978	1050	872	782
	6	1430	1100	1140	956	1020	851	763
	7	1420	1090	1130	948	1010	844	757
	8	1410	1070	1120	939	1000	835	749
	9	1400	1060	1110	928	990	826	741
	10	1380	1040	1100	917	978	816	732
	11	1370	1020	1080	905	964	804	721
	12	1360	994	1070	892	949	792	710
	13	1340	970	1050	877	933	779	699
	14	1330	946	1030	862	917	765	686
	15	1310	920	1010	846	899	750	673
	16	1290	893	992	830	880	735	659
	17	1270	865	971	812	861	719	645
	18	1250	837	949	794	841	703	630
	19	1230	808	926	775	820	685	615
	20	1210	778	903	756	799	668	599
	22	1160	718	855	716	755	632	566
	24	1120	657	806	675	710	594	533
	26	1070	597	755	633	664	556	498
	28	1020	539	704	591	617	517	464
	30	964	482	653	548	571	479	429
	32	911	427	602	506	525	441	395
	34	858	379	553	465	481	404	362
	36	805	338	505	425	438	368	330
	38	753	303	459	386	395	333	298
	40	702	273	414	349	357	300	269

Properties

	16×12	16×8	14×10		12×10		
r_{my} (in.)	4.84	3.30	4.02	4.08	3.94	4.00	4.03
r_{mx} / r_{my}	1.25	1.72	1.30	1.29	1.15	1.15	1.15
$\phi_b M_{nx}$ (kip-ft)	497	390	362	281	290	225	191
$\phi_b M_{ny}$ (kip-ft)	407	240	288	224	256	199	168
P_{ex} $(K_x L_x)^2$ / 10^4 (kip-ft^2)	245	174	150	125	101	85.3	76.4
P_{ey} $(K_y L_y)^2$ / 10^4 (kip-ft^2)	157	58.7	88.8	74.8	76.6	64.4	57.8

Note: Heavy line indicates Kl / r of 200.

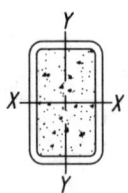

$F_y = 46$ ksi

COMPOSITE COLUMNS
Rectangular structural tubing
$f_c' = 3.5$ ksi
Axial design strength in kips

Steel Tube	Nominal Size		12×8				12×6		
	Thickness (in.)	$5/8$	$1/2$	$3/8$	$5/16$	$5/8$	$1/2$	$3/8$	$5/16$
	Wt./ft	76.33	62.46	47.90	40.35	67.82	55.66	42.79	36.10
	F_y				46 ksi				
Effective length KL (ft) with respect to least radius of gyration, r_{my}	0	1060	914	757	677	906	780	642	569
	6	1020	881	731	653	850	732	604	536
	7	1010	869	721	645	830	716	591	524
	8	991	856	711	635	808	697	576	511
	9	973	842	699	625	783	677	559	497
	10	954	826	686	613	757	655	542	481
	11	934	808	671	600	729	631	523	464
	12	912	790	656	587	699	606	503	447
	13	888	770	640	572	668	580	482	429
	14	864	749	623	557	637	553	460	410
	15	838	728	606	541	604	526	438	390
	16	812	705	587	525	571	498	416	370
	17	785	682	568	508	538	470	393	351
	18	757	658	549	491	506	443	371	331
	19	728	634	529	473	473	415	348	311
	20	699	610	509	455	441	387	326	291
	22	641	560	468	418	379	335	283	253
	24	583	510	427	382	321	284	241	216
	26	525	461	387	346	273	242	206	184
	28	470	413	348	310	236	209	177	159
	30	416	368	310	277	205	182	154	138
	32	366	324	273	244	180	160	136	122
	34	324	287	242	216	160	142	120	108
	36	289	256	216	193	142	126	107	96
	38	260	230	194	173	128	113	96	86
	40	234	207	175	156		102	87	78

Properties

r_{my} (in.)	3.14	3.20	3.26	3.28	2.37	2.42	2.48	2.51
r_{mx} / r_{my}	1.38	1.37	1.37	1.37	1.73	1.73	1.72	1.71
$\phi_b M_{nx}$ (kip-ft)	300	250	195	165	252	210	165	140
$\phi_b M_{ny}$ (kip-ft)	226	189	147	125	154	129	101	86.6
$P_{ex} (K_x L_x)^2 / 10^4$ (kip-ft^2)	95.2	83.7	69.9	62.8	74.5	65.8	55.0	49.1
$P_{ey} (K_y L_y)^2 / 10^4$ (kip-ft^2)	50.3	44.5	37.5	33.5	24.8	22.0	18.6	16.7

Note: Heavy line indicates Kl / r of 200.

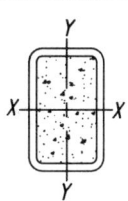

COMPOSITE COLUMNS
Rectangular structural tubing
$f_c' = 3.5$ ksi
Axial design strength in kips

F_y = 46 ksi

Steel Tube	Nominal Size		10×8				10×6			
	Thickness (in.)	½	⅜	5⁄16	¼	½	⅜	5⁄16	¼
	Wt./ft	55.66	42.79	36.10	29.23	48.85	37.69	31.84	25.82
	F_y				46 ksi				

Effective length KL (ft) with respect to least radius of gyration, r_{my}		10×8				10×6			
	0	800	662	589	516	676	556	493	429
	6	770	638	568	497	633	522	463	403
	7	759	629	560	490	619	511	453	394
	8	747	619	551	483	602	497	441	384
	9	734	608	542	475	584	483	428	373
	10	720	597	531	465	564	467	415	361
	11	704	584	520	455	543	450	400	348
	12	687	570	508	445	521	432	384	335
	13	669	555	495	433	497	413	368	320
	14	650	540	481	421	474	394	351	306
	15	630	524	467	409	449	375	334	291
	16	610	507	452	396	425	355	316	276
	17	589	490	437	383	400	335	299	261
	18	568	473	422	369	375	315	281	245
	19	546	455	406	355	351	295	263	230
	20	524	437	390	341	327	275	246	215
	22	479	400	357	313	281	238	213	186
	24	435	364	325	285	237	202	181	159
	26	391	328	293	257	202	172	154	135
	28	349	293	262	230	174	148	133	116
	30	308	260	233	204	152	129	116	101
	32	271	228	204	179	133	113	102	89
	34	240	202	181	159	118	101	90	79
	36	214	181	162	142	105	90	80	70
	38	192	162	145	127	95	80	72	63
	40	173	146	131	115		73	65	57

Properties									
r_{my} (in.)	3.12	3.18	3.21	3.24	2.37	2.43	2.46	2.49
r_{mx}/r_{my}	1.19	1.19	1.19	1.19	1.50	1.49	1.48	1.48
$\phi_b M_{nx}$ (kip-ft)	190	149	127	104	157	124	106	86.6
$\phi_b M_{ny}$ (kip-ft)	163	128	109	89.0	110	86.9	74.2	61.1
$P_{ex}\ (K_x L_x)^2/10^4$ (kip-ft²)	52.9	44.3	39.6	34.6	41.1	34.6	30.8	26.9
$P_{ey}\ (K_y L_y)^2/10^4$ (kip-ft²)	37.2	31.4	28.1	24.6	18.3	15.6	14.0	12.3

Note: Heavy line indicates Kl/r of 200.

COMPOSITE DESIGN

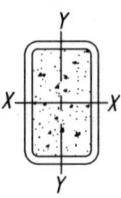

			$F_y = 46$ ksi

COMPOSITE COLUMNS
Rectangular structural tubing
$f_c' = 3.5$ ksi
Axial design strength in kips

Steel Tube	Nominal Size	10×5			8×6			
	Thickness (in.)	³⁄₈	⁵⁄₁₆	¹⁄₄	¹⁄₂	³⁄₈	⁵⁄₁₆	¹⁄₄
	Wt./ft	35.13	29.72	24.12	42.05	32.58	27.59	22.42
	F_y				46 ksi			

Effective length KL (ft) with respect to least radius of gyration, r_{my}								
	0	502	445	385	573	471	417	362
	6	459	408	353	535	440	391	339
	7	445	395	342	522	430	381	331
	8	428	381	330	507	418	371	322
	9	411	365	317	491	405	360	312
	10	392	349	302	473	391	347	302
	11	372	331	287	455	376	334	291
	12	351	313	272	435	360	321	279
	13	330	295	256	415	344	306	266
	14	309	276	240	394	327	291	254
	15	288	257	223	373	310	276	241
	16	266	239	207	352	293	261	228
	17	245	220	191	330	275	246	214
	18	225	202	176	309	258	231	201
	19	205	185	161	288	241	216	188
	20	186	168	146	267	224	201	175
	22	154	139	121	228	192	172	151
	24	129	117	101	192	162	145	127
	26	110	99	86	163	138	124	108
	28	95	86	74	141	119	107	93
	30	83	75	65	123	104	93	81
	32	73	66	57	108	91	82	72
	34	64	58	50	96	81	72	63
	36				85	72	65	57
	38				76	65	58	51
	40							46

Properties								
r_{my} (in.)		2.04	2.07	2.09	2.31	2.36	2.39	2.42
r_{mx} / r_{my}		1.72	1.71	1.72	1.25	1.25	1.25	1.25
$\phi_b M_{nx}$ (kip-ft)		111	95.2	78.3	111	88.3	75.6	62.1
$\phi_b M_{ny}$ (kip-ft)		68.7	58.7	48.3	91.1	72.5	62.1	51.1
$P_{ex} (K_x L_x)^2 / 10^4$ (kip-ft²)		29.5	26.5	23.1	23.2	19.7	17.6	15.3
$P_{ey} (K_y L_y)^2 / 10^4$ (kip-ft²)		9.98	9.01	7.83	14.8	12.5	11.2	9.83

Note: Heavy line indicates Kl / r of 200.

F_y = 46 ksi

COMPOSITE COLUMNS
Rectangular structural tubing
$f_c' = 3.5$ ksi
Axial design strength in kips

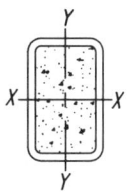

Steel Tube	Nominal Size		8×4					7×5				
	Thickness (in.)	$\frac{5}{8}$	$\frac{1}{2}$	$\frac{3}{8}$	$\frac{5}{16}$	$\frac{1}{4}$	$\frac{1}{2}$	$\frac{3}{8}$	$\frac{5}{16}$	$\frac{1}{4}$	$\frac{3}{16}$
	Wt./ft	42.30	35.24	27.48	23.34	19.02	35.24	27.48	23.34	19.02	14.53
	F_y					46 ksi					

Effective length KL (ft) with respect to least radius of gyration, r_{my}											
0	532	459	375	331	285	467	383	339	292	244
6	453	394	325	287	248	422	347	308	266	223
7	427	373	309	273	236	407	336	298	257	215
8	399	350	291	258	223	390	322	286	248	207
9	370	326	272	241	209	372	308	273	237	198
10	340	301	252	224	194	353	293	260	226	189
11	309	275	232	206	179	333	276	246	214	179
12	279	250	212	188	164	312	260	232	201	169
13	250	225	192	171	149	291	243	217	189	158
14	221	200	172	154	134	270	226	202	176	148
15	194	177	153	137	120	249	209	187	163	137
16	170	156	135	121	107	228	192	172	150	126
17	151	138	120	107	94	208	176	158	138	116
18	135	123	107	96	84	189	160	144	126	106
19	121	110	96	86	76	170	145	131	115	96
20	109	100	87	78	68	153	131	118	103	87
22	90	82	72	64	56	127	108	97	85	72
24	76	69	60	54	47	107	91	82	72	60
26			51	46	40	91	77	70	61	52
28						78	67	60	53	44
30						68	58	52	46	39
32							51	46	40	34
34										30

Properties										
r_{my} (in.)	1.49	1.54	1.60	1.62	1.65	1.90	1.95	1.98	2.01	2.04
r_{mx} / r_{my}	1.76	1.75	1.73	1.73	1.72	1.31	1.30	1.30	1.30	1.29
$\phi_b M_{nx}$ (kip-ft)	99.3	85.2	68.7	59.0	48.6	79.7	63.8	54.9	45.5	35.2
$\phi_b M_{ny}$ (kip-ft)	59.8	51.8	42.1	36.2	30.1	62.8	50.4	43.5	35.9	27.9
$P_{ex} (K_x L_x)^2 / 10^4$ (kip-ft^2)	18.1	16.3	13.9	12.4	10.9	14.0	11.9	10.7	9.35	7.83
$P_{ey} (K_y L_y)^2 / 10^4$ (kip-ft^2)	5.85	5.34	4.65	4.16	3.66	8.23	7.01	6.32	5.55	4.67

Note: Heavy line indicates Kl / r of 200.

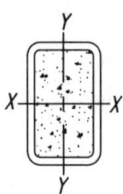

	F_y = 46 ksi
COMPOSITE COLUMNS	
Rectangular structural tubing	
f_c' = 3.5 ksi	
Axial design strength in kips	

	Nominal Size		7×4				6×4			
Steel Tube	Thickness (in.)	3/8	5/16	1/4	3/16	1/2	3/8	5/16	1/4	3/16
	Wt./ft	24.93	21.21	17.32	13.25	28.43	22.37	19.08	15.62	11.97
	F_y					**46 ksi**				

Effective length KL (ft) with respect to least radius of gyration, r_{my}										
	0	338	298	256	213	364	300	265	228	189
	6	291	258	223	185	309	257	228	197	164
	7	276	245	212	176	291	243	216	187	156
	8	259	230	200	166	272	228	203	176	147
	9	242	215	187	156	252	212	189	164	137
	10	224	200	173	145	231	196	175	152	127
	11	205	183	160	134	210	179	160	140	117
	12	187	167	146	122	189	162	146	127	107
	13	168	151	132	111	168	146	131	115	97
	14	151	136	119	100	149	130	118	103	87
	15	134	121	106	90	130	115	104	92	77
	16	118	107	94	79	114	101	92	81	68
	17	104	94	83	70	101	89	81	72	60
	18	93	84	74	63	90	80	72	64	54
	19	83	76	67	56	81	71	65	57	48
	20	75	68	60	51	73	65	59	52	44
	22	62	56	50	42	61	53	48	43	36
	24	52	47	42	35	51	45	41	36	30
	26	45	40	36	30			35	31	26

Properties										
r_{my} (in.)		1.57	1.60	1.63	1.66	1.48	1.54	1.57	1.60	1.63
r_{mx} / r_{my}		1.56	1.56	1.55	1.54	1.39	1.38	1.38	1.37	1.37
$\phi_b M_{nx}$ (kip-ft)		55.2	47.6	39.7	30.7	53.1	43.1	37.6	31.3	24.4
$\phi_b M_{ny}$ (kip-ft)		37.3	32.3	26.8	20.9	39.7	32.6	28.3	23.6	18.4
$P_{ex} (K_x L_x)^2 / 10^4$ (kip-ft^2)		9.83	8.87	7.72	6.42	7.61	6.62	5.96	5.20	4.39
$P_{ey} (K_y L_y)^2 / 10^4$ (kip-ft^2)		4.04	3.66	3.23	2.72	3.93	3.46	3.15	2.78	2.34

Note: Heavy line indicates Kl / r of 200.

F_y = 46 ksi

COMPOSITE COLUMNS
Rectangular structural tubing
$f_c' = 3.5$ ksi
Axial design strength in kips

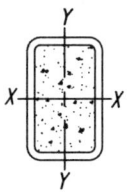

Steel Tube	Nominal Size		5×4			
	Thickness (in.)		⅜	⁵⁄₁₆	¼	³⁄₁₆
	Wt./ft		19.82	16.96	13.91	10.70
	F_y		**46 ksi**			
		0	252	216	177	136
Effective length KL (ft) with respect to least radius of gyration, r_{my}		6	215	185	153	118
		7	203	175	145	112
		8	190	164	136	106
		9	176	153	127	99
		10	162	141	118	92
		11	147	129	108	85
		12	133	117	98	77
		13	119	105	89	70
		14	106	93	79	63
		15	93	82	70	56
		16	81	73	62	49
		17	72	64	55	44
		18	64	57	49	39
		19	58	51	44	35
		20	52	46	40	32
		22	43	38	33	26
		24	36	32	28	22
		26			23	19

Properties				
r_{my} (in.)	1.50	1.53	1.56	1.59
r_{mx} / r_{my}	1.19	1.20	1.19	1.19
$\phi_b M_{nx}$ (kip-ft)	32.6	28.4	23.8	18.6
$\phi_b M_{ny}$ (kip-ft)	27.9	24.3	20.4	16.0
$P_{ex} (K_x L_x)^2 / 10^4$ (kip-ft²)	3.98	3.56	3.03	2.40
$P_{ey} (K_y L_y)^2 / 10^4$ (kip-ft²)	2.80	2.49	2.13	1.70
Note: Heavy line indicates Kl / r of 200.				

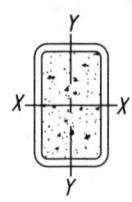

	Fy = 46 ksi

COMPOSITE COLUMNS
Rectangular structural tubing
$f_c' = 5$ ksi
Axial design strength in kips

Steel Tube	Nominal Size	16×12	16×8	14×10		12×10		
	Thickness (in.)	½	½	½	⅜	½	⅜	5⁄16
	Wt./ft	89.68	76.07	76.07	58.10	69.27	53.00	44.60
	F_y				46 ksi			

Effective length KL (ft) with respect to least radius of gyration, r_{my}		16×12	16×8	14×10 (½)	14×10 (⅜)	12×10 (½)	12×10 (⅜)	12×10 (5⁄16)
	0	1630	1250	1300	1110	1150	984	897
	6	1600	1210	1270	1080	1120	959	874
	7	1590	1190	1250	1070	1110	950	866
	8	1580	1180	1240	1060	1100	940	857
	9	1560	1150	1230	1050	1090	929	846
	10	1550	1130	1210	1040	1070	916	835
	11	1530	1110	1190	1020	1060	902	822
	12	1520	1080	1170	1000	1040	888	808
	13	1500	1060	1150	987	1020	872	794
	14	1480	1030	1130	969	1000	855	778
	15	1460	1000	1110	949	981	837	762
	16	1440	967	1090	929	960	819	745
	17	1410	935	1060	908	937	800	727
	18	1390	902	1040	886	914	780	709
	19	1360	869	1010	863	890	759	690
	20	1340	835	983	840	866	738	671
	22	1290	766	928	792	815	695	631
	24	1230	698	870	743	763	650	591
	26	1170	630	812	693	711	605	549
	28	1110	565	753	643	658	560	508
	30	1050	502	695	593	605	515	467
	32	987	442	638	544	554	471	426
	34	926	391	582	496	504	428	387
	36	865	349	528	450	456	387	350
	38	804	313	476	405	409	348	314
	40	745	283	429	365	369	314	283

Properties								
r_{my} (in.)		4.84	3.30	4.02	4.08	3.94	4.00	4.03
r_{mx} / r_{my}		1.25	1.72	1.30	1.29	1.15	1.15	1.15
$\phi_b M_{nx}$ (kip-ft)		497	390	362	281	290	225	191
$\phi_b M_{ny}$ (kip-ft)		407	240	288	224	256	199	168
$P_{ex} (K_x L_x)^2 / 10^4$ (kip-ft^2)		256	180	155	131	105	89.2	80.4
$P_{ey} (K_y L_y)^2 / 10^4$ (kip-ft^2)		164	60.7	92.1	78.4	79.3	67.3	60.8

Note: Heavy line indicates Kl / r of 200.

$F_y = 46$ ksi

COMPOSITE COLUMNS
Rectangular structural tubing
$f'_c = 5$ ksi
Axial design strength in kips

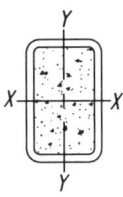

	Nominal Size	12×8				12×6			
Steel Tube	Thickness (in.)	$\frac{5}{8}$	$\frac{1}{2}$	$\frac{3}{8}$	$\frac{5}{16}$	$\frac{5}{8}$	$\frac{1}{2}$	$\frac{3}{8}$	$\frac{5}{16}$
	Wt./ft	76.33	62.46	47.90	40.35	67.82	55.66	42.79	36.10
	F_y	46 ksi							
Effective length KL (ft) with respect to least radius of gyration, r_{my}	0	1140	997	845	768	961	839	705	635
	6	1090	959	814	739	899	785	661	595
	7	1080	946	802	728	877	767	646	581
	8	1060	931	790	717	853	746	629	566
	9	1040	914	775	704	826	723	610	549
	10	1020	896	760	689	797	698	589	531
	11	997	876	743	674	766	672	567	511
	12	972	854	725	658	734	644	544	490
	13	946	832	706	640	701	615	520	469
	14	919	808	686	622	666	586	496	446
	15	890	783	665	603	631	555	470	424
	16	861	758	643	583	595	525	445	401
	17	831	732	621	562	559	494	419	378
	18	800	705	598	542	524	463	394	355
	19	768	677	575	520	489	433	368	332
	20	736	650	552	499	454	403	343	309
	22	672	594	505	456	388	345	295	266
	24	608	538	457	413	327	291	250	225
	26	545	483	411	371	279	248	213	192
	28	485	431	366	330	240	214	183	165
	30	427	380	323	291	209	187	160	144
	32	375	334	284	255	184	164	140	127
	34	332	296	252	226	163	145	124	112
	36	296	264	225	202	145	130	111	100
	38	266	237	202	181	130	116	100	90
	40	240	214	182	163		105	90	81
	Properties								
	r_{my} (in.)	3.14	3.20	3.26	3.28	2.37	2.42	2.48	2.51
	r_{mx} / r_{my}	1.38	1.37	1.37	1.37	1.73	1.73	1.72	1.71
	$\phi_b M_{nx}$ (kip-ft)	300	250	195	165	252	210	165	140
	$\phi_b M_{ny}$ (kip-ft)	226	189	147	125	154	129	101	86.6
	$P_{ex} (K_x L_x)^2 / 10^4$ (kip-ft²)	97.5	86.3	72.8	65.8	76.0	67.5	56.9	51.0
	$P_{ey} (K_y L_y)^2 / 10^4$ (kip-ft²)	51.5	45.9	39.0	35.1	25.3	22.5	19.3	17.4

Note: Heavy line indicates Kl / r of 200.

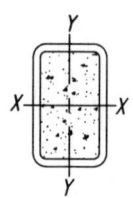

			F_y = 46 ksi
COMPOSITE COLUMNS			
Rectangular structural tubing			
f'_c = 5 ksi			
Axial design strength in kips			

	Nominal Size		10×8				10×6			
Steel Tube	Thickness (in.)		½	⅜	5/16	¼	½	⅜	5/16	¼
	Wt./ft		55.66	42.79	36.10	29.23	48.85	37.69	31.84	25.82
	F_y		46 ksi							
Effective length KL (ft) with respect to least radius of gyration, r_{my}		0	868	734	664	593	725	609	548	485
		6	834	706	638	569	677	569	512	454
		7	822	696	629	561	660	556	500	443
		8	808	684	618	552	642	540	486	431
		9	793	671	607	541	621	524	471	417
		10	776	657	594	530	599	505	455	403
		11	758	642	580	517	576	486	437	387
		12	739	626	565	504	551	466	419	371
		13	719	609	550	490	526	444	400	354
		14	697	591	534	476	499	423	381	337
		15	675	572	517	460	472	400	361	319
		16	652	553	499	444	445	378	340	301
		17	628	533	481	428	418	355	320	283
		18	604	512	463	412	391	333	300	265
		19	580	492	444	395	365	311	280	248
		20	555	471	425	378	339	289	261	230
		22	506	429	387	344	289	247	223	197
		24	456	387	349	310	243	208	188	166
		26	408	347	313	277	207	177	160	141
		28	362	308	277	245	178	153	138	122
		30	317	270	243	215	155	133	120	106
		32	279	237	213	189	137	117	106	93
		34	247	210	189	167	121	104	94	83
		36	220	187	169	149	108	93	84	74
		38	198	168	151	134	97	83	75	66
		40	178	152	137	121		75	68	60

Properties									
r_{my} (in.)		3.12	3.18	3.21	3.24	2.37	2.43	2.46	2.49
r_{mx} / r_{my}		1.19	1.19	1.19	1.19	1.50	1.49	1.48	1.48
$\phi_b M_{nx}$ (kip-ft)		190	149	127	104	157	124	106	86.6
$\phi_b M_{ny}$ (kip-ft)		163	128	109	89.0	110	86.9	74.2	61.1
P_{ex} $(K_x L_x)^2$ / 10^4 (kip-ft²)		54.4	46.0	41.3	36.4	42.1	35.7	32.0	28.1
P_{ey} $(K_y L_y)^2$ / 10^4 (kip-ft²)		38.3	32.6	29.3	25.9	18.8	16.1	14.5	12.8

Note: Heavy line indicates Kl / r of 200.

F_y = 46 ksi

COMPOSITE COLUMNS
Rectangular structural tubing
$f_c' = 5$ ksi
Axial design strength in kips

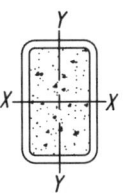

Steel Tube	Nominal Size		10×5			8×6			
	Thickness (in.)		$\frac{3}{8}$	$\frac{5}{16}$	$\frac{1}{4}$	$\frac{1}{2}$	$\frac{3}{8}$	$\frac{5}{16}$	$\frac{1}{4}$
	Wt./ft		35.13	29.72	24.12	42.05	32.58	27.59	22.42
	F_y					46 ksi			

Effective length KL (ft) with respect to least radius of gyration, r_{my}									
		0	544	489	431	610	512	460	406
		6	496	446	393	568	477	429	379
		7	479	431	380	554	465	418	369
		8	461	415	365	538	451	406	359
		9	441	397	349	520	437	393	347
		10	419	378	332	500	421	379	334
		11	397	358	315	480	404	363	321
		12	374	337	296	459	386	347	307
		13	350	316	278	436	367	331	292
		14	326	295	259	414	349	314	277
		15	303	273	240	390	329	297	262
		16	279	252	221	367	310	279	247
		17	256	232	203	344	291	262	231
		18	234	212	185	321	271	245	216
		19	212	192	168	298	252	228	201
		20	191	174	152	276	234	211	186
		22	158	143	126	233	198	179	158
		24	133	121	106	196	167	150	133
		26	113	103	90	167	142	128	113
		28	98	89	78	144	122	111	97
		30	85	77	68	125	107	96	85
		32	75	68	59	110	94	85	75
		34	66	60	53	98	83	75	66
		36				87	74	67	59
		38				78	66	60	53
		40							48

Properties								
r_{my} (in.)	2.04	2.07	2.09	2.31	2.36	2.39	2.42
r_{mx} / r_{my}	1.72	1.71	1.72	1.25	1.25	1.25	1.25
$\phi_b M_{nx}$ (kip-ft)	111	95.2	78.3	111	88.3	75.6	62.1
$\phi_b M_{ny}$ (kip-ft)	68.7	58.7	48.3	91.1	72.5	62.1	51.1
P_{ex} $(K_x L_x)^2 / 10^4$ (kip-ft^2)	30.4	27.4	24.1	23.7	20.2	18.2	16.0
P_{ey} $(K_y L_y)^2 / 10^4$ (kip-ft^2)	10.3	9.31	8.15	15.1	12.9	11.6	10.3

Note: Heavy line indicates Kl/r of 200.

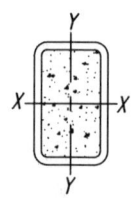

COMPOSITE COLUMNS
Rectangular structural tubing
$$f_c' = 5 \text{ ksi}$$
Axial design strength in kips

$F_y = 46$ ksi

Steel Tube	Nominal Size		8×4					7×5				
	Thickness (in.)		⅝	½	⅜	⁵⁄₁₆	¼	½	⅜	⁵⁄₁₆	¼	³⁄₁₆
	Wt./ft		42.30	35.24	27.48	23.34	19.02	35.24	27.48	23.34	19.02	14.53
	F_y						46 ksi					
Effective length KL (ft) with respect to least radius of gyration, r_{my}	0		552	482	401	358	313	492	411	369	324	278
	6		468	411	345	308	271	444	372	333	293	251
	7		441	389	327	292	257	427	358	322	283	242
	8		411	364	307	275	241	409	343	308	271	232
	9		380	338	286	256	225	390	327	294	259	221
	10		349	311	265	237	208	369	310	279	246	210
	11		317	284	242	217	191	347	292	263	232	198
	12		285	257	220	198	174	325	274	247	217	186
	13		254	230	199	178	157	302	255	230	203	173
	14		224	205	178	160	141	279	237	213	188	161
	15		196	180	157	142	125	257	218	197	174	148
	16		172	158	138	124	110	235	200	181	159	136
	17		153	140	123	110	98	213	182	165	145	124
	18		136	125	109	98	87	193	165	149	132	112
	19		122	112	98	88	78	173	148	134	119	101
	20		110	101	89	80	71	156	134	121	107	91
	22		91	84	73	66	58	129	111	100	89	75
	24		77	70	62	55	49	108	93	84	74	63
	26				52	47	42	92	79	72	63	54
	28							80	68	62	55	47
	30							69	59	54	48	41
	32								52	47	42	36
	34											32
Properties												
r_{my} (in.)			1.49	1.54	1.60	1.62	1.65	1.90	1.95	1.98	2.01	2.04
r_{mx} / r_{my}			1.76	1.75	1.73	1.73	1.72	1.31	1.30	1.30	1.30	1.29
$\phi_b M_{nx}$ (kip-ft)			99.3	85.2	68.7	59.0	48.6	79.7	63.8	54.9	45.5	35.2
$\phi_b M_{ny}$ (kip-ft)			59.8	51.8	42.1	36.2	30.1	62.8	50.4	43.5	35.9	27.9
$P_{ex} (K_x L_x)^2 / 10^4$ (kip-ft²)			18.3	16.6	14.2	12.8	11.2	14.3	12.2	11.1	9.70	8.20
$P_{ey} (K_y L_y)^2 / 10^4$ (kip-ft²)			5.92	5.43	4.75	4.28	3.79	8.38	7.18	6.51	5.75	4.90

Note: Heavy line indicates Kl / r of 200.

F_y = 46 ksi

COMPOSITE COLUMNS
Rectangular structural tubing
f_c' = 5 ksi
Axial design strength in kips

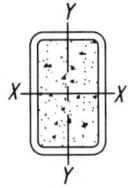

Steel Tube	Nominal Size	7×4				6×4				
	Thickness (in.)	3/8	5/16	1/4	3/16	1/2	3/8	5/16	1/4	3/16
	Wt./ft	24.93	21.21	17.32	13.25	28.43	22.37	19.08	15.62	11.97
	F_y	46 ksi								
Effective length KL (ft) with respect to least radius of gyration, r_{my}	0	360	321	281	239	380	318	285	249	211
	6	308	276	242	206	321	271	243	213	181
	7	291	261	229	195	302	256	230	202	171
	8	273	245	215	183	282	240	215	189	161
	9	254	228	201	171	260	222	200	176	149
	10	234	211	186	158	238	204	184	162	138
	11	214	193	170	145	215	186	168	148	126
	12	194	175	155	131	193	168	152	134	114
	13	174	158	139	119	172	150	137	121	103
	14	155	141	125	106	151	133	121	108	92
	15	137	124	110	94	132	117	107	95	81
	16	120	109	97	83	116	103	94	83	71
	17	106	97	86	73	103	91	83	74	63
	18	95	86	77	65	92	81	74	66	56
	19	85	78	69	59	82	73	67	59	50
	20	77	70	62	53	74	66	60	53	45
	22	64	58	51	44	61	54	50	44	38
	24	53	49	43	37	52	46	42	37	32
	26	45	41	37	31			36	32	27

Properties									
r_{my} (in.)	1.57	1.60	1.63	1.66	1.48	1.54	1.57	1.60	1.63
r_{mx} / r_{my}	1.56	1.56	1.55	1.54	1.39	1.38	1.38	1.37	1.37
$\phi_b M_{nx}$ (kip-ft)	55.2	47.6	39.7	30.7	53.1	43.1	37.6	31.3	24.4
$\phi_b M_{ny}$ (kip-ft)	37.3	32.3	26.8	20.9	39.7	32.6	28.3	23.6	18.4
$P_{ex} (K_x L_x)^2 / 10^4$ (kip-ft^2)	10.0	9.10	7.97	6.70	7.72	6.76	6.10	5.36	4.56
$P_{ey} (K_y L_y)^2 / 10^4$ (kip-ft^2)	4.12	3.76	3.33	2.84	3.99	3.53	3.22	2.86	2.44

Note: Heavy line indicates Kl / r of 200.

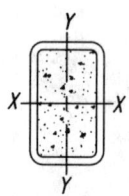

		F_y = 46 ksi

COMPOSITE COLUMNS
Rectangular structural tubing
$f_c' = 5$ ksi
Axial design strength in kips

Steel Tube	Nominal Size		5×4			
	Thickness (in.)		⅜	⁵⁄₁₆	¼	³⁄₁₆
	Wt./ft		19.82	16.96	13.91	10.70
	F_y		46 ksi			
Effective length KL (ft) with respect to least radius of gyration, r_{my}		0	262	224	185	142
		6	222	192	159	123
		7	210	181	150	117
		8	196	170	141	110
		9	181	157	131	102
		10	166	145	121	95
		11	151	132	111	87
		12	136	119	101	79
		13	121	107	91	72
		14	107	95	81	64
		15	94	84	71	57
		16	83	73	63	50
		17	73	65	56	44
		18	65	58	50	40
		19	59	52	45	36
		20	53	47	40	32
		22	44	39	33	27
		24	37	33	28	22
		26			24	19

Properties

	⅜	⁵⁄₁₆	¼	³⁄₁₆
r_{my} (in.)	1.50	1.53	1.56	1.59
r_{mx} / r_{my}	1.19	1.20	1.19	1.19
$\phi_b M_{nx}$ (kip-ft)	32.6	28.4	23.8	18.6
$\phi_b M_{ny}$ (kip-ft)	27.9	24.3	20.4	16.0
$P_{ex} (K_x L_x)^2 / 10^4$ (kip-ft²)	4.04	3.61	3.07	2.44
$P_{ey} (K_y L_y)^2 / 10^4$ (kip-ft²)	2.83	2.52	2.16	1.72

Note: Heavy line indicates Kl / r of 200.

REFERENCES

Allison, H., 1991, *Low- and Medium-Rise Steel Buildings,* AISC Steel Design Guide Series No. 5, American Institute of Steel Construction, Chicago, IL.

Fisher, J. M. and M. A. West, 1990, *Serviceability Design Considerations for Low-Rise Buildings,* AISC Steel Design Guide Series No. 3, AISC, Chicago.

Griffis, L. G., 1992, *Load and Resistance Factor Design of W-Shapes Encased in Concrete,* AISC Steel Design Guide Series No. 6, AISC, Chicago.

PART 6

SPECIFICATIONS AND CODES

Load and Resistance Factor Design Specification for Structural Steel Buildings

December 1, 1993

AMERICAN INSTITUTE OF STEEL CONSTRUCTION, INC.
**One East Wacker Drive, Suite 3100
Chicago, Illinois 60601-2001**

PREFACE

The AISC *Load and Resistance Factor Design (LRFD) Specification for Structural Steel Buildings* is based on reliability theory. As have all AISC Specifications, this LRFD Specification has been based upon past successful usage, advances in the state of knowledge, and changes in design practice. The LRFD Specification has been developed as a consensus document to provide a uniform practice in the design of steel-framed buildings. The intention is to provide design criteria for routine use and not to provide specific criteria for infrequently encountered problems which occur in the full range of structural design. Providing definitive provisions to cover all cases would make the LRFD Specification too cumbersome for routine design usage.

The LRFD Specification is the result of the deliberations of a committee of structural engineers with wide experience and high professional standing, representing a wide geographical distribution throughout the U.S. The committee includes approximately equal numbers of engineers in private practice and code agencies, engineers involved in research and teaching, and engineers employed by steel fabricating and producing companies.

In order to avoid reference to proprietary steels which may have limited availability, only those steels which can be identified by ASTM specifications are approved under this Specification. However, some steels covered by ASTM specifications, but subject to more costly manufacturing and inspection techniques than deemed essential for structures covered by this Specification, are not listed, even though they may provide all the necessary characteristics for reliable usage in structural applications. Approval of such steels in lieu of less expensive steels is left to the owner's representative.

The Appendices to this Specification are an integral part of the Specification.

A non-mandatory Commentary has been prepared to provide background for the Specification provisions and the user is encouraged to consult it.

The principal changes incorporated in this edition of the Specification include:

• Updated web crippling design provisions.
• Recommendations for the use of heavy rolled shapes and welded members made up of thick plates.
• Updated provisions for slender web girders and unsymmetric members.
• Revised provisions for built-up compression members.
• Improved C_b equation.
• Provisions for slip-critical joints designed at factored loads.
• Reorganization and expansion of material on stability of unbraced frames.
• Reorganization and expansion of Chapters F and K.

- Alternative fillet-weld design strength.
- Addition of beam-web opening provisions.

The reader is cautioned that professional judgment must be exercised when data or recommendations in the Specification are applied. The publication of the material contained herein is not intended as a representation or warranty on the part of the American Institute of Steel Construction, Inc.—or any other person named herein— that this information is suitable for general or particular use, or freedom from infringement of any patent or patents. Anyone making use of this information assumes all liability arising from such use. The design and detailing of steel structures is within the expertise of professional individuals who are competent by virtue of education, training, and experience for the application of engineering principles and the provisions of this specification to the design and/or detailing of a particular structure.

By the Committee,

Arthur P. Arndt, Chairman	Jay W. Larson
William E. Moore, II, Vice Chairman	William J. LeMessurier
Horatio Allison	H. S. Lew
Reidar Bjorhovde	Bill Lindley, II
Roger L. Brockenbrough	Stanley D. Lindsey
W. F. Chen	Richard W. Marshall
Andrew K. Courtney	Lisa McCasland
Robert O. Disque	Robert McCluer
Joseph Dudek	William A. Milek
Duane S. Ellifritt	Duane K. Miller
Bruce Ellingwood	Walter P. Moore, Jr.
Shu-Jin Fang	Thomas M. Murray
Steven J. Fenves	Gary G. Nichols
Roger E. Ferch	Clarkson W. Pinkham
James M. Fisher	Egor P. Popov
John W. Fisher	Donald R. Sherman
Theodore V. Galambos	Frank Sowokinos
Geerhard Haaijer	William A. Thornton
Richard B. Haws	Raymond H. R. Tide
Mark Holland	Ivan M. Viest
Ira Hooper	Lyle L. Wilson
Donald L. Johnson	Joseph A. Yura
L. A. Kloiber	Nestor R. Iwankiw, Secretary

TABLE OF CONTENTS

Symbols

The section number in parentheses after the definition of a symbol refers to the section where the symbol is first defined.

A	Cross-sectional area, in.^{2} (F1.2)
A_B	Loaded area of concrete, in.^{2} (I2.4)
A_b	Nominal body area of a fastener, in.^{2} (J3.7)
A_c	Area of concrete, in.^{2} (I2.2)
A_c	Area of concrete slab within effective width, in.^{2} (I5.2)
A_D	Area of an upset rod based on the major diameter of its threads, in.^{2} (J3.6)
A_e	Effective net area, in.^{2} (B3)
A_f	Area of flange, in.^{2} (Appendix F3)
A_{fe}	Effective tension flange area, in.^{2} (B10)
A_{fg}	Gross area of flange, in.^{2} (B10)
A_{fn}	Net area of flange, in.^{2} (B10)
A_g	Gross area, in.^{2} (A5)
A_{gt}	Gross area subject to tension, in.^{2} (J4.3)
A_{gv}	Gross area subject to shear, in.^{2} (J4.3)
A_n	Net area, in.^{2} (B2)
A_{nt}	Net area subject to tension, in.^{2} (J4.2)
A_{nv}	Net area subject to shear, in.^{2} (J4.1)
A_{pb}	Projected bearing area, in.^{2} (J8.1)
A_r	Area of reinforcing bars, in.^{2} (I2.2)
A_s	Area of steel cross section, in.^{2} (I2.2)
A_{sc}	Cross-sectional area of stud shear connector, in.^{2} (I5.3)
A_{sf}	Shear area on the failure path, in.^{2} (D3)
A_w	Web area, in.^{2} (F2.1)
A_1	Area of steel bearing concentrically on a concrete support, in.^{2} (J9)
A_2	Total cross-sectional area of a concrete support, in.^{2} (J9)
B	Factor for bending stress in tees and double angles (F1.2)
B	Factor for bending stress in web-tapered members, in., defined by Equations A-F3-8 through A-F3-11 (Appendix F3)
B_1, B_2	Factors used in determining M_u for combined bending and axial forces when first-order analysis is employed (C1)
C_{PG}	Plate-girder coefficient (Appendix G2)
C_b	Bending coefficient dependent on moment gradient (F1.2a)
C_m	Coefficient applied to bending term in interaction formula for prismatic members and dependent on column curvature caused by applied moments (C1)

C'_m	Coefficient applied to bending term in interaction formula for tapered members and dependent on axial stress at the small end of the member (Appendix F3)
C_p	Ponding flexibility coefficient for primary member in a flat roof (K2)
C_s	Ponding flexibility coefficient for secondary member in a flat roof (K2)
C_v	Ratio of "critical" web stress, according to linear buckling theory, to the shear yield stress of web material (Appendix G3)
C_w	Warping constant, in.^{6} (F1.2)
D	Outside diameter of circular hollow section, in. (Appendix B5.3)
D	Dead load due to the weight of the structural elements and permanent features on the structure (A4.1)
D	Factor used in Equation A-G4-2, dependent on the type of transverse stiffeners used in a plate girder (Appendix G4)
E	Modulus of elasticity of steel ($E = 29,000$ ksi) (E2)
E	Earthquake load (A4.1)
E_c	Modulus of elasticity of concrete, ksi (I2.2)
E_m	Modified modulus of elasticity, ksi (I2.2)
F_{BM}	Nominal strength of the base material to be welded, ksi (J2.4)
F_{EXX}	Classification number of weld metal (minimum specified strength), ksi (J2.4)
F_L	Smaller of $(F_{yf} - F_r)$ or F_{yw}, ksi (F1.2)
$F_{b\gamma}$	Flexural stress for tapered members defined by Equations A-F3-4 and A-F3-5 (Appendix F3)
F_{cr}	Critical stress, ksi (E2)
$F_{crft}, F_{cry}, F_{crz}$	Flexural-torsional buckling stresses for double-angle and tee-shaped compression members, ksi (E3)
F_e	Elastic buckling stress, ksi (Appendix E3)
F_{ex}	Elastic flexural buckling stress about the major axis, ksi (Appendix E3)
F_{ey}	Elastic flexural buckling stress about the minor axis, ksi (Appendix E3)
F_{ez}	Elastic torsional buckling stress, ksi (Appendix E3)
F_{my}	Modified yield stress for composite columns, ksi (I2.2)
F_n	Nominal shear rupture strength, ksi (J4)
F_r	Compressive residual stress in flange (10 ksi for rolled; 16.5 ksi for welded), ksi (Table B5.1)
$F_{s\gamma}$	Stress for tapered members defined by Equation A-F3-6, ksi (Appendix F3)
F_u	Specified minimum tensile strength of the type of steel being used, ksi (B10)
F_w	Nominal strength of the weld electrode material, ksi (J2.4)
$F_{w\gamma}$	Stress for tapered members defined by Equation A-F3-7, ksi (Appendix F3)
F_y	Specified minimum yield stress of the type of steel being used, ksi. As used in this Specification, "yield stress" denotes either the specified minimum yield point (for those steels that have a yield point) or specified yield strength (for those steels that do not have a yield point) (A5)
F_{yf}	Specified minimum yield stress of the flange, ksi (Table B5.1)
F_{yr}	Specified minimum yield stress of reinforcing bars, ksi (I2.2)
F_{yst}	Specified minimum yield stress of the stiffener material, ksi (Appendix G4)

F_{yw}	Specified minimum yield stress of the web, ksi (Table B5.1)
G	Shear modulus of elasticity of steel, ksi ($G = 11,200$) (F1.2)
H	Horizontal force, kips (C1)
H	Flexural constant (E3)
H_s	Length of stud connector after welding, in. (I3.5)
I	Moment of inertia, in.^{4} (F1.2)
I_d	Moment of inertia of the steel deck supported on secondary members, in.^{4} (K2)
I_p	Moment of inertia of primary members, in.^{4} (K2)
I_s	Moment of inertia of secondary members, in.^{4} (K2)
I_{st}	Moment of inertia of a transverse stiffener, in.^{4} (Appendix G4)
I_{yc}	Moment of inertia about y axis referred to compression flange, or if reverse curvature bending referred to smaller flange, in.^{4} (Appendix F1)
J	Torsional constant for a section, in.^{4} (F1.2)
K	Effective length factor for prismatic member (B7)
K_z	Effective length factor for torsional buckling (Appendix E3)
K_γ	Effective length factor for a tapered member (Appendix F3)
L	Story height, in. (C1)
L	Length of connection in the direction of loading, in. (B3)
L	Live load due to occupancy and moveable equipment (A4.1)
L_b	Laterally unbraced length; length between points which are either braced against lateral displacement of compression flange or braced against twist of the cross section, in. (F1.2)
L_c	Length of channel shear connector, in. (I5.4)
L_e	Edge distance, in. (J3.10)
L_p	Limiting laterally unbraced length for full plastic bending capacity, uniform moment case ($C_b = 1.0$), in. (F1.2)
L_p	Column spacing in direction of girder, ft (K2)
L_{pd}	Limiting laterally unbraced length for plastic analysis, in. (F1.2)
L_r	Limiting laterally unbraced length for inelastic lateral-torsional buckling, in. (F1.2)
L_r	Roof live load (A4.1)
L_s	Column spacing perpendicular to direction of girder, ft (K2)
M_A	Absolute value of moment at quarter point of the unbraced beam segment, kip-in. (F1.2)
M_B	Absolute value of moment at centerline of the unbraced beam segment, kip-in. (F1.2)
M_C	Absolute value of moment at three-quarter point of the unbraced beam segment, kip-in. (F1.2)
M_{cr}	Elastic buckling moment, kip-in. (F1.2)
M_{lt}	Required flexural strength in member due to lateral frame translation only, kip-in. (C1)
M_{max}	Absolute value of maximum moment in the unbraced beam segment, kip-in. (F1.2)
M_n	Nominal flexural strength, kip-in. (F1.1)
M'_{nx}, M'_{ny}	Flexural strength defined in Equations A-H3-7 and A-H3-8 for use in alternate interaction equations for combined bending and axial force, kip-in. (Appendix H3)

M_{nt}	Required flexural strength in member assuming there is no lateral translation of the frame, kip-in. (C1)
M_p	Plastic bending moment, kip-in. (F1.1)
M_p'	Moment defined in Equations A-H3-5 and A-H3-6, for use in alternate interaction equations for combined bending and axial force, kip-in. (Appendix H3)
M_r	Limiting buckling moment, M_{cr}, when $\lambda = \lambda_r$ and $C_b = 1.0$, kip-in. (F1.2)
M_u	Required flexural strength, kip-in. (C1)
M_y	Moment corresponding to onset of yielding at the extreme fiber from an elastic stress distribution ($= F_y S$ for homogeneous sections), kip-in. (F1.1)
M_1	Smaller moment at end of unbraced length of beam or beam-column, kip-in.
M_2	Larger moment at end of unbraced length of beam or beam-column, kip-in.
N	Length of bearing, in. (K1.3)
N_r	Number of stud connectors in one rib at a beam intersection (I3.5)
P_{e1}, P_{e2}	Elastic Euler buckling load for braced and unbraced frame, respectively, kips (C1)
P_n	Nominal axial strength (tension or compression), kips (D1)
P_p	Bearing load on concrete, kips (J9)
P_u	Required axial strength (tension or compression), kips (Table B5.1)
P_y	Yield strength, kips (Table B5.1)
Q	Full reduction factor for slender compression elements (Appendix E3)
Q_a	Reduction factor for slender stiffened compression elements (Appendix B5)
Q_n	Nominal strength of one stud shear connector, kips (I5)
Q_s	Reduction factor for slender unstiffened compression elements (Appendix B5.3)
R	Load due to initial rainwater or ice exclusive of the ponding contribution (A4.1)
R_{PG}	Plate girder bending strength reduction factor (Appendix G)
R_e	Hybrid girder factor (Appendix F1)
R_n	Nominal strength (A5.3)
R_v	Web shear strength, kips (K1.7)
S	Elastic section modulus, in.^{3} (F1.2)
S	Spacing of secondary members, ft (K2)
S	Snow load (A4.1)
S_x'	Elastic section modulus of larger end of tapered member about its major axis, in.^{3} (Appendix F3)
S_{eff}	Effective section modulus about major axis, in.^{3} (Appendix F1)
S_{xt}, S_{xc}	Elastic section modulus referred to tension and compression flanges, respectively, in.^{3} (Appendix F1)
T	Tension force due to service loads, kips (J3.9)
T_b	Specified pretension load in high-strength bolt, kips (J3.9)
T_u	Required tensile strength due to factored loads, kips (Appendix J3.9b)
U	Reduction coefficient, used in calculating effective net area (B3)
V_n	Nominal shear strength, kips (F2.2)

V_u	Required shear strength, kips (Appendix G4)
W	Wind load (A4.1)
X_1	Beam buckling factor defined by Equation F1-8 (F1.2)
X_2	Beam buckling factor defined by Equation F1-9 (F1.2)
Z	Plastic section modulus, in.^{3} (F1.1)
a	Clear distance between transverse stiffeners, in. (Appendix F2.2)
a	Distance between connectors in a built-up member, in. (E4)
a	Shortest distance from edge of pin hole to edge of member measured parallel to direction of force, in. (D3)
a_r	Ratio of web area to compression flange area (Appendix G2)
a'	Weld length, in. (B10)
b	Compression element width, in. (B5.1)
b_e	Reduced effective width for slender compression elements, in. (Appendix B5.3)
b_{eff}	Effective edge distance, in. (D3)
b_f	Flange width, in. (B5.1)
c_1, c_2, c_3	Numerical coefficients (I2.2)
d	Nominal fastener diameter, in. (J3.3)
d	Overall depth of member, in. (B5.1)
d	Pin diameter, in. (D3)
d	Roller diameter, in. (J8.2)
d_L	Depth at larger end of unbraced tapered segment, in. (Appendix F3)
d_b	Beam depth, in. (K1.7)
d_c	Column depth, in. (K1.7)
d_o	Depth at smaller end of unbraced tapered segment, in. (Appendix F3)
e	Base of natural logarithm = 2.71828...
f	Computed compressive stress in the stiffened element, ksi (Appendix B5.3)
f_{b1}	Smallest computed bending stress at one end of a tapered segment, ksi (Appendix F3)
f_{b2}	Largest computed bending stress at one end of a tapered segment, ksi (Appendix F3)
f_c'	Specified compressive strength of concrete, ksi (I2.2)
f_o	Stress due to $1.2D + 1.2R$, ksi (Appendix K2)
f_{un}	Required normal stress, ksi (H2)
f_{uv}	Required shear stress, ksi (H2)
f_v	Required shear stress due to factored loads in bolts or rivets, ksi (J3.7)
g	Transverse center-to-center spacing (gage) between fastener gage lines, in. (B2)
h	Clear distance between flanges less the fillet or corner radius for rolled shapes; and for built-up sections, the distance between adjacent lines of fasteners or the clear distance between flanges when welds are used, in. (B5.1)
h	Distance between centroids of individual components perpendicular to the member axis of buckling, in. (E4)
h_c	Twice the distance from the centroid to the following: the inside face of the compression flange less the fillet or corner radius, for rolled shapes; the nearest line of fasteners at the compression flange or the inside faces

	of the compression flange when welds are used, for built-up sections, in. (B5.1)
h_r	Nominal rib height, in. (I3.5)
h_s	Factor used in Equation A-F3-6 for web-tapered members (Appendix F3)
h_w	Factor used in Equation A-F3-7 for web-tapered members (Appendix F3)
j	Factor defined by Equation A-F2-4 for minimum moment of inertia for a transverse stiffener (Appendix F2.3)
k	Distance from outer face of flange to web toe of fillet, in. (K1.3)
k_v	Web plate buckling coefficient (Appendix F2.2)
l	Laterally unbraced length of member at the point of load, in. (B7)
l	Length of bearing, in. (J8.2)
l	Length of connection in the direction of loading, in. (B3)
l	Length of weld, in. (B3)
m	Ratio of web to flange yield stress or critical stress in hybrid beams (Appendix G2)
r	Governing radius of gyration, in. (B7)
r_{To}	For the smaller end of a tapered member, the radius of gyration, considering only the compression flange plus one-third of the compression web area, taken about an axis in the plane of the web, in. (Appendix F3.4)
r_i	Minimum radius of gyration of individual component in a built-up member, in. (E4)
r_{ib}	Radius of gyration of individual component relative to centroidal axis parallel to member axis of buckling, in. (E4)
r_m	Radius of gyration of the steel shape, pipe, or tubing in composite columns. For steel shapes it may not be less than 0.3 times the overall thickness of the composite section, in. (I2)
$\bar{r}_o$	Polar radius of gyration about the shear center, in. (E3)
r_{ox} , r_{oy}	Radius of gyration about x and y axes at the smaller end of a tapered member, respectively, in. (Appendix F3.3)
r_x, r_y	Radius of gyration about x and y axes, respectively, in. (E3)
r_{yc}	Radius of gyration about y axis referred to compression flange, or if reverse curvature bending, referred to smaller flange, in. (Appendix F1)
s	Longitudinal center-to-center spacing (pitch) of any two consecutive holes, in. (B2)
t	Thickness of connected part, in. (D3)
t_f	Flange thickness, in. (B5.1)
t_f	Flange thickness of channel shear connector, in. (I5.4)
t_w	Web thickness of channel shear connector, in. (I5.4)
t_w	Web thickness, in. (B5.3)
w	Plate width; distance between welds, in. (B3)
w	Unit weight of concrete, lbs/cu ft. (I2)
w_r	Average width of concrete rib or haunch, in. (I3.5)
x	Subscript relating symbol to strong axis bending
x_o, y_o	Coordinates of the shear center with respect to the centroid, in. (E3)
$\bar{x}$	Connection eccentricity, in. (B3)
y	Subscript relating symbol to weak axis bending
z	Distance from the smaller end of tapered member used in Equation A-F3-1 for the variation in depth, in. (Appendix F3)

α Separation ratio for built-up compression members $= \dfrac{h}{2r_{ib}}$ (E4)

Δ_{oh} Translation deflection of the story under consideration, in. (C1)

γ Depth tapering ratio (Appendix F3). Subscript for tapered members (Appendix F3)

ζ Exponent for alternate beam-column interaction equation (Appendix H3)

η Exponent for alternate beam-column interaction equation (Appendix H3)

λ_c Column slenderness parameter (C1)

λ_e Equivalent slenderness parameter (Appendix E3)

λ_{eff} Effective slenderness ratio defined by Equation A-F3-2 (Appendix F3)

λ_p Limiting slenderness parameter for compact element (B5.1)

λ_r Limiting slenderness parameter for noncompact element (B5.1)

ϕ Resistance factor (A5.3)

ϕ_b Resistance factor for flexure (F1)

ϕ_c Resistance factor for compression (A5)

ϕ_c Resistance factor for axially loaded composite columns (I2.2)

ϕ_{sf} Resistance factor for shear on the failure path (D3)

ϕ_t Resistance factor for tension (D1)

ϕ_v Resistance factor for shear (F2.2)

CHAPTER A

GENERAL PROVISIONS

A1. SCOPE

The *Load and Resistance Factor Design Specification for Structural Steel Buildings* governs the design, fabrication, and erection of steel-framed buildings. As an alternative, the AISC *Specification for Structural Steel Buildings, Allowable Stress Design and Plastic Design* is permitted.

A2. LIMITS OF APPLICABILITY

1. Structural Steel Defined

As used in this Specification, the term *structural steel* refers to the steel elements of the structural steel frame essential to the support of the required loads. Such elements are enumerated in Section 2.1 of the AISC *Code of Standard Practice for Steel Buildings and Bridges*. For the design of cold-formed steel structural members, whose profiles contain rounded corners and slender flat elements, the provisions of the American Iron and Steel Institute *Load and Resistance Factor Design Specification for the Design of Cold-Formed Steel Structural Members* are recommended.

2. Types of Construction

Two basic types of construction and associated design assumptions are permissible under the conditions stated herein, and each will govern in a specific manner the strength of members and the types and strength of their connections.

Type FR (fully restrained), commonly designated as "rigid-frame" (continuous frame), assumes that connections have sufficient rigidity to maintain the angles between intersecting members.

Type PR (partially restrained) assumes that connections have insufficient rigidity to maintain the angles between intersecting members.

The type of construction assumed in the design shall be indicated on the design documents. The design of all connections shall be consistent with the assumption.

Type PR construction under this Specification depends upon a predictable proportion of full end restraint. When a portion of the full end restraint of members is used in the design for strength of the connected members or for the stability of the structure as a whole, the capacity of the connections to provide the needed restraint shall be documented in the technical literature or established by analytical or empirical means.

When the connection restraint is ignored, commonly designated "simple framing," it is assumed that for the transmission of gravity loads the ends of the beams and girders are connected for shear only and are free to rotate. For "simple framing" the following requirements apply:

(1) The connections and connected members shall be adequate to resist the factored gravity loads as "simple beams."

(2) The connections and connected members shall be adequate to resist the factored lateral loads.

(3) The connections shall have sufficient inelastic rotation capacity to avoid overload of fasteners or welds under combined factored gravity and lateral loading.

Type PR construction may necessitate some inelastic, but self-limiting, deformation of a structural steel part.

A3. MATERIAL

1. Structural Steel

1a. ASTM Designations

Material conforming to one of the following standard specifications is approved for use under this Specification:

Structural Steel, ASTM A36

Pipe, Steel, Black and Hot-Dipped, Zinc-Coated Welded and Seamless ASTM A53, Gr. B

High-Strength Low-Alloy Structural Steel, ASTM A242

Cold-Formed Welded and Seamless Carbon Steel Structural Tubing in Rounds and Shapes, ASTM A500

Hot-Formed Welded and Seamless Carbon Steel Structural Tubing, ASTM A501

High-Yield-Strength, Quenched and Tempered Alloy Steel Plate, Suitable for Welding, ASTM A514

High-Strength Carbon-Manganese Steel of Structural Quality, ASTM A529

Steel, Sheet and Strip, Carbon, Hot-Rolled, Structural Quality, ASTM A570, Gr. 40, 45, and 50

High-Strength Low-Alloy Columbium-Vanadium Steels of Structural Quality, ASTM A572

High-Strength Low-Alloy Structural Steel with 50 ksi Minimum Yield Point to 4-in. Thick, ASTM A588

Steel, Sheet and Strip, High-Strength, Low-Alloy, Hot-Rolled and Cold-Rolled, with Improved Atmospheric Corrosion Resistance, ASTM A606

Steel, Sheet and Strip, High-Strength, Low-Alloy, Columbium or Vanadium, or Both, Hot-Rolled and Cold-Rolled, ASTM A607

Hot-Formed Welded and Seamless High-Strength Low-Alloy Structural Tubing, ASTM A618

Structural Steel for Bridges, ASTM A709

Quenched and Tempered Low-Alloy Structural Steel Plate with 70 ksi Minimum Yield Strength to 4-in. Thick, ASTM A852

Certified mill test reports or certified reports of tests made by the fabricator or a testing laboratory in accordance with ASTM A6 or A568, as applicable, shall constitute sufficient evidence of conformity with one of the above ASTM standards. If requested, the fabricator shall provide an affidavit stating that the structural steel furnished meets the requirements of the grade specified.

1b. Unidentified Steel

Unidentified steel, if surface conditions are acceptable according to criteria contained in ASTM A6, may be used for unimportant members or details, where the precise physical properties and weldability of the steel would not affect the strength of the structure.

1c. Heavy Shapes

For ASTM A6 Group 4 and 5 rolled shapes to be used as members subject to primary tensile stresses due to tension or flexure, toughness need not be specified if splices are made by bolting. If such members are spliced using complete-joint penetration welds, the steel shall be specified in the contract documents to be supplied with Charpy V-Notch testing in accordance with ASTM A6, Supplementary Requirement S5. The impact test shall meet a minimum average value of 20 ft-lbs. absorbed energy at +70°F and shall be conducted in accordance with ASTM A673 with the following exceptions:

(1) The center longitudinal axis of the specimens shall be located as near as practical to midway between the inner flange surface and the center of the flange thickness at the intersection with the web mid-thickness.

(2) Tests shall be conducted by the producer on material selected from a location representing the top of each ingot or part of an ingot used to produce the product represented by these tests.

For plates exceeding 2-in. thick used for built-up cross-sections with bolted splices and subject to primary tensile stresses due to tension or flexure, material toughness need not be specified. If such cross-sections are spliced using complete-joint penetration welds, the steel shall be specified in the contract documents to be supplied with Charpy V-Notch testing in accordance with ASTM A6, Supplementary Requirement S5. The impact test shall be conducted by the producer in accordance with ASTM A673, Frequency P, and shall meet a minimum average value of 20 ft-lbs. absorbed energy at +70°F.

The above supplementary requirements also apply when complete-joint penetration welded joints through the thickness of ASTM A6 Group 4 and 5 shapes and built-up cross sections with thickness exceeding two inches are used in connections subjected to primary tensile stress due to tension or flexure of such members. The requirements need not apply to ASTM A6 Group 4 and 5 shapes and built-up members with thickness exceeding two inches to which members other than ASTM A6 Group 4 and 5 shapes and built-up members are connected by complete-joint penetration welded joints through the thickness of the thinner material to the face of the heavy material.

Additional requirements for joints in heavy rolled and built-up members are given in Sections J1.5, J1.6, J2, and M2.2.

2.   Steel Castings and Forgings

Cast steel shall conform to one of the following standard specifications:

> Mild-to-Medium-Strength Carbon-Steel Castings for General Applications, ASTM A27, Gr. 65-35
> High-Strength Steel Castings for Structural Purposes, ASTM A148 Gr. 80-50

Steel forgings shall conform to the following standard specification:

> Steel Forgings Carbon and Alloy for General Industrial Use, ASTM A668

Certified test reports shall constitute sufficient evidence of conformity with standards.

3.   Bolts, Washers, and Nuts

Steel bolts, washers, and nuts shall conform to one of the following standard specifications:

> Carbon and Alloy Steel Nuts for Bolts for High-Pressure and High-Temperature Service, ASTM A194
> Carbon Steel Bolts and Studs, 60,000 psi Tensile Strength, ASTM A307
> Structural Bolts, Steel, Heat-Treated, 120/105 ksi Minimum Tensile Strength, ASTM A325
> Quenched and Tempered Steel Bolts and Studs, ASTM A449
> Heat-Treated Steel Structural Bolts, 150 ksi Min. Tensile Strength, ASTM A490
> Carbon and Alloy Steel Nuts, ASTM A563
> Hardened Steel Washers, ASTM F436

A449 bolts are permitted to be used only in connections requiring bolt diameters greater than $1\frac{1}{2}$-in. and shall not be used in slip-critical connections.

Manufacturer's certification shall constitute sufficient evidence of conformity with the standards.

4.   Anchor Bolts and Threaded Rods

Anchor bolt and threaded rod steel shall conform to one of the following standard specifications:

> Structural Steel, ASTM A36
> Alloy Steel and Stainless Steel Bolting Materials for High-Temperature Service, ASTM A193
> Quenched and Tempered Alloy Steel Bolts, Studs and Other Externally Threaded Fasteners, ASTM A354
> High-Strength Low-Alloy Columbium-Vanadium Steels of Structural Quality, ASTM A572
> High-Strength Low-Alloy Structural Steel with 50,000 psi Minimum Yield Point to 4-in. Thick, ASTM A588
> High-Strength Nonheaded Steel Bolts and Studs, ASTM A687

Threads on bolts and rods shall conform to the Unified Standard Series of ANSI B18.1 and shall have Class 2A tolerances.

Steel bolts conforming to other provisions of Section A3.3 are permitted as anchor bolts. A449 material is acceptable for high-strength anchor bolts and threaded rods of any diameter.

Manufacturer's certification shall constitute sufficient evidence of conformity with the standards.

5.   Filler Metal and Flux for Welding

Welding electrodes and fluxes shall conform to one of the following specifications of the American Welding Society:

Specification for Carbon Steel Electrodes for Shield Metal Arc Welding, AWS A5.1

Specification for Low-Alloy Steel Covered Arc Welding Electrodes, AWS A5.5

Specification for Carbon Steel Electrodes and Fluxes for Submerged Arc Welding, AWS A5.17

Specification for Carbon Steel Filler Metals for Gas Shielded Arc Welding, AWS A5.18

Specification for Carbon Steel Electrodes for Flux Cored Arc Welding, AWS A5.20

Specification for Low-Alloy Steel Electrodes and Fluxes for Submerged Arc Welding, AWS A5.23

Specification for Low-Alloy Steel Filler Metals for Gas Shielded Arc Welding, AWS A5.28

Specification for Low-Alloy Steel Electrodes for Flux Cored Arc Welding, AWS A5.29

Manufacturer's certification shall constitute sufficient evidence of conformity with the standards. Electrodes (filler metals) that are suitable for the intended application shall be selected. Weld metal notch toughness is generally not critical for building construction.

6.   Stud Shear Connectors

Steel stud shear connectors shall conform to the requirements of *Structural Welding Code—Steel*, AWS D1.1.

Manufacturer's certification shall constitute sufficient evidence of conformity with the code.

A4.   LOADS AND LOAD COMBINATIONS

The nominal loads shall be the minimum design loads stipulated by the applicable code under which the structure is designed or dictated by the conditions involved. In the absence of a code, the loads and load combinations shall be those stipulated in the American Society of Civil Engineers Standard *Minimum Design Loads for Buildings and Other Structures*, ASCE 7. For design purposes, the loads stipulated by the applicable code shall be taken as nominal loads. For ease of reference, the more common ASCE load combinations are listed in the following section.

Seismic design of buildings assigned to the higher risk Seismic Performance Categories defined in the AISC *Seismic Provisions for Structural Steel Buildings* shall comply with that document. Seismic design not covered by the AISC

Seismic Provisions for Structural Steel Buildings shall be in accordance with this Specification.

1. Loads, Load Factors, and Load Combinations

The following nominal loads are to be considered:

D : dead load due to the weight of the structural elements and the permanent features on the structure

L : live load due to occupancy and moveable equipment

L_r : roof live load

W : wind load

S : snow load

E : earthquake load determined in accordance with Part I of the AISC *Seismic Provisions for Structural Steel Buildings*

R : load due to initial rainwater or ice exclusive of the ponding contribution

The required strength of the structure and its elements must be determined from the appropriate critical combination of factored loads. The most critical effect may occur when one or more loads are not acting. The following load combinations and the corresponding load factors shall be investigated:

$$1.4D \tag{A4-1}$$

$$1.2D + 1.6L + 0.5(L_r \text{ or } S \text{ or } R) \tag{A4-2}$$

$$1.2D + 1.6(L_r \text{ or } S \text{ or } R) + (0.5L \text{ or } 0.8W) \tag{A4-3}$$

$$1.2D + 1.3W + 0.5L + 0.5(L_r \text{ or } S \text{ or } R) \tag{A4-4}$$

$$1.2D \pm 1.0E + 0.5L + 0.2S \tag{A4-5}$$

$$0.9D \pm (1.3W \text{ or } 1.0E) \tag{A4-6}$$

Exception: The load factor on L in combinations A4-3, A4-4, and A4-5 shall equal 1.0 for garages, areas occupied as places of public assembly, and all areas where the live load is greater than 100 psf.

2. Impact

For structures carrying live loads which induce impact, the assumed nominal live load shall be increased to provide for this impact in combinations A4-2 and A4-3.

If not otherwise specified, the increase shall be:

For supports of elevators and elevator machinery 100%

For supports of light machinery, shaft or motor driven, not less than 20%

For supports of reciprocating machinery or power driven units,
    not less than. 50%

For hangers supporting floors and balconies 33%

For cab-operated traveling crane support girders and their
    connections 25%

For pendant-operated traveling crane support girders and their
    connections 10%

3. **Crane Runway Horizontal Forces**

The nominal lateral force on crane runways to provide for the effect of moving crane trolleys shall be a minimum of 20 percent of the sum of weights of the lifted load and of the crane trolley, but exclusive of other parts of the crane. The force shall be assumed to be applied at the top of the rails, acting in either direction normal to the runway rails, and shall be distributed with due regard for lateral stiffness of the structure supporting the rails.

The nominal longitudinal force shall be a minimum of 10 percent of the maximum wheel loads of the crane applied at the top of the rail, unless otherwise specified.

A5. DESIGN BASIS

1. **Required Strength at Factored Loads**

The required strength of structural members and connections shall be determined by structural analysis for the appropriate factored load combinations given in Section A4.

Design by either elastic or plastic analysis is permitted, except that design by plastic analysis is permitted only for steels with specified yield stresses not exceeding 65 ksi and is subject to provisions of Sections B5.2, C2, E1.2, F1.2d, H1, and I1.

Beams and girders composed of compact sections, as defined in Section B5.1, and satisfying the unbraced length requirements of Section F1.2d (including composite members) which are continuous over supports or are rigidly framed to columns may be proportioned for nine-tenths of the negative moments produced by gravity loading at points of support, provided that the maximum positive moment is increased by one-tenth of the average negative moments. This reduction is not permitted for hybrid beams, members of A514 steel, or moments produced by loading on cantilevers. If the negative moment is resisted by a column rigidly framed to the beam or girder, the one-tenth reduction may be used in proportioning the column for combined axial force and flexure, provided that the axial force does not exceed ϕ_c times $0.15A_gF_y$,

where

A_g = gross area, in.^{2}
F_y = specified minimum yield stress, ksi
ϕ_c = resistance factor for compression

2. **Limit States**

LRFD is a method of proportioning structures so that no applicable limit state is exceeded when the structure is subjected to all appropriate factored load combinations.

Strength limit states are related to safety and concern maximum load carrying capacity. Serviceability limit states are related to performance under normal service conditions. The term "resistance" includes both strength limit states and serviceability limit states.

3. Design for Strength

The design strength of each structural component or assemblage must equal or exceed the required strength based on the factored loads. The design strength ϕR_n for each applicable limit state is calculated as the nominal strength R_n multiplied by a resistance factor ϕ.

The required strength is determined for each applicable load combination as stipulated in Section A4.

Nominal strengths R_n and resistance factors ϕ are given in Chapters D through K.

4. Design for Serviceability and Other Considerations

The overall structure and the individual members, connections, and connectors shall be checked for serviceability. Provisions for design for serviceability are given in Chapter L.

A6. REFERENCED CODES AND STANDARDS

The following documents are referenced in this Specification:

American National Standards Institute
ANSI B18.1-72

American Society of Civil Engineers
ASCE 7-88

American Society for Testing and Materials

ASTM A6-91b	ASTM A27-87	ASTM A36-91
ASTM A53-88	ASTM A148-84	ASTM A193-91
ASTM A194-91	ASTM A242-91a	ASTM A307-91
ASTM A325-91c	ASTM A354-91	ASTM A449-91a
ASTM A490-91	ASTM A500-90a	ASTM A501-89
ASTM A502-91	ASTM A514-91	ASTM A529-89
ASTM A563-91c	ASTM A570-91	ASTM A572-91
ASTM A588-91a	ASTM A606-91a	ASTM A607-91
ASTM A618-90a	ASTM A668-85a	ASTM A687-89
ASTM A709-91	ASTM A852-91	ASTM C33-90
ASTM C330-89	ASTM F436-91	

American Welding Society

AWS D1.1-92	AWS A5.1-91	AWS A5.5-81
AWS A5.17-89	AWS A5.18-79	AWS A5.20-79
AWS A5.23-90	AWS A5.28-79	AWS A5.29-80

Research Council on Structural Connections
Load and Resistance Factor Design Specification for Structural Joints Using ASTM A325 or A490 Bolts, 1988

American Iron and Steel Institute
Load and Resistance Factor Design Specification for Cold-Formed Steel Members, 1991

American Institute of Steel Construction, Inc.
Code of Standard Practice for Steel Buildings and Bridges, 1992
Seismic Provisions for Structural Steel Buildings, 1992
Specification for Load and Resistance Factor Design of Single-Angle Members,
1993

A7. DESIGN DOCUMENTS

1. Plans

The design plans shall show a complete design with sizes, sections, and relative locations of the various members. Floor levels, column centers and offsets shall be dimensioned. Drawings shall be drawn to a scale large enough to show the information clearly.

Design documents shall indicate the type or types of construction as defined in Section A2.2 and include the required strengths (moments and forces) if necessary for preparation of shop drawings.

Where joints are to be assembled with high-strength bolts, the design documents shall indicate the connection type (i.e., snug-tight bearing, fully-tensioned bearing, direct tension, or slip-critical).

Camber of trusses, beams, and girders, if required, shall be specified in the design documents. The requirements for stiffeners and bracing shall be shown in the design documents.

2. Standard Symbols and Nomenclature

Welding and inspection symbols used on plans and shop drawings shall be the American Welding Society symbols. Welding symbols for special requirements not covered by AWS is permitted to be used provided a complete explanation thereof is shown in the design documents.

3. Notation for Welding

Weld lengths called for in the design documents and on the shop drawings shall be the net effective lengths.

CHAPTER B

DESIGN REQUIREMENTS

This chapter contains provisions which are common to the Specification as a whole.

B1. GROSS AREA

The gross area A_g of a member at any point is the sum of the products of the thickness and the gross width of each element measured normal to the axis of the member. For angles, the gross width is the sum of the widths of the legs less the thickness.

B2. NET AREA

The net area A_n of a member is the sum of the products of the thickness and the net width of each element computed as follows:

In computing net area for tension and shear, the width of a bolt hole shall be taken as $\frac{1}{16}$-in. greater than the nominal dimension of the hole.

For a chain of holes extending across a part in any diagonal or zigzag line, the net width of the part shall be obtained by deducting from the gross width the sum of the diameters or slot dimensions as provided in Section J3.2, of all holes in the chain, and adding, for each gage space in the chain, the quantity $s^2 / 4g$

where

s = longitudinal center-to-center spacing (pitch) of any two consecutive holes, in.

g = transverse center-to-center spacing (gage) between fastener gage lines, in.

For angles, the gage for holes in opposite adjacent legs shall be the sum of the gages from the back of the angles less the thickness.

In determining the net area across plug or slot welds, the weld metal shall not be considered as adding to the net area.

B3. EFFECTIVE NET AREA FOR TENSION MEMBERS

The effective net area for tension members shall be determined as follows:

1. When a tension load is transmitted directly to each of the cross-sectional elements by fasteners or welds, the effective net area A_e is equal to the net area A_n.

2. When a tension load is transmitted by fasteners or welds through some but not

all of the cross-sectional elements of the member, the effective net area A_e shall be computed as:

$$A_e = AU \qquad (B3\text{-}1)$$

where

A = area as defined below
U = reduction coefficient
$\quad$ = $1 - (\bar{x}/L) \le 0.9$ or as defined in B3c or B3d $\qquad$ (B3-2)
$\bar{x}$ = connection eccentricity, in.
L = length of connection in the directions of loading, in.

Larger values of U are permitted to be used when justified by tests or other rational criteria.

(a) When the tension load is transmitted only by bolts or rivets:

A = A_n
$\quad$ = net area of member, in.^{2}

(b) When the tension load is transmitted only by longitudinal welds to other than a plate member or by longitudinal welds in combination with transverse welds:

A = A_g
$\quad$ = gross area of member, in.^{2}

(c) When the tension load is transmitted only by transverse welds:

A = area of directly connected elements, in.^{2}
U = 1.0

(d) When the tension load is transmitted to a plate by longitudinal welds along both edges at the end of the plate for $l \ge w$:

A = area of plate, in.^{2}

For $l \ge 2w$ .... $U = 1.00$
For $2w > l \ge 1.5w$ .... $U = 0.87$
For $1.5w > l \ge w$ .... $U = 0.75$

where

l = length of weld, in.
w = plate width (distance between welds), in.

For effective area of connecting elements, see Section J5.2.

B4. STABILITY

General stability shall be provided for the structure as a whole and for each of its elements.

Consideration shall be given to the significant effects of the loads on the deflected shape of the structure and its individual elements.

B5.   LOCAL BUCKLING

1.   Classification of Steel Sections

Steel sections are classified as compact, noncompact, or slender-element sections. For a section to qualify as compact, its flanges must be continuously connected to the web or webs and the width-thickness ratios of its compression elements must not exceed the limiting width-thickness ratios λ_p from Table B5.1. If the width-thickness ratio of one or more compression elements exceeds λ_p, but does not exceed λ_r, the section is noncompact. If the width-thickness ratio of any element exceeds λ_r from Table B5.1, the section is referred to as a slender-element compression section.

For unstiffened elements which are supported along only one edge parallel to the direction of the compression force, the width shall be taken as follows:

(a) For flanges of I-shaped members and tees, the width b is half the full-flange width, b_f.

(b) For legs of angles and flanges of channels and zees, the width b is the full nominal dimension.

(c) For plates, the width b is the distance from the free edge to the first row of fasteners or line of welds.

(d) For stems of tees, d is taken as the full nominal depth.

For stiffened elements which are supported along two edges parallel to the direction of the compression force, the width shall be taken as follows:

(a) For webs of rolled or formed sections, h is the clear distance between flanges less the fillet or corner radius at each flange; h_c is twice the distance from the centroid to the inside face of the compression flange less the fillet or corner radius.

(b) For webs of built-up sections, h is the distance between adjacent lines of fasteners or the clear distance between flanges when welds are used, and h_c is twice the distance from the centroid to the nearest line of fasteners at the compression flange or the inside face of the compression flange when welds are used.

(c) For flange or diaphragm plates in built-up sections, the width b is the distance between adjacent lines of fasteners or lines of welds.

(d) For flanges of rectangular hollow structural sections, the width b is the clear distance between webs less the inside corner radius on each side. If the corner radius is not known, the width may be taken as the total section width minus three times the thickness.

For tapered flanges of rolled sections, the thickness is the nominal value halfway between the free edge and the corresponding face of the web.

2.   Design by Plastic Analysis

Design by plastic analysis is permitted when flanges subject to compression involving hinge rotation and all webs have a width-thickness ratio less than or equal to the limiting λ_p from Table B5.1. For circular hollow sections see Footnote d of Table B5.1.

Design by plastic analysis is subject to the limitations in Section A5.1.

3.   Slender-Element Compression Sections

For the flexural design of I-shaped sections, channels and rectangular or circular sections with slender flange elements, see Appendix F1. For other shapes in flexure or members in axial compression that have slender compression elements, see Appendix B5.3. For plate girders with slender web elements, see Appendix G.

B6.   BRACING AT SUPPORTS

At points of support for beams, girders and trusses, restraint against rotation about their longitudinal axis shall be provided.

B7.   LIMITING SLENDERNESS RATIOS

For members in which the design is based on compression, the slenderness ratio Kl/r preferably should not exceed 200.

For members in which the design is based on tension, the slenderness ratio l/r preferably should not exceed 300. The above limitation does not apply to rods in tension. Members in which the design is dictated by tension loading, but which may be subject to some compression under other load conditions, need not satisfy the compression slenderness limit.

B8.   SIMPLE SPANS

Beams, girders and trusses designed on the basis of simple spans shall have an effective length equal to the distance between centers of gravity of the members to which they deliver their end reactions.

B9.   END RESTRAINT

When designed on the assumption of full or partial end restraint due to continuous, semicontinuous, or cantilever action, the beams, girders, and trusses, as well as the sections of the members to which they connect, shall be designed to carry the factored forces and moments so introduced, as well as all other factored forces, without exceeding the design strengths prescribed in Chapters D through K, except that some inelastic but self-limiting deformation of a part of the connection is permitted.

B10. PROPORTIONS OF BEAMS AND GIRDERS

Rolled or welded shapes, plate girders and cover-plated beams shall, in general, be proportioned by the moment of inertia of the gross section. No deduction shall be made for bolt or rivet holes in either flange provided that

$$0.75F_u A_{fn} \geq 0.9F_y A_{fg} \tag{B10-1}$$

where A_{fg} is the gross flange area and A_{fn} is the net flange area calculated in

TABLE B5.1
Limiting Width-Thickness Ratios for Compression Elements

	Description of Element	Width Thickness Ratio	Limiting Width-Thickness Ratios	
			λ_p (compact)	λ_r (non compact)
Unstiffened Elements	Flanges of I-shaped rolled beams and channels in flexure	b/t	$65/\sqrt{F_y}$ [c]	$141/\sqrt{F_y-10}$
	Flanges of I-shaped hybrid or welded beams in flexure	b/t	$65/\sqrt{F_{yf}}$	$\dfrac{162}{\sqrt{(F_{yf}-16.5)/k_c}}$ [f]
	Flanges projecting from built-up compression members	b/t	NA	$109/\sqrt{F_y/k_c}$ [f]
	Outstanding legs of pairs of angles in continuous contact, flanges of I-shaped members and channels in axial compression; angles and plates projecting from beams or compression members	b/t	NA	$95/\sqrt{F_y}$
	Legs of single angle struts; legs of double angle struts with separators; unstiffened elements, i.e., supported along one edge	b/t	NA	$76/\sqrt{F_y}$
	Stems of tees	d/t	NA	$127/\sqrt{F_y}$

accordance with the provisions of Sections B1 and B2 and F_u is the specified minimum tensile strength.

If

$$0.75F_u A_{fn} < 0.9F_y A_{fg} \tag{B10-2}$$

the member flexural properties shall be based on an effective tension flange area A_{fe}

$$A_{fe} = \frac{5}{6}\frac{F_u}{F_y}A_{fn} \tag{B10-3}$$

Hybrid girders may be proportioned by the moment of inertia of their gross section, subject to the applicable provisions in Appendix G1, provided they are not required to resist an axial force greater than ϕ_b times $0.15F_{yf}A_g$, where F_{yf} is the specified yield stress of the flange material and A_g is the gross area. No limit is placed on the web stresses produced by the applied bending moment for which a hybrid girder is designed, except as provided in Section K3 and Appendix K3. To qualify as hybrid girders, the flanges at any given section shall have the same cross-sectional area and be made of the same grade of steel.

TABLE B5.1 (cont.)
Limiting Width-Thickness Ratios for Compression Elements

	Description of Element	Width Thick-ness Ratio	Limiting Width-Thickness Ratios	
			λ_p (compact)	λ_r (non compact)
Stiffened Elements	Flanges of square and rectangular box and hollow structural sections of uniform thickness subject to bending or compression; flange cover plates and diaphragm plates between lines of fasteners or welds	b/t	$190/\sqrt{F_y}$	$238/\sqrt{F_y}$
	Unsupported width of cover plates perforated with a succession of access holes [b]	b/t	NA	$317/\sqrt{F_y}$
	Webs in flexural compression [a]	h/t_w	$640/\sqrt{F_y}$ [c]	$970/\sqrt{F_y}$ [g]
	Webs in combined flexural and axial compression	h/t_w	for $P_u/\phi_b P_y \leq 0.125$ [c] $\dfrac{640}{\sqrt{F_y}}\left(1 - \dfrac{2.75P_u}{\phi_b P_y}\right)$ for $P_u/\phi_b P_y > 0.125$ [c] $\dfrac{191}{\sqrt{F_y}}\left(2.33 - \dfrac{P_u}{\phi_b P_y}\right) \geq \dfrac{253}{\sqrt{F_y}}$	[g] $\dfrac{970}{\sqrt{F_y}}\left(1 - 0.74\dfrac{P_u}{\phi_b P_y}\right)$
	All other uniformly compressed stiffened elements, i.e., supported along two edges	b/t h/t_w	NA	$253/\sqrt{F_y}$
	Circular hollow sections In axial compression In flexure	D/t	[d] NA $2,070/F_y$	$3,300/F_y$ $8,970/F_y$

[a] For hybrid beams, use the yield strength of the flange F_{yf} instead of F_y

[b] Assumes net area of plate at widest hole.

[c] Assumes an inelastic rotation capacity of 3. For structures in zones of high seismicity, a greater rotation capacity may be required.

[d] For plastic design use $1,300/F_y$.

[e] F_r = compressive residual stress in flange
= 10 ksi for rolled shapes
= 16.5 ksi for welded shapes

[f] $k_c = \dfrac{4}{\sqrt{h/t_w}}$ but not less than $0.35 \leq k_c \leq 0.763$

[g] For members with unequal flanges, see Appendix B5.1. F_y is the specified minimum yield stress of the type of steel being used.

Flanges of welded beams or girders may be varied in thickness or width by splicing a series of plates or by the use of cover plates.

The total cross-sectional area of cover plates of bolted or riveted girders shall not exceed 70 percent of the total flange area.

High-strength bolts, rivets, or welds connecting flange to web, or cover plate to flange, shall be proportioned to resist the total horizontal shear resulting from the bending forces on the girder. The longitudinal distribution of these bolts, rivets, or intermittent welds shall be in proportion to the intensity of the shear. However, the longitudinal spacing shall not exceed the maximum permitted for compression or tension members in Section E4 or D2, respectively. Bolts, rivets, or welds connecting flange to web shall also be proportioned to transmit to the web any loads applied directly to the flange, unless provision is made to transmit such loads by direct bearing.

Partial length cover plates shall be extended beyond the theoretical cutoff point and the extended portion shall be attached to the beam or girder by high-strength bolts in a slip-critical connection, rivets, or fillet welds. The attachment shall be adequate, at the applicable design strength given in Sections J2.2, J3.8, or K3 to develop the cover plate's portion of the flexural design strength in the beam or girder at the theoretical cutoff point.

For welded cover plates, the welds connecting the cover plate termination to the beam or girder in the length a', defined below, shall be adequate, at the applicable design strength, to develop the cover plate's portion of the design strength in the beam or girder at the distance a' from the end of the cover plate. The length a', measured from the end of the cover plate, shall be:

(a) A distance equal to the width of the cover plate when there is a continuous weld equal to or larger than three-fourths of the plate thickness across the end of the plate and continuous welds along both edges of the cover plate in the length a'.

(b) A distance equal to one and one-half times the width of the cover plate when there is a continuous weld smaller than three-fourths of the plate thickness across the end of the plate and continuous welds along both edges of the cover plate in the length a'.

(c) A distance equal to two times the width of the cover plate when there is no weld across the end of the plate, but continuous welds along both edges of the cover plate in the length a'.

CHAPTER C

FRAMES AND OTHER STRUCTURES

This chapter contains general requirements for stability of the structure as a whole.

C1. SECOND ORDER EFFECTS

Second order ($P\Delta$) effects shall be considered in the design of frames.

In structures designed on the basis of plastic analysis, the required flexural strength M_u shall be determined from a second-order plastic analysis that satisfies the requirements of Section C2. In structures designed on the basis of elastic analysis, M_u for beam-columns, connections, and connected members shall be determined from a second-order elastic analysis or from the following approximate second-order analysis procedure:

$$M_u = B_1 M_{nt} + B_2 M_{lt} \tag{C1-1}$$

where

M_{nt} = required flexural strength in member assuming there is no lateral translation of the frame, kip-in.

M_{lt} = required flexural strength in member as a result of lateral translation of the frame only, kip-in.

$$B_1 = \frac{C_m}{(1 - P_u / P_{e1})} \geq 1 \tag{C1-2}$$

$P_{e1} = A_g F_y / \lambda_c^2$ where λ_c is the slenderness parameter, in which the effective length factor K in the plane of bending shall be determined in accordance with Section C2.1, for the braced frame.

$$\lambda_c = \frac{Kl}{r\pi} \sqrt{\frac{F_y}{E}}$$

P_u = required axial compressive strength for the member under consideration, kips

C_m = a coefficient based on elastic first-order analysis assuming no lateral translation of the frame whose value shall be taken as follows:

(a) For compression members not subject to transverse loading between their supports in the plane of bending,

$$C_m = 0.6 - 0.4(M_1 / M_2) \tag{C1-3}$$

where M_1 / M_2 is the ratio of the smaller to larger moments at the ends of that portion of the member unbraced in the plane of bending under consideration. M_1 / M_2 is positive when the member is bent in reverse curvature, negative when bent in single curvature.

(b) For compression members subjected to transverse loading between their supports, the value of C_m shall be determined either by rational analysis or by the use of the following values:

For members whose ends are restrained. . . . . $C_m = 0.85$
For members whose ends are unrestrained. . . . . $C_m = 1.00$

$$B_2 = \frac{1}{1 - \Sigma P_u \left(\dfrac{\Delta_{oh}}{\Sigma HL} \right)} \qquad\qquad \text{(C1-4)}$$

or

$$B_2 = \frac{1}{1 - \dfrac{\Sigma P_u}{\Sigma P_{e2}}} \qquad\qquad \text{(C1-5)}$$

ΣP_u = required axial strength of all columns in a story, kips
Δ_{oh} = lateral inter-story deflection, in.
ΣH = sum of all story horizontal forces producing Δ_{oh}, kips
L = story height, in.
P_{e2} = $A_g F_y / \lambda_c^2$, kips, where λ_c is the slenderness parameter, in which the effective length factor K in the plane of bending shall be determined in accordance with Section C2.2, for the unbraced frame.

C2. FRAME STABILITY

1. Braced Frames

In trusses and frames where lateral stability is provided by diagonal bracing, shear walls, or equivalent means, the effective length factor K for compression members shall be taken as unity, unless structural analysis shows that a smaller value may be used.

The vertical bracing system for a braced multistory frame shall be determined by structural analysis to be adequate to prevent buckling of the structure and to maintain the lateral stability of the structure, including the overturning effects of drift, under the factored loads given in Section A4.

The vertical bracing system for a multistory frame may be considered to function together with in-plane shear-resisting exterior and interior walls, floor slabs, and roof decks, which are properly secured to the structural frames. The columns, girders, beams, and diagonal members, when used as the vertical bracing system, may be considered to comprise a vertically cantilevered simply con-nected truss in the analyses for frame buckling and lateral stability. Axial deformation of all members in the vertical bracing system shall be included in the lateral stability analysis.

In structures designed on the basis of plastic analysis, the axial force in these members caused by factored gravity plus factored horizontal loads shall not exceed $0.85\phi_c$ times $A_g F_y$.

Girders and beams included in the vertical bracing system of a braced multistory frame shall be proportioned for axial force and moment caused by concurrent factored horizontal and gravity loads.

2. Unbraced Frames

In frames where lateral stability depends upon the bending stiffness of rigidly connected beams and columns, the effective length factor K of compression members shall be determined by structural analysis. The destabilizing effects of gravity loaded columns whose simple connections to the frame do not provide resistance to lateral loads shall be included in the design of the moment-frame columns. Stiffness reduction adjustment due to column inelasticity is permitted.

Analysis of the required strength of unbraced multistory frames shall include the effects of frame instability and column axial deformation under the factored loads given in Section A4.

In structures designed on the basis of plastic analysis, the axial force in the columns caused by factored gravity plus factored horizontal loads shall not exceed $0.75\phi_c$ times $A_g F_y$.

CHAPTER D

TENSION MEMBERS

This chapter applies to prismatic members subject to axial tension caused by static forces acting through the centroidal axis. For members subject to combined axial tension and flexure, see Section H1.1. For threaded rods, see Section J3. For block shear rupture strength at end connections of tension members, see Section J4.3. For the design tensile strength of connecting elements, see Section J5.2. For members subject to fatigue, see Section K3.

D1. DESIGN TENSILE STRENGTH

The design strength of tension members $\phi_t P_n$ shall be the lower value obtained according to the limit states of yielding in the gross section and fracture in the net section.

(a) For yielding in the gross section:

$$\phi_t = 0.90$$
$$P_n = F_y A_g \tag{D1-1}$$

(b) For fracture in the net section:

$$\phi_t = 0.75$$
$$P_n = F_u A_e \tag{D1-2}$$

where

A_e = effective net area, in.^{2}
A_g = gross area of member, in.^{2}
F_y = specified minimum yield stress, ksi
F_u = specified minimum tensile strength, ksi
P_n = nominal axial strength, kips

When members without holes are fully connected by welds, the effective net section used in Equation D1-2 shall be as defined in Section B3. When holes are present in a member with welded-end connections, or at the welded connection in the case of plug or slot welds, the net section through the holes shall be used in Equation D1-2.

D2. BUILT-UP MEMBERS

For limitations on the longitudinal spacing of connectors between elements in continuous contact consisting of a plate and a shape or two plates, see Section J3.5.

The longitudinal spacing of connectors between components should preferably limit the slenderness ratio in any component between the connectors to 300.

Either perforated cover plates or tie plates without lacing are permitted to be used on the open sides of built-up tension members. Tie plates shall have a length not less than two-thirds the distance between the lines of welds or fasteners connecting them to the components of the member. The thickness of such tie plates shall not be less than one-fiftieth of the distance between these lines. The longitudinal spacing of intermittent welds or fasteners at tie plates shall not exceed six inches. The spacing of tie plates shall be such that the slenderness ratio of any component in the length between tie plates should preferably not exceed 300.

D3. PIN-CONNECTED MEMBERS AND EYEBARS

The pin diameter shall not be less than seven-eighths times the eyebar body width.

The pin-hole diameter shall not be more than $\frac{1}{32}$-in. greater than the pin diameter.

For steels having a yield stress greater than 70 ksi, the hole diameter shall not exceed five times the plate thickness and the width of the eyebar body shall be reduced accordingly.

In pin-connected members, the pin hole shall be located midway between the edges of the member in the direction normal to the applied force. For pin-connected members in which the pin is expected to provide for relative movement between connected parts while under full load, the diameter of pin hole shall not be more than $\frac{1}{32}$-in. greater than the diameter of the pin. The width of the plate beyond the pin hole shall be not less than the effective width on either side of the pin hole.

In pin-connected plates other than eyebars, the minimum net area beyond the bearing end of the pin hole, parallel to the axis of the member, shall not be less than two-thirds of the net area required for strength across the pin hole.

The design strength of a pin-connected member ϕP_n shall be the lowest value of the following limit states:

(a) Tension on the net effective area:

$$\phi = \phi_t = 0.75 \qquad (D3\text{-}1)$$
$$P_n = 2tb_{eff}F_u$$

(b) Shear on the effective area:

$$\phi_{sf} = 0.75 \qquad (D3\text{-}2)$$
$$P_n = 0.6A_{sf}F_u$$

(c) For bearing on the projected area of the pin, see Section J8.

(d) For yielding in the gross section, use Equation D1-1.

where

  a = shortest distance from edge of the pin hole to the edge of the member measured parallel to the direction of the force, in.

  $A_{sf} = 2t(a + d/2)$, in.^{2}

  $b_{eff} = 2t + 0.63$, but not more than the actual distance from the edge of the hole to the edge of the part measured in the direction normal to the applied force, in.

  d = pin diameter, in.

  t = thickness of plate, in.

The corners beyond the pin hole are permitted to be cut at 45° to the axis of the member, provided the net area beyond the pin hole, on a plane perpendicular to the cut, is not less than that required beyond the pin hole parallel to the axis of the member.

The design strength of eyebars shall be determined in accordance with Section D1 with A_g taken as the cross-sectional area of the body.

Eyebars shall be of uniform thickness, without reinforcement at the pin holes, and have circular heads whose periphery is concentric with the pin hole.

The radius of transition between the circular head and the eyebar body shall be not less than the head diameter.

The width of the body of the eyebars shall not exceed eight times its thickness.

The thickness of less than ½-in. is permissible only if external nuts are provided to tighten pin plates and filler plates into snug contact. The width b from the hole edge to the plate edge perpendicular to the direction of applied load shall be greater than two-thirds and, for the purpose of calculation, not more than three-fourths times the eyebar body width.

CHAPTER E

COLUMNS AND OTHER COMPRESSION MEMBERS

This chapter applies to compact and non-compact prismatic members subject to axial compression through the centroidal axis. For members subject to combined axial compression and flexure, see Section H1.2. For members with slender compression elements, see Appendix B5.3. For tapered members, see Appendix F3. For single-angle members, see AISC *Specification for Load and Resistance Design of Single-Angle Members*.

E1. EFFECTIVE LENGTH AND SLENDERNESS LIMITATIONS

1. Effective Length

The effective length factor K shall be determined in accordance with Section C2.

2. Design by Plastic Analysis

Design by plastic analysis, as limited in Section A5.1, is permitted if the column slenderness parameter λ_c does not exceed $1.5K$.

E2. DESIGN COMPRESSIVE STRENGTH FOR FLEXURAL BUCKLING

The design strength for flexural buckling of compression members whose elements have width-thickness ratios less than λ_r from Section B5.1 is $\phi_c P_n$:

$$\phi_c = 0.85$$
$$P_n = A_g F_{cr} \tag{E2-1}$$

(a) For $\lambda_c \leq 1.5$

$$F_{cr} = (0.658^{\lambda_c^2}) F_y \tag{E2-2}$$

(b) For $\lambda_c > 1.5$

$$F_{cr} = \left[\frac{0.877}{\lambda_c^2} \right] F_y \tag{E2-3}$$

where

$$\lambda_c = \frac{Kl}{r\pi} \sqrt{\frac{F_y}{E}} \tag{E2-4}$$

A_g = gross area of member, in.^{2}
F_y = specified yield stress, ksi
E = modulus of elasticity, ksi

K = effective length factor
l = laterally unbraced length of member, in.
r = governing radius of gyration about the axis of buckling, in.

For members whose elements do not meet the requirements of Section B5.1, see Appendix B5.3.

E3. DESIGN COMPRESSIVE STRENGTH FOR FLEXURAL-TORSIONAL BUCKLING

The design strength for flexural-torsional buckling of double-angle and tee-shaped compression members whose elements have width-thickness ratios less than λ_r from Section B5.1 is $\phi_c P_n$:

$$\phi_c = 0.85$$
$$P_n = A_g F_{crft}$$

$$F_{crft} = \left(\frac{F_{cry} + F_{crz}}{2H}\right)\left[1 - \sqrt{1 - \frac{4F_{cry}F_{crz}H}{(F_{cry} + F_{crz})^2}}\right] \qquad \text{(E3-1)}$$

where:

$$F_{crz} = \frac{GJ}{A\bar{r}_o^2}$$

$\bar{r}_o$ = polar radius of gyration about shear center, in. (see Equation A-E3-8)

$$H = 1 - \left(\frac{x_o^2 + y_o^2}{\bar{r}_o^2}\right)$$

x_o, y_o = coordinate of shear center with respect to the centroid, in.
x_o = 0 for double-angle and tee-shaped members (y-axis of symmetry)

F_{cry} is determined according to Section E2 for flexural buckling about the y-axis

of symmetry for $\lambda_c = \dfrac{Kl}{r_y \pi}\sqrt{\dfrac{F_y}{E}}$.

For double-angle and tee-shaped members whose elements do not meet the requirements of Section B5.1, see Appendix B5.3 to determine F_{cry} for use in Equation E3-1.

Other singly symmetric and unsymmetric columns, and doubly symmetric columns, such as cruciform or built-up columns, with very thin walls shall be designed for the limit states of flexural-torsional and torsional buckling in accordance with Appendix E3.

E4. BUILT-UP MEMBERS

At the ends of built-up compression members bearing on base plates or milled surfaces, all components in contact with one another shall be connected by a weld having a length not less than the maximum width of the member or by bolts spaced longitudinally not more than four diameters apart for a distance equal to 1½ times the maximum width of the member.

Along the length of built-up compression members between the end connections required above, longitudinal spacing for intermittent welds, bolts, or rivets shall be adequate to provide for the transfer of the required forces. For limitations on the longitudinal spacing of connectors between elements in continuous contact consisting of a plate and a shape or two plates, see Section J3.5. Where a component of a built-up compression member consists of an outside plate, the maximum spacing shall not exceed the thickness of the thinner outside plate times $127/\sqrt{F_y}$, nor 12 inches, when intermittent welds are provided along the edges of the components or when fasteners are provided on all gage lines at each section. When fasteners are staggered, the maximum spacing on each gage line shall not exceed the thickness of the thinner outside plate times $190/\sqrt{F_y}$ nor 18 inches.

Individual components of compression members composed of two or more shapes shall be connected to one another at intervals, a, such that the effective slenderness ratio Ka/r_i of each of the component shapes, between the connectors, does not exceed three-fourths times the governing slenderness ratio of the built-up member. The least radius of gyration r_i shall be used in computing the slenderness ratio of each component part. The end connection shall be welded or fully tensioned bolted with clean mill scale or blasted cleaned faying surfaces with Class A coatings.

The design strength of built-up members composed of two or more shapes shall be determined in accordance with Section E2 and Section E3 subject to the following modification. If the buckling mode involves relative deformations that produce shear forces in the connectors between individual shapes, Kl/r is replaced by $(Kl/r)_m$ determined as follows:

(a) For intermediate connectors that are snug-tight bolted:

$$\left(\frac{Kl}{r}\right)_m = \sqrt{\left(\frac{Kl}{r}\right)_o^2 + \left(\frac{a}{r_i}\right)^2} \qquad (E4\text{-}1)$$

(b) For intermediate connectors that are welded or fully-tensioned bolted:

$$\left(\frac{Kl}{r}\right)_m = \sqrt{\left(\frac{Kl}{r}\right)_o^2 + 0.82\frac{\alpha^2}{(1+\alpha^2)}\left(\frac{a}{r_{ib}}\right)^2} \qquad (E4\text{-}2)$$

where

$\left(\dfrac{Kl}{r}\right)_o$ = column slenderness of built-up member acting as a unit

$\left(\dfrac{Kl}{r}\right)_m$ = modified column slenderness of built-up member

$\dfrac{a}{r_i}$ = largest column slenderness of individual components

$\dfrac{a}{r_{ib}}$ = column slenderness of individual components relative to its centroidal axis parallel to axis of buckling

a = distance between connectors, in.

r_i = minimum radius of gyration of individual component, in.

r_{ib} = radius of gyration of individual component relative to its centroidal axis parallel to member axis of buckling, in.

α = separation ratio = $h / 2r_{ib}$

h = distance between centroids of individual components perpendicular to the member axis of buckling, in.

Open sides of compression members built up from plates or shapes shall be provided with continuous cover plates perforated with a succession of access holes. The unsupported width of such plates at access holes, as defined in Section B5.1, is assumed to contribute to the design strength provided that:

(1) The width-thickness ratio conforms to the limitations of Section B5.1.

(2) The ratio of length (in direction of stress) to width of hole shall not exceed two.

(3) The clear distance between holes in the direction of stress shall be not less than the transverse distance between nearest lines of connecting fasteners or welds.

(4) The periphery of the holes at all points shall have a minimum radius of $1\frac{1}{2}$-in.

As an alternative to perforated cover plates, lacing with tie plates is permitted at each end and at intermediate points if the lacing is interrupted. Tie plates shall be as near the ends as practicable. In main members providing design strength, the end tie plates shall have a length of not less than the distance between the lines of fasteners or welds connecting them to the components of the member. Intermediate tie plates shall have a length not less than one-half of this distance. The thickness of tie plates shall be not less than one-fiftieth of the distance between lines of welds or fasteners connecting them to the segments of the members. In welded construction, the welding on each line connecting a tie plate shall aggregate not less than one-third the length of the plate. In bolted and riveted construction, the spacing in the direction of stress in tie plates shall be not more than six diameters and the tie plates shall be connected to each segment by at least three fasteners.

Lacing, including flat bars, angles, channels, or other shapes employed as lacing, shall be so spaced that l / r of the flange included between their connections shall not exceed the governing slenderness ratio for the member as a whole. Lacing shall be proportioned to provide a shearing strength normal to the axis of the member equal to two percent of the compressive design strength of the member. The l / r ratio for lacing bars arranged in single systems shall not exceed 140. For double lacing this ratio shall not exceed 200. Double lacing bars shall be joined at the intersections. For lacing bars in compression, l is permitted to be taken as the unsupported length of the lacing bar between welds or fasteners connecting it to the components of the built-up member for single lacing, and 70 percent of that distance for double lacing. The inclination of lacing bars to the axis of the member shall preferably be not less than 60° for single lacing and

45° for double lacing. When the distance between the lines of welds or fasteners in the flanges is more than 15 inches, the lacing shall preferably be double or be made of angles.

For additional spacing requirements, see Section J3.

E5. PIN-CONNECTED COMPRESSION MEMBERS

Pin connections of pin-connected compression members shall conform to the requirements of Section D3 except Equations D3-1 and D3-2 do not apply.

CHAPTER F

BEAMS AND OTHER FLEXURAL MEMBERS

This chapter applies to compact and noncompact prismatic members subject to flexure and shear. For members subject to combined flexure and axial force, see Section H1. For members subject to fatigue, see Section K4. For members with slender compression elements, see Appendix B5. For web-tapered members, see Appendix F3. For members with slender web elements (plate girders), see Appendix G. For single-angle members, the AISC *Specification for Load and Resistance Factor Design of Single-Angle Members* is applicable.

F1. DESIGN FOR FLEXURE

The nominal flexural strength M_n is the lowest value obtained according to the limit states of: (a) yielding; (b) lateral-torsional buckling; (c) flange local buckling; and (d) web local buckling. For laterally braced compact beams with $L_b \le L_p$, only the limit state of yielding is applicable. For unbraced compact beams and noncompact tees and double angles, only the limit states of yielding and lateral-torsional buckling are applicable. The lateral-torsional buckling limit state is not applicable to members subject to bending about the minor axis, or to square or circular shapes.

This section applies to homogeneous and hybrid shapes with at least one axis of symmetry and which are subject to simple bending about one principal axis. For simple bending, the beam is loaded in a plane parallel to a principal axis that passes through the shear center or the beam is restrained against twisting at load points and supports. Only the limit states of yielding and lateral-torsional buckling are considered in this section. The lateral-torsional buckling provisions are limited to doubly symmetric shapes, channels, double angles, and tees. For lateral-torsional buckling of other singly symmetric shapes and for the limit states of flange local buckling and web local buckling of noncompact or slender-element sections, see Appendix F1. For unsymmetric shapes and beams subject to torsion combined with flexure, see Section H2. For biaxial bending, see Section H1.

1. Yielding

The flexural design strength of beams, determined by the limit state of yielding, is $\phi_b M_n$:

$$\phi_b = 0.90$$
$$M_n = M_p \tag{F1-1}$$

where

M_p = plastic moment (= $F_y Z \leq 1.5 M_y$ for homogeneous sections), kip-in.
M_y = moment corresponding to onset of yielding at the extreme fiber from an elastic stress distribution (= $F_y S$ for homogeneous section and $F_{yf} S$ for hybrid sections), kip-in.

2.   Lateral-Torsional Buckling

This limit state is only applicable to members subject to major axis bending. The flexural design strength, determined by the limit state of lateral-torsional buckling, is $\phi_b M_n$:

ϕ_b = 0.90
M_n = nominal strength determined as follows:

2a.   Doubly Symmetric Shapes and Channels with $L_b \leq L_r$

The nominal flexural strength is:

$$M_n = C_b \left[M_p - (M_p - M_r) \left(\frac{L_b - L_p}{L_r - L_p} \right) \right] \leq M_p \qquad \text{(F1-2)}$$

where:

L_b = distance between points braced against lateral displacement of the compression flange, or between points braced to prevent twist of the cross section, in.

In the above equation, C_b is a modification factor for non-uniform moment diagrams where, when both ends of the beam segment are braced:

$$C_b = \frac{12.5 M_{max}}{2.5 M_{max} + 3 M_A + 4 M_B + 3 M_C} \qquad \text{(F1-3)}$$

where

M_{max} = absolute value of maximum moment in the unbraced segment, kip-in.
M_A = absolute value of moment at quarter point of the unbraced segment
M_B = absolute value of moment at centerline of the unbraced beam segment
M_C = absolute value of moment at three-quarter point of the unbraced beam segment

C_b is permitted to be conservatively taken as 1.0 for all cases. For cantilevers or overhangs where the free end is unbraced, $C_b = 1.0$.

The limiting unbraced length for full plastic bending capacity, L_p, shall be determined as follows.

(a) For I-shaped members including hybrid sections and channels:

$$L_p = \frac{300 r_y}{\sqrt{F_{yf}}} \qquad \text{(F1-4)}$$

(b) For solid rectangular bars and box sections:

$$L_p = \frac{3,750r_y}{M_p}\sqrt{JA}$$

(F1-5)

where

A = cross-sectional area, in.^{2}
J = torsional constant, in.^{4}

The limiting laterally unbraced length L_r and the corresponding buckling moment M_r shall be determined as follows:

(a) For doubly symmetric I-shaped members and channels:

$$L_r = \frac{r_y X_1}{F_L} \sqrt{1 + \sqrt{1 + X_2 F_L^2}}$$

(F1-6)

$$M_r = F_L S_x$$

(F1-7)

where

$$X_1 = \frac{\pi}{S_x}\sqrt{\frac{EGJA}{2}}$$

(F1-8)

$$X_2 = 4\frac{C_w}{I_y}\left(\frac{S_x}{GJ}\right)^2$$

(F1-9)

S_x = section modulus about major axis, in.^{3}
E = modulus of elasticity of steel (29,000 ksi)
G = shear modulus of elasticity of steel (11,200 ksi)
F_L = smaller of $(F_{yf} - F_r)$ or F_{yw}
F_r = compressive residual stress in flange; 10 ksi for rolled shapes, 16.5 ksi for welded shapes
F_{yf} = yield stress of flange, ksi
F_{yw} = yield stress of web, ksi
I_y = moment of inertia about y-axis, in.^{4}
C_w = warping constant, in.^{6}

Equations F1-4 and F1-6 are conservatively based on $C_b = 1.0$.

(b) For solid rectangular bars and box sections:

$$L_r = \frac{57,000r_y\sqrt{JA}}{M_r}$$

(F1-10)

$$M_r = F_{yf} S_x$$

(F1-11)

2b. Doubly Symmetric Shapes and Channels with $L_b > L_r$

The nominal flexural strength is:

$$M_n = M_{cr} \le M_p$$

(F1-12)

where M_{cr} is the critical elastic moment, determined as follows:

(a) For doubly symmetric I-shaped members and channels:

$$M_{cr} = C_b \frac{\pi}{L_b} \sqrt{EI_yGJ + \left(\frac{\pi E}{L_b}\right)^2 I_y C_w}$$

(F1-13)

$$= \frac{C_b S_x X_1 \sqrt{2}}{L_b / r_y} \sqrt{1 + \frac{X_1^2 X_2}{2(L_b / r_y)^2}}$$

(b) For solid rectangular bars and symmetric box sections:

$$M_{cr} = \frac{57{,}000 \, C_b \sqrt{JA}}{L_b / r_y}$$

(F1-14)

2c. Tees and Double Angles

For tees and double-angle beams loaded in the plane of symmetry:

$$M_n = M_{cr} = \frac{\pi \sqrt{EI_yGJ}}{L_b} [B + \sqrt{1 + B^2}]$$

(F1-15)

where

$M_n \le 1.5 M_y$ for stems in tension
$M_n \le 1.0 M_y$ for stems in compression
$B = \pm 2.3(d / L_b) \sqrt{I_y / J}$ (F1-16)

The plus sign for B applies when the stem is in tension and the minus sign applies when the stem is in compression. If the tip of the stem is in compression anywhere along the unbraced length, use the negative value of B.

2d. Unbraced Length for Design by Plastic Analysis

Design by plastic analysis, as limited in Section A5.1, is permitted for a compact section member bent about the major axis when the laterally unbraced length L_b of the compression flange adjacent to plastic hinge locations associated with the failure mechanism does not exceed L_{pd}, determined as follows:

(a) For doubly symmetric and singly symmetric I-shaped members with the compression flange equal to or larger than the tension flange (including hybrid members) loaded in the plane of the web

$$L_{pd} = \frac{[3{,}600 + 2{,}200 \, (M_1 / M_2)] \, r_y}{F_y}$$

(F1-17)

where

F_y = specified minimum yield stress of the compression flange, ksi
M_1 = smaller moment at end of unbraced length of beam, kip-in.
M_2 = larger moment at end of unbraced length of beam, kip-in.
r_y = radius of gyration about minor axis, in.
(M_1 / M_2) is positive when moments cause reverse curvature and negative for single curvature

(b) For solid rectangular bars and symmetric box beams

$$L_{pd} = \frac{5,000 + 3,000 \, (M_1 / M_2)}{F_y} \, r_y \geq 3,000 r_y / F_y \qquad \text{(F1-18)}$$

There is no limit on L_b for members with circular or square cross sections nor for any beam bent about its minor axis.

In the region of the last hinge to form, and in regions not adjacent to a plastic hinge, the flexural design strength shall be determined in accordance with Section F1.2.

F2.   DESIGN FOR SHEAR

This section applies to unstiffened webs of singly or doubly symmetric beams, including hybrid beams, and channels subject to shear in the plane of the web. For the design shear strength of webs with stiffeners, see Appendix F2 or Appendix G3. For shear in the weak direction of the shapes above, pipes, and unsymmetric sections, see Section H2. For web panels subject to high shear, see Section K1.7. For shear strength at connections, see Sections J4 and J5.

1.   Web Area Determination

The web area A_w shall be taken as the overall depth d times the web thickness t_w.

2.   Design Shear Strength

The design shear strength of unstiffened webs, with $h / t_w \leq 260$, is $\phi_v V_n$, where

$\phi_v = 0.90$
V_n = nominal shear strength defined as follows

For $h / t_w \leq 418 / \sqrt{F_{yw}}$

$$V_n = 0.6 F_{yw} A_w \qquad \text{(F2-1)}$$

For $418 / \sqrt{F_{yw}} < h / t_w \leq 523 / \sqrt{F_{yw}}$

$$V_n = 0.6 F_{yw} A_w (418 / \sqrt{F_{yw}}) / (h / t_w) \qquad \text{(F2-2)}$$

For $523 / \sqrt{F_{yw}} < h / t_w \leq 260$

$$V_n = (132,000 A_w) / (h / t_w)^2 \qquad \text{(F2-3)}$$

The general design shear strength of webs with or without stiffeners is given in Appendix F2.2 and an alternative method utilizing tension field action is given in Appendix G3.

3.   Transverse Stiffeners

See Appendix F2.3.

F3.   WEB-TAPERED MEMBERS

See Appendix F3.

F4. BEAMS AND GIRDERS WITH WEB OPENINGS

The effect of all web openings on the design strength of steel and composite beams shall be determined. Adequate reinforcement shall be provided when the required strength exceeds the net strength of the member at the opening.

CHAPTER G

PLATE GIRDERS

I-shaped plate girders shall be distinguished from I-shaped beams on the basis of the web slenderness ratio h / t_w . When this value is greater than λ_r, the provisions of Appendices G1 and G2 shall apply for design flexural strength. For $h / t_w \leq \lambda_r$, the provisions of Chapter F or Appendix F shall apply for design flexural strength. For girders with unequal flanges, see Appendix B5.1.

The design shear strength and transverse stiffener design shall be based on either Section F2 (without tension-field action) or Appendix G3 (with tension-field action). For girders with unequal flanges, see Appendix B5.1.

CHAPTER H

MEMBERS UNDER COMBINED FORCES AND TORSION

This chapter applies to prismatic members subject to axial force and flexure about one or both axes of symmetry, with or without torsion, and torsion only. For web-tapered members, see Appendix F3.

H1. SYMMETRIC MEMBERS SUBJECT TO BENDING AND AXIAL FORCE

1. Doubly and Singly Symmetric Members in Flexure and Tension

The interaction of flexure and tension in symmetric shapes shall be limited by Equations H1-1a and H1-1b.

(a) For $\dfrac{P_u}{\phi P_n} \geq 0.2$

$$\frac{P_u}{\phi P_n} + \frac{8}{9}\left(\frac{M_{ux}}{\phi_b M_{nx}} + \frac{M_{uy}}{\phi_b M_{ny}}\right) \leq 1.0 \qquad \text{(H1-1a)}$$

(b) For $\dfrac{P_u}{\phi P_n} < 0.2$

$$\frac{P_u}{2\phi P_n} + \left(\frac{M_{ux}}{\phi_b M_{nx}} + \frac{M_{uy}}{\phi_b M_{ny}}\right) \leq 1.0 \qquad \text{(H1-1b)}$$

where

P_u = required tensile strength, kips
P_n = nominal tensile strength determined in accordance with Section D1, kips
M_u = required flexural strength determined in accordance with Section C1, kip-in.
M_n = nominal flexural strength determined in accordance with Section F1, kip-in.
x = subscript relating symbol to strong axis bending
y = subscript relating symbol to weak axis bending.
ϕ = ϕ_t = resistance factor for tension (see Section D1)
ϕ_b = resistance factor for flexure = 0.90

A more detailed analysis of the interaction of flexure and tension is permitted in lieu of Equations H1-1a and H1-1b.

2.   **Doubly and Singly Symmetric Members in Flexure and Compression**

The interaction of flexure and compression in symmetric shapes shall be limited by Equations H1-1a and H1-1b where

P_u = required compressive strength, kips

P_n = nominal compressive strength determined in accordance with Section E2, kips

M_u = required flexural strength determined in accordance with Section C1 kip-in.

M_n = nominal flexural strength determined in accordance with Section F1, kip-in.

x = subscript relating symbol to strong axis bending

y = subscript relating symbol to weak axis bending.

ϕ = ϕ_c = resistance factor for compression, = 0.85 (see Section E2)

ϕ_b = resistance factor for flexure = 0.90

H2. UNSYMMETRIC MEMBERS AND MEMBERS UNDER TORSION AND COMBINED TORSION, FLEXURE, SHEAR, AND/OR AXIAL FORCE

The design strength ϕF_y of the member shall equal or exceed the required strength expressed in terms of the normal stress f_{un} or the shear stress f_{uv}, determined by elastic analysis for the factored loads:

(a) For the limit state of yielding under normal stress:

$$f_{un} \le \phi F_y \tag{H2-1}$$
$$\phi = 0.90$$

(b) For the limit state of yielding under shear stress:

$$f_{uv} \le 0.6\phi F_y \tag{H2-2}$$
$$\phi = 0.90$$

(c) For the limit state of buckling:

$$f_{un} \text{ or } f_{uv} \le \phi_c F_{cr}, \text{ as applicable} \tag{H2-3}$$
$$\phi_c = 0.85$$

Some constrained local yielding is permitted adjacent to areas which remain elastic.

H3. ALTERNATIVE INTERACTION EQUATIONS FOR MEMBERS UNDER COMBINED STRESS

See Appendix H3.

CHAPTER I

COMPOSITE MEMBERS

This chapter applies to composite columns composed of rolled or built-up structural steel shapes, pipe or tubing, and structural concrete acting together and to steel beams supporting a reinforced concrete slab so interconnected that the beams and the slab act together to resist bending. Simple and continuous composite beams with shear connectors and concrete-encased beams, constructed with or without temporary shores, are included.

I1.   DESIGN ASSUMPTIONS

Force Determination. In determining forces in members and connections of a structure that includes composite beams, consideration shall be given to the effective sections at the time each increment of load is applied.

Elastic Analysis. For an elastic analysis of continuous composite beams without haunched ends, it is permissible to assume that the stiffness of a beam is uniform throughout the beam length. The stiffness is permitted to be computed using the moment of inertia of the composite transformed section in the positive moment region.

Plastic Analysis. When plastic analysis is used, the strength of flexural composite members shall be determined from plastic stress distributions.

Plastic Stress Distribution for Positive Moment. If the slab in the positive moment region is connected to the steel beam with shear connectors, a concrete stress of $0.85f_c'$ is permitted to be assumed uniformly distributed throughout the effective compression zone. Concrete tensile strength shall be neglected. A uniformly distributed steel stress of F_y shall be assumed throughout the tension zone and throughout the compression zone in the structural steel section. The net tensile force in the steel section shall be equal to the compressive force in the concrete slab.

Plastic Stress Distribution for Negative Moment. If the slab in the negative moment region is connected to the steel beam with shear connectors, a tensile stress of F_{yr} shall be assumed in all adequately developed longitudinal reinforcing bars within the effective width of the concrete slab. Concrete tensile strength shall be neglected. A uniformly distributed steel stress of F_y shall be assumed throughout the tension zone and throughout the compression zone in the structural steel section. The net compressive force in the steel section shall be equal to the total tensile force in the reinforcing steel.

Elastic Stress Distribution. When a determination of elastic stress distribution is required, strains in steel and concrete shall be assumed directly proportional to the distance from the neutral axis. The stress shall equal strain times modulus of elasticity for steel, E, or modulus of elasticity for concrete, E_c. Concrete tensile strength shall be neglected. Maximum stress in the steel shall not exceed F_y. Maximum compressive stress in the concrete shall not exceed $0.85f_c'$ where f_c' is the specified compressive strength of the concrete. In composite hybrid beams, the maximum stress in the steel flange shall not exceed F_{yf} but the strain in the web may exceed the yield strain; the stress shall be taken as F_{yw} at such locations.

Fully Composite Beam. Shear connectors are provided in sufficient numbers to develop the maximum flexural strength of the composite beam. For elastic stress distribution it shall be assumed that no slip occurs.

Partially Composite Beam. The shear strength of shear connectors governs the flexural strength of the partially composite beam. Elastic computations such as those for deflections, fatigue, and vibrations shall include the effect of slip.

Concrete-Encased Beam. A beam totally encased in concrete cast integrally with the slab may be assumed to be interconnected to the concrete by natural bond, without additional anchorage, provided that: (1) concrete cover over beam sides and soffit is at least two inches; (2) the top of the beam is at least 1½-in. below the top and two inches above the bottom of the slab; and (3) concrete encasement contains adequate mesh or other reinforcing steel to prevent spalling of concrete.

Composite Column. A steel column fabricated from rolled or built-up steel shapes and encased in structural concrete or fabricated from steel pipe or tubing and filled with structural concrete shall be designed in accordance with Section I2.

I2.   COMPRESSION MEMBERS

1.   Limitations

To qualify as a composite column, the following limitations shall be met:

(1) The cross-sectional area of the steel shape, pipe, or tubing shall comprise at least four percent of the total composite cross section.

(2) Concrete encasement of a steel core shall be reinforced with longitudinal load carrying bars, longitudinal bars to restrain concrete, and lateral ties. Longitudinal load carrying bars shall be continuous at framed levels; longitudinal restraining bars may be interrupted at framed levels. The spacing of ties shall be not greater than two-thirds of the least dimension of the composite cross section. The cross-sectional area of the transverse and longitudinal reinforcement shall be at least 0.007 sq. in. per inch of bar spacing. The encasement shall provide at least 1½-in. of clear cover outside of both transverse and longitudinal reinforcement.

(3) Concrete shall have a specified compressive strength f_c' of not less than 3 ksi nor more than 8 ksi for normal weight concrete and not less than 4 ksi for light weight concrete.

(4) The specified minimum yield stress of structural steel and reinforcing bars

used in calculating the strength of a composite column shall not exceed 55 ksi.

(5) The minimum wall thickness of structural steel pipe or tubing filled with concrete shall be equal to $b\sqrt{F_y/3E}$ for each face of width b in rectangular sections and $D\sqrt{F_y/8E}$ for circular sections of outside diameter D.

2. Design Strength

The design strength of axially loaded composite columns is $\phi_c P_n$,

where

ϕ_c = 0.85
P_n = nominal axial compressive strength determined from Equations E2-1 through E2-4 with the following modifications:

(1) A_s = gross area of steel shape, pipe, or tubing, in.^{2} (replaces A_g)

r_m = radius of gyration of the steel shape, pipe, or tubing except that for steel shapes it shall not be less than 0.3 times the overall thickness of the composite cross section in the plane of buckling, in. (replaces r)

(2) Replace F_y with modified yield stress F_{my} from Equation I2-1 and replace E with modified modulus of elasticity E_m from Equation I2-2.

$$F_{my} = F_y + c_1 F_{yr}(A_r/A_s) + c_2 f_c'(A_c/A_s) \qquad \text{(I2-1)}$$

$$E_m = E + c_3 E_c (A_c/A_s) \qquad \text{(I2-2)}$$

where

A_c = area of concrete, in.^{2}
A_r = area of longitudinal reinforcing bars, in.^{2}
A_s = area of steel, in.^{2}
E = modulus of elasticity of steel, ksi
E_c = modulus of elasticity of concrete. E_c is permitted to be computed from $E_c = w^{1.5}\sqrt{f_c'}$ where w, the unit weight of concrete, is expressed in lbs./cu. ft and f_c' is expressed in ksi.
F_y = specified minimum yield stress of steel shape, pipe, or tubing, ksi
F_{yr} = specified minimum yield stress of longitudinal reinforcing bars, ksi
f_c' = specified compressive strength of concrete, ksi
c_1, c_2, c_3 = numerical coefficients. For concrete-filled pipe and tubing: $c_1 = 1.0$, $c_2 = 0.85$, and $c_3 = 0.4$; for concrete encased shapes $c_1 = 0.7$, $c_2 = 0.6$, and $c_3 = 0.2$

3. Columns with Multiple Steel Shapes

If the composite cross section includes two or more steel shapes, the shapes shall be interconnected with lacing, tie plates, or batten plates to prevent buckling of individual shapes before hardening of concrete.

4. Load Transfer

The portion of the design strength of axially loaded composite columns resisted by concrete shall be developed by direct bearing at connections. When the supporting concrete area is wider than the loaded area on one or more sides and otherwise restrained against lateral expansion on the remaining sides, the maximum design strength of concrete shall be $1.7\phi_c f_c' A_B$,

where

$\phi_c = 0.60$
A_B = loaded area

I3. FLEXURAL MEMBERS

1. Effective Width

The effective width of the concrete slab on each side of the beam center-line shall not exceed:

(a) one-eighth of the beam span, center to center of supports;

(b) one-half the distance to the center-line of the adjacent beam; or

(c) the distance to the edge of the slab.

2. Strength of Beams with Shear Connectors

The positive design flexural strength $\phi_b M_n$ shall be determined as follows:

(a) For $h/t_w \leq 640/\sqrt{F_{yf}}$:

$\phi_b = 0.85$; M_n shall be determined from the plastic stress distribution on the composite section.

(b) For $h/t_w > 640/\sqrt{F_{yf}}$:

$\phi_b = 0.90$; M_n shall be determined from the superposition of elastic stresses, considering the effects of shoring.

The negative design flexural strength $\phi_b M_n$ shall be determined for the steel section alone, in accordance with the requirements of Chapter F.

Alternatively, the negative design flexural strength $\phi_b M_n$ shall be computed with: $\phi_b = 0.85$ and M_n determined from the plastic stress distribution on the composite section, provided that:

(1) Steel beam is an adequately braced compact section, as defined in Section B5.

(2) Shear connectors connect the slab to the steel beam in the negative moment region.

(3) Slab reinforcement parallel to the steel beam, within the effective width of the slab, is properly developed.

3. Strength of Concrete-Encased Beams

The design flexural strength $\phi_b M_n$ shall be computed with $\phi_b = 0.90$ and M_n determined from the superposition of elastic stresses, considering the effects of shoring.

Alternatively, the design flexural strength $\phi_b M_n$ shall be computed with $\phi_b = 0.90$ and M_n determined from the plastic stress distribution on the steel section alone.

4. Strength During Construction

When temporary shores are not used during construction, the steel section alone shall have adequate strength to support all loads applied prior to the concrete attaining 75 percent of its specified strength f_c'. The design flexural strength of the steel section shall be determined in accordance with the requirements of Section F1.

5. Formed Steel Deck

5a. General

The design flexural strength $\phi_b M_n$ of composite construction consisting of concrete slabs on formed steel deck connected to steel beams shall be determined by the applicable portions of Section I3.2, with the following modifications.

This section is applicable to decks with nominal rib height not greater than three inches. The average width of concrete rib or haunch w_r shall be not less than two inches, but shall not be taken in calculations as more than the minimum clear width near the top of the steel deck. See Section I3.5c for additional restrictions.

The concrete slab shall be connected to the steel beam with welded stud shear connectors $\frac{3}{4}$-in. or less in diameter (AWS D1.1). Studs shall be welded either through the deck or directly to the steel beam. Stud shear connectors, after installation, shall extend not less than $1\frac{1}{2}$-in. above the top of the steel deck.

The slab thickness above the steel deck shall be not less than two inches.

5b. Deck Ribs Oriented Perpendicular to Steel Beam

Concrete below the top of the steel deck shall be neglected in determining section properties and in calculating A_c for deck ribs oriented perpendicular to the steel beams.

The spacing of stud shear connectors along the length of a supporting beam shall not exceed 36 inches.

The nominal strength of a stud shear connector shall be the value stipulated in Section I5 multiplied by the following reduction factor:

$$\frac{0.85}{\sqrt{N_r}} (w_r / h_r) [(H_s / h_r) - 1.0] \leq 1.0 \qquad (I3-1)$$

where

h_r = nominal rib height, in.
H_s = length of stud connector after welding, in., not to exceed the value $(h_r + 3)$ in computations, although actual length may be greater
N_r = number of stud connectors in one rib at a beam intersection, not to exceed three in computations, although more than three studs may be installed

w_r = average width of concrete rib or haunch (as defined in Section I3.5a), in.

To resist uplift, steel deck shall be anchored to all supporting members at a spacing not to exceed 18 inches. Such anchorage shall be provided by stud connectors, a combination of stud connectors and arc spot (puddle) welds, or other devices specified by the designer.

5c. Deck Ribs Oriented Parallel to Steel Beam

Concrete below the top of the steel deck may be included in determining section properties and shall be included in calculating A_c in Section I5.

Steel deck ribs over supporting beams may be split longitudinally and separated to form a concrete haunch.

When the nominal depth of steel deck is 1½-in. or greater, the average width w_r of the supported haunch or rib shall be not less than two inches for the first stud in the transverse row plus four stud diameters for each additional stud.

The nominal strength of a stud shear connector shall be the value stipulated in Section I5, except that when w_r / h_r is less than 1.5, the value from Section I5 shall be multiplied by the following reduction factor:

$$0.6(w_r / h_r)[(H_s / h_r) - 1.0] \leq 1.0 \qquad (I3\text{-}2)$$

where h_r and H_s are as defined in Section I3.5b and w_r is the average width of concrete rib or haunch as defined in Section I3.5a.

6. Design Shear Strength

The design shear strength of composite beams shall be determined by the shear strength of the steel web, in accordance with Section F2.

I4. COMBINED COMPRESSION AND FLEXURE

The interaction of axial compression and flexure in the plane of symmetry on composite members shall be limited by Section H1.2 with the following modifications:

M_n = nominal flexural strength determined from plastic stress distribution on the composite cross section except as provided below, kip-in.

P_{e1}, P_{e2} = $A_s F_{my} / \lambda_c^2$ elastic buckling load, kips

F_{my} = modified yield stress, ksi, see Section I2

ϕ_b = resistance factor for flexure from Section I3

ϕ_c = resistance factor for compression = 0.85

λ_c = column slenderness parameter defined by Equation E2-4 as modified in Section I2.2

When the axial term in Equations H1-1a and H1-1b is less than 0.3, the nominal flexural strength M_n shall be determined by straight line transition between the nominal flexural strength determined from the plastic distribution on the composite cross sections at $(P_u / \phi_c P_n) = 0.3$ and the flexural strength at $P_u = 0$ as determined in Section I3. If shear connectors are required at $P_u = 0$, they shall be provided whenever $P_u / \phi_c P_n$ is less than 0.3.

I5. SHEAR CONNECTORS

This section applies to the design of stud and channel shear connectors. For connectors of other types, see Section I6.

1. Materials

Shear connectors shall be headed steel studs not less than four stud diameters in length after installation, or hot rolled steel channels. The stud connectors shall conform to the requirements of Section A3.6. The channel connectors shall conform to the requirements of Section A3. Shear connectors shall be embedded in concrete slabs made with ASTM C33 aggregate or with rotary kiln produced aggregates conforming to ASTM C330, with concrete unit weight not less than 90 pcf.

2. Horizontal Shear Force

Except for concrete-encased beams as defined in Section I1, the entire horizontal shear at the interface between the steel beam and the concrete slab shall be assumed to be transferred by shear connectors. For composite action with concrete subject to flexural compression, the total horizontal shear force between the point of maximum positive moment and the point of zero moment shall be taken as the smallest of the following: (1) $0.85f_c' A_c$; (2) $A_s F_y$; and (3) ΣQ_n;

where

f_c' = specified compressive strength of concrete, ksi
A_c = area of concrete slab within effective width, in.^{2}
A_s = area of steel cross section, in.^{2}
F_y = minimum specified yield stress, ksi
ΣQ_n = sum of nominal strengths of shear connectors between the point of maximum positive moment and the point of zero moment, kips

For hybrid beams, the yield force shall be computed separately for each component of the cross section; $A_s F_y$ of the entire cross section is the sum of the component yield forces.

In continuous composite beams where longitudinal reinforcing steel in the negative moment regions is considered to act compositely with the steel beam, the total horizontal shear force between the point of maximum negative moment and the point of zero moment shall be taken as the smaller of $A_r F_{yr}$ and ΣQ_n;

where

A_r = area of adequately developed longitudinal reinforcing steel within the effective width of the concrete slab, in.^{2}
F_{yr} = minimum specified yield stress of the reinforcing steel, ksi
ΣQ_n = sum of nominal strengths of shear connectors between the point of maximum negative moment and the point of zero moment, kips

3. Strength of Stud Shear Connectors

The nominal strength of one stud shear connector embedded in a solid concrete slab is

$$Q_n = 0.5A_{sc}\sqrt{f_c'E_c} \leq A_{sc}F_u \tag{I5-1}$$

where

A_{sc} = cross-sectional area of a stud shear connector, in.^{2}
f_c' = specified compressive strength of concrete, ksi
F_u = minimum specified tensile strength of a stud shear connector, ksi
E_c = modulus of elasticity of concrete, ksi

For stud shear connector embedded in a slab on a formed steel deck, refer to Section I3 for reduction factors given by Equations I3-1 and I3-2 as applicable. The reduction factors apply only to $0.5A_{sc}\sqrt{f_c'E_c}$ term in Equation I5-1.

4. Strength of Channel Shear Connectors

The nominal strength of one channel shear connector embedded in a solid concrete slab is

$$Q_n = 0.3(t_f + 0.5t_w)L_c\sqrt{f_c'E_c} \tag{I5-2}$$

where

t_f = flange thickness of channel shear connector, in.
t_w = web thickness of channel shear connector, in.
L_c = length of channel shear connector, in.

5. Required Number of Shear Connectors

The number of shear connectors required between the section of maximum bending moment, positive or negative, and the adjacent section of zero moment shall be equal to the horizontal shear force as determined in Section I5.2 divided by the nominal strength of one shear connector as determined from Section I5.3 or Section I5.4.

6. Shear Connector Placement and Spacing

Unless otherwise specified shear connectors required each side of the point of maximum bending moment, positive or negative, shall be distributed uniformly between that point and the adjacent points of zero moment. However, the number of shear connectors placed between any concentrated load and the nearest point of zero moment shall be sufficient to develop the maximum moment required at the concentrated load point.

Except for connectors installed in the ribs of formed steel decks, shear connectors shall have at least one inch of lateral concrete cover. Unless located over the web, the diameter of studs shall not be greater than 2.5 times the thickness of the flange to which they are welded. The minimum center-to-center spacing of stud connectors shall be six diameters along the longitudinal axis of the supporting composite beam and four diameters transverse to the longitudinal axis of the supporting composite beam, except that within the ribs of formed steel decks the center-to-center spacing may be as small as four diameters in any direction. The maximum center-to-center spacing of shear connectors shall not exceed eight times the total slab thickness. Also see Section I3.5b.

I6.   SPECIAL CASES

When composite construction does not conform to the requirements of Section I1 through Section I5, the strength of shear connectors and details of construction shall be established by a suitable test program.

CHAPTER J

CONNECTIONS, JOINTS, AND FASTENERS

This chapter applies to connecting elements, connectors, and the affected elements of the connected members subject to static loads. For connections subject to fatigue, see Appendix K3.

J1. GENERAL PROVISIONS

1. Design Basis

Connections consist of affected elements of connected members (e.g. beam webs), connecting elements (e.g., gussets, angles, brackets), and connectors (welds, bolts, rivets). These components shall be proportioned so that their design strength equals or exceeds the required strength determined by structural analysis for factored loads acting on the structure or a specified proportion of the strength of the connected members, whichever is appropriate.

2. Simple Connections

Except as otherwise indicated in the design documents, connections of beams, girders, or trusses shall be designed as flexible, and are permitted to ordinarily be proportioned for the reaction shears only. Flexible beam connections shall accommodate end rotations of unrestrained (simple) beams. To accomplish this, some inelastic but self-limiting deformation in the connection is permitted.

3. Moment Connections

End connections of restrained beams, girders, and trusses shall be designed for the combined effect of forces resulting from moment and shear induced by the rigidity of the connections.

4. Compression Members with Bearing Joints

When columns bear on bearing plates or are finished to bear at splices, there shall be sufficient connectors to hold all parts securely in place.

When compression members other than columns are finished to bear, the splice material and its connectors shall be arranged to hold all parts in line and shall be proportioned for 50 percent of the required strength of the member.

All compression joints shall be proportioned to resist any tension developed by the factored loads specified by load combination A4-6.

5. Splices in Heavy Sections

This paragraph applies to ASTM A6 Group 4 and 5 rolled shapes, or shapes built-up by welding plates more than two inches thick together to form the cross section, and where the cross section is to be spliced and subject to primary tensile stresses due to tension or flexure. When the individual elements of the cross section are spliced prior to being joined to form the cross section in accordance with AWS D1.1, Article 3.4.6, the applicable provisions of AWS D1.1 apply in lieu of the requirements of this section. When tensile forces in these sections are to be transmitted through splices by complete-joint-penetration groove welds, material notch-toughness requirements as given in Section A3.1c, weld access hole details as given in Section J1.6, welding preheat requirements as given in Section J2.8, and thermal-cut surface preparation and inspection requirements as given in Section M2.2 apply.

At tension splices in ASTM A6 Group 4 and 5 shapes and built-up members of material more than two inches thick, weld tabs and backing shall be removed and the surfaces ground smooth.

When splicing ASTM A6 Group 4 and 5 rolled shapes or shapes built-up by welding plates more than two inches thick to form a cross section, and where the section is to be used as a primary compression member, all weld access holes required to facilitate groove welding operations shall satisfy the provisions of Section J1.6.

Alternatively, splicing of such members subject to compression, including members which are subject to tension due to wind or seismic loads, shall be accomplished using splice details which do not induce large weld shrinkage strains; for example partial-joint-penetration flange groove welds with fillet-welded surface lap plate splices on the web, bolted lap plate splices, or combination bolted/fillet-welded lap plate splices.

6. Beam Copes and Weld Access Holes

All weld access holes required to facilitate welding operations shall have a length from the toe of the weld preparation not less than 1½ times the thickness of the material in which the hole is made. The height of the access hole shall be adequate for deposition of sound weld metal in the adjacent plates and provide clearance for weld tabs for the weld in the material in which the hole is made, but not less than the thickness of the material. In hot-rolled shapes and built-up shapes, all beam copes and weld access holes shall be shaped free of notches and sharp re-entrant corners except, when fillet web-to-flange welds are used in built-up shapes, access holes are permitted to terminate perpendicular to the flange.

For ASTM A6 Group 4 and 5 shapes and built-up shapes of material more than two inches thick, the thermally cut surfaces of beam copes and weld access holes shall be ground to bright metal and inspected by either magnetic particle or dye penetrant methods prior to deposition of splice welds. If the curved transition portion of weld access holes and beam copes are formed by predrilled or sawed holes, that portion of the access hole or cope need not be ground. Weld access holes and beam copes in other shapes need not be ground nor inspected by dye penetrant or magnetic particle methods.

7. Minimum Strength of Connections

Except for lacing, sag rods, or girts, connections providing design strength shall be designed to support a factored load not less than 10 kips.

8.   Placement of Welds and Bolts

Groups of welds or bolts at the ends of any member which transmit axial force into that member shall be sized so that the center of gravity of the group coincides with the center of gravity of the member, unless provision is made for the eccentricity. The foregoing provision is not applicable to end connections of statically-loaded single angle, double angle, and similar members.

9.   Bolts in Combination with Welds

In new work, A307 bolts or high-strength bolts proportioned as bearing-type connections shall not be considered as sharing the load in combination with welds. Welds, if used, shall be proportioned for the entire force in the connection. In slip-critical connections, high-strength bolts are permitted to be considered as sharing the load with the welds.

In making welded alterations to structures, existing rivets and high-strength bolts tightened to the requirements for slip-critical connections are permitted to be utilized for carrying loads present at the time of alteration and the welding need only provide the additional design strength required.

10.   High-Strength Bolts in Combination with Rivets

In both new work and alterations, in connections designed as slip-critical connections in accordance with the provisions of Section J3, high-strength bolts are permitted to be considered as sharing the load with rivets.

11.   Limitations on Bolted and Welded Connections

Fully tensioned high-strength bolts (see Table J3.1) or welds shall be used for the following connections:

Column splices in all tier structures 200 ft or more in height.

Column splices in tier structures 100 to 200 ft in height, if the least horizontal dimension is less than 40 percent of the height.

Column splices in tier structures less than 100 ft in height, if the least horizontal dimension is less than 25 percent of the height.

Connections of all beams and girders to columns and of any other beams and girders on which the bracing of columns is dependent, in structures over 125 ft in height.

In all structures carrying cranes of over five-ton capacity: roof-truss splices and connections of trusses to columns, column splices, column bracing, knee braces, and crane supports.

Connections for supports of running machinery, or of other live loads which produce impact or reversal of stress.

Any other connections stipulated on the design plans.

In all other cases connections are permitted to be made with A307 bolts or snug-tight high-strength bolts.

For the purpose of this section, the height of a tier structure shall be taken as the vertical distance from the curb level to the highest point of the roof beams in the case of flat roofs, or to the mean height of the gable in the case of roofs having a rise of more than $2\frac{2}{3}$ in 12. Where the curb level has not been established, or where the structure does not adjoin a street, the mean level of the adjoining land shall be used instead of curb level. It is permissible to exclude penthouses in computing the height of structure.

J2.   WELDS

All provisions of the American Welding Society *Structural Welding Code Steel*, AWS D1.1, apply under this specification, except Chapter 10—Tubular Structures, which is outside the scope of this specification, and except that the provisions of the listed AISC LRFD Specification Sections apply under this Specification in lieu of the cited AWS Code provisions as follows:

AISC Section J1.5 and J1.6 in lieu of AWS Section 3.2.5
AISC Section J2.2 in lieu of AWS Section 2.3.2.4
AISC Table J2.5 in lieu of AWS Table 8.1
AISC Table A-K3.2 in lieu of AWS Section 2.5
AISC Section K3 and Appendix K3 in lieu of AWS Chapter 9
AISC Section M2.2 in lieu of AWS Section 3.2.2

1.   Groove Welds

1a.   Effective Area

The effective area of groove welds shall be considered as the effective length of the welds times the effective throat thickness.

The effective length of a groove weld shall be the width of the part joined.

The effective throat thickness of a complete-joint-penetration groove weld shall be the thickness of the thinner part joined.

The effective throat thickness of a partial-joint-penetration groove weld shall be as shown in Table J2.1.

The effective throat thickness of flare groove weld when flush to the surface of a bar or 90° bend in formed section shall be as shown in Table J2.2. Random sections of production welds for each welding procedure, or such test sections as may be required by design documents, shall be used to verify that the effective throat is consistently obtained.

Larger effective throat thicknesses than those in Table J2.2 are permitted, provided the fabricator can establish by qualification the consistent production of such larger effective throat thicknesses. Qualification shall consist of sectioning the weld normal to its axis, at mid-length and terminal ends. Such sectioning shall be made on a number of combinations of material sizes representative of the range to be used in the fabrication or as required by the designer.

1b.   Limitations

The minimum effective throat thickness of a partial-joint-penetration groove weld shall be as shown in Table J2.3. Weld size is determined by the thicker of the two parts joined, except that the weld size need not exceed the thickness of the thinnest part joined when a larger size is required by calculated strength. For this exception, particular care shall be taken to provide sufficient preheat for soundness of the weld.

TABLE J2.1
Effective Throat Thickness of
Partial-Penetration Groove Welds

Welding Process	Welding Position	Included Angle at Root of Groove	Effective Throat Thickness
Shielded metal arc Submerged arc		J or U joint	Depth of chamfer
Gas metal arc	All	Bevel or V joint $\geq 60^\circ$	
Flux-cored arc		Bevel or V joint $< 60^\circ$ but $\geq 45^\circ$	Depth of chamfer minus $\frac{1}{8}$-in.

TABLE J2.2
Effective Throat Thickness of Flare Groove Welds

Type of Weld	Radius (R) of Bar or Bend	Effective Throat Thickness
Flare bevel groove	All	$\frac{5}{16}R$
Flare V-groove	All	$\frac{1}{2}R$

[a] Use $\frac{3}{8}R$ for Gas Metal Arc Welding (except short circuiting transfer process) when $R \geq 1$ in.

TABLE J2.3
Minimum Effective Throat Thickness of
Partial-Joint-Penentration Groove Welds

Material Thickness of Thicker Part Joined (in.)	Minimum Effective Throat Thickness[a] (in.)
To $\frac{1}{4}$ inclusive	$\frac{1}{8}$
Over $\frac{1}{4}$ to $\frac{1}{2}$	$\frac{3}{16}$
Over $\frac{1}{2}$ to $\frac{3}{4}$	$\frac{1}{4}$
Over $\frac{3}{4}$ to $1\frac{1}{2}$	$\frac{5}{16}$
Over $1\frac{1}{2}$ to $2\frac{1}{4}$	$\frac{3}{8}$
Over $2\frac{1}{4}$ to 6	$\frac{1}{2}$
Over 6	$\frac{5}{8}$

[a] See Section J2.

2. Fillet Welds

2a. Effective Area

The effective area of fillet welds shall be as defined in American Welding Society Code D1.1 Article 2.3.2, except 2.3.2.4. The effective throat thickness of a fillet weld shall be the shortest distance from the root of the joint to the face of the diagrammatic weld, except that for the fillet welds made by the submerged

TABLE J2.4
Minimum Size of Fillet Welds[b]

Material Thickness of Thicker Part Joined (in.)	Minimum Size of Fillet Weld[a] (in.)
To 1/4 inclusive	1/8
Over 1/4 to 1/2	3/16
Over 1/2 to 3/4	1/4
Over 3/4	5/16

[a] Leg dimension of fillet welds. Single pass welds must be used.
[b] See Section J2.2b for maximum size of fillet welds.

arc process, the effective throat thickness shall be taken equal to the leg size for 3/8-in. and smaller fillet welds, and equal to the theoretical throat plus 0.11-in. for fillet welds over 3/8-in.

For fillet welds in holes and slots, the effective length shall be the length of the centerline of the weld along the center of the plane through the throat. In the case of overlapping fillets, the effective area shall not exceed the nominal cross-sectional area of the hole or slot, in the plane of the faying surface.

2b. Limitations

The *minimum size of fillet welds* shall be not less than the size required to transmit calculated forces nor the size as shown in Table J2.4 which is based upon experiences and provides some margin for uncalculated stress encountered during fabrication, handling, transportation, and erection. These provisions do not apply to fillet weld reinforcements of partial- or complete-joint-penetration welds.

The *maximum size of fillet welds* of connected parts shall be:

(a) Along edges of material less than 1/4-in. thick, not greater than the thickness of the material.

(b) Along edges of material 1/4-in. or more in thickness, not greater than the thickness of the material minus 1/16-in., unless the weld is especially designated on the drawings to be built out to obtain full-throat thickness. In the as-welded condition, the distance between the edge of the base metal and the toe of the weld is permitted to be less than 1/16-in. provided the weld size is clearly verifiable.

(c) For flange-web welds and similar connections, the actual weld size need not be larger than that required to develop the web capacity, and the requirements of Table J2.4 need not apply.

The *minimum effective length of fillet welds* designed on the basis of strength shall be not less than four times the nominal size, or else the size of the weld shall be considered not to exceed 1/4 of its effective length. If longitudinal fillet welds are used alone in end connections of flat-bar tension members, the length of each fillet weld shall be not less than the perpendicular distance between them. The transverse spacing of longitudinal fillet welds used in end connections of tension members shall comply with Section B3.

The *maximum effective length of fillet welds* loaded by forces parallel to the weld, such as lap splices, shall not exceed 70 times the fillet weld leg. A uniform stress distribution may be assumed throughout the maximum effective length.

Intermittent fillet welds may be used to transfer calculated stress across a joint or faying surfaces when the strength required is less than that developed by a continuous fillet weld of the smallest permitted size, and to join components of built-up members. The effective length of any segment of intermittent fillet welding shall be not less than four times the weld size, with a minimum of $1\frac{1}{2}$-in.

In lap joints, the minimum amount of lap shall be five times the thickness of the thinner part joined, but not less than one inch. Lap joints joining plates or bars subjected to axial stress shall be fillet welded along the end of both lapped parts, except where the deflection of the lapped parts is sufficiently restrained to prevent opening of the joint under maximum loading.

Fillet welds terminations shall not be at the extreme ends or sides of parts or members. They shall be *either* returned continuously around the ends or sides, respectively for a distance of not less than two times the nominal weld size *or* shall terminate not less than the nominal weld size from the sides or ends except as follows. For details and structural elements such as brackets, beam seats, framing angles, and simple end plates which are subject to cyclic (fatigue) out-of-plane forces and/or moments of frequency and magnitude that would tend to initiate a progressive failure of the weld, fillet welds *shall be returned* around the side or end for a distance not less than two times the nominal weld size. For framing angles and simple end-plate connections which depend upon flexibility of the outstanding legs for connection flexibility, if end returns are used, their length shall not exceed four times the nominal size of the weld. Fillet welds which occur on opposite sides of a common plane shall be interrupted at the corner common to both welds. End returns shall be indicated on the design and detail drawings.

Fillet welds in holes or slots may be used to transmit shear in lap joints or to prevent the buckling or separation of lapped parts and to join components of built-up members. Such fillet welds may overlap, subject to the provisions of Section J2. Fillet welds in holes or slots are not to be considered plug or slot welds.

3.　Plug and Slot Welds

3a.　Effective Area

The effective shearing area of plug and slot welds shall be considered as the nominal cross-sectional area of the hole or slot in the plane of the faying surface.

3b.　Limitations

Plug or slot welds are permitted to be used to transmit shear in lap joints or to prevent buckling of lapped parts and to join component parts of built-up members.

The diameter of the holes for a plug weld shall not be less than the thickness of the part containing it plus $\frac{5}{16}$-in., rounded to the next larger odd $\frac{1}{16}$-in., nor greater than the minimum diameter plus $\frac{1}{8}$-in. or $2\frac{1}{4}$ times the thickness of the weld.

The minimum center-to-center spacing of plug welds shall be four times the diameter of the hole.

The length of slot for a slot weld shall not exceed 10 times the thickness of the weld. The width of the slot shall be not less than the thickness of the part containing it plus $\frac{5}{16}$-in. rounded to the next larger odd $\frac{1}{16}$-in., nor shall it be larger than $2\frac{1}{4}$ times the thickness of the weld. The ends of the slot shall be semicircular or shall have the corners rounded to a radius of not less than the thickness of the part containing it, except those ends which extend to the edge of the part.

The minimum spacing of lines of slot welds in a direction transverse to their length shall be four times the width of the slot. The minimum center-to-center spacing in a longitudinal direction on any line shall be two times the length of the slot.

The thickness of plug or slot welds in material $\frac{5}{8}$-in. or less in thickness shall be equal to the thickness of the material. In material over $\frac{5}{8}$-in. thick, the thickness of the weld shall be at least one-half the thickness of the material but not less than $\frac{5}{8}$-in.

4. Design Strength

The design strength of welds shall be the lower value of $\phi F_{BM} A_{BM}$ and $\phi F_w A_w$, when applicable. The values of ϕ, F_{BM} , and F_w and limitations thereon are given in Table J2.5,

where

F_{BM} = nominal strength of the base material, ksi
F_w = nominal strength of the weld electrode, ksi
A_{BM} = cross-sectional area of the base material, in.^{2}
A_w = effective cross-sectional area of the weld, in.^{2}
ϕ = resistance factor

Alternatively, fillet welds loaded in-plane are permitted to be designed in accordance with Appendix J2.4.

5. Combination of Welds

If two or more of the general types of welds (groove, fillet, plug, slot) are combined in a single joint, the design strength of each shall be separately computed with reference to the axis of the group in order to determine the design strength of the combination.

6. Matching Weld Metal

The choice of electrode for use with complete-joint-penetration groove welds subject to tension normal to the effective area shall comply with the requirements for matching weld metals given in AWS D1.1.

TABLE J2.5
Design Strength of Welds

Types of Weld and Stress [a]	Material	Resistance Factor ϕ	Nominal Strength F_{BM} or F_w	Required Weld Strength Level [b,c]
Complete-Joint-Penetration Groove Weld				
Tension normal to effective area	Base	0.90	F_y	Matching weld must be used.
Compression normal to effective area	Base	0.90	F_y	Weld metal with a strength level equal to or less than matching weld metal is permitted to be used.
Tension or compression parallel to axis of weld				
Shear on effective area	Base Weld electrode	0.90 0.80	$0.60F_y$ $0.60F_{EXX}$	
Partial-Joint-Penetration Groove Weld				
Compression normal to effective area	Base	0.90	F_y	Weld metal with a strength level equal to or less than matching weld metal is permitted to be used.
Tension or compression parallel to axis of weld [d]				
Shear parallel to axis of weld	Base Weld electrode	0.75	[e] $0.60F_{EXX}$	
Tension normal to effective area	Base Weld electrode	0.90 0.80	F_y $0.60F_{EXX}$	
Fillet Welds				
Shear on effective area	Base Weld electrode	0.75	[e] $0.60F_{EXX}$ [f]	Weld metal with a strength level equal to or less than matching weld metal is permitted to be used.
Tension or compression parallel to axis of weld [d]	Base	0.90	F_y	
Plug or Slot Welds				
Shear parallel to faying surfaces (on effective area)	Base Weld electrode	0.75	[e] $0.60F_{EXX}$	Weld metal with a strength level equal to or less than matching weld metal is permitted to be used.

[a] For definition of effective area, see Section J2.
[b] For matching weld metal, see Table 4.1, AWS D1.1.
[c] Weld metal one strength level stronger than matching weld metal is permitted.
[d] Fillet welds and partial-joint-penetration groove welds joining component elements of built-up members, such as flange-to-web connections, are not required to be designed with regard to the tensile or compressive stress in these elements parallel to the axis of the welds.
[e] The design of connected material is governed by Sections J4 and J5.
[f] For alternative design strength, see Appendix J2.4.

7.   Mixed Weld Metal

When notch-toughness is specified, the process consumables for all weld metal, tack welds, root pass, and subsequent passes deposited in a joint shall be compatible to assure notch-tough composite weld metal.

8.   Preheat for Heavy Shapes

For ASTM A6 Group 4 and 5 shapes and welded built-up members made of plates more than two inches thick, a preheat equal to or greater than 350°F shall be used when making groove-weld splices.

J3.   BOLTS AND THREADED PARTS

1.   High-Strength Bolts

Use of high-strength bolts shall conform to the provisions of the *Load and Resistance Factor Design Specification for Structural Joints Using ASTM A325 or A490 Bolts*, as approved by the Research Council on Structural Connections, except as otherwise provided in this Specification.

If required to be tightened to more than 50 percent of their minimum specified tensile strength, A449 bolts in tension and bearing-type shear connections shall have an ASTM F436 hardened washer installed under the bolt head, and the nuts shall meet the requirements of ASTM A563. When assembled, all joint surfaces, including those adjacent to the washers, shall be free of scale, except tight mill scale. Except as noted below, all A325 and A490 bolts shall be tightened to a bolt tension not less than that given in Table J3.1. Tightening shall be done by any of the following methods: turn-of-nut method, a direct tension indicator, calibrated wrench, or alternative design bolt.

Bolts in connections not subject to tension loads, where slip can be permitted and where loosening or fatigue due to vibration or load fluctuations are not design considerations, need only to be tightened to the snug-tight condition. The snug-tight condition is defined as the tightness attained by either a few impacts of an impact wrench or the full effort of a worker with an ordinary spud wrench that brings the connected plies into firm contact. The nominal strength value given in Table J3.2 for bearing-type connections shall be used for bolts tightened to the snug-tight condition. Bolts tightened only to the snug-tight condition shall be clearly identified on the design and erection drawings.

When A490 bolts over one inch in diameter are used in slotted or oversize holes in external plies, a single hardened washer conforming to ASTM F436, except with $\frac{5}{16}$-in. minimum thickness, shall be used in lieu of the standard washer.

In slip-critical connections in which the direction of loading is toward an edge of a connected part, adequate bearing strength at factored load shall be provided based upon the applicable requirements of Section J3.10.

2.   Size and Use of Holes

In slip-critical connections in which the direction of loading is toward the edge of the connected part, adequate bearing capacity at factored load shall be provided based upon the applicable requirements of Section J3.10.

The *maximum sizes* of holes for rivets and bolts are given in Table J3.3, except

TABLE J3.1
Minimum Bolt Tension, kips*

Bolt Size, in.	A325 Bolts	A490 Bolts
$\frac{1}{2}$	12	15
$\frac{5}{8}$	19	24
$\frac{3}{4}$	28	35
$\frac{7}{8}$	39	49
1	51	64
$1\frac{1}{8}$	56	80
$1\frac{1}{4}$	71	102
$1\frac{3}{8}$	85	121
$1\frac{1}{2}$	103	148

* Equal to 0.70 of minimum tensile strength of bolts, rounded off to nearest kip, as specified in ASTM specifications for A325 and A490 bolts with UNC threads.

that larger holes, required for tolerance on location of anchor bolts in concrete foundations, are allowed in column base details.

Standard holes shall be provided in member-to-member connections, unless oversized, short-slotted, or long-slotted holes in bolted connections are approved by the designer. Finger shims up to $\frac{1}{4}$-in. are permitted in slip-critical connections designed on the basis of standard holes without reducing the nominal shear strength of the fastener to that specified for slotted holes.

Oversized holes are allowed in any or all plies of slip-critical connections, but they shall not be used in bearing-type connections. Hardened washers shall be installed over oversized holes in an outer ply.

Short-slotted holes are allowed in any or all plies of slip-critical or bearing-type connections. The slots are permitted to be used without regard to direction of loading in slip-critical connections, but the length shall be normal to the direction of the load in bearing-type connections. Washers shall be installed over short-slotted holes in an outer ply; when high-strength bolts are used, such washers shall be hardened.

Long-slotted holes are allowed in only one of the connected parts of either a slip-critical or bearing-type connection at an individual faying surface. Long-slotted holes are permitted to be used without regard to direction of loading in slip-critical connections, but shall be normal to the direction of load in bearing-type connections. Where long-slotted holes are used in an outer ply, plate washers, or a continuous bar with standard holes, having a size sufficient to completely cover the slot after installation, shall be provided. In high-strength bolted connections, such plate washers or continuous bars shall be not less than $\frac{5}{16}$-in. thick and shall be of structural grade material, but need not be hardened. If hardened washers are required for use of high-strength bolts, the hardened washers shall be placed over the outer surface of the plate washer or bar.

3.   Minimum Spacing

The distance between centers of standard, oversized, or slotted holes, shall not

TABLE J3.2
Design Strength of Fasteners

Description of Fasteners	Tensile Strength		Shear Strength in Bearing-type Connections	
	Resistance Factor ϕ	Nominal Strength, ksi	Resistance Factor ϕ	Nominal Strength, ksi
A307 bolts		45 [a]		24 [b,e]
A325 bolts, when threads are not excluded from shear planes		90 [d]		48 [e]
A325 bolts, when threads are excluded from shear planes		90 [d]		60 [e]
A490 bolts, when threads are not excluded from shear planes		113 [d]		60 [e]
A490 bolts, when threads are excluded from shear planes	0.75	113 [d]	0.75	75 [e]
Threaded parts meeting the requirements of Sect. A3, when threads are not excluded from shear planes		$0.75F_u$ [a,c]		$0.40F_u$
Threaded parts meeting the requirements of Sect. A3, when threads are excluded from shear planes		$0.75F_u$ [a,c]		$0.50F_u$ [a,c]
A502, Gr. 1, hot-driven rivets		45 [a]		25 [e]
A502, Gr. 2 & 3, hot-driven rivets		60 [a]		33 [e]

[a] Static loading only.
[b] Threads permitted in shear planes.
[c] The nominal tensile strength of the threaded portion of an upset rod, based upon the cross-sectional area at its major thread diameter, A_D shall be larger than the nominal body area of the rod before upsetting times F_y.
[d] For A325 and A490 bolts subject to tensile fatigue loading, see Appendix K3.
[e] When bearing-type connections used to splice tension members have a fastener pattern whose length, measured parallel to the line of force, exceeds 50 in., tabulated values shall be reduced by 20 percent.

be less than $2\frac{2}{3}$ times the nominal diameter of the fastener; a distance of $3d$ is preferred. Refer to Section J3.10 for bearing strength requirement.

4. Minimum Edge Distance

The distance from the center of a standard hole to an edge of a connected part shall not be less than either the applicable value from Table J3.4, or as required

TABLE J3.3
Nominal Hole Dimensions

Bolt Diameter	Standard (Dia.)	Oversize (Dia.)	Short-slot (Width × Length)	Long-slot Width × Length
		Hole Dimensions		
$\frac{1}{2}$	$\frac{9}{16}$	$\frac{5}{8}$	$\frac{9}{16} \times \frac{11}{16}$	$\frac{9}{16} \times 1\frac{1}{4}$
$\frac{5}{8}$	$\frac{11}{16}$	$\frac{13}{16}$	$\frac{11}{16} \times \frac{7}{8}$	$\frac{11}{16} \times 1\frac{9}{16}$
$\frac{3}{4}$	$\frac{13}{16}$	$\frac{15}{16}$	$\frac{13}{16} \times 1$	$\frac{13}{16} \times 1\frac{7}{8}$
$\frac{7}{8}$	$\frac{15}{16}$	$1\frac{1}{16}$	$\frac{15}{16} \times 1\frac{1}{8}$	$\frac{15}{16} \times 2\frac{3}{16}$
1	$1\frac{1}{16}$	$1\frac{1}{4}$	$1\frac{1}{16} \times 1\frac{5}{16}$	$1\frac{1}{16} \times 2\frac{1}{2}$
$\geq 1\frac{1}{8}$	$d + \frac{1}{16}$	$d + \frac{5}{16}$	$(d + \frac{1}{16}) \times (d + \frac{3}{8})$	$(d + \frac{1}{16}) \times (2.5 \times d)$

TABLE J3.4
Minimum Edge Distance,[a] in.
(Center of Standard Hole[b] to Edge of Connected Part)

Nominal Rivet or Bolt Diameter (in.)	At Sheared Edges	At Rolled Edges of Plates, Shapes or Bars, or Gas Cut Edges [c]
$\frac{1}{2}$	$\frac{7}{8}$	$\frac{3}{4}$
$\frac{5}{8}$	$1\frac{1}{8}$	$\frac{7}{8}$
$\frac{3}{4}$	$1\frac{1}{4}$	1
$\frac{7}{8}$	$1\frac{1}{2}$ [d]	$1\frac{1}{8}$
1	$1\frac{3}{4}$ [d]	$1\frac{1}{4}$
$1\frac{1}{8}$	2	$1\frac{1}{2}$
$1\frac{1}{4}$	$2\frac{1}{4}$	$1\frac{5}{8}$
Over $1\frac{1}{4}$	$1\frac{3}{4} \times$ Diameter	$1\frac{1}{4} \times$ Diameter

[a] Lesser edge distances are permitted to be used provided Equations from J3.10, as appropriate, are satisfied.
[b] For oversized or slotted holes, see Table J3.8.
[c] All edge distances in this column are permitted to be reduced $\frac{1}{8}$-in. when the hole is at a point where stress does not exceed 25 percent of the maximum design strength in the element.
[d] These are permitted to be $1\frac{1}{4}$-in. at the ends of beam connection angles and shear end plates.

in Section J3.10. The distance from the center of an oversized or slotted hole to an edge of a connected part shall be not less than that required for a standard hole to an edge of a connected part plus the applicable increment C_2 from Table J3.8. Refer to Section J3.10 for bearing strength requirement.

5. **Maximum Spacing and Edge Distance**

The maximum distance from the center of any bolt or rivet to the nearest edge of parts in contact shall be 12 times the thickness of the connected part under consideration, but shall not exceed six inches. The longitudinal spacing of connectors between elements in continuous contact consisting of a plate and a shape or two plates shall be as follows:

(a) For painted members or unpainted members not subject to corrosion, the spacing shall not exceed 24 times the thickness of the thinner plate or 12 inches.

(b) For unpainted members of weathering steel subject to atmospheric corrosion, the spacing shall not exceed 14 times the thickness of the thinner plate or seven inches.

6. Design Tension or Shear Strength

The design tension or shear strength of a high-strength bolt or threaded part is

$$\phi F_n A_b$$

where

ϕ = resistance factor tabulated in Table J3.2

F_n = nominal tensile strength F_t, or shear strength, F_v, tabulated in Table J3.2, ksi

A_b = nominal unthreaded body area of bolt or threaded part (for upset rods, see Footnote c, Table J3.2), in.^{2}

The applied load shall be the sum of the factored loads and any tension resulting from prying action produced by deformation of the connected parts.

7. Combined Tension and Shear in Bearing-Type Connections

The design strength of a bolt or rivet subject to combined tension and shear is $\phi F_t A_b$, where ϕ is 0.75 and the nominal tension stress F_t shall be computed from the equations in Table J3.5 as a function of f_v, the required shear stress produced by the factored loads. The design shear strength ϕF_v, tabulated in Table J3.2, shall equal or exceed the shear stress, f_v.

8. High-Strength Bolts in Slip-Critical Connections

The design for shear of high-strength bolts in slip-critical connections shall be in accordance with either Section J3.8a or J3.8b and checked for bearing in accordance with J3.2 and J3.10.

8a. Slip-Critical Connections Designed at Service Loads

The design resistance to shear of a bolt in a slip-critical connection is $\phi F_v A_b$,

where

ϕ = 1.0 for standard, oversized, short-slotted, and long-slotted holes when the long slot is perpendicular to the line of force

ϕ = 0.85 for long-slotted holes when the long slot is parallel to the line of force

F_v = nominal slip-critical shear resistance tabulated in Table J3.6, ksi

The design resistance to shear shall equal or exceed the shear on the bolt due to service loads. When the loading combination includes wind loads in addition to dead and live loads, the total shear on the bolt due to combined load effects, at service load, may be multiplied by 0.75.

The values for F_v in Table J3.6 are based on Class A (slip coefficient 0.33), clean mill scale and blast cleaned surfaces with class A coatings. When specified by

TABLE J3.5
Tension Stress Limit (F_t), ksi
Fasteners in Bearing-type Connections

Description of Fasteners	Threads Included in the Shear Plane	Threads Excluded from the Shear Plane
A307 bolts	$59 - 1.9f_v \leq 45$	
A325 bolts	$117 - 1.9f_v \leq 90$	$117 - 1.5f_v \leq 90$
A490 bolts	$147 - 1.9f_v \leq 113$	$147 - 1.5f_v \leq 113$
Threaded parts A449 bolts over 1½ diameter	$0.98F_u - 1.9f_v \leq 0.75F_u$	$0.98F_u - 1.5f_v \leq 0.75F_u$
A502 Gr.1 rivets	$59 - 1.8f_v \leq 45$	
A502 Gr.2 rivets	$78 - 1.8f_v \leq 60$	

TABLE J3.6
Slip-Critical Nominal Resistance to Shear, ksi,
of High-Strength Bolts[a]

Type of Bolt	Nominal Resistance to Shear		
	Standard Size Holes	Oversized and Short-slotted Holes	Long-slotted Holes
A325	17	15	12
A490	21	18	15

[a] For each shear plane.

the designer, the nominal slip resistance for connections having special faying surface conditions are permitted to be adjusted to the applicable values in the RCSC Load and Resistance Factor Design Specification.

Finger shims up to ¼-in. are permitted to be introduced into slip-critical connections designed on the basis of standard holes without reducing the design shear stress of the fastener to that specified for slotted holes.

8b.   **Slip-Critical Connections Designed at Factored Loads**

See Appendix J3.8b.

9.   **Combined Tension and Shear in Slip-Critical Connections**

The design of slip-critical connections subject to tensile forces shall be in accordance with either Sections J3.9a and J3.8a or Sections J3.9b and J3.8b.

9a.   **Slip-Critical Connections Designed at Service Loads**

The design resistance to shear of a bolt in a slip-critical connection subject to a tensile force T due to service loads shall be computed according to Section J3.8a multiplied by the following reduction factor,

$$\left(1 - \frac{T}{T_b}\right)$$

where

T_b = minimum bolt pre-tension from Table J3.1

9b.   Slip-Critical Connections Designed at Factored Loads

See Appendix J3.9b.

10.   Bearing Strength at Bolt Holes

The design bearing strength at bolt holes is ϕR_n, where

ϕ = 0.75
R_n = nominal bearing strength

Bearing strength shall be checked for both bearing-type and slip-critical connections. The use of oversize holes and short- and long-slotted holes parallel to the line of force is restricted to slip-critical connections per Section J3.2.

In the following sections:

L_e = distance (in.) along the line of force from the edge of the connected part to the center of a standard hole or the center of a short- and long-slotted hole perpendicular to the line of force. For oversize holes and short- and long-slotted holes parallel to the line of force, L_e shall be increased by the increment C_2 of Table J3.8.

s = distance (in.) along the line of force between centers of standard holes, or between centers of short- and long-slotted holes perpendicular to the line of force. For oversize holes and short- and long-slotted holes parallel to the line of force, s shall be increased by the spacing increment C_1 of Table J3.7.

d = diameter of bolt, in.
F_u = specified minimum tensile strength of the critical part, ksi
t = thickness of the critical connected part, in. For countersunk bolts and rivets, deduct one-half the depth of the countersink.

(a) When $L_e \geq 1.5d$ and $s \geq 3d$ and there are two or more bolts in line of force:

For standard holes; short and long-slotted holes perpendicular to the line of force; oversize holes in slip-critical connections; and long and short-slotted holes in slip-critical connections when the line of force is parallel to the axis of the hole:

When deformation around the bolt holes is a design consideration

$$R_n = 2.4dtF_u \qquad\qquad\text{(J3-1a)}$$

When deformation around the bolt holes is not a design consideration, for the bolt nearest the edge

$$R_n = L_e tF_u \leq 3.0dtF_u \qquad\qquad\text{(J3-1b)}$$

TABLE J3.7
Values of Spacing Increment C_1, in.

Nominal Diameter of Fastener	Oversize Holes	Slotted Holes		
		Perpendicular to Line of Force	Parallel to Line of Force	
			Short-slots	Long-slots [a]
$\leq 7/8$	$1/8$	0	$3/16$	$1\frac{1}{2}d - 1/16$
1	$3/16$	0	$1/4$	$1 7/16$
$\geq 1\frac{1}{8}$	$1/4$	0	$5/16$	$1\frac{1}{2}d - 1/16$

[a] When length of slot is less than maximum allowed in Table J3.5, C_1 are permitted to be reduced by the difference between the maximum and actual slot lengths.

TABLE J3.8
Values of Edge Distance Increment C_2, in.

Nominal Diameter of Fastener (in.)	Oversized Holes	Slotted Holes		
		Long Axis Perpendicular to Edge		Long Axis Parallel to Edge
		Short Slots	Long Slots [a]	
$\leq 7/8$	$1/16$	$1/8$		
1	$1/8$	$1/8$	$3/4 d$	0
$\geq 1\frac{1}{8}$	$1/8$	$3/16$		

[a] When length of slot is less than maximum allowable (see Table J3.5), C_2 are permitted to be reduced by one-half the difference between the maximum and actual slot lengths.

and for the remaining bolts

$$R_n = (s - d/2)tF_u \leq 3.0dtF_u \qquad (J3\text{-}1c)$$

For long-slotted bolt holes perpendicular to the line of force:

$$R_n = 2.0dtF_u \qquad (J3\text{-}1d)$$

(b) When $L_e < 1.5d$ or $s < 3d$ or for a single bolt in the line of force:

For standard holes; short and long-slotted holes perpendicular to the line of force; oversize holes in slip-critical connections; and long and short-slotted holes in slip-critical connections when the line of force is parallel to the axis of the hole:

For a single bolt hole or the bolt hole nearest the edge when there are two or more bolt holes in the line of force

$$R_n = L_e tF_u \leq 2.4dtF_u \qquad (J3\text{-}2a)$$

For the remaining bolt holes

$$R_n = (s - d/2)tF_u \leq 2.4dtF_u \qquad \text{(J3-2b)}$$

For long-slotted bolt holes perpendicular to the line of force:

For a single bolt hole or the bolt hole nearest the edge where there are two or more bolt holes in the line of force

$$R_n = L_e tF_u \leq 2.0dtF_u \qquad \text{(J3-2c)}$$

For the remaining bolt holes

$$R_n = (s - d/2)tF_u \leq 2.0dtF_u \qquad \text{(J3-2d)}$$

11. Long Grips

A307 bolts providing design strength, and for which the grip exceeds five diameters, shall have their number increased one percent for each additional $1/16$-in. in the grip.

J4. DESIGN RUPTURE STRENGTH

1. Shear Rupture Strength

The design strength for the limit state of rupture along a shear failure path in the affected elements of connected members shall be taken as ϕR_n

where

$\phi = 0.75$
$R_n = 0.6F_u A_{nv}$ \qquad (J4-1)
$A_{nv} = $ net area subject to shear, in.^{2}

2. Tension Rupture Strength

The design strength for the limit state of rupture along a tension path in the affected elements of connected members shall be taken as ϕR_n

where

$\phi = 0.75$
$R_n = F_u A_{nt}$ \qquad (J4-2)
$A_{nt} = $ net area subject to tension, in.^{2}

3. Block Shear Rupture Strength

Block shear is a limit state in which the resistance is determined by the sum of the shear strength on a failure path(s) and the tensile strength on a perpendicular segment. It shall be checked at beam end connections where the top flange is coped and in similar situations, such as tension members and gusset plates. When ultimate rupture strength on the net section is used to determine the resistance on one segment, yielding on the gross section shall be used on the perpendicular segment. The block shear rupture design strength, ϕR_n, shall be determined as follows:

(a) When $F_u A_{nt} \geq 0.6F_u A_{nv}$:

$$\phi R_n = \phi[0.6F_y A_{gv} + F_u A_{nt}] \qquad \text{(J4-3a)}$$

(b) When $0.6F_u A_{nv} > F_u A_{nt}$:

$$\phi R_n = \phi[0.6F_u A_{nv} + F_y A_{gt}] \qquad \text{(J4-3b)}$$

where

$\phi\ \ = 0.75$
A_{gv} = gross area subject to shear, in.^{2}
A_{gt} = gross area subject to tension, in.^{2}
A_{nv} = net area subjected to shear, in.^{2}
A_{nt} = net area subjected to tension, in.^{2}

J5.   CONNECTING ELEMENTS

This section applies to the design of connecting elements, such as plates, gussets, angles, brackets, and the panel zones of beam-to-column connections.

1.   Eccentric Connections

Intersecting axially stressed members shall have their gravity axis intersect at one point, if practicable; if not, provision shall be made for bending and shearing stresses due to the eccentricity. Also see Section J1.8.

2.   Design Strength of Connecting Elements in Tension

The design strength, ϕR_n, of welded, bolted, and riveted connecting elements statically loaded in tension (e.g., splice and gusset plates) shall be the lower value obtained according to limit states of yielding, rupture of the connecting element, and block shear rupture.

(a) For tension yielding of the connecting element:

$$\phi = 0.90$$
$$R_n = A_g F_y \qquad \text{(J5-1)}$$

(b) For tension rupture of the connecting element:

$$\phi = 0.75$$
$$R_n = A_n F_u \qquad \text{(J5-2)}$$

where A_n is the net area, not to exceed $0.85A_g$.

(c) For block shear rupture of connecting elements, see Section J4.3.

3.   Other Connecting Elements

For all other connecting elements, the design strength, ϕR_n, shall be determined for the applicable limit state to ensure that the design strength is equal to or greater than the required strength, where R_n is the nominal strength appropriate to the geometry and type of loading on the connecting element. For shear yielding of the connecting element:

$$\phi = 0.90$$
$$R_n = 0.60A_g F_y \qquad \text{(J5-3)}$$

If the connecting element is in compression an appropriate limit state analysis shall be made.

J6. FILLERS

In welded construction, any filler $\frac{1}{4}$-in. or more in thickness shall extend beyond the edges of the splice plate and shall be welded to the part on which it is fitted with sufficient weld to transmit the splice plate load, applied at the surface of the filler. The welds joining the splice plate to the filler shall be sufficient to transmit the splice plate load and shall be long enough to avoid overloading the filler along the toe of the weld. Any filler less than $\frac{1}{4}$-in. thick shall have its edges made flush with the edges of the splice plate and the weld size shall be the sum of the size necessary to carry the splice plus the thickness of the filler plate.

When bolts or rivets carrying loads pass through fillers thicker than $\frac{1}{4}$-in., except in connections designed as slip-critical connections, the fillers shall be extended beyond the splice material and the filler extension shall be secured by enough bolts or rivets to distribute the total stress in the member uniformly over the combined section of the member and the filler, or an equivalent number of fasteners shall be included in the connection. Fillers between $\frac{1}{4}$-in. and $\frac{3}{4}$-in. thick, inclusive, need not be extended and developed, provided the design shear strength of the bolts is reduced by the factor, $0.4(t - 0.25)$, where t is the total thickness of the fillers, up to $\frac{3}{4}$-in.

J7. SPLICES

Groove-welded splices in plate girders and beams shall develop the full strength of the smaller spliced section. Other types of splices in cross sections of plate girders and beams shall develop the strength required by the forces at the point of splice.

J8. BEARING STRENGTH

The strength of surfaces in bearing is ϕR_n, where

$\phi = 0.75$
R_n is defined below for the various types of bearing

(a) For milled surfaces, pins in reamed, drilled, or bored holes, and ends of fitted bearing stiffeners,

$$R_n = 1.8 F_y A_{pb} \qquad\qquad (J8\text{-}1)$$

where

F_y = specified minimum yield stress, ksi
A_{pb} = projected bearing area, in.^{2}

(b) For expansion rollers and rockers,

If $d \le 25$ in.,

$$R_n = 1.2(F_y - 13)ld / 20 \qquad\qquad (J8\text{-}2)$$

If $d > 25$ in.,

$$R_n = 6.0(F_y - 13)l\sqrt{d} / 20 \tag{J8-3}$$

where

d = diameter, in.
l = length of bearing, in.

J9. COLUMN BASES AND BEARING ON CONCRETE

Proper provision shall be made to transfer the column loads and moments to the footings and foundations.

In the absence of code regulations, design bearing loads on concrete may be taken as $\phi_c P_p$:

(a) On the full area of a concrete support

$$P_p = 0.85f_c'A_1 \tag{J9-1}$$

(b) On less than the full area of a concrete support

$$P_p = 0.85f_c'A_1\sqrt{A_2/A_1} \tag{J9-2}$$

where

$\phi_c = 0.60$
A_1 = area of steel concentrically bearing on a concrete support, in.^{2}
A_2 = maximum area of the portion of the supporting surface that is geometrically similar to and concentric with the loaded area, in.^{2}
$\sqrt{A_2/A_1} \leq 2$

J10. ANCHOR BOLTS AND EMBEDMENTS

Anchor bolts and embedments shall be designed in accordance with American Concrete Institute or Prestressed Concrete Institute criteria. If the load factors and combinations given in Section A4.1 are used, a reduction in the ϕ factors specified by ACI shall be made based on the ratio of load factors given in Section A4.1 and in ACI.

CHAPTER K

CONCENTRATED FORCES, PONDING, AND FATIGUE

This chapter covers member strength design considerations pertaining to concentrated forces, ponding, and fatigue.

K1. FLANGES AND WEBS WITH CONCENTRATED FORCES

1. Design Basis

Sections K1.2 through K1.7 apply to single and double concentrated forces as indicated in each Section. A single concentrated force is tensile or compressive. Double concentrated forces, one tensile and one compressive, form a couple on the same side of the loaded member.

Transverse stiffeners are required at locations of concentrated tensile forces in accordance with Section K1.2 for the flange limit state of local bending, and at unframed ends of beams and girders in accordance with Section K1.8. Transverse stiffeners or doubler plates are required at locations of concentrated forces in accordance with Sections K1.3 through K1.6 for the web limit states of yielding, crippling, sidesway buckling, and compression buckling. Doubler plates or diagonal stiffeners are required in accordance with Section K1.7 for the web limit state of panel-zone shear.

Transverse stiffeners and diagonal stiffeners required by Sections K1.2 through K1.8 shall also meet the requirements of Section K1.9. Doubler plates required by Sections K1.3 through K1.6 shall also meet the requirements of Section K1.10.

2. Local Flange Bending

This Section applies to both tensile single-concentrated forces and the tensile component of double-concentrated forces.

A pair of transverse stiffeners extending at least one-half the depth of the web shall be provided adjacent to a concentrated tensile force centrally applied across the flange when the required strength of the flange exceeds ϕR_n, where

$$\phi = 0.90$$
$$R_n = 6.25 t_f^2 F_{yf} \qquad \text{(K1-1)}$$

where

F_{yf} = specified minimum yield stress of the flange, ksi
t_f = thickness of the loaded flange, in.

If the length of loading measured across the member flange is less than $0.15b$, where b is the member flange width, Equation K1-1 need not be checked.

When the concentrated force to be resisted is applied at a distance from the member end that is less than $10t_f$, R_n shall be reduced by 50 percent.

When transverse stiffeners are required, they shall be welded to the loaded flange to develop the welded portion of the stiffener. The weld connecting transverse stiffeners to the web shall be sized to transmit the unbalanced force in the stiffener to the web. Also, see Section K1.9.

3.   Local Web Yielding

This Section applies to single-concentrated forces and both components of double-concentrated forces.

Either a pair of transverse stiffeners or a doubler plate, extending at least one-half the depth of the web, shall be provided adjacent to a concentrated tensile or compressive force when the required strength of the web at the toe of the fillet exceeds ϕR_n, where

$$\phi = 1.0$$

and R_n is determined as follows:

(a) When the concentrated force to be resisted is applied at a distance from the member end that is greater than the depth of the member d,

$$R_n = (5k + N)F_{yw}\,t_w \qquad (K1\text{-}2)$$

(b) When the concentrated force to be resisted is applied at a distance from the member end that is less than or equal to the depth of the member d,

$$R_n = (2.5k + N)F_{yw}t_w \qquad (K1\text{-}3)$$

In Equations K1-2 and K1-3, the following definitions apply:

   F_{yw} = specified minimum yield stress of the web, ksi
   N = length of bearing (not less than k for end beam reactions), in.
   k = distance from outer face of the flange to the web toe of the fillet, in.
   t_w = web thickness, in.

When required for a tensile force normal to the flange, transverse stiffeners shall be welded to the loaded flange to develop the connected portion of the stiffener. When required for a compressive force normal to the flange, transverse stiffeners shall either bear on or be welded to the loaded flange to develop the force transmitted to the stiffener. The weld connecting transverse stiffeners to the web shall be sized to transmit the unbalanced force in the stiffener to the web. Also, see Section K1.9.

Alternatively, when doubler plates are required, see Section K1.10.

4.   Web Crippling

This Section applies to both compressive single-concentrated forces and the compressive component of double-concentrated forces.

Either a transverse stiffener, a pair of transverse stiffeners, or a doubler plate, extending at least one-half the depth of the web, shall be provided adjacent to a concentrated compressive force when the required strength of the web exceeds ϕR_n, where

$$\phi = 0.75$$

and R_n is determined as follows:

(a) When the concentrated compressive force to be resisted is applied at a distance from the member end that is greater than or equal to $d/2$,

$$R_n = 135t_w^2 \left[1 + 3 \left(\frac{N}{d} \right) \left(\frac{t_w}{t_f} \right)^{1.5} \right] \sqrt{\frac{F_{yw}t_f}{t_w}} \qquad \text{(K1-4)}$$

(b) When the concentrated compressive force to be resisted is applied at a distance from the member end that is less than $d/2$,

For $N/d \le 0.2$,

$$R_n = 68t_w^2 \left[1 + 3 \left(\frac{N}{d} \right) \left(\frac{t_w}{t_f} \right)^{1.5} \right] \sqrt{\frac{F_{yw}t_f}{t_w}} \qquad \text{(K1-5a)}$$

For $N/d > 0.2$,

$$R_n = 68t_w^2 \left[1 + \left(\frac{4N}{d} - 0.2 \right) \left(\frac{t_w}{t_f} \right)^{1.5} \right] \sqrt{\frac{F_{yw}t_f}{t_w}} \qquad \text{(K1-5b)}$$

In Equations K1-4 and K1-5, the following definitions apply:

d = overall depth of the member, in.
t_f = flange thickness, in.

When transverse stiffeners are required, they shall either bear on or be welded to the loaded flange to develop the force transmitted to the stiffener. The weld connecting transverse stiffeners to the web shall be sized to transmit the unbalanced force in the stiffener to the web. Also, see Section K1.9.

Alternatively, when doubler plates are required, see Section K1.10.

5.   Sidesway Web Buckling

This Section applies only to compressive single-concentrated forces applied to members where relative lateral movement between the loaded compression flange and the tension flange is not restrained at the point of application of the concentrated force.

The design strength of the web is ϕR_n, where

$$\phi = 0.85$$

and R_n is determined as follows:

(a) If the compression flange is restrained against rotation:

for $(h/t_w)/(l/b_f) \leq 2.3$,

$$R_n = \frac{C_r t_w^3 t_f}{h^2}\left[1 + 0.4\left(\frac{h/t_w}{l/b_f}\right)^3\right]$$ (K1-6)

for $(h/t_w)/(l/b_f) > 2.3$, the limit state of sidesway web buckling does not apply.

When the required strength of the web exceeds ϕR_n, local lateral bracing shall be provided at the tension flange or either a pair of transverse stiffeners or a doubler plate, extending at least one-half the depth of the web, shall be provided adjacent to the concentrated compressive force.

When transverse stiffeners are required, they shall either bear on or be welded to the loaded flange to develop the full applied force. The weld connecting transverse stiffeners to the web shall be sized to transmit the force in the stiffener to the web. Also, see Section K1.9.

Alternatively, when doubler plates are required, they shall be sized to develop the full applied force. Also, see Section K1.10.

(b) If the compression flange is *not* restrained against rotation:

for $(h/t_w)/(l/b_f) \leq 1.7$,

$$R_n = \frac{C_r t_w^3 t_f}{h^2}\left[0.4\left(\frac{h/t_w}{l/b_f}\right)^3\right]$$ (K1-7)

for $(h/t_w)/(l/b_f) > 1.7$, the limit state of sidesway web buckling does not apply.

When the required strength of the web exceeds ϕR_n, local lateral bracing shall be provided at both flanges at the point of application of the concentrated forces.

In Equations K1-6 and K1-7, the following definitions apply:

l = largest laterally unbraced length along either flange at the point of load, in.
b_f = flange width, in.
t_w = web thickness, in.
h = clear distance between flanges less the fillet or corner radius for rolled shapes; distance between adjacent lines of fasteners or the clear distance between flanges when welds are used for built-up shapes, in.
C_r = 960,000 when $M_u < M_y$ at the location of the force, ksi
   = 480,000 when $M_u \geq M_y$ at the location of the force, ksi

6. Compression Buckling of the Web

This Section applies to a pair of compressive single-concentrated forces or the compressive components in a pair of double-concentrated forces, applied at both flanges of a member at the same location.

Either a single transverse stiffener, or pair of transverse stiffeners, or a doubler plate, extending the full depth of the web, shall be provided adjacent to concentrated compressive forces at both flanges when the required strength of the web exceeds ϕR_n, where

$$\phi = 0.90$$

and

$$R_n = \frac{4,100 t_w^3 \sqrt{F_{yw}}}{h} \qquad \text{(K1-8)}$$

When the pair of concentrated compressive forces to be resisted is applied at a distance from the member end that is less than $d/2$, R_n shall be reduced by 50 percent.

When transverse stiffeners are required, they shall either bear on or be welded to the loaded flange to develop the force transmitted to the stiffener. The weld connecting transverse stiffeners to the web shall be sized to transmit the unbalanced force in the stiffener to the web. Also, see Section K1.9.

Alternatively, when doubler plates are required, see Section K1.10.

7. Panel-Zone Web Shear

Either doubler plates or diagonal stiffeners shall be provided within the boundaries of the rigid connection of members whose webs lie in a common plane when the required strength exceeds ϕR_v, where

$$\phi = 0.90$$

and R_v is determined as follows:

(a) When the effect of panel-zone deformation on frame stability is *not* considered in the analysis,

For $P_u \leq 0.4 P_y$

$$R_v = 0.60 F_y d_c t_w \qquad \text{(K1-9)}$$

For $P_u > 0.4 P_y$

$$R_v = 0.60 F_y d_c t_w \left(1.4 - \frac{P_u}{P_y} \right) \qquad \text{(K1-10)}$$

(b) When frame stability, including plastic panel-zone deformation, is considered in the analysis:

For $P_u \leq 0.75 P_y$

$$R_v = 0.60 F_y d_c t_w \left(1 + \frac{3 b_{cf} t_{cf}^2}{d_b d_c t_w} \right) \qquad \text{(K1-11)}$$

For $P_u > 0.75 P_y$

$$R_v = 0.60 F_y d_c t_w \left(1 + \frac{3 b_{cf} t_{cf}^2}{d_b d_c t_w} \right) \left(1.9 - \frac{1.2 P_u}{P_y} \right) \qquad \text{(K1-12)}$$

In Equations K1-9 through K1-12, the following definitions apply:

t_w = column web thickness, in.
b_{cf} = width of column flange, in.
t_{cf} = thickness of the column flange, in.
d_b = beam depth, in.
d_c = column depth, in.
F_y = yield strength of the column web, ksi
P_y = $F_y A$, axial yield strength of the column, kips
A = column cross-sectional area, in.^{2}

When doubler plates are required, they shall meet the criteria of Section F2 and shall be welded to develop the proportion of the total shear force which is to be carried.

Alternatively, when diagonal stiffeners are required, the weld connecting diagonal stiffeners to the web shall be sized to transmit the stiffener force caused by unbalanced moments to the web. Also, see Section K1.9.

8.   Unframed Ends of Beams and Girders

At unframed ends of beams and girders not otherwise restrained against rotation about their longitudinal axes, a pair of transverse stiffeners, extending the full depth of the web, shall be provided. Also, see Section K1.9.

9.   Additional Stiffener Requirements for Concentrated Forces

Transverse and diagonal stiffeners shall also comply with the following criteria:

(1) The width of each stiffener plus one-half the thickness of the column web shall not be less than one-third of the width of the flange or moment connection plate delivering the concentrated force.

(2) The thickness of a stiffener shall not be less than one-half the thickness of the flange or moment connection plate delivering the concentrated load, and not less than its width times $\sqrt{F_y} / 95$.

Full depth transverse stiffeners for compressive forces applied to a beam or plate girder flange shall be designed as axially compressed members (columns) in accordance with the requirements of Section E2, with an effective length of $0.75h$, a cross section composed of two stiffeners and a strip of the web having a width of $25t_w$ at interior stiffeners and $12t_w$ at the ends of members.

The weld connecting bearing stiffeners to the web shall be sized to transmit the excess web shear force to the stiffener. For fitted bearing stiffeners, see Section J8.1.

10.   Additional Doubler Plate Requirements for Concentrated Forces

Doubler plates required by Sections K1.3 through K1.6 shall also comply with the following criteria:

(1) The thickness and extent of the doubler plate shall provide the additional material necessary to equal or exceed the strength requirements.

(2) The doubler plate shall be welded to develop the proportion of the total force transmitted to the doubler plate.

K2. PONDING

The roof system shall be investigated by structural analysis to assure adequate strength and stability under ponding conditions, unless the roof surface is provided with sufficient slope toward points of free drainage or adequate individual drains to prevent the accumulation of rainwater.

The roof system shall be considered stable and no further investigation is needed if:

$$C_p + 0.9C_s \le 0.25 \qquad (K2\text{-}1)$$

$$I_d \ge 25(S^4)10^{-6} \qquad (K2\text{-}2)$$

where

$$C_p = \frac{32L_sL_p^4}{10^7 I_p}$$

$$C_s = \frac{32SL_s^4}{10^7 I_s}$$

L_p = column spacing in direction of girder (length of primary members), ft
L_s = column spacing perpendicular to direction of girder (length of secondary members), ft
S = spacing of secondary members, ft
I_p = moment of inertia of primary members, in.^{4}
I_s = moment of inertia of secondary members, in.^{4}
I_d = moment of inertia of the steel deck supported on secondary members, in.^{4} per ft

For trusses and steel joists, the moment of inertia I_s shall be decreased 15 percent when used in the above equation. A steel deck shall be considered a secondary member when it is directly supported by the primary members.

See Appendix K2 for an alternate determination of flat roof framing stiffness.

K3. FATIGUE

Few members or connections in conventional buildings need to be designed for fatigue, since most load changes in such structures occur only a small number of times or produce only minor stress fluctuations. The occurrence of full design wind or earthquake loads is too infrequent to warrant consideration in fatigue design. However, crane runways and supporting structures for machinery and equipment are often subject to fatigue loading conditions.

Members and their connections subject to fatigue loading shall be proportioned in accordance with the provisions of Appendix K3 for service loads.

CHAPTER L

SERVICEABILITY DESIGN CONSIDERATIONS

This chapter is intended to provide design guidance for serviceability considerations.

Serviceability is a state in which the function of a building, its appearance, maintainability, durability, and comfort of its occupants are preserved under normal usage. The general design requirement for serviceability is given in Section A5.4. Limiting values of structural behavior to ensure serviceability (e.g., maximum deflections, accelerations, etc.) shall be chosen with due regard to the intended function of the structure. Where necessary, serviceability shall be checked using realistic loads for the appropriate serviceability limit state.

L1.   CAMBER

If any special camber requirements are necessary to bring a loaded member into proper relation with the work of other trades, as for the attachment of runs of sash, the requirements shall be set forth in the design documents.

Beams and trusses detailed without specified camber shall be fabricated so that after erection any camber due to rolling or shop assembly shall be upward. If camber involves the erection of any member under a preload, this shall be noted in the design documents.

L2.   EXPANSION AND CONTRACTION

Adequate provision shall be made for expansion and contraction appropriate to the service conditions of the structure.

L3.   DEFLECTIONS, VIBRATION, AND DRIFT

1.   Deflections

Deformations in structural members and structural systems due to service loads shall not impair the serviceability of the structure.

2.   Floor Vibration

Vibration shall be considered in designing beams and girders supporting large areas free of partitions or other sources of damping where excessive vibration due to pedestrian traffic or other sources within the building is not acceptable.

3.   Drift

Lateral deflection or drift of structures due to code-specified wind or seismic loads shall not cause collision with adjacent structures nor exceed the limiting values of such drifts which may be specified or appropriate.

L4. CONNECTION SLIP

For the design of slip-critical connections see Sections J3.8 and J3.9.

L5. CORROSION

When appropriate, structural components shall be designed to tolerate corrosion or shall be protected against corrosion that may impair the strength or serviceability of the structure.

CHAPTER M

FABRICATION, ERECTION, AND QUALITY CONTROL

This chapter provides requirements for shop drawings, fabrication, shop painting, erection, and quality control.

M1. SHOP DRAWINGS

Shop drawings giving complete information necessary for the fabrication of the component parts of the structure, including the location, type, and size of all welds, bolts, and rivets, shall be prepared in advance of the actual fabrication. These drawings shall clearly distinguish between shop and field welds and bolts and shall clearly identify slip-critical high-strength bolted connections.

Shop drawings shall be made in conformity with good practice and with due regard to speed and economy in fabrication and erection.

M2. FABRICATION

1. Cambering, Curving, and Straightening

Local application of heat or mechanical means is permitted to be used to introduce or correct camber, curvature, and straightness. The temperature of heated areas, as measured by approved methods, shall not exceed 1,100°F for A514 and A852 steel nor 1,200°F for other steels.

2. Thermal Cutting

Thermally cut edges shall meet the requirements of AWS 3.2.2 with the exception that thermally cut free edges which will be subject to calculated static tensile stress shall be free of round bottom gouges greater than $\frac{3}{16}$-in. deep and sharp V-shaped notches. Gouges greater than $\frac{3}{16}$-in. deep and notches shall be removed by grinding or repaired by welding.

Re-entrant corners, except re-entrant corners of beam copes and weld access holes, shall meet the requirements of AWS 3.2.4. If other specified contour is required it must be shown on the contract documents.

Beam copes and weld access holes shall meet the geometrical requirements of Section J1.6. For beam copes and weld access holes in ASTM A6 Group 4 and 5 shapes and welded built-up shapes with material thickness greater than two inches, a preheat temperature of not less than 150°F shall be applied prior to thermal cutting.

3. Planing of Edges

Planing or finishing of sheared or thermally cut edges of plates or shapes is not required unless specifically called for in the design documents or included in a stipulated edge preparation for welding.

4.   Welded Construction

The technique of welding, the workmanship, appearance, and quality of welds and the methods used in correcting nonconforming work shall be in accordance with AWS D1.1 except as modified in Section J2.

5.   Bolted Construction

All parts of bolted members shall be pinned or bolted and rigidly held together during assembly. Use of a drift pin in bolt holes during assembly shall not distort the metal or enlarge the holes. Poor matching of holes shall be cause for rejection.

If the thickness of the material is not greater than the nominal diameter of the bolt plus $\frac{1}{8}$-in., the holes are permitted to be punched. If the thickness of the material is greater than the nominal diameter of the bolt plus $\frac{1}{8}$-in., the holes shall be either drilled or sub-punched and reamed. The die for all sub-punched holes, and the drill for all sub-drilled holes, shall be at least $\frac{1}{16}$-in. smaller than the nominal diameter of the bolt. Holes in A514 steel plates over $\frac{1}{2}$-in. thick shall be drilled.

Fully inserted finger shims, with a total thickness of not more than $\frac{1}{4}$-in. within a joint, are permitted in joints without changing the design strength (based upon hole type) for the design of connections. The orientation of such shims is independent of the direction of application of the load.

The use of high-strength bolts shall conform to the requirements of the RCSC *Load and Resistance Factor Design Specification for Structural Joints Using ASTM A325 or A490 Bolts.*

6.   Compression Joints

Compression joints which depend on contact bearing as part of the splice strength shall have the bearing surfaces of individual fabricated pieces prepared by milling, sawing, or other suitable means.

7.   Dimensional Tolerances

Dimensional tolerances shall be in accordance with the AISC *Code of Standard Practice.*

8.   Finish of Column Bases

Column bases and base plates shall be finished in accordance with the following requirements:

(1) Steel bearing plates two inches or less in thickness are permitted without milling, provided a satisfactory contact bearing is obtained. Steel bearing plates over two inches but not over four inches in thickness are permitted to be straightened by pressing or, if presses are not available, by milling for all bearing surfaces (except as noted in subparagraphs 2 and 3 of this section), to obtain a satisfactory contact bearing. Steel bearing plates over four inches

in thickness shall be milled for all bearing surfaces (except as noted in subparagraphs 2 and 3 of this section).

(2) Bottom surfaces of bearing plates and column bases which are grouted to ensure full bearing contact on foundations need not be milled.

(3) Top surfaces of bearing plates need not be milled when full-penetration welds are provided between the column and the bearing plate.

M3. SHOP PAINTING

1. General Requirements

Shop painting and surface preparation shall be in accordance with the provisions of the AISC *Code of Standard Practice*.

Shop paint is not required unless specified by the contract documents.

2. Inaccessible Surfaces

Except for contact surfaces, surfaces inaccessible after shop assembly shall be cleaned and painted prior to assembly, if required by the design documents.

3. Contact Surfaces

Paint is permitted unconditionally in bearing-type connections. For slip-critical connections, the faying surface requirements shall be in accordance with the RCSC *Specification for Structural Joints Using ASTM A325 or A490 Bolts*, paragraph 3(b).

4. Finished Surfaces

Machine-finished surfaces shall be protected against corrosion by a rust-inhibitive coating that can be removed prior to erection, or which has characteristics that make removal prior to erection unnecessary.

5. Surfaces Adjacent to Field Welds

Unless otherwise specified in the design documents, surfaces within two inches of any field weld location shall be free of materials that would prevent proper welding or produce objectionable fumes during welding.

M4. ERECTION

1. Alignment of Column Bases

Column bases shall be set level and to correct elevation with full bearing on concrete or masonry.

2. Bracing

The frame of steel skeleton buildings shall be carried up true and plumb within the limits defined in the AISC *Code of Standard Practice*. Temporary bracing shall be provided, in accordance with the requirements of the *Code of Standard Practice*, wherever necessary to support all loads to which the structure may be subjected, including equipment and the operation of same. Such bracing shall be left in place as long as required for safety.

3. Alignment

No permanent bolting or welding shall be performed until the adjacent affected portions of the structure have been properly aligned.

4. Fit of Column Compression Joints and Base Plates

Lack of contact bearing not exceeding a gap of $\frac{1}{16}$-in., regardless of the type of splice used (partial-joint-penetration groove welded, or bolted), is permitted. If the gap exceeds $\frac{1}{16}$-in., but is less than $\frac{1}{4}$-in., and if an engineering investigation shows that sufficient contact area does not exist, the gap shall be packed out with non-tapered steel shims. Shims need not be other than mild steel, regardless of the grade of the main material.

5. Field Welding

Shop paint on surfaces adjacent to joints to be field welded shall be wire brushed if necessary to assure weld quality.

Field welding of attachments to installed embedments in contact with concrete shall be done in such a manner as to avoid excessive thermal expansion of the embedment which could result in spalling or cracking of the concrete or excessive stress in the embedment anchors.

6. Field Painting

Responsibility for touch-up painting, cleaning, and field painting shall be allocated in accordance with accepted local practices, and this allocation shall be set forth explicitly in the design documents.

7. Field Connections

As erection progresses, the structure shall be securely bolted or welded to support all dead, wind, and erection loads.

M5. QUALITY CONTROL

The fabricator shall provide quality control procedures to the extent that the fabricator deems necessary to assure that all work is performed in accordance with this Specification. In addition to the fabricator's quality control procedures, material and workmanship at all times may be subject to inspection by qualified inspectors representing the purchaser. If such inspection by representatives of the purchaser will be required, it shall be so stated in the design documents.

1. Cooperation

As far as possible, all inspection by representatives of the purchaser shall be made at the fabricator's plant. The fabricator shall cooperate with the inspector, permitting access for inspection to all places where work is being done. The purchaser's inspector shall schedule this work for minimum interruption to the work of the fabricator.

2. Rejections

Material or workmanship not in reasonable conformance with the provisions of this Specification may be rejected at any time during the progress of the work.

The fabricator shall receive copies of all reports furnished to the purchaser by the inspection agency.

3. Inspection of Welding

The inspection of welding shall be performed in accordance with the provisions of AWS D1.1 except as modified in Section J2.

When visual inspection is required to be performed by AWS certified welding inspectors, it shall be so specified in the design documents.

When nondestructive testing is required, the process, extent, and standards of acceptance shall be clearly defined in the design documents.

4. Inspection of Slip-Critical High-Strength Bolted Connections

The inspection of slip-critical high-strength bolted connections shall be in accordance with the provisions of the RCSC *Load and Resistance Factor Design Specification for Structural Joints Using ASTM A325 or A490 Bolts.*

5. Identification of Steel

The fabricator shall be able to demonstrate by a written procedure and by actual practice a method of material application and identification, visible at least through the "fit-up" operation, of the main structural elements of a shipping piece.

The identification method shall be capable of verifying proper material application as it relates to:

(1) Material specification designation

(2) Heat number, if required

(3) Material test reports for special requirements.

APPENDIX B

DESIGN REQUIREMENTS

Appendix B5.1 provides an expanded definition of limiting width-thickness ratio for webs in combined flexure and axial compression. Appendix B5.3 applies to the design of members containing slender compression elements.

B5.   LOCAL BUCKLING

1.   Classification of Steel Sections

For members with unequal flanges and with webs in combined flexural and axial compression, λ_r for the limit state of web local buckling is

$$\lambda_r = \frac{253}{\sqrt{F_y}}\left[1 + 2.83\left(\frac{h}{h_c}\right)\left(1 - \frac{P_u}{\phi_b P_y}\right)\right] \qquad (A\text{-}B5\text{-}1)$$

$$\frac{3}{4} \le \frac{h}{h_c} \le \frac{3}{2}$$

For members with unequal flanges with webs subjected to flexure only, λ_r for the limit state of web local buckling is

$$\lambda_r = \frac{253}{\sqrt{F_y}}\left[1 + 2.83\left(\frac{h}{h_c}\right)\right] \qquad (A\text{-}B5\text{-}2)$$

$$\frac{3}{4} \le \frac{h}{h_c} \le \frac{3}{2}$$

where λ_r, h, and h_c are as defined in Section B5.1.

These substitutions shall be made in Appendices F and G when applied to members with unequal flanges. If the compression flange is larger than the tension flange, λ_r shall be determined using Equation A-B5-1, A-B5-2, or Table B5.1.

3.   Slender-Element Compression Sections

Axially loaded members containing elements subject to compression which have a width-thickness ratio in excess of the applicable λ_r as stipulated in Section B5.1 shall be proportioned according to this Appendix. Flexural members with slender compression elements shall be designed in accordance with Appendices F and G. Flexural members with proportions not covered by Appendix F1 shall be designed in accordance with this Appendix.

3a. Unstiffened Compression Elements

The design strength of unstiffened compression elements whose width-thickness ratio exceeds the applicable limit λ_r as stipulated in Section B5.1 shall be subject to a reduction factor Q_s. The value of Q_s shall be determined by Equations A-B5-3 through A-B5-10, as applicable. When such elements comprise the compression flange of a flexural member, the maximum required bending stress shall not exceed $\phi_b F_y Q_s$, where $\phi_b = 0.90$. The design strength of axially loaded compression members shall be modified by the appropriate reduction factor Q, as provided in Appendix B5.3c.

(a) For single angles:

when $76.0/\sqrt{F_y} < b/t < 155/\sqrt{F_y}$:

$$Q_s = 1.340 - 0.00447(b/t)\sqrt{F_y} \qquad \text{(A-B5-3)}$$

when $b/t > 155/\sqrt{F_y}$:

$$Q_s = 15{,}500/[F_y(b/t)^2] \qquad \text{(A-B5-4)}$$

(b) For flanges, angles, and plates projecting from rolled beams or columns or other compression members:

when $95.0/\sqrt{F_y} < b/t < 176/\sqrt{F_y}$:

$$Q_s = 1.415 - 0.00437(b/t)\sqrt{F_y} \qquad \text{(A-B5-5)}$$

when $b/t \geq 176/\sqrt{F_y}$:

$$Q_s = 20{,}000/[F_y(b/t)^2] \qquad \text{(A-B5-6)}$$

(c) For flanges, angles and plates projecting from built-up columns or other compression members:

when $109/\sqrt{F_y/k_c} < b/t < 200/\sqrt{F_y/k_c}$:

$$Q_s = 1.415 - 0.00381(b/t)\sqrt{F_y/k_c} \qquad \text{(A-B5-7)}$$

when $b/t \geq 200/\sqrt{F_y/k_c}$:

$$Q_s = 26{,}200k_c/[F_y(b/t)^2] \qquad \text{(A-B5-8)}$$

The coefficient, k_c, shall be computed as follows:

(a) For I-shaped sections:

$$k_c = \frac{4}{\sqrt{h/t_w}}, \quad 0.35 \leq k_c \leq 0.763$$

where:

h = depth of web, in.
t_w = thickness of web, in.

(b) For other sections:

$k_c = 0.763$

(d) For stems of tees:

when $127/\sqrt{F_y} < b/t < 176/\sqrt{F_y}$:

$$Q_s = 1.908 - 0.00715(b/t)\sqrt{F_y} \qquad \text{(A-B5-9)}$$

when $b/t \geq 176/\sqrt{F_y}$:

$$Q_s = 20,000/[F_y(b/t)^2] \qquad \text{(A-B5-10)}$$

where

   b = width of unstiffened compression element as defined in Section B5.1, in.
   t = thickness of unstiffened element, in.
   F_y = specified minimum yield stress, ksi

3b.   Stiffened Compression Elements

When the width-thickness ratio of uniformly compressed stiffened elements (except perforated cover plates) exceeds the limit λ_r stipulated in Section B5.1, a reduced effective width b_e shall be used in computing the design properties of the section containing the element.

(a) For flanges of square and rectangular sections of uniform thickness:

when $\dfrac{b}{t} \geq \dfrac{238}{\sqrt{f}}$:

$$b_e = \frac{326t}{\sqrt{f}}\left[1 - \frac{64.9}{(b/t)\sqrt{f}}\right] \qquad \text{(A-B5-11)}$$

otherwise $b_e = b$.

(b) For other uniformly compressed elements:

when $\dfrac{b}{t} \geq \dfrac{253}{\sqrt{f}}$:

$$b_e = \frac{326t}{\sqrt{f}}\left[1 - \frac{57.2}{(b/t)\sqrt{f}}\right] \qquad \text{(A-B5-12)}$$

otherwise $b_e = b$

where

   b = actual width of a stiffened compression element, as defined in Section B5.1, in.
   b_e = reduced effective width, in.
   t = element thickness, in.
   f = computed elastic compressive stress in the stiffened elements, based on the design properties as specified in Appendix B5.3c, ksi. If unstiffened elements are included in the total cross section, f for the stiffened element must be such that the maximum compressive stress in the unstiffened element does not exceed $\phi_c F_{cr}$ as defined in Appendix B5.3c with $Q = Q_s$ and $\phi_c = 0.85$, or $\phi_b F_y Q_s$ with $\phi_b = 0.90$, as applicable.

(c) For axially loaded circular sections with diameter-to-thickness ratio D/t greater than $3,300/F_y$ but less than $13,000/F_y$

$$Q = Q_a = \frac{1,100}{F_y(D/t)} + \frac{2}{3} \qquad \text{(A-B5-13)}$$

where

D = outside diameter, in.
t = wall thickness, in.

3c. Design Properties

Properties of sections shall be determined using the full cross section, except as follows:

In computing the moment of inertia and elastic section modulus of flexural members, the effective width of uniformly compressed stiffened elements b_e, as determined in Appendix B5.3b, shall be used in determining effective cross-sectional properties.

For unstiffened elements of the cross section, Q_s is determined from Appendix B5.3a. For stiffened elements of the cross section

$$Q_a = \frac{\text{effective area}}{\text{actual area}} \qquad \text{(A-B5-14)}$$

where the effective area is equal to the summation of the effective areas of the cross section.

3d. Design Strength

For axially loaded compression members the gross cross-sectional area and the radius of gyration r shall be computed on the basis of the actual cross section. The critical stress F_{cr} shall be determined as follows:

(a) For $\lambda_c\sqrt{Q} \le 1.5$:

$$F_{cr} = Q(0.658^{Q\lambda_c^2})F_y \qquad \text{(A-B5-15)}$$

(a) For $\lambda_c\sqrt{Q} > 1.5$:

$$F_{cr} = \left[\frac{0.877}{\lambda_c^2}\right]F_y \qquad \text{(A-B5-16)}$$

where

$$Q = Q_sQ_a \qquad \text{(A-B5-17)}$$

Cross sections comprised of only unstiffened elements, $Q = Q_s$, ($Q_a = 1.0$)

Cross sections comprised of only stiffened elements, $Q = Q_a$, ($Q_s = 1.0$)

Cross sections comprised of both stiffened and unstiffened elements, $Q = Q_sQ_a$

APPENDIX E

COLUMNS AND OTHER COMPRESSION MEMBERS

This Appendix applies to the strength of doubly symmetric columns with thin plate elements, singly symmetric and unsymmetric columns for the limit states of flexural-torsional and torsional buckling.

E3. DESIGN COMPRESSIVE STRENGTH FOR FLEXURAL-TORSIONAL BUCKLING

The strength of compression members determined by the limit states of torsional and flexural-torsional buckling is $\phi_c P_n$,

where

$\phi_c = 0.85$
P_n = nominal resistance in compression, kips
$\quad = A_g F_{cr}$ $\hfill$ (A-E3-1)
A_g = gross area of cross section, in.^{2}

The nominal critical stress F_{cr} is determined as follows:

(a) For $\lambda_e\sqrt{Q} \leq 1.5$:

$$F_{cr} = Q(0.658^{Q\lambda_e^2})F_y \hfill \text{(A-E3-2)}$$

(b) For $\lambda_e\sqrt{Q} > 1.5$:

$$F_{cr} = \left[\frac{0.877}{\lambda_e^2}\right]F_y \hfill \text{(A-E3-3)}$$

where

$$\lambda_e = \sqrt{F_y / F_e} \hfill \text{(A-E3-4)}$$

F_y = specified minimum yield stress of steel, ksi
Q = 1.0 for elements meeting the width-thickness ratios λ_r of Section B5.1
$\quad = Q_s Q_a$ for elements not meeting the width-thickness ratios λ_r of Section B5.1 and determined in accordance with the provisions of Appendix B5.3

The critical torsional or flexural-torsional elastic buckling stress F_e is determined as follows:

(a) For doubly symmetric shapes:

$$F_e = \left[\frac{\pi^2 E C_w}{(K_z l)^2} + GJ\right]\frac{1}{I_x + I_y}$$

(A-E3-5)

(b) For singly symmetric shapes where y is the axis of symmetry:

$$F_e = \frac{F_{ey} + F_{ez}}{2H}\left(1 - \sqrt{1 - \frac{4F_{ey}F_{ez}H}{(F_{ey} + F_{ez})^2}}\right)$$

(A-E3-6)

(c) For unsymmetric shapes, the critical flexural-torsional elastic buckling stress F_e is the lowest root of the cubic equation

$$(F_e - F_{ex})(F_e - F_{ey})(F_e - F_{ez}) - F_e^2(F_e - F_{ey})\left(\frac{x_o}{r_o}\right)^2 - F_e^2(F_e - F_{ex})\left(\frac{y_o}{r_o}\right)^2 = 0 \quad \text{(A-E3-7)}$$

where

K_z = effective length factor for torsional buckling
E = modulus of elasticity, ksi
G = shear modulus, ksi
C_w = warping constant, in.^{6}
J = torsional constant, in.^{4}
I_x, I_y = moment of inertia about the principal axes, in.^{4}
x_o, y_o = coordinates of shear center with respect to the centroid, in.

$$\bar{r}_o^2 = x_o^2 + y_o^2 + \frac{I_x + I_y}{A}$$

(A-E3-8)

$$H = 1 - \left(\frac{x_o^2 + y_o^2}{\bar{r}_o^2}\right)$$

(A-E3-9)

$$F_{ex} = \frac{\pi^2 E}{(K_x l / r_x)^2}$$

(A-E3-10)

$$F_{ey} = \frac{\pi^2 E}{(K_y l / r_y)^2}$$

(A-E3-11)

$$F_{ez} = \left(\frac{\pi^2 E C_w}{(K_z l)^2} + GJ\right)\frac{1}{A\bar{r}_o^2}$$

(A-E3-12)

where

A = cross-sectional area of member, in.^{2}
l = unbraced length, in.
K_x, K_y = effective length factors in x and y directions
r_x, r_y = radii of gyration about the principal axes, in.
$\bar{r}_o$ = polar radius of gyration about the shear center, in.

APPENDIX F

BEAMS AND OTHER FLEXURAL MEMBERS

Appendix F1 provides the design flexural strength of beams and girders. Appendix F2 provides the design shear strength of webs with and without stiffeners and requirements on transverse stiffeners. Appendix F3 applies to web-tapered members.

F1. DESIGN FOR FLEXURE

The design strength for flexural members is $\phi_b M_n$ where $\phi_b = 0.90$ and M_n is the nominal strength.

Table A-F1.1 provides a tabular summary of Equations F1-1 through F1-15 for determining the nominal flexural strength of beams and girders. For slenderness parameters of cross sections not included in Table A-F1.1, see Appendix B5.3. For flexural members with unequal flanges see Appendix B5.1 for the determination of λ_r for the limit state of web local buckling.

The nominal flexural strength M_n is the lowest value obtained according to the limit states of yielding: lateral-torsional buckling (LTB); flange local buckling (FLB); and web local buckling (WLB).

The nominal flexural strength M_n shall be determined as follows for each limit state:

(a) For $\lambda \le \lambda_p$:

$$M_n = M_p \qquad \text{(A-F1-1)}$$

(b) For $\lambda_p < \lambda \le \lambda_r$:

For the limit state of lateral-torsional buckling:

$$M_n = C_b \left[M_p - (M_p - M_r)\left(\frac{\lambda - \lambda_p}{\lambda_r - \lambda_p}\right)\right] \le M_p \qquad \text{(A-F1-2)}$$

For the limit states of flange and web local buckling:

$$M_n = M_p - (M_p - M_r)\left(\frac{\lambda - \lambda_p}{\lambda_r - \lambda_p}\right) \qquad \text{(A-F1-3)}$$

(c) For $\lambda > \lambda_r$:

For the limit state of lateral-torsional buckling and flange local buckling:

$$M_n = M_{cr} = SF_{cr} \le M_p \qquad \text{(A-F1-4)}$$

For design of girders with slender webs, the limit state of web local buckling is not applicable. See Appendix G2.

For λ of the flange $> \lambda_r$ in shapes not included in Table A-F1.1, see Appendix B5.3.

For λ of the web $> \lambda_r$, see Appendix G.

The terms used in the above equations are:

M_n = nominal flexural strength, kip-in.

M_p = $F_y Z$, plastic moment $\leq 1.5 F_y S$, kip-in.

M_{cr} = buckling moment, kip-in.

M_r = limiting buckling moment (equal to M_{cr} when $\lambda = \lambda_r$), kip-in.

λ = controlling slenderness parameter

    = minor axis slenderness ratio L_b / r_y for lateral-torsional buckling

    = flange width-thickness ratio b / t for flange local buckling as defined in Section B5.1

    = web depth-thickness ratio h / t_w for web local buckling as defined in Section B5.1

λ_p = largest value of λ for which $M_n = M_p$

λ_r = largest value of λ for which buckling is inelastic

F_{cr} = critical stress, ksi

C_b = Bending coefficient dependent on moment gradient, see Section F1.2a, Equation F1-3

S = section modulus, in.^{3}

L_b = laterally unbraced length, in.

r_y = radius of gyration about minor axis, in.

The applicable limit states and equations for M_p, M_r, F_{cr}, λ, λ_p, and λ_r are given in Table A-F1.1 for shapes covered in this Appendix. The terms used in the table are:

A = cross-sectional area, in.^{2}

F_L = smaller of $(F_{yf} - F_r)$ or F_{yw}, ksi

F_r = compressive residual stress in flange

    = 10 ksi for rolled shapes

    = 16.5 ksi for welded shapes

F_y = specified minimum yield strength, ksi

F_{yf} = yield strength of the flange, ksi

F_{yw} = yield strength of the web, ksi

I_{yc} = moment of inertia of compression flange about y axis or if reverse curvature bending, moment of inertia of smaller flange, in.^{4}

J = torsional constant, in.^{4}

R_e = see Appendix G2

S_{eff} = effective section modulus about major axis, in.^{3}

S_{xc} = section modulus of the outside fiber of the compression flange, in.^{3}

S_{xt} = section modulus of the outside fiber of the tension flange, in.^{3}

Z = plastic section modulus, in.^{3}

b = flange width, in.

d = overall depth, in.

f = computed compressive stress in the stiffened element, ksi
h = clear distance between flanges less the fillet or corner radius at each flange, in.
r_{yc} = radius of gyration of compression flange about y axis or if reverse curvature bending, smaller flange, in.
t_f = flange thickness, in.
t_w = web thickness, in.

F2. DESIGN FOR SHEAR

2. Design Shear Strength

The design shear strength of stiffened or unstiffened webs is $\phi_v V_n$,

where

$\phi_v = 0.90$
V_n = nominal shear strength defined as follows:

For $h / t_w \leq 187\sqrt{k_v / F_{yw}}$:

$$V_n = 0.6 F_{yw} A_w \tag{A-F2-1}$$

For $187\sqrt{k_v / F_{yw}} < h / t_w \leq 234\sqrt{k_v / F_{yw}}$:

$$V_n = 0.6 F_{yw} A_w (187\sqrt{k_v / F_{yw}}) / (h / t_w) \tag{A-F2-2}$$

For $h / t_w > 234\sqrt{k_v / F_{yw}}$:

$$V_n = A_w (26{,}400 k_v) / (h / t_w)^2 \tag{A-F2-3}$$

where

$k_v = 5 + 5 / (a / h)^2$
    $= 5$ when $a / h > 3$ or $a / h > [260 / (h / t)]^2$
a = distance between transverse stiffeners, in.
h = for rolled shapes, the clear distance between flanges less the fillet or corner radius, in.
   = for built-up welded sections, the clear distance between flanges, in.
   = for built-up bolted or riveted sections, the distance between fastener lines, in.

3. Transverse Stiffeners

Transverse stiffeners are not required in plate girders where $h / t_w \leq 418 / \sqrt{F_{yw}}$, or where the required shear, V_u, as determined by structural analysis for the factored loads, is less than or equal to $0.6\phi_v A_w F_{yw} C_v$, where C_v is determined for $k_v = 5$ and $\phi_v = 0.90$.

Transverse stiffeners used to develop the web design shear strength as provided in Appendix F2.2 shall have a moment of inertia about an axis in the web center for stiffener pairs or about the face in contact with the web plate for single stiffeners, which shall not be less than $a t_w^3 j$, where

$$j = 2.5 / (a / h)^2 - 2 \geq 0.5 \tag{A-F2-4}$$

TABLE A-F1.1
Nominal Strength Parameters

Shape	Plastic Moment M_p	Limit State of Buckling	Limiting Buckling Moment M_r
Channels and doubly and singly symmetric I-shaped beams (including hybrid beams) bent about major axis [a]	$F_y Z_x$ [b]	LTB doubly symmetric members and channels	$F_L S_x$
		LTB singly symmetric members	$F_L S_{xc} \le F_{yt} S_{xt}$
		FLB	$F_L S_x$
		WLB	$R_e F_{yt} S_x$
Channels and doubly and singly symmetric I-shaped members bent about minor axis [a]	$F_y Z_y$	FLB	$F_y S_y$

NOTE: LTB applies only for strong axis bending.
[a] Excluding double angles and tees.
[b] Computed from fully plastic stress distribution for hybrid sections.

[c] $X_1 = \dfrac{\pi}{S_x} \sqrt{\dfrac{EGJA}{2}}$ $X_2 = 4 \dfrac{C_w}{I_y} \left(\dfrac{S_x}{GJ} \right)^2$

[d] $\lambda_r = \dfrac{X_1}{F_L} \sqrt{1 + \sqrt{1 + X_2 F_L^2}}$

[e] $F_{cr} = \dfrac{M_{cr}}{S_{xc}}$, where $M_{cr} = \dfrac{57,000 C_b}{L_b} \sqrt{I_y J} \, [B_1 + \sqrt{(1 + B_2 + B_1^2)}] \le M_p$

where
$B_1 = 2.25[2(I_{yc}/I_y) - 1](h/L_b)\sqrt{(I_y/J)}$
$B_2 = 25(1 - I_{yc}/I_y)(I_{yc}/J)(h/L_b)^2$
$C_b = 1.0$ if $I_{yc}/I_y < 0.1$ or $I_{yc}/I_y > 0.9$.

TABLE A-F1.1 (cont'd)
Nominal Strength Parameters

Critical Stress F_{cr}	Slenderness Parameters			Limitations
	λ	λ_p	λ_r	
$\dfrac{C_b X_1 \sqrt{2}}{\lambda} \sqrt{1 + \dfrac{X_1^2 X_2}{2\lambda^2}}$	$\dfrac{L_b}{r_y}$	$\dfrac{300}{\sqrt{F_{yf}}}$	[c, d]	Applicable for I-shaped members if $h/t_w \le \lambda_r$ when $h/t_w > \lambda_r$ See Appendix G.
[e]	$\dfrac{L_b}{r_{yc}}$	$\dfrac{300}{\sqrt{F_{yf}}}$	Value of λ for which $M_{cr}(C_b = 1) = M_r$	
[f]	$\dfrac{b}{t}$	$\dfrac{65}{\sqrt{F_{yf}}}$	[g]	
Not applicable	$\dfrac{h}{t_w}$	$\dfrac{640}{\sqrt{F_{yf}}}$	λ_r as defined in Section B5.1	
Same as for major axis				

[f] $F_{cr} = \dfrac{20{,}000}{\lambda^2}$ for rolled shapes

$F_{cr} = \dfrac{26{,}200 k_c}{\lambda^2}$ for welded shapes

where
$k_c = 4 / \sqrt{h/t_w}$ and $0.35 \le k_c \le 0.763$

[g] $\lambda_r = \dfrac{141}{\sqrt{F_L}}$ for rolled shapes

$\lambda_r = \dfrac{162}{\sqrt{F_L / k_c}}$ for welded shapes

TABLE A-F1.1 (cont'd)
Nominal Strength Parameters

Shape	Plastic Moment M_p	Limit State of Buckling	Limiting Buckling Moment M_r
Solid symmetric shapes, except rectangular bars, bent about major axis	$F_y Z_x$		Not applicable
Solid rectangular bars bent about major axis	$F_y Z_x$	LTB	$F_y S_x$
Symmetric box sections loaded in a plane of symmetry	$F_y Z$	LTB	$F_{yf} S_{eff}$
		FLB	$F_L S_{eff}$
		WLB	Same as for I-shape
Circular tubes	$F_y Z$	LTB	Not applicable
		FLB	$M_n = \left(\dfrac{600}{D/t} + F_y \right) S$ [h]
		WLB	Not applicable

[h] This equation is to be used in place of Equation A-F1-4.

TABLE A-F1.1 (cont'd)
Nominal Strength Parameters

Critical Stress F_{cr}	Slenderness Parameters			Limitations
	λ	λ_p	λ_r	
Not applicable				
$\dfrac{57,000 C_b \sqrt{JA}}{\lambda S_x}$	$\dfrac{L_b}{r_y}$	$\dfrac{3,750\sqrt{JA}}{M_p}$	$\dfrac{57,000\sqrt{JA}}{M_{r,}}$	
$\dfrac{57,000 C_b \sqrt{JA}}{\lambda S_x}$	$\dfrac{L_b}{r_y}$	$\dfrac{3,750\sqrt{JA}}{M_p}$	$\dfrac{57,000\sqrt{JA}}{M_r}$	Applicable if $h\,/\,t_w \leq 970\,/\sqrt{F_{yf}}$
$\dfrac{S_{eff}}{S_x} F_y$ [i]	$\dfrac{b}{t}$	$\dfrac{190}{\sqrt{F_y}}$	$\dfrac{238}{\sqrt{F_y}}$	
Same as for I-shape				
Not applicable				
$\dfrac{9,570}{D\,/\,t}$	$D\,/\,t$	$\dfrac{2,070}{F_y}$	$\dfrac{8,970}{F_y}$	$D\,/\,t < \dfrac{13,000}{F_y}$
Not applicable				

[i] S_{eff} is the effective section modulus for the section with a compression flange b_e defined in
Appendix B5.3b

Intermediate stiffeners are permitted to be stopped short of the tension flange, provided bearing is not needed to transmit a concentrated load or reaction. The weld by which intermediate stiffeners are attached to the web shall be terminated not less than four times nor more than six times the web thickness from the near toe of the web-to-flange weld. When single stiffeners are used, they shall be attached to the compression flange, if it consists of a rectangular plate, to resist any uplift tendency due to torsion in the flange. When lateral bracing is attached to a stiffener, or a pair of stiffeners, these, in turn, shall be connected to the compression flange to transmit one percent of the total flange stress, unless the flange is composed only of angles.

Bolts connecting stiffeners to the girder web shall be spaced not more than 12 in. on center. If intermittent fillet welds are used, the clear distance between welds shall not be more than 16 times the web thickness nor more than 10 in.

F3.　WEB-TAPERED MEMBERS

The design of tapered members meeting the requirements of this section shall be governed by the provisions of Chapters D through H, except as modified by this Appendix.

1.　General Requirements

In order to qualify under this Specification, a tapered member shall meet the following requirements:

(1) It shall possess at least one axis of symmetry which shall be perpendicular to the plane of bending if moments are present.

(2) The flanges shall be of equal and constant area.

(3) The depth shall vary linearly as

$$d = d_o \left(1 + \gamma \frac{z}{L} \right) \tag{A-F3-1}$$

where

d_o = depth at smaller end of member, in.
d_L = depth at larger end of member, in.
γ = $(d_L - d_o) / d_o \leq$ the smaller of $0.268(L / d_o)$ or 6.0
z = distance from the smaller end of member, in.
L = unbraced length of member measured between the center of gravity of the bracing members, in.

2.　Design Tensile Strength

The design strength of tapered tension members shall be determined in accordance with Section D1.

3.　Design Compressive Strength

The design strength of tapered compression members shall be determined in accordance with Section E2, using an effective slenderness parameter λ_{eff} computed as follows:

$$\lambda_{eff} = \frac{S}{\pi} \sqrt{\frac{QF_y}{E}} \qquad \text{(A-F3-2)}$$

where

S = KL / r_{oy} for weak axis buckling and $K_\gamma L / r_{ox}$ for strong axis buckling
K = effective length factor for a prismatic member
K_γ = effective length factor for a tapered member as determined by a rational analysis
r_{ox} = strong axis radius of gyration at the smaller end of a tapered member, in.
r_{oy} = weak axis radius of gyration at the smaller end of a tapered member, in.
F_y = specified minimum yield stress, ksi
Q = reduction factor
     = 1.0 if all elements meet the limiting width-thickness ratios λ_r of Section B5.1
     = $Q_s Q_a$, determined in accordance with Appendix B5.3, if any stiffened and/or unstiffened elements exceed the ratios λ_r of Section B5.1
E = modulus of elasticity for steel, ksi

The smallest area of the tapered member shall be used for A_g in Equation E2-1.

4. **Design Flexural Strength**

The design flexural strength of tapered flexural members for the limit state of lateral-torsional buckling is $\phi_b M_n$, where $\phi_b = 0.90$ and the nominal strength is

$$M_n = (5/3)S'_x F_{b\gamma} \qquad \text{(A-F3-3)}$$

where

S'_x = the section modulus of the critical section of the unbraced beam length under consideration

$$F_{b\gamma} = \frac{2}{3}\left[1.0 - \frac{F_y}{6B\sqrt{F_{s\gamma}^2 + F_{w\gamma}^2}}\right] F_y \leq 0.60F_y \qquad \text{(A-F3-4)}$$

unless $F_{b\gamma} \leq F_y / 3$, in which case

$$F_{b\gamma} = B\sqrt{F_{s\gamma}^2 + F_{w\gamma}^2} \qquad \text{(A-F3-5)}$$

In the preceding equations,

$$F_{s\gamma} = \frac{12 \times 10^3}{h_s L d_o / A_f} \qquad \text{(A-F3-6)}$$

$$F_{w\gamma} = \frac{170 \times 10^3}{(h_w L / r_{To})^2} \qquad \text{(A-F3-7)}$$

where

h_s = factor equal to $1.0 + 0.0230\gamma\sqrt{Ld_o / A_f}$
h_w = factor equal to $1.0 + 0.00385\gamma\sqrt{L / r_{To}}$
r_{To} = radius of gyration of a section at the smaller end, considering only the

compression flange plus one-third of the compression web area, taken about an axis in the plane of the web, in.

A_f = area of the compression flange, in.^{2}

and where B is determined as follows:

(a) When the maximum moment M_2 in three adjacent segments of approximately equal unbraced length is located within the central segment and M_1 is the larger moment at one end of the three-segment portion of a member:

$$B = 1.0 + 0.37\left(1.0 + \frac{M_1}{M_2}\right) + 0.50\gamma\left(1.0 + \frac{M_1}{M_2}\right) \geq 1.0 \qquad \text{(A-F3-8)}$$

(b) When the largest computed bending stress f_{b2} occurs at the larger end of two adjacent segments of approximately equal unbraced lengths and f_{b1} is the computed bending stress at the smaller end of the two-segment portion of a member:

$$B = 1.0 + 0.58\left(1.0 + \frac{f_{b1}}{f_{b2}}\right) - 0.70\gamma\left(1.0 + \frac{f_{b1}}{f_{b2}}\right) \geq 1.0 \qquad \text{(A-F3-9)}$$

(c) When the largest computed bending stress f_{b2} occurs at the smaller end of two adjacent segments of approximately equal unbraced length and f_{b1} is the computed bending stress at the larger end of the two-segment portion of a member:

$$B = 1.0 + 0.55\left(1.0 + \frac{f_{b1}}{f_{b2}}\right) + 2.20\gamma\left(1.0 + \frac{f_{b1}}{f_{b2}}\right) \geq 1.0 \qquad \text{(A-F3-10)}$$

In the foregoing, $\gamma = (d_L - d_o)/d_o$ is calculated for the unbraced length that contains the maximum computed bending stress. M_1/M_2 is considered as negative when producing single curvature. In the rare case where M_1/M_2 is positive, it is recommended that it be taken as zero. f_{b1}/f_{b2} is considered as negative when producing single curvature. If a point of contraflexure occurs in one of two adjacent unbraced segments, f_{b1}/f_{b2} is considered as positive. The ratio $f_{b1}/f_{b2} \neq 0$.

(d) When the computed bending stress at the smaller end of a tapered member or segment thereof is equal to zero:

$$B = \frac{1.75}{1.0 + 0.25\sqrt{\gamma}} \qquad \text{(A-F3-11)}$$

where $\gamma = (d_L - d_o)/d_o$ is calculated for the unbraced length adjacent to the point of zero bending stress.

5.   **Design Shear Strength**

The design shear strength of tapered flexural members shall be determined in accordance with Section F2.

6.   **Combined Flexure and Axial Force**

For tapered members with a single web taper subject to compression and bending about the major axis, Equation H1-1 applies, with the following modifications: P_n and P_{ex} shall be determined for the properties of the smaller end, using appropriate effective length factors. M_{nx}, M_u, and M_{px} shall be determined for the larger end; $M_{nx} = (5/3) S_x' F_{b\gamma}$, where S_x' is the elastic section modulus of the larger end, and $F_{b\gamma}$ is the design flexural stress of tapered members. C_{mx} is replaced by C_m', determined as follows:

(a) When the member is subjected to end moments which cause single curvature bending and approximately equal computed moments at the ends:

$$C_m' = 1.0 + 0.1 \left(\frac{P_u}{\phi_b P_{ex}} \right) + 0.3 \left(\frac{P_u}{\phi_b P_{ex}} \right)^2 \qquad \text{(A-F3-12)}$$

(b) When the computed bending moment at the smaller end of the unbraced length is equal to zero:

$$C_m' = 1.0 - 0.9 \left(\frac{P_u}{\phi_b P_{ex}} \right) + 0.6 \left(\frac{P_u}{\phi_b P_{ex}} \right)^2 \qquad \text{(A-F3-13)}$$

When the effective slenderness parameter $\lambda_{eff} \geq 1.5$ and combined stress is checked incrementally along the length, the actual area and the actual section modulus at the section under investigation is permitted to be used.

APPENDIX G

PLATE GIRDERS

This appendix applies to I-shaped plate girders with slender webs.

G1. LIMITATIONS

Doubly and singly symmetric single-web non-hybrid and hybrid plate girders loaded in the plane of the web shall be proportioned according to the provisions of this Appendix or Section F2, provided that the following limits are satisfied:

(a) For $\dfrac{a}{h} \le 1.5$:

$$\frac{h}{t_w} \le \frac{2,000}{\sqrt{F_{yf}}} \qquad \text{(A-G1-1)}$$

(b) For $\dfrac{a}{h} > 1.5$:

$$\frac{h}{t_w} \le \frac{14,000}{\sqrt{F_{yf}\,(F_{yf}+16.5)}} \qquad \text{(A-G1-2)}$$

where

a = clear distance between transverse stiffeners, in.
h = clear distance between flanges less the fillet or corner radius for rolled shapes; and for built-up sections, the distance between adjacent lines of fasteners or the clear distance between flanges when welds are used, in.
t_w = web thickness, in.
F_{yf} = specified minimum yield stress of a flange, ksi

In unstiffened girders h/t_w shall not exceed 260.

G2. DESIGN FLEXURAL STRENGTH

The design flexural strength for plate girders with slender webs shall be $\phi_b M_n$, where $\phi_b = 0.90$ and M_n is the lower value obtained according to the limit states of tension-flange yield and compression-flange buckling. For girders with unequal flanges, see Appendix B5.1 for the determination of λ_r for the limit state of web local buckling.

(a) For tension-flange yield:

$$M_n = S_{xt} R_e F_{yt} \qquad \text{(A-G2-1)}$$

(b) For compression-flange buckling:

$$M_n = S_{xc}R_{PG}R_eF_{cr} \tag{A-G2-2}$$

where

$$R_{PG} = 1 - \frac{a_r}{1,200 + 300a_r}\left(\frac{h_c}{t_w} - \frac{970}{\sqrt{F_{cr}}}\right) \le 1.0 \tag{A-G2-3}$$

R_e = hybrid girder factor

$$= \frac{12 + a_r(3m - m^3)}{12 + 2a_r} \le 1.0 \text{ (for non-hybrid girders, } R_e = 1.0)$$

a_r = ratio of web area to compression flange area (≤ 10)
m = ratio of web yield stress to flange yield stress or to F_{cr}
F_{cr} = critical compression flange stress, ksi
F_{yt} = yield stress of tension flange, ksi
S_{xc} = section modulus referred to compression flange, in.^{3}
S_{xt} = section modulus referred to tension flange, in.^{3}
h_c = twice the distance from the centroid to the nearest line of fasteners at the compression flange or the inside of the face of the compression flange when welds are used

The critical stress F_{cr} to be used is dependent upon the slenderness parameters λ, λ_p, λ_r, and C_{PG} as follows:

For $\lambda \le \lambda_p$:

$$F_{cr} = F_{yf} \tag{A-G2-4}$$

For $\lambda_p < \lambda \le \lambda_r$:

$$F_{cr} = C_bF_{yf}\left[1 - \frac{1}{2}\left(\frac{\lambda - \lambda_p}{\lambda_r - \lambda_p}\right)\right] \le F_{yf} \tag{A-G2-5}$$

For $\lambda > \lambda_r$:

$$F_{cr} = \frac{C_{PG}}{\lambda^2} \tag{A-G2-6}$$

In the foregoing, the slenderness parameter shall be determined for both the limit state of lateral-torsional buckling and the limit state of flange local buckling; the slenderness parameter which results in the lowest value of F_{cr} governs.

(a) For the limit state of lateral-torsional buckling:

$$\lambda = \frac{L_b}{r_T} \tag{A-G2-7}$$

$$\lambda_p = \frac{300}{\sqrt{F_{yf}}} \tag{A-G2-8}$$

$$\lambda_r = \frac{756}{\sqrt{F_{yf}}} \tag{A-G2-9}$$

$$C_{PG} = 286,000C_b \tag{A-G2-10}$$

where

C_b = see Section F1.2, Equation F1-3

r_T = radius of gyration of compression flange plus one-third of the compression portion of the web, in.

(b) For the limit state of flange local buckling:

$$\lambda = \frac{b_f}{2t_f} \tag{A-G2-11}$$

$$\lambda_p = \frac{65}{\sqrt{F_{yf}}} \tag{A-G2-12}$$

$$\lambda_r = \frac{230}{\sqrt{F_{yf}/k_c}} \tag{A-G2-13}$$

$$C_{PG} = 26{,}200k_c \tag{A-G2-14}$$

$$C_b = 1.0$$

where $k_c = 4/\sqrt{h/t_w}$ and $0.35 \le k_c \le 0.763$.

The limit state of flexural web local buckling is not applicable.

G3. DESIGN SHEAR STRENGTH WITH TENSION FIELD ACTION

The design shear strength with tension field action shall be $\phi_v V_n$, kips, where $\phi_v = 0.90$ and V_n is determined as follows:

(a) For $h/t_w \le 187\sqrt{k_v/F_{yw}}$:

$$V_n = 0.6A_w F_{yw} \tag{A-G3-1}$$

(b) For $h/t_w > 187\sqrt{k_v/F_{yw}}$:

$$V_n = 0.6A_w F_{yw}\left(C_v + \frac{1-C_v}{1.15\sqrt{1+(a/h)^2}}\right) \tag{A-G3-2}$$

where

C_v = ratio of "critical" web stress, according to linear buckling theory, to the shear yield stress of web material

Also see Appendix G4 and G5.

For end-panels in non-hybrid plate girders, all panels in hybrid and web-tapered plate girders, and when a/h exceeds 3.0 or $[260/(h/t_w)]^2$, tension field action is not permitted and

$$V_n = 0.6A_w F_{yw} C_v \tag{A-G3-3}$$

The web plate buckling coefficient k_v is given as

$$k_v = 5 + \frac{5}{(a/h)^2} \tag{A-G3-4}$$

except that k_v shall be taken as 5.0 if a/h exceeds 3.0 or $[260/(h/t_w)]^2$.

The shear coefficient C_v is determined as follows:

(a) For $187\sqrt{\dfrac{k_v}{F_{yw}}} \le \dfrac{h}{t_w} \le 234\sqrt{\dfrac{k_v}{F_{yw}}}$:

$$C_v = \frac{187\sqrt{k_v/F_{yw}}}{h/t_w}$$

(A-G3-5)

(b) For $\dfrac{h}{t_w} > 234\sqrt{\dfrac{k_v}{F_{yw}}}$:

$$C_v = \frac{44{,}000 k_v}{(h/t_w)^2 F_{yw}}$$

(A-G3-6)

G4. TRANSVERSE STIFFENERS

Transverse stiffeners are not required in plate girders where $h/t_w \le 418/\sqrt{F_{yw}}$, or where the required shear V_u, as determined by structural analysis for the factored loads, is less than or equal to $0.6\phi_v A_w F_{yw} C_v$, where C_v is determined for $k_v = 5$ and $\phi_v = 0.90$. Stiffeners may be required in certain portions of a plate girder to develop the required shear or to satisfy the limitations given in Appendix G1. Transverse stiffeners shall satisfy the requirements of Appendix F2.3.

When designing for tension field action, the stiffener area A_{st} shall not be less than

$$\frac{F_{yw}}{F_{yst}}\left[0.15 D h t_w (1 - C_v)\frac{V_u}{\phi_v V_n} - 18 t_w^2\right] \ge 0$$

(A-G4-1)

where

  F_{yst} = specified yield stress of the stiffener material, ksi
  D = 1 for stiffeners in pairs
      = 1.8 for single angle stiffeners
      = 2.4 for single plate stiffeners

C_v and V_n are defined in Appendix G3, and V_u is the required shear at the location of the stiffener.

G5. FLEXURE-SHEAR INTERACTION

For $0.6\phi V_n \le V_u \le \phi V_n$ ($\phi = 0.90$) and $0.75\phi M_n \le M_u \le \phi M_n$ ($\phi = 0.90$), plate girders with webs designed for tension field action shall satisfy the additional flexure-shear interaction criteria:

$$\frac{M_u}{\phi M_n} + 0.625 \frac{V_u}{\phi V_n} \leq 1.375 \qquad \text{(A-G5-1)}$$

where M_n is the nominal flexural strength of plate girders from Appendix G2 or Section F1, $\phi = 0.90$, and V_n is the nominal shear strength from Appendix G3.

APPENDIX H

MEMBERS UNDER COMBINED FORCES AND TORSION

This appendix provides alternative interaction equations for biaxially loaded I-shaped members with $b_f / d \leq 1.0$ and box-shaped members.

H3. ALTERNATIVE INTERACTION EQUATIONS FOR MEMBERS UNDER COMBINED STRESS

For biaxially loaded I-shaped members with $b_f / d \leq 1.0$ and box-shaped members in braced frames only, the use of the following interaction equations in lieu of Equations H1-1a and H1-1b is permitted. Both Equations A-H3-1 and A-H3-2 shall be satisfied.

$$\left(\frac{M_{ux}}{\phi_b M'_{px}} \right)^\zeta + \left(\frac{M_{uy}}{\phi_b M'_{py}} \right)^\zeta \leq 1.0 \qquad \text{(A-H3-1)}$$

$$\left(\frac{C_{mx} M_{ux}}{\phi_b M'_{nx}} \right)^\eta + \left(\frac{C_{my} M_{uy}}{\phi_b M'_{ny}} \right)^\eta \leq 1.0 \qquad \text{(A-H3-2)}$$

The terms in Equations A-H3-1 and A-H3-2 are determined as follows:

(a) For I-shaped members:

For $b_f / d < 0.5$:

$$\zeta = 1.0$$

For $0.5 \leq b_f / d \leq 1.0$:

$$\zeta = 1.6 - \frac{P_u / P_y}{2[\ln(P_u / P_y)]} \qquad \text{(A-H3-3)}$$

For $b_f / d < 0.3$:

$$\eta = 1.0$$

For $0.3 \leq b_f / d \leq 1.0$:

$$\eta = 0.4 + \frac{P_u}{P_y} + \frac{b_f}{d} \geq 1.0 \qquad \text{(A-H3-4)}$$

where

b_f = flange width, in.
d = member depth, in.
C_m = coefficient applied to the bending term in interaction equation for

prismatic members and dependent on column curvature caused by applied moments, see Section C1.

$$M'_{px} = 1.2M_{px}[1 - (P_u / P_y)] \le M_{px} \qquad \text{(A-H3-5)}$$

$$M'_{py} = 1.2M_{py}[1 - (P_u / P_y)^2] \le M_{py} \qquad \text{(A-H3-6)}$$

$$M'_{nx} = M_{nx}\left(1 - \frac{P_u}{\phi_c P_n}\right)\left(1 - \frac{P_u}{P_{ex}}\right) \qquad \text{(A-H3-7)}$$

$$M'_{ny} = M_{ny}\left(1 - \frac{P_u}{\phi_c P_n}\right)\left(1 - \frac{P_u}{P_{ey}}\right) \qquad \text{(A-H3-8)}$$

(b) For box-section members:

$$\zeta = 1.7 - \frac{P_u / P_y}{\ln(P_u / P_y)} \qquad \text{(A-H3-9)}$$

$$\eta = 1.7 - \frac{P_u / P_y}{\ln(P_u / P_y)} - a\lambda_x\left(\frac{P_u}{P_y}\right)^b > 1.1 \qquad \text{(A-H3-10)}$$

For $P_u / P_y \le 0.4$, $a = 0.06$, and $b = 1.0$;

For $P_u / P_y > 0.4$, $a = 0.15$, and $b = 2.0$;

$$M'_{px} = 1.2M_{px}[1 - P_u / P_y] \le M_{px} \qquad \text{(A-H3-11a)}$$

$$M'_{py} = 1.2M_{py}[1 - P_u / P_y] \le M_{py} \qquad \text{(A-H3-11b)}$$

$$M'_{nx} = M_{nx}\left(1 - \frac{P_u}{\phi_c P_n}\right)\left(1 - \frac{P_u}{P_{ex}}\frac{1.25}{(B/H)^{1/3}}\right) \qquad \text{(A-H3-12)}$$

$$M'_{ny} = M_{ny}\left(1 - \frac{P_u}{\phi_c P_n}\right)\left(1 - \frac{P_u}{P_{ey}}\frac{1.25}{(B/H)^{1/2}}\right) \qquad \text{(A-H3-13)}$$

where

P_n = nominal compressive strength determined in accordance with Section E2, kips

P_u = required axial strength, kips

P_y = compressive yield strength $A_g F_y$, kips

ϕ_b = resistance factor for flexure = 0.90

ϕ_c = resistance factor for compression = 0.85

P_e = Euler buckling strength $A_g F_y / \lambda_c^2$, where λ_c is the column slenderness parameter defined by Equation E2-4, kips

M_u = required flexural strength, kip-in.

M_n = nominal flexural strength, determined in accordance with Section F1, kip-in.

M_p = plastic moment $\le 1.5 F_y S$, kip-in.

B = outside width of box section parallel to major principal axis x, in.

H = outside depth of box section perpendicular to major principal axis x, in.

APPENDIX J

CONNECTIONS, JOINTS, AND FASTENERS

Appendix J2.4 provides the alternative design strength for fillet welds. Appendices J3.8 and J3.9 pertain to the design of slip-critical connections using factored loads.

J2. WELDS

4. Design Strength

In lieu of the constant design strength for fillet welds given in Table J2.5, the following procedure is permitted.

(a) The design strength of a linear weld group loaded in-plane through the center of gravity is $\phi F_w A_w$:

$$F_w = 0.60 F_{EXX}(1.0 + 0.50 \sin^{1.5}\theta)$$

where

ϕ = 0.75
F_w = nominal stress, ksi
F_{EXX} = electrode classification number, i.e., minimum specified strength, ksi
θ = angle of loading measured from the weld longitudinal axis, degrees
A_w = effective area of weld throat, in.^{2}

(b) The design strength of weld elements within a weld group that are loaded in-plane and analyzed using an instantaneous center of rotation method to maintain deformation compatibility and non-linear load deformation behavior of variable angle loaded welds is $\phi F_{wx} A_w$ and $\phi F_{wy} A_w$:

where

F_{wx} = ΣF_{wix}
F_{wy} = ΣF_{wiy}
F_{wi} = $0.60 F_{EXX}(1.0 + 0.50 \sin^{1.5}\theta) f(p)$
$f(p)$ = $[p(1.9 - 0.9p)]^{0.3}$
ϕ = 0.75
F_{wi} = nominal stress in any ith weld element, ksi
F_{wix} = x component of stress F_{wi}
F_{wiy} = y component of stress F_{wi}
p = Δ_i / Δ_m, ratio of element i deformation to its deformation at maximum stress
Δ_m = $0.209(\theta + 2)^{-0.32}D$, deformation of weld element at maximum stress, in.

Δ_i = deformation of weld elements at intermediate stress levels, linearly proportioned to the critical deformation based on distance from the instantaneous center of rotation, r_i, in.

     = $r_i \Delta_u / r_{crit}$

Δ_u = $1.087(\theta + 6)^{-0.65} D \leq 0.17D$, deformation of weld element at ultimate stress (fracture), usually in element furthest from instantaneous center of rotation, in.

D = leg size of the fillet weld, in.

r_{crit} = distance from instantaneous center of rotation to weld element with minimum Δ_u / r_i ratio

J3. BOLTS AND THREADED PARTS

8. High-Strength Bolts in Slip-Critical Connections

8b. Slip-Critical Connections Designed at Factored Loads

It is permissible to proportion slip-critical connections at factored loads. The design slip resistance for use at factored loads, ϕR_{str}, shall equal or exceed the required force due to the factored loads, where:

$$R_{str} = 1.13 \mu T_m N_b N_s \qquad \text{(A-J3-1)}$$

where:

T_m = minimum fastener tension given in Table J3.1, kips
N_b = number of bolts in the joint
N_s = number of slip planes
μ = mean slip coefficient for Class A, B, or C surfaces, as applicable, or as established by tests

    (a) For Class A surfaces (unpainted clean mill scale steel surfaces or surfaces with Class A coating on blast-cleaned steel), $\mu = 0.33$

    (b) For Class B surfaces (unpainted blast-cleaned steel surfaces or surfaces with Class B coatings on blast-cleaned steel), $\mu = 0.50$

    (c) For Class C surfaces (hot-dip galvanized and roughened surfaces), $\mu = 0.40$

ϕ = resistance factor

    (a) For standard holes, $\phi = 1.0$

    (b) For oversize and short-slotted holes, $\phi = 0.85$

    (c) For long-slotted holes transverse to the direction of load, $\phi = 0.70$

    (d) For long-slotted holes parallel to the direction of load, $\phi = 0.60$

9. Combined Tension and Shear in Slip-Critical Connections

9b. Slip-Critical Connections Designed at Factored Loads

When using factored loads as the basis for design of slip-critical connections subject to applied tension, T, that reduces the net clamping force, the slip resistance ϕR_{str} according to Appendix J3.8b shall be multiplied by the following factor in which T_u is the required tensile strength at factored loads:

$$[1 - T_u / (1.13 T_m N_b)] \qquad \text{(A-J3-2)}$$

APPENDIX K

CONCENTRATED FORCES, PONDING, AND FATIGUE

Appendix K2 provides an alternative determination of roof stiffness. Appendix K4 pertains to members and connections subjected to fatigue loading.

K2. PONDING

The provisions of this Appendix are permitted to be used when a more exact determination of flat roof framing stiffness is needed than that given by the provision of Section K2 that $C_p + 0.9C_s \leq 0.25$.

For any combination of primary and secondary framing, the stress index is computed as

$$U_p = \left(\frac{F_y - f_o}{f_o}\right)_p \text{ for the primary member} \tag{A-K2-3}$$

$$U_s = \left(\frac{F_y - f_o}{f_o}\right)_s \text{ for the secondary member} \tag{A-K2-4}$$

where

f_o = the stress due to $1.2D + 1.2R$ (D = nominal dead load, R = nominal load due to rain water or ice exclusive of the ponding contribution)*

Enter Figure A-K2.1 at the level of the computed stress index U_p determined for the primary beam; move horizontally to the computed C_s value of the secondary beams and then downward to the abscissa scale. The combined stiffness of the primary and secondary framing is sufficient to prevent ponding if the flexibility constant read from this latter scale is more than the value of C_p computed for the given primary member; if not, a stiffer primary or secondary beam, or combination of both, is required. In the above,

$$C_p = \frac{32L_sL_p^4}{10^7I_p}$$

$$C_s = \frac{32SL_s^4}{10^7I_s}$$

---

* Depending upon geographic location, this loading should include such amount of snow as might also be present, although ponding failures have occurred more frequently during torrential summer rains when the rate of precipitation exceeded the rate of drainage runoff and the resulting hydraulic gradient over large roof areas caused substantial accumulation of water some distance from the eaves. A load factor of 1.2 shall be used for loads resulting from these phenomena.

where

L_p = column spacing in direction of girder (length of primary members), ft
L_s = column spacing perpendicular to direction of girder (length of secondary members), ft
S = spacing of secondary members, ft
I_p = moment of inertia of primary members, in.^{4}
I_s = moment of inertia of secondary members, in.^{4}

A similar procedure must be followed using Figure A-K2.2.

Roof framing consisting of a series of equally spaced wall-bearing beams is considered as consisting of secondary members supported on an infinitely stiff

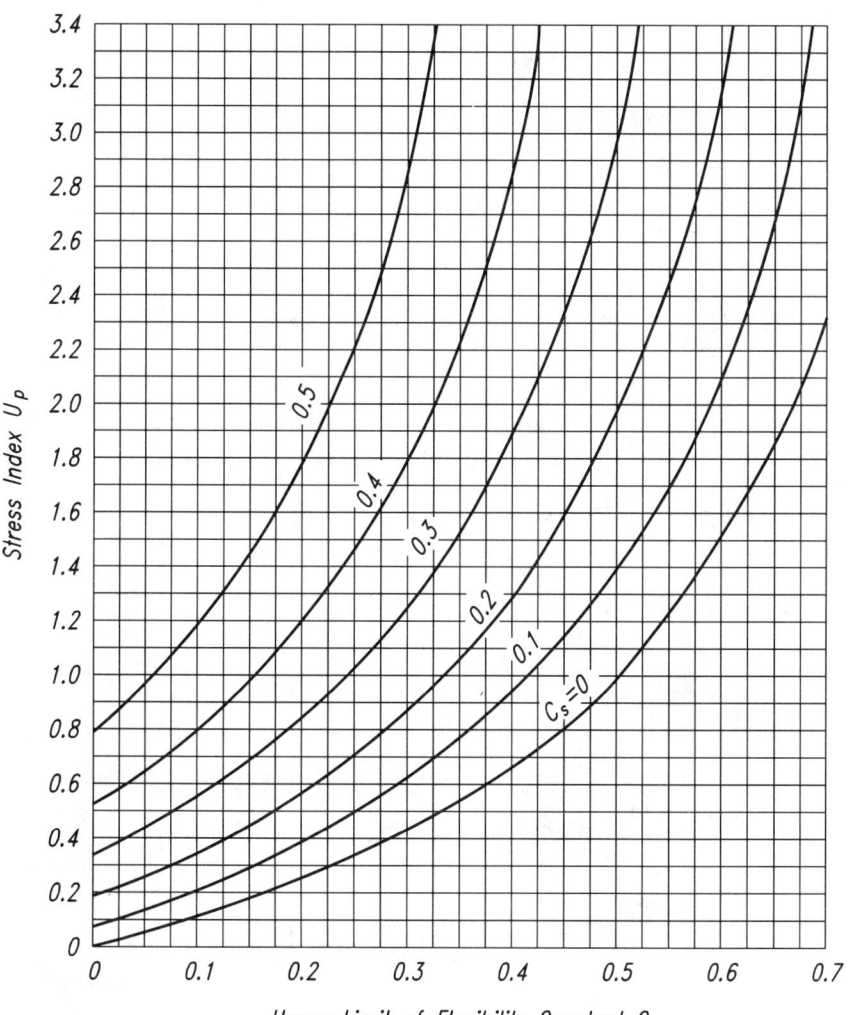

Fig. A-K2.1. Limiting flexibility coefficient for the primary systems.

primary member. For this case, enter Figure A-K2.2 with the computed stress index U_s. The limiting value of C_s is determined by the intercept of a horizontal line representing the U_s value and the curve for $C_p = 0$.

The ponding deflection contributed by a metal deck is usually such a small part of the total ponding deflection of a roof panel that it is sufficient merely to limit its moment of inertia (per foot of width normal to its span) to 0.000025 times the fourth power of its span length. However, the stability against ponding of a roof consisting of a metal roof deck of relatively slender depth-span ratio, spanning between beams supported directly on columns, may need to be checked. This can be done using Figure A-K2.1 or A-K2.2 using as C_s the flexibility constant for a one-foot width of the roof deck ($S = 1.0$).

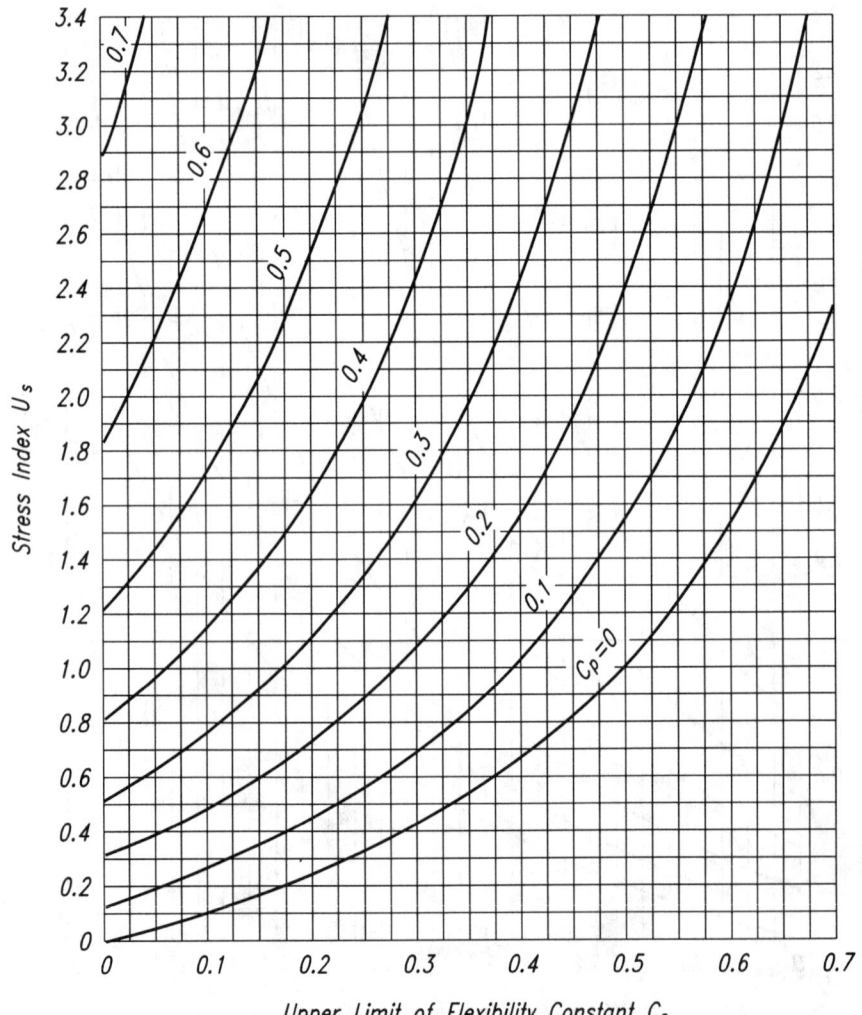

Fig. A-K2.2. Limiting flexibility coefficient for the secondary systems.

TABLE A-K3.1
Number of Loading Cycles

Loading Condition	From	To
Loading Condition	From	To
1	20,000 [a]	100,000 [b]
2	100,000	500,000 [c]
3	500,000	2,000,000 [d]
4	Over 2,000,000	

[a] Approximately equivalent to two applications every day for 25 years.
[b] Approximately equivalent to 10 applications every day for 25 years.
[c] Approximately equivalent to 50 applications every day for 25 years.
[d] Approximately equivalent to 200 applications every day for 25 years.

Since the shear rigidity of the web system of steel joists and trusses is less than that of a solid plate, their moment of inertia shall be taken as 85 percent of their chords.

K3. FATIGUE

Members and connections subject to fatigue loading shall be proportioned in accordance with the provisions of this Appendix.

Fatigue, as used in this Specification, is defined as the damage that may result in fracture after a sufficient number of fluctuations of stress. Stress range is defined as the magnitude of these fluctuations. In the case of a stress reversal, the stress range shall be computed as the numerical sum of maximum repeated tensile and compressive stresses or the sum of maximum shearing stresses of opposite direction at a given point, resulting from differing arrangement of live load.

1. Loading Conditions; Type and Location of Material

In the design of members and connections subject to repeated variation of live load, consideration shall be given to the number of stress cycles, the expected range of stress, and the type and location of member or detail.

Loading conditions shall be classified according to Table A-K3.1.

The type and location of material shall be categorized according to Table A-K3.2.

2. Design Stress Range

The maximum range of stress at service loads shall not exceed the design stress range specified in Table A-K3.3.

3. Design Strength of Bolts in Tension

When subject to tensile fatigue loading, fully tensioned A325 or A490 bolts shall be designed for the combined tensile design strength due to combined external and prying forces in accordance with Table A-K3.4.

TABLE A-K3.2
Type and Location of Material

General Condition	Situation	Kind of Stress [a]	Stress Category (see Table A-K3.3)	Illustrative Example Nos. (see Fig. A-K3.1) [b]
Plain Material	Base metal with rolled or cleaned surface. Flame-cut edges with ANSI smoothness of 1,000 or less	T or Rev.	A	1,2
Built-up Members	Base metal and weld metal in members without attachments, built-up plates or shapes connected by continuous full-penetration groove welds or by continuous fillet welds parallel to the direction of applied stress	T or Rev.	B	3,4,5,6
	Base metal and weld metal in members without attachments, built-up plates, or shapes connected by full-penetration groove welds with backing bars not removed, or by partial-penetration groove welds parallel to the direction of applied stress	T or Rev.	B′	3,4,5,6
	Base metal at toe of welds on girder webs or flanges adjacent to welded transverse stiffeners	T or Rev.	C	7
	Base metal at ends of partial length welded coverplates narrower than the flange having square or tapered ends, with or without welds across the ends or wider than flange with welds across the ends			
	Flange thickness ≤ 0.8 in.	T or Rev.	E	5
	Flange thickness > 0.8 in.	T or Rev.	E′	5
	Base metal at end of partial length welded coverplates wider than the flange without welds across the ends		E′	5

[a] "T" signifies range in tensile stress only; "Rev." signifies a range involving reversal of tensile or compressive stress; "S" signifies range in shear, including shear stress reversal.

[b] These examples are provided as guidelines and are not intended to exclude other reasonably similar situations.

[c] Allowable fatigue stress range for transverse partial-penetration and transverse fillet welds is a function of the effective throat, depth of penetration, and plate thickness. See Frank and Fisher, *Journal of the Structural Division*, Vol. 105 No. ST9, Sept. 1979.

TABLE A-K3.2 (cont'd)
Type and Location of Material

General Condition	Situation	Kind of Stress [a]	Stress Category (see Table A-K3.3)	Illustrative Example Nos. (see Fig. A-K3.1) [b]
Groove Welds	Base metal and weld metal at full-penetration groove welded splices of parts of similar cross section ground flush, with grinding in the direction of applied stress and with weld soundness established by radiographic or ultrasonic inspection in accordance with the requirements of 9.25.2 or 9.25.3 of AWS D1.1	T or Rev.	B	10,11
	Base metal and weld metal at full-penetration groove welded splices at transitions in width or thickness, with welds ground to provide slopes no steeper than 1 to 2½ with grinding in the direction of applied stress, and with weld soundness established by radiographic or ultrasonic inspection in accordance with the requirements of 9.25.2 or 9.25.3 of AWS D1.1			
	A514 base metal	T or Rev.	B′	12,13
	Other base metals	T or Rev.	B	12,13
	Base metal and weld metal at full-penetration groove welded splices, with or without transitions having slopes no greater than 1 to 2½ when reinforcement is not removed but weld soundness is established by radiographic or ultrasonic inspection in accordance with requirements of 9.25.2 or 9.25.3 of AWS D1.1	T or Rev.	C	10,11,12,13
Partial-Penetration Groove Welds	Weld metal of partial-penetration transverse groove welds, based on effective throat area of the weld or welds	T or Rev.	F [c]	
Fillet-welded Connections	Base metal at intermittent fillet welds	T or Rev.	E	

TABLE A-K3.2 (cont'd)
Type and Location of Material

General Condition	Situation	Kind of Stress [a]	Stress Category (see Table A-K3.3)	Illustrative Example Nos. (see Fig. A-K3.1) [b]
Fillet-welded Connections (Continued)	Base metal at junction of axially loaded members with fillet-welded end connections. Welds shall be disposed about the axis of the member so as to balance weld stresses			
	$b \leq 1$ in.	T or Rev.	E	17,18
	$b > 1$ in.	T or Rev.	E′	17,18
	Base metal at members connected with transverse fillet welds			
	$b \leq \frac{1}{2}$-in.	T or Rev.	C	20,21
	$b > \frac{1}{2}$-in.		See Note	
Fillet Welds	Weld metal of continuous or intermittent longitudinal or transverse fillet welds	S	F [c]	15,17,18 20,21
Plug or Slot Welds	Base metal at plug or slot welds	T or Rev.	E	27
	Shear on plug or slot welds	S	F	27
Mechanically Fastened Connections	Base metal at gross section of high-strength bolted slip-critical connections, except axially loaded joints which induce out-of-plane bending in connected material	T or Rev.	B	8
	Base metal at net section of other mechanically fastened joints	T or Rev.	D	8,9
	Base metal at net section of fully tensioned high-strength, bolted-bearing connections	T or Rev.	B	8,9
Eyebar or Pin Plates	Base metal at net section of eyebar head or pin plate	T or Rev.	E	28,29
Attachments	Base metal at details attached by full-penetration groove welds subject to longitudinal and/or transverse loading when the detail embodies a transition radius R with the weld termination ground smooth and for transverse loading, the weld soundness established by radiographic or ultrasonic inspection in accordance with 9.25.2 or 9.25.3 of AWS D1.1			

TABLE A-K3.2 (cont'd)
Type and Location of Material

General Condition	Situation	Kind of Stress [a]	Stress Category (see Table A-K3.3)	Illustrative Example Nos. (see Fig. A-K3.1) [b]
Attachments (Continued)	Longitudinal loading			
	$R >$ 24 in.	T or Rev.	B	14
	24 in. $> R >$ 6 in.	T or Rev.	C	14
	6 in. $> R >$ 2 in.	T or Rev.	D	14
	2 in. $> R$	T or Rev.	E	14
	Detail base metal for transverse loading: equal thickness and reinforcement removed			
	$R >$ 24 in.	T or Rev.	B	14
	24 in. $> R >$ 6 in.	T or Rev.	C	14
	6 in. $> R >$ 2 in.	T or Rev.	D	14
	2 in. $> R$	T or Rev.	E	14,15
	Detail base metal for transverse loading: equal thickness and reinforcement not removed			
	$R >$ 24 in.	T or Rev.	C	14
	24 in. $> R >$ 6 in.	T or Rev.	C	14
	6 in. $> R >$ 2 in.	T or Rev.	D	14
	2 in. $> R$	T or Rev.	E	14,15
	Detail base metal for transverse loading: unequal thickness and reinforcement removed			
	$R >$ 2 in.	T or Rev.	D	14
	2 in. $> R$	T or Rev.	E	14,15
	Detail base metal for transverse loading: Unequal thickness and reinforcement not removed			
	All R	T or Rev.	E	14,15
	Detail base metal for transverse loading			
	$R >$ 6 in.	T or Rev.	C	19
	6 in. $> R >$ 2 in.	T or Rev.	D	19
	2 in. $> R$	T or Rev.	E	19
	Base metal at detail attached by full-penetration groove welds subject to longitudinal loading			
	$2 < a < 12b$ or 4 in.	T or Rev.	D	15
	$a > 12b$ or 4 in. when $b \leq 1$ in.	T or Rev.	E	15
	$a > 12b$ or 4 in. when $b > 1$ in.	T or Rev.	E'	15

TABLE A-K3.2 (cont'd)
Type and Location of Material

General Condition	Situation	Kind of Stress [a]	Stress Category (see Table A-K3.3)	Illustrative Example Nos. (see Fig. A-K3.1) [b]
Attachments (Continued)	Base metal at detail attached by fillet welds or partial-penetration groove welds subject to longitudinal loading			
	$a < 2$ in.	T or Rev.	C	15,23,24 25,26
	2 in. $< a < 12b$ or 4 in.	T or Rev.	D	15,23,24,26
	$a > 12b$ or 4 in. when $b \le 1$ in.	T or Rev.	E	15,23,24,26
	$a > 12b$ or 4 in. when $b > 1$ in.	T or Rev.	E'	15,23,24,26
	Base metal attached by fillet welds or partial-penetration groove welds subjected to longitudinal loading when the weld termination embodies a transition radius with the weld termination ground smooth			
	$R > 2$ in. $R \le 2$ in.	T or Rev. T or Rev.	D E	19 19
	Fillet-welded attachments where the weld termination embodies a transition radius, weld termination ground smooth, and main material subject to longitudinal loading Detail base metal for transverse loading:			
	$R > 2$ in. $R < 2$ in.	T or Rev. T or Rev.	D E	19 19
	Base metal at stud-type shear connector attached by fillet weld or automatic end weld	T or Rev.	C	22
	Shear stress on nominal area of stud-type shear connectors	S	F	

TABLE A-K 3.3
Design Stress Range, ksi

Category (From Table A-K3.2)	Loading Condition 1	Loading Condition 2	Loading Condition 3	Loading Condition 4
A	63	37	24	24
B	49	29	18	16
B′	39	23	15	12
C	35	21	13	10 [a]
D	28	16	10	7
E	22	13	8	4.5
E′	16	9.2	5.8	2.6
F	15	12	9	8

[a] Flexural stress range of 12 ksi permitted at toe of stiffener welds or flanges.

TABLE A-K3.4
Design Strength of A325 or A490 Bolts
Subject to Tension

Number of cycles	Design strength
Not more than 20,000	As specified in Section J3
From 20,000 to 500,000	$0.30\, A_b F_u$ [a]
More than 500,000	$0.25\, A_b F_u$ [a]

[a] At service loads.

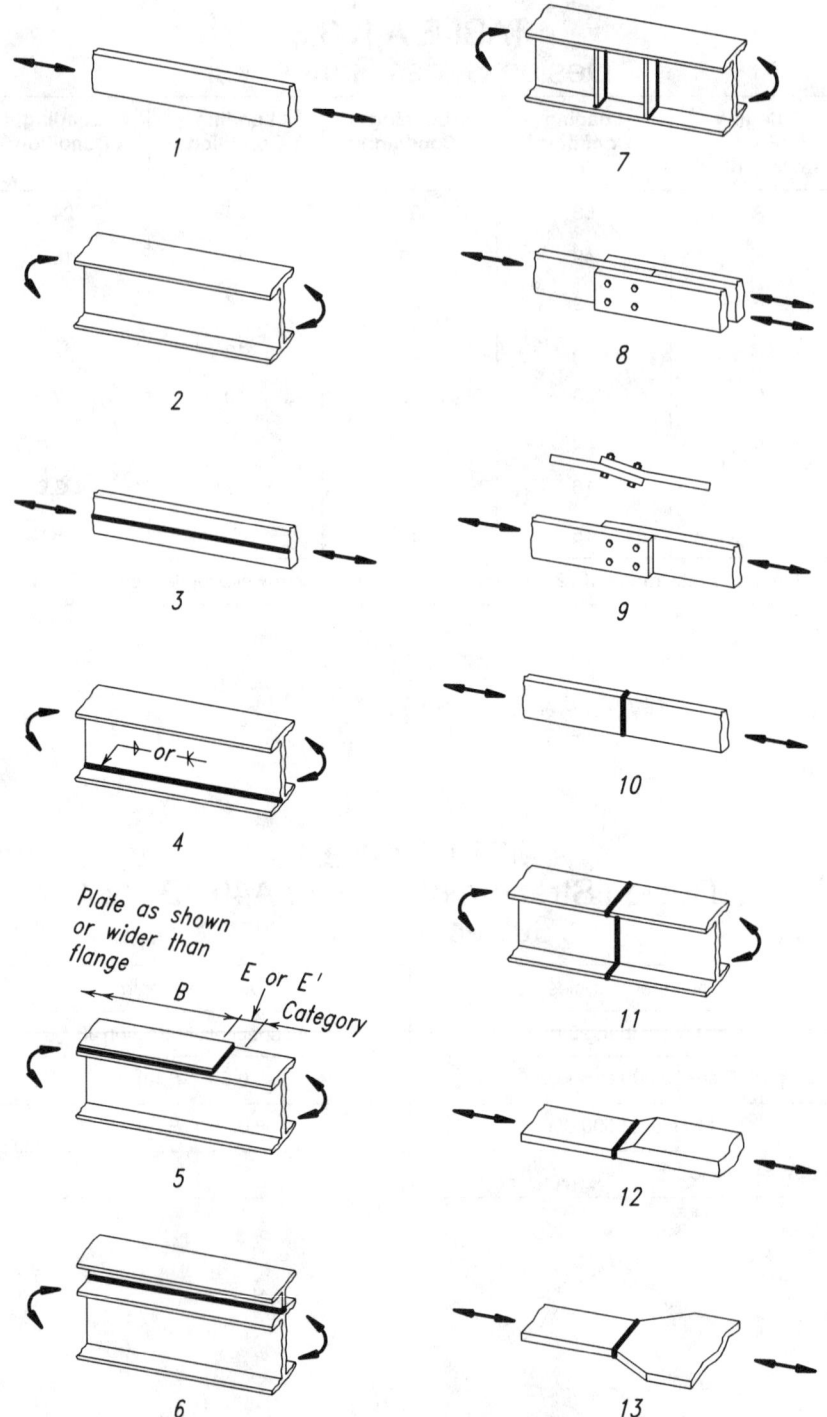

Fig. A-K3.1. Illustrative examples.

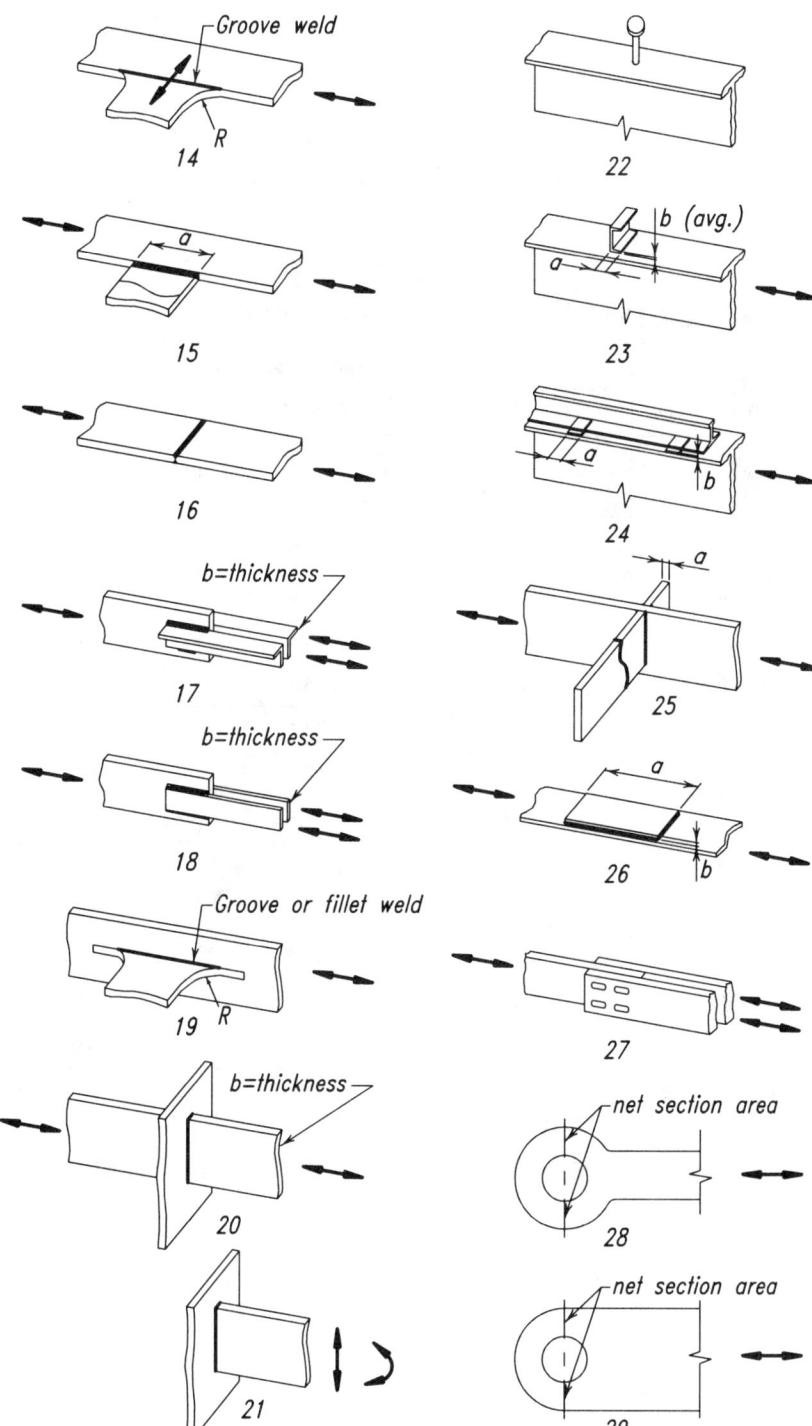

Fig. A-K3.1. Illustrative examples (cont.).

NUMERICAL VALUES

TABLE 1
Design Strength as a Function of F_y

F_y (ksi)	Design Stress (ksi)		
	$0.54F_y$ [a]	$0.85F_y$ [b]	$0.90F_y$ [c]
33	17.8	28.1	29.7
35	18.9	29.8	31.5
36	19.4	30.6	32.4
40	21.6	34.0	36.0
42	22.7	35.7	37.8
45	24.3	38.3	40.5
46	24.8	39.1	41.4
50	27.0	42.5	45.0
55	29.7	46.8	49.5
60	32.4	51.0	54.0
65	35.1	55.3	58.5
70	37.8	59.5	63.0
90	48.6	76.5	81.0
100	54.0	85.0	90.0

[a] See Section F2, Equation F2-1
[b] See Section E2, Equation E2-1
[c] See Section D1, Equation D1-1

TABLE 2
Design Strength as a Function of F_y

Item	ASTM Desig- nation	F_y (ksi)	F_u (ksi)	Design Strength (ksi)				
				Connection Part of Designated Steel		Bolt or Threaded Part of Designated Steel		
				Tension 0.75 × F_u [a]	Bearing 0.75 × 2.4F_u [b]	Tension 0.75 × 0.75F_u [c]	Shear 0.75 × 0.40F_u [d]	Shear 0.75 × 0.50F_u [e]
Shapes, Plates, Bars, Sheet and Tubing or Threaded parts	A36	36	58-80	43.5	104	32.6	17.4	21.8
	A53	35	60	45.0	108	—	—	—
	A242⌉ A588⌋	50 42 40	70 63 60	52.5 47.3 45.0	126 113 108	39.4 35.4 33.8	21.0 18.9 18.0	26.3 23.2 22.5
	A500	33/39 [f] 42/46 [f] 46/50 [f]	45 58 62	33.8 43.5 46.5	81 104 112	— — —	— — —	— — —
	A501	36	58	43.5	104	—	—	—
	A529	42	60-85	45.0	108	33.8	18.0	22.5
	A570	40 42	55 58	41.3 43.5	99 104	— —	— —	— —
	A572	42 50 60 65	60 65 75 80	45.0 48.8 56.3 60.0	108 117 135 144	33.8 36.6 42.2 45.0	18.0 19.5 22.5 24.0	22.5 24.4 28.1 30.0
	A514	100 90	110-130 100-130	82.5 75.0	198 180	61.9 56.3	33.0 30.0	41.3 37.5
	A606	45 50	65 70	48.8 52.5	117 126	— —	— —	— —
	A607	45 50 55 60 65 70	60 65 70 75 80 85	45.0 48.8 52.5 56.3 60.0 63.8	108 117 126 135 144 153	— — — — — —	— — — — — —	— — — — — —
	A618	50 50	70 65	52.5 48.8	126 117	— —	— —	— —

TABLE 2 (cont'd)
Design Strength as a Function of F_u

Item				Design Strength (ksi)				
				Connection Part of Designated Steel		**Bolt of Threaded Part of Designated Steel**		
	ASTM Desig- nation	F_y **(ksi)**	F_u **(ksi)**	**Tension** $0.75 \times$ F_u [a]	**Bearing** $0.75 \times$ $2.4F_u$ [b]	**Tension** $0.75 \times$ $0.75F_u$ [c]	**Shear** $0.75 \times$ $0.40F_u$ [d]	**Shear** $0.75 \times$ $0.50F_u$ [e]
Bolts	A449	92	120	—	—	67.5	36.0	45.0
		81	105	—	—	59.1	31.5	39.4
		58	90	—	—	50.6	27.0	33.8

[a] On effective net area, see Sections D1, J5.2.
[b] Produced by fastener in shear, see Section J3.10. Note that smaller maximum design bearing stresses, as a function of hole type spacing, are given.
[c] On nominal body area, see Table J3.2.
[d] Threads not excluded from shear plane, see Table J3.2.
[e] Threads excluded from shear plane, see Table J3.2.
[f] Smaller value for circular shapes, larger for square or rectangular shapes.
Note: For dimensional and size limitations, see the appropriate ASTM Specification.

TABLE 3-36
Design Stress for Compression Members of 36 ksi Specified Yield Stress Steel, $\phi_c = 0.85$[a]

$\frac{Kl}{r}$	$\phi_c F_{cr}$ ksi	$\frac{Kl}{r}$	$\phi_c F_{cr}$ ksi	$\frac{Kl}{r}$	$\phi_c F_{cr}$ ksi	$\frac{Kl}{r}$	$\phi_c F_{cr}$ ksi	$\frac{Kl}{r}$	$\phi_c F_{cr}$ ksi
1	30.60	41	28.01	81	21.66	121	14.16	161	8.23
2	30.59	42	27.89	82	21.48	122	13.98	162	8.13
3	30.59	43	27.76	83	21.29	123	13.80	163	8.03
4	30.57	44	27.64	84	21.11	124	13.62	164	7.93
5	30.56	45	27.51	85	20.92	125	13.44	165	7.84
6	30.54	46	27.37	86	20.73	126	13.27	166	7.74
7	30.52	47	27.24	87	20.54	127	13.09	167	7.65
8	30.50	48	27.11	88	20.36	128	12.92	168	7.56
9	30.47	49	26.97	89	20.17	129	12.74	169	7.47
10	30.44	50	26.83	90	19.98	130	12.57	170	7.38
11	30.41	51	26.68	91	19.79	131	12.40	171	7.30
12	30.37	52	26.54	92	19.60	132	12.23	172	7.21
13	30.33	53	26.39	93	19.41	133	12.06	173	7.13
14	30.29	54	26.25	94	19.22	134	11.88	174	7.05
15	30.24	55	26.10	95	19.03	135	11.71	175	6.97
16	30.19	56	25.94	96	18.84	136	11.54	176	6.89
17	30.14	57	25.79	97	18.65	137	11.37	177	6.81
18	30.08	58	25.63	98	18.46	138	11.20	178	6.73
19	30.02	59	25.48	99	18.27	139	11.04	179	6.66
20	29.96	60	25.32	100	18.08	140	10.89	180	6.59
21	29.90	61	25.16	101	17.89	141	10.73	181	6.51
22	29.83	62	24.99	102	17.70	142	10.58	182	6.44
23	29.76	63	24.83	103	17.51	143	10.43	183	6.37
24	29.69	64	24.67	104	17.32	144	10.29	184	6.30
25	29.61	65	24.50	105	17.13	145	10.15	185	6.23
26	29.53	66	24.33	106	16.94	146	10.01	186	6.17
27	29.45	67	24.16	107	16.75	147	9.87	187	6.10
28	29.36	68	23.99	108	16.56	148	9.74	188	6.04
29	29.28	69	23.82	109	16.37	149	9.61	189	5.97
30	29.18	70	23.64	110	16.19	150	9.48	190	5.91
31	29.09	71	23.47	111	16.00	151	9.36	191	5.85
32	28.99	72	23.29	112	15.81	152	9.23	192	5.79
33	28.90	73	23.12	113	15.63	153	9.11	193	5.73
34	28.79	74	22.94	114	15.44	154	9.00	194	5.67
35	28.69	75	22.76	115	15.26	155	8.88	195	5.61
36	28.58	76	22.58	116	15.07	156	8.77	196	5.55
37	28.47	77	22.40	117	14.89	157	8.66	197	5.50
38	28.36	78	22.22	118	14.70	158	8.55	198	5.44
39	28.25	79	22.03	119	14.52	159	8.44	199	5.39
40	28.13	80	21.85	120	14.34	160	8.33	200	5.33

[a] When element width-to-thickness ratio exceeds λ_r, see Appendix B5.3.

TABLE 3-50
Design Stress for Compression Members of 50 ksi Specified Yield Stress Steel, $\phi_c = 0.85$[a]

$\frac{Kl}{r}$	$\phi_c F_{cr}$ ksi	$\frac{Kl}{r}$	$\phi_c F_{cr}$ ksi	$\frac{Kl}{r}$	$\phi_c F_{cr}$ ksi	$\frac{Kl}{r}$	$\phi_c F_{cr}$ ksi	$\frac{Kl}{r}$	$\phi_c F_{cr}$ ksi
1	42.50	41	37.59	81	26.31	121	14.57	161	8.23
2	42.49	42	37.36	82	26.00	122	14.33	162	8.13
3	42.47	43	37.13	83	25.68	123	14.10	163	8.03
4	42.45	44	36.89	84	25.37	124	13.88	164	7.93
5	42.42	45	36.65	85	25.06	125	13.66	165	7.84
6	42.39	46	36.41	86	24.75	126	13.44	166	7.74
7	42.35	47	36.16	87	24.44	127	13.23	167	7.65
8	42.30	48	35.91	88	24.13	128	13.02	168	7.56
9	42.25	49	35.66	89	23.82	129	12.82	169	7.47
10	42.19	50	35.40	90	23.51	130	12.62	170	7.38
11	42.13	51	35.14	91	23.20	131	12.43	171	7.30
12	42.05	52	34.88	92	22.89	132	12.25	172	7.21
13	41.98	53	34.61	93	22.58	133	12.06	173	7.13
14	41.90	54	34.34	94	22.28	134	11.88	174	7.05
15	41.81	55	34.07	95	21.97	135	11.71	175	6.97
16	41.71	56	33.79	96	21.67	136	11.54	176	6.89
17	41.61	57	33.51	97	21.36	137	11.37	177	6.81
18	41.51	58	33.23	98	21.06	138	11.20	178	6.73
19	41.39	59	32.95	99	20.76	139	11.04	179	6.66
20	41.28	60	32.67	100	20.46	140	10.89	180	6.59
21	41.15	61	32.38	101	20.16	141	10.73	181	6.51
22	41.02	62	32.09	102	19.86	142	10.58	182	6.44
23	40.89	63	31.80	103	19.57	143	10.43	183	6.37
24	40.75	64	31.50	104	19.28	144	10.29	184	6.30
25	40.60	65	31.21	105	18.98	145	10.15	185	6.23
26	40.45	66	30.91	106	18.69	146	10.01	186	6.17
27	40.29	67	30.61	107	18.40	147	9.87	187	6.10
28	40.13	68	30.31	108	18.12	148	9.74	188	6.04
29	39.97	69	30.01	109	17.83	149	9.61	189	5.97
30	39.79	70	29.70	110	17.55	150	9.48	190	5.91
31	39.62	71	29.40	111	17.27	151	9.36	191	5.85
32	39.43	72	29.11	112	16.99	152	9.23	192	5.79
33	39.25	73	28.79	113	16.71	153	9.11	193	5.73
34	39.06	74	28.48	114	16.42	154	9.00	194	5.67
35	38.86	75	28.17	115	16.13	155	8.88	195	5.61
36	38.66	76	27.86	116	15.86	156	8.77	196	5.55
37	38.45	77	27.55	117	15.59	157	8.66	197	5.50
38	38.24	78	27.24	118	15.32	158	8.55	198	5.44
39	38.03	79	26.93	119	15.07	159	8.44	199	5.39
40	37.81	80	26.62	120	14.82	160	8.33	200	5.33

[a] When element width-to-thickness ratio exceeds λ_r, see Appendix B5.3.

TABLE 4
Values of $\phi_c F_{cr}/F_y$, $\phi_c = 0.85$
for Determining Design Stress for Compression
Members for Steel of Any Yield Stress[a]

λ_c	$\phi_c F_{cr}/F_y$	λ_c	$\phi_c F_{cr}/F_y$	λ_c	$\phi_c F_{cr}/F_y$	λ_c	$\phi_c F_{cr}/F_y$
0.02	0.850	0.82	0.641	1.62	0.284	2.42	0.127
0.04	0.849	0.84	0.632	1.64	0.277	2.44	0.125
0.06	0.849	0.86	0.623	1.66	0.271	2.46	0.123
0.08	0.848	0.88	0.614	1.68	0.264	2.48	0.121
0.10	0.846	0.90	0.605	1.70	0.258	2.50	0.119
0.12	0.845	0.92	0.596	1.72	0.252	2.52	0.117
0.14	0.843	0.94	0.587	1.74	0.246	2.54	0.116
0.16	0.841	0.96	0.578	1.76	0.241	2.56	0.114
0.18	0.839	0.98	0.568	1.78	0.235	2.58	0.112
0.20	0.836	1.00	0.559	1.80	0.230	2.60	0.110
0.22	0.833	1.02	0.550	1.82	0.225	2.62	0.109
0.24	0.830	1.04	0.540	1.84	0.220	2.64	0.107
0.26	0.826	1.06	0.531	1.86	0.215	2.66	0.105
0.28	0.823	1.08	0.521	1.88	0.211	2.68	0.104
0.30	0.819	1.10	0.512	1.90	0.206	2.70	0.102
0.32	0.814	1.12	0.503	1.92	0.202	2.72	0.101
0.34	0.810	1.14	0.493	1.94	0.198	2.74	0.099
0.36	0.805	1.16	0.484	1.96	0.194	2.76	0.098
0.38	0.800	1.18	0.474	1.98	0.190	2.78	0.096
0.40	0.795	1.20	0.465	2.00	0.186	2.80	0.095
0.42	0.789	1.22	0.456	2.02	0.183	2.82	0.094
0.44	0.784	1.24	0.446	2.04	0.179	2.84	0.092
0.46	0.778	1.26	0.437	2.06	0.176	2.86	0.091
0.48	0.772	1.28	0.428	2.08	0.172	2.88	0.090
0.50	0.765	1.30	0.419	2.10	0.169	2.90	0.089
0.52	0.759	1.32	0.410	2.12	0.166	2.92	0.087
0.54	0.752	1.34	0.401	2.14	0.163	2.94	0.086
0.56	0.745	1.36	0.392	2.16	0.160	2.96	0.085
0.58	0.738	1.38	0.383	2.18	0.157	2.98	0.084
0.60	0.731	1.40	0.374	2.20	0.154	3.00	0.083
0.62	0.724	1.42	0.365	2.22	0.151	3.02	0.082
0.64	0.716	1.44	0.357	2.24	0.149	3.04	0.081
0.66	0.708	1.46	0.348	2.26	0.146	3.06	0.080
0.68	0.700	1.48	0.339	2.28	0.143	3.08	0.079
0.70	0.692	1.50	0.331	2.30	0.141	3.10	0.078
0.72	0.684	1.52	0.323	2.32	0.138	3.12	0.077
0.74	0.676	1.54	0.314	2.34	0.136	3.14	0.076
0.76	0.667	1.56	0.306	2.36	0.134	3.16	0.075
0.78	0.659	1.58	0.299	2.38	0.132	3.18	0.074
0.80	0.650	1.60	0.291	2.40	0.129	3.20	0.073

[a] When element width-to-thickness ratios exceed λ_r, see Appendix B5.3.
Values of $\lambda_c > 2.24$ exceed Kl/r of 200 for $F_y = 36$
Values of $\lambda_c > 2.64$ exceed Kl/r of 200 for $F_y = 50$

NUMERICAL VALUES

TABLE 5
Values of Kl / r for $F_y = 36$ and 50 ksi

λ_c	Kl/r $F_y = 36$	Kl/r $F_y = 50$	λ_c	Kl/r $F_y = 36$	Kl/r $F_y = 50$
0.02	1.8	1.5	0.82	73.1	62.0
0.04	3.6	3.0	0.84	74.9	63.6
0.06	4.3	4.5	0.86	76.7	65.1
0.08	7.1	6.1	0.88	78.5	66.6
0.10	8.9	7.6	0.90	80.2	68.1
0.12	10.7	9.1	0.92	82.0	69.6
0.14	12.5	10.6	0.94	83.8	71.1
0.16	14.3	12.1	0.96	85.6	72.6
0.18	16.0	13.6	0.98	87.4	74.1
0.20	17.8	15.1	1.00	89.2	75.7
0.22	19.6	16.6	1.02	90.9	77.2
0.24	21.4	18.2	1.04	92.7	78.7
0.26	23.2	19.7	1.06	94.5	80.2
0.28	25.0	21.2	1.08	96.3	81.7
0.30	26.7	22.7	1.10	98.1	83.2
0.32	28.5	24.2	1.12	99.9	84.7
0.34	30.3	25.7	1.14	101.6	86.3
0.36	32.1	27.2	1.16	103.4	87.8
0.38	33.9	28.8	1.18	105.2	89.3
0.40	35.7	30.3	1.20	107.0	90.8
0.42	37.4	31.8	1.22	108.8	92.3
0.44	39.2	33.3	1.24	110.6	93.8
0.46	41.0	34.8	1.26	112.3	95.3
0.48	42.8	36.3	1.28	114.1	96.8
0.50	44.6	37.8	1.30	115.9	98.4
0.52	46.4	39.3	1.32	117.7	99.9
0.54	48.1	40.9	1.34	119.5	101.4
0.56	49.9	42.4	1.36	121.3	102.9
0.58	51.7	43.9	1.38	123.0	104.4
0.60	53.5	45.4	1.40	124.8	105.9
0.62	55.3	46.9	1.42	126.6	107.4
0.64	57.1	48.4	1.44	128.4	108.9
0.66	58.8	49.9	1.46	130.2	110.5
0.68	60.6	51.4	1.48	132.0	112.0
0.70	62.4	53.0	1.50	133.7	113.5
0.72	64.2	54.5	1.52	135.5	115.0
0.74	66.0	56.0	1.54	137.3	116.5
0.76	67.8	57.5	1.56	139.1	118.0
0.78	69.5	59.0	1.58	140.9	119.5
0.80	71.3	60.5	1.60	142.7	121.1

TABLE 5 (cont'd)
Values of Kl / r for $F_y = 36$ and 50 ksi

λ_c	Kl/r $F_y = 36$	Kl/r $F_y = 50$	λ_c	Kl/r $F_y = 50$
1.62	144.4	122.6	2.42	183.1
1.64	146.2	124.1	2.44	184.6
1.66	148.0	125.6	2.46	186.1
1.68	149.8	127.1	2.48	187.6
1.70	151.6	128.6	2.50	189.1
1.72	153.4	130.1	2.52	190.7
1.74	155.1	131.6	2.54	192.2
1.76	156.9	133.2	2.56	193.7
1.78	158.7	134.7	2.58	195.2
1.80	160.5	136.2	2.60	196.7
1.82	162.3	137.7	2.62	198.2
1.84	164.1	139.2	2.64	199.7
1.86	165.8	140.7		
1.88	167.6	142.2		
1.90	169.4	143.8		
1.92	171.2	145.3		
1.94	173.0	146.8		
1.96	174.8	148.3		
1.98	176.5	149.8		
2.00	178.3	151.3		
2.02	180.1	152.8		
2.04	181.9	154.3		
2.06	183.7	155.9		
2.08	185.5	157.4		
2.10	187.2	158.9		
2.12	189.0	160.4		
2.14	190.8	161.9		
2.16	192.6	163.4		
2.18	194.4	164.9		
2.20	196.2	166.5		
2.22	197.9	168.0		
2.24	199.7	169.5		
2.26		171.0		
2.28		172.5		
2.30		174.0		
2.32		175.5		
2.34		177.0		
2.36		178.6		
2.38		180.1		
2.40		181.6		

Heavy line indicates Kl/r of 200.

NUMERICAL VALUES

TABLE 6
Slenderness Ratios of Elements as a Function of F_y
From Table B5.1

Ratio	F_y (ksi)					
	36	42	46	50	60	65
$65/\sqrt{F_y}$	10.8	10.0	9.6	9.2	8.4	8.1
$76/\sqrt{F_y}$	12.7	11.7	11.2	10.7	9.8	9.4
$95/\sqrt{F_y}$	15.8	14.7	14.0	13.4	12.3	11.8
$127/\sqrt{F_y}$	21.2	19.6	18.7	18.0	16.4	15.8
$141/\sqrt{F_y-10}$	27.7	24.9	23.5	22.3	19.9	19.0
$190/\sqrt{F_y}$	31.7	29.3	28.0	26.9	24.5	23.6
$238/\sqrt{F_y}$	39.7	36.7	35.1	33.7	30.7	29.5
$253/\sqrt{F_y}$	42.2	39.0	37.3	35.8	32.7	31.4
$317/\sqrt{F_y}$	52.8	48.9	46.7	44.8	40.9	39.3
$640/\sqrt{F_y}$	107.0	98.8	94.4	90.5	82.6	79.4
$970/\sqrt{F_y}$	162.0	150.0	143.0	137.0	125.0	120.0
$1,300/F_y$	36.1	31.0	28.3	26.0	21.7	20.0
$2,070/F_y$	57.5	49.3	45.0	41.4	34.5	31.8
$3,300/F_y$	91.7	78.6	71.7	66.0	55.0	50.8
$8,970/F_y$	249.0	214.0	195.0	179.0	150.0	138.0

TABLE 7
Values of C_m
for Use in Section C1

$\dfrac{M_1}{M_2}$	C_m	$\dfrac{M_1}{M_2}$	C_m	$\dfrac{M_1}{M_2}$	C_m
−1.00	1.00	−0.45	0.78	0.10	0.56
−0.95	0.98	−0.40	0.76	0.15	0.54
−0.90	0.96	−0.35	0.74	0.20	0.52
−0.85	0.94	−0.30	0.72	0.25	0.50
−0.80	0.92	−0.25	0.70	0.30	0.48
−0.75	0.90	−0.20	0.68	0.35	0.46
−0.70	0.88	−0.15	0.66	0.40	0.44
−0.65	0.86	−0.10	0.64	0.45	0.42
−0.60	0.84	−0.05	0.62	0.50	0.40
				0.60	0.36
−0.55	0.82	0	0.60	0.80	0.28
−0.50	0.80	0.05	0.58	1.00	0.20

Note 1: $C_m = 0.6 - 0.4(M_1 / M_2)$.
Note 2: M_1 / M_2 is positive for reverse curvature and negative for single curvature. $|M_1| \leq |M_2|$

TABLE 8

Values of P_e / A_g
for Use in Section C1 for Steel of Any Yield Stress

$\frac{Kl}{r}$	P_e/A_g (ksi)	$\frac{Kl}{r}$	P_e/A_g (ksi)	$\frac{Kl}{r}$	P_e/A_g (ksi)	$\frac{Kl}{r}$	P_e/A_g (ksi)	$\frac{Kl}{r}$	P_e/A_g (ksi)	$\frac{Kl}{r}$	P_e/A_g (ksi)
21	649.02	51	110.04	81	43.62	111	23.23	141	14.40	171	9.79
22	591.36	52	105.85	82	42.57	112	22.82	142	14.19	172	9.67
23	541.06	53	101.89	83	41.55	113	22.42	143	14.00	173	9.56
24	496.91	54	98.15	84	40.56	114	22.02	144	13.80	174	9.45
25	457.95	55	94.62	85	39.62	115	21.64	145	13.61	175	9.35
26	423.40	56	91.27	86	38.70	116	21.27	146	13.43	176	9.24
27	392.62	57	88.09	87	37.81	117	20.91	147	13.25	177	9.14
28	365.07	58	85.08	88	36.96	118	20.56	148	13.07	178	9.03
29	340.33	59	82.22	89	36.13	119	20.21	149	12.89	179	8.93
30	318.02	60	79.51	90	35.34	120	19.88	150	12.72	180	8.83
31	297.83	61	76.92	91	34.56	121	19.55	151	12.55	181	8.74
32	279.51	62	74.46	92	33.82	122	19.23	152	12.39	182	8.64
33	262.83	63	72.11	93	33.09	123	18.92	153	12.23	183	8.55
34	247.59	64	69.88	94	32.39	124	18.61	154	12.07	184	8.45
35	233.65	65	67.74	95	31.71	125	18.32	155	11.91	185	8.36
36	220.85	66	65.71	96	31.06	126	18.03	156	11.76	186	8.27
37	209.07	67	63.76	97	30.42	127	17.75	157	11.61	187	8.18
38	198.21	68	61.90	98	29.80	128	17.47	158	11.47	188	8.10
39	188.18	69	60.12	99	29.20	129	17.20	159	11.32	189	8.01
40	178.89	70	58.41	100	28.62	130	16.94	160	11.18	190	7.93
41	170.27	71	56.78	101	28.06	131	16.68	161	11.04	191	7.85
42	162.26	72	55.21	102	27.51	132	16.43	162	10.91	192	7.76
43	154.80	73	53.71	103	26.98	133	16.18	163	10.77	193	7.68
44	147.84	74	52.57	104	26.46	134	15.94	164	10.64	194	7.60
45	141.34	75	50.88	105	25.96	135	15.70	165	10.51	195	7.53
46	135.26	76	49.55	106	25.47	136	15.47	166	10.39	196	7.45
47	129.57	77	48.27	107	25.00	137	15.25	167	10.26	197	7.38
48	124.23	78	47.04	108	24.54	138	15.03	168	10.14	198	7.30
49	119.21	79	45.86	109	24.09	139	14.81	169	10.02	199	7.23
50	114.49	80	44.72	110	23.65	140	14.60	170	9.90	200	7.16

Note: $P_e/A_g = \dfrac{\pi^2 E}{(Kl/r)^2}$, use for both P_{e1} and P_{e2}.

TABLE 9-36
$\dfrac{\phi_v V_n}{A_w}$ (ksi) for Plate Girders by Appendix F2 for 36 ksi Yield Stress Steel, Tension Field Action Not Included

$\dfrac{h}{t_w}$	Aspect ratio a/h: Stiffener Spacing to Web Depth													
	0.5	0.6	0.7	0.8	0.9	1.0	1.2	1.4	1.6	1.8	2.0	2.5	3.0	Over 3.0
60	19.4	19.4	19.4	19.4	19.4	19.4	19.4	19.4	19.4	19.4	19.4	19.4	19.4	19.4
70	19.4	19.4	19.4	19.4	19.4	19.4	19.4	19.4	19.4	19.4	19.4	19.4	19.4	19.4
80	19.4	19.4	19.4	19.4	19.4	19.4	19.4	19.4	19.4	19.4	18.9	18.2	17.9	16.9
90	19.4	19.4	19.4	19.4	19.4	19.4	19.4	18.5	17.8	17.2	16.8	16.2	15.9	14.7
100	19.4	19.4	19.4	19.4	19.4	19.2	17.6	16.6	16.0	15.5	14.9	13.8	13.2	11.9
110	19.4	19.4	19.4	19.4	18.4	17.4	16.0	14.8	13.7	12.8	12.3	11.4	10.9	9.8
120	19.4	19.4	19.4	18.1	16.9	16.0	14.0	12.5	11.5	10.8	10.3	9.6	9.2	8.3
130	19.4	19.4	18.2	16.7	15.6	14.1	11.9	10.6	9.8	9.2	8.8	8.2	7.8	7.0
140	19.4	18.8	16.9	15.5	13.5	12.1	10.3	9.2	8.4	7.9	7.6	7.0	6.7	6.1
150	19.4	17.6	15.7	13.5	11.8	10.6	8.9	8.0	7.3	6.9	6.6	6.1	5.9	5.3
160	18.9	16.5	14.1	11.9	10.4	9.3	7.9	7.0	6.5	6.1	5.8	5.4		4.6
170	17.8	15.5	12.5	10.5	9.2	8.2	7.0	6.2	5.7	5.4	5.1			4.1
180	16.8	13.9	11.1	9.4	8.2	7.3	6.2	5.5	5.1	4.8	4.6			3.7
200	14.9	11.2	9.0	7.6	6.6	5.9	5.0	4.5	4.1					3.0
220	12.3	9.3	7.5	6.3	5.5	4.9	4.2							2.5
240	10.3	7.8	6.3	5.3	4.6	4.1								2.1
260	8.8	6.6	5.3	4.5	3.9	3.5								1.8
280	7.6	5.7	4.6	3.9										
300	6.6	5.0	4.0											
320	5.8	4.4												

TABLE 9-50
$\dfrac{\phi_v V_n}{A_w}$ (ksi) for Plate Girders by Appendix F2 for 50 ksi Yield Stress Steel, Tension Field Action Not Included

$\dfrac{h}{t_w}$	Aspect ratio a/h: Stiffener Spacing to Web Depth													
	0.5	0.6	0.7	0.8	0.9	1.0	1.2	1.4	1.6	1.8	2.0	2.5	3.0	Over 3.0
60	27.0	27.0	27.0	27.0	27.0	27.0	27.0	27.0	27.0	27.0	27.0	27.0	27.0	26.6
70	27.0	27.0	27.0	27.0	27.0	27.0	27.0	27.0	26.9	26.1	25.5	24.6	24.0	22.8
80	27.0	27.0	27.0	27.0	27.0	27.0	26.0	24.5	23.5	22.8	22.3	21.5	20.6	18.6
90	27.0	27.0	27.0	27.0	26.5	25.1	23.1	21.8	20.4	19.2	18.3	17.0	16.3	14.7
100	27.0	27.0	27.0	25.6	23.9	22.6	20.1	17.9	16.5	15.5	14.9	13.8	13.2	11.9
110	27.0	27.0	25.3	23.2	21.7	19.6	16.6	14.8	13.7	12.8	12.3	11.4	10.9	9.8
120	27.0	25.9	23.2	21.1	18.4	16.5	14.0	12.5	11.5	10.8	10.3	9.6	9.2	8.3
130	27.0	23.9	21.4	18.0	15.7	14.1	11.9	10.6	9.8	9.2	8.8	8.2	7.8	7.0
140	25.5	22.2	18.4	15.5	13.5	12.1	10.3	9.2	8.4	7.9	7.6	7.0	6.7	6.1
150	23.8	19.9	16.1	13.5	11.8	10.6	8.9	8.0	7.3	6.9	6.6	6.1	5.9	5.3
160	22.3	17.5	14.1	11.9	10.4	9.3	7.9	7.0	6.5	6.1	5.8	5.4		4.6
170	20.6	15.5	12.5	10.5	9.2	8.2	7.0	6.2	5.7	5.4	5.1			4.1
180	18.3	13.9	11.1	9.4	8.2	7.3	6.2	5.5	5.1	4.8	4.6			3.7
200	14.9	11.2	9.0	7.6	6.6	5.9	5.0	4.5	4.1					3.0
220	12.3	9.3	7.5	6.3	5.5	4.9	4.2							2.5
240	10.3	7.8	6.3	5.3	4.6	4.1								2.1
260	8.8	6.6	5.3	4.5	3.9	3.5								1.8
280	7.6	5.7	4.6	3.9										

TABLE 10-36
$\dfrac{\phi_v V_n}{A_w}$ (ksi) for Plate Girders by Appendix G for 36 ksi Yield Stress Steel, Tension Field Action Included[b]

(Italic values indicate gross area,
as percent of $(h \times t_w)$ required for pairs of
intermediate stiffeners of 36 ksi yield stress
steel with $V_u / \phi V_n = 1.0$) [a]

$\dfrac{h}{t_w}$	Aspect ratio a/h: Stiffener Spacing to Web Depth													
	0.5	0.6	0.7	0.8	0.9	1.0	1.2	1.4	1.6	1.8	2.0	2.5	3.0	Over 3.0 [c]
60	19.4	19.4	19.4	19.4	19.4	19.4	19.4	19.4	19.4	19.4	19.4	19.4	19.4	19.4
70	19.4	19.4	19.4	19.4	19.4	19.4	19.4	19.4	19.4	19.4	19.4	19.4	19.4	19.4
80	19.4	19.4	19.4	19.4	19.4	19.4	19.4	19.4	19.4	19.4	19.1	18.6	18.3	16.9
90	19.4	19.4	19.4	19.4	19.4	19.4	19.4	19.0	18.5	18.2	17.8	17.3	16.8	14.7
100	19.4	19.4	19.4	19.4	19.4	19.3	18.6	18.1	17.6	17.2	16.6	15.6	14.9	11.9
110	19.4	19.4	19.4	19.4	19.1	18.7	17.9	17.2	16.3	15.6	15.1	14.0	13.3	9.8
120	19.4	19.4	19.4	19.0	18.5	18.1	17.0	16.0	15.1	14.4	13.9	12.8	12.0	8.3
130	19.4	19.4	19.1	18.6	18.1	17.4	16.1	15.1	14.2	13.5	12.9	11.8	11.0	7.0
140	19.4	19.3	18.7	18.2	17.4	16.6	15.4	14.4	13.5	12.8	12.2	11.0	10.2	6.1
150	19.4	19.0	18.4	17.5	16.7	16.0	14.8	13.8	12.9	12.2	11.6	10.4	9.6	5.3
160	19.3	18.7	17.9	17.0	16.2	15.5	14.3	13.3	12.4	11.7	11.1	9.9		4.6
170	19.1	18.4	17.4	16.6	15.8	15.1	13.9	12.9	12.0	11.3	10.7			4.1
										0.3	*0.4*			
180	18.9	18.0	17.1	16.2	15.5	14.8	13.6	12.6	11.7	11.0	10.4			3.7
							0.2	*0.7*	*1.1*	*1.3*	*1.5*			
200	18.4	17.3	16.4	15.6	14.9	14.2	13.1	12.0	11.2					3.0
				0.1	*0.9*	*1.4*	*2.1*	*2.5*	*2.8*					
220	17.8	16.9	16.0	15.2	14.5	13.8	12.7							2.5
		1.1	*2.0*	*2.6*	*3.0*	*3.6*								
240	17.4	16.5	15.7	14.9	14.2	13.5								2.1
		1.5	*2.7*	*3.4*	*3.9*	*4.3*								
260	17.1	16.2	15.4	14.6	14.0	13.3								1.8
	1.3	*3.0*	*4.0*	*4.6*	*5.0*	*5.4*								
280	16.8	16.0	15.2	14.4										
	2.7	*4.2*	*5.0*	*5.6*										

NUMERICAL VALUES

TABLE 10-36 (cont'd)
$\dfrac{\phi_v V_n}{A_w}$ (ksi) for Plate Girders by Appendix G for 36 ksi Yield Stress Steel, Tension Field Action Included[b]

(Italic values indicate gross area,
as percent of ($h \times t_w$) required for pairs of
intermediate stiffeners of 36 ksi yield stress
steel with $V_u / \phi V_n = 1.0$)[a]

$\dfrac{h}{t_w}$	Aspect ratio a / h: Stiffener Spacing to Web Depth													
	0.5	0.6	0.7	0.8	0.9	1.0	1.2	1.4	1.6	1.8	2.0	2.5	3.0	Over 3.0 [c]
300	16.6	15.8	15.0											
	3.9	*5.2*	*5.9*											
320	16.4	15.6												
	4.9	*6.0*												

[a] For area of single-angle and single-plate stiffeners, or when $V_u / \phi V_n < 1.0$, see Equation A-G4-1.
[b] For end-panels and all panels in hybrid and web-tapered plate girders, use Table 9-36.
[c] Same as for Table 9-36.
Note: Girders so proportioned that the computed shear is less than that given in right-hand column do not require intermediate stiffeners.

TABLE 10-50
$\dfrac{\phi_v V_n}{A_w}$ (ksi) for Plate Girders by Appendix G for 50 ksi Yield Stress Steel, Tension Field Action Included[b]

(Italic values indicate gross area, as percent of $(h \times t_w)$ required for pairs of intermediate stiffeners of 50 ksi yield stress steel with $V_u / \phi V_n = 1.0$)[a]

$\dfrac{h}{t_w}$	Aspect ratio a/h: Stiffener Spacing to Web Depth													
	0.5	0.6	0.7	0.8	0.9	1.0	1.2	1.4	1.6	1.8	2.0	2.5	3.0	Over 3.0 [c]
60	27.0	27.0	27.0	27.0	27.0	27.0	27.0	27.0	27.0	27.0	27.0	27.0	27.0	26.6
70	27.0	27.0	27.0	27.0	27.0	27.0	27.0	27.0	26.9	26.5	26.1	25.4	24.9	22.8
80	27.0	27.0	27.0	27.0	27.0	27.0	26.5	25.8	25.1	24.6	24.1	23.3	22.4	18.6
90	27.0	27.0	27.0	27.0	26.8	26.3	25.3	24.4	23.4	22.5	21.7	20.2	19.2	14.7
100	27.0	27.0	27.0	26.5	25.9	25.3	24.0	22.5	21.4	20.4	19.6	18.0	17.0	11.9
110	27.0	27.0	26.5	25.8	25.1	24.2	22.4	21.0	19.8	18.8	18.0	16.4	15.3	9.8
120	27.0	26.7	25.9	25.1	24.0	23.0	21.2	19.8	18.6	17.6	16.8	15.2	14.1	8.3
130	27.0	26.2	25.4	24.1	23.0	22.0	20.3	18.9	17.7	16.7	15.9	14.2	13.1	7.0
140	26.7	25.8	24.5	23.3	22.2	21.3	19.6	18.2	17.0	16.0	15.1	13.5	12.3	6.1
150	26.3	25.2	23.9	22.7	21.6	20.7	19.0	17.6	16.4	15.4	14.5	12.9	11.7	5.3
160	26.0	24.6	23.3	22.2	21.1	20.2	18.5	17.1	15.9	14.9	14.0	12.4		4.6
									0.2	*0.4*	*0.5*	*0.8*		
170	25.6	24.1	22.8	21.7	20.7	19.8	18.1	16.7	15.2	14.5	13.6			4.1
							0.5	*1.0*	*1.2*	*1.4*	*1.6*			
180	25.1	23.7	22.4	21.3	20.3	19.4	17.8	16.4	15.2	14.2	13.3			3.7
					0.4	*0.9*	*1.5*	*1.9*	*2.2*	*2.3*	*2.5*			
200	24.3	23.0	21.8	20.8	19.8	18.9	17.3	15.9	14.7					3.0
			1.0	*1.8*	*2.3*	*2.7*	*3.2*	*3.5*	*3.7*					
220	23.7	22.5	21.4	20.4	19.4	18.5	16.9							2.5
		1.7	*2.7*	*3.3*	*3.8*	*4.1*	*4.5*							
240	23.2	22.1	21.0	20.0	19.1	18.2								2.1
	1.8	*3.2*	*4.0*	*4.6*	*4.9*	*5.2*								
260	23.0	21.8	20.8	19.8	18.8	18.0								
	3.2	*4.4*	*5.1*	*5.6*	*5.9*	*6.1*								
280	22.7	21.6	20.6	19.6										
	4.4	*5.4*	*6.0*	*6.4*										

[a] For area of single-angle and single-plate stiffeners, or when $V_u / \phi V_n < 1.0$, see Equation A-G4-1.
[b] For end-panels and all panels in hybrid and web-tapered plate girders, use Table 9-50.
[c] Same as for Table 9-50.
Note: Girders so proportioned that the computed shear is less than that given in right-hand column do not require intermediate stiffeners.

TABLE 11
Nominal Horizontal Shear Load for One Connector Q_n, kips[a]
From Equations I5-1 and I5-2

Connector [b]	Specified Compressive Strength of Concrete, f_c', ksi [d]		
	3.0	3.5	4.0
½-in. dia. × 2-in. hooked or headed stud	9.4	10.5	11.6
⅝-in. dia. × 2½-in. hooked or headed stud	14.6	16.4	18.1
¾-in. dia. × 3-in. hooked or headed stud	21.0	23.6	26.1
⅞-in. dia. × 3½-in. hooked or headed stud	28.6	32.1	35.5
Channel C3 × 4.1	10.2 L_c [c]	11.5 L_c [c]	12.7 L_c [c]
Channel C4 × 5.4	11.1 L_c [c]	12.4 L_c [c]	13.8 L_c [c]
Channel C5 × 6.7	11.9 L_c [c]	13.3 L_c [c]	14.7 L_c [c]

[a] Applicable only to concrete made with ASTM C33 aggregates.
[b] The nominal horizontal loads tabulated may also be used for studs longer than shown.
[c] L_c = length of channel, inches.
[d] $F_u > 0.5(f_c' w)^{0.75}$, w = 145 lbs./cu. ft.

COMMENTARY
on the Load and Resistance Factor Design Specification for Structural Steel Buildings

December 1, 1993

INTRODUCTION

The Specification is intended to be complete for normal design usage.

The Commentary furnishes background information and references for the benefit of the engineer seeking further understanding of the derivations and limits of the specification.

The Specification and Commentary are intended for use by design professionals with demonstrated engineering competence.

CHAPTER A

GENERAL PROVISIONS

A1. SCOPE

Load and Resistance Factor Design (LRFD) is an improved approach to the design of structural steel for buildings. It involves explicit consideration of limit states, multiple load factors, and resistance factors, and implicit probabilistic determination of reliability. The designation LRFD reflects the concept of factoring both loads and resistance. This type of factoring differs from the AISC allowable stress design (ASD) Specification (AISC, 1989), where only the resistance is divided by a factor of safety (to obtain allowable stress) and from the plastic design portion of that Specification, where only the loads are multiplied by a common load factor. The LRFD method was devised to offer the designer greater flexibility, more rationality, and possible overall economy.

The format of using resistance factors and multiple load factors is not new, as several such design codes are in effect [the ACI-318 Strength Design for Reinforced Concrete (ACI, 1989) and the AASHTO Load Factor Design for Bridges (AASHTO, 1989)]. Nor should the new LRFD method give designs radically different from the older methods, since it was tuned, or "calibrated," to typical representative designs of the earlier methods. The principal new ingredient is the use of a probabilistic mathematical model in the development of the load and resistance factors, which made it possible to give proper weight to the accuracy with which the various loads and resistances can be determined. Also, it provides a rational methodology for transference of test results into design provisions. A more rational design procedure leading to more uniform reliability is the practical result.

A2. LIMITS OF APPLICABILITY

2. Types of Construction

The provisions for these types of construction have been revised to provide for a truer recognition of the actual degree of connection restraint in the structural design. All connections provide some restraint. Depending on the amount of restraint offered, connections are classified as either Type FR or PR. This classification renames the Type I connection of the AISC ASD Specification to Type FR and includes both Type II and Type III of that Specification under a new, more general classification of Type PR.

Just as in the allowable stress design (ASD) provisions, construction utilizing Type FR connections may be designed in LRFD using either elastic or plastic analysis provided the appropriate Specification provisions are satisfied.

For Type PR construction which uses the "simple framing" approach, the restraint of the connection is ignored, provided the given conditions are met. This is no change from the ASD provisions. Where there is evidence of the actual moment rotation capability of a given type of connection, the use of designs incorporating the connection restraint is permitted just as in ASD. The designer should, when incorporating connection restraint into the design, take into account the reduced connection stiffness on the stability of the structure and its effect on the magnitude of second order effects.

A3.   MATERIAL

1.   Structural Steel

1a.   ASTM Designations

The grades of structural steel approved for use under the LRFD Specification, covered by ASTM standard specifications, extend to a yield stress of 100 ksi. Some of these ASTM standards specify a minimum yield point, while others specify a minimum yield strength. The term "yield stress" is used in the Specification as a generic term to denote either the yield point or the yield strength.

It is important to be aware of limitations of availability that may exist for some combinations of strength and size. Not all structural section sizes are included in the various material specifications. For example, the 60 ksi yield strength steel in the A572 specification includes plate only up to $1\frac{1}{4}$-in. in thickness. Another limitation on availability is that even when a product is included in the specifications, it may be only infrequently produced by the mills. Specifying these products may result in procurement delays or require ordering large quantities directly from the producing mills. Consequently, it is prudent to check availability before completing the details of a design.

Properties in the direction of rolling are of principal interest in the design of steel structures. Hence, yield stress as determined by the standard tensile test is the principal mechanical property recognized in the selection of the steels approved for use under the Specification. It must be recognized that other mechanical and physical properties of rolled steel, such as anisotropy, ductility, notch toughness, formability, corrosion resistance, etc., may also be important to the satisfactory performance of a structure.

It is not possible to incorporate in the Commentary adequate information to impart full understanding of all factors which might merit consideration in the selection and specification of materials for unique or especially demanding applications. In such a situation the user of the Specification is advised to make use of reference material contained in the literature on the specific properties of concern and to specify supplementary material production or quality requirements as provided for in ASTM material specifications. One such case is the design of highly restrained welded connections (AISC, 1973). Rolled steel is anisotropic, especially insofar as ductility is concerned; therefore, weld contraction strains in the region of highly restrained welded connections may exceed the capabilities of the material if special attention is not given to material selection, details, workmanship, and inspection.

Another special situation is that of fracture control design for certain types of service conditions (AASHTO, 1989). The relatively warm temperatures of steel in buildings, the essentially static strain rates, the stress intensity, and the number of cycles of full design stress make the probability of fracture in building structures extremely remote. Good workmanship and good design details incorporating joint geometry that avoids severe stress concentrations are generally the most effective means of providing fracture-resistant construction. However, for especially demanding service conditions such as low temperatures with impact loading, the specification of steels with superior notch toughness may be warranted.

1c. Heavy Shapes

The web-to-flange intersection and the web center of heavy hot-rolled shapes as well as the interior portions of heavy plates may contain a coarser grain structure and/or lower toughness material than other areas of these products. This is probably caused by ingot segregation, as well as somewhat less deformation during hot rolling, higher finishing temperature, and a slower cooling rate after rolling for these heavy sections. This characteristic is not detrimental to suitability for service for compression members, or for non-welded members.

However, when heavy cross sections are joined by splices or connections using complete-joint penetration welds which extend through the coarser and/or lower notch-tough interior portions, tensile strains induced by weld shrinkage may result in cracking, for example a complete-joint penetration welded connection of a heavy cross section beam to any column section. When members of lesser thickness are joined by complete-joint penetration welds, which induce smaller weld shrinkage strains, to the finer grained and/or more notch-tough surface material of ASTM A6 Group 4 and 5 shapes and heavy built-up cross sections, the potential for cracking is significantly lower, for example a complete penetration groove welded connection of a non-heavy cross-section beam to a heavy cross-section column.

For critical applications such as primary tension members, material should be specified to provide adequate toughness at service temperatures. Because of differences in the strain rate between the Charpy V-Notch (CVN) impact test and the strain rate experienced in actual structures, the CVN test is conducted at a temperature higher than the anticipated service temperature for the structure. The location of the CVN test is shown in Figure C-A3.1.

The toughness requirements of A3.1c are intended only to provide material of reasonable toughness for ordinary service application. For unusual applications and/or low temperature service, more restrictive requirements and/or toughness requirements for other section sizes and thicknesses may be appropriate.

To minimize the potential for fracture, the notch toughness requirements of A3.1c must be used in conjunction with good design and fabrication procedures. Specific requirements are given in Sections J1.5, J1.6, J2.3, and M2.2.

3. Bolts, Washers, and Nuts

The ASTM standard for A307 bolts covers two grades of fasteners. Either grade may be used under the LRFD Specification; however, it should be noted that Gr. B is intended for pipe flange bolting and Gr. A is the quality long in use for structural applications.

4.   **Anchor Bolts and Threaded Rods**

Since there is a limit on the maximum available length of A325 and A490, the use of these bolts for anchor bolts with design lengths longer than the maximum available lengths has presented problems in the past. The inclusion of A687 material in this Specification allows the use of higher strength material for bolts longer than A325 and A490 bolts. The designer should be aware that pretensioning anchor bolts is not recommended due to relaxation and stress corrosion after pretensioning.

The designer should specify the appropriate thread and SAE fit for threaded rods used as load-carrying members.

5.   **Filler Metal and Flux for Welding**

The filler metal specifications issued by the American Welding Society are general specifications which include filler metals suitable for building construction, as well as consumables that would not be suitable for building construction. For example, some electrodes covered by the specifications are specifically limited to single pass applications, while others are restricted to sheet metal applications. Many of the filler metals listed are "low hydrogen", that is, they deposit weld metal with low levels of diffusable hydrogen. Other materials are not. Filler metals listed under the various A5 specifications may or may not have required impact toughness, depending on the specific electrode classification. Notch toughness is generally not critical for weld metal used in building construction. However, on structures subject to dynamic loading, the engineer may require the filler metals used to deliver notch-tough weld deposits. Filler metals may be classified in either the as welded or post weld heat treated (stress relieved) condition. Since most structural applications will not involve stress relief, it is important to utilize filler materials that are classified in conditions similar to those experienced by the actual structure.

When specifying filler metal and/or flux by AWS designation, the applicable standard specifications should be carefully reviewed to assure a complete understanding of the designation reference. This is necessary because the AWS designation systems are not consistent. For example, in the case of electrodes for shielded metal arc welding (AWS A5.1), the first two or three digits indicate the nominal tensile strength classification, in ksi, of the weld metal and the final two digits indicate the type of coating; however, in the case of mild steel

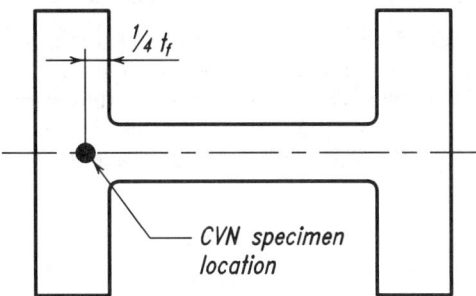

Fig. C-A3.1. Location from which Charpy impact specimen shall be taken.

electrodes for submerged arc welding (AWS A5.17), the first one or two digits times 10 indicate the nominal tensile strength classification, while the final digit or digits times 10 indicate the testing temperature in degrees F, for weld metal impact tests. In the case of low-alloy steel covered arc welding electrodes (AWS A5.5), certain portions of the designation indicate a requirement for stress relief, while others indicate no stress relief requirement.

Engineers do not, in general, specify the exact filler metal to be employed on a particular structure. Rather, the decision as to which welding process and which filler metal is to be utilized is usually left with the fabricator. To ensure that the proper filler metals are used, codes restrict the usage of certain filler materials, or impose qualification testing to prove the suitability of the specific electrode.

A4.   LOADS AND LOAD COMBINATIONS

1.   Loads, Load Factors, and Load Combinations

The load factors and load combinations given in Section A4.1 were developed to be used with the recommended minimum loads given in ASCE 7 *Minimum Design Loads for Buildings and Other Structures* (ASCE, 1988). The load factors and load combinations are developed in Ellingwood et al. (1982). The target reliability indices β underlying the load factors are approximately 3.0 for combinations with gravity loads only (dead, snow, and live loads), 2.5 for combinations with wind included, and 1.75 for combinations with earthquake loads. See Commentary A5.3 for definition of β.

The load factors and load combinations recognize that when several loads act in combination with the dead load (e.g., dead plus live plus wind), only one of these takes on its maximum lifetime value, while the other load is at its "arbitrary point-in-time value" (i.e., at a value which can be expected to be on the structure at any time). For example, under dead, live, and wind loads the following combinations are appropriate:

$$\gamma_D D + \gamma_L L \qquad\qquad\qquad (\text{C-A4-1})$$

$$\gamma_D D + \gamma_{L_a} L_a + \gamma_w W \qquad\qquad (\text{C-A4-2})$$

$$\gamma_D D + \gamma_L L + \gamma_{w_a} W_a \qquad\qquad (\text{C-A4-3})$$

where γ is the appropriate load factor as designated by the subscript symbol. Subscript a refers to an "arbitrary point-in-time" value.

The mean value of arbitrary point-in-time live load L_a is on the order of 0.24 to 0.4 times the mean maximum lifetime live load L for many occupancies, but its dispersion is far greater. The arbitrary point-in-time wind load W_a, acting in conjunction with the maximum lifetime live load, is the maximum daily wind. It turns out that $\gamma_{w_a} W_a$ is a negligible quantity so only two load combinations remain:

$$1.2D + 1.6L \qquad\qquad\qquad (\text{C-A4-4})$$

$$1.2D + 0.5L + 1.3W \qquad\qquad (\text{C-A4-5})$$

The load factor 0.5 assigned to L in the second equation reflects the statistical properties of L_a, but to avoid having to calculate yet another load, it is reduced so it can be combined with the maximum lifetime wind load.

The nominal loads D, L, W, E, and S are the code loads or the loads given in ASCE 7. The new specified earthquake loads are based on post-elastic energy dissipation in the structure, and are higher than those traditionally specified for allowable stress design (NEHRP, 1992). The new edition of ASCE Standard 7 on structural loads expected to be released in 1993 has adopted the new seismic design recommendations, as has the AISC *Seismic Provisions for Structural Steel Buildings* (AISC, 1992). The load factors on E in Load Combinations A4-5 and A4-6 have been reduced from 1.5 to 1.0 to be consistent with the specification of earthquake force in these new documents. The reader is referred to the commentaries to these documents for an expanded discussion on seismic loads, load factors, and seismic design of steel buildings.

2. Impact

A mass of the total moving load (wheel load) is used as the basis for impact loads on crane runway girders, because maximum impact load results when cranes travel while supporting lifted loads.

The increase in load, in recognition of random impacts, is not required to be applied to supporting columns because the impact load effects (increase in eccentricities or increases in out-of-straightness) will not develop or will be negligible during the short duration of impact. For additional information on crane girder design criteria see AISE Technical Report No. 13.

A5. DESIGN BASIS

1. Required Strength at Factored Loads

LRFD permits the use of both elastic and plastic structural analyses. LRFD provisions result in essentially the same methodology for, and end product of, plastic design as included in the AISC ASD Specification (AISC, 1989), except that the LRFD provisions tend to be more liberal, reflecting added experience and the results of further research. The 10 percent redistribution permitted is consistent with that in the AISC ASD Specification (AISC, 1989).

2. Limit States

A limit state is a condition which represents the limit of structural usefulness. Limit states may be dictated by functional requirements, such as maximum deflections or drift; they may be conceptual, such as plastic hinge or mechanism formation; or they may represent the actual collapse of the whole or part of the structure, such as fracture or instability. Design criteria ensure that a limit state is violated only with an acceptably small probability by selecting the load and resistance factors and nominal load and resistance values which will never be exceeded under the design assumptions.

Two kinds of limit states apply for structures: limit states of strength which define safety against the extreme loads during the intended life of the structure, and limit states of serviceability which define the functional requirements. The LRFD Specification, like other structural codes, focuses on the limit states of strength because of overriding considerations of public safety for the life, limb, and property of human beings. This does not mean that limit states of serviceability are not important to the designer, who must equally ensure functional performance and economy of design. However, these latter considerations permit more exercise of judgment on the part of designers. Minimum considerations of public safety, on the other hand, are not matters of individual judgment and, therefore, specifications dwell more on the limit states of strength than on the limit states of serviceability.

Limit states of strength vary from member to member, and several limit states may apply to a given member. The following limit states of strength are the most common: onset of yielding, formation of a plastic hinge, formation of a plastic mechanism, overall frame or member instability, lateral-torsional buckling, local buckling, tensile fracture, development of fatigue cracks, deflection instability, alternating plasticity, and excessive deformation.

The most common serviceability limit states include unacceptable elastic deflections and drift, unacceptable vibrations, and permanent deformations.

3.　Design for Strength

The general format of the LRFD Specification is given by the formula:

$$\Sigma \gamma_i Q_i \leq \phi R_n \qquad\qquad \text{(C-A5-1)}$$

where

Σ　　　= summation
i　　　= type of load, i.e., dead load, live load, wind, etc.
Q_i　　= nominal load effect
γ_i　　= load factor corresponding to Q_i
$\Sigma \gamma_i Q_i$ = required resistance
R_n　　= nominal resistance
ϕ　　　= resistance factor corresponding to R_n
ϕR_n　= design strength

The left side of Equation C-A5-1 represents the required resistance computed by structural analysis based upon assumed loads, and the right side of Equation C-A5-1 represents a limiting structural capacity provided by the selected members. In LRFD, the designer compares the effect of factored loads to the strength actually provided. The term design strength refers to the resistance or strength ϕR_n that must be provided by the selected member. The load factors γ and the resistance factors ϕ reflect the fact that loads, load effects (the computed forces and moments in the structural elements), and the resistances can be determined only to imperfect degrees of accuracy. The resistance factor ϕ is equal to or less than 1.0 because there is always a chance for the actual resistance to be less than the nominal value R_n computed by the equations given in Chapters D through K. Similarly, the load factors γ reflect the fact that the actual load effects may deviate from the nominal values of Q_i computed from the specified nominal loads. These factors account for unavoidable inaccuracies in the theory, variations in the material properties and dimensions, and uncertainties in the determination of loads. They provide a margin of reliability to account for unexpected loads. They do not account for gross error or negligence.

The LRFD Specification is based on (1) probabilistic models of loads and resistance, (2) a calibration of the LRFD criteria to the 1978 edition of the AISC ASD Specification for selected members, and (3) the evaluation of the resulting criteria by judgment and past experience aided by comparative design office studies of representative structures.

The following is a brief probabilistic basis for LRFD (Ravindra and Galambos, 1978, and Ellingwood et al., 1982). The load effects Q and the resistance factor R are assumed to be statistically independent random variables. In Figure C-A5.1, frequency distributions for Q and R are portrayed as separate curves on a common plot for a hypothetical case. As long as the resistance R is greater than (to the right of) the effects of the loads Q, a margin of safety for the particular limit state exists. However, because Q and R are random variables, there is some small probability that R may be less than Q, $(R < Q)$. This limit state probability is related to the degree of overlap of the frequency distributions in Figure C-A5.1, which depends on their relative positioning (R_m vs. Q_m) and their dispersions.

An equivalent situation may be represented as in Figure C-A5.2. If the expression $R < Q$ is divided by Q and the result expressed logarithmically, the result will be a single frequency distribution curve combining the uncertainties of both R and Q. The probability of attaining a limit state $(R < Q)$ is equal to the probability that $\ln(R/Q) < 0$ and is represented by the shaded area in the diagram.

The shaded area may be reduced and thus reliability increased in either of two ways: (1) by moving the mean of $\ln(R/Q)$ to the right, or (2) by reducing the spread of the curve for a given position of the mean relative to the origin. A convenient way of combining these two approaches is by defining the position of the mean using the standard deviation of $\ln(R/Q)$, as the unit of measure. Thus, the distance from the origin to the mean is measured as the number of standard deviations of the function $\ln(R/Q)$. As shown in Figure C-A5.2, this

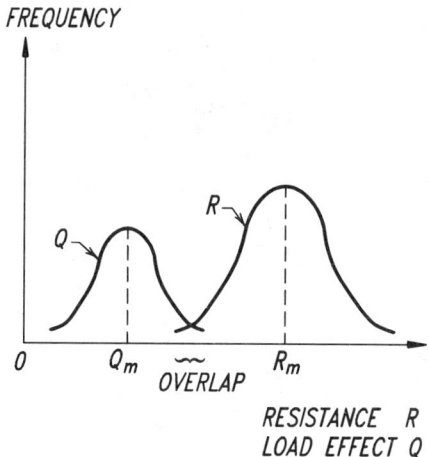

Fig. C-A5.1. Frequency distribution of load effect Q and resistance R.

is stated as β times $\sigma_{\ln(R/Q)}$, the standard deviation of $\ln(R/Q)$. The factor β therefore is called the "reliability index."

If the actual shape of the distribution of $\ln(R/Q)$ were known, and if an acceptable value of the probability of reaching the limit state could be agreed upon, one could establish a completely probability-based set of design criteria. Unfortunately, this much information frequently is not known. The distribution shape of each of the many variables (material, loads, etc.) has an influence on the shape of the distribution of $\ln(R/Q)$. Often only the means and the standard deviations of the many variables involved in the makeup of the resistance and the load effect can be estimated. However, this information is enough to build an approximate design criterion which is independent of the knowledge of the distribution, by stipulating the following design condition:

$$\beta\sigma_{\ln(R/Q)} \approx \beta\sqrt{V_R^2 + V_Q^2} \leq \ln(R_m/Q_m) \qquad \text{(C-A5-2)}$$

In this formula, the standard deviation has been replaced by the approximation $\sqrt{V_R^2 + V_Q^2}$, where $V_R = \sigma_R/R_m$ and $V_Q = \sigma_Q/Q_m$ (σ_R and σ_Q are the standard deviations, R_m and Q_m are the mean values, V_R and V_Q are the coefficients of variation, respectively, of the resistance R and the load effect Q). For structural elements and the usual loadings R_m, Q_m, and the coefficients of variation, V_R and V_Q, can be estimated, so a calculation of

$$\beta = \frac{\ln(R_m/Q_m)}{\sqrt{V_R^2 + V_Q^2}} \qquad \text{(C-A5-3)}$$

will give a comparative value of the measure of reliability of a structure or component.

The description of the determination of β as given above is a simple way of defining the probabilistic method used in the development of LRFD. A more refined method, which can accommodate more complex design situations (such as the beam-column interaction equation) and include probabilistic distributions other than the lognormal distribution used to derive Equation C-A5-3, has been developed since the publication of Ravindra and Galambos (1978), and is fully described in Galambos, et al. (1982). This latter method has been used in the development of the recommended load factors (see Section A4). The two methods give essentially the same β values for most steel structural members and connections.

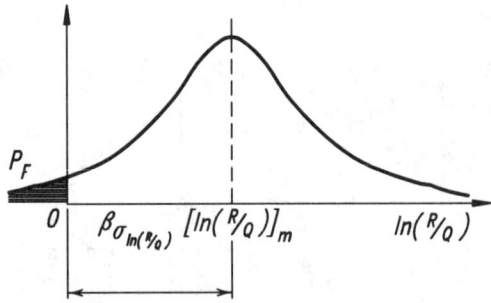

Fig. C-A5.2. Definition of reliability index.

Statistical properties (mean values and coefficients of variations) are presented for the basic material properties and for steel beams, columns, composite beams, plate girders, beam-columns, and connection elements in a series of eight articles in the September 1978 issue of the *Journal of the Structural Division of ASCE* (Vol. 104, ST9). The corresponding load statistics are given in Galambos, et al. (1982). Based on these statistics, the values of β inherent in the 1978 edition of the AISC ASD Specification were evaluated under different load combinations (live/dead, wind/dead, etc.), and for various tributary areas for typical members (beams, columns, beam-columns, structural components, etc.). As might be expected, there was a considerable variation in the range of β values. Examination of the many β values associated with ASD revealed certain trends. For example, compact rolled beams (flexure) and tension members (yielding) had β values that decreased from about 3.1 at $L/D = 0.50$ to 2.4 at $L/D = 4$. This decrease is a result of ASD applying the same factor to dead load, which is relatively predictable, and live load, which is more variable. For bolted or welded connections, β was on the order of 4 to 5. Reliability indices for load combinations involving wind and earthquake loads tended to be lower. Based on a thorough assessment of implied reliabilities in existing acceptable design practice, common load factors for various structural materials (steel, reinforced concrete, etc.) were developed in Ellingwood et al. (1982).

One of the features of the probability-based method used in the development of LRFD is that the variations of β values can be reduced by specifying several "target" β values and selecting multiple load and resistance factors to meet these targets. The Committee on Specifications set the point at which LRFD is calibrated to ASD at $L/D = 3.0$ for braced compact beams in flexure and tension members at yield. The resistance factor, ϕ, for these limit states is 0.90, and the implied β is approximately 2.6 for members and 4.0 for connections; this larger β value for connections reflects the fact that connections are expected to be stronger than the members that they connect. Limit states for other members are handled consistently.

Computer methods as well as charts are given in Ellingwood et al. (1982) for the use of specification writers to determine the resistance factors ϕ. These factors can also be approximately determined by the following:

$$\phi = (R_m/R_n) \exp(-0.55\beta V_r) \qquad \text{(C-A5-4)*}$$

where

R_m = mean resistance
R_n = nominal resistance according to the equations in Chapters D through K
V_r = coefficient of variation of the resistance

## 4.	Design for Serviceability and Other Considerations

Nominally, serviceability should be checked at the unfactored loads. For combinations of gravity and wind or seismic loads some additional reduction factor may be warranted.

---

* Note that exp (x) is identical to the more familiar e^x.

CHAPTER B

DESIGN REQUIREMENTS

B2. NET AREA

Critical net area is based on net width and load transfer at a particular chain.

B3. EFFECTIVE NET AREA FOR TENSION MEMBERS

Section B3 deals with the effect of shear lag, which is applicable to both welded and bolted tension members. The reduction coefficient U is applied to the net area A_n of bolted members and to the gross area A_g of welded members. As the length of connection l is increased, the shear lag effect is diminished. This concept is expressed empirically by Equation B3-2. Munse and Chesson (1963) have shown, using this expression to compute an effective net area, with few exceptions, the estimated strength of some 1,000 bolted and riveted connection test specimens correlated with observed test results within a scatterband of ±10 percent. Newer research (Easterling and Gonzales, 1993) provides further justification for current provisions.

For any given profile and connected elements, $\bar{x}$ is a fixed geometric property. It is illustrated as the distance from the connection plane, or face of the member, to the centroid of the member section resisting the connection force. See Figure C-B3.1. Length l is dependent upon the number of fasteners or equivalent length of weld required to develop the given tensile force, and this in turn is dependent upon the mechanical properties of the member and the capacity of the fasteners or weld used. The length l is illustrated as the distance, parallel to the line of force, between the first and last fasteners in a line for bolted connections. The number of bolts in a line, for the purpose of the determination of l, is determined by the line with the maximum number of bolts in the connection. For staggered bolts, the out-to-out dimension is used for l. See Figure C-B3.2. There is insufficient data to establish a value of U if all lines have only one bolt, but it is probably conservative to use A_e equal to the net area of the connected element. For welded connections, l is the length of the member parallel to the line of force that is welded. For combinations of longitudinal and transverse welds (see Figure C-B3.3), l is the length of longitudinal weld because the transverse weld has little or no effect on the shear lag problem, i.e., it does little to get the load into the unattached portions of the member.

Previous issues of this Specification have presented values for U for bolted or riveted connections of W, M, and S shapes, tees cut from these shapes, and other shapes. These values are acceptable for use in lieu of calculated values from Equation B3-2 and are retained here for the convenience of designers.

For bolted or riveted connections the following values of U may be used:

(a) W, M, or S shapes with flange widths not less than two-thirds the depth, and structural tees cut from these shapes, provided the connection is to the flanges and has no fewer than three fasteners per line in the direction of stress, $U = 0.90$

(b) W, M, or S shapes not meeting the conditions of subparagraph a, structural tees cut from these shapes, and all other shapes including built-up cross

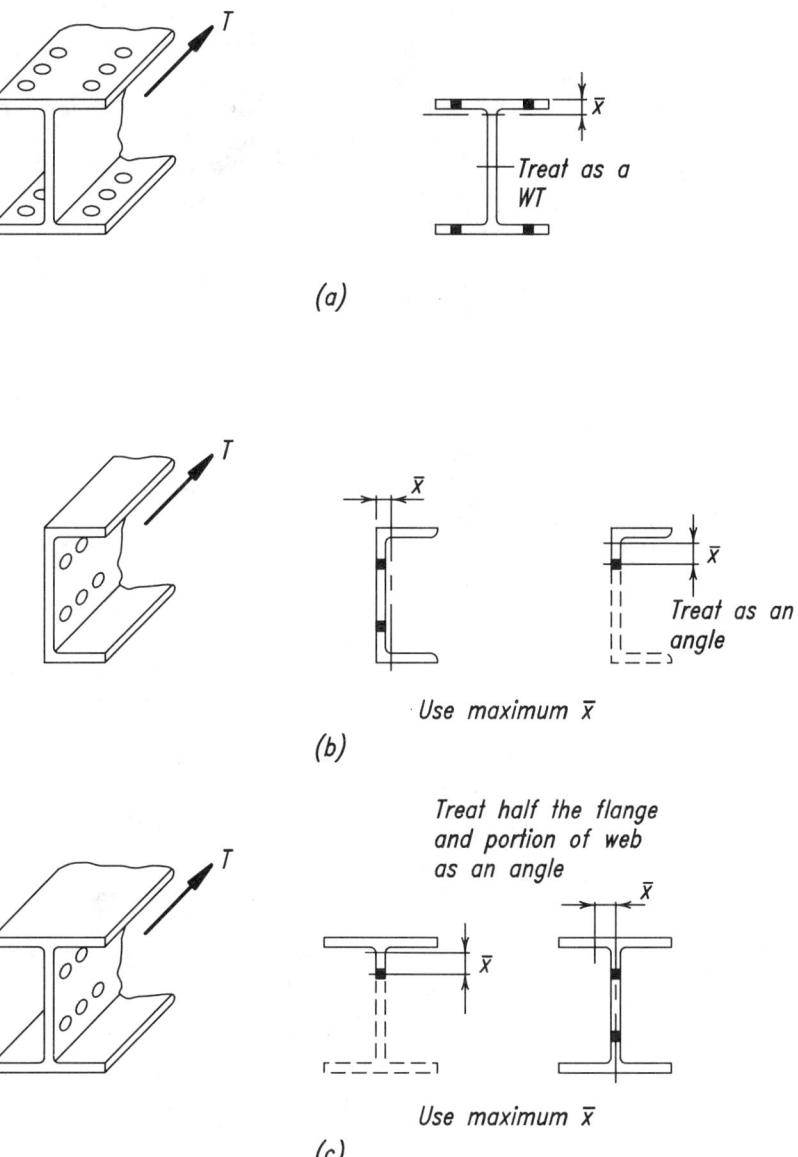

Fig. C-B3.1. Determination of $\bar{x}$ for U.

sections, provided the connection has no fewer than three fasteners per line in the direction of stress, $U = 0.85$

(c) All members having only two fasteners per line in the direction of stress, $U = 0.75$

When a tension load is transmitted by fillet welds to some but not all elements of a cross section, the weld strength will control.

B5.   LOCAL BUCKLING

For the purposes of this Specification, steel sections are divided into compact sections, noncompact sections, and sections with slender compression elements. Compact sections are capable of developing a *fully plastic* stress distribution and they possess a rotational capacity of approximately 3 before the onset of local buckling (Yura et al., 1978). Noncompact sections can develop the yield stress in compression elements before local buckling occurs, but will not resist inelastic local buckling at the strain levels required for a fully plastic stress distribution. Slender compression elements buckle elastically before the yield stress is achieved.

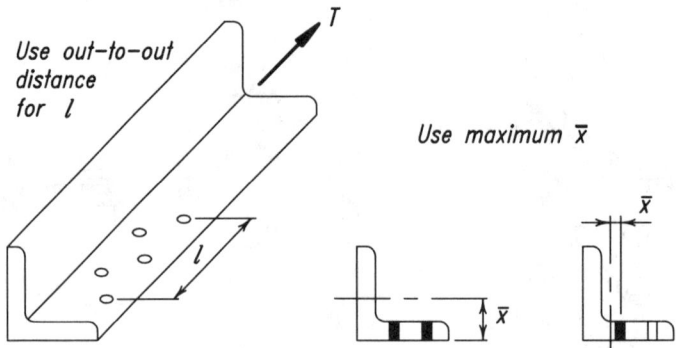

Fig. C-B3.2. Staggered holes.

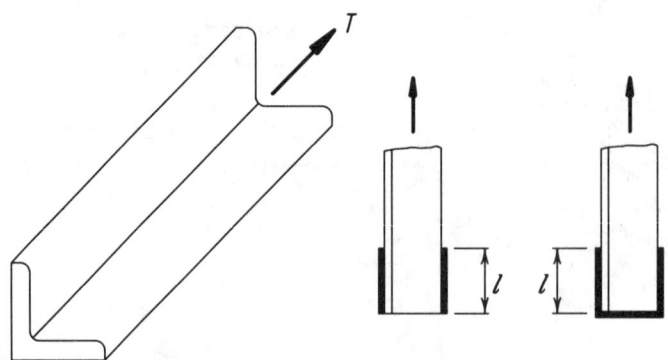

Fig. C-B3.3. Longitudinal and transverse welds.

TABLE C-B5.1
Limiting Width-Thickness Ratios for Compression Elements

Description of Element	Width-Thickness Ratio	Limiting Width-thickness Ratios λ_p	
		Non-seismic	Seismic
Flanges of I-shaped sections (including hybrid sections) and channels in flexure [a]	b/t	$65/\sqrt{F_y}$	$52/\sqrt{F_y}$
Webs in combined flexural and axial compression	h/t_w	For $P_u/\phi_b P_y \leq 0.125$	
		$\dfrac{640}{\sqrt{F_y}}\left(1-\dfrac{2.75P_u}{\phi_b P_y}\right)$	$\dfrac{520}{\sqrt{F_y}}\left(1-\dfrac{1.54P_u}{\phi_b P_y}\right)$
		For $P_u/\phi_b P_y > 0.125$	
		$\dfrac{191}{\sqrt{F_y}}\left(2.33-\dfrac{P_u}{\phi_b P_y}\right)\geq\dfrac{253}{\sqrt{F_y}}$	

[a] For hybrid beams use F_{yf} in place of F_y.

The dividing line between compact and noncompact sections is the limiting width-thickness ratio λ_p. For a section to be compact, all of its compression elements must have width-thickness ratios smaller than the limiting λ_p.

A greater inelastic rotation capacity than provided by the limiting values λ_p given in Table C-B5.1 may be required for some structures in areas of high seismicity. It has been suggested that in order to develop a ductility of from 3 to 5 in a structural member, ductility factors for elements would have to lie in the range of 5 to 15. Thus, in this case it is prudent to provide for an inelastic rotation of 7 to 9 times the elastic rotation (Chopra and Newmark, 1980). In order to provide for this rotation capacity, the limits λ_p for local flange and web buckling would be as shown in Table C-B5.1 (Galambos, 1976).

More information on seismic design is contained in the AISC *Seismic Provisions for Structural Steel Buildings*.

Another limiting width-thickness ratio is λ_r, representing the distinction between noncompact sections and sections with slender compression elements. As long as the width-thickness ratio of a compression element does not exceed the limiting value λ_r, local elastic buckling will not govern its strength. However, for those cases where the width-thickness ratios exceed λ_r, elastic buckling strength must be considered. A design procedure for such slender-element compression sections, based on elastic buckling of plates, is given in Appendix B5.3. The effective width Equation A-B5-12 applies strictly to stiffened elements under uniform compression. It does not apply to cases where the compression element is under stress gradient. A method of dealing with the stress gradient in a compression element is provided in Section B2 of the AISI *Design Specifications for Cold-Formed Steel Structural Members,* 1986 and Addendum, 1989. An exception is plate girders with slender webs. Such plate girders are capable of developing postbuckling strength in excess of the elastic buckling load. A design procedure for plate girders including tension field action is given in Appendix G.

The values of the limiting ratios λ_p and λ_r specified in Table B5.1 are similar to those in AISC (1989) and Table 2.3.3.3 of Galambos (1976), except that: (1) $\lambda_p = 65 / \sqrt{F_y}$, limited in Galambos (1976) to indeterminate beams when moments are determined by elastic analysis and to determinate beams, was adopted for all conditions on the basis of Yura et al. (1978); and (2) $\lambda_p = 1{,}300 / F_y$ for circular hollow sections was obtained from Sherman (1976).

The high shape factor for circular hollow sections makes it impractical to use the same slenderness limits to define the regions of behavior for different types of loading. In Table B5.1, the values of λ_p for a compact shape that can achieve the plastic moment, and λ_r for bending, are based on an analysis of test data from several projects involving the bending of pipes in a region of constant moment (Sherman and Tanavde, 1984 and Galambos, 1988). The same analysis produced the equation for the inelastic moment capacity in Table A-F1.1 in Appendix F1. However, a more restrictive value of λ_p is required to prevent inelastic local buckling from limiting the plastic hinge rotation capacity needed to develop a mechanism in a circular hollow beam section (Sherman, 1976).

The values of λ_r for axial compression and for bending are both based on test data. The former value has been used in building specifications since 1968 (Winter, 1970). Appendices B5 and F1 also limit the diameter-to-thickness ratio for any circular section to $13{,}000 / F_y$. Beyond this, the local buckling strength decreases rapidly, making it impractical to use these sections in building construction.

Following the SSRC recommendations (Galambos, 1988) and the approach used for other shapes with slender compression elements, a Q factor is used for circular sections to account for interaction between local and column buckling. The Q factor is the ratio between the local buckling stress and the yield stress. The local buckling stress for the circular section is taken from the inelastic AISI criteria (Winter, 1970) and is based on tests conducted on fabricated and manufactured cylinders. Subsequent tests on fabricated cylinders (Galambos, 1988) confirm that this equation is conservative.

The definitions of the width and thickness of compression elements agree with the 1978 AISC ASD Specification with minor modifications. Their applicability extends to sections formed by bending and to unsymmetrical and hybrid sections.

For built-up I-shaped sections under axial compression, modifications have been made to the flange local buckling criterion to include web-flange interaction. The k_c in the λ_r limit, in Equations A-B5-7 and A-B5-8 and the elastic buckling Equation A-B5-8 are the same that are used for flexural members. Theory indicates that the web-flange interaction in axial compression is at least as severe as in flexure. Rolled shapes are excluded from this criteria because there are no standard sections with proportions where the interaction would occur. In built-up sections where the interaction causes a reduction in the flange local buckling strength, it is likely that the web is also a thin stiffened element.

The k_c factor accounts for the interaction of flange and web local buckling demonstrated in experiments conducted by Johnson (1985). The maximum limit of 0.763 corresponds to $F_{cr} = 20,000 / \lambda^2$ which was used as the local buckling strength in earlier editions of both the ASD and LRFD Specifications. An $h / t_w = 27.5$ is required to reach $k_c = 0.763$. Fully fixed restraint for an un-stiffened compression element corresponds to $k_c = 1.3$ while zero restraint gives $k_c = 0.42$. Because of web-flange interactions it is possible to get $k_c < 0.42$ from the new k_c formula. If $h / t_w > 970 / \sqrt{F_y}$ use $h / t_w = 970 / \sqrt{F_y}$ in the k_c equation, which corresponds to the 0.35 limit.

Illustrations of some of the requirements of Table B5.1 are shown in Figure C-B5.1.

B7. LIMITING SLENDERNESS RATIOS

Chapters D and E provide reliable criteria for resistance of axially loaded members based on theory and confirmed by test for all significant parameters including slenderness. The advisory upper limits on slenderness contained in Section B7 are based on professional judgment and practical considerations of economics, ease of handling, and care required to minimize inadvertent damage during fabrication, transport, and erection. Out-of-straightness within reason-able tolerances does not affect the strength of tension members, and the effect of out-of-straightness within specified tolerances on the strength of compression members is accounted for in formulas for resistance. Applied tension tends to reduce, whereas compression tends to amplify, out-of-straightness. Therefore, more liberal criteria are suggested for tension members, including those subject to small compressive forces resulting from transient loads such as earthquake and wind. For members with slenderness ratios greater than 200, these compres-sive forces correspond to stresses less than 2.6 ksi.

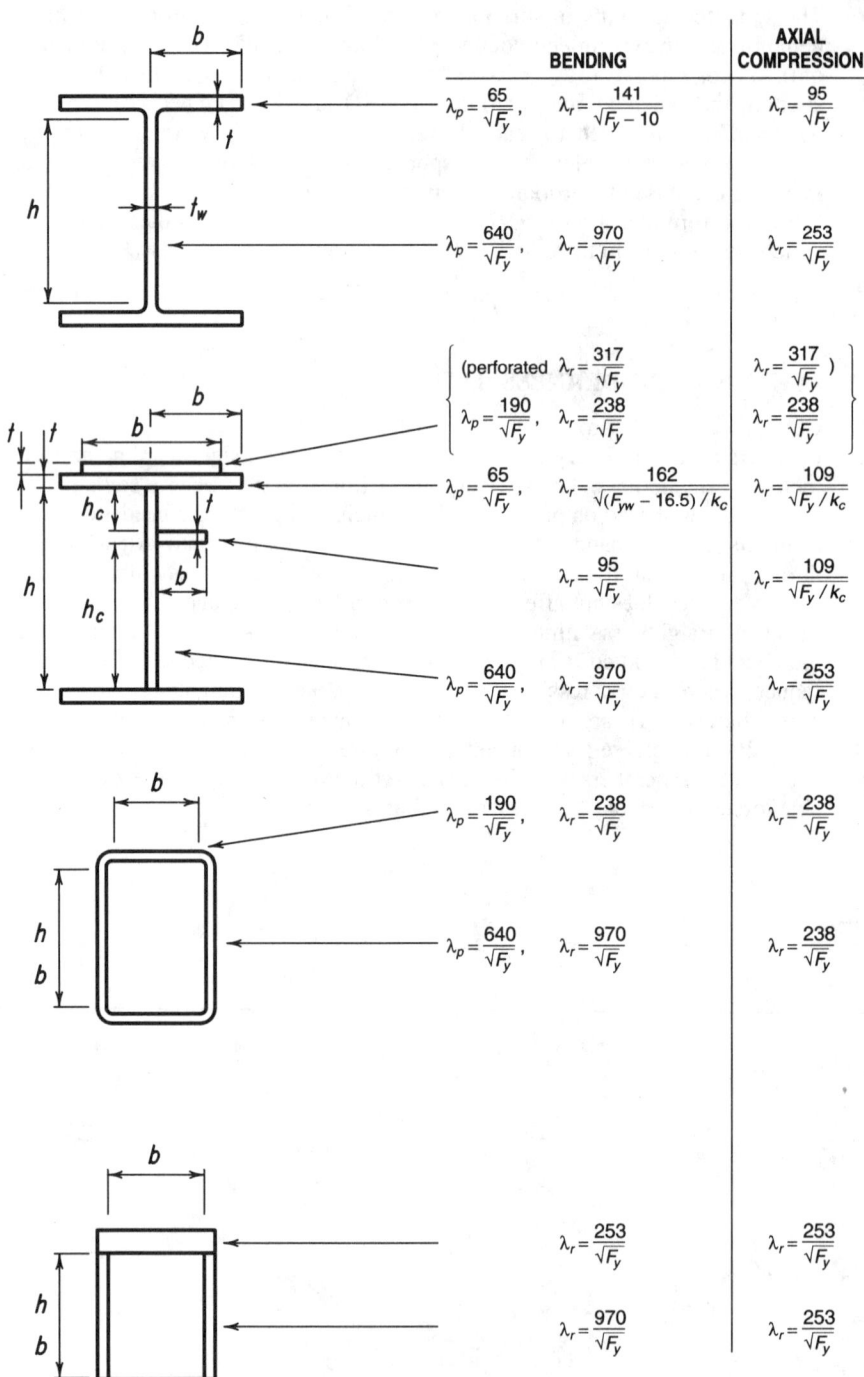

Fig. C-B5.1. Selected examples of Table B5.1 requirements.

CHAPTER C

FRAMES AND OTHER STRUCTURES

C1. SECOND ORDER EFFECTS

While resistance to wind and seismic loading can be provided in certain buildings by means of shear walls, which also provide for overall frame stability at factored gravity loading, other building frames must provide this resistance by frame action. This resistance can be achieved in several ways, e.g., by a system of bracing, by a moment-resisting frame, or by any combination of lateral force-resisting elements.

For frames under combined gravity and lateral loads, drift (horizontal deflection caused by applied loads) occurs at the start of loading. At a given value of the applied loads, the frame has a definite amount of drift Δ. In unbraced frames, additional secondary bending moments, known as the $P\Delta$ moments, may be developed in the columns and beams of the lateral load-resisting systems in each story. P is the total gravity load above the story and Δ is the story drift. As the applied load increases, the $P\Delta$ moments also increase. Therefore, the $P\Delta$ effect must often be accounted for in frame design. Similarly, in braced frames, increases in axial forces occur in the members of the bracing systems; however, such effects are usually less significant. The designer should consider these effects for all types of frames and determine if they are significant. Since $P\Delta$ effects can cause frame drifts to be larger than those calculated by ignoring them, they should also be included in the service load drift analysis when they are significant.

In unbraced frames designed by plastic analysis, the limit of $0.75\phi_c P_y$ on column axial loads has also been retained to help ensure stability.

The designer may use second-order elastic analysis to compute the maximum factored forces and moments in a member. These represent the required strength. Alternatively, for structures designed on the basis of elastic analysis, the designer may use first order analysis and the amplification factors B_1 and B_2.

In the general case, a member may have first order moments not associated with sidesway which are multiplied by B_1, and first order moments produced by forces causing sidesway which are multiplied by B_2.

The factor B_2 applies only to moments caused by forces producing sidesway and is calculated for an entire story. In building frames designed to limit Δ_{oh}/L to a predetermined value, the factor B_2 may be found in advance of designing individual members.

Drift limits may also be set for design of various categories of buildings so that the effect of secondary bending can be insignificant (Kanchanalai and Lu, 1979;

ATC, 1978). It is conservative to use the B_2 factor with the sum of the sway and the no-sway moments, i.e., with $M_{lt} + M_{nt}$.

The two kinds of first order moment M_{nt} and M_{lt} may both occur in sidesway frames from gravity loads. M_{nt} is defined as a moment developed in a member with frame sidesway prevented. If a significant restraining force is necessary to prevent sidesway of an unsymmetrical structure (or an unsymmetrically loaded symmetrical structure), the moments induced by releasing the restraining force will be M_{lt} moments, to be multiplied by B_2. In most reasonably symmetric frames, this effect will be small. If such a moment $B_2 M_{lt}$ is added algebraically to the $B_1 M_{nt}$ moment developed with sidesway prevented, a fairly accurate value of M_u will result. End moments produced in sidesway frames by lateral loads from wind or earthquake will always be M_{lt} moments to be multiplied by B_2.

When first order end moments in members subjected to axial compression are magnified by B_1 and B_2 factors, equilibrium requires that they be balanced by moments in connected members (Figure C-C1.1). This can generally be accomplished satisfactorily by distributing the difference between the magnified moment and the first order moment to any other moment-resisting members attached to the compressed member (or members) in proportion to the relative stiffness of the uncompressed members. Minor imbalances may be neglected in the judgment of the engineer. However, complex conditions, such as occur when there is significant magnification in several members meeting at a joint, may require a second order elastic analysis. Connections shall also be designed to resist the magnified end moments.

For compression members in braced frames, B_1 is determined from C_m values which are similar to the values in the AISC ASD Specification. A significant

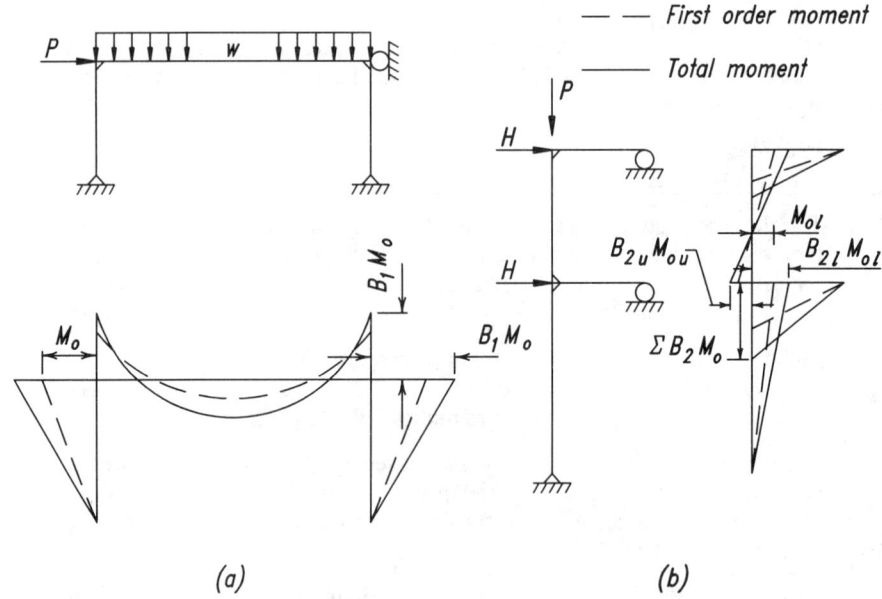

(a)   (b)

Fig. C-C1.1. Moment amplification.

difference, however, is that B_1 is never less than 1. When $C_m = 1$ for a compression member loaded between its supports, the factors of $\frac{8}{9}$ and $\frac{1}{2}$ make the new equations more liberal than Equation H1-1 of the AISC ASD Specification. For $C_m \leq 1$ (for members with unequal end moments), the new equations will be slightly more conservative than the AISC ASD Specification for a very slender member with low C_m. For the entire range of l / r and C_m, the equations compare very closely to exact inelastic solutions of braced members.

The center-to-center member length is usually used in the structural analysis. In braced and unbraced frames, P_n is governed by the maximum slenderness ratio regardless of the plane of bending. However, P_{e1} and P_{e2} are always calculated using the slenderness ratio in the plane of bending. Thus, when flexure is about the strong axis only, two different values of slenderness ratio may be involved in solving a given problem.

When second order analysis is used, it must account for the interaction of the factored load effects, that is, combinations of factored loads must be used in analysis. Superposition of forces obtained from separate analyses is not adequate.

When bending occurs about both the x and the y axes, the required flexural strength calculated about each axis is adjusted by the value of C_m and P_{e1} or P_{e2} corresponding to the distribution of moment and the slenderness ratio in its plane of bending, and is then taken as a fraction of the design bending strength, $\phi_b M_n$, about that axis, with due regard to the unbraced length of the compression flange where this is a factor.

Equations C1-2 and C1-3 approximate the maximum second order moments in compression members with no relative joint translation and no transverse loads between the ends of the member. This approximation is compared to an exact solution (Ketter, 1961) in Figure C-C1.2. For single curvature, Equation C1-3 is slightly unconservative, for a zero end moment it is almost exact, and for double curvature it is conservative. The 1978 AISC ASD Specification imposed the limit $C_m \geq 0.4$ which corresponds to a M_1 / M_2 ratio of 0.5. However, Figure C-C1.2 shows that if, for example, $M_1 / M_2 = 0.8$, the $C_m = 0.28$ is already very conservative, so the limit has been removed. The limit was originally adopted from Austin (1961), which was intended to apply to lateral-torsional buckling, not second-order in-plane bending strength. The AISC Specifications, both in the 1989 ASD and LRFD, use a modification factor C_b as given in Equation F1-3 for lateral-torsional buckling. C_b is approximately the inverse of C_m as presented in Austin (1961) with a 0.4 limit. In Zandonini (1985) it was pointed out that Equation C1-3 could be used for in-plane second order moments if the 0.4 limit was eliminated. Unfortunately, Austin (1961) was misinterpreted and a lateral-torsional buckling solution was used for an in-plane second-order analysis. This oversight has now been corrected.

For beam columns with transverse loadings, the second-order moment can be approximated by using the following equation

$$C_m = 1 + \psi P_u / P_{e1}$$

For simply supported members

where

$$\psi = \frac{\pi^2 \delta_0 EI}{M_0 L^2} - 1$$

δ_0 = maximum deflection due to transverse loading, in.

M_0 = maximum factored design moment between supports due to transverse loading, kip-in.

For restrained ends some limiting cases (Iwankiw, 1984) are given in Table C-C1.1 together with two cases of simply supported beam-columns. These values of C_m are always used with the maximum moment in the member. For the restrained-end cases, the values of B_1 will be most accurate if values of $K < 1.0$ corresponding to the end boundary conditions are used in calculating P_{e1}. In lieu of using the equations above, $C_m = 1.0$ can be used conservatively for transversely loaded members with unrestrained ends and 0.85 for restrained ends.

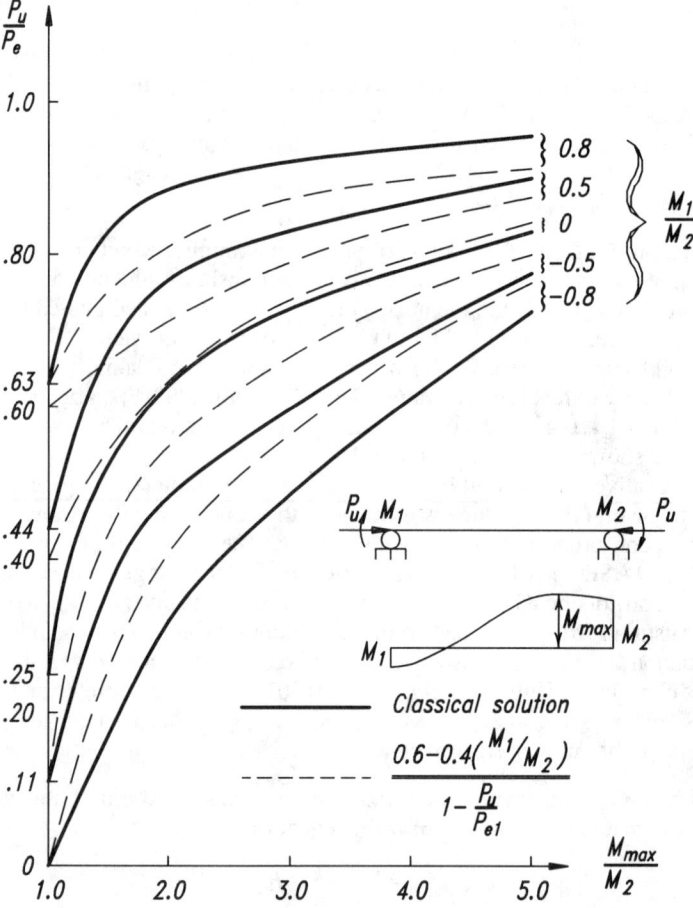

Fig. C-C1.2. Second-order moments for braced beam-column.

TABLE C-C1.1
Amplification Factors ψ and C_m

Case	ψ	C_m
	0	1.0
	−0.4	$1 - 0.4\dfrac{P_u}{P_{e1}}$
	−0.4	$1 - 0.4\dfrac{P_u}{P_{e1}}$
	−0.2	$1 - 0.2\dfrac{P_u}{P_{e1}}$
	−0.3	$1 - 0.3\dfrac{P_u}{P_{e1}}$
	−0.2	$1 - 0.2\dfrac{P_u}{P_{e1}}$

If, as in the case of a derrick boom, a beam-column is subject to transverse (gravity) load and a calculable amount of end moment, the value δ_0 should include the deflection between supports produced by this moment.

Stiffness reduction adjustment due to column inelasticity is permitted.

C2. FRAME STABILITY

The stability of structures must be considered from the standpoint of the structure as a whole, including not only the compression members, but also the beams, bracing system, and connections. The stability of individual elements must also be provided. Considerable attention has been given in the technical literature to this subject, and various methods of analysis are available to assure stability. The SSRC *Guide to Stability Design Criteria for Metal Structures* (Galambos, 1988) devotes several chapters to the stability of different types of members considered as individual elements, and then considers the effects of individual elements on the stability of the structure as a whole.

The effective length concept is one method of estimating the interaction effects of the total frame on a compression element being considered. This concept uses

TABLE C-C2.1
K Values for Columns

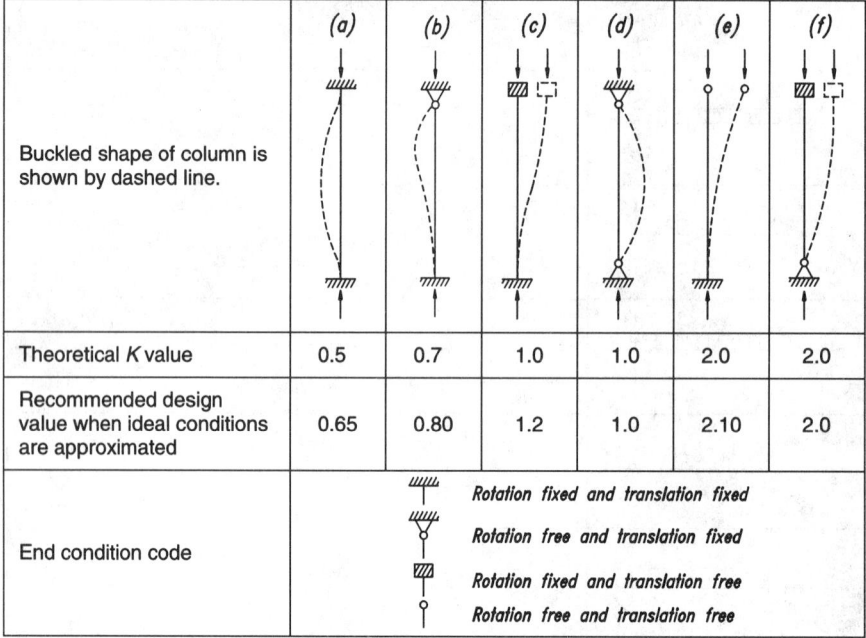

	(a)	(b)	(c)	(d)	(e)	(f)
Buckled shape of column is shown by dashed line.						
Theoretical K value	0.5	0.7	1.0	1.0	2.0	2.0
Recommended design value when ideal conditions are approximated	0.65	0.80	1.2	1.0	2.10	2.0
End condition code		Rotation fixed and translation fixed				
		Rotation free and translation fixed				
		Rotation fixed and translation free				
		Rotation free and translation free				

K factors to equate the strength of a framed compression element of length *L* to an equivalent pin-ended member of length *KL* subject to axial load only. Other rational methods are available for evaluating the stability of frames subject to gravity and side loading and individual compression members subject to axial load and moments. However, the effective-length concept is the only tool currently available for handling several cases which occur in practically all structures, and it is an essential part of many analysis procedures. Although the concept is completely valid for ideal structures, its practical implementation involves several assumptions of idealized conditions which will be mentioned later.

Two conditions, opposite in their effect upon column strength under axial loading, must be considered. If enough axial load is applied to the columns in an unbraced frame dependent entirely on its own bending stiffness for resistance to lateral deflection of the tops of the columns with respect to their bases (see Figure C-C2.1), the effective length of these columns will exceed the actual length. On the other hand, if the same frame were braced to resist such lateral movement, the effective length would be less than the actual length, due to the restraint (resistance to joint translation) provided by the bracing or other lateral support. The ratio *K*, effective column length to actual unbraced length, may be greater or less than 1.0.

The theoretical *K* values for six idealized conditions in which joint rotation and translation are either fully realized or nonexistent are tabulated in Table C-C2.1.

Also shown are suggested design values recommended by the Structural Stability Research Council (formerly the Column Research Council) for use when these conditions are approximated in actual design. In general, these suggested values are slightly higher than their theoretical equivalents, since joint fixity is seldom fully realized.

If the column base in Case f of Table C-C2.1 were truly pinned, K would actually exceed 2.0 for a frame such as that pictured in Figure C-C2.1, because the flexibility of the horizontal member would prevent realization of full fixity at the top of the column. On the other hand, it has been shown (Galambos, 1960) that the restraining influence of foundations, even where these footings are designed only for vertical load, can be very substantial in the case of flat-ended column base details with ordinary anchorage. For this condition, a design K value of 1.5 would generally be conservative in Case f.

While in some cases masonry walls provide enough lateral support for building frames to control lateral deflection, light curtain wall construction and wide column spacing can create a situation where only the bending stiffness of the frame provides this support. In this case the effective length factor K for an unbraced length of column L is dependent upon the bending stiffness provided by the other in-plane members entering the joint at each end of the unbraced segment. If the combined stiffness provided by the beams is sufficiently small, relative to that of the unbraced column segments, KL could exceed two or more story heights (Bleich, 1952).

Several rational methods are available to estimate the effective length of the columns in an unbraced frame with sufficient accuracy. These range from simple interpolation between the idealized cases shown in Table C-C2.1 to very complex analytical procedures. Once a trial selection of framing members has been made, the use of the alignment chart in Figure C-C2.2 affords a fairly rapid

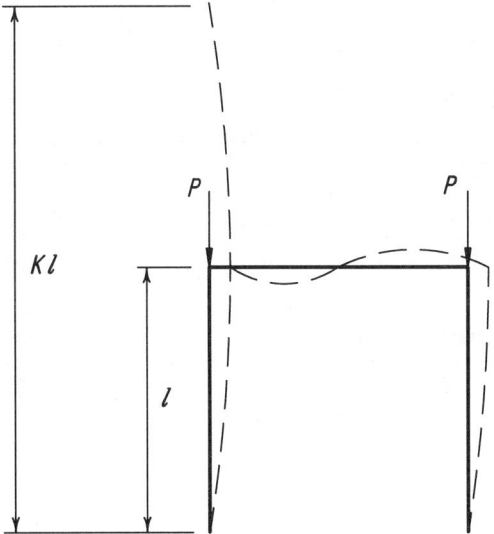

Fig. C-C2.1. Column effective length.

method for determining adequate K values. However, it should be noted that this alignment chart is based upon assumptions of idealized conditions which seldom exist in real structures (Galambos, 1988). These assumptions are as follows:

(1) Behavior is purely elastic.

(2) All members have constant cross section.

(3) All joints are rigid.

(4) For braced frames, rotations at opposite ends of beams are equal in magnitude, producing single-curvature bending.

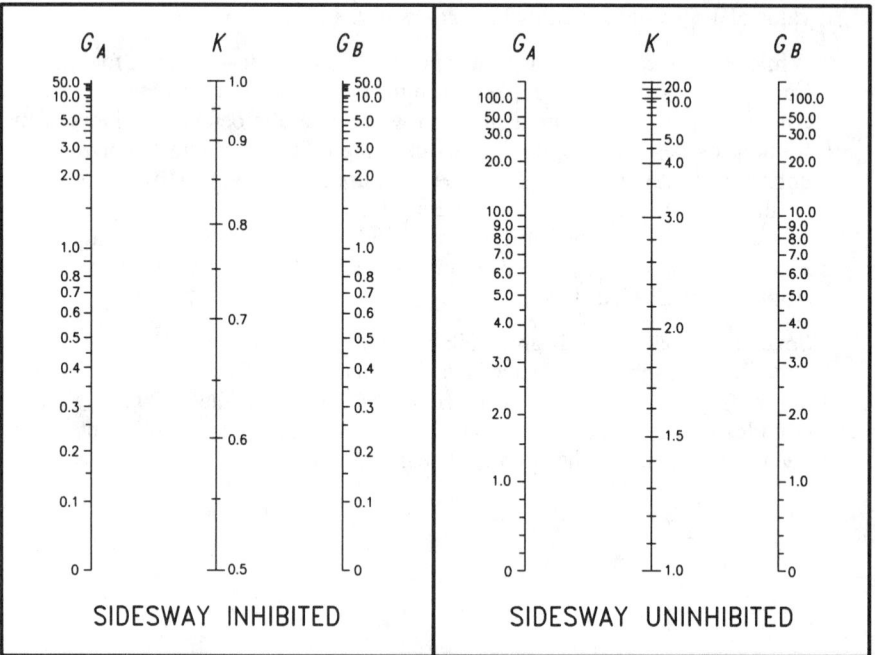

The subscripts A and B refer to the joints at the two ends of the column section being considered. G is defined as

$$G = \frac{\Sigma(I_c / L_c)}{\Sigma(I_g / L_g)}$$

in which Σ indicates a summation of all members rigidly connected to that joint and lying on the plane in which buckling of the column is being considered. I_c is the moment of inertia and L_c the unsupported length of a column section, and I_g is the moment of inertia and L_g the unsupported length of a girder or other restraining member. I_c and I_g are taken about axes perpendicular to the plane of buckling being considered.

For column ends supported by but not rigidly connected to a footing or foundation, G is theoretically infinity, but, unless actually designed as a true friction-free pin, may be taken as "10" for practical designs. If the column end is rigidly attached to a properly designed footing, G may be taken as 1.0. Smaller values may be used if justified by analysis.

Fig. C-C2.2. Alignment chart for effective length of columns in continuous frames.

(5) For unbraced frames, rotations at opposite ends of the restraining beams are equal in magnitude, producing reverse-curvature bending.

(6) The stiffness parameters $L\sqrt{P/EI}$ of all columns are equal.

(7) Joint restraint is distributed to the column above and below the joint in proportion to I/L of the two columns.

(8) All columns buckle simultaneously.

(9) No significant axial compression force exists in the girders.

Where the actual conditions differ from these assumptions, unrealistic designs may result. There are design procedures available which may be used in the calculation of G for use in Figure C-C2.2 to give results that better reflect the conditions in real structures (Yura, 1971; Disque, 1973; Bjorhovde, 1984; Davison et al., 1988).

Leaning columns (sized for gravity loads only, based on an assumed K of 1.0) may be used in unbraced frames provided that the destabilizing effects due to their lack of lateral stiffness from simple connections to the frame ($K = \infty$) is included in the design of the moment frame columns. A stabilizing column in one direction may be a leaning column in the transverse direction if it is rigidly connected in only one plane. LeMessurier (1977) presented an overall discussion of this problem and recommended a general solution for unbraced frames. In lieu of this and more exact analyses, the following design approximations are suggested.

When unbraced moment-resisting frames are the only source of lateral rigidity for a given direction of a story, the upper bound of sidesway stiffness in that direction, measured, in shear force per radian of drift, is $\Sigma P_L = \Sigma HL/\Delta_{oh}$. This force may be found from a first-order lateral load analysis, without gravity loads, where ΣH is the total story shear. (The calculation of B_2 using interstory drift as in LRFD Equation C1-4 also uses the term $\Sigma HL/\Delta_{oh}$). Since most of the moment-resisting columns in the frame will directly support axial loads, the bending stiffness of the columns will be reduced, lowering the sidesway stiffness ΣP_L.

An estimate of the reduced sidesway stiffness of the frame may be found by calculating P_{e2} for each moment-resisting column, in the direction under consideration, by using the nomograph for sidesway K based on local boundary conditions measured by G_A and G_B. G is normally assumed as:

$$G = \frac{\Sigma \dfrac{I_c}{L_c}}{\Sigma \dfrac{I_g}{L_g}}$$

This definition of G is based on the assumption that girders restraining columns have equal moments (same clockwise direction) at each end determined by an analysis for lateral loads only. When this assumption is violated, a significant overestimate of ΣP_{e2} may occur. Accurate G values may be found from an examination of girder end moments from such an analysis. The correct L_g should be taken as $L_g' = L_g \left[2 - \dfrac{M_F}{M_N} \right]$ where M_F is the moment at the far end of the girder under consideration and M_N is the moment at the near end. When $\dfrac{M_F}{M_N} > 2$, L_g' becomes negative which, although real, will result in negative values of G. Negative values of G are beyond the scope of the nomograph but are valid for use in Equation C-C2-2.

The reduced total stiffness of the whole story, when each rigidly connected column is loaded with its maximum load P_{e2}, is ΣP_{e2}. ΣP_{e2} calculated in this way will always satisfy:

$$.82 \Sigma P_L < \Sigma P_{e2} < \Sigma P_L$$

Many common framing arrangements include within a story loaded columns designed, in a particular direction, with $K = 1$. Such columns, often called leaning columns, receive lateral stability from the stiffness of columns with rigid moment-resisting connections. The required axial compressive strength of such leaning columns is called P_{uo} where the subscript implies no shear resistance to lateral loads. The ratio of the loads on all leaning columns in a story to the total of all loads on the story is:

$$\frac{\Sigma P_{uo}}{\Sigma P_u} = R_L$$

The ratio of all story loads to the loads on columns providing sidesway is:

$$N = \frac{1}{1 - R_L} = \frac{\Sigma P_u}{\Sigma P_u - \Sigma P_{uo}}$$

If the story stiffness ΣP_{e2} is calculated from the nomograph K values, the net stiffness available to stabilize the rigid column is:

$$\Sigma P_{e2} (1 - R_L) = \frac{\Sigma P_{e2}}{N}$$

If there is no redistribution of $\Sigma P_{e2} / N$ among the rigid columns, the modified capacity of an individual column is, conservatively:

$$P_{e2}' = \frac{P_{e2}}{N}$$

It follows that a modified K_i' including leaning effects is:

$$K_i' = \sqrt{N} \times K_i \qquad \text{(C-C2-1)}$$

A more exact value of K' to account for loss of stiffness to leaning columns can be found from an iterative solution of:

$$6 \left[\frac{\dfrac{\pi}{K_i'}}{\tan \dfrac{\pi}{K'}} - R_L \right] [G_A + G_B] - \left(\frac{\pi}{K'} \right)^2 G_A G_B + 36 \left[1 - R_L \left(\frac{\tan \dfrac{\pi}{2K'}}{\dfrac{\pi}{2K'}} \right) \right] = 0$$

$$\text{(C-C2-2)}$$

When $R_L = 0$, this equation reduces to the equation solved by the sidesway uninhibited nomograph.

The 1993 LRFD Specification no longer limits K to unity in sidesway frames and redistribution of stiffness between members of a frame may be advantageous. There are several ways of doing this. Based on the assumption that ΣP_{e2} is constant, regardless of loading distribution, an adjusted distribution of stiffness to the ith column of a story is:

$$P_{ei}' = \frac{P_{ui}}{\Sigma P_u}[\Sigma P_{e2}] = \left(\frac{\pi}{K_i'}\right)^2 \frac{EI_i}{L^2} \qquad \text{(C-C2-3a)}$$

except

$$P_{ei}' < 1.6 P_{ei} \qquad \text{(C-C2-3b)}$$

or in terms of K directly with E and L^2 constant,

$$K_i' = \sqrt{\frac{\Sigma P_u}{P_{ui}} \times \frac{I_i}{\Sigma\left(\dfrac{I_i}{K_i^2}\right)}} \qquad \text{(C-C2-4a)}$$

except

$$K_i' \geq \sqrt{\frac{5}{8}} K_i \qquad \text{(C-C2-4b)}$$

where

  K_i' = effective length factor with story stability effect for ith rigid column
  I_i = moment of inertia in plane of bending for ith rigid column
  K_i = effective length for ith rigid-column factor based on alignment chart for unbraced frame
  P_{ui} = required axial compressive strength for ith rigid column
  ΣP_u = required axial compressive strength of all columns in a story

These expressions include consideration of leaning effects but, in addition, allow concentration of lateral stiffness on relatively weak columns. To limit the error involved with the assumption that ΣP_{e2} is constant and to avoid the possibility of failure of a weak column in the sidesway prevented mode, the modified P_{ei}' for a member should not exceed 1.6 times the P_{ei} for the member included in the sum ΣP_{e2}.

An alternate formulation which is simple to use but may give lower design values than the expressions above when leaning effects are minimal is:

$$P_{ei}' = \frac{P_{ui}}{\Sigma P_u}[\Sigma P_L][.85 + .15R_L] = \left(\frac{\pi}{K_i'}\right)^2 \frac{EI_i}{L^2} \qquad \text{(C-C2-5a)}$$

except

$$P_{ei}' \le 1.7P_{Li} \qquad \text{(C-C2-5b)}$$

where $P_{Li} = \dfrac{H_i L}{\Delta_{oh}}$ and H_i is the shear in the ith column included in ΣH.

The limits set by Equations C-C2-3b, C-C2-4b, and C-C2-5b have been chosen to avoid unconservative error exceeding five percent in extreme cases.

Although K_i' may be found, it is an unnecessary step since the key parameter $\lambda_c^2 = AF_y / P_e'$ is the only one required by Chapter E to find the design capacity $\phi_c P_n$. Design values may be found directly from:

$$\phi_c P_n = \phi_c\,.658^{\left[\frac{AF_y}{P_e'}\right]} AF_y \text{ when } P_e' > \frac{4}{9}AF_y \qquad \text{(C-C2-6a)}$$

$$\phi_c P_n = \phi_c\,.877 P_e' \text{ when } P_e' \le \frac{4}{9}AF_y \qquad \text{(C-C2-6b)}$$

Because frames that use partially restrained (PR) connections violate the condition that all joints are rigid, special attention should be paid to calculation of the proper G value (Barakat and Chen, 1991).

If roof decks or floor slabs, anchored to shear walls or vertical plane bracing systems, are counted upon to provide lateral support for individual columns in a building frame, due consideration must be given to their stiffness when functioning as horizontal diaphragms (Winter, 1958).

Translation of the joints in the plane of a truss is inhibited and, due to end restraint, the effective length of compression members might be assumed to be less than the distance between panel points. However, it is usual practice to take K as equal to 1.0 (Galambos, 1988); if all members of the truss reached their ultimate load capacity simultaneously, the restraints at the ends of the compression members would be greatly reduced.

CHAPTER D

TENSION MEMBERS

D1. DESIGN TENSILE STRENGTH

Due to strain hardening, a ductile steel bar loaded in axial tension can resist, without fracture, a force greater than the product of its gross area and its coupon yield stress. However, excessive elongation of a tension member due to uncontrolled yielding of its gross area not only marks the limit of its usefulness, but can precipitate failure of the structural system of which it is a part. On the other hand, depending upon the reduction of area and other mechanical properties of the steel, the member can fail by fracture of the net area at a load smaller than required to yield the gross area. Hence, general yielding of the gross area and fracture of the net area both constitute failure limit states. The relative values of ϕ_t given for yielding and fracture reflect the same basic difference in factor of safety as between design of members and design of connections in the AISC ASD Specification.

The length of the member in the net area is negligible relative to the total length of the member. As a result, the strain hardening condition is quickly reached and yielding of the net area at fastener holes does not constitute a limit state of practical significance.

D2. BUILT-UP MEMBERS

The slenderness ratio L/r of tension members other than rods, tubes, or straps should preferably not exceed the limiting value of 300. This slenderness limit recommended for tension members is not essential to the structural integrity of such members; it merely assures a degree of stiffness such that undesirable lateral movement ("slapping" or vibration) will be unlikely.

See Section B7 and Commentary Section E4.

D3. PIN-CONNECTED MEMBERS AND EYEBARS

Forged eyebars have generally been replaced by pin-connected plates or eyebars thermally cut from plates. Provisions for the proportioning of eyebars contained in the LRFD Specification are based upon standards evolved from long experience with forged eyebars. Through extensive destructive testing, eyebars have been found to provide balanced designs when they are thermally cut instead of forged. The somewhat more conservative rules for pin-connected members of nonuniform cross section and those not having enlarged "circular" heads are likewise based on the results of experimental research (Johnston, 1939).

Somewhat stockier proportions are provided for eyebars and pin-connected members fabricated from steel having a yield stress greater than 70 ksi, in order to eliminate any possibility of their "dishing" under the higher design stress.

CHAPTER E

COLUMNS AND OTHER COMPRESSION MEMBERS

E1. EFFECTIVE LENGTH AND SLENDERNESS LIMITATIONS

1. Effective Length

The Commentary on Section C2 regarding frame stability and effective length factors applies here. Further analytic methods, formulas, charts, and references for the determination of effective length are provided in Chapter 15 of the SSRC Guide (Galambos, 1988).

2. Design by Plastic Analysis

The limitation on λ_c is essentially the same as that for l/r in Chapter N of the 1989 AISC Specification—Allowable Stress Design and Plastic Design.

E2. DESIGN COMPRESSIVE STRENGTH FOR FLEXURAL BUCKLING*

Equations E2-2 and E2-3 are based on a reasonable conversion of research data into design equations. Conversion of the allowable stress design (ASD) equations which was based on the CRC—Column Research Council—curve (Galambos, 1988) was found to be cumbersome for two reasons. The first was the nature of the ASD variable safety factor. Secondly, the difference in philosophical origins of the two design procedures requires an assumption of a live load-to-dead load ratio (L/D).

Since all L/D ratios could not be considered, a value of approximately 1.1 at λ equal to 1.0 was used to calibrate the exponential equation for columns with the lower range of λ against the appropriate ASD provision. The coefficient with the Euler equation was obtained by equating the ASD and LRFD expressions at λ of 1.5.

Equations E2-2 and E2-3 are essentially the same curve as column-strength curve 2P of the Structural Stability Research Council which is based on an initial out-of-straightness curve of $l/1,500$ (Bjorhovde, 1972 and 1988; Galambos, 1988; Tide, 1985).

It should be noted that this set of column equations has a range of reliability (β) values. At low- and high-column slenderness, β values exceeding 3.0 and 3.3 respectively are obtained compared to β of 2.60 at L/D of 1.1. This is considered satisfactory, since the limits of out-of-straightness combined with residual stress have not been clearly established. Furthermore, there has been

---

* For tapered members see Commentary Appendix F3.

no history of unacceptable behavior of columns designed using the ASD procedure. This includes cases with L/D ratios greater than 1.1.

Equations E2-2 and E2-3 can be restated in terms of the more familiar slenderness ratio Kl/r. First, Equation E2-2 is expressed in exponential form,

$$F_{cr} = [\exp(-0.419\lambda_c^2)]F_y \qquad \text{(C-E2-1)}$$

Note that $\exp(x)$ is identical to e^x. Substitution of λ_c according to definition of λ_c in Section E2 gives,

For $\dfrac{Kl}{r} \leq 4.71\sqrt{\dfrac{E}{F_y}}$

$$F_{cr} = \left\{\exp\left[-0.0424\frac{F_y}{E}\left(\frac{Kl}{r}\right)^2\right]\right\}F_y \qquad \text{(C-E2-2)}$$

For $\dfrac{Kl}{r} > 4.71\sqrt{\dfrac{E}{F_y}}$

$$F_{cr} = \frac{0.877\pi^2 E}{\left(\dfrac{Kl}{r}\right)^2} \qquad \text{(C-E2-3)}$$

E3. DESIGN COMPRESSIVE STRENGTH FOR FLEXURAL-TORSIONAL BUCKLING

Torsional buckling of symmetric shapes and flexural-torsional buckling of unsymmetric shapes are failure modes usually not considered in the design of hot-rolled columns. They generally do not govern, or the critical load differs very little from the weak axis planar buckling load. Such buckling loads may, however, control the capacity of symmetric columns made from relatively thin plate elements and unsymmetric columns. Design equations for determining the strength of such columns are given in Appendix E3.

Tees that conform to the limits in Table C-E3.1 need not be checked for flexural-torsional buckling.

A simpler and more accurate design strength for the special case of tees and double-angles is based on Galambos (1991) wherein the y-axis of symmetry flexural-buckling strength component is determined directly from the column formulas.

The separate AISC *Specification for Load and Resistance Factor Design of Single-Angle Members* contains detailed provisions not only for the limit state of compression, but also for tension, shear, flexure, and combined forces.

TABLE C-E3.1
Limiting Proportions for Tees

Shape	Ratio of Full Flange Width to Profile Depth	Ratio of Flange Thickness to Web or Stem Thickness
Built-up tees	≥ 0.50	≥ 1.25
Rolled tees	≥ 0.50	≥ 1.10

E4. BUILT-UP MEMBERS

Requirements for detailing and design of built-up members, which cannot be stated in terms of calculated stress, are based upon judgment and experience.

The longitudinal spacing of connectors connecting components of built-up compression members must be such that the slenderness ratio l / r of individual shapes does not exceed three-fourths of the slenderness ratio of the entire member. Additional requirements are imposed for built-up members consisting of angles. However, these minimum requirements do not necessarily ensure that the effective slenderness ratio of the built-up member is equal to that for the built-up member acting as a single unit. Section E4 gives formulas for modified slenderness ratios that are based on research and take into account the effect of shear deformation in the connectors (Zandonini, 1985). Equation E4-1 for snug-tight intermediate connectors is empirically based on test results (Zandonini, 1985). The new Equation E4-2 is derived from theory and verified by test data. In both cases the end connection must be welded or slip-critical bolted (Aslani and Goel, 1991). The connectors must be designed to resist the shear forces which develop in the buckled member. The shear stresses are highest where the slope of the buckled shape is maximum (Bleich, 1952).

Maximum fastener spacing less than that required for strength may be needed to ensure a close fit over the entire faying surface of components in continuous contact. Specific requirements are given for weathering steel members exposed to atmospheric corrosion (Brockenbrough, 1983).

The provisions governing the proportioning of perforated cover plates are based upon extensive experimental research (Stang and Jaffe, 1948).

CHAPTER F

BEAMS AND OTHER FLEXURAL MEMBERS

F1. DESIGN FOR FLEXURE

1. Yielding

The bending strength of a laterally braced compact section is the plastic moment M_p. If the shape has a large shape factor (ratio of plastic moment to the moment corresponding to the onset of yielding at the extreme fiber), significant inelastic deformation may occur at service load if the section is permitted to reach M_p at factored load. The limit of $1.5M_y$ at factored load will control the amount of inelastic deformation for sections with shape factors greater than 1.5. This provision is not intended to limit the plastic moment of a hybrid section with a web yield stress lower than the flange yield stress. Yielding in the web does not result in significant inelastic deformations. In hybrid sections, $M_y = F_{yf}S$.

Lateral-torsional buckling cannot occur if the moment of inertia about the bending axis is equal to or less than the moment of inertia out of plane. Thus, for shapes bent about the minor axis and shapes with $I_x = I_y$, such as square or circular shapes, the limit state of lateral-torsional buckling is not applicable and yielding controls if the section is compact.

2. Lateral-Torsional Buckling

2a. Doubly Symmetric Shapes and Channels with $L_b \leq L_r$

The basic relationship between nominal moment M_n and unbraced length L_b is shown in Figure C-F1.1 for a compact section with $C_b = 1.0$. There are four principal zones defined on the basic curve by L_{pd}, L_p, and L_r. Equation F1-4 defines the maximum unbraced length L_p to reach M_p with uniform moment. Elastic lateral-torsional buckling will occur when the unbraced length is greater than L_r given by Equation F1-6. Equation F1-2 defines the inelastic lateral-torsional buckling as a straight line between the defined limits L_p and L_r. Buckling strength in the elastic region $L_b > L_r$ is given by Equation F1-14 for I-shaped members.

For other moment diagrams, the lateral buckling strength is obtained by multiplying the basic strength by C_b as shown in Figure C-F1.1. The maximum M_n, however, is limited to M_p. Note that L_p given by Equation F1-4 is merely a definition which has physical meaning when $C_b = 1.0$. For C_b greater than 1.0, larger unbraced lengths are permitted to reach M_p as shown by the curve for $C_b > 1.0$. For design, this length could be calculated by setting Equation F1-2 equal to M_p and solving this equation for L_b using the desired C_b value.

The equation

$$C_b = 1.75 + 1.05(M_1/M_2) + 0.3(M_1/M_2)^2 \le 2.3 \qquad \text{(C-F1-1)}$$

has been used since 1961 to adjust the flexural-torsional buckling equation for variations in the moment diagram within the unbraced length. This equation is applicable only to moment diagrams that are straight lines between braced points. The equation provides a lower bound fit to the solutions developed by Salvadori (1956) which are shown in Figure C-F1.2. Another equation

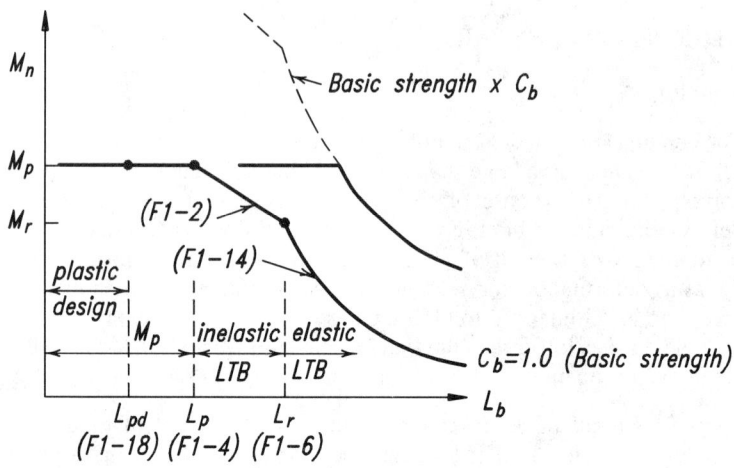

Fig. C-F1.1. Nominal moment as a function of unbraced length and moment gradient.

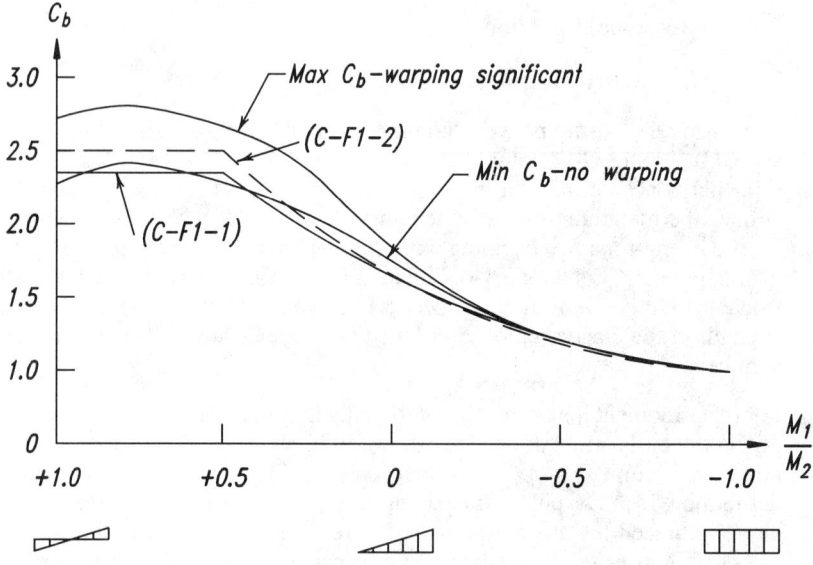

Fig. C-F1.2. Moment modifier C_b for beams.

$$C_b = \frac{1}{0.6 - 0.4\dfrac{M_1}{M_2}} \leq 2.5 \qquad \text{(C-F1-2)}$$

fits the average value theoretical solutions when the beams are bent in reverse curvature and also provides a reasonable fit to the theory. If the maximum moment within the unbraced segment is equal to or larger than the end moment, $C_b = 1.0$ is used.

The equations above can be easily misinterpreted and misapplied to moment diagrams that are not straight within the unbraced segment. Kirby and Nethercot (1979) presented an equation which applies to various shapes of moment diagrams within the unbraced segment. Their equation has been adjusted slightly to the following

$$C_b = \frac{12.5 M_{\max}}{2.5 M_{\max} + 3 M_A + 4 M_B + 3 M_C} \qquad \text{(C-F1-3)}$$

This equation gives more accurate solutions for fixed-end beams, and the adjusted equation reduces exactly to Equation C-F1-2 for a straight line moment diagram in single curvature. The new C_b equation is shown in Figure C-F1.3 for straight line moment diagrams. Other moment diagrams along with exact theoretical solutions in the SSRC Guide (Galambos, 1988) show good comparison with the new equation. The absolute value of the three interior quarter-point moments plus the maximum moment, regardless of its location are used in the equation. The maximum moment in the unbraced segment is always used for comparison with the resistance. The length between braces, not the distance to inflection points, and C_b is used in the resistance equation.

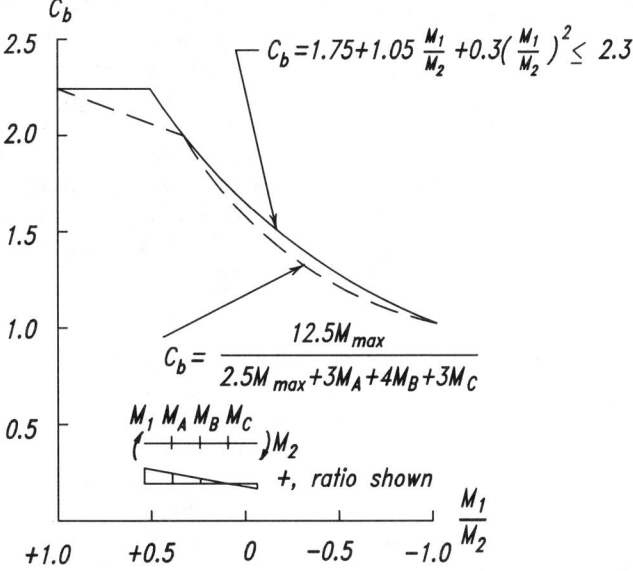

Fig. C-F1.3. C_b for a straight line moment diagram—prismatic beam.

It is still satisfactory to use the former C_b factor, Equation C-F1-1, for straight line moment diagrams within the unbraced length.

The elastic strength of hybrid beams is identical to homogeneous beams. The strength advantage of hybrid sections becomes evident only in the inelastic and plastic slenderness ranges.

2b. Doubly Symmetric Shapes and Channels with $L_b > L_r$

The equation given in the Specification assumes that the loading is applied along the beam centroidal axis. If the load is placed on the top flange and the flange is not braced, there is a tipping effect that reduces the critical moment; conversely, if the load is suspended from the bottom flange and is not braced, there is a stabilizing effect which increases the critical moment (Galambos, 1988). For unbraced top flange loading, the reduced critical moment may be conservatively approximated by setting the warping buckling factor X_2 to zero.

An effective length factor of unity is implied in these critical moment equations to represent a worst case pinned-pinned unbraced segment. Including consideration of any end restraint of the adjacent segments on the critical segment can increase its buckling capacity. The effects of beam continuity on lateral-torsional buckling have been studied and a simple and conservative design method, based on the analogy of end-restrained nonsway columns with an effective length factor less than one, has been proposed (Galambos, 1988).

2c. Tees and Double-Angles

The lateral-torsional buckling strength (LTS) of singly symmetric tee beams is given by a fairly complex formula (Galambos, 1988). Equation F1-15 is a simplified formulation based on Kitipornchai and Trahair (1980). See also Ellifritt, et al., 1992.

The C_b used for I-shaped beams is unconservative for tee beams with the stem in compression. For such cases $C_b = 1.0$ is appropriate. When beams are bent in reverse curvature, the portion with the stem in compression may control the LTB resistance even though the moments may be small relative to other portions of the unbraced length with $C_b \approx 1.0$. This is because the LTB strength of a tee with the stem in compression may be only about one-fourth of the capacity for the stem in tension. Since the buckling strength is sensitive to the moment diagram, C_b has been conservatively taken as 1.0. In cases where the stem is in tension, connection details should be designed to minimize any end restraining moments which might cause the stem to be in compression.

2d. Unbraced Length for Plastic Analysis

In the AISC ASD Specification, Chapter N, the unbraced length of a beam that permits the attainment of plastic moments, and ensures sufficient rotation capacity to redistribute moments, is given by two formulas which depend on the moment ratio at the ends of the unbraced length. One length is permitted for $M_1 / M_2 < -0.5$ (almost uniform moment), and a substantially larger length for $M_1 / M_2 > -0.5$. These two equations are replaced by Equation F1-18 to provide a continuous function between unbraced length and end moment ratio so there is no abrupt change for a slight change in moment ratio near -0.5. At $M_1 / M_2 = -0.5$ (uniform moment) the maximum unbraced length is almost the same as that in the AISC ASD Specification. There is a substantial increase in unbraced length for positive moment ratios (reverse curvature) because the yielding is confined to zones close to the brace points (Yura, et al., 1978).

Equation F1-18 is an equation in similar form for solid rectangular bars and symmetric box beams. Equations F1-17 and F1-18 assume that the moment diagram within the unbraced length next to plastic hinge locations is reasonably linear. For nonlinear diagrams between braces, judgment should be used in choosing a representative ratio.

Equations F1-17 and F1-18 were developed to provide rotation capacities of at least 3.0, which are sufficient for most applications (Yura, et al., 1978). When inelastic rotations of 7 to 9 are deemed appropriate in areas of high seismicity, as discussed in Commentary Section B5, Equation F1-17 would become:

$$L_{pd} = \frac{2500 r_y}{F_y} \qquad \text{(C-F1-3)}$$

F2.   DESIGN FOR SHEAR

For unstiffened webs $k_v = 5.0$,

therefore $187\sqrt{k_v / F_{yw}} = 418 / \sqrt{F_{yw}}$, and $234\sqrt{k_v / F_{yw}} = 523 / \sqrt{F_{yw}}$.

For webs with $h / t_w \le 187\sqrt{k_v / F_{yw}}$, the nominal shear strength V_n is based on shear yielding of the web, Equation F2-1 and Equation A-F2-1. This h / t_w limit was determined by setting the critical stress causing shear buckling F_{cr} equal to the yield stress of the web F_{yw} in Equation 35 of Cooper et al. (1978) and Timoshenko and Gere (1961). When $h / t_w > 187\sqrt{k_v / F_{yw}}$, the web shear strength is based on buckling. Basler (1961) suggested taking the proportional limit as 80 percent of the yield stress of the web. This corresponds to $h / t_w = (187/0.8)$ $(\sqrt{k_v / F_{yw}})$. Thus, when $h / t_w > 234(\sqrt{k_v / F_{yw}})$, the web strength is determined from the elastic buckling stress given by Equation 6 of Cooper et al., (1978) and Timoshenko and Gere (1961):

$$F_{cr} = \frac{\pi^2 E k_v}{12(1 - v^2)(h / t_w)^2} \qquad \text{(C-F2-1)}$$

The nominal shear strength, given by Equation F2-3 and A-F2-3, was obtained by multiplying F_{cr} by the web area and using $E = 29{,}000$ ksi and $v = 0.3$. A straight line transition, Equation F2-2 and AF2-2, is used between the limits $187(\sqrt{k_v / F_{yw}})$ and $234(\sqrt{k_v / F_{yw}})$.

The shear strength of flexural members follows the approach used in the AISC ASD Specification, except for two simplifications. First, the expression for the plate buckling coefficient k_v has been simplified; it corresponds to that given by AASHTO *Standard Specification for Highway Bridges* (1989). The earlier expression for k_v was a curve fit to the exact expression; the new expression is just as accurate. Second, the alternate method (tension field action) for web shear strength is placed in Appendix G because it was desired that only one method appear in the main body of the Specification with alternate methods given in the Appendix. When designing plate girders, thicker unstiffened webs will frequently be less costly than lighter stiffened web designs because of the additional fabrication. If a stiffened girder design has economic advantages, the tension field method in Appendix G will require fewer stiffeners.

The equations in this section were established assuming monotonically increasing loads. If a flexural member is subjected to load reversals causing cyclic yielding over large portions of a web, such as may occur during a major earthquake, special design considerations may apply (Popov, 1980).

F4. BEAMS AND GIRDERS WITH WEB OPENINGS

Web openings in structural floor members may be necessary to accommodate various mechanical, electrical, and other systems. Strength limit states, including local buckling of the compression flange, web, and tee-shaped compression zone above or below the opening, lateral buckling and moment-shear interaction, or serviceability may control the design of a flexural member with web openings. The location, size, and number of openings are important and empirical limits for them have been identified. One general procedure for assessing these effects and the design of any needed reinforcement for both steel and composite beams is given in Darwin (1990) and in ASCE (1992, 1992a).

CHAPTER H

MEMBERS UNDER COMBINED FORCES AND TORSION

H1. SYMMETRIC MEMBERS SUBJECT TO BENDING AND AXIAL FORCE

Equations H1-1a and H1-1b are simplifications and clarifications of similar equations used in the AISC ASD Specification since 1961. Previously, both equations had to be checked. In the new formulation the applicable equation is governed by the value of the first term, $P_u / \phi P_n$. For bending about one axis only, the equations have the form shown in Figure C-H1.1.

The first term $P_u / \phi P_n$ has the same significance as the axial load term f_a / F_a in Equations H1-1 of the AISC ASD Specification. This means that for members in compression P_n must be based on the largest effective slenderness ratio Kl / r. In the development of Equations H1-1a and H1-1b, a number of alternative formulations were compared to the exact inelastic solutions of 82 sidesway cases reported in Kanchanalai (1977). In particular, the possibility of using Kl / r as the actual column length ($K = 1$) in determining P_n, combined with an elastic second order moment M_u, was studied. In those cases where the true P_n based on Kl / r, with $K = 1.0$, was in the inelastic range, the errors proved to be unacceptably large without the additional check that $P_u \le \phi_c P_n$, P_n being based on effective length. Although deviations from exact solutions were reduced, they still remained high.

In summary, it is not possible to formulate a safe general interaction equation

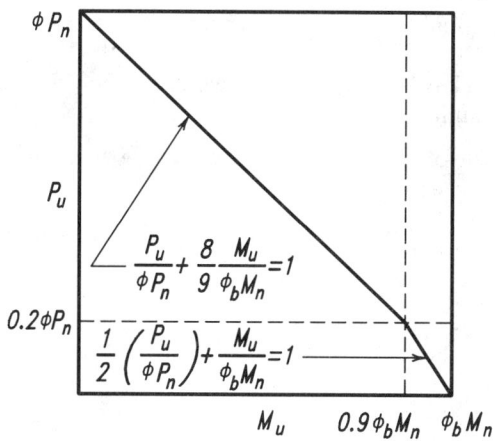

Fig. C-H1.1. Beam-column interaction equations.

for compression without considering effective length directly (or indirectly by a second equation). Therefore, the requirement that the nominal compressive strength P_n be based on the effective length KL in the general equation is continued in the LRFD Specification as it has been in the AISC ASD Specification since 1961. It is not intended that these provisions be applicable to limit nonlinear secondary flexure that might be encountered in large amplitude earthquake stability design (ATC, 1978).

The defined term M_u is the maximum moment in a member. In the calculation of this moment, inclusion of beneficial second order effects of tension is optional. But consideration of detrimental second order effects of axial compression and translation of gravity loads is required. Provisions for calculation of these effects are given in Chapter C.

The interaction equations in Appendix H3 have been recommended for biaxially loaded H and wide flange shapes in Galambos (1988) and Springfield (1975). These equations which can be used only in braced frames represent a considerable liberalization over the provisions given in Section H1; it is, therefore, also necessary to check yielding under service loads, using the appropriate load and resistance factors for the serviceability limit state in Equation H1-1a or H1-1b with $M_{ux} = S_x F_y$ and $M_{uy} = S_y F_y$. Appendix H3 also provides interaction equations for rectangular box-shaped beam-columns. These equations are taken from Zhou and Chen (1985).

H2.   UNSYMMETRIC MEMBERS AND MEMBERS UNDER TORSION AND COMBINED TORSION, FLEXURE, SHEAR, AND/OR AXIAL FORCE

This section deals with types of cross sections and loadings not covered in Section H1, especially where torsion is a consideration. For such cases it is recommended to perform an elastic analysis based on the theoretical numerical methods available from the literature for the determination of the maximum normal and shear stresses, or for the elastic buckling stresses. In the buckling calculations an equivalent slenderness parameter is determined for use in Equation E2-2 or E2-3, as follows:

$$\lambda_e = \sqrt{F_y / F_e}$$

where F_e is the elastic buckling stress determined from a stability analysis. This procedure is similar to that of Appendix E3.

For the analysis of members with open sections under torsion refer to AISC (1983).

CHAPTER I

COMPOSITE MEMBERS

I1. DESIGN ASSUMPTIONS

Force Determination. Loads applied to an unshored beam before the concrete has hardened are resisted by the steel section alone, and only loads applied after the concrete has hardened are considered as resisted by the composite section. It is usually assumed for design purposes that concrete has hardened when it attains 75 percent of its design strength. In beams properly shored during construction, all loads may be assumed as resisted by the composite cross section. Loads applied to a continuous composite beam with shear connectors throughout its length, after the slab is cracked in the negative moment region, are resisted in that region by the steel section and by properly anchored longitudinal slab reinforcement.

For purposes of plastic analysis all loads are considered resisted by the composite cross section, since a fully plastic strength is reached only after considerable yielding at the locations of plastic hinges.

Elastic Analysis. The use of constant stiffness in elastic analyses of continuous beams is analogous to the practice in reinforced concrete design.

Plastic Analysis. For composite beams with shear connectors, plastic analysis may be used only when the steel section in the positive moment region has a compact web, i.e., $h/t_w \leq 640/\sqrt{F_{yf}}$, and when the steel section in the negative moment region is compact, as required for steel beams alone. No compactness limitations are placed on encased beams, but plastic analysis is permitted only if the direct contribution of concrete to the strength of sections is neglected; the concrete is relied upon only to prevent buckling.

Plastic Stress Distribution for Positive Moment. Plastic stress distributions are described in Commentary Section I3, and a discussion of the composite participation of slab reinforcement is presented.

Plastic Stress Distribution for Negative Moment. Plastic stress distributions are described in Commentary Section I3.

Elastic Stress Distribution. The strain distribution at any cross section of a composite beam is related to slip between the structural steel and concrete elements. Prior to slip, strain in both steel and concrete is proportional to the distance from the neutral axis for the elastic transformed section. After slip, the strain distribution is discontinuous, with a jump at the top of the steel shape. The strains in steel and concrete are proportional to distances from separate neutral axes, one for steel and the other for concrete.

Fully Composite Beam. Either tensile yield strength of the steel section or the compressive stress of the concrete slab governs the maximum flexural strength of a fully composite beam subjected to a positive moment. The tensile yield strength of the longitudinal reinforcing bars in the slab governs the maximum flexural strength of a fully composite beam subjected to a negative moment. When shear connectors are provided in sufficient numbers to fully develop this maximum flexural strength, any slip that occurs prior to yielding is minor and has negligible influence both on stresses and stiffness.

Partially Composite Beam. The effects of slip on elastic properties of a partially composite beam can be significant and should be accounted for in calculations of deflections and stresses at service loads. Approximate elastic properties of partially composite beams are given in Commentary Section I3. For simplified design methods, see Hansell, et al. (1978).

Concrete-Encased Beam. When the dimensions of a concrete slab supported on steel beams are such that the slab can effectively serve as the flange of a composite T-beam, and the concrete and steel are adequately tied together so as to act as a unit, the beam can be proportioned on the assumption of composite action.

Two cases are recognized: fully encased steel beams, which depend upon natural bond for interaction with the concrete, and those with mechanical anchorage to the slab (shear connectors), which do not have to be encased.

I2.   COMPRESSION MEMBERS

1.   Limitations

(a) The lower limit of four percent on the cross-sectional area of structural steel differentiates between composite and reinforced concrete columns. If the area is less than four percent, a column with a structural steel core should be designed as a reinforced concrete column.

(b) The specified minimum quantity of transverse and longitudinal reinforcement in the encasement should be adequate to prevent severe spalling of the surface concrete during fires.

(c) Very little of the supporting test data involved concrete strengths in excess of 6 ksi, even though the cylinder strength for one group of four columns was 9.6 ksi. Normal weight concrete is believed to have been used in all tests. Thus, the upper limit of concrete strength is specified as 8 ksi for normal weight concrete. A lower limit of 3 ksi is specified for normal weight concrete and 4 ksi for lightweight concrete to encourage the use of good quality, yet readily available, grades of structural concrete.

(d) Encased steel shapes and longitudinal reinforcing bars are restrained from buckling as long as the concrete remains sound. A limit strain of 0.0018, at which unconfined concrete remains unspalled and stable, serves analytically to define a failure condition for composite cross sections under uniform axial strain. The limit strain of 0.0018 corresponds approximately to 55 ksi.

(e) The specified minimum wall thicknesses are identical to those in the 1989 ACI Building Code (1989). The purpose of this provision is to prevent buckling of the steel pipe or tubing before yielding.

2. Design Strength

The procedure adopted for the design of axially loaded composite columns is described in detail in Galambos and Chapuis (1980). It is based on the equation for the strength of a short column derived in Galambos and Chapuis (1980), and the same reductions for slenderness as those specified for steel columns in Section E2. The design follows the same path as the design of steel columns, except that the yield stress of structural steel, the modulus of elasticity of steel, and the radius of gyration of the steel section, are modified to account for the effect of concrete and longitudinal reinforcing bars. A detailed explanation of the origin of these modifications may be found in SSRC Task Group 20 (1979). Galambos and Chapuis (1980) includes comparisons of the design procedure with 48 tests of axially loaded stub columns, 96 tests of concrete-filled pipes or tubing, and 26 tests of concrete-encased steel shapes. The mean ratio of the test failure loads to the predicted strengths was 1.18 for all 170 tests, and the corresponding coefficient of variation was 0.19.

3. Columns with Multiple Steel Shapes

This limitation is based on Australian research reported in Bridge and Roderick (1978), which demonstrated that after hardening of concrete the composite column will respond to loading as a unit even without lacing, tie plates, or batten plates connecting the individual steel sections.

4. Load Transfer

To avoid overstressing either the structural steel section or the concrete at connections, a transfer of load to concrete by direct bearing is required.

When a supporting concrete area is wider on all sides than the loaded area, the maximum design strength of concrete is specified by ACI (1989) as $1.7\phi_B f_c' A_B$ where $\phi_B = 0.7$ is the strength reduction factor in bearing on concrete and A_B is the loaded area. Because the AISC LRFD Specification is based on the lower ASCE 7 load factors (ASCE, 1988), $\phi_B = 0.60$ in the AISC LRFD Specification. The portion of the design load of an axially loaded column ϕP_n resisted by the concrete may be expressed as $(c_2 f_c' A_c / A_s F_{my})\phi_B P_n$.

Accordingly,

$$A_B \geq \frac{\phi_B\, c_2\, A_c\, P_n}{\phi_B\, 1.7 A_s F_{my}} = \frac{c_2\, A_c\, P_n}{1.7 A_s F_{my}} \qquad \text{(C-I2-1)}$$

I3. FLEXURAL MEMBERS

1. Effective Width

LRFD provisions for effective width omit any limit based on slab thickness, in accordance with both theoretical and experimental studies, as well as current composite beam codes in other countries (ASCE, 1979). The same effective width rules apply to composite beams with a slab on either one side or both sides of the beam. To simplify design, effective width is based on the full span, center-to-center of supports, for both simple and continuous beams.

2. Strength of Beams with Shear Connectors

This section applies to simple and continuous composite beams with shear connectors, constructed with or without temporary shores.

Positive Flexural Design Strength. Flexural strength of a composite beam in the positive moment region may be limited by the plastic strength of the steel section, the concrete slab, or shear connectors. In addition, web buckling may limit flexural strength if the web is slender and a significantly large portion of the web is in compression.

According to Table B5.1, local web buckling does not reduce the plastic strength of a bare steel beam if the beam depth-to-web thickness ratio is not larger than $640 / \sqrt{F_y}$. In the absence of web buckling research on composite beams, the same ratio is conservatively applied to composite beams. Furthermore, for more slender webs, the LRFD Specification conservatively adopts first yield as the flexural strength limit. In this case, stresses on the steel section from permanent loads applied to unshored beams before the concrete has hardened must be superimposed on stresses on the composite section from loads applied to the beams after hardening of concrete. In this superposition, all permanent loads should be multiplied by the dead load factor and all live loads should be multiplied by the live load factor. For shored beams, all loads may be assumed as resisted by the composite section.

When first yield is the flexural strength limit, the elastic transformed section is used to calculate stresses on the composite section. The modular ratio $n = E / E_c$ used to determine the transformed section depends on the specified unit weight and strength of concrete. Note that this procedure for compact beams differs from the requirements of Section I2 of the 1989 AISC ASD Specification.

Plastic Stress Distribution for Positive Moment. When flexural strength is determined from the plastic stress distribution shown in Figure C-I3.1, compression force C in the concrete slab is the smallest of:

$$C = A_{sw}F_{yw} + 2A_{sf}F_{yf} \tag{C-I3-1}$$

$$C = 0.85f_c'A_c \tag{C-I3-2}$$

$$C = \Sigma Q_n \tag{C-I3-3}$$

For a non-hybrid steel section, Equation C-I3-1 becomes $C = A_s F_y$

where

f_c' = specified compressive strength of concrete, ksi
A_c = area of concrete slab within effective width, in.^{2}
A_s = area of steel cross section, in.^{2}
A_{sw} = area of steel web, in.^{2}
A_{sf} = area of steel flange, in.^{2}
F_y = minimum specified yield stress of steel, ksi
F_{yw} = minimum specified yield stress of web steel, ksi
F_{yf} = minimum specified yield stress of flange steel, ksi
ΣQ_n = sum of nominal strengths of shear connectors between the point of

maximum positive moment and the point of zero moment to either side, kips

Longitudinal slab reinforcement makes a negligible contribution to the compression force, except when Equation C-I3-2 governs. In this case, the area of longitudinal reinforcement within the effective width of the concrete slab times the yield stress of the reinforcement may be added in determining C.

The depth of the compression block is

$$a = \frac{C}{0.85f_c'b} \qquad \text{(C-I3-4)}$$

where

b = effective width of concrete slab, in.

A fully composite beam corresponds to the case of C governed by the yield strength of the steel beam or the compressive strength of the concrete slab, as in Equation C-I3-1 or C-I3-2. The number and strength of shear connectors govern C for a partially composite beam as in Equation C-I3-3.

The plastic stress distribution may have the plastic neutral axis (PNA) in the web, in the top flange of the steel section or in the slab, depending on the value of C.

The nominal plastic moment resistance of a composite section in positive bending is given by the following equation and Figure C-I3.1:

$$M_n = C(d_1 + d_2) + P_y(d_3 - d_2) \qquad \text{(C-I3-5)}$$

where

P_y = tensile strength of the steel section; for a non-hybrid steel section $P_y = A_s F_y$, kips
d_1 = distance from the centroid of the compression force C in concrete to the top of the steel section, in.
d_2 = distance from the centroid of the compression force in the steel section to the top of the steel section, in. For the case of no compression in the steel section $d_2 = 0$.

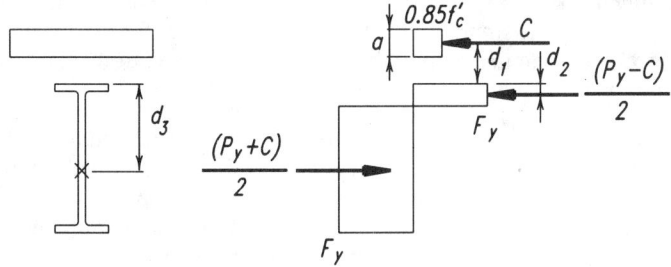

Fig. C-I3.1. Plastic stress distribution for positive moment in composite beams.

d_3 = distance from P_y to the top of the steel section, in.

Equation C-I3-5 is generally applicable including both non-hybrid and hybrid steel sections symmetrical about one or two axes.

Approximate Elastic Properties of Partially Composite Beams. Elastic calculations for stress and deflection of partially composite beams should include the effects of slip.

The effective moment of inertia I_{eff} for a partially composite beam is approximated by

$$I_{eff} = I_s + \sqrt{(\Sigma Q_n / C_f)}(I_{tr} - I_s) \tag{C-I3-6}$$

where

I_s = moment of inertia for the structural steel section, in.[4]

I_{tr} = moment of inertia for the fully composite uncracked transformed section, in.[4]

ΣQ_n = strength of shear connectors between the point of maximum positive moment and the point of zero moment to either side, kips

C_f = compression force in concrete slab for fully composite beam; smaller of Equations C-I3-1 and C-I3-2, kips

The effective section modulus S_{eff}, referred to the tension flange of the steel section for a partially composite beam, is approximated by

$$S_{eff} = S_s + \sqrt{(\Sigma Q_n / C_f)}(S_{tr} - S_s) \tag{C-I3-7}$$

where

S_s = section modulus for the structural steel section, referred to the tension flange, in.[3]

S_{tr} = section modulus for the fully composite uncracked transformed section, referred to the tension flange of the steel section, in.[3]

Equations C-I3-6 and C-I3-7 should not be used for ratios $\Sigma Q_n / C_f$ less than 0.25. This restriction is to prevent excessive slip, as well as substantial loss in beam stiffness. Studies indicate that Equations C-I3-6 and C-I3-7 adequately reflect the reduction in beam stiffness and strength, respectively, when fewer connectors are used than required for full composite action (Grant et al., 1977).

Negative Flexural Design Strength. The flexural strength in the negative moment region is the strength of the steel beam alone or the plastic strength of the composite section made up of the longitudinal slab reinforcement and the steel section.

Plastic Stress Distribution for Negative Moment. When an adequately braced compact steel section and adequately developed longitudinal reinforcing bars act compositely in the negative moment region, the nominal flexural strength is determined from the plastic stress distributions as shown in Figure C-I3.2. The tensile force T in the reinforcing bars is the smaller of:

$$T = A_r F_{yr} \tag{C-I3-8}$$

$$T = \Sigma Q_n \qquad \text{(C-I3-9)}$$

where

A_r = area of properly developed slab reinforcement parallel to the steel beam and within the effective width of the slab, in.[2]

F_{yr} = specified yield stress of the slab reinforcement, ksi

ΣQ_n = sum of the nominal strengths of shear connectors between the point of maximum negative moment and the point of zero moment to either side, kips

A third theoretical limit on T is the product of the area and yield stress of the steel section. However, this limit is redundant in view of practical limitations on slab reinforcement.

The nominal plastic moment resistance of a composite section in negative bending is given by the following equation:

$$M_n = T (d_1 + d_2) + P_{yc} (d_3 - d_2) \qquad \text{(C-I3-10)}$$

where

P_{yc} = the compressive strength of the steel section; for a non-hybrid section $P_{yc} = A_s F_y$, kips

d_1 = distance from the centroid of the longitudinal slab reinforcement to the top of the steel section, in.

d_2 = distance from the centroid of the tension force in the steel section to the top of the steel section, in.

d_3 = distance from P_{yc} to the top of the steel section, in.

Transverse Reinforcement for the Slab. Where experience has shown that longitudinal cracking detrimental to serviceability is likely to occur, the slab should be reinforced in the direction transverse to the supporting steel section. It is recommended that the area of such reinforcement should be at least 0.002 times the concrete area in the longitudinal direction of the beam and should be uniformly distributed.

3. Strength of Concrete-Encased Beams

Tests of concrete-encased beams demonstrated that (1) the encasement drastically reduces the possibility of lateral-torsional instability and prevents local buckling of the encased steel, (2) the restrictions imposed on the encasement

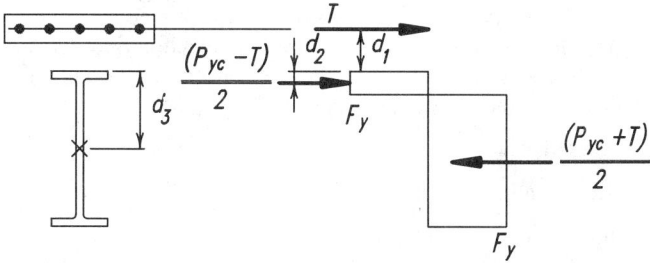

Fig. C-I3.2. Plastic stress distribution for negative moment.

practically prevent bond failure prior to first yielding of the steel section, and (3) bond failure does not necessarily limit the moment capacity of an encased steel beam (ASCE, 1979). Accordingly, the LRFD Specification permits two alternate design methods: one based on the first yield in the tension flange of the composite section and the other based on the plastic moment capacity of the steel beam alone. No limitations are placed on the slenderness of either the composite beam or the elements of the steel section, since the encasement effectively inhibits both local and lateral buckling.

In the method based on first yield, stresses on the steel section from permanent loads applied to unshored beams before the concrete has hardened must be superimposed on stresses on the composite section from loads applied to the beams after hardening of the concrete. In this superposition, all permanent loads should be multiplied by the dead load factor and all live loads should be multiplied by the live load factor. For shored beams, all loads may be assumed as resisted by the composite section. Complete interaction (no slip) between the concrete and steel is assumed.

The contribution of concrete to the strength of the composite section is ordinarily larger in positive moment regions than in negative moment regions. Accordingly, design based on the composite section is more advantageous in the regions of positive moments.

4.   Strength During Construction

When temporary shores are not used during construction, the steel beam alone must resist all loads applied before the concrete has hardened enough to provide composite action. Unshored beam deflection caused by wet concrete tends to increase slab thickness and dead load. For longer spans this may lead to instability analogous to roof ponding. An excessive increase of slab thickness may be avoided by beam camber.

When forms are not attached to the top flange, lateral bracing of the steel beam during construction may not be continuous and the unbraced length may control flexural strength, as defined in Section F1.

The LRFD Specification does not include special requirements for a margin against yield during construction. According to Section F1, maximum factored moment during construction is $0.90F_yZ$ where F_yZ is the plastic moment $(0.90F_yZ \approx 0.90 \times 1.1F_yS)$. This is equivalent to approximately the yield moment, F_yS. Hence, required flexural strength during construction prevents moment in excess of the yield moment.

Load factors for construction loads should be determined for individual projects according to local conditions, with the factors listed in Section A4 as a guide. Once the concrete has hardened, slab weight becomes a permanent dead load and the dead load factor applies to any load combinations.

5.   Formed Steel Deck

Figure C-I3.3 is a graphic presentation of the terminology used in Section I3.5.

When studs are used on beams with formed steel deck, they may be welded directly through the deck or through prepunched or cut-in-place holes in the deck. The usual procedure is to install studs by welding directly through the deck; however, when the deck thickness is greater than 16 gage for single thickness, or 18 gage for each sheet of double thickness, or when the total thickness of galvanized coating is greater than 1.25 ounces/sq. ft, special precautions and procedures recommended by the stud manufacturer should be followed.

The design rules for composite construction with formed steel deck are based upon a study (Grant, et al., 1977) of the then available test results. The limiting parameters listed in Section I3.5 were established to keep composite construction with formed steel deck within the available research data.

Seventeen full size composite beams with concrete slab on formed steel deck

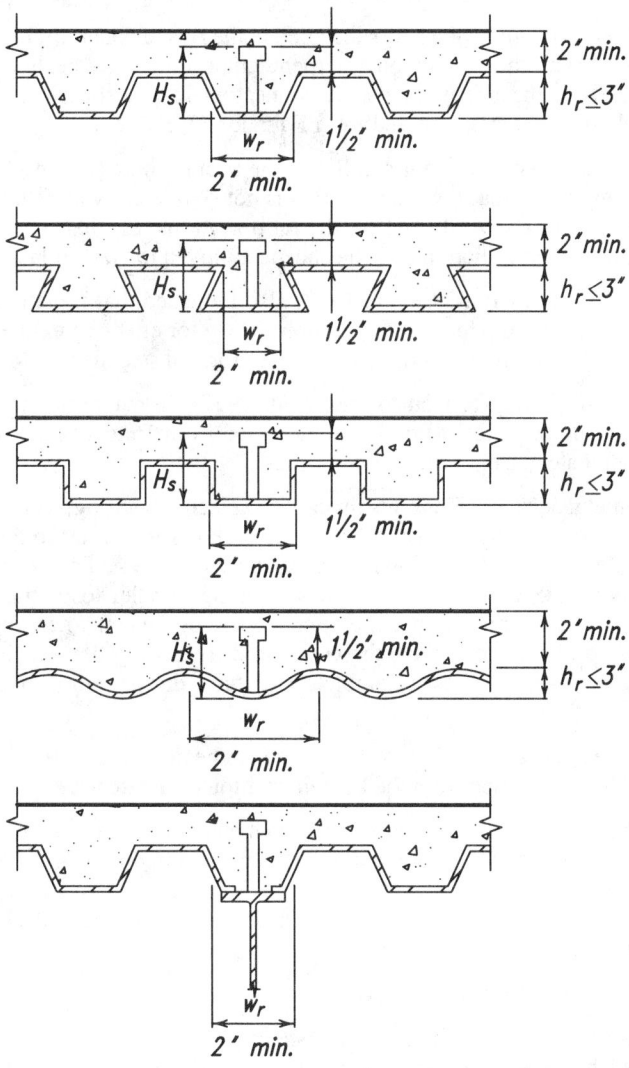

Fig. C-I3.3. Steel deck limits.

were tested at Lehigh University and the results supplemented by the results of 58 tests performed elsewhere. The range of stud and steel deck dimensions encompassed by the 75 tests were limited to:

(1) Stud dimensions: $\frac{3}{4}$-in. dia. $\times$ 3.00 to 7.00 in.

(2) Rib width: 1.94 in. to 7.25 in.

(3) Rib height: 0.88 in. to 3.00 in.

(4) Ratio w_r / h_r: 1.30 to 3.33

(5) Ratio H_s / h_r: 1.50 to 3.41

(6) Number of studs in any one rib: 1, 2, or 3

The strength of stud connectors installed in the ribs of concrete slabs on formed steel deck with the ribs oriented perpendicular to the steel beam is reasonably estimated by the strength of stud connectors in flat soffit composite slabs multiplied by values computed from Equation I3-1.

For the case where ribs run parallel to the beam, limited testing (Grant et al., 1977) has shown that shear connection is not significantly affected by the ribs. However, for narrow ribs, where the ratio w_r / h_r is less than 1.5, a shear stud reduction factor, Equation I3-2, has been employed in view of lack of test data.

The Lehigh study (Grant et al., 1977) also indicated that Equation C-I3-7 for effective section modulus and Equation C-I3-6 for effective moment of inertia were valid for composite construction with formed steel deck.

Based on the Lehigh test data (Grant, et al., 1977), the maximum spacing of steel deck anchorage to resist uplift was increased from 16 to 18 inches in order to accommodate current production profiles.

When metal deck includes units for carrying electrical wiring, crossover headers are commonly installed over the cellular deck perpendicular to the ribs. They create trenches which completely or partially replace sections of the concrete slab above the deck. These trenches, running parallel to or transverse to a composite beam, may reduce the effectiveness of the concrete flange. Without special provisions to replace the concrete displaced by the trench, the trench should be considered as a complete structural discontinuity in the concrete flange.

When trenches are parallel to the composite beam, the effective flange width should be determined from the known position of the trench.

Trenches oriented transverse to composite beams should, if possible, be located in areas of low bending moment and the full required number of studs should be placed between the trench and the point of maximum positive moment. Where the trench cannot be located in an area of low moment, the beam should be designed as non-composite.

6.   Design Shear Strength

A conservative approach to vertical shear provisions for composite beams is adopted by assigning all shear to the steel section web. This neglects any concrete slab contribution and serves to simplify the design.

I4.　COMBINED COMPRESSION AND FLEXURE

The procedure adopted for the design of beam-columns is described and supported by comparisons with test data in Galambos and Chapuis (1980). The basic approach is identical to that specified for steel columns in Section H1.

The nominal axial strength of a beam-column is obtained from Section I2.2, while the nominal flexural strength is determined from the plastic stress distribution on the composite section. An approximate formula for this plastic moment resistance of a composite column is given in Galambos and Chapuis (1980).

$$M_n = M_p = ZF_y + \tfrac{1}{3}(h_2 - 2c_r)A_rF_{yr} + \left(\frac{h_2}{2} - \frac{A_wF_y}{1.7f_c'h_1}\right)A_wF_y \qquad \text{(C-I4-1)}$$

where

A_w = web area of encased steel shape; for concrete-filled tubes, $A_w = 0$, in.^{2}

Z = plastic section modulus of the steel section, in.^{3}

c_r = average of distance from compression face to longitudinal reinforcement in that face and distance from tension face to longitudinal reinforcement in that face, in.

h_1 = width of composite cross section perpendicular to the plane of bending, in.

h_2 = width of composite cross section parallel to the plane of bending, in.

The supporting comparisons with beam-column tests included 48 concrete-filled pipes or tubing and 44 concrete-encased steel shapes (Galambos and Chapuis, 1980). The overall mean test-to-prediction ratio was 1.23 and the coefficient of variation 0.21.

The last paragraph in Section I4 provides a transition from beam-columns to beams. It involves bond between the steel section and concrete. Section I3 for beams requires either shear connectors or full, properly reinforced encasement of the steel section. Furthermore, even with full encasement, it is assumed that bond is capable of developing only the moment at first yielding in the steel of the composite section. No test data are available on the loss of bond in composite beam-columns. However, consideration of tensile cracking of concrete suggests $P_u / \phi_c P_n = 0.3$ as a conservative limit. It is assumed that when $P_u / \phi_c P_n$ is less than 0.3, the nominal flexural strength is reduced below that indicated by plastic stress distribution on the composite cross section unless the transfer of shear from the concrete to the steel is provided for by shear connectors.

I5.　SHEAR CONNECTORS

1.　Materials

Tests (Ollgaard et al., 1971) have shown that fully composite beams with concrete meeting the requirements of Part 3, Chapter 4, "Concrete Quality," of ACI (1989), made with ASTM C33 or rotary-kiln produced C330 aggregates, develop full flexural capacity.

2. Horizontal Shear Force

Composite beams in which the longitudinal spacing of shear connectors was varied according to the intensity of statical shear, and duplicate beams in which the connectors were uniformly spaced, exhibited the same ultimate strength and the same amount of deflection at normal working loads. Only a slight deformation in the concrete and the more heavily stressed connectors is needed to redistribute the horizontal shear to other less heavily stressed connectors. The important consideration is that the total number of connectors be sufficient to develop the shear V_h on either side of the point of maximum moment. The provisions of the LRFD Specification are based upon this concept of composite action.

In computing the design flexural strength at points of maximum negative bending, reinforcement parallel to the steel beam within the effective width of the slab may be included, provided such reinforcement is properly anchored beyond the region of negative moment. However, enough shear connectors are required to transfer, from the slab to the steel beam, the ultimate tensile force in the reinforcement.

3. Strength of Stud Shear Connectors

Studies have defined stud shear connector strength in terms of normal weight and lightweight aggregate concretes as a function of both concrete modulus of elasticity and concrete strength as given by Equation I5-1.

Equation I5-1, obtained from Ollgaard, et al. (1971), corresponds to Tables I4.1 and I4.2 in Section I4 of the 1989 AISC ASD Specification. Note that an upper bound on stud shear strength is the product of the cross-sectional area of the stud times its ultimate tensile strength.

The LRFD Specification does not specify a resistance factor for shear connector strength. The resistance factor for the flexural strength of a composite beam accounts for all sources of variability, including those associated with the shear connectors.

4. Strength of Channel Shear Connectors

Equation I5-2 is a modified form of the formula for the strength of channel connectors developed by Slutter and Driscoll (1965). The modification has extended its use to lightweight concrete.

6. Shear Connector Placement and Spacing

Uniform spacing of shear connectors is permitted except in the presence of heavy concentrated loads.

When stud shear connectors are installed on beams with formed steel deck, concrete cover at the sides of studs adjacent to sides of steel ribs is not critical. Tests have shown that studs installed as close as is permitted to accomplish welding of studs does not reduce the composite beam capacity.

Studs not located directly over the web of a beam tend to tear out of a thin flange before attaining full shear-resisting capacity. To guard against this contingency, the size of a stud not located over the beam web is limited to 2½ times the flange thickness (Goble, 1968).

The minimum spacing of connectors along the length of the beam, in both flat soffit concrete slabs and in formed steel deck with ribs parallel to the beam, is six diameters; this spacing reflects development of shear planes in the concrete slab (Ollgaard et al., 1971). Since most test data are based on the minimum transverse spacing of four diameters, this transverse spacing was set as the minimum permitted. If the steel beam flange is narrow, this spacing requirement may be achieved by staggering the studs with a minimum transverse spacing of three diameters between the staggered row of studs. The reduction in connector capacity in the ribs of formed steel decks is provided by the factor $0.85 / \sqrt{N_r}$, which accounts for the reduced capacity of multiple connectors, including the effect of spacing. When deck ribs are parallel to the beam and the design requires more studs than can be placed in the rib, the deck may be split so that adequate spacing is available for stud installation. Figure C-I5.1 shows possible connector arrangements.

I6.   SPECIAL CASES

Tests are required for construction that falls outside the limits given in the Specification. Different types of shear connectors may require different spacing and other detailing than stud and channel connectors.

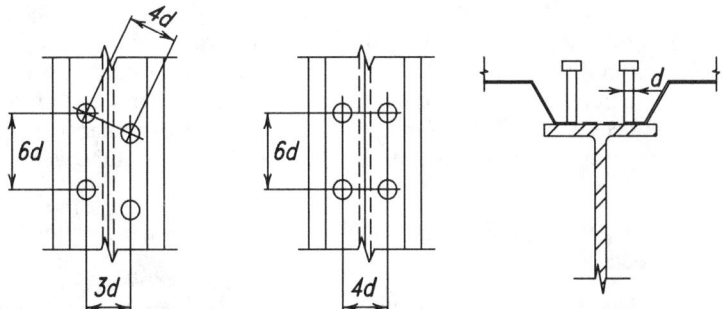

Fig. C-I5.1. Shear connector arrangements.

CHAPTER J

CONNECTIONS, JOINTS, AND FASTENERS

J1. GENERAL PROVISIONS

5. Splices in Heavy Sections

Solidified but still-hot weld metal contracts significantly as it cools to ambient temperature. Shrinkage of large welds between elements which are not free to move to accommodate the shrinkage, causes strains in the material adjacent to the weld that can exceed the yield point strain. In thick material, the weld shrinkage is restrained in the thickness direction as well as in the width and length directions, causing triaxial stresses to develop that may inhibit the ability of ductile steel to deform in a ductile manner. Under these conditions, the possibility of brittle fracture increases.

When splicing ASTM A6 Group 4 and 5 rolled sections or heavy welded built-up members, the potentially harmful weld shrinkage strains can be avoided by using bolted splices or fillet-welded lap splices or splices that combine a welded and bolted detail (see Figure C-J1.1). Details and techniques that perform well for materials of modest thickness usually must be changed or supplemented by more demanding requirements when welding thick material. Also, the provisions of the *Structural Welding Code,* AWS D1.1, are minimum requirements that apply to most structural welding situations; however, when designing and fabricating welded splices of ASTM A6 Group 4 and 5 shapes and similar built-up cross sections, special consideration must be given to all aspects of the welded splice detail.

- Notch-toughness requirements should be specified for tension members. See Commentary A3.

- Generously sized weld access holes, Figure C-J1.2, are required to provide increased relief from concentrated weld shrinkage strains, to avoid close juncture of welds in orthogonal directions, and to provide adequate clearance for the exercise of high quality workmanship in hole preparation, welding, and ease of inspection.

- Preheating for thermal cutting is required to minimize the formation of a hard surface layer.

- Grinding to bright metal and inspection using magnetic particle or dye-penetrant methods is required to remove the hard surface layer and to assure smooth transitions free of notches or cracks.

In addition to tension splices of truss chord members and tension flanges of flexural members, other joints fabricated of heavy sections subject to tension should be given special consideration during design and fabrication.

8.   Placement of Welds and Bolts

Slight eccentricities between the gravity axis of single and double angle members and the center of gravity of connecting rivets or bolts have long been ignored as having negligible effect on the static strength of such members. Tests (Gibson and Wake, 1942) have shown that similar practice is warranted in the case of welded members in statically loaded structures.

However, the fatigue life of eccentrically loaded welded angles has been shown to be very short (Kloppel and Seeger, 1964). Notches at the roots of fillet welds are harmful when alternating tensile stresses are normal to the axis of the weld, as could occur due to bending when axial cyclic loading is applied to angles with end welds not balanced about the neutral axis. Accordingly, balanced welds are indicated when such members are subjected to cyclic loading (see Figure C-J1.3).

9.   Bolts in Combination with Welds

Welds will not share the load equally with mechanical fasteners in bearing-type connections. Before ultimate loading occurs, the fastener will slip and the weld will carry an indeterminately larger share of the load.

Accordingly, the sharing of load between welds and A307 bolts or high-strength bolts in a bearing-type connection is not recommended. For similar reasons, A307 bolts and rivets should not be assumed to share loads in a single group of fasteners.

For high-strength bolts in slip-critical connections to share the load with welds it is advisable to fully tension the bolts before the weld is made. If the weld is placed first, angular distortion from the heat of the weld might prevent the faying action required for development of the slip-critical force. When the bolts are fully tensioned before the weld is made, the slip-critical bolts and the weld may be assumed to share the load on a common shear plane (Kulak, et al., 1987). The heat of welding near bolts will not alter the mechanical properties of the bolts.

In making alterations to existing structures, it is assumed that whatever slip is

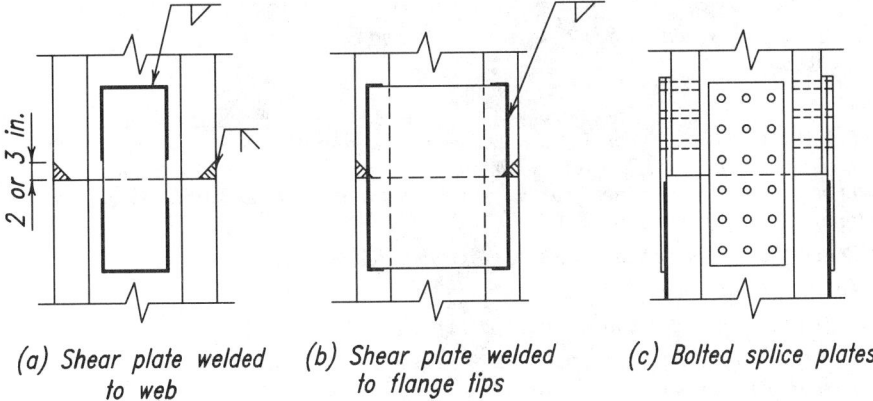

(a) Shear plate welded to web   *(b) Shear plate welded to flange tips*   *(c) Bolted splice plates*

Fig. C-J1.1. Alternative splices that minimize weld restraint tensile stresses.

likely to occur in high-strength bolted bearing-type connections or riveted connections will have already taken place. Hence, in such cases the use of welding to resist all stresses, other than those produced by existing dead load present at the time of making the alteration, is permitted.

It should be noted that combinations of fasteners as defined herein does not refer to connections such as shear plates for beam-to-column connections which are welded to the column and bolted to the beam flange or web (Kulak, et al., 1987) and other comparable connections.

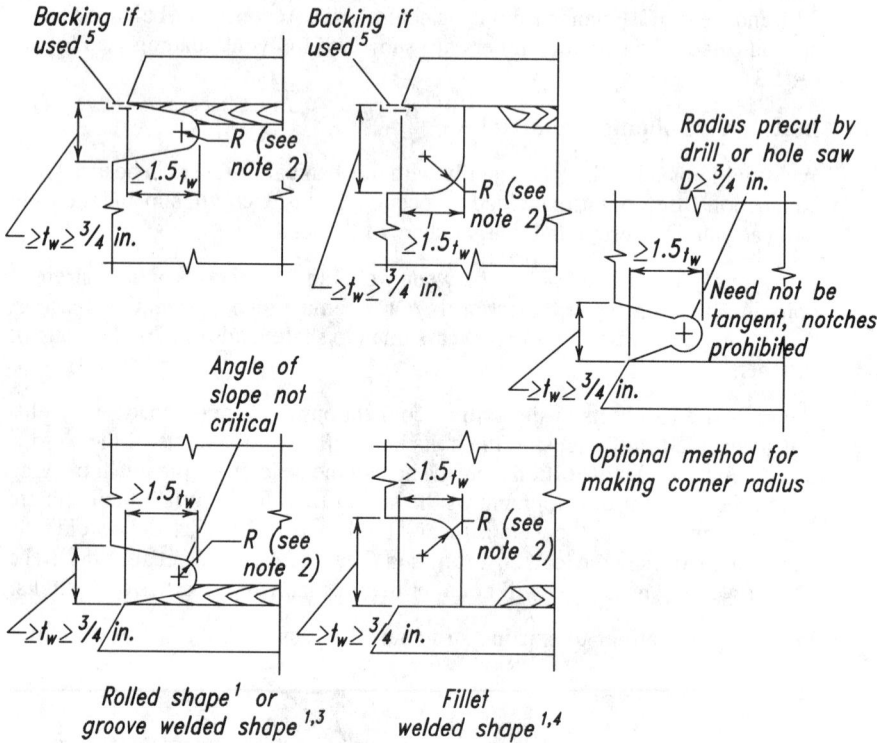

Fig. C-J1.2. Weld access hole and beam cope geometry.

Notes:

1. *For ASTM A6 Group 4 and 5 shapes and welded built-up shapes with plate thickness more than 2 in., preheat to 150°F prior to thermal cutting, grind and inspect thermally cut edges of access hole using magnetic particle or dye penetration methods prior to making web and flange splice groove welds.*

2. *Radius shall provide smooth notch-free transition; $R \geq \frac{3}{8}$-in. (typical $\frac{1}{2}$-in.)*

3. *Access opening made after welding web to flange.*

4. *Access opening made before welding web to flange.*

5. *These are typical details for joints welded from one side against steel backing. Alternative joint designs should be considered.*

10.   High-Strength Bolts in Combination with Rivets

When high-strength bolts are used in combination with rivets, the ductility of the rivets permits the direct addition of the strengths of both fastener types.

J2.   WELDS

1.   Groove Welds

The engineer preparing contract design drawings cannot specify the depth of groove without knowing the welding process and the position of welding. Accordingly, only the effective throat for partial joint-penetration groove welds should be specified on design drawings, allowing the fabricator to produce this effective throat with his own choice of welding process and position.

The weld reinforcement is not used in determining the effective throat thickness of a groove weld (see Table J2.1).

2.   Fillet Welds

2a.   Effective Area

The effective throat of a fillet weld is based upon the root of the joint and the face of the diagrammatic weld, hence this definition gives no credit for weld penetration or reinforcement at the weld face. If the fillet weld is made by the submerged arc welding process, some credit for penetration is made. If the leg size of the resulting fillet weld exceeds ⅜-in., then 0.11 in. is added to the theoretical throat. This increased weld throat is allowed because the submerged arc process produces deep penetration of welds of consistent quality. However, it is necessary to run a short length of fillet weld to be assured that this increased penetration is obtained. In practice, this is usually done initially by cross-sectioning the runoff plates of the joint. Once this is done, no further testing is required, as long as the welding procedure is not changed.

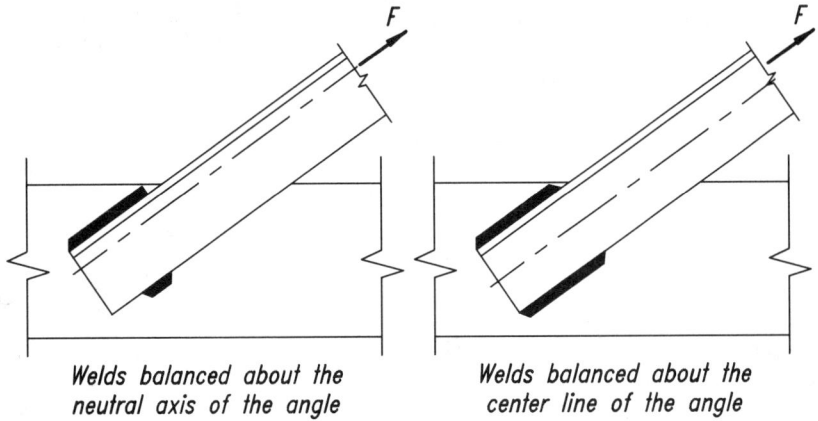

Welds balanced about the neutral axis of the angle   *Welds balanced about the center line of the angle*

Figure C-J1.3

2b.   Limitations

Table J2.4 provides a minimum size of fillet weld for a given thickness of the thicker part joined.

The requirements are not based upon strength considerations, but upon the quench effect of thick material on small welds. Very rapid cooling of weld metal may result in a loss of ductility. Further, the restraint to weld-metal shrinkage provided by thick material may result in weld cracking. Because a $\frac{5}{16}$-in. fillet weld is the largest that can be deposited in a single pass by SMAW process, $\frac{5}{16}$-in. applies to all material $\frac{3}{4}$-in. and greater in thickness, but minimum preheat and interpass temperature are required by AWS D1.1.* Both the design engineer and the shop welder must be governed by the requirements.

Table J2.3 gives the minimum effective throat of a partial joint-penetration groove weld. Notice that Table J2.3 for partial joint-penetration groove welds goes up to a plate thickness of over 6 in. and a minimum weld throat of $\frac{5}{8}$-in., whereas, for fillet welds Table J2.4 goes up to a plate thickness of over $\frac{3}{4}$-in. and a minimum leg size of fillet weld of only $\frac{5}{16}$-in. The additional thickness for partial-penetration welds is to provide for reasonable proportionality between weld and material thickness.

For plates of $\frac{1}{4}$-in. or more in thickness, it is necessary that the inspector be able to identify the edge of the plate to position the weld gage. This is assured if the weld is kept back at least $\frac{1}{16}$-in. from the edge, as shown in Figure C-J2.1.

Where longitudinal fillet welds are used alone in a connection (see Figure C-J2.2), Section J2.2b requires the length of each weld to be at least equal to the width of the connecting material because of shear lag (Fisher, et al., 1978).

By providing a minimum lap of five times the thickness of the thinner part of a lap joint, the resulting rotation of the joint when pulled will not be excessive, as shown in Figure C-J2.3. Fillet welded lap joints under tension tend to open and

---

* See Table J2.4.

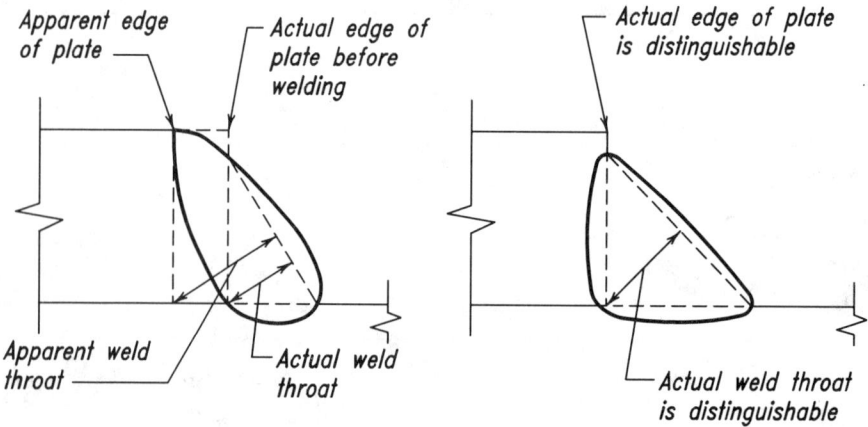

Fig. C-J2.1. Identification of plate edge.

apply a tearing action at the root of the weld as shown in Figure C-J2.4b, unless restrained by a force F as shown in Figure C-J2.4a.

End returns are not essential for developing the capacity of fillet welded connections and have a negligible effect on their strength. Their use has been encouraged to insure that the weld size is maintained over the length of the weld, to enhance the fatigue resistance of cyclically loaded flexible end connections, and to increase the plastic deformation capability of such connections.

The weld capacity database on which the specifications were developed had no end returns. This includes the study by Higgins and Preece (1968), seat angle tests by Lyse and Schreiner (1935), the seat and top angle tests by Lyse and Gibson (1937), beam webs welded directly to column or girder by fillet welds by Johnston and Deits (1941), and the eccentrically loaded welded connections reported by Butler, Pal, and Kulak (1972). Hence, the current design-resistance values and joint-capacity models do not require end returns, when the required weld size is provided. Johnston and Green (1940) noted that movement consistent with the design assumption of no end restraint (i.e., joint flexibility) was

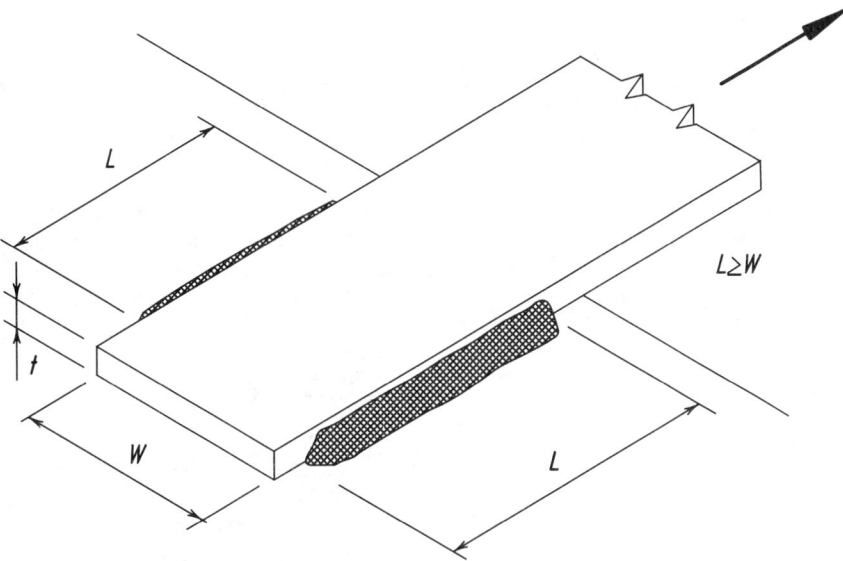

Fig. C-J2.2. Longitudinal fillet welds.

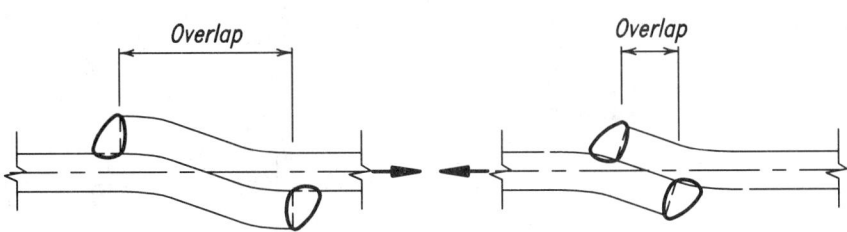

Fig. C-J2.3. Minimum lap.

enhanced without end returns. They also verified that greater plastic deformation of the connection was achieved when end returns existed, although the strength was not significantly different.

There are numerous welded joints where it is not possible to provide end returns and where it is also possible to provide the desired weld size. These joints as well as the seat angle and the web angle connections cited earlier do not require end returns when the weld size is adequate and fatigue is not a design consideration.

4. Design Strength

The strength of welds is governed by the strength of either the base material or the deposited weld metal. Table J2.5 contains the resistance factors and nominal weld strengths, as well as a number of limitations.

It should be noted that in Table J2.5 the nominal strength of fillet welds is determined from the effective throat area, whereas the strength of the connected parts is governed by their respective thicknesses. Figure C-J2.5 illustrates the shear planes for fillet welds and base material:

(a) Plane 1-1, in which the resistance is governed by the shear strength for material A.

(b) Plane 2-2, in which the resistance is governed by the shear strength of the weld metal.

(c) Plane 3-3, in which the resistance is governed by the shear strength of the material B.

The resistance of the welded joint is the lowest of the resistance calculated in each plane of shear transfer. Note that planes 1-1 and 3-3 are positioned away from the fusion areas between the weld and the base material. Tests have

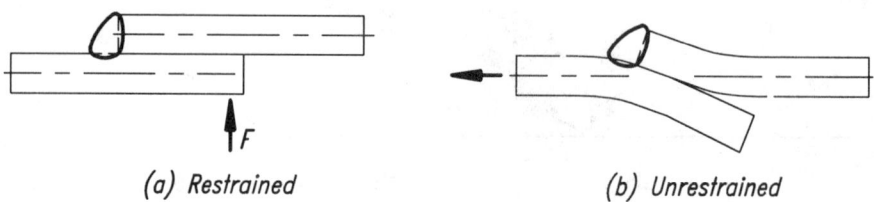

(a) Restrained　　　　　　　　　　　　　(b) Unrestrained

Fig. C-J2.4. Restraint of lap joints.

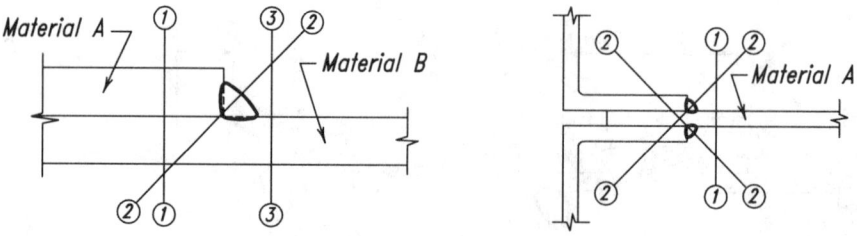

Fig. C-J2.5. Shear planes for fillet welds loaded in longitudinal shear.

demonstrated that the stress on this fusion area is not critical in determining the shear strength of fillet welds (Preece, 1968).

The shear planes for plug and partial penetration groove welds are shown in Figure C-J2.6 for the weld and base metal. Generally the base metal will govern the shear strength.

5. Combination of Welds

This method of adding weld strengths does not apply to a welded joint using a partial-penetration single bevel groove weld with a superimposed fillet weld. In this case, the effective throat of the combined joint must be determined and the design strength based upon this throat area.

7. Mixed Weld Metal

Problems can occur when incompatible weld metals are used in combination and notch-tough composite weld metal is required. For instance, tack welds deposited using a self-shielded process with aluminum deoxidizers in the electrodes and subsequently covered by SAW weld passes can result in composite weld metal with low notch-toughness, despite the fact that each process by itself could provide notch-tough weld metal.

J3. BOLTS AND THREADED PARTS

1. High-Strength Bolts

In general, the use of high-strength bolts is required to conform to the provisions of the *Load and Resistance Factor Design Specification for Structural Joints Using ASTM A325 or A490 Bolts* (RCSC, 1988) as approved by the Research Council on Structural Connections.

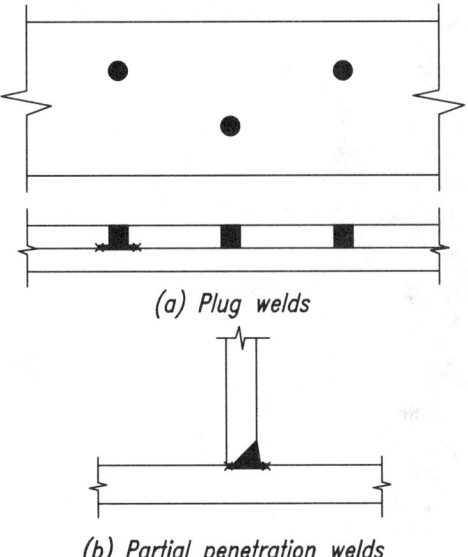

(a) Plug welds

(b) Partial penetration welds

Fig. C-J2.6. Shear planes for plug and partial-penetration welds.

Occasionally the need arises for the use of high-strength bolts of diameters and lengths in excess of those available for A325 and A490 bolts, as for example, anchor bolts for fastening machine bases. For this situation Section A3.3 permits the use of A449 bolts and A354 threaded rods.

2. Size and Use of Holes

To provide some latitude for adjustment in plumbing up a frame during erection, three types of enlarged holes are permitted, subject to the approval of the designer. The nominal maximum sizes of these holes are given in Table J3.3.

The use of these enlarged holes is restricted to connections assembled with bolts and is subject to the provisions of Sections J3.3 and J3.4.

3. Minimum Spacing

The *maximum* factored strength R_n at a bolt or rivet hole in bearing requires that the distance between the centerline of the first fastener and the edge of a plate toward which the force is directed should not be less than $1\frac{1}{2}d$, where d is the fastener diameter (Kulak et al., 1987). By similar reasoning the distance measured in the line of force, from the centerline of any fastener to the nearest edge of an adjacent hole, should not be less than $3d$, to ensure maximum design strength in bearing. Plotting of numerous test results indicates that the critical bearing strength is directly proportional to the above defined distances up to a maximum value of $3d$, above which no additional bearing strength is achieved (Kulak et al., 1987). Table J3.7 lists the increments that must be added to adjust the spacing upward to compensate for an increase in hole dimension parallel to the line of force. Section J3.10 gives the bearing strength criteria as a function of spacing.

4. Minimum Edge Distance

Critical bearing stress is a function of the material tensile strength, the spacing of fasteners, and the distance from the edge of the part to the center line of the nearest fastener. Tests have shown (Kulak et al., 1987) that a linear relationship exists between the ratio of critical bearing stress to tensile strength (of the connected material) and the ratio of fastener spacing (in the line of force) to fastener diameter. The following equation affords a good lower bound to published test data for single-fastener connections with standard holes, and is conservative for adequately spaced multi-fastener connections:

$$\frac{F_{pcr}}{F_u} = \frac{l_e}{d} \qquad \text{(C-J3-1)}$$

where

F_{pcr} = critical bearing stress, ksi
F_u = tensile strength of the connected material, ksi
l_e = distance, along a line of transmitted force, from the center of a fastener to the nearest edge of an adjacent fastener or to the free edge of a connected part (in the direction of stress), in.
d = diameter of fastener, in.

The provisions of Section J3.3 are concerned with l_e as hole spacing, whereas Section J3.4 is concerned with l_e as edge distance in the direction of stress.

Section J3.10 establishes a maximum bearing strength. Spacing and/or edge distance may be increased to provide for a required bearing strength, or bearing force may be reduced to satisfy a spacing and/or edge distance limitation.

It has long been known that the critical bearing stress of a single fastener connection is more dependent upon a given edge distance than multi-fastener connections (Jones, 1940). For this reason, longer edge distances (in the direction of force) are required for connections with one fastener in the line of transmitted force than required for those having two or more.

The recommended minimum distance transverse to the direction of load is primarily a workmanship tolerance. It has little, if any, effect on the strength of the member.

5.   Maximum Spacing and Edge Distance

Limiting the edge distance to not more than 12 times the thickness of an outside connected part, but not more than six inches, is intended to provide for the exclusion of moisture in the event of paint failure, thus preventing corrosion between the parts which might accumulate and force these parts to separate. More restrictive limitations are required for connected parts of unpainted weathering steel exposed to atmospheric corrosion.

6.   Design Tension or Shear Strength

Tension loading of fasteners is usually accompanied by some bending due to the deformation of the connected parts. Hence, the resistance factor ϕ, by which R_n is multiplied to obtain the design tensile strength of fasteners, is relatively low. The nominal tensile strength values in Table J3.2 were obtained from the equation

$$R_n = 0.75 A_b F_u \qquad\qquad (C\text{-}J3\text{-}2)$$

While the equation was developed for bolted connections (Kulak et al., 1987), it was also conservatively applied to threaded parts and to rivets. The nominal strength of A307 bolts was discounted by 5 ksi.

In connections consisting of only a few fasteners, the effects of strain on the shear in bearing fasteners is negligible (Kulak et al., 1987; Fisher et al., 1978). In longer joints, the differential strain produces an uneven distribution between fasteners (those near the end taking a disproportionate part of the total load), so that the maximum strength per fastener is reduced. The AISC ASD Specification permits connections up to 50 in. in length without a reduction in maximum shear stress. With this in mind the resistance factor ϕ for shear in bearing-type connections has been selected to accommodate the same range of connections.

The values of nominal shear strength in Table J3.2 were obtained from the equation

$$R_n / m A_b = 0.50 F_u \qquad\qquad (C\text{-}J3\text{-}3)$$

when threads are excluded from the shear planes and

$$R_n / m A_b = 0.40 F_u \qquad\qquad (C\text{-}J3\text{-}4)$$

when threads are not excluded from the shear plane, where m is the number of shear planes (Kulak et al., 1987). While developed for bolted connections, the equations were also conservatively applied to threaded parts and rivets. The value given for A307 bolts was obtained from Equation C-J3-4 but is specified for all cases regardless of the position of threads. For A325 bolts, no distinction is made between small and large diameters, even though the minimum tensile strength F_u is lower for bolts with diameters in excess of one inch. It was felt that such a refinement of design was not justified, particularly in view of the low resistance factor ϕ, increasing ratio of tensile area to gross area and other compensating factors.

7. Combined Tension and Shear in Bearing-Type Connections

Tests have shown that the strength of bearing fasteners subject to combined shear and tension resulting from externally applied forces can be closely defined by an ellipse (Kulak et al., 1987). Such a curve can be replaced, with only minor deviations, by three straight lines as shown in Figure C-J3.1. This latter representation offers the advantage that no modification of either type stress is required in the presence of fairly large magnitudes of other types. This linear representation was adopted for Table J3.5, giving a limiting tensile stress F_t as a function of the shearing stress f_v for bearing-type connections.

8. High-Strength Bolts in Slip-Critical Connections

Connections classified as slip-critical include those cases where slip could theoretically exceed an amount deemed by the Engineer of Record to affect the suitability for service of the structure by excessive distortion or reduction in strength or stability, even though the nominal strength of the connection may be adequate. Also included are those cases where slip of any magnitude must be prevented, for example, joints subject to fatigue, connectors between elements of built-up members at their ends (Sections D2 and E4), and bolts in combination with welds (Section J1.9).

The onset of slipping in a high-strength bolted, slip-critical connection is not an

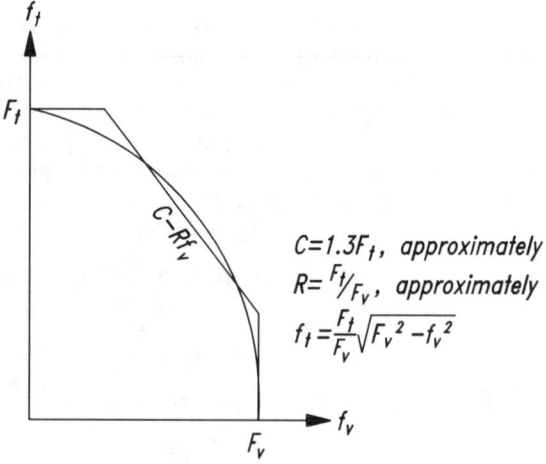

$$C = 1.3F_t, \text{ approximately}$$
$$R = \frac{F_t}{F_v}, \text{ approximately}$$
$$f_t = \frac{F_t}{F_v}\sqrt{F_v{}^2 - f_v{}^2}$$

Figure C-J3.1.

indication that maximum capacity of the connection has been reached. Its occurrence may be only a serviceability limit state. In the case of bolts in holes with only small clearance, such as standard holes and slotted holes loaded transverse to the axis of the slot in practical connections, the freedom to slip generally does not exist because one or more bolts are in bearing even before load is applied due to normal fabrication tolerances and erection procedures. Further, the consequences of slip, if it can occur at all, are trivial except for a few situations as noted above.

Slip of slip-critical connections is likely to occur at approximately 1.4 to 1.5 times the service loads. For standard holes, oversized holes, and short slotted holes the connection can be designed either at service loads (Section J3.8a) or at factored loads (Appendix J3.8b). The nominal loads and ϕ factors have been adjusted accordingly. The number of connectors will be essentially the same for the two procedures because they have been calibrated to give similar results. Slight differences will occur because of variation in the ratio of live load to dead load.

In connections containing long slots that are parallel to the direction of the applied load, slip of the connection prior to attainment of the factored load might be large enough to alter the usual assumption of analysis that the undeformed structure can be used to obtain the internal forces. To guard against this occurring, the design slip resistance is further reduced by 0.85 when designing at service load (Section J3.8a) and by setting ϕ to 0.60 in conjunction with factored loads (Appendix J3.8b).

While the possibility of a slip-critical connection slipping into bearing under anticipated service conditions is small, such connections must comply with the provisions of Section J3.10 in order to prevent connection failure at the maximum load condition.

10.   Bearing Strength at Bolt Holes

The recommended bearing stress on pins is not the same as for bolts as explained in Section J8.

Bearing values are not provided as a protection to the fastener, because it needs no such protection. Therefore, the same bearing value applies to joints assembled by bolts, regardless of fastener shear strength or the presence or absence of threads in the bearing area.

Tests (Frank and Yura, 1981) have demonstrated that hole elongation greater than 0.25 in. will begin to develop as the bearing stress is increased beyond the values given in Equations J3-1a and J3-1d, especially if it is combined with high tensile stress on the net section, even though rupture does not occur. Equations J3-1b and J3-1c consider the effect of hole ovalization (deformation greater than 0.25 in.) whenever the upper design limit ($3.0dtF_u$) is deemed acceptable. These latter equations also establish the design limit for a single bolt, or two or more bolts, whenever the bolt arrangement results in each bolt singly in line with the direction of the applied force. Because two separate limit states are considered (deformation and strength) with both limit states equated to a bearing stress ($2.4F_u$ or $2.0F_u$ and $3.0F_u$, respectively) conflicting design strengths may result, either acceptable, when intermediate edge distance and bolt spacing values are considered.

11.   Long Grips

Provisions requiring a decrease in calculated stress for A307 bolts having long grips (by arbitrarily increasing the required number in proportion to the grip length) are not required for high-strength bolts. Tests (Bendigo et al., 1963) have demonstrated that the ultimate shearing strength of high-strength bolts having a grip of eight or nine diameters is no less than that of similar bolts with much shorter grips.

J4.   DESIGN RUPTURE STRENGTH

Tests (Birkemoe and Gilmor, 1978) on coped beams indicated that a tearing failure mode (rupture) can occur along the perimeter of the bolt holes as shown in Figure C-J4.1. This block shear mode combines tensile strength on one plane and shear strength on a perpendicular plane. The failure path is defined by the center lines of the bolt holes. The block shear failure mode is not limited to the coped ends of beams. Other examples are shown in Figure CJ4.1 and C-J4.2.

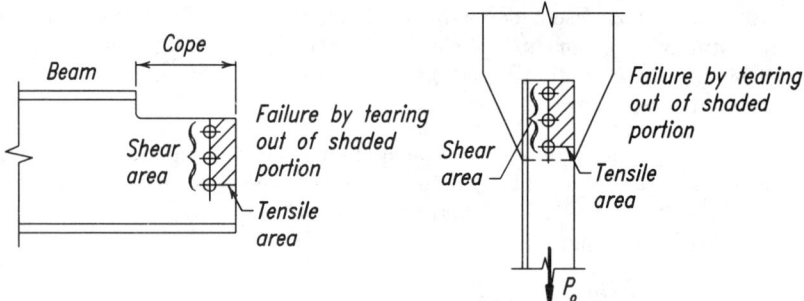

Fig. C-J4.1. Failure surface for block shear rupture limit state.

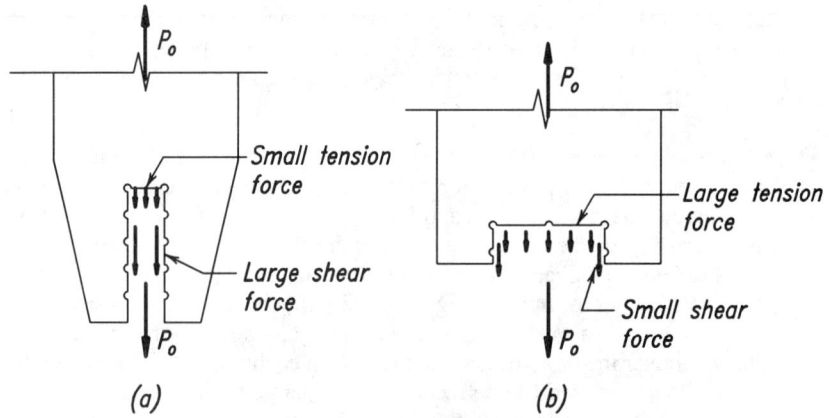

Fig. C-J4.2 Block shear rupture in tension.

The block shear failure mode should also be checked around the periphery of welded connections. Welded connection block shear is determined using $\phi = 0.75$ in conjunction with the area of both the fracture and yielding planes (Yura, 1988).

The LRFD Specification has adopted a conservative model to predict block shear strength. Test results suggest that it is reasonable to add the yield strength on one plane to the rupture strength of the perpendicular plane (Ricles and Yura, 1983 and Hardash and Bjorhovde, 1985). Therefore, two possible block shear strengths can be calculated; rupture strength F_u on the net tensile section along with shear yielding $0.6F_y$ on the gross section on the shear plane(s), or rupture $0.6F_u$ on the net shear area(s) combined with yielding F_y on the gross tensile area. This is the basis of Equations J4-3 and J4-4.

These equations are consistent with the philosophy in Chapter D for tension members, where gross area is used for the limit state of yielding and net area is used for rupture. The controlling equation is the one that produces the *larger* rupture force. This can be explained by the two extreme examples given in Figure C-J4.2. In Case a, the total force is resisted primarily by shear, so shear rupture, not shear yielding, should control the block shear tearing mode; therefore, use Equation J4-4. For Case b, block shear cannot occur until the tension area ruptures as given by Equation J4-3. If Equation J4-4 (shear rupture on the small area and yielding on the large tension area) is checked for Case b, a smaller P_o will result. In fact, as the shear area gets smaller and approaches zero, the use of Equation J4-4 for Case b would give a block shear strength based totally on *yielding* of the gross tensile area. Block shear is a rupture or tearing phenomenon not a yielding limit state. Therefore, the proper equation to use is the one with the larger rupture term.

J5.   CONNECTING ELEMENTS

2.   Design Strength of Connecting Elements in Tension

Tests have shown that yield will occur on the gross section area before the tensile capacity of the net section is reached, if the ratio $A_n / A_g \leq 0.85$ (Kulak et al., 1987). Since the length of connecting elements is small compared to the member length, inelastic deformation of the gross section is limited. Hence, the effective net area A_n of the connecting element is limited to $0.85A_g$ in recognition of the limited inelastic deformation and to provide a reserve capacity.

J6.   FILLERS

The practice of securing fillers by means of additional fasteners, so that they are, in effect, an integral part of a shear-connected component, is not required where a connection is designed to be a slip-critical connection using high-strength bolts. In such connections, the resistance to slip between filler and either connected part is comparable to that which would exist between the connected parts if no fill were present.

Filler plates may be used in lap joints of welded connections that splice parts of different thickness, or where there may be an offset in the joint.

J8. BEARING STRENGTH

The LRFD Specification provisions for bearing on milled surfaces, Section J8, follow the same philosophy of earlier AISC ASD Specifications. In general, the design is governed by a deformation limit state at service loads resulting in stresses nominally at $9/10$ of yield. Adequate safety is provided by post-yield strength as deformation increases. Tests on pin connections (Johnston, 1939) and on rockers (Wilson, 1934) have confirmed this behavior.

As used throughout the LRFD Specification, the terms "milled surface," "milled," and "milling" are intended to include surfaces which have been accurately sawed or finished to a true plane by any suitable means.

J9. COLUMN BASES AND BEARING ON CONCRETE

The equations for resistance of concrete in bearing are the same as ACI 318-89 except that AISC equations use $\phi = 0.60$ while ACI uses $\phi = 0.70$, since ACI specifies larger load factors than the ASCE load factors specified by AISC.

J10. ANCHOR BOLTS AND EMBEDMENTS

ACI 318 and 349 Appendix B and the PCI Handbook include recommended procedures for the design of anchor bolts and embedments.

CHAPTER K

CONCENTRATED FORCES, PONDING, AND FATIGUE

K1. FLANGES AND WEBS WITH CONCENTRATED FORCES

1. Design Basis

The LRFD Specification separates flange and web strength requirements into distinct categories representing different limit state criteria, i.e., local flange bending (Section K1.2), local web yielding (Section K1.3), web crippling (Section K1.4), sidesway web buckling (Section K1.5), compression buckling of the web (Section K1.6), and panel zone web shear (Section K1.7).

These criteria are applied to two distinct types of concentrated forces which act on member flanges. *Single concentrated forces* may be tensile, such as those delivered by tension hangers, or compressive, such as those delivered by bearing plates at beam interior positions, reactions at beam ends, and other *bearing* connections. *Double concentrated forces,* one tensile and one compressive, form a couple on the same side of the loaded member, such as that delivered to column flanges through welded and bolted *moment* connections.

2. Local Flange Bending

Where a tensile force is applied through a plate welded across a flange, that flange must be sufficiently rigid to prevent deformation of the flange and the corresponding high-stress concentration in the weld in line with the web.

The effective column flange length for local flange bending is $12t_f$ (Graham, et al., 1959). Thus, it is assumed that yield lines form in the flange at $6t_f$ in each direction from the point of the applied concentrated force. To develop the fixed edge consistent with the assumptions of this model, an additional $4t_f$ and therefore a total of $10t_f$, is required for the full flange-bending strength given by Equation K1-1. In the absence of applicable research, a 50 percent reduction has been introduced for cases wherein the applied concentrated force is less than $10t_f$ from the member end.

This criterion given by Equation K1-1 was originally developed for *moment* connections, but it also applies to *single concentrated forces* such as tension hangers consisting of a plate welded to the bottom flange of a beam and transverse to the beam web.

3. Local Web Yielding

The web strength criteria have been established to limit the stress in the web of a member into which a force is being transmitted. It should matter little whether the member receiving the force is a beam or a column; however, Galambos (1976) and AISC (1978), references upon which the LRFD Specification is based, did make such a distinction. For beams, a 2:1 stress gradient through the flange was used, whereas the gradient through column flanges was 2½:1. In Section K1.3, the 2½:1 gradient is used for both cases.

This criterion applies to both *bearing* and *moment* connections.

4. Web Crippling

The expression for resistance to web crippling at a concentrated force is a departure from previous specifications (IABSE, 1968; Bergfelt, 1971; Hoglund, 1971; and Elgaaly, 1983). Equations K1-4 and K1-5 are based on research by Roberts (1981). The increase in Equation K1-5b for $N/d > 0.2$ was developed after additional testing (Elgaaly, 1991) to better represent the effect of longer bearing lengths at ends of members. All tests were conducted on bare steel beams without the expected beneficial contributions of any connection or floor attachments. Thus, the resulting criteria are considered conservative for such applications.

These equations were developed for *bearing* connections, but are also generally applicable to *moment* connections. However, for the rolled shapes listed in Part 1 of the LRFD Manual with F_y not greater than 50 ksi, the web crippling criterion will never control the design in a *moment* connection except for a W12×50 or W10×33 column.

The web crippling phenomenon has been observed to occur in the web adjacent to the load flange. For this reason, a half-depth stiffener (or stiffeners) or a half-depth doubler plate is expected to eliminate this limit state.

5. Sidesway Web Buckling

The sidesway web buckling criterion was developed after observing several unexpected failures in tested beams (Summers and Yura, 1982). In those tests the compression flanges were braced at the concentrated load, the web was squeezed into compression, and the tension flange buckled (see Figure C-K1.1).

Sidesway web buckling will not occur in the following cases. For flanges restrained against rotation:

$$\frac{h/t_w}{l/b_f} > 2.3 \tag{C-K1-1}$$

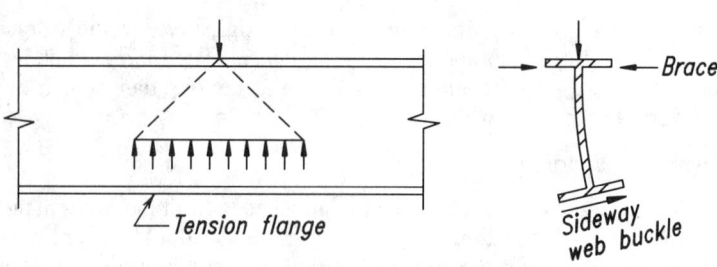

Fig. C-K1.1. Sidesway web buckling.

For flanges *not* restrained against rotation:

$$\frac{h/t_w}{l/b_f} > 1.7 \qquad \text{(C-K-1-2)}$$

where l is as shown in Figure C-K1.2.

Sidesway web buckling can also be prevented by the proper design of lateral bracing or stiffeners at the load point. It is suggested that local bracing at both flanges be designed for one percent of the concentrated force applied at that point. Stiffeners must extend from the load point through at least one-half the beam or girder depth. In addition, the pair of stiffeners should be designed to carry the full load. If flange rotation is permitted at the loaded flange, neither stiffeners nor doubler plates will be effective.

In the 1st Edition LRFD Manual, the sidesway web buckling equations were based on the assumption that $h/t_f = 40$, a convenient assumption which is generally true for economy beams. This assumption has been removed so that the equations will be applicable to all sections.

These equations were developed only for *bearing* connections and do not apply to *moment* connections.

6. **Compression Buckling of the Web**

When compressive forces are applied to both flanges of a member at the same

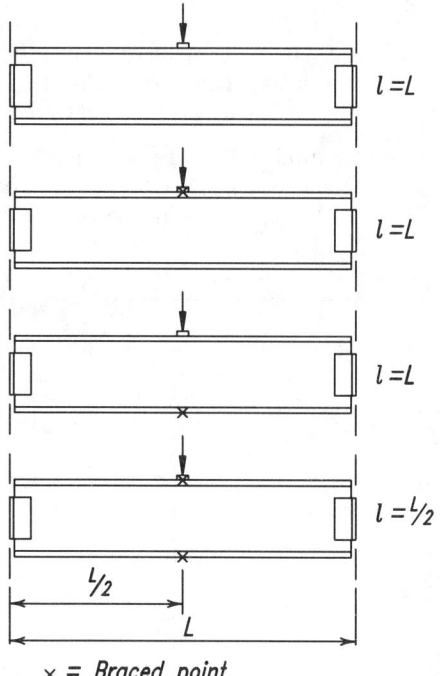

$\times$ = *Braced point*

Fig. C-K1.2. Unbraced flange length.

location, as by *moment* connections at both flanges of a column, the member web must have its slenderness ratio limited to avoid the possibility of buckling. This is done in the LRFD Specification with Equation K1-8, which is a modified form of a similar equation used in the ASD Specification. This equation is applicable to a pair of *moment* connections, and to other pairs of compressive forces applied at both flanges of a member, for which N/d is small (<1). When N/d is not small, the member web should be designed as a compression member in accordance with Chapter E.

Equation K1-8 is predicated on an interior member loading condition. In the absence of applicable research, a 50 percent reduction has been introduced for cases wherein the compressive forces are close to the member end.

Equation K1-8 has also traditionally been applied when there is a *moment* connection to only one flange of the column and compressive force is applied to only one flange. Its use in this case is conservative.

7. **Panel Zone Web Shear**

The column web shear stresses may be high within the boundaries of the rigid connection of two or more members whose webs lie in a common plane. Such webs should be reinforced when the calculated factored force ΣF along plane A-A in Figure C-K1.3 exceeds the column web design strength ϕR_v, where

$$\Sigma F_u = \frac{M_{u1}}{d_{m1}} + \frac{M_{u2}}{d_{m2}} - V_u \qquad \text{(C-K1-3)}$$

and

$M_{u1} = M_{u1L} + M_{u1G}$ = the sum of the moments due to the factored lateral load M_{u1L} and the moments due to factored gravity load M_{u1G} on the leeward side of the connection, kip-in.

$M_{u2} = M_{u2L} - M_{u2G}$ = the difference between the moments due to the factored lateral load M_{u2L} and the moments due to factored gravity load M_{u2G} on the windward side of the connection, kip-in.

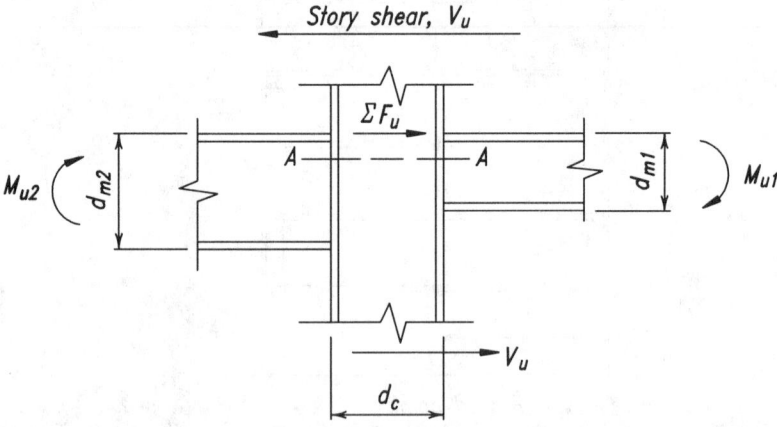

Fig. C-K1.3. Forces in panel zone.

d_{m1}, d_{m2} = distance between flange forces in a moment connection, in.

Conservatively, 0.95 times the beam depth has been used for d_m in the past.

If $\Sigma F_u \leq \phi R_v$, no reinforcement is necessary, i.e., $t_{req} \leq t_w$, where t_w is the column web thickness.

Consistent with elastic first order analysis, Equations K1-9 and K1-10 limit panel-zone behavior to the elastic range. While such connection panels possess large reserve capacity beyond initial general shear yielding, the corresponding inelastic joint deformations may adversely affect the strength and stability of the frame or story (Fielding and Huang, 1971 and Fielding and Chen, 1973). Panel-zone shear yielding affects the overall frame stiffness and, therefore, the ultimate-strength second-order effects may be significant. The shear/axial inter-action expression of Equation K1-10, as shown in Figure C-K1.4, is chosen to ensure elastic panel behavior.

If adequate connection ductility is provided and the frame analysis considers the inelastic panel-zone deformations, then the additional inelastic shear strength is recognized in Equations K1-11 and K1-12 by the factor

$$\left(1 + \frac{3 b_{cf}\, t_{cf}^2}{d_b\, d_c\, t_w}\right)$$

This inelastic shear strength has been most often utilized for design of frames in high seismic zones and should be used when the panel zone is to be designed to match the strength of the members from which it is formed.

The shear/axial interaction expression incorporated in Equation K1-12 (see Figure C-K1.5) is similar to that contained in the previous issue of this specifi-cation and recognizes the observed fact that when the panel-zone web has completely yielded in shear, the axial column load is carried in the flanges.

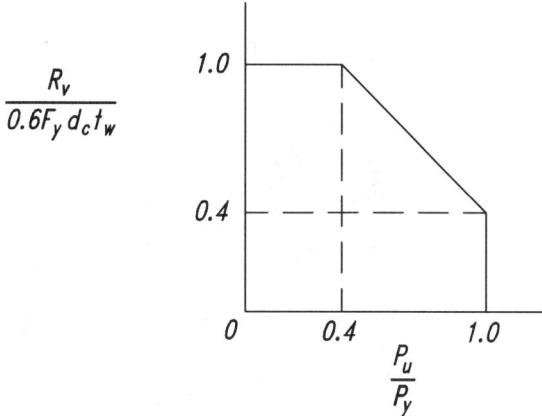

Fig. C-K1.4. Interaction of shear and axial force—elastic.

K2. PONDING

As used in the LRFD Specification, *ponding* refers to the retention of water due solely to the deflection of flat roof framing. The amount of this water is dependent upon the flexibility of the framing. Lacking sufficient framing stiffness, its accumulated weight can result in collapse of the roof if a strength evaluation is not made (ASCE, 1990).

Representing the deflected shape of the primary and critical secondary member as a half-sine wave, the weight and distribution of the ponded water can be estimated and, from this, the contribution that the deflection each of these members makes to the total ponding deflection can be expressed (Marino, 1966):

For the primary member:

$$\Delta_w = \frac{\alpha_p \Delta_o \left[1 + 0.25\pi\alpha_s + 0.25\pi\rho(1 + \alpha_s)\right]}{1 - 0.25\pi\alpha_p\alpha_s}$$

For the secondary member:

$$\delta_w = \frac{\alpha_s \delta_o \left[1 + \frac{\pi^3}{32}\alpha_p + \frac{\pi^2}{8\rho}(1 + \alpha_p) + 0.185\alpha_s\alpha_p\right]}{1 - 0.25\pi\alpha_p\alpha_s}$$

In these expressions Δ_o and δ_o are, respectively, the primary and secondary beam deflections due to loading present at the initiation of ponding, $\alpha_p = C_p / (1 - C_p)$, $\alpha_s = C_s / (1 - C_s)$, and $\rho = \delta_o / \Delta_o = C_s / C_p$.

Using the above expressions for Δ_w and δ_w, the ratios Δ_w / Δ_o and δ_w / δ_o can be computed for any given combination of primary and secondary beam framing using, respectively, the computed value of parameters C_p and C_s defined in the LRFD Specification.

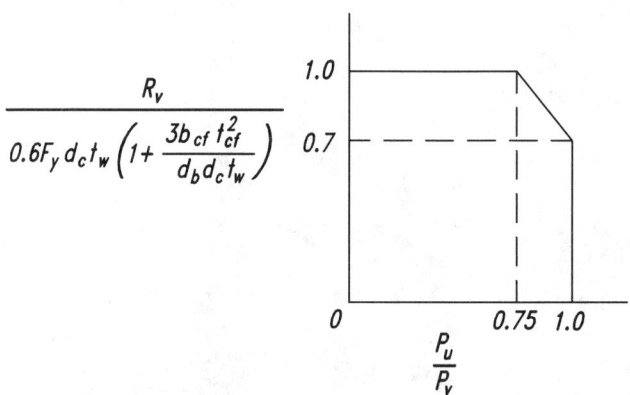

Fig. C-K1.5. Interaction of shear and axial force—inelastic.

Even on the basis of unlimited elastic behavior, it is seen that the ponding deflections would become infinitely large unless

$$\left(\frac{C_p}{1-C_p}\right)\left(\frac{C_s}{1-C_s}\right) < \frac{4}{\pi}$$

Since elastic behavior is not unlimited, the effective bending strength available in each member to resist the stress caused by ponding action is restricted to the difference between the yield stress of the member and the stress f_o produced by the total load supported by it before consideration of ponding is included.

Note that elastic deflection is directly proportional to stress. The admissible amount of ponding in either the primary or critical (midspan) secondary member, in terms of the applicable ratio Δ_w/Δ_o and δ_w/δ_o, can be represented as $(F_y-f_o)/f_o$. Substituting this expression for Δ_w/Δ_o and δ_w/δ_o, and combining with the foregoing expressions for Δ_w and δ_w, the relationship between critical values for C_p and C_s and the available elastic bending strength to resist ponding is obtained. The curves presented in Figures A-K2.1 and A-K2.2 are based upon this relationship. They constitute a design aid for use when a more exact determination of required flat roof framing stiffness is needed than given by the LRFD Specification provision that $C_p + 0.9C_s \leq 0.25$.

Given any combination of primary and secondary framing, the stress index is computed as

$$U_p = \left(\frac{F_y - f_o}{f_o}\right)_p \quad \text{for the primary member}$$

$$U_s = \left(\frac{F_y - f_o}{f_o}\right)_s \quad \text{for the secondary member}$$

where f_o, in each case, is the computed bending stress, ksi, in the member due to the supported loading, neglecting ponding effect. Depending upon geographic location, this loading should include such amount of snow as might also be present, although ponding failures have occurred more frequently during torrential summer rains when the rate of precipitation exceeded the rate of drainage runoff and the resulting hydraulic gradient over large roof areas caused substantial accumulation of water some distance from the eaves.

Given the size, spacing, and span of a tentatively selected combination of primary and secondary beams, for example, one may enter Figure A-K2.1 at the level of the computed stress index U_p, determined for the primary beam; move horizontally to the computed C_s value of the secondary beams; then move downward to the abscissa scale. The combined stiffness of the primary and secondary framing is sufficient to prevent ponding if the flexibility constant read from this latter scale is more than the value of C_p computed for the given primary member; if not, a stiffer primary or secondary beam, or combination of both, is required.

If the roof framing consists of a series of equally-spaced wall-bearing beams, they would be considered as secondary members, supported on an infinitely stiff primary member. For this case, one would use Figure A-K2.2. The limiting value of C_s would be determined by the intercept of a horizontal line representing the U_s value and the curve for $C_p = 0$.

The ponding deflection contributed by a metal deck is usually such a small part of the total ponding deflection of a roof panel that it is sufficient merely to limit its moment of inertia (in.^{4} per foot of width normal to its span) to 0.000025 times the fourth power of its span length, as provided in the LRFD Specification. However, the stability against ponding of a roof consisting of a metal roof deck of relatively slender depth-span ratio, spanning between beams supported directly on columns, may need to be checked. This can be done using Figure A-K2.1 or A-K2.2 with the following computed values:

U_p= stress index for the supporting beam
U_s= stress index for the roof deck
C_p= flexibility constant for the supporting beams
C_s= flexibility constant for one foot width of the roof deck ($S = 1.0$)

Since the shear rigidity of the web system is less than that of a solid plate, the moment of inertia of steel joists and trusses should be taken as somewhat less than that of their chords.

CHAPTER L

SERVICEABILITY DESIGN CONSIDERATIONS

Serviceability criteria are formulated to prevent disruptions of the functional use and damage to the structure during its normal everyday use. While malfunctions may not result in the collapse of a structure or in loss of life or injury, they can seriously impair the usefulness of the structure and lead to costly repairs. Neglect of serviceability may result in unacceptably flexible structures.

There are essentially three types of structural behavior which may impair serviceability:

(1) Excessive local damage (local yielding, buckling, slip, or cracking) that may require excessive maintenance or lead to corrosion.

(2) Excessive deflection or rotation that may affect the appearance, function, or drainage of the structure, or may cause damage to nonstructural components and their attachments.

(3) Excessive vibrations induced by wind or transient live loads which affect the comfort of occupants of the structure or the operation of mechanical equipment.

In allowable stress design, the AISC Specification accounts for possible local damage with factors of safety included in the allowable stresses, while deflection and vibration are controlled, directly or indirectly, by limiting deflections and span-depth ratios. In the past, these rules have led to satisfactory performance of structures, with perhaps the exception of large open floor areas without partitions. In LRFD the serviceability checks should consider the appropriate loads, the response of the structure, and the reaction of the occupants to the structural response.

Examples of loads that may require consideration of serviceability include permanent live loads, wind, and earthquake; effects of human activities such as walking, dancing, etc.; temperature fluctuations; and vibrations induced by traffic near the building or by the operation of mechanical equipment within the building.

Serviceability checks are concerned with adequate performance under the appropriate load conditions. Elastic behavior can usually be assumed. However, some structural elements may have to be examined with respect to their long-term behavior under load.

It is difficult to specify limiting values of structural performance based on serviceability considerations because these depend to a great extent on the type of structure, its intended use, and subjective physiological reaction. For example, acceptable structural motion in a hospital clearly would be much less than in an ordinary industrial building. It should be noted that humans perceive levels of structural motion that are far less than motions that would cause any structural damage. Serviceability limits must be determined through careful consideration by the designer and client.

L1. CAMBER

The engineer should consider camber when deflections at the appropriate load level present a serviceability problem.

L2. EXPANSION AND CONTRACTION

As in the case of deflections, the satisfactory control of expansion cannot be reduced to a few simple rules, but must depend largely upon the good judgment of qualified engineers.

The problem is more serious in buildings with masonry walls than with prefabricated units. Complete separation of the framing, at widely spaced expansion joints, is generally more satisfactory than more frequently located devices dependent upon the sliding of parts in bearing, and usually less expensive than rocker or roller expansion bearings.

Creep and shrinkage of concrete and yielding of steel are among the causes, other than temperature, for dimensional changes.

L3. DEFLECTIONS, VIBRATION, AND DRIFT

1. Deflections

Excessive transverse deflections or lateral drift may lead to permanent damage to building elements, separation of cladding, or loss of weathertightness, damaging transfer of load to non-load-supporting elements, disruption of operation of building service systems, objectionable changes in appearance of portions of the buildings, and discomfort of occupants.

The LRFD Specification does not provide specific limiting deflections for individual members or structural assemblies. Such limits would depend on the function of the structure (ASCE, 1979; CSA, 1989; Ad Hoc Committee, 1986). Provisions that limit deflections to a percentage of span may not be adequate for certain long-span floor systems; a limit on maximum deflection that is independent of span length may also be necessary to minimize the possibility of damage to adjoining or connecting nonstructural elements.

2. Floor Vibration

The increasing use of high-strength materials and efficient structural schemes leads to longer spans and more flexible floor systems. Even though the use of a deflection limit related to span length generally precluded vibration problems in the past, some floor systems may require explicit consideration of the dynamic, as well as the static, characteristics of the floor system.

The dynamic response of structures or structural assemblies may be difficult to analyze because of difficulties in defining the actual mass, stiffness, and damping characteristics. Moreover, different load sources cause varying responses. For example, a steel beam-concrete slab floor system may respond to live loading as a non-composite system, but to transient excitation from human activity as an orthotropic composite plate. Nonstructural partitions, cladding, and built-in furniture significantly increase the stiffness and damping of the structure and frequently eliminate potential vibration problems. The damping can also depend on the amplitude of excitation.

The general objective in minimizing problems associated with excessive structural motion is to limit accelerations, velocities, and displacements to levels that would not be disturbing to the building occupants. Generally, occupants of a building find sustained vibrations more objectionable than transient vibrations.

The levels of peak acceleration that people find annoying depend on frequency of response. Thresholds of annoyance for transient vibrations are somewhat higher and depend on the amount of damping in the floor system. These levels depend on the individual and the activity at the time of excitation (ASCE, 1979; ISO, 1974; CSA, 1989; Murray, 1991; and Ad Hoc Committee, 1986).

The most effective way to reduce effects of continuous vibrations is through vibration isolation devices. Care should be taken to avoid resonance, where the frequency of steady-state excitation is close to the fundamental frequency of the system. Transient vibrations are reduced most effectively by increasing the damping in the structural assembly. Mechanical equipment which can produce objectionable vibrations in any portion of a structure should be adequately isolated to reduce the transmission of such vibrations to critical elements of the structure.

3.   Drift

The LRFD Specification does not provide specific limiting values for lateral drift. If a drift analysis is desired, the stiffening effect of non-load-supporting elements such as partitions and infilled walls may be included in the analysis of drift.

Some irrecoverable inelastic deformations may occur at given load levels in certain types of construction. The effect of such deformations may be negligible or serious, depending on the function of the structure, and should be considered by the designer on a case by case basis.

The deformation limits should apply to structural assemblies as a whole. Reasonable tolerance should also be provided for creep. Where load cycling occurs, consideration should be given to the possibility of increases in residual deformation that may lead to incremental failure.

L5.   CORROSION

Steel members may deteriorate in particular service environments. This deterioration may appear either in external corrosion, which would be visible upon inspection, or in undetected changes that would reduce its strength. The designer should recognize these problems by either factoring a specific amount of damage tolerance into the design or providing adequate protection systems (e.g., coatings, cathodic protection) and/or planned maintenance programs so that such problems do not occur.

CHAPTER M

FABRICATION, ERECTION, AND QUALITY CONTROL

M2. FABRICATION

1. Cambering, Curving, and Straightening

The use of heat for straightening or cambering members is permitted for A514 and A852 steel, as it is for other steels. However, the maximum temperature permitted is 1,100°F compared to 1,200°F for other steels.

The cambering of flexural members, when required by the contract documents, may be accomplished in various ways. In the case of trusses and girders, the desired curvature can be built in during assembly of the component parts. Within limits, rolled beams can be cold-cambered at the producing mills.

The local application of heat has come into common use as a means of straightening or cambering beams and girders. The method depends upon an ultimate shortening of the heat-affected zones. A number of such zones, on the side of the member that would be subject to compression during cold-cambering or "gagging," are heated enough to be "upset" by the restraint provided by surrounding unheated areas. Shortening takes place upon cooling.

While the final curvature or camber can be controlled by these methods, it must be realized that some deviation, due to workmanship error and permanent change due to handling, is inevitable.

2. Thermal Cutting

Preferably thermal cutting shall be done by machine. The requirement for a positive preheat of 150°F minimum when thermal cutting beam copes and weld access holes in ASTM A6 Group 4 and 5 shapes, and in built-up shapes made of material more than two inches thick, tends to minimize the hard surface layer and the initiation of cracks.

5. Bolted Construction

In the past, it has been required to tighten to a specified tension all ASTM A325 and A490 bolts in both slip-critical and bearing-type connections. The requirement was changed in 1985 to permit most bearing-type connections to be tightened to a snug-tight condition.

In a snug-tight bearing connection, the bolts cannot be subjected to tension loads, slip can be permitted, and loosening or fatigue due to vibration or load fluctuations are not design considerations.

It is suggested that snug-tight bearing-type connections be used in applications where A307 bolts would be permitted.

This section provides rules for the use of oversized and slotted holes paralleling the provisions which have been in the RCSC Specification (RCSC, 1988) since 1972, extended to include A307 bolts which are outside the scope of the high-strength bolt specifications.

M3. SHOP PAINTING

The surface condition of steel framing disclosed by the demolition of long-standing buildings has been found to be unchanged from the time of its erection, except at isolated spots where leakage may have occurred. Even in the presence of leakage, the shop coat is of minor influence (Bigos et al., 1954).

The LRFD Specification does not define the type of paint to be used when a shop coat is required. Conditions of exposure and individual preference with regard to finish paint are factors which bear on the selection of the proper primer. Hence, a single formulation would not suffice. For a comprehensive treatment of the subject, see SSPC (1989).

5. Surfaces Adjacent to Field Welds

The Specification allows for welding through surface materials, including appropriate shop coatings, that do not adversely affect weld quality nor create objectionable fumes.

M4. ERECTION

4. Fit of Column Compression Joints and Base Plates

Tests at the University of California-Berkeley (Popov and Stephen, 1977) on spliced full-size columns with joints that had been intentionally milled out-of-square, relative to either strong or weak axis, demonstrated that the load-carrying capacity was the same as that for a similar unspliced column. In the tests, gaps of $\frac{1}{16}$-in. were not shimmed; gaps of $\frac{1}{4}$-in. were shimmed with non-tapered mild steel shims. Minimum size partial-penetration welds were used in all tests. No tests were performed on specimens with gaps greater than $\frac{1}{4}$-in.

5. Field Welding

The purpose of wire brushing shop paint, on surfaces adjacent to joints to be field welded, is to reduce the possibility of porosity and cracking and also to reduce any environmental hazard. Although there are limited tests which indicate that painted surfaces result in sound welds without wire brushing, other studies have resulted in excessive porosity and/or cracking when welding coated surfaces. Wire brushing to reduce the paint film thickness minimizes rejectable welds. Grinding or other procedures beyond wire brushing is not necessary.

APPENDIX B

DESIGN REQUIREMENTS

B5. LOCAL BUCKLING

1. Classification of Steel Sections

The limiting width-thickness λ_p and λ_r ratios for webs in pure flexure $(P_u / \phi_b P_y = 0)$ and with axial compression have been revised in terms of (h / t) rather than (h_c / t). The simplified formulation in Table B5.1 for λ_r based on double symmetry with equal flanges $(h / h_c = 1)$ is unconservative when the compression flange is smaller than the tension flange, and conservative if the reverse is true. The more accurate limit is given in Appendix B5.1 as a function of h_c. Figure C-A-B5.1 illustrates the λ_r variation for axial compression and flange asymmetry effects.

The $\frac{3}{4}$ minimum and $\frac{3}{2}$ maximum restrictions on h / h_c in Equations A-B5-1 and A-B5-2 approximately correspond to the 0.1 and 0.9 range of I_{yc} / I_y for a member to be considered a singly symmetric I shape. Otherwise, when the flange areas differ by more than a factor of two, the member should be conservatively designed as a tee section.

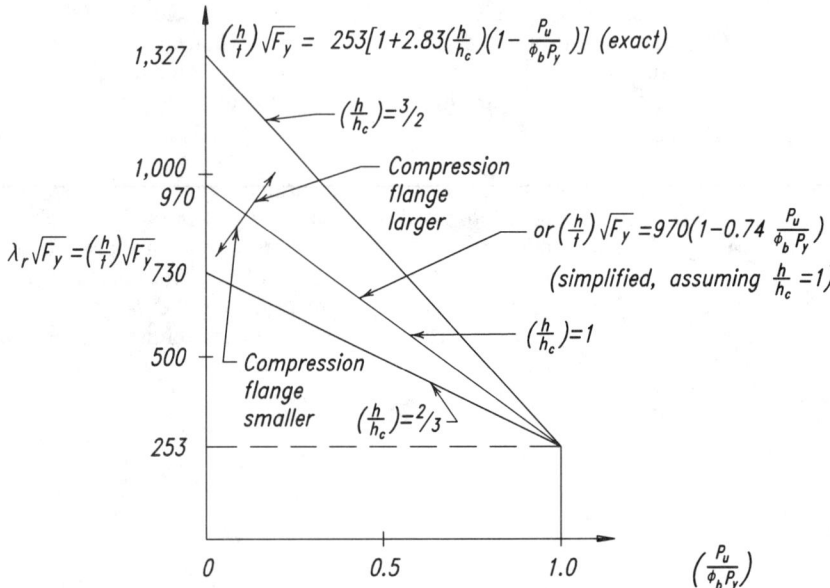

Fig. C-A-B5.1. Local web buckling for I-shaped members.

APPENDIX E

COLUMNS AND OTHER COMPRESSION MEMBERS

E3. DESIGN COMPRESSIVE STRENGTH FOR FLEXURAL-TORSIONAL BUCKLING

The equations in Appendix E3 for determining the flexural-torsional elastic buckling loads of columns are derived in texts on structural stability (Timoshenko and Gere (1961), Bleich (1952), Galambos (1968), and Chen and Atsuta (1977), for example). Since these equations for flexural-torsional buckling apply only to elastic buckling, they must be modified for inelastic buckling when $F_{cr} > 0.5F_y$. This is accomplished through the use of the equivalent slenderness factor $\lambda_e = \sqrt{F_y / F_e}$.

APPENDIX F

BEAMS AND OTHER FLEXURAL MEMBERS

F1. DESIGN FOR FLEXURE

Three limit states must be investigated to determine the moment capacity of flexural members: lateral-torsional buckling (LTB), local buckling of the compression flange (FLB), and local buckling of the web (WLB). These limit states depend, respectively, on the beam slenderness ratio L_b / r_y, the width-thickness ratio b / t of the compression flange and the width-thickness ratio h / t_w of the web. For convenience, all three measures of slenderness are denoted by λ.

Variations in M_n with L_b are shown in Figure C-A-F1.1. The discussion of plastic, inelastic, and elastic buckling in Commentary Section F1 with reference to lateral-torsional buckling applies here except for an important difference in the significance of λ_p for lateral-torsional buckling and local buckling. Values of λ_p for FLB and WLB produce a compact section with a rotation capacity of about three (after reaching M_p) before the onset of local buckling, and therefore meet the requirements for plastic analysis of load effects (Commentary Section B5). On the other hand, values of λ_p for LTB do not allow plastic analysis because they do not provide rotation capacity beyond that needed to develop M_p. Instead $L_b \leq L_{pd}$ (Section F1.2d) must be satisfied.

Analyses to include restraint effects of adjoining elements are discussed in Galambos (1988). Analysis of the lateral stability of members with shapes not covered in this appendix must be performed according to the available literature (Galambos, 1988).

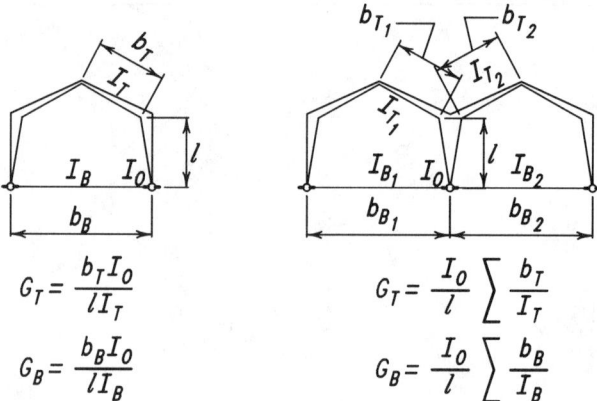

$$G_T = \frac{b_T I_0}{l I_T}$$

$$G_B = \frac{b_B I_0}{l I_B}$$

$$G_T = \frac{I_0}{l} \sum \frac{b_T}{I_T}$$

$$G_B = \frac{I_0}{l} \sum \frac{b_B}{I_B}$$

Figure C-A-F1.1

See the Commentary for Section B5 for the discussion of the equation regarding the bending capacity of circular sections.

F3.   WEB-TAPERED MEMBERS

1.   General Requirements

The provision contained in Appendix F3 covers only those aspects of the design of tapered members that are unique to tapered members. For other criteria of design not specifically covered in Appendix F3, see the appropriate portions of this Specification and Commentary.

The design of wide-flange columns with a single web taper and constant flanges follows the same procedure as for uniform columns according to Section E2, except the column slenderness parameter λ_c for major axis buckling is determined for a slenderness ratio $K_\gamma L / r_{ox}$, and for minor axis buckling for KL / r_{oy}, where K_γ is an effective length factor for tapered members, K is the effective length factor for prismatic members, and r_{ox} and r_{oy} are the radii of gyration about the x and the y axes, respectively, taken at the smaller end of the tapered member.

For stepped columns or columns with other than a single web taper, the elastic critical stress is determined by analysis or from data in reference texts or research reports (Chapters 11 and 13 in Timoshenko and Gere (1961) and Bleich (1952) and Kitipornchai and Trahair (1980)), and then the same procedure of using λ_{eff} is utilized in calculating the factored resistance.

This same approach is recommended for open section built-up columns (columns with perforated cover plates, lacing, and battens) where the elastic critical buckling stress determination must include a reduction for the effect of shear. Methods for calculating the elastic buckling strength of such columns are given in Chapter 12 of the SSRC Guide (Galambos, 1988) and in Timoshenko and Gere (1961) and Bleich (1952).

3.   Design Compressive Strength

The approach in formulating $F_{a\gamma}$ of tapered columns is based on the concept that the critical stress for an axially loaded tapered column is equal to that of a prismatic column of different length, but of the same cross section as the smaller end of the tapered column. This has resulted in an equivalent effective length factor K_γ for a tapered member subjected to axial compression (Lee et al., 1972). This factor, which is used to determine the value of S in Equations A-F3-2 and λ_c in Equation E2-3, can be determined accurately for a symmetrical rectangular rigid frame comprised of prismatic beams and tapered columns.

With modifying assumptions, such a frame can be used as a mathematical model to determine with sufficient accuracy the influence of the stiffness $\Sigma(I / b)_g$ of beams and rafters which afford restraint at the ends of a tapered column in other cases such as those shown in Figure C-A-F1.1. From Equations A-F3-2 and E2-3, the critical load P_{cr} can be expressed as $\pi^2 EI_o / (K_\gamma l)^2$. The value of K_γ can be obtained by interpolation, using the appropriate chart from Lee et al. (1972) and restraint modifiers G_T and G_B. In each of these modifiers the tapered column, treated as a prismatic member having a moment of inertia I_o, computed at the smaller end, and its actual length l, is assigned the stiffness I_o / l, which is then divided by the stiffness of the restraining members at the end of the tapered column under consideration.

4.   Design Flexural Strength

The development of the design bending stress for tapered beams follows closely with that for prismatic beams. The basic concept is to replace a tapered beam by an equivalent prismatic beam with a different length, but with a cross section identical to that of the smaller end of the tapered beam (Lee et al., 1972). This has led to the modified length factors h_s and h_w in Equations A-F3-6 and A-F3-7.

Equations A-F3-6 and A-F3-7 are based on total resistance to lateral buckling, using both St. Venant and warping resistance. The factor B modifies the basic $F_{b\gamma}$ to members which are continuous past lateral supports. Categories a, b, and c of Appendix F3.4 usually apply; however, it is to be noted that they apply only when the axial force is small and adjacent unbraced segments are approximately equal in length. For a single member, or segments which do not fall into category a, b, c, or d, the recommended value of B is unity. The value of B should also be taken as unity when computing the value of $F_{b\gamma}$ to obtain M_n to be used in Equations H1-1 and C1-1, since the effect of moment gradient is provided for by the factor C_m. The background material is given in WRC Bulletin No. 192 (Morrell and Lee, 1974).

APPENDIX G

PLATE GIRDERS

Appendix G is taken from AISI Bulletin 27 (Galambos, 1978). Comparable provisions are included in the AISC ASD Specification. The provisions are presented in an appendix as they are seldom used and produce designs which are often less economical than plate girders designed without tension-field action.

The web slenderness ratio $h/t_w = 970/\sqrt{F_{yf}}$ that distinguishes plate girders from beams is written in terms of the flange yield stress, because for hybrid girders inelastic buckling of the web due to bending depends on the flange strain.

The equation for R_e used in the 1986 LRFD Specification was the same as that used in the AASHTO *Standard Specification for Highway Bridges*. In this edition, the equation for R_e, used in the AISC ASD Specification since 1969, is used because its derivation is published (Gaylord and Gaylord, 1992 and ASCE-AASHTO, 1968) and it is more accurate than the AASHTO equation.

G2. DESIGN FLEXURAL STRENGTH

In previous versions of the AISC Specification a coefficient of $0.0005a_r$ was used in R_{PG} based on the work of Basler (1961). This value is valid for $a_r \leq 2$. In that same paper, Basler developed a more general coefficient, applicable to all ratios of A_w/A_f which has now been adopted because application of the previous equation to sections with large a_r values gives unreasonable results. An arbitrary limit of $a_r \leq 10$ is imposed so that the R_{PG} expression is not applied to sections approaching a tee shape.

APPENDIX H

MEMBERS UNDER COMBINED FORCES AND TORSION

H3. ALTERNATIVE INTERACTION EQUATIONS FOR MEMBERS UNDER COMBINED STRESS

In the case of members not subject to flexural buckling, i.e., $L_b < L_{pd}$, the use of somewhat more liberal interaction Equations A-H3-5 and A-H3-6 is acceptable as an alternative when the flexure is about one axis only.

The alternative interaction Equations A-H3-1 and A-H3-2 for biaxially loaded H and wide-flange column shapes were taken from Galambos (1988), Springfield (1975), and Tebedge and Chen (1974).

For I-shaped members with $b_f / d > 1.0$, use of Section H1 is recommended, because no additional research is available for this case.

APPENDIX J

CONNECTIONS, JOINTS, AND FASTENERS

J2. WELDS

4. Design Strength

When weld groups are loaded in shear by an external load that does not act through the center of gravity of the group, the load is eccentric and will tend to cause a relative rotation and translation between the parts connected by the weld. The point about which rotation tends to take place is called the instantaneous center of rotation. Its location is dependent upon the load eccentricity, geometry of the weld group, and deformation of the weld at different angles of the resultant elemental force relative to the weld axis.

The individual resistance force of each unit weld element can be assumed to act on a line perpendicular to a ray passing through the instantaneous center and that element's location (see Figure C-A-J2.1).

The ultimate shear strength of weld groups can be obtained from the load deformation relationship of a single-unit weld element. This relationship was originally given by Butler (1972) for E60 electrodes. Curves for E70 electrodes used in the Appendix were obtained by Lesik (1990).

Unlike the load-deformation relationship for bolts, strength and deformation performance in welds are dependent on the angle θ that the resultant elemental force makes with the axis of the weld element (see Figure C-A-J2.1). The actual load deformation relationship for welds is given in Figure C-A-J2.2, taken from Lesik and Kennedy (1990). Conversion of the SI equation to foot-pound units results in the following weld strength equation for R_n:

$$R_n = 0.852(1.0 + 0.50 \sin^{1.5}\theta)F_{EXX} A_w$$

Because the maximum strength is limited to $0.60F_{EXX}$ for longitudinally loaded welds ($\theta = 0°$), the LRFD Specification provision provides, in the reduced equation coefficient, a reasonable margin for any variation in welding techniques and procedures. To eliminate possible computational difficulties, the maximum deformation in the weld elements is limited to $0.17D$. For design convenience, a simple elliptical formula is used for $f(p)$ to closely approximate the empirically derived polynomial in Lesik (1990).

The total resistance of all the weld elements combine to resist the eccentric ultimate load, and when the correct location of the instantaneous center has been selected, the three in-plane equations of statics (ΣF_x, ΣF_y, ΣM) will be satisfied. Numerical techniques, such as those given by Brandt (1982), have been developed to locate the instantaneous center of rotation subject to convergence tolerances.

Earlier editions of the AISC *Manual of Steel Construction* (AISC, 1980, 1986, 1989) took advantage of the inelastic redistribution of stresses that is inherent in the Appendix J2.4 procedure. However, in each of the utilized computational techniques the resulting coefficients were factored down so that the maximum

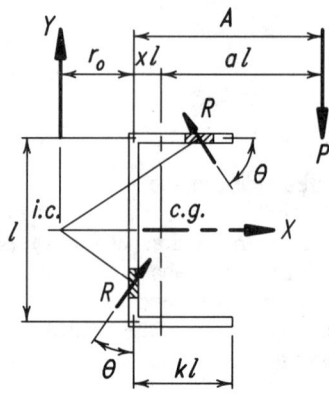

Figure C-A-J2.1

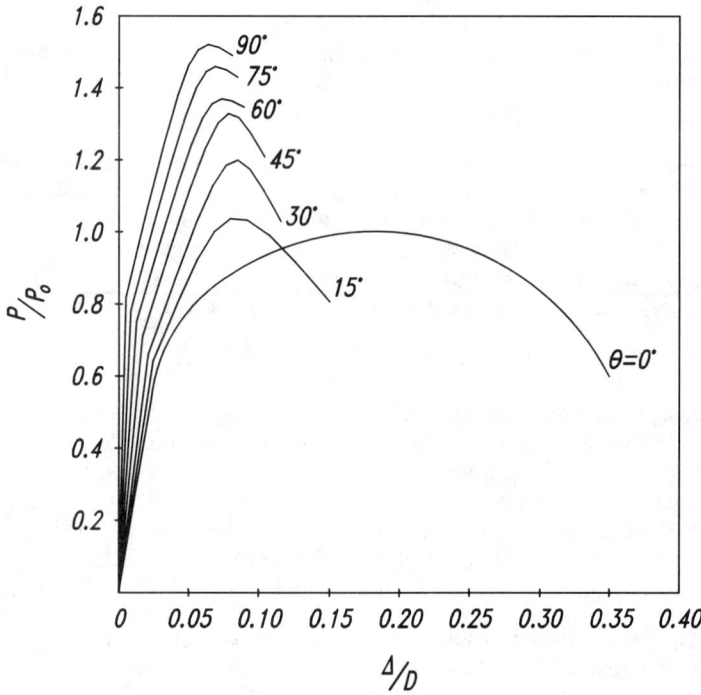

Figure C-A-J2.2

stress, at any point in the weld group, did not exceed the limiting value specified by either the Allowable Stress Design or LRFD Specifications, $0.3F_u$ or $0.6F_u$, respectively. As a result, the tabulated weld-capacity data shown in the appropriate referenced manual tables will be found to be conservative relative to the data obtained using the computational procedure presented in Appendix J2.4.

APPENDIX K

CONCENTRATED FORCES, PONDING, AND FATIGUE

K3. FATIGUE

Because most members in building frames are not subject to a large enough number of cycles of full design stress application to require design for fatigue, the provisions covering such designs have been placed in Appendix K3.

When fatigue is a design consideration, its severity is most significantly affected by the number of load applications, the magnitude of the stress range, and the severity of the stress concentrations associated with the particular details. These factors are not encountered in normal building designs; however, when encountered and when fatigue is of concern, all provisions of Appendix K3 must be satisfied.

Members or connections subject to less than 20,000 cycles of loading will not involve a fatigue condition, except in the case of repeated loading involving large ranges of stress. For such conditions, the admissible range of stress can conservatively be taken as one and one-half times the applicable value given in Table A-K3.3 for "Loading Condition 1."

Fluctuation in stress which does not involve tensile stress does not cause crack propagation and is not considered to be a fatigue situation. On the other hand, in elements of members subject solely to calculated compression stress, fatigue cracks may initiate in regions of high tensile residual stress. In such situations, the cracks generally do not propagate beyond the region of the residual tensile stress, because the residual stress is relieved by the crack. For this reason stress ranges that are completely in compression are not included in the column headed by "Kind of Stress" in Table A-K3.2. This is also true of comparable tables of the current AASHTO and AREA specifications.

When fabrication details involving more than one category occur at the same location in a member, the stress range at that location must be limited to that of the most restrictive category. By locating notch-producing fabrication details in regions subject to a small range of stress, the need for a member larger than required by static loading will often be eliminated.

Extensive test programs (Fisher et al., 1970; and Fisher et al., 1974) using full size specimens, substantiated by theoretical stress analysis, have confirmed the following general conclusions:

(1) Stress range and notch severity are the dominant stress variables for welded details and beams.

(2) Other variables such as minimum stress, mean stress, and maximum stress are not significant for design purposes.

(3) Structural steels with yield points of 36 to 100 ksi do not exhibit significantly different fatigue strength for given welded details fabricated in the same manner.

Allowable stress ranges can be read directly from Table A-K3.3 for a particular category and loading condition. The values are based on extensive research (Keating and Fisher, 1985).

Provisions for bolts subjected to tension are given in Table A-K3.4. Tests have uncovered dramatic differences in fatigue life, not completely predictable from the various published equations for estimating the actual magnitude of prying force (Kulak et al., 1987). To limit the uncertainties regarding prying action on the fatigue behavior of these bolts, the tensile stresses given in Table J3.2 are approved for use under extended cyclic loading only if the prying force, included in the design tensile force, is small. When this cannot be assured, the design tensile stress is drastically reduced to cover any conceivable prying effect.

The use of other types of mechanical fasteners to resist applied cyclic loading in tension is not recommended. Lacking a high degree of assured pretension, the range of stress is generally too great to resist such loading for long.

However, all types of mechanical fasteners survive unharmed when subject to cyclic shear stresses sufficient to fracture the connected parts, which is provided for elsewhere in Appendix K3.

References

Ad Hoc Committee on Serviceability Research (1986), "Structural Serviceability: A Critical Appraisal and Research Needs," *Journal of Structural Engineering,* ASCE, Vol. 112, No. 12, 1986.

American Association of State Highway and Transportation Officials (1989), *Standard Specification for Highway Bridges,* 1989.

American Concrete Institute (1989), *Building Code Requirements for Reinforced Concrete,* ACI 318-89, Detroit, MI, 1989.

American Institute of Steel Construction, Inc. (1973), "Commentary on Highly Restrained Welded Connections," *Engineering Journal,* AISC, Vol. 10, No. 3, 3rd Quarter, 1973.

American Institute of Steel Construction, Inc. (1978), *Specification for the Design, Fabrication, and Erection of Structural Steel for Buildings,* Chicago, IL, 1978.

American Institute of Steel Construction, Inc. (1983), *Torsional Analysis of Steel Members,* Chicago, IL, 1983.

American Institute of Steel Construction, Inc. (1986), *Load and Resistance Factor Design Specification for Structural Steel Buildings,* Chicago, IL, 1986.

American Institute of Steel Construction, Inc. (1989), *Specification for Structural Steel Buildings—Allowable Stress Design and Plastic Design,* Chicago, IL, 1989.

American Institute of Steel Construction, Inc. (1992), *Seismic Provisions of Structural Steel Buildings,* AISC, Chicago, IL, 1992.

American Society of Civil Engineers (1971), "Plastic Design in Steel," *ASCE Manual of Engineering Practice,* No. 41, 2nd Edition, New York, 1971.

American Society of Civil Engineers (1979), *Structural Design of Tall Steel Buildings,* New York, 1979.

American Society of Civil Engineers (1988), *Minimum Design Loads for Buildings and Other Structures,* ASCE7-88, New York.

ASCE Task Committee on Design Criteria for Composite Structures in Steel and Concrete (1992), "Proposed Specification for Structural Steel Beams with Web Openings," D. Darwin, Chmn., *Journal of Structural Engineering,* ASCE, Vol. 118, No. ST12, December 1992.

ASCE Task Committee on Design Criteria for Composite Structures in Steel and Concrete (1992a), "Commentary on Proposed Specification for Structural Steel

Beams with Web Openings," D. Darwin, Chmn., *Journal of Structural Engineering,* ASCE, Vol. 118, No. ST12, December 1992.

Joint ASCE-AASHTO Committee on Flexural Members (1968), "Design of Hybrid Steel Beams," Report of Subcommitte 1, *Journal of the Structural Division,* ASCE, Vol. 94, No. ST6, June 1968.

Aslani, F. and S. C. Goel (1991), "An Analytical Criteria for Buckling Strength of Built-Up Compression Members," *Engineering Journal,* AISC, 4th Quarter, 1991.

ATC (1978), *Tentative Provisions for the Development of Seismic Regulations for Buildings,* ATC Publication 3-06, June 1978.

American Welding Bureau (1931), *Report of Structural Welding Committee,* 1931.

Austin, W. J. (1961), "Strength and Design of Metal Beam-Columns," *Journal of the Structural Division,* ASCE, Vol. 87, No. ST4, April 1961.

Basler, K. (1961), "Strength of Plate Girders in Shear," *Journal of the Structural Division,* ASCE, Vol. 104, No. ST9, October 1961.

Basler, K. and B. Thurlimann (1963), "Strength of Plate Girders in Bending," *Journal of the Structural Division,* ASCE, Vol. 89, No. ST4, August 1963.

Barakat, Munzer and Wai-Fah Chen (1991), "Design Analysis of Semi-Rigid Frames: Evaluation and Implementation," *Engineering Journal,* AISC, Vol. 28, No. 2, 2nd Quarter, 1991.

Beedle, L. S and L. Tall (1960), "Basic Column Strength," *Journal of the Structural Division,* ASCE, Vol. 86, No. ST7, July 1960.

Bendigo, R. A., R. M. Hansen, and J. L. Rumpf (1963), "Long Bolted Joints," *Journal of the Structural Division,* ASCE, Vol. 89, No. ST6, December 1963.

Bergfelt, A. (1971), *Studies and Tests on Slender Plate Girders Without Stiffeners,* March 1971.

Bigos, J., G. W. Smith, E. F. Ball, and P. J. Foehl (1954), *Shop Paint and Painting Practice,* 1954 Proceedings, AISC National Engineering Conference, Milwaukee, WI.

Birkemoe, P. C. and M. I. Gilmor (1978), "Behavior of Bearing-Critical Double-Angle Beam Connections," *Engineering Journal,* AISC, Vol. 15, No. 4, 4th Quarter, 1978.

Bjorhovde, Reidar (1972), *Deterministic and Probabilistic Approaches to the Strength of Steel Columns,* Ph.D. Dissertation, Lehigh University, Bethlehem, PA, May.

Bjorhovde, Reidar (1984), "Effect of End Restraint on Column Strength—Practical Applications," *Engineering Journal,* AISC, Vol. 21, No. 1, 1st Quarter, 1984.

Bjorhovde, Reidar (1988), "Columns: From Theory to Practice," *Engineering Journal,* AISC, Vol. 25, No. 1, 1st Quarter 1988.

Bjorhovde, Reidar, T. V. Galambos, and M. K. Ravindra (1978), "LRFD Criteria for Steel Beam-Columns," *Journal of the Structural Division,* ASCE, Vol. 104, No. ST9, September.

Bleich, F. (1952), *Buckling Strength of Metal Structures,* McGraw-Hill Book Co., New York, NY, 1952.

Brandt, G. D. (1982), "A General Solution for Eccentric Loads on Weld Groups," *Engineering Journal,* AISC, Vol. 19, No. 3, 3rd Quarter, 1982.

Bridge, P. Q. and J. W. Roderick (1978), "Behavior of Built-Up Composite Columns," *Journal of the Structural Division,* ASCE, Vol. 104, No. ST7, July 1978, pp. 1141–1165.

Brockenbrough, R. L. (1983), "Considerations in the Design of Bolted Joints for Weathering Steel," *Engineering Journal,* AISC, Vol. 20, No. 1, 1st Quarter, 1983.

Butler, Lorne J., Shubendu Pal, and Geoffrey L. Kulak (1972), "Eccentrically Loaded Welded Connections," *Journal of the Structural Division,* ASCE, Vol. 98, No. ST5, May 1972.

Canadian Standards Association (1989), *Limit States Design of Steel Structures,* Appendices G, H, and I, CSA S16.1-M89, Rexdale, Ontario, Canada, 1989.

Chen, W. F. and T. Atsuta (1976), *Theory of Beam Columns, Volume I: In-Plane Behavior and Design,* McGraw-Hill, New York, 1976.

Chen, W. F. and T. Atsuta (1977), *Theory of Beam Columns, Volume II: Space Behavior and Design,* McGraw-Hill, New York, 1977.

Cheong-Siat Moy, F., E. Ozer, and L. W. Lu (1977), "Strength of Steel Frames Under Gravity Loads," Journal of the Structural Division, ASCE, Vol. 103, No. ST6, June 1977.

Chopra, A. K. and N. M. Newmark (1980), *Design of Earthquake Resistant Structures,* John Wiley and Sons, Inc., New York, NY, 1980.

Cooper, P. B., T. V. Galambos, and M. K. Ravindra (1978), "LRFD Criteria for Plate Girders," *Journal of the Structural Division,* ASCE, Vol. 104, No. ST9, September 1978.

Daniels, J. H. and L. W. Lu (1972), "Plastic Subassemblage Analysis for Unbraced Frames," *Journal of the Structural Division,* ASCE, Vol. 98, No. ST8, August 1972.

Darwin, David (1990), *Steel and Composite Beams with Web Openings,* AISC Steel Design Guide Series No. 2, AISC, Chicago, IL, 1990.

Davison, J. B., P. A. Kirby, and D. A. Nethercot (1988), "Semi-Rigid Connections in Isolation and in Frames," in *Connections in Steel Structures: Behavior, Strength and Design,* Elsevier Applied Science Publishers, London.

Disque, R. O. (1973), "Inelastic K-Factor in Design," *Engineering Journal,* AISC, Vol. 10, No. 2, 2nd Quarter, 1973.

Easterling, W. S. and L. Gonzales (1993), "Shear Lag Effects in Steel Tension Members," *Engineering Journal,* AISC, Vol. 30, No. 2, 2nd Quarter, 1993.

Elgaaly, M. (1983), "Web Design Under Compressive Edge Loads," *Engineering Journal,* AISC, Vol. 20, No. 4, 4th Quarter, 1983.

Elgaaly, M. and R. Salkar (1991), "Web Crippling Under Edge Loading," *Proceedings of AISC National Steel Construction Conference,* Washington, D.C., 1991.

Ellifritt, D. S., G. Wine, T. Sputo, and S. Samuel (1992), "Flexural Strength of WT Sections," *Engineering Journal,* AISC, Vol. 29, No. 2, 2nd Quarter, 1992.

Ellingwood, B. E., J. G. MacGregor, T. V. Galambos, and C. A. Cornell (1982), "Probability-Based Load Criteria: Load Factors and Load Combinations," *Journal of the Structural Division,* ASCE, Vol. 108, No. 5, 1982.

Fielding, D. J. and J. S. Huang (1971), "Shear in Steel Beam-to-Column Connections," *Welding Journal,* AWS, Vol. 50, No. 7, Research Supplement 1971, pp. 313–326.

Fielding, D. J. and W. F. Chen (1973), "Steel Frame Analysis and Connection Shear Deformation," *Journal of the Structural Division,* ASCE, Vol. 99, No. ST1, January 1973.

Fisher, J. W., P. A. Albrecht, B. T. Yen, D. J. Klingerman, and B. M. McNamee (1974), *Fatigue Strength of Steel Beams With Welded Stiffeners and Attachments,* National Cooperative Highway Research Program, Report 147, Washington, D.C., 1974.

Fisher, J. W., K. H. Frank, M. A. Hirt, and B. M. McNamee (1970), *Effect of Weldments on the Fatigue Strength of Beams,* National Cooperative Highway Research Program, Report 102, Washington, D.C., 1970.

Fisher, J. W., T. V. Galambos, G. L. Kulak, and M. K. Ravindra (1978), "Load and Resistance Factor Design Criteria for Connectors," *Journal of the Structural Division,* ASCE, Vol. 104, No. ST9, September 1978.

Frank, K. H. and J. A. Yura (1981), *An Experimental Study of Bolted Shear Connections,* FHWA/RD-81/148, December 1981.

Freeman, F. R. (1930), *The Strength of Arc-Welded Joints,* Proc. Inst. Civil Engineers, Vol. 231, London, England, 1930.

Galambos, T. V. (1960), "Influence of Partial Base Fixity on Frame Stability," *Journal of the Structural Division,* ASCE, Vol. 86, No. ST5, May 1960.

Galambos, T. V. (1968), *Structural Members and Frames,* Prentice-Hall, Englewood Cliffs, NJ, 1968.

Galambos, T. V. (1976), *Proposed Criteria for Load Resistance Factor Design of Steel Building Structures,* Research Report No. 45, Civil Engineering Dept., Washington University, St. Louis, MO, May 1976.

Galambos, T. V. (1978), *Bulletin No. 27,* American Iron and Steel Institute, Washington, D.C., January 1978.

Galambos, T. V. (1980), *Reliability of Axially Loaded Columns,* Washington University, Department of Civil Engineering, St. Louis, MO, December 1980.

Galambos, T. V. (ed.) (1988), *Guide to Stability Design Criteria for Metal Structures,* Structural Stability Research Council, 4th Edition, John Wiley & Sons, 1988.

Galambos, T. V. (1991), "Design of Axially Loaded Compressed Angles," *Structural Stability Research Council Annual Technical Session Proceedings,* 1991.

Galambos, T.V. and J. Chapuis (1980), *LRFD Criteria for Composite Columns and Beam-Columns,* Revised Draft, Washington University, Dept. of Civil Engineering, St. Louis, MO, December 1980.

Galambos, T. V., B. Ellingwood, J. G. MacGregor, and C. A. Cornell (1982), "Probability-Based Load Criteria: Assessment of Current Design Practice," *Journal of the Structural Division,* ASCE, Vol. 108, No. ST5, May 1982.

Galambos, T. V. and M. K. Ravindra (1973), *Tentative Load and Resistance Factor Design Criteria for Steel Buildings,* Research Report No. 18, Washington University, Dept. of Civil Engineering, St. Louis, MO, September 1973.

Galambos, T. V. and M. K. Ravindra (1976), *Load and Resistance Factor Design Criteria for Steel Beams,* Research Report No. 27, Washington University, Dept. of Civil Engineering, St. Louis, MO, February 1976.

Galambos, T. V. and M. K. Ravindra (1978), "Properties of Steel for Use in LRFD," *Journal of the Structural Division,* ASCE, Vol. 104, No. ST9, September 1978.

Gaylord, Edwin H. Jr., Charles N. Gaylord, and James E. Stallmeyer (1992), *Design of Steel Structures,* 3rd Edition, McGraw-Hill Book Co., New York, 1992.

Gibson, G. T. and B. T. Wake (1942), "An Investigation of Welded Connections for Angle Tension Members," *The Welding Journal,* American Welding Society, January 1942.

Goble, G. G. (1968), "Shear Strength of Thin Flange Composite Specimens," *Engineering Journal,* AISC, Vol. 5., No. 2, 2nd Quarter, 1968.

Grant, J. A., J. W. Fisher, and R. G. Slutter (1977), "Composite Beams with Formed Steel Deck," *Engineering Journal,* AISC, Vol. 14, No. 1, 1st Quarter, 1977.

Hall, D. H. (1981), "Proposed Steel Column Strength Criteria," *Journal of the Structural Division,* ASCE, Vol. 107, No. ST4, April 1981.

Hansell, W. C., T. V. Galambos, M. K. Ravindra, and I. M. Viest (1978), "Composite Beam Criteria in LRFD," *Journal of the Structural Division,* ASCE, Vol. 104, No. ST9, September 1978.

Hardash, S. G. and R. Bjorhovde (1985), "New Design Criteria for Gusset Plates in Tension," *Engineering Journal,* AISC, Vol. 22, No. 2, 2nd Quarter, 1985.

Hoglund, T. (1971), *Simply Supported Long Thin Plate I-Girders Without Web Stiffeners, Subjected to Distributed Transverse Load,* Dept. of Building Statics and Structural Engineering of the Royal Institute of Technology, Stockholm, Sweden, 1971.

International Association of Bridge and Structural Engineering (1968), *Final Report of the Eighth Congress,* Zurich, September 1968.

International Organization for Standardization (1974), *Guide for the Evaluation of Human Exposure to Whole-Body Vibration,* Document ISO 2631, September 1974.

Iwankiw, N. (1984), "Note on Beam-Column Moment Amplification Factor," *Engineering Journal,* AISC, Vol. 21, No. 1, 1st Quarter, 1984.

Johnson, D. L. (1985), *An Investigation into the Interaction of Flanges and Webs in Wide-Flange Shapes,* 1985 Proceeding SSRC Annual Technical Session, Cleveland, OH, Structural Stability Research Council, Lehigh University, Bethlehem, PA.

Johnston, B. G. (1939), *Pin-Connected Plate Links,* 1939 ASCE Transactions.

Johnston, B. G. and Deits, "Tests of Miscellaneous Welded Building Connections," *Welding Journal,* November 1941.

Johnston, B. G. and L. F. Green (1940), "Flexible Welded Angle Connections," *The Welding Journal*, October 1940.

Jones, J. (1940), "Static Tests on Riveted Joints," *Civil Engineering*, May 1940.

Kanchanalai, T. (1977), *The Design and Behavior of Beam-Columns in Unbraced Steel Frames*, AISI Project No. 189, Report No. 2, Civil Engineering/Structures Research Lab, University of Texas-Austin, October 1977.

Kanchanalai, T. and L. W. Lu (1979), "Analysis and Design of Framed Columns Under Minor Axis Bending," *Engineering Journal*, AISC, Vol. 16, No. 2, 2nd Quarter, 1979.

Keating, P. B. and J. W. Fisher (1985), *Review of Fatigue Tests and Design Criteria on Welded Details*, NCHRP Project 12-15(50), October 1985, Washington, D.C.

Ketter, R. L. (1961), "Further Studies of the Strength of Beam-Columns," *Journal of the Structural Division*, ASCE, Vol. 87, No. ST6, August 1961.

Kirby, P. A. and D. A. Nethercot (1979), *Design for Structural Stability*, John Wiley and Sons, Inc., New York, NY, 1979.

Kitipornchai, S. and N. S. Trahair (1980), "Buckling Properties of Monosymmetric I-Beams," *Journal of the Structural Division*, ASCE, Vol. 109, No. ST5, May 1980.

Kloppel, K. and T. Seeger (1964), "Dauerversuche Mit Einsohnittigen Hv-Verbindurgen Aus ST37," *Der Stahlbau*, Vol. 33, No. 8, August 1964, pp. 225–245 and Vol. 33, No. 11, November 1964, pp. 335–346.

Kotecki, D. S. and R. A. Moll (1970), "A Toughness Study of Steel Weld Metal from Self-Shielded, Flux-Cored Electrodes, Part 1," *Welding Journal*, Vol. 49, April 1970.

Kotecki, D. S. and R. A. Moll (1972), "A Toughness Study of Steel Weld Metal from Self-Shielded, Flux-Cored Electrodes, Part 2," *Welding Journal*, Vol. 51, March 1972.

Kulak, G. L., J. W. Fisher, and J. H. A. Struik (1987), *Guide to Design Criteria for Bolted and Riveted Joints*, 2nd Edition, John Wiley & Sons, New York, NY, 1987.

Lee, G. D., M. L. Morrell, and R. L. Ketter (1972), "Design of Tapered Members," *WRC Bulletin*, No. 173, June 1972.

LeMessurier, W. J. (1976), "A Practical Method of Second Order Analysis, Part 1— Pin-Jointed Frames," *Engineering Journal*, AISC, Vol. 13, No. 4, 4th Quarter, 1976.

LeMessurier, W. J. (1977), "A Practical Method of Second Order Analysis, Part 2— Rigid Frames," *Engineering Journal*, AISC, Vol. 14, No. 2, 2nd Quarter, 1977.

LeMessurier, W. J., R. J. McNamara, and J. C. Scrivener (1974), "Approximate Analytical Model for Multi-Story Frames," *Engineering Journal*, AISC, Vol. 11, No. 4, 4th Quarter, 1974.

Lesik, D. F. and D. J. L. Kennedy (1990), "Ultimate Strength of Fillet Welded Connections Loaded in Plane," *Canadian Journal of Civil Engineering*, National Research Council of Canada, Ottawa, Canada, Vol. 17, No. 1, 1990.

Liapunov, S. (1974), "Ultimate Load Studies of Plane Multi-Story Steel Rigid

Frames," *Journal of the Structural Division,* ASCE, Vol. 100, No. ST8, Proc. Paper 10750, August 1974.

Lim, L. C. and L. W. Lu (1970), *The Strength and Behavior of Laterally Unsupported Columns,* Fritz Engineering Laboratory Report No. 329.5, Lehigh University, Bethlehem, PA, June 1970.

Lu, L. W. (1967), "Design of Braced Multi-Story Frames by the Plastic Method," *Engineering Journal,* AISC, Vol. 4, No. 1, 1st Quarter, 1967.

Lu, L. W., E. Ozer, J. H. Daniels, O. S. Okten, and S. Morino (1977), "Strength and Drift Characteristics of Steel Frames," *Journal of the Structural Division,* ASCE, Vol. 103, No. ST11, November 1977.

Lyse, I. and Schreiner, "An Investigation of Welded Seat Angle Connections," *Welding Journal,* February 1935.

Lyse, I. and G. J. Gibson, "Effect of Welded Top Angles on Beam-Column Connections," *Welding Journal,* October 1937.

Marino, F. J. (1966), "Ponding of Two-Way Roof Systems," *Engineering Journal,* AISC, Vol. 3, No. 3, 3rd Quarter, 1966.

Morrell, M. L. and G. C. Lee (1974), *Allowable Stress for Web-Tapered Members,* WRC Bulletin 192, Welding Research Council, New York, February 1974.

Munse, W. H. and E. Chesson, Jr. (1963), "Riveted and Bolted Joints: Net Section Design," *Journal of the Structural Division,* ASCE, Vol. 89, No. ST1, February 1963.

Murray, Thomas M. (1991), "Building Floor Vibrations," *Engineering Journal,* AISC, Vol. 28, No. 3, 3rd Quarter 1991.

NEHRP (1992), *NEHRP Recommended Provisions for the Development of Seismic Regulations for New Buildings,* Federal Emergency Management Agency Report, FEMA 222, January 1992.

Ollgaard, J. G., R. G. Slutter, and J. W. Fisher (1971), "Shear Strength of Stud Shear Connections in Lightweight and Normal Weight Concrete," *Engineering Journal,* AISC, Vol. 8, No. 2, 2nd Quarter, 1971.

Popov, E. P. (1980), "An Update on Eccentric Seismic Bracing," *Engineering Journal,* AISC, Vol. 17, No. 3, 3rd Quarter, 1980.

Popov, E. P. and R. M. Stephen (1977), "Capacity of Columns with Splice Imperfections," *Engineering Journal,* AISC, Vol. 14, No. 1, 1st Quarter, 1977.

Preece, F. R. (1968), *AWS-AISC Fillet Weld Study—Longitudinal and Transverse Shear Tests,* Testing Engineers, Inc., Los Angeles, May 31, 1968.

Rao, N. R. N., M. Lohrmann, and L. Tall (1966), "Effect of Strain Rate on the Yield Stress of Structural Steels," *Journal of Materials,* Vol. 1, No. 1, ASTM, March 1966.

Ravindra, M. K. and T. V. Galambos (1978), "Load and Resistance Factor Design for Steel," *Journal of the Structural Division,* ASCE, Vol. 104, No. ST9, September 1978.

Research Council on Structural Connections (1988), *Load and Resistance Factor*

Design Specification for Structural Joints Using ASTM A325 or A490 Bolts, 1988, AISC, Chicago, IL.

Ricles, J. M. and J. A. Yura (1983), "Strength of Double-Row Bolted Web Connections," *Journal of the Structural Division,* ASCE, Vol. 109, No. ST1, January 1983.

Roberts, T. M. (1981), "Slender Plate Girders Subjected to Edge Loading," *Proceedings, Institution of Civil Engineers,* Part 2, 71, September 1981, London.

Ross, D. A. and W. F. Chen (1976), "Design Criteria for Steel I-Columns Under Axial Load and Biaxial Bending," *Canadian Journal of Civil Engineering,* Vol. 3, No. 3, 1976.

Salvadori, M. (1956), "Lateral Buckling of Eccentrically Loaded I-Columns," *1956 ASCE Transactions,* Vol. 122-1.

Sherman, D. R. (1976), *Tentative Criteria for Structural Applications of Steel Tubing and Pipe,* American Iron and Steel Institute, Washington, D.C., August 1976.

Sherman, D. R. and A. S. Tanavde (1984), *Comparative Study of Flexural Capacity of Pipes,* Civil Engineering Department Report, University of Wisconsin-Milwaukee, March 1984.

Slutter, R. G. and G. C. Driscoll (1965), "Flexural Strength of Steel-Concrete Composite Beams," *Journal of the Structural Division,* ASCE, Vol. 91, No. ST2, April 1965.

Springfield, J. (1975), "Design of Column Subject to Biaxial Bending," *Engineering Journal,* AISC, Vol. 12, No. 3, 3rd Quarter, 1975.

Springfield, J. and P. F. Adams (1972), "Aspects of Column Design in Tall Steel Buildings," *Journal of the Structural Division,* ASCE, Vol. 9, No. ST5, May 1972.

Stang, A. H. and B. S. Jaffe (1948), *Perforated Cover Plates for Steel Columns,* Research Paper RP1861, National Bureau of Standards, 1948, Washington, D.C.

Steel Structures Painting Council (1989), *Steel Structures Painting Manual, Vol. 2, Systems and Specifications,* SSPC, Pittsburgh, PA, 1982.

Structural Stability Research Council Task Group 20 (1979), "A Specification for the Design of Steel-Concrete Composite Columns," *Engineering Journal,* AISC, Vol. 16, No. 4, 4th Quarter, 1979.

Summers, Paul A. and Joseph A. Yura (1982), *The Behavior of Beams Subjected to Concentrated Loads,* Phil M. Ferguson Structural Engineering Laboratory Report No. 82-5, University of Texas, Austin, TX, August 1982.

Tebedge, N. and W. F. Chen (1974), "Design Criteria for H-Columns Under Biaxial Loading," *ASCE Journal of the Structural Division,* Vol. 100, ST3, 1974.

Terashima, H. and P. H. M. Hart (1984), "Effect of Aluminum on Carbon, Manganese, Niobium Steel Submerged Arc Weld Metal Properties," *Welding Journal,* Vol. 63, June 1984.

Tide, R. H. R. (1985), *Reasonable Column Design Equations,* Annual Technical Session of Structural Stability Research Council, April 16–17, 1985, Cleveland, OH, SSRC, Lehigh University, Bethlehem, PA.

Timoshenko, S. P. and J. M. Gere (1961), *Theory of Elastic Stability,* McGraw-Hill Book Company, 1961.

Wilson, W. M. (1934), *The Bearing Value of Rollers,* Bulletin No. 263, University of Illinois Engineering Experiment Station, Urbana, IL, 1934.

Winter, G. (1958), "Lateral Bracing of Columns and Beams," *Journal of the Structural Division,* ASCE, Vol. 84, No. ST2, March 1958.

Winter, G. (1970), *Commentary on the 1968 Edition of Light Gage Cold-Formed Steel Design Manual,* American Iron and Steel Institute, Washington, D.C., 1970.

Wood, B. R., D. Beaulieu, and P. F. Adams (1976), "Column Design by P-Delta Method," *Journal of the Structural Division,* ASCE, Vol. 102, No. ST2, February 1976.

Yura, J. A. (1971), "The Effective Length of Columns in Unbraced Frames," *Engineering Journal,* AISC, Vol. 8, No. 2, April 1971.

Yura, J. A., et al. (1978), "The Bending Resistance of Steel Beams," *Journal of the Structural Division,* ASCE, Vol. 104, No. ST9, September 1978.

Yura, Joseph A. (1988), *Elements for Teaching Load & Resistance Factor Design,* AISC, Chicago, IL, April 1988.

Zandonini, R. (1985), "Stability of Compact Built-Up Struts: Experimental Investigation and Numerical Simulation," (in Italian) *Construzioni Metalliche,* No. 4, 1985.

Zhou, S. P. and W. F. Chen (1985), "Design Criteria for Box Columns Under Biaxial Loading," *Journal of Structural Engineering,* ASCE, Vol. 111, No. ST12, December 1985.

Glossary

Alignment chart for columns. A nomograph for determining the effective length factor *K* for some types of columns

Amplification factor. A multiplier of the value of moment or deflection in the unbraced length of an axially loaded member to reflect the secondary values generated by the eccentricity of the applied axial load within the member

Aspect ratio. In any rectangular configuration, the ratio of the lengths of the sides

Batten plate. A plate element used to join two parallel components of a built-up column, girder, or strut rigidly connected to the parallel components and designed to transmit shear between them

Beam. A structural member whose primary function is to carry loads transverse to its longitudinal axis

Beam-column. A structural member whose primary function is to carry loads both transverse and parallel to its longitudinal axis

Bent. A plane framework of beam or truss members which support loads and the columns which support these members

Biaxial bending. Simultaneous bending of a member about two perpendicular axes

Bifurcation. The phenomenon whereby a perfectly straight member under compression may either assume a deflected position or may remain undeflected, or a beam under flexure may either deflect and twist out of plane or remain in its in-plane deflected position

Braced frame. A frame in which the resistance to lateral load or frame instability is primarily provided by a diagonal, a *K* brace, or other auxiliary system of bracing

Brittle fracture. Abrupt cleavage with little or no prior ductile deformation

Buckling load. The load at which a perfectly straight member under compression assumes a deflected position

Built-up member. A member made of structural metal elements that are welded, bolted, or riveted together

Cladding. The exterior covering of the structural components of a building

Cold-formed members. Structural members formed from steel without the application of heat

Column. A structural member whose primary function is to carry loads parallel to its longitudinal axis

Column curve. A curve expressing the relationship between an axial column strength and slenderness ratio

Combined mechanism. A mechanism determined by plastic analysis procedure which combines elementary beam, panel, and joint mechanisms

Compact section. Compact sections are capable of developing a fully plastic stress distribution and possess rotation capacity of approximately three before the onset of local buckling

Composite beam. A steel beam structurally connected to a concrete slab so that the beam and slab respond to loads as a unit. See also *Concrete-encased beam*

Concrete-encased beam. A beam totally encased in concrete cast integrally with the slab

Connection. Combination of joints used to transmit forces between two or more members. Categorized by the type and amount of force transferred (moment, shear, end reaction). See also *Splices*

Critical load. The load at which bifurcation occurs as determined by a theoretical stability analysis

Curvature. The rotation per unit length due to bending

Design documents. See *Structural design documents*

Design strength. Resistance (force, moment, stress, as appropriate) provided by element or connection; the product of the nominal strength and the resistance factor

Diagonal bracing. Inclined structural members carrying primarily axial load employed to enable a structural frame to act as a truss to resist horizontal loads

Diaphragm. Floor slab, metal wall, or roof panel possessing a large in-plane shear stiffness and strength adequate to transmit horizontal forces to resisting systems

Diaphragm action. The in-plane action of a floor system (also roofs and walls) such that all columns framing into the floor from above and below are maintained in their same position relative to each other

Double concentrated forces. Two equal and opposite forces which form a couple on the same side of the loaded member

Double curvature. A bending condition in which end moments on a member cause the member to assume an S shape

Drift. Lateral deflection of a building

Drift index. The ratio of lateral deflection to the height of the building

Ductility factor. The ratio of the total deformation at maximum load to the elastic-limit deformation

Effective length. The equivalent length KL used in compression formulas and determined by a bifurcation analysis

Effective length factor K. The ratio between the effective length and the unbraced length of the member measured between the centers of gravity of the bracing members

Effective moment of inertia. The moment of inertia of the cross section of a member that remains elastic when partial plastification of the cross section takes place, usually under the combination of residual stress and applied stress. Also, the moment of inertia based on effective widths of elements that buckle locally. Also, the moment of inertia used in the design of partially composite members

Effective stiffness. The stiffness of a member computed using the effective moment of inertia of its cross section

Effective width. The reduced width of a plate or slab which, with an assumed uniform stress distribution, produces the same effect on the behavior of a structural member as the actual plate width with its nonuniform stress distribution

Elastic analysis. Determination of load effects (force, moment, stress, as appropriate) on members and connections based on the assumption that material deformation disappears on removal of the force that produced it

Elastic-perfectly plastic. A material which has an idealized stress-strain curve that varies linearly from the point of zero strain and zero stress up to the yield point of the material, and then increases in strain at the value of the yield stress without any further increases in stress

Embedment. A steel component cast in a concrete structure which is used to transmit externally applied loads to the concrete structure by means of bearing, shear, bond, friction, or any combination thereof. The embedment may be fabricated of structural-steel plates, shapes, bars, bolts, pipe, studs, concrete reinforcing bars, shear connectors, or any combination thereof

Encased steel structure. A steel-framed structure in which all of the individual frame members are completely encased in cast-in-place concrete

Euler formula. The mathematical relationship expressing the value of the Euler load in terms of the modulus of elasticity, the moment of inertia of the cross section, and the length of a column

Euler load. The critical load of a perfectly straight, centrally loaded pin-ended column

Eyebar. A particular type of pin-connected tension member of uniform thickness with forged or flame cut head of greater width than the body proportioned to provide approximately equal strength in the head and body

Factored load. The product of the nominal load and a load factor

Fastener. Generic term for welds, bolts, rivets, or other connecting device

Fatigue. A fracture phenomenon resulting from a fluctuating stress cycle

First-order analysis. Analysis based on first-order deformations in which equilibrium conditions are formulated on the undeformed structure

Flame-cut plate. A plate in which the longitudinal edges have been prepared by oxygen cutting from a larger plate

Flat width. For a rectangular tube, the nominal width minus twice the outside corner radius. In absence of knowledge of the corner radius, the flat width may be taken as the total section width minus three times the thickness

Flexible connection. A connection permitting a portion, but not all, of the simple beam rotation of a member end

Floor system. The system of structural components separating the stories of a building

Force. Resultant of distribution of stress over a prescribed area. A reaction that develops in a member as a result of load (formerly called total stress or stress). Generic term signifying axial loads, bending moment, torques, and shears

Fracture toughness. Measurement of the ability to absorb energy without fracture. Generally determined by impact loading of specimens containing a notch having a prescribed geometry

Frame buckling. A condition under which bifurcation may occur in a frame

Frame instability. A condition under which a frame deforms with increasing lateral deflection under a system of increasing applied monotonic loads until a maximum value of the load called the stability limit is reached, after which the frame will continue to deflect without further increase in load

Fully composite beam. A composite beam with sufficient shear connectors to develop the full flexural strength of the composite section

High-cycle fatigue. Failure resulting from more than 20,000 applications of cyclic stress

Hybrid beam. A fabricated steel beam composed of flanges with a greater yield strength than that of the web. Whenever the maximum flange stress is less than or equal to the web yield stress the girder is considered homogeneous

Hysteresis loop. A plot of force versus displacement of a structure or member subjected to reversed, repeated load into the inelastic range, in which the path followed during release and removal of load is different from the path for the addition of load over the same range of displacement

Inclusions. Nonmetallic material entrapped in otherwise sound metal

Incomplete fusion. Lack of union by melting of filler and base metal over entire prescribed area

Inelastic action. Material deformation that does not disappear on removal of the force that produced it

Instability. A condition reached in the loading of an element or structure in which continued deformation results in a decrease of load-resisting capacity

Joint. Area where two or more ends, surfaces, or edges are attached. Categorized by type of fastener or weld used and method of force transfer

K bracing. A system of struts used in a braced frame in which the pattern of the struts resembles the letter K, either normal or on its side

Lamellar tearing. Separation in highly restrained base metal caused by through-thickness strains induced by shrinkage of adjacent weld metal

Lateral bracing member. A member utilized individually or as a component of a lateral bracing system to prevent buckling of members or elements and/or to resist lateral loads

Lateral (or lateral-torsional) buckling. Buckling of a member involving lateral deflection and twist

Leaning column. Gravity-loaded column where connections to the frame (simple connections) do not provide resistance to lateral loads

Limit state. A condition in which a structure or component becomes unfit for service and is judged either to be no longer useful for its intended function (*serviceability limit state*) or to be unsafe (*strength limit state*)

Limit states. Limits of structural usefulness, such as brittle fracture, plastic collapse, excessive deformation, durability, fatigue, instability, and serviceability

Load factor. A factor that accounts for unavoidable deviations of the actual load from the nominal value and for uncertainties in the analysis that transforms the load into a load effect

Loads. Forces or other actions that arise on structural systems from the weight of all permanent construction, occupants and their possessions, environmental effects, differential settlement, and restrained dimensional changes. *Permanent* loads are those loads in which variations in time are rare or of small magnitude. All other loads are *variable* loads. See *Nominal loads*

LRFD (Load and Resistance Factor Design). A method of proportioning structural components (members, connectors, connecting elements, and assemblages) such that no applicable limit state is exceeded when the structure is subjected to all appropriate load combinations

Local buckling. The buckling of a compression element which may precipitate the failure of the whole member

Low-cycle fatigue. Fracture resulting from a relatively high-stress range resulting in a relatively small number of cycles to failure

Lower bound load. A load computed on the basis of an assumed equilibrium moment diagram in which the moments are not greater than M_p that is less than or at best equal to the true ultimate load

Mechanism. An articulated system able to deform without an increase in load, used in the special sense that the linkage may include real hinges or plastic hinges, or both

Mechanism method. A method of plastic analysis in which equilibrium between external forces and internal plastic hinges is calculated on the basis of an assumed mechanism. The failure load so determined is an upper bound

Nominal loads. The magnitudes of the loads specified by the applicable code

Nominal strength. The capacity of a structure or component to resist the effects of loads, as determined by computations using specified material strengths and dimensions and formulas derived from accepted principles of structural mechanics or by field tests or laboratory tests of scaled models, allowing for modeling effects and differences between laboratory and field conditions

Noncompact section. Noncompact sections can develop the yield stress in compression elements before local buckling occurs, but will not resist inelastic local buckling at strain levels required for a fully plastic stress distribution

P-Delta effect. Secondary effect of column axial loads and lateral deflection on the moments in members

Panel zone. The zone in a beam-to-column connection that transmits moment by a shear panel

Partially composite beam. A composite beam for which the shear strength of shear connectors governs the flexural strength

Plane frame. A structural system assumed for the purpose of analysis and design to be two-dimensional

Plastic analysis. Determination of load effects (force, moment, stress, as appropriate) on members and connections based on the assumption of rigid-plastic behavior, i.e., that equilibrium is satisfied throughout the structure and yield is not exceeded anywhere. Second order effects may need to be considered

Plastic design section. The cross section of a member which can maintain a full plastic moment through large rotations so that a mechanism can develop; the section suitable for plastic design

Plastic hinge. A yielded zone which forms in a structural member when the plastic moment is attained. The beam is assumed to rotate as if hinged, except that it is restrained by the plastic moment M_p

Plastic-limit load. The maximum load that is attained when a sufficient number of yield zones have formed to permit the structure to deform plastically without further increase in load. It is the largest load a structure will support, when perfect plasticity is assumed and when such factors as instability, second-order effects, strain hardening, and fracture are neglected

Plastic mechanism. See *Mechanism*

Plastic modulus. The section modulus of resistance to bending of a completely yielded cross section. It is the combined static moment about the neutral axis of the cross-sectional areas above and below that axis

Plastic moment. The resisting moment of a fully yielded cross section

Plastic strain. The difference between total strain and elastic strain

Plastic zone. The yielded region of a member

Plastification. The process of successive yielding of fibers in the cross section of a member as bending moment is increased

Plate girder. A built-up structural beam

Post-buckling strength. The load that can be carried by an element, member, or frame after buckling

Primary stress. A primary stress is any normal stress or shear stress developed by an imposed loading which is necessary to satisfy the laws of equilibrium of external and internal forces, moments, and torques. A primary stress is not self-limiting.

Redistribution of moment. A process which results in the successive formation of plastic hinges so that less highly stressed portions of a structure may carry increased moments

Required strength. Load effect (force, moment, stress, as appropriate) acting on element or connection determined by structural analysis from the factored loads (using most appropriate critical load combinations)

Residual stress. The stresses that remain in an unloaded member after it has been formed into a finished product. (Examples of such stresses include, but are not limited to, those induced by cold bending, cooling after rolling, or welding.)

Resistance. The capacity of a structure or component to resist the effects of loads. It is determined by computations using specified material strengths, dimensions and formulas derived from accepted principles of structural mechanics, or by field tests or laboratory tests of scaled models, allowing for modeling effects and differences between laboratory and field conditions. Resistance is a generic term that includes both strength and serviceability limit states

Resistance factor. A factor that accounts for unavoidable deviations of the actual strength from the nominal value and the manner and consequences of failure

Rigid frame. A structure in which connections maintain the angular relationship between beam and column members under load

Root of the flange. Location on the web of the corner radius termination point or the toe of the flange-to-web weld. Measured as the k distance from the far side of the flange

Rotation capacity. The incremental angular rotation that a given shape can accept prior to local failure defined as $R = (\theta_u / \theta_p) - 1$ where θ_u is the overall rotation attained at the factored load state and θ_p is the idealized rotation corresponding to elastic theory applied to the case of $M = M_p$

St. Venant torsion. That portion of the torsion in a member that induces only shear stresses in the member

Second-order analysis. Analysis based on second-order deformations, in which equilibrium conditions are formulated on the deformed structure

Service load. Load expected to be supported by the structure under normal usage; often taken as the nominal load

Serviceability limit state. Limiting condition affecting the ability of a structure to preserve its appearance, maintainability, durability, or the comfort of its occupants or function of machinery under normal usage

Shape factor. The ratio of the plastic moment to the yield moment, or the ratio of the plastic modulus to the section modulus for a cross section

Shear friction. Friction between the embedment and the concrete that transmits shear loads. The relative displacement in the plane of the shear load is considered to be resisted by shear-friction anchors located perpendicular to the plane of the shear load

Shear lugs. Plates, welded studs, bolts, and other steel shapes that are embedded in the concrete and located transverse to the direction of the shear force and that transmit shear loads, introduced into the concrete by local bearing at the shear lug-concrete interface

Shear wall. A wall that in its own plane resists shear forces resulting from applied wind, earthquake, or other transverse loads or provides frame stability. Also called a structural wall

Sidesway. The lateral movement of a structure under the action of lateral loads, unsymmetrical vertical loads, or unsymmetrical properties of the structure

Sidesway buckling. The buckling mode of a multistory frame precipitated by the relative lateral displacements of joints, leading to failure by sidesway of the frame

Simple plastic theory. See *Plastic design*

Single curvature. A deformed shape of a member having one smooth continuous arc, as opposed to double curvature which contains a reversal

Slender-element section. The cross section of a member which will experience local buckling in the elastic range

Slenderness ratio. The ratio of the effective length of a column to the radius of gyration of the column, both with respect to the same axis of bending

Slip-critical joint. A bolted joint in which the slip resistance of the connection is required

Space frame. A three-dimensional structural framework (as contrasted to a plane frame)

Splice. The connection between two structural elements joined at their ends to form a single, longer element

Stability-limit load. Maximum (theoretical) load a structure can support when second-order instability effects are included

Stepped column. A column with changes from one cross section to another occurring at abrupt points within the length of the column

Stiffener. A member, usually an angle or plate, attached to a plate or web of a beam or girder to distribute load, to transfer shear, or to prevent buckling of the member to which it is attached

Stiffness. The resistance to deformation of a member or structure measured by the ratio of the applied force to the corresponding displacement

Story drift. The difference in horizontal deflection at the top and bottom of a story

Strain hardening. Phenomenon wherein ductile steel, after undergoing considerable deformation at or just above yield point, exhibits the capacity to resist substantially higher loading than that which caused initial yielding

Strain-hardening strain. For structural steels that have a flat (plastic) region in the stress-strain relationship, the value of the strain at the onset of strain hardening

Strength design. A method of proportioning structural members using load factors and resistance factors such that no applicable limit state is exceeded (also called load and resistance factor design)

Strength limit state. Limiting condition affecting the safety of the structure, in which the ultimate load-carrying capacity is reached

Stress. Force per unit area

Stress concentration. Localized stress considerably higher than average (even in uniformly loaded cross sections of uniform thickness) due to abrupt changes in geometry or localized loading

Strong axis. The major principal axis of a cross section

Structural design documents. Documents prepared by the designer (plans, design details, and job specifications)

Structural system. An assemblage of load-carrying components which are joined together to provide regular interaction or interdependence

Stub column. A short compression-test specimen, long enough for use in measuring the stress-strain relationship for the complete cross section, but short enough to avoid buckling as a column in the elastic and plastic ranges

Subassemblage. A truncated portion of a structural frame

Supported frame. A frame which depends upon adjacent braced or unbraced frames for resistance to lateral load or frame instability. (This transfer of load is frequently provided by the floor or roof system through diaphragm action or by horizontal cross bracing in the roof.)

Tangent modulus. At any given stress level, the slope of the stress-strain curve of a material in the inelastic range as determined by the compression test of a small specimen under controlled conditions

Temporary structure. A general term for anything that is built or constructed (usually to carry construction loads) that will eventually be removed before or after completion of construction and does not become part of the permanent structural system

Tensile strength. The maximum tensile stress that a material is capable of sustaining

Tension field action. The behavior of a plate girder panel under shear force in which diagonal tensile stresses develop in the web and compressive forces develop in the transverse stiffeners in a manner analogous to a Pratt truss

Toe of the fillet. Termination point of fillet weld or of rolled section fillet

Torque-tension relationship. Term applied to the wrench torque required to produce specified pre-tension in high-strength bolts

Turn-of-nut method. Procedure whereby the specified pre-tension in high-strength bolts is controlled by rotation of the wrench a predetermined amount after the nut has been tightened to a snug fit

Unbraced frame. A frame in which the resistance to lateral load is provided by the bending resistance of frame members and their connections

Unbraced length. The distance between braced points of a member, measured between the centers of gravity of the bracing members

Undercut. A notch resulting from the melting and removal of base metal at the edge of a weld

Universal-mill plate. A plate in which the longitudinal edges have been formed by a rolling process during manufacture. Often abbreviated as UM plate

Upper bound load. A load computed on the basis of an assumed mechanism which will always be at best equal to or greater than the true ultimate load

Vertical bracing system. A system of shear walls, braced frames, or both, extending through one or more floors of a building

Von Mises yield criterion. A theory which states that inelastic action at any point in a body under any combination of stresses begins only when the strain energy of distortion per unit volume absorbed at the point is equal to the strain energy of distortion absorbed per unit volume at any point in a simple tensile bar stressed to the elastic limit under a state of uniaxial stress. It is often called the maximum strain-energy-of-distortion theory. Accordingly, shear yield occurs at 0.58 times the yield strength

Warping torsion. That portion of the total resistance to torsion that is provided by resistance to warping of the cross section

Weak axis. The minor principal axis of a cross section

Weathering steel. A type of high-strength, low-alloy steel which can be used in normal environments (not marine) and outdoor exposures without protective paint covering. This steel develops a tight adherent rust at a decreasing rate with respect to time

Web buckling. The buckling of a web plate

Web crippling. The local failure of a web plate in the immediate vicinity of a concentrated load or reaction

Working load. Also called service load. The actual load assumed to be acting on the structure

Yield moment. In a member subjected to bending, the moment at which an outer fiber first attains the yield stress

Yield plateau. The portion of the stress-strain curve for uniaxial tension or compression in which the stress remains essentially constant during a period of substantially increased strain

Yield point. The first stress in a material at which an increase in strain occurs without an increase in stress, the yield point less than the maximum attainable stress

Yield strength. The stress at which a material exhibits a specified limiting deviation from the proportionality of stress to strain. Deviation expressed in terms of strain

Yield stress. Yield point, yield strength, or yield stress level as defined

Yield-stress level. The average stress during yielding in the plastic range, the stress determined in a tension test when the strain reaches 0.005 in. per in.

Specification for Load and Resistance Factor Design of Single-Angle Members

December 1, 1993

AMERICAN INSTITUTE OF STEEL CONSTRUCTION, INC.
One East Wacker Drive, Suite 3100, Chicago, IL 60601-2001

PREFACE

The intention of the AISC Specification is to cover the common everyday design criteria in routine design office usage. It is not feasible to also cover the many special and unique problems encountered within the full range of structural design practice. This separate Specification and Commentary addresses one such topic—single-angle members—to provide needed design guidance for this more complex structural shape under various load and support conditions.

The single-angle design criteria were developed through a consensus process by the AISC Task Committee 116 on Single-Angle Members:

Donald R. Sherman, Chairman
Hansraj G. Ashar
Wai-Fah Chen
Raymond D. Ciatto
Mohamed Elgaaly
Theodore V. Galambos
Thomas G. Longlais
LeRoy A. Lutz
William A. Milek
Raymond H. R. Tide
Nestor R. Iwankiw, Secretary

The assistance of the Structural Stability Research Council Task Group on Single Angles in the preparation and review of this document is acknowledged.

The full AISC Committee on Specifications has reviewed and endorsed this Specification.

A non-mandatory Commentary provides background for the Specification provisions and the user is encouraged to consult it.

The principal changes in this edition include:

- establishing upper limit of single-angle flexural strength at 1.25 of the yield moment

- increasing resistance factor for compression to 0.90

- removing flexural-torsional buckling consideration for compression members

- considering the sense of flexural stresses in the combined force interaction check

The reader is cautioned that professional judgment must be exercised when data or recommendations in this Specification are applied. The publication of the material contained herein is not intended as a representation or warranty on the part of the American Institute of Steel Construction, Inc.—or any other person named herein—that this information is suitable for general or particular use, or freedom from infringement of any patent or patents. Anyone making use of this information assumes all liability arising from such use. The design of structures is within the scope of expertise of a competent licensed structural engineer, architect, or other licensed professional for the application of principles to a particular structure.

Specification for Load and Resistance Factor Design of Single-Angle Members

December 1, 1993

1. SCOPE

This document contains Load and Resistance Factor Design (LRFD) criteria for hot-rolled, single-angle members with equal and unequal legs in tension, shear, compression, flexure, and for combined forces. It is intended to be compatible with, and a supplement to, the 1993 AISC Specification for Structural Steel Buildings—Load and Resistance Factor Design (AISC LRFD) and repeats some common criteria for ease of reference. For design purposes, the conservative simplifications and approximations in the Specification provisions for single angles are permitted to be refined through a more precise analysis. As an alternative to this Specification, the 1989 AISC *Specification for Allowable Stress Design of Single-Angle Members* is permitted.

The Specification for single-angle design supersedes any comparable but more general requirements of the AISC LRFD. All other design, fabrication, and erection provisions not directly covered by this document shall be in compliance with the AISC LRFD. In the absence of a governing building code, the factored load combinations in AISC LRFD Section A4 shall be used to determine the required strength. For design of slender, cold-formed steel angles, the current AISI *LRFD Specification for the Design of Cold-Formed Steel Structural Members* is applicable.

2. TENSION

The tensile design strength $\phi_t P_n$ shall be the lower value obtained according to the limit states of yielding, $\phi_t = 0.9$, $P_n = F_y A_g$, and fracture, $\phi_t = 0.75$, $P_n = F_u A_e$.

a. For members connected by bolting, the net area and effective net area shall be determined from AISC LRFD Specification Sections B1 to B3 inclusive.

b. When the load is transmitted by longitudinal welds only or a combination of longitudinal and transverse welds through just one leg of the angle, the effective net area A_e shall be:

$$A_e = A_g U \tag{2-1}$$

where

A_g = gross area of member

$$U = \left(1 - \frac{\bar{x}}{l}\right) \le 0.9$$

$\bar{x}$ = connection eccentricity
l = length of connection in the direction of loading

c. When a load is transmitted by transverse weld through just one leg of the angle, A_e is the area of the connected leg and $U = 1$.

For members whose design is based on tension, the slenderness ratio l/r preferably should not exceed 300. Members in which the design is dictated by tension loading, but which may be subject to some compression under other load conditions, need not satisfy the compression slenderness limits.

3.   SHEAR

For the limit state of yielding in shear, the shear stress, f_{uv}, due to flexure and torsion shall not exceed:

$$f_{uv} \le \phi_v 0.6 F_y \tag{3-1}$$
$$\phi_v = 0.9$$

4.   COMPRESSION

The design strength of compression members shall be $\phi_c P_n$

where

$\phi_c = 0.90$
$P_n = A_g F_{cr}$

a. For $\lambda_c \sqrt{Q} \le 1.5$:

$$F_{cr} = Q\left(0.658^{Q\lambda_c^2}\right) F_y \tag{4-1}$$

b. For $\lambda_c \sqrt{Q} \ge 1.5$:

$$F_{cr} = \left[\frac{0.877}{\lambda_c^2}\right] F_y \tag{4-2}$$

$$\lambda_c = \frac{Kl}{r\pi}\sqrt{\frac{F_y}{E}}$$

F_y = specified minimum yield stress of steel
Q = reduction factor for local buckling

The reduction factor Q shall be:

when $\dfrac{b}{t} \le 0.446 \sqrt{\dfrac{E}{F_y}}$:

$$Q = 1.0 \qquad\qquad (4\text{-}3a)$$

when $0.446 \sqrt{\dfrac{E}{F_y}} < \dfrac{b}{t} < 0.910 \sqrt{\dfrac{E}{F_y}}$:

$$Q = 1.34 - 0.761 \dfrac{b}{t} \sqrt{\dfrac{F_y}{E}} \qquad\qquad (4\text{-}3b)$$

when $\dfrac{b}{t} \ge 0.910 \sqrt{\dfrac{E}{F_y}}$:

$$Q = \dfrac{0.534E}{F_y \left(\dfrac{b}{t}\right)^2} \qquad\qquad (4\text{-}3c)$$

b = full width of longest angle leg
t = thickness of angle

For members whose design is based on compressive force, the largest effective slenderness ratio preferably should not exceed 200.

5. FLEXURE

The flexure design strengths of Section 5.1 shall be used as indicated in Sections 5.2 and 5.3

5.1. Flexural Design Strength

The flexural design strength shall be limited to the minimum value $\phi_b M_n$ determined from Sections 5.1.1, 5.1.2, and 5.1.3, as applicable, with $\phi_b = 0.9$.

5.1.1. For the limit state of local buckling when the tip of an angle leg is in compression:

when $\dfrac{b}{t} \le 0.382 \sqrt{\dfrac{E}{F_y}}$:

$$M_n = 1.25 F_y S_c \qquad\qquad (5\text{-}1a)$$

when $0.382 \sqrt{\dfrac{E}{F_y}} < \dfrac{b}{t} \le 0.446 \sqrt{\dfrac{E}{F_y}}$:

$$M_n = F_y S_c \left[1.25 - 1.49 \left(\frac{b/t}{0.382 \sqrt{\dfrac{E}{F_y}}} - 1 \right) \right] \qquad \text{(5-1b)}$$

when $\dfrac{b}{t} > 0.446 \sqrt{\dfrac{E}{F_y}}$:

$$M_n = Q F_y S_c \qquad \text{(5-1c)}$$

where

b = full width of angle leg with tip in compression
Q = reduction factor per Equations 4-3a, b, and c
S_c = elastic section modulus to the tip in compression relative to axis of bending
E = modulus of elasticity

5.1.2. For the limit state of yielding when the tip of an angle leg is in tension

$$M_n = 1.25 M_y \qquad \text{(5-2)}$$

where

M_y = yield moment about the axis of bending

5.1.3. For the limit state of lateral-torsional buckling:

when $M_{ob} \le M_y$:

$$M_n = [0.92 - 0.17 M_{ob}/M_y] M_{ob} \qquad \text{(5-3a)}$$

when $M_{ob} > M_y$:

$$M_n = [1.58 - 0.83 \sqrt{M_y/M_{ob}}] M_y \le 1.25 M_y \qquad \text{(5-3b)}$$

where

M_{ob} = elastic lateral-torsional buckling moment, from Section 5.2 or 5.3 as applicable

5.2. Bending about Geometric Axes

5.2.1. a. Angle bending members with lateral-torsion restraint along the length shall be designed on the basis of geometric axis bending with the nominal flexural strength M_n limited to the provisions of Sections 5.1.1 and 5.1.2.

b. For equal-leg angles if the lateral-torsional restraint is only at the point of maximum moment, the required moment shall be limited to $\phi_b M_n$ per Section 5.1. M_y shall be computed using the geometric axis section modulus and M_{ob} shall be substituted by using 1.25 times M_{ob} computed from Equation 5-4.

5.2.2. Equal-leg angle members without lateral-torsional restraint subjected to flexure applied about one of the geometric axes are permitted to be designed considering only geometric axis bending provided:

a. The yield moment shall be based on use of 0.80 of the geometric axis section modulus.

b. For the angle-leg tips in compression, the nominal flexural strength M_n shall be determined by the provisions in Section 5.1.1 and in Section 5.1.3,

where

$$M_{ob} = \frac{0.66Eb^4tC_b}{l^2}[\sqrt{1 + 0.78(lt/b^2)^2} - 1] \qquad (5\text{-}4)$$

l = unbraced length

$$C_b = \frac{12.5M_{max}}{2.5M_{max} + 3M_A + 4M_B + 3M_C} \leq 1.5$$

where

M_{max} = absolute value of maximum moment in the unbraced beam segment

M_A = absolute value of moment at quarter point of the unbraced beam segment

M_B = absolute value of moment at centerline of the unbraced beam segment

M_C = absolute value of moment at three-quarter point of the unbraced beam segment

c. For the angle-leg tips in tension, the nominal flexural strength shall be determined according to Section 5.1.2.

5.2.3. Unequal-leg angle members without lateral-torsional restraint subjected to bending about one of the geometric axes shall be designed using Section 5.3.

5.3. Bending about Principal Axes

Angles without lateral-torsional restraint shall be designed considering principal-axis bending, except for the alternative of Section 5.2.2, if appropriate. Bending about both of the principal axes shall be evaluated as required in Section 6.

5.3.1. Equal-leg angles:

a. Major-axis bending:

The nominal flexural strength M_n about the major principal axis shall be determined by the provisions in Section 5.1.1 and in Section 5.1.3,

where

$$M_{ob} = C_b \frac{0.46Eb^2t^2}{l} \qquad (5\text{-}5)$$

b. Minor-axis bending:

The nominal design strength M_n about the minor principal axis shall be determined by Section 5.1.1 when the leg tips are in compression, and by Section 5.1.2 when the leg tips are in tension.

5.3.2. Unequal-leg angles:

  a. Major-axis bending:

     The nominal flexural strength M_n about the major principal axis shall be determined by the provisions in Section 5.1.1 for the compression leg and in Section 5.1.3,

     where

$$M_{ob} = 4.9E \frac{I_z}{l^2} C_b \left[\sqrt{\beta_w^2 + 0.052(lt/r_z)^2} + \beta_w \right] \tag{5-6}$$

     I_z = minor principal axis moment of inertia

     r_z = radius of gyration for minor principal axis

     $\beta_w = \left[\dfrac{1}{I_w} \displaystyle\int_A z(w^2 + z^2)dA \right] - 2z_o$, special section property for unequal-leg angles, positive for short leg in compression and negative for long leg in compression (see Commentary for values for common angle sizes). If the long leg is in compression anywhere along the unbraced length of the member, the negative value of β_w shall be used.

     z_o = coordinate along z axis of the shear center with respect to centroid

     I_w = moment of inertia for major principal axis

  b. Minor-axis bending:

     The nominal design strength M_n about the minor principal axis shall be determined by Section 5.1.1 when leg tips are in compression and by Section 5.1.2 when leg tips are in tension.

6. COMBINED FORCES

The interaction equation shall be evaluated for the principal bending axes either by addition of all the maximum axial and flexural terms, or by considering the sense of the associated flexural stresses at the critical points of the cross section, the flexural terms are either added to or subtracted from the axial load term.

6.1. Members in Flexure and Axial Compression

6.1.1. The interaction of flexure and axial compression applicable to specific locations on the cross section shall be limited by Equations 6-1a and 6-1b:

For $\dfrac{P_u}{\phi P_n} \geq 0.2$

$$\left| \frac{P_u}{\phi P_n} + \frac{8}{9}\left(\frac{M_{uw}}{\phi_b M_{nw}} + \frac{M_{uz}}{\phi_b M_{nz}} \right) \right| \leq 1.0 \tag{6-1a}$$

For $\dfrac{P_u}{\phi P_n} \le 0.2$

$$\left| \frac{P_u}{2\phi P_n} + \left(\frac{M_{uw}}{\phi_b M_{nw}} + \frac{M_{uz}}{\phi_b M_{nz}} \right) \right| \le 1.0 \qquad (6\text{-}1b)$$

P_u = required compressive strength

P_n = nominal compressive strength determined in accordance with Section 4

M_u = required flexural strength

M_n = nominal flexural strength for tension or compression in accordance with Section 5, as appropriate. Use section modulus for specific location in the cross section and consider the type of stress.

ϕ = ϕ_c = resistance factor for compression = 0.90

ϕ_b = resistance factor for flexure = 0.90

w = subscript relating symbol to major-axis bending

z = subscript relating symbol to minor-axis bending

In Equations 6-1a and 6-1b when M_n represents the flexural strength of the compression side, the corresponding M_u shall be multiplied by B_1.

$$B_1 = \frac{C_m}{1 - \dfrac{P_u}{P_{e1}}} \ge 1.0 \qquad (6\text{-}2)$$

C_m = bending coefficient defined in AISC LRFD

P_{e1} = elastic buckling load for the braced frame defined in AISC LRFD

6.1.2. For members constrained to bend about a geometric axis with nominal flexural strength determined per Section 5.2.1, the radius of gyration r for P_{e1} shall be taken as the geometric axis value. The bending terms for the principal axes in Equations 6-1a and 6-1b shall be replaced by a single geometric axis term.

6.1.3. Alternatively, for equal-leg angles without lateral-torsional restraint along the length and with bending applied about one of the geometric axes, the provisions of Section 5.2.2 are permitted for the required and design bending strength. If Section 5.2.2 is used for M_n, the radius of gyration about the axis of bending r for P_{e1} shall be taken as the geometric axis value of r divided by 1.35 in the absence of a more detailed analysis. The bending terms for the principal axes in Equations 6-1a and 6-1b shall be replaced by a single geometric axis term.

6.2. Members in Flexure and Axial Tension

The interaction of flexure and axial tension shall be limited by Equations 6-1a and 6-1b where

P_u = required tensile strength

P_n = nominal tensile strength determined in accordance with Section 2

M_u = required flexural strength

M_n = nominal flexural strength for tension or compression in accordance with Section 5, as appropriate. Use section modulus for specific location in the cross section and consider the type of stress.

ϕ = ϕ_t = resistance factor for tension = 0.90

ϕ_b = resistance factor for flexure = 0.90

For members subject to bending about a geometric axis, the required bending strength evaluation shall be in accordance with Sections 6.1.2 and 6.1.3. Second-order effects due to axial tension and bending interaction are permitted to be considered in the determination of M_u for use in Formulas 6-1a and 6-1b. In lieu of using Formulas 6-1a and 6-1b, a more detailed analysis of the interaction of flexure and tension is permitted.

Commentary on the Specification for Load and Resistance Factor Design of Single-Angle Members

December 1, 1993

INTRODUCTION

This Specification is intended to be complete for normal design usage in conjunction with the main 1993 AISC LRFD Specification and Commentary.

This Commentary furnishes background information and references for the benefit of the engineer seeking further understanding of the derivation and limits of the specification.

The Specification and Commentary are intended for use by design professionals with demonstrated engineering competence.

C2. TENSION

The criteria for the design of tension members in AISC LRFD Specification Section D1 have been adopted for angles with bolted connections. However, recognizing the effect of shear lag when the connection is welded, the criteria in Section B3 of the AISC LRFD Specification have been applied.

The advisory upper slenderness limits are not due to strength considerations but are based on professional judgment and practical considerations of economics, ease of handling, and transportability. The radius of gyration about the z axis will produce the maximum l/r and, except for very unusual support conditions, the maximum Kl/r. Since the advisory slenderness limit for compression members is less than for tension members, an accommodation has been made for members with $Kl/r > 200$ that are always in tension, except for unusual load conditions which produce a small compression force.

C3. SHEAR

Shear stress due to factored loads in a single-angle member are the result of the gradient in the bending moment along the length (flexural shear) and the torsional moment.

The maximum elastic stress due to flexural shear may be computed by

$$f_v = \frac{1.5 V_b}{bt} \qquad \text{(C3-1)}$$

where

V_b = component of the shear force parallel to the angle leg with length b and thickness t, kips

The stress, which is constant through the thickness, should be determined for both legs to determine the maximum.

The 1.5 factor is the calculated elastic value for equal-leg angles loaded along one of the principal axes. For equal-leg angles loaded along one of the geometric axes (laterally braced or unbraced) the factor is 1.35. Constants between these limits may be calculated conservatively from $V_b Q / I$ to determine the maximum stress at the neutral axis.

Alternatively, if only flexural shear is considered, a uniform flexural shear stress in the leg of V_b / bt may be used due to inelastic material behavior and stress redistribution.

If the angle is not laterally braced against twist, a torsional moment is produced equal to the applied transverse load times the perpendicular distance e to the shear center, which is at the heel of the angle cross section. Torsional moments are resisted by two types of shear behavior: pure torsion (St. Venant) and warping torsion (AISC, 1983). If the boundary conditions are such that the cross section is free to warp, the applied torsional moment M_T is resisted by pure shear stresses as shown in Figure C3.1a. Except near the ends of the legs, these stresses are constant along the length of the leg, and the maximum value can be approximated by

$$f_v = M_T t / J = \frac{3 M_T}{At} \qquad \text{(C3-2)}$$

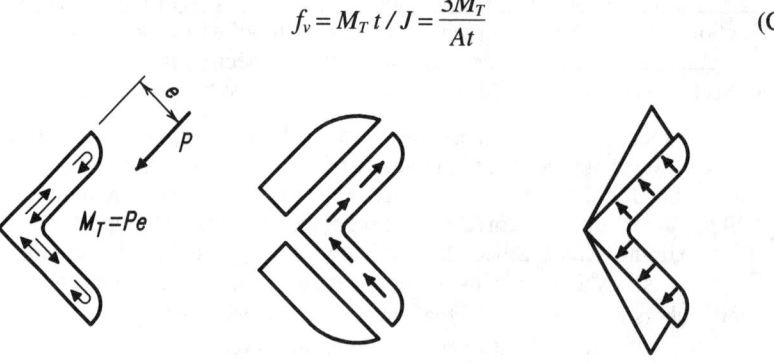

(a) Pure torsion   (b) In-plane warping   (c) Across-thickness warping

Fig. C.3.1. Shear stresses due to torsion.

where

J = torsional constant (approximated by $\Sigma bt^3 / 3$ when precomputed value unavailable)

A = angle cross-sectional area

At section where warping is restrained, the torsional moment is resisted by warping shear stresses of two types (Gjelsvik, 1981). One type is in-plane (contour) as shown in Figure C3.1b, which varies from zero at the toe to a maximum at the heel of the angle. The other type is across the thickness and is sometimes referred to as secondary warping shear. As indicated in Figure C3.1c, it varies from zero at the heel to a maximum at the toe.

In an angle with typical boundary conditions and unrestrained load point, the torsional moment produces all three types of shear stresses (pure, in-plane warping, and secondary warping) in varying proportions along its length. The total applied moment is resisted by a combination of three types of internal moments that differ in relative proportions according to the distance from the boundary condition. Using typical angle dimensions, it can be shown that the two warping shears are approximately the same order of magnitude and are less than 20 percent of the pure shear stress for the same torsional moment. Therefore, it is conservative to compute the torsional shear stress using the pure shear equation and total applied torsional moment M_T as if no warping restraint were present. This stress is added directly to the flexural shear stress to produce a maximum surface shear stress near the mid-length of a leg. Since this sum is a local maximum that does not extend through the thickness, applying the limit of $\phi_v 0.6 F_y$ adds another degree of conservatism relative to the design of other structural shapes.

In general, torsional moments from laterally unrestrained transverse loads also produce warping normal stresses that are superimposed on bending stresses. However, since the warping strength for a single angle is relatively small, this additional bending effect is negligible and often ignored in design practice.

C4.   COMPRESSION

The provisions for the critical compression stress account for the three possible limit states that may occur in an angle column depending on its proportions: general column flexural buckling, local buckling of thin legs, and flexural-torsional buckling of the member. The Q-factor in the equation for critical stress accounts for the local buckling, and the expressions for Q are nondimensionalized from AISC LRFD Specification (AISC, 1993) Appendix B5. Flexural-torsional buckling is covered in Appendix E of the AISC LRFD Specification (AISC, 1993). This strength limit state is approximated by the Q-factor reduction for slender-angle legs. For non-slender sections where $Q = 1$, flexural-torsional buckling is relevant for relatively short columns, but it was shown by Galambos (1991) that the error of neglecting this effect is not significant. For this reason no explicit consideration of this effect is required in these single-angle specifications. The provisions of Appendix E of AISC LRFD may be conservatively used to directly consider flexural-torsional buckling for single-angle members.

The effective length factors for angle columns may be determined by consulting the paper by Lutz (1992).

The resistance factor ϕ was increased from 0.85 in AISC LRFD for all cross sections to 0.90 for single angles only because it was shown that a ϕ of 0.90 provides an equivalent degree of reliability (Galambos, 1992).

C5. FLEXURE

Flexural strength limits are established for yielding, local buckling, and lateral-torsional buckling. In addition to addressing the general case of unequal-leg single angles, the equal-leg angle is treated as a special case. Furthermore, bending of equal-leg angles about a geometric axis, an axis parallel to one of the legs, is addressed separately as it is a very common situation.

The tips of an angle refer to the free edges of the two legs. In most cases of unrestrained bending, the flexural stresses at the two tips will have the same sign (tension or compression). For constrained bending about a geometric axis, the tip stresses will differ in sign. Criteria for both tension and compression at the tip should be checked as appropriate, but in most cases it will be evident which controls.

Appropriate serviceability limits for single-angle beams need also to be considered. In particular, for longer members subjected to unrestrained bending, deflections are likely to control rather than lateral-torsional or local buckling strength.

C5.1.1. These provisions follow the LRFD format for nominal flexural resistance. There is a region of full yielding, a linear transition to the yield moment, and a region of local buckling. The strength at full yielding is limited to a shape factor of 1.25, which is less than that corresponding to the plastic moment of an angle. The factor of 1.25 corresponds to an allowable stress of $0.75F_y$, which has traditionally been used for rectangular shapes and for weak axis bending. It is used for angles due to uncertainties in developing the full plastic moment and to limit the large distortion of sections with large shape factors.

The b/t limits and the criteria for local buckling follow typical AISC criteria for single angles under uniform compression. They are conservative when the leg is subjected to non-uniform compression due to flexure.

C5.1.2. Since the shape factor for angles is in excess of 1.5, the nominal design strength $M_n = 1.25M_y$ for compact members is justified provided that instability does not control.

C5.1.3. Lateral-torsional instability may limit the flexural strength of an un-braced single-angle beam. As illustrated in Figure C5.1, Equation 5-3a represents the elastic buckling portion with the nominal flexural strength, M_n, varying from 75 percent to 92 percent of the theoretical buckling moment, M_{ob}. Equation 5-3b represents the inelastic buckling transition expression between $0.75M_y$ and $1.25M_y$. At M_{ob} greater than approximately $6M_y$, the unbraced length is adequate to develop the maximum beam flexural strength of $M_n = 1.25M_y$. These formulas were based on Australian research on single angles in flexure and on an analytical model consisting of two rectangular elements of length equal to the actual angle leg width minus one-half the thickness (Leigh and Lay, 1984; Australian Institute of Steel Construction, 1975; Leigh and Lay, 1978; Madugula and Kennedy, 1985). Figure C5.1 reflects the higher nominal moment strength than was implied by the $0.66F_y$ allowable stress in the ASD version.

A new and more general C_b moment gradient formula consistent with the 1993 AISC LRFD Specification is used to correct lateral-torsional stability equations from the assumed most severe case of uniform moment throughout the unbraced length ($C_b = 1.0$). The equation for C_b used in the ASD version is applicable only to moment diagrams that are straight lines between brace points. In lieu of a more detailed analysis, the reduced maximum limit of 1.5 is imposed for single-angle beams to represent conservatively the lower envelope of this cross section's non-uniform bending response.

C5.2.1. An angle beam loaded parallel to one leg will deflect and bend about that leg only if the angle is restrained laterally along the length. In this case simple bending occurs without any torsional rotation or lateral

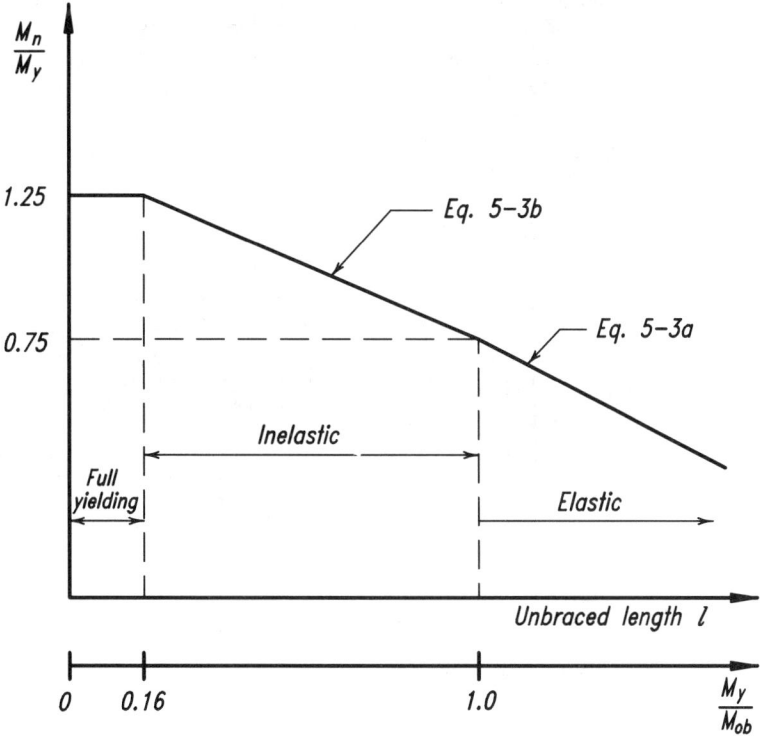

Fig. C5.1. *Lateral-torsional buckling of a single-angle beam.*

deflection and the geometric axis section properties should be used in the evaluation of the flexural design strength and deflection. If only the point of maximum moment is laterally braced, lateral-torsional buckling of the unbraced length under simple bending must also be checked, as outlined in Section 5.2.1b.

C5.2.2. When bending is applied about one leg of a laterally unrestrained single angle, it will deflect laterally as well as in the bending direction. Its behavior can be evaluated by resolving the load and/or moments into principal axis components and determining the sum of these principal axis flexural effects. Section 5.2.2 is provided to simplify and expedite the design calculations for this common situation with equal-leg angles.

For such unrestrained bending of an equal-leg angle, the resulting maximum normal stress at the angle tip (in the direction of bending) will be approximately 25 percent greater than calculated using the geometric axis section modulus. The value of M_{ob} in Equation 5-4 and the evaluation of M_y using 0.80 of the geometric axis section modulus reflect bending about the inclined axis shown in Figure C5.2.

The deflection calculated using the geometric axis moment of inertia has to be increased 82 percent to approximate the total deflection. Deflection has two components, a vertical component (in the direction of applied load) 1.56 times the calculated value and a horizontal component of 0.94 of the calculated value. The resultant total deflection is in the general direction of the weak principal axis bending of the angle (see Figure C5.2). These unrestrained bending deflections

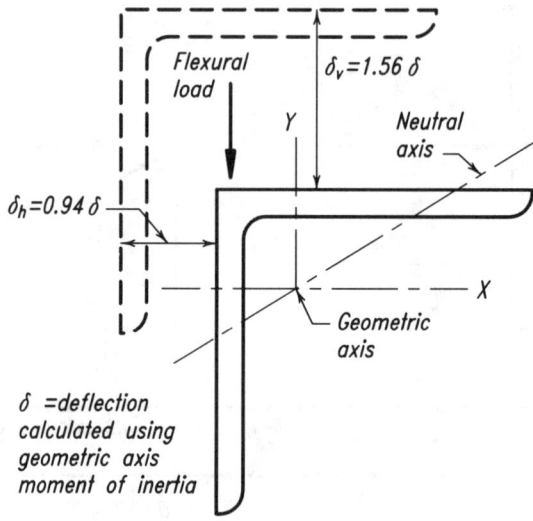

Fig. C5.2. Geometric axis bending of laterally unrestrained
equal-leg angles.

should be considered in evaluating serviceability and will often control the design over lateral-torsional buckling.

The horizontal component of deflection being approximately 60 percent of the vertical deflection means that the lateral restraining force required to achieve purely vertical deflection (Section 5.2.1) must be 60 percent of the applied load value (or produce a moment 60 percent of the applied value) which is very significant.

Lateral-torsional buckling is limited by M_{ob} (Leigh and Lay, 1984 and 1978) in Equation 5-4, which is based on

$$M_{cr} = \frac{2.33Eb^4t}{(1 + 3\cos^2\theta)(Kl)^2} \times$$

$$\left[\sqrt{\sin^2\theta + \frac{0.156(1 + 3\cos^2\theta)(Kl)^2t^2}{b^4}} + \sin\theta \right] \qquad (C5-1)$$

(the general expression for the critical moment of an equal-leg angle) with $\theta = -45°$ which is the most severe condition with the angle heel (shear center) in tension. Flexural loading which produces angle-heel compression can be conservatively designed by Equation 5-4 or more exactly by using the above general M_{cr} equation with $\theta = 45°$ (see Figure C5.3). With the angle heel in compression, Equation C5-1 will slightly exceed the yield moment limit of $1.25(0.8S_xF_y)$ only for relatively few high slenderness cases. For pure bending situations, deflections would be unreasonably large under these conditions. However, considering the interaction of flexure and compression in an angle with $F_y = 50$ ksi, b/t equal to 16 and the largest l/r of 200, Equation C5-1 will produce results eight percent less than the modified yield moment. This situation could arise in a compression angle where the load is transferred by end gusset plates attached to one leg only. In this case the flexure term in the interaction is about 0.5 which reduces the effect

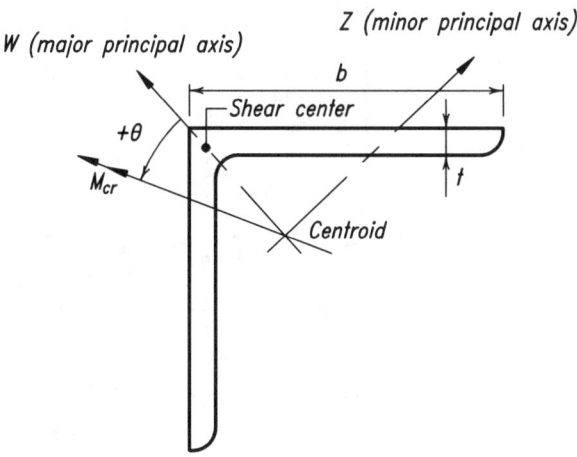

Fig. C5.3. Equal-leg angle with general moment loading.

to less than four percent and the end restraints provide an unknown increase in the lateral-torsional buckling strength. Consequently only the yield limit is required to be checked in Section 5.2.2 when the leg tips are in tension.

Lateral-torsional buckling will reduce the nominal bending strength only when l/b is relatively large. If the lt/b^2 parameter (which is a ratio of l/b over b/t) is small (less than approximately 2.5 with $C_b = 1$), there is no need to check lateral-torsional stability inasmuch as local buckling provisions of Section 5.1.1 will control the nominal bending strength.

Lateral-torsional buckling will produce $M_n < 1.25M_y$ for equal-leg angles only if M_{ob} by Equation 5-4 is less than about $6M_y$, for $C_b = 1.0$. Limits for l/b as a function of b/t are shown graphically in Figure C5.4. Local buckling and deflections must be checked separately.

Stress at the tip of the angle leg parallel to the applied bending axis is of the same sign as the maximum stress at the tip of the other leg when the single angle is unrestrained. For an equal-leg angle this stress is about one-third of the maximum stress. It is only necessary to check the nominal bending strength based on the tip of the angle leg with the maximum stress when evaluating such an angle. Since this maximum moment per Section 5.2.2 represents combined principal axis moments and Equation 5-4 represents the design limit for these combined

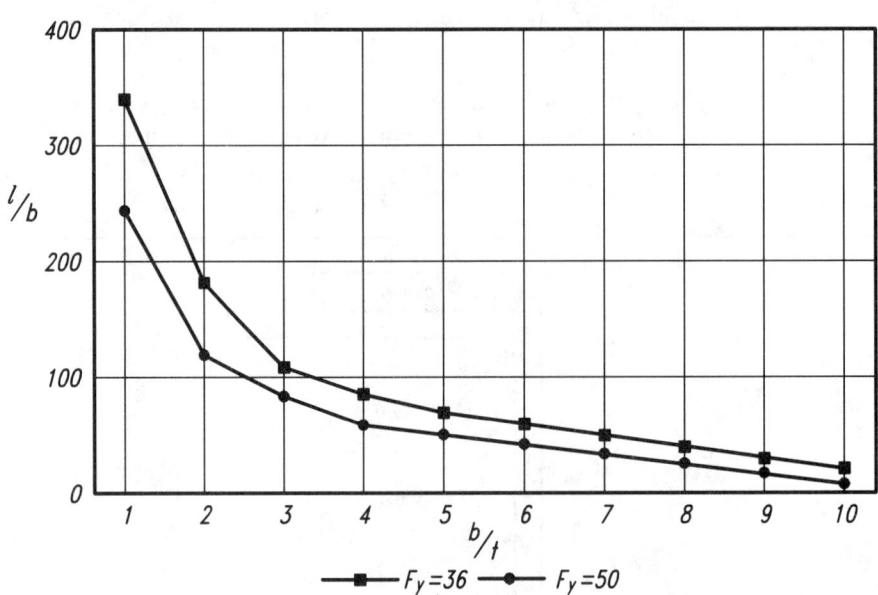

Fig. C5.4. Equal leg single-angle lateral buckling limits for $M_n = 1.25M_y$ about geometric axis.

flexural moments, only a single flexural term needs to be considered when evaluating combined flexural and axial effects.

C5.2.3. For unequal-leg angles without lateral-torsional restraint the applied load or moment must be resolved into components along the two principal axis in all cases and designed for biaxial bending using the interaction equation.

C5.3.1. Under major axis bending of equal-leg angles Equation 5-5 in combination with 5-3a or 5-3b controls the nominal design moment against overall lateral-torsional buckling of the angle. This is based on M_{cr}, given earlier with $\theta = 0$.

Lateral-torsional buckling for this case will reduce the stress below $1.25M_y$ only for $l/t \geq 4800/F_y$ or $0.160E/F_y$ ($M_{ob} = 6M_y$). If the lt/b^2 parameter is small (less than approximately $1.5C_b$ for this case), local buckling will control the nominal design moment and M_n based on lateral-torsional buckling need not be evaluated. Local buckling must be checked using Section 5.1.1.

C5.3.2. Lateral-torsional buckling about the major principal W axis of an unequal-leg angle is controlled by M_{ob} in Equation 5-6. Section property β_w reflects the location of the shear center relative to the principal axis of the section and the bending direction under uniform bending. Positive β_w and maximum M_{ob} occurs when the shear center is in flexural compression while negative β_w and minimum M_{ob} occurs when the shear center is in flexural tension (see Figure C5.5). This β_w effect is consistent with behavior of singly symmetric I-shaped beams which are more stable when the compression flange is larger than the tension flange. For principal W-axis bending of equal-leg angles, β_w is equal to zero due to symmetry and Equation 5-6 reduces to Equation 5-5 for this special case.

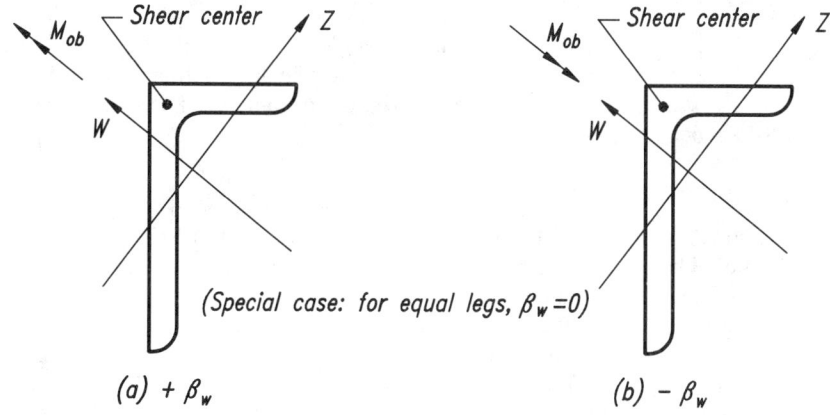

Fig. C5.5. Unequal-leg angle in bending.

TABLE C5.1
β_w Values for Angles

Angle Size (in.)	β_w (in.)*
9 × 4	6.54
8 × 6	3.31
8 × 4	5.48
7 × 4	4.37
6 × 4	3.14
6 × 3.5	3.69
5 × 3.5	2.40
5 × 3	2.99
4 × 3.5	0.87
4 × 3	1.65
3.5 × 3	0.87
3.5 × 2.5	1.62
3 × 2.5	0.86
3 × 2	1.56
2.5 × 2	0.85
Equal legs	0.00

* Has positive or negative value depending on direction of bending (see Figure C5.5).

For reverse curvature bending, part of the unbraced length has positive β_w, while the remainder negative β_w, and conservatively, the negative value is assigned for that entire unbraced segment.

β_w is essentially independent of angle thickness (less than one percent variation from mean value) and is primarily a function of the leg widths. The average values shown in Table C5.1 may be used for design.

C6.   COMBINED STRESSES

The stability and strength interaction equations of AISC LRFD Specification Chapter H have been adopted with modifications to account for various conditions of bending that may be encountered. Bending will usually accompany axial loading in a single-angle member since the axial load and connection along the legs are eccentric to the centroid of the cross section. Unless the situation conforms to Section 5.2.1 or 5.2.2 in that Section 6.1.2 or 6.1.3 may be used, the applied moment should be resolved about the principal axes for the interaction check.

For the non-symmetric and singly symmetric single angles, the interaction expression related to stresses at a particular location on the cross section is the most accurate due to lack of double symmetry. At a particular location, it is possible to have stresses of different sign from the various components such that a combination of tensile and compressive stress will represent a critical condition. The absolute value of the combined terms must be checked at the angle-leg tips and heel and compared with 1.0.

When using the combined force expressions for single angles, M_{uw} and M_{uz} are positive as customary. The evaluation of M_n in Section 5.1 is dependent on the location on the cross section being examined by using the appropriate value of section modulus, S. Since the sign of the stress is important in using Equations 6-1a and 6-1b, M_n is considered either positive or negative by assigning a sign to S to reflect the stress condition as adding to, or subtracting from, the axial load effect. A designer may choose to use any consistent sign convention.

It is conservative to ignore this refinement and simply use positive critical M_n values in the bending terms and add the absolute values of all terms (Elgaaly, Davids, and Dagher, 1992 and Adluri and Madugula, 1992).

Alternative special interaction equations for single angles have recently been published (Adluri and Madugula, 1992).

C6.1.3. When the total maximum flexural stress is evaluated for a laterally unrestrained length of angle per Section 5.2, the bending axis is the inclined axis shown in Figure C5.2. The radius of gyration modification for the moment amplification about this axis is equal to $\sqrt{1.82} = 1.35$ to account for the increased unrestrained bending deflection relative to that about the geometric axis for the laterally unrestrained length. The 1.35 factor is retained for angles braced only at the point of maximum moment to maintain a conservative calculation for this case. If the brace exhibits any flexibility permitting lateral movement of the angle, use of $r = r_x$ would not be conservative.

List of References

Alduri, S. M. and Madugula, M. K. S. (1992), "Eccentrically Loaded Steel Single-Angle Struts," AISC *Engineering Journal,* 2nd Quarter.

American Institute of Steel Construction, Inc. (1983), *Torsional Analysis of Steel Members,* Chicago, IL.

American Institute of Steel Construction, Inc. (1993), *Load and Resistance Factor Design Specification for Structural Steel Buildings,* Chicago, IL.

American Institute of Steel Construction, Inc. (1989), *Specification for Allowable Stress Design of Single-Angle Members,* Chicago, IL.

Australian Institute of Steel Construction (1975), *Australian Standard AS1250,* 1975.

Elgaaly, M., Davids, W. and Dagher, H. (1992), "Non-Slender Single-Angle Struts," AISC *Engineering Journal,* 2nd Quarter.

Galambos, T. V. (1991), "Stability of Axially Loaded Compressed Angles," Structural Stability Research Council, *Annual Technical Session Proceedings,* Apr. 15–17, 1991, Chicago, IL.

Gjelsvik, A. (1981), *The Theory of Thin-walled Bars,* John Wiley and Sons, New York.

Leigh, J. M. and M. G. Lay (1978), "Laterally Unsupported Angles with Equal and Unequal Legs," Report MRL 22/2 July 1978, Melbourne Research Laboratories, Clayton.

Leigh, J. M. and M. G. Lay (1984), "The Design of Laterally Unsupported Angles," in *Steel Design Current Practice,* Section 2, Bending Members, American Institute of Steel Construction, Inc., January 1984.

Lutz, L. A. (1992), "Critical Slenderness of Compression Members with Effective Lengths About Nonprincipal Axes," Structural Stability Research Council, Annual Technical Session Proceedings, Apr. 6–7, 1992, Pittsburgh, PA.

Madugula, M. K. S. and J. B. Kennedy (1985), *Single and Compound Angle Members,* Elsevier Applied Science, New York.

Seismic Provisions for Structural Steel Buildings

June 15, 1992

AMERICAN INSTITUTE OF STEEL CONSTRUCTION, INC.
One East Wacker Drive, Suite 3100, Chicago, IL 60601-2001

PREFACE

The intention of the main AISC *Specification* is to cover the common everyday design criteria in routine office usage. It is not feasible to also cover the many special and unique problems encountered within the full range of structural design practice. This document is a separate Specification which addresses one such topic, steel seismic provisions. It contains its own list of Symbols, a Glossary and a non-mandatory Commentary which has been included to provide background for the provisions.

The AISC Specification Task Committee 113 on Seismic Provisions to supplement the current Load and Resistance Factor Design (LRFD) and Allowable Stress Design (ASD) Specification for Structural Steel Buildings acknowledges the various contributions of several groups to the completion of this document: the Structural Engineers Association of California (SEAOC), the National Science Foundation, and the Building Seismic Safety Council. The main AISC Committee on Specification enhanced these provisions by careful scrutiny, discussions, suggestions for improvements, and endorsement. The members of this Task Committee, as principal authors of the AISC Seismic Provisions, are most grateful to all of the above groups and people. Special recognition must also be given to the leadership expertise, and perseverance of Task Committee Chairman Egor Popov and Technical Secretary Clarkson Pinkham.

The principal changes in this edition of the Seismic Provisions are the conversion to the loads and design format recommended by the 1991 National Earthquake Hazards Reduction Program (NEHRP) document.

The reader is cautioned that professional judgment must be exercised when data or recommendations in this Specification are applied. The publication of the material contained herein is not intended as a representation or warranty on the part of the American Institute of Steel Construction, Inc.—or any other person named herein—that this information is suitable for general or particular use, or freedom from infringement of any patent or patents. Anyone making use of this information assumes all liability arising from such use. The design of structures is within the scope of expertise of a competent licensed structural engineer, architect, or other licensed professional for the application of principles to a particular structure.

By the AISC Subcommittee,

E. P. Popov, Chair	S. D. Lindsey
R. Becker	H. W. Martin
G. G. Deierlein	C. M. Saunders
M. D. Engelhardt	J. B. Shantz
S. J. Fang	I. M. Viest
R. E. Ferch	N. F. G. Youssef
R. D. Hanson	C. W. Pinkham, Technical Secretary
J. R. Harris	N. Iwankiw, Recording Secretary
K. Kasai	

May 22, 1992

Symbols

The section numbers in parentheses after the definition of a symbol refers to the section where the symbol is first used.

A_e Effective net area, in.^{2} (9)
A_f Flange area of member, in.^{2} (6)
A_g Gross area, in.^{2} (8)
A_{st} Area of link stiffener, in.^{2} (10)
A_v Seismic coefficient representing the effective peak velocity-related acceleration. (2)
A_w Effective area of weld, in.^{2} (6)
A_w Link web area, in.^{2} (10)
C_s Response factor related to the fundamental period of the building. (3)
D Dead load due to the self-weight of the structure and the permanent elements on the structure, kips. (3)
E Earthquake load. (3)
F_{BM} Nominal strength of the base material to be welded, ksi. (6)
F_{EXX} Classification strength of weld metal, ksi. (6)
F_w Nominal strength of the weld electrode material, ksi. (6)
F_y Specified minimum yield strength of the type of steel being used, ksi. (8)
F_{yb} F_y of a beam, ksi. (8)
F_{yc} F_y of a column, ksi. (6)
H Average story height above and below a beam-to-column connection., in. (8)
L Live load due to occupancy and moveable equipment, kips. (3)
L Unbraced length of compression or bracing member, in. (8)
L_r Roof live load, kips. (3)
M_n Nominal moment strength of a member or joint, kip-in. (8)
M_p Plastic bending moment, kip-in. (8)
M_{pa} Plastic bending moment modified by axial load ratio, kip-in. (10)
M_u Required flexural strength on a member or joint, kip-in. (8)
P_D Required axial strength on a column resulting from application of dead load, D, kips. (6)
P_E Required axial strength on a column resulting from application of the specified earthquake load, E, kips. (6)
P_L Required axial strength on a column resulting from application of live load, L, kips. (6)
P_u Required axial strength on a column or a link, kips. (10)
P_n Nominal axial strength of a column, kips. (6)
P_u^* Required axial strength on a brace, kips. (9)
P_{uc} Required axial strength on a column based on load combination with seismic loads, kips. (8)
P_y Nominal yield axial strength of a member = $F_y A_g$, kips. (10)

R	Response modification factor. (3)
R'	Load due to initial rainwater or ice exclusive of the ponding contribution, kips. (Symbol R is used in the *Specification*). (3)
R_n	Nominal strength of a member. (8)
S	Snow load, kips. (3)
V	Base shear due to earthquake load, kips. (3)
V_n	Nominal shear strength of a member, kips. (8)
V_u	Required shear strength on a member, kips. (8)
V_p	Nominal shear strength of an active link, kips. (10)
V_{pa}	Nominal shear strength of an active link modified by the axial load magnitude, kips. (10)
W	Wind load, kips. (3)
W_g	Total weight of the building, kips. (3)
Z_b	Plastic section modulus of a beam, in.^{3} (8)
Z_c	Plastic section modulus of a column, in.^{3} (8)
b	Width of compression element, in. (Table 8-1)
b_f	Flange width, in. (8)
b_{cf}	Column flange width, in. (8)
d_b	Overall beam depth, in. (8)
d_c	Overall column depth, in. (8)
d_z	Overall panel zone depth between continuity plates, in. (8)
e	EBF link length, in. (10)
h	Assumed web depth for stability, in. (Table 8-1)
r	Governing radius of gyration, in. (9)
r_y	Radius of gyration about y axis, in. (8)
t_{bf}	Thickness of beam flange, in. (8)
t_{cf}	Thickness of column flange, in. (8)
t_f	Thickness of flange, in. (8)
t_p	Thickness of panel zone including doubler plates, in. (8)
t_w	Thickness of web, in. (8)
t_z	Thickness of panel zone (doubler plates not necessarily included), in. (8)
w_z	Width of panel zone between column flanges, in. (8)
α	Fraction of member force transferred across a particular net section. (9)
ρ	Ratio of required axial force P_u to required shear strength V_u of a link. (10)
k	Slenderness parameter. (9)
k_p	Limiting slenderness parameter for compact element. (8)
k_r	Limiting slenderness parameter for non-compact element. (9)
ϕ	Resistance factor. (6,10)
ϕ_b	Resistance factor for beams. (6)
ϕ_c	Resistance factor for columns in compression. (6,10)
ϕ_t	Resistance factor for columns in tension. (6)
ϕ_v	Resistance factor for shear strength of panel zone of beam-to-column connections. (8)
ϕ_w	Resistance factor for welds. (6)

Glossary

Beam. A structural member whose primary function is to carry loads transverse to its longitudinal axis, usually a horizontal member in a seismic frame system.

Braced Frame. An essentially vertical truss system of concentric or eccentric type that resists lateral forces on the structural system.

Concentrically Braced Frame (CBF). A braced frame in which all members of the bracing system are subjected primarily to axial forces. The CBF shall meet the requirements of Sect. 9.

Connection. Combination of joints used to transmit forces between two or more members. Categorized by the type and amount of force transferred (moment, shear, end reaction).

Continuity Plates. Column stiffeners at top and bottom of the panel zone.

Design strength. Resistance (force, moment, stress, as appropriate) provided by element or connection; the product of the nominal strength and the resistance factor.

Diagonal Bracing. Inclined structural members carrying primarily axial load employed to enable a structural frame to act as a truss to resist horizontal loads.

Dual System. A dual system is a structural system with the following features:

- An essentially complete space frame which provides support for gravity loads.
- Resistance to lateral load is provided by moment resisting frames (SMF) or (OMF) which is capable of resisting at least 25 percent of the base shear and concrete or steel shear walls, steel eccentrically (EBF) or concentrically (CBF) braced frames.
- Each system shall be also designed to resist the total lateral load in proportion to its relative rigidity.

Eccentrically Braced Frame (EBF). A diagonal braced frame in which at least one end of each bracing member connects to a beam a short distance from a beam-to-column connection or from another beam-to-brace connection. The EBF shall meet the requirements of Sect. 10.

Essential Facilities. Those facilities defined as essential in the applicable code under which the structure is designed. In the absence of such a code, see ASCE 7-92.

Joint. Area where two or more ends, surfaces, or edges are attached. Categorized by type of fastener or weld used and method of force transfer.

K Braced Frame. A concentric braced frame (CBF) in which a pair of diagonal braces located on one side of a column is connected to a single point within the clear column height.

Lateral Support Member. Member designed to inhibit lateral buckling or lateral-torsional buckling of primary frame members.

Link. In EBF, the segment of a beam which extends from column to column, located between the end of a diagonal brace and a column or between the ends of two diagonal braces of the EBF. The length of the link is defined as the clear distance between the diagonal brace and the column face or between the ends of two diagonal braces.

Link Intermediate Web Stiffeners. Vertical web stiffeners placed within the link.

Link Rotation Angle. The link rotation angle is the plastic angle between the link and the beam outside of the link when the total story drift is E' / E times the drift derived using the specified base shear, V.

Link Shear Design Strength. The lesser of ϕV_p or $2\phi M_p/e$, where $\phi = 0.9$, $V_p = 0.55 F_y\, dt_w$ and e = the link length except as modified by Sect. S9.2.f.

LRFD. (Load and Resistance Factor Design). A method of proportioning structural components (members, connectors, connecting elements, and assemblages) such that no applicable limit state is exceeded when the structure is subjected to all design load combinations.

Moment Frame. A building frame system in which seismic shear forces are resisted by shear and flexure in members and joints of the frame.

Nominal loads. The magnitudes of the loads specified by the applicable code.

Nominal strength. The capacity of a structure or component to resist the effects of loads, as determined by computations using specified material strengths and dimensions and formulas derived from accepted principles of structural mechanics or by field tests or laboratory tests of scaled models, allowing for modeling effects, and differences between laboratory and field conditions.

Ordinary Moment Frame (OMF). A moment frame system which meets the requirements of Sect. 7.

P - Delta effect. Secondary effect of column axial loads and lateral deflection on the shears and moments in members.

Panel Zone. Area of beam-to-column connection delineated by beam and column flanges.

Required Strength. Load effect (force, moment, stress, as appropriate) acting on element of connection determined by structural analysis from the factored loads (using most appropriate critical load combinations).

Resistance Factor. A factor that accounts for unavoidable deviations of the actual strength from the nominal value and the manner and consequences of failure.

Slip-Critical Joint. A bolted joint in which slip resistance of the connection is required.

Special Moment Frame (SMF). A moment frame system which meets the requirements of Sect. 8.

Structural System. An assemblage of load-carrying components which are joined together to provide regular interaction or interdependence.

V Braced Frame. A concentrically braced frame (CBF) in which a pair of diagonal braces located either above or below a beam is connected to a single point within the clear beam span. Where the diagonal braces are below the beam, the system is also referred to as an Inverted V Braced Frame.

X Braced Frame. A concentrically braced frame (CBF) in which a pair of diagonal braces crosses near mid-length of the braces.

Y Braced Frame. An eccentrically braced frame (EBF) in which the stem of the Y is the link of the EBF system.

Seismic Provisions for Structural Steel Buildings

June 15, 1992

Part I—Load and Resistance Factor Design (LRFD)

1. SCOPE

These special seismic requirements are to be applied in conjunction with the AISC *Load and Resistance Factor Design Specification for Structural Steel Buildings* (LRFD), 1986; hereinafter referred to as the *Specification*. They are intended for the design and construction of structural steel members and connections in buildings for which the design forces resulting from earthquake motions have been determined on the basis of energy dissipation in the non-linear range of response.

Seismic provisions and the nominal loads for each Seismic Performance Category, Seismic Hazard Exposure Group, or Seismic Zone shall be as specified by the applicable code under which the structure is designed or where no code applies, as dictated by the conditions involved. In the absence of a code, the Performance Categories, Seismic Hazard Exposure Groups, loads and load combinations shall be as given herein.

2. SEISMIC PERFORMANCE CATEGORIES

Seismic Performance Categories vary with the Seismic Hazard Exposure Group shown in Table 2-1, the Effective Peak Velocity Related Acceleration, A_v, and the Seismic Hazard Exposure Group shown in Table 2-2.

In addition to the general requirements assigned to the various Seismic Performance Categories in the applicable building code for all types of construction, the following requirements apply to fabricated steel construction for buildings and structures with similar structural characteristics.

2.1. Seismic Performance Categories A, B, and C

Buildings assigned to Categories A, B, and C, except Category C in Seismic Hazard Exposure Group III where the value of $A_v \geq 0.10$, shall be designed either in accordance with solely the *Specification* or in accordance with the *Specification* and these provisions.

TABLE 2-1
Seismic Hazard Exposure Groups

Group III	Buildings having essential facilities that are necessary for post-earthquake recovery and requiring special requirements for access and functionality.
Group II	Buildings that constitute a substantial public hazard because of occupancy or use.
Group I	All buildings not classified in Groups II and III.

2.2. Seismic Performance Category C

Buildings assigned to Category C in Seismic Hazard Exposure Group III where the value of $A_v \geq 0.10$ shall be designed in accordance with the *Specification* as modified by the additional provisions of this section.

2.2.a. Steel used in seismic resisting systems shall be limited by the provisions of Sect. 5.

2.2.b. Columns in seismic resisting systems shall be designed in accordance with Sect. 6.

2.2.c. Ordinary Moment Frames (OMF) shall be designed in accordance with the provisions of Sect. 7.

2.2.d. Special Moment Frames (SMF) are required to conform only to the requirements of Sects. 8.2, 8.7, and 8.8.

2.2.e. Braced framed systems shall conform to the requirements of Sects. 9 or 10 when used alone or in combination with the moment frames of the seismic resisting system.

2.2.f. A quality assurance plan shall be submitted to the regulatory agency for the seismic force resisting system of the building.

2.3. Seismic Performance Categories D and E

Buildings assigned to Categories D and E shall be designed in accordance with the *Specification* as modified by the additional provisions of this section.

2.3.a. Steel used in seismic resisting systems shall be limited by the provisions of Sect. 5.

2.3.b. Columns in seismic resisting systems shall be designed in accordance with Sect. 6.

2.3.c. Ordinary Moment Frames (OMF) shall be designed in accordance with the provisions of Sect. 7.

2.3.d. Special Moment Frames (SMF) shall be designed in accordance with the provisions of Sect. 8.

2.3.e. Braced framed systems shall conform to the requirements of Sects. 9.

TABLE 2-2
Seismic Performance Categories

Value of A_v	Seismic Hazard Exposure Group		
	I	II	III
$0.20 \leq A_v$	D	D	E
$0.15 \leq A_v < 0.20$	C	D	D
$0.10 \leq A_v < 0.15$	C	C	C
$0.05 \leq A_v < 0.10$	B	B	C
$A_v < 0.05$	A	A	A

(CBF) or 10. (EBF) when used alone or in combination with the moment frames of the seismic resisting system.

The use of K-bracing systems shall not be permitted as part of the seismic resisting system except as permitted by Sect. 9.5. (Low Buildings)

2.3.f. A quality assurance plan shall be submitted to the regulatory agency for the seismic force resisting system of the building.

3. LOADS, LOAD COMBINATIONS, AND NOMINAL STRENGTHS

3.1. Loads and Load Combinations

The following specified loads and their effects on the structure shall be taken into account:

D : dead load due to the weight of the structural elements and the permanent features on the structure.
L : live load due to occupancy and moveable equipment.
L_r : roof live load.
W : wind load.
S : snow load.
E : earthquake load (where the horizontal component is derived from base shear Formula $V = C_s W_g$).
R' : load due to initial rainwater or ice exclusive of the ponding contribution.

In the Formula $V = C_s W_g$ for base shear:

C_s = Seismic design coefficient
W_g = Total weight of the building, see the applicable code.

For the nominal loads as defined above, see the applicable code.

The required strength of the structure and its elements shall be determined from the appropriate critical combination of factored loads. The following Load Combinations and corresponding load factors shall be investigated:

$$1.4D \tag{3-1}$$

$$1.2D + 1.6L + 0.5(L_r \text{ or } S \text{ or } R') \tag{3-2}$$

$$1.2D + 1.6(L_r \text{ or } S \text{ or } R') + (0.5L \text{ or } 0.8W) \tag{3-3}$$

$$1.2D + 1.3W + 0.5L + 0.5(L_r \text{ or } S \text{ or } R') \qquad (3\text{-}4)$$

$$1.2D \pm 1.0E + 0.5L + 0.2S \qquad (3\text{-}5)$$

$$0.9D \pm (1.0E \text{ or } 1.3W) \qquad (3\text{-}6)$$

Exception: The load factor on L in Load Combinations 3-3, 3-4, and 3-5 shall equal 1.0 for garages, areas occupied as places of public assembly, and all areas where the live load is greater than 100 psf.

Other special load combinations are included with specific design requirements throughout these provisions.

Orthogonal earthquake effects shall be included in the analysis unless noted specifically otherwise in the governing building code.

Where required by these provisions, an amplified horizontal earthquake load of $0.4R \times E$ (where the term $0.4R$ is greater or equal to 1.0) shall be applied in lieu of the horizontal component of earthquake load E in the load combinations above. The term R is the earthquake response modification coefficient contained in the applicable code. The additional load combinations using the amplified horizontal earthquake load are:

$$1.2D + 0.5L + 0.2S \pm 0.4R \times E \qquad (3\text{-}7)$$

$$0.9D \pm 0.4R \times E \qquad (3\text{-}8)$$

Exception: The load factor on L in Load Combinations 3-7 shall equal 1.0 for garages, areas occupied as places of public assembly and all areas where the live load is greater than 100 psf.

The term $0.4R$ in Load Combinations 3-7 and 3-8 shall be greater or equal to 1.0.

Where the amplified load is required, orthogonal effects are not required to be included.

3.2. Nominal Strengths

The nominal strengths shall be as provided in the *Specification.*

4.   STORY DRIFT

Story drift shall be calculated using the appropriate load effects consistent with the structural system and the method of analysis. Limits on story drift shall be in accordance with the governing code and shall not impair the stability of the structure.

5.   MATERIAL SPECIFICATIONS

Steel used in seismic force resisting systems shall be as listed in Sect. A3.1 of the *Specification,* except for buildings over one story in height. The steel used in seismic resisting systems described in Sections 8, 9, and 10 shall be limited to the following ASTM Specifications: A36, A500 (Grades B and C), A501, A572 (Grades 42 and 50), and A588. The steel used for base plates shall meet one of the preceding ASTM Specifications or ASTM A283 Grade D.

6. COLUMN REQUIREMENTS

6.1. Column Strength

When $P_u / \phi P_n > 0.5$, columns in seismic resisting frames, in addition to complying with the *Specification*, shall be limited by the following requirements:

6.1.a. Axial compression loads:

$$1.2P_D + 0.5P_L + 0.2P_S + 0.4R \times P_E \leq \phi_c P_n \qquad (6\text{-}1)$$

where the term $0.4R$ is greater or equal to 1.0.

Exception: The load factor on P_L in Load Combination 6-1 shall equal 1.0 for garages, areas occupied as places of public assembly, and all areas where the live load is greater than 100 psf.

6.1.b. Axial tension loads:

$$0.9P_D - 0.4R \times P_E \leq \phi_t P_n \qquad (6\text{-}2)$$

where the term $0.4R$ is greater or equal to 1.0.

6.1.c. The axial Load Combinations 6-1 and 6-2 are not required to exceed either of the following:

1. The maximum loads transferred to the column, considering 1.25 times the design strengths of the connecting beam or brace elements of the structure.

2. The limit as determined by the foundation capacity to resist overturning uplift.

6.2. Column Splices

Column splices shall have a design strength to develop the column axial loads given in Sect. 6.1.a, b, and c as well as the Load Combinations 3-1 to 3-6.

6.2.a. In column splices using either complete or partial penetration welded joints, beveled transitions are not required when changes in thickness and width of flanges and webs occur.

6.2.b. Splices using partial penetration welded joints shall not be within 3 ft of the beam-to-column connection. Column splices that are subject to net tension forces shall comply with the more critical of the following:

1. The design strength of partial penetration welded joints, the lesser of $\phi_w F_w A_w$ or $\phi_w F_{BM} A_w$, shall be at least 150 percent of the required strength, where $\phi_w = 0.8$ and $F_w = 0.6 F_{EXX}$.

2. The design strength of welds shall not be less than $0.5 F_{yc} A_f$, where F_{yc} is the yield strength of the column material and A_f is the flange area of the smaller column connected.

7. REQUIREMENTS FOR ORDINARY MOMENT FRAMES (OMF)

7.1. Scope

Ordinary Moment Frames (OMF) shall have a design strength as provided in the *Specification* to resist the Load Combinations 3-1 through 3-6 as modified by the following added provisions:

7.2. Joint Requirements

All beam-to-column and column to beam connections in OMF which resist seismic forces shall meet one of the following requirements:

7.2.a. FR (fully restrained) connections conforming with Sect. 8.2, except that the required flexural strength, M_u, of a column-to-beam joint is not required to exceed the nominal plastic flexural strength of the connection.

7.2.b. FR connections with design strengths of the connections meeting the requirements of Sect. 7.1 using the Load Combinations 3-7 and 3-8.

7.2.c. Either FR or PR (partially restrained) connections shall meet the following:

     1. The design strengths of the members and connections meet the requirements of Sect. 7.1.

     2. The connections have been demonstrated by cyclic tests to have adequate rotation capacity at a story drift calculated at a horizontal load of $0.4R \times E$, (where the term $0.4R$ is equal to or greater than 1.0).

     3. The additional drift due to PR connections shall be considered in design.

     FR and PR connections are described in detail in Sect. A2 of the *Specification.*

8. REQUIREMENTS FOR SPECIAL MOMENT FRAMES (SMF)

8.1. Scope

Special Moment Frames (SMF) shall have a design strength as provided in the *Specification* to resist the Load Combinations 3-1 through 3-6 as modified by the following added provisions:

8.2. Beam-to-Column Joints

8.2.a. The required flexural strength, M_u, of each beam-to-column joint shall be the lesser of the following quantities:

     1. The plastic bending moment, M_p, of the beam.

     2. The moment resulting from the panel zone nominal shear strength, V_n, as determined using Equation 8-1.

     The joint is not required to develop either of the strengths defined above if it is shown that under an amplified frame deformation produced by Load Combinations 3-7 and 3-8, the design strength of the members at the connection is adequate to support the vertical loads, and the required lateral force resistance is provided by other means.

8.2.b. The required shear strength, V_u, of a beam-to-column joint shall be determined using the Load Combination $1.2D + 0.5L + 0.2S$ plus the shear resulting from M_u, as defined in Sect. 8.2.a., on each end of the beam. Alternatively, V_u shall be justified by a rational analysis. The required shear strength is not required to exceed the shear resulting from Load Combination 3-7.

8.2.c. The design strength, ϕR_n, of a beam-to-column joint shall be considered adequate to develop the required flexural strength, M_u, of the beam if it conforms to the following:

1. The beam flanges are welded to the column using complete penetration welded joints.

2. The beam web joint has a design shear strength ϕV_n greater than the required shear, V_u, and conforms to either:

   a. Where the nominal flexural strength of the beam, M_n, considering only the flanges is greater than 70 percent of the nominal flexural strength of the entire beam section [i.e., $b_f\, t_f\, (d{-}t_f)F_{yf} \geq 0.7M_p$]; the web joint shall be made by means of welding or slip-critical high strength bolting, or;

   b. Where $b_f\, t_f\, (d{-}t_f)F_{yf} < 0.7M_p$, the web joint shall be made by means of welding the web to the column directly or through shear tabs. That welding shall have a design strength of at least 20 percent of the nominal flexural strength of the beam web. The required beam shear, V_u, shall be resisted by further welding or by slip-critical high-strength bolting or both.

8.2.d. Alternate Joint Configurations: For joint configurations utilizing welds or high-strength bolts, but not conforming to Sect. 8.2.c, the design strength shall be determined by test or calculations to meet the criteria of Sect. 8.2.a. Where conformance is shown by calculation, the design strength of the joint shall be 125 percent of the design strengths of the connected elements.

8.3. Panel Zone of Beam-to-Column Connections (Beam web parallel to column web)

8.3.a. Shear Strength: The required shear strength, V_u, of the panel zone shall be based on beam bending moments determined from the Load Combinations 3-5 and 3-6. However, V_u is not required to exceed the shear forces determined from $0.9\Sigma\phi_b M_p$ of the beams framing into the column flanges at the connection. The design shear strength, $\phi_v V_n$, of the panel zone shall be determined by the following formula:

$$\phi_v V_n = 0.6\phi_v F_y d_c\, t_p \left[1 + \frac{3b_{cf}\, t_{cf}^2}{d_b\, d_c\, t_p} \right] \text{ where for this case } \phi_v = 0.75. \qquad (8\text{-}1)$$

where:

t_p = Total thickness of panel zone including doubler plates, in.
d_c = Overall column section depth, in.
b_{cf} = Width of the column flange, in.

t_{cf} = Thickness of the column flange, in.
d_b = Overall beam depth, in.
F_y = Specified yield strength of the panel zone steel, ksi.

8.3.b. Panel Zone Thickness: The panel zone thickness, t_z, shall conform to the following:

$$t_z \geq (d_z + w_z) / 90 \qquad (8\text{-}2)$$

where:

d_z = the panel zone depth between continuity plates, in.
w_z = the panel zone width between column flanges, in.

For this purpose, t_z shall not include any doubler plate thickness unless the doubler plate is connected to the web with plug welds adequate to prevent local buckling of the plate.

Where a doubler plate is used without plug welds to the column web, the doubler plate shall conform to Eq. 8-2.

8.3.c. Panel Zone Doubler Plates: Doubler plates provided to increase the design strength of the panel zone or to reduce the web depth thickness ratio shall be placed next to the column web and welded across the plate width along the top and bottom with at least a minimum fillet weld. The doubler plates shall be fastened to the column flanges using either butt or fillet welded joints to develop the design shear strength of the doubler plate.

8.4. Beam and Column Limitations

8.4.a. Beam Flange Area: There shall be no abrupt changes in beam flange areas in plastic hinge regions.

8.4.b. Width-Thickness Ratios: Beams and columns shall comply with λ_p in Table 8-1 in lieu of those in Table B5.1 of the *Specification*.

8.5. Continuity Plates

Continuity plates shall be provided if required by the provisions in the *Specification* for webs and flanges with concentrated forces and if the nominal column local flange bending strength R_n is less than $1.8 F_{yb} b_f t_{bf}$, where:

$R_n = 6.25(t_{cf})^2 F_{yf}$, and
F_{yb} = Specified minimum yield strength of beam, ksi.
F_{yf} = Specified minimum yield strength of column flange, ksi.
b_f = Beam flange width, in.
t_{bf} = Beam flange thickness, in.
t_{cf} = Column flange thickness, in.

Continuity plates shall be fastened by welds to both the column flanges and either the column webs or doubler plates.

8.6. Column-Beam Moment Ratio

At any beam-to-column connection, one of the following relationships shall be satisfied:

TABLE 8-1
Limiting Width Thickness Ratios λ_p
for Compression Elements

Description of Element	Width-Thickness Ratio	Limiting Width-Thickness Ratios λ_p
Flanges of I-shaped non-hybrid sections and channels in flexure. Flanges of I-shaped hybrid beams in flexure.	b/t	$52/\sqrt{F_y}$
Webs in combined flexural and axial compression.	h/t_w	For $P_u/\phi_b P_y \leq 0.125$
		$\dfrac{520}{\sqrt{F_y}}\left[1-\dfrac{1.54P_u}{\phi_b P_y}\right]$
		For $P_u/\phi_b P_y > 0.125$
		$\dfrac{191}{\sqrt{F_y}}\left[2.33-\dfrac{P_u}{\phi_b P_y}\right] \geq \dfrac{253}{\sqrt{F_y}}$

$$\frac{\Sigma Z_c(F_{yc}-P_{uc}/A_g)}{\Sigma Z_b F_{yb}} \geq 1.0, \tag{8-3}$$

$$\frac{\Sigma Z_c(F_{yc}-P_{uc}/A_g)}{V_n d_b H/(H-d_b)} \geq 1.0, \tag{8-4}$$

where:

A_g = Gross area of a column, in.^{2}
F_{yb} = Specified minimum yield strength of a beam, ksi.
F_{yc} = Specified minimum yield strength of a column, ksi.
H = Average of the story heights above and below the joint, in.
P_{uc} = Required axial strength in the column (in compression) ≥ 0
V_n = Nominal strength of the panel zone as determined from Equation 8-1, ksi.
Z_b = Plastic section modulus of a beam, in.^{3}
Z_c = Plastic section modulus of a column, in.^{3}
d_b = Average overall depth of beams framing into the connection, in.

These requirements do not apply in any of the following cases, provided the columns conform to the requirements of Sect. 8.4:

8.6.a. Columns with $P_{uc} < 0.3F_{yc}A_g$.

8.6.b. Columns in any story that has a ratio of design shear strength to design force 50 percent greater than the story above.

8.6.c. Any column not included in the design to resist the required seismic shears, but included in the design to resist axial overturning forces.

8.7. Beam-to-Column Connection Restraint

8.7.a. Restrained Connection:

1. Column flanges at a beam-to-column connection require lateral support only at the level of the top flanges of the beams when a column is shown to remain elastic outside of the panel zone, using one of the following conditions:

   a. Ratios calculated using Eqs. 8-3 or 8-4 are greater than 1.25.

   b. Column remains elastic when loaded with Load Combination 3-7.

2. When a column cannot be shown to remain elastic outside of the panel zone, the following provisions apply:

   a. The column flanges shall be laterally supported at the levels of both top and bottom beam flanges.

   b. Each column flange lateral support shall be designed for a required strength equal to 2.0 percent of the nominal beam flange strength $(F_y b_f t_f)$.

   c. Column flanges shall be laterally supported either directly, or indirectly, by means of the column web or beam flanges.

8.7.b. Unrestrained Connections: A column containing a beam-to-column connection with no lateral support transverse to the seismic frame at the connection shall be designed using the distance between adjacent lateral supports as the column height for buckling transverse to the seismic frame and conform to Sect. H of the *Specification* except that:

1. The required column strength shall be determined from the Load Combination 3-5 where E is the least of:

   a. The amplified earthquake force $0.4R \times E$ (where the term $0.4R$ shall be equal to or greater than 1.0).

   b. 125 percent of the frame design strength based on either beam or panel zone design strengths.

2. The L/r for these columns shall not exceed 60.

3. The required column moment transverse to the seismic frame shall include that caused by the beam flange force specified in Sect. 8.7.a.2.b plus the added second order moment due to the resulting column displacement in this direction.

8.8. Lateral Support of Beams

Both flanges of beams shall be laterally supported directly or indirectly. The unbraced length between lateral supports shall not exceed 2,500 r_y/F_y. In addition, lateral supports shall be placed at concentrated loads where an analysis indicates a hinge will be formed during inelastic deformations of the SMF.

9. REQUIREMENTS FOR CONCENTRICALLY BRACED (CBF) BUILDINGS

9.1. Scope

Concentrically Braced Frames (CBF) are braced systems whose worklines essentially intersect at points. Minor eccentricities, where the worklines intersect within the width of the bracing members, are acceptable if accounted for in the design. CBF shall have a design strength as provided in the *Specification* to resist the Load Combinations 3-1 through 3-6 as modified by the following added provisions:

9.2. Bracing Members

9.2.a. Slenderness: Bracing members shall have an $\frac{L}{r} \leq \frac{720}{\sqrt{F_y}}$ except as permitted in Sect. 9.5.

9.2.b. Compressive Design Strength: The design strength of a bracing member in axial compression shall not exceed $0.8\phi_c P_n$.

9.2.c. Lateral Force Distribution: Along any line of bracing, braces shall be deployed in alternate directions such that, for either direction of force parallel to the bracing, at least 30 percent but no more than 70 percent of the total horizontal force shall be resisted by tension braces, unless the nominal strength, P_n, of each brace in compression is larger than the required strength, P_u, resulting from the application of the Load Combinations 3-7 or 3-8. A line of bracing, for the purpose of this provision, is defined as a single line or parallel lines whose plan offset is 10 percent or less of the building dimension perpendicular to the line of bracing.

9.2.d. Width-Thickness Ratios: Width-thickness ratios of stiffened and unstiffened compression elements in braces shall comply with Sect. B5 in the *Specification*. Braces shall be compact or non-compact, but not slender (i.e., $\lambda < \lambda_r$). Circular sections shall have an outside diameter to wall thickness ratio not exceeding $1,300 / F_y$; rectangular tubes shall have a flat-width to wall thickness not exceeding $110 / \sqrt{F_y}$, unless the circular section or tube walls are stiffened.

9.2.e. Built-up Member Stitches: For all built-up braces, the first bolted or welded stitch on each side of the midlength of a built up member shall be designed to transmit a force equal to 50 percent of the nominal strength of one element to the adjacent element. Not less than two stitches shall be equally spaced about the member centerline.

9.3. Bracing Connections

9.3.a. Forces: The required strength of bracing joints (including beam-to-column joints if part of the bracing system) shall be the least of the following:

1. The design axial tension strength of the bracing member.

2. The force in the brace resulting from the Load Combinations 3-7 or 3-8.

3. The maximum force, indicated by an analysis, that is transferred to the brace by the system.

9.3.b. Net Area: In bolted brace joints, the minimum ratio of effective net section area to gross section area shall be limited by:

$$\frac{A_e}{A_g} \geq \frac{1.2\alpha P_u^*}{\phi_t P_n} \qquad (9\text{-}1)$$

where:

A_e = Effective net area as defined in Equation B3-1 of the *Specification*.
P_u^* = Required strength on the brace as determined in Sect. 9.3.a.
P_n = Nominal tension strength as specified in Chapter D of the *Specification*.
ϕ_t = Special resistance factor for tension = 0.75.
α = Fraction of the member force from Sect. 9.3.a that is transferred across a particular net section.

9.3.c. Gusset Plates:

1. Where analysis indicates that braces buckle in the plane of the gusset plates, the gusset and other parts of the connection shall have a design strength equal to or greater than the in-plane nominal bending strength of the brace.

2. Where the critical buckling strength is out-of-plane of the gusset plate, the brace shall terminate on the gusset a minimum of two times the gusset thickness from the theoretical line of bending which is unrestrained by the column or beam joints. The gusset plate shall have a required compressive strength to resist the compressive design strength of the brace member without local buckling of the gusset plate. For braces designed for axial load only, the bolts or welds shall be designed to transmit the brace forces along the centroids of the brace elements.

9.4. Special Bracing Configuration Requirements

9.4.a. V and Inverted V Type Bracing:

1. The design strength of the brace members shall be at least 1.5 times the required strength using Load Combinations 3-5 and 3-6.

2. The beam intersected by braces shall be continuous between columns.

3. A beam intersected by V braces shall be capable of supporting all tributary dead and live loads assuming the bracing is not present.

4. The top and bottom flanges of the beam at the point of intersection of V braces shall be designed to support a lateral force equal to 1.5 percent of the nominal beam flange strength ($F_y b_f t_f$).

9.4.b. K bracing, where permitted:

1. The design strength of K brace members shall be at least 1.5 times the required strength using Load Combinations 3-5 and 3-6.

2. A column intersected by K braces shall be continuous between beams.

3. A column intersected by K braces shall be capable of supporting all dead and live loads assuming the bracing is not present.

4. Both flanges of the column at the point of intersection of K braces shall be designed to support a lateral force equal to 1.5 percent of the nominal column flange strength ($F_y b_f t_f$).

9.5. Low Buildings

Braced frames not meeting the requirements of Sect. 9.2 through 9.4 shall only be used in buildings not over two stories and in roof structures if Load Combinations 3-7 and 3-8 are used for determining the required strength of the members and connections.

10. REQUIREMENTS FOR ECCENTRICALLY BRACED FRAMES (EBF)

10.1. Scope

Eccentrically braced frames shall be designed so that under inelastic earthquake deformations, yielding will occur in the links. The diagonal braces, the columns, and the beam segments outside of the links shall be designed to remain elastic under the maximum forces that will be generated by the fully yielded and strain hardened links, except where permitted by this section.

10.2. Links

10.2.a. Beams with links shall comply with the width-thickness ratios in Table 8-1.

10.2.b. The specified minimum yield stress of steel used for links shall not exceed $F_y = 50$ ksi.

10.2.c. The web of a link shall be single thickness without doubler plate reinforcement and without openings.

10.2.d. Except as limited by Sect. 10.2.f., the required shear strength of the link, V_u, shall not exceed the design shear strength of the link, ϕV_n, where:

ϕV_n = Link design shear strength of the link = the lesser of ϕV_p or $2\phi M_p / e$, kips.
$V_p = 0.6 F_y (d - 2t_f) t_w$, kips.
$\phi = 0.9$.
e = link length, in.

10.2.e. If the required axial strength, P_u, in a link is equal to or less than $0.15P_y$, where $P_y = A_g F_y$, the effect of axial force on the link design shear strength need not be considered.

10.2.f. If the required axial strength, P_u, in a link exceeds $0.15P_y$, the following additional limitations shall be required:

1. The link design shear strength shall be the lesser of ϕV_{pa} or $2\phi M_{pa} / e$, where:

$$V_{pa} = V_p \sqrt{1 - (P_u / P_y)^2}$$

$$M_{pa} = 1.18M_p[1 - (P_u / P_y)]$$
$$\phi = 0.9$$

2. The length of the link shall not exceed:

$$[1.15 - 0.5\rho(A_w / A_g)]1.6M_p / V_p \text{ for } \rho(A_w / A_g) \geq 0.3 \text{ and}$$

$$1.6M_p / V_p \text{ for } \rho(A_w / A_g) < 0.3, \text{ where:}$$

$$A_w = (d - 2t_f) \, t_w$$
$$\rho = P_u / V_u$$

10.2.g. The link rotation angle is the plastic angle between the link and the beam outside of the link when the total story drift is $0.4R$ times the drift determined using the specified base shear V. The term $0.4R$ shall be equal to or greater than 1.0. Except as noted in Sect. 10.4.d, the link rotation angle shall not exceed the following values:

1. 0.09 radians for links of length $1.6M_p / V_p$ or less.

2. 0.03 radians for links of length $2.6M_p / V_p$ or greater.

3. Linear interpolation shall be used for links of length between $1.6M_p / V_p$ and $2.6M_p / V_p$.

10.2.h. Alternatively, the top story of an EBF building having over five stories shall be a CBF.

10.3. Link Stiffeners

10.3.a. Full depth web stiffeners shall be provided on both sides of the link web at the diagonal brace ends of the link. These stiffeners shall have a combined width not less than $(b_f - 2t_w)$ and a thickness not less than $0.75t_w$ or ⅜-in., whichever is larger, where b_f and t_w are the link flange width and link web thickness, respectively.

10.3.b. Links shall be provided with intermediate web stiffeners as follows:

1. Links of lengths $1.6M_p / V_p$ or less shall be provided with intermediate web stiffeners spaced at intervals not exceeding $(30t_w - d / 5)$ for a link rotation angle of 0.09 radians or $(52t_w - d / 5)$ for link rotation angles of 0.03 radians or less. Linear interpolation shall be used for values between 0.03 and 0.09 radians.

2. Links of length greater than $2.6M_p / V_p$ and less than $5M_p / V_p$ shall be provided with intermediate web stiffeners placed at a distance of $1.5b_f$ from each end of the link.

3. Links of length between $1.6M_p / V_p$ and $2.6M_p / V_p$ shall be provided with intermediate web stiffeners meeting the requirements of 1 and 2 above.

4. No intermediate web stiffeners are required in links of lengths greater than $5M_p / V_p$.

5. Intermediate link web stiffeners shall be full depth. For links less than 25 inches in depth, stiffeners are required on only one side of the link web. The thickness of one-sided stiffeners shall not be less than t_w or

⅜-in., whichever is larger, and the width shall be not less than $(b_f / 2) - t_w$. For links 25 inches in depth or greater, similar intermediate stiffeners are required on both sides of the web.

10.3.c. Fillet welds connecting link stiffener to the link web shall have a design strength adequate to resist a force of $A_{st}F_y$, in which A_{st} = area of the stiffener. The design strength of fillet welds fastening the stiffener to the flanges shall be adequate to resist a force of $A_{st}F_y / 4$.

10.4. Link-to-Column Connections

Where a link is connected to a column, the following additional requirements shall be met:

10.4.a. The length of links connected to columns shall not exceed $1.6M_p / V_p$ unless it is demonstrated that the link-to-column connection is adequate to develop the required inelastic rotation of the link.

10.4.b. The link flanges shall have complete penetration welded joints to the column. The joint of the link web to the column shall be welded. The required strength of the welded joint shall be at least the nominal axial, shear, and flexural strengths of the link web.

10.4.c. The need for continuity plates shall be determined according to the requirements of Sect. 8.5.

10.4.d. Where the link is connected to the column web, the link flanges shall have complete penetration welded joints to plates and the web joint shall be welded. The required strength of the link web shall be at least the nominal axial, shear, and flexural strength of the link web. The link rotation angle shall not exceed 0.015 radians for any link length.

10.5. Lateral Support of Link

Lateral supports shall be provided at both the top and bottom flanges of link at the ends of the link. End lateral supports of links shall have a design strength of 6 percent of the link flange nominal strength computed as $F_y b_f t_f$.

10.6. Diagonal Brace and Beam Outside of Link

10.6.a. The required combined axial and moment strength of the diagonal brace shall be the axial forces and moments generated by 1.25 times the nominal shear strength of the link as defined in Sect. 10.2. The design strengths of the diagonal brace, as determined by Sect. H (including Appendix H) of the *Specification*, shall exceed the required strengths as defined above.

10.6.b. The required strength of the beam outside of the link shall be the forces generated by at least 1.25 times the nominal shear strength of the link and shall be provided with lateral support to maintain the stability of the beam. Lateral supports shall be provided at both top and bottom flanges of the beam and each shall have a design strength to resist at least 1.5 percent of the beam flange nominal strength computed as $F_y b_f t_f$.

10.6.c. At the connection between the diagonal brace and the beam at the link end of the brace, the intersection of the brace and beam centerlines shall be at the end of the link or in the link. The beam shall not be spliced within or adjacent to the connection between the beam and the brace.

10.6.d. The required strength of the diagonal brace-to-beam connection at the link end of the brace shall be at least the nominal strength of the brace. No part of this connection shall extend over the link length. If the brace resists a portion of the link end moment, the connection shall be designed as Type FR (Fully Restrained).

10.6.e. The width-thickness ratio of brace shall satisfy λ_p of Table B5.1 of the *Specification*.

10.7. Beam-to-Column Connections

Beam-to-column connections away from links are permitted to be designed as a pin in the plane of the web. The connection shall have a design strength to resist torsion about the longitudinal axis of the beam based on two equal and opposite forces of at least 1.5 percent of the beam flange nominal strength computed as $F_y b_f t_f$ acting laterally on the beam flanges.

10.8. Required Column Strength

The required strength of columns shall be determined by Load Combinations 3-5 and 3-6 except that the moments and axial loads introduced into the column at the connection of a link or brace shall not be less than those generated by 1.25 times the nominal strength of the link.

11.　QUALITY ASSURANCE

The general requirements and responsibilities for performance of a quality assurance plan shall be in accordance with the requirements of the regulatory agency and specifications by the design engineer.

The special inspections and special tests needed to establish that the construction is in conformance with these provisions shall be included in a quality assurance plan.

The minimum special inspection and testing contained in the quality assurance plan beyond that required by the *Specification* shall be as follows:

Groove welded joints subjected to net tensile forces which are part of the seismic force resisting systems of Sects. 8, 9, and 10 shall be tested 100 percent either by ultrasonic testing or by other approved equivalent methods conforming to AWS D1.1.

Exception: The nondestructive testing rate for an individual welder shall be reduced to 25 percent with the concurrence of the person responsible for structural design, provided the reject rate is demonstrated to be 5 percent or less of the welds tested for the welder.

Part II—Allowable Stress Design (ASD) Alternative

As an alternative to the LRFD seismic design procedures for structural steel design given in PART I, the design procedures in the *Specification for Structural Steel Buildings—Allowable Stress Design and Plastic Design,* AISC 1989 are permitted as modified by PART II of these provisions. When using ASD, the provisions of PART I of these seismic provisions shall apply except the following sections shall be substituted for, or added to, the appropriate sections as indicated:

1. SCOPE

Revise the first paragraph of PART I, Sect. 1 to read as follows:

These special requirements are to be applied in conjunction with the AISC *Specification for Structural Steel Buildings—Allowable Stress Design and Plastic Design* hereinafter referred to as *Specification.* They are intended for the design and construction of structural steel members and connections in buildings for which the design forces resulting from earthquake motions have been determined on the basis of energy dissipation in the nonlinear range of response.

3. LOADS, LOAD COMBINATIONS AND NOMINAL STRENGTHS

Substitute the following for Section 3.2 in PART I:

3.2. Nominal Strengths

The nominal strengths of members shall be determined as follows:

3.2.a. Replace Sect. A5.2 of the *Specification* to read: "The nominal strength of structural steel members for resisting seismic forces acting alone or in combination with dead and live loads shall be determined by multiplying 1.7 times the allowable stresses in Sect. D, E, F, G, J, and K."

3.2.b. Amend the first paragraph of Sect. N1 of the *Specification* by deleting "or earthquake" and adding: "The nominal strength of members shall be determined by the requirements contained herein. Except as modified by these rules, all pertinent provisions of Chapters A through M shall govern."

3.2.c. In Sect H1 of the *Specification* the definition of F_e' shall read as follows:

$$F_e' = \frac{\pi^2 E}{(Kl_b / r_b)^2}$$

where:

l_b = the actual length in the plane of bending.
r_b = the corresponding radius of gyration.
K = the effective length factor in the plane of bending.

Add the following section to PART I:

3.3. Design Strengths

3.3.a. The design strengths of structural steel members and connections subjected to seismic forces in combination with other prescribed loads shall be determined by converting allowable stresses into nominal strengths and multiplying such nominal strengths by the resistance factors herein.

3.3.b. Resistance factors, ϕ, for use in Part II shall be as follows:

Flexure	$\phi_b = 0.90$
Compression and axially loaded composite members	$\phi_c = 0.85$

Eyebars and pin connected members:
Shear of the effective area	$\phi_{sf} = 0.75$
Tension on net effective area	$\phi_t = 0.75$
Bearing on the project area of pin	$\phi_t = 1.0$

Tension members:
Yielding on gross section	$\phi_t = 0.90$
Fracture in the net section	$\phi_t = 0.75$

Shear	$\phi_v = 0.90$

Connections:
Base plates that develop the strength of the members or structural systems	$\phi = 0.90$
Welded connections that do not develop the strength of the member or structural system, including connection of base plates and anchor bolts	$\phi = 0.67$
Partial Penetration welds in columns when subjected to tension stresses	$\phi = 0.80$

High strength bolts (A325 and A490) and rivets:
Tensile strength	$\phi = 0.75$
Shear strength in bearing-type joints	$\phi = 0.65$
Slip-critical joints	$\phi = 1.0$

A307 bolts:
Tensile strength	$\phi = 0.75$
Shear strength in bearing-type joints	$\phi = 0.60$

Substitute the following for Section 7 in PART I in its entirety:

7.   REQUIREMENTS FOR ORDINARY MOMENT FRAMES (OMF)

7.1. Scope

Ordinary Moment Frames (OMF) shall have a design strength as provided in the *Specification* to resist the Load Combinations 3-5 and 3-6 as modified by the following added provisions:

7.2. Joint Requirements

All beam-to-column and column to beam connections in OMF which resist seismic forces shall meet one of the following requirements:

7.2.a. Type 1 connections conforming with Sect. 8.2, except that the required flexural strength, M_u, of a column-to-beam joint are not required to exceed that required to develop the nominal plastic flexural strength of the connection.

7.2.b. Type 1 connections capable of inelastic deformation and the design strengths of the connections meeting the requirements of Sect. 7.1 using the Load Combinations 3-7 and 3-8.

7.2.c. Either Type 1 or Type 3 connections are permitted provided:

1. The design strengths of the members and connections meet the requirements of Sect. 7.1.

2. The connections have been demonstrated by cyclic tests to have adequate rotation capacity at a story drift calculated at a horizontal load of $0.4R \times E$ (where the term $0.4R$ is equal to or greater than 1.0).

3. The additional drift due to Type 3 connections shall be considered in design.

Type 1 and Type 3 connections are described in detail in Sect. A2 of the *Specification*.

Substitute the following in Sections 10.6.a and 10.6.d in PART I:

10.6.a. Delete reference to Appendix H.

10.6.d. The last sentence shall read: "If the brace resists a portion of the link end moment as described above, the connection shall be designed as a Type 1 connection."

Commentary on the Seismic Provisions for Structural Steel Buildings

June 15, 1992

Part I—LRFD Provisions

1. SCOPE

Load and Resistance Factor Design (LRFD) is an improved approach to the design of structural steel for buildings. The method involves explicit consideration of limit states, multiple load and resistance factors, and implicit probabilistic determination of reliability. The designation LRFD reflects the concept of factoring both loads and resistance. The LRFD method was devised to offer the designer greater flexibility, more rationality and possible overall economy.

The First Edition of the LRFD *Specification* was published and distributed in 1986.[1] It did not contain the special requirements necessary in the design and construction of steel buildings which are required to respond to high earthquake input by deformations into the nonlinear range. The seismic design forces specified in the building codes have been set with consideration given to the energy dissipation generated during the non-linear response.

The provisions contained in this document are to be used in conjunction with the AISC LRFD *Specification* in the design of buildings in the areas of moderate and high seismicity.

The load provisions have been modified from those contained in the *Specification* to be consistent with the load provisions contained in the soon to be published BOCA and SBCCI building codes and the ASCE 7-93, Minimum Design Loads for Buildings and Other Structures.[2] All these new seismic load provisions are modeled on the the 1991 NEHRP[3] earthquake provisions.

2. SEISMIC PERFORMANCE CATEGORIES

Buildings are classified into three types depending on the occupancy and use of each as related to the special hazards resulting from earthquake environment.

The Seismic Hazard Exposure Groups listed in Table 2-1 are defined in detail, with examples of buildings in each type, in ASCE 7-93. The Seismic Performance Category to be used in the design of a specific building is defined by the seismic coefficient representing the peak velocity-related acceleration of the building site, A_v, and the Seismic Hazard Exposure Group related to the occupancy and use of the building. The five categories, A through E, given in Table 2-2 specify design and detail requirements that would be required for the seismic design of the building. These categories establish the level of requirements to be used in items such as detailing limitations, quality assurance, method of analyses, orthogonal effects, and change of building use. The general requirements for each of the categories are given in ASCE 7-93. The differences related specifically to structural steel design are repeated in this *Specification*.

3. LOADS AND LOAD COMBINATIONS

The most frequently used load factors and load combinations given in Sect. A4.1 of the *Specification* are repeated in this Section to reduce the amount of cross-referencing to other documents. They have been modified to be consistent with the anticipated ASCE 7-93. The most notable modification is the reduction of the load factor on E to 1.0. This results from the limit states load model used in ASCE 7-93. For design of structures subjected to impact loads, see the *Specification*.

The earthquake load and load effects E in ASCE 7-93 are composed of two parts. E is the sum of the seismic horizontal load effects and one half of A_v times the dead load effects. The second part adds an effect simulating vertical accelerations concurrent to the usual horizontal earthquake effects.

The load factors and load combinations reflect the fact that when several loads act in combination with the dead load, e.g., dead plus live plus earthquake loads, only one of these takes on its maximum lifetime value, while the other load is at its "arbitrary point-in-time value," at a value which can be expected to be on the structure at any time. The most critical effect may occur when one or more load types are not acting.

The basic requirements for dual systems are given in the Glossary to clarify the use of the EBF in a dual system and to indicate that steel moment frames can also be used as part of a dual system with concrete shear walls.

An amplification factor to earthquake load E of $0.4R$ is prescribed for limited use in this set of provisions. It is used as an amplification of the deflections determined using the earthquake forces specified in ASCE 7-93. It was derived by assuming that deflections due to large earthquake response would be the same regardless of the reductions in applied forces due to the inelastic response of the type of lateral force resisting system.[56] The amount of this amplification was assumed to be two times the deflections generated by forces specified for a buildings with $R = 5$. This amplification factor is thus $2R / 5$ or $0.4R$. However, with $R = 2.5$ or less it is felt that the amplification factor should not be less than 1.0. The load combinations to be used with the amplification factor are given by formulas 3-7 and 3-8. Specific values of R are not needed for determination of the amplified load because R is cancelled out when substituted in the formula for the horizontal seismic base shear, V. The added complication that would be required to consider orthogonal effects with the amplified force is not deemed to be necessary.

The ASCE 7-93 provisions are detailed earthquake load provisions in which two methods of analysis are provided. The first is frequently referred to as the "Static Force Procedure" or "Equivalent Lateral Force Procedure." The second method is the "Modal Analysis Procedure." In both methods a linearly elastic model is assumed. Other "Dynamic Analysis Procedures" are permitted both with linearly elastic or non-linear models as long as the internal forces and deformations in the members are determined using a model consistent with the procedure adopted. Guidelines for use of these other methods of analyses are provided in the Commentary to ASCE 7-93.

These earthquake provisions refer to the load provisions of ASCE 7-93. By changing the load combination portion of Section 3, these provisions can be made compatible with other sets of load provisions.[4] For instance, the following changes can be made to the provisions in Section 3 to make them compatible with the following document: (S_n is used in this Commentary for snow loads to distinguish them from site effects that use the symbol S).

1991 UNIFORM BUILDING CODE:[5]
(SEAOC seismic provisions are similar)[6]

The required strength on the structure and its elements must be determined from the appropriate critical combination of factored loads. The most critical effect may occur when one or more loads are not acting. The following load combinations and corresponding load factors shall be investigated:

$$1.4D \tag{3-1}$$

$$1.2D + 1.6L + 0.5(Lr \text{ or } S \text{ or } R') \tag{3-2}$$

$$1.2D + 1.6(Lr \text{ or } S \text{ or } R') + (0.5L \text{ or } 0.8W) \tag{3-3}$$

$$1.2D + 1.3W + 0.5L + 0.5(Lr \text{ or } S \text{ or } R') \tag{3-4}$$

$$1.2D + 1.5E + 0.5L + 0.2S \tag{3-5}$$

$$0.9D - (1.3W \text{ or } 1.5E) \tag{3-6}$$

Exception: The load factor on L in Load Combinations 3-3, 3-4, and 3-5 shall equal 1.0 for garages, areas occupied as places of public assembly, and all areas where the live load is greater than 100 psf.

Other special load combinations are included with specific design requirements throughout these provisions.

Where required by these provisions, an amplification factor is applied on the earthquake load $E = \frac{3}{8}R_w$, where R_w is a response factor similar to the factor R except used to reduce the earthquake load to a working stress design level. Earthquake loads are similar to those found in ASCE 7-93 except for the R_w factor.

Earthquake loads are defined in detail in Section 2334 of 1991 UBC. The revised load combinations are:

$$1.2D + (3R_w / 8)E + 0.5L + 0.2S \qquad (3\text{-}7)$$

$$0.9D - (3R_w / 8)E \qquad (3\text{-}8)$$

Exception: The load factor on L in Load Combination 3-7 shall equal 1.0 for garages, areas occupied as places of public assembly, and all areas where the live load is greater than 100 psf.

The amplification factor was derived by using the similar assumptions that were used in deriving the factor for ASCE 7-93. The same type of building with $R = 5$ in ASCE 7-93 has a Structure System Coefficient $R_w = 8$ in 1991 UBC. The deflection determined by this R_w was used as the value to be amplified by 3. Thus $(3R_w / 8)E$.

Where the use of the amplification factor to load E is required, orthogonal effects need not be included.

The 1991 UBC outlines in detail many of the requirements for "Dynamic Lateral Force Procedure." The following is a summary of these requirements:

The ground motion may be defined in one of five ways:

1. The plot of a normalized response spectra may be used.

2. A site specific response spectra based on geologic, tectonic seismologic, and soil characteristics of the site may be used. As per SEAOC the damping ratio shall be 5 percent unless another value is shown to be consistent with the structural behavior of the building.

3. Site specific time histories are to be representative of actual earthquake motions. Spectra developed from these time histories would follow item (2) above.

4. Site specific response spectra and time histories developed for sites with a profile having more than 40 feet of soft clay per SEAOC shall be based on ground motion having a 10 percent probability of exceedance in 50 years; the effects of lengthening of the structural period on response amplification due to soil-structure resonance shall be included; and the design base shear shall be determined by dividing by a factor not greater than R_w for the structure.

5. A two-thirds factor shall be used on horizontal motions to determine the vertical component of ground motion unless specifically determined otherwise for the site. The mathematical model shall represent the actual structure adequately for the calculation of all the significant features of the dynamic response. Three dimensional models shall be used for highly irregular plan configurations if a rigid or semi-rigid diaphragm is used.

A response spectrum analysis shall be an elastic dynamic analysis of all the significant peak modal responses combined in a statistical manner to obtain an approximate total structural response. When the base shear is less than that determined from the Static Lateral Force Procedure, it shall be increased to 100 percent of the static base shear for irregular structures, shall be taken as 90 percent of the static base shear for regular structures where the fundamental period is determined using the structural characteristics of the building system, and shall be set at 80 percent for regular structures. Accidental torsion shall be accounted for by appropriate adjustments in the model. Where a dual system is used, the combined system shall be accounted for in the modelling; the backup Special Moment Frame (SMF) shall be capable of resisting 25 percent of the base shear used for the design of the total system. The analysis of the backup SMF may either use the Static of Dynamic Lateral Force Procedures.

A time history analysis shall be an elastic or inelastic dynamic analysis of a model of a structure subjected to specified time history of ground motion. The time dependent dynamic response of the structure to these motions is obtained through numerical integration of its equations of motion. These analyses shall be based on established principles of mechanics.

Scaling of base shear determined by a response spectrum analysis results in making the Load Combinations 3-1 through 3-6 as well as 3-7 and 3-8 applicable to this method of analysis in the 1991 UBC. No scaling effect is specified for the results of time history dynamic analysis (either elastic or inelastic). In this case, it is necessary to define the specified time histories which will result in the structure responding to the limit of essentially elastic response. This would be the level to determine the required resistance of the system. In order to determine the deformations corresponding to the specified drift limits, the force level shall be divided by a factor of 1.5.

4. STORY DRIFT

Deflection limits are commonly used in design to assure the serviceability of the structure. These serviceability limit states are variable, since they depend upon the structural usage and contents. The *Specification* does not specify these serviceability limits, since they are regarded as a matter of engineering judgment, rather than general design limits.[54]

Like deflection limits, drift limits for both wind and seismic design are excluded from these Seismic Provisions. Research has shown that seismic drift control provides a function beyond assuring the serviceability of the structure. The added strength and stiffness which drift limits often provide in moment frames improves the performance of structures during earthquakes. Model codes, load standards, and resource documents contain specific seismic drift limits but there are major differences among them. There is neither uniform agreement regarding appropriate code specified drift limits nor how they should be applied. Further it is difficult to estimate the actual story drift of moment frames with panel zone yielding. Nevertheless, drift control is important to serviceability and stability of the structure. It is recommended the designer review drift limits in the appropriate code and use those applicable for the serviceability and stability of the structure under consideration.

The story drift limitations of ASCE 7-93 are applied to an amplified story drift that estimates the story drift that would occur during a large earthquake. The story drift is defined as the difference of deflection between the top and bottom of the story under consideration. For determining the story drift the deflection determined using the earthquake forces E is amplified by a deflection amplification factor, C_d, which is dependent on the type of building system. The story drifts when determined by an elastic analysis, including the P-Δ effect when

TABLE C-4.1
Tentative Allowable Story Drift

Building	Seismic Hazard Exposure Group		
	I	II	III
Single story buildings without equipment attached to the structural resisting system and with interior walls, partitions, ceilings, and exterior wall system that have been designed to accommodate the story drifts.	No limit	$0.020h_{sx}$	$0.015h_{sx}$
Buildings with 4 stories or less with interior walls, partitions ceilings, and exterior wall system that have been designed to accommodate the story drifts.	$0.025h_{sx}$	$0.020h_{sx}$	$0.015h_{sx}$
All other buildings.	$0.020h_{sx}$	$0.015h_{sx}$	$0.010h_{sx}$

Where h_{sx} is the story height of the story drift calculated.

applicable, have limits depending on the Seismic Hazard Exposure Group of the building as shown in Table C-4.1.

In calculating the elastic drift, the forces may be based on the fundamental period of the building without the arbitrary limit specified for determining the seismic design forces in the framing members. ASCE 7-93 does not prescribe explicit requirements for building separations. An admonition is included, however, that all portions of the building shall be designed and constructed as an integral unit in resisting seismic forces unless separated structurally by a sufficient distance to avoid damaging contact between components under amplified deformations. The latter are determined by multiplying the elastic deflection by a deflection amplification factor, C_d, which is based on the type and materials of the seismic resisting system. If the effects of hammering between segments can be shown not to be detrimental, separations could be reduced.

1991 UBC Requirements:

In order to comply with the 1991 UBC requirements, the story drift shall be calculated including the translational and torsional deflections resulting from the application of unfactored lateral forces. Story drift is defined as the displacement of one level relative to the level above or below. The calculated story drift shall not exceed $0.04 / R_w$ nor 0.005 times the story height for structures with fundamental periods of less than 0.7 seconds and shall not exceed $0.03 / R_w$ nor 0.004 times the story height for structures with fundamental periods of 0.7 seconds or greater. For the purpose of this limit the fundamental period is the same as that used for determining the base shear.

For calculating the drift, the lateral forces may be calculated using a base shear V defined as:

$$V = \frac{ZIC}{R_w} W, \text{ in which}$$

Z = The seismic zone coefficient.
I = An importance factor.
R_w = A numerical coefficient related to the type of construction.

$$C = \frac{1.25S}{T^{2/3}}, \text{ in which}$$

S = Site coefficient.
T = Fundamental period of vibration, which may be determined using the structural properties and deformational characteristics of the resisting elements of the lateral force resisting framing. Method A of determining T need not be applied for drift determination. The lower bound of 0.075 on the ratio C/R_w may also be neglected.
W = Dead load used to calculate seismic loads.

The story drift limits need not be applied if it is demonstrated that greater drift can be tolerated without affecting life safety by damage to either structural and non-structural elements.

There are no drift limits on single story steel framed structures with low occupancies. This would generally apply to buildings such as warehouses, parking garages, aircraft hangers, factories, workshops and agricultural buildings. These buildings are not allowed to have brittle finishes and are not allowed to have equipment attached to the structural frame unless the finish or equipment attachment is detailed to accommodate the additional drift.

5.   MATERIAL SPECIFICATIONS

The list of structural steels for use in designing to earthquake motion has been chosen with consideration given to the inelastic properties of the steels and their weldability. In general, the steels selected possess the following characteristics:

- Ratio of tensile strength to yield strength between 1.2 to 1.8.

- Pronounced stress-strain plateau at yield strength.

- Large inelastic strain capability.

- Tension elongation of 20 percent or greater in a 2-in. gage length.

- Good weldability for inelastic behavior.

Other steels including those with a specified yield point greater than 50 ksi should not be used without demonstrating that equivalent inelastic behavior can be attained.

6.   COLUMN REQUIREMENTS

6.1. Column Strength

During the maximum probable earthquake expected at any site, axial forces calculated using the specified design earthquake may be exceeded. This is a result of the reduction in lateral force for use in analysis of an elastic model of the structure, the underestimation of the overturning forces in this analysis, and the concurrent vertical accelerations which are not explicitly specified as a required design load. The amplifications required in this section provide an approximation of these actions by providing a limit to the required axial force. The two special Load Combinations 6-1 and 6-2 account for these effects; one as a minimum required column compressive required strength and the other on the minimum required tensile strength. They are to be applied without consideration of any concurrent flexure on the members.

The exceptions provided for these limits are self limiting conditions stating that the required axial strengths need not exceed the limits based on the design strength of the overall system to transfer axial loads to the column. For instance, if pile foundations are used, the design strength of the piles in tension may be much larger than the required strength because the size of the foundation may depend on the required strength in compression.

6.2. Column Splices

Column splices are required to have design strengths adequate to join column elements together not only to resist the axial, flexural, and shear forces required at the splice location by the usual load combinations 3-1 through 3-6 but also the forces specified in 6.1.

Butt weld splices in columns where it is anticipated that potential dynamic loading consists only of wind or earthquake forces are not required by these specifications to provide the transition of thicknesses given in Section 9.20 of AWS D1.1.[7] If other types of frequent, high cycle dynamic loadings are also present, the transition requirements should be met.

Partial penetration welds in thick members, such as occur in column flange splices, are very brittle under tensile loading, showing virtually no ductility.[8-9] Recognizing this behavior in seismic design, the location of column splices is moved away from the beam-to-column connection to reduce bending and a 50 percent increase is stipulated in required strength of the splice.

The possibility for developing tensile stresses in such welds during a maximum probable seismic event should be considered. If there is probability of such a condition developing, the use of splice plates welded to the lower part of the column and bolted to the upper part is suggested. If for the noted adverse condition, the suggested detail is not practical, the possibility of fracture in partial penetration welded joints should be recognized, and some restraint from uncontrolled relative movement at the splice be provided. This can be achieved, for example, by having wide splice plates on both sides of the column web to maintain alignment. Shake table experiments have shown that if some columns, unattached at the base, reseat themselves after lifting, the performance of a steel frame remains tolerable.[10]

These provisions apply for common frame configurations. The designer should review the conditions found in columns in tall stories, large changes in column sizes at the splice, or where the possibility of a single curvature exists on a column over multiple stories to determine if special design strength or special detailing is necessary at the splice.

7. REQUIREMENTS FOR ORDINARY MOMENT FRAMES (OMF)

7.1. Scope

Ordinary moment frames of structural steel are moment frames which do not meet the requirements for special design and detailing contained in Section 8. OMF of structural steel do exist and are being built in all areas of seismic activity. Experience has shown that in most instances the buildings of this type have responded without significant structural damage. In recent years advances in analytical procedures have minimized the natural margins of safety normally found in buildings that were designed by approximate methods. Thus it is prudent to require that the design of the beam-to-column connection be adequate to develop the strength of the members framing into the connection as is specified in Sect. 8.2 unless the connection has a design strength significantly larger than the required strengths required by Load Combinations 3-5 and 3-6. Thus unless the connection can develop the full strength of members framing into it, the Load Combinations 3-7 and 3-8 should be used to provide the required strength on the connection.

7.2. Joint Requirements

Although for OMF it is not required to meet most of the special detailing requirements given in Sect. 8, consideration should be given to using as many of the requirements as practical, particularly in those locations where good engineering judgment would suggest that the use of the special detailing requirements would provide improved system and member ductility and stability. The provision requiring a demonstration of rotation capacity is included to permit the use of connections not permitted under the provisions of Sect. 8, such as top and bottom angle joints, in areas where the added drift is acceptable.

8. REQUIREMENTS FOR SPECIAL MOMENT FRAMES (SMF)

8.1. Scope

The requirements in this Section are for those buildings whose lateral force resisting systems are moment frames in the higher seismic zones. The special provisions, when reasonably applied, provide SMF with reliable ductile systems. Non-ductile behavior is inhibited so that nonlinear response to large earthquake motions can occur in components of the frames having a capability of ductile behavior. The concepts are not new but the provisions are supported by tests and analyses.[11–17] SMF systems when properly designed have, in general, resulted in reliable ductile structural systems that respond well to high earthquake motions for both low and high rise buildings.

Inelastic energy absorption through ductile behavior of members of SMF can occur at three places usually adjacent to the beam-to-column connection. Flexural hinges can form in the beams and columns and shear yielding can occur in the area of the panel zone. Within limits and specific restraints, inelastic yielding is permitted in each or in combinations of these three areas. The primary concern when designing the frame for inelastic behavior is to prevent brittle fracture and severe buckling in and adjacent to the zone of inelasticity. The final selection of the appropriate zones of inelasticity is left to the design engineer. Different problems are presented to the design engineer depending on which of the three areas is chosen to have the lowest inelastic threshold.

Yielding in columns is permitted but is considered by many design engineers to be the least desirable. Special limitations are provided for this type of yielding by the provisions in Sect. 8.6. and the bracing required in Sect. 8.7. If the first inelastic mode is chosen to be shear yielding of the panel zone, the limitations of Sect. 8.3 would be required. This usually results in the flexibility of the panel zone being a significant contributor to the total story drift and consideration of this flexibility should be included in analyses.[18] If the designer chooses to avoid inelastic behavior at the above two locations, the yield hinge will form in the beam. This requires the critical design items to be the beam-to-column connection and the beam stability.

8.2.　Beam-to-Column Joints

The special limitations provided for these joints are intended to assure that inelastic hinging that may occur in the connection during the response to high seismic activity will not take place at the joinery but in one of the two adjoining locations, namely in the beam or in the panel zone.[19–24] Some of the more common beam-to-column connections are illustrated in Fig. C-8.1. Beam-to-column connections are not only designed to meet the loads prescribed by the Load Combinations 3-1 through 3-6 but also designed to resist the requirements based on the nominal strengths of the members actually used in the framing system. Frequently the frame member sizes may be sized to limit drift or to meet requirements of load combinations other than those containing seismic loads. Thus to provide frames having the capability of deforming into the nonlinear range without having a connection failure, the required strength on the connection is most frequently based on the design strength of the members actually used.

An exception is provided for joints that are not designed to contribute to the lateral force resisting system. In order to demonstrate that the joint will be capable of undergoing large deformation, the elastic or inelastic joint rotations that would be induced by deforming the frame into an amplified displacement of $0.4R$ times that under Load Combinations 3-5 and 3-6 are required. The term $0.4R$ should not be less than 1.0. If the "non-moment resisting" web connection were to be a shear tab joined to the column flange by welding and bolted to the beam web, the connection should be proportioned to either yield in the tab or by use of horizontally slotted holes for the bolts. Fracture should not occur in the welded joint to the column. (See Fig. C-8.2.)

The required shear strength, V_u, of the beam-to-column joint is defined as the summation of the factored gravity loads and the shear resulting from the required moments on the two ends of the beam. The easy method is to assume that M_p occurs at each end of the beam. However, when Load Combination 3-7 is used in which one end only of the beam reaches M_p, or the panel zone nominal shear reaches V_n as defined in Sect. 8.2.a, the shear resulting from hinging at both ends of the beam need not be used.

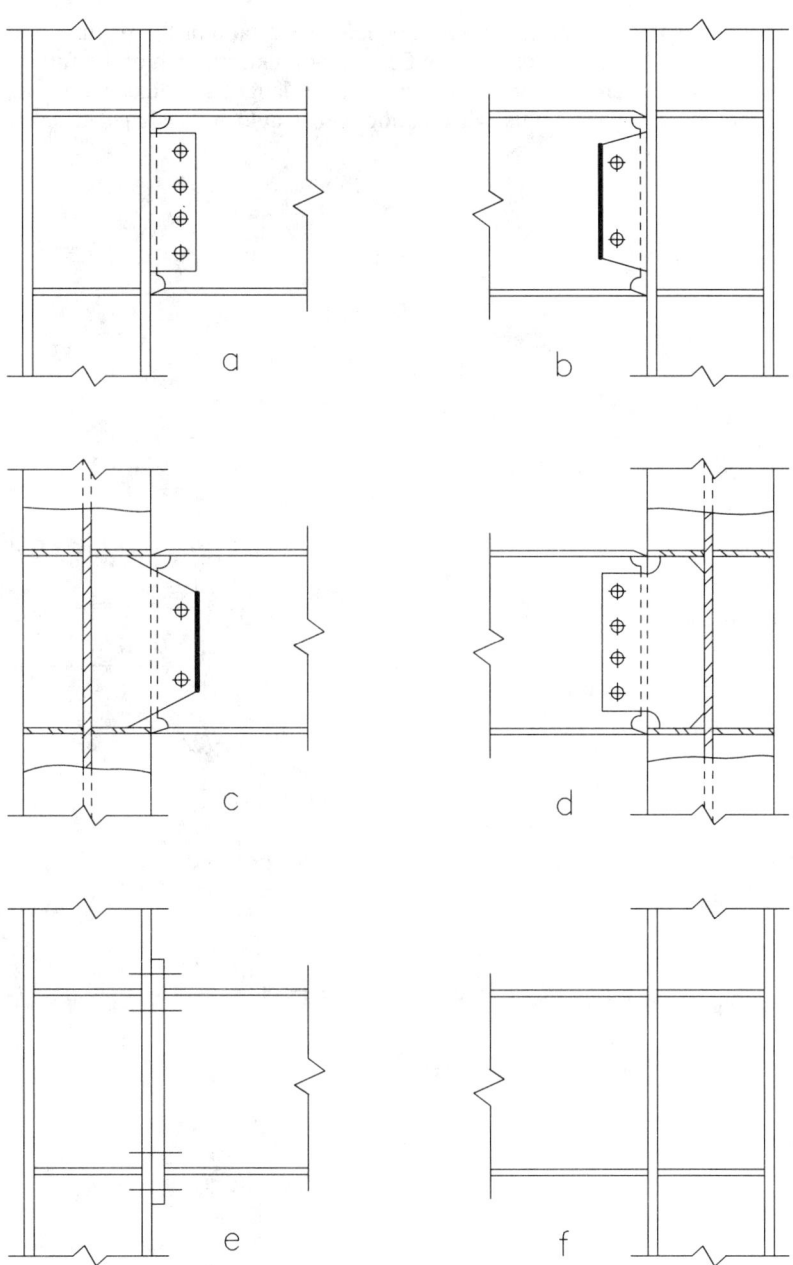

Fig. C-8.1. Beam-to-column connections.

When the required flexural strength of the joint is M_p of the beam, the type of joint is prescribed to be one of three types:

First is the joint where both flanges and web are fully welded to develop their portions of the moment and shear strength of the beam. (See Fig. C-8.3.)

Second is the joint of those beams which have a ratio of the flexural nominal strength of the flanges only to the flexural nominal strength of the full section of at least 70 percent. For this connection, the flanges are joined with complete penetration welded joints whereas the web would be designed to carry the

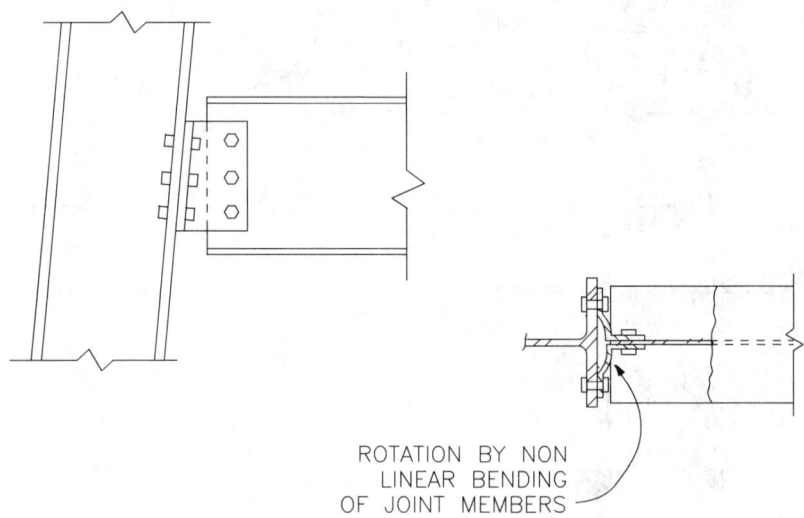

ROTATION BY NON
LINEAR BENDING
OF JOINT MEMBERS

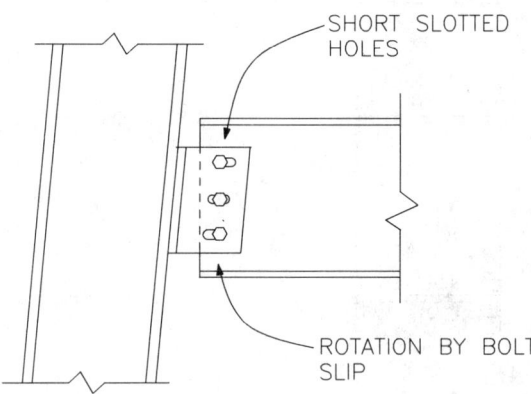

SHORT SLOTTED
HOLES

ROTATION BY BOLT
SLIP

Fig. C-8.2. Simple connections.

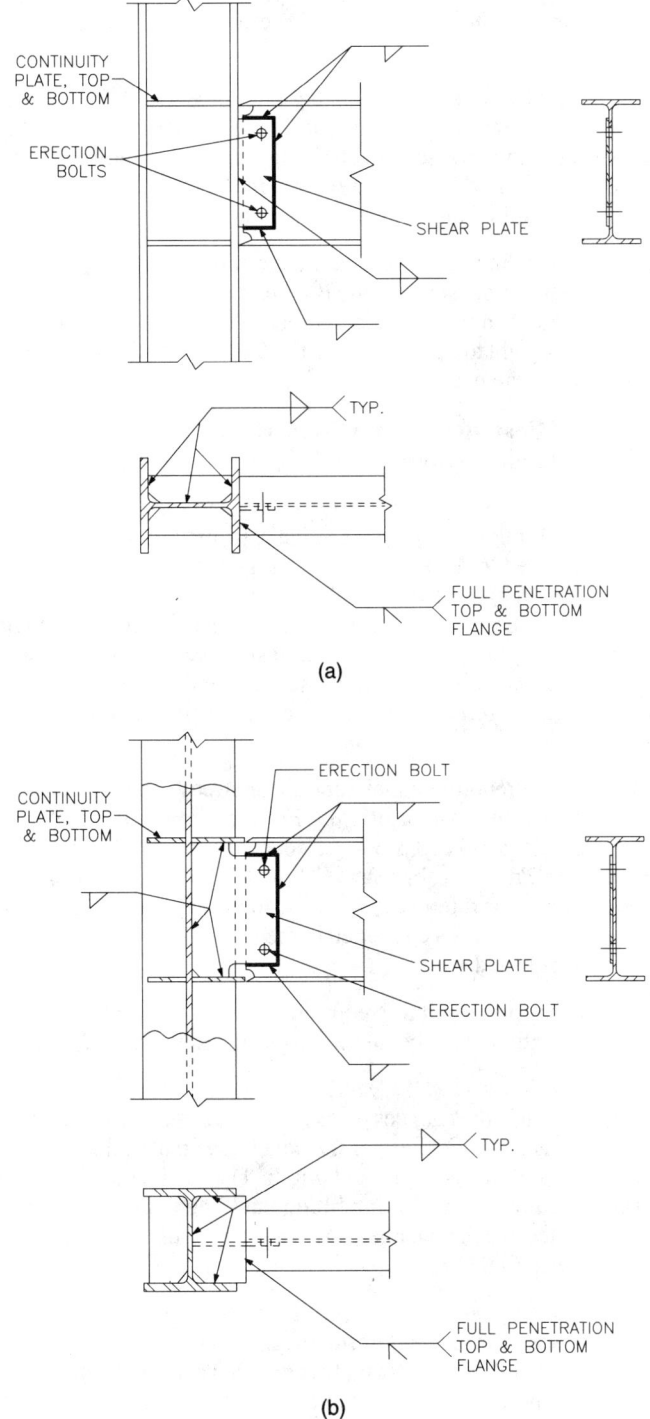

CONTINUITY PLATE, TOP & BOTTOM

ERECTION BOLTS

SHEAR PLATE

TYP.

FULL PENETRATION TOP & BOTTOM FLANGE

(a)

ERECTION BOLT

CONTINUITY PLATE, TOP & BOTTOM

SHEAR PLATE

ERECTION BOLT

TYP.

FULL PENETRATION TOP & BOTTOM FLANGE

(b)

Fig. C-8.3. Beam-column joint.

required shear by either welds or by slip-critical high strength bolts. (See Fig. C-8.4.)

Third is the joint of beams not meeting the 70 percent criteria. This would be similar to the second joint except that the beam web is required to be welded directly or through shear tabs even though the web is bolted to the shear tab. The welds are required to have a nominal moment strength at least equal to 20 percent of the nominal moment strength of the full beam web. (See Fig. C-8.5.)

Other joints than the ones specified are permitted to be used but the adequacy of the joint requires substantiation either by tests or by calculations. Where the adequacy is demonstrated by calculations, additional conservatism is provided by requiring the joint to develop at least 125 percent of the nominal moment and shear strength of the beam.

8.3. Panel Zone of Beam-to-Column Connection (Beam web parallel to column web)

During recent years many cyclic tests have shown the ductility of shear yielding in panel zones through many cycles of inelastic distortions.[17,25-28] Thus the panel zone does not need to develop beam hinging and a method of determining the nominal shear strength of the panel zone is needed. The usual assumption of Von Mises shear limit of $0.577F_y dt$ did not predict the actual behavior of many of the tests. Many panel zone and beam tests have shown that strain hardening and other phenomena have enabled shear strengths in excess of $1.0F_y dt$ to be developed. Eq. 8-1 reflects the significant strength provided by thick column flanges.

In calculating the required panel zone shear strength the UBC 1991 magnifies the specified load by a factor of 1.85. For the LRFD specification, the typical Load Combinations 3-5 and 3-6 are used and the nominal web shear strength is defined as $0.6F_y dt$, rather than $0.55F_y dt$ which had been used in plastic design and in some previous references. In order to provide the same level of safety as determined by tests and as contained in the UBC 1991, a lower resistance factor $\phi_v = 0.75$ was selected.

An upper limit is placed on the required shear strength of the panel zone of 0.9 times the summation of the beam design plastic moments $\phi_b M_p$ framing into the connection.

In order to minimize the chances of shear buckling during inelastic deformations of the panel zone, the thickness of the panel zone material is limited to not less than $\frac{1}{90}$ of the sum of its depth and width. The thickness of any doubler plate used is assumed ineffective in inhibiting buckling unless it is connected to the panel zone plate in such a manner, such as plug welds, to prevent local buckling of the plate. (See Fig. C-8.6.)

Whenever doubler plates are used (i.e., increased strength, compliance with Eq. 8-2, or to reduce panel zone deformations), the plates are required to be close to the column web. The doubler plates are to have at least minimum fillet welds across the top and bottom and to have either butt or fillet welds to the column flanges. These details are provided to closely simulate the joints that have been found to perform satisfactorily in the cyclic tests that have been performed. Fillet

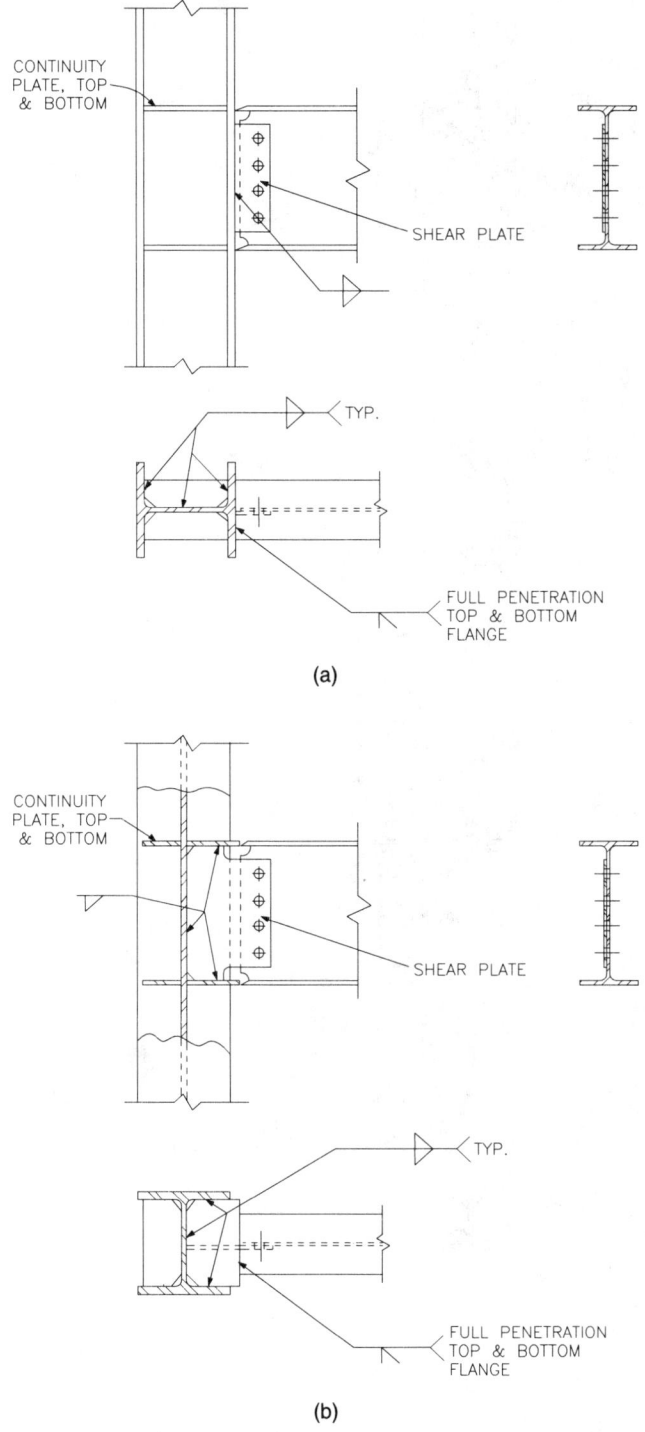

Fig. C-8.4. Beam-column joint, $b_f t_f (d_b - t_f)F_y \geq 0.7F_yZ_x$.

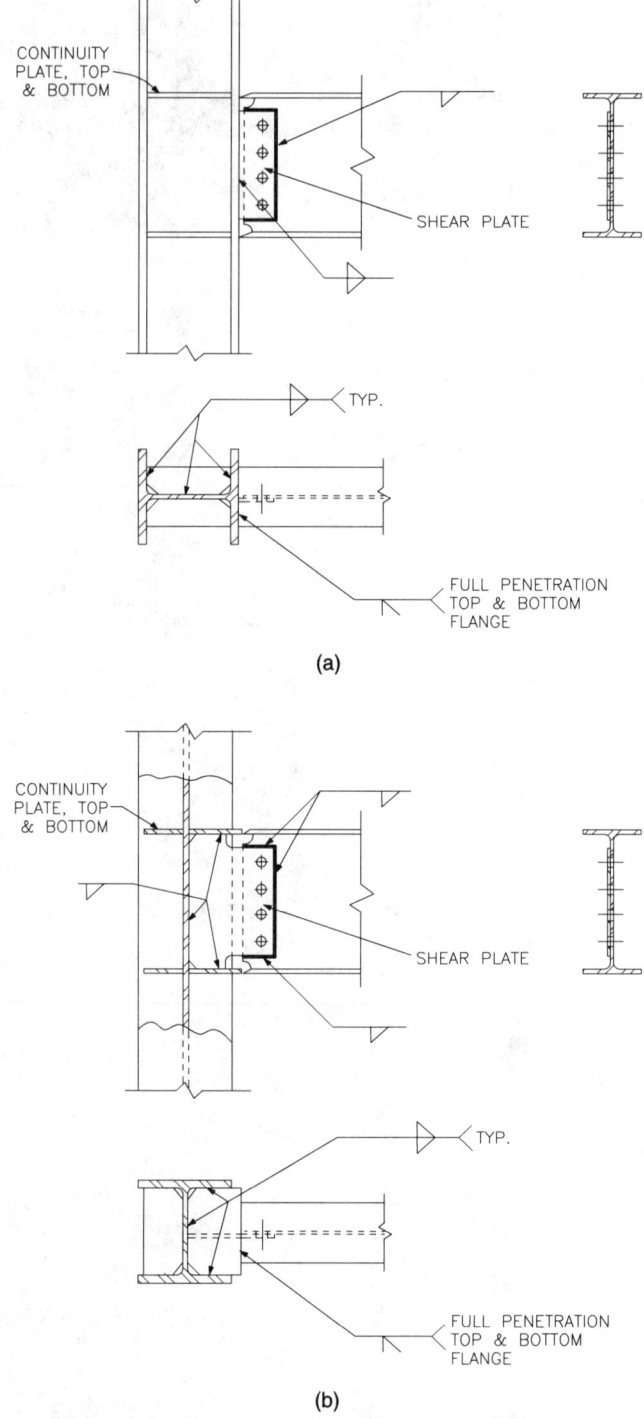

Fig. C-8.5. Beam-column joint, $b_f t_f (d_b - t_f)F_y < 0.7F_y Z_x$.

welding is encouraged to assist in minimizing the built-in weld stresses and the cost of welding.

Doubler plates may be designed to extend between continuity plates which are welded directly to the column web or they may extend above and below the continuity plates which are welded to the doubler plate. For the latter case, the horizontal welds at the top and bottom of the doubler plate should be sized to transfer all loads imposed by the design system. In particular, the welds to the column web should be designed to transfer load from the doubler plate to the column web for their portion of load from the continuity plate.

For the fillet or butt welds of the doubler plate to the column flanges, the following items should be considered:

- The vertical shear and bending loads of beams or girders framing perpendicular to the column web and supported by the doubler plate.

- The compression or tension load delivered to the column web and doubler plate by the flanges of the girders framing into the column flanges.

For examples of doubler plate connections, see Ref. 55 and Fig. C-8.6.

The use of diagonal stiffeners for strengthening and stiffening of the panel zone has not been adequately tested for low cycle reversed loading into the inelastic range. Thus no specific recommendations are made at this time for special seismic requirements for this detail.

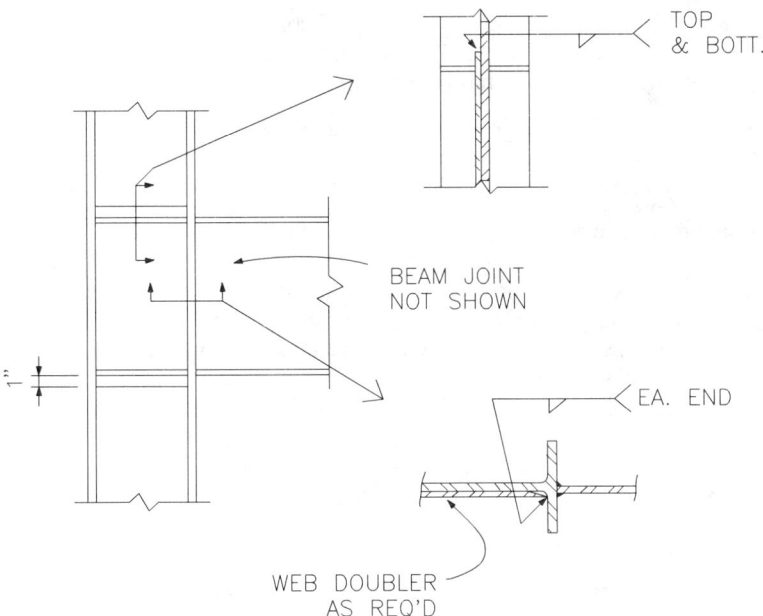

TOP
& BOTT.

BEAM JOINT
NOT SHOWN

EA. END

WEB DOUBLER
AS REQ'D

Fig. C-8.6. Panel zone detail (with doubler).

8.4. Beam Limitations

In order to minimize the cost of connections it has occasionally been suggested that the beam cross-section be reduced immediately adjacent to the column. This type of assemblage can result in a very brittle mode of failure. Detailing that results in a concentration of stress in an area where inelastic deformations are anticipated under large seismic response is discouraged.

The width thickness ratio of projecting elements should be within those which provide the cross-section with stability against local buckling.

The limits given in Table 8-1 are deemed adequate by the Committee for ductilities to 6 or 7 based on the tests performed to date.[29-32] Further testing may result in some modifications of these limits.

8.5. Continuity Plates

Sect. K1 of the *Specification* gives the design requirements for webs and flanges with concentrated forces. Sect. K1.2 gives the design strength in local buckling in a flange under the action of a tensile force. When the design strength is inadequate, column web stiffeners are required. In moment resisting frames, an interior beam-to-column connection has tension on one flange and compression on the opposite side. When stiffeners are required, it is normal to place a full depth stiffener on each side of the column web. As this stiffener provides a load path for the flanges on both sides of the column, it is commonly called a continuity plate. The stiffener not only provides resistance to local flange buckling but also provides a boundary to the very highly stressed panel zone. When it is anticipated that there could be a plastic hinge adjacent to the column, the required force to determine whether a continuity plate is required is not the design earthquake force given by the load combinations 3-1 through 3-6. It is the force exerted by the beam connection when the full plastic moment with possible strain hardening has been formed.

Tests have shown that hinging occurs due to local flange buckling when a compact section is strain hardened to about $1.3F_y$.[20] At the joint, the flanges of the beam can be strain-hardened to a force of $1.8F_y\, b_f\, t_f$.

Using this force as the required strength on the continuity plate is conservative as there is only a small moment strength contributed by the bolted web connection. Since the flange continuity plate is needed to protect the weld at the joint of the beam flange to column flange, consideration should be given to their use in connections where the calculations indicate they may not be required. Continuity plates have been used in almost all cyclic joint tests that have performed well.[17] When tests have been performed on specimens not meeting the requirements of Sect. K1.2, the joints have performed poorly. For the actual design of the continuity plates, Sect. K1.8 of the LRFD specification would apply.

8.6. Column-Beam Moment Ratio

Tests have shown that moment frame subassemblages in which yielding of columns occurred did not exhibit any loss of lateral force resistance at displacements representative of maximum expected earthquake response.[33] Most engineers believe, however, that the performance of seismic moment frames is more predictable if columns outside of the panel zone do not yield. The tests necessary to formulate truly appropriate criteria have not been conducted. In the past, many frames have been designed with the assumption that the first hinging occurs in the columns and until recently no code provisions for this behavior have been enforced. There have not been any documented failures in past earthquakes directly attributable to column hinging. Design situations do occur where elimination of the "strong beam-weak column" connection type would be grossly impractical.

The committee feels that some interim provisions are appropriate. Thus Eqs. 8-3 and 8-4 are introduced. These formulas require that the initial potential for yielding at a beam-to-column connection be in the beam or panel zone rather than in the column.

The exceptions to the "strong column-weak beam" connection type require that the column be a compact section and include one of the following characteristics:

a. Have a low required axial strength.

b. Be a column in a story which has a significantly stronger design story shear strength than the story above.

c. Be a column that is not part of the lateral force resisting system except to support the axial load from the overturning moment of the building as a whole.

Wherever possible the committee recommends that the hinging conform to the requirements of Sect. 8.2.

8.7. Beam-to-Column Connection Restraint

In order to function properly, particularly if inelastic behavior in or adjacent to the beam-to-column connection occurs during high seismic activity, the column needs to be braced to prevent rotation out of the plane of the moment frame.

8.7.a. Restrained Connections: Beam-to-column connections are usually restrained laterally by roof or floor framing. For these cases, lateral support of the connection is required only at the level of the top flanges of the beams as long as the column can be shown to remain elastic. The two criteria to demonstrate that the column remains elastic are arbitrary but appear to be reasonable assumptions until otherwise demonstrated by test.

If the column cannot be demonstrated to remain elastic, a hinge would be potentially forming and the column should be laterally supported at the levels of both the top and bottom flanges of the beam.

The lateral support provided at the beam-to-column connection is to be designed using a required strength of 2 percent of the nominal beam strength. It is recognized from the limited test data available that the lateral support provided should also be rigid enough to inhibit lateral movement of the column flanges.[32] Designers should carefully design the lateral support member to be composed of reasonably rigid elements and be anchored to rigid supports.

The lateral support provided the beam-to-column connection is not required to be a separate member at the connection in all cases. It may be shown that the lateral support force can be adequately carried by the column web or the beam flanges.

8.7.b. Unrestrained Connection: Unrestrained connections can occur in special cases as in two story frames, at mechanical floors or for architectural space layout. When this does occur, special care should be provided to minimize the potential of out-of-plane buckling at the connection. Three arbitrary provisions are given for the columns to assure that this buckling does not occur.

8.8. Lateral Support of Beams

The lateral support for beams is defined in Chapter F in the LRFD design specifications. In moment resisting frames, the beams are nearly always in double curvature between columns unless one end is pinned. If the formula for plastic design were used as a guide and assuming M_p at one end and pinned at the other, formula F1-1 would yield $3,600r_y / F_y$. With $F_y = 36$ ksi, $L_{pd} = 100r_y$. The 1991 UBC has $96r_y$ for this limitation. Due to the low cycle oscillating motion of the frames under earthquake loading and the uncertainty of the locations of hinging under the various loading combinations, a more conservative approach was appropriate and set the maximum limit of the spacing of lateral support for frame beams at $2,500r_y / F_y$ for both top and bottom flanges.

9.   REQUIREMENTS FOR CONCENTRICALLY BRACED FRAMES (CBF)

9.1 Scope

The provisions contained in the Section are for braced frame systems of Building Categories C, D, and E where the braces are designed to carry all the lateral force shears or are used in combination with a moment resisting frame. If used in combination with a moment frame system, the moment frames should follow the requirements of Sects. 7 or 8 as required by the local Building Code. In a Concentrically Braced Frame (CBF), the bracing members are so arranged that the brace members primarily act with axial loading. CBF usually are in one of the following five types. (See Figs. C-9.1 through C-9.5).

Ductility of CBF systems producing a pattern of reasonably stable reversible distortions provides justification for basing seismic design on reduced displacements that can be expected during a strong earthquake. CBF systems, by the fact that the primary forces in the bracing system are axial tension and compression, are very limited in reversible inelastic distortions. Tests have shown that after buckling, an axially loaded member rapidly loses strength with repeated inelastic load reversals and does not return to its original straight position.[34] For this reason in high seismic areas, CBF systems have not been permitted by codes for tall or special buildings without being combined with a moment resisting frame. Codes also have required significantly higher levels of design force so that the possibility of large uncontrolled inelastic deformations will not occur. For instance, ASCE 7-88 in Sect. 9.9.5 requires that CBF be designed to a force 1.25 times the normal design force given in Sect. 9.4 for the system involved.

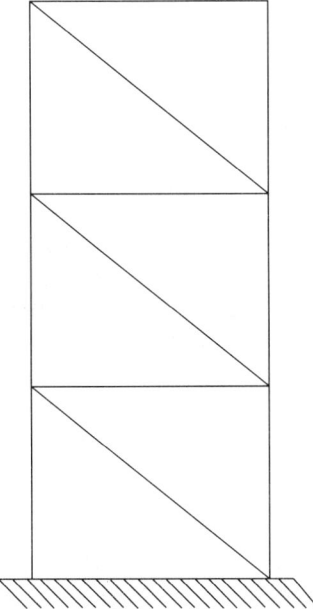

Fig. C-9.1. Diagonal braced frame.

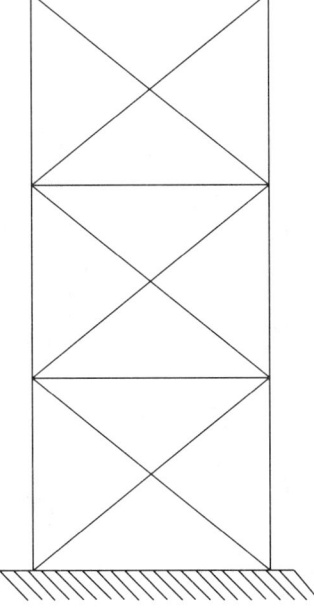

Fig. C-9.2. X-braced frame.

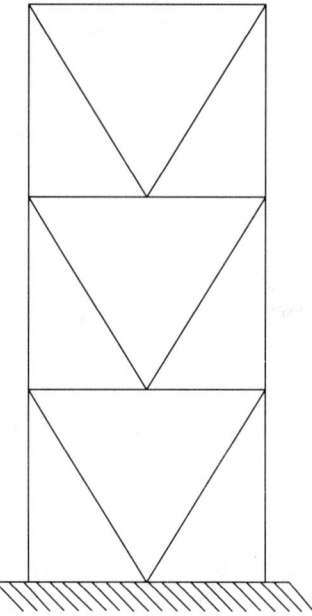

Fig. C-9.3. V-braced frame.

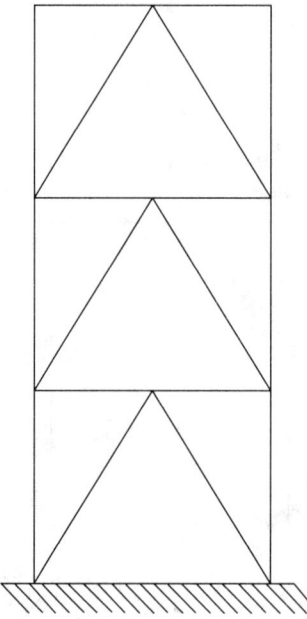

Fig. C-9.4. Inverted V-braced frame.

In Sect. 9.4 of this specification, for Special Configurations, this higher force factor is raised to 1.5.

The performance of CBF systems in earthquakes is acceptable as long as they retain stable configuration. The emphasis of these provisions is on raising the level of stable behavior and protecting against brittle failures.

When an axially loaded brace buckles in compression, several developments take place:

a. When buckling occurs, additional load is transferred to the tension brace increasing the force it must carry.

b. The buckling of the brace may cause excessive rotation at the brace ends and local connection failure.

c. The buckling can cause local or torsional buckling to occur near mid span.

d. If the buckling causes the brace to bow out of plane of the braced frame, non-structural encasement of the frame system can be destroyed.

e. Brace buckling can occur non-symmetrically which would induce large torsional response.

f. Excessive buckling can affect non-structural systems which are attached to the frame.[6]

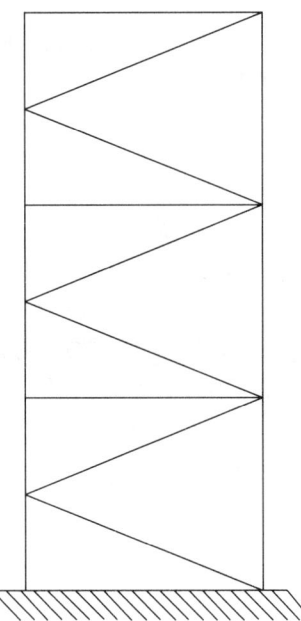

Fig. C-9.5. K-braced frame.

9.2. Bracing Members

9.2.a. Slenderness: Except for low buildings using the required strength given in Sect. 9.5, the slenderness (L/r) of members of CBF systems is limited. In the post-buckling range, the compressive nominal axial strength deteriorates.[34] Hysteresis loops of tested assemblies take on a severely pinched shape. (See Fig. C-9.6.)

Braces with small L/r dissipate more energy because in the post-buckling range they undergo cyclic inelastic bending which slender braces cannot. Very slender braces have almost no stiffness in a buckled configuration. On a load reversal, the brace quickly assumes a straightened configuration and very rapidly picks up a tensile force. This rapid increase in the brace force may cause impact loading and may lead to a brittle failure of the connection.

The curvatures associated with cyclic inelastic bending of braces may be large and local buckling can develop. This local buckling may be so severe as to result in localized kinking of the brace or the connection plate elements causing crack propagation and fracture. Such fractures have been obseved rather early in tests of tubular bracing members.[35] This characteristic is more prevalent in rectangular and square tube braces. Consideration should be given to using composite tubes with concrete fill to inhibit buckling.[36]

9.2.b. Compressive Design Strength: Due to the cyclic nature of seismic response, the compressive design strength of bracing members is reduced to 80 percent of the value given in the *Specification*, Chapter E.

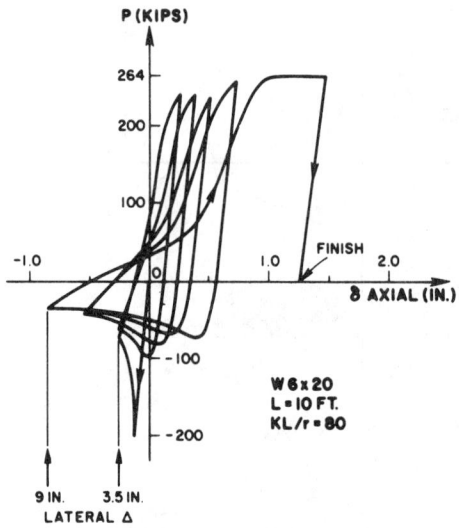

Fig. C-9.6. P–δ diagram for a strut.

This reduction factor is a simplified value from the factor proposed by others which varies with KL/r.[6] When evaluating the nominal strength of the bracing system for the purpose of determining the maximum load the bracing will impose on other systems (such as Eq. 6-1), the reduction for cyclic behavior should not be used for design as it would underestimate the nominal strength of the bracing system during the early cycles of seismic response.

9.2.c.   Lateral Force Distribution: This provision attempts to balance the tensile and compressive resistances across the width and breadth of the building since at large loads the capacity of buckled compression braces may be substantially less than that of tension braces. An exception is provided for the case where the bracing members were sufficiently oversized to provide essentially elastic seismic response.

9.2.d.   Width-Thickness Ratios: In Sect. B5 of the *Specification*, definitions are given to three types of sections. The compact section is one which has elements with width-thickness ratios, λ, less than λ_p. Non-compact sections are those with elements $\lambda_p \leq \lambda \leq \lambda_r$. Slender compression sections are those which have at least one element for which λ is greater than λ_r. The latter sections are prone to local buckling and are not to be used for the bracing members covered in this Section. The circular section wall thickness limitation was chosen to be the same as for Plastic Design in the *Specification*. Due to the repetitive nature of cyclic loading for rectangular tubular sections, a more stringent requirement on the b/t ratios is specified based on tests.[35-36] Filling of tubing with lean concrete has been shown to effectively stiffen the tube walls.

9.2.e.   Built-up Member Stitches: The special requirements for built-up member stitches were chosen from test data.[37] They are intended for members built up from double angles and channels, and may not be appropriate for markedly different shapes.

9.3. Bracing Connections

9.3.a.   In CBF systems, the bracing members normally carry most of the seismic story shear, particularly if a dual system is not used. The required strength on brace connections should be adequate so that failure by out-of-plane buckling of gussets or brittle fracture of the connection are not the critical failure mechanism.

The minimum of the three criteria, (i.e., the design axial tension strength of the bracing member, the force generated by the amplified load combinations of 3-7 and 3-8, and the maximum force that could be generated by the overall system) determine the required strength on both the brace connection and the beam-to-column connection if it is part of the bracing system. The latter criterion is intended to cover the possibility that the shear could be limited by the amount of overturning that could be developed.

9.3.b.   Net Area: Eq. 9-1 extends the concepts of LRFD Sect. B3 to the forces given in Section 9.2.a above.

9.3.c.   Gusset Plates: Gusset plates in CBF systems are frequently the critical design element in a system required to deform into the inelastic range. The increased force required for design of CBF tends to reduce the inelastic demand but may be insufficient to totally eliminate the problem. If the critical buckling mode of the braced member is in the plane of the CBF, the gussets and their joints should have a design strength capable to resist the nominal strength of the brace in that direction. If the critical buckling mode is out of the plane of the CBF, each gusset shall be detailed to permit the formation of a hinge line in the gusset. (See Fig. C-9.7.)

9.4. Special Bracing Configuration Requirements

In addition to the general requirements for bracing members and their connections given above, special limitations are applied to V and K types of CBF systems due to their special configurations.

9.4.a. V and Inverted V Type Bracing: If one diagonal of a V type brace were to buckle in compression, the force in the tension brace would become larger than the force in the buckled brace. The vertical resultant of these two forces could then impose a large vertical deformation on the horizontal member of the bracing system. (See Fig. C-9.8.) If the connection at the point of the V tip were pinned, there would be no resistance to this deformation. If a continuous horizontal member survives and undergoes a deformation reversal, the previously buckled diagonal member would

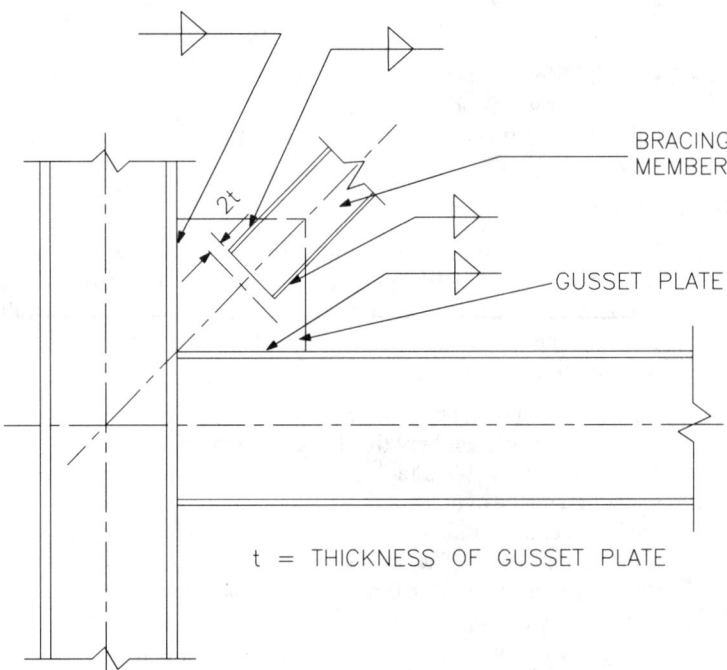

Fig. C-9.7. Brace-to-gusset plate requirement for buckling out-of-plane bracing system.

not return to its original alignment and the diagonal member which was in tension could exceed its capacity in compression. In this manner both diagonal members would be in a buckled condition. This behavior would cause the post buckling strength of the braced system to deteriorate rapidly.[38] (See Fig. C-9.9.)

Near the tip point of the V is a zone where inelastic rotations are likely to occur, members should be braced against out-of-plane buckling. Several options were considered for CBF systems using the V type bracing. One was to prohibit its use, a second was to impose stringent limitations on the slenderness ratios of the bracing members, and a third was to provide a larger axial load capacity for the diagonal members. The latter option was adopted by providing a design axial strength 1.5 times the required axial strength in lieu of the 1.25 normally required for other CBF. It is also required that the beam be continuous throughout the bay and that this beam be designed to carry the tributary vertical gravity loads without considering the support provided by the diagonal members of the V.

A review of more recent testing of V braced systems may in future editions be able to modify some of the current limitations.

9.4.b. K Bracing: In areas of high seismicity where it is envisioned that inelastic response to large motions will be required, the K type of CBF system is not a desirable method for seismic resistance. The same behavior discussed in the V type bracing occurs, but in the case of the K system a buckled brace causes the column to deform horizontally. Potentially this could cause column buckling and subsequent collapse.

In buildings of Categories A, B, and a portion of C, the K system is permitted unrestricted by these provisions. For the remainder of Category C as per

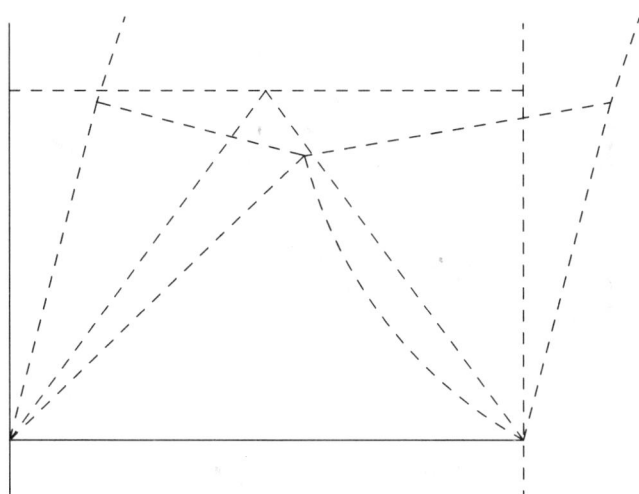

Fig. C-9.8. Failure mechanism of inverted V-braced frame.

Sect. 2.2, however, K braces shall meet the requirements Sects. 9.4.b and 9.5. This requires a 50 percent increase in design axial load for the braces and a continuous column though at the story mid-height. It is recommended that K type bracing not be used even where permitted for seismic resistance unless other configurations are impractical.

9.5. Low Buildings

One of the few problem areas observed in the seismic performance of smaller steel buildings using the CBF system pertain to the size and type of member connections used. Quite frequently the critical design horizontal load is wind rather than seismic. In these cases, the sizing of bracing members is larger than would be required if seismic loads were the only design horizontal loads. Thus for smaller buildings and roof structures, the special provisions for CBF systems have been waived if the seismic resisting system has been designed using the amplified loads given in Load Combinations 3-7 and 3-8. This waiver would permit, for instance, an X braced or diagonal braced system in which the bracing members would be assumed to be in tension only.

10.   REQUIREMENTS FOR ECCENTRICALLY BRACED FRAMES (EBF)

10.1. Scope

Research[39–49] has shown that buildings using the EBF system possess the ability to combine high stiffness in the elastic range together with excellent ductility and energy dissipation capacity in the inelastic range. In the elastic range, the lateral stiffness of an EBF system is comparable to that of a CBF system, particularly when short link lengths are used. In the inelastic range, EBF systems

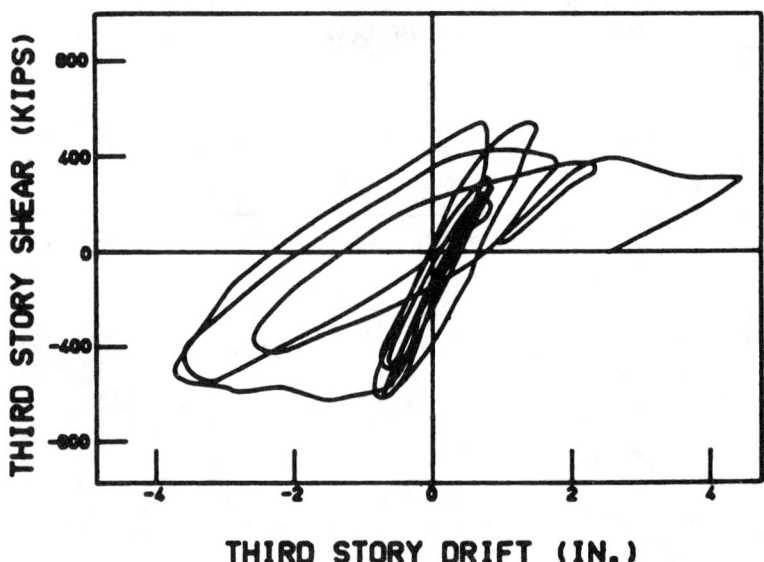

Fig. C-9.9. Story shear–story drift diagram for frame with inverted V-bracing.

provide stable, ductile behavior under severe cyclic loading, comparable to that of a SMF system. The EBF is composed of columns, beams, and braces in which at least one end of each bracing member connects to a beam at a short distance from a beam-to- column connection or from an adjacent beam-to-brace connection. (See Fig. C-10.1.) The short distance of the beam between the brace connection and the column or between brace connections is called the link. The design purpose of an EBF system creates a system that will yield primarily in the links. The special provisions for EBF systems are intended to satisfy this criterion and to ensure that cyclic yielding in the links can occur in a stable manner. The yielding in the links is accomplished by ensuring that the diagonal braces, the columns, and the portion of the beam outside of the links remain essentially elastic under forces that can be generated by fully yielding and strain hardened links.

Arrangements of braces can be made in which links may not be fully effective. One such arrangement is the one shown on Fig. C-10.2 in which links are provided at each end of the brace. If the upper link has significantly lower design shear strength than the story below, the upper link deforms inelastically and limits the force that can be delivered to the brace to deform the lower link inelastically. When this condition occurs the upper link is termed an active link, whereas the lower link is an inactive link. Having potentially inactive links in the EBF system increases the difficulty of analysis. The plastic analyses show that in some cases the lower link yields due to the combined effect of D, L, and E loads, and the frame capacity becomes smaller than expected.[50] It also increases the cost of the structure by requiring full link details on the inactive links even though the brace would be sized by the strength of the active link and the brace connection at an inactive link could be designed as a pin. Thus it is best to arrange a system that contains only active links as those shown in Fig. C-10.1. Design suggestions have been compiled in Ref. 48.

In Sect. 10.1 in conformity with the strong column–weak beam concept, plastic hinges should not develop in columns at floor beam levels in EBF. The occurrence of such plastic hinges, together with those forming in the links, could result in a soft story and must be prevented. There are two important code provisions intended to prevent this from happening. First, according to Sect. 6.1, the required axial column strength includes P_E, based on application of the amplified earthquake load $0.4R_E$. Second, per Sect. 10.8, the required strength of columns due to the forces introduced at the connection of a link and/or brace is based on these forces multiplied by a factor of 1.25. Note that for a severe earthquake the formation of plastic hinges at column bases is generally unavoidable.

10.2. Links

The general provisions for links to ensure that stable yielding occurs are included under this heading.

10.2.a. Beams with links are required to be compact shapes following the same criteria as SMF systems (Table 8-1).

10.2.b. In order to provide steel with proven ductile behavior the yield stress of steel is limited to 50 ksi.

10.2.c. Doubler plates on the link web are not permitted as they do not perform as intended in inelastic deformations. Openings are not permitted as they adversely affect the yielding of the link web.

10.2.d. The link design shear strength ϕV_n is the lesser of that determined from the yield shear or twice the plastic moment strength divided by the link length. This ϕV_n should be greater than or equal to the required shear determined from the Load Combinations 3-5 or 3-6.

10.2.e. If the required axial load on the link is less than $0.15P_y$, the effects of the axial load can be ignored, In general, the axial load is negligible because the horizontal component of the brace load is transmitted to the beam outside of the link. However, due to a particular arrangement of the framing, substantial axial forces can develop in the link. For such cases, the limitations given in f. apply, and the design shear strength and link lengths are required to be reduced to ensure stable yielding.

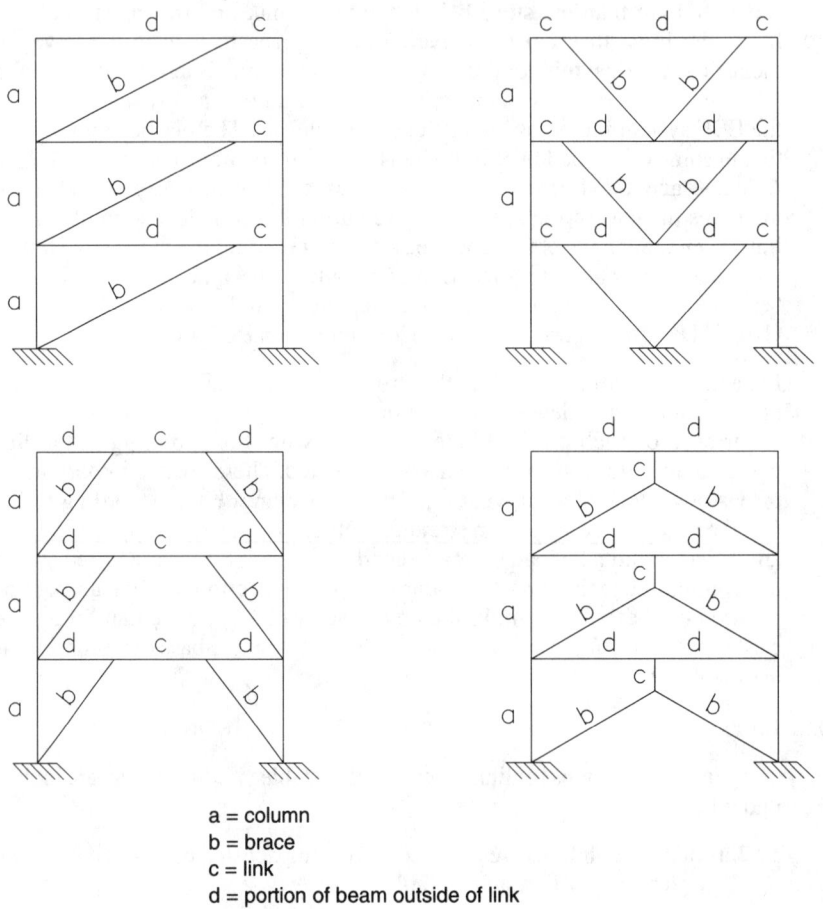

a = column
b = brace
c = link
d = portion of beam outside of link

Fig. C-10.1. Common types of eccentric braced frames.

10.2.f. See Commentary 10.2.e.

10.2.g. The link rotation angle is defined in the Specifications as the plastic angle between the link and the beam outside the link when the total story drift Δ_t, calculated using amplified earthquake forces $0.4R \times E$. The plastic link rotation can be conservatively determined assuming that the EBF bay will deform in a rigid-plastic mechanism. Several such mechanisms are illustrated for various EBF configurations in Fig. C-10.3. The plastic angle is determined using a story drift $\Delta_p = \Delta_t - \Delta_e$, where Δ_e the elastic story drift can conservatively be assumed to be zero. The plastic story drift angle $\theta_p = \Delta_p / h$ follows from geometry. The actual plastic link rotation angle can be determined by non-linear elastic-plastic analyses if a more explicit definition of the angle is desired.

An inverted Y system is shown on Fig. C-10.1. In this system the precise definition given in the Glossary for the link rotation angle does not apply but the concept is the same as in the other systems, as shown on Fig. C-10.3. As usual both ends of the link are required to be laterally supported.

The link length of $1.6M_p / V_p$ indicates the limit chosen for the link to act primarily in shear. The link length $2.6M_p / V_p$ is the lower limit of a flexural link. Straight line interpolation is used for the intermediate link lengths.

It has been demonstrated experimentally[51-52] as well as analytically[48] that the first floor links usually experience the largest plastic deformation. In extreme cases this may result in a tendency to develop a soft story. The

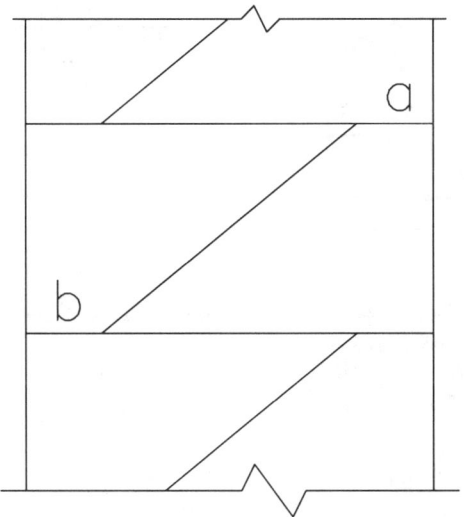

ϕV_n – link a (active link) $< \phi V_n$ – link b (inactive link)

Fig. C-10.2. EBF—active and inactive link.

plastic link rotations tend to attenuate at higher floors, and decrease with the increasing frame periods. Therefore for severe seismic applications a conservative design for the links in the first two or three floors is recommended. This can be achieved by increasing the minimum design shear strengths of these links on the order of 10 percent over that specified in Sect. 10.2.d. An even more conservative approach would be to have vertical connecting members at the ends of the links in a few lower floors.

The use of the framing shown in Fig. C-10.1 can be advantageous where

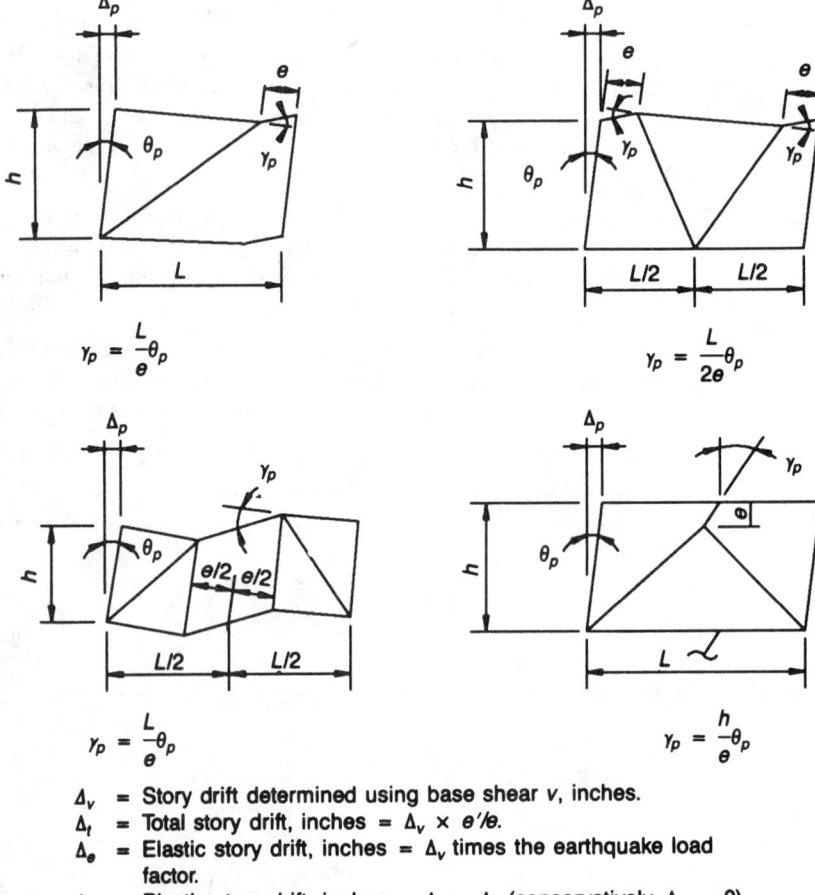

$$\gamma_p = \frac{L}{e}\theta_p \qquad \gamma_p = \frac{L}{2e}\theta_p$$

$$\gamma_p = \frac{L}{e}\theta_p \qquad \gamma_p = \frac{h}{e}\theta_p$$

Δ_v = Story drift determined using base shear v, inches.
Δ_t = Total story drift, inches = $\Delta_v \times e'/e$.
Δ_e = Elastic story drift, inches = Δ_v times the earthquake load factor.
Δ_p = Plastic story drift, inches = $\Delta_t - \Delta_e$ (conservatively, $\Delta_e = 0$).
e = Link length, inches.
h = Story height, inches.
L = Column to column distance, inches.
θ_p = Plastic story drift angle, radians = Δ_p/h.
γ_p = Link rotation angle, radians.

Fig. C-10.3. Link rotation angle.

the beam-column-brace connections can be designed as simple connections. Welds of the link flanges are avoided in this kind of framing.

By changing the link lengths the stiffness of an EBF can be modified. In this manner the frame periods can be optimized.

10.2.h. The intent of this provision is to permit a CBF on the top floor of an EBF building over five stories tall with application of an earthquake response modification coefficient R appropriate for an EBF.

10.3. Link Stiffeners

Properly detailed and restrained webs can provide stable, ductile, and predictable behavior under severe cyclic loading. The design of the EBF link requires close attention to the detailing of the link web thickness and stiffeners.

10.3.a. Full depth stiffeners are required at the end of all EBF links and serve to transfer the link shears to the reacting elements as well as restraining the link web against buckling.

10.3.b. In shear links, the spacing of intermediate web stiffeners is varied depending on the magnitude of the link rotation angle.[45] The closer spacing is provided for the system with the greatest angle. Flexural links having lengths greater than $2.6M_p / V_p$ but less than $5M_p / V_p$ are required to have an intermediate stiffener at a distance from the link end equal to 1.5 times the beam flange width. Links between shear and flexural limits would have intermediate stiffeners meeting the requirement of both shear and flexural links. When the link length is greater than $5M_p / V_p$, no intermediate stiffeners are required. Intermediate stiffeners are required to be full depth in order to effectively react against shear buckling. Intermediate stiffeners are required on both sides of the web for links 25 inches in depth or greater. For links less than 25 in. deep, the stiffener need be on one side only.

10.3.c. All link stiffeners are required to be fillet welded to the link web. These welds shall have a required strength equal to the nominal vertical tensile strength of the stiffener. The connection to the link flanges should be similar.

10.4. Link-to-Column Connections

There are special connection requirements for the connections of links to columns. The intent is to provide connections which can transfer not only the shear and moment forces of the links but also torsion due to flange buckling. The *Specification* does not explicitly address the column panel zone design requirements at link-column connections, as little research is available on this issue. However, from research on panel zones for SMF systems, it is believed that limited yielding of panel zones in EBF systems would not be detrimental. Pending future research on this topic, a suggested design approach is as follows:

Compute the required shear strength of the panel zone based on the bending moment at the column end of the link, as given by the equations in Sect. 10.6.a in the commentary of these provisions. The corresponding panel zone design shear strength should then be computed according to Eq. 8-1 of these provisions.

10.5. Lateral Support of the Link

One of the essential items to ensure stable inelastic behavior of the EBF system is to restrain the ends of the link from twisting out of plane. The 6 percent of the nominal strength of the beam flange defines the required strength on the lateral support member and its connections.

10.6. Diagonal Brace and Beam Outside of Links

10.6.a. A basic requirement of EBF design is that yielding be restricted primarily to the links. Accordingly, the diagonal brace and the beam segment outside of the link should be designed to resist the maximum forces that can be generated by the link, accounting for the sources of link overstrength. Link overstrength can be attributed primarily to strain hardening, effects of composite floor systems, and the actual yield strength of the link exceeding the specified yield strength. In EBF research literature, for design of the brace and the beam, an overstrength factor of 1.5 has generally been applied to the nominal strength of the link. Using this overstrength factor, the brace and beam segments were checked using their nominal strength, i.e., using $\phi = 1.0$. This approach considers that designing for an overstrength factor of 1.5 represents an extreme loading condition for the beam and brace, and therefore a relaxation of the ϕ factor was appropriate to avoid an overly conservative design.[49] Sect. 10.6.a specifies that the design strength of the beam and diagonal brace exceed the forces generated by 1.25 times the nominal link shear strength, maintaining approximately the same basic design approach for the diagonal brace and beam. That is, based on a ϕ factor of 0.85 on axial compression in the beam or brace, the effective overstrength factor becomes 1.25 / 0.85, or about 1.5. For bending moments in the beam or diagonal brace, for which ϕ is 0.9, the overstrength factor becomes 1.25 / 0.9, or about 1.4, representing a slight relaxation from the test criterion.

Based on a link overstrength factor of 1.25, the required strength of the diagonal brace and beam segment outside of the link can be taken as the forces generated by the following values of link shear and link end moment:

For $e \le 2M_p / V_p$,	link shear	$= 1.25V_p$
	link end moment	$= e(1.25V_p)/2$
For $e > 2M_p / V_p$,	link shear	$= 2(1.25M_p)/e$
	link end moment	$= 1.25M_p$

The above equations are based on the assumption that link end moments will be equal when the link achieves its limit strength. For links of length $e \le 1.3M_p / V_p$ attached to columns, experiments have shown that link end moments do not equalize.[44] For this situation, link shear and link end moments can be taken as:

For $e \le 1.3M_p / V_p$ next to column,

link shear $= 1.25V_p$

moment at column end of link $= 0.8M_p$
moment at brace end of link $= e(1.25V_p) - 0.8M_p$

The link shear force will generate axial force in the diagonal brace, and for most EBF configurations, will also generate substantial axial force in the beam segment outside of the link. The ratio of beam or brace axial force to link shear force is controlled primarily by the geometry of the EBF and is therefore not affected by inelastic activity within the EBF.[47] Therefore, this ratio can be taken from an elastic frame analysis and used to scale up the beam and brace axial force to a level corresponding to the link shear force specified in the above equations. At the brace end of the link, the link end moment will be transferred to the brace and to the beam. If the diagonal brace and its connection remains elastic, based on link overstrength design considerations, some minor inelastic rotation can be tolerated in the beam outside of the link.

10.6.b. Typically in EBF design, the intersection of the brace and beam center-lines is located at the end of the link. However, as permitted by Sect. 10.6.b, the brace connection should be designed with an eccentricity so that the brace and beam centerlines intersect inside of the link. This eccentricity in the connection generates a moment that is opposite in sign to the link end moment. Consequently, the value of link end moment

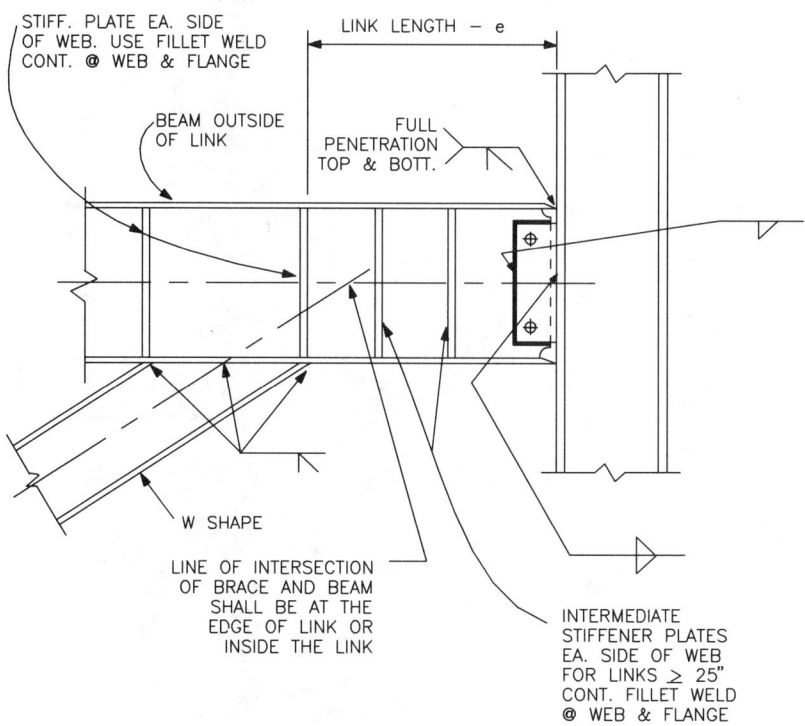

Fig. C-10.4

given above can be reduced by the moment generated by this brace connection eccentricity. This may substantially reduce the moment that will be required to be resisted by the beam and brace, and may be advantageous in design. The intersection of the brace and beam center-lines should not be located outside of the link, as this increases the bending moment generated in the beam and brace. See Figs. C-10.4 and C-10.5.

10.6.c. If the brace connection at the link is designed as a pin, the beam by itself shall be adequate to resist the entire link end moment. This condition normally would occur only on EBF with short links. If the brace is considered to resist a portion of the link end moment, then the brace connection at the link should be designed as fully restrained, as required by Sect. 10.6.c. Test results on several brace connection details subject to axial force and bending moment are reported in Ref. 47.

10.6.d. When checking the requirements of Sect. 10.6, both the beam and diagonal brace should, in general, be treated as beam-columns in strength and stability computations. Unlike CBF, the brace of an EBF may be subject to significant bending moments. For the beam segment outside of the link, adequate lateral bracing should be provided to maintain its stability under the axial force and bending moment gener-

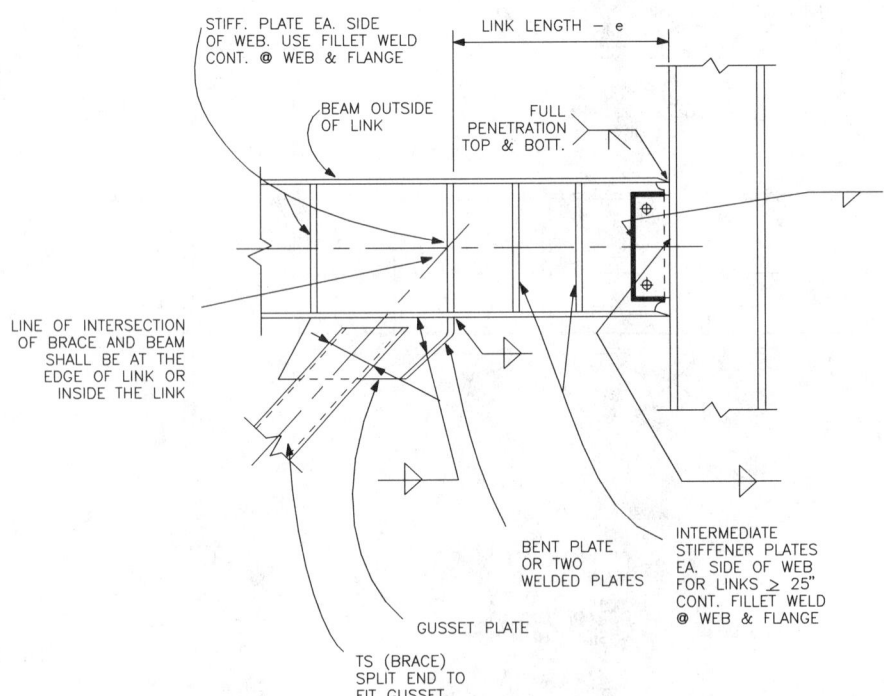

Fig. C-10.5

ated by the link, as required by Sect. 10.6.d. If the stability of the beam is provided by adequate lateral support, tests have shown that limited yielding of the beam segment is not detrimental to EBF performance, and for some EBF configurations may be unavoidable.[47] However, the combined flexural strength of the beam and the brace, reduced for the presence of axial force, should be adequate to resist the link end moment. For EBF geometries with very small angles between the beam and the brace and/or for EBF with long links, satisfying the requirements of Sect. 10.6.e. may require very heavy braces, and in extreme cases, may require cover plates on the beams. EBF with relatively steep braces, e.g., brace-beam angles greater than about 40 degrees, combined with short links are preferable for avoiding design problems with the brace and beam segment outside of the link. A general discussion on design issues related to the beams and braces of an EBF is provided in Ref. 49, with further details provided in Ref. 47.

10.7. Beam-to-Column Connection

If the arrangement of the EBF system is such that a link is not adjacent to a column, a simple pinned connection is considered to be adequate if the connection provides some restraint against torsion in the beam. The magnitude of torsion is calculated by considering perpendicular forces equal to 1.5 percent of the nominal axial flange tensile strength applied in opposite directions on each flange.

10.8. Required Column Strength

As the shear strength of the adjoining critical link is potentially greater than the nominal strength due to strain hardening, the required column strength is required to be designed for the increased moment and axial load due to the load from the adjacent link or brace.

11.   QUALITY ASSURANCE

As the behavior of all steel framing during a major earthquake is dependent on the workmanship of the fabricator in providing sound joints, the design engineer is advised to provide for adequate assurance control, particularly on the tension groove welded joints of the seismic resisting system. ASCE 7-92 provides special requirements for inspection and testing based on the Seismic Performance Category of the building to be built. The special requirements for structural steel construction are in general those that would normally be required for construction in all areas of seismic activity.

Part II—ASD Provisions

1.   SCOPE

As noted in PART I, the special seismic requirements are collateral provisions related to the AISC *Load and Resistance Factor Design Specification*. As that document was first published in 1986, the references to earthquake load were not current. The provisions in PART I use limit state load models derived from the 1991 NEHRP[3] and the soon to be published ASCE 7-93.[2]

The provisions in PART II allow a designer to apply AISC *Allowable Stress Design Specification for Structural Steel Buildings* (ASD)[53] in the design of the seismic lateral force resisting system based upon limit state loads. If the user wishes to use ASD in the design of the seismic lateral force resisting system where the loads are based upon service loads, the loads need to be converted to factored levels consistent with those in PART I. The PART II provisions are intended to be used in conjunction with PART I by either adding to or substituting to the provisions of Part I.

3.2   Nominal Strengths, and

3.3   Design Strengths

These provisions modify PART I to convert allowable stresses into equivalent nominal strengths by multiplying allowable stresses by 1.7 as noted. Design strengths are determined by multiplying ϕ times the nominal strengths.

7.2, 10.6.a, and 10.6.d

These modifications to PART I requirements change FR and PR connections to Type 1 and Type 3 connections consistent with ASD nomenclature.

List of References

1. AISC, *Load and Resistance Factor Design Specification,* American Institute of Steel Construction, Inc., Chicago, IL, 1986.

2. ASCE 7-93, *Minimum Design Loads for Buildings and Other Structures,* American Society of Civil Engineers, New York, NY, 1993 (to be published).

3. BSSC, NEHRP (National Earthquake Hazards Reduction Program) *Recommended Provisions for the Development of Seismic Regulations for New Buildings,* Building Seismic Safety Council, Federal Emergency Management Agency, Washington, DC, 1992.

4. Luft, R. W., "Comparison Among Earthquake Codes," *Earthquake Spectra,* Earthquake Engineering Research Institute, Vol. 5, No. 4, November 1989.

5. ICBO, *Uniform Building Code,* International Conference of Building Officials, Whittier, CA, 1991.

6. SEAOC, *Recommended Lateral Force Requirements,* Seismology Committee, Structural Engineers Association of California, Sacramento/San Francisco/Los Angeles, CA, 1988.

7. AWS, D1.1-92, *Structural Welding Code,* American Welding Society, Inc., Miami, FL, 1992.

8. Popov, E. P., Stephen, R. M., "Tensile Capacity of Partial Penetration Welds," *Journal of the Structural Division,* American Society of Civil Engineers, Vol. 103, No. ST9, September 1977.

9. Bruneau, M., Mahin, S. A., and Popov, E. P., *Ultimate Behavior of Butt Welded Splices in Heavy Rolled Steel Sections,* Report No. UCB/EERC-87/10, Earthquake Engineering Research Center, University of California, Berkeley, CA, 1987.

10. Huckelbridge, A. A., Clough, R. W., *Earthquake Simulator Tests of Nine-Story Steel Frame with Columns Allowed to Uplift,* Report No. UCB/EERC-77/23, Earthquake Engineering Research Center, University of California, Berkeley, CA, 1977.

11. Carpenter, L. D., Lu, L-W., *Reversed and Repeated Load Tests of Full Scale Steel Frames,* Fritz Engineering Laboratory Report No. 332.7, Lehigh University, Bethlehem, PA, 1972.

12. Galambos, T. V., *Deformation and Energy Absorption Capacity of Steel Structures in the Inelastic Range,* Bulletin No. 8, American Iron and Steel Institute, New York, NY, 1968.

13. Krawinkler, H., Bertero, V. V., and Popov, E. P., *Inelastic Behavior of Steel Beam-to-Column Subassemblages,* Report No. UCB/EERC-71/7, Earthquake Engineering Research Center, University of California, Berkeley, CA, 1971.

14. Bertero, V. V., Popov, E. P., and Krawinkler, H., *Further Studies on Seismic Behavior of Steel Beam-Column Subassemblages,* Report No. UCB/EERC-73/27, Earthquake Engineering Research Center, University of California, Berkeley, CA, 1973.

15. Popov, E. P., "Seismic Behavior of Structural Assemblages," *Journal of the Structural Division,* American Society of Civil Engineers, Vol. 106, No. ST7, July 1980.

16. *Tall Building Systems and Concepts,* Monograph on Planning and Design of Tall Buildings, Council on Tall Buildings and Urban Habitat, American Society of Civil Engineers, New York, NY, 1980.

17. Popov, E. P., Amin, N. R., Louie, J. J., and Stephen, R. M., "Cyclic Behavior of Large Beam Column Assemblies," *Earthquake Spectra,* Professional Journal of the Earthquake Engineering Research Institute, Vol. 1, No. 2, February 1985.

18. Tsai, K. C. and Popov, E. P., "Seismic Panel Zone Design Effect on Elastic Story Drift in Steel Moment Resisting Frames," *Journal of Structural Division,* in press.

19. Nicoletti, J. P., Pinkham, C. W., Saunders, C. M., Teal, E. J., *A Synthesis of Steel Research for Code Development,* Structural Steel Educational Council, San Francisco, CA, 1984.

20. Popov, E. P. and Pinkney, R. B., *Behavior of Steel Building Connections Subjected to Inelastic Strain Reversals—Experimental Data,* Bulletin No. 14, American Iron and Steel Institute, November 1968.

21. Popov, E. P. and Stephen, R. M., *Cyclic Loading of Full-Size Steel Connections,* Bulletin No 21, American Iron and Steel Institute, February 1972.

22. Driscoll, G. C. and Beedle, L. S., "Suggestions for Avoiding Beam-to-Column Web Connection Failure," *Engineering Journal,* American Institute of Steel Construction, Chicago, IL, 1st Qtr., 1982.

23. Tsai, K. C. and Popov, E. P., *Two Beam-to-Column Web Connections,* Report No. UCB/EERC-86/05, Earthquake Engineering Research Center, University of California, Berkeley, CA, 1986.

24. Popov, E. P. and Tsai, K. C., *Performance of Large Seismic Steel Moment Connections Under Cyclic Loads,* Proceedings; Structural Engineers Association of California Convention, San Diego, CA, October 1987.

25. Slutter, R., *Tests of Panel Zone Behavior in Beam Column Connections,* Lehigh University, Report No. 200.81.403.1, Bethlehem, PA.

26. Becker, E. R., *Panel Zone Effect on the Strength of Rigid Steel Frames,* University of Southern California Structural Mechanics Laboratory, USCOE 001, June 1971.

27. Fielding, D. J., Huang, J. S., "Shear in Steel Beam-to-Column Connections," *Welding Journal,* Vol. 50, No. 7, Research Supplement, 1971.

28. Krawinkler, H., "Shear in Beam-Column Joints in Seismic Design of Steel

Frames," *Engineering Journal,* American Institute of Steel Construction, Chicago, IL, Vol. 15, 1978.

29. Sawyer, H. A., "Post-Elastic Behavior of Wide-Flange Steel Beams," *Journal of the Structural Division,* Vol. 87, No. ST8, American Society of Civil Engineers, December 1961.

30. Lay, M. G., "Flange Local Buckling in Wide-Flange Shapes," *Journal of the Structural Division,* Vol. 91, No. ST6, American Society of Civil Engineers, December 1965.

31. Kemp, A. R., "Factors Affecting the Rotation Capacity of Plastically Designed Members," *The Structural Engineer,* Vol. 64B, No. 2, June 1986.

32. Bansal, J. P., *The Lateral Instability of Continuous Steel Beams,* CESRL Dissertation No. 71-1, University of Texas, Austin, TX, 1971.

33. Krawinkler, H., Bertero, V. V., Popov, E. P., *Hysteresis Behavior of Steel Columns,* Report No. UCB/EERC-75/11, Earthquake Engineering Research Center, University of California, Berkeley, CA, 1975.

34. Black, R. C., Wenger, W. A., Popov, E. P., *Inelastic Buckling of Steel Struts Under Cyclic Load Reversals,* Report No. UCB/EERC-80/40, Earthquake Engineering Research Center, University of California, Berkeley, CA, 1980.

35. Tang, X., Goel, S. C., *Seismic Analysis and Design Considerations of Braced Steel Structures,* UMCE Report 87-4, University of Michigan, Ann Arbor, MI, 1987.

36. Uang, C-M., and Bertero, V. V., *Earthquake Simulation Tests and Associated Studies of 0.3-Scale Model of a Six-Story Concentrically Braced Steel Structure,* Report No. UCB/EERC—86/10, EERC, Berkeley, CA, December 1986.

37. Liu, Z., Goel, S. C., *Investigation of Concrete Filled Steel Tubes under Cyclic Bending and Buckling,* UMCE Report 87-3, University of Michigan, Ann Arbor, MI, 1987.

38. Astaneh, A., Goel, S. C., Hanson, R, D., "Earthquake-Resistant Design of Double Angle Bracings," *Engineering Journal,* American Institute of Steel Construction, Chicago, IL, Vol. 23, No. 4, 1986.

39. Roeder, C. W. and Popov, E. P., "Eccentrically Braced Frames for Earthquakes," *Journal of the Structural Division,* Vol. 104, No. 3, American Society of Civil Engineers, March 1978.

40. Libby, J. R., "Eccentrically Braced Frame Construction—A Case History," *Engineering Journal,* American Institute of Steel Construction, Chicago, IL, Vol. 18, No. 4, 1981.

41. Merovich, A. T., Nicoletti, J. P. and Hartle, E., "Eccentric Bracing in Tall Buildings," *Journal of the Structural Division,* Vol. 108, No. 9, American Society of Civil Engineers, September 1982.

42. Hjelmstad, K. D. and Popov, E. P., "Cyclic Behavior and Design of Link Beams," *Journal of the Structural Division,* Vol. 109, No. 10, American Society of Civil Engineers, October 1983.

43. Malley, J. O. and Popov, E. P., "Shear Links in Eccentrically Braced Frames,"

Journal of the Structural Division, Vol. 110, No. 9, American Society of Civil Engineers, September 1984.

44. Kasai, K. and Popov, E. P., "General Behavior of WF Steel Shear Link Beams," *Journal of the Structural Division,* Vol. 112, No. 2, American Society of Civil Engineers, February 1986.

45. Kasai, K. and Popov, E. P., "Cyclic Web Buckling Control for Shear Link Beams," *Journal of the Structural Division,* Vol. 112, No. 3, American Society of Civil Engineers, March 1986.

46. Ricles, J. M. and Popov, E. P., *Dynamic Analysis of Seismically Resistant Eccentrically Braced Frames,* Report No. UCB/EERC-87/107, Earthquake Engineering Research Center, University of California, Berkeley, CA, 1987.

47. Engelhardt, M. D. and Popov, E. P., *Behavior of Long Links in Eccentrically Braced Frames,* Report No. UCB/EERC-89/01, Earthquake Engineering Research Center, University of California, Berkeley, CA, 1989.

48. Popov, E. P., Engelhardt, M. D. and Ricles, J. M., "Eccentrically Brace Frames: U. S. Practice," *Engineering Journal,* American Institute of Steel Construction, Chicago, IL., Vol. 26, No. 2, 1989. p. 66–80.

49. Engelhardt, M. D. and Popov, E. P., "On Design of Eccentrically Braced Frames," Earthquake Engineering Research Institute, El Cerrito, CA, *Earthquake Spectra,* Vol. 5, No.3, August 1989.

50. Kasai, K. and Popov, E. P., *On Seismic Design of Eccentrically Braced Steel Frames,* Proceedings, 8th World Conference on Earthquake Engineering, July 1984, San Francisco, CA., Vol. 5, pp.387–394.

51. Whittaker, A. S., Uang, C-M., and Bertero, V. V., *Earthquake Simulation Tests and Associated studies of a 0.3-Scale Model of a Six-Story Eccentrically Braced Steel Structure,* Report No. UBC/EERC-87/02, EERC, Berkeley, CA., 1987.

52. Foutch, D. A., "Seismic Behavior of Eccentrically Braced Steel Building," ASCE *Journal of Structural Engineering,* Vol. 115, No. 8, August 1989, pp 1857–1876.

53. AISC, *Allowable Stress Design Specification,* American Institute of Steel Construction, Inc., Chicago, IL, 1989.

54. AISC, *Serviceability Design Considerations for Low-Rise Buildings,* Steel Design Guide Series 3, American Institute of Steel Construction, Inc.

55. Structural Steel Education Council, *Steel Connections/Details and Relative Costs,* Moraga, CA., 1986.

56. Uang, C. M., "Establishing R (or R_w) and C_d Factors for Building Seismic Provisions," *Journal of Structural Engineering,* Vol. 117, No. 1, American Society of Civil Engineers, Jan. 1991.

LOAD AND RESISTANCE FACTOR DESIGN

Specification for Structural Joints Using ASTM A325 or A490 Bolts

Approved by Research Council on Structural Connections
of the Engineering Foundation, **June 8, 1988**.
Endorsed by American Institute of Steel Construction
Endorsed by Industrial Fasteners Institute

AMERICAN INSTITUTE OF STEEL CONSTRUCTION, INC.
One East Wacker Drive, Suite 3100, Chicago, IL 60601-2001

PREFACE

The purpose of the Research Council on Structural Connections is to stimulate and support such investigation as may be deemed necessary and valuable to determine the suitability and capacity of various types of structural connections, to promote the knowledge of economical and efficient practices relating to such structural connections, and to prepare and publish related standards and such other documents as necessary to achieving its purpose.

The Council membership consists of qualified structural engineers from the academic and research institutions, practicing design engineers, suppliers, and manufacturers of threaded fasteners, fabricators and erectors and code writing authorities. Each version of the Specification is based upon deliberations and letter ballot of the full Council membership.

The first *Specification for Assembly of Structural Joints Using High Tensile Steel Bolts* approved by the Council was published in January 1951. Since that time the Council has published 12 succeeding editions each based upon past successful usage, advances in the state of knowledge and changes in engineering design practice. This version of the Council's *Load and Resistance Factor Design Specification* is significantly reorganized and revised from earlier versions.

The intention of the Specifications is to cover the design criteria and normal usage and practices involved in the everyday use of high-strength bolts in steel-to-steel structural connections. It is not intended to cover the full range of structural connections using threaded fasteners nor the use of high-strength bolts other than those included in ASTM A325 or ASTM A490 Specifications nor the use of ASTM A325 or A490 bolts in connections with material other than steel within the grip.

A Commentary has been prepared to accompany these Specifications to provide background and aid the user to better understand and apply the provisions.

The user is cautioned that independent professional judgment must be exercised when data or recommendations set forth in these Specifications are applied. The design and the proper installation and inspection of bolts in structural connections is within the scope of expertise of a competent licensed architect, structural engineer or other licensed professional for the application of the principles to a particular case.

LOAD AND RESISTANCE FACTOR DESIGN

Specification for Structural Joints Using ASTM A325 or A490 Bolts

Approved by Research Council on Structural Connections of the
Engineering Foundation, June 8, 1988.
Endorsed by American Institute of Steel Construction
Endorsed by Industrial Fasteners Institute

1. Scope

This Specification relates to the load and resistance factor design of structural joints using ASTM A325 high-strength bolts, ASTM A490 high-strength bolts or equivalent fasteners, and for the installation of such bolts in connections of structural steel members. The Specification relates only to those aspects of the connected materials that bear upon the performance of the fasteners.

Design and construction shall conform to an applicable load and resistance factor design code or specification for structures of carbon, high-strength low alloy steel or quenched and tempered structural steel. Load and resistance factor design is a method of proportioning structural components such that no applicable limit state is exceeded when the structure is subject to all appropriate load combinations. When a structure or component ceases to fulfill the intended purpose in some way, it is said to have exceeded a limit state. Strength limit states concern maximum load carrying capacity, and thus generally are related to safety. Serviceability limit states are usually related to performance under normal service conditions, and thus usually are not related to strength or safety. (See Commentary.) The term "resistance" includes both strength limit states and serviceability limit states.

The design strength, ϕR_n (nominal strength multiplied by a resistance factor), of each structural component or assemblage must equal or exceed the effect of the factored loads (nominal loads multiplied by load factors, with due recognition for load combinations). Thus, both the load factor and the resistance factor must be known to determine the reliability of the design, identified in load and resistance factor design as the "safety index." Although the load factors are not stated in this Specification, load criteria contained in American National Standard "Building Code Requirements for Minimum Design Loads in Buildings and Other Structures," ANSI A58.1-1982, were used as the basis for determining the resistance factors. For construction governed by other design load criteria, appropriate adjustment of resistance factors may be required.

The attached Commentary provides background information in order that the user may better understand the provisions of the Specification.

2. Bolts, Nuts, Washers and Paint

(a) Bolt Specifications. Bolts shall conform to the requirements of the current edition of the American Society for Testing and Materials' "Specification for High-Strength Bolts for Structural Steel Joints," ASTM A325, or "Specification for Heat Treated, Steel Structural Bolts, 150 ksi Tensile Strength," ASTM A490, except as provided in paragraph (d) of this section. The Engineer of Record shall specify the type of bolts to be used.

(b) Bolt Geometry. Bolt dimensions shall conform to the current American National Standards Institute's standard, "Heavy Hex Structural Bolts," ANSI Standard B18.2.1, except as provided in paragraph (d) of this section. The length of bolts shall be such that the end of the bolt will be flush with or project beyond the face of the nut when properly installed.

(c) Nut Specifications. Nuts shall conform to the current chemical and mechanical requirements of the American Society for Testing and Materials' "Specification for Carbon and Alloy Steel Nuts," ASTM A563, or "Specification for Carbon and Alloy Steel Nuts for Bolts for High-Pressure and High-Temperature Service," ASTM A194. The grade and surface finish of nuts for each type shall be as follows:

A325 Bolt Type	Nut Specification, Grade and Finish
1 and 2, plain (uncoated)	A563 C, C3, D, D3 and DH3 or A194 2 and 2H; plain
1 and 2, galvanized	A563 DH or A194 2H; galvanized and lubricated
3, plain	A563 C3 and DH3; plain

A490 Bolt Type	Nut Specification, Grade and Finish
1 and 2, plain	A563 DH and DH3 or A194 2H; plain
3, plain	A563 DH3; plain

Nut dimensions shall conform to the current American National Standards Institute's standard, "Heavy Hex Nuts," ANSI Standard B18.2.2., except as provided in paragraph (d) of this section.

(d) Alternative Fastener Designs. Other fasteners or fastener assemblies which meet the materials, manufacturing and chemical composition requirements of ASTM A325 or ASTM A490, as applicable, and which meet the mechanical property requirements of the same specifications in full-size tests, and which have a body diameter and bearing areas under the head and nut not less than those provided by a bolt and nut of the same nominal dimensions prescribed by paragraphs 2(b) and 2(c), may be used subject to the approval of the Engineer of Record. Such alternative fasteners may differ in other dimensions from those of the specified bolts and nuts. Their installation procedure and inspection may differ from procedures specified for regular high-strength bolts in Sections 8 and 9. When a different installation procedure or inspection is used, it shall be detailed in a supplemental specification applying to the alternative fastener, and that specification must be approved by the Engineer of Record.

(e) **Washers.** Flat circular washers and square or rectangular beveled washers shall conform to the current requirements of the American Society for Testing and Materials' "Specification for Hardened Steel Washers," ASTM F436.

(f) **Load Indicating Devices.** Load indicating devices may be used in conjunction with bolts, nuts and washers specified in 2(a) through 2(e). Load indicating devices shall conform to the requirements of American Society for Testing and Materials' "Specification for Compressible-Washer-Type Direct Tension Indicators for Use with Structural Fasteners," ASTM F959. Subject to the approval of the Engineer of Record, direct tension indicating devices different from those meeting the requirements of ASTM F959 may be used provided they satisfy the requirements of 8(d)(4). If their installation procedure and inspection are not identical to that specified in 8(d)(4), they shall be detailed in supplemental specifications provided by the manufacturer and subject to the approval of the Engineer of Record.

(g) **Faying Surface Coatings.** Paint, if used on faying surfaces of connections which are not specified to be slip critical, may be of any formulation. Paint, used on the faying surfaces of connections specified to be slip critical, shall be qualified by test in accordance with "Test Method to Determine the Slip Coefficient for Coatings Used in Bolted Joints" as published by the Research Council on Structural Connections. (See Appendix A.) Manufacturer's certification shall include a certified copy of the test report.

3. Bolted Parts

(a) **Connected Material.** All material within the grip of the bolt shall be steel. There shall be no compressible material such as gaskets or insulation within the grip. Bolted steel parts shall fit solidly together after the bolts are tightened, and may be coated or noncoated. The slope of the surfaces of parts in contact with the bolt head or nut shall not exceed 1:20 with respect to a plane normal to the bolt axis.

(b) **Surface Conditions.** When assembled, all joint surfaces, including surfaces adjacent to the bolt head and nut, shall be free of scale, except tight mill scale, and shall be free of dirt or other foreign material. Burrs that would prevent solid seating of the connected parts in the snug tight condition shall be removed.

Paint is permitted unconditionally on the faying surfaces in connections except in slip-critical connections as defined in Section 5(a).

The faying surfaces of slip-critical connections shall meet the requirements of the following paragraphs, as applicable.

(1) In noncoated joints, paint, including any inadvertent overspray, shall be excluded from areas closer than one bolt diameter but not less

than one inch from the edge of any hole and all areas within the bolt pattern.

(2) Joints specified to have painted faying surfaces shall be blast cleaned and coated with a paint which has been qualified as Class A or B in accordance with the requirements of paragraph 2(g), except as provided in 3(b)3.

(3) Subject to the approval of the Engineer of Record, coatings providing a slip coefficient less than 0.33 may be used provided the mean slip coefficient is established by test in accordance with the requirements of paragraph 2(g), and the design slip resistance, ϕR_s, calculated in accordance with the formula in Section 5(b) or 5(c).

(4) Coated joints shall not be assembled before the coatings have cured for the minimum time used in the qualifying test.

(5) Faying surfaces specified to be galvanized shall be hot-dip galvanized in accordance with American Society for Testing and Materials' "Specification for Zinc (Hot-Galvanized) Coatings on Products Fabricated from Rolled, Pressed, and Forged Steel Shapes, Plates, Bars, and Strip," ASTM A123 and shall subsequently be roughened by means of hand wire brushing. Power wire brushing is not permitted.

(c) Hole Types. Hole types recognized under this specification are standard holes, oversize holes, short slotted holes and long slotted holes. The nominal dimensions for each type hole shall be not greater than those shown in Table 1. Holes not more than $\frac{1}{32}$ inch larger in diameter than the true decimal equivalent of the nominal diameter that may result from a drill or reamer of the nominal diameter are considered acceptable. The slightly conical hole that naturally results from punching operations is considered acceptable. The width of slotted holes which are produced by flame cutting or a combination of drilling or punching and flame cutting shall generally be not more than $\frac{1}{32}$ inch greater than the nominal width except that gouges not more than $\frac{1}{16}$ inch deep shall be permitted. For statically loaded connections, the flame cut surface need not be ground. For dynamically loaded connections, the flame cut surface shall be ground smooth.

4. Design of Bolted Connections

Expressions for design strengths, ϕR_n, of bolts subject to axial tension, shear and combined shear and tension are given in 4(a) and 4(b). They are to be compared to the effect of the factored loads. The design resistances of bolts subject to cyclic application of axial tension are given in 4(e). They are to be compared to effect of cyclically applied nominal (service) loads.

(a) Tension and Shear Strength Limit States. The design strength in axial tension for A325 and A490 bolts which are tightened to the minimum fastener tension specified in Table 4 is ϕR_n. The design strength in shear for A325 and A490 bolts, independent of the installed bolt pretension, is ϕR_n where:

$$R_n = F_n A_b \qquad \text{(LRFD 4.1)}$$

Table 1. Nominal Hole Dimensions

Bolt Dia.	Hole Dimensions			
	Standard (Dia.)	Oversize (Dia.)	Short Slot (Width × Length)	Long Slot (Width × Length)
$\frac{1}{2}$	$\frac{9}{16}$	$\frac{5}{8}$	$\frac{9}{16} \times \frac{11}{16}$	$\frac{9}{16} \times 1\frac{1}{4}$
$\frac{5}{8}$	$\frac{11}{16}$	$\frac{13}{16}$	$\frac{11}{16} \times \frac{7}{8}$	$\frac{11}{16} \times 1\frac{9}{16}$
$\frac{3}{4}$	$\frac{13}{16}$	$\frac{15}{16}$	$\frac{13}{16} \times 1$	$\frac{13}{16} \times 1\frac{7}{8}$
$\frac{7}{8}$	$\frac{15}{16}$	$1\frac{1}{16}$	$\frac{15}{16} \times 1\frac{1}{8}$	$\frac{15}{16} \times 2\frac{3}{16}$
1	$1\frac{1}{16}$	$1\frac{1}{4}$	$1\frac{1}{16} \times 1\frac{5}{16}$	$1\frac{1}{16} \times 2\frac{1}{2}$
$\geq 1\frac{1}{8}$	$d + \frac{1}{16}$	$d + \frac{5}{16}$	$(d + \frac{1}{16}) \times (d + \frac{3}{8})$	$(d + \frac{1}{16}) \times (2.5 \times d)$

In this expression:

R_n = nominal strength of a bolt subject to axial tension or shear, kips

F_n = nominal strength from Table 2 for appropriate kind of load, ksi

A_b = area of bolt corresponding to nominal diameter, in.^{2}

ϕ = resistance factor from Table 2.

(b) Combined Tension and Shear Strength Limit State. In bearing connections in which the applied shear force is greater than $\frac{1}{3}$ the design shear strength according to 4(a), the design strength in axial tension for A325 and A490 bolts is ϕR_n where:

$$R_n = F_{nt} A_b \qquad \text{(LRFD 4.2)}$$

Where

R_n = nominal tension strength of a bolt subject to concurrent shear, kips

F_{nt} = nominal tension strength of a bolt as calculated by formulas in Table 3, ksi

A_b = area of bolt corresponding to nominal diameter, in.^{2}

ϕ = resistance factor equal to 0.75

In Table 3, f_v equals the shear force on the bolt in ksi.

(c) Bearing Strength Limit State. The design bearing strength on the connected material for all bolts in a connection with two or more bolts in the line of force in standard, oversize, or short slotted holes when the edge distance in direction of force is not less than $1\frac{1}{2}d$ and the distance center to center of bolts is not less than $3d$ is ϕR_n where:

$$R_n = 2.4 dt F_u \qquad \text{(LRFD 4.3)}$$

The design bearing strength on the connected material for all bolts in a connection with two or more bolts in the line of force in long slotted holes perpendicular to the direction of force when the edge distance, L, is not less than $1\frac{1}{2}d$ and the distance center to center of bolts is not less than 3d is ϕR_n where:

$$R_n = 2.0 dt F_u \qquad \text{(LRFD 4.4)}$$

The design bearing strength on the connected material for the bolt nearest to the free edge in the direction of force when two or more bolts are in the line of force in standard, oversize, or short slotted holes but with the

Table 2. Nominal Strength of Fasteners

Load Condition	Nominal Strength (ksi)		Resistance Factor, ϕ
	A325	A490	
Applied Static Tension.[a,b,c]	90	113	0.75
Shear on bolt with threads in shear plane.	48[d]	60[d]	0.75
Shear on bolt without threads in shear plane.	60[d]	75[d]	0.75

a. Bolts must be tensioned to requirements of Table 4.
b. See 4(e) for bolts subject to tensile fatigue.
c. Except as required by 4(b).
d. In shear connections transmitting axial force whose length between extreme fasteners measured parallel to the line of force exceeds 50 inches, tabulated values shall be reduced 20 percent.

Table 3. Nominal Tension Strength for Bolts in Bearing Connections (Nominal Tensile Strength, F_{nt}, ksi.)

Fastener Grade	Threads Not Excluded from Shear Plane	Threads Excluded from Shear Plane
ASTM A325	$(90^2 - 3.52f_v^2)^{0.5}$	$(90^2 - 2.25f_v^2)^{0.5}$
ASTM A490	$(113^2 - 3.54f_v^2)^{0.5}$	$(113^2 - 2.27f_v^2)^{0.5}$

edge distance less than $1\frac{1}{2}d$ and for a single bolt in the line of force is ϕR_n where:

$$R_n = LtF_u \leq 3.0dtF_u \qquad \text{(LRFD 4.5)}$$

When two or more bolts are in the line of force in standard, oversize, or short slotted holes and if deformation around the bolt holes is not a design consideration, the design strength in bearing for the individual bolts of a connection may be taken as ϕR_n where:

$$R_n = LtF_u \leq 3.0dtF_u \qquad \text{(LRFD 4.6)}$$

In the foregoing:
R_n = nominal bearing strength of connected material, kips
F_u = specified minimum tensile strength of the connected part, ksi
L = distance in the direction of the force from the center of a standard hole or transverse slotted hole to the edge of the connected part or the distance center to center of standard holes or transverse slots, as applicable, in.
d = nominal diameter of bolt, in.
t = thickness of connected material, in.
ϕ = resistance factor = 0.75

(d) **Prying Action.** The force in bolts required to support loads by means of direct tension shall be calculated considering the effects of the external load and any tension resulting from prying action produced by deformation of the connected parts.

(e) **Tensile Fatigue.** When subject to tensile fatigue loading, the tensile stress in the bolt due to the nominal (service) load plus the prying force resulting from cyclic application of nominal load shall not exceed the following design resistances in kips per square inch. The nominal diameter of the bolt shall be used in calculating the bolt stress. In no case shall the calculated prying force exceed 60 percent of the externally applied load.

Number of Cycles	A325	A490
Not more than 20,000	44	54
From 20,000 to 500,000	40	49
More than 500,000	31	38

Bolts subject to tensile fatigue loading must be tensioned to requirements of Table 4.

5. Design Check for Slip Resistance

(a) **Slip-Critical Joints.** Joints in which, in the judgment of the Engineer of Record, slip would be detrimental to the behavior of the joint, are defined as slip-critical. As discussed in the Commentary, these include but are not necessarily limited to joints subject to fatigue or significant load reversal, joints with bolts in oversize holes or slotted holes with the applied force approximately in the direction of the long dimension of the slots and joints in which welds and bolts share in transmitting shear loads at a common faying surface. Slip-critical joints shall be checked for slip resistance. At the option of the Engineer of Record, the required check may be based upon either nominal loads or factored loads. When serviceability at the nominal (service) load is the design criterion, the design slip resistance specified in Section 5(b) shall be compared with the effects of the nominal loads. When slip of the joint at the factored load level would affect the ability of the structure to support the factored load, the design slip resistance specified in Section 5(c) shall be compared to the effects of the factored loads.

Slip-critical joints shall also be checked to ensure that the ultimate strength of the joint as a bearing joint is equal to or greater than the effect of the factored loads.

Slip-critical joints must be designated on the contract plans and in the specifications. Bolts used in slip-critical joints shall be installed in accordance with the provisions of Section 8(d).

(b) **Slip-Critical Joints Designed at the Nominal Load Level.** Slip-critical joints for which nominal loads are the design criterion shall, in addition to meeting the requirements of Section 4, be proportioned so that the force due to nominal (service) loads does not exceed the design slip resistance for use at nominal loads (service) loads, ϕR_s, where:

$$R_s = D\mu T_m N_b N_s \qquad \text{(LRFD 5.1)}$$

Where:
R_s = nominal slip resistance of a bolt for use at nominal loads, kips
T_m = minimum fastener tension given in Table 4, kips
N_b = number of bolts in the joint
N_s = number of slip planes
D = slip probability factor*
    = 0.81 for μ equal to 0.33
    = 0.86 for μ equal to 0.40

$\quad$ = 0.86 for μ equal to 0.50

μ = mean slip coefficient for Class A, B or C surfaces,† as applicable, or as established by tests

$\quad$ = 0.33 for Class A surfaces (unpainted clean mill scale steel surfaces or surfaces with Class A coating on blast-cleaned steel)

$\quad$ = 0.50 for Class B surfaces (unpainted blast-cleaned steel surfaces or surfaces with Class B coatings on blast-cleaned steel)

$\quad$ = 0.40 for Class C surfaces (hot-dip galvanized and roughened surfaces)

ϕ = 1.0 for standard holes

$\quad$ = 0.85 for oversize and short slotted holes

$\quad$ = 0.70 for long slotted holes transverse to the direction of load

$\quad$ = 0.60 for long slotted holes parallel to the direction of load

* D is a multiplier that reflects the distribution of actual slip coefficient values about the mean, the ratio of measured bolt tensile strength to the specified minimum values, and a slip probability level. Use of other values of D (see Commentary) must be approved by the Engineer of Record.

† Coatings classified as Class A or Class B includes those coatings which provide a mean slip coefficient not less than 0.33 or 0.50, respectively, as determined by "Test Method to Determine the Slip Coefficient for Coatings Used in Bolted Connections."

Table 4. Fastener Tension Required for Slip-Critical Connections and Connections Subject to Direct Tension

Nominal Bolt Size, Inches	Minimum Tension[a] in 1,000s of Pounds (kips)	
	A325 Bolts	A490 Bolts
½	12	15
⅝	19	24
¾	28	35
⅞	39	49
1	51	64
1⅛	56	80
1¼	71	102
1⅜	85	121
1½	103	148

a. Equal to 70 percent of specified minimum tensile strengths of bolts (as specified in ASTM Specifications for tests of full size A325 and A490 bolts with UNC threads loaded in axial tension) rounded to the nearest kip.

$\quad$ When using nominal loads as the basis for design of slip-critical connections subject to applied tension, T, that reduces the net clamping force, the slip resistance (ϕR_s) shall be multiplied by the following factor in which T is the applied tensile force at nominal loads

$$[1 - T/(0.82T_m N_b)] \qquad \text{(LRFD 5.2)}$$

(c) Slip-Critical Joints Designed at Factored Load Level. Slip-critical joints for which factored loads are the design criterion shall, in addition to meeting the requirements of Section 4, be proportioned so that the force due to the factored loads shall not exceed the design slip resistance for use at factored loads, ϕR_{str}, where:

$$R_{str} = 1.13\mu T_m N_b N_s \qquad \text{(LRFD 5.3)}$$

Where terms in Formula (LRFD 5.3) are as defined in 5(b).

When using factored loads as the basis for design of slip-critical connections subject to applied tension, T, that reduces the net clamping force, the slip resistance (ϕR_s) shall be multiplied by the following factor in which T is the applied tensile force at nominal loads

$$[1 - T/(1.13 T_m N_b)] \qquad \text{(LRFD 5.4)}$$

6. Loads in Combination

When the reduced probabilities of maximum loads acting concurrently are accounted for by load combination factors, the resistances given in this Specification shall not be increased.

7. Design Details of Bolted Connections

(a) **Standard Holes.** In the absence of approval by the Engineer of Record for use of other hole types, standard holes shall be used in high-strength bolted connections.

(b) **Oversize and Slotted Holes.** When approved by the Engineer of Record, oversize holes, short slotted holes or long slotted holes may be used subject to the following joint detail requirements:

    (1) Oversize holes may be used in all plies of connections in which the design slip resistance of the connection is greater than the factored nominal load.

    (2) Short slotted holes may be used in any or all plies of connections in which the design strength (Section 4(a)) is greater than the factored nominal load provided the load is applied approximately normal (between 80 and 100 degrees) to the axis of the slot. Short slotted holes may be used without regard for the direction of applied load in any or all plies of connections in which the design slip resistance (Section 5(b)) is greater than the factored nominal load.

    (3) Long slotted holes may be used in one of the connected parts at any individual faying surface in connections in which the design strength (Section 4(a)) is greater than the factored nominal load provided the load is applied approximately normal (between 80 and 100 degrees) to the axis of the slot. Long slotted holes may be used in one of the connected parts at any individual faying surface without regard for the direction of applied load on connections in which the design slip resistance (Section 5(b)) is greater than the factored nominal load.

    (4) Fully inserted finger shims between the faying surfaces of load transmitting elements of connections are not to be considered a long slot element of a connection.

(c) **Washer Requirements.** Design details shall provide for washers in high-strength bolted connections as follows:

    (1) Where the outer face of the bolted parts has a slope greater than 1:20 with respect to a plane normal to the bolt axis, a hardened beveled washer shall be used to compensate for the lack of parallelism.

(2) Hardened washers are not required for connections using A325 and A490 bolts except as required in paragraphs 7(c)(3) through 7(c)(7) for slip-critical connections and connections subject to direct tension or as required by paragraph 8(c) for shear/bearing connections.

(3) Hardened washers shall be used under the element turned in tightening when the tightening is to be performed by calibrated wrench method.

(4) Irrespective of the tightening method, hardened washers shall be used under both the head and the nut when A490 bolts are to be installed and tightened to the tension specified in Table 4 in material having a specified yield point less than 40 ksi.

(5) Where A325 bolts of any diameter or A490 bolts equal to or less than 1 inch in diameter are to be installed and tightened in an oversize or short slotted hole in an outer ply, a hardened washer conforming to ASTM F436 shall be used.

(6) When A490 bolts over 1 inch in diameter are to be installed and tightened in an oversize or short slotted hole in an outer ply, hardened washers conforming to ASTM F436 except with $\frac{5}{16}$ inch minimum thickness shall be used under both the head and the nut in lieu of standard thickness hardened washers. Multiple hardened washers with combined thickness equal to or greater than $\frac{5}{16}$ inch do not satisfy this requirement.

(7) Where A325 bolts of any diameter or A490 bolts equal to or less than 1 inch in diameter are to be installed and tightened in a long slotted hole in an outer ply, a plate washer or continuous bar of at least $\frac{5}{16}$ inch thickness with standard holes shall be provided. These washers or bars shall have a size sufficient to completely cover the slot after installation and shall be of structural grade material, but need not be hardened except as follows. When A490 bolts over 1 inch in diameter are to be used in long slotted holes in external plies, a single hardened washer conforming to ASTM F436 but with $\frac{5}{16}$ inch minimum thickness shall be used in lieu of washers or bars of structural grade material. Multiple hardened washers with combined thickness equal to or greater than $\frac{5}{16}$ inch do not satisfy this requirement.

(8) Alternative design fasteners meeting the requirements of 2(d) with a geometry which provides a bearing circle on the head or nut with a diameter equal to or greater than the diameter of hardened washers meeting the requirements ASTM F436 satisfy the requirements for washers specified in paragraphs 7(c)(4) and 7(c)(5).

8. Installation and Tightening

(a) **Handling and Storage of Fasteners.** Fasteners shall be protected from dirt and moisture at the job site. Only as many fasteners as are anticipated to be installed and tightened during a work shift shall be taken from protected storage. Fasteners not used shall be returned to protected storage at the end of the shift. Fasteners shall not be cleaned of lubricant that is present in as-delivered condition. Fasteners which accumulate rust or dirt resulting from job site conditions shall be cleaned and relubricated prior to installation.

(b) Tension Calibrator. A tension measuring device shall be at all job sites where bolts in slip-critical joints or connections subject to direct tension are being installed and tightened. The tension measuring device shall be used to confirm (1) the suitability of the complete fastener assembly and method of tightening, including lubrication, if required to satisfy the requirements of Table 4, (2) to calibrate the wrenches, if applicable, and (3) to confirm the understanding and proper use by the bolting crew of the method to be used. The frequency of confirmation testing, the number of tests to be performed, and the test procedure shall be as specified in 8(d), as applicable. The accuracy of the tension measuring device shall be confirmed through calibration by an approved testing agency at least annually.

(c) Joint Assembly and Tightening of Shear/Bearing Connections.

  (1) Snug Tightened Bolts. Bolts in connections not within the slip-critical category as defined in Section 5(a) nor subject to tension loads nor required to be pretensioned bearing connections in accordance with 8(c)(2) shall be installed in properly aligned holes, but need only be tightened to the snug tight condition. The snug tight condition is defined as the tightness that exists when all plies in a joint are in firm contact. (See Commentary.) If a slotted hole occurs in an outer ply, a flat hardened washer or common plate washer shall be installed over the slot.

  (2) Tensioned Shear/Bearing Connections. The Engineer of Record may designate certain shear/bearing connections to be tightened to pretension in excess of snug tight. When so designated and identified on the contract drawings, the bolts in such connections shall be installed and tightened in accordance with one of the methods described in Subsections 8(d)(1) through 8(d)(4), but shall not be subject to the requirements for faying surface conditions of slip-critical connections contained in 3(b). The bolts need not be subject to inspection testing to determine the actual level of bolt pretension unless required by the Engineer of Record.

(d) Joint Assembly and Tightening of Slip-Critical and Direct Tension Connections. In slip-critical connections and connections subject to direct tension, fasteners together with washers of size and quality specified, located as required by Section 7(c), shall be installed in properly aligned holes and tightened by any of the methods described in Subsections 8(d)(1) through 8(d)(4) to at least the minimum tension specified in Table 4 when all the fasteners are tight. Tightening may be done by turning the bolt while the nut is prevented from rotating when it is impractical to turn the nut. Impact wrenches, if used, shall be of adequate capacity and sufficiently supplied with air to perform the required tightening of each bolt in approximately 10 seconds. Slip-critical connections and connections subject to direct tension shall be clearly identified on the drawings.

  (1) Turn-of-Nut Tightening. When turn-of-nut tightening is used, hardened washers are not required except as may be specified in 7(c).

   A representative sample of not less than three bolt and nut assemblies of each diameter, length, grade and lot to be used in the work shall be checked at the start of work in a device capable of indicating bolt tension. The test shall demonstrate that the method for estimating the snug tight condition and controlling the turns from snug tight to be used by the bolting crew develops a tension not less than 5 percent greater than the tension required by Table 4.

Bolts shall be installed in all holes of the connection and brought to a "snug tight" condition. Snug tight is defined as the tightness that exists when the plies of the joint are in firm contact. Snug tightening shall progress systematically from the most rigid part of the connection to the free edges until all bolts are simultaneously snug tight and the connection is fully compacted. In some cases, proper tensioning of the bolts may require more than a single cycle of systematic tightening to produce a uniform snug tight condition. Following this initial operation, all bolts in the connection shall be tightened further by application of the rotation specified in Table 5. During the tightening operation, there shall be no rotation of the part not turned by the wrench. Tightening shall progress systematically from the most rigid part of the joint to its free edges.

(2) **Calibrated Wrench Tightening:** Calibrated wrench tightening may be used only when installation procedures are calibrated on a daily basis and when a hardened washer is used under the element turned in tightening. (See the Commentary to this Section.) This specification does not recognize standard torques determined from tables or from formulas which are assumed to relate torque to tension.

When calibrated wrenches are used for installation, they shall be set to provide a tension not less than 5 percent in excess of the minimum tension specified in Table 4. The installation procedures shall be calibrated at least once each working day by tightening representative sample fastener assemblies in a device capable of indicating actual bolt tension. The representative fastener assemblies shall consist of three bolts from each lot, diameter, length and grade with nuts from each lot, diameter and grade and with a hardened washer from the washers being used in the work under the element turned in tightening. Wrenches shall be recalibrated when significant difference is noted in the surface condition of the bolts' threads, nuts or washers. It shall be verified during actual installation in the assembled steelwork that the wrench adjustment selected by the calibration does not produce a nut or bolt head rotation from snug tight greater than that permitted in Table 5. If manual torque wrenches are used, nuts shall be turned in the tightening direction when torque is measured.

When calibrated wrenches are used to install and tension bolts in a connection, bolts shall be installed with hardened washers under the element turned in tightening bolts in all holes of the connection and brought to a snug tight condition. Snug tightening shall progress systematically from the most rigid part of the connections to the free edges until bolts are uniformly snug tight and the plies of the joint are in firm contact. Following this initial tightening operation, the connection shall be tightened using the calibrated wrench. Tightening shall progress systematically from the most rigid part

Table 5. Nut Rotation from Snug Tight Condition[a,b]

Bolt length (underside of head to end of bolt)	Disposition of Outer Face of Bolted Parts		
	Both faces normal to bolt axis	One face normal to bolt axis and other sloped not more than 1:20 (beveled washer not used)	Both faces sloped not more than 1:20 from normal to the bolt axis (beveled washer not used)
Up to and including 4 diameters	1/3 turn	1/2 turn	2/3 turn
Over 4 diameters but not exceeding 8 dia.	1/2 turn	2/3 turn	5/6 turn
Over 8 diameters but not exceeding 12 dia.[c]	2/3 turn	5/6 turn	1 turn

a. Nut rotation is relative to bolt regardless of the element (nut or bolt) being turned. For bolts installed by 1/2 turn and less, the tolerance should be plus or minus 30 degrees; for bolts installed by 2/3 turn and more, the tolerance should be plus or minus 45 degrees.
b. Applicable only to connections in which all material within the grip of the bolt is steel.
c. No research has been performed by the Council to establish the turn-of-nut procedure for bolt lengths exceeding 12 diameters. Therefore, the required rotation must be determined by actual test in a suitable tension measuring device which simulates conditions of solidly fitted steel.

of the joint to its free edges. During snugging and final tightening the element not turned in tightening shall be held to prevent rotation which will damage threads. In some cases, proper tensioning of the bolts may require more than a single cycle of systematic tightening to ensure all bolts are tightened to at least the prescribed amount.

(3) **Installation of Alternative Design Bolts.** When fasteners which incorporate a design feature intended to indicate a predetermined tension or torque has been applied or to control bolt installation tension or torque, and which have been qualified under Section 2(d) are to be installed, a representative sample of not less than three bolts of each diameter, length and grade shall be checked at the job site in a device capable of indicating bolt tension. The test assembly shall include flat hardened washers, if required in the actual connection, arranged as in the actual connections to be tensioned. The calibration test shall demonstrate that each bolt develops a tension not less than 5 percent greater than the tension required by Table 4. Manufacturer's installation procedure as required by Section 2(d) shall be followed for installation of bolts in the calibration device and in all connections.

When alternative design fasteners are used in the work, bolts shall be installed in all holes of the connection and initially tightened sufficiently to bring all plies of the joint into firm contact with the bolts uniformly tight but without yielding or fracturing the control or indicator element of the fasteners. In some cases, proper tensioning of the bolts may require more than a single cycle of systematic partial tightening. After all plies of the joint are in firm contact, all fasteners shall be further tightened, progressing systematically from the most rigid part of the connection to the free edges in a manner that will minimize relaxation of previously tightened fasteners. In some cases, proper tensioning of the bolts may require more than a single cycle of systematic partial tightening prior to final yielding or fracture of the control or indicator element of individual fasteners.

(4) **Direct Tension Indicator Tightening.** When bolts are to be installed using direct tension indicator devices to indicate bolt tension, a representative sample of not less than three devices for each diameter and grade of fastener shall be tested with three typical bolts in a calibration device capable of indicating bolt tension. The test assembly shall include flat hardened washers, if required in the actual connection, arranged as those in the actual connections to be tensioned. The calibration test shall demonstrate that the device indicates a tension not less than 5 percent greater than that required by Table 4.

When bolts are installed in the work using direct tension indicators meeting the requirements of ASTM F959, bolts shall be installed in all holes of the connection and tightened until all plies of the joint are in firm contact and fasteners are uniformly snug tight. Snug tight is indicated by partial compression of the direct tension indicator protrusions. All fasteners shall then be tightened, progressing systematically from the most rigid part of the connection to the free edges in a manner that will minimize relaxation of previously tightened fasteners. In some cases, proper tensioning of the bolts may require more than a single cycle of systematic partial tightening prior to final tightening to deform the protrusion to the specified gap.

Special attention shall be given to proper installation of flat hardened washers when direct tension indicator devices are used with bolts installed in oversize or slotted holes and when the load indicating devices are used under the turned element.

If direct tension indicators different from those meeting the requirements of ASTM F959 are used, manufacturer's installation procedure as required by Section 2(f), shall be followed for installation of bolts in the calibration device and in all connections, and in addition the general requirements for use of direct tension indicators meeting the requirements of ASTM F959 shall be met.

(e) **Identification of Tightening Requirements.** Bolts in slip-critical connections or bolts subject to axial tension which are to be installed and tightened in accordance by one of the methods in 8(d) and which require inspection to ensure that requirements of Table 4 are satisfied shall be clearly identified on the contract drawings. Shear/bearing connections which are to be installed by one of the methods in 8(d) but which need not be inspected to ensure bolt tensions specified in Table 4 are met shall be clearly identified on the contract drawings.

(f) **Reuse of Bolts.** A490 bolts and galvanized A325 bolts shall not be reused. Other A325 bolts may be reused if approved by the Engineer of Record. Touching up or retightening previously snug tightened bolts which may have been loosened by the snugging of adjacent bolts shall not be considered to be a reuse.

9. Inspection

(a) **Inspector Responsibility.** When inspection is required by the contract documents, the Inspector shall determine while the work is in progress that the requirements of Sections 2, 3 and 8, as appropriate, of this Specification are met in the work. All connections shall be inspected to ensure that the plies of the connected elements have been brought into firm contact.

Bolts in connections not identified as being slip-critical nor subject to direct tension nor as tensioned bearing connections as provided in 8(c)(2) should not be inspected for bolt tension. For connections identified to be installed in accordance with 8(c)(2), the Inspector shall monitor installation and tightening of bolts to ensure that bolts are tightened in accordance with one of the methods of 8(d), but should not test the bolts for actual installed pretension.

For all connections specified to be slip critical or subject to axial tension, the Inspector shall observe the demonstration testing, and calibration procedures when such calibration is required, and shall monitor the installation of bolts to determine that all plies of the material have been drawn together and that the selected procedure has been used to tighten all bolts to ensure that the specified procedure was followed to achieve the pretension specified in Table 4. Bolts installed by procedures in Section 8(d) may reach tensions substantially greater than values given in Table 4, but this shall not be cause for rejection.

(b) **Arbitration Inspection.** When high-strength bolts in slip-critical connections and connections subject to direct tension have been installed by any of the tightening methods in Section 8(d) and inspected in accordance with Section 9(a) and a disagreement exists as to the minimum tension of the installed bolts, the following arbitration procedure may be used. Other methods for arbitration inspection may be used if approved by the Engineer of Record.

(1) The Inspector shall use a manual torque wrench which indicates torque by means of a dial or which may be adjusted to give an indication that the job inspecting torque has been reached.

(2) This Specification does not recognize standard torques determined from tables or from formulas which are assumed to relate torque to tension. Testing using such standard torques shall not be considered valid.

(3) A representative sample of five bolts from the diameter, length and grade of the bolts being inspected shall be tightened in the tension measuring device by any convenient means to an initial condition equal to approximately 15 percent of the required fastener tension and then to the minimum tension specified in Table 4. Material under the turned element in the tension measuring device shall be the same as in the actual installation, that is, structural steel or

hardened washer. Tightening beyond the initial condition must not produce greater nut rotation than 1½ times that permitted in Table 5. The job inspecting torque shall be taken as the average of three values thus determined after rejecting the high and low values. The inspecting wrench shall then be applied to the tightened bolts in the work and the torque necessary to turn the nut or head 5 degrees (approximately 1 inch at 12 inch radius) in the tightening direction shall be determined.

(4) Bolts represented by the sample in the foregoing paragraph which have been tightened in the structure shall be inspected by applying, in the tightening direction, the inspecting wrench and its job torque to 10 percent of the bolts, but not less than 2 bolts, selected at random in each connection in question. If no nut or bolt head is turned by application of the job inspecting torque, the connection shall be accepted as properly tightened. If any nut or bolt is turned by the application of the job inspecting torque, all bolts in the connection shall be tested, and all bolts whose nut or head is turned by the job inspecting torque shall be tightened and reinspected. Alternatively, the fabricator or erector, at his option, may retighten all of the bolts in the connection and then resubmit the connection for the specified inspection.

(c) **Delayed Verification Inspection.** The procedures specified in Sections 9(a) and (b) are intended for inspection of bolted connections and verification of pretension at the time of tensioning the joint. If verification of bolt tension is required after a passage of a period of time and exposure of the completed joints, the procedure of Section 9(b) will provide indication of bolt tension which is of questionable accuracy. Procedures appropriate to the specific situation should be used for verification of bolt tension. This might involve use of the arbitration inspection procedure contained herein, or might require the development and use of alternate procedures. (See Commentary.)

APPENDIX A

Testing Method to Determine the Slip Coefficient for Coatings Used in Bolted Joints

Reprinted from *Engineering Journal*
American Institute of Steel Construction, Third Quarter, 1985.

JOSEPH A. YURA and KARL H. FRANK

In 1975, the Steel Structures Painting Council (SSPC) contacted the Research Council on Riveted and Bolted Structural Joints (RCRBSJ), now the Research Council on Structural Connections (RCSC), regarding the difficulties and costs which steel fabricators encounter with restrictions on coatings of contact surfaces for friction-type structural joints. The SSPC also expressed the need for a "standardized test which can be conducted by any certified testing agency at the initiative and expense of any interested party, including the paint manufacturer." And finally, the RCSC was requested to "prepare and promulgate a specification for the conduct of such a standard test for slip coefficients."

The following Testing Method is the answer of Research Council on Structural Connections to the SSPC request. The test method was developed by Professors Joseph A. Yura and Karl H. Frank of the University of Texas at Austin under a grant from the Federal Highway Administration. The Testing Method was approved by the RCSC on June 14, 1984.

1.0 GENERAL PROVISIONS

1.1 Purpose and Scope

The purpose of the testing procedure is to determine the slip coefficient of a coating for use in high-strength bolted connections. The testing specification ensures that the creep deformation of the coating due to both the clamping force of the bolt and the service load joint shear are such that the coating will provide satisfactory performance under sustained loading.

Joseph A. Yura, M. ASCE, is Warren S. Bellows Centennial Professor in Civil Engineering, University of Texas at Austin, Austin, Texas.

Karl H. Frank, A.M. ASCE, is Associate Professor, Department of Civil Engineering, University of Texas at Austin, Austin, Texas.

1.2 Definition of Essential Variables

Essential variables mean those variables which, if changed, will require retesting of the coating to determine its slip coefficient. The essential variables are given below. The relationship of these variables to the limitation of application of the coating for structural joints is also given.

The *time interval* between application of the coating and the time of testing is an essential variable. The time interval must be recorded in hours and any special curing procedures detailed. Curing according to published manufacturer's recommendations would not be considered a special curing procedure. The coatings are qualified for use in structural connections which are assembled after coating for a time equal to or greater than the interval used in the test specimens. Special curing conditions used in the test specimens will also apply to the use of the coating in the structural connections.

The *coating thickness* is an essential variable. The maximum average coating thickness allowed on the bolted structure will be the average thickness, rounded to the nearest whole mil, of the coating used on the creep test specimens minus 2 mils.

The *composition of the coating,* including the thinners used, and its method of manufacture are essential variables. Any change will require retesting of the coating.

1.3 Retesting

A coating which fails to meet the creep or the post-creep slip test requirements given in Sect. 4 may be retested in accordance with methods in Sect. 4 at a lower slip coefficient, without repeating the static short-term tests specified in Sect. 3. Essential variables must remain unchanged in the retest.

2.0 TEST PLATES AND COATING OF THE SPECIMENS

2.1 Test Plates

The test specimen plates for the short-term static tests are shown in Fig. 1. The plates are 4×4 in. plates, ⅝-in. thick, with a 1-in. dia. hole drilled 1½ in. ± ¹⁄₁₆ in. from one edge. The specimen plates for the creep specimen are shown in Fig. 2. The plates are 4×7 in., ⅝-in. thick, with two 1-in. holes, 1½ in. ± ¹⁄₁₆ in. from each end. The edges of the plates may be milled, as rolled or saw cut. Flame cut edges

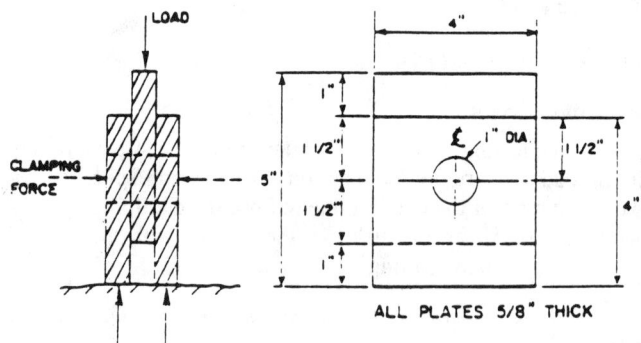

Fig. 1. Compression test specimen

are not permitted. The plates should be flat enough to ensure they will be in reasonably full contact over the faying surface. Any burrs, lips or rough edges should be filed or milled flat. The arrangement of the specimen plates for the testing is shown in Figs. 2 and 3. The plates are to be fabricated from a steel with a minimum yield strength between 36 to 50 ksi.

If specimens with more than one bolt are desired, the contact surface per bolt should be 4×3 in. as shown for the single bolt specimen in Fig. 1.

2.2 Specimen Coating

The coatings are to be applied to the specimens in a manner consistent with the actual intended structural application. The method of applying the coating and the surface preparation should be given in the test report. The specimens are to be coated to an average thickness 2 mils (0.05 mm) greater than average thickness to be used in the structure. The thickness of the total coating and the primer, if used, shall be measured on the contact surface of the specimens. The thickness should be measured in accordance with the Steel Structures Painting Council specification SSPC-PA2, Measurement of Dry Paint Thickness with Magnetic Gages.[1] Two spot readings (six gage readings) should be made for each contact surface. The overall average thickness from the three plates comprising a specimen is the average thickness for the specimen. This value should be reported for each specimen. The average coating thickness of the three creep specimens will be calculated and reported. The average thickness of the creep specimen minus two mils rounded to the nearest whole mil is the maximum average thickness of the coating to be used in the faying surface of a structure.

The time between painting and specimen assembly is to be the same for all specimens within ±4 hours. The average time is to be calculated and reported. The two coating applications required in Sect. 3 are to use the same equipment and procedures.

3.0 SLIP TESTS

The methods and procedures described herein are used to determine experimentally the slip coefficient (sometimes called the coefficient of friction) under short-term static loading for high-strength bolted connections. The slip coefficient will be determined by testing two sets of five specimens. The two sets are to be coated at different times at least one week apart.

3.1 Compression Test Setup

The test setup shown in Fig. 3 has two major loading components, one to apply a clamping force to the specimen plates and another to apply a compressive load to the specimen so that the load is transferred across the faying surfaces by friction.

Clamping Force System. The clamping force system consists of a ⅞-in. dia. threaded rod which passes through the specimen and a centerhole compression ram. A 2H nut is used at both ends of the rod, and a hardened washer is used at each side of the test specimen. Between the ram and the specimen is a specially fabricated ⅞-in. 2H nut in which the threads have been drilled out so that it will slide with little resistance along the rod. When oil is pumped into the centerhole ram,

---

1. Steel Structures Painting Council, *Steel Structures Painting Manual*, Vols. 1 and 2. Pittsburgh, Pa., 1982.

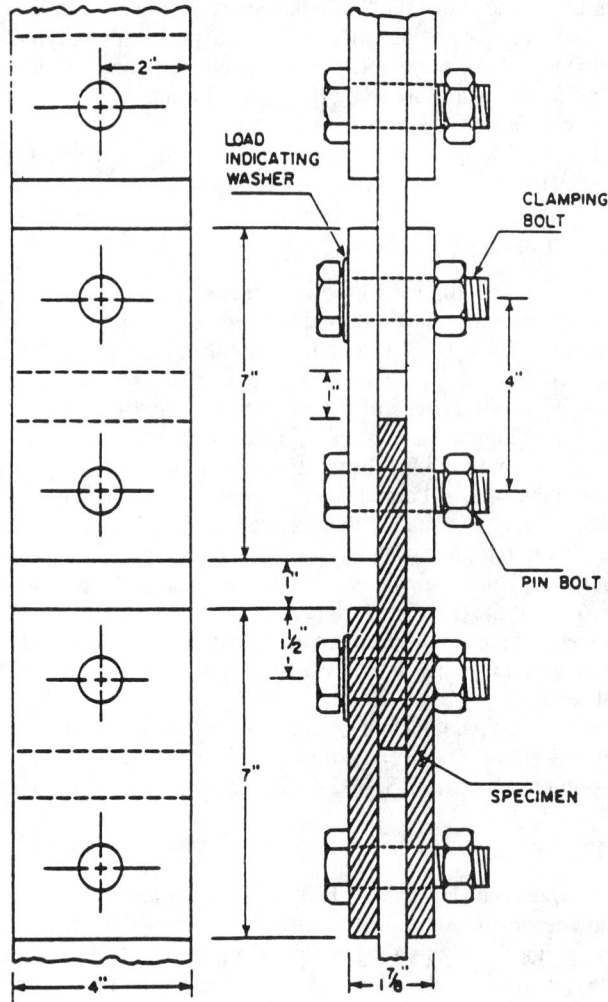

Fig. 2.   Creep test specimens

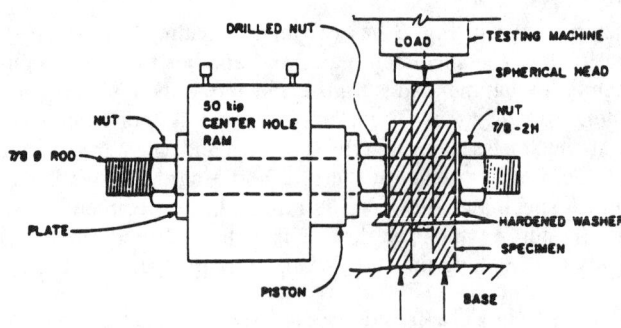

Fig. 3.   Test setup

the piston rod extends, thus forcing the special nut against one of the outside plates of the specimen. This action puts tension in the threaded rod and applies a clamping force to the specimen which simulates the effect of a tightened bolt. If the diameter of the centerhole ram is greater than 1 in., additional plate washers will be necessary at the ends of the ram. The clamping force system must have a capability to apply a load of at least 49 kips and maintain this load during the test with an accuracy of ±1%.

Compressive Load System. A compressive load is applied to the specimen until slip occurs. This compressive load can be applied by a compression test machine or compression ram. The machine, ram and the necessary supporting elements should be able to support a force of 90 kips.

The compression loading system should have an accuracy of 1.0% of the slip load.

3.2 Instrumentation

Clamping Force. The clamping force must be measured within 0.5 kips. This may be accomplished by measuring the pressure in the calibrated ram or placing a load cell in series with the ram.

Compression Load. The compression load must be measured during the test. This may be accomplished by direct reading from a compression testing machine, a load cell in series with the specimen and the compression loading device, or pressure readings on a calibrated compression ram.

Slip Deformation. The relative displacement of the center plate and the two outside plates must be measured. This displacement, called slip for simplicity, should be the average which occurs at the centerline of the specimen. This can be accomplished by using the average of two gages placed on the two exposed edges of the specimen or by monitoring the movement of the loading head relative to the base. If the latter method is used, due regard must be taken for any slack that may be present in the loading system prior to application of the load. Deflections can be measured by dial gages or any other calibrated device which has an accuracy of 0.001 in.

3.3 Test Procedure

The specimen is installed in the test setup as shown in Fig. 3. Before the hydraulic clamping force is applied, the individual plates should be positioned so that they are in, or are close to, bearing contact with the ⅞-in. threaded rod in a direction opposite to the planned compressive loading to ensure obvious slip deformation. Care should be taken in positioning the two outside plates so that the specimen will be straight and both plates are in contact with the base.

After the plates are positioned, the centerhold ram is engaged to produce a clamping force of 49 kips. The applied clamping force should be maintained within ±0.5 kips during the test until slip occurs.

The spherical head of the compression loading machine should be brought in contact with the center plate of the specimen after the clamping force is applied. The spherical head or other appropriate device ensures uniform contact along the edge of the plate, thus eliminating eccentric loading. When 1 kip or less of compressive load is applied, the slip gages should be engaged or attached. The purpose of engaging the deflection gage(s), after a slight load is applied, is to eliminate initial specimen settling deformation from the slip readings.

When the slip gages are in place, the compression load is applied at a rate not exceeding 25 kips (109 kN) per minute, or 0.003 in. of slip displacement per minute until the slip load is reached. The test should be terminated when a slip of 0.05 in. or greater is recorded. The load-slip relationship should preferably be monitored continuously on an X-Y plotter throughout the test, but in lieu of continuous data, sufficient load-slip data must be recorded to evaluate the slip load defined below.

3.4 Slip Load

Typical load-slip response is shown in Fig. 4. Three types of curves are usually observed and the slip load associated with each type is defined as follows:

Curve (a). Slip load is the maximum load, provided this maximum occurs before a slip of 0.02 in. is recorded.

Curve (b). Slip load is the load at which the slip rate increases suddenly.

Curve (c). Slip load is the load corresponding to a deformation of 0.02 in. This definition applies when the load vs. slip curves show a gradual change in response.

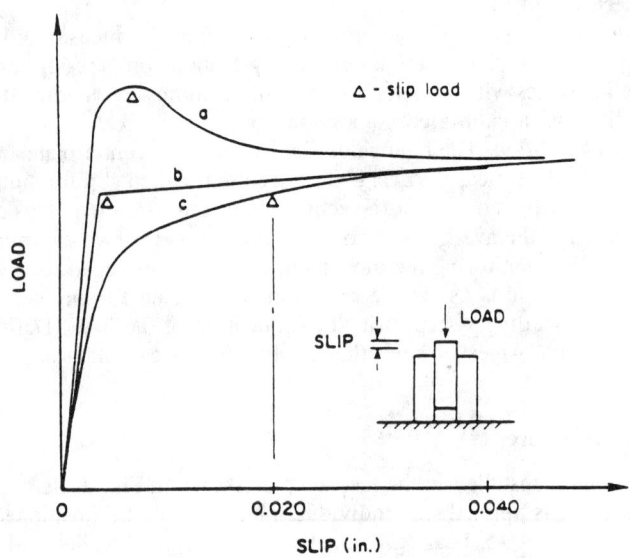

Fig. 4. Definition of slip load

3.5 Coefficient of Slip

The slip coefficient k_s for an individual specimen is calculated as follows:

$$k_s = \frac{\text{slip load}}{2 \times \text{clamping force}}$$

The mean slip coefficient for both sets of five specimens must be compared. If the two means differ by more than 25%, using the smaller mean as the base, a third five-specimen set must be tested. The mean and standard deviation of the data from all specimens tested define the slip coefficient of the coating.

3.6 Alternate Test Methods

Other test methods to determine slip may be used provided the accuracy of load measurement and clamping satisfies the conditions presented in the previous sections. For example, the slip load may be determined from a tension-type test setup rather than the compression-type as long as the contact surface area per fastener of the test specimen is the same as shown in Fig. 1. The clamping force of at least 49 kips may be applied by any means provided the force can be established within $\pm 1\%$. Strain-gaged bolts can usually provide the desired accuracy. However, bolts installed by turn-of-nut method, tension indicating fasteners and load indicator washers usually show too much variation to be used in the slip test.

4.0 TENSION CREEP TESTS

The test method outlined is intended to ensure the coating will not undergo significant creep deformation under service loading. The test also determines the loss in clamping force in the fastener due to the compression or creep of the paint. Three replicate specimens are to be tested.

4.1 Test Setup

Tension-type specimens, as shown in Fig. 2, are to be used. The replicate specimens are to be linked together in a single chain-like arrangement, using loose pin bolts, so the same load is applied to all specimens. The specimens shall be assembled so the specimen plates are bearing against the bolt in a direction opposite to the applied tension loading. Care should be taken in the assembly of the specimens to ensure the centerline of the holes used to accept the pin bolts is in line with the bolts used to assemble the joint. The load level, specified in Sect. 4.2, shall be maintained constant within $\pm 1\%$ by springs, load maintainers, servo controllers, dead weights or other suitable equipment. The bolts used to clamp the specimens together shall be $\frac{7}{8}$-in. dia. A490 bolts. All bolts should come from the same lot.

The clamping force in the bolts should be a minimum of 49 kips. The clamping force is to be determined by calibrating the bolt force with bolt elongation, if standard bolts are used. Special fasteners which control the clamping force by other means such as bolt torque or strain gages may be used. A minimum of three bolt calibrations must be performed using the technique selected for bolt force determination. The average of the three-bolt calibration is to be calculated and reported. The method of measuring bolt force must ensure the clamping force is within ± 2 kips (9 kN) of the average value.

The relative slip between the outside plates and the center plates shall be measured to an accuracy of 0.001 in. (0.02 mm). This is to be measured on both sides of each specimen.

4.2 Test Procedure

The load to be placed on the creep specimens is the service load permitted for $\frac{7}{8}$-in. A490 bolts in slip-critical connections by the latest edition of the *Specification for Structural Joints Using ASTM A325 or A490 Bolts*[2] for the particular slip coefficient category under consideration. The load is to be placed on the specimen and held

---

2. Research Council on Structural Connections, *Specification for Structural Joints Using ASTM A325 or A490 Bolts*, American Institute of Steel Construction, Inc., Chicago, November 1985.

for 1,000 hours. The creep deformation of a specimen is calculated using the average reading of the two displacements on each side of the specimen. The difference between the average after 1,000 hours and the initial average reading taken within one-half hour after loading the specimens is defined as the creep deformation of the specimen. This value is to be reported for each specimen. If the creep deformation of any specimen exceeds 0.005 in. (0.12 mm), the coating has failed the test for the slip coefficient used. The coating may be retested using new specimens in accordance with this section at a load corresponding to a lower value of slip coefficient.

If the value of creep deformation is less than 0.005 in. (0.12 mm) for all specimens, the specimens are to be loaded in tension to a load calculated as

$$P_u = \text{average clamping force} \times \text{design slip coefficient} \times 2$$

since there are two slip planes. The average slip deformation which occurs at this load must be less than 0.015 in. (0.38 mm) for the three specimens. If the deformation is greater than this value, the coating is considered to have failed to meet the requirements for the particular slip coefficient used. The value of deformation for each specimen is to be reported.

COMMENTARY

The slip coefficient under short-term static loading has been found to be independent of clamping force, paint thickness and hole diameter.[3] The slip coefficient can be easily determined using the hydraulic bolt test setup included in this specification. The slip load measured in this setup yields the slip coefficient directly since the clamping force is controlled. The slip coefficient k, is given by

$$k_s = \frac{\text{slip load}}{2 \times \text{clamping force}}$$

The resulting slip coefficient has been found to correlate with both tension and compression tests of bolted specimens. However, tests of bolted specimens revealed that the clamping force may not be constant but decreases with time due to the compressive creep of the coating on the faying surfaces and under the nut and bolt head. The reduction of the clamping force can be considerable for joints with high clamping force and thick coatings, as much as a 20% loss. This reduction in clamping force causes a corresponding reduction in the slip load. The resulting reduction in slip load must be considered in the procedure used to determine the design allowable slip loads for the coating.

The loss in clamping force is a characteristic of the coating. Consequently, it cannot be accounted for by an increase in the factor of safety or a reduction in the clamping force used for design without unduly penalizing coatings which do not exhibit this behavior.

The creep deformation of the bolted joint under the applied shear loading is also an important characteristic and a function of the coating applied. Thicker coatings tend to creep more than thinner coatings. Rate of creep deformation increases as the applied load approaches the slip load. Extensive testing has shown the rate of creep is not constant with time, rather it decreases with time. After 1,000 hours of loading, the additional creep deformation is negligible.

3. Frank, K. H., and J. A. Yura, *An Experimental Study of Bolted Shear Connections*, FHWA/RD-81-148, Federal Highway Administration, Washington, D.C., December 1981.

The proposed test methods are designed to provide the necessary information to evaluate the suitability of a coating for slip critical bolted connections and to determine the slip coefficient to be used in the design of the connections. The initial testing of the compression specimens provides a measure of the scatter of the slip coefficient. In order to get better statistical information, a third set of specimens must be tested whenever the means of the initial two sets differ by more than 25%.

The creep tests are designed to measure the paint's creep behavior under the service loads determined by the paint's slip coefficient based on the compression test results. The slip test conducted at the conclusion of the creep test is to ensure the loss of clamping force in the bolt does not reduce the slip load below that associated with the design slip coefficient. A490 bolts are specified, since the loss of clamping force is larger for these bolts than A325 bolts. Qualifying of the paint for use in a structure at an average thickness of 2 mils less than the test specimen is to ensure that a casual buildup of paint due to overspray, etc., does not jeopardize the coating's performance.

The use of 1-in. (25 mm) holes in the specimens is to ensure that adequate clearance is available for slip. Fabrication tolerances, coating buildup on the holes and assembly tolerances reduce the apparent clearances.

Commentary on Specifications for Structural Joints Using ASTM A325 or A490 Bolts

June 8, 1988.

Historical Notes

When first approved by the Research Council on Structural Connections of the Engineering Foundation, January 1951, the "Specification for Assembly of Structural Joints Using High-Strength Bolts" merely permitted the substitution of a like number of A325 high-strength bolts for hot driven ASTM A141 (presently identified as A502, Grade 1) steel rivets of the same nominal diameter. It was required that all contact surfaces be free of paint. As revised in 1954, the omission of paint was required to apply only to "joints subject to stress reversal, impact or vibration, or to cases where stress redistribution due to joint slippage would be undesirable." This relaxation of the earlier provision recognized the fact that, in a great many cases, movement of the connected parts that brings the bolts into bearing against the sides of their holes is in no way detrimental.

In the first edition of the Specification published in 1951, a table of torque to tension relationships for bolts of various diameters was included. It was soon demonstrated in research that a variation in the torque to tension relationship of as high as plus or minus 40 percent must be anticipated unless the relationship is established individually for each bolt lot, diameter and fastener condition. Hence, by the 1954 edition of the Specification, recognition of standard torque to tension relationships in the form of tabulated values or formulas was withdrawn. Recognition of the calibrated wrench method of tightening was retained, however, until 1980, but with the requirement that the torque required for installation or inspection be determined specifically for the bolts being installed on a daily basis. Recognition of the method was withdrawn in 1980 because of continuing controversy resulting from failure of users to adhere to the detailed requirements for valid use of the method both during installation and inspection. With the 1985 version of the Specification, the calibrated wrench method was reinstated, but with more detailed requirements which should be carefully followed.

The increasing use of high-strength steels created the need for bolts substantially stronger than A325 in order to resist the much greater forces they support without resort to very large connections. To meet this need, a new ASTM specification, A490, was developed. When provisions for the use of these bolts were included in this Specification in 1964, it was required that they be tightened to their specified proof load, as was required for the installation of A325 bolts. However, the ratio of proof load to specified minimum tensile strength is approximately 0.7 for A325 bolts, whereas it is 0.8 for A490 bolts. Calibration studies have shown that high-strength bolts have ultimate load capacities in torqued tension which vary from about

80 to 90 percent of the pure-tension tensile strength.[1] Hence, if minimum strength A490 bolts were supplied and they experienced the maximum reduction due to torque required to induce the tension, there is a possibility that these bolts could not be tightened to proof load by any method of installation. Also, statistical studies have shown that tightening to the 0.8 times tensile strength under calibrated wrench control may result in some "twist-off" bolt failures during installation or in some cases a slight amount of under-tightening.[2] Therefore, the required installed tension for A490 bolts was reduced to 70 percent of the specified minimum tensile strength. For consistency, but with only minor change, the initial tension required for A325 bolts was also set at 70 percent of their specified minimum tensile strength and, at the same time, the values for minimum required pretension were rounded off to the nearest kip.

C1 Scope

This Specification deals only with two types of high-strength bolts, namely, ASTM A325 and A490, and to their installation in structural steel joints. The provisions may not be relied upon for high-strength fasteners of other chemical composition or mechanical properties or size. The provisions do not apply to ASTM A325 or A490 fasteners when material other than steel is included in the grip. The provisions do not apply to high-strength anchor bolts.

The Specification relates only to the performance of fasteners in structural steel connections and those few aspects of the connected material that affect the performance of the fasteners in connections. Many other aspects of connection design and fabrication are of equal importance and must not be overlooked. For information on questions of design of connected material, not covered herein, the user is directed to standard textbooks on design of structural steel and also to Kulak, G. L., J. W. Fisher, and J. H. A. Struik, *Guide to Design Criteria for Bolted and Riveted Joints*, 2nd ed., New York: John Wiley & Sons, 1987. (Hereinafter referred to as the *Guide*.)

C2 Bolts, Nuts, Washers and Paint

Complete familiarity with the referenced ASTM Specification requirements is necessary for the proper application of this Specification. Discussion of referenced specifications in this Commentary is limited to only a few frequently overlooked or little-understood items.

In this Specification, a single style of fastener (heavy hex structural bolts with heavy hex nuts) available in two strength grades (A325 and A490) is specified as a principal style, but conditions for acceptance of other types of fasteners are provided.

Bolt Specifications. ASTM A325 and A490 bolts are manufactured to dimensions as specified in ANSI Standard B18.2.1 for Heavy Hex Structural Bolts. The basic dimensions, as defined in Fig. C1, are shown in Table C1. The principal geometric features of heavy hex structural bolts that distinguish them from bolts for general application are the size of the head and the body length. The head of the heavy hex

---

1. Christopher, R. J., G. L. Kulak, and J. W. Fisher, "Calibration of Alloy Steel Bolts," ASCE *Journal of the Structural Division*, Vol. 92, No. ST2, Proc. Paper 4768, April, 1966, pp. 19–40.
2. Gill, P. J., "Specifications of Minimum Preloads for Structural Bolts," Memorandum 30, G.K.N. Group Research Laboratory, England, 1966 (Unpublished Report).

Table C1

Nominal Bolt Size, Inches D	Bolt Dimensions, Inches Heavy Hex Structural Bolts			Nut Dimensions, Inches Heavy Hex nuts	
	Width across flats, F	Height H	Thread length	Width across flats, W	Height H
$\frac{1}{2}$	$\frac{7}{8}$	$\frac{5}{16}$	1	$\frac{7}{8}$	$\frac{31}{64}$
$\frac{5}{8}$	$1\frac{1}{16}$	$\frac{25}{64}$	$1\frac{1}{4}$	$1\frac{1}{16}$	$\frac{39}{64}$
$\frac{3}{4}$	$1\frac{1}{4}$	$\frac{15}{32}$	$1\frac{3}{8}$	$1\frac{1}{4}$	$\frac{47}{64}$
$\frac{7}{8}$	$1\frac{7}{16}$	$\frac{35}{64}$	$1\frac{1}{2}$	$1\frac{7}{16}$	$\frac{55}{64}$
1	$1\frac{5}{8}$	$\frac{39}{64}$	$1\frac{3}{4}$	$1\frac{5}{8}$	$\frac{64}{64}$
$1\frac{1}{8}$	$1\frac{13}{16}$	$\frac{11}{16}$	2	$1\frac{13}{16}$	$1\frac{7}{64}$
$1\frac{1}{4}$	2	$\frac{25}{32}$	2	2	$1\frac{7}{32}$
$1\frac{3}{8}$	$2\frac{3}{16}$	$\frac{27}{32}$	$2\frac{1}{4}$	$2\frac{3}{16}$	$1\frac{11}{32}$
$1\frac{1}{2}$	$2\frac{3}{8}$	$\frac{15}{16}$	$2\frac{1}{4}$	$2\frac{3}{8}$	$1\frac{15}{32}$

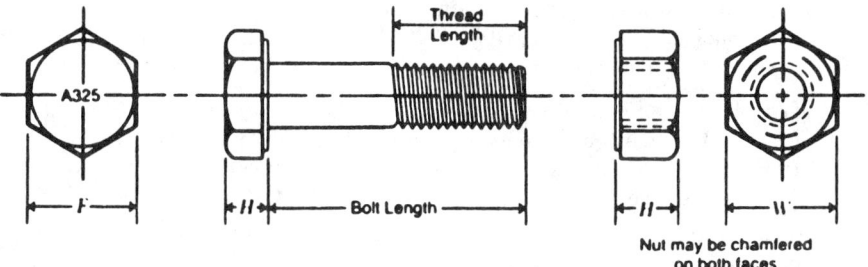

Nut may be chamfered on both faces

Fig. C1. Heavy hex structural bolt and heavy hex nut

structural bolt is specified to be the same size as a heavy hex nut of the same nominal diameter in order that the ironworker may use a single size wrench or socket on both the bolt head and the nut. Heavy hex structural bolts have shorter thread length than bolts for general application. By making the body length of the bolt the control dimension, it has been possible to exclude the thread from all shear planes, except in the case of thin outside parts adjacent to the nut. Depending upon the amount of bolt length added to adjust for incremental stock lengths, the full thread may extend into the grip by as much as $\frac{3}{8}$ inch for $\frac{1}{2}$, $\frac{5}{8}$, $\frac{3}{4}$, $\frac{7}{8}$, $1\frac{1}{4}$, and $1\frac{1}{2}$ in. diameter bolts and as much as $\frac{1}{2}$ inch for 1, $1\frac{1}{8}$ and $1\frac{3}{8}$ in. diameter bolts. Inclusion of some thread run-out in the plane of shear is permissible. Of equal or even greater importance is exercise of care to provide sufficient thread for nut tightening to keep the nut threads from jamming into the thread run-out. When the thickness of an outside part is less than the amount the threads may extend into the grip tabulated above, it may be necessary to call for the next increment of bolt length together with sufficient flat washers to ensure full tightening of the nut without jamming nut threads on the thread run-out.

There is an exception to the short thread length requirements for ASTM A325 bolts discussed in the foregoing. Beginning with ASTM A325-83, supplementary requirements have been added to the ASTM A325 Specification which permit the purchaser, when the bolt length is equal to or shorter than four times the nominal diameter, to specify that the bolt be threaded for the full length of the shank. This exception to the requirements for thread length of heavy hex structural bolts was provided in the Specification in order to increase economy through simplified ordering and inventory control in the fabrication and erection of structures using relatively thin materials where strength of the connection is not dependent upon shear strength of the bolt, whether threads are in the shear plane or not. The Specification requires that bolts ordered to such supplementary requirements be marked with the symbol A325T.

In order to determine the required bolt length, the value shown in Table C2 should be added to the grip (i.e., the total thickness of all connected material, exclusive of washers). For each hardened flat washer that is used, add $\frac{5}{32}$ inch, and for each beveled washer add $\frac{5}{16}$ inch. The tabulated values provide appropriate allowances for manufacturing tolerances, and also provide for full thread engagement (defined as having the end of the bolt at least flush with the face of the nut) with an installed heavy hex nut. The length determined by the use of Table C2 should be adjusted to the next longer $\frac{1}{4}$ inch length.

ASTM A325 and ASTM A490 currently provide for three types (according to metallurgical classification) of high-strength structural bolts, supplied in sizes $\frac{1}{2}$ inch to $1\frac{1}{2}$ inch inclusive except for A490 Type 2 bolts which are available in diameters from $\frac{1}{2}$ inch to 1 inch inclusive:

Type 1. Medium carbon steel for A325 bolts, alloy steel for A490 bolts.
Type 2. Low carbon martensitic steel for both A325 and A490 bolts.
Type 3. Bolts having improved atmospheric corrosion resistance and weathering characteristics for both A325 and A490 bolts.

When the bolt type is not specified, either Type 1, Type 2 or Type 3 may be supplied at the option of the manufacturer. Special attention is called to the requirement in ASTM A325 that, where elevated temperature applications are involved, Type 1 bolts shall be specified by the purchaser. This is because the chemistry of Type 2 bolts permits heat treatment at sufficiently low temperatures that subsequent heating to elevated temperatures may affect the mechanical properties.

Heavy Hex Nuts. Heavy hex nuts for use with A325 bolts may be manufactured to the requirements of ASTM A194 for grades 2 or 2H or the requirements of ASTM A563 for grades DH, except that nuts to be galvanized for use with galvanized bolts must be hardened nuts meeting the requirements for ASTM A563 grade DH.

The heavy hex nuts for use with A490 bolts may be manufactured to the requirements of ASTM A194 for grade 2H or the requirements of ASTM A563 for grade DH.

Galvanized High-Strength Bolts. Galvanized high-strength bolts and nuts must be considered as a manufactured matched assembly; hence, comments relative to them have not been included in the foregoing paragraphs where bolts and nuts have been considered separately. Insofar as the hot-dip galvanized bolt and nut assembly, per se, is concerned, four principal factors need be discussed in order that the provisions of the Specification may be understood and properly applied. These are (1) the effect of the hot-dip galvanizing process on the mechanical properties of high-strength steels, (2) the effect of hot-dip galvanized coatings on the nut stripping strength, (3) the effect of galvanizing upon the torque involved in the tightening operation, and (4) shipping requirements.

The ASTM Specifications for galvanized A325 high-strength bolts recognize

Table C2

Nominal Bolt Size, Inches	To Determine Required Bolt Length Add to Grip, in Inches
½	$^{11}/_{16}$
⅝	⅞
¾	1
⅞	1⅛
1	1¼
1⅛	1½
1¼	1⅝
1⅜	1¾
1½	1⅞

both the hot-dip galvanizing process and the mechanical galvanizing process. The effects of the two processes upon the performance characteristics and requirements for proper installation are distinctly different; therefore, distinction between the two must be noted in the comments which follow. ASTM A325 Specifications require that all components of a fastener assembly (nuts, bolts and washers) shall have been coated by the same process and that the supplier's option is limited to one process per item with no mixed processes in a lot. Mixing a bolt galvanized by one process with a nut galvanized by the other may result in a unworkable assembly.

Effect of Hot-Dip Galvanizing on the Strength of Steels. Steels in the 200 ksi and higher tensile strength range are subject to embrittlement if hydrogen is permitted to remain in the steel and the steel is subjected to high tensile stress. The minimum tensile strength of A325 bolts is 105 or 120 ksi, depending upon the size, comfortably below the critical range. The required maximum tensile strength for A490 bolts was set at 170 ksi in order to provide a little more than a 10 percent margin below 200 ksi; however, because manufacturers must target their production slightly higher than the required minimum, A490 bolts close to the critical range of tensile strength must be anticipated. For black bolts this is not a cause for concern, but, if the bolt is hot-dip galvanized, a hazard of delayed brittle fracture in service exists because of the real possibility of introduction of hydrogen into the steel during the pickling operation of the hot-dip galvanizing process and the subsequent "sealing-in" of the hydrogen by the zinc coating. There also exists the possibility of cathodic hydrogen adsorption arising from corrosion process in service in aggressive environments. ASTM Specifications provide for the galvanizing of A325 bolts but not A490 bolts. Galvanizing of A490 bolts is not permitted. Because pickling and emersion in molten zinc is not involved, galvanizing by the mechanical process essentially avoids potential for hydrogen embrittlement.

The heat treatment temperatures for Type 2 ASTM A325 bolts are in the range of the molten zinc temperatures for hot-dip galvanizing; therefore there is a potential for diminishing the heat treated mechanical properties of Type 2 A325 bolts by the hot-dip galvanizing process. For this reason, the current Specifications require that only mechanical galvanizing shall be used on Type 2 ASTM A325 bolts.

Nut Stripping Strength. Hot-dip galvanizing affects the stripping strength of the nut-bolt assembly primarily because, to accommodate the relatively thick zinc coatings of non-uniform thickness on bolt threads, it is usual practice to hot-dip galvanize the blank nut and then to tap the nut oversize after galvanizing. This overtapping results in a reduction in the amount of engagement between the steel portions of the male and female threads with a consequent approximately 25 percent reduction in the stripping strength. Only the stronger hardened nuts have adequate strength to meet specification requirements even with the reduction due to overtapping; therefore, ASTM A325 specifies that only Grades DH and 2H be used for galvanized nuts. This requirement should not be overlooked if non-galvanized nuts are purchased and then sent to a local galvanizer for hot-dip galvanizing. Because the mechanical galvanizing process results in a more uniformly distributed and smooth zinc coating, nuts may be tapped oversize before galvanizing by an amount less than required for the hot-dip process before galvanizing. This results in a better bolt-nut fit with zinc coating on the internal threads of the nut.

Effect of Galvanizing Upon Torque Involved in Tightening. Research[3] has shown that, in the as-galvanized condition, galvanizing both increases the friction between the bolt and nut threads and also makes the torque induced tension much more variable. Lower torque and more consistent results are provided if the nuts are lubricated; thus, ASTM A325 requires that a galvanized bolt and lubricated galvanized nut shall be assembled in a steel joint with a galvanized washer and tested in accordance with ASTM A563 by the manufacturer prior to shipment to ensure that the galvanized nut with the lubricant provided may be rotated from the snug tight condition well in excess of the rotation required for full tensioning of the bolts without stripping. The requirement applies to both hot-dip and mechanical galvanized fasteners.

Shipping Requirements for Galvanized Bolts and Nuts. The above requirements clearly indicate (1) that galvanized bolts and nuts are to be treated as a matched assembly, (2) that the seller must supply nuts which have been lubricated and tested with the supplied bolts, and (3) that nuts and bolts must be shipped together in the same shipping container. Purchase of galvanized bolts and galvanized nuts from separate sources is not in accordance with the intent of the ASTM Specifications because the control of overtapping and the testing and application of lubricant would be lost. Because some of the lubricants used to meet the requirements of ASTM Specifications are water soluble, it is advisable that galvanized bolts and nuts be shipped and stored in plastic bags in wood or metal containers.

Washers. The primary function of washers is to provide a hardened non-galling surface under the element turned in tightening, particularly for those installation procedures which depend upon torque for control or inspection. Circular hardened washers meeting the requirements of ASTM A436 provide an increase in bearing area of 45 to 55 percent over the area provided by a heavy hex bolt head or nut; however, tests have shown that standard thickness washers play only a minor role in distributing the pressure induced by the bolt pretension, except where oversize or short slotted holes are used. Hence, consideration is given to this function only in the case of oversize and short slotted holes. The requirement for standard thickness hardened washers, when such washers are specified as an aid in the distribu-

---

3. Birkemoe, P. C., and D. C. Herrschaft, "Bolted Galvanized Bridges—Engineering Acceptance Near," ASCE *Civil Engineering,* April 1970.

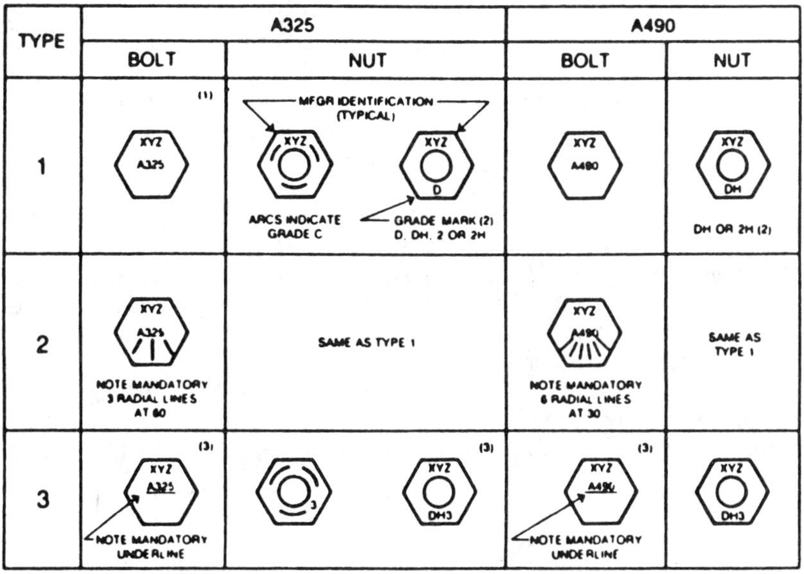

TYPE	A325		A490	
	BOLT	**NUT**	**BOLT**	**NUT**
1	(1) XYZ A325	MFGR IDENTIFICATION (TYPICAL) — ARCS INDICATE GRADE C — GRADE MARK (2) D, DH, 2 OR 2H	XYZ A490	XYZ DH — DH OR 2H (2)
2	XYZ A325 — NOTE MANDATORY 3 RADIAL LINES AT 60	SAME AS TYPE 1	XYZ A490 — NOTE MANDATORY 6 RADIAL LINES AT 30	SAME AS TYPE 1
3	(3) XYZ A325 — NOTE MANDATORY UNDERLINE	(3) 3 — DH3	(3) XYZ A490 — NOTE MANDATORY UNDERLINE	XYZ DH3

(1) ADDITIONAL OPTIONAL 3 RADIAL LINES AT 120° MAY BE ADDED
(2) TYPE 3 ALSO ACCEPTABLE
(3) ADDITIONAL OPTIONAL MARK INDICATING WEATHERING GRADE MAY BE ADDED

Fig. C2. Required marking for acceptable bolt and nut assemblies

tion of pressure, is waived for alternative design fasteners which incorporate a bearing surface under the head of the same diameter as the hardened washer; however, the requirements for hardened washers to satisfy the principal requirement of providing a non-galling surface under the element turned in tightening is not waived. The maximum thickness is the same for all standard washers up to and including 1½ inch bolt diameter in order that washers may be produced from a single stock of material.

The requirement that heat-treated washers not less than $\frac{5}{16}$ inch thick be used to cover oversize and slotted holes in external plies, when A490 bolts of 1⅛ inch or larger diameter are used, was found necessary to distribute the high clamping pressure so as to prevent collapse of the hole perimeter and enable development of the desired clamping force. Preliminary investigation has shown that a similar but less severe deformation occurs when oversize or slotted holes are in the interior plies. The reduction in clamping force may be offset by "keying," which tends to increase the resistance to slip. These effects are accentuated in joints of thin plies.

Marking. Heavy hex structural bolts and heavy hex nuts are required by ASTM Specifications to be distinctively marked. Certain markings are mandatory. In addition to the mandatory markings, the manufacturer may apply additional distinguishing marking. The mandatory and optional markings are shown in Figure C2.

Paint. In the previous edition of the Specification, generic names for paints applied to faying surfaces was the basis for categories of allowable working stresses in "friction" type connections. Research[4] completed since the adoption of the 1980 Specification has demonstrated that the slip coefficients for paints described by a generic type are not single values but depend also upon the type of vehicle used. Small differences in formulation from manufacturer to manufacturer or from lot to lot with a single manufacturer significantly affect slip coefficients if certain essential variables within a generic type are changed. It is unrealistic to assign paints to categories with relatively small incremental differences between categories based solely upon a generic description. As a result of the research, a test method was developed and adopted by the Council titled "Test Method to Determine the Slip Coefficient for Coatings Used in Bolted Joints." A copy of this document is appended to this Specification as Appendix A. The method, which requires requalification if an essential variable is changed, is the sole basis for qualification of any coating to be used under this Specification. Further, normally only two categories of slip coefficient for paints to be used in slip-critical joints are recognized: Class A for coatings which do not reduce the slip coefficient below that provided by clean mill scale, and Class B for paints which do not reduce the slip coefficient below that of blast-cleaned steel surfaces.

The research cited in the preceding paragraph also investigated the effect of varying the time from coating the faying surfaces to assembly of the connection and tightening the bolts. The purpose was to ascertain if partially cured paint continued to cure within the assembled joint over a period of time. It was learned that all curing ceased at the time the joint was assembled and tightened and that paint coatings that were not fully cured acted much as a lubricant would; thus, the slip resistance of the joint was severely reduced from that which was provided by faying surfaces that were fully cured prior to assembly.

C3 Bolted Parts

Material Within the Grip. The Specification is intended to apply to structural joints in which all of the material within the grip of the bolt is steel.

Surface Conditions. The Test Method to Determine the Slip Coefficient for Coatings Used in Bolted Joints includes long-term creep test requirements to ensure reliable performance for qualified paint coatings. However, it must be recognized that in the case of hot-dip galvanized coatings, especially if the joint consists of many plies of thickly coated material, relaxation of bolt tension may be significant and may require retensioning of the bolts subsequent to the initial tightening. Research[5] has shown that a loss of pretension of approximately 6.5 percent occurred for galvanized plates and bolts due to relaxation as compared with 2.5 percent for uncoated joints. This loss of bolt tension occurred in five days with negligible loss recorded thereafter. This loss can be allowed for in design or pretension may be brought back to the prescribed level by retightening the bolts after an initial period of "settling-in."

Since it was first published, this Specification has permitted the use of bolt holes $\frac{1}{16}$ inch larger than the bolts installed in them. Research[6] has shown that, where

4. Frank, Karl H. and J. A. Yura, "An Experimental Study of Bolted Shear Connections." FHWA/RD-81/148, December 1981.

5. Munse, W. H., "Structural Behavior of Hot Galvanized Bolted Connections," 8th International Conference on Hot-dip Galvanizing, London, England, June 1967.

6. Allen, R. N. and J. W. Fisher, "Bolted Joints With Oversize or Slotted Holes," ASCE *Journal of the Structural Division*, Vol. 94, No. ST9, September, 1968.

greater latitude is needed in meeting dimensional tolerances during erection, somewhat larger holes can be permitted for bolts ⅝ inch diameter and larger without adversely affecting the performance of shear connections assembled with high-strength bolts. The oversize and slotted hole provisions of this Specification are based upon these findings. Because an increase in hole size generally reduces the net area of a connected part, the use of oversize holes is subject to approval by the Engineer of Record.

Burrs. Based upon tests[7] which demonstrated that the slip resistance of joints was unchanged or slightly improved by the presence of burrs, burrs which do not prevent solid seating of the connected parts in the snug tight condition need not be removed. On the other hand, parallel tests in the same program demonstrated that large burrs can cause a small increase in the required turns from snug tight condition to achieve specified pretension with turn-of-nut method of tightening.

Unqualified Paint on Faying Surfaces. An extension to the research on the slip resistance of shear connections cited in footnote 4 investigated the effect of ordinary paint coatings on limited portions of the contact area within joints and the effect of overspray over the total contact area. The tests[8] demonstrated that the effective area for transfer of shear by friction between contact surfaces was concentrated in an annular ring around and close to the bolts. Paint on the contact surfaces approximately one inch but not less than the bolt diameter away from the edge of the hole did not reduce the slip resistance. On the other hand, in recognition of the fact that, in connections of thick material involving a number of bolts on multiple gage lines, bolt pretension might not be adequate to completely flatten and pull thick material into tight contact around every bolt, the Specification requires that all areas between bolts also be free of paint. (See Figure C3.) The new requirements have a potential for increased economy because the paint-free area may easily be protected using masking tape located relative to the hole pattern, and, further, the narrow paint strip around the perimeter of the faying surface will minimize uncoated material outside the connection requiring field touch-up.

This research also investigated the effect of various degrees of inadvertent overspray on slip resistance. It was found that even the smallest amount of overspray of ordinary paint (that is, not qualified as Class A) within the specified paint-free area on clean mill scale reduced the slip resistance significantly. On blast-cleaned surfaces, the presence of a small amount of overspray was not as detrimental. For simplicity, the Specification prohibits any overspray from areas required to be free of paint in slip-critical joints regardless of whether the surface is clean mill scale or blast cleaned.

Galvanized Faying Surfaces. The slip factor for initial slip with clean hot-dip galvanized surfaces is of the order of 0.19 as compared with a factor of about 0.35 for clean mill scale. However, research (see note 3) has shown that the slip factor of galvanized surfaces is significantly improved by treatments such as hand wire brushing or light "brush-off" grit blasting. In either case, the treatment must be controlled in order to achieve the necessary roughening or scoring. Power wire brushing is unsatisfactory because it tends to polish rather than roughen the surface.

---

7. Polyzois, D. and J. A. Yura, "Effect of Burrs on Bolted Friction Connections," AISC *Engineering Journal*, 22 (No. 3) Third Quarter 1985.

8. Polyzois, D. and K. Frank, "Effect of Overspray and Incomplete Masking of Faying Surfaces on the Slip Resistance of Bolted Connections," AISC *Engineering Journal*, 23 (No. 2), 2nd Quarter 1986.

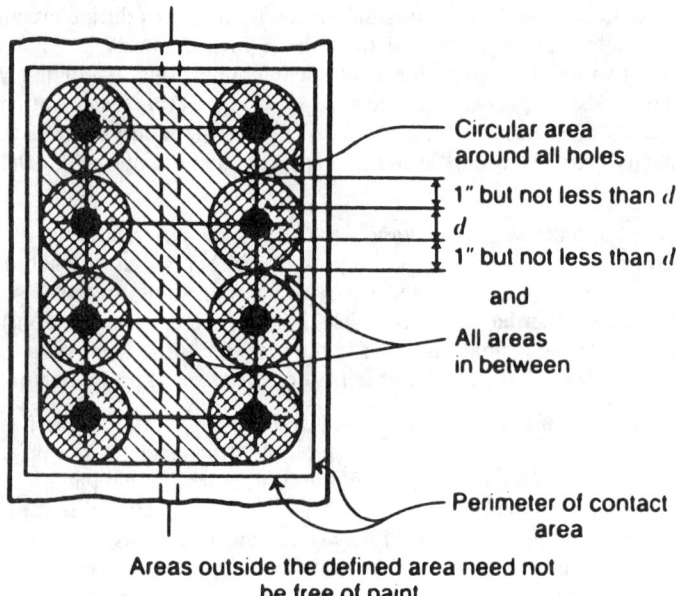

Circular area
around all holes

1" but not less than d

d

1" but not less than d

and

All areas
in between

Perimeter of contact
area

**Areas outside the defined area need not
be free of paint.**

Fig. C3. Areas outside the defined area need not be free of paint

Field experience and test results have indicated that galvanized members may have a tendency to continue to slip under sustained loading.[9] Tests of hot-dip galvanized joints subject to sustained loading show a creep-type behavior. Treatments to the galvanized faying surfaces prior to assembly of the joint which caused an increase in the slip resistance under short duration loads did not significantly improve the slip behavior under sustained loading.

C4 Design for Strength of Bolted Connections

Background for Design Stresses. With the 1985 edition of the Specification, the arbitrary designations "friction type" and "bearing type" connections used in former editions, and which were frequently misinterpreted as implying an actual difference in the manner of performance or strength of the two types of connection, were discontinued in order to focus attention more upon the real manner of performance of bolted connections.

In bolted connections subject to shear-type loading, the load is transferred between the connected parts by friction up to a certain level of force which is dependent upon the total clamping force on the faying surfaces and the coefficient of friction of the faying surfaces. The connectors are not subject to shear, nor is the connected material subject to bearing stress. As loading is increased to a level in excess of the frictional resistance between the faying surfaces, slip occurs, but failure in the sense of rupture does not occur. As even higher levels of load are applied, the load is resisted by shear in the fastener and bearing upon the connected material plus some uncertain amount of friction between the faying surfaces. The final failure will be by shear failure of the connectors, or by tear out of the connected

9. Kulak, G. L., J. W. Fisher, and J. H. A. Struik, "Guide to Design Criteria for Bolted and Riveted Joints," 2nd ed., New York: John Wiley & Sons, 1987, p. 208. (Hereinafter referred to as the *Guide.*)

material, or by unacceptable ovalization of the holes. Final failure load is independent of the clamping force provided by the bolts.[10]

The design of high-strength bolted connections under this Specification begins with consideration of strength required to prevent premature failure by shear of the connectors or bearing failure of the connected material. Next, for connections which are defined as "slip-critical," the resistance to slip load is checked. Because the combined effect of frictional resistance with shear or bearing has not been systematically studied and is uncertain, any potential greater resistance due to combined effect is ignored.

Connection Slip. There are practical cases in the design of structures where slip of the connection is desirable in order to permit rotation in a joint or to minimize the transfer of moment. Additionally there are cases where, because of the number of fasteners in a joint, the probability of slip is extremely small or where, if slip did occur, it would not be detrimental to the serviceability of the structure. In order to provide for such cases while at the same time making use of the higher shear strength of high-strength bolts, as contrasted to ASTM A307 bolts, the Specification now permits joints tightened only to the snug tight condition.

The maximum amount of slip that can occur in connections that are not classified as slip-critical is, theoretically, an amount equal to two hole clearances. In practical terms, it is observed to be much less than this. In laboratory tests it is usually about one-half a hole clearance. This is because the acceptable inaccuracies in the location of holes within a pattern of bolts would usually cause one or more bolts to be in bearing in the initial unloaded condition. Further, in statically loaded structures, even with perfectly positioned holes, the usual method of erection would cause the weight of the connected elements to put the bolts into direct bearing at the time the member is supported on loose bolts and the lifting crane is unhooked. Subsequent additional gravity loading could not cause additional connection slip.

Connections classified as slip-critical include those cases where slip could theoretically exceed an amount deemed by the Engineer of Record to affect the suitability for service of the structure by excessive distortion or reduction in strength or stability, even though the resistance to fracture of the connection, per se, may be adequate. Also included are those cases where slip of any magnitude must be prevented, for example, joints subject to load reversal.

Shear and Bearing on Fasteners. Several interrelated parameters influence the shear and bearing strength of connections. These include such geometric parameters as the net-to-gross-area ratio of the connected parts, the ratio of the net area of the connected parts to the total shear-resisting area of the fasteners, and the ratio of transverse fastener spacing to fastener diameter and to the connected part thickness. In addition, the ratio of yield strength to tensile strength of the steel comprising the connected parts, as well as the total distance between extreme fasteners, measured parallel to the line of direct tensile force, play a part.

In the past, a balanced design concept had been sought in developing criteria for mechanically fastened joints to resist shear between connected parts by means of bearing of the fasteners against the sides of the holes. This philosophy resulted in wide variations in the factor of safety for the fasteners, because the ratio of yield to tensile strength increases significantly with increasingly stronger grades of steel. It had no application at all in the case of very long joints used to transfer direct

---

10. Ibid., pp. 49–52.

tension, because the end fasteners "unbutton" before the plate can attain its full strength or before the interior fasteners can be loaded to their rated shear capacity.

By means of a mathematical model it was possible to study the interrelationship of the previously mentioned parameters.[11,12] It has been shown that the factor of safety against shear failure ranged from 3.3 for compact (short) joints to approximately 2.0 for joints with an overall length in excess of 50 inches. It is of interest to note that the longest (and often the most important) joints had the lowest factor, indicating that a factor of safety of 2.0 has proven satisfactory in service.

The absence of design strength provisions specifically for the case where a bolt in double shear has a non-threaded shank in one shear plane and a threaded section in the other shear plane is because of the uncertainty of manner of sharing the load between the two different shear areas. It also recognizes that knowledge as to the bolt placement (which might leave both shear planes in the threaded section) is not ordinarily available to the detailer. If threads occur in one shear plane, the conservative assumption is made that threads are in all shear planes.

The nominal strength and resistance factors for fasteners subject to applied tension or shear are given in Table 2. The values are based upon the research and recommendations reported in the *Guide*. With the wealth of data available, it was possible through statistical analyses to adjust resistances to provide more uniform reliability for all loading and joint types. The design resistances provide designs approximately equivalent to the designs provided by the allowable stresses in the 1980 edition of the Specification. The design of connections is more conservative than that of the connected members of buildings and bridges by a substantial margin, in the sense that the failure load of the fasteners is substantially in excess of the maximum serviceability limit (yield) of the connected material.

Design for Tension. The nominal strengths specified for applied tension[13] are intended to apply to the external bolt load plus any tension resulting from prying action produced by deformation of the connected parts. The recommended design strength is approximately equal to the initial tightening force; thus, when loaded to the nominal (service) load, high-strength bolts will experience little if any actual change in stress. For this reason, bolts in connections in which the applied loads subject the bolts to axial tension are required to be fully tensioned, even though the connection may not be subject to fatigue loading nor classified as slip-critical.

Properly tightened A325 and A490 bolts are not adversely affected by repeated application of the recommended service load tensile stress, provided the fitting material is sufficiently stiff, so that the prying force is a relatively small part of the applied tension.[14] The provisions covering bolt tensile fatigue are based upon study of test reports of bolts that were subjected to repeated tensile load to failure.

Design for Shear. The nominal strength in shear is based upon the observation that the shear strength of a single high-strength bolt is about 0.62 times the tensile strength of that bolt.[15] However, in shear connections with more than two bolts in the line of force, deformation of the connected material causes nonuniform bolt shear force distribution so that the strength of the connection in terms of the average bolt strength

11. Fisher, J. W. and L. S. Beedle, "Analysis of Bolted Butt Joints," ASCE *Journal of the Structural Division*, 91 (No. ST5), October 1965.
12. *Guide*, pp. 89–116; 126–132.
13. Ibid., pp. 263–286.
14. Ibid., pp. 272.
15. Ibid., pp. 44–50.

goes down as the joint length increases.[16] Rather than provide a function that reflects this decrease in average fastener strength with joint length, a single reduction factor of 0.80 was applied to the 0.62 multiplier. The result will accommodate bolts in all joints up to about 50 inches in length without seriously affecting the economy of very short joints. As noted in the footnotes to Table 2, bolts in joints longer than 50 inches in length must be further discounted by an additional 20 percent.

The average value of the nominal strength for bolts with threads in the shear plane has been determined by a series of tests[17] to be 0.833 F_u with a standard deviation of 0.03. A value of 0.80 was taken as a factor to account for the shear strength of a bolt with threads in the shear plane based upon the area corresponding to the nominal body area of the bolt.

The shear strength of bolts is not affected by pretension in the fasteners provided the connected material is in contact at the faying surfaces.

The design shear strength equals the nominal shear strength multiplied by a resistance factor of 0.75.

Combined Tension and Shear. The nominal strength of fasteners subject to combined tension and shear is provided by elliptical interaction curves in Table 3 which account for the connection length effect on bolts loaded in shear, the ratio of shear strength to tension strength of threaded fasteners, and the ratios of root area to nominal body area and tensile stress area to nominal body area.[18] No reduction in the design shear strength is required when applied tensile stress is equal to or less than the design tensile strength. Although the elliptical interaction curve provides the best estimate of the strength of bolts subject to combined shear and tension and thus is used in this Specification, it would be within the intent of the Specification for invoking specifications to use a three straight line approximation of the ellipse.

Design for Bearing. Bearing stress produced by a high-strength bolt pressing against the side of the hole in a connected part is important only as an index to behavior of the connected part. It is of no significance to the bolt. The critical value can be derived from the case of a single bolt at the end of a tension member.

It has been shown,[19] using finger-tight bolts, that a connected plate will not fail by tearing through the free edge of the material if the distance L, measured parallel to the line of applied force from a single bolt to the free edge of the member toward which the force is directed, is not less than the diameter of the bolt multiplied by the ratio of the bearing stress to the tensile strength of the connected part.

The criterion for nominal bearing strength is

$$L/d \geq R_n/F_u$$

where
R_n = nominal bearing pressure
F_u = specified minimum tensile strength of the connected part.

As a practical consideration, a lower limit of 1.5 is placed on the ratio L/d and an upper limit of 1.5 on the ratio F_p/F_u and an upper limit of 3.0 on the ratio R_n/F_u.

The foregoing leads to the rules governing bearing strength in the specification.

---

16. Ibid., pp. 99–104.
17. Yura, J. A., K. H. Frank, and D. Polyois, "High Strength Bolts for Bridges." *PMFSEL Report No. 87-3*, May 1987, Phil M. Ferguson Structural Engineering Laboratory, University of Texas at Austin.
18. *Guide*, pp. 50–51.
19. Ibid., pp. 141–143.

The bearing pressure permitted in the 1980 Specification and the current provisions are fully justifiable from the standpoint of strength of the connected material. However, even though rupture does not occur, recent tests have demonstrated that ovalization of the hole will begin to develop as the bearing stress is increased beyond the previously permitted stress, especially if it is combined with high tensile stress on the net section. Furthermore, when high bearing stress is combined with high tensile stress on the net section and the effect of exterior versus interior plies, lower ultimate strengths than previously reported result in addition to the hole ovalization.

Recognizing that initiation of hole ovalization occurs well below the ultimate strength, and to facilitate standardization in detailing and fabrication, sufficiently conservative simplified criteria have been provided in a formula format for usual applications. The more accurate formula in which the strength is related to the distance L may be used for special cases such as those with very large bolts or very thin material.

For connections with more than a single bolt in the direction of force, the resistance may be taken as the sum of the resistances of the individual bolts.

C5 Design Check for Slip Resistance

The Specification recognizes that, for a number of cases, slip of a joint would be undesirable or must be precluded. Such joints are termed "slip-critical" joints. This is somewhat different from the previous term "friction type" connection. The new terminology was adopted in order to focus attention on the fact that all tightened high-strength bolted joints resist load by friction between the faying surfaces up to the slip load and subsequently are able to resist even greater loads by shear and bearing. The strength of the joint is not related to the slip load. The Specification requires that the two different resistances be considered separately.

The consequences of slip into bearing varies from application to application; hence the determination of which connections shall be designed and installed as slip-critical is best left to judgment and a conscious decision on the part of the Engineer of Record. Also, the determination of whether the potential slippage of a joint is critical at nominal load level as a serviceability consideration or whether slippage could result in distortions of the frame such that the ability of the frame to resist factored loads would be reduced can be determined only by the Engineer of Record. The following comments reflect the collective thinking of the Council as developed during numerous meetings and reviews of drafts of the Specification and Commentary. They are provided as guidance and an indication of the intent of the Specification.

In the case of bolts in holes with only small clearance, such as standard holes and slotted holes loaded transverse to the axis of the slot in practical connections, the freedom to slip generally does not exist because one or more bolts are in bearing even before load is applied due to normal fabrication tolerances and erection procedures. Further, the consequences of slip, if it can occur at all, are trivial except for a few situations. If for some reason it is deemed critical, design should probably be on the basis of nominal loads (Section 5(b)).

In connections containing long slots that are parallel to the direction of the applied load, slip of the connection prior to attainment of the factored load might be large enough to alter the usual assumption of analysis that the undeformed structure can be used to obtain the internal forces. The Specification allows the designer two alternatives in this case. If the connection is designed so that it will not slip under the effect of the nominal loads, then the effect of the factored loads acting on the deformed structure (deformed by the maximum amount of slip in the long slots at all locations) must be included in the structural analysis. Alternatively the connection can be designed so that it will not slip at loads up to the factored load level. These requirements are noted in Clause 7(b)(3).

Joints subject to full reverse cyclic loading are clearly slip-critical joints since slip would permit back-and-forth movement of the joint and early fatigue failure. However, for joints subject to pulsating load that does not involve reversal of load direction, proper fatigue design could be provided either as a slip-critical joint on the basis of stress on the gross section or as a non-slip-critical joint on the basis of stress on the net section. Because fatigue results from repeated application, the service load rather than the overload load design should be based upon nominal load criteria (Section 5(b)).

For high-strength bolts in combination with welds in statically loaded conditions and considering new work only, the nominal strength may be taken as the sum of two contributions.[20] One results from the slip resistance of the bolted parts and may be determined in accordance with Section 5(c). The second results from the resistance of the welds as provided by applicable welding specifications. If one type of connector is already loaded when the second type of connector is introduced, the nominal strength cannot be obtained by adding the two resistances. The *Guide* should be consulted in these cases.

From the definition of the term "coefficient of slip" (friction), the expression for nominal slip resistances for bolts in standard holes is apparent and needs no explanation. The mean value of slip coefficients from many tests on clean mill scale, blast-cleaned steel surfaces and galvanized and roughened surfaces were taken as the basis for the three classes of surfaces.

In the 1978 edition of the Specification, nine classes of faying surface conditions were introduced, and significant increases were made in the recommended allowable stresses for proportioning connections which function by transfer of shear between connected parts by friction. These classes and stresses were adopted on the basis of statistical evaluation of the information then available.

Extensive data developed through research sponsored by the Council and others during the past ten years has been statistically analyzed to provide improved information on slip probability of connections in which the bolts have been preloaded to the requirements of Table 4. Two principal variables—coefficient of friction of the faying surfaces and bolt pretension—were found to dominate the slip resistance of connections.

An examination of the slip (friction) coefficient data for a wide range of surface conditions indicates that the data are distributed normally, and the standard deviation is essentially the same for each surface condition class. This means that different reduction factors should be applied to classes of surfaces with different mean values of coefficients of friction—the smaller the mean value of the coefficient of friction, the smaller (more severe) the appropriate reduction factor—in order to provide equivalent reliability of slip resistance.

The bolt clamping force data indicate that bolt tensions are distributed normally for each method of tightening. However, the data also indicate that the mean values of the bolt tensions are different for each method. If the calibrated wrench method is used to tighten ASTM A325 bolts, the mean value of bolt tension is about 1.13

---

20. Ibid., pp. 238–40.

times the minimum specified tension in Table 4. If the turn-of-nut method is used, the mean value of tension is about 1.35 times the minimum specified preload for A325 bolts and about 1.27 for A490 bolts.

The combined effects of the variability of coefficient of friction and bolt tension have been accounted for in the slip probability factor, D, of the formula for nominal slip resistance in Section 5(b). The values of the slip probability factor, D, given by 5(b) imply a 90 percent reliability that slip will not occur if the calibrated wrench method of installation is used. If the turn-of-nut method is used, a reliability of about 95 percent will be provided.

Reference is made to *Guide to Design Criteria for Bolted and Riveted Joints* (2nd ed., New York: John Wiley and Sons, 1987 p. 135) for tables of values of D appropriate for other mean slip coefficients and slip probabilities and suitable for direct substitution into the formula for slip resistance in Section 5(b).

The frequency distribution and mean value of clamping force for bolts tightened by turn-of-nut method are higher than calibrated wrench installation because of the elimination of variables which affect torque-tension ratios and due to higher-than-specified minimum strength of production bolts. Because properly applied turn-of-nut installation induces yield point strain in the bolt, the higher-than-specified yield strength of production bolts will be mobilized and result in higher clamping force by the method. On the other hand, the calibrated wrench method, which is dependent upon the calibration of wrenches to slightly more than Table 4 tensions, independent of the actual bolt properties, will not mobilize any additional strength of production bolts. High clamping force might be achieved by the calibrated wrench method if the wrench was set to a higher torque value. However, this would require more attention to the degrees of rotation to prevent excessive deformation of the bolt or torsional bolt failure.

Because of the effects of oversize and slotted holes on the induced tension in bolts using any of the specified installation methods, lower values are provided for bolts in these hole types. In the case of bolts in long slotted holes, even though the slip load is the same for bolts loaded transverse or parallel to the axis of the slot, the values for bolts loaded parallel to the axis has been further reduced based upon judgment in recognition of the greater consequences of slip.

Attention is called to the fact that the criteria for slip resistance are for the case of connections subject to a coaxial load. For cases in which the load tends to rotate the connection in the plane of the faying surface, a modified formula accounting for the placement of bolts relative to the center of rotation should be used.[21]

Connections of the type shown in Figure C4(a), in which some of the bolts (A) lose a part of their clamping force due to applied tension, suffer no overall loss of frictional resistance. The bolt tension produced by the moment is coupled with a compensating compressive force (C) on the other side of the axis of bending. In a connection of the type shown in Fig. C4(b), however, all fasteners (B) receive applied tension which reduces the initial compression force at the contact surface. If slip under load cannot be tolerated, the design slip-load value of the bolts in shear should be reduced in proportion to the ratio of residual axial force to initial tension. If slip of the joint can be tolerated, the bolt shear stress should be reduced according to the tension-shear interaction as outlined in the *Guide*, page 71. Because the bolts are subject to applied axial tension, they are required to be pretensioned in either case.

---

21. Ibid., pp. 217–30.

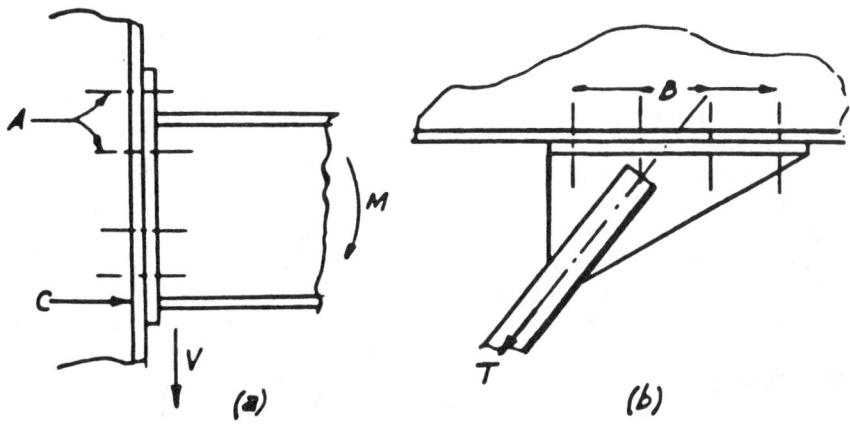

Fig. C4

While connections with bolts pretensioned to the levels specified in Table 4 do not ordinarily slip into bearing when subject to anticipated loads, it is required that they meet the requirements of Section 5 in order to maintain the factor of safety of 2 against fracture in the event that the bolts do slip into bearing as a result of large unforeseen loads.

To cover those cases where a coefficient of friction less than 0.33 might be adequate for a given situation, the Specification provides that, subject to the approval of the Engineer of Record, and provided the mean slip coefficient is determined by the specified test procedure and the appropriate slip probability factor, *D,* is selected from the literature, faying surface coatings providing lower slip resistance than Class A coating may be used.

It should be noted that both Class A and Class B coatings are required to be applied to blast-cleaned steel.

High-Strength Bolts in Combination with Welds or Rivets. For high-strength bolts in combination with welds in statically loaded conditions and considering new work only, the nominal strength may be taken as the sum of the two contributions. If one type of connector is already loaded when the second type of connector is introduced, the nominal strength cannot be obtained by sum of the two resistances. The *Guide* should be consulted in these cases.

For high-strength bolts in combination with welds in fatigue loaded applications, available data are not sufficient to develop general design recommendations at this time. High-strength bolts in combination with rivets are rarely encountered in modern practice. If need arises, guidance may be found in the *Guide.*

C7 Design Details of Bolted Connections

A new section has been added with this edition of the Specification in order to bring together a number of requirements for proper design and detailing of high-strength bolted connections. The material covered in the Specification, and in Section 7 in particular, is not intended to provide comprehensive coverage of the design of high-strength bolted connections. For example, other design considerations of importance to the satisfactory performance of the connected material such as block shear, shear lag, prying action, connection stiffness, effect on the performance of the structure and others are beyond the scope of this Specification and Commentary.

Proper location of hardened washers is as important as other elements of a detail to the performance of the fasteners. Drawings and details should clearly reflect the number and disposition of washers, especially the thick hardened washers that are required for several slotted hole applications. Location of washers is a design consideration that should not be left to the experience of the iron worker.

While hardened washers are not required with some methods of installation, their use will overcome the effects of galling under the element turned in tightening.

Finger shims are a necessary device or tool of the trade to permit adjusting alignment and plumbing of structures. When these devices are fully and properly inserted, they do not have the same effect on bolt tension relaxation or the connection performance as do long slotted holes in an outer ply. When fully inserted, the shim provides support around approximately 75 percent of the perimeter of the bolt in contrast to the greatly reduced area that exists with a bolt centered in a long slot. Further, finger shims would always be enclosed on both sides by the connected material which would be fully effective in bridging the space between the fingers.

C8 Installation and Tightening

Several methods for installation and tensioning of high-strength bolts, when tensioning is required, are provided without preference in the Specification. Each method recognized in Section 8, when properly used as specified, may be relied upon to provide satisfactory results. All methods may be misused or abused.

At the expense of redundancy, the provisions stipulating the manner in which each method is intended to be used are set forth in complete detail in order that the rules for each method may stand alone without need for footnotes or reference to other sections. If the methods are conscientiously implemented, good results should be routinely achieved.

Connections Not Requiring Full Tensioning. In the Commentary, Section C6 of the previous edition of the Specification, it was pointed out that "bearing" type connections need not be tested to ensure that the specified pretension in the bolts had been provided, but specific provision permitting relaxation of the tensioning requirement was not contained in the body of the Specification. In the present edition of the Specification, separate installation procedures are provided for bolts that are not within the slip-critical or direct tension category. The intent in making this change is to improve the quality of bolted steel construction and reduce the frequency of costly controversies by focusing attention, both during the installation and tensioning phase and during inspection, on the true slip-critical connections, rather than diluting the effort through the requirement for costly tensioning and tension testing of the great many connections where such effort serves no useful purpose. The requirement for identification of connections on the drawings may be satisfied either by identifying the slip-critical and direct tension connections which must be fully tightened and inspected or by identifying the connections which need be tightened only to the snug tight condition.

Under the provisions of some other specifications, certain shear/bearing connections are required to be tightened well beyond the snug tight conditions;[22] how-

---

22. For example, American Institute of Steel Construction, "Specification for Design Fabrication and

ever, because the joints are in bearing, prevention of slip of the joint is not a concern in these connections. Because they are not slip-critical joints, they should not be subject to the same requirements as slip-critical joints, especially the requirements for faying surface coatings and conditions. To ensure proper tightness of the connections, they should be tightened by one of the four methods in 8(d); however, inspection should be limited to monitoring the work to confirm that the bolt tightening procedure is properly applied. Inspection should not include testing to ensure that any specific level of tension has been achieved.

In the Specification, snug tight is defined as the tightness that exists when all plies are in firm contact. This may usually be attained by a few impacts of an impact wrench or the full effort of a man using an ordinary spud wrench. In actuality, snug tight is a degree of tightness which will vary from joint to joint depending upon the thickness, flatness and degree of parallelism of the connected material. In most joints, the plies will pull together at snug tight; however, in some joints in thick material, it may not be possible to have continuous contact throughout the faying surface area. In such joints, the slip resistance of the completed joints will not be reduced because compressive forces between the faying surfaces, however distributed, must be in equilibrium with the total of the tensile forces in all bolts.

Tension Calibrating Devices. At the present time, there is no known economical means for determining the tension in a bolt that has previously been installed in a connection. The actual tension in a bolt installed in a tension calibrator (hydraulic tension indicating device) is directly indicated by the dial of the device, provided the device is properly calibrated. Such a device is an economical and valuable tool that should be readily available whenever high-strength bolts are to be installed in either slip-critical or shear/bearing connections. The testing of as-recieved bolts and nuts at the job site is a requirement of the Specification because instances of counterfeit under strength fasteners not meeting the requirements of the ASTM Specification have not infrequently occurred. Job site testing provides a practical means for ensuring that nonconforming fasteners are not incorporated in the work. Further, although the several elements of a fastener assembly may conform to the minimum requirements of their separate ASTM Specifications, their compatibility in an assembly or the need for lubrication can only be ensured by testing of the assembly. Hence, such devices are important for testing the complete fastener assembly as it will be used with the method of tightening to be used to ensure the suitability of bolts and nuts (probably produced by different manufacturers), other elements, and the adequacy of impact wrenches and/or air pressure to provide the specified tension using the selected method. Testing before start of installation of fasteners in the work will also identify potential sources of problems, such as the need for lubrication to prevent failure of bolts by combined high torque with tension, under-strength assemblies due to excessive overtapping of hot-dip galvanized nuts, and to clarify for the bolting crews and inspectors the proper implementation of the selected installation method to be used. Such devices are essential to the confirmation testing of alternative design fasteners, direct tension indicators, and to verify the proper use of the turn-of-nut procedure. They are also essential to the specified procedure for the calibrated wrench method of installation, and for the specified procedure for determining a valid testing torque when such inspection by a torque method is required.

Erection of Structural Steel for Buildings," Section 1.15.12, stipulates several cases where high-strength bolts in bearing connections are to be fully tensioned independent of whether potential slip is a concern or not.

They are the only known economically available tool for field use for determining realistic torque to tension relationships for given fastener assemblies.

Experience on many projects has shown that bolts and/or nuts not meeting the requirements of the applicable ASTM specification would have been identified prior to installation if they had been tested as an assembly in a tension calibrator. The controversy and great expense of replacing bolts installed in the structure when the nonconforming bolts were discovered at a later date would have been avoided.

Hydraulic tension calibrating devices capable of indicating bolt tension undergo a slight deformation under load. Hence, the nut rotation corresponding to a given tension reading may be somewhat larger than it would be if the same bolt were tightened against a solid steel abutment. Stated differently, the reading of the calibrating device tends to underestimate the tension which a given rotation of the turned element would induce in a bolt in an actual joint. This should be borne in mind when using such devices to establish a tension-rotation relationship.

Slip-critical Connections and Connections Subject to Direct Tension. Four methods for joint assembly and tightening are provided for slip-critical and direct tension connections. It has repeatedly been demonstrated in the laboratory that each of the four installation methods provides the specified pretension when used properly with specified fasteners in good condition, but improperly applied methods or under-strength fasteners or fasteners in poor condition provide uncertain pretensions. Therefore, regardless of the method used and prior to the commencement of work, it is required to be demonstrated by installation of a representative sample of the fastener assemblies in the tension calibrator that the specified pretension can be achieved using the procedure to be used with the fasteners to be used by the crews who will be doing the work.

With any of the four described tensioning methods, it is important to install bolts in all holes of the connection and bring them to an intermediate level of tension generally corresponding to snug tight in order to compact the joint. Even after being fully tightened, some thick parts with uneven surfaces may not be in contact over the entire faying surface. In itself, this is not detrimental to the performance of the joint. As long as the specified bolt tension is present in all bolts of the completed connection, the clamping force equal to the total of the tensions in all bolts will be transferred at the locations that are in contact and be fully effective in resisting slip through friction. If however, individual bolts are installed and tightened in a single continous operation, bolts which are tightened first will be subsequently relaxed by the tightening of the adjacent bolts. The total of the forces in all bolts will be reduced, which will reduce the slip load whether there is uninterrupted contact between the surfaces or not.

With all methods, tightening should begin at the most rigidly fixed or stiffest point and progress toward the free edges, both in the initial snugging up and in the final tightening.

Turn-of-Nut-Tightening. When properly implemented, turn-of-nut method provides more uniform tension in the bolts than does torque controlled tensioning methods because it is primarily dependent upon bolt elongation into the inelastic range.

Consistency and reliability method is dependent upon ensuring that the joint is well compacted and all bolts are uniformly tight at a snug tight condition prior to application of the final required partial turn. Under-tightened bolts will result if this starting condition is not achieved because subsequent turning of the nut will first close the gap before meaningful elongation of the bolt occurs as would be the case with solid steel in the grip. Reliability is also dependent upon ensuring that the turn that is applied is relative between the bolt and nut; thus the element not turned in tightening should be prevented from rotating while the required degree of turn is applied to the turned element. Reliability and inspectability of the method may be improved by having the outer face of the nut match-marked to the protruding end of the bolt after the joint has been snug tightened but prior to final tightening. Such marks may be applied by the wrench operator using a crayon or dab of paint. Such marks in their relatively displaced position after tightening will afford the inspector a means for noting the rotation that was applied.

Problems with turn-of-nut tightening have been encountered with hot-dip galvanized bolts. In some cases, the problems have been attributed to especially effective lubricant applied by the manufacturer to ensure that bolts from stock will meet the ASTM Specification requirements without the need for relubricating and retesting. Job site tests in the tension indicating device demonstrated the lubricant reduced the coefficient of friction between the bolt and nut to the degree that "the full effort of a man using an ordinary spud wrench" to snug tighten the joint actually induced the full required tension. Also, because the nuts could be removed by an ordinary spud wrench they were erroneously judged improperly tightened by the inspector. Research (see note 3) confirms that lubricated high-strength bolts may require only one-half as much torque to induce the specified tension. In other cases of problems with hot-dip galvanized bolts, the absence of lubrication or lack of proper overtapping caused seizing of the nut and bolt threads which resulted in twist failure of the bolt at less than specified tension. For such situations, use of a tension indicating device and the fasteners being installed may be helpful in establishing either the need for lubrication or alternate criteria for snug tight at about one-half the tension required by Table 4.

Because reliability of the method is independent of the presence or absence of washers, washers are not required except for oversize and slotted holes in an outer ply. In the absence of washers, testing after the fact using a torque wrench method is highly unreliable.

That is, the turn-of-nut method of installation, properly applied, is more reliable and consistent than the testing method. The best method for inspection of the method is for the Inspector to observe the required job site confirmation testing of the fasteners and the method to be used, followed by monitoring of the work in progress to ensure that the method is routinely properly applied.

Calibrated Wrench Method. Research has demonstrated that scatter in induced tension is to be expected when torque is used as an indirect indicator of tension. Numerous variables, which are not related to tension, affect torque. For example, the finish and tolerance on bolt threads, the finish and tolerance on the nut threads, the fact that the bolt and nut may not be produced by the same manufacturer, the degree of lubrication, the job site conditions contributing to dust and dirt or corrosion on the threads, the friction that exists to varying degree between the turned element and the supporting surface, the variability of the air pressure on the torque wrenches due to length of air lines or number of wrenches operating from the same source, the condition and lubrication of the wrench which may change within a work shift, and other factors all bear upon the effectiveness of the calibrated torque wrench to induce tension.

Recognition of the calibrated wrench method of tightening was removed from the Specification with the 1980 edition. This action was taken because it is the least reliable of all methods of installation and many costly controversies had occured. It is suspected that shortcut procedures in the use of the calibrated wrench method of installation, not in accordance with the Specification provisions, were probably being used. Further, torque controlled inspection procedures based upon "standard" or calculated inspection torques rather than torques determined as required by the Specification were being routinely used. These incorrect procedures plus others had a compounding effect upon the uncertainty of the installed bolt tension, and were responsible for many of the controversies.

It is recognized, however, that if the calibrated wrench method is implemented without shortcuts as intended by the Specification, that there will be a 90 percent assurance that the tensions specified in Table 4 will be equaled or exceeded. Because the Specification should not prohibit any method which will give acceptable results when used as specified, the calibrated wrench method of installation was reinstated in the 1985 edition of the Council Specification. However, to improve upon the previous situation, the 1985 version of the Specification was modified to require better control. Wrenches must be calibrated daily for each diameter and grade of bolt. Hardened washers must be used. Fasteners must be protected from dirt and moisture at the job site. Additionally, to achieve reliable results attention should be given to the control, insofar as it is practical, of those controllable factors which contribute to variability. For example, bolts and nuts should be purchased from reliable manufacturers with a record of good quality control to minimize the variabilty of the fit. Bolts and nuts should be adequately and uniformly lubricated. Water soluble lubricants should be avoided.

Installation of Alternative Design Fasteners. It is the policy of the Council to recognize only fasteners covered by ASTM Specifications; however, it cannot be denied that a general type of alternative design fastener, produced by several manufacturers, is used on a significant number of projects as permitted by Section2(d). The bolts referred to involve a splined end extending beyond the threaded portion of the bolt which is gripped by a specially designed wrench chuck which provides a means for turning the nut relative to the bolt. While such bolts are subject to many of the variables affecting torque mentioned in the preceding section, they are produced and shipped by the manufacturers as a nut-bolt assembly under good quality control, which apparently minimizes some of the negative aspects of the torque controlled process.

While these alternative design fasteners have been demonstrated to consistently provide tension in the fastener meeting the requirements of Table 5 in controlled tests in tension indicating devices, it must be recognized that the fastener may be misused and provide results as unreliable as those with other methods. They must be used in the as-delivered clean lubricated condition. The requirements of this Specification and the installation requirements of the manufacturer's specification required by Section 2(d) must be adhered to.

As with other methods, a representative sample of the bolts to be used should be tested to ensure that, when used in accordance with the manufacturer's instructions, they do, in fact, provide tension, as specified in Table 5. In the actual joints, bolts must be installed in all holes of a connection and all fasteners tightened to an intermediate level of tension adequate to pull all material into contact. Only after this has been accomplished should the fasteners be fully tensioned in a systematic manner and the splined end sheared off. The sheared off splined end merely signifies that at some time the bolt has been subjected to a torque adequate to cause the shearing. If the fasteners are installed and tensioned in a single continuous operation, they will give a misleading indication to the Inspector that the bolts are properly tightened. Therefore, the only way to inspect these fasteners with assurance is to observe the job site testing of the fasteners and installation procedure and then monitor the work while in progress to ensure that the specified procedure is routinely followed.

Direct Tension Indicator Tightening. This Specification recognizes load indicating devices covered by the American Society for Testing and Materials' "Specification for Compressible-Washer Type Direct Tension Indicators For Use With Structural Fasteners," ASTM F959, in Section 2(f). The referenced device is a hardened washer incorporating several small formed arches which are designed to deform in a controlled manner when subjected to load. These load indicator washers are the sole type of device known which is directly dependent upon the tension load in the bolt, rather than upon some indirect parameter, to indicate the tension in a bolt.

As with the alternative design load indicating bolts, load indicating washers are dependent upon the quality control of the producer and proper use in accordance with the manufacturer's installation procedures and these Specifications. If the load indicator washers delivered for use in a specific application are tested at the job site to demonstrate that all components of the assembly do provide a proper indication of bolt tension, they are reliable if they are properly used by the bolting crews. Direct tension indicators meeting the requirements of ASTM F959 depend upon tension in the fastener to cause inelastic deformation of the formed arches. Bolts together with the load indicator washer plus any other washers required by Specification should be installed in all holes of the connection and the bolts tightened to approximately one-half the specified tension (deformation of the formed arches by about one-half the amount required to compress them to the specified gap) to ensure that plies of the joint have been brought into firm contact. Only after this initial tightening operation should the bolts be fully tensioned in a systematic manner. If the bolts are installed and tensioned in a single continuous operation, the load indicator washers will give the inspector a misleading indication that bolts are uniformly tensioned to the specified tension. Therefore, the only way to inspect fasteners with which load indicator washers are used with assurance is to observe the job site testing of the devices and installation procedure and then routinely monitor the work while in progress to ensure that the specified procedure is followed.

Use of direct tension indicators provides a reliable means for tensioning galvanized fasteners because it avoids the factors which affect other methods.

During installation, care must be taken to ensure that the indicator nubs are oriented to bear against the hardened bearing surface of the bolt head or against a hardened flat washer if used under the nut.

C9 Inspection

It is apparent from the commentary on installation procedures that the inspection procedures giving the best assurance that bolts are properly installed and tensioned is provided by Inspector observation of the calibration testing of the fasteners using the selected installation procedure followed by monitoring of the work in progress to ensure that the procedure that was demonstrated to provide the specified tension is routinely adhered to. When such a program is followed, no further evidence of proper bolt tension is required.

If testing for bolt tension using torque wrenches is conducted subsequent to the time the work of installation and tightening of bolts performed, the test procedure is subject to all of the uncertainties of torque controlled calibrated wrench installation. Additionally, the absence of many of the controls necessary to minimize variability of the torque to tension relationship, which are unnecessary for the other methods of bolt installation, such as use of hardened washers, careful attention to lubrication and the uncertainty of the effect of passage of time and exposure in the installed condition all reduce the reliability of the arbitration inspection results. The fact that in many cases it may have to be based upon a job test torque determined by using bolts only assumed to be representative of the bolts in the actual job, or using bolts removed from completed joints, makes the test procedure less reliable than a properly implemented installation procedure it is used to verify. Verification inspection using ultrasonic extensometers is accurate but costly and time-consuming, and requires that each tested bolt must be loosened to zero tension for calibration. Therefore, extensometers should be used for inspection only in the most critical cases. The arbitration inspection procedure contained in the Specification is provided, in spite of its limitations, as the most feasible available at this time.

Code of Standard Practice

for Steel Buildings and Bridges

Adopted Effective June 10, 1992
American Institute of Steel Construction, Inc.

AMERICAN INSTITUTE OF STEEL CONSTRUCTION, INC.
One East Wacker Drive, Suite 3100, Chicago, IL 60601-2001

PREFACE

When contractual documents do not contain specific provisions to the contrary, existing trade practices are considered to be incorporated into the relationships between the parties to a contract. As in any industry, trade practices have developed among those involved in the purchase, design, fabrication and erection of structural steel. The American Institute of Steel Construction has continuously surveyed the structural steel fabrication industry to determine standard practices and, commencing in 1924, published its *Code of Standard Practice*. Since that date, the Code has been periodically updated to reflect new and changing technology and practices of the industry.

It is the Institute's intention to provide to owners, architects, engineers, contractors and others associated with construction, a useful framework for a common understanding of acceptable standards when contracting for structural steel construction.

This edition is the fourth complete revision of the Code since it was first published. It includes a number of new sections covering new subjects not included in the previous Code, but which are an integral part of the relationship of the parties to a contract.

The Institute acknowledges the valuable information and suggestions provided by trade associations and other organizations associated with construction and the fabricating industry in developing this current *Code of Standard Practice*.

While every precaution has been taken to insure that all data and information presented is as accurate as possible, the Institute cannot assume responsibility for errors or oversights in the information published herein, or the use of the information published or incorporation of such information in the preparation of detailed engineering plans. The Code should not replace the judgment of an experienced architect or engineer who has the responsibility of design for a specific structure.

Code of Standard Practice
for Steel Buildings and Bridges

Adopted Effective June 10, 1992
American Institute of Steel Construction, Inc.

SECTION 1. GENERAL PROVISIONS

1.1. Scope

The practices defined herein have been adopted by the AISC as the commonly accepted standards of the structural steel fabricating industry. In the absence of other instructions in the contract documents, the trade practices defined in this *Code of Standard Practice*, as revised to date, govern the fabrication and erection of structural steel.

1.2. Definitions

AISC Specification–The *Specification for the Design, Fabrication and Erection of Structural Steel for Buildings* as adopted by the American Institute of Steel Construction.

ANSI–American National Standards Institute.

Architect/Engineer–The owner's designated representative with full responsibility for the design and integrity of the structure. (The EOR)

ASTM–The material standard of the American Society for Testing and Materials.

AWS Code–The *Structural Welding Code* of the American Welding Society.

Code–The *Code of Standard Practice* as adopted by the American Institute of Steel Construction.

Contract Documents–The documents which define the responsibilities of the parties involved in bidding, purchasing, supplying and erecting structural steel. Such documents normally consist of a contract, plans and specifications.

Drawings–Shop and field erection drawings prepared by the fabricator and erector for the performance of the work.

Erector–The party responsible for the erection of the structural steel.

Fabricator–The party responsible for furnishing fabricated structural steel.

General Contractor–The owner's designated representative with full responsibility for the construction of the structure.

MBMA–Metal Building Manufacturers Association.

Mill Material–Steel mill products ordered expressly for the requirements of a specific project.

Owner–The owner of the proposed structure or his designated representatives, who may be the architect, engineer, general contractor, construction manager, public authority or others.

Owner's Authorized Representative – That person designated by the owner to have the responsibility for the approval of shop drawings. This is usually the structural engineer of record for the project.

Plans–Design drawings furnished by the party responsible for the design of the structure.

Release for Construction–The release by the owner permitting the fabricator to commence work under the contract, including ordering material and preparing shop drawings.

SSPC–The Steel Structures Painting Council, publishers of the *Steel Structures Painting Manual*, Vol. 2, "Systems and Specifications."

Tier–The word Tier used in Section 7.11 is defined as a column shipping piece.

1.3. Design Criteria for Buildings and Similar Type Structures

In the absence of other instructions, the provisions of the AISC Specification govern the design of the structural steel.

1.4. Design for Bridges

In the absence of other instructions, the following provisions govern, as applicable:
Standard Specifications for Highway Bridges of the American Association of State Highway and Transportation Officials
Specifications for Steel Railway Bridges of the American Railway Engineering Association
Structural Welding Code of the American Welding Society

1.5. Responsibility for Design

1.5.1. When the owner provides the design, plans and specifications, the fabricator and erector are not responsible for the suitability, adequacy or legality of the design. The fabricator is not responsible for the safety of erection if the structure is erected by others.

1.5.2. When the owner enters into a direct contract with the fabricator to both design and fabricate an entire, completed steel structure, the fabricator is responsible for the structural adequacy of the design. The fabricator is not responsible for the safety of erection if the structure is erected by others.

1.6. Patented Devices

Except when the contract documents call for the design to be furnished by the fabricator or erector, the fabricator and erector assume that all necessary patent rights have been obtained by the owner and that the fabricator or erector will be fully protected in the use of patented designs, devices or parts required by the contract documents.

SECTION 2.0. CLASSIFICATION OF MATERIALS

2.1. Definition of Structural Steel

"Structural Steel," as used to define the scope of work in the contract documents, consists of the steel elements of the structural steel frame essential to support the design loads. Unless otherwise specified in the contract documents, these elements consist of material as shown on the structural steel plans and described as:

Anchor bolts for structural steel
Base or bearing plates
Beams, girders, purlins and girts
Bearings of steel for girders, trusses or bridges
Bracing
Columns, posts
Connecting materials for framing structural steel to structural steel
Crane rails, splices, stops, bolts and clamps
Door frames constituting part of the structural steel frame
Expansion joints connected to the structural steel frame
Fasteners for connecting structural steel items:
    Shop rivets
    Permanent shop bolts
    Shop bolts for shipment
    Field rivets for permanent connections
    Field bolts for permanent connections
    Permanent pins
Floor plates (checkered or plain) attached to the structural steel frame
Grillage beams and girders
Hangers essential to the structural steel frame
Leveling plates, wedges, shims & leveling screws
Lintels, if attached to the structural steel frame
Marquee or canopy framing
Machinery foundations of rolled steel sections and/or plate attached to the
    structural frame

Monorail elements of standard structural shapes when attached to the structural frame
Roof frames of standard structural shapes
Shear connectors–if specified to be shop attached
Struts, tie rods and sag rods forming part of the structural steel frame
Trusses.

2.2. Other Steel or Metal Items

The classification "Structural Steel," does not include steel, iron or other metal items not generally described in Section 2.1, even when such items are shown on the structural steel plans or are attached to the structural frame. These items include but are not limited to:

Cables for permanent bracing or suspension systems
Chutes and hoppers
Cold-formed steel products
Concrete or masonry reinforcing steel
Door and corner guards
Embedded steel parts in precast or poured concrete
Flagpole support steel
Floor plates (checkered or plain) not attached to the structural steel frame
Grating and metal deck
Items required for the assembly or erection of materials supplied by trades other than structural steel fabricators or erectors
Ladders and safety cages
Lintels over wall recesses
Miscellaneous metal
Non-steel bearings
Open-web, long-span joists and joist girders
Ornamental metal framing
Shear connectors – if specified to be field installed
Stacks, tanks and pressure vessels
Stairs, catwalks, handrail and toeplates
Trench or pit covers.

SECTION 3. PLANS AND SPECIFICATIONS

3.1. Structural Steel

In order to insure adequate and complete bids, and to enable the timely preparation of shop drawings and timely fabrication, the fabricator must be able to rely upon the completeness of the contract documents. The contract documents can be assumed to provide complete structural steel design plans clearly showing the work to be performed and giving the size, section, material grade and the location of all members, floor levels, column centers and offsets, and camber of members, with sufficient dimensions to convey accurately the quantity and nature of the structural steel to be furnished. Structural steel specifications include any special requirements controlling the fabrication and erection of the structural steel. Contract drawings, specifications and addenda must be numbered and dated for purposes of identification.

3.1.1. Wind bracing, connections, column stiffeners, column web doubler plates, bearing stiffeners on beams and girders, web reinforcement, openings for other trades, and other special details where required are shown in sufficient detail so that they may be readily understood.

3.1.2. Plans include sufficient data concerning assumed loads, shears, moments and axial forces to be resisted by the individual members and their connections, as may be required for the development of connection details on the shop drawings. Unless otherwise indicated in the contract documents, the plans are based upon consideration of the loads and forces to be resisted by the steel frame in the completed and fully connected condition. See Section 7.9.

3.1.3. Where connections are not shown, the connections are to be in accordance with the requirements of the AISC Specification.

3.1.4. When loose lintels and leveling plates are required to be furnished as part of the contract requirements, the plans and specifications show the size, section and location of all pieces.

3.1.5. Whenever steel frames, in the completely erected and fully connected state, require interaction with other elements not classified as structural steel (see Section 2) to provide stability and strength to resist loads for which the frame is designed, the non-self-supporting frame and the major elements not classified as structural steel, such as diaphragms, masonry and/or concrete shear walls, shall be identified in the contract documents. See Section 7.9.3.

3.1.6. When camber is required for cantilevered members, the magnitude and direction of camber are shown.

3.1.7.   The contract documents specify all the painting requirements, including the identification of members to be painted, surface preparation, paint specifications, manufacturer's product identification and the required minimum and maximum dry film thickness, in mils, of the shop coat. Contract documents must clearly indicate all individual members which are to be left unpainted so as to receive concrete, sprayed on fireproofing or for other reasons.

3.2. Architectural, Electrical and Mechanical

Architectural, electrical and mechanical plans may be used as a supplement to the structural steel plans to define detail configurations and construction information, provided all requirements for the quantities and locations of structural steel are noted on the structural steel plans.

3.3. Discrepancies

In case of discrepancies between plans and specifications for buildings, the specifications govern. In case of discrepancies between plans and specifications for bridges, the plans govern. In case of discrepancies between scale dimensions on the plans and figures written on them, the figures govern. In case of discrepancies between the structural steel plans and the architectural plans or plans for other trades, the structural steel plans govern.

3.4. Legibility of Plans

Plans are clearly legible and made to a scale not less than $1/_8$ in. to the foot. More complex information is furnished to an adequate scale to convey the information clearly.

3.5. Special Conditions

When it is required that a project be advertised for bidding before the requirements of Section 3.1 can be met, the owner must provide sufficient information in the form of scope, drawings, weights, outline specifications, and other descriptive data to enable the fabricator and erector to prepare a knowledgeable bid.

SECTION 4. SHOP AND ERECTION DRAWINGS

4.1. Owner Responsibility

To enable the fabricator and erector to properly and expeditiously proceed with the work, the owner must furnish, in a timely manner and in accordance with the contract documents, complete structural steel plans and specifications released for construction. "Released for construction" plans and specifications are required by the fabricator for ordering mill material and the preparation and completion of shop and erection drawings. Plans provided as part of a contract bid package are considered to be "released for construction" unless otherwise noted.

4.2. Approval

When shop drawings are made by the fabricator, prints thereof are submitted to the owner for his examination and approval. The fabricator includes a maximum allowance of fourteen (14) calendar days in his schedule for the return of shop drawings. Return of shop drawings is noted with the owner's approval, or approval subject to corrections as noted. The fabricator makes the corrections and furnishes corrected prints to the owner. Approval of shop drawings, approval "subject to corrections noted," or similar approvals, constitute the owner's release for the fabricator to begin fabrication. The fabricator retains flexibility to determine the fabrication schedule necessary to meet the project's requirements.

4.2.1. Approval by the owner's authorized representative of shop drawings prepared by the fabricator indicates that the fabricator has correctly interpreted the contract requirements, and may rely upon these drawings in the fabrication process. Where the fabricator must select or complete connection details, this approval constitutes acceptance by the owner's authorized representative of design responsibility for the structural adequacy of such connections. If a fabricator wishes to change a connection that is fully detailed in the contract documents, the fabricator shall submit the change for review by the owner's authorized representative in a manner that clearly indicates that a change is being requested. Approval of this submittal constitutes acceptance by the owner's authorized representative of design responsibility for the structural adequacy of the changed detail. Approval under any of the circumstances described in this Section does not relieve the fabricator of the responsibility for accuracy of detailed dimensions on shop drawings, nor the general fit-up of parts to be assembled in the field.

4.2.2. Unless specifically stated to the contrary, any additions, deletions or changes indicated on the approval of shop and erection drawings are authorizations by the owner to release the additions, deletions or revisions for construction.

4.3. Drawings Furnished by Owner

When the shop drawings are furnished by the owner, he must deliver them to the fabricator in time to permit material procurement and fabrication to proceed in an orderly manner in accordance with the prescribed time schedule. The owner prepares these shop drawings, insofar as practicable, in accordance with the shop and drafting room standards of the fabricator. The owner is responsible for the completeness and accuracy of shop drawings so furnished.

SECTION 5. MATERIALS

5.1. Mill Materials

When the fabricator receives "released for construction" plans and specifications, the fabricator may immediately place orders for the materials necessary for fabrication. The contract documents must note any material or areas which should not be ordered due to a design which is incomplete or subject to revision.

5.1.1.   Mill tests are performed to demonstrate material conformance to ASTM specifications in accordance with the contract requirements.   Unless special requirements are included in the contract documents, mill testing is limited to those tests required by the applicable ASTM material specifications. Mill test reports are furnished by the fabricator only if requested by the owner, either in the contract documents or in separate written instructions prior to the time the fabricator places his material orders with the mill.

5.1.2.   When material received from the mill does not satisfy ASTM A6 tolerances for camber, profile, flatness or sweep, the fabricator is permitted to perform corrective work by the use of controlled heating and mechanical straightening, subject to the limitations of the AISC Specification.

5.1.3.   Corrective procedures described in ASTM A6 for reconditioning the surface of structural steel plates and shapes before shipment from the producing mill may also be performed by the fabricator, at the fabricator's option, when variations described in ASTM A6 are discovered or occur after receipt of the steel from the producing mill.

5.1.4.   When special requirements demand tolerances more restrictive than allowed by ASTM A6, such requirements are defined in the contract documents and the fabricator has the option of corrective measures as described above.

5.2. Stock Materials

5.2.1. Many fabricators maintain stocks of steel products for use in their fabricating operations. Materials taken from stock by the fabricator to be used for structural purposes must be of a quality at least equal to that required by the ASTM specifications applicable to the classification covering the intended use.

5.2.2. Mill test reports are accepted as sufficient record of the quality of materials carried in stock by the fabricator. The fabricator reviews and retains the mill test reports covering the materials he purchases for stock, but the fabricator does not maintain records that identify individual pieces of stock material against individual mill test reports. Such records are not required if the fabricator purchases for stock under established specifications as to grade and quality.

5.2.3. Stock materials purchased under no particular specifications or under specifications less rigid than those mentioned above, or stock materials which have not been subject to mill or other recognized test reports, are not used without the express approval of the owner, except where the quality of the material could not affect the integrity of the structure.

SECTION 6. FABRICATION AND DELIVERY

6.1. Identification of Material

6.1.1. High strength steel and steel ordered to special requirements is marked by the supplier, in accordance with ASTM A6 requirements, prior to delivery to the fabricator's shop or other point of use.

6.1.2. High strength steel and steel ordered to special requirements that has not been marked by the supplier in accordance with Section 6.1.1 is not used until its identification is established by means of tests as specified in Section A3.1 of the AISC Specification, and until a fabricator's identification mark, as described in Section 6.1.3, has been applied.

6.1.3. During fabrication, up to the point of assembling members, each piece of high strength steel and steel ordered to special requirements carries a fabricator's identification mark or an original supplier's identification mark. The fabricator's identification mark is in accordance with the fabricator's established identification system, which is on record and available for the information of the owner or his representative, the building commissioner and the inspector, prior to the start of fabrication.

6.1.4. Members made of high strength steel and steel ordered to special requirements are not given the same assembling or erecting mark as members made of other steel, even though they are of identical dimensions and detail.

6.2. Preparation of Material

6.2.1. Thermal cutting of structural steel may be performed by hand or mechanically guided means.

6.2.2. Surfaces noted as "finished" on the drawings are defined as having a maximum ANSI roughness height value of 500. Any fabricating technique, such as friction sawing, cold sawing, milling, etc., that produces such a finish may be used.

6.3. Fitting and Fastening

6.3.1. Projecting elements of connection attachments need not be straightened in the connecting plane if it can be demonstrated that installation of the connectors or fitting aids will provide reasonable contact between faying surfaces.

6.3.2. Runoff tabs are often required to produce sound welds. The fabricator or erector does not remove them unless specified in the contract documents. When their removal is required, they may be hand flame-cut close to the edge of the finished member with no further finishing required, unless other finishing is specifically called for in the contract documents.

6.3.3. All high-strength bolts for shop attached connection material are to be installed in the shop in accordance with the *Specification for Structural Joints Using A325 or A490 Bolts*, unless otherwise noted on the shop drawings.

6.4. Dimensional Tolerances

6.4.1. A variation of $1/32$ in. is permissible in the overall length of members with both ends finished for contact bearing as defined in Section 6.2.2.

6.4.2. Members without ends finished for contact bearing, which are to be framed to other steel parts of the structure, may have a variation from the detailed length not greater than $1/16$ in. for members 30 ft or less in length, and not greater than $1/8$ in. for members over 30 ft in length.

6.4.3. Unless otherwise specified, structural members, whether of a single-rolled shape or built-up, may vary from straightness within the tolerances allowed for wide-flange shapes by ASTM Specification A6, except that the tolerance on deviation from straightness of compression members is $\frac{1}{1000}$ of the axial length between points which are to be laterally supported.

Completed members should be free from twists, bends and open joints. Sharp kinks or bends are cause for rejection of material.

6.4.4.   Beams and trusses detailed without specified camber are fabricated so that after erection any camber due to rolling or shop fabrication is upward.

6.4.5.   When members are specified on the contract documents as requiring camber, the shop fabrication tolerance shall be minus zero / plus $\frac{1}{2}$ in. for members 50 ft and less in length, or minus zero / plus ($\frac{1}{2}$ in. plus $\frac{1}{8}$ in. for each 10 ft or fraction thereof in excess of 50 ft in length) for members over 50 ft. Members received from the rolling mill with 75% of the specified camber require no further cambering. For purposes of inspection, camber must be measured in the fabricator's shop in the unstressed condition.

6.4.6.   Any permissible deviation in depths of girders may result in abrupt changes in depth at splices. Any such difference in depth at a bolted joint, within the prescribed tolerances, is taken up by fill plates. At welded joints the weld profile may be adjusted to conform to the variation in depth, provided that the minimum cross section of required weld is furnished and that the slope of the weld surface meets AWS Code requirements.

6.5. Shop Painting (See also Section 3.1.7.)

6.5.1.   The shop coat of paint is the prime coat of the protective system. It protects the steel for only a short period of exposure in ordinary atmospheric conditions, and is considered a temporary and provisional coating. The fabricator does not assume responsibility for deterioration of the prime coat that may result from exposure to ordinary atmospheric conditions, nor from exposure to corrosive conditions more severe than ordinary atmospheric conditions.

6.5.2.   In the absence of other requirements in the contract documents, the fabricator hand cleans the steel of loose rust, loose mill scale, dirt and other foreign matter, prior to painting, by means of wire brushing or by other methods elected by the fabricator, to meet the requirements of SSPC-SP2. The fabricator's workmanship on surface preparation is considered accepted by the owner unless specifically disapproved prior to paint application.

6.5.3.   Unless specifically excluded, paint is applied by brush, spray, roller coating, flow coating or dipping, at the election of the fabricator. When the term "shop coat" or "shop paint" is used with no paint system specified, the fabricator's standard paint shall be applied to a minimum dry film thickness of one mil.

6.5.4. Steel not requiring shop paint is cleaned of oil or grease by solvent cleaners and cleaned of dirt and other foreign material by sweeping with a fiber brush or other suitable means.

6.5.5. Abrasions caused by handling after painting are to be expected. Touch-up of these blemished areas is the responsibility of the contractor performing field touch-up or field painting.

6.6. Marking and Shipping of Materials

6.6.1. Erection marks are applied to the structural steel members by painting or other suitable means, unless otherwise specified in the contract documents.

6.6.2. Rivets and bolts are commonly shipped in separate containers according to length and diameter; loose nuts and washers are shipped in separate containers according to sizes. Pins and other small parts, and packages of rivets, bolts, nuts and washers are usually shipped in boxes, crates, kegs or barrels. A list and description of the material usually appears on the outside of each closed container.

6.7. Delivery of Materials

6.7.1 Fabricated structural steel is delivered in such sequence as will permit the most efficient and economical performance of both shop fabrication and erection. If the owner wishes to prescribe or control the sequence of delivery of materials, the owner reserves such right and defines the requirements in the contract documents. If the owner contracts separately for delivery and erection, the owner must coordinate planning between contractors.

6.7.2. Anchor bolts, washers and other anchorage or grillage materials to be built into masonry should be shipped so that they will be on hand when needed. The owner must allow the fabricator sufficient time to fabricate and ship such materials before they are needed.

6.7.3. The quantities of material shown by the shipping statement are customarily accepted by the owner, fabricator and erector as correct. If any shortage is claimed, the owner or erector should immediately notify the carrier and the fabricator in order that the claim may be investigated.

6.7.4. The size and weight of structural steel assemblies may be limited by shop capabilities, the permissible weight and clearance dimensions of available transportation and the job site conditions. The fabricator limits the number of field splices to those consistent with minimum project cost.

6.7.5. If material arrives at its destination in damaged condition, it is the responsibility of the receiving party to promptly notify the fabricator and carrier prior to unloading the material, or immediately upon discovery.

SECTION 7. ERECTION

7.1. Method of Erection

When the owner wishes to control the method and sequence of erection, or when certain members cannot be erected in their normal sequence, the owner so specifies in the contract documents. In the absence of such restrictions, the erector will proceed using the most efficient and economical method and sequence available to the erector consistent with the contract documents. When the owner contracts separately for fabrication and erection services, the owner is responsible for coordinating planning between contractors.

7.2. Site Conditions

The owner provides and maintains adequate access roads into and through the site for the safe delivery and movement of derricks, cranes, trucks, other necessary equipment, and the material to be erected. The owner affords the erector a firm, properly graded, drained, convenient and adequate space at the site for the operation of the erector's equipment, and removes all overhead obstructions such as power lines, telephone lines, etc., in order to provide a safe working area for erection of the steelwork. The erector provides and installs the safety protection required for his own work. Any protection for other trades not essential to the steel erection activity is the responsibility of the owner. When safety protection provided by the erector is left remaining in an area to be used by other trades after the steel erection activity is completed, the owner shall be responsible for accepting and maintaining this protection, assuring that it is adequate for the protection of all other affected trades, assuring that it complies with all applicable safety regulations when being used by other trades, indemnifying the erector from any damages incurred as a result of the safety protection's use by other trades, removing the safety equipment when no longer required, and returning it to the erector in the same condition as it was received. When the structure does not occupy the full available site, the owner provides adequate storage space to enable the fabricator and erector to operate at maximum practicable speed.

7.3. Foundations, Piers and Abutments

The accurate location, strength, suitability and access to all foundations, piers and abutments is the sole responsibility of the owner.

7.4. Building Lines and Bench Marks

The owner is responsible for accurate location of building lines and bench marks at the site of the structure, and for furnishing the erector with a plan containing all such information. At each level the owner establishes offset building lines and reference elevations for the use of the erector in the positioning of adjustable construction elements.

7.5. Installation of Anchor Bolts and Embedded Items

7.5.1.   Anchor bolts and foundation bolts are set by the owner in accordance with an approved drawing. They must not vary from the dimensions shown on the erection drawings by more than the following:

(a) $1/8$ in. center to center of any two bolts within an anchor bolt group, where an anchor bolt group is defined as the set of anchor bolts which receive a single fabricated steel shipping piece.

(b) $1/4$ in. center to center of adjacent anchor bolt groups.

(c)   Elevation of the top of anchor bolts $\pm 1/2$ in.

(d)   Maximum accumulation of $1/4$ in. per hundred ft along the established column line of multiple anchor bolt groups, but not to exceed a total of 1 in., where the established column line is the actual field line most representative of the centers of the as-built anchor bolt groups along a line of columns.

(e) $1/4$ in. from the center of any anchor bolt group to the established column line through that group.

(f)   The tolerances of paragraphs b, c and d apply to offset dimensions shown on the plans, measured parallel and perpendicular to the nearest established column line for individual columns shown on the plans to be offset from established column lines.

7.5.2.   Unless shown otherwise, anchor bolts are set perpendicular to the theoretical bearing surface.

7.5.3.   Other embedded items or connection materials between the structural steel and the work of other trades are located and set by the owner in accordance with approved location or erection drawings. Accuracy of these items must satisfy the erection tolerance requirements of Section 7.11.3.

7.5.4.   All work performed by the owner is completed so as not to delay or interfere with the erection of the structural steel.

7.6. Bearing Devices

The owner sets to line and grade all leveling plates, leveling nuts and loose bearing plates which can be handled without a derrick or crane. All other bearing devices supporting structural steel are set and wedged, shimmed or adjusted with leveling screws by the erector to lines and grades established by the owner. The fabricator provides the wedges, shims or leveling screws that are required, and clearly scribes the bearing devices with working lines to facilitate proper alignment. Promptly after the setting of any bearing devices, the owner checks lines and grades, and grouts as required. The final location and proper grouting of bearing devices are the responsibility of the owner. Tolerance on elevation relative to established grades of bearing devices, whether set by the owner or by the erector, is $\pm$ $^1/_8$ in.

7.7. Field Connection Material

7.7.1. The fabricator provides field connection details consistent with the requirements of the contract documents which will, in the fabricator's opinion, result in the most economical fabrication and erection cost.

7.7.2. When the fabricator erects the structural steel, the fabricator supplies all materials required for temporary and permanent connection of the component parts of the structural steel.

7.7.3. When the erection of the structural steel is performed by someone other than the fabricator, the fabricator furnishes the following field connection material:

(a) Bolts of required size and in sufficient quantity for all field connections of steel to steel which are to be permanently bolted. Unless high-strength bolts or other special types of bolts and washers are specified, common bolts are furnished. An extra 2 percent of each bolt size (diameter and length) is furnished.

(b) Rivets of required size and in sufficient quantity for all field connections of steel to steel which are to be riveted field connections. An extra 10 percent of each rivet size is furnished.

(c) Shims shown as necessary for make-up of permanent connections of steel to steel.

(d) Back-up bars or run-off tabs that may be required for field welding.

7.7.4. When the erection of the structural steel is performed by someone other than the fabricator, the erector furnishes all welding electrodes, fit-up bolts and drift pins used for erection of the structural steel.

7.7.5. Field-installed shear connectors are supplied by the shear connector applicator.

7.7.6. Metal deck support angles are the responsibility of the metal deck supplier.

7.8. Loose Material

Loose items of structural steel not connected to the structural frame are set by the owner without assistance from the erector, unless otherwise specified in the contract documents.

7.9. Temporary Support of Structural Steel Frames

7.9.1. General

Temporary supports, such as temporary guys, braces, falsework, cribbing or other elements required for the erection operation will be determined and furnished and installed by the erector. These temporary supports will secure the steel framing, or any partly assembled steel framing, against loads comparable in intensity to those for which the structure was designed, resulting from wind, seismic forces and erection operations, but not the loads resulting from the performance of work by or the acts of others, nor such unpredictable loads as those due to tornado, explosion or collision.

7.9.2. Self-supporting Steel Frames

A self-supporting steel frame is one that provides the required stability and resistance to gravity loads and design wind and seismic forces without interaction with other elements of the structure. The erector furnishes and installs only those temporary supports that are necessary to secure any element or elements of the steel framing until they are made stable without external support.

Special erection sequences or other considerations which are required to provide stability during the erection process must be set out in the contract documents in detail.

7.9.3. Non-Self-supporting Steel Frames

A non-self-supporting steel frame is one that, when fully assembled and connected, requires interaction with other elements not classified as Structural Steel to provide stability and strength to resist loads for which the frame is designed. Such frames shall be clearly designated as "non-self-supporting." The major elements not classified as structural steel, such as steel deck diaphragms, masonry and/or concrete shear walls, shall be identified in the contract documents.

When elements not classified as structural steel interact with the structural steel elements to provide stability and/or strength to resist loads, the owner is responsible for the installation, structural adequacy during installation, and timely completion of all such elements. The contract documents must specify the sequence and schedule of placement of such elements and the effects of the loads imposed on the structural steel frame by partially or completely installed interacting elements. The erector furnishes and installs temporary support as necessary in accordance with this information but does not thereby assume responsibility for the appropriateness of the sequence specified.

7.9.4. Special Erection Conditions

When the design concept of a structure is dependent upon the use of shores, jacks or loads which must be adjusted as erection progresses to set or maintain camber or prestress, such requirement is specifically stated in the contract documents.

7.9.5. Removal of Temporary Supports

The temporary guys, braces, falsework, cribbing and other elements required for the erection operation, which are furnished and installed by the erector, are not the property of the owner.

In *self-supporting structures*, temporary supports are not required after the structural steel for a self-supporting element is located and finally fastened within the required tolerances. After such final fastening, the erector is no longer responsible for temporary support of the self-supporting element and may remove the temporary supports.

In *non-self-supporting structures*, the erector may remove temporary supports when the necessary non-structural steel elements are complete. Temporary supports are not to be removed without the consent of the erector. At completion of steel erection, any temporary supports that are required to be left in place are removed by the owner and returned to the erector in good condition.

7.9.6. Temporary Supports for Other Work

Should temporary supports beyond those defined as the responsibility of the erector in Sections 7.9.1, 7.9.2 and 7.9.3 be required, either during or after the erection of the structural steel, responsibility for the supply and installation of such supports rests with the owner.

7.10. Temporary Floors and Handrails for Buildings

The erector provides floor coverings, handrails and walkways as required by law and applicable safety regulations for protection of his own personnel. As work progresses, the erector removes such facilities from units where the erection operations are completed, unless other arrangements are included in the contract documents. The owner is responsible for all protection necessary for the work of other trades. When permanent steel decking is used for protective flooring and is installed by the owner, all such work is performed so as not to delay or interfere with erection progress and is scheduled by the owner and installed in a sequence adequate to meet all safety regulations. (See Section 7.2)

7.11. Frame Tolerances

7.11.1. Overall Dimensions

Some variation is to be expected in the finished overall dimensions of structural steel frames. Such variations are deemed to be within the limits of good practice when they do not exceed the cumulative effect of rolling tolerances, fabricating tolerances and erection tolerances.

7.11.2. Working Points and Working Lines

Erection tolerances are defined relative to member working points and working lines as follows:

(a) For members other than horizontal members, the member work point is the actual center of the member at each end of the shipping piece.
(b) For horizontal members, the working point is the actual center line of the top flange or top surface at each end.
(c) Other working points may be substituted for ease of reference, providing they are based upon these definitions.
(d) The member working line is a straight line connecting the member working points.

7.11.3. Position and Alignment

The tolerances on position and alignment of member working points and working lines are as follows:

7.11.3.1. Columns

Individual column shipping pieces are considered plumb if the deviation of the working line from a plumb line does not exceed 1:500, subject to the following limitations:

(a) The member working points of column shipping pieces adjacent to elevator shafts may be displaced no more than 1 in. from the established column line in the first 20 stories; above this level, the displacement may be increased $1/_{32}$ in. for each additional story up to a maximum of 2 in.

(b) The member working points of exterior column shipping pieces may be displaced from the established column line no more than 1 in. toward nor 2 in. away from the building line in the first 20 stories; above the 20th story, the displacement may be increased $1/_{16}$ in. for each additional story, but may not exceed a total displacement of 2 in. toward nor 3 in. away from the building line.

(c) The member working points of exterior column shipping pieces at any splice level for multi-tier buildings and at the tops of columns for single tier buildings may not fall outside a horizontal envelope, parallel to the building line, $1^{1}/_{2}$ in. wide for buildings up to 300 ft in length. The width of the envelope may be increased by $1/_{2}$ in. for each additional 100 ft in length, but may not exceed 3 in.

(d) The member working points of exterior column shipping pieces may be displaced from the established column line, in a direction parallel to the building line, no more than 2 in. in the first 20 stories; above the 20th story, the displacement may be increased $1/_{16}$ in. for each additional story, but may not exceed a total displacement of 3 in. parallel to the building line.

7.11.3.2. Members Other Than Columns

(a) Alignment of members which consist of a single straight shipping piece containing no field splices, except cantilevered members, is considered acceptable if the variation in alignment is caused solely by the variation of column alignment and/or primary supporting member alignment within the permissible limits for fabrication and erection of such members.

(b) The elevation of members connecting to columns is considered acceptable if the distance from the member working point to the upper milled splice line of the column does not deviate more than plus $3/_{16}$ in. or minus $5/_{16}$ in. from the distance specified on the drawings.

(c) The elevation of members which consist of a single shipping piece, other than members connected to columns, is considered acceptable if the variation in actual elevation is caused solely by the variation in elevation of the supporting members which are within permissible limits for fabrication and erection of such members.

(d) Individual shipping pieces which are segments of field assembled units containing field splices between points of support are considered plumb, level and aligned if the angular variation of the working line of each shipping piece relative to the plan alignment does not exceed 1:500.

(e) The elevation and alignment of cantilevered members shall be considered plumb, level and aligned if the angular variation of the working line from a straight line extended in the plan direction from the working point at its supported end does not exceed 1:500.

(f) The elevation and alignment of members which are of irregular shape shall be considered plumb, level and aligned if the fabricated member is within its tolerance and its supporting member or members are within the tolerances specified in this Code.

7.11.3.3. Adjustable Items

The alignment of lintels, wall supports, curb angles, mullions and similar supporting members for the use of other trades, requiring limits closer than the foregoing tolerances, cannot be assured unless the owner's plans call for adjustable connections of these members to the supporting structural frame. The fabricator may provide nonadjustable connections unless the contract documents specifically show or specify them as adjustable.When adjustable connections are specified, the owner's plans must provide for the total adjustment required to accommodate the tolerances on the steel frame for the proper alignment of these supports for other trades. The tolerances on position and alignment of such adjustable items are as follows:

(a) Adjustable items are considered to be properly located in their vertical position when their location is within $3/_8$ in. of the location established from the upper milled splice line of the nearest column to the support location as specified on the drawings.

(b) Adjustable items are considered to be properly located in their horizontal position when their location is within $3/_8$ in. of the proper location relative to the established finish line at any particular floor.

(c) The ends of adjustable items which abut are considered to be properly located when aligned to within $3/_{16}$ in. of each other both vertically and horizontally.

7.11.4. Responsibility for Clearances

In the design of steel structures, the owner is responsible for providing clearances and adjustments of material furnished by other trades to accommodate all of the foregoing tolerances of the structural steel frame.

7.11.5. Acceptance of Position and Alignment

Prior to placing or applying any other materials, the owner is responsible for determining that the location of the structural steel is acceptable for plumbness, level and alignment within tolerances. The erector is given timely notice of acceptance by the owner or a listing of specific items to be corrected in order to obtain acceptance.

Such notice is rendered immediately upon completion of any part of the work and prior to the start of work by other trades that may be supported, attached or applied to the structural steelwork.

7.12. Correction of Errors

Normal erection operations include the correction of minor misfits by moderate amounts of reaming, chipping, welding or cutting, and the drawing of elements into line through the use of drift pins. Errors which cannot be corrected by the foregoing means, or which require major changes in member configuration, are reported immediately to the owner and fabricator by the erector, to enable whoever is responsible either to correct the error or to approve the most efficient and economic method of correction to be used by others.

7.13. Cuts, Alterations and Holes for Other Trades

Neither the fabricator nor the erector will cut, drill or otherwise alter his work, or the work of other trades, to accommodate other trades, unless such work is clearly specified in the contract documents. Whenever such work is specified, the owner is responsible for furnishing complete information as to materials, size, location and number of alterations in a timely manner so that the preparation of shop drawings will not be delayed.

7.14. Handling and Storage

The erector takes reasonable care in the proper handling and storage of steel during erection operations to avoid accumulation of excess dirt and foreign matter. The erector is not responsible for removal from the steel of dust, dirt or other foreign matter which accumulates during the erection period as the result of site conditions or exposure to the elements.

7.15. Field Painting

The erector does not paint field bolt heads and nuts, field rivet heads and field welds, nor touch up abrasions of the shop coat, nor perform any other field painting.

7.16. Final Cleaning Up

Upon completion of erection and before final acceptance, the erector removes all of his falsework, rubbish and temporary buildings.

SECTION 8. QUALITY CONTROL

8.1. General

8.1.1. The fabricator maintains a quality control program to the extent deemed necessary so that the work is performed in accordance with this Code, the AISC Specification, and contract documents. The fabricator has the option to use the AISC Quality Certification Program in establishing and administering the quality control program.

8.1.2. The erector maintains a quality control program to the extent the erector deems necessary so that all of the work is performed in accordance with this Code, the AISC Specification and the contract documents. The erector shall be capable of performing the erection of the structural steel, and shall provide the equipment, personnel and management for the scope, magnitude and required quality of each project.

8.1.3. When the owner requires more extensive quality control or independent inspection by qualified personnel, or requires the fabricator to be certified by the AISC Quality Certification Program, this shall be clearly stated in the contract documents, including a definition of the scope of such inspection.

8.2. Mill Material Inspection

The fabricator customarily makes a visual inspection, but does not perform any material tests, depending upon mill reports to signify that the mill product satisfies material order requirements. The owner relies on mill tests required by contract and on such additional tests as he orders the fabricator to have made at the owner's expense. If mill inspection operations are to be monitored, or if tests other than mill tests are desired, the owner so specifies in the contract documents and should arrange for such testing through the fabricator to assure coordination.

8.3. Non-Destructive Testing

When non-destructive testing is required, the process, extent, technique and standards of acceptance are clearly defined in the contract documents.

8.4. Surface Preparation and Shop Painting Inspection

Surface preparation and shop painting inspection must be planned for acceptance of each operation as completed by the fabricator. Inspection of the paint system, including material and thickness, is made promptly upon completion of the paint application. When wet film thickness is inspected, it must be measured during the application.

8.5. Independent Inspection

When contract documents specify inspection by other than the fabricator's and erector's own personnel, both parties to the contract incur obligations relative to the performance of the inspection.

8.5.1. The fabricator and erector provide the inspector with access to all places where work is being done. A minimum of 24 hours notification is given prior to commencement of work.

8.5.2. Inspection of shop work by the owner or his representative is performed in the fabricator's shop to the fullest extent possible. Such inspections should be in sequence, timely, and performed in such a manner as will not disrupt fabrication operations and will permit repair of non-conforming work prior to any required painting while the material is still in process in the fabrication shop.

8.5.3. Inspection of field work must be completed promptly so that corrections can be made without delaying the progress of the work.

8.5.4. Rejection of material or workmanship not in conformance with the contract documents may be made at any time during the progress of the work. However, this provision does not relieve the owner of his obligation for timely, in-sequence inspections.

8.5.5. Copies of all reports prepared by the owner's inspection representative must be given to the fabricator and erector immediately after the inspection to allow any necessary corrective work to be performed in a timely manner.

8.5.6. The owner's inspection representative may not suggest, direct, or approve the fabricator or erector to deviate from the contract documents or approved shop drawings, or approve such deviation, without the express written approval of the engineer of record or the person designated as the owner's authorized representative.

SECTION 9. CONTRACTS

9.1. Types of Contracts

9.1.1. For contracts stipulating a lump sum price, the work required to be performed by the fabricator and erector is completely defined by the contract documents.

9.1.2. For contracts stipulating a price per pound, the scope of work, type of materials, character of fabrication, and conditions of erection are based upon the contract documents which must be representative of the work to be performed.

9.1.3. For contracts stipulating a price per item, the work required to be performed by the fabricator and erector is based upon the quantity and the character of items described in the contract documents.

9.1.4. For contracts stipulating unit prices for various categories of structural steel, the scope of the work required to be performed by the fabricator and erector is based upon the quantity, character and complexity of the items in each category as described in the contract documents. The contract documents must be representative of the work to be done in each category.

9.2. Calculation of Weights

Unless otherwise set forth in the contract, on contracts stipulating a price per pound for fabricated structural steel delivered and/or erected, the quantities of materials for payment are determined by the calculation of gross weight of materials as shown on the shop drawings.

9.2.1. The unit weight of steel is assumed to be 490 pounds per cubic ft. The unit weight of other materials is in accordance with the manufacturer's published data for the specific product.

9.2.2. The weights of shapes, plates, bars, steel pipe and structural tubing are calculated on the basis of shop drawings showing actual quantities and dimensions of material furnished, as follows:

  (a) The weight of all structural shapes, steel pipe and structural tubing is calculated using the nominal weight per ft and the detailed overall length.
  (b) The weight of plates and bars is calculated using the detailed overall rectangular dimensions.

(c)   When parts can be economically cut in multiples from material of larger dimensions, the weight is calculated on the basis of the theoretical rectangular dimensions of the material from which the parts are cut.

(d)   When parts are cut from structural shapes, leaving a non-standard section not useable on the same contract, the weight is calculated on the basis of the nominal unit weight of the section from which the parts are cut.

(e)   No deductions are to be made for material removed by cuts, copes, clips, blocks, drilling, punching, boring, slot milling, planing or weld joint preparation.

9.2.3.   The calculated weights of castings are determined from the shop drawings of the pieces. An allowance of 10 percent is added for fillets and overrun. Scale weights of rough castings may be used if available.

9.2.4.   The items for which weights are shown in tables in the *AISC Manual of Steel Construction* are calculated on the basis of tabulated unit weights.

9.2.5.   The weight of items not included in the tables in the *AISC Manual of Steel Construction* shall be taken from the manufacturers' catalog and the manufacturers' shipping weight shall be used.

9.2.6.   The weight of shop or field weld metal and protective coatings is not included in the calculated weight for pay purposes.

9.3. Revisions to Contract Documents

9.3.1.   Revisions to the contract are made by issuance of new documents or re-issuance of existing documents. In either case, all revisions, including revisions communicated by annotation of shop or erection drawings, must be clearly and individually indicated and the documents dated and identified by revision number. All contract drawings shall be identified by the same drawing number throughout the duration of the job regardless of the revision. The engineer of record is responsible for reviewing the overall structural design to identify all components which will be affected by a change to any individual component.

9.3.2.   A revision to the requirements of the contract documents is made by change order, extra work order, or notations on the shop and erection drawings when returned upon approval.

9.3.3.   Unless specifically stated to the contrary, the issuance of a revision is authorization by the owner to release these documents for construction.

9.4. Contract Price Adjustment

9.4.1.   When the scope of work and responsibilities of the fabricator and erector are changed from those previously established by the contract documents, an appropriate modification of the contract price is made. In computing the contract price adjustment, the fabricator and erector consider the quantity of work added or deleted, modifications in the character of the work, and the timeliness of the change with respect to the status of material ordering, detailing, fabrication and erection operations.

9.4.2.   Requests for contract price adjustments are presented by the fabricator and erector in a timely manner and are accompanied by a description of the change in sufficient detail to permit evaluation and timely approval by the owner.

9.4.3.   Price per pound and price per item contracts generally provide for additions or deletions to the quantity of work prior to the time work is released for construction. Changes to the character of the work, at any time, or additions and/or deletions to the quantity of the work after it is released for detailing, fabrication, or erection, may require a contract price adjustment.

9.5. Scheduling

9.5.1.   The contract documents specify the schedule for the performance of the work. This schedule states when the "released for construction" plans will be issued and when the job site, foundations, piers and abutments will be ready, free from obstructions and accessible to the erector, so that erection can start at the designated time and continue without interference or delay caused by the owner or other trades.

9.5.2.   The fabricator and erector have the responsibility to advise the owner, in a timely manner, of the effect any revision has on the contract schedule.

9.5.3.   If the fabrication or erection is significantly delayed due to design revisions, or for other reasons which are the owner's responsibility, the fabricator and erector are compensated for additional costs incurred.

9.6. Terms of Payment

The terms of payment for the contract shall be outlined in the contract documents.

SECTION 10. ARCHITECTURALLY EXPOSED STRUCTURAL STEEL

10.1. Scope

This section of the Code defines additional requirements which apply only to members specifically designated by the contract documents as "Architecturally Exposed Structural Steel" (AESS). All provisions of Sections 1 through 9 of the Code apply unless specifically modified in this section. AESS members or components are fabricated and erected with the care and dimensional tolerances indicated in this section.

10.2. Additional Information Required in Contract Documents

(a) Specific identification of members or components which are to be AESS.
(b) Fabrication and erection tolerances which are more restrictive than provided for in this section.
(c) Requirements, if any, of a test panel or components for inspection and acceptance standards prior to the start of fabrication.

10.3. Fabrication

10.3.1. Rolled Shapes

Permissible tolerances for out-of-square or out-of-parallel, depth, width and symmetry of rolled shapes are as specified in ASTM Specification A6. No attempt to match abutting cross-sectional configurations is made unless specifically required by the contract documents. The as-fabricated straightness tolerances of members are one-half of the standard camber and sweep tolerances in ASTM A6.

10.3.2. Built-up Members

The tolerances on overall profile dimensions of members made up from a series of plates, bars and shapes by welding are limited to the accumulation of permissible tolerances of the component parts as provided by ASTM Specification A6. The as-fabricated straightness tolerances for the member as a whole are one-half the standard camber and sweep tolerances for rolled shapes in ASTM A6.

10.3.3. Weld Show-through

It is recognized that the degree of weld show-through, which is any visual indication of the presence of a weld or welds on the side away from the viewer, is a function of weld size and material thickness. The members or components will be acceptable as produced unless specific acceptance criteria for weld show-through are included in the contract documents.

10.3.4. Joints

All copes, miters and butt cuts in surfaces exposed to view are made with uniform gaps of $1/8$ in. if shown to be open joints, or in reasonable contact if shown without gap.

10.3.5. Welding

Reasonably smooth and uniform as-welded surfaces are acceptable on all welds exposed to view. Butt and plug welds do not project more than $1/16$ in. above the exposed surface. No finishing or grinding is required except where clearances or fit of other components may necessitate, or when specifically required by the contract documents.

10.3.6. Weathering Steel

Members fabricated of weathering steel which are to be AESS shall not have erection marks or other painted marks on surfaces that are to be exposed in the completed structure. If cleaning other than SSPC-SP6 is required, these requirements shall be defined in the contract documents.

10.4.   Delivery of Materials

The fabricator uses special care to avoid bending, twisting or otherwise distorting individual members.

10.5.   Erection

10.5.1. General

The erector uses special care in unloading, handling and erecting the steel to avoid marking or distorting the steel members. Care is also taken to minimize damage to any shop paint. If temporary braces or erection clips are used, care is taken to avoid unsightly surfaces upon removal. Tack welds are ground smooth and holes are filled with weld metal or body solder and smoothed by grinding or filing. The erector plans and executes all operations in such a manner that the close fit and neat appearance of the structure will not be impaired.

10.5.2. Erection Tolerances

Unless otherwise specifically designated in the contract documents, members and components are plumbed, leveled and aligned to a tolerance not to exceed one-half the amount permitted for structural steel. These erection tolerances for AESS require that the owner's plans specify adjustable connections between AESS and the structural steel frame or the masonry or concrete supports, in order to provide the erector with means for adjustment.

10.5.3. Components with Concrete Backing

When AESS is backed with concrete, it is the general contractor's responsibility to provide sufficient shores, ties and strongbacks to assure against sagging, bulging, etc., of the AESS resulting from the weight and pressure of the wet concrete.

Commentary on the

Code of Standard Practice

for Steel Buildings and Bridges

Adopted Effective June 10, 1992
American Institute of Steel Construction, Inc.

PREFACE

This Commentary has been prepared to assist those who use the *Code of Standard Practice* in understanding the background, basis and intent of its provisions.

Each section in the Commentary is referenced by corresponding sections in the Code. Not all sections of the Code are discussed; sections are covered only if it is believed that additional explanation may be helpful.

While every precaution has been taken to insure that all data and information presented is as accurate as possible, the Institute cannot assume responsibility for errors or oversights in the information published herein or the use of the information published or incorporating such information in the preparation of detailed engineering plans. The figures are for illustrative purposes only and are not intended to be applicable to any actual design. The information should not replace the judgment of an experienced architect or engineer who has the responsibility of design for a specific structure.

Commentary on the Code of Standard Practice for Steel Buildings and Bridges

Adopted Effective June 10, 1992
American Institute of Steel Construction, Inc.

SECTION 1. GENERAL PROVISIONS

1.1. Scope

This Code is not applicable to metal building systems, which are the subject of standards published by the Metal Building Manufacturers Association in their *Metal Building Systems Manual*. AISC has not participated in the development of the MBMA code and, therefore, takes no position and is not responsible for any of its provisions.

This Code is not applicable to standard steel joists, which are the subject of *Recommended Code of Standard Practice for Steel Joists*, published by the Steel Joist Institute. AISC has not participated in the development of the SJI code and, therefore, takes no position and is not responsible for any of its provisions.

SECTION 2. CLASSIFICATION OF MATERIALS

2.2. Other Steel or Metal Items

These items include materials which may be supplied by the steel fabricator which require coordination between other material suppliers and trades. If they are to be supplied by the fabricator, they must be specifically called for and detailed in contract documents.

SECTION 3. PLANS AND SPECIFICATIONS

3.1. Structural Steel

Project specifications vary greatly in complexity and completeness. There is a benefit to the owner if the specifications leave the contractor reasonable latitude in performing his work. However, critical requirements affecting the integrity of the structure, or necessary to protect the owner's interest, must be covered in the contract documents. The following checklist is included for reference:

Standard codes and specifications governing structural steelwork
Material specifications
Mill test reports
Welded joint configuration
Weld procedure qualification
Bolting specifications
Special requirements for work of other trades
Runoff tabs
Wind bracing
Connections or data for connection development
Column stiffeners
Column web doubler plates
Bearing stiffeners on beams and girders
Web reinforcement
Openings for other trades
Surface preparation and shop painting
Shop inspection
Field inspection
Non-destructive testing, including acceptance criteria
Special requirements on delivery
Special erection limitations
Temporary bracing for non-self-supporting structures
Special fabrication and erection tolerances for AESS
Special pay weight provisions

The structural steel plans must provide the elevations of all members as well as the dimensions to the centerline of all members (or the backs of angles or channels) relative to the grid lines, column centerline or other nearby members unless the locations of those members must be coordinated by the general contractor with the requirements of another trade. When the necessary dimensions are not given, the fabricator is not in a position to order material or start shop drawings in a timely manner and may be delayed while attempting to get the information.

SECTION 4. SHOP AND ERECTION DRAWINGS

4.1. Owner's Responsibility

The owner's responsibility for the proper planning of the work and the com-munication of all facts of his particular project is a requirement of the Code, not only at the time of bidding, but also throughout the term of any project. The contract documents, including the plans and specification, are for the purpose of communication. It is the owner's responsibility to properly define the scope of work, and to define information or items required and outlined in the plans and specifications. When the owner releases plans and specifications for construction, the fabricator and erector rely on the fact that these are the owner's requirements for his project.

The Code defines the owner as including a designated representative such as the architect, engineer or project manager, and when these representatives direct specific action to be taken, they are acting as and for the owner.

On phased construction projects, to insure the orderly flow of material procure-ment, detailing, fabrication and erection activities, it is essential that designs are not continuously revised after progressive releases for construction are made. In essence, once a portion of a design is released for construction, the essential elements of that design should be "frozen" to assure adherence to the construction schedule or all parties should reach an understanding on the effects of future changes as they affect scheduled deliveries and added costs, if any.

4.2. Approval

4.2.1. From the inception of the *Code of Standard Practice*, AISC and the industry in general have recognized that the engineer of record is the only individual who has all the information necessary to evaluate the total impact of connection details on the overall structural design of the project. This authority has traditionally been exercised during the approval process for shop and erection drawings. The EOR has retained the final and total responsibility for the adequacy and safety of the entire structure since at least the 1927 edition of the *Code of Standard Practice*. In those instances where a fabricator develops the detailed configuration of connections during the preparation of shop drawings, the fabricator does not thereby become responsible for the structural integrity of that part of the overall structure.

In the first issue of the Code, as printed in the first AISC Manual in 1927, this was stated as "Shop Drawings prepared by the Seller and approved by a representative of the Buyer shall be deemed the correct interpretation of the work to be done, but does not relieve the Seller of responsibility for the accuracy of details." This statement was modified in the 1952 revision of the Code to read "...the owner must return one set of prints to the fabricator with a notation of the owner's outright approval or approval subject to corrections as noted." In 1972 the Code stated "Approval by the owner of shop drawings prepared by the fabricator indicates that the fabricator has correctly interpreted the contract requirements, and that any connections designed by the fabricator are of adequate capacity for the design requirements." The Code was again modified in 1976 saying "Approval by the owner of shop drawings prepared by the fabricator indicates that the fabricator has correctly interpreted the contract requirements. This approval constitutes the owner's acceptance of all responsibility for the design adequacy of any connection designed by the fabricator as a part of his preparation of these shop drawings." This statement was not changed in the 1986 revision of the Code.

The current revision of Paragraph 4.2.1 of the Code is intended to clarify the use of the word "Owner." Consequently, the term "owner" has been replaced by "owner's authorized representative," usually meaning the engineer of record. The continuing concept that the structural engineer of record is the sole individual who can best assure the safety of the completed structure has not been modified. This system has worked well for at least the past 65 years, and has achieved a commendable safety record where its principles have been steadfastly applied.

In the preparation of contract drawings, the engineer of record (EOR) has two basic choices in the showing of connection details. The EOR may fully design and detail connections for all conditions. However, in order to allow the owner to benefit from the economies inherent in allowing the fabricator to choose the most efficient connections for the fabricator's shop and erection processes, the EOR may allow the fabricator to select the types of connection and show them in complete detail on the shop drawings for the EOR's approval. In either case, the approval of the shop drawings by the owner's authorized representative constitutes acceptance by the owner's authorized representative of design responsibility for the structural adequacy of the connections shown on the shop drawings. Contracts attempting to share or allocate design responsibility are strongly discouraged. Individual state codes and licensing requirements may vary widely in allowing such allocation of responsibility.

Should the engineer of record elect to fully design and detail connections on the contract documents, the EOR has the obligation to show all fastener sizes, arrangement, quantities and grades, as well as all connection material and weld types, sizes and lengths for each individual member or part to be joined. All requirements for bracing details, stiffeners, doublers, web or cope reinforcement or similar items necessary for the completeness of the design must be sized and shown in complete detail. The fabricator is responsible for correctly reflecting this information in the preparation of shop drawings. Should the fabricator wish to deviate from these specific details or call a problem to the attention of the engineer of record, the fabricator must either do so in writing prior to the preparation of shop drawings, or clearly note the deviation on the drawings submitted for approval. This requirement is not intended in any way to negate the responsibility of the owner's authorized representative to review completely each shop drawing for structural adequacy during the approval process.

If the engineer of record does not show fully designed and detailed connections on the contract documents and allows the fabricator to select connection types when detailing shop drawings, the contract documents must give all reactions, moments, or other forces required for each individual member of parts to be joined so that when preparing shop drawings, the fabricator's detailers and checkers may determine the appropriate connection either by selection from tables shown in AISC publications or by simple calculation. The fabricator can assume that the reactions, moments or other forces given by the engineer are appropriate for the loads to be applied to the structure. All requirements for bracing details, stiffeners, doublers, web or cope reinforcement or similar items necessary for the completeness of the design must be shown in sufficient detail so as to allow the fabricator to submit an accurate estimate of cost at the time of bid.

It is suggested that highly complex connections be fully designed on the contract documents or developed in a timely manner by the EOR after consulting with the fabricator regarding accepted, current and standard practices for fabrication and erection so that the detailing and fabricating processes will not be delayed. In the latter case, a pre-detailing meeting between the EOR and the fabricator may be appropriate to facilitate this exchange of information. In the event that design loads or other information necessary for development of connections is not shown on the contract documents, this information must be furnished to the fabricator in a timely manner.

If the engineer of record elects to utilize typical details which must be interpreted or modified by the fabricator to meet conditions occurring in a structure, such interpretation is forwarded to the engineer of record for review and approval by way of detail or shop drawing submittals.

Where state codes and licensing requirements allow fabricators to design and fabricate complete steel structures, and a fabricator has contracted to provide such services, submittals to the owner or applicable public reviewing authority will normally include only those documents customarily submitted by licensed design professionals on comparable projects within the same licensing jurisdiction.

SECTION 5. MATERIALS

5.1. Mill Materials

The fabricator may purchase materials in stock lengths, exact lengths or multiples of exact lengths to suit the dimensions shown on the contract drawings. Such purchases will normally be job-specific in nature and may not be capable of being utilized on other projects or returned for full credit if subsequent design changes make these materials unsuitable for their originally intended use. The fabricator should be paid for these materials upon delivery from the mill, subject to appropriate additional payment or credit if subsequent unanticipated modification or reorder is required. Purchasing materials to exact lengths is not considered fabrication.

5.1.2. Mill dimensional tolerances are completely set forth as part of ASTM A6. Variation in cross sectional geometry of rolled members must be recognized by the designer, the fabricator and erector (see Fig. 1). Such tolerances are mandatory because roll wear, thermal distortions of the hot cross section immediately after leaving the forming rolls, and differential cooling distortions that take place on the cooling beds are economically beyond precise control. Absolute perfection of cross sectional geometry is not of structural significance and, if the tolerances are recognized and provided for, also not of architectural significance. ASTM A6 also stipulates straightness and camber tolerances which are adequate for most conventional construction. However, these characteristics may be controlled or corrected to closer tolerances during the fabrication process when the unique demands of a particular project justify the added cost.

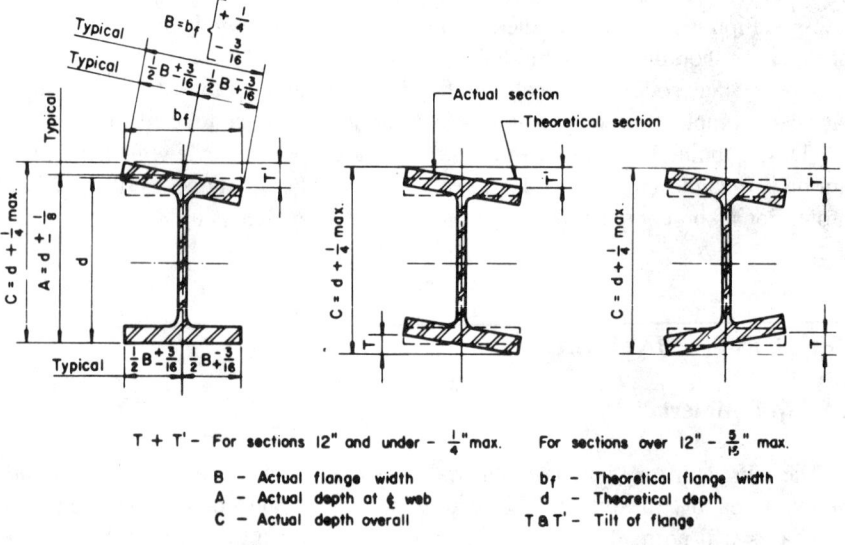

Fig. 1 Mill tolerances on cross-section dimensions

SECTION 6. FABRICATION AND DELIVERY

6.4. Dimensional Tolerances

Fabrication tolerances are stipulated in several specification documents, each applicable to a special area of construction. Basic fabrication tolerances are stipulated in Sections 6.4 and 10 of the Code and Section M2.7 of the AISC Specification. Other specifications and codes frequently incorporated by reference in the contract documents are the *AWS Structural Welding Code and AASHTO Standard Specifications for Highway Bridges.*

6.4.5. Due to the release of stresses, there is no known way to verify camber once members are received in the field. Camber may only be measured in the fabrication shop in the unstressed condition and does not take into account the dead weight of the member, the restraint caused by the end connections in the erected state or any dead load which may be intended to be applied.

6.5. Shop Painting

6.5.2., 6.5.3. The selection of a paint system is a design decision involving many factors including owner's preference, service life of the structure, severity of environmental exposure, cost of both initial application and future renewals, and compatibility of the various components comprising the paint system, i.e., surface preparation, prime coat and subsequent coats.

Because inspection of shop painting needs to be concerned with workmanship at each stage of the operation, the fabricator provides notice of the schedule of operations and affords access to the work site to inspectors. Inspection must be coordinated with that schedule in such a way as to avoid delay of the scheduled operations.

Acceptance of the prepared surface must be made prior to application of the prime coat because the degree of surface preparation cannot be readily verified after painting. Time delay between surface preparation and application of the prime coat can result in unacceptable deterioration of a properly prepared surface, necessitating a repetition of surface preparation. This is especially true with blast-cleaned surfaces. Therefore, to avoid potential deterioration of the surface it is assumed that surface preparation is accepted unless it is inspected and rejected prior to the scheduled application of the prime coat.

The prime coat in any paint system is designed to maximize the wetting and adherence characteristics of the paint, usually at the expense of its weathering capabilities. Deterioration of the shop paint normally begins immediately after exposure to the elements and worsens as the duration of exposure is extended. Consequently, extended exposure of the prime coat to weather or to a corrosive atmosphere will lead to its deterioration and may necessitate repair, possibly including repetition of surface preparation and primer application in limited areas. With the introduction of high performance paint systems, delay in the application of the prime coat has become more critical. High performance paint systems generally require a greater degree of surface preparation, as well as early application of weathering protection for the prime coat.

Since the fabricator does not control the selection of the paint system, the compatibility of the various components of the total paint system, nor the length of exposure of the prime coat, he cannot guarantee the performance of the prime coat or any other part of the system. Rather, the fabricator is responsible only for accomplishing the specified surface preparation and for applying the shop coat or coats in accordance with the contract documents.

Section 6.5.2 stipulates cleaning the steel to the requirements of SSPC-SP2. This section is not meant as an exclusive cleaning level, but rather that level of surface preparation which will be furnished if the steel is to be painted and if the job specifications are silent or do not require more stringent surface preparation requirements.

Further information regarding shop painting is available in *A Guide to Shop Painting of Structural Steel*, published jointly by the Steel Structures Painting Council and the American Institute of Steel Construction.

6.5.4. Extended exposure of unpainted steel which has been cleaned for subsequent fire protection material application can be detrimental to the fabricated product. Most levels of cleaning require the removal of all loose mill scale, but permit some amount of "tightly adhering mill scale." When a piece of structural steel which has been cleaned to an acceptable level is left exposed to a normal environment, moisture can penetrate behind the scale, and some "lifting" of the scale by the oxidation products is to be expected. Cleanup of "lifted" mill scale is not the responsibility of the fabricator, but is assigned by contract requirement to an appropriate contractor.

Section 6.5.4 of the Code is not applicable to weathering steel, for which special cleaning specifications are always required in the contract documents.

SECTION 7. ERECTION

7.5. Installation of Anchor Bolts and Embedded Items

7.5.1. While the general contractor must make every effort to set anchor bolts accurately to theoretical drawing dimensions, minor deviations may occur. The tolerances set forth in this section were compiled from data collected from general contractors and erectors. They can be attained by using reasonable care and will ordinarily allow the steel to be erected and plumbed to required tolerances. If special conditions require closer tolerances, the contractor responsible for setting the anchor bolts should be so informed by the contract documents. When anchor bolts are set in sleeves, the adjustment provided may be used to satisfy the required anchor bolt setting tolerances.

The tolerances established in this section of the Code have been selected to be compatible with oversize holes in base plates, as recommended in the AISC textbook *Detailing for Steel Construction*.

An *anchor bolt group* is the set of anchor bolts which receive a single fabricated steel shipping piece.

The *established column line* is the actual field line most representative of the centers of the as-built anchor bolt groups along a line of columns. It must be straight or curved as shown on the plans.

7.6. Bearing Devices

The $\frac{1}{8}$ in. tolerance on elevation of bearing devices relative to established grades is provided to permit some variation in setting bearing devices and to account for attainable accuracy with standard surveying instruments. The use of leveling plates larger than 22 in. x 22 in. is discouraged and grouting is recommended with larger sizes. For purposes of erection stability, the use of leveling nuts is discouraged when base plates have less than four (4) anchor bolts.

7.9.3. Non Self-Supporting Steel Frames

To rationally provide temporary supports and/or bracing, the erector must be informed by the owner of the sequence of installation and the effect of loads imposed by such elements at various stages during the sequence until they become effective. The overall strength and stability of a non self-supporting steel frame may be dependent upon the installation of non-structural steel elements such as concrete floor diaphragms, concrete or masonry shear walls, precast concrete facade pieces, etc. The requirement for these elements to be in place to provide overall strength and stability for the structural steel frame should be made clear in the contract documents in order that the need for temporary support may be clearly understood. For example, precast tilt-up slabs or channel slab facia elements which depend upon attachment to the steel frame for stability against overturning due to eccentricity of their gravity load may induce significant unbalanced lateral forces on the bare steel frame when partially installed.

7.11.   Framing Tolerances

The erection tolerances defined in this section of the Code have been developed through long-standing usage as practical criteria for the erection of structural steel. Erection tolerances were first defined by AISC in its *Code of Standard Practice* of

October, 1924 in Section 7 (f), "Plumbing Up." With the changes that took place in the types and use of materials in building construction after World War II, and the increasing demand by architects and owners for more specific tolerances, AISC adopted new standards for erection tolerances in Section 7 (h) of the March 15, 1959 edition of the Code. Experience has proven that those tolerances can be economically obtained.

The current requirements were first published in the October 1,1972 edition of the Code. They provide an expanded set of criteria over earlier Code editions. The basic premise that the final accuracy of location of any specific point in a structural steel frame results from the combined mill, fabrication and erection tolerances, rather than from the erection tolerances alone, remains unchanged in this edition of the Code. However, to improve clarity, pertinent standard fabrication tolerances are now stipulated in Section 7.11, rather than by reference to the AISC Specification as in previous editions. Additionally, expanded coverage has been given to the definition of working points and working lines governing measurements of the actual steel location. Illustrations for defining and applying the applicable Code tolerances are provided in this Commentary.

The recent trend in building work is away from built-in-place construction wherein compatibility of the frame and the facade or other collateral materials is automatically provided for by the routine procedures of the crafts. Building construction today frequently incorporates prefabricated components wherein large units are developed with machine-like precision to dimensions that are theoretically correct for a perfectly aligned steel frame with ideal member cross sections. This type of construction has made the magnitude of the tolerances allowed for structural steel building frames increasingly of concern to owners, architects and engineers. This has led to the inclusion in job specifications of unrealistically small tolerances, which indicate a general lack of recognition of the effects of the accumulation of dead load, temperature effects and mill, fabrication and erection tolerances. Such tolerances are not economically feasible and do not measurably increase the structure's functional value. This edition of the Code incorporates tolerances previously found to be practical and presents them in a precise and clear manner. Actual application methods have been considered and the application of the tolerance limitations to the actual structure have been defined.

7.11.3. Position and Alignment

The limitations described in Section 7.11.3.1 and illustrated in Figs. 2 and 3 make it possible to maintain built-in-place or prefabricated facades in a true vertical plane up to the 20th story, if connections which provide for 3 in. adjustment are used. Above the 20th story, the facade may be maintained within $\frac{1}{16}$ in. per story with a maximum total deviation of 1 in. from a true vertical plane, if the 3 in. adjustment is provided.

Section 7.11.3.1(c) limits the position of exterior column working points at any given splice elevation to a narrow horizontal envelope parallel to the building line (see Fig. 4). This envelope is limited to a width of $1\frac{1}{2}$ in., normal to the building line, in up to 300 ft of building length. The horizontal location of this envelope is not necessarily directly above or below the corresponding envelope at the adjacent splice elevations, but should be within the limitation of the 1:500 allowable tolerance in plumbness of the controlling columns (see Fig. 3).

Connections permitting adjustments of plus 2 in. to minus 3 in. (5 in. total) will be necessary in cases where the architect or owner insists upon attempting to construct the facade to a true vertical plane above the 20th story.

Usually there is a differential shortening of the internal versus the external columns during construction, due to non-uniform rate of accumulation of dead load stresses (see Fig. 5). The amount of such differential shortening is indeterminate because it varies dependent upon construction sequence from day to day as the construction progresses, and does not reach its maximum shortening until the building is in service. When floor concrete is placed while columns are supporting different percentages of their full design loads, the floor must be finished to slopes established by measurements from the tops of beams at column connections. The effects of

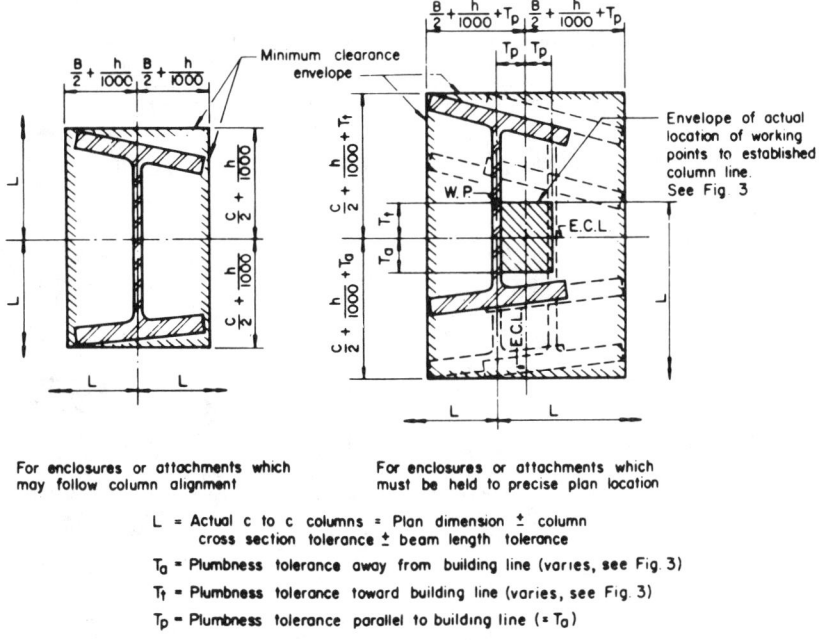

Fig. 2. Clearance required to accommodate accumulated column tolerances

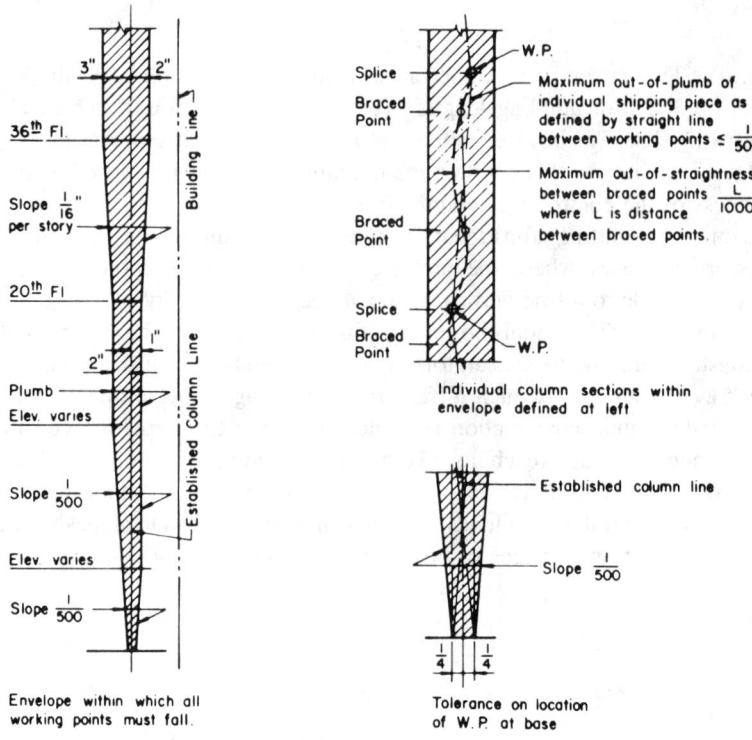

Fig. 3. *Exterior column plumbness tolerances normal to building line*

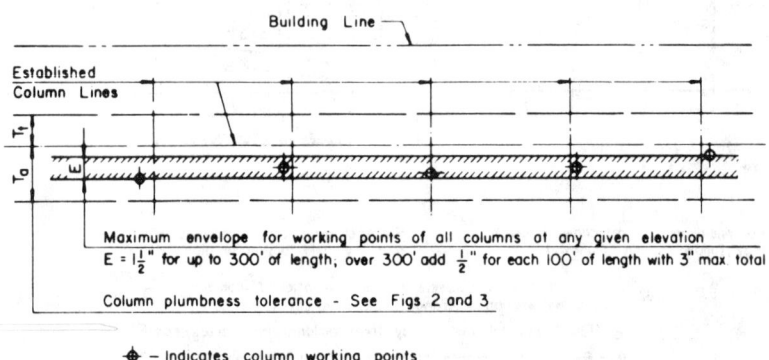

Fig. 4. *Tolerances in plan at any splice elevation of exterior columns*

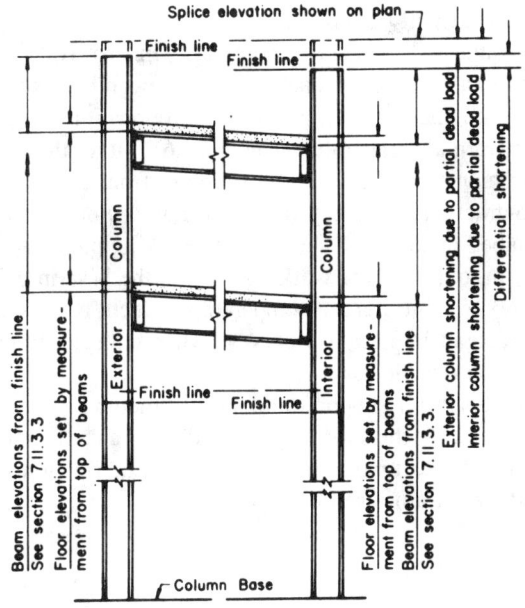

On a particular date during the erection of structural steel and placement of other material, (floor concrete, facade etc.) the interior columns will be carrying a higher percentage of their final loads than the exterior columns Therefore, for equal design unit stresses, the actual stress on that date for interior columns will be greater than the actual stresses on exterior columns.
When all dead loads have been applied, stresses and shortening in all columns will be approximately equal.

Fig. 5. Effect of differential column shortening

differential shortening, plus mill camber and deflections, become very important when there is little cover over the steel, when there are electrical fittings mounted on the steel flooring whose tops are supposed to be flush with the finished floor, when there is small clearance between bottom of beams and top of door frames, etc., and when there is little clearance around ductwork. To finish floors to a precisely level plane, for example by the use of laser leveling techniques, can result in significant differential floor thicknesses, different increases above design dead loads for individual columns and, thus, permanent differential column shortening and out-of-level completed floors.

Similar considerations make it unfeasible to attempt to set the elevation of a given floor in a multistory building by reference to a bench mark at the base of the structure. Columns are fabricated to a length tolerance of ± $\frac{1}{32}$ in. while under a zero state of stress. As dead loads accumulate, the column shortening which takes place is negligible within individual stories and in low buildings, but will accumulate to significant magnitude in tall buildings. Thus, the upper floors of tall buildings will be excessively thick and the lower floors will be below the initial finish elevation if floor elevations are established relative to a ground level bench mark.

If foundations and base plates are accurately set to grade and the lengths of individual column sections are checked for accuracy prior to erection, and if floor elevations are established by reference to the elevation of the top of beams, the effect of column shortening due to dead load will be minimized.

Since a long unencased steel frame will expand or contract $\frac{1}{8}$ in. per 100 ft for each change of 15°F in temperature, and since the change in length can be assumed to act about the center of rigidity, the end columns anchored to foundations will be plumb only when the steel is at normal temperature (see Fig. 6). It is therefore necessary to correct field measurements of offsets to the structure from established baselines for the expansion or contraction of the exposed steel frame. For example, a building 200-ft long that is plumbed up at 100°F should have working points at the tops of end columns positioned $\frac{1}{2}$ in. out from the working point at the base in order for the column to be plumb at 60°F. Differential temperature effects on column length should also be taken into account in plumbing surveys when tall steel frames are subject to strong sun exposure on one side.

The alignment of lintels, spandrels, wall supports and similar members used to connect other building construction units to the steel frame should have an adjustment of sufficient magnitude to allow for the accumulative effect of mill, fabrication and erection tolerances on the erected steel frame (see Fig. 7).

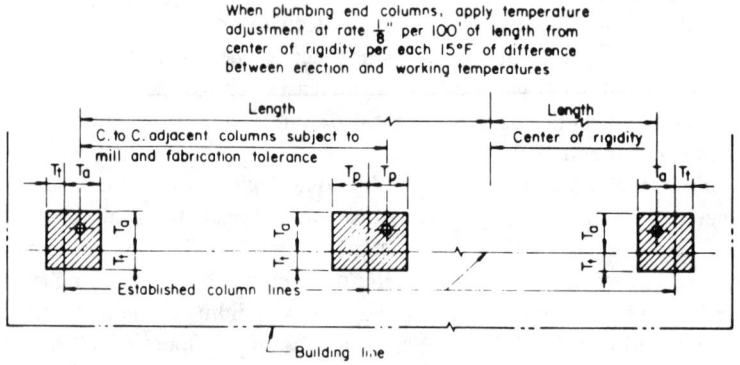

Fig. 6. Tolerances in plan location of columns

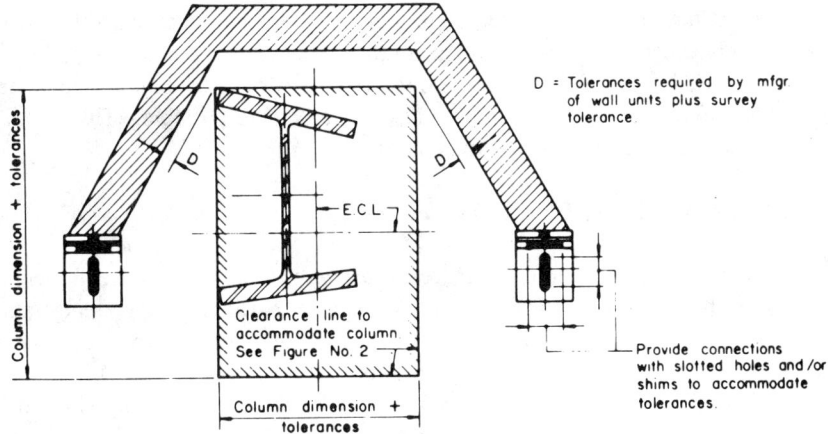

D = Tolerances required by mfgr. of wall units plus survey tolerance.

Provide connections with slotted holes and/or shims to accommodate tolerances.

If fascia joints are set from nearest column finish line, allow $\pm\frac{5}{8}$" for vertical adjustment. Owners plans for fascia details must allow for progressive shortening of steel columns.

Fig. 7. Clearance required to accommodate fascia

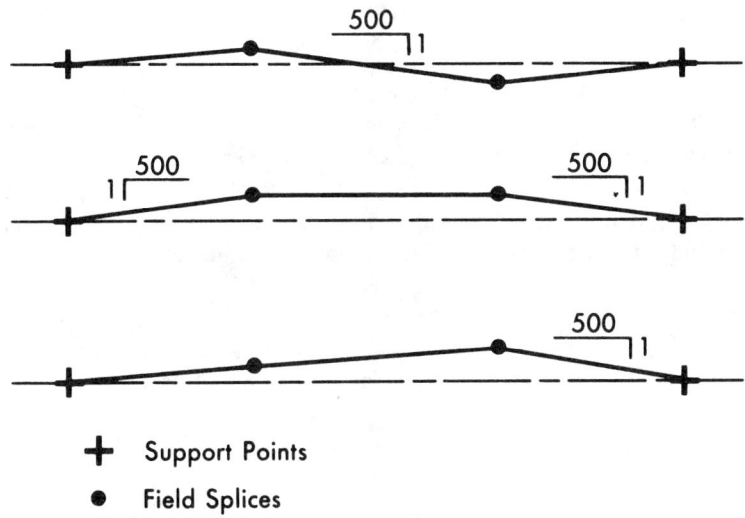

✛ Support Points

● Field Splices

Fig. 8. Alignment tolerances for members with field splices

7.11.3.2. Alignment Tolerance for Members with Field Splices

The angular misalignment of the working line of all fabricated shipping pieces relative to the line between support points of the member as a whole in erected position must not exceed 1 in 500. Note that the tolerance is not stated in terms of a linear displacement at any point and is not to be taken as the overall length between supports divided by 500. Typical examples are shown in Fig. 8. Numerous conditions within tolerance for these and other cases are possible. This condition applies to both plan and elevation tolerances.

7.11.4. Responsibility for Clearances

In spite of all efforts to minimize inaccuracies, deviations will still exist; therefore, in addition, the designs of prefabricated wall panels, partition panels, fenestrations, floor-to-ceiling door frames and similar elements must provide for clearance and details for adjustment as described in Section 7.11.4. Designs must provide for adjustment in the vertical dimension of prefabricated facade panels supported by the steel frame because the accumulation of shortening of stressed steel columns will result in the unstressed facade supported at each floor level being higher than the steel frame connections to which it must be attached. Observations in the field have shown that where a heavy facade is erected to a greater height on one side of a multistory building than on the other, the steel framing will be pulled out of alignment. Facades should be erected at a relatively uniform rate around the perimeter of the structure.

7.14. Handling and Storage

Handling Painted Steel

During storage, loading, transport, unloading and erection, blemish marks caused by slings, chains, blocking, tie-downs, etc., occur in varying degrees. Abrasions caused by handling or cartage after painting are to be expected. The owner/engineer must recognize that any shop applied coating, no matter how carefully protected, will require touch-up in the field. Touch-up of these blemished areas is the responsibility of the contractor performing the field touch-up of field painting.

Cleaning After Erection

The responsibility for proper storage and handling of fabricated steel at the construction site during erection is properly the erector's. Shop-painted steel stored in the field pending erection should be kept free of the ground and so positioned as to minimize water-holding pockets. The owner or general contractor is responsible for providing suitable site conditions and proper access so that the fabricator/erector may perform its work.

Site conditions are frequently muddy, sandy or dusty, or a combination of all three, during the erection period. Under such conditions it may be impossible to store and handle the steel in such a way as to completely avoid accumulation of mud, dirt or sand on the surface of the steel, even though the fabricator/erector manages to proceed with the work.

Repairs of damage to painted surfaces and/or removal of foreign materials due to adverse site conditions are outside the scope of responsibility of the fabricator/erector when reasonable attempts at proper handling and storage have been made.

SECTION 8. QUALITY CONTROL

8.1.1. The AISC Quality Certification Program confirms to the construction industry that a certified structural steel fabricating plant has the capability by reason of commitment, personnel, organization, experience, procedures, knowledge and equipment to produce fabricated structural steel of the required quality for a given category of structural steelwork. The AISC Quality Certification Program is not intended to involve inspection and/or judgment of product quality on individual projects. Neither is it intended to guarantee the quality of specific fabricated steel products.

SECTION 9. CONTRACTS

9.2. Calculation of Weights

The standard procedure for calculation of weights that is described in the Code meets the need for a universally acceptable system for defining "pay weights" in contracts based on the weight of delivered and/or erected materials. This procedure permits owners to easily and accurately evaluate price per pound proposals from potential suppliers and enables both parties to a contract to have a clear understanding of the basis for payment.

The Code procedure affords a simple, readily understood method of calculation which will produce pay weights which are consistent throughout the industry and which may be easily verified by the owner. While this procedure does not produce actual weights, it can be used by purchasers and suppliers to define a widely accepted basis for bidding and contracting for structural steel. However, any other system can be used as the basis for a contractual agreement. When other systems are used, both supplier and purchaser should clearly understand how the alternate procedure is handled.

9.3. Revisions to Contract Documents

9.3.1.   Revisions to the Contract are implemented by issuance of new documents or re-issuance of existing documents. Individual revisions must be noted where they occur and documents must be dated with latest issue date and the reasons for issuance must be identified.

9.3.2.   Revisions to the Contract are also implemented by change order, extra work order, or notations on the shop and erection drawings when returned from approval. However, revisions implemented in this manner must be incorporated subsequently as revisions to the plans and/or specifications and re-issued in accordance with Section 9.3.1.

9.3.3.   The issuance of revisions authorizes the fabricator and erector to incorporate the revisions into the work. This authorization obligates the owner to pay the fabricator and erector for costs associated with changed and/or additional work.

9.6. Terms of Payment

These terms include such items as progress payments for material, fabrication, erection, retainage, performance and payment bonds and final payment. If a performance or payment bond, paid for by the owner, is required by contract, then no retainage shall be required.

SECTION 10. ARCHITECTURALLY EXPOSED STRUCTURAL STEEL

The rapidly increasing use of exposed structural steel as a medium of architectural expression has given rise to a demand for closer dimensional tolerances and smoother finished surfaces than required for ordinary structural steel framing.

This section of the Code establishes standards for these requirements which take into account both the desired finished appearance and the abilities of the fabrication shop to produce the desired product. These requirements were previously contained in the AISC *Specification for Architecturally Exposed Structural Steel* which architects and engineers have specified in the past. It should be pointed out that the term "Architecturally Exposed Structural Steel" (AESS), as covered in this section, must be specified in the contract documents if the fabricator is required to meet the fabricating standards of Section 10, and applies only to that portion of the structural steel so identified. In order to avoid misunderstandings and to hold costs to a minimum, only those steel surfaces and connections which will remain exposed and subject to normal view by pedestrians or occupants of the completed structure should be designated as AESS.

AISC
Quality Certification
Program

AMERICAN INSTITUTE OF STEEL CONSTRUCTION, INC.

AISC
Quality Certification
Program

In recent years, the quality of construction methods and materials has become the subject of increasing concern to building officials, highway officials, and designers. One result of this concern has been the enactment of ever more demanding inspection requirements intended to ensure product quality. In many cases, however, these more demanding inspection requirements have not been based upon demonstrated unsatisfactory performance of structures in service. Rather, they have been based upon the capacity of sophisticated test equipment, or upon standards developed for nuclear construction rather than conventional construction. Adding to the problem, arbitrary interpretation of specifications by inspectors has too often been made without rational consideration of the type of construction involved. The result has been spiraling increases in the costs of fabrication of structural steel and of inspection, which must be paid by owners without necessarily assuring that the product quality required has been improved.

Product inspection, although it has a valid place in the construction process, is not the most logical or practical way to assure that structural steelwork will conform to the requirements of contract documents and satisfy the intended use. A better solution can be found in the exercise of good quality control and quality assurance by the fabricator *throughout the entire production process*.

Recognizing this fact, and seeking some valid, objective method whereby a fabricator's capability for assuring a quality product could be evaluated, a number of code authorities have, in recent years, instituted steps to establish fabricator registration programs. However, these independent efforts resulted in extremely inconsistent criteria. They were developed primarily by inspectors or inspection agencies who were experienced in testing, but were not familiar with the complexities of the many steps, procedures, techniques, and controls required to assure quality throughout the fabricating process. Neither were these inspection agencies qualified to determine the various levels of quality required to assure satisfactory performance in meeting the service requirements of the many different types of steel structures.

Recognizing the need for a comprehensive national standard for fabricator certification, and concerned by the trend toward costly inspection requirements that could not be justified by rational quality standards, the American Institute of Steel Construction has developed and implemented a voluntary Quality Certification Program, whereby any structural steel fabricating plant—whether a member of AISC or not—can have its capability for assuring quality production evaluated on a fair and impartial basis.

THE AISC PROGRAM

The AISC Quality Certification Program does not involve inspection and/or judgment of product quality on individual projects. Neither does it guarantee the quality of specific fabricated steel products. Rather, the purpose of the AISC Quality Certification Program is to confirm to the construction industry that a Certified structural steel fabricating plant *has the personnel, organization, experience, procedures, knowledge, equipment, capability and commitment to produce fabricated steel of the required quality for a given category of structural steelwork.*

The AISC Quality Certification Program was developed by a group of highly qualified shop operation personnel from large, medium, and small structural steel fabricating plants throughout the United States. These individuals all had extensive experience and were fully aware of where and how problems can arise during the production process and of the steps and procedures that must be followed during fabrication to assure that the finished product meets the quality requirements of the contract.

The program was reviewed and strongly endorsed by an Independent Board of Review comprised of 17 prominent structural engineers from throughout the United States, who were not associated with the steel fabricating industry, but were well qualified in matters of quality requirements for reliable service of all types of steel structures.

CATEGORIES OF CERTIFICATION

A fabricator may apply for certification of a plant in one of the following categories of structural steelwork:

I: **Conventional Steel Structures** — Small Public Service and Institutional Buildings, (Schools, etc.), Shopping Centers, Light Manufacturing Plants, Miscellaneous and Ornamental Iron Work, Warehouses, Sign Structures, Low Rise, Truss Beam/Column Structures, Simple Rolled Beam Bridges.

II: **Complex Steel Building Structures** — Large Public Service and Institutional Buildings, Heavy Manufacturing Plants, Powerhouses (fossil, non-nuclear), Metal Producing/Rolling Facilities, Crane Bridge Girders, Bunkers and Bins, Stadia, Auditoriums, High Rise Buildings, Chemical Processing Plants, Petroleum Processing Plants.

III: **Major Steel Bridges** — All bridge structures other than simple rolled beam bridges.

MB: **Metal Building Systems** — Pre-engineered Metal Building Structures.

Supplement: **Auxiliary and Support Structures for Nuclear Power Plants** — This supplement, applicable to nuclear plant structures designed under the AISC Specification, but not to pressure-retaining structures, offers utility companies and designers of nuclear power plants a certification program that will eliminate the need for many of the more costly, conflicting programs now in use. A fabricator must hold certification in either Category I, II or III prior to application for certification in this category.

Certification in Category II automatically includes Category I. Certification in Category III automatically includes Categories I and II. Certification in Category MB is not transferable to any other Category.

INSPECTION-EVALUATION PROCEDURE

An outside, experienced, professional organization, ABS Quality Evaluations, Inc. (a subsidiary of American Bureau of Shipping) has been retained by AISC to perform the plant Inspection-Evaluation in accordance with a standard check list and rating procedure established by AISC for each certification category in the program. Upon completion of this Inspection-Evaluation, ABS Quality Evaluations, Inc. (commonly known as ABS-QE) will recommend to AISC that a fabricator be approved or disapproved for certification. ABS-QE's Inspection-Evaluation is totally independent of the fabricator's and AISC's influence, and their evaluation is not subject to review by AISC.

At a time mutually agreed upon by the fabricator, AISC, and ABS-QE, the Inspection-Evaluation team visits the plant to investigate and rate the following basic plant functions directly and indirectly affecting quality assurance: General Management, Engineering and Drafting, Procurement, Shop Operations, and Quality Control. The Inspection-Evaluation team will perform the following:

1. Confirm data submitted with the Application for Certification.
2. Interview key supervisory personnel and subordinate employees.
3. Observe and rate the organization in operation, including procedures used in functions affecting quality assurance.
4. Inspect and rate equipment and facilities.
5. At an "exit interview," review with plant management the completed check list observations and evaluation scoring, including discussions of deficiencies and omissions, if any.

The number of days required for Inspection-Evaluation varies according to the size and complexity of the plant, but usually requires two to five days.

CERTIFICATION

Following recommendation for Certification by the Inspection-Evaluation team, AISC will issue a certificate identifying the fabricator, the plant, and the Category of Certification. The certificate is valid for a three year period, subject to annual review in the form of unannounced inspections early in the second and third year periods. The certificate is endorsed annually, provided there is successful completion of the unannounced second and third year inspection.

An annual self-audit, based on the standard check list, must be made by plant management during the 11th and 23rd months after initial Certification. This self-audit must be retained at the plant and made available to the Inspection-Evaluation team during the unannounced second and third year inspections.

At the end of the third year, the cycle begins again with a complete prescheduled Inspection-Evaluation and the issuance of a new certificate.

Part 7

MISCELLANEOUS DATA AND MATHEMATICAL INFORMATION

Table 7-1.
WIRE AND SHEET METAL GAGES
Equivalent thickness in decimals of an inch

Gage No.	U.S. Standard Gage for Uncoated Hot & Cold-Rolled Sheets[b]	Galvanized Sheet Gage for Hot-Dipped Zinc Coated Sheets[b]	USA Steel Wire Gage	Gage No.	U.S. Standard Gage for Uncoated Hot & Cold-Rolled Sheets[b]	Galvanized Sheet Gage for Hot-Dipped Zinc Coated Sheets[b]	USA Steel Wire Gage
7/0	—	—	.490	13	.0897	.0934	.092[a]
6/0	—	—	.462[a]	14	.0747	.0785	.080
5/0	—	—	.430[a]	15	.0673	.0710	.072
4/0	—	—	.394[a]	16	.0598	.0635	.062[a]
3/0	—	—	.362[a]	17	.0538	.0575	.054
2/0	—	—	.331	18	.0478	.0516	.048[a]
1/0	—	—	.306	19	.0418	.0456	.041
1	—	—	.283	20	.0359	.0396	.035[a]
2	—	—	.262[a]	21	.0329	.0366	—
3	.2391	—	.244[a]	22	.0299	.0336	—
4	.2242	—	.225[a]	23	.0269	.0306	—
5	.2092	—	.207	24	.0239	.0276	—
6	.1943	—	.192	25	.0209	.0247	—
7	.1793	—	.177	26	.0179	.0217	—
8	.1644	.1681	.162	27	.0164	.0202	—
9	.1495	.1532	.148[a]	28	.0149	.0187	—
10	.1345	.1382	.135	29	—	.0172	—
11	.1196	.1233	.120[a]	30	—	.0157	—
12	.1046	.1084	.106[a]				

[a]Rounded value. The steel wire gage has been taken from ASTM A510 "General Requirements for Wire Rods and Coarse Round Wire, Carbon Steel." Sizes originally quoted to four decimal equivalent places have been rounded to three decimal places in accordance with rounding procedures of ASTM "Recommended Practice" E29.
[b]The equivalent thicknesses are for information only. The product is commonly specified to decimal thickness, not to gage number.

Table 7-2.
AISI STANDARD NOMENCLATURE FOR FLAT ROLLED CARBON STEEL

Thickness (Inches)	Width (Inches)					
	To 3½ incl.	Over 3½ To 6	Over 6 To 8	Over 8 To 12	Over 12 To 48	Over 48
0.2300 & thicker	Bar	Bar	Bar	Plate	Plate	Plate
0.2299 to 0.2031	Bar	Bar	Strip	Strip	Sheet	Plate
0.2030 to 0.1800	Strip	Strip	Strip	Strip	Sheet	Plate
0.1799 to 0.0449	Strip	Strip	Strip	Strip	Sheet	Sheet
0.0448 to 0.0344	Strip	Strip				
0.0343 to 0.0255	Strip	Hot-rolled sheet and strip not generally produced in these widths and thicknesses				
0.0254 & thinner						

Table 7-3.
COEFFICIENTS OF EXPANSION

The coefficient of linear expansion (ε) is the change in length, per unit, for a change of one degree of temperature. The coefficient of surface expansion is approximately two times the linear coefficient, and the coefficient of volume expansion, for solids, is approximately three times the linear coefficient.

A bar, free to move, will increase in length with an increase in temperature and will decrease in length with a decrease in temperature. The change in length will be $\varepsilon t l$, where ε is the coefficient of linear expansion, t the change in temperature and l the length. If the ends of a bar are fixed, a change in temperature (t) will cause a change in the unit stress of $E\varepsilon t$, and in force of $AE\varepsilon t$, where A is the cross-sectional area of the bar and E the modulus of elasticity.

The following table gives the coefficient of linear expansion for 100°, or 100 times the value indicated above.

Example: A piece of medium steel is exactly 40 ft long at 60°F. Find the length at 90°F assuming the ends are free to move.

$$\text{change of length} = \varepsilon t l = \frac{.00065 \times 30 \times 40}{100} = .0078 \text{ ft}$$

The length at 90° is 40.0078 ft

Example: A piece of medium carbon steel is exactly 40 ft long and the ends are fixed. If the temperature increases 30°F, what is the resulting change in the unit stress?

$$\text{change in unit stress} = E\varepsilon t = \frac{29,000 \times .00065 \times 30}{100} = 5.7 \text{ ksi}$$

COEFFICIENTS OF EXPANSION FOR 100 DEGREES = 100ε

Materials	Linear Expansion Centigrade	Linear Expansion Fahrenheit	Materials	Linear Expansion Centigrade	Linear Expansion Fahrenheit
METALS AND ALLOYS			**STONE AND MASONRY**		
Aluminum, wrought	.00231	.00128	Ashlar masonry	.00063	.00035
Brass	.00188	.00104	Brick Masonry	.00061	.00034
Bronze	.00181	.00101	Cement, portland	.00126	.00070
Copper	.00168	.00093	Concrete	.00099	.00055
Iron, cast, gray	.00106	.00059	Granite	.00080	.00044
Iron, wrought	.00120	.00067	Limestone	.00076	.00042
Iron, wire	.00124	.00069	Marble	.00081	.00045
Lead	.00286	.00159	Plaster	.00166	.00092
Magnesium, various alloys	.0029	.0016	Rubble masonry	.00063	.00035
Nickel	.00126	.00070	Sandstone	.00097	.00054
Steel , mild	.00117	.00065	Slate	.00080	.00044
Steel, stainless, 18-8	.00178	.00099			
Zinc, rolled	.00311	.00173			
TIMBER			**TIMBER**		
Fir ⎫	.00037	.00021	Fir ⎫	.0058	.0032
Maple ⎬ parallel to fiber	.00064	.00036	Maple ⎬ perpendicular to	.0048	.0027
Oak ⎬	.00049	.00027	Oak ⎬ fiber	.0054	.0030
Pine ⎭	.00054	.00030	Pine ⎭	.0034	.0019

EXPANSION OF WATER

Maximum Density = 1

C°	Volume	C°	Volume	C°	Volume	C°	Volume	C°	Volume	C°	Volume
0	1.000126	10	1.000257	30	1.004234	50	1.011877	70	1.022384	90	1.035829
4	1.000000	20	1.001732	40	1.007627	60	1.016954	80	1.029003	100	1.043116

Table 7-4.
WEIGHTS AND SPECIFIC GRAVITIES

Substance	Weight lb per cu ft	Specific Gravity	Substance	Weight lb per cu ft	Specific Gravity
ASHLAR, MASONRY			**MINERALS**		
Granite, syenite, gneiss	165	2.3–3.0	Asbestos	153	2.1–2.8
Limestone, marble	160	2.3–2.8	Barytes	281	4.50
Sandstone, bluestone	140	2.1–2.4	Basalt	184	2.7–3.2
			Bauxite	159	2.55
MORTAR RUBBLE			Borax	109	1.7–1.8
MASONRY			Chalk	137	1.8–2.6
Granite, syenite, gneiss	155	2.2–2.8	Clay, marl	137	1.8–2.6
Limestone, marble	150	2.2–2.6	Dolomite	181	2.9
Sandstone, bluestone	130	2.0–2.2	Feldspar, orthoclase	159	2.5–2.6
			Gneiss, serpentine	159	2.4–2.7
DRY RUBBLE MASONRY			Granite, syenite	175	2.5–3.1
Granite, syenite, gneiss	130	1.9–2.3	Greenstone, trap	187	2.8–3.2
Limestone, marble	125	1.9–2.1	Gypsum, alabaster	159	2.3–2.8
Sandstone, bluestone	110	1.8–1.9	Hornblende	187	3.0
			Limestone, marble	165	2.5–2.8
BRICK MASONRY			Magnesite	187	3.0
Pressed brick	140	2.2–2.3	Phosphate rock, apatite	200	3.2
Common brick	120	1.8–2.0	Porphyry	172	2.6–2.9
Soft brick	100	1.5–1.7	Pumice, natural	40	0.37–0.90
			Quartz, flint	165	2.5–2.8
CONCRETE MASONRY			Sandstone, bluestone	147	2.2–2.5
Cement, stone, sand	144	2.2–2.4	Shale, slate	175	2.7–2.9
Cement, slag. etc.	130	1.9–2.3	Soapstone, talc	169	2.6–2.8
Cement, cinder, etc.	100	1.5–1.7			
VARIOUS BUILDING					
MATERIALS			**STONE, QUARRIED, PILED**		
Ashes, cinders	40–45	—	Basalt, granite, gneiss	96	—
Cement, portland, loose	90	—	Limestone, marble, quartz . . .	95	—
Cement, portland, set	183	2.7–3.2	Sandstone	82	—
Lime, gypsum, loose	53–64	—	Shale	92	—
Mortar, set	103	1.4–1.9	Greenstone, hornblende . . .	107	—
Slags, bank slag	67–72	—			
Slags, bank screenings	98–117	—			
Slags, machine slag	96	—	**BITUMINOUS SUBSTANCES**		
Slags, slag sand	49–55	—	Asphaltum	81	1.1–1.5
			Coal, anthracite	97	1.4–1.7
EARTH, ETC., EXCAVATED			Coal, bituminous	84	1.2–1.5
Clay, dry	63	—	Coal, lignite	78	1.1–1.4
Clay, damp, plastic	110	—	Coal, peat, turf, dry	47	0.65–0.85
Clay and gravel, dry	100	—	Coal, charcoal, pine	23	0.28–0.44
Earth, dry, loose	76	—	Coal, charcoal, oak	33	0.47–0.57
Earth, dry, packed	95	—	Coal, coke	75	1.0–1.4
Earth, moist, loose	78	—	Graphite	131	1.9–2.3
Earth, moist, packed	96	—	Paraffine	56	0.87–0.91
Earth, mud, flowing	108	—	Petroleum	54	0.87
Earth, mud, packed	115	—	Petroleum, refined	50	0.79–0.82
Riprap, limestone	80–85	—	Petroleum, benzine	46	0.73–0.75
Riprap, sandstone	90	—	Petroleum, gasoline	42	0.66–0.69
Riprap, shale	105	—	Pitch	69	1.07–1.15
Sand, gravel, dry, loose	90–105	—	Tar, bituminous	75	1.20
Sand, gravel, dry, packed . . .	100–120	—			
Sand, gravel, wet	118–120	—			
EXCAVATIONS IN WATER					
Sand or gravel	60	—	**COAL AND COKE, PILED**		
Sand or gravel and clay	65	—	Coal, anthracite	47–58	—
Clay	80	—	Coal, bituminous, lignite	40–54	—
River mud	90	—	Coal, peat, turf	20–26	—
Soil	70	—	Coal charcoal	10–14	—
Stone riprap	65	—	Coal coke	23–32	—

The specific gravities of solids and liquids refer to water at 4°C, those of gases to air at 0°C and 760 mm pressure. The weights per cubic foot are derived from average specific gravities, except where stated that weights are for bulk, heaped, or loose material, etc.

Table 7-4 (cont.).
WEIGHTS AND SPECIFIC GRAVITIES

Substance	Weight lb per cu ft	Specific Gravity	Substance	Weight lb per cu ft	Specific Gravity
METALS, ALLOYS, ORES			**TIMBER, U.S. SEASONED**		
Aluminum, cast, hammered . .	165	2.55–2.75	Moisture content by weight:		
Brass, cast, rolled	534	8.4–8.7	Seasoned timber 15 to 20%		
Bronze, 7.9 to 14% Sn	509	7.4–8.9	Green timber up to 50%		
Bronze, aluminum	481	7.7	Ash, white, red	40	0.62–0.65
Copper, cast, rolled	556	8.8–9.0	Cedar, white, red	22	0.32–.038
Copper ore, pyrites	262	4.1–4.3	Chestnut	41	0.66
Gold, cast, hammered	1205	19.25–19.3	Cypress	30	0.48
Iron, cast, pig	450	7.2	Fir, Douglas spruce	32	0.51
Iron, wrought	485	7.6–7.9	Fir, eastern	25	0.40
Iron, speigel-eisen	468	7.5	Elm, white	45	0.72
Iron, ferro-silicon	437	6.7–7.3	Hemlock	29	0.42–0.52
Iron ore, hematite	325	5.2	Hickory	49	0.74–0.84
Iron ore, hematite in bank . .	160–180	—	Locust	46	0.73
Iron ore, hematite loose . . .	130–160	—	Maple, hard	43	0.68
Iron ore, limonite	237	3.6–4.0	Maple, white	33	0.53
Iron ore, magnetite	315	4.9–5.2	Oak, chestnut	54	0.86
Iron slag	172	2.5–3.0	Oak, live	59	0.95
Lead	710	11.37	Oak, red, black	41	0.65
Lead ore, galena	465	7.3–7.6	Oak, white	46	0.74
Magnesium, alloys	112	1.74–1.83	Pine, Oregon	32	0.51
Manganese	475	7.2–8.0	Pine, red	30	0.48
Manganese ore, pyrolusite . .	259	3.7–4.6	Pine, white	26	0.41
Mercury	849	13.6	Pine, yellow, long-leaf	44	0.70
Monel Metal	556	8.8–9.0	Pine, yellow, short-leaf	38	0.61
Nickel	565	8.9–9.2	Poplar	30	0.48
Platinum, cast, hammered . . .	1330	21.1–21.5	Redwood, California	26	0.42
Silver, cast, hammered	656	10.4–10.6	Spruce, white, black	27	0.40–0.46
Steel, rolled	490	7.85	Walnut, black	38	0.61
Tin, cast, hammered	459	7.2–7.5	Walnut, white	26	0.41
Tin ore, cassiterite	418	6.4–7.0			
Zinc, cast, rolled	440	6.9–7.2			
Zinc ore, blende	253	3.9–4.2			
			VARIOUS LIQUIDS		
			Alcohol, 100%	49	0.79
			Acids, muriatic 40%	75	1.20
			Acids, nitric 91%	94	1.50
VARIOUS SOLIDS			Acids, sulphuric 87%	112	1.80
Cereals, oats . . . . . bulk	32	—	Lye, soda 66%	106	1.70
Cereals, barley . . . . bulk	39	—	Oils, vegetable	58	0.91–0.94
Cereals, corn, rye . . . . bulk	48	—	Oils, mineral, lubricants	57	0.90–0.93
Cereals, wheat . . . . bulk	48	—	Water, 4°C max. density . . .	62.428	1.0
Hay and Straw . . . . . bales	20	—	Water, 100°C	59.830	0.9584
Cotton, Flax, Hemp	93	1.47–1.50	Water, ice	56	0.88–0.92
Fats	58	0.90–0.97	Water, snow, fresh fallen . . .	8	.125
Flour, loose	28	0.40–0.50	Water, sea water	64	1.02–1.03
Flour, pressed	47	0.70–0.80			
Glass, common	156	2.40–2.60			
Glass, plate or crown	161	2.45–2.72			
Glass, crystal	184	2.90–3.00			
Leather	59	0.86–1.02	**GASES**		
Paper	58	0.70–1.15	Air, 0°C 760 mm	.08071	1.0
Potatoes, piled	42	—	Ammonia	.0478	0.5920
Rubber, caoutchouc	59	0.92–0.96	Carbon dioxide	.1234	1.5291
Rubber goods	94	1.0–2.0	Carbon monoxide	.0781	0.9673
Salt, granulated, piled	48	—	Gas, illuminating	.028–.036	0.35–0.45
Saltpeter	67	—	Gas, natural	.038–.039	0.47–0.48
Starch	96	1.53	Hydrogen	.00559	0.0693
Sulphur	125	1.93–2.07	Nitrogen	.0784	0.9714
Wool	82	1.32	Oxygen	.0892	1.1056

The specific gravities of solids and liquids refer to water at 4°C, those of gases to air at 0°C and 760 mm pressure. The weights per cubic foot are derived from average specific gravities, except where stated that weights are for bulk, heaped, or loose material, etc.

Table 7-5.
WEIGHTS OF BUILDING MATERIALS

Materials	Weight lb per sq ft	Materials	Weight lb per sq ft
CEILINGS		**PARTITIONS**	
Channel suspended system	1	Clay Tile	
Lathing and plastering	See Partitions	3 in.	17
Acoustical fiber tile	1	4 in.	18
		6 in.	28
		8 in.	34
		10 in.	40
FLOORS		Gypsum Block	
Steel Deck	See Manufacturer	2 in.	$9\frac{1}{2}$
		3 in.	$10\frac{1}{2}$
Concrete-Reinforced 1 in.		4 in.	$12\frac{1}{2}$
Stone	$12\frac{1}{2}$	5 in.	14
Slag	$11\frac{1}{2}$	6 in.	$18\frac{1}{2}$
Lightweight	6 to 10	Wood Studs 2×4	
		12–16 in. o.c.	2
Concrete-Plain 1 in.		Steel partitions	4
Stone	12	Plaster 1 inch	
Slag	11	Cement	10
Lightweight	3 to 9	Gypsum	5
		Lathing	
Fills 1 inch		Metal	$\frac{1}{2}$
Gypsum	6	Gypsum Board $\frac{1}{2}$-in.	2
Sand	8		
Cinders	4		
Finishes			
Terrazzo 1 in.	13		
Ceramic or Quarry Tile $\frac{3}{4}$-in.	10		
Linoleum $\frac{1}{4}$-in.	1	**WALLS**	
Mastic $\frac{3}{4}$-in.	9	Brick	
Hardwood $\frac{7}{8}$-in.	4	4 in.	40
Softwood $\frac{3}{4}$-in.	$2\frac{1}{2}$	8 in.	80
		12 in.	120
		Hollow Concrete Block	
		(Heavy Aggregate)	
ROOFS		4 in.	30
Copper or tin	1	6 in.	43
Corrugated steel	See Manufactuer	8 in.	55
3-ply ready roofing	1	$12\frac{1}{2}$-in.	80
3-ply felt and gravel	$5\frac{1}{2}$	Hollow Concrete Block	
5-ply felt and gravel	6	(Light Aggregate)	
		4 in.	21
Shingles		6 in.	30
Wood	2	8 in.	38
Asphalt	3	12 in.	55
Clay tile	9 to 14	Clay tile (Load Bearing)	
Slate $\frac{1}{4}$	10	4 in.	25
		6 in.	30
Sheathing		8 in.	33
Wood $\frac{3}{4}$-in.	3	12 in.	45
Gypsum 1 in.	4	Stone 4 in.	55
		Glass Block 4 in.	18
Insulation 1 in.		Window, Glass, Frame, & Sash	8
Loose	$\frac{1}{2}$	Curtain Walls	See Manufacturer
Poured	2	Structural Glass 1 in.	15
Rigid	$1\frac{1}{2}$	Corrugated Cement Asbestos $\frac{1}{4}$-in.	3

For weights of other materials used in building construction, see Table 7-4.

SI UNITS FOR STRUCTURAL STEEL DESIGN

Although there are seven metric base units in the SI system, only four are currently used by AISC in structural steel design. These base units are listed in Table 7-6.

Table 7-6. Base SI Units for Steel Design

Quantity	Unit	Symbol
Length	meter	m
mass	kilogram	kg
time	second	s
temperature	celcius	°C

Similarly, of the numerous decimal prefixes included in the SI system, only three are used in steel design; see Table 7-7.

Table 7-7. SI Prefixes for Steel Design

Prefix	Symbol	Order of Magnitude	Expression
mega	M	10^6	1,000,000 (one million)
kilo	k	10^3	1,000 (one thousand)
milli	m	10^{-3}	0.001 (one thousandth)

In addition, three derived units are applicable to the present conversion. They are shown in Table 7-8.

Table 7-8. Derived SI Units for Steel Design

Quantity	Name	Symbol	Expression
force	newton	N	$N = kg \times m/s^2$
stress	pascal	Pa	$Pa = N/m^2$
energy	joule	J	$J = N \times m$

Although specified in SI, the pascal is not universally accepted as the unit of stress. Because section properties are expressed in millimeters, it is more convenient to express stress in newtons per square millimeter ($1 \text{ N/mm}^2 = 1 \text{ MPa}$). This is the practice followed in recent international structural design standards. It should be noted that the joule, as the unit of energy, is used to express energy absorption requirements for impact tests. Moments are expressed in terms of N×m.

A summary of the conversion factors relating traditional U.S. units of measurement to the corresponding SI units is given in Table 7-9.

Table 7-9. Summary of SI Conversion Factors

Multiply	by:	to obtain:
inch (in.)	25.4	millimeters (mm)
foot (ft)	305	millimeters (mm)
pound-mass (lb)	0.454	kilogram (kg)
pound-force (lbf)	4.448	newton (N)
ksi	6.895	N/mm^2
ft-lbf	1.356	joule (J)
psf	47.88	N/m^2
plf	14.59	N/m

Note that fractions resulting from metric conversion should be rounded to whole millimeters. Common fractions of inches and their metric equivalent are in Table 7-10.

Table 7-10. SI Equivalents of Fractions of an Inch

Fraction, in.	Exact conversion, mm	Rounded to: (mm)
$\frac{1}{16}$	1.5875	2
$\frac{1}{8}$	3.175	3
$\frac{3}{16}$	4.7625	5
$\frac{1}{4}$	6.35	6
$\frac{5}{16}$	7.9375	8
$\frac{3}{8}$	9.525	10
$\frac{7}{16}$	11.1125	11
$\frac{1}{2}$	12.7	13
$\frac{5}{8}$	15.875	16
$\frac{3}{4}$	19.05	19
$\frac{7}{8}$	22.225	22
1	25.4	25

Bolt diameters are taken directly from the ASTM Specifications A325M and A490M rather than converting the diameters of bolts dimensioned in inches. The metric bolt designations are in Table 7-11.

Table 7-11. SI Bolt Designation

Designation	Diameter, mm	Diameter, in.
M16	16	0.63
M20	20	0.79
M22	22	0.87
M24	24	0.94
M27	27	1.06
M30	30	1.18
M36	36	1.42

The yield strengths of structural steels are taken from the metric ASTM Specifications. It should be noted that the yield points are slightly different from the traditional values. See Table 7-12. The modulus of elasticity of steel E is taken as 200,000 N/mm². The shear modulus of elasticity of steel G is 77,000 N/mm².

Table 7-12. SI Steel Yield Stresses

ASTM Designation	Yield stress, N/mm²	Yield stress, ksi
A36M	250	36.26
A572M Gr. 345 A588M	345	50.04
A852M	485	70.34
A514M	690	100.07

Table 7-13.
WEIGHTS AND MEASURES
International System of Units (SI)[a]
(Metric practice)

BASE UNITS

Quantity	Unit	Symbol
length	meter	m
mass	kilogram	kg
time	second	s
electric current	ampere	A
thermodynamic temperature	kelvin	K
amount of substance	mole	mol
luminous intensity	candela	cd

SUPPLEMENTARY UNITS

Symbol	Unit	Symbol
plane angle	radian	rad
solid angle	steradian	sr

DERIVED UNITS (WITH SPECIAL NAMES)

Quantity	Unit	Symbol	Formula
force	newton	N	$kg\text{-}m/s^2$
pressure, stress	pascal	Pa	N/m^2
energy, work, quantity of heat	joule	J	N-m
power	watt	W	J/s

DERIVED UNITS (WITHOUT SPECIAL NAMES)

Quantity	Unit	Formula
area	square meter	m^2
volume	cubic meter	m^3
velocity	meter per second	m/s
acceleration	meter per second squared	m/s^2
specific volume	cubic meter per kilogram	m^3/kg
density	kilogram per cubic meter	kg/m^3

SI PREFIXES

Multiplication Factor	Prefix	Symbol
$1\ 000\ 000\ 000\ 000\ 000\ 000 = 10^{18}$	exa	E
$1\ 000\ 000\ 000\ 000\ 000 = 10^{15}$	peta	P
$1\ 000\ 000\ 000\ 000 = 10^{12}$	tera	T
$1\ 000\ 000\ 000 = 10^{9}$	giga	G
$1\ 000\ 000 = 10^{6}$	mega	M
$1\ 000 = 10^{3}$	kilo	k
$100 = 10^{2}$	hecto[b]	h
$10 = 10^{1}$	deka[b]	da
$0.1 = 10^{-1}$	deci[b]	d
$0.01 = 10^{-2}$	centi[b]	c
$0.001 = 10^{-3}$	milli	m
$0.000\ 001 = 10^{-6}$	micro	μ
$0.000\ 000\ 001 = 10^{-9}$	nano	n
$0.000\ 000\ 000\ 001 = 10^{-12}$	pico	p
$0.000\ 000\ 000\ 000\ 001 = 10^{-15}$	femto	f
$0.000\ 000\ 000\ 000\ 000\ 001 = 10^{-18}$	atto	a

[a]Refer to ASTM E380 for more complete information on SI.
[b]Use is not recommended.

Table 7-14.
WEIGHTS AND MEASURES
United States System

LINEAR MEASURE

Inches	Feet	Yards	Rods	Furlongs	Miles
1.0 =	.08333 =	.02778 =	.0050505 =	.00012626 =	.00001578
12.0 =	1.0 =	.33333 =	.0606061 =	.00151515 =	.00018939
36.0 =	3.0 =	1.0 =	.1818182 =	.00454545 =	.00056818
198.0 =	16.5 =	5.5 =	1.0 =	.025 =	.003125
7,920.0 =	660.0 =	220.0 =	40.0 =	1.0 =	.125
63,360.0 =	5,280.0 =	1,760.0 =	320.0 =	8.0 =	1.0

SQUARE AND LAND MEASURE

Sq. Inches	Square Feet	Square Yards	Square Rods	Acres	Sq. Miles
1.0 =	.006944 =	.000772			
144.0 =	1.0 =	.111111			
1,296.0 =	9.0 =	1.0 =	.03306 =	.000207	
39,204.0 =	272.25 =	30.25 =	1.0 =	.00625 =	.0000098
	43,560.0 =	4,840.0 =	160.0 =	1.0 =	.0015625
		3,097,600.0 =	102,400.0 =	640.0 =	1.0

AVOIRDUPOIS WEIGHTS

Grains	Drams	Ounces	Pounds	Tons
1.0 =	.03657 =	.002286 =	.000143 =	.0000000714
27.34375 =	1.0 =	.0625 =	.003906 =	.00000195
437.5 =	16.0 =	1.0 =	.0625 =	.00003125
7,000.0 =	256.0 =	16.0 =	1.0 =	.0005
14,000,000.0 =	512,000.0 =	32,000.0 =	2,000.0 =	1.0

DRY MEASURE

Pints	Quarts	Pecks	Cubic Feet	Bushels
1.0 =	.5 =	.0625 =	.01945 =	.01563
2.0 =	1.0 =	.125 =	.03891 =	.03125
16.0 =	8.0 =	1.0 =	.31112 =	.25
51.42627 =	25.71314 =	3.21414 =	1.0 =	.80354
64.0 =	32.0 =	4.0 =	1.2445 =	1.0

LIQUID MEASURE

Gills	Pints	Quarts	U.S. Gallons	Cubic Feet
1.0 =	.25 =	.125 =	.03125 =	.00418
4.0 =	1.0 =	.5 =	.125 =	.01671
8.0 =	2.0 =	1.0 =	.250 =	.03342
32.0 =	8.0 =	4.0 =	1.0 =	.1337
			7.48052 =	1.0

Table 7-15.
SI CONVERSION FACTORS[a]

Quantity	Multiply	by	to obtain	
Length	inch	25.400	millimeter	mm
	foot	0.305	meter	m
	yard	0.914	meter	m
	mile (U.S. Statute)	1.609	kilometer	km
	millimeter	39.370×10^{-3}	inch	in
	meter	3.281	foot	ft
	meter	1.094	yard	yd
	kilometer	0.621	mile	mi
Area	square inch	0.645×10^3	square millimeter	mm^2
	square foot	0.093	square meter	m^2
	square yard	0.836	square meter	m^2
	square mile (U.S. Statute)	2.590	square kilometer	km^2
	acre	4.047×10^3	square meter	m^2
	acre	0.405	hectare	
	square millimeter	1.550×10^{-3}	square inch	in^2
	square meter	10.764	square foot	ft^2
	square meter	1.196	square yard	yd^2
	square kilometer	0.386	square mile	mi^2
	square meter	0.247×10^{-3}	acre	
	hectare	2.471	acre	
Volume	cubic inch	16.387×10^3	cubic millimeter	mm^3
	cubic foot	28.317×10^{-3}	cubic meter	m^3
	cubic yard	0.765	cubic meter	m^3
	gallon (U.S. liquid)	3.785	liter	L
	quart (U.S. liquid)	0.946	liter	L
	cubic millimeter	61.024×10^{-6}	cubic inch	in^3
	cubic meter	35.315	cubic foot	ft^3
	cubic meter	1.308	cubic yard	yd^3
	liter	0.264	gallon (U.S. liquid)	gal
	liter	1.057	quart (U.S. liquid)	qt
Mass	ounce (avoirdupois)	28.350	gram	g
	pound (avoirdupois)	0.454	kilogram	kg
	short ton	0.907×10^3	kilogram	kg
	gram	35.274×10^{-3}	ounce (avoirdupois)	oz av
	kilogram	2.205	pound (avoirdupois)	lb av
	kilogram	1.102×10^{-3}	short ton	

[a]Refer to ASTM E380 for more complete information on SI.
The conversion factors tabulated herein have been rounded.

Table 7-15 (cont.).
SI CONVERSION FACTORS[a]

Quantity	Multiply	by	to obtain	
Force	ounce-force	0.278	newton	N
	pound-force	4.448	newton	N
	newton	3.597	ounce-force	
	newton	0.225	pound-force	lbf
Bending Moment	pound-force-inch	0.113	newton-meter	N-m
	pound-force-foot	1.356	newton-meter	N-m
	newton-meter	8.851	pound-force-inch	lbf-in
	newton-meter	0.738	pound-force-foot	lbf-ft
Pressure, Stress	pound-force per square inch	6.895	kilopascal	kPa
	foot of water (39.2 F)	2.989	kilopascal	kPa
	inch of mercury (32 F)	3.386	kilopascal	kPa
	kilopascal	0.145	pound-force per square inch	lbf/in^2
	kilopascal	0.335	foot of water (39.2 F)	
	kilopascal	0.295	inch of mercury (32 F)	
Energy, Work, Heat	foot-pound-force	1.356	joule	J
	[b]British thermal unit	1.055×10^3	joule	J
	[b]calorie	4.187		J
	kilowatt hour	3.600×10^6	joule	J
	joule	0.738	foot-pound-force	ft-lbf
	joule	0.948×10^{-3}	[b]British thermal unit	Btu
	joule	0.239	[b]calorie	
	joule	0.278×10^{-6}	kilowatt hour	kW-h
Power	foot-pound-force/second	1.356	watt	W
	[b]British thermal unit per hour	0.293	watt	W
	horsepower (550 ft lbf/s)	0.746	kilowatt	kW
	watt	0.738	foot-pound-force/ second	ft-lbf/s
	watt	3.412	[b]British thermal unit per hour	Btu/h
	kilowatt	1.341	horsepower (550 ft-lbf/s)	hp
Angle	degree	17.453×10^{-3}	radian	rad
	radian	57.296	degree	
Temperature	degree Fahrenheit	$t°C = (t°F - 32)/1.8$	degree Celsius	
	degree Celsius	$t°F = 1.8 \times t°C + 32$	degree Fahrenheit	

[a]Refer to ASTM E380 for more complete information on SI.
[b]International Table.
The conversion factors tabulated herein have been rounded.

BRACING FORMULAS

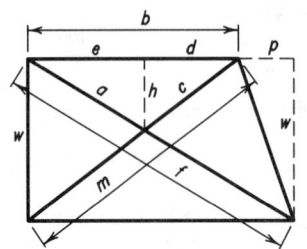 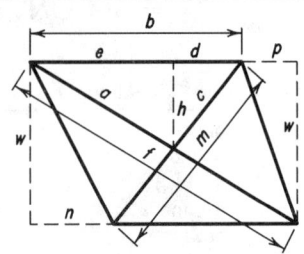

Given	To Find	Formula
bpw	f	$\sqrt{(b+p)^2 + w^2}$
bw	m	$\sqrt{b^2 + w^2}$
bp	d	$b^2 \div (2b+p)$
bp	e	$b(b+p) \div (2b+p)$
bfp	a	$bf \div (2b+p)$
bmp	c	$bm \div (2b+p)$
bpw	h	$bw \div (2b+p)$
afw	h	$aw \div f$
cmw	h	$cw \div m$

Given	To Find	Formula
bpw	f	$\sqrt{(b+p)^2 + w^2}$
bnw	m	$\sqrt{(b-n)^2 + w^2}$
bnp	d	$b(b-n) \div (2b+p-n)$
bnp	e	$b(b+p) \div (2b+p-n)$
bfnp	a	$bf \div (2b+p-n)$
bmnp	c	$bm \div (2b+p-n)$
bnpw	h	$bw \div (2b+p-n)$
afw	h	$aw \div f$
cmw	h	$cw \div m$

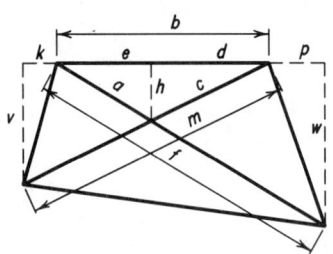

Given	To Find	Formula
bpw	f	$\sqrt{(b+p)^2 + w^2}$
bkv	m	$\sqrt{(b+k)^2 + v^2}$
bkpvw	d	$bw(b+k) \div [v(b+p) + w(b+k)]$
bkpvw	e	$bv(b+p) \div [v(b+p) + w(b+k)]$
bfkpvw	a	$fbv \div [v(b+p) + w(b+k)]$
bkmpvw	c	$bmw \div [v(b+p) + w(b+k)]$
bkpvw	h	$bvw \div [v(b+p) + w(b+k)]$
afw	h	$aw \div f$
cmv	h	$cv \div m$

PARALLEL BRACING

$k = (\log B - \log T) \div$ no. of panels. Constant k plus the logarithm of any line equals the log of the corresponding line in the next panel below.

$$a = TH \div (T + e + p)$$

$$b = Th \div (T + e + p)$$

$$c = \sqrt{(\tfrac{1}{2}T + \tfrac{1}{2}e)^2 + a^2}$$

$$d = ce \div (T + e)$$

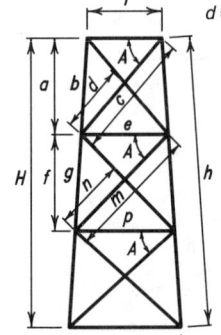

$\log e = k + \log T$
$\log f = k + \log a$
$\log g = k + \log b$
$\log m = k + \log c$
$\log n = k + \log d$
$\log p = k + \log e$

The above method can be used for any number of panels. In the formulas for "a" and "b" the sum in parenthesis, which in the case shown is (T + e + p), is always composed of all the horizontal distances except the base.

PROPERTIES OF PARABOLA AND ELLIPSE

PARABOLA

if $\dfrac{H}{B} \le 0.1$, Approx. ½ perimeter $= \sqrt{B^2 + \dfrac{4}{3}H^2}$

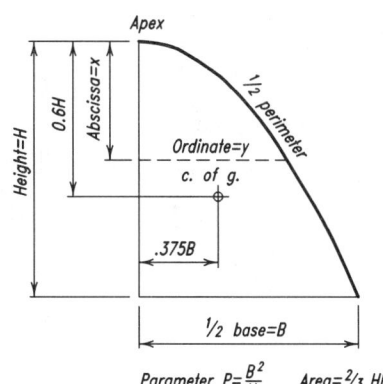

Parameter $P = \dfrac{B^2}{H}$ Area $= \dfrac{2}{3} HB$

$x = \dfrac{y^2}{P}$

$y = \sqrt{xP}$

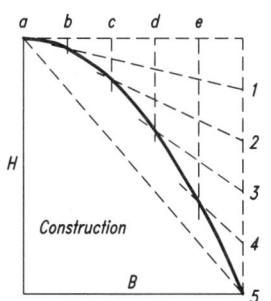

Construction

ELLIPSE

Approx. ¼ perimeter $= \dfrac{\pi}{4}\sqrt{2(H^2 + B^2)}$

$(x^2 \div H^2) + (y^2 \div B^2) = 1$

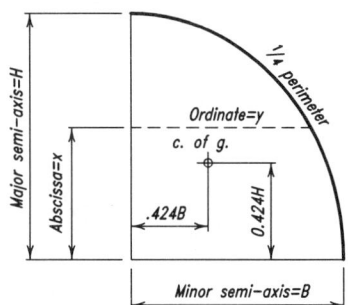

Area $= .7854Dd$

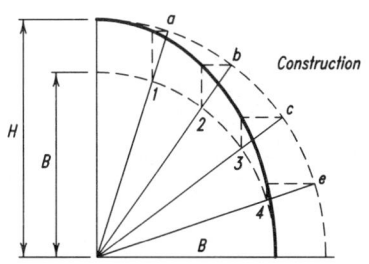

Construction

AREA BETWEEN PARABOLIC CURVE AND SECANT

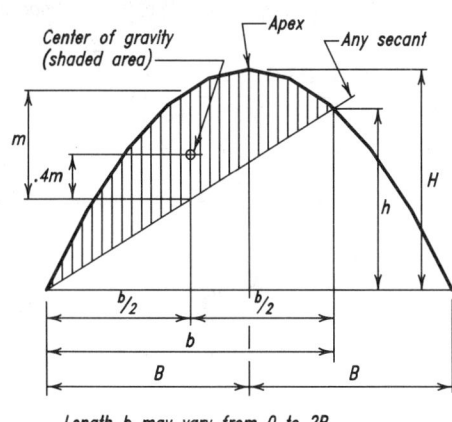

Center of gravity (shaded area)

Apex

Any secant

Length b may vary from 0 to 2B

$h = Hb\left(\dfrac{2B - b}{B^2}\right)$

$m = \dfrac{Hb^2}{4B^2}$

Shaded area

$= \dfrac{2}{3}bm$

$= \dfrac{Hb^3}{6B^2}$

PROPERTIES OF THE CIRCLE

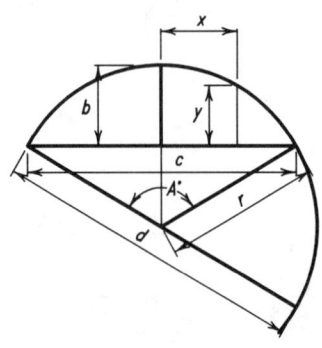

Circumference = 6.28318 r = 3.14159 d

Diameter = 0.31831 circumference

Area = 3.14159r^2

Arc $a = \dfrac{\pi r A^\circ}{180^\circ} = 0.017453\ rA^\circ$

Angle $A^\circ = \dfrac{180^\circ a}{\pi r} = 57.29578\ \dfrac{a}{r}$

Radius $r = \dfrac{4b^2 + c^2}{8b}$

Chord $c = 2\sqrt{2br - b^2} = 2r \sin\dfrac{A}{2}$

Rise $b = r - \dfrac{1}{2}\sqrt{4r^2 - c^2} = \dfrac{c}{2}\tan\dfrac{A}{4}$

$= 2r \sin^2 \dfrac{A}{4} = r + y - \sqrt{r^2 - x^2}$

$y = b - r + \sqrt{r^2 - x^2}$

$x = \sqrt{r^2 - (r + y - b)^2}$

Diameter of circle of equal periphery as square = 1.27324 side of square
Side of square of equal periphery as circle   = 0.78540 diameter of circle
Diameter of circle circumscribed about square = 1.41421 side of square
Side of square inscribed in circle   = 0.70711 diameter of circle

CIRCULAR SECTOR

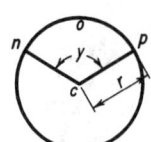

r = radius of circle y = angle ncp in degrees
Area of Sector $ncpo$ = ½ (length of arc nop x r)

$= $ Area of Circle $\times \dfrac{y}{360}$

$= 0.0087266 \times r^2 \times y$

CIRCULAR SEGMENT

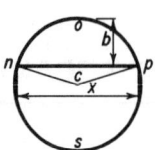

r = radius of circle x = chord b = rise
Area of Segment nop = Area of Sector $ncpo$ – Area of triangle ncp

$= \dfrac{(\text{Length of arc } nop \times r) - x(r - b)}{2}$

Area of Segment nsp = Area of Circle – Area of Segment nop

VALUES FOR FUNCTIONS OF π

$\pi = 3.14159265359$, log = 0.4971499

$\pi^2 = 9.8696044$, log = 0.9942997	$\dfrac{1}{\pi} = 0.3183099$, log = $\overline{1}.5028501$	$\sqrt{\dfrac{1}{\pi}} = 0.5641896$, log = $\overline{1}.7514251$
$\pi^3 = 31.0062767$, log = 1.4914496	$\dfrac{1}{\pi^2} = 0.1013212$, log = $\overline{1}.0057003$	$\dfrac{\pi}{180} = 0.0174533$, log = $\overline{2}.2418774$
$\sqrt{\pi} = 1.7724539$, log = 0.2485749	$\dfrac{1}{\pi^3} = 0.0322515$, log = $\overline{2}.5085504$	$\dfrac{180}{\pi} = 57.2957795$, log = 1.7581226

Note: Logs of fractions such as $\overline{1}.5028501$ and $\overline{2}.5085500$ may also be written 9.5028501 – 10 and 8.5085500 – 10 respectively

PROPERTIES OF GEOMETRIC SECTIONS

SQUARE
Axis of moments through center

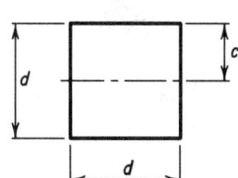

$$A = d^2$$
$$c = \frac{d}{2}$$
$$I = \frac{d^4}{12}$$
$$S = \frac{d^3}{6}$$
$$r = \frac{d}{\sqrt{12}} = .288675\, d$$
$$Z = \frac{d^3}{4}$$

SQUARE
Axis of moments on base

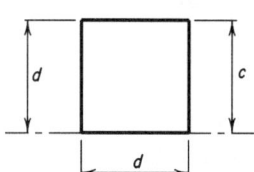

$$A = d^2$$
$$c = d$$
$$I = \frac{d^4}{3}$$
$$S = \frac{d^3}{3}$$
$$r = \frac{d}{\sqrt{3}} = .577350\, d$$

SQUARE
Axis of moments on diagonal

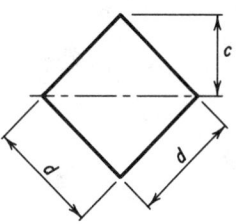

$$A = d^2$$
$$c = \frac{d}{\sqrt{2}} = .707107\, d$$
$$I = \frac{d^4}{12}$$
$$S = \frac{d^3}{6\sqrt{2}} = .117851\, d^3$$
$$r = \frac{d}{\sqrt{12}} = .288675\, d$$
$$Z = \frac{2c^3}{3} = \frac{d^3}{3\sqrt{2}} = .235702\, d^3$$

RECTANGLE
Axis of moments through center

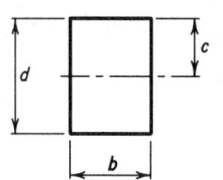

$$A = bd$$
$$c = \frac{d}{2}$$
$$I = \frac{bd^3}{12}$$
$$S = \frac{bd^2}{6}$$
$$r = \frac{d}{\sqrt{12}} = .288675\, d$$
$$Z = \frac{bd^2}{4}$$

PROPERTIES OF GEOMETRIC SECTIONS (cont.)

RECTANGLE
Axis of moments on base

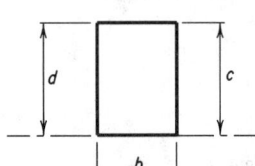

$A = bd$

$c = d$

$I = \dfrac{bd^3}{3}$

$S = \dfrac{bd^2}{3}$

$r = \dfrac{d}{\sqrt{3}} = .577350\,d$

RECTANGLE
Axis of moments on diagonal

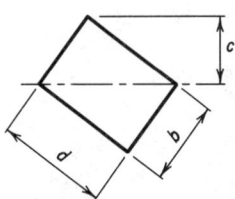

$A = bd$

$c = \dfrac{bd}{\sqrt{b^2 + d^2}}$

$I = \dfrac{b^3 d^3}{6(b^2 + d^2)}$

$S = \dfrac{b^2 d^2}{6\sqrt{b^2 + d^2}}$

$r = \dfrac{bd}{\sqrt{6(b^2 + d^2)}}$

RECTANGLE
Axis of moments any line
through center of gravity

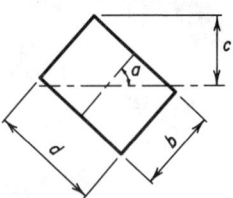

$A = bd$

$c = \dfrac{b \sin a + d \cos a}{2}$

$I = \dfrac{bd\,(b^2 \sin^2 a + d^2 \cos^2 a)}{12}$

$S = \dfrac{bd\,(b^2 \sin^2 a + d^2 \cos^2 a)}{6\,(b \sin a + d \cos a)}$

$r = \sqrt{\dfrac{b^2 \sin^2 a + d^2 \cos^2 a}{12}}$

HOLLOW RECTANGLE
Axis of moments through center

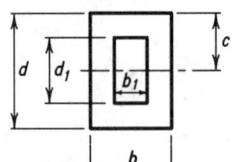

$A = bd - b_1 d_1$

$c = \dfrac{d}{2}$

$I = \dfrac{bd^3 - b_1 d_1^3}{12}$

$S = \dfrac{bd^3 - b_1 d_1^3}{6d}$

$r = \sqrt{\dfrac{bd^3 - b_1 d_1^3}{12A}}$

$Z = \dfrac{bd^2}{4} - \dfrac{b_1 d_1^2}{4}$

PROPERTIES OF GEOMETRIC SECTIONS (cont.)

EQUAL RECTANGLES
Axis of moments through center of gravity

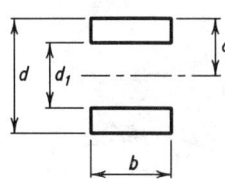

$$A = b(d - d_1)$$

$$c = \frac{d}{2}$$

$$I = \frac{b(d^3 - d_1^3)}{12}$$

$$S = \frac{b(d^3 - d_1^3)}{6d}$$

$$r = \sqrt{\frac{d^3 - d_1^3}{12(d - d_1)}}$$

$$Z = \frac{b}{4}(d^2 - d_1^2)$$

UNEQUAL RECTANGLES
Axis of moments through center of gravity

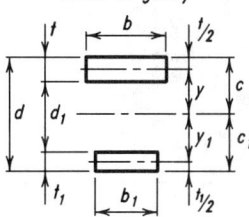

$$A = bt + b_1 t_1$$

$$c = \frac{\frac{1}{2} bt^2 + b_1 t_1 (d - \frac{1}{2} t_1)}{A}$$

$$I = \frac{bt^3}{12} + bty^2 + \frac{b_1 t_1^3}{12} + b_1 t_1 y_1^2$$

$$S = \frac{I}{c} \qquad S_1 = \frac{I}{c_1}$$

$$r = \sqrt{\frac{I}{A}}$$

$$Z = \frac{A}{2} \left[d - \left(\frac{t + t_1}{2} \right) \right]$$

TRIANGLE
Axis of moments through center of gravity

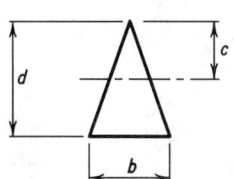

$$A = \frac{bd}{2}$$

$$c = \frac{2d}{3}$$

$$I = \frac{bd^3}{36}$$

$$S = \frac{bd^2}{24}$$

$$r = \frac{d}{\sqrt{18}} = .235702\, d$$

TRIANGLE
Axis of moments on base

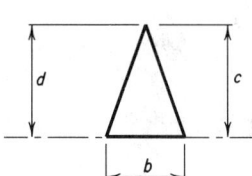

$$A = \frac{bd}{2}$$

$$c = d$$

$$I = \frac{bd^3}{12}$$

$$S = \frac{bd^2}{12}$$

$$r = \frac{d}{\sqrt{6}} = .408248\, d$$

PROPERTIES OF GEOMETRIC SECTIONS (cont.)

TRAPEZOID
Axis of moments through center of gravity

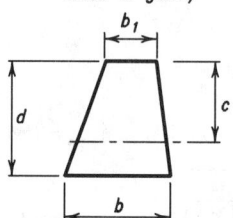

$$A = \frac{d(b + b_1)}{2}$$

$$c = \frac{d(2b + b_1)}{3(b + b_1)}$$

$$I = \frac{d^3(b^2 + 4bb_1 + b_1^2)}{36(b + b_1)}$$

$$S = \frac{d^2(b^2 + 4bb_1 + b_1^2)}{12(2b + b_1)}$$

$$r = \frac{d}{6(b + b_1)}\sqrt{2(b^2 + 4bb_1 + b_1^2)}$$

CIRCLE
Axis of moments through center

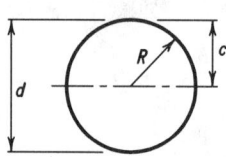

$$A = \frac{\pi d^2}{4} = \pi R^2 = .785398\, d^2 = 3.141593\, R^2$$

$$c = \frac{d}{2} = R$$

$$I = \frac{\pi d^4}{64} = \frac{\pi R^4}{4} = .049087\, d^4 = .785398\, R^4$$

$$S = \frac{\pi d^3}{32} = \frac{\pi R^3}{4} = .098175\, d^3 = .785398\, R^3$$

$$r = \frac{d}{4} = \frac{R}{2}$$

$$Z = \frac{d^3}{6}$$

HOLLOW CIRCLE
Axis of moments through center

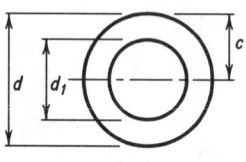

$$A = \frac{\pi(d^2 - d_1^2)}{4} = .785398\,(d^2 - d_1^2)$$

$$c = \frac{d}{2}$$

$$I = \frac{\pi(d^4 - d_1^4)}{64} = .049087\,(d^4 - d_1^4)$$

$$S = \frac{\pi(d^4 - d_1^4)}{32\,d} = .098175\,\frac{d^4 - d_1^4}{d}$$

$$r = \frac{\sqrt{d^2 + d_1^2}}{4}$$

$$Z = \frac{d^3}{6} - \frac{d_1^3}{6}$$

HALF CIRCLE
Axis of moments through center of gravity

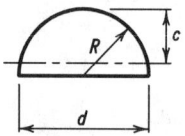

$$A = \frac{\pi R^2}{2} = 1.570796\, R^2$$

$$C = R\left(1 - \frac{4}{3\pi}\right) = .575587\, R$$

$$I = R^4\left(\frac{\pi}{8} - \frac{8}{9\pi}\right) = .109757\, R^4$$

$$S = \frac{R^3}{24}\frac{(9\pi^2 - 64)}{(3\pi - 4)} = .190687\, R^3$$

$$r = R\frac{\sqrt{9\pi^2 - 64}}{6\pi} = .264336\, R$$

PROPERTIES OF GEOMETRIC SECTIONS (cont.)

PARABOLA

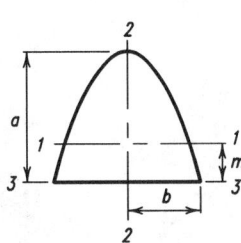

$$A = \frac{4}{3} ab$$

$$m = \frac{2}{5} a$$

$$I_1 = \frac{16}{175} a^3 b$$

$$I_2 = \frac{4}{15} ab^3$$

$$I_3 = \frac{32}{105} a^3 b$$

HALF PARABOLA

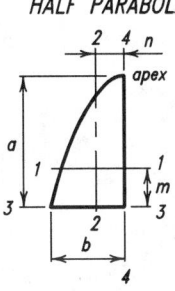

$$A = \frac{2}{3} ab$$

$$m = \frac{2}{5} a$$

$$n = \frac{3}{8} b$$

$$I_1 = \frac{8}{175} a^3 b$$

$$I_2 = \frac{19}{480} ab^3$$

$$I_3 = \frac{16}{105} a^3 b$$

$$I_4 = \frac{2}{15} ab^3$$

COMPLEMENT OF HALF PARABOLA

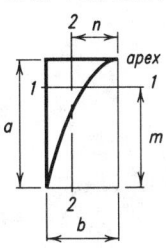

$$A = \frac{1}{3} ab$$

$$m = \frac{7}{10} a$$

$$n = \frac{3}{4} b$$

$$I_1 = \frac{37}{2,100} a^3 b$$

$$I_2 = \frac{1}{80} ab^3$$

PARABOLIC FILLET IN RIGHT ANGLE

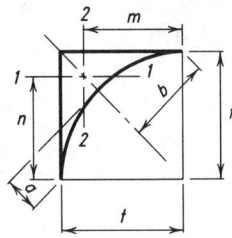

$$a = \frac{t}{2\sqrt{2}}$$

$$b = \frac{t}{\sqrt{2}}$$

$$A = \frac{1}{6} t^2$$

$$m = n = \frac{4}{5} t$$

$$I_1 = I_2 = \frac{11}{2,100} t^4$$

PROPERTIES OF GEOMETRIC SECTIONS (cont.)

* HALF ELLIPSE

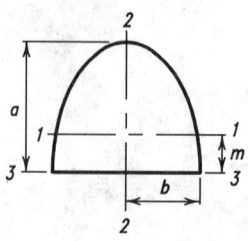

$$A = \frac{1}{2}\pi ab$$

$$m = \frac{4a}{3\pi}$$

$$I_1 = a^3b\left(\frac{\pi}{8} - \frac{8}{9\pi}\right)$$

$$I_2 = \frac{1}{8}\pi ab^3$$

$$I_3 = \frac{1}{8}\pi a^3b$$

* QUARTER ELLIPSE

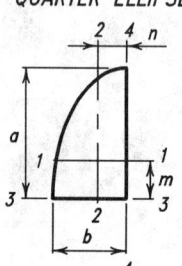

$$A = \frac{1}{4}\pi ab$$

$$m = \frac{4a}{3\pi}$$

$$n = \frac{4b}{3\pi}$$

$$I_1 = a^3b\left(\frac{\pi}{16} - \frac{4}{9\pi}\right)$$

$$I_2 = ab^3\left(\frac{\pi}{16} - \frac{4}{9\pi}\right)$$

$$I_3 = \frac{1}{16}\pi a^3b$$

$$I_4 = \frac{1}{16}\pi ab^3$$

* ELLIPTIC COMPLEMENT

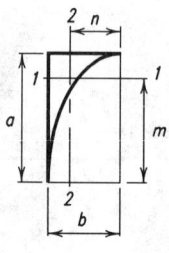

$$A = ab\left(1 - \frac{\pi}{4}\right)$$

$$m = \frac{a}{6\left(1 - \frac{\pi}{4}\right)}$$

$$n = \frac{b}{6\left(1 - \frac{\pi}{4}\right)}$$

$$I_1 = a^3b\left(\frac{1}{3} - \frac{\pi}{16} - \frac{1}{36\left(1 - \frac{\pi}{4}\right)}\right)$$

$$I_2 = ab^3\left(\frac{1}{3} - \frac{\pi}{16} - \frac{1}{36\left(1 - \frac{\pi}{4}\right)}\right)$$

*To obtain properties of half circle, quarter circle, and circular complement, substitute $a = b = R$.

PROPERTIES OF GEOMETRIC SECTIONS (cont.)

REGULAR POLYGON

Axis of moments through center

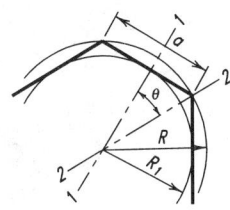

n = Number of sides

$$\theta = \frac{180°}{n}$$

$$a = 2\sqrt{R^2 - R_1^2}$$

$$R = \frac{a}{2\sin\theta}$$

$$R_1 = \frac{a}{2\tan\theta}$$

$$A = \frac{1}{4} na^2 \cot\theta = \frac{1}{2} nR^2 \sin 2\theta = nR_1^2 \tan\theta$$

$$I_1 = I_2 = \frac{A(6R^2 - a^2)}{24} = \frac{A(12R_1^2 + a^2)}{48}$$

$$r_1 = r_2 = \sqrt{\frac{6R^2 - a^2}{24}} = \sqrt{\frac{12R_1^2 + a^2}{48}}$$

ANGLE

Axis of moments through center of gravity

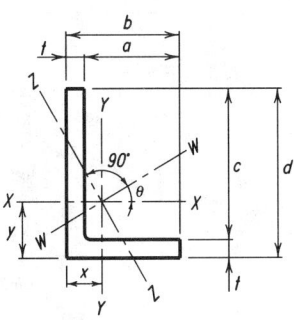

$$\tan 2\theta = \frac{2K}{I_y - I_x}$$

$$A = t(b + c) \quad x = \frac{b^2 + ct}{2(b + c)} \quad y = \frac{d^2 + at}{2(b + c)}$$

K = Product of Inertia about X-X and Y-Y

$$= \mp \frac{abcdt}{4(b + c)}$$

$$I_x = \frac{1}{3}(t(d - y)^3 + by^3 - a(y - t)^3)$$

$$I_Y = \frac{1}{3}(t(b - x)^3 + dx^3 - c(x - t)^3)$$

$$I_z = I_x \sin^2\theta + I_Y \cos^2\theta + K \sin 2\theta$$

$$I_w = I_x \cos^2\theta + I_Y \sin^2\theta - K \sin 2\theta$$

K is negative when heel of angle, with respect to center of gravity, is in 1st or 3rd quadrant, positive when in 2nd or 4th quadrant.

BEAMS AND CHANNELS

Transverse force oblique through center of gravity

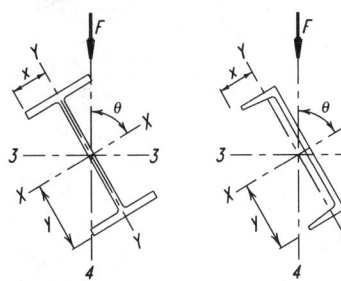

$$I_3 = I_x \sin^2\theta + I_Y \cos^2\theta$$

$$I_4 = I_x \cos^2\theta + I_Y \sin^2\theta$$

$$f_b = M\left(\frac{y}{I_x}\sin\theta + \frac{x}{I_Y}\cos\theta\right)$$

where M is bending moment due to force F.

TRIGONOMETRIC FORMULAS

TRIGOMETRIC FUNCTIONS

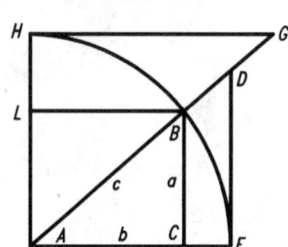

Radius $AF = 1$

$$= \sin^2 A + \cos^2 A = \sin A \, \text{cosec} \, A$$
$$= \cos A \sec A = \tan A \cot A$$

$$\sin A = \frac{\cos A}{\cot A} = \frac{1}{\text{cosec} \, A} = \cos A \tan A = \sqrt{1 - \cos^2 A} = BC$$

$$\cos A = \frac{\sin A}{\tan A} = \frac{1}{\sec A} = \sin A \cot A = \sqrt{1 - \sin^2 A} = AC$$

$$\tan A = \frac{\sin A}{\cos A} = \frac{1}{\cot A} = \sin A \sec A = FD$$

$$\cot A = \frac{\cos A}{\sin A} = \frac{1}{\tan A} = \cos A \, \text{cosec} \, A = HG$$

$$\sec A = \frac{\tan A}{\sin A} = \frac{1}{\cos A} = AD$$

$$\text{cosec} \, A = \frac{\cot A}{\cos A} = \frac{1}{\sin A} = AG$$

RIGHT ANGLED TRIANGLES

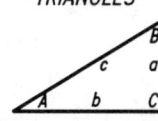

$$a^2 = c^2 - b^2$$
$$b^2 = c^2 - a^2$$
$$c^2 = a^2 + b^2$$

	Required					
Known	**A**	**B**	**a**	**b**	**c**	**Area**
a, b	$\tan A = \dfrac{a}{b}$	$\tan B = \dfrac{b}{a}$			$\sqrt{a^2 + b^2}$	$\dfrac{ab}{2}$
a, c	$\sin A = \dfrac{a}{c}$	$\cos B = \dfrac{a}{c}$		$\sqrt{c^2 - a^2}$		$\dfrac{a\sqrt{c^2 - a^2}}{2}$
A, a		$90° - A$		$a \cot A$	$\dfrac{a}{\sin A}$	$\dfrac{a^2 \cot A}{2}$
A, b		$90° - A$	$b \tan A$		$\dfrac{b}{\cos A}$	$\dfrac{b^2 \tan A}{2}$
A, c		$90° - A$	$c \sin A$	$c \cos A$		$\dfrac{c^2 \sin 2A}{4}$

OBLIQUE ANGLED TRIANGLES

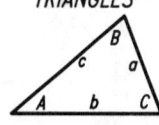

$$s = \frac{a + b + c}{2}$$

$$K = \sqrt{\frac{(s - a)(s - b)(s - c)}{s}}$$

$$a^2 = b^2 + c^2 - 2bc \cos A$$
$$b^2 = a^2 + c^2 - 2ac \cos B$$
$$c^2 = a^2 + b^2 - 2ab \cos C$$

	Required					
Known	**A**	**B**	**C**	**b**	**c**	**Area**
a, b, c	$\tan \dfrac{1}{2} A = \dfrac{K}{s - a}$	$\tan \dfrac{1}{2} B = \dfrac{K}{s - b}$	$\tan \dfrac{1}{2} C = \dfrac{K}{s - c}$			$\sqrt{s(s - a)(s - b)(s - c)}$
a, A, B			$180° - (A + B)$	$\dfrac{a \sin B}{\sin A}$	$\dfrac{a \sin C}{\sin A}$	
a, b, A		$\sin B = \dfrac{b \sin A}{a}$			$\dfrac{b \sin C}{\sin B}$	
a, b, C	$\tan A = \dfrac{a \sin C}{b - a \cos C}$				$\sqrt{a^2 + b^2 - 2ab \cos C}$	$\dfrac{ab \sin C}{2}$

DECIMALS OF AN INCH
For each 64th of an inch
With millimeter equivalents

Fractions	1/64ths	Decimal	Millimeters (Approx.)	Fractions	1/64ths	Decimal	Millimeters (Approx.)
—	1	.015625	0.397	—	33	.515625	13.097
1/32	2	.03125	0.794	17/32	34	.53125	13.494
—	3	.046875	1.191	—	35	.546875	13.891
1/16	4	.0625	1.588	9/16	36	.5625	14.288
—	5	.078125	1.984	—	37	.578125	14.684
3/32	6	.09375	2.381	19/32	38	.59375	15.081
—	7	.109375	2.778	—	39	.609375	15.478
1/8	8	.125	3.175	5/8	40	.625	15.875
—	9	.140625	3.572	—	41	.640625	16.272
5/32	10	.15625	3.969	21/32	42	.65625	16.669
—	11	.171875	4.366	—	43	.671875	17.066
3/16	12	.1875	4.763	11/16	44	.6875	17.463
—	13	.203125	5.159	—	45	.703125	17.859
7/32	14	.21875	5.556	23/32	46	.71875	18.256
—	15	.234375	5.953	—	47	.734375	18.653
1/4	16	.250	6.350	3/4	48	.750	19.050
—	17	.265625	6.747	—	49	.765625	19.447
9/32	18	.28125	7.144	25/32	50	.78125	19.844
—	19	.296875	7.541	—	51	.796875	20.241
5/16	20	.3125	7.938	13/16	52	.8125	20.638
—	21	.328125	8.334	—	53	.828125	21.034
11/32	22	.34375	8.731	27/32	54	.84375	21.431
—	23	.359375	9.128	—	55	.859375	21.828
3/8	24	.375	9.525	7/8	56	.875	22.225
—	25	.390625	9.922	—	57	.890625	22.622
13/32	26	.40625	10.319	29/32	58	.90625	23.019
—	27	.421875	10.716	—	59	.921875	23.416
7/16	28	.4375	11.113	15/16	60	.9375	23.813
—	29	.453125	11.509	—	61	.953125	24.209
15/32	30	.46875	11.906	31/32	62	.96875	24.606
—	31	.484375	12.303	—	63	.984375	25.003
1/2	32	.500	12.700	1	64	1.000	25.400

DECIMALS OF A FOOT
For each 32nd of an inch

Inch	0	1	2	3	4	5
0	0	.0833	.1667	.2500	.3333	.4167
1/32	.0026	.0859	.1693	.2526	.3359	.4193
1/16	.0052	.0885	.1719	.2552	.3385	.4219
3/32	.0078	.0911	.1745	.2578	.3411	.4245
1/8	.0104	.0938	.1771	.2604	.3438	.4271
5/32	.0130	.0964	.1797	.2630	.3464	.4297
3/16	.0156	.0990	.1823	.2656	.3490	.4323
7/32	.0182	.1016	.1849	.2682	.3516	.4349
1/4	.0208	.1042	.1875	.2708	.3542	.4375
9/32	.0234	.1068	.1901	.2734	.3568	.4401
5/16	.0260	.1094	.1927	.2760	.3594	.4427
11/32	.0286	.1120	.1953	.2786	.3620	.4453
3/8	.0313	.1146	.1979	.2812	.3646	.4479
13/32	.0339	.1172	.2005	.2839	.3672	.4505
7/16	.0365	.1198	.2031	.2865	.3698	.4531
15/32	.0391	.1224	.2057	.2891	.3724	.4557
1/2	.0417	.1250	.2083	.2917	.3750	.4583
17/32	.0443	.1276	.2109	.2943	.3776	.4609
9/16	.0469	.1302	.2135	.2969	.3802	.4635
19/32	.0495	.1328	.2161	.2995	.3828	.4661
5/8	.0521	.1354	.2188	.3021	.3854	.4688
21/32	.0547	.1380	.2214	.3047	.3880	.4714
11/16	.0573	.1406	.2240	.3073	.3906	.4740
23/32	.0599	.1432	.2266	.3099	.3932	.4766
3/4	.0625	.1458	.2292	.3125	.3958	.4792
25/32	.0651	.1484	.2318	.3151	.3984	.4818
13/16	.0677	.1510	.2344	.3177	.4010	.4844
27/32	.0703	.1536	.2370	.3203	.4036	.4870
7/8	.0729	.1563	.2396	.3229	.4063	.4896
29/32	.0755	.1589	.2422	.3255	.4089	.4922
15/16	.0781	.1615	.2448	.3281	.4115	.4948
31/32	.0807	.1641	.2472	.3307	.4141	.4974

DECIMALS OF A FOOT (cont.)
For each 32nd of an inch

Inch	6	7	8	9	10	11
0	.5000	.5833	.6667	.7500	.8333	.9167
1/32	.5026	.5859	.6693	.7526	.8359	.9193
1/16	.5052	.5885	.6719	.7552	.8385	.9219
3/32	.5078	.5911	.6745	.7578	.8411	.9245
1/8	.5104	.5938	.6771	.7604	.8438	.9271
5/32	.5130	.5964	.6797	.7630	.8464	.9297
3/16	.5156	.5990	.6823	.7656	.8490	.9323
7/32	.5182	.6016	.6849	.7682	.8516	.9349
1/4	.5208	.6042	.6875	.7708	.8542	.9375
9/32	.5234	.6068	.6901	.7734	.8568	.9401
5/16	.5260	.6094	.6927	.7760	.8594	.9427
11/32	.5286	.6120	.6953	.7786	.8620	.9453
3/8	.5313	.6146	.6979	.7813	.8646	.9479
13/32	.5339	.6172	.7005	.7839	.8672	.9505
7/16	.5365	.6198	.7031	.7865	.8698	.9531
15/32	.5391	.6224	.7057	.7891	.8724	.9557
1/2	.5417	.6250	.7083	.7917	.8750	.9583
17/32	.5443	.6276	.7109	.7943	.8776	.9609
9/16	.5469	.6302	.7135	.7969	.8802	.9635
19/32	.5495	.6328	.7161	.7995	.8828	.9661
5/8	.5521	.6354	.7188	.8021	.8854	.9688
21/32	.5547	.6380	.7214	.8047	.8880	.9714
11/16	.5573	.6406	.7240	.8073	.8906	.9740
23/32	.5599	.6432	.7266	.8099	.8932	.9766
3/4	.5625	.6458	.7292	.8125	.8958	.9792
25/32	.5651	.6484	.7318	.8151	.8984	.9818
13/16	.5677	.6510	.7344	.8177	.9010	.9844
27/32	.5703	.6536	.7370	.8203	.9036	.9870
7/8	.5729	.6563	.7396	.8229	.9063	.9896
29/32	.5755	.6589	.7422	.8255	.9089	.9922
15/16	.5781	.6615	.7448	.8281	.9115	.9948
31/32	.5807	.6641	.7474	.8307	.9141	.9974

GENERAL NOMENCLATURE

A	Cross-sectional area, in.^{2}
A	Horizontal distance from end panel point to mid-span of a truss, ft.
A	Minimum side dimension for square or rectangular beveled washer, in.
A_B	Loaded area of concrete, in.^{2}
A_{BM}	Cross-sectional area of base metal for a welded joint, in.^{2}
A_b	Nominal body area of a fastener, in.^{2}
A_b	Nominal bolt area, in.^{2}
A_c	Area of concrete in a composite column, in.^{2}
A_c	Area of concrete slab within effective width, in.^{2}
A_{cp}	Projected surface area of concrete cone surrounding headed anchor rods, in.^{2}
A_e	Effective net area, in.^{2}
A_f	Area of flange, in.^{2}
A_{fe}	Effective tension flange area, in.^{2}
A_{fg}	Gross area of flange, in.^{2}
A_{fn}	Net area of flange, in.^{2}
A_g	Gross area, in.^{2}
A_{gt}	Gross area subject to tension, in.^{2}
A_{gv}	Gross area subject to shear, in.^{2}
A_n	Net area, in.^{2}
A_{nt}	Net area subject to tension, in.^{2}
A_{nv}	Net area subject to shear, in.^{2}
A_{pb}	Projected bearing area, in.^{2}
A_r	Area of reinforcing bars, in.^{2}
A_s	Area of steel cross section, in.^{2}
A_{sc}	Cross-sectional area of stud shear connector, in.^{2}
A_{sf}	Shear area on the failure path, in.^{2}
A_{st}	Cross-sectional area of stiffener or pair of stiffeners, in.^{2}
A_v	Seismic coefficient representing the effective peak velocity-related acceleration
A_w	Area of web clear of flanges, in.^{2}
A_w	Effective area of weld, in.^{2}
A_0	Initial amplitude of a floor system due to a heel-drop excitation, in.
A_1	Area of steel bearing concentrically on a concrete support, in.^{2}
A_2	Total cross-sectional area of a concrete support, in.^{2}
B	Factor for bending stress in tees and double angles, defined by LRFD Specification Equation F1-16
B	Factor for bending stress in web-tapered members, defined by LRFD Specification Equations A-F3-8 through A-F3-11, in.
B	Horizontal distance from mid-span of a truss to a given panel point, ft.
B	Base plate width, in.
B_1, B_2	Factors used in determining M_u for combined bending and axial forces when elastic, first order analysis is employed

INDEX

BF	A factor that can be used to calculate the flexural strength for unbraced length L_b between L_p and L_r, defined in Part 4
C	Required mid-span camber, in.
C	Width across points of square or hex bolt head or nut, or maximum diameter of countersunk bolt head, in.
C	Coefficient for eccentrically loaded bolt and weld groups
C_{PG}	Plate girder coefficient
C_{Tot}	Sum of compressive forces in a composite beam, kips
C_a, C_b	Coefficients used in extended end-plate connection design
C_b	Bending coefficient dependent upon moment gradient
C_c	Beam reaction coefficient (Part 5)
C_{con}	Effective concrete flange force for a composite beam, kips
C_m	Coefficient applied to bending term in interaction formula for prismatic members and dependent upon column curvature caused by applied moments
$C_m{}'$	Coefficient applied to bending term in interaction formula for tapered members and dependent upon axial stress at the small end of the member
C_p	Ponding flexibility coefficient for primary member in a flat roof
C_s	Ponding flexibility coefficient for secondary member in a flat roof
C_s	Seismic response factor related to the fundamental period of the building
C_{stl}	Compressive force in steel in a composite beam, kips
C_v	Ratio of "critical" web stress, according to linear buckling theory, to the shear yield stress of web material
C_w	Warping constant, in.[6]
C_1	Loading constant used in deflection calculations (Part 4)
C_1	Clearance for tightening, in. (see Tables 8-4 and 8-5)
C_1	Electrode coefficient for relative strength of electrodes where, for E70 electrodes, $C_1 = 1.00$ (see Table 8-37)
C_2	Clearance for entering, in. (see Tables 8-4 and 8-5)
C_3	Clearance for fillet based on one standard hardened washer, in. (see Tables 8-4 and 8-5)
CG	Center of gravity
D	Outside diameter of circular hollow section, in.
D	Dead load, due to the weight of the structural elements and permanent features on the structure
D	Factor used in LRFD Specification Equation A-G4-2, dependent on the type of transverse stiffeners used in a plate girder
D	Offset from the base line at a panel point of a truss, in.
D	Damping in percent of critical
D	Slip probability factor for bolts
D	Number of sixteenths-of-an-inch in the weld size
DLF	Dynamic load factor
E	Modulus of elasticity of steel (29,000 ksi)
E	Earthquake load
E	Minimum edge distance for clipped washer, in.

E	Minimum effective throat thickness for partial-joint-penetration groove weld, in.
E_c	Modulus of elasticity of concrete, ksi
E_m	Modified modulus of elasticity for the design of composite columns, ksi
EBF	Eccentrically braced frame (Seismic Specification)
ENA	Elastic neutral axis
F	Width across flats of bolt head, in.
F	Clearance for tightening staggered bolts, in. (see Tables 8-4 and 8-5)
F_{BM}	Nominal strength of the base material to be welded, ksi
F_{EXX}	Classification strength of weld metal, ksi
F_L	Smaller of $(F_{yf} - F_r)$ or F_{yw}, ksi
$F_{b\gamma}$	Flexural stress for tapered members defined by LRFD Specification Equations A-F4-4 and A-F4-5, ksi
F_{cr}	Critical stress, ksi
$F_{crft}, F_{cry}, F_{crz}$	Flexural-torsional buckling stresses for double-angle and tee-shaped compression members, ksi
F_e	Elastic buckling stress, ksi
F_{ex}	Elastic flexural buckling stress about the major axis, ksi
F_{ey}	Elastic flexural buckling stress about the minor axis, ksi
F_{ez}	Elastic torsional buckling stress, ksi
F_{my}	Modified yield stress for the design of composite columns, ksi
F_n	Nominal shear rupture strength, ksi
F_n, F_{nt}	Nominal strength of bolt, ksi
F_p	Nominal bearing stress on fastener, ksi
F_r	Compressive residual stress in flange, ksi
$F_{s\gamma}$	Stress for tapered members defined by LRFD Specification Equation A-F3-6, ksi
F_t	Nominal tensile strength of bolt from LRFD Specification Table J3.2, ksi
F_u	Specified minimum tensile strength of the type of steel being used, ksi
F_v	Nominal shear strength of bolt from LRFD Specification Table J3.2, ksi
F_w	Nominal strength of the weld electrode material, ksi
$F_{w\gamma}$	Stress for tapered members defined by LRFD Specification Equation A-F3-7, ksi
F_y	Specified minimum yield stress of the type of steel being used, ksi. As used in the LRFD Specification, "yield stress" denotes either the specified minimum yield point (for steels that have a yield point) or specified yield strength (for steels that do not have a yield point)
F_y'''	The theoretical maximum yield stress (ksi) based on the web depth-thickness ratio (h/t_w) above which the web of a column is considered a "slender element" (See LRFD Specification Table B5.1)

$$= \left[\frac{253}{h/t_w} \right]^2$$

Note: In the tables, — indicates $F_y''' > 65$ ksi.

F_{yb}	F_y of a beam, ksi
F_{yc}	F_y of a column, ksi

F_{yf}	Specified minimum yield stress of the flange, ksi
F_{yr}	Specified minimum yield stress of the longitudinal reinforcing bars, ksi
$F_{y\,st}$	Specified minimum yield stress of the stiffener material, ksi
F_{yw}	Specified minimum yield stress of the web, ksi
G	Shear modulus of elasticity of steel (11,200 ksi)
G	Ratio of the total column stiffness framing into a joint to that of the stiffening members framing into the same joint
H	Horizontal force, kips
H	Flexural constant in LRFD Specification Equation E3-1
H	Average story height
H	Height of bolt head or nut, in.
H	Theoretical thread height, in. (see Table 8-7)
H_s	Length of stud connector after welding, in.
H_1	Height of bolt head, in. (see Tables 8-4 and 8-5)
H_2	Maximum bolt shank extension based on one standard hardened washer, in. (see Tables 8-4 and 8-5)
I	Moment of inertia, in.^{4}
I_{LB}	Lower bound moment of inertia for composite section, in.^{4}
I_c	Moment of inertia of column section about axis perpendicular to plane of buckling, in.^{4}
I_d	Moment of inertia of the steel deck supported on secondary members, in.^{4}
I_g	Moment of inertia of girder about axis perpendicular to plane of buckling, in.^{4}
I_p	Moment of inertia of primary member in flat roof framing, in.^{4}
I_p	Polar moment of inertia of bolt and weld groups ($= I_x + I_y$), in.^{4} per in.^{2}
I_s	Moment of inertia of secondary member in flat roof framing, in.^{4}
I_{st}	Moment of inertia of a transverse stiffener, in.^{4}
I_t	Transformed moment of inertia of the composite section, in.^{4}
I_x	Moment of inertia of bolt and weld groups about X-axis, in.^{4} per in.^{2}
I_y	Moment of inertia of bolt and weld groups about Y-axis, in.^{4} per in.^{2}
I_{yc}	Moment of inertia of compression flange about y axis or if reverse curvature bending, moment of inertia of smaller flange, in.^{4}
IC	Instantaneous center of rotation
ID	Nominal inside diameter of flat circular washer, in.
J	Torsional constant for a section, in.^{4}
K	Effective length factor for a prismatic member
K	Coefficient for estimating the natural frequency of a beam (Part 4)
K	Minimum root diameter of threaded fastener, in. (see Table 8-7)
K_{area}	An idealized area representing the contribution of the fillet to the steel beam area, as defined in the composite beam model of Part 5, in.^{2}
K_{dep}	Fillet depth, $(k - t_f)$, in.
K_i'	Modified effective length factor of a column
K_z	Effective length factor for torsional buckling
K_γ	Effective length factor for a tapered member
L	Unbraced length of member measured between the centers of gravity of the bracing members, in. or ft, as indicated

L	Span length, ft
L	Length of connection in the direction of loading, in.
L	Story height, in. or ft, as indicated
L	Live load due to occupancy and moveable equipment
L	Edge distance or center-to-center distance for holes, in.
L'	Total live load
L_b	Laterally unbraced length; length between points which are either braced against lateral displacement of the compression flange or braced against twist of the cross section, in. or ft, as indicated
L_c	Length of channel shear connector, in.
L_c	Unsupported length of a column section, ft
L_e	Edge distance, in.
L_{eh}	Horizontal edge distance, in.
L_{ev}	Vertical edge distance, in.
L_g	Unsupported length of a girder or other restraining member, ft
L_h	Hook length for hooked anchor rods, in.
L_m	Limiting laterally unbraced length for full plastic flexural strength, in. or ft, as indicated
L_m'	Limiting laterally unbraced length for the maximum design flexural strength for noncompact shapes, in. or ft, as indicated
L_p	Column spacing in direction of girder, ft
L_p	Limiting laterally unbraced length for full plastic flexural strength, uniform moment case ($C_b = 1.0$), in. or ft, as indicated
L_p'	Limiting laterally unbraced length for the maximum design flexural strength for noncompact shapes, uniform moment case ($C_b = 1.0$), in. or ft, as indicated
L_{pd}	Limiting laterally unbraced length for plastic analysis, in. or ft, as indicated
L_r	Limiting laterally unbraced length for inelastic lateral-torsional buckling, in. or ft, as indicated
L_r	Roof live load
L_s	Column spacing perpendicular to direction of girder, ft
M	Beam bending moment, kip-in. or kip-ft, as indicated
M_A	Absolute value of moment at quarter point of the unbraced beam segment, kip-in.
M_B	Absolute value of moment at centerline of the unbraced beam segment, kip-in.
M_C	Absolute value of moment at three-quarter point of the unbraced beam segment, kip-in.
M_{LL}	Beam moment due to live load, kip-in. or kip-ft, as indicated
M_T	Applied torsional moment, kip-in.
M_{cr}	Elastic buckling moment, kip-in. or kip-ft, as indicated
M_{eu}	Required flexural strength for extended end-plate connections, kip-in.
M_{lt}	Required flexural strength in member due to lateral frame translation, kip-in.
M_{max}	Maximum bending moment, kip-in. or kip-ft, as indicated

$M_{\max}$	Absolute value of maximum moment in the unbraced beam segment, kip-in.
M_n	Nominal flexural strength, kip-in. or kip-ft, as indicated
M_n'	Maximum design flexural strength for noncompact shapes, when $L_b \leq L_m'$, kip-in. or kip-ft, as indicated
M_{nt}	Required flexural strength in member assuming there is no lateral translation of the frame, kip-in.
M_{nx}', M_{ny}'	Flexural strength defined in LRFD Specification Equations A-H3-7 and A-H3-8, for use in the alternate interaction equations for combined bending and axial force, kip-in. or kip-ft, as indicated
M_p	Plastic bending moment, kip-in. or kip-ft, as indicated
M_p'	Moment defined in LRFD Specification Equations A-H3-5 and A-H3-6, for use in the alternate interaction formulas for combined bending and axial force, kip-in. or kip-ft, as indicated
M_{pa}	Plastic bending moment modified by axial load ratio, kip-in.
M_r	Limiting buckling moment, M_{cr}, when $\lambda = \lambda_r$ and $C_b = 1.0$, kip-in. or kip-ft, as indicated
M_u	Required flexural strength, kip-in. or kip-ft, as indicated
M_y	Initial yield bending moment, kip-in.
M_{ob}	Elastic lateral-torsional buckling moment, kip-in. or kip-ft, as indicated
M_1	Smaller moment at end of unbraced length of beam or beam-column, kip-in.
M_2	Larger moment at end of unbraced length of beam or beam-column, kip-in.
N	Length of bearing, in.
N	Ratio of the factored gravity load supported by all columns in a story to that supported by the columns in the rigid frame
N	Length of base plate, in.
N_b	Number of bolts in a joint
N_{eff}	Number of beams effective in resisting floor vibration (Part 4)
N_r	Number of stud connectors in one rib at a beam intersection, not to exceed 3 in calculations
N_s	Number of slip planes
OD	Nominal outside diameter of flat circular washer, in.
P	Concentrated load, kips
P	Bolt stagger, in.
P	Thread pitch, in. (see Table 8-7)
P_D	Unfactored dead load, kips
P_E	Unfactored earthquake load, kips
P_L	Unfactored live load, kips
P_S	Unfactored snow load, kips
P_{bf}	Applied factored beam flange force in moment connections, kips
P_e, P_{e1}, P_{e2}	Euler buckling strengths, kips
P_{fb}	Resistance to local flange bending per LRFD Specification Equation K1-1 (used to check need for column web stiffeners), kips
P_n	Nominal axial strength (tension or compression), kips
P_p	Bearing load on concrete, kips
P_u	Factored concentrated beam load, kips

P_u	Required axial strength (tension or compression), kips
$P_u e$	Induced moment due to eccentricity e in an eccentrically loaded bolt or weld group, kip-in.
P_{uf}	Factored beam flange force, tensile or compressive, kips
P_{wb}	Resistance to compression buckling of the web per LRFD Specification Equation K1-8 (used to check need for column web stiffening), kips
P_{wi}	A factor consisting of terms from the second portion of LRFD Specification Equation K1-2 (used in a column web stiffener check for local web yielding), kips/in.
P_{wo}	A factor consisting of the first portion of LRFD Specification Equation K1-2 (used in a column web stiffener check for local web yielding), kips
P_y	Yield strength, kips
PNA	Plastic neutral axis
Q	Full reduction factor for slender compression elements
Q_a	Reduction factor for slender stiffened compression elements
Q_f	Statical moment for a point in the flange directly above the vertical edge of the web, in.^{3}
Q_i	Load effects
Q_n	Nominal strength of one stud shear connector, kips
Q_s	Reduction factor for slender unstiffened compression elements
Q_w	Statical moment at mid-depth of the section, in.^{3}
R	Nominal load due to initial rainwater or ice exclusive of the ponding contribution
R	Nominal reaction, kips
R	Earthquake response modification coefficient
R_{PG}	Plate girder bending strength reduction factor
R_e	Hybrid girder factor
R_n	Nominal resistance or strength, kips
R_s	Nominal slip resistance of a bolt, kips
R_u	Required strength determined from factored loads; must be less than or equal to design strength ϕR_n
$R_{u\,st}$	Required strength for transverse stiffener (factored force delivered to stiffener), kips
R_v	Web shear strength, kips
R_1	An expression consisting of the first portion of LRFD Specification Equation K1-3, kips
R_2	An expression consisting of terms from the second portion of LRFD Specification Equation K1-3, kips/in.
R_3	An expression consisting of the first portion of LRFD Specification Equation K1-5a, kips
R_4	An expression consisting of terms from the second portion of LRFD Specification Equation K1-5a, kips/in.
R_5	An expression consisting of terms from LRFD Specification Equation K1-5b, kips
R_6	An expression consisting of terms from LRFD Specification Equation K1-5b, kips/in.

S	Elastic section modulus, in.^{3}
S	Spacing, in. or ft, as indicated
S	Snow load
S	Groove depth for partial-joint-penetration groove welds, in.
S'	Additional elastic section modulus corresponding to $\frac{1}{16}$-in. increase in web thickness for built-up wide flange sections, in.^{3}
S_c	Elastic section modulus to the tip of the angle in compression, in.^{3}
S_{eff}	Effective section modulus about major axis, in.^{3}
S_{net}	Net elastic section modulus, in.^{3}
S_w	Warping statical moment at a point on the cross section, in.^{4}
S_x	Elastic section modulus about major axis, in.^{3}
S_x'	Elastic section modulus of larger end of tapered member about its major axis, in.^{3}
S_{xt}, S_{xc}	Elastic section modulus referred to tension and compression flanges, respectively, in.^{3}
SRF	Stiffness reduction factors (Table 3-1), for use with the alignment charts (Figure 3-1) in the determination of effective length factors K for columns
T	Distance between web toes of fillets at top and at bottom of web, in. $= d - 2k$
T	Tension force due to service loads, kips
T	Thickness of flat circular washer or mean thickness of square or rectangular beveled washer, in.
T	Unfactored tensile force on slip-critical connections designed at service loads, kips
T_b, T_m	Minimum bolt tension for fully tensioned bolts from LRFD Specification Table J3.1, kips
T_{stl}	Tensile force in steel in a composite beam, kips
T_{Tot}	Sum of tensile forces in a composite beam, kips
T_u	Factored tensile force, kips
U	Reduction coefficient, used in calculating effective net area
V	Shear force, kips
V_b	Shear force component, kips
V_h	Total horizontal force transferred by the shear connections, kips
V_n	Nominal shear strength, kips
V_u	Required shear strength, kips
W	Wind load
W	Uniformly distributed load, kips
W	Weight, lbs or kips, as indicated
W	Width across flats of nut, in.
W_c	Uniform load constant for beams, kip-ft
W_{no}	Normalized warping function at a point at the flange edge, in.^{2}
W_u	Total factored uniformly distributed load, kips
X_1	Beam buckling factor defined by LRFD Specification Equation F1-8
X_2	Beam buckling factor defined by LRFD Specification Equation F1-9
Y_{ENA}	Distance from bottom of steel beam to elastic neutral axis, in.
Y_{con}	Distance from top of steel beam to top of concrete, in.

$Y1$	Distance from top of steel beam to the plastic neutral axis, in.
$Y2$	Distance from top of steel beam to the concrete flange force in a composite beam, in.
Z	Plastic section modulus, in.^{3}
Z'	Additional plastic section modulus corresponding to $\frac{1}{16}$-inch increase in web thickness for built-up wide flange section, in.^{3}
Z_e	Effective plastic section modulus, in.^{3}
a	Clear distance between transverse stiffeners, in.
a	Distance between connectors in a built-up member, in.
a	Effective concrete flange thickness of a composite beam, in.
a	Shortest distance from edge of pinhole to edge of member measured parallel to direction of force, in.
a	Coefficient for eccentrically loaded weld group
a	Distance from bolt centerline to edge of fitting subjected to prying action, but not greater than $1.25b$, in.
a	Depth of bracket plate, in.
a_r	Ratio of web area to compression flange area
b	Compression element width, in.
b	Effective concrete flange width in a composite beam, in.
b	Width of composite column section, in.
b	Minimum shelf dimension for deposition of fillet weld, in.
b	Width of bracket plate, in.
b	Distance from bolt centerline to face of fitting subjected to prying action, in.
b_e	Reduced effective width for slender compression elements, in.
b_{eff}	Effective edge distance, in.
b_f	Flange width of rolled beam or plate girder, in.
b_s	Width of transverse stiffener, in.
b_s	Width of extended end-plate, in.
c	Distance from the neutral axis to the extreme fiber of the cross section, in.
c	Cope length, in.
c_1, c_2, c_3	Numerical coefficients used in the calculation of the modified yield stress and modulus of elasticity for composite columns
d	Nominal fastener diameter, in.
d	Overall depth of member, in.
d	Pin diameter, in.
d	Roller diameter, in.
d_L	Depth at larger end of unbraced tapered segment, in.
d_b	Nominal bolt diameter, in.
d_c	Column depth, in.
d_c	Cope depth, in.
d_{ct}	Top-flange cope depth, in.
d_{cb}	Bottom-flange cope depth, in.
d_h	Hole diameter, in.
d_m	Moment arm between resultant tensile and compressive forces due to a moment or eccentric force, in.

d_z	Overall panel-zone depth, in.
d_0	Depth at smaller end of unbraced tapered segment, in.
e	Eccentricity, in.
e	Base of natural logarithms = 2.71828...
e	Link length in eccentrically braced frame (EBF), in.
e_o	Horizontal distance from the outer edge of a channel web to its shear center, in.
f	Computed compressive stress in the stiffened element, ksi
f	Natural frequency, hz
f	Plate buckling model adjustment factor for beams coped at top flange only
f_b	Maximum bending stress, ksi
f_{b1}	Smallest computed bending stress at one end of a tapered segment, ksi
f_{b2}	Largest computed bending stress at one end of a tapered segment, ksi
f_c'	Specified compressive strength of concrete, ksi
f_d	Adjustment factor for beams coped at both flanges
f_{un}	Required normal stress, ksi
f_{uv}	Required shear stress, ksi
f_v	Computed shear stress, ksi
f_o	Stress due to $1.2D + 1.2R$, ksi
g	Transverse center-to-center spacing (gage) between fastener gage lines, in.
g	Acceleration due to gravity = 32.2 ft/sec^2 = 386 in./sec^2
h	Clear distance between flanges less the fillet or corner radius for rolled shapes; and for built-up sections, the distance between adjacent lines of fasteners or the clear distance between flanges when welds are used, in.
h	Depth of composite column section, in.
h_c	Twice the distance from the centroid to the following: the inside face of the compression flange less the fillet or corner radius, for rolled shapes; the nearest line of fasteners at the compression flange or the inside face of the compression flange when welds are used, for built-up sections, in.
h_r	Nominal rib height, in.
h_s	Factor used in LRFD Specification Equation A-F3-6 for web-tapered members
h_w	Factor used in LRFD Specification Equation A-F3-7 for web-tapered members
h_o	Remaining web depth of coped beam, in.
j	Factor defined by LRFD Specification Equations A-F2-4 for minimum moment of inertia for a transverse stiffener
k	Distance from outer face of flange to web toe of fillet, in.
k	Slenderness parameter
k	Plate buckling coefficient for beams coped at top flange only
k_s	Bolt slip coefficient
k_v	Web plate buckling coefficient
k_1	Distance from web center line to flange toe of fillet, in.
l	Unbraced length of member, in.
l	Span length, in.
l	Length of bearing, in.

l	Length of connection in the direction of loading, in.
l	Length of weld, in.
l	Characteristic length of weld group (see Tables 8-38 through 8-45), in.
l_o	Distance from center of gravity (CG) to instantaneous center of rotation (IC) of bolt or weld group, in.
m	Ratio of web to flange yield stress or critical stress in hybrid beams
m	Coefficient for converting bending to an approximate equivalent axial load in beam-columns (Part 3)
m	Cantilever dimension for base plate (see Part 11), in.
n	Number of shear connectors between point of maximum positive moment and the point of zero moment to each side
n	Number of bolts in a vertical row
n	Number of threads per inch on threaded fasteners
n	Cantilever dimension for base plate (see Part 11), in.
n'	Number of bolts above the neutral axis (in tension)
p	Length of supporting flange parallel to stem or leg of hanger tributary to each bolt in determinimg prying action, in.
p_e	Effective span used to compute M_{eu} for extended end-plate connections, in.
p_f	Distance from centerline of bolt to nearer surface of tension flange in extended end-plate connections, in.
q_u	Additional tension per bolt resulting from prying action produced by deformation of the connected parts, kips/bolt
r	Governing radius of gyration, in.
r_T	Radius of gyration of compression flange plus one third of the compression portion of the web taken about an axis in the plane of the web, in.
r_{To}	Radius of gyration, r_T, for the smaller end of a tapered member, in.
r_i	Minimum radius of gyration of individual component in a built-up member, in.
r_{ib}	Radius of gyration of individual component relative to centroidal axis parallel to member axis of buckling, in.
r_m	Radius of gyration of steel shape, pipe, or tubing in composite columns. For steel shapes it may not be less than 0.3 times the overall thickness of the composite section, in.
r_n	Nominal strength per bolt from LRFD Specification
r_x, r_y	Radius of gyration about x and y axes respectively, in.
r_{ut}	Required tensile strength per bolt or per inch of weld (factored force per bolt or per inch of weld due to a tensile force), kips/bolt
r_{ut}	Required shear strength per bolt or per inch of weld (factored force per bolt or per inch of weld due to a shear force), kips/bolt
r_{yc}	Radius of gyration about y axis referred to compression flange, or if reverse curvature bending, referred to smaller flange, in.
$\bar{r}_o$	Polar radius of gyration about the shear center, in.
r_{ox}, r_{oy}	Radius of gyration about x and y axes at the smaller end of a tapered member respectively, in.
s	Longitudinal center-to-center spacing (pitch) of any two consecutive holes, in.
s	Bolt spacing, in.

t	Thickness, in.
t	Change in temperature, degrees Fahrenheit or Celsius, as indicated
t_b	Thickness of beam flange or connection plate delivering concentrated force, in.
t_c	Flange or angle thickness required to develop design tensile strength of bolts with no prying action, in.
t_e	Total required effective thickness of column web with doubler plate, in.
t_f	Flange thickness, in.
t_f	Flange thickness of channel shear connector, in.
t_p	Thickness of base plate, in.
t_p	Panel zone thickness including doubler plates, in.
$t_{p\,req}$	Required doubler plate thickness, in.
t_s	Extended end-plate thickness, in.
t_w	Web thickness, in.
t_w	Web thickness of channel shear connector, in.
t_{wb}	Beam web thickness, in.
t_{wc}	Column web thickness, in.
t_z	Panel zone thickness, in.
u	Factor for approximate design of beam-columns (Part 3)
w	Uniformly distributed load per unit of length, kips/in.
w	Fillet weld size, in.
w	Plate width; distance between welds, in.
w	Subscript relating symbol to strong principal axis of angle
w	Unit weight of concrete, lbs/ft^3
w_r	Average width of concrete rib or haunch, in.
w_z	Panel zone width, in.
x	Subscript relating symbol to strong axis bending
x	Horizontal distance, in.
$\bar{x}$	Horizontal distance from the outer edge of a channel web to its centroid, in.
$\bar{x}$	Connection eccentricity, in.
x_p	Horizontal distance from the designated edge of member to its plastic neutral axis, in.
x_o	Horizontal distance, in.
x_o, y_o	Coordinates of the shear center with respect to the centroid, in.
y	Moment arm between centroid of tensile forces and compressive forces, in.
y	Subscript relating symbol to weak axis bending
y_p	Vertical distance from the designated edge of member to its plastic neutral axis, in.
y_1, y_2	Vertical distance from designated edge of member to center of gravity, in.
z	Distance from the smaller end of tapered member used in LRFD Specification Equation A-F3-1 for the variation in depth, in.
z	Subscript relating symbol to weak principal axis of angle
z	Coefficient for buckling of triangular-shaped bracket plate
Δ	Deflection, in.
Δ_{LL}	Live load deflection, in.

Δ_{oh}	Translation deflection of the story under consideration, in.
α	Separation ratio for built-up compression members, LRFD Specification Equation E4
α	Fraction of member force transferred across a particular net section
α	Ratio of moment at bolt line to moment at stem line for determining prying action in hanger connections
α	Ideal distance from face of column flange or web to centroid of gusset-to-beam connection for bracing connections and uniform force method, in.
$\bar{\alpha}$	Actual distance from face of column flange or web to centroid of gusset-to-beam connection for bracing connections and uniform force method, in.
α_m	Coefficient for calculating M_{eu} for extended end-plate connections
β	Ideal distance from face of beam flange to centroid of gusset-to-column connection for bracing connections and uniform force method, in.
$\bar{\beta}$	Actual distance from face of beam flange to centroid of gusset-to-column connection for bracing connections and uniform force method, in.
β_w	Special section property for unequal-leg angles (Single Angle Specification)
γ	Depth tapering ratio
γ	Subscript relating symbol to tapered members
γ_i	Load factor
δ	Deflection, in.
δ	Ratio of net area at bolt line to gross area at face of stem or angle leg used to determine prying action for hanger connections
ε	Coefficient of linear expansion, with units as indicated
ζ	Exponent for alternate beam-column interaction equation
η	Exponent for alternate beam-column interaction equation
λ	Slenderness parameter
λ_c	Column slenderness parameter
λ_e	Equivalent slenderness parameter
λ_{eff}	Effective slenderness ratio defined by LRFD Specification Equation A-F3-2
λ_p	Limiting slenderness parameter for compact element
λ_r	Limiting slenderness parameter for noncompact element
μ	Coefficient of friction; mean slip coefficient for bolts
ρ	Ratio of P_u to V_u of a link in an eccentrically braced frame (EBF)
ϕ	Resistance factor
ϕ_b	Resistance factor for flexure
ϕ_c	Resistance factor for compression
ϕ_c	Resistance factor for axially loaded composite columns
ϕ_r	Resistance factor for compression, used in web crippling equations
ϕ_{sf}	Resistance factor for shear on the failure path
ϕ_t	Resistance factor for tension
ϕ_v	Resistance factor for shear
ϕ_w	Resistance factor for welds
ϕF_{bc}	Design buckling stress for coped beams, ksi

ϕR_n Design strength from LRFD Specification; must equal or exceed required strength R_u

ϕr_n Design strength per bolt or per inch of weld from LRFD Specification; must equal or exceed required strength per bolt or per inch of weld r_u

kip   1,000 pounds

ksi   Stress, kips/in.^{2}

INDEX

Notes

Notes

<u>Notes</u>

<u>Notes</u>

Notes

Notes

<u>Notes</u>

Notes

Notes

<u>Notes</u>